MW00837121

Electricity for HVACR

JOSEPH MORAVEK
Nance Universal HVACR Technical School, Beaumont, Texas

PEARSON

Boston Columbus Indianapolis New York San Francisco Upper Saddle River
Amsterdam Cape Town Dubai London Madrid Milan Munich Paris Montreal Toronto
Delhi Mexico City São Paulo Sydney Hong Kong Seoul Singapore Taipei Tokyo

Editorial Director: Vernon R. Anthony
Senior Acquisitions Editor: Lindsey Gill
Editorial Assistant: Nancy Kesterson
Director of Marketing: David Gesell
Senior Marketing Coordinator: Alicia Wozniak
Marketing Assistant: Crystal Gonzalez
Program Manager: Maren L. Miller
Operations Specialist: Deidra Skahill
Development Manager: Robin C. Bonner for Aptara®, Inc.
Development Editor: Leslie Lahr
Senior Art Director: Diane Y. Ernsberger
Art Director: Jayne Conte
Cover Designer: Suzanne Behnke
Permissions Researcher: Maria Siriano
Image Permission Coordinator: Mike Lackey
Cover Art: Fieldpiece Instruments
Lead Media Project Manager: April Cleland
Full-Service Project Management: Peggy Kellar for Aptara®, Inc.
Composition: Aptara®, Inc.
Printer/Binder: LSC Communications, Kendallville
Cover Printer: LSC Communications, Kendallville
Text Font: Minion

Credits and acknowledgments borrowed from other sources and reproduced, with permission, in this textbook appear on the pages 573–577.

Library of Congress Cataloging-in-Publication Data

Moravek, Joseph.
Electricity for HVACR/Joseph Moravek.—First edition.
pages cm
Includes bibliographical references and index.
ISBN-13: 978-0-13-512534-2
ISBN-10: 0-13-512534-0
1. Air conditioning—Electric equipment. 2. Heating—Electric equipment. 3. Ventilation—Electric equipment. 4. Refrigeration and refrigerating machinery—Electric equipment. 5. Electric circuits. I. Title.
TK4035.A35M67 2013
621.319'24—dc23

2013009166

10 2021

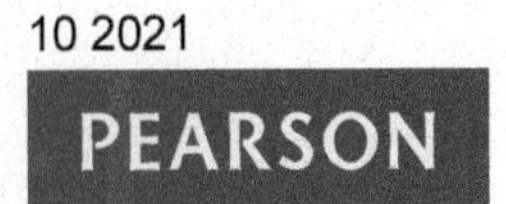

ISBN 13: 978-0-13-512534-2
ISBN 10: 0-13-512534-0

This book is dedicated to my wife,

Martha E. Moravek,

who tolerated my absence for the period while I was working on this project

Preface

This book was written to help a person with no electrical experience or training to understand the operation of HVACR electrical circuits. This material does not get into the theory of electrical circuits; instead, it provides the fundamental information needed to understand and repair an HVACR system. Therein lies the real goal of this book: to teach the reader how to correctly diagnosis and solve electrical problems. As an aspiring HVACR technician, you will never know it all. There is always something new and interesting to learn. If you are a lifetime learner, you will like our profession.

Let's look at some of the features and benefits of *Electricity for HVACR.*

HOW THE TEXT IS ORGANIZED

Electricity for HVACR is divided into 28 progressive units. It is designed with the spiral learning concept, with each new unit building on content learned from the previous unit. The first units begin with fundamentals such as defining terms used in our industry. A unit on the safe use of instruments addresses how to use diagnostic tools and instruments safely. This includes information on purchasing a quality and safe voltage-measuring product. The first half of the book includes many units that discuss the common components found in HVACR systems. These units describe the operation of the electrical components and how to troubleshoot them.

Next, the middle content discusses the symbols and components that make up an electrical diagram. The reader is slowly taken through the process of understanding electrical diagrams. Many examples are used to explain the operating sequence of diagrams.

Unit 16 covers the "green" electronically commutated motor (ECM). This motor represents a great advance in technology. Because this new, advanced technology is not well understood, the goal of the unit is to help the learner understand the operation and troubleshooting of ECMs.

Units 21 through 24 describe the sequence of operation for air conditioning, gas heat, electric heat, and heat pump systems, respectively. These units apply what was learned in the preceding 20 units.

The final four units describe the all-important troubleshooting process. Unit 25 discusses how to get started. It offers some basic techniques to help the reader apply troubleshooting skills at the beginning of the search for the problem. Ultimately, this is the goal of this book: learning how to troubleshoot and repair. If a student-tech cannot find and repair an HVACR problem, then this book will help him or her "figure it out."

IMPORTANT FEATURES OF THE TEXT

What You Need to Know

Each unit opens with learning objectives entitled "What You Need to Know." These objectives serve as a unit outline and, more importantly, provide the student with attainable learning goals.

Summary

Each unit concludes with a Summary that recaps and highlights the unit content.

Review Questions

Review Questions found at the end of each unit challenge readers to test their understanding of the reading. The questions cover each important aspect of the unit's content. Some questions require identification of HVACR components, use of troubleshooting skills, and tracing of wiring diagrams.

Tech Tips

Tech Tips provide hands-on hints based on the author's years of experience in the field. These tips offer the new (and experienced) technician practical insights into key HVACR components, their installation, and troubleshooting techniques.

Safety Tips

Safety Tips offer crucial, and sometimes lifesaving, pointers for the technician in the field. A long productive career in HVACR requires stringent safety practices. Accident and injury prevention is tantamount to good workmanship—look after yourself and those around you.

Green Tips

Green Tips highlight new technologies and components being used in the field that have been designed to save electricity. These tips offer the new technician a chance to become familiar with some of the latest advancements in the HVACR industry.

Sequence of Operation Boxes

Sequence of Operation boxes describe, step by step, the operation of specific pieces of equipment. The numbered steps are tied to a numbered diagram, which creates a visual learning experience for the student.

Troubleshooting Boxes

Troubleshooting boxes offer instruction on how to troubleshoot an electrical problem, one step at a time. To aid student comprehension of the troubleshooting process, each numbered step of instruction is matched to a numbered photograph of actual equipment or a numbered diagram.

Examples

Examples are offered to help students work with the formulas and calculations used in the HVACR field. Examples present a real-world problem and then walk the student through the steps of the solution.

Service Ticket

The Service Ticket presented at the end of each unit puts the student in an on-the-job situation. Each scenario involves a problem that requires the reader to use the concepts and techniques presented in the unit to arrive at a satisfactory resolution. The scenario includes an evaluation and assessment of the problem, the most probable cause, and a solution that satisfies both the customer and the technician's employer.

Electrical Diagrams

Electricity for HVACR has many special features, but the 14 enlarged electrical diagrams that are included with the book make it stand out from other HVACR books. These are referred to as Electrical Diagram ED-1, ED-2, etc. It can be difficult to read textbook diagrams because they are limited by the size of the page. Printing a two-page diagram on facing pages is possible, but the break between the pages makes the diagrams difficult to follow, which creates a problem for the user. For these reasons, we have included 14 enlarged electrical diagrams that are much easier for the learner to follow during discussions by the instructor or when working with the detailed information in each unit.

The electrical diagrams are actual diagrams one might come across in the field. The diagrams cover the gamut of the HVACR field, including gas heat, electric heat, and heat pumps; and residential, commercial, and industrial air conditioning systems, including chilled water systems. The electrical diagrams were selected to represent all areas of HVACR, yet they are basic enough to be understood for the entry-level technician.

COMPREHENSIVE SUPPLEMENT PACKAGE

Download Instructor Resources from the Instructor Resource Center

To access supplementary materials online, instructors need to request an instructor access code. Go to www.pearsonhighered.com/irc to register for an instructor access code. Within 48 hours of registering, you will receive a confirming e-mail including an instructor access code. Once you have received your code, locate your text in the online catalog and click on the Instructor Resources button on the left side of the catalog product page. Select a supplement, and a login page will appear. Once you have logged in, you can access instructor material for all Pearson textbooks. If you have any difficulties accessing the site or downloading a supplement, please contact Customer Service at http://247pearsoned.custhelp.com.

Instructor's Solutions Manual, ISBN-10: 0-13-512537-5

Includes solutions to the end-of-unit questions and the lab manual.

PowerPoint Slides, ISBN-10: 0-13-302725-2

Includes a comprehensive, colorful image bank and PowerPoint presentations for all units.

Lab Manual, ISBN-10: 0-13-512536-7

The printed lab manual offers basic coverage of all of the major lessons in each unit.

MyTest, ISBN-10: 0-13-512576-6

MyTest is a comprehensive set of test questions that match the key objectives for each unit.

Acknowledgments

Without the help of those listed here, the quality of the product you are about to explore would have suffered. The HVACR professional must thank everyone listed here for their support of our industry. Their time, insights, and suggestions helped make this book better and stronger.

Vernon Anthony
Pearson

Dan Trudden
Pearson

Larry Giroux
Giroux Air Conditioning Training

Roger Stuksa

Curtis McGuirt

Jeff Zinsmayer
Zinsmayer Design and Construction

Thanks to all of the HVACR contractors who answered my polls and offered suggestions to make this a premium product.

Many reviewers improved this product with ideas, suggestions, and content. They are:

Timothy Andera
South Dakota State University

Douglas Broughman
Augusta Technical College

Edward A. Burns
Harrisburg Area Community College

Danny B. Burris
Eastfield College of Dallas County
Community College District

Thomas Bush
South Florida Community College

Victor Cafarchia
El Camino College

Michael D. Covington
Sullivan College of Technology

Dave DeRoche
St. Clair College Windsor, Ontario

Patrick F. Duschl
Fortis College Cincinnati

Michael W. Falvey
Copiah-Lincoln Community College

Luther Gardner
Somerset County Technology Center

Nicholas Griewahn
Northern Michigan University

Patrick Heeb
Long Beach City College

John Hohman
EDEMPCO

James Janich
College of DuPage

Rick Lutz
Red Rock Community College

Christopher Mohalley
Genteq

Michael Mutarelli
Lehigh Carbon Community College

Paul Oppenheim
University of Florida

Thomas E. Owen, Jr.
Sullivan College of Technology and Design

Jacky Skelton
Arkansas Northeastern College

Samuel Shane Todd
Ogeechee Technical College

Mark Van Doren
Ivy Tech Community College

Freddie Williams
Lanier Technical College

I would like to thank the following people for their contributions toward making this work a professional benefit to the HVACR profession.

First, I want to thank Pearson for selecting me to complete this important project.

I want to thank Dr. John Hohman, Ph.D., for his contribution to writing several high-performance units. Dr. Hohman also offered good advice and support prior to starting the project.

My wife, Martha Moravek, developed and organized the important glossary.

Others who offered continued support and mentoring were Larry Giroux of Giroux Training, and Chris Mohalley of Genteq and Roger Stuksa, an independent HVACR contractor.

I want to thank the Aptara, Inc., Production Department for making this product so appealing and useful to the student.

SPECIAL THANK YOU

I want to express special appreciation to Leslie Lahr of Aptara, Inc., for her talents and perseverance in helping me to the goal line. It was difficult and she was the biggest supporter and motivator in bringing this work to fruition. Her editing skills brought out the best in this project. Leslie does not have an air conditioning background, which helped. If she could understand what was being conveyed, the learner should be able to do the same. Many of her suggestions, ideas, concepts, and improvements are seen in every unit. The success of this project can be largely contributed to Leslie. Great job, Leslie—it was a pleasure to have worked with you. I hope we can do another project together.

Maren Miller managed to keep all of the pieces together with remarkable skill. She orchestrated not only the process of building the book, but also the people needed to get the job done.

Initially, Robin Bonner of Aptara Productions kept the project rolling and directed in a positive manner. Thank you, Robin.

The final manuscript and art content were managed by Peggy Kellar, Aptara, Inc., and Lorretta Palagi, Quantum Publishing Services, Inc. Their sharp editing skills offered final improvements so that we can offer the finest published product possible.

It is difficult to obtain permission to use HVACR images. Maria Siriano of Redline Ink met the challenge.

I want to recognize the contributions of the following companies. They provided images and granted permission to use pictures, tables, and diagrams of their fine products. These images are used throughout the book and supporting supplements.

Air-Conditioning, Heating, and Refrigeration Institute (AHRI)

Alco Products

Allen Bradley

Allied Air Enterprises

American Standard

AmRad Engineering

ARESCO

Bolt Depot

Carrier Corporation

Copeland Compressors

Don Crawshaw, HVAC/R Productions

Diversitech

Edison Electric Institute

Emerson Climate Technologies

Fasco Motors

Fieldpiece Instruments

Fluke Corporation

Genteq

Giroux Air Conditioning Training

Goodman Manufacturing

Hampden Engineering

Dr. John Hohman

ICM Controls

Ideal Industries

Johnson Supply

National Fire Protection Agency

OSHA

Pearson Production Department

Refrigeration Basics

Reliant Energy

Ritchie Manufacturing

Rockwell Automation

Sealed Unit Parts Company (SUPCO)

Siemens

Sporlan Corporation

Square D

Tecumseh Corporation

Thermostat Recycling Corporation

3M

Trane Corporation

U.S. Department of Energy

U.S. Information Administration

Venstar

WEG

www.hobbyteam@hobby-hour.com

www.onsemi.com

www.quest-comp.com

York

Zebra Instruments

Author contact information:

Joseph Moravek
zmoravek@aol.com

About the Author

Joe Moravek has been in the air conditioning profession since 1975 when he went to work doing heat load calculations on low-income homes in Houston, Texas. From there he was employed by the University of Houston's College of Technology where he was involved with the Texas Energy Extension Service offering energy-saving training and consulting for homeowners and small commercial businesses. Other related jobs included working for the City of Houston in the Energy Conservation Office, working as an HVACR technician in the Parks and Recreation Department, and working as a mechanical inspector in the Occupancy Department. During his time with the City of Houston, he taught as an adjunct instructor for San Jacinto College and Houston Community College. Joe found his love for teaching and left the municipal environment to become a full-time HVACR instructor. Joe received his master's degree in education from the University of Houston in 1980.

Next, he was the lead HVACR instructor for 14 years at Lee College, in Baytown, Texas, helping to improve the quality of the courses the department offered. The next challenge was at the corporate level. Joe was the first corporate training manager for Hunton Trane in Houston, Texas. He developed curriculum, taught classes, and improved the delivery quality of training material. His current job is as Training Director at Nance Universal HVACR Technical School in Beaumont, Texas. This was a new position for Nance School and for the past five years, Joe has worked to attract quality instructors and offer an exceptional curriculum.

Joe has published several books, reviewed more than two dozen training works, and has published many HVACR book reviews. Finally, the author is kept busy with his consulting business, Mechanical Training Services. This work gets him involved in training, developing training, and helping solve contractors' technical problems on the job. He is a licensed air conditioning and refrigeration contractor in Texas and conducts required continuing education classes for his fellow contractors. He welcomes comments on this book and can be contacted at zmoravek@aol.com.

Contents

UNIT 1

What You Need to Know to Understand Electricity

WHAT YOU NEED TO KNOW

After studying this unit, you will be able to:

1. Describe the components of electricity.
2. Explain how alternating current (AC) and direct current (DC) electricity work.
3. Define important electrical terms and symbols used in HVACR.

This unit provides you with the most basic electrical information required to understand how the electricity in our equipment works. As you progress you will notice that this book does not go into deep theory applications such as how atoms work or electrical theory. We are not trying to turn you into an electrical engineer. You do, however, need to understand HVACR electric circuits so that you can troubleshoot and repair systems. Our goal is to provide you with knowledge about the basics of electricity and to offer practical information that you can use on your job as a professional technician. This unit is important because electrical defects in HVACR equipment are the most common problems you will encounter.

To begin, you need an understanding of the terms used in the HVACR profession. This will help you communicate with other HVACR professionals as well as your clients. Electrical terms and definitions will be used throughout this textbook. You will notice in HVACR electricity that some terms have the same definition. For example, the word *volt* can be used to refer to potential difference, electrical pressure, electrical force, or the abbreviations **E** or **V**. Also, consider the term *electrical diagrams*. Different names may be used for electrical diagrams, but they are essentially the same thing. Electrical diagrams are also called wiring diagrams, schematic diagrams, ladder diagrams, and connections diagrams. Throughout this book, we will present you with all of the different terms used in the electrical side of our profession. Our intention is not to confuse you, but to show that different terms are used to represent the same thing.

1.1 WHAT IS ELECTRICITY?

Electricity is the flow of electrons between atoms in common wire materials such as copper and aluminum. The flow of electrons is known as **current flow**. Electron flow is illustrated in Figure 1-1 by the movement of electrons in a wire. Current flow creates a magnetic field in the wire. The magnetism in the wire can be designed to move motors, operate switches, and do other work in an HVACR system.

Figure 1-2 shows that **voltage** is the pressure pushing the current flow. The top illustration compares water pressure to voltage pressure. Water pressure can be compared to voltage (electrical pressure) and water flow can be compared to electron flow. The greater the pressure developed by the chilled water pump, the greater the pressure in the piping system. The middle and bottom diagrams show voltage pressure. The higher the voltage, the greater the voltage pressure. The greater the electrical pressure, the higher the voltage in the wire. If the voltage pressure is too low, electron flow will not occur.

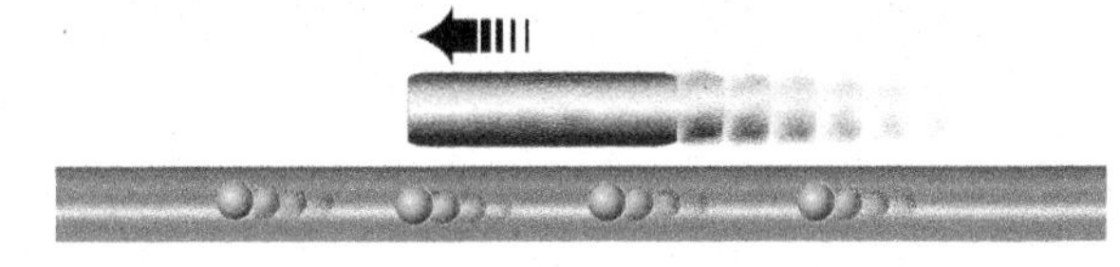

Figure 1-1 The flow of electrons is known as current flow.

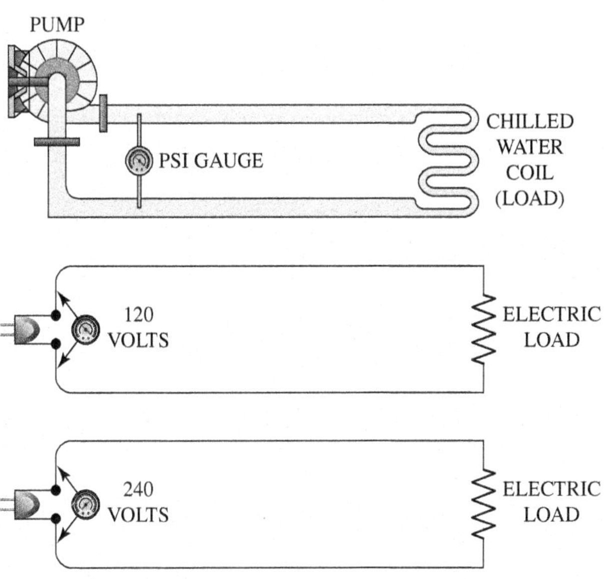

Figure 1-2 Water pressure compared to voltage pressure. The gauges in the middle and lower images show that the higher the voltage, the greater the voltage pressure.

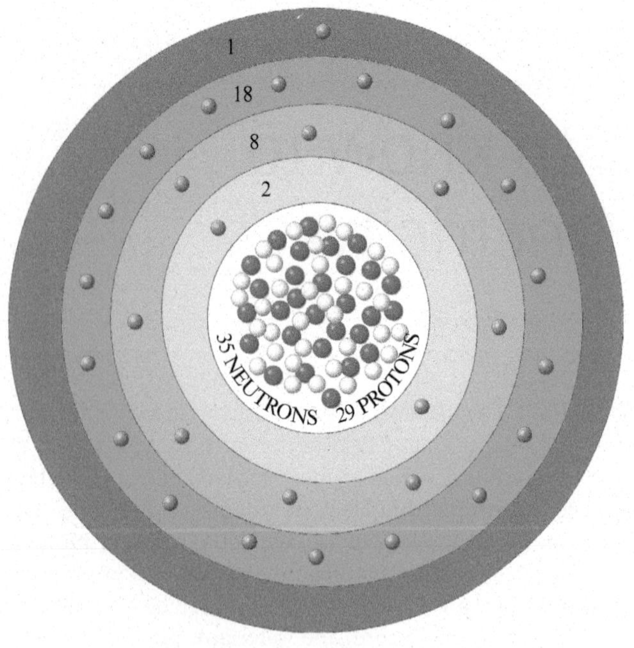

Figure 1-3 This illustration of a copper atom shows that the first three inner electron energy levels of the copper atom are full, but the outermost level has only one electron. The single electron is the free electron that will be transferred between atoms to create what we know as electricity.

The atom is constructed of neutrons, protons, and electrons as illustrated in Figure 1-3.

The neutrons and protons are stable components and stay linked within the core of the atom. Free **electrons** can be found around the outer shell. A good conductor of electricity will have one or more free electrons in the outer shell. These free electrons move between atoms and create the current flow, as shown in Figure 1-4.

Electron flow occurs as a chain reaction. Notice how the electrons are transferred in the outer shell of this atom. The greater the number of free electrons, the greater the current flow. This transfer of electrons results in what we call electricity. Materials known as **conductors** have many free electrons in the atom's orbit that are easily transferred between atoms. Conductors used in HVACR equipment are normally copper wire. **Insulators** have few or no free electrons to share; therefore, they resist electron or current flow. Rubber and glass are good insulators.

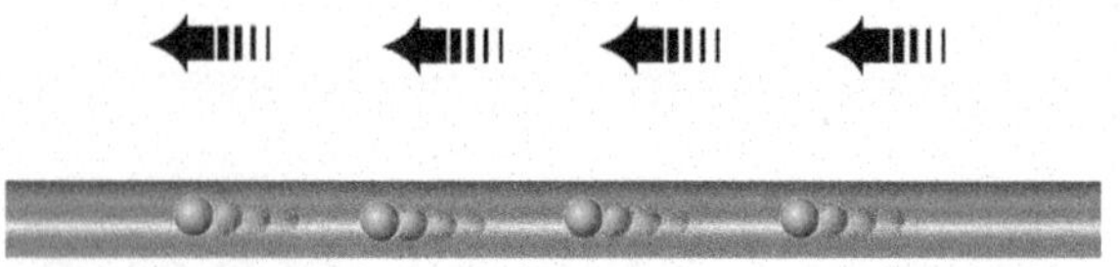

Figure 1-5 Electrons flow in only one direction with direct current. The DC can either be positive or negative.

1.2 TYPES OF ELECTRICITY

There are two types of electricity: **direct current** (**DC**) and **alternating current** (**AC**). An example of direct current is battery voltage. Direct current reaches its level of voltage and stays at that point as illustrated in Figure 1-5. It flows in only one direction and can be positive or negative, depending on how the DC voltage is hooked in to the circuit.

As seen in Figure 1-6, alternating current varies as positive and negative around a zero center voltage.

The alternations occur 60 times a second. The complete positive and negative cycle is called the **hertz (Hz)** or **cycle.** Hertz determines the speed of a motor. The higher the hertz, the faster the motor's revolutions per minute (RPM). Note that 50-Hz alternations are used in some countries. Alternating current is the most common power source found in HVACR equipment. Direct current circuits may be found on circuit boards, but troubleshooting individual components on a circuit board is not usually practiced. Checking DC voltage is most likely done to check battery voltage.

SAFETY TIP

Knowing the correct hertz for a motor or other electrical component is important. Values of 60 and 50 Hz are commonly found around the world. Sixty hertz is normally found in the Western Hemisphere in North and South America. Installing a motor with the wrong hertz rating may damage a motor, so be aware that 50 Hz motors are made in the United States and sometimes find their way into the supply stream. Hertz or cycles is one of several electrical characteristics you should check when changing an electrical device.

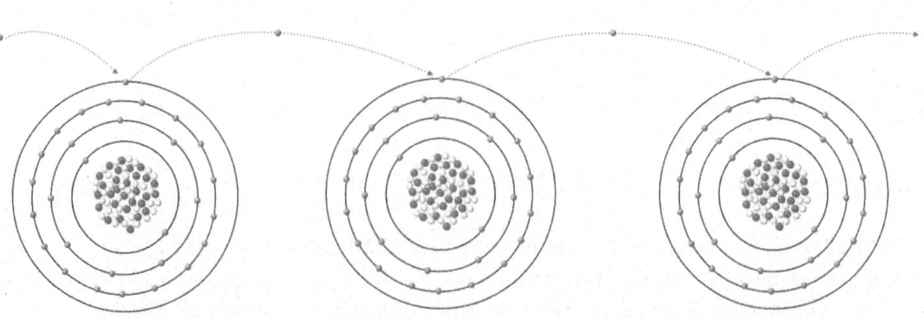

Figure 1-4 Electron flow occurs as a chain reaction.

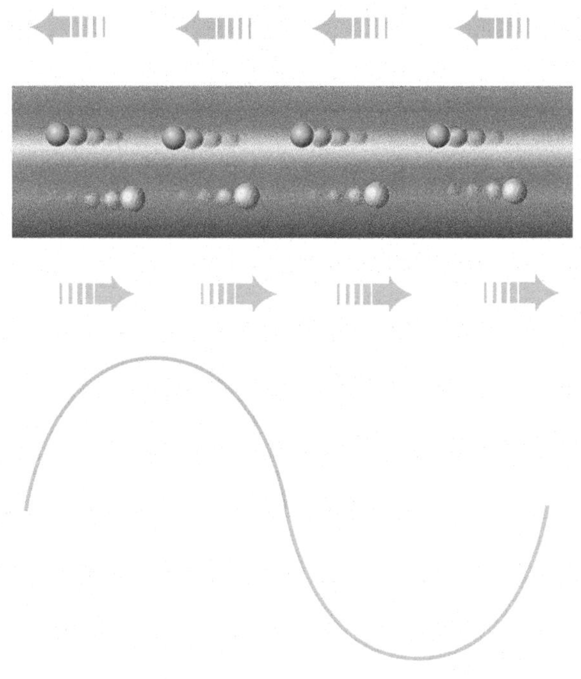

Figure 1-6 In alternating current the direction of electron flow reverses at regular intervals. These variations are used to change the magnetic field in a component, which helps move motors and switches.

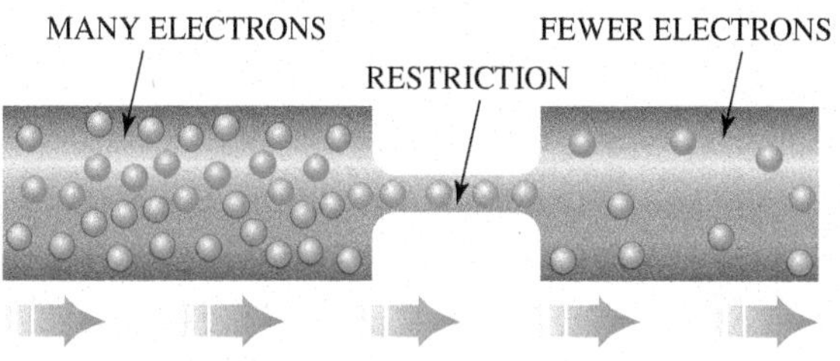

Figure 1-7 Current flow or electron exchange is slowed down as resistance is increased. Note that many electrons are available on the left side of the restriction or load. The restriction or load reduces the electron flow as evidenced by fewer electrons flowing on the right side.

1.3 ELECTRICAL TERMS USED IN HVACR

To understand the terms used in our profession, we must define them in simple terms. This section defines each term and give an example that relates to the definition. It is imperative for you to know these terms so that you can communicate problems and issues with the customer and your supervisor. Using the wrong term can create a misunderstanding that may lead to confusion and time lost on the job. Learn the following terms and use them as part of your professional language:

Electron: An electron is the part of an atom that changes position from atom to atom, which creates current flow. To be specific, the movement of electrons is current flow.

Volt: The term *volt* or *voltage* refers to the pressure that pushes electrons through a circuit. The greater the pressure, the higher the voltage. The term *volt* is also stated as potential difference, electromotive force (EMF), **V**, or **E**. Common voltages used in HVACR are 24, 120, 208, 240, and 480 V. A much higher voltage is possible on larger systems.

Control voltage: Control voltage is the electrical section of the HVACR system that manages the on and off cycles. A thermostat is usually placed in the control voltage circuit. The thermostat turns the system on and off to satisfy the customer's desired temperature setting. The control voltage in an air conditioning unit is normally lower than the system supply voltage. Common air conditioning control voltages use 24 or 120 VAC. Some air conditioning systems also use low-voltage DC signals.

Ampere: An ampere or amp is a measure of electron movement in a circuit. The greater the electron movement, the higher the amperage in a circuit. The term *ampere* is also stated as amp, current flow, and intensity and is abbreviated as **A** or ***I***. Amperage may be whole numbers or decimals such as 0.1, 0.01, 0.001, or 0.0001 A. The lower amperage measurements are known as milliamps or microamps. One milliamp is equal to one one-thousandth of an amp. One microamp is equal to one one-millionth of an amp. Very low amperage is used in gas heating and other control circuits.

Resistance: Resistance is the force that slows down the flow of electrons in a circuit, as illustrated in Figure 1-7. Resistance may be a matter of an undersized wire or a wire that does not conduct electrons easily. The greater the resistance, the lower the electron or amp flow in a circuit. The term *resistance* is also stated in ohms and is abbreviated with the letter ***R*** or the Greek letter omega, Ω. Glass, rubber, and plastic are good insulators and stop current flow.

Watt: The term *watt* is a reference to the power consumed by a circuit. The watt is the rate of doing work. A watt, or *wattage*, is calculated by multiplying the voltage by the amperage as follows:

$$\text{Watt} = V \times I \quad \text{or} \quad \text{Watt} = E \times I$$

The watt is used by utility companies to bill for electricity and is used by technicians to determine if the correct wattage is being drawn by a circuit, especially an electric heat circuit. The term *watt* or *wattage* is also known as power and is abbreviated **W** or **P**.

Load: A load is a component that has resistance to electron flow. Examples of loads in HVACR equipment are a motor or electric heating element. Figures 1-8A and 1-8B show common loads found in our profession.

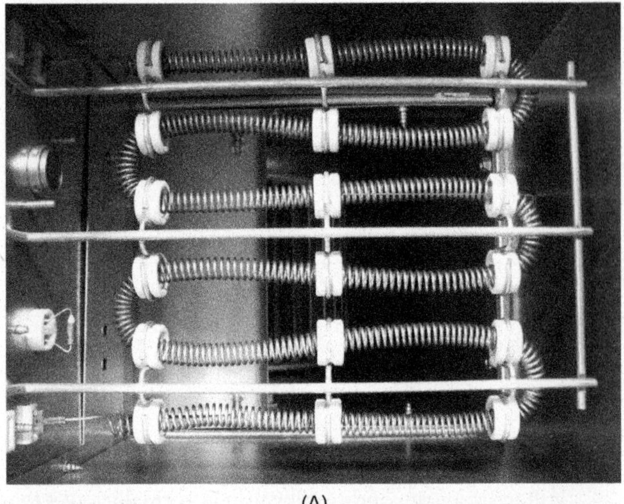

(A)

(B)

Figure 1-8 (A) An electric strip is a load in a heating system. The wire coil is made of resistive wire that heats up when electricity is applied. This is similar to the heating system in an electric stove or countertop toaster. (B) This motor is considered a load that is used in an air handler or furnace to move air through the duct system.

Complete circuit: A complete circuit is a path through which electricity flows. It is also known as a closed circuit. The simplest complete circuit includes a power supply, a load or resistance, and interconnecting wires. Figure 1-9 illustrates a complete circuit.

Open circuit: An open circuit is one that has a break or "opening" in the wire or load. The break stops the flow of electrons. Figure 1-10 shows an open circuit.

Short circuit: A short circuit occurs when the power goes to ground or when the load has very low resistance. Figure 1-11 is a diagram of a short circuit.

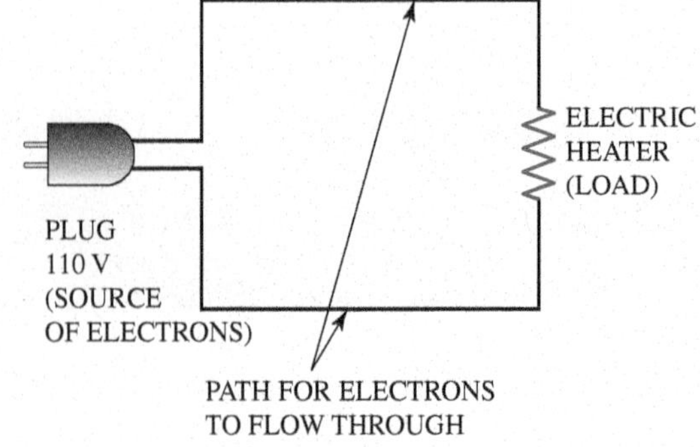

Figure 1-9 A complete circuit is a path through which electricity flows. In this diagram the load is the electric heater.

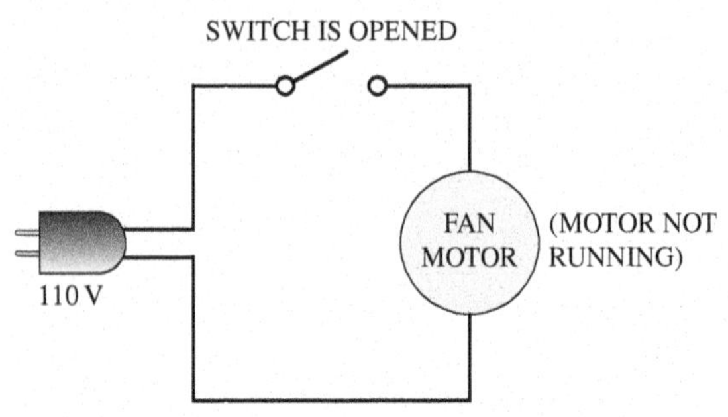

Figure 1-10 This is an open circuit with no current flow. The switch is open, which blocks the current flow.

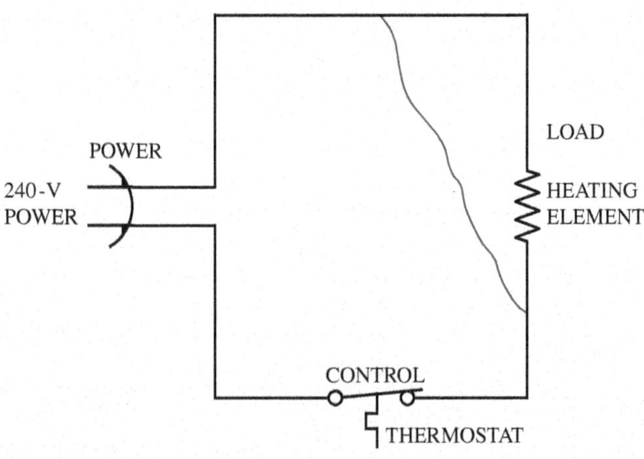

Figure 1-11 This is a short circuit because the power supply is bypassing the load. A red wire is causing the power supplied to be shorted out. The power supply or wire will be damaged in this shorted condition.

Series circuit: A series circuit is a circuit that has only one path for current flow. The same current flows through each component in the circuit, as shown in Figure 1-12. A series circuit can have one load or a number of loads. Figure 1-12 is a series circuit with four heaters as loads. The total amp draw is 5 amps. All points in this series circuit have a 5-A current flow.

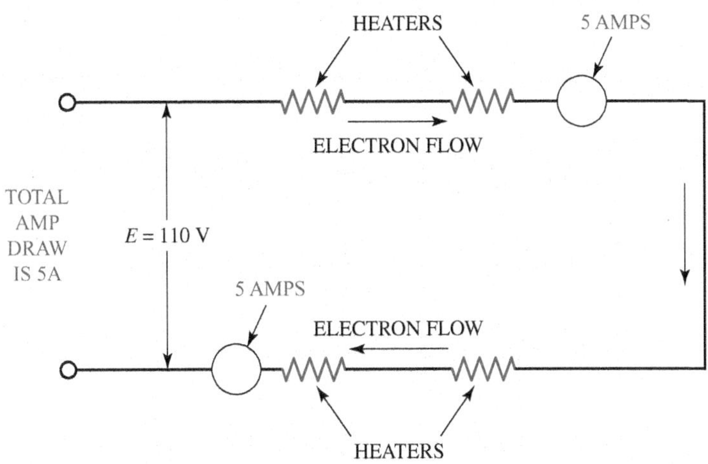

Figure 1-12 This is a series circuit with four heaters as loads. The total amp draw is 5 amps. All points in this series circuit have a 5-A current flow.

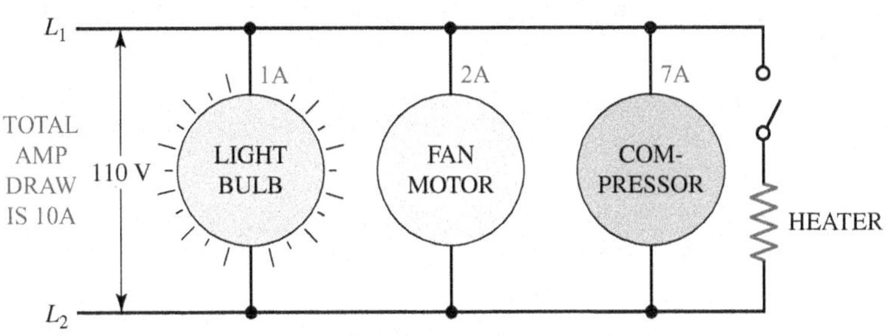

Figure 1-13 This is a parallel circuit with four loads. The 110-V power supply is applied across each load. The total amperage would be 10 amps if the amp draw in the light bulb is 1 A, the amp draw of the fan motor is 2 A, and the amp draw of the compressor is 7 A.

Parallel circuit: A parallel circuit is a circuit with parallel paths for current flow as shown in Figure 1-13. The voltage is the same across each of the loads. However, the current through each load varies. The figure shows a parallel circuit with four loads. The 110-V power supply is applied across each load. The total amperage provided by the power supply is equal to the combined amperage in each of the branch loads. For example, the total amperage would be 10 amps:

- The amp draw of the light bulb is 1 A.
- The amp draw of the fan motor is 2 A.
- The amp draw of the compressor is 7 A.

Combination circuit: A combination circuit is a circuit that combines series and parallel circuits, as illustrated in Figure 1-14. This is the most common type of circuit found in HVACR equipment. Part of the circuit is in series with the load and part is in parallel with the series circuit. The R_1 and R_2 resistors are in parallel with each other. This parallel circuit is in series with the R_3 resistor. It is essential to understand the various types of circuits found in HVACR equipment. Identifying the type of circuit assists the technician in understanding and troubleshooting equipment.

Voltmeter: A voltmeter is an electrical instrument used to measure voltage. The voltmeter is usually part of an electrical instrument called a multimeter. As the name implies, the multimeter in Figure 1-15 contains several electrical meters in one case. The meter should measure:

- DC volts
- AC volts
- Amperage
- Resistance.

Ammeter: An ammeter is an electrical instrument used to measure amperage. The ammeter can be part of a multimeter, as previously shown in Figure 1-15, or it can be a stand-alone meter. One popular amperage meter uses a clamp device that opens around a wire to measure amperage. Figure 1-16 shows a common clamp-on type of ammeter used in our industry. Current flow creates magnetism in the wire. The clamp-on meter converts

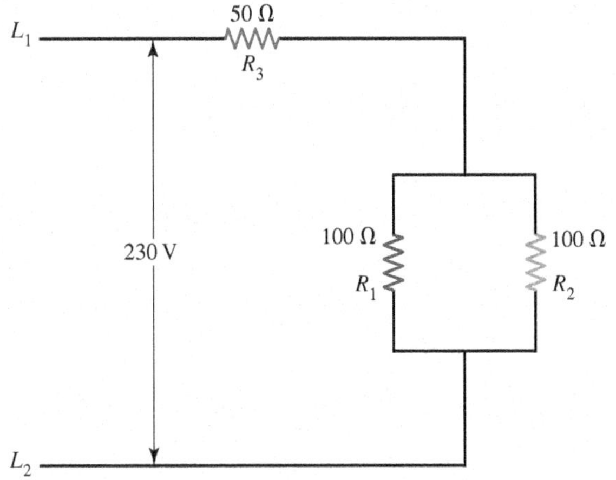

Figure 1-14 This is a series-parallel or combination circuit. The R_1 and R_2 resistors are in parallel with each other. This parallel circuit is in series with the R_3 resistor.

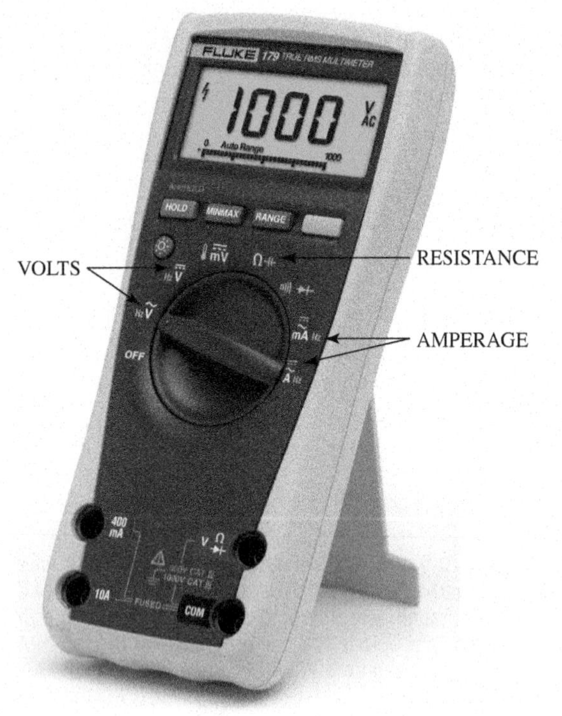

Figure 1-15 This multimeter includes options for measuring voltage, resistance, and amperage. A voltmeter is usually part of a multimeter.

Figure 1-16 An ammeter is clamped on to a wire so amperage can be measured.

that magnetic field into an amperage reading. The greater the current flow, the greater the magnetic field developed by the wire and, hence, the higher the amperage.

Ohmmeter: An ohmmeter is used to measure resistance in a load such as a motor or relay coil. The ohmmeter is usually part of a multimeter, as previously shown in Figure 1-15.

1.4 COMBINING VOLTS, AMPS, AND RESISTANCE

Figure 1-17 shows a simple circuit that combines voltage, amperage (current), and resistance into one series circuit. This is a DC voltage circuit that powers a resistor or simple heater. As you will soon learn, the electrical symbol for a resistor and heater are the same. Pressure from the DC voltage pushes electrons through the resistor.

Figure 1-18 illustrates the Ohm's law formula, which will be discussed in detail in Unit 2. Notice that the formula uses voltage, current, and resistance, which can be measured with a multimeter. Ohm's law can be used only with resistive circuits such as a resistor or heater.

Electrical symbols: Electrical symbols are representative images of electric components in a wiring diagram. Generally these symbols do not look like the electrical

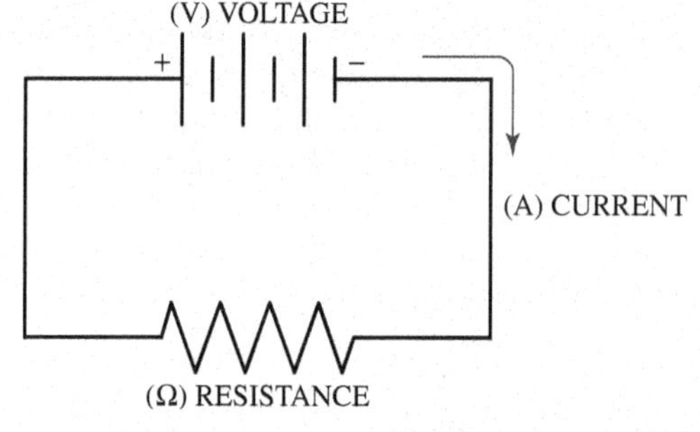

Figure 1-17 A simple circuit that combines voltage, amperage (current), and resistance into one series circuit. Pressure from the DC voltage pushes electrons through the resistor.

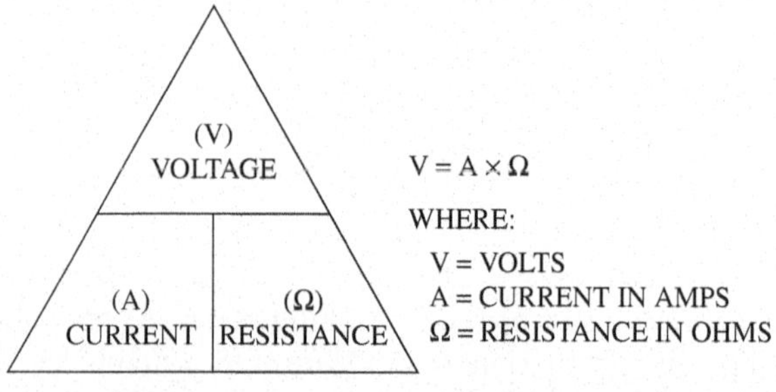

Figure 1-18 Formula for Ohm's law. Ohm's law can be used only with resistive circuits such as a resistor or heater.

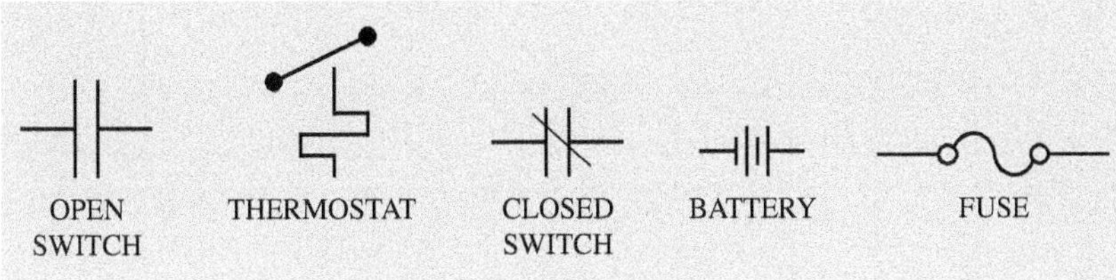

Figure 1-19 Here are a few symbols that may be used in understanding an electrical diagram. The symbols are joined to form an electrical diagram, which is like a road map for the HVACR system. See the back cover for a complete list of common electrical symbols used in our profession.

components they represent. Symbols are like the images that are seen on road maps. Symbols are covered in detail in Unit 17. Figure 1-19 illustrates some common symbols used in heating, air conditioning, ventilation, and refrigeration. See the back cover for a complete list of common electrical symbols used in the HVACR industry.

Electrical legend: An electrical legend is comparable to a legend found on a road map. The legend shows what each symbol represents. A typical legend for an air conditioning condenser is shown in the accompanying table. The components are abbreviated because there is limited space on an electrical diagram for spelling out all of the component names. For example, "CONT" is the abbreviation for the contactor. A contactor is an electromechanical switch used to provide power to the compressor and condenser fan motor. A list of abbreviations and acronyms is provided in Appendix A at the back of this book. There is no universal standard for legends or symbols, so you will find variations among manufacturers.

Wiring Diagram Legend for an Air Conditioning Condenser

CAP	Dual-Run Capacitor
CH	Crankcase Heater
CHS	Crankcase Heater Switch
COMP	Compressor
CONT	Contactor
CTD	Compressor Time Delay
DTS	Discharge Temperature Switch
HPS	High-Pressure Switch
IFR	Indoor Fan Relay
LLS	Liquid Line Solenoid Valve
LPS	Low-Pressure Switch
OFM	Outdoor Fan Motor
OPS	Oil Pressure Safety
SC	Start Capacitor
SR	Start Relay
ST	Start Thermistor

Wiring diagram: A wiring diagram or schematic is the road map of the electrical system. Figure 1-20 shows a typical wiring diagram with symbols and wiring connections for a residential condensing unit. Such diagrams help technicians find their way around a circuit. It is an aid for troubleshooting the system. Various types of wiring diagrams are discussed in detail in later units.

Circuit boards: As our knowledge of electronics has advanced, additional circuit boards are being used in air conditioning and heating equipment. See Figure 1-21. Circuit boards, sometimes called solid-state circuit boards, are used to reduce and miniaturize the number of components in HVACR equipment. The term *solid state* means that the components are not mechanical and do not have moving parts. Some circuit boards have microprocessor circuits that assist in troubleshooting the equipment. This topic will be discussed in Unit 19.

TECH TIP

Keep an open mind! Electrical **troubleshooting** requires experience and an open mind. Troubleshooting means that you are going to find the problem and repair it. At first you will need help. Experience comes with time and working on the job doing various HVACR activities. Even working around properly functioning equipment and cleaning equipment provide valuable experience. It is important to know how a system is supposed to operate when everything is functioning correctly.

Keeping an open mind is most important until the new technician gains experience. Being open minded means checking all options and exploring the unknown. Many times in our profession we will not totally understand every piece of equipment that we have to troubleshoot. The technician must be able to determine what is working in order to troubleshoot what is not working. This is done by carefully checking and eliminating sections that are not creating the problem.

SERVICE TICKET

Using electrical terminology correctly is important when you communicate with and get assistance from your supervisor. Let's use the following example to see why.

As an air conditioning apprentice, you are sent to survey a problem with a small commercial refrigeration job. The customer says the cooler is getting warm and the outdoor section is not operating. You recently purchased and learned to operate some of the basic electrical instruments such as the voltmeter and clamp-on ammeter. You are not expected to do the final troubleshooting, but to report back to the office with an initial finding since the experienced technician will soon

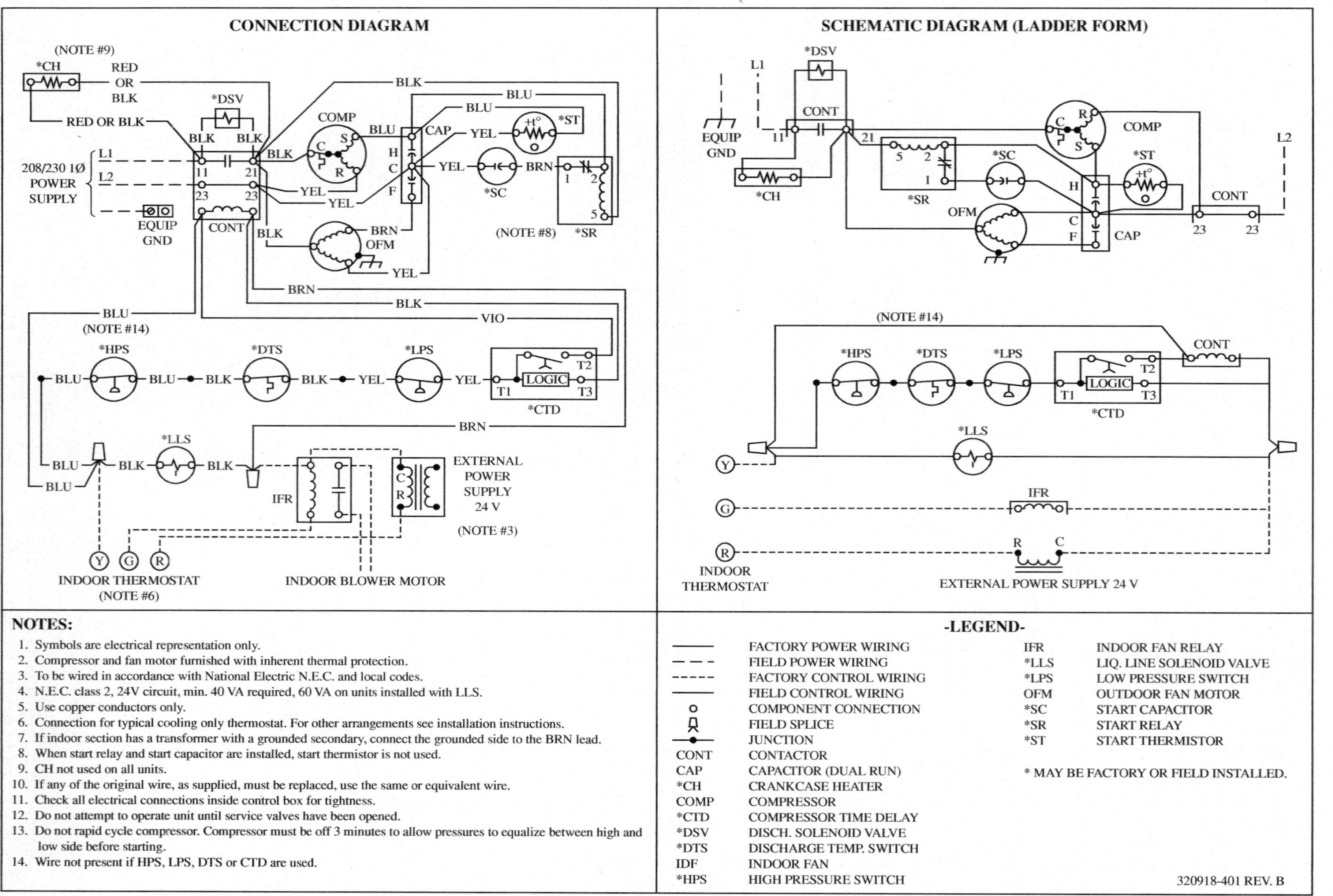

Figure 1-20 Wiring diagram used for troubleshooting. Optional components are marked with an asterisk (*). The optional components on the lower half of this diagram are the HPS (high-pressure switch), DTS (discharge thermostat switch), LPS (low-pressure switch), LLS (liquid line solenoid valve), and CTD (compressor time delay).

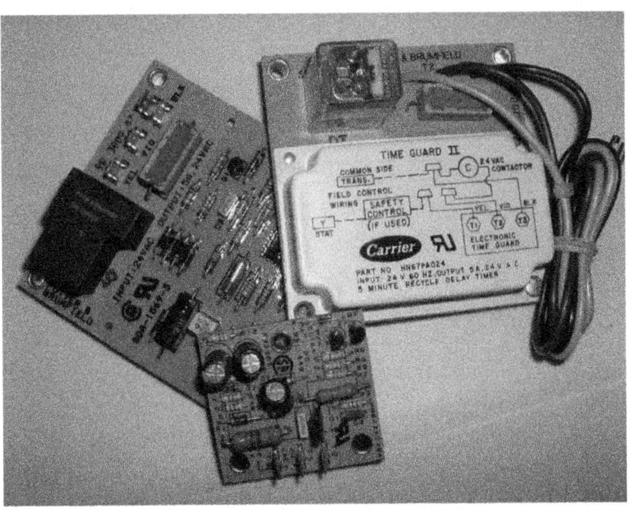

Figure 1-21 Several types of circuit boards are found in air conditioning units.

follow. This is a chance to show the company the "new tech" knows a little something.

You notice that the condensing unit is not working. The evaporator blower is operating. You decide to do a few checks on the condensing unit. The voltage measurement at the **disconnect** and the condenser is 120 volts. The disconnect is the power switch. You decide to call the office with this information and seek advice on how to proceed. You talk with a supervisor and state that you measured 120 amps on the condensing unit. The supervisor repeats in a loud voice, "120 amps!" The supervisor tells you that the compressor might be locked up or there could be a problem with the starting components on the compressor and gives you a few tips on what to check. The compressor is not the problem. Unfortunately, you reported the 120-volt measurement as 120 amps, which changed the way the system was approached for troubleshooting.

Knowing the correct terminology is important to save time troubleshooting and prevent embarrassment.

SUMMARY

This unit was an introductory section that introduces some basic terms used when working with HVACR equipment. It is essential to understand these basic terms in order to communicate with your supervisor, the manufacturer's tech support people, and customers. These terms will be discussed and applied in more detail in the following units. Some of these terms are easily misunderstood or confusing. You saw in the Service Ticket scenario that using the wrong term can change the outcome when communicating with a supervisor or other technician. Yes, there are many new terms to learn and understand. Refer back to these terms or the terms in the glossary at the end of the book to drill down to the fundamental terms needed to become a good technician.

REVIEW QUESTIONS

1. What is electricity?
2. What is current flow or amperage?
3. What creates a magnetic field in a conductor: voltage or current flow?
4. What is voltage?
5. What is the difference between a conductor and an insulator?
6. What are the two common types of electricity?
7. What is control voltage?
8. What is resistance?
9. What is a watt?
10. How are watts calculated?
11. What is a meant by a load?
12. What are the differences among a complete circuit, an open circuit, and a short circuit?
13. What is a series circuit?
14. What is a parallel circuit?
15. What is a combination circuit?
16. What is the most common type of circuit found in HVACR equipment?
17. What is an electrical legend?
18. Looking at a diagram legend, what does LPS mean? What does HPS mean?
19. How are optional components marked in Figure 1-20?
20. Why are circuit boards used in HVACR equipment?

UNIT 2

Ohm's Law and Circuit Operation

WHAT YOU NEED TO KNOW

After studying this unit, you will be able to:

1. Calculate volts, amperage draw, and resistance using Ohm's law.
2. Identify series circuits.
3. Apply the series circuit laws to electrical circuits.
4. Identify parallel circuits.
5. Apply the parallel circuit laws to electrical circuits.
6. Identify combination circuits.
7. Calculate volts, amperage draw, and resistance using the watts formula.

Learning how to use Ohm's law will help you understand how current and voltage flow in a circuit. In this unit we discuss how resistance affects circuit operation. Resistance changes the current flow and has an influence on the voltage drops in a circuit. We will see that voltage, current, and resistance have an influence on each other.

First, we review the types of circuits. They are:

- **Series circuits**
- **Parallel circuits**
- **Combination circuits.**

These circuits were discussed in Unit 1. The discussion here will be more detailed. It is important to understand these circuits to solve **Ohm's law** problems, but more importantly to learn how current and voltage flow vary in a circuit. This concept is fundamental to troubleshooting. The technician has to understand the relationship of voltage, current, and resistance to be able to troubleshoot. The relationship is known as Ohm's law. If you do not learn this concept, you will not be able to troubleshoot HVACR equipment beyond the obvious burned-out component, which is just a small part of the overall troubleshooting process. This unit shows you how to solve Ohm's law and, more importantly, how to use it to understand circuits and troubleshooting.

This unit also explores the use of the **watts formula.** Some call it Watt's law, but that is incorrect. It is a variation of the formula for Ohm's law. The watts formula compares the voltage and current flow in a circuit to determine how it affects the component wattage. The watts formula is also referred to as the **power formula** or *power calculation.*

Finally, Ohm's law and the watts formula are only accurate with resistive loads. **Resistive loads** are electric heaters, crankcase heaters, resistors, or incandescent bulbs. The other types of loads, inductive or capacitive loads, use a more complex formula to determine current and voltage conditions. The resistance in an inductive or capacitive circuit is called *impedance.* The **impedance formula** includes resistance plus the effects of magnetism or capacitance. Inductive loads are devices that use magnetism to operate a component. For example, a motor, transformer, or relay coil is an **inductive load.**

An inductive load in the form of a motor and transformer is shown in Figures 2-1A and 2-1B, respectively.

A **capacitive load** has a capacitor in the circuit. A dual-run capacitor is shown in Figure 2-2. The resistance the capacitor creates is called **capacitive reactance.** Many loads have a combination of resistance, induction, and capacitance. These loads will be discussed throughout the book, but learning the impedance formula is beyond the scope of this book and really not necessary to do troubleshooting. Learning and understanding Ohm's law and the watts formula, however, will aid in the troubleshooting process.

Figure 2-1A A motor works on the principle of changing magnetic fields. This is an inductive load.

Figure 2-1B The motor and transformer are inductive load components. Inductive components work on the principle of magnetism to create motor motion or induce voltage; such is the case with this control transformer.

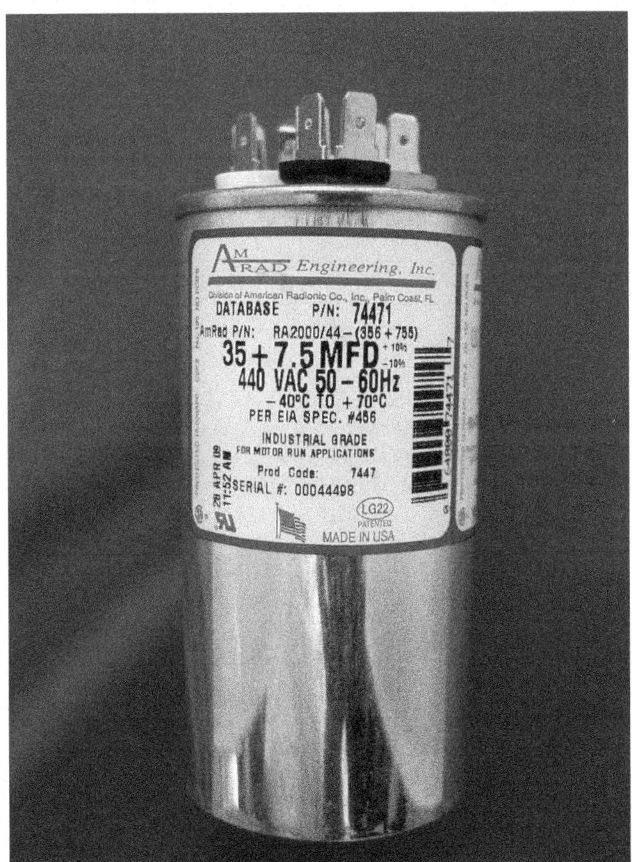

Figure 2-2 This dual-run capacitor can be used to operate two different motors. There are two capacitors in one shell.

First we introduce Ohm's law. Then we review the various circuit types, starting with the series circuit. Following this sequence, Ohm's law is examined using practical examples and applied troubleshooting problems.

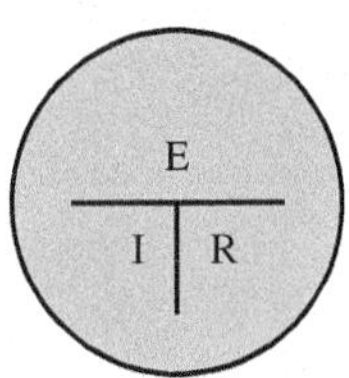

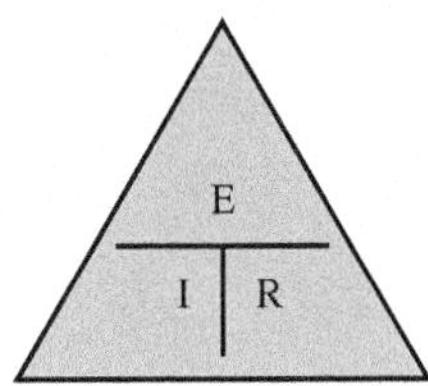

Figure 2-3 Two symbols are used to represent Ohm's law, $E = I \times R$.

2.1 TECHNICAL REVIEW

The terms *voltage*, *ampere*, *resistance*, and *watt* will be used many times in this unit. Here is a review of those terms:

Volt: The term *volt* or *voltage* refers to the pressure that pushes electrons through a circuit.

Ampere: An ampere or amp is a measure of electron movement in a circuit.

Resistance: Resistance is the force that slows down the flow of electrons in a circuit.

Watt: The term *watt* or *wattage* is a reference to the power consumed by a circuit. The watt is the rate of doing work.

Complete circuit: The voltage that leaves the power supply will need to return to the other side of the power supply for electrons to flow and the circuit to work.

The Ohm's law formula $E = I \times R$ is expressed in a circle or triangle as shown in Figure 2-3. This formula is discussed in the next section.

2.2 OHM'S LAW

Ohm's law is

$$E = I \times R$$

where

E = voltage
I = current
R = resistance.

The formula can also be changed to find current (I) and resistance (R). Here are the formulas for those options:

$$I = \frac{E}{R}$$

$$R = \frac{E}{I}$$

EXAMPLE 2.1 CALCULATING VOLTS

What is the voltage supplied to a compressor crankcase heater that draws 4 A and measures 60 Ω of resistance?

SOLUTION

$$E = I \times R$$
$$E = 4A \times 60\ \Omega$$
$$E = 240\ V$$

EXAMPLE 2.2 CALCULATING AMPERAGE DRAW

Calculate the amp draw of the simple series circuit shown in Figure 2-4.

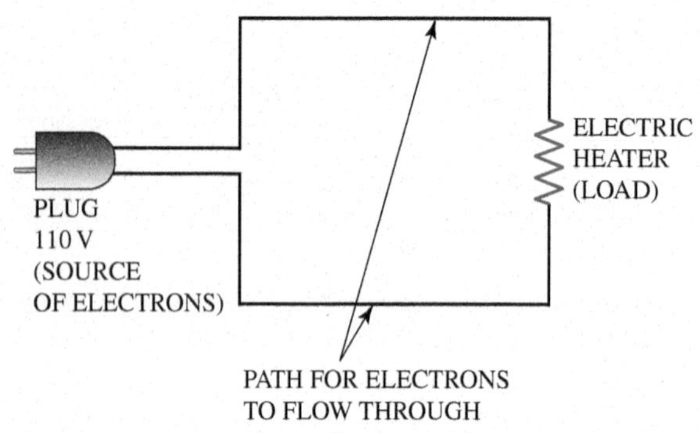

Figure 2-4 This is the simplest type of series circuit. It has a 110-V power supply, a path for electron flow, and an electric heater as the load.

SOLUTION

As measured by an ohmmeter, the resistance of the heater in Figure 2-4 is 11 Ω. The voltage measured is 110 V. We will use Ohm's law to calculate the amp draw of this simple series circuit.

$$I = \frac{E}{R}$$
$$I = \frac{110\ V}{11\ \Omega}$$
$$I = 10\ A$$

Every point in the circuit has 10 amps flowing through it when power is supplied, even the middle of the heat strip.

EXAMPLE 2.3 CALCULATING RESISTANCE

You have two identical 120-V electric heaters. One heater is functional and the second heater has an open heating element. You need to select the correct replacement heating element using the resistance specifications. You remember Ohm's law plus you have a voltmeter and ammeter (or *amp meter*) to determine the resistance.

SOLUTION

Turning on the good electric heater, the voltage measures 115 V with an amperage draw of 10 A. Using the Ohm's law formula, the resistance is calculated as follows:

$$R = \frac{E}{I}$$
$$R = \frac{115\ V}{10\ A}$$
$$R = 11.5\ \Omega$$

You select the replacement heating element that is closest to 11.5 Ω.

2.3 HOW *E*, *I*, AND *R* INFLUENCE EACH OTHER

Figure 2-5 represents a balance between current, resistance, and voltage potential. This is written as

$$E = I \times R$$

which should be recognizable by now as the familiar Ohm's law.

Figure 2-6 illustrates an imbalance. While keeping the voltage the same, the resistance is increased. Increased resistance reduces current flow. The increased resistance causes the current to drop, as shown by the lowered point on the left side of the seesaw.

Finally, Figure 2-7 illustrates an imbalance. While keeping the voltage level the same, the resistance is decreased. Decreased resistance increases the current flow. The reduced resistance causes the current flow to increase, as shown by the lowered point on the right side of the seesaw.

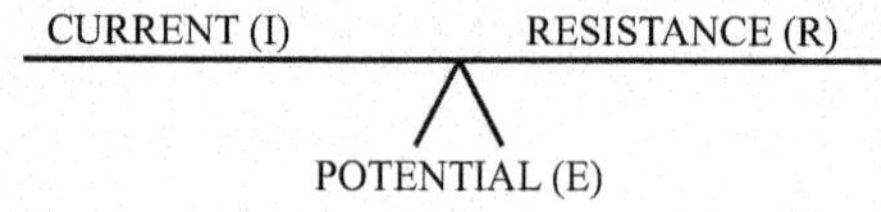

Figure 2-5 This seesaw balance is a starting point that is used in the later figures. Multiplying current and resistance gives you the supplied voltage. In this figure, the seesaw is balanced, meaning the circuit is in equilibrium.

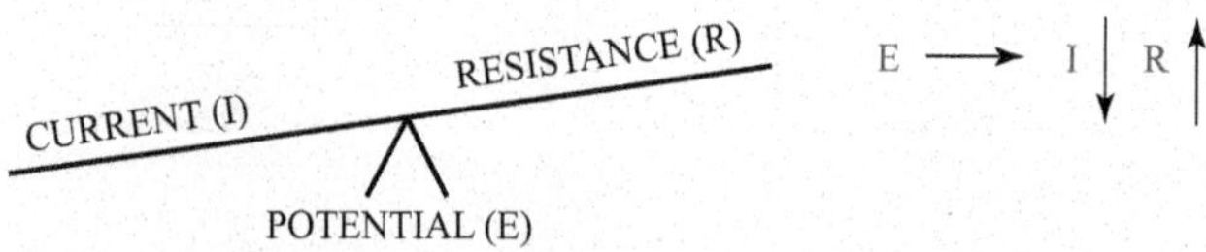

Figure 2-6 There is an imbalance in this equation. While keeping the voltage the same, the resistance is increased. This rise in resistance reduces the electron flow, which causes the current to drop (note the lowered point on the left side of the seesaw).

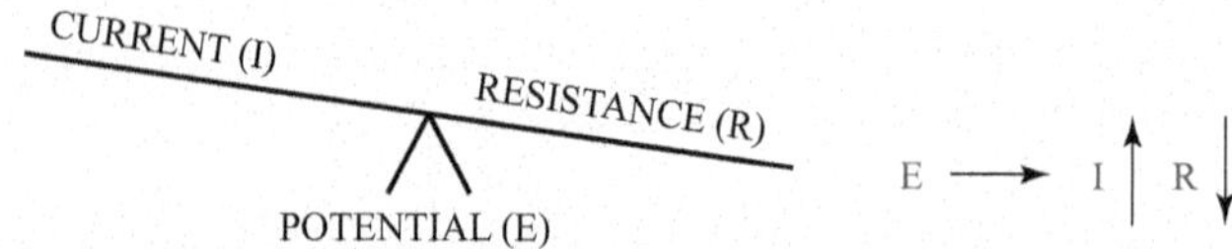

Figure 2-7 There is an imbalance in this equation. By keeping the voltage the same, but decreasing the resistance, the current will increase (note the lowered point on the right side of the seesaw).

The use of a balance can be used to track the current and voltage variations as well. You may have noticed that the resistance has the greatest impact on the Ohm's law formula. The supply voltage does not usually vary much in a circuit. The current flow depends on the resistance in the circuit. Even though Ohm's law works well with totally resistive loads, having inductive loads like a motor or capacitive loads from capacitors, any increases and decreases in current flow have similar effects in a circuit. *Reminder:* Ohm's law formula cannot be used to directly calculate inductive or capacitive loads or the voltage and current flow in these circuits.

EXAMPLE 2.4 DETERMINING WIRE SIZING

Let's apply the information we have learned to a problem with a burned wire used to feed an electric heater. We could use this information to size wire that can handle 240 V and a heater resistance of 12 Ω.

SOLUTION

$$I = \frac{E}{R} = \frac{240\text{ V}}{12\ \Omega} = 20\text{ A}$$

Using Ohm's law we have determined that the replacement wire for the electric heater needs to handle at least 10 amps or it is considered undersized.

Undersized wire will overheat and burn out. Another indication of undersized wire is a warm wire. The wire should be the surrounding, or *ambient*, temperature. This is only one small way learning Ohm's law can help you troubleshoot.

2.4 SERIES CIRCUITS

The series circuit shown in Figure 2-4 has one component in line with the power supply. A more complex series circuit can be seen in Figure 2-8.

Certain rules govern the operation of a series circuit. If you learn the following rules you will be able to build a good troubleshooting foundation:

- The current is the same throughout a series circuit.
- There is a voltage drop across each of the loads in a circuit. For example, in Figure 2-9 if each heater has the same resistance, the voltage drop across each heater will be the same.
- Current flow will stop if there is a break in a series circuit. Applied voltage will be present on either side of an open wire or open component even though the current is not flowing.
- The lower the resistance, the higher the current flow.
- The higher the resistance, the lower the current flow.
- The resistance is additive in a series circuit. To find the total resistance, simply add up the value of all resistances.
- Remember that when voltage leaves a power supply it must return in order to have a complete circuit.

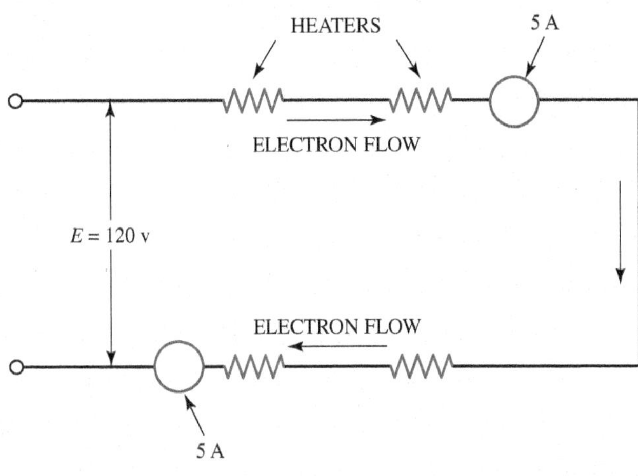

Figure 2-9 This series circuit has four heat strips as loads. If the heaters have the same resistance, the voltage drop across each heater will be the same. In this case, each heater will have 30 V applied across them (120 V/4 = 30 V). The amperage throughout the series circuit is the same; in this case, 5 A.

Some examples of components you will find in series with loads are **thermostats, high-pressure switches, low-pressure switches**, and **safety devices.**

2.5 PARALLEL CIRCUITS

Figure 2-10 shows a parallel circuit. It is not a practical circuit, but it shows four parallel loads across the L_1 and L_2 power source. One load can open and the remaining loads will continue to function. *Parallel* means that each circuit

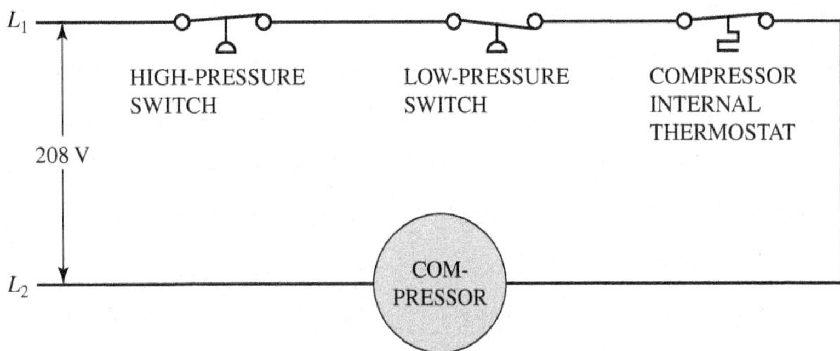

Figure 2-8 This more complex series circuit has a compressor motor load and three safety switches. The safety switches do not drop voltage and are not a load.

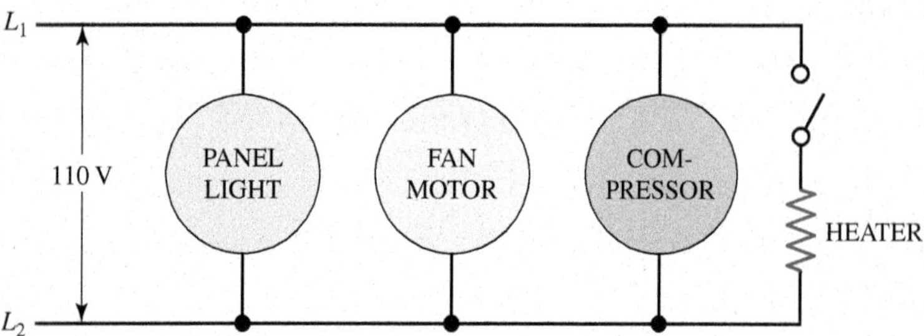

Figure 2-10 This parallel circuit has four loads across the L_1 and L_2 power source. The loads are the panel light, fan motor, compressor, and crankcase heater.

has two voltage legs running parallel with the loads between them. A parallel circuit is really a ladder diagram with the rails of the ladder being the voltage source, and the rungs the loads with control switches.

Here are the characteristics of a parallel circuit:

- The voltage is equal across each line load.
- The total current is divided among all loads.
- The total current is equal to the sum of the current in each line load.
- An open load will not stop the voltage to the other parallel circuits.
- Total resistance in a parallel circuit is less than the resistance of the branch with the least resistance. For example, the total resistance in a parallel circuit with resistances of 5, 10, and 15 Ω will be less than 5 Ω. If you compare this to parallel water piping, the resistance in two or more parallel pipes will be less than the resistance in one pipe.

EXAMPLE 2.5 USING OHM'S LAW TO DETERMINE AMPERAGE DRAW

There is more than one way to determine the amp draw of electric heat strips. One way is to measure the amperage with a clamp-on ammeter. But what if your ammeter is in the shop for calibration? What can you do then?

You can determine the amperage draw of a set of parallel heat strips by using Ohm's law. With all of the heat strips in operation, measure the applied voltage. For this example, you measured 240 V. Quickly turn off and remove the voltage to the heat strips and measure the resistance. The resistance will vary somewhat as the heat strips cool; therefore, to obtain the most accurate reading, measure the resistance as soon as possible after shutdown and when the voltage is disconnected. The total resistance measured was 6 Ω. This is the resistance of the heat strips in parallel, not an individual heat strip. Use Ohm's law to determine the approximate amp draw.

SOLUTION

We'll use the Ohm's law formula I = E/R to determine the approximate amperage draw.

We know from our measurements that E = 240 V and R = 6 Ω, so

$$I = \frac{240\text{ V}}{6\ \Omega}$$

$$I = 40\text{ A}$$

The heat strips should be drawing approximately 40 A. The wire should be sized to handle at least 40 amps or more. Wire sizing will be discussed in the unit on the National Electrical Code, Unit 6. A single electric heat strip draws about 15 to 20 A at 240 V. As you can see, at lower voltages, let's say, 120 V, the amperage draw will be less.

You will find when we study the watts formula later in this unit that fewer volts and amps means less heat output. As you can see from the preceding example, you can roughly determine the amperage draw of a resistive load without a clamp-on ammeter.

2.6 COMBINATION CIRCUITS

Combination circuits, also known as series-parallel circuits, are the most common circuits found in our industry. As the name indicates, these circuits are a combination of series and parallel circuits. Figure 2-11 is a simple series-parallel circuit.

Here are the characteristics of a combination circuit:

- The current flow in each circuit depends on its location in the circuit. For example, in Figure 2-11 the current flow through R_3 is the same as the combined current flow through R_1 and R_2. Because R_2 and R_3 have the same resistance, the current flow through R_1 will be evenly split between the parallel circuits.
- The voltage will be split between the loads.
- The voltage across parallel components will be the same.

Refer again to Figure 2-12. When thermostat **THER** is closed, the main fuse blows. By using Ohm's law we understand that a load is required or excessive amperage will be drawn.

EXAMPLE 2.6 DETERMINING THE CAUSE OF A BLOWN MAIN FUSE

Troubleshoot the following problem in the combination circuit shown in Figure 2-12.

Let's assume that this circuit is designed to draw no more than 25 amps. The 25 amps will be unevenly divided between the four parallel lines. There is a 25-A fuse to protect against short-circuit conditions. Motors and other magnetic loads draw very high amperage the second they start, so the fuse is larger than the full-load amperage so that it does not blow on motor start-up.

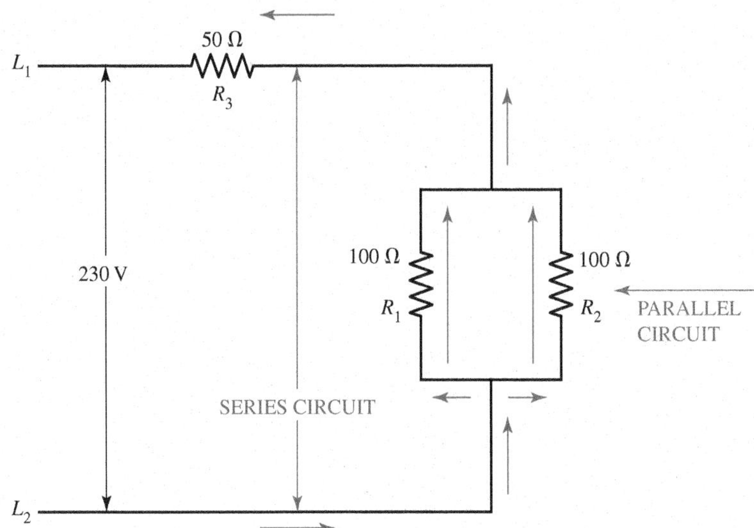

Figure 2-11 This is a simple combination circuit, also known as a series-parallel circuit. Current flows in series with a parallel circuit.

SEQUENCE OF OPERATION 2.1: SERIES AND PARALLEL CIRCUITS

A basic, but functional series-parallel circuit is shown in Figure 2-12. This is a condensing unit. The indoor blower section is not shown.

Here is how the circuit operates:

1. The thermostat, **THER**, closes and completes a circuit to Ⓒ, the contactor coil. At this time we are assuming no problems; therefore, the **LPS** and **HPS** will be closed. This is a series circuit since the same current flows through all switches and contactor coil Ⓒ.
2. The energized contactor coil Ⓒ closes the double-pole contactor at the bottom of the diagram and starts the operation of the **COMP** and **CFM1**. The condensing unit is in full operation. **COMP** and **CFM1** are in parallel with each other.
3. **CFM2 THER** will be closed if the outdoor temperature rises above a temperature setpoint. For this example, we will use 80°F as the temperature setpoint.
4. The **CFM2 THER** contacts will close and start the operation of the second condenser fan motor, **CFM2**. This is a series circuit.

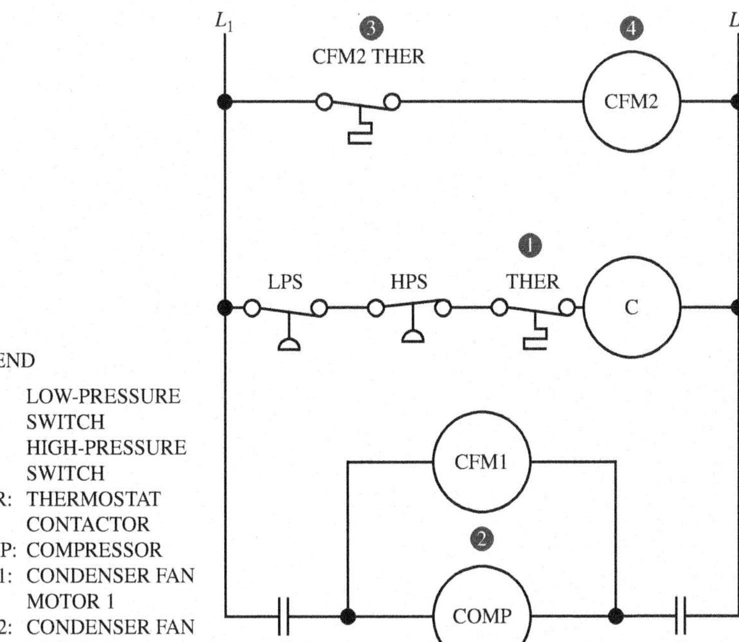

Figure 2-12 This is a simple series-parallel circuit.

Here are some reasonable amperage numbers for the loads in Figure 2-12:

- Condenser fan motor **CFM2** draws 3 amps.
- Contactor coil **C** draws 0.5 amps.
- Condenser fan motor **CFM1** draws 4 amps.
- Compressor motor **COMP** draws 13 amps.

The total rated load amperage when all components are operating is 20.5 A. The load resistance must drop dramatically to blow a 25-A fuse. How does that look in the Ohm's law formula?

$$I = \frac{E}{R}$$

In this formula, when resistance R decreases, current I in the circuit increases. The question is this: Which one of the components in Figure 2-12 has reduced resistance such that the amperage draw increases high enough to blow the fuse?

SOLUTION

There are a couple ways to solve this problem.

First method: One way is to remove the power from the circuit. Next, remove the wires feeding power to the load. Check the whole circuit with an ohmmeter. A shorted component with a very low resistance (less than 5 Ω) may cause the fuse to open. Next, check the resistance on each component until you locate the component that has very low resistance.

Second method: Another way to solve the problem is as follows: Disconnect the power to the circuit. Remove power to one side of each of the loads. Reconnect the power. Does the fuse blow with the loads disconnected? If so, it is not the load; the problem is a short to ground from the power wires feeding the loads. If the disconnected loads do not blow the fuse, disconnect the power and reconnect one load. Apply power. Does the fuse blow? If not, use the same steps until you come to the load that, when connected to power, blows the fuse. That will be the circuit that needs to be checked for a short circuit.

Note to reader: We are using Ohm's law in an inductive circuit. This is not the correct way to use the Ohm's law formula, but for general troubleshooting it assists in figuring out the problem. Even in inductive circuits the voltage, current, and resistance vary in the same way they do in a resistive circuit, but you cannot use Ohm's law for the exact calculation of *E*, *I*, and *R*.

2.7 USING THE OHM'S LAW WHEEL

The Ohm's law formula $E = I \times R$ and variations on this formula can be remembered by using the following easy steps:

1. Draw the Ohm's law formula within a circle as shown in Figure 2-13.
2. Cover up what you are looking for. In Figure 2-14, cover up *E* since we are looking for voltage. The remaining formula, which is $I \times R$, is exposed.
3. Next, find current or *I*. Covering *I* exposes the formula $\frac{E}{R}$. See Figure 2-15.
4. Finally, find resistance *R*. Covering *R* exposes the formula $\frac{E}{I}$. See Figure 2-16.

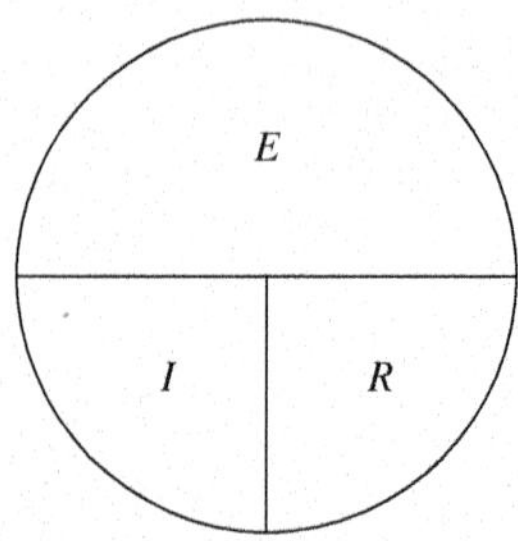
E
I
R

Figure 2-13 Draw an Ohm's law pie chart as shown here. This will assist in using the Ohm's law formula. (Note that it can be drawn without the circle with the same results.)

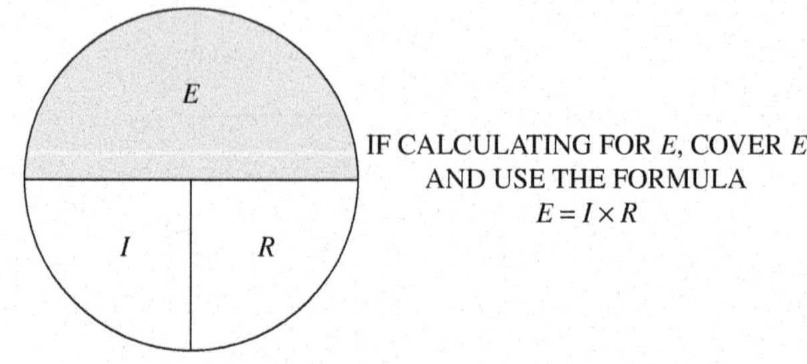
E
I
R
IF CALCULATING FOR E, COVER E
AND USE THE FORMULA
E = I × R

Figure 2-14 After drawing the Ohm's law pie chart, cover up the section you are trying to solve for. In this case we are trying to find the voltage; therefore, the *E* is shaded (or place your finger over the shaded area). The solution to finding voltage is exposed: $I \times R$.

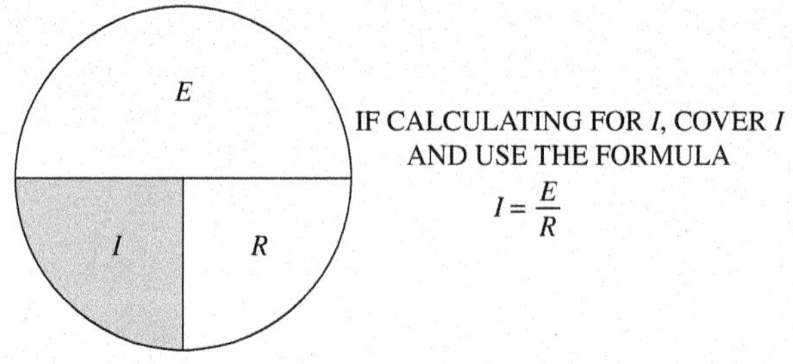
E
I
R
IF CALCULATING FOR I, COVER I
AND USE THE FORMULA
I = E/R

Figure 2-15 In this case we are looking for the current; therefore, the *I* is shaded (or place your finger over the shaded area). The solution to finding current is exposed: $\frac{E}{R}$.

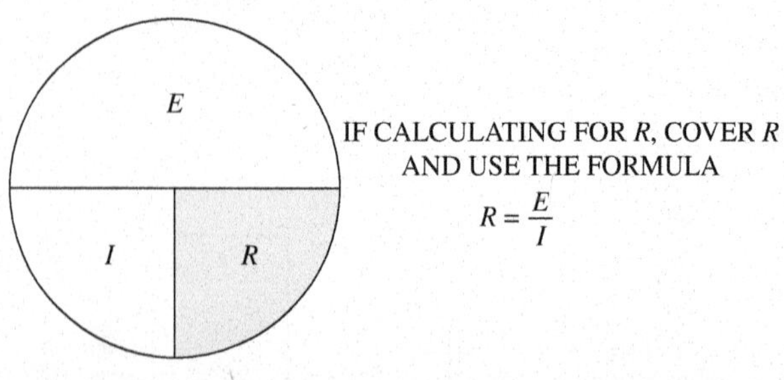
E
I
R
IF CALCULATING FOR R, COVER R
AND USE THE FORMULA
R = E/I

Figure 2-16 In this case we are looking for the resistance; therefore, the *R* is shaded (or place your finger over the shaded area). The solution to finding resistance is exposed: $\frac{E}{I}$.

2.8 USING THE WATTS FORMULA

The word *watt* is used to express a unit of electrical power. The capital letter W is commonly used as the abbreviation for the term watt(s). It is also sometimes called power, in which case its abbreviation is a capital letter P.

Watts are the product of volts and amps, which is stated as $W = E \times I$ or $P = E \times I$, where E is the voltage and I is the amps, or current. Volts or amps can be found when the watts and the voltage or current variable are known. Here are the formulas for finding voltage or current:

$$E = \frac{W}{I}$$

$$I = \frac{W}{E}$$

The watts formula $W = E \times I$ and versions of this formula can be remembered by using the following easy steps: Draw the watts formula as shown in Figure 2-17.

In Figure 2-18, cover up W since we are looking for voltage. The remaining formula, $E \times I$, is exposed.

Next, find voltage or E. Covering E exposes the formula $\frac{W}{I}$. See Figure 2-19.

Finally, find amperage draw or I. Covering I exposes the formula $\frac{W}{E}$. See Figure 2-20.

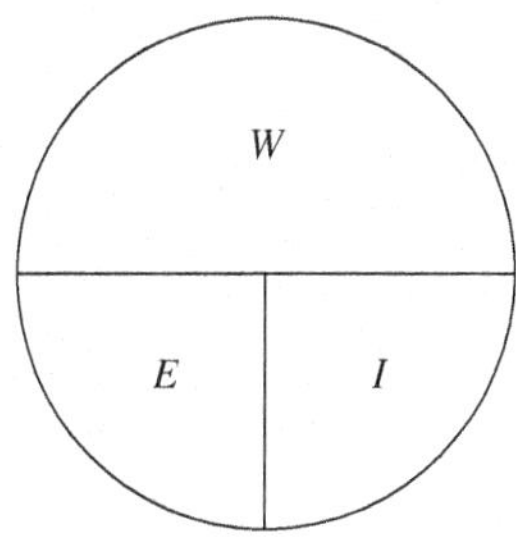

Figure 2-17 Draw a watts formula pie chart as shown here. This will assist in using the watts formula. (Note that it can be drawn without the circle with the same results.)

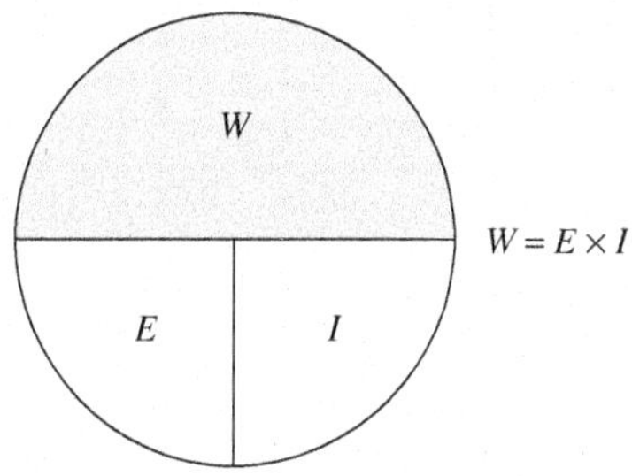

Figure 2-18 After drawing the watts formula pie chart, cover up the section you are trying to solve. In this case we are looking for the watts; therefore, the W is shaded (or place your finger over the shaded area). The solution to finding wattage, which is $E \times I$, is exposed.

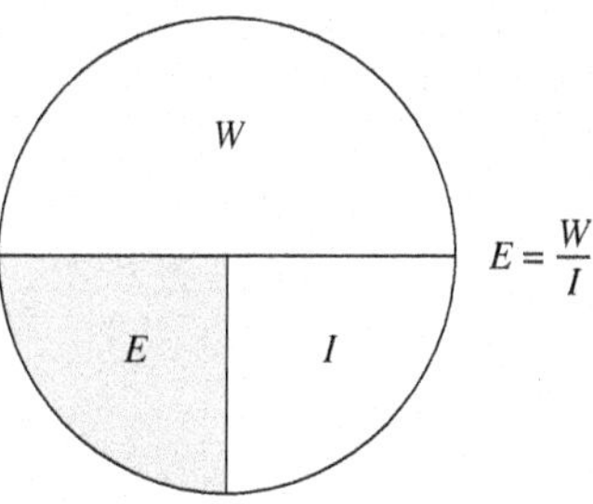

Figure 2-19 In this case we are looking for the volts; therefore, the E is shaded (or place your finger over the shaded area). The solution to finding voltage is exposed: $\frac{W}{I}$.

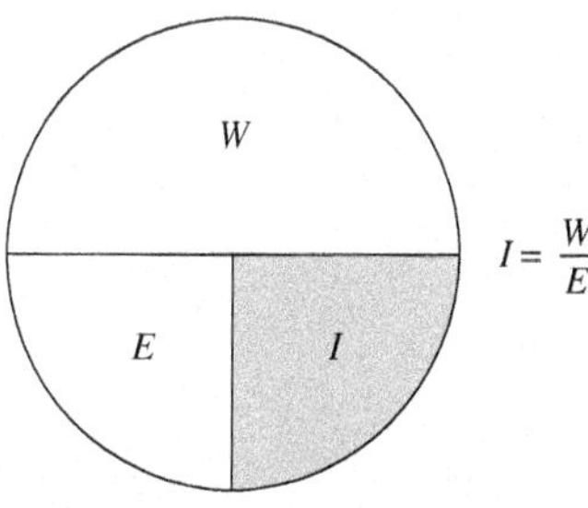

Figure 2-20 In this case we are looking for the amperage draw; therefore, the I is shaded (or place your finger over the shaded area). The solution to finding voltage is exposed: $\frac{W}{E}$.

As you can see, the watts formula wheel is similar to the Ohm's law wheel and not difficult to use.

EXAMPLE 2.7 USING THE WATTS FORMULA TO DETERMINE TOTAL WATTAGE

One single-phase electric strip heater is rated at 5,000 watts or 5 kW. The abbreviation *kW* means kilowatts. A kilowatt is 1,000 watts because *kilo* means 1,000. Some heat strips have a lower kilowatt rating such as 4 or 4.5 kW, but 5 kW is fairly standard. The higher the number of kilowatts, the greater the heat output at the rated voltage input. If the input voltage is lower, the heat output will be less.

The problem we need to troubleshoot is inadequate heat from a heater that uses three electric strip heaters, rated at 5 kW each. The 5-kW heat strips are in parallel with each other; therefore, the furnace has three heat strips in parallel with each other. Is the problem with the heat strips or is the heating system too small to heat the space? How do we determine if the heating system is putting out 5 kW of heat?

SOLUTION

The heating system is turned on and the thermostat turned up several degrees above the room temperature. The heating system is allowed to operate for several minutes to ensure that all of the heat strips are energized. Here are the voltage and total amperage readings after the system has been operating for a few minutes:

- 240 V
- 42 A.

Using the watts formula we can see that the total wattage is less than the 15-kW rating (3 strips × 5 kW). How many watts do we have with the voltage and amperage readings?

$$W = E \times I$$
$$= 240\text{ V} \times 42\text{ A}$$
$$W = 10{,}080\text{ W}$$

Measured power is 10.08 kW.

We can see from this calculation that we are approximately 5,000 watts short of heat. This is 33% of the total heating capacity. Here are possible reasons for the inadequate heat:

- One of the heat strips is not working, or a heat strip may be open.
- An open heater thermal overload is controlling power to the heat strips.
- A connecting wire is open.
- The relay or contactor feeding the heat strip is defective.

You can troubleshoot these common problems by using an ohmmeter with the power disconnected and the component wiring removed.

2.9 COMBINED OHM'S LAW AND WATTS FORMULA WHEEL

The Ohm's law and watts formulas can be combined into one formula wheel that is easy to use. See Figure 2-21.

Figure 2-22 shows the formula wheel broken down into four quarters. Each section has three different formulas to assist in finding amps, volts, resistance, or watts.

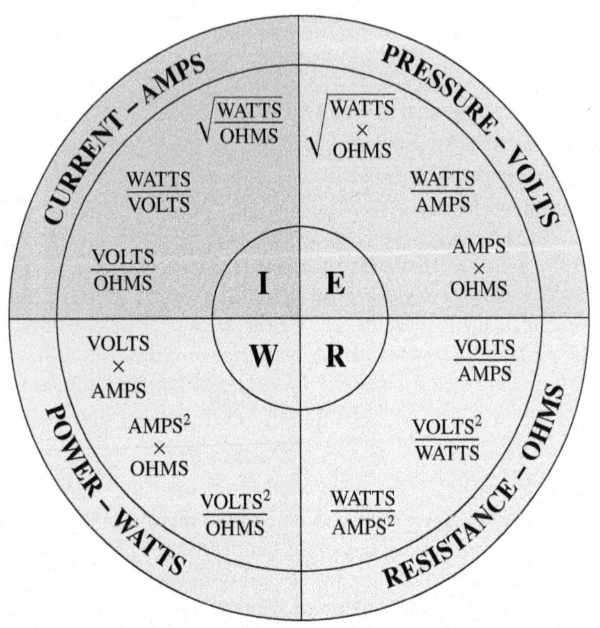

Figure 2-21 This pie chart combines the Ohm's law and watts formula into one easy-to-use wheel. This wheel also has other formulas that have not been discussed, but are useful.

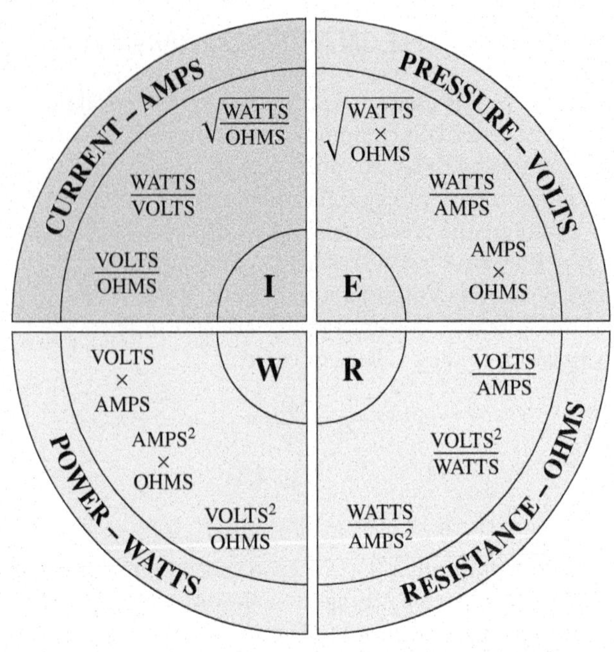

Figure 2-22 The pie chart is dissected into four quarters. Each quarter has a root formula. The root formulas are divided into amps (*I*), volts (*E*), ohms (*R*), and watts (*W*).

Here is a breakdown of each quarter section:

Current-Amps

$$I = \sqrt{\frac{\text{watts}}{\text{ohms}}} = \frac{\text{watts}}{\text{volts}} = \frac{\text{volts}}{\text{ohms}} = I$$

Pressure-Volts

$$E = \sqrt{\text{watts} \times \text{ohms}} = \frac{\text{watts}}{\text{amps}} = \text{amps} \times \text{ohms} = E$$

Resistance-Ohms

$$R = \frac{\text{volts}}{\text{amps}} = \frac{\text{volts}^2}{\text{watts}} = \frac{\text{watts}}{\text{amps}^2} = R$$

Power-Watts

$$W = \text{volts} \times \text{amps} = \text{amps}^2 \times \text{ohms} = \frac{\text{volts}^2}{\text{ohms}} = W$$

TECH TIP

The abbreviation *Btu* refers to British thermal unit. The Btu is a measurement of heat. It can also be expressed as British thermal units per hour, or Btuh. In HVACR applications Btuh may not always be expressed, but it is understood as the amount of heat added or removed per hour:

$$1\text{ W} = 3.4\text{ Btu}$$

It is easy to find the heat output of electric heat. Calculate the wattage when all of the heat strips are energized using the watts formula: $W = E \times I$. Then multiply the total wattage by 3.4 to determine the heat output in Btu's. For example, a 5-kW heat strip will produce 17,000 Btuh of heat:

$$5{,}000\text{ W} \times 3.4 = 17{,}000\text{ Btuh}$$

EXAMPLE 2.8 USING THE COMBINED FORMULA WHEEL TO CHECK HEAT OUTPUT

Let's look at how to use a couple of these formulas from the formula wheel in Figure 2-21 or 2-22.

A technician wants to check heat output, but her ammeter is in the repair shop. What steps should she follow?

SOLUTION

1. The tech measures the resistance and operating voltage of the heat strip. Using the pie chart, the tech finds a formula that will give her the answer:

$$W = \frac{E^2}{R} \text{ or } \frac{E \times E}{R}$$

2. Next, the tech wants to estimate the amperage draw before starting a system that has four single-phase heat strips wired in parallel. The tech realizes that the wire size will depend on the amperage in the circuit. The formula wheel has the solution to aid in this estimation.
3. The tech will need to know the kilowatts of the heat strip. Kilowatt ratings are readily available and many times listed on a tag attached to the heat strip housing. In this case, these are 4-kW heat strips.
4. The tech measures the resistance of the heat strip. The resistance measures 18 Ω. The tech uses the following formula to determine the amperage draw of one heat strip:

$$I = \sqrt{\frac{W}{R}}$$

$$I = \sqrt{\frac{4{,}000\text{ W}}{18\ \Omega}}$$

$$I = \sqrt{222}$$

$$I = 14.9\text{ A (round to 15 A)}$$

Four heat strips will draw 4 × 15 A = 60 A (estimate).

5. The tech individually measures 18 Ω across each heat strip. They are all good. Because the furnace is not energized, the other heat strips will not be hooked in parallel. When loads are hooked in parallel, the resistance is lower than that of the smallest resistor.
6. She calculates the heat output when all four heat strips are energized by the following formula:

$$\text{Btuh} = \text{watts} \times 3.4$$

$$\text{Btuh} = 16{,}000\ W \times 3.4$$

$$\text{Btuh} = 54{,}400$$

2.10 CALCULATING RESISTANCE

Calculating resistance is an extension of using Ohm's law. In the following sections we discuss resistance in one of these forms:

- **Electric heat strips**
- **Crankcase heaters**
- **Carbon or wire-wound resistors**
- **Incandescent light bulbs** (the old-style light bulbs with a filament).

Even though inductive loads, like motors and transformers, have resistance, the formula we are going to discuss here will not be accurate if used with inductive loads. Special impedance formulas are used to calculate the resistance of loads that are controlled by magnetism. Ohm's law cannot accurately calculate the true resistance of an inductive or capacitive circuit. In the following sections, we calculate the resistance in series and parallel. We will again see how voltage and current vary with resistance.

2.11 RESISTANCE IN SERIES CIRCUITS

Figure 2-23 shows a series circuit with four resistive loads. These loads could be heaters or simple resistors. This is not a practical circuit, but it is useful in helping you understand Ohm's law.

Finding Total Resistance

As you would expect, resistance in a series is cumulative; that is, the resistance adds up. The formula for this can be expressed as follows, where R_T stands for total resistance:

$$R_T = R_1 + R_2 + R_3 + R_4 + \text{etc.}$$

$$R_T = 4\ \Omega + 10\ \Omega + 12\ \Omega + 14\ \Omega$$

$$R_T = 40\ \Omega$$

Calculating Current Flow in a Series Circuit

Now let's find the current flow in this circuit using the total resistance we just calculated:

$$I = \frac{E}{R}$$

$$I = \frac{120\text{ V}}{40\ \Omega}$$

$$I = 3\text{ A}$$

One of the laws of a series circuit is that the current is the same everywhere in the circuit. In Figure 2-23, 3 amps are present at all three circled points in this complete circuit when the voltage is supplied.

Calculating Voltage Drop in a Series Circuit

Determine the voltage drop across each resistor in the series circuit in Figure 2-24.The voltage is divided up across each resistive load. Multiply the amperage draw by the resistance of each load. We will use the following formula:

$$E = I \times R$$

Load 1: $E_1 = I_1 \times R_1$
$E_1 = 3\text{ A} \times 4\ \Omega$
$E_1 = 12\text{ V}$

Load 2: $E_2 = I_2 \times R_2$
$E_2 = 3\text{ A} \times 10\ \Omega$
$E_2 = 30\text{ V}$

Figure 2-23 This series circuit has four resistors of 4, 10, 12, and 14 Ω. The total resistance is 40 Ω.

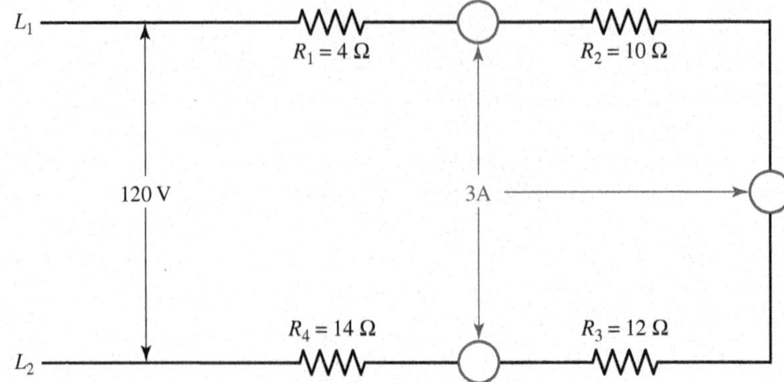

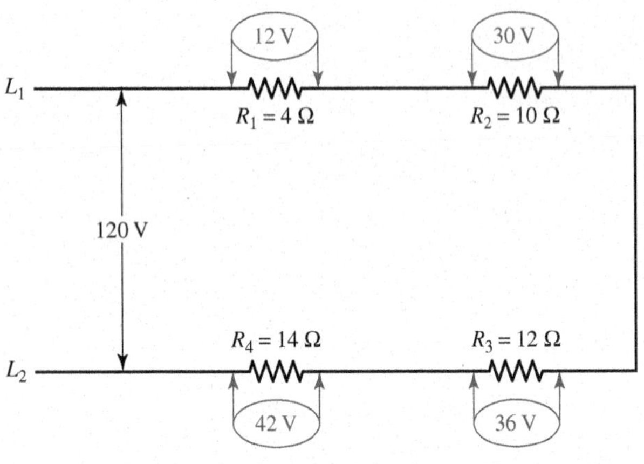

Figure 2-24 The voltage drop across the load in a series circuit depends on the size of the resistance. The larger the resistance, the greater the voltage drop. The total voltage drop of all loads will equal the supply voltage.

Load 3: $E_3 = I_3 \times R_3$
$E_3 = 3\text{ A} \times 12\ \Omega$
$E_3 = 36\text{ V}$

Load 4: $E_4 = I_4 \times R_4$
$E_4 = 3\text{ A} \times 14\ \Omega$
$E_4 = 42\text{ V}$

Verify Voltage Calculation

One way to verify that your calculation is correct is to add the voltage. The sum should equal the supplied voltage, E_T:

$$E_T = E_1 + E_2 + E_3 + E_4$$
$$E_T = 12\text{ V} + 30\text{ V} + 36\text{ V} + 42\text{ V} = 120\text{ V}$$

You can measure the voltage drop across each of these resistances when power is applied. Under the conditions stated here, you would measure these voltages: 12 V + 30 V + 36 V + 42 V.

TECH TIP

Table 2-1 can be used to understand the meaning of different prefixes and symbols. As an example, the term *kilowatt* refers to a 1,000 watts. We will use *kilo* and the other prefixes shown in this table throughout this book.

Table 2-1 Common Prefixes and Their Meanings

Prefix	Symbol	Number	Exponent
mega	M	1,000,000	10^6
kilo	k	1,000	10^3
milli	m	0.001	10^{-3}
micro	μ	0.000001	10^{-6}

2.12 RESISTANCE IN PARALLEL CIRCUITS

Determining resistance in a parallel circuit is different from determining resistance in a series circuit. We can compare parallel circuits to water piping. The resistance of a simple water piping circuit (series circuit) is based on the size and length of the pipe. The larger the diameter and the shorter the length of the water pipe, the lower the resistance to water flow. Figure 2-25 represents a series water circuit.

The size of the pipe can be compared to the wire size in a series circuit, and the water flow rate can be equated to the electron flow rate in wire. The larger the wire size, the lower the resistance to electron flow and, therefore, the greater the current flow in the circuit.

Figure 2-26 provides an example of a parallel piping circuit. We see that each water branch has 3 GPM of water flow.

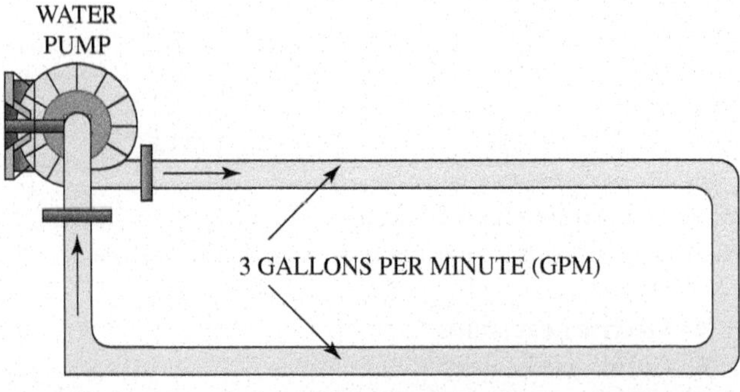

Figure 2-25 This series water circuit has a water flow rate of 3 gallons per minute, normally stated as 3 GPM.

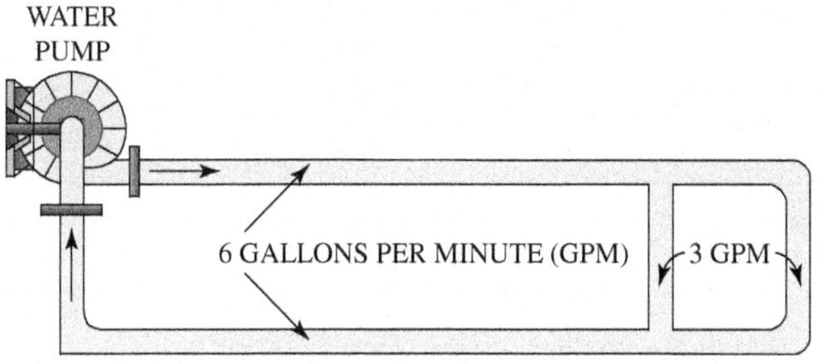

Figure 2-26 This parallel piping circuit has less resistance to water flow than a series circuit because the water path is split between two waterways.

The pump is providing 6 GPM of pump capacity. This would be the same in a parallel wiring circuit. As you can see, a parallel circuit has less resistance than a series circuit because there are more paths through which the electricity can flow.

Three methods are used to calculate the resistance of a parallel circuit:

- **Product over the sum method**
- **Reciprocal method**
- **Equal resistance method.**

Next we will learn how to use these formulas to calculate resistance in parallel circuits.

Product Over the Sum Method

The product over the sum method is one of the easier ways to find the resistance of a parallel circuit. This method can only be applied to two parallel branches at a time. If a parallel circuit has more than two branches, then this method must be repeated more than once, using only two resistors at a time. We start with a simple two-resistor parallel circuit as shown in Figure 2-27.

The formula for the product over the sum method is:

$$R_T = \frac{R_1 \times R_2}{R_1 + R_2}$$

Applied to the circuit in Figure 2-27, we have

$$R_1 = 100\ \Omega$$
$$R_2 = 100\ \Omega$$

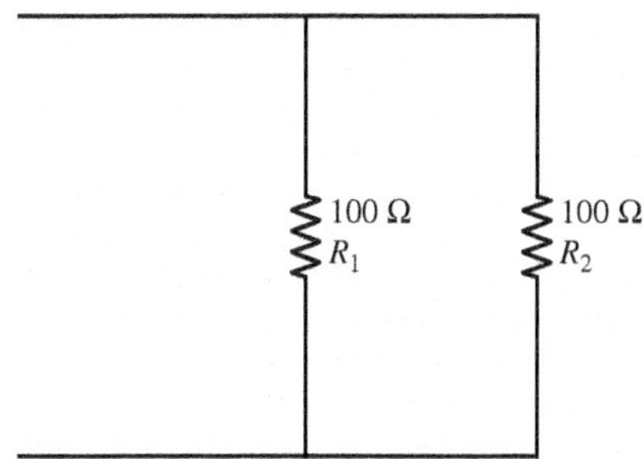

Figure 2-27 This parallel circuit has two 100-Ω resistors in parallel. Using the product over the sum formula, we find the resistance to be 50 Ω. The resistance in a parallel circuit is lower than or equal to that of the smallest resistor.

$$R_T = \frac{100 \times 100}{100 + 100}$$

$$R_T = \frac{10{,}000}{200}$$

$$R_T = 50\ \Omega$$

As mentioned, this formula can be used with more than two resistors, but you can only work with two resistors at a time. To calculate resistance for additional resistors, take the R_T value found from the calculation of the first pair of resistors and use it in the next calculation as R_1, then introduce the new resistor as R_2, and so on, as demonstrated in Example 2.9.

EXAMPLE 2.9 FINDING THE TOTAL RESISTANCE OF THREE RESISTORS IN PARALLEL

Use the product over the sum method to find the total resistance of the three parallel resistors shown in Figure 2-28.

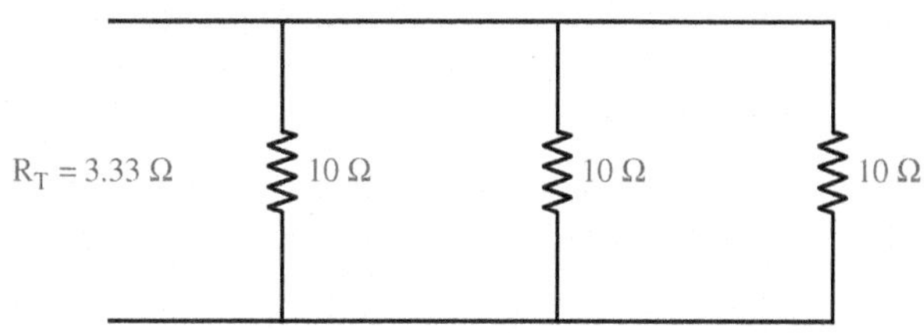

Figure 2-28 This parallel circuit has three parallel resistors. Use the product over the sum formula to find the total resistance.

SOLUTION

First, select any two resistors. Because all of the resistances are 10 Ω, the calculations for this circuit will be easy:

$$R_T = \frac{R_1 \times R_2}{R_1 + R_2}$$

$$R_T = \frac{10 \times 10}{10 + 10}$$

$$R_T = \frac{100}{20}$$

$$R_T = 5\ \Omega$$

Next, take the 5-Ω solution for the first two resistors and use the formula one more time in the calculation for the final 10-Ω resistor:

$$R_T = \frac{R_1 \times R_2}{R_1 + R_2}$$

$$R_T = \frac{5 \times 10}{5 + 10}$$

$$R_T = \frac{50}{15}$$

$$R_T = 3.33\ \Omega$$

Reciprocal Method

The reciprocal method can be used to calculate resistance for more than two resistors at a time. Here is the reciprocal method formula:

$$\frac{1}{R_T} = \frac{1}{R_1} + \frac{1}{R_2} + \frac{1}{R_3} + \frac{1}{R_4}, \ldots, \text{etc.}$$

Using Figure 2-28 again, let's calculate resistance using the reciprocal formula:

$$\frac{1}{R_T} = \frac{1}{R_1} + \frac{1}{R_2} + \frac{1}{R_3}$$

$$\frac{1}{R_T} = \frac{1}{10\ \Omega} + \frac{1}{10\ \Omega} + \frac{1}{10\ \Omega}$$

Now, 1 divided by 10 is 0.1. So we have

$$\frac{1}{R_T} = 0.1 + 0.1 + 0.1$$

Next, add the answers on the right side of the equation:

$$\frac{1}{R_T} = 0.3$$

Convert $\frac{1}{R_T}$ to R_T by dividing each side of the equation into 1:

$$\frac{1}{R_T} = \frac{1}{0.3}$$

$$R_T = 3.33\ \Omega$$

As you can see, this formula needs to be carefully followed. There is room for error, which will change the outcome. Many students prefer the product over the sum method since it is easier to understand. The wise student will use both of these formulas to verify the correct answer when calculating resistance in parallel circuits.

Equal Resistance Method

The equal resistance method of calculating resistance is used for circuits that contain equal branch resistance values. For example, in the parallel circuit in Figure 2-28 there are three 10-Ω resistors. The resistance can be easily calculated with the following formula:

$$R_T = \frac{R}{\text{number of resistors}}$$

$$R_T = \frac{10\ \Omega}{3\ \text{resistors}}$$

$$R_T = 3.33\ \Omega$$

Remember, this formula can only be used for circuits in which all of the resistors have the same value.

Basic Rules of Thumb for Calculating Parallel Resistance

Here are some basic rules of thumb that will assist you in determining if your parallel resistance calculation is correct. These rules only apply to parallel resistors.

- The total resistance of a parallel circuit will be less than that of the smallest resistor. Here is an extreme example: A circuit has two resistors in parallel. The resistances are 1 Ω and 1 MΩ. From the rule of thumb, we know the total resistance will be less than 1 Ω.
- When the resistances are equal, divide the resistance by the number of resistors. This is the equal resistance method. It can be used for any resistors in a parallel circuit as long as the resistors all have the same resistance. For example, consider a parallel circuit with four resistors. Two resistors are 10 Ω each and two are 20 Ω. Using this rule of thumb, the two 10-Ω resistors equal 5 Ω (10 Ω ÷ 2 = 5 Ω) and the two 20-Ω resistors equal 10 Ω (20 Ω ÷ 2 = 10 Ω).

2.13 RESISTANCE IN SERIES-PARALLEL CIRCUITS

Resistors in a combination circuit follow the same rules as series or parallel circuits. Each series or parallel part of the circuit is treated as if it were a separate circuit. Figure 2-29 is a series-parallel circuit with a 50-Ω resistor in series with two 100-Ω parallel resistors.

You may be able to calculate the total circuit resistance in your head. First calculate the resistance of the two parallel resistors. This is 50 Ω. Add the 50-Ω resistance of the parallel resistors to the 50-Ω resistance of the series resistor to get a total circuit resistance of 100 Ω.

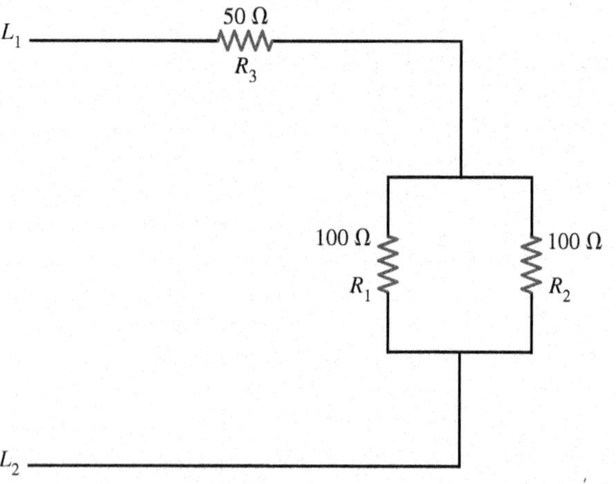

Figure 2-29 This series-parallel circuit has a 50-Ω resistor in series with two 100-Ω parallel resistors. The total resistance is 100 Ω. The resistance of the 50-Ω series resistor is added to the resistance of the two 100-Ω parallel resistors, which is 50 Ω. This will equal 50 Ω + 50 Ω = 100 Ω.

SERVICE TICKET

You are assigned to answer a service call for a heating unit. The customer states that there is not enough heat on very cold days. He says that the electric heater runs continuously, but the temperature in the house still drops a few degrees when it is cold, particularly at night. What do you think is the problem?

Some of the common problems related to reduced heat output include the following:

- One or more heat strips are open.
- The overload device is open.
- Airflow is low, causing the overload to cut out the heat strips.
- Heating capacity is inadequate.

The operational check includes taking measurements of running voltage and amperage. Removing the heater panel, you notice that the unit is wired with four heat strips. The heating system is running. You raise the thermostat temperature a few degrees to ensure that the system stays on during the troubleshooting process.

The following measurements and observations are recorded on the service ticket:

- Voltage: 205 V.
- Total amperage draw: 64 A. This is an average of 16 A per heat strip. Each heat strip is checked, and each measures about 16 A per heat strip.
- The air filter is clean.
- Airflow is good.
- The temperature rise between the return and the supply is 50°F, which is good for electric heat.

You monitor the heating system for 15 minutes. All four heat strips are checked with a clamp-on ammeter and found to be operating continuously. What is the problem?

If the system is left to run with a change in outlet temperature and the heat strips do not cycle off, then the problem is most likely an inadequate heating system. Increasing the voltage to 240 V would increase the wattage output. Let's do the math using the watts formula, $W = E \times I$.

Existing condition:

$$W = E \times I$$

$$13{,}120\ \text{W} = 205\ \text{V} \times 64\ \text{A}$$

Condition with increased voltage:

$$W = E \times I$$

$$15{,}360\ \text{W} = 240\ \text{V} \times 64\ \text{A}$$

To continue this example, it is helpful to know how to calculate British thermal units per hour (Btuh) from watts. As mentioned earlier, Btuh is a measurement of heat. This is the heat generated by the watts of electric energy created by a space heater or used by a crankcase heater to keep compressor oil separate from its refrigerant. The formula is simply:

$$\text{Btuh} = \text{total watts} \times 3.4$$

In the example just discussed, the difference is an increase of 2,100 watts, which is 7,140 Btuh (2,100 W × 3.4 = 7,140 Btuh).

The reasons why a heating system may be experiencing reduced voltage are many. There may be **voltage drops** in the electrical distribution system such as the breaker box or junction boxes, or the unit's wire may be undersized. In addition, the electric company may actually be supplying the lower voltage. A higher voltage supply could be requested.

If voltage drops are not found or if the electric company will not provide a higher voltage service, it is up to you to make additional recommendations. Think about it—what would you recommend?

You can recommend a larger or supplemental heating system. Even though energy efficiency improvements may not be your job, you could recommend an energy upgrade. The homeowner could improve the thermal efficiency of the building envelope. Thermal efficiency is achieved by reducing heat transfer in the winter and summer. Improvements such as installing attic, wall, or floor insulation will save energy, especially if the amount of existing insulation is limited or nonexistent. Reducing air filtrations is the least expensive option for saving energy and reducing operating time. Investments in caulking and weather stripping pay for themselves in 1 year. These types of improvements would save energy during both the winter and summer.

The solutions are many:

- Increase the voltage to the heat strips.
- Install more heat capacity or heat strips.
- Improve the thermal efficiency of the structure by adding insulation or energy-efficient windows and by reducing infiltration.
- A combination of these options.

SUMMARY

The purpose of learning Ohm's law and the watts formula is to help you understand the relationship of volts, amps, resistance, and power. Understanding this relationship will help you figure out problems and generate solutions when troubleshooting. When technicians first learn these formulas, they simply think of them as formulas that they need to learn to pass a course. These formulas, however, are the first step in figuring things out.

When you are solving problems, think of the relationships between volts, amps, resistance, and watts. Initially, you may not see the relationships. Keep trying to do so, however, because once you do, the information will become automatic and useful. You will eventually use these formulas to comprehend and decipher problems without thinking directly about the formulas themselves. Problem solving is not learned overnight, but flourishes with practice and an understanding of systems and their component parts.

This is not an easy unit to comprehend, but the material is the foundation to understanding circuit operation and troubleshooting techniques.

This unit presented many new terms, ideas, and concepts. Reread sections that are not clear. Work the Review Questions, which have been designed to help you apply the information you have just learned.

REVIEW QUESTIONS

1. What is the voltage in a circuit that draws 10 A and has a resistance of 12 Ω?
2. What is the current flow in a circuit that measures 120 V and has a resistance of 12 Ω?
3. What is the resistance in a circuit that measures 120 V and has a 20-A flow through it?
4. What is the power of a circuit that draws 10 A and measures 120 V?
5. What is the amperage draw in a circuit that uses 1,200 W and measures 240 V?
6. What is the voltage in a circuit that uses 1,200 W and draws 10 A?
7. How many watts does a heater produce that uses 240 V and a total resistance of 10 Ω?
8. How many Btu's of heat are generated by a 5-kW heat strip?
9. How many amps will be drawn when a heat strip uses 10,000 W and measures a total resistance of 10 Ω?
10. What is the wattage of a heat strip if the amperage draw is 20 A and the strip has a resistance of 3 Ω?
11. What is the resistance of a series circuit with the following resistances: 10, 20, and 30 Ω?
12. What is the resistance of a parallel circuit that has two 10-Ω heat strips?

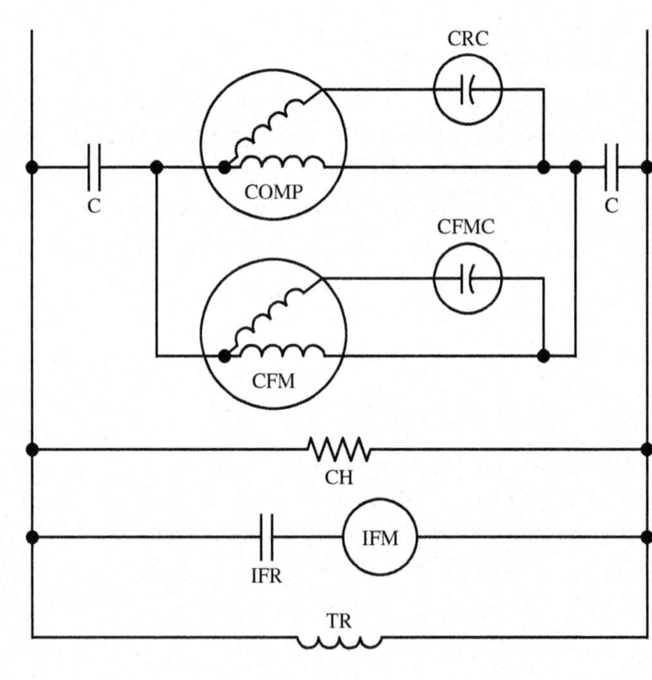

LEGEND

C:	CONTACTOR
COMP:	COMPRESSOR
CRC:	COMPRESSOR RUNNING CAPACITOR
CFM:	CONDENSER FAN MOTOR
CFMC:	CONDENSER FAN MOTOR CAPACITOR
IFR:	INDOOR FAN RELAY
IFM:	INDOOR FAN MOTOR
HP:	HIGH-PRESSURE SWITCH
LP:	LOW-PRESSURE SWITCH
CH:	CRANKCASE HEATER
TR:	TRANSFORMER

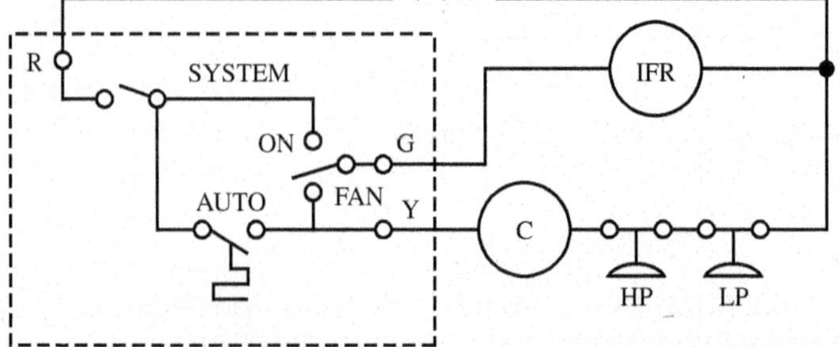

Figure 2-30 Use this circuit to answer Review Questions 20 through 23.

13. What is the resistance of a parallel circuit that has three 10-Ω heat strips?

14. What is the resistance of a parallel circuit that has four 10-Ω heat strips?

15. The clamp-on ammeter measures 20 A (240 V and 12 Ω) on the right side of a heat strip that is in a series circuit. What amperage will the ammeter measure on the left side of the heat strip?

Applying Ohm's Law and the Watts Formula to Resistive Loads

16. In an unusual design, a compressor manufacturer has two 120-V crankcase heaters wired in series. Each heater measures about 20 Ω with 240 V applied across the series heaters. Each crankcase heater must have the same heat output to be effective in keeping liquid refrigerant out of the crankcase. The tech is required to measure the operating voltage across each heater. What voltage reading should the technician expect across each heater if everything is functioning properly?

17. The heating requirements for a training area are 100,000 Btuh. The heating system for that area consists of three 5-kW heat strips wired in parallel. Will this provide enough heat for the training area?

18. Three 5-kW heat strips are wired in parallel. The supplied voltage is 240 V. The heat strips are labeled H_1, H_2, and H_3. The tech measures 241 V across H_3. What voltage would be expected across H_1?

19. You are asked to go to a job site and replace a shorted-out crankcase heater. You notice that the tag on the defective heater indicates the voltage is 240 V and 100 W. To ensure that the heater is working when the new crankcase heater is installed, what will the amperage on the heater be when it is powered up?

20. Refer to Figure 2-30.
    a. Is CFM in series or parallel to the C contacts?
    b. Is CFM and COMP in series or parallel with each other?
    c. Is this a series, parallel, or combination circuit?

21. Refer to Figure 2-30.
    a. Is the crankcase heater in series or parallel with the transformer?
    b. Is the low-pressure switch in series or parallel with the Ⓒ contactor coil?

22. Refer to Figure 2-30. What is the amp draw of the crankcase heater if the supply voltage is 240 V and the resistance is 100 Ω?

23. Refer to Figure 2-30. How many watts will be generated by the crankcase heater in Review Question 22?

UNIT 3

Safe Use of Electrical Instruments

WHAT YOU NEED TO KNOW

After studying this unit, you will be able to:

1. Identify the features and benefits of a multimeter.
2. List the options and functions found on a clamp-on ammeter.
3. Describe how to measure low and high amperage conditions.
4. Select a multimeter and clamp-on meter that will be useful in HVACR work.
5. Describe the safety considerations that take place when selecting a multimeter.
6. List the hazards associated with using electrical instruments.

Understanding how to use electrical instruments is required to evaluate and repair a HVACR problem. Because electricity cannot be seen, our instruments are our "eyes" into the system. In a way, an electrical meter is like an x-ray. An x-ray machine uses invisible technology to produce visual images of, say, the human body for diagnostics and troubleshooting. An electrical meter converts unseen voltage and current into a digital readout that helps technicians diagnose the HVACR patient. The meter can also be used to detect a complete circuit or determine if a component has a complete path for electricity to follow. This unit explores various instrument options and teaches you how to select the best meter for a particular job. The most important information in this unit, however, covers and stresses electrical safety when using meters and working around electricity.

Safety cannot be overemphasized, so we open this unit with a list to consider when working on electrical systems. Thanks go to Jim White with Shermco Industries for creating this list and providing some real-life insight into our profession.

Ten Mistakes People Make When Working on Electrical Systems *by Jim White, Shermco Industries.*

1. **Thinking that it's "only" 120 volts or 208 volts or 480 volts or....**
   "It's only low voltage." Okay, I'll admit that you can have an open casket with a low-voltage hit, but you'll still be dead. The only difference between low and high voltage is how fast it can kill you. High voltage kills instantly; low voltage may take a little longer.

   Dr. A.G. Soto, consulting physician to Ontario Power Generation, presented a paper at the 2007 IEEE Electrical Safety Workshop discussing low-voltage shock exposures. In that paper, he stated that a 120-V shock can kill up to 48 hours later. He also stated that many emergency department physicians are unfamiliar with the effects of an electric shock and that an EKG may not show a problem. The injury to the heart muscle tends to spread over time and cannot always be identified using EKGs.

2. **Working on energized systems or equipment when it can be de-energized.**
   When I was working in a power plant (back in the 1970s), we never de-energized anything, whether it could be or not. My boss had great contempt for anyone who would actually ask to de-energize equipment before working on it. He would tell anyone foolish enough to suggest turning it off, "You're an electrician, work it hot! That's what you're trained to do!" His other favorite saying was, "If you want to be here tomorrow, you'll get this done today." Can you feel the love?

   De-energizing is the only way to eliminate hazards. The use of **arc flash** personal protective equipment (PPE) just increases your chances of survival; it doesn't guarantee it. Just be aware that until equipment and systems are placed in an electrically safe work condition, proper PPE and procedures must be used to protect the worker. See Article 120 in the National Fire Protection Association (NFPA) Standard 70E 2009.

3. **Not wearing PPE.**
   This could go into number 2 above, but people really don't like wearing rubber insulating gloves or arc flash PPE. It's hot, uncomfortable, restricts movement, and slows the entire work process down—not only while wearing it, but when selecting the correct PPE and putting it on and taking it off. *But* it will save your life. One of the most likely times people neglect to wear their PPE is during troubleshooting. The rationale seems to be, "I'm not really working on it; I'm just testing it." Yet, National Institute for Occupational Safety and Health studies have found that 24% of electrical accidents are caused by troubleshooting, voltage testing, and similar activities. We have a tendency to ignore hazards associated with tasks we consider "safe."

   In a previous job, when I was surveying a 480-V, 250-A molded-case circuit breaker, the worker I was with pushed his bifocals up on his forehead so he could read the label on the breaker. Immediately after he had dropped his glasses back onto his nose, the breaker blew up! Luckily, he only had some red spots on his face and some singed hair, because he was backing his head out when it let go. Metal droplets were imbedded into the lenses of his glasses, but because of them, he wasn't seriously injured. We investigated why that breaker might have failed and never found a good reason; it was just time for it to fail. Carbon buildup from earlier fault interruption, eroded contact material that gets sprayed up into the arc chutes, a weakened dielectric due to the extreme heating of arc interruption—all of these weaken circuit breakers and could have caused what seemed like a perfectly good breaker to fail suddenly. You never know.

4. **Going to sleep during safety training.**
Nothing like a good nap to get you ready for a hard day's work! Every Monday morning Shermco does a 1-hour safety meeting for all technicians. We call it the "Monday Moaner," because the technicians really want to be at their job sites, not getting "preached to." All of us like to do what we are comfortable with, even if there's a better way to do things. Add that to the fact that wearing PPE and filling out forms are part of the required steps, and fuggedaboutit!

The other side of the coin is that a lot of safety training is so boring! I've been to some sessions where, by the end, I'm praying for a mercy killing—either me or the instructor, I don't care which! Safety training has to be focused, concise, and interesting; otherwise everyone tunes it out.

5. **Using outdated or defective test equipment to troubleshoot.**
When the leads are frayed or the meter's dodgy, it's time to replace it. I worked with a technician who had used the same Wiggy solenoid tester for 7 years. You couldn't read the faceplate, the coil was so weak that it didn't even vibrate, and the leads had been pulled loose from the bottom. Almost every time he used it he got nailed! One day, right after he was shocked (for the kazillionth time), I said, "Hey, let me see your Wiggy." He handed it down and I twirled it around my head and smacked it into a concrete column. The coil came springing out and he charged down the ladder like an enraged bull! I handed him my new Wiggy and said, "Take this new one—that one's going to get you killed," to which he replied, "I've had that since I was an apprentice!" Don't get emotionally attached to inanimate objects. If you really love your old voltage tester, take it home and make a little shrine to it—just don't bring it to work.

The NFPA committee was concerned enough to mandate a requirement that only portable electric tools and test equipment that were properly rated could be used. The National Electrical Code (NEC), Section 110.9(A)(1) regulation, *Use of Equipment, Rating*, states, "Test instruments, equipment and their accessories shall be rated for the circuits to which they will be connected."

This statement is followed by a reference to ANSI/ISA Standard 61010-1, *Safety Requirements for Electrical Equipment for Measurement, Control, and Laboratory Use—Part 1: General Requirements*: "for rating and design requirements for voltage measurement and test instruments intended for use on electrical systems 1000V and below."

6. **Not wearing the right PPE.**
No, I'm not repeating myself. Some people think that if they wear anything by way of PPE, that should be enough. Although it is true that the injuries you would sustain probably wouldn't be quite as severe as if you hadn't worn any PPE at all, there's a high probability that if the right type of PPE had been worn, you wouldn't have had an injury at all. This could also probably go under number 4, because if you aren't paying attention during safety training, you probably can't choose the right PPE either. Do you know how to interpret arc flash labels? What do you do if the electrical power equipment you're working on does not have an arc flash label? Do you know how to use the tables in the NFPA 70E handbook? Do you refer to the notes when you use the tables? If you answer "no" to any of these questions, you aren't choosing the right PPE. As a matter of fact, you probably would not be considered qualified by the Occupational Safety and Health Administration (OSHA). Your company has the responsibility to provide training so you meet OSHA's definition of a qualified electrical worker, but you are the one who will be exposed to the hazard, and you're the one who will get burned. You need to do the homework to protect yourself!

7. **Trusting your safety to someone else.**
An OSHA compliance officer I know investigated an arc flash incident in which two electricians had been working together for years. The one who was injured had asked his buddy if the circuit had been checked and was dead, to which his buddy replied, "Yeah." The tech who got injured really didn't think that the circuit had been checked, but he didn't want to offend his partner, so he didn't pursue the question. When he started working on the circuit, it blew up, causing severe arc flash burns. He stated, "If I had to do it over again I would have checked it myself and not worried about so-and-so's feelings." Actually, those weren't his words, but they won't allow me to print what he really did say. You get the idea, though.

Sometimes relationships cause us to not follow through when we should. Either we don't want to offend someone, as in the preceding example, or we don't want to look incompetent to our coworkers. However you want to put it ("Nothing personal, I'd just like to make sure I don't get my face blown off"), don't neglect to prove systems dead personally.

8. **Not performing required maintenance of power system equipment.**
Too often companies look at maintenance costs as an overhead expense. Nothing could be further from the truth. The problem is that it is difficult to determine the

cost savings for things that don't happen: unscheduled outages, loss of production, buying equipment at premium prices, overtime, disposing of the cratered equipment, and so forth. Those of us who have been through the maintenance wars have seen the costs associated with neglect, but newer managers and those in accounting may find that difficult to appreciate. Liken it to automobile maintenance. You go out and buy that new ZR1 and then do no maintenance for 100,000 miles. What condition do you think it will be in?

9. **Not carrying your gloves with you.**
During my safety training classes, I like to ask how many people actually carry their rubber insulating gloves with them. Maybe one or two will raise their hands. Well, guess what? If you don't carry them, you aren't using them. This might go along with thinking that low voltage won't hurt you. We get buzzed and it's no big deal. At the beginning of 2008 in Athens, Texas, three TXU Energy workers were working on a 120/208-V transformer. One of the workers stood and said, "Well, boys. Looks like I got bit again," took three steps, and died. Carry your gloves and use them—always.

10. **Not using an Energized Electrical Work Permit system.**
People tend to hate paperwork, including myself. The Energized Electrical Work Permit is one great exception. OSHA wants us to plan each job, gather the right tools and equipment to do the job safely, and follow our work plan. How do we document hazards and risks or do a PPE assessment? The OSHA field safety compliance officers I know all tell me the same thing: If it's not documented, you can't prove that you did it. The Energized Electrical Work Permit provides the means to plan the work, assess the hazard and the risk, choose the proper PPE for the job, and document it.

There's always something else that could be included in this list, but these ten get you thinking. We go through life making small mistake after small mistake and nothing happens, until we happen to get the wrong alignment of small mistakes and now an accident has occurred. Once the accident starts, we have no control over it, so the best thing to do is to avoid the small mistakes and tighten up the way we work.

Credit Due: Jim White is the training director for Shermco Industries in Irving, Texas, and a Level IV International Electrical Testing Association (NETA) technician. Jim represents NETA on the NFPA 70E and B committees as well as the Arc Flash Hazard Work Group. He chaired the 2008 IEEE Electrical Safety Workshop.

3.1 READ THE DIRECTIONS!

Reading the directions and understanding how to use any meter you purchase is the most important step to becoming a good troubleshooter. Most meters have numerous helpful functions as well as limitations. It is important to learn about the functions so you can get the most use out of the instrument. It is also essential to know its limitations so that you do not get false readings when troubleshooting. For example, the amount of ohms or resistance a meter can measure is limited in some meters. If the limit is 1,000 Ω and you are measuring a relay coil with 1,200 Ω of resistance, the meter will read "infinity" or "open." Even though the coil may be good, your meter will lead you to believe that the coil is open or defective. This is an operator error.

All instruments have limits. Know your instrument's limits! Read the directions. After using the instrument for awhile, read the directions again. You will be surprised about the features and benefits that you missed during the first reading.

TECH TIP

Have extra batteries available for all electrical instruments used in your service work. Most service jobs are pressed for time and having to stop to purchase batteries is not a professional practice. Think ahead and always plan for the worst. Having a backup meter is another good professional practice.

3.2 DIGITAL MULTIMETERS

The **digital multimeter**, or **DMM**, is an electronic tape measure for taking electrical measurements. It uses a digital display to show, for example, volts, ohms, and amperes. Other features such as a capacitor checker, temperature tester, and diode tester are available on some models. A common DMM is shown in Figure 3-1. Wearing gloves gives the user an extra layer of protection against shock. Note, however, that gloves will not prevent severe shock.

Figure 3-1 The digital multimeter, also known as a DMM, measures voltage, resistance, and amperage. This one is reading 199.98 volts AC.

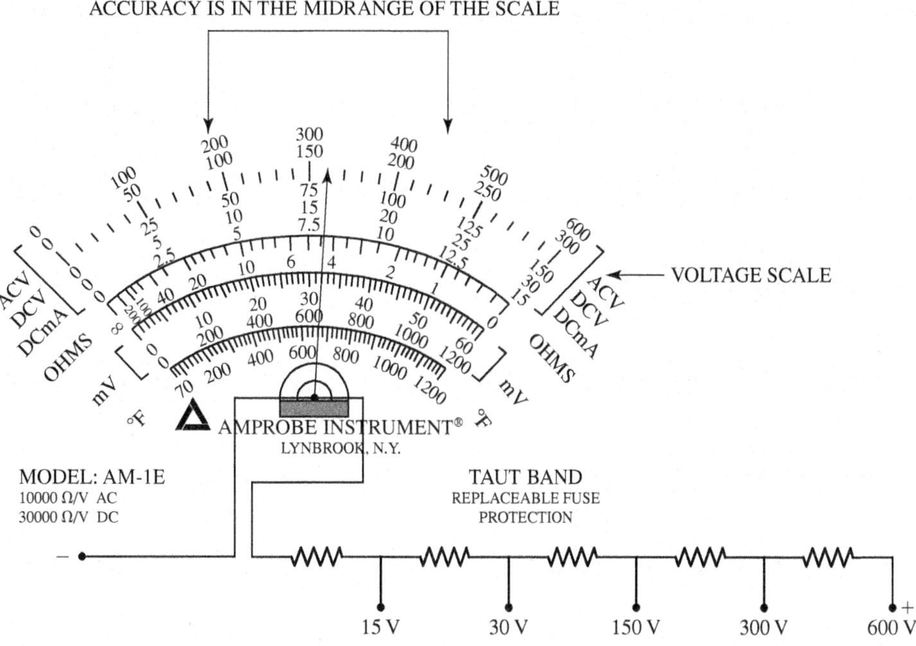

Figure 3-2 The analog voltmeter has its best accuracy in the middle part of the scale. Accuracy drops off dramatically at the lower and upper scale readings. At midrange the analog scale is ±2% accurate.

A meter that uses a swinging needle to display a reading is called an **analog multimeter**. The first multimeters were analog or meter movement types. We will concentrate on the DMM since it is the predominant instrument used in our profession today.

Choosing the right meter for the job can be challenging unless you know what features to look for and what the features do. This section is designed to assist you in selecting the multimeter that will work best for you. When selecting a good quality meter it is important to answer the following questions:

- What are the safety aspects of the multimeter?
- What features do you need?
- What is the easiest way to get the most out of your meter?
- Which meter is best suited to the environment in which you will be working?

Using and understanding the standard features of the instrument will allow you to derive the benefits of quick and accurate troubleshooting and problem solving.

First we will look at the importance of a DMM's accuracy, display, resolution, range, and RMS features. After that, we explore DMM options such as measuring voltage, current, and resistance and other important readings found on a quality meter.

DMM Accuracy

Accuracy refers to how close the DMM measurement is to the actual value of the signal being measured. Accuracy is expressed as a percentage (%) of the reading. The typical accuracy of a quality DMM meter is ±1% or less. This means if the true voltage is 120 V, the display will be within ±1.2 V of the real voltage and show a reading of 118.8 to 121.2 V. This level of accuracy is good for HVACR work. Analog multimeters, the meters with a needle-type movement, are less accurate. For example, the typical accuracy of a good analog multimeter is ±2%. This accuracy is at the midrange or middle portion of the analog scale as shown in Figure 3-2.

The accuracy on the lower and upper ends of the analog scale drops dramatically. At one-tenth of full scale, the level of accuracy drops by 20%. Analog multimeters can be used in our profession as long as the user is aware of the inaccuracy of the readings on its lower and upper scales. The analog meter is most accurate in the middle 50% of the scale.

DMM Display

A digital display is easier for a technician to use and understand than an analog meter. The analog meter requires interpretation of various scales and estimation of the values that fall between the lines. In contrast, with a DMM a technician can choose to display readings to several decimal places. This enhances the accuracy of digital meters.

Quality DMMs may have a bar graph that shows changes and trends in the measured signal. Such a bar graph is shown in Figure 3-3, below the voltage reading on the digital display. At a full scale reading of 1,000 V, the bar below the digital readout fully populates the bar graph. The DMM display is durable and less prone to damage when dropped or when the meter range is exceeded when compared to instruments with a needle-type movement.

DMM Resolution and Range

Resolution refers to how accurate or fine a measurement a meter can take. By knowing the resolution of a meter, the technician can determine if it is possible to see a change in the measured signal. For example, in the 150-V range, the meter can measure voltage in 1-V increments. A specific meter would need to be selected if half-volt increments were required. It is important to have a DMM that can measure volts, amps, or ohms in one-digit increments. You will find

Figure 3-3 At a full scale reading of 1,000 V, the bar below the digital readout stretches the full length of the bar graph. A quarter bar would indicate a voltage reading of 250 V.

that most meters will measure better than this, down to the tenths, such as 120.2 V or 5.5 Ω. Finer measurements are not normally required in the HVACR industry; in fact, even decimal readings are not normally required and sometimes confuse the novice user of these instruments.

In summary, resolution as it relates to a DMM is the smallest change in measured value to which the instrument will respond.

As the range increases, the resolution decreases. Unlike high-definition television, resolution has *nothing* to do with the how clear or sharp the digital display is presented on the meter screen.

To determine the range and resolution of your DMM, turn the meter to the volt, alternating current range (ACV) and press the **Range** button. The **Auto** feature on the display will disappear. Depending on the selected range, the resolution will vary. Here is an example of the ranges and resolutions of a leading manufacturer's digital meter:

Range	Resolution
600.0 mV	0.1 mV (= 1/10 mV)
6.000 V	0.001 V (= 1 mV)
60.00 V	0.01 V (= 10 mV)
600.0 V	0.1 V (= 100 mV)
1,000 V	1 V (= 1,000 mV)

For maximum resolution, choose the lowest possible range. To exit the manual **Range** mode, hold the **Range** button for 2 seconds. (*Note:* The procedure for exiting the manual **Range** mode varies among manufacturers.)

TECH TIP

One of the fundamental steps of electrical troubleshooting is voltage measurement. After surveying the job site and testing for proper supply, voltage is usually the first step taken when troubleshooting. If there is no voltage present, or if the voltage is too low or too high, the problem will need to be corrected before continuing the troubleshooting process.

3.3 RMS

RMS means **root mean square.** It relates to alternating current, sawtooth, square, or ripple voltage patterns. It does not relate to true DC voltage, which has a flat pattern and does not vary. A visual representation of RMS voltage is provided in Figure 3-4. This figure shows a full-wave, alternating current sine wave. The part of the wave above the line is positive; the part below the line is negative. It is important to use a voltmeter that measures true RMS. A true RMS meter displays the actual usable voltage and is an accurate way of measuring voltage. The RMS value is the effective DC value of the AC voltage or 0.707 times the peak voltage.

All DMM meters are accurate when measuring a pure alternating current sine wave as illustrated on the left side of Figure 3-5.

Non-RMS or average-measuring rated meters will provide voltage measurements that are higher or lower than the actual voltage when measuring sine waves that are irregular, sawtooth, square, or ripple shaped. These nonsinusoidal voltages are found in circuit boards and other electronic controls. Selecting a quality DMM and **clamp-on ammeter**

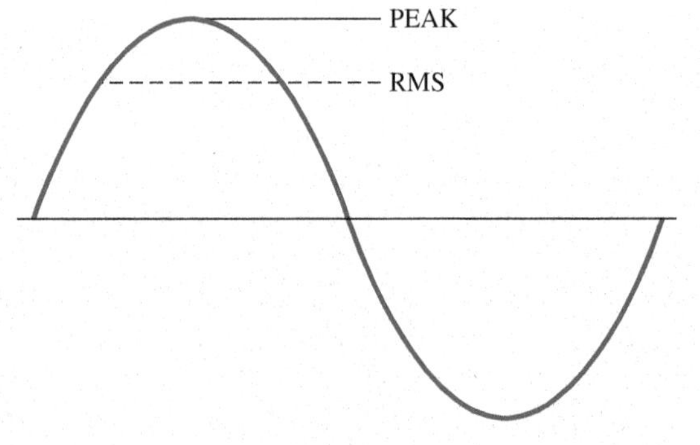

Figure 3-4 This is a full-wave, alternating current sine wave.

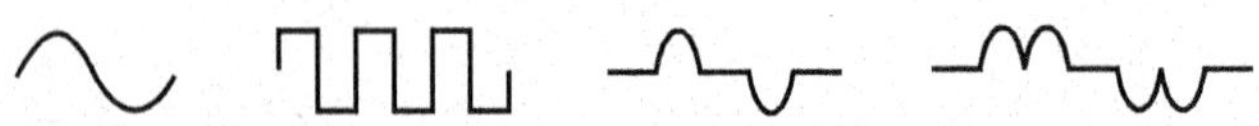

Figure 3-5 This table compares the average voltage reading versus true RMS voltage readings. The average responds correctly to a true sinusoidal wave. Other voltages will be too high or too low. A true RMS meter is the best selection for all HVACR applications.

Figure 3-6 The same wire with two different amp readings. Which one do you trust?

that measure true RMS voltage is an important step toward proper troubleshooting so that the technician does not get false voltage readings.

Figure 3-6 shows the readings of two different amperage-measuring instruments. The amp meter on the right measures true RMS and is therefore more accurate when measuring irregular current flow sine waves. The branch circuit above feeds a nonlinear load with distorted current. The true RMS clamp-on meter on the right reads amps accurately. The average-measuring clamp-on meter on the right reads low by about 32%.

SAFETY TIP

When measuring inline amperage make sure that the power is off before cutting or breaking a circuit to insert the DMM for current measurements. Even the smallest current can be dangerous.

3.4 WHAT DO ALL THOSE SYMBOLS MEAN?

Let's take a look now at the common **symbols** found on multimeters. It is important to recognize their meaning so that the meter is not misread or misapplied, which could give a false reading or damage the meter.

Symbol	Meaning
V ⎓	VDC
V ~	VAC
mV	millivolts (0.001 V or 1/1,000 V)
A	amps
mA	milliamps (0.001 A or 1/1,000 A)
μA	microamps (0.000001 A or 1/1,000,000 A)
Ω	resistance (ohms)
kΩ	kilohms
MΩ	megohms
)))	continuity beeper
—|(—	capacitor checker (μF or microfarads)
—▶|—	diode tester
Hz	hertz (cycles/sec)
dB	decibels (sound)
Range	manual measurement ranging
Hold, TouchHold/ AutoHOLD	last stable reading
MIN-MAX	highest or lowest recorded readings
⚡	dangerous voltage levels
⚠	caution; see manual

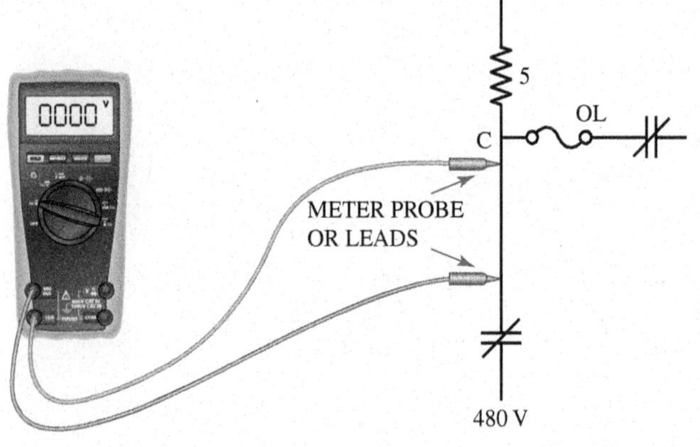

Figure 3-7 A voltmeter measures potential difference, not the voltage in a line. This meter is measuring 0 V because it is not measuring a potential or voltage difference on the wire with 480 V applied.

3.5 VOLTMETER FUNCTION

The voltmeter function is the most common feature used by the HVACR technician. It is important to know where voltage is lost in a circuit and if the voltage is correct.

The **voltmeter** measures potential difference. Potential difference refers to the fact that there must be a voltage difference between the two meter probes for the meter to be able to measure a voltage. For example, a single bare power line with a potential of 480 V will measure 0 V if the two meter probes are placed on the same line, as shown in Figure 3-7.

Figure 3-8 shows a voltmeter measuring potential difference. This illustrates that 230 V will be measured if one meter probe is placed on the wire of the transformer and the other probe placed on a different wire on the transformer volt source. Figure 3-9 shows that 115 V will be measured from the transformer to ground. In both cases a potential difference must exist in order for the voltmeter to register a voltage.

Some instruments use a light to detect a voltage as shown in Figure 3-10.

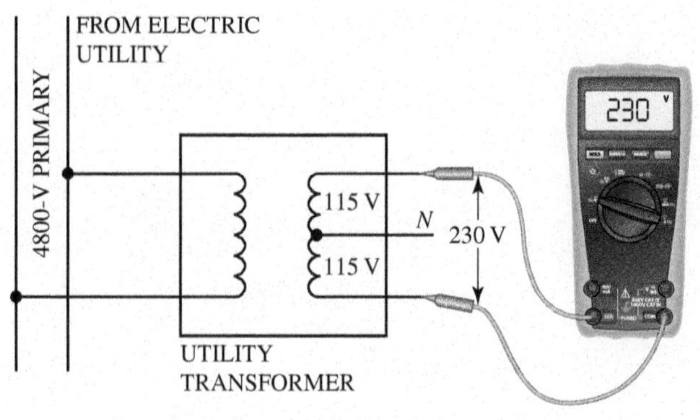

Figure 3-8 This shows that the voltmeter measures 230 V across two hot lines of a transformer. There is a voltage difference between the two lines.

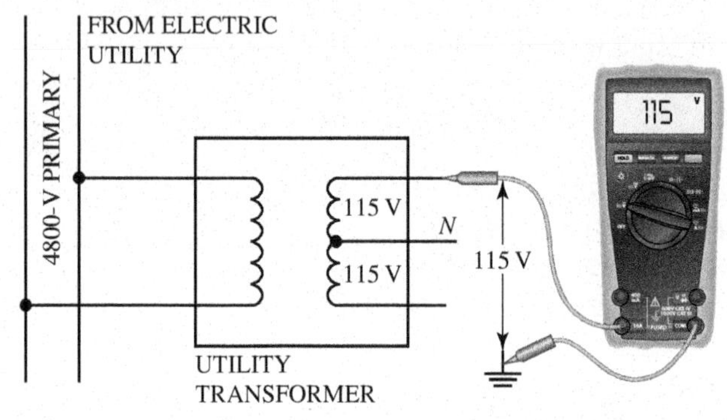

Figure 3-9 This shows that the voltmeter measures 115 V to ground. There is a voltage difference or potential between the top of the transformer to ground.

These types of instruments are known as **noncontact voltage detectors**. To determine the reliability of this meter option, check it by measuring a known, live voltage prior to its first use. Recheck with a voltmeter to be certain.

These light instruments will detect the existence of voltage, but will not indicate the exact amount of voltage present.

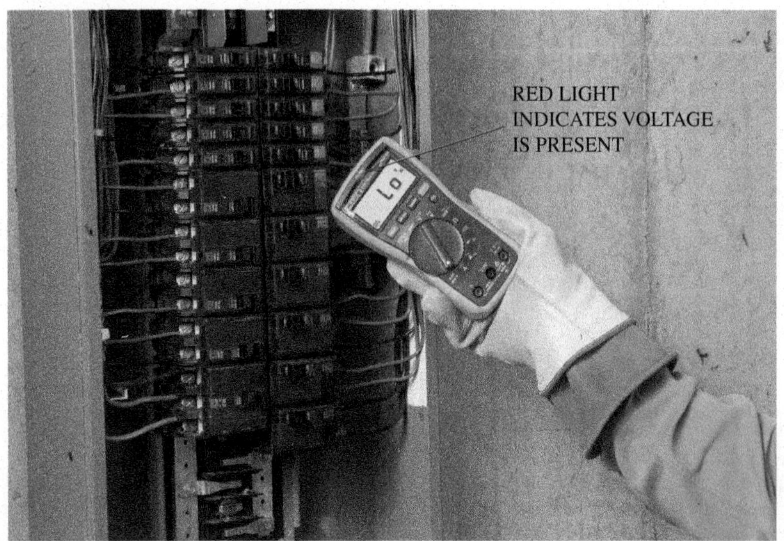

Figure 3-10 This shows a noncontact voltmeter option. The red light at the top of the meter indicates that voltage is present. The meter also shows the voltage level of **Lo**, but not the exact voltage present. This type of meter can be used as a safety check when the technician thinks that the circuit is dead.

These are good devices for determining if it is safe to touch a circuit with your fingers. This device should **not** be used, however, to make the final determination about whether there is voltage in a circuit. Like any instrument, this voltage detector can become defective at any time. Use other methods to ensure that a circuit is dead. The final test before touching a potential voltage should be shorting the "dead circuit" to ground.

3.6 MEASURING AC AND DC VOLTAGE

Many DMM voltmeters require the technician to select the voltage type: AC or DC. (Note, however, that some voltmeters automatically determine the voltage type for you.) The AC voltage selection may be indicated as follows:

∼ VAC

This is the symbol for DC:

⎓ VDC

If the voltmeter is set to measure the wrong type of voltage, the technician will get an incorrect voltage reading. This can mislead the technician and confuse the troubleshooting process.

When measuring AC it does not matter which probe goes on which line being measured. When measuring DC the polarity, positive or negative, is generally not important for DMM instruments. Reverse polarity is indicated as a negative (−) reading on the meter, but the reading will be accurate. For example, a reverse polarity reading on a 9-V battery will read −9 V (negative 9 volts).

3.7 MEASURING RESISTANCE

The resistance feature on a digital multimeter is the second most common feature used by the HVACR professional. This is known as the **ohmmeter** function. The meter may have an auto ranging feature or a resistance range selection feature. The auto ranging feature allows the technician to select one setting, for example, resistance. The meter determines the resistance range and gives the technician a digital reading.

Some meters provide various resistance ranges from which to select. This is indicated on the face of the meter as follows:

R × 1

R × 10

R × 100

R × 1,000

R × 10,000

R × 100,000

How do you read resistance? Digital multimeters handle resistance readings in various ways. Here we discuss one of the most common ways, but keep in mind that it is always best to refer to the owner's manual to determine the specifics for your meter.

If you select a low resistance range such as R × 1, the meter will only read a resistance within the low resistance range, which may top out at 1,000 Ω. Therefore, a resistance higher than 1,000 Ω will be indicated as over the limit by showing **OL** on the display. This can be confusing to the technician. Does this mean that the component is open? Or is the reading outside the lower range of 1,000 Ω? The technician should increase the range setting to the next higher resistance level until a resistance reading is provided. If no resistance reading is given when the meter is set to the highest resistance range, the component being tested is open.

Note: Many poor quality meters have low resistance ranges and will give false indications for high-resistance coils such as potential relays, gas valves, or motor shorts that may be greater than a couple of thousand ohms. If the meter's resistance range is only 1,000 Ω and the coil resistance is 1,001 Ω, the meter will read **OL** or an open circuit. Because the meter has a limited high resistance range, the **OL** reading will mislead the technician into believing the circuit being tested is open.

3.8 READING AMPERAGE

The second most common electrical instrument used by the HVACR technician is the **clamp-on ammeter.** Most clamp-on meters are the type shown in Figure 3-11. (A glove is worn as a safety practice when using an ammeter. Note, however, that the glove pictured in this figure is not an electrically rated glove.) The clamp-on ammeter measures the magnetic field in a wire. The higher the current flow in the wire, the greater the

Figure 3-11 This is a self-contained clamp-on ammeter. It is measuring 8.4 A on one of the breaker wires.

(A)

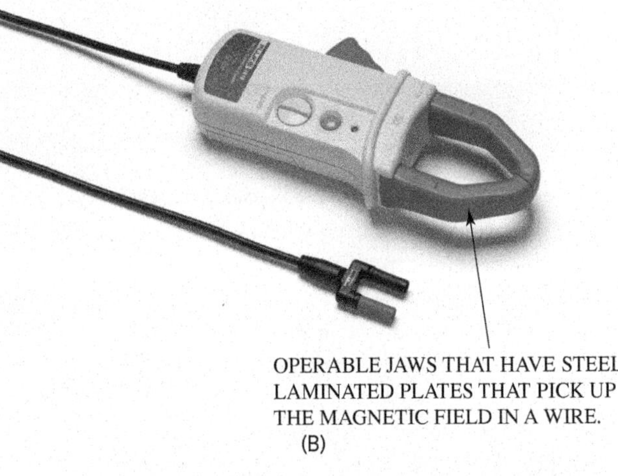
OPERABLE JAWS THAT HAVE STEEL
LAMINATED PLATES THAT PICK UP
THE MAGNETIC FIELD IN A WIRE.

(B)

Figure 3-12 (A) A current-measuring accessory is used with a DMM to monitor load current. The clamp-on jaw accessory is connected via a lead to the DMM, which provides an amperage readout. (B) Close-up view of a clamp-on jaw accessory used with a DMM. This device expands the use of the DMM.

Figure 3-12B shows a close-up view of the clamp-on jaw accessory used with a DMM. This device expands the use of the DMM. Under the plastic jaws there are layers of laminated steel plates that pick up the magnetic field generated by current passing through the wire. The higher the amperage draw, the greater the magnetic field generated by the wire. The magnetic field is converted to an amperage reading on the meter's display.

Let's return to our discussion of the self-contained clamp-on ammeter. As mentioned, the jaws of a clamp-on ammeter open and surround one wire. The jaws sample the magnitude of the magnetism created by the current passing through the wire and changes it into a current reading on the instrument. The model shown in Figure 3-13 has options to measure voltage and limited resistance. The steel-laminated jaws of the meter should be tightly closed and the wire should be placed near the center of the jaws. Loosely closed jaws will not complete the circuit. The meter will give a false low reading if the jaws are not clean and tightly closed.

Low current flow (between 0.1 and 2.0 A) can also be accurately measured by wrapping the jaws of a clamp-on meter with the wire of the circuit to be measured, as shown in Figure 3-14. Wrapping the wire around the jaws increases the magnetic flow induced into the steel-laminated plates inside the jaws. The amperage reading on the clamp-on meter is divided by the number of wire wraps around the jaws. For example,

Figure 3-13 The clamp-on ammeter opens and surrounds one wire. The jaws need to be tightly closed to get an accurate reading.

magnetic field. Only one wire should be measured at a time. Placing the clamp around two or more wires at the same time will give a false amperage reading since the magnetic fields are going in opposite directions and cancelling each other out.

As discussed earlier, some DMMs have amperage-measuring options. One option is to use a clamp-on jaw with the DMM to measure amperage. The amperage jaw has a lead that plugs into the DMM as shown in Figure 3-12A. This type of clamp-on amp jaw is normally an optional accessory for a DMM.

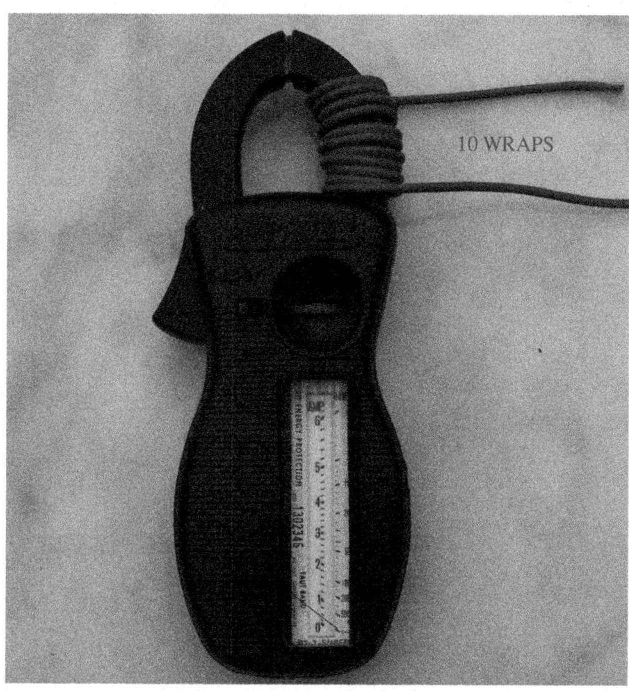

Figure 3-14 This ammeter is set up to measure low current with an analog clamp-on ammeter. The jaws of the clamp-on ammeter are wrapped 10 times with the wire being measured. The reading on the meter scale will be divided by the number of wraps, 10, to determine the true current reading.

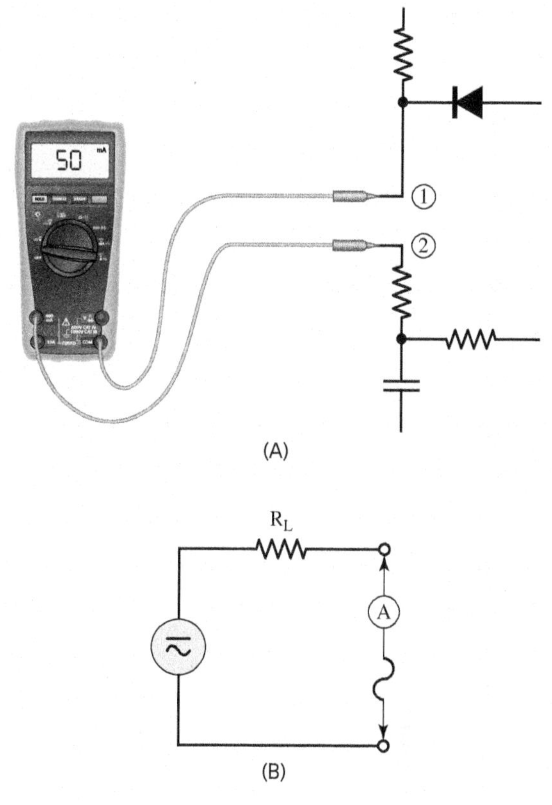

Figure 3-15 (A) A DMM can be used to measure amperage in series through the meter. (B) This is a simplified schematic view of Figure 3-15A. The *A* represents the ammeter in series with the load R_L.

the jaws of the clamp-on ammeter in Figure 3-14 are wrapped 10 times. The reading on the meter scale will be divided by 10 to determine the true current reading. If the reading is 10 A, then 10 A ÷ 10 = 1 A. The circuit is drawing 1 amp.

Some ammeters are analog (with a needle-type movement). Because the analog meter scale can be misread or misinterpreted, digital clamp-on ammeters are recommended for their accurate visual display. Many digital clamp-on ammeters have a peak amperage hold option. This allows the technician the opportunity to measure peak amps such as those found during motor-starting conditions.

Many clamp-on ammeters have options such as extra probes to measure voltage and resistance. The resistance range is low, usually less than 2,000 Ω. It is important to be aware of the resistance limits since some relay coils have higher resistance. For example, a coil with a 5,000 Ω resistance will read open if a clamp-on meter's resistance limit is 2,000 Ω. This type of low-resistance device is not recommended for checking motor resistance to ground. Instead, a meter than measures megohms should be used to measure grounded conditions. Again, knowing the operating limits of any instrument is important.

Measuring Amperage in Series with a DMM

Digital multimeter amperage readings can be taken with the meter placed in series with the current flow as illustrated in Figures 3-15A and 3-15B.

Figure 3-15A illustrates how to use a DMM to measure amperage in series. Using a DMM to measure inline amperage requires the technician to kill the power to the circuit. The circuit must be safely interrupted and the meter leads placed in series with the current flow being measured. It is important to have an idea of the current being measured. Connect the DMM in series with the load to measure amperage.

Most DMMs have an amperage limit of 10 amps or less; therefore, the circuit should not exceed this maximum amperage reading. Accidentally connecting the inline amp feature of the DMM across a load will blow the meter's fuse or damage the meter. Many meters are not protected against this type of misuse and have been known to explode. If you are not certain of the current draw, use a clamp-on ammeter first. If the current is less than 10 amps, turn off the power and hook up the inline ammeter.

Finally, these meters may also have a lower amperage range and may be able to measure milliamps (mA) and microamps (μA) AC and/or DC. Check the setting on the meter before measuring amperage, milliamps, and microamps. Start measuring on the highest amperage scale if you are uncertain of the current draw. You can always step down to the milliamp or microamp scales if you do not get a reading. Turn the power off each time you switch amperage scales.

In summary, there are two ways to measure amperage: the clamp-on amp meter method and the series, inline amperage hook-up. The clamp-on method is recommended for higher amperages.

SAFETY TIP

Measuring voltage while selecting the amperage range will damage your meter and possibly injure you. DMMs are not protected against misapplications. Accidentally connecting the inline amp feature of the DMM across a live voltage load will blow the meter's fuse or damage the meter. Many meters are not protected against this type of misuse and have been known to explode.

3.9 OPTIONAL FEATURES

Some DMMs have additional features or options. These options may include a temperature tester, capacitor tester, and diode tester. Of these features the temperature tester and capacitor tester are the most valuable. The temperature tester can be used to check air temperatures, superheat, and subcooling (see Figure 3-16).

The capacitor option is valuable for troubleshooting. A weak capacitor can prevent a motor from starting or can create a high current running condition. The high current running condition will cause the motor to trip out on its overcurrent protection. The capacitor checker will determine if the capacitor is within ±5% of the rated capacitance value. A meter is not normally selected for these features, but it is handy to have them in one instrument.

Another option is the MIN-MAX feature (see Figure 3-17). This feature displays the minimum or maximum reading

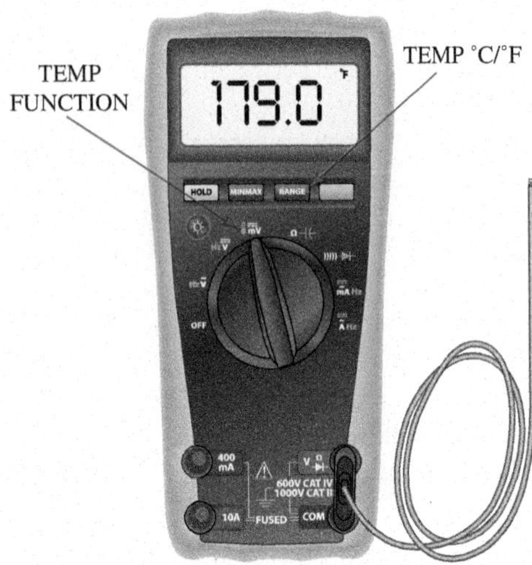

Figure 3-16 This DMM shows the temperature measurement function. It is part of the microvolt (mV) function. The temperature tester can be used to measure air temperature or be taped to a refrigerant line to measure superheating or subcooling.

Figure 3-17 The MIN-MAX feature shows the technician the lowest and highest ranges of what is being measured.

recorded during that meter use. This can be used to determine if there is a large voltage fluctuation in the circuit being measured.

Figure 3-18 shows another accessory device used to measure temperature on a suction line, discharge line, or liquid line. The accessory clamps on the refrigerant line like a jumper cable clamp. The clamp surface has a temperature sensor that is used to detect the surface temperature of the pipe. This allows measurement of superheat or subcooling.

Other accessories may include airflow-measuring devices, pressure-measuring devices, and high voltage–measuring adapters. Figures 3-19 through 3-24 illustrate some of the common accessories available for digital multimeters. Not all accessories are cross-compatible between meter model numbers or between manufacturers.

SAFETY TIP

Do not measure voltage with the test probe in the current jack. Meters are not designed to protect against this misapplication. The meter could explode! Figure 3-25 illustrates what can happen to a meter when misapplied. Read the meter's directions, warnings, and cautions prior to use. Misapplications are not covered under manufacturers' warranties.

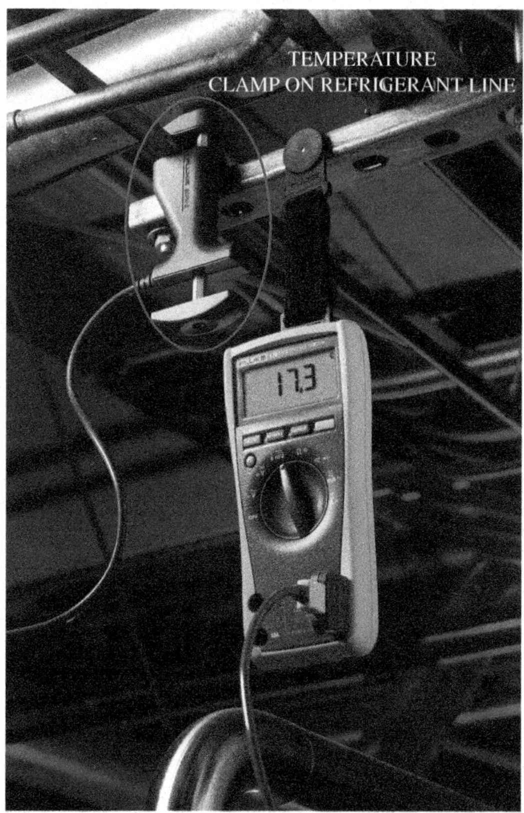

Figure 3-18 The temperature accessory for this DMM clamps onto a refrigerant line to measure superheat or subcooling. Notice that the meter is supported by a strap that has a strong magnet so that the tech does not need to hold the meter. The temperature is reading 17.3°C.

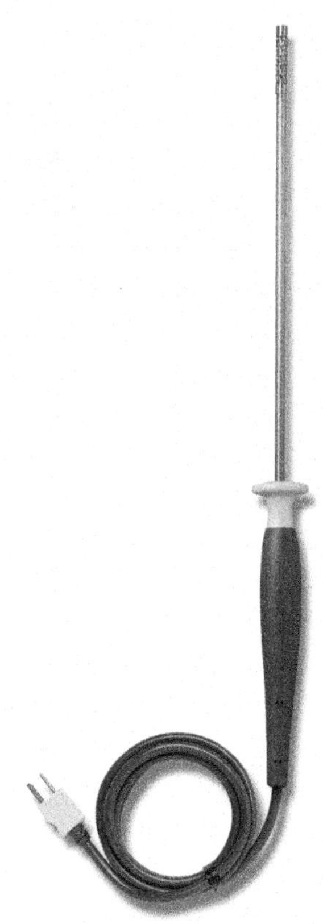

Figure 3-20 This air-measuring probe accessory can be used to measure return air and supply air temperatures. This will not give accurate surface or liquid measurements.

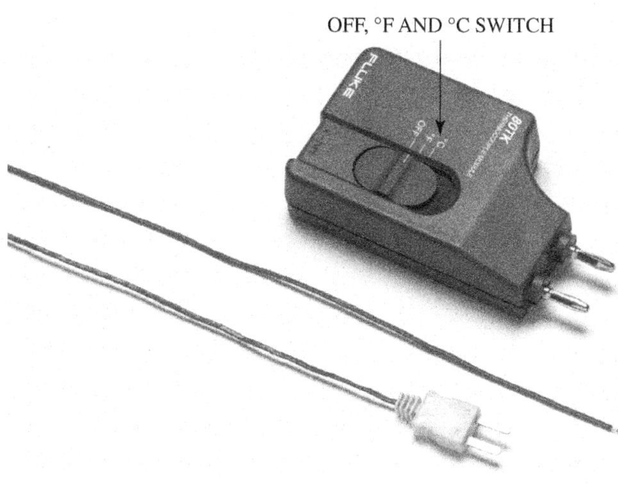

Figure 3-19 This temperature probe accessory can measure Fahrenheit or centigrade temperatures.

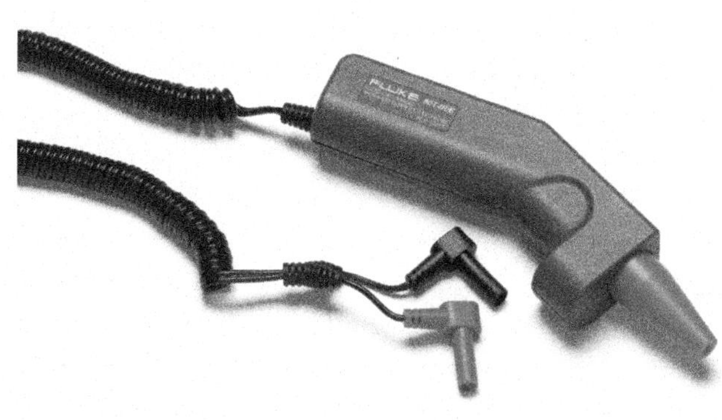

Figure 3-21 The infrared temperature measuring device is only accurate on dull, black surfaces. This will not give accurate readings for measuring superheating or subcooling. It provides a good indication of the general temperature of return and supply grilles.

Figure 3-22 Temperature can be measured using a clamp device. The clamp temperature device is good for measuring superheat and subcooling.

In summary, select a quality meter with the following features:

- Resolution that can display at least one-tenth of each scale
- A large visual display and bar graph reading
- True RMS readings.

Most importantly, select meters based on safety standards!

TECH TIP

The DMM is one of the mostly commonly used instruments for troubleshooting, so you should always have a backup meter available should your primary meter fail or show inaccurate reading. You can use a multipurpose meter such as a clamp-on ammeter with voltage and limited resistance functions as a backup device.

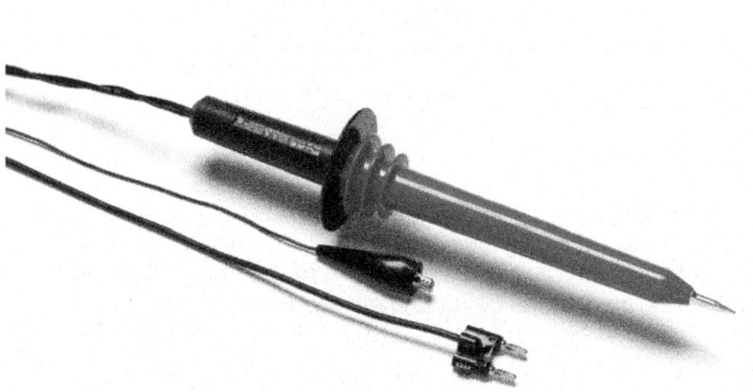

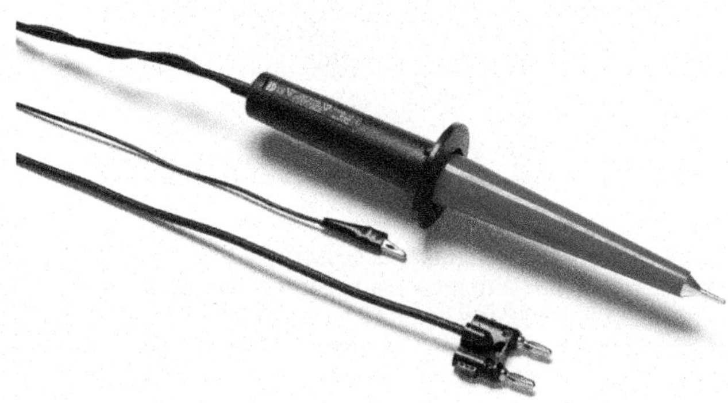

Figure 3-23 Special probes can be used to measure high voltage, low current in circuits like those found in electronic air filters. The electronic air filter may generate 10,000 V at a very low current flow.

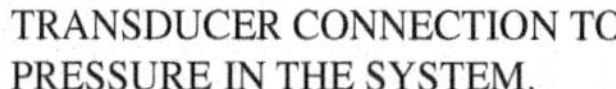

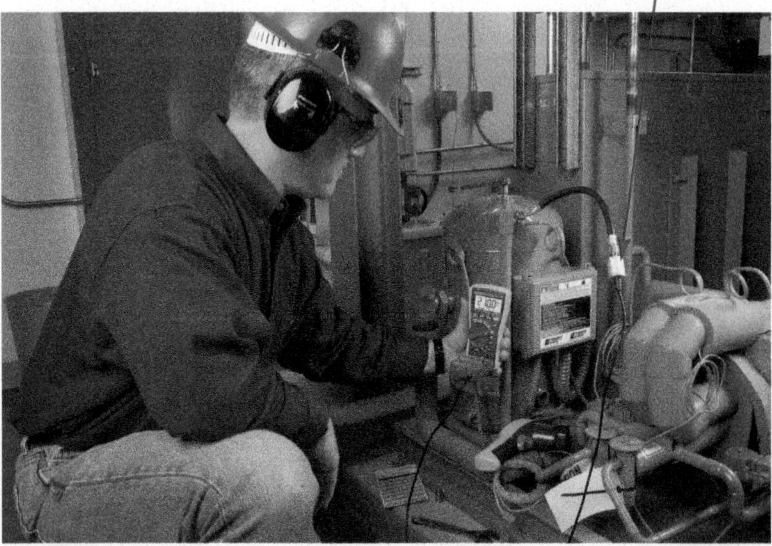

Figure 3-24 This accessory is a pressure transducer. The transducer converts pressure to a digital readout on the DMM. This can be used in lieu of a refrigeration gauge.

Figure 3-25 Misapplication of a meter can end in a damaged meter and an injured technician. Good quality meters protect against misuse, but sometimes the best engineered protection can still result in damage.

3.10 BEWARE OF GHOST VOLTAGE

Ghost voltage is a false voltage reading on a digital voltmeter. The ghost voltage is caused by capacitive coupling between energized and adjacent, unused wiring in a circuit. Even though there is enough voltage present to provide a sensitive meter reading, the circuit does not have enough current-carrying capacity to operate a component. An example of this is illustrated in Figure 3-26.

Analog voltmeters have low-impedance (Lo Z or 10-kΩ) circuits, which cause ghost voltage to dissipate and register a true 0-V reading. Impedance of **Lo Z** is a term used to express resistance in a capacitor or coil circuit. The disadvantage of an analog meter is that it loads down an electronic circuit and can change the operating characteristics of the solid-state device being tested.

When taking a voltage reading, the DMM has high-impedance (Hi Z or 1-MΩ) circuits, which do not interfere with the operation of circuit boards and other solid-state devices. The high resistance of a DMM circuit is similar to placing a high resistance in parallel with an existing circuit. The high resistance has little current flow; therefore, it does not affect the operation of the circuit it is paralleling.

The solution to reading ghost voltage is to use a digital multimeter with a low-impedance option. This is also known as a dual-impedance multimeter. It has the normal high-impedance feature and the low-impedance (Lo Z) option. Most ghost voltages are about 80% of the normally supplied voltage. For example, if the normal circuit voltage is 120 V, the ghost voltage would be around 95 V. This creates a puzzle for the technician. The technician will wonder if 95 V is the true voltage or a ghost, inducted voltage. Setting the DMM to the Lo Z setting will reduce the ghost voltage to zero. If it is a true low voltage rather than a normal voltage condition, it will continue to register 95 V. The DMM should *not* be placed on the Lo Z setting when measuring solid-state circuits. As mentioned, the low impedance can affect the operation of the circuit, especially solid-state circuit boards.

SAFETY TIP

According to OSHA, an average of one worker is electrocuted on the job every day. The four main types of electrical injuries are:

- Electrocution (death due to electrical shock)
- Electrical shock
- Burns from electrical shock
- Falls from electrical shock.

Don't be one of these statistics.

TECH TIP

Use the correct fuse (Figure 3-27).

A fuse has an amperage and a voltage rating. Matching both ratings is important. Installing a higher amperage fuse will not protect the meter from dangerous overcurrent conditions. Installing a lower voltage fuse will not allow the fuse element to separate enough to stop electricity from arcing across an open fuse element. The fuses used in our industry are time-delay or HVACR-type fuses. This means they will tolerate a short burst of high current before opening. This is required because many motor loads in HVACR equipment start with a momentary high current draw. If a time-delay fuse design were not used, this overcurrent protection would blow on start-up or at least more often than necessary, creating what is called a nuisance fuse trip.

3.11 SELECTING A SAFE MULTIMETER

DMM devices provide varying levels of inherent protection from misuse or dangerous conditions. Price is an indicator of safety features and other options. It costs the manufacturer of these products to make them safe and have them tested by a certified laboratory. The International Electrotechnical Commission (IEC) has published safety standards for working on electrical systems. Use meters that meet the IEC standards, which include voltage ratings and work environment. For example, if the measured voltage is around 480 V, the meter should have a Category III 600-V or 1,000-V rating. Sometimes Category III is abbreviated CAT III. In addition to the IEC rating, the meter should also have additional certifications from one or more of the following safety testing organizations: UL, CSA, VDE, or TUV. Figure 3-28 shows the

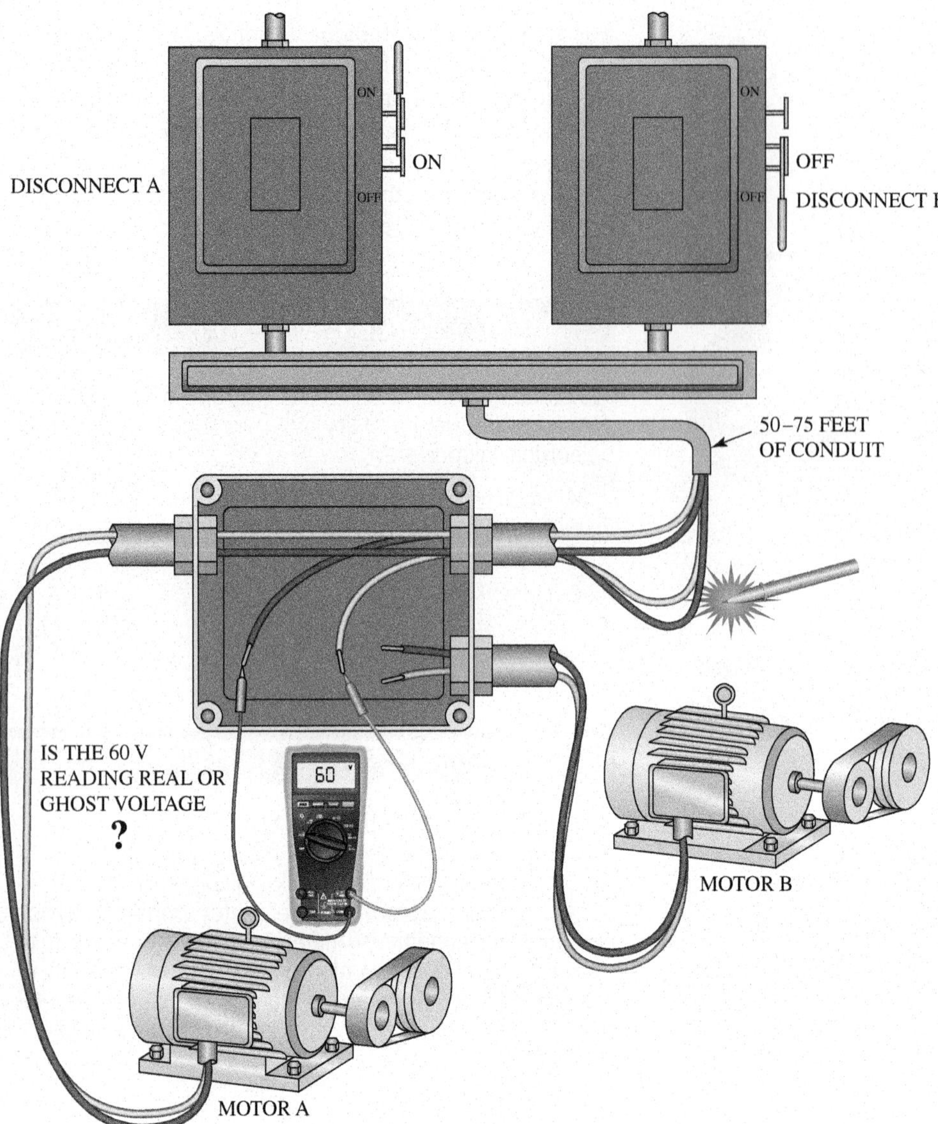

Figure 3-26 Here is an example of how ghost voltages can occur and where they can confuse voltage measurements. Both the voltmeter and the noncontact voltage indicator show a voltage measurement created by a ghost voltage.

approval symbols for these organizations. A manufacturer is allowed to imprint these symbols on meters that have been tested and found safe by these organizations. This will ensure that the multimeter you choose was actually tested for safety. Do not get a false sense of security, however—even a safe meter can be dangerous to the user if misapplied.

The following points are important when shopping for a safe multimeter. Visually check for:

- The category rating (e.g., "1000V CAT III" or "600V CAT IV") imprinted on the case
- Double-insulated meter case
- Shrouded connectors and finger guards as shown in Figure 3-29
- Insulation on probe leads that is not melted, cut, cracked, or damaged
- Connectors on probes that are not damaged (e.g., no insulation has pulled away from end connectors)
- Probe tips that are not burned, loose, or broken.

Figure 3-30 (page 42) is checklist for selecting a safe multimeter. Use it to help ensure your safety.

3.12 HOW MUCH ELECTRICITY IS TOO MUCH?

Electricity is the deadly, invisible killer. Even if an electrical shock does not kill you, it can injure you for life. The following points show you how just a small amount of electricity can hurt and harm:

- Currents greater than 75 milliamps (75 mA = 0.0075 amp) can cause ventricular fibrillation (rapid, ineffective heartbeat).
- Currents greater than 75 mA will cause death in a few minutes unless a defibrillator is used.
- 75 mA is not much current—a regular power drill uses 30 times as much.

Figure 3-27 Fuses have an amperage rating and a voltage rating. It is important not to exceed the amperage rating. The voltage rating of the replacement fuse should be equal to or higher than that of the blown fuse.

Figure 3-28 How can you tell if you are purchasing a CAT II or CAT III meter? Look for the symbol and category listing number of an independent testing lab such as UL, CSA, TUV, or other recognized approval agency.

SAFETY TIP

The following standards relate to electrical safety. Understanding their importance will lead to a long and successful career in the HVACR profession:

- NFPA 70E-1
- National Electrical Code
- OSHA 29-CFR 1910
- IEEE Standard 1584-2002.

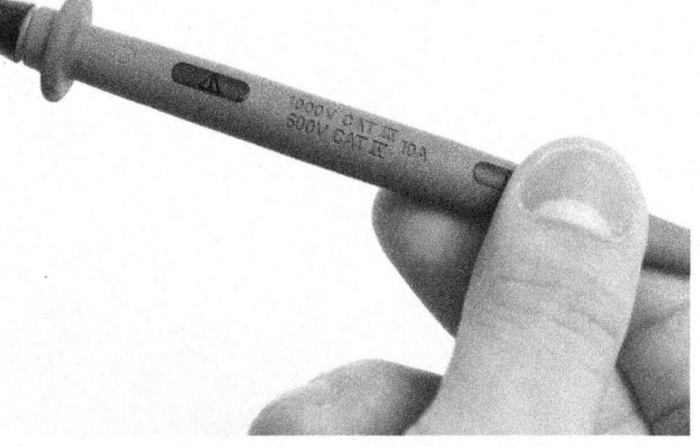

Figure 3-29 Test leads should be double insulated, recessed/shrouded, with finger guards. Note the category imprints. Replace when damaged.

Figure 3-30 Meter safety selection standards. Your meter should pass all of these standards in order to qualify as a "safe meter."

METER SCORECARD		
TESTED & CERTIFIED BY TWO OR MORE INDEPENDENT LABS	❑ PASS	❑ FAIL
VISUAL INSPECTION OF TESTER FOR CRACKS OR FADED DISPLAY	❑ PASS	❑ FAIL
VISUAL INSPECTION OF TEST LEADS FOR CRACKS, CAT RATING, ETC.	❑ PASS	❑ FAIL
TEST LEAD CONTINUITY	❑ PASS	❑ FAIL
RATED CAT III 600 OR 1000 VOLT OR CAT IV 600 VOLT	❑ PASS	❑ FAIL
TESTER IS DOUBLE INSULATED	❑ PASS	❑ FAIL
OHMS AND CONTINUITY CIRCUIT PROTECTION	❑ PASS	❑ FAIL
TESTER HAS APPROPRIATE FUSES AND THEY ARE WORKING	❑ PASS	❑ FAIL

*BASED ON NFPA 70E, IEC 61010 AND ANSI S82.02 STANDARDS.

3.13 ARC FLASH

This section is not meant to be the final word on arc flash; it is an introduction to the safety aspects of this possibly dangerous condition.

Work around any amount of voltage requires a minimum of safety gear as shown in Figure 3-31. The gloves in Figure 3-31 may not necessarily be shock-resistant gloves, but any protection is better than exposed, bare skin. When using general-purpose gloves around electricity, it is a safe practice to act as if you have no gloves on at all. Do not allow gloves to touch bare wires. The use of safety glasses is a very important part of this safety mix. The technician in Figure 3-31 is checking the electrical panel for voltage that feeds a condensing unit. For standard electrical measurements, high-impedance meters are preferred, unless ghost voltage is present.

An arc flash is a voltage across the resistance of air that results in an arc or voltage flash. An arc flash is an unexpected voltage to ground or a lower voltage between two voltage probes, as shown in Figure 3-32. Arc flash voltages are usually 480 V or greater. Technicians working around high voltages require special **personal protective equipment** known as **PPE.**

Figure 3-33shows the correct way to measure high voltage with insulated protective gloves and the lockout and tagout procedures. Higher voltages will require additional distance and PPE. Check the **National Fire Protection Association (NFPA)** arc flash recommendations when working with voltages over 480 V. Even voltages of less than 480 V are deadly.

Systems that operate on 600 V or less have an arc flash zone of up to 4 feet. The arc flash zone on voltages above 600 V is based on a recommended guideline and is greater than 4 feet.

Electric arcs produce some of the highest temperatures known to occur on earth—up to 35,000°F. This is hotter than the sun's surface. All known materials are vaporized at this temperature. When materials vaporize they expand in volume (copper expands by 67,000 times; water by 1,670 times). The air blast created by an arc flash can spread molten metal

Figure 3-31 The safe technician will wear gloves and safety glasses when measuring voltage. This technician is reading 227 VAC.

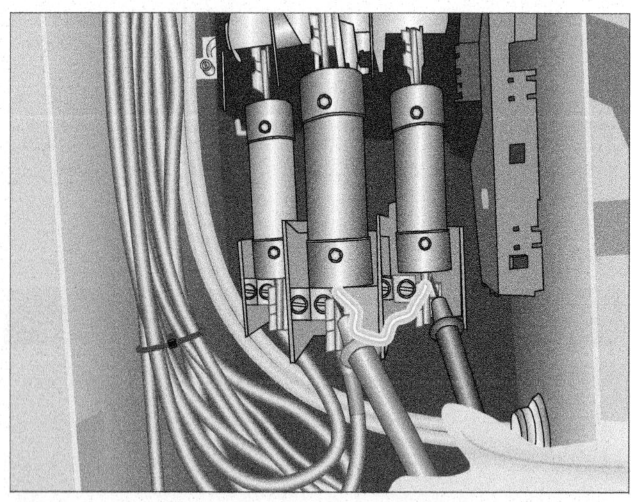

Figure 3-32 An arc flash can occur between high-voltage and low-voltage sources. Working around high voltages requires the technician to wear PPE.

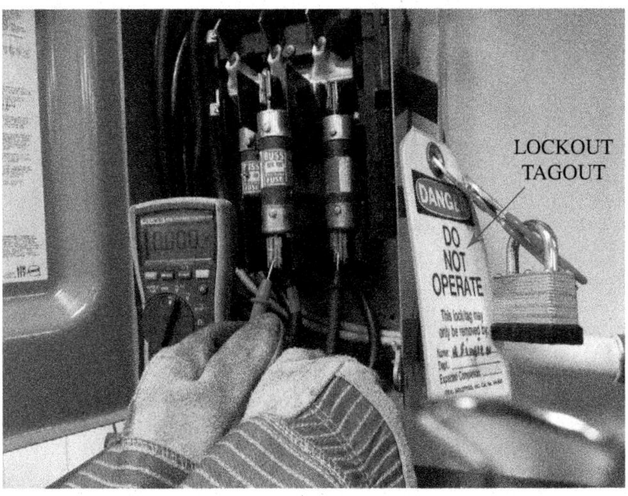

Figure 3-33 The technician uses heavy, insulated rubber gloves when measuring high voltage. Notice on the right side the lockout and tagout safety action tags used to protect the technician against accidental energizing of the part of the circuit that is assumed dead. The voltmeter at left is reading 0 V.

to great distances with force. The results of arc flash are rapidly expanding gases, extreme pressure and sound waves, and molten metal and metal plasma.

Possible Effects of Arc Flash on Humans

The pressure developed by a blast's pressure waves can throw workers across a room. Pressure on the chest can be higher than 2,000 lb/ft^2, blowing the clothes off the body. Clothing can be ignited from several feet away. Clothed areas can be burned more severely than exposed skin if clothing melts.

Hearing loss can occur from an arc flash sound blast. The sound can have a magnitude as high as 140 dB at a distance of 2 feet from the arc.

Arcs in enclosures, such as a motor control center (MCC) or a panel board, magnify blasts and the energy transmitted as the blast is forced to the open side of the enclosure and toward the worker. This situation is referred to as an *arc-in-the-box* event.

Possible Causes of Arc Flash

Dust and impurities contribute to an arc flash. Dust and impurities on insulating surfaces can provide a path for current, allowing it to flash over and create an arc discharge across the surface.

Corrosion can also contribute to arc flash. Corrosion of equipment creates impurities on insulating and conducting surfaces. Corrosion can also weaken the contact between conductor terminals, increasing the contact resistance through oxidation or other contamination. Heat is generated on the contacts when they are closed due to current flowing though now resistive contacts than once had no resistance. Sparks may be produced when the contacts close or open. This can lead to arcing faults closest to a lower voltage potential source. In other words, the spark will jump from the contacts to ground or a lower voltage source.

Condensation or water vapor can drip, causing tracking on the surface of insulating materials. This can create a flashover to ground.

Spark discharge is another way an arc flash can occur. Accidental contact, dropping of tools, and voltage arcing over to meter probes can cause a spark discharge.

Overvoltage across narrow gaps is a sign of arc flash. Overvoltage can occur if the power supply fails to stay within its range. Lightning can strike the power supply many miles away and create an arc flash on the equipment on which you are working.

Failure of insulating materials causes arc flash.

Note: For a low-voltage system (480/277 V), a 3- to 4-inch arc can become "stabilized" and persist for an extended period of time.

3.14 WHAT TO WEAR WHEN MEASURING VOLTAGE

Personal protective equipment (PPE) provides protection against an arc flash or minor electrical sparks. PPE includes the following clothing and equipment safety considerations:

- Voltage-rated gloves
- Voltage-rated tools
- Doubled-layered switching hood and hearing protection
- Untreated natural fiber, including T-shirts and long pants (No synthetic fibers are allowed, alone or in blends; must use all-cotton or fire-rated shirts and pants.)
- Fire-resistant clothing: long-sleeve shirt, pants and coveralls
- Fire-resistant protective equipment: hard hat, safety glasses, leather gloves and leather work shoes.

3.15 SAFETY TIPS FROM CANADA

The Construction Safety Association of Ontario, Canada, states the following:

> Momentary high-voltage transients or spikes can travel through a multimeter at any time and without warning. Motors, capacitors, lighting, and power conversion equipment such as variable-speed drives are all possible sources of spikes. Important tips to know in the safe use of multimeters include the following:
>
> - Ensure that the meter's voltage is appropriate for the work being done.
> - Use PPE such as eye protection, flame-resistant clothing, long-sleeve shirt, dielectric safety boots, rubber gloves with leather protectors, mats, blankets, and shields.
> - Check the manufacturer's manual for specific cautions. Moisture and cold affect the performance of your meter.

- Wipe clean the multimeter and test leads to remove any surface contamination prior to use.
- Start with high ranges of the multimeter, then move to lower ranges when the values to be measured are uncertain.
- Connect to ground first and disconnect to ground last.
- Test the multimeter on a known power source to verify the meter's proper function before and after testing the suspect circuit.

SAFETY TIP

Safety standard 70E, established by the National Fire Protection Association, sets electrical safety standards for the workplace. The standard requires that voltage above 50 volts be de-energized before working on the equipment. This also requires lockout/tagout procedures. The standard realizes that there are tasks in HVACR that cannot be done without the equipment being energized. Standard 70E requires that PPE be used to protect the technician from three hazards:

- *Electric shock:* any part of the body making contact with electricity
- *Arc flash:* intense heat and light energy that is generated by an electrical short circuit condition in the area where the technician is working
- *Arc blast:* an explosion resulting from the expanding gas generated by a short circuit; the explosion may send flying parts and molten metal through the air.

3.16 SAFETY TIPS WHEN USING METERS

Most voltages you will be working with are less than 600 V. This requires a meter that will meet a Category III 1,000-V minimum rating. This means that the meter has the proper circuit fuse protection and insulating leads to be used safely in most HVACR applications.

Even though you are using a good quality meter that meets the highest safety standards, the meter can still be damaged and even explode. This could occur if the user changes scales while the meter is connected to a power supply. Damage can also occur if the power source exceeds the rated value of the meter or the meter is used on the wrong option. For example, reading voltage when the meter is set to the amperage reading scale will damage the meter. This type of misuse is not protected and can be dangerous to the technician.

3.17 LOCKOUT/TAGOUT PROCEDURE

The **lockout/tagout** procedure, also known as **LOTO**, is a way of ensuring that power that a technician has turned off will not be turned on until the technician has completed his work. Injuries and deaths attributed to electric shock can be avoided by implementing a comprehensive LOTO program, complete with periodic safety training, clearly written procedures, and proper lockout/tagout equipment. Various types of required lockout/tagout items are shown in Figure 3-34.

Figure 3-34 Here are a variety of lockout and tagout items necessary to be safe on the job. At a minimum you will need the one key lock, tag, and the multiple locking device.

Figure 3-35 shows an example of a correct LOTO installation. The disconnect is open or in the down position. The locking mechanism is over the disconnect loop. Each locking mechanism has only one key, so once a technician has locked out the disconnect, only that technician has access because the tech has the only key. The name of the tech who

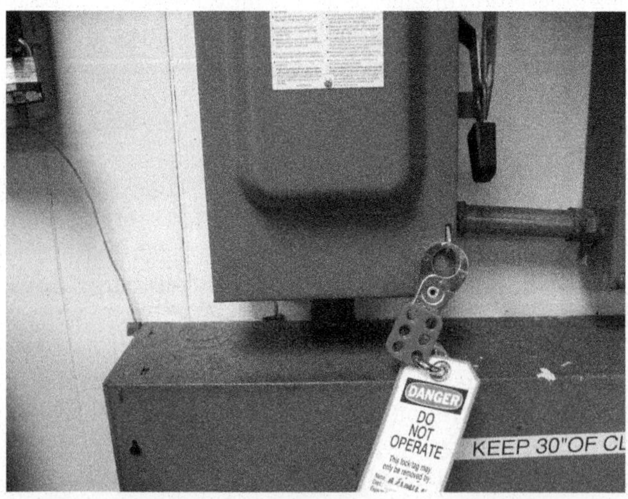

Figure 3-35 This is an example of a correct LOTO installation.

has locked out the equipment is on the tag. Additional locks and tags may be added to the disconnect if other techs are working on the system and want to be assured that no power is applied without their knowledge.

For a full description of a LOTO program, view the web page of the Environmental Safety and Health Group of OSHA at www.OSHA.gov.

SERVICE TICKET

As part of a weekly technicians' meeting, the service manager asks her staff to bring in their multimeters and the instructions for that meter. As the meeting opens, the manager has everyone read the directions on how to use their meter and its features and benefits. After 20 minutes of the meter instruction review, the manager asks each technician to state what feature or benefit he or she learned about that he or she was unaware of before the service meeting. Every technician had learned one or more new things about the meter. Many attending were excited and wanted to try to use what they had learned. It was a good meeting. *Remember:* Read your meter's instructions prior to using it and after you have used the meter for awhile. Often, meters have numerous functions that never get used because the user does not realize how and when to use them.

SUMMARY

This unit discussed the features and benefits of digital multimeters and clamp-on ammeters. It is important to select electrical instruments based on their safety, quality, and the features it has that you require to do a good job. Price is usually the first consideration that influences a technician's decision to purchase a meter—but price should be the last consideration. Consider what you pay for a quality instrument on an annual basis and you will notice that the yearly cost is not that much. A good quality meter will last more than 10 years. Divide the cost of the meter by 10 to determine the annual cost of having a safe and accurate instrument to use. For example, if you purchase a DMM with a life expectancy of 10 years for $400, that would equal an investment of $40 per year or about $3 a month. Most technicians would be willing to rent a quality meter for $3 a month. *Remember:* These are your professional tools and you need them to do a good job.

We close this unit as we opened it and include a list of what *not* to do when taking electrical measurements. The following information is presented by Fluke Instruments as a service to the HVACR industry.

Ten Dumb Things Smart People Do When Testing Electricity

Anyone who makes a living by working with electricity quickly develops a healthy respect for anything with even a remote chance of being "live." Yet the pressures of getting a job done on time or getting a mission-critical piece of equipment back on line can result in carelessness and uncharacteristic mistakes by even the most seasoned electrician. The list below was developed as a quick reminder of what not to do when taking electrical measurements.

1. Do not **replace the original fuse with a cheaper one.** If your digital multimeter meets today's safety standards, that fuse is a special safety fuse designed to pop before an overload hits your hands. When you change your DMM fuse, be sure to replace it with an authorized fuse.
2. Do not **use a bit of wire or metal to get around the fuse all together.** That may seem like a quick fix if you're caught without an extra fuse, but that fuse could be all that ends up between you and a spike headed your way.
3. Do not **use the wrong test tool for the job.** It's important to match your DMM to the work ahead. Make sure your test tool holds the correct CAT rating for each job you do, even if it means switching DMMs throughout the day.
4. Do not **grab the cheapest DMM on the rack.** You can upgrade later, right? Maybe not, if you end up a victim of a safety accident because that cheap test tool didn't actually contain the safety features it advertised. Look for independent laboratory testing.
5. Do not **leave your safety glasses in your shirt pocket.** Take them out. Put them on. It's important. Ditto insulated gloves and flame-resistant clothing.
6. Do not **work on a live circuit.** De-energize the circuit whenever possible. If the situation requires you to work on a live circuit, use properly insulated tools, wear safety glasses or a face shield and insulated gloves, remove watches or other jewelry, stand on an insulated mat, and wear flame-resistant clothing, not regular work clothes.
7. Do not **fail to use proper lockout/tagout procedures.**
8. Do not **keep both hands on the test probes.** Don't! When working with live circuits, remember the old electrician's trick. Keep one hand in your pocket. That lessens the chance of a closed circuit across your chest and through your heart. Hang or rest the meter if possible. Try to avoid holding it with your hands to minimize personal exposure to the effects of transients to the meter.
9. Do not **neglect your leads.** Test leads are an important component of DMM safety. Make sure your leads match the CAT level of your job as well. Look for test leads with double insulation, shrouded input connectors, finger guards, and a nonslip surface.
10. Do not **hang onto your old test tool forever.** Today's test tools contain safety features unheard of even a few years ago, features that are worth the cost of an equipment upgrade and a lot less expensive than an emergency room visit.

REVIEW QUESTIONS

1. What is the purpose of reading the operating instructions for a multimeter?
2. What does the abbreviation DMM mean?
3. What three basic measurements should a DMM be able to measure?
4. What is the typical accuracy of a DMM?
5. What does resolution mean when referring to a DMM?
6. What does meter range mean when referring to a DMM?
7. Why is it important to have a voltmeter that can measure true RMS?
8. What safety organizations should your meter have listed?
9. What do the following symbols found on multimeters mean?

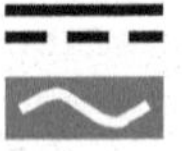

mV

A

mA

μA

Ω

kΩ, MΩ

))))

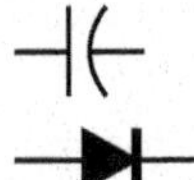

Hz

dB

Hold

MIN- MAX

10. The voltmeter can only measure ______________ difference.
11. Low resistance is measured on which scale?
12. How many wires should be surrounded or clamped on when measuring amperage using a clamp-on ammeter?
13. How is the DMM set up to measure amperage?
14. What low range of low amperage can a DMM measure?
15. What is a *ghost voltage*?
16. Is the digital or analog meter most likely to measure ghost voltage?
17. List five features of a safe multimeter.
18. What is an arc flash?
19. What PPE is required to reduce dangers from arc flash or minor electrical sparks?
20. What is the importance of lockout/tagout procedures?
21. What six meter accessories are available on some DMMs?
22. How can you measure low-amperage conditions with a clamp-on ammeter?
23. The clamp-on ammeter measures amperage. What other measuring options does a clamp-on ammeter have?
24. Review the "Ten Mistakes People Make When Working on Electrical Systems," found at the beginning of this unit. Select the point that is most important to you and discuss why it is most important to you.
25. Name five things that can be measured with the DMM shown in Figure 3-36.
26. Can the meter shown in Figure 3-36 be used as an inline or clamp-on amperage reading device?

Figure 3-36 Use this meter to answer Review Questions 25 and 26.

UNIT 4
Electrical Fasteners

WHAT YOU NEED TO KNOW

After studying this unit, you will be able to:

1. Identify the color of a wire nut required to connect specific wire sizes.
2. Describe the difference between a bolt and screw.
3. List the types and sizes of nails.
4. Describe how to attach a wire to a quick disconnect.
5. Describe where to use butt, fork, spade, and ring connectors.
6. Identify the color of the quick connector used with specific wire sizes.
7. List the tools used with fasteners.
8. Describe the approved electrical tape.

A **fastener** is a hardware device that mechanically joins or affixes two or more objects together. Some examples of fasteners we use in our industry are nylon ties, screws, bolts, tapes, and mastics. As a professional it is important to know the names and applications of the many types of fasteners used in our industry. You will learn the names of the various types of fasteners as you gain experience. Some of the names used for a fastener are specific to an area of the country. For this reason, you should always try to use the generic or common name for a fastener so that you will be understood no matter where you are working.

Fasteners are used daily in HVACR work. They are used to securely close condensing unit and furnace cabinets, to join wires, and to seal ductwork and fabricate support for HVACR equipment. Many types of fasteners are used throughout the world. This unit is limited to fasteners used in the HVACR trade.

This unit also discusses some of the tools used in conjunction with fasteners. This discussion includes nut drivers, nylon tensioning tools, and wire crimper/strippers.

4.1 WIRE NUTS

Wire nuts are used to mechanically join two or more wires. (Note that Wire-Nut® is a registered trademark of Ideal Industries, Inc.) They are also known as cone or thimble connectors.

Twist-on wire connectors are available in a variety of sizes and shapes and are typically made from plastic, with a tapered, conducting metal coiled insert that threads onto the wires to hold the wires secure. Many of the newer wire nuts do not include the metal spring insert. Twist-on wire connectors are intended to be installed by hand only, and on the surface may include molded grooves, wings, or blades intended to assist in hand installation.

Twist-on wire connectors are commonly color coded to indicate the nut's size. They are often used as an alternative to terminal blocks or soldering of conductors because they are cleaner, faster, and easier to remove for future rework.

Twist-on wire connectors are generally not recommended for use with aluminum wire.

As a safety tip, cut the power when possible from the wires that will be connected or disconnected under a wire nut. The wire nuts can be installed in one of two ways. The wires to be connected are stripped back 7/16 in. for 16-AWG wire and 3/8 in. for all other gauge wires. One method as shown in Figure 4-1 is to place the wires side by side, making sure the ends of the wire are even. When fastening solid and stranded wire together, align the stranded wire slightly ahead of the solid wire. Place the appropriate size wire nut over the wires and twist clockwise.

Some technicians like to twist the wires together first then place the wire nut over the wires and twist to tighten. When using this practice, it is recommended that you expose more wire when you strip it back since the pretwist shortens the wire bundle. Sometimes when the wire is pretwisted, its diameter increases and makes it difficult to install into the wire nut shell. If this happens, simply untwist and straighten the wires and place the wire nut over the wires and then twist.

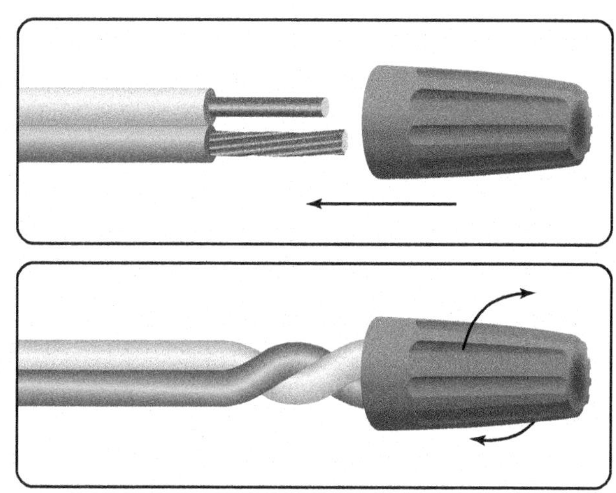

Figure 4-1 The correct way to use a wire fastener.

Table 4-1 Wire Nut Table

This table helps the technician decide what size wire nuts are required for the number and size of the conductor wires. This table is for pressure-type wire connectors. It can be used on solid or stranded copper-to-copper (Cu/Cu) combinations. This is a copper-only connector.

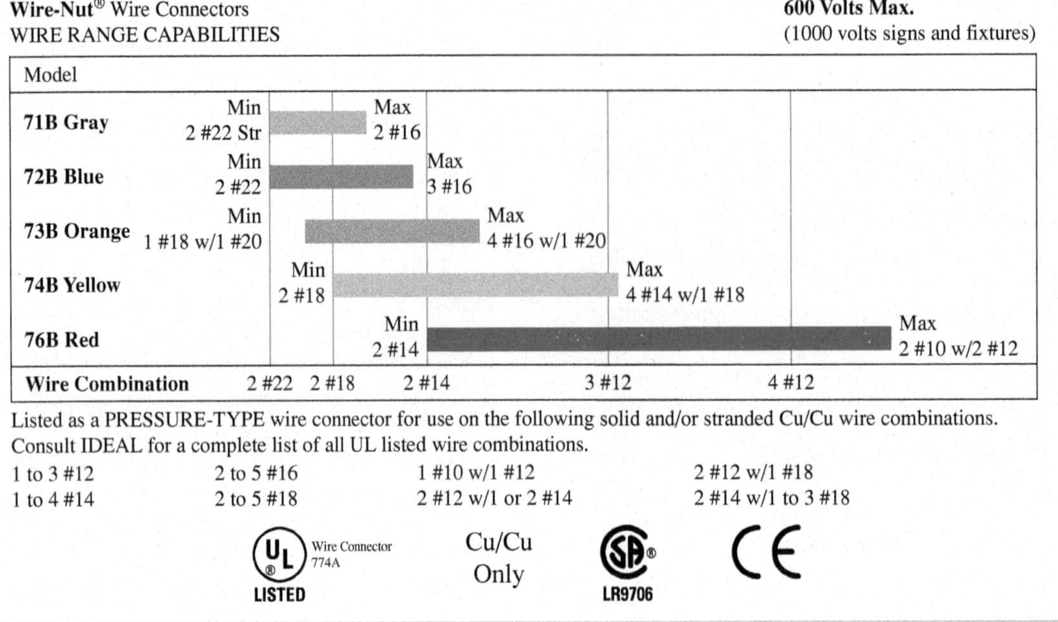

(Images provided courtesy Ideal Industries, Inc.)

Most plastic wire nuts have a 221°F temperature limit and a 600-volt rating. Higher temperature applications may require the use of ceramic wire nuts and higher temperature wire insulation.

The color of a wire nut indicates what **wire gauge** and how many wires a wire nut can accommodate. The larger the **gauge number**, the smaller the wire diameter. For example, Table 4-1 indicates that a gray wire nut will accommodate two 22-gauge or two 16-gauge wires. Notice that minimum and maximum wire sizes are given. Using an undersized wire will not provide a tight fit when twisting the wire nut on the wires. If you must use an undersized wire with a wire nut, double the wire by bending it back on itself to double its diameter. This will make the fit with the wire nut tighter.

Reviewing Table 4-1, a blue wire nut will handle two 22-gauge wires or up to three 16-gauge wires. An orange wire nut will handle one 18-gauge wire with one 20-gauge wire or four 16-gauge wires with one 20-gauge wire.

As you can see from the table, the larger wire nuts have more wire options in addition to being able to handle larger gauge wires.

Table 4-2 compares the wire nut color to the matching wire pairs that can go into the wire nut. You will notice that many of the color codes are repeated. For example, the table shows a big blue and a little blue.

In summary, wire nuts are common electrical fasteners. The professional tech will have available a selection of each type of wire nut listed in Table 4-1. There are larger wire nuts, but these are the most commonly used sizes found in our industry. The wire nut is a convenient way to connect and disconnect wiring.

Table 4-2 Wire Nut Colors and Wire Sizes

Color	Size
Big blue	#10 and larger
Gray	#14 and larger
Red	#14 and larger
Big tan	#14 and larger
Yellow	#18 and larger
Big orange	#18 to #14
Little orange	#22 to #14
Little blue	#22 to #16
Small tan	#22 to #18

4.2 SCREWS

Screws are used in many applications in our profession. They are used to keep equipment together, mount thermostats, mechanically fasten metal ducts together, and a hundred and one other uses. Typical types of screws are **sheetmetal screws**, **wood screws**, and **machine screws**. Sheetmetal screws are the most common type found in HVACR equipment. **Hex head** (six-sided) **sheetmetal screws** are commonly found in residential and commercial equipment.

The hex head screw shown in Figure 4-2 is **self-drilling**. The tip is designed to bore into the sheet metal and does not

Figure 4-2 This hex head sheetmetal screw is self-drilling. A pilot hole is not needed in most sheetmetal applications.

Figure 4-3 Hex head sheetmetal screw with slot head. This screw can be driven with a hex driver or slot screwdriver. This sheetmetal screw is known as a self-starting screw.

require the predrilling of a pilot hole. It is like having a drill bit on the tip of the screw.

Many hex screws have a slotted head for a screwdriver option as shown in Figure 4-3. This screw is self-starting and is sharp enough to drill into light-gauge metal such as that found in most metal ductwork. Thicker gauge sheet metal will require a pilot hole created by drilling a small hole.

Common hex head sizes used on HVACR units are 1/4, 5/16, and 3/8 in. **Hex screwdrivers**, also called **nut drivers**, are used when working with these types of screws (see Figure 4-4).

The color of the handle of the screwdriver determines the size that is used to fit the hex head. Larger screws and nut drivers are used on large tonnage systems that require strong sheetmetal security.

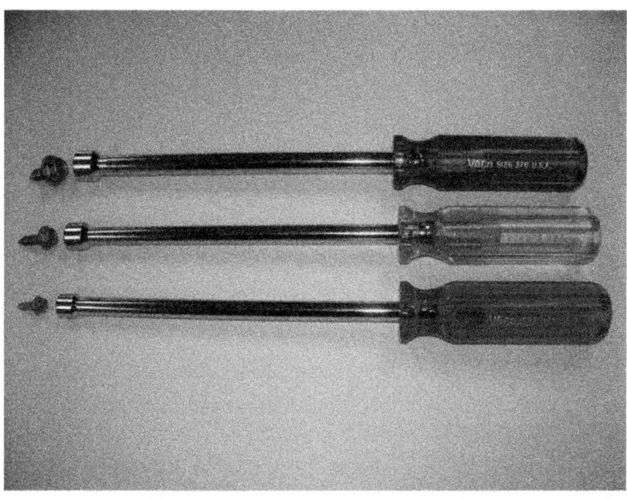

Figure 4-4 Three common nut or hex screwdrivers. The red-handled nut driver fits a ¼-in. hex head screw or nut. The gold one fits a 5/16-in. head, and the blue a 3/8-in. head.

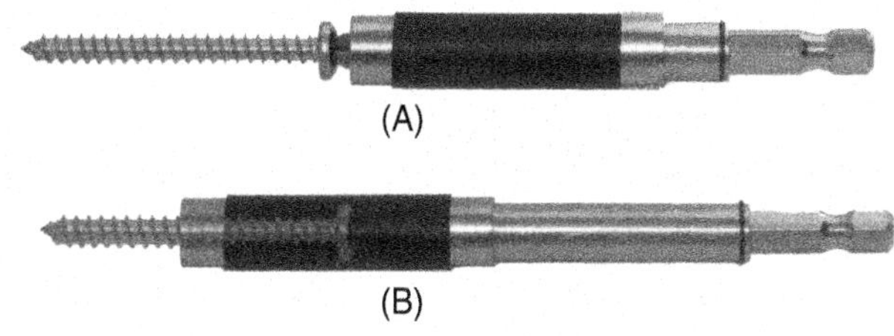

Figure 4-5 (A) A screwdriver magnetic chuck is used to hold the screw in the Phillips point. (B) The holder is retracted around the screw to keep it in place, which helps when trying to reach out with the screw to start it in the hole. A magnetic chuck is also available for hex head screws.

A **magnetic chuck** is useful when using a hex head fitting on a variable-speed drill. As shown in Figure 4-5, the magnetic adapter holds the screw in place when trying to reach and insert a screw into a hole.

A number of different screw head designs are commonly found inside and outside HVACR equipment. Figure 4-6 shows four common types:

- Six-sided hex head screw (Figure 4-6A)
- Round head screw (Figure 4-6B)
- Pan head screw (Figure 4-6C)
- Flat head screw (Figure 4-6D).

Round head and pan head screws may be found in electrical boxes and electrical terminal strips. The rounded head prevents the wire insulation damage that could occur if the insulation were to come in contact with the sharper edge of a hex head screw. Round, pan, and flat head screws have a straight slot or Phillips slot in the head. If a hex head screw has a slot, it will be a straight slot.

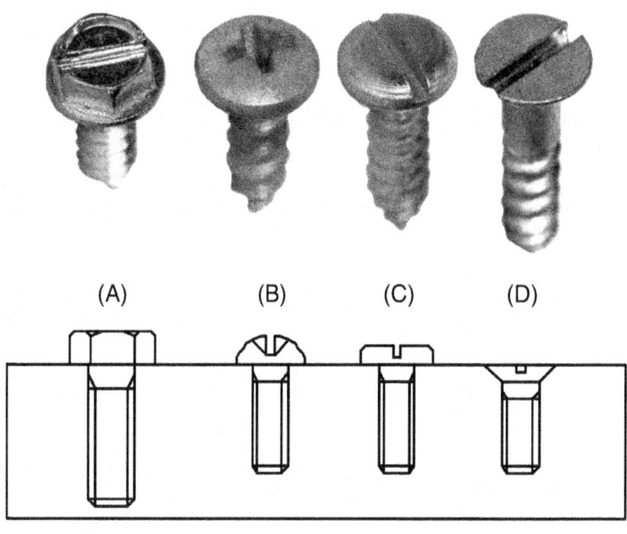

Figure 4-6 Common screws from left to right are the (A) hex head, (B) round head, (C) pan head, and (D) flat head. The bottom part of this figure shows how each screw is represented in a drawing.

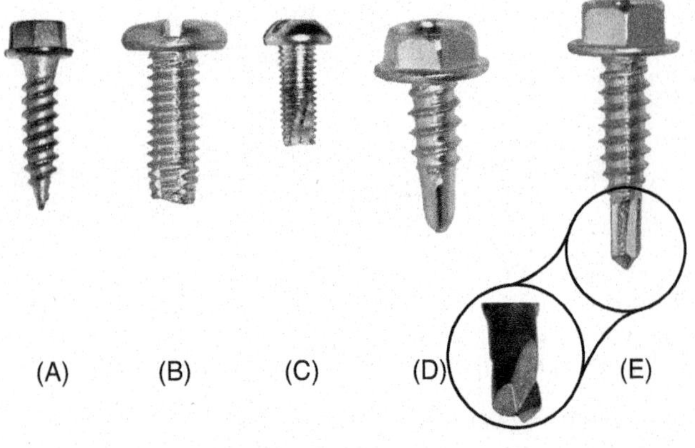

Figure 4-7 Common types of screws found in HVACR work include the (A) self-starting screw, (B) self-threading screw, (C) self-tapping screw, (D) round-tipped screw, and (E) self-drilling screw.

Figure 4-7 compares five types of screw tips:

- **Self-starting screw** (Figure 4-7A)
- **Self-threading screw** (Figure 4-7B)
- **Self-tapping screw** (Figure 4-7C)
- **Round-tipped screw** (Figure 4-7D)
- **Self-drilling screw** (Figure 4-7E)

These types of screws are commonly found in HVACR work. **Wood screws** and **sheetrock screws** are used to support ducts and site fabricate equipment stands and return air plenums. Figure 4-8 compares these two types of screws. Wood screws are usually shiny and tapered, with the thread stopping a distance from the head. The sheetrock screw is black or gray and normally has a full thread from tip to head. Wood screws are stronger than drywall screws. For the same length, the wood screw has a tapered body and larger diameter.

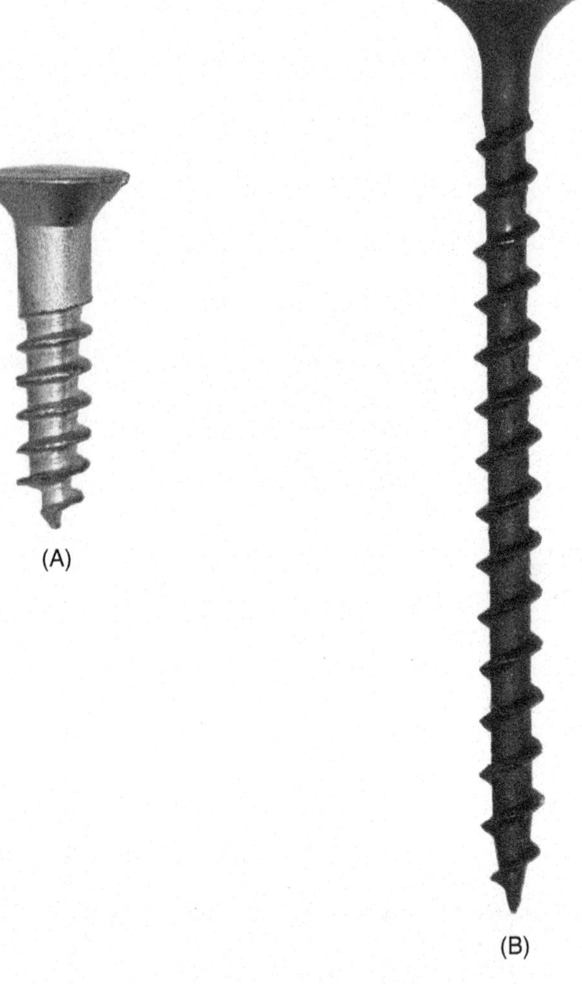

Figure 4-8 A wood screw is shown on the left (A), a drywall screw on the right (B). Wood screws are stronger than drywall screws. For the same length, a wood screw is tapered and has a larger diameter.

4.3 BOLTS, NUTS, AND WASHERS

A **bolt** is a type of fastener characterized by a helical ridge, known as a *thread*, that is wrapped around a cylinder. A common bolt is shown in Figure 4-9.

Threads are designed to mate with a complementary thread, known as an internal thread, often in the form of a nut or other object that has the internal thread formed into it. The most common uses of screws are to hold objects together and to position objects.

There is a debate about the difference between a screw and a bolt. For this book we will keep it simple. A screw will have some type of tip, whereas a bolt will be flat on the end that goes into an inside threaded hole or nut.

Screws often have a head, which is a specially formed section on one end of the screw that allows it to be turned, or driven. Common tools for driving screws include screwdrivers, sockets, and wrenches. The head is usually larger than the body of the bolt, which keeps the bolt from being driven deeper than its length and also provides a bearing surface. There are exceptions; for instance, **carriage bolts** have a domed head that is not designed to be driven; **set screws** have a head that is smaller than the outer diameter of the screw; and **J-bolts** do not have a head and are not designed to be driven. The cylindrical portion of the bolt or screw from the underside of the head to the tip is known as the shank; it may be fully threaded or partially threaded.

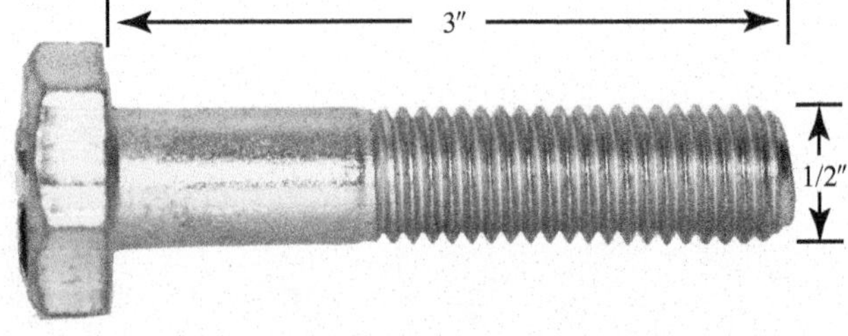

Figure 4-9 The length of a bolt is measured from the bottom of the head to the end of the tip. In this example, the length is 3 in. The diameter is measured on the outside of the course threads.

The majority of screws and bolts are tightened by clockwise rotation, which is termed a **right-hand thread**. Screws and bolts with left-hand threads are used in exceptional cases. For example, when the screw will be subject to anticlockwise forces (which would work to undo a right-hand thread), a left-hand-threaded screw would be an appropriate choice.

SAE International, a professional organization for engineers, publishes the standard **SAE J429**, which defines the bolt *grades* for inch-system sized bolts and screws. Grades indicate the strength of a bolt and range from 0 to 8, with 8 being the strongest. Higher grades do not exist within the specification. SAE grades 5 and 8 are the most common. Grade numbers are often stamped on the head of the bolt.

High-strength steel bolts usually have a hexagonal head with an International Organization for Standardization (ISO) strength rating, called its property class, stamped on the head. The absence of a marking/number indicates a lower grade bolt with low strength. The property classes most often used are 5.8, 8.8, and 10.9. The number before the decimal point is the tensile ultimate strength. The number after the point is 10 times the ratio of the tensile yield strength to the tensile ultimate strength. For example, a property class 5.8 bolt has a nominal (minimum) tensile ultimate strength of 500 MPa, and a tensile yield strength of 0.8 times the tensile ultimate strength or 0.8(500) = 400 MPa.

Tensile ultimate strength is the stress at which a bolt will fail. Tensile yield strength is the stress at which the bolt will experience a permanent set (an elongation from which it will not recover when the force is removed) of 0.2% offset strain. When a fastener elongates prior to reaching its yield point, the fastener is said to be operating in the elastic region; elongation beyond the yield point is referred to as operating in the plastic region, because at that point the fastener has suffered permanent plastic deformation.

Mild steel bolts have a property class of 4.6. High-strength steel bolts have a property class of 8.8 or above.

The same type of screw or bolt can be made in many different grades of material. For critical high-tensile-strength applications, low-grade bolts may fail, resulting in damage or injury. On standard SAE bolts, a distinctive pattern of marking is impressed on the heads to allow inspection and validation of the strength of the bolt. However, low-cost counterfeit fasteners may be found with actual strengths that are far less than that indicated by the markings. Such inferior fasteners are a danger to life and property when used in aircraft, automobiles, heavy trucks, and similar critical applications.

Like screws, a bolt's head has many different configurations. It can be hex, slotted, Phillips, or any of more than a dozen other designs. Figure 4-10 shows two bolt identifier templates.

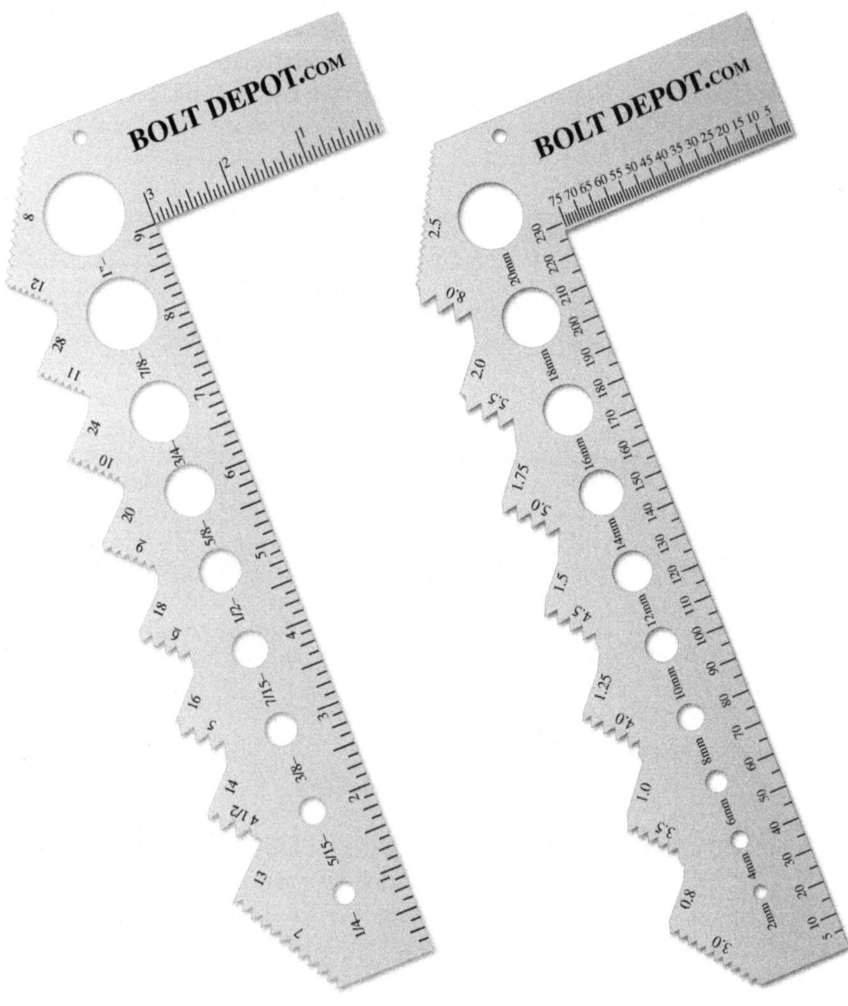

Figure 4-10 The bolt identification template on the left is used for standard SAE bolts. The template on the right is for metric bolts.

4.4 NAILS

Nails are occasionally used to hang duct strap, build equipment stands, or make return air plenums for residential equipment to sit on. Today's nails are typically made of steel and are often dipped or coated to prevent corrosion in harsh conditions or improve adhesion. Ordinary nails for wood are usually of a soft, low-carbon or "mild" steel, about 0.1% carbon. Nails for concrete are harder, with a higher percentage of carbon, to reduce bending while being driven into a hard material.

Nails are typically driven into wood by a hammer, a pneumatic nail gun, or a small explosive charge or primer. A nail holds materials together by friction in the axial direction and shear strength laterally. The point of the nail is also sometimes bent over or clinched after driving to prevent loosening.

Nails are made in a great variety of forms for specialized purposes. The most common is a wire nail. Other types of nails include pins, tacks, brads, and spikes.

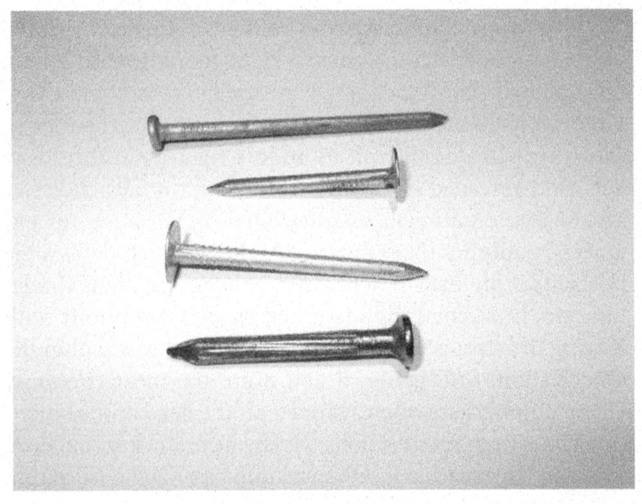

Figure 4-11 The head of a common nail is larger than the diameter of the shaft. A finishing nail is used when you want to hide the nail head.

U.S. Nail Sizes

In the United States, the length of a nail is designated by its penny size, written with a number and the abbreviation ***d*** for **penny**; for example, *10d* for a ten-penny nail. The larger the number, the longer the nail (see Table 4-3). Nails under 1¼ in. long, often called brads, are sold mostly in small packages with only a length designation or with length and wire gauge designations; for example, 1-in. 18-gauge or 3/4-in. 16-gauge nails.

Penny sizes originally referred to the price for a hundred nails in England in the 15th century: the larger the nail, the higher the cost per hundred. The system remained in use in England into the 20th century, but is obsolete there today. The *d* is an abbreviation for *denarius*, a Roman coin similar to a penny; this was the abbreviation for a penny in the United Kingdom before converting to the decimal system. Nail sizes used in HVACR projects are 6d, 8d, and 10d. The size will vary depending on the application.

Table 4-3 U.S. Nail Sizing Table

The larger the number in the left column, the longer the nail shaft. Common nail sizes include 6d, 8d, and 10d.

Penny Size	Length (in.)	Length (nearest mm)
2d	1	25
3d	1¼	32
4d	1½	38
6d	2	51
7d	2¼	57
8d	2½	65
9d	2¾	70
10d	3	76
12d	3¼	83
16d	3½	89
20d	4	102
30d	4½	115
40d	5	127
50d	5½	140
60d	6	152

TECH TIP

Let's take a look at some of the nails you may use in our profession. The usual type of nail that you will see is called the *common nail*. The head of the common nail is three to four times the diameter of the shaft. The *box* nail has a smaller shank when compared to a common nail. *Casing* nails have a slightly larger head than a finishing nail. Casing nails are used for crating or making wooden cases for equipment. The large head provides a larger capping surface. Finishing nails are used when you want the nail flush or countersunk with the wood surface. Finishing nails are used in trim such as a door frame or baseboard. *Galvanized* nails are used when moisture exposure is a problem. See Figure 4-11 for a comparison of two different types of nails.

4.5 NYLON STRAPS

Nylon straps or **nylon ties** are used to seal flex duct and bundle wires. They are handy fasteners that can be used for various applications on an HVACR job. Figure 4-12 shows how a nylon tie is used to seal flex duct on a duct collar. The tie is placed through the eye of the strap and hand tightened. A

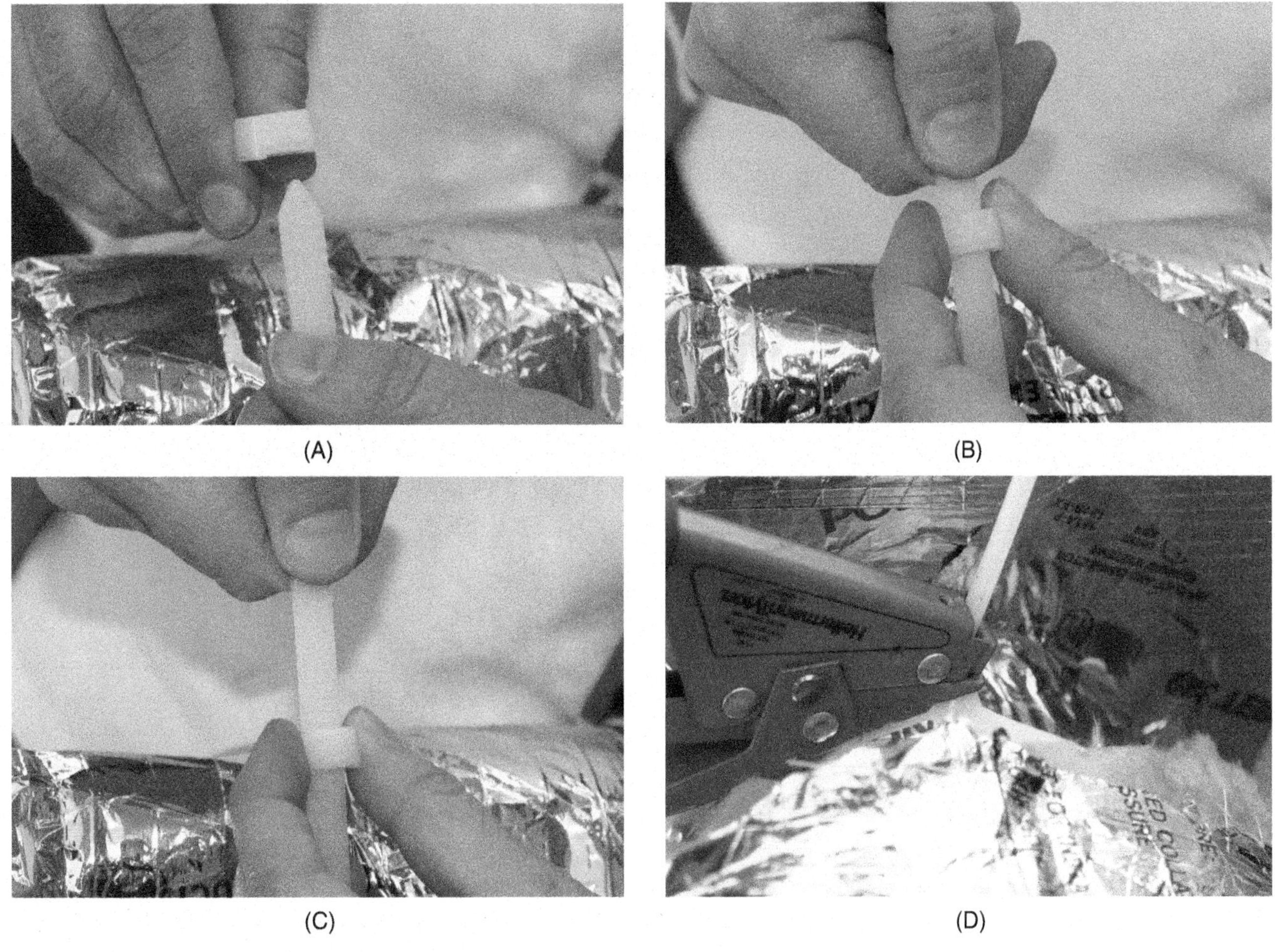

Figure 4-12 (A) Insert the end of the nylon tie strap through the ratchet mechanism. (B) Use your hand to take the slack out of the strap. (C) Pull the strap as tight as you can. (D) Use a duct strap tensioning tool to finish tightening and to cut the end of the tie. Nylon ties can be used to gather a group of wires for a neat installation and to keep wires out of the way of moving parts.

tensioning tool, as shown in Figure 4-12D, is used to tighten and cut the tie.

Figure 4-13 shows how to bundle wire for a neat installation. In this case you do not want to overtighten the strap because you may damage the wire or wire insulation. Excess strap is snipped off with a wire cutter to preserve a neat and professional installation.

TECH TIP

Figure 4-14 shows how to release a nylon strap. This is easier to accomplish on larger ties because the ratchet clip is larger and easier to release.

Finally, cable ties or nylon ties come in various packages. Figure 4-15 shows a variety pack that contains different sizes of ties. Larger sizes of ties tend to be packaged by themselves. It is good practice to have a variety of cable ties available to make your work look professional. They are also handy for tying down a load or loop handle.

4.6 QUICK CONNECTORS

Quick connectors are used to conveniently attach and remove wires from a circuit. A wire(s) is placed in the round end of the connector and crimped closed. It is best to use a **crimping tool** as shown in Figure 4-16 to mechanically bond the quick connect to the wire. When the crimp is completed, tug on the wire to ensure it has a good mechanical connection. This tool doubles as a **wire cutter** and **insulation stripping tool**—a very handy multipurpose tool.

There are several common types of connectors. A **female, slip-on connector** is shown in Figure 4-17. This is the most common type of quick connector found on HVACR equipment.

Figure 4-18 illustrates a **male, slip-on connector**. The male spade connector will slip into a female connector.

The insulated color at the base of the connector indicates the wire gauge that will go into the connector:

- Red handles 22 to 18 AWG (gauge wire size).
- Blue handles 16 to 14 AWG.
- Yellow handles 12 to 10 AWG.

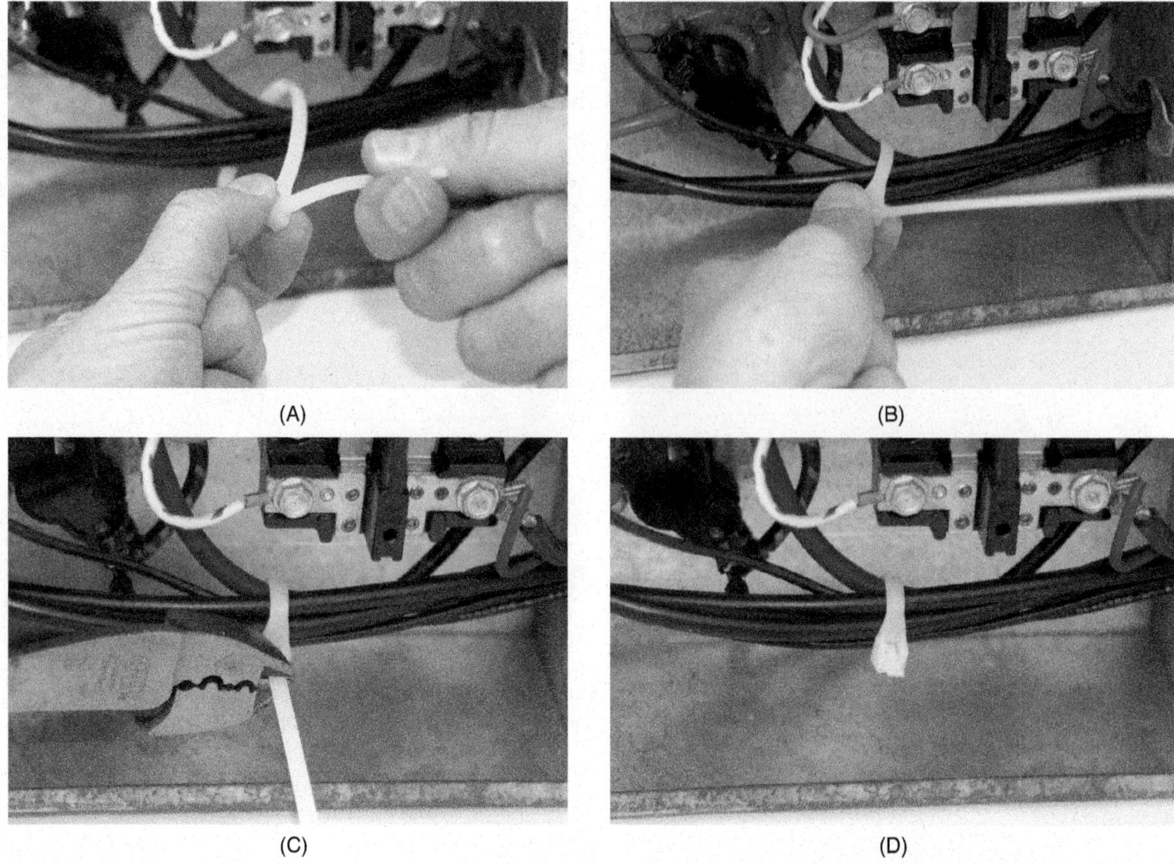

Figure 4-13 (A) Use a small wire tie to bundle the wires. (B) Pull the tie tight by hand. (C) The excess amount of tie is removed with a wire cutter. (D) The final product is a neat, professional installation.

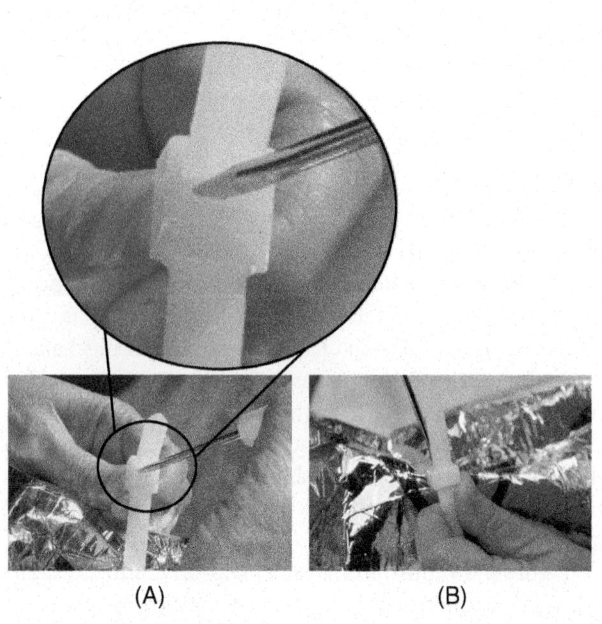

Figure 4-14 Nylon straps that have not been drawn totally tight may be released by opening the ratchet tab and pulling the tie strap in a reverse direction to loosen it. A very small, flat-blade screwdriver is used to open up the ratchet mechanism and free the strap.

Figure 4-15 Cable ties are packaged one size to a pack or as a variety pack, as shown here.

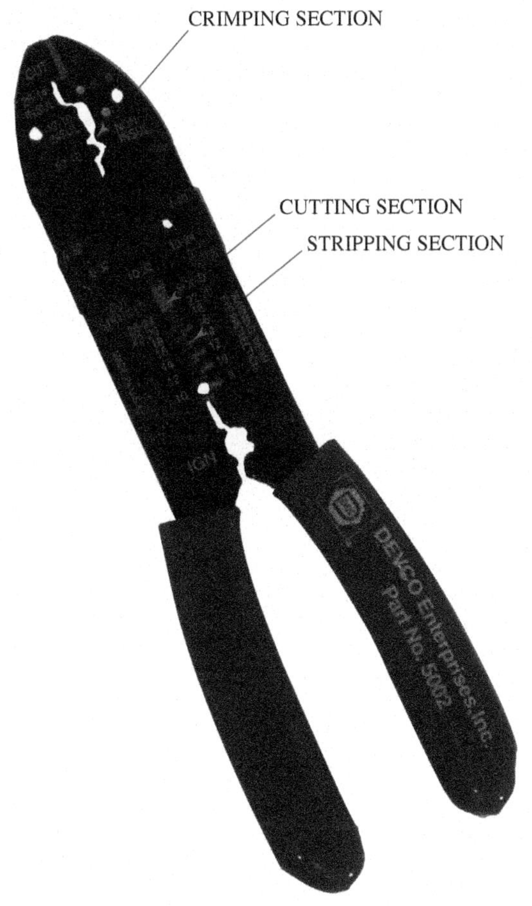

Figure 4-16 A crimping, wire stripping, and cutting tool is used to cut a wire, strip back insulation, and crimp or mechanically secure a wire in a connector. After crimping the connection, tug on the wire to check for a good mechanical bond.

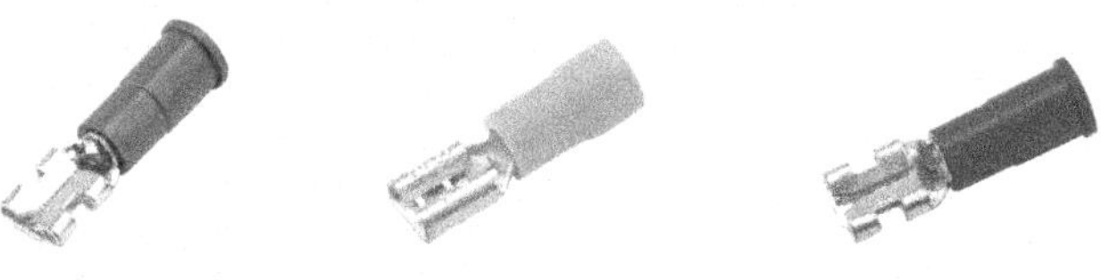

Figure 4-17 On these female quick connectors, the color of the insulation dictates the size of wire that can go into the connector end.

Larger connector sizes use the same colors and handle the following wire gauges:

- Larger red handles 8 AWG.
- Larger blue handles 6 AWG.
- Larger yellow handles 4 AWG.

The technician can also purchase connectors without the insulation. In most cases, however, it is good practice to use insulated connectors to reduce the possibility of a shorted wire or a shock to the technician. Two wires may be placed into a connector opening and crimped. In such a case, the connector size will need to be larger than the single-wire connector gauges listed in the preceding color chart.

Figure 4-18 The male connector or spade is another type of quick connector. It is not as commonly used as the female connector.

Figure 4-19 The flag quick connector is a female quick connector with the wire connection on the side instead of on the end of the connector. Flag connectors are not insulated.

The **flag connector** (see Figure 4-19) uses a side connector instead of a straight-on connector.

The **ring connector** (see Figure 4-20) is used mostly by manufacturers.

The **fork connector** (see Figure 4-21) is more popular with technicians, since this allows the technician to complete a wiring hook-up without totally removing the terminal screw.

The **butt connector** (see Figure 4-22) is used to connect two wires.

Figure 4-20 Ring connectors are used by some manufacturers for wire hook-ups. The terminal screw must be removed in order to install or remove a wire using a ring connector.

Figure 4-21 A fork connector is sometimes used in place of a ring connector. The fork connector requires loosening, but not total removal of the terminal screw.

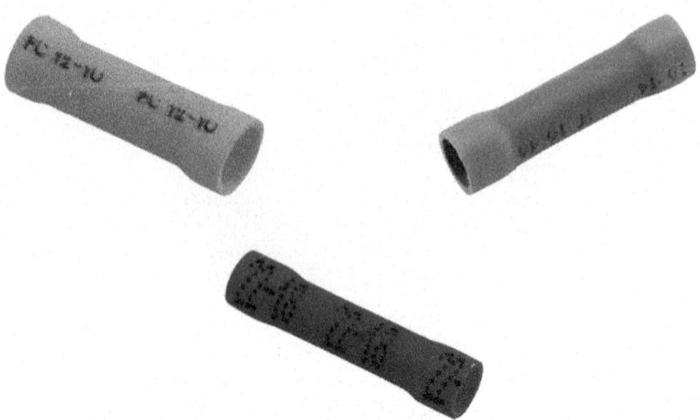

Figure 4-22 Butt connectors are used to join two pieces of wire.

Another type of connector is the **double-male, single-female connector**, as shown in Figure 4-23.

All of these connectors have a purpose and make our work a little easier. Prior to having these connectors, the system would need terminal blocks as illustrated in Figure 4-24.

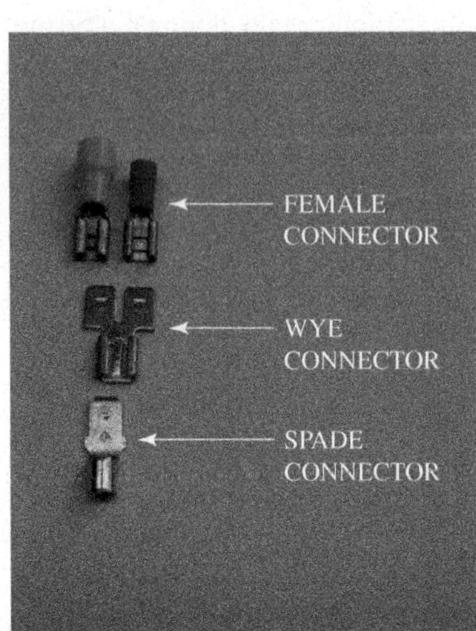

Figure 4-23 This is a double-male, single-female connector. his ties two connectors into one junction point.

Figure 4-24 A terminal block is used as a central point to tie wires together. Quick connectors have been used to reduce the use of a terminal block.

The advent of the quick connector saves time, both for wiring hook-ups and troubleshooting.

TECH TIP

Using a knife or box cutter to strip the insulation off wire may cut some of the strands in the wire bundle. This will lower the current-carrying capabilities of the wire and create a hot spot where the broken wires are trying to move current across the cut surface. You should instead use a wire stripper and select the wire gauge being stripped, as shown in Figure 4-25.

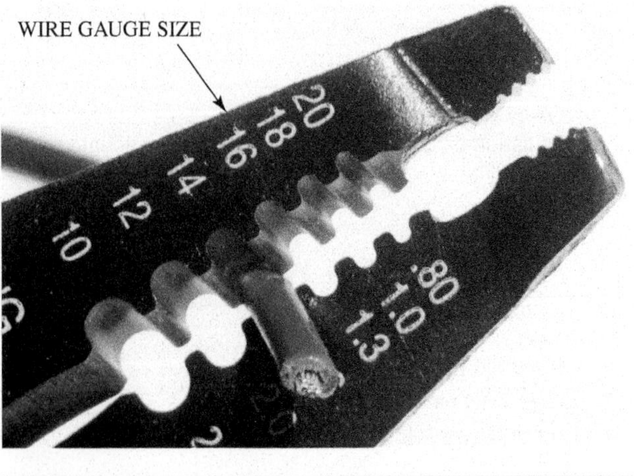

Figure 4-25 A wire stripper tool reduces cutting and nicking of the wire conductor. Select the correct wire gauge when stripping a wire. If you are unsure about the gauge size, start with the larger gauge first, then step down the gauge size to prevent conductor damage.

Figure 4-26 Electrical tape comes in various colors; the most common color is black.

4.7 ELECTRICAL TAPE

Electrical tape is used to isolate bare wires from shorting to ground or unintended electrical pathways. Think of electrical tape as wire insulation for exposed wire connections. Electrical tape, as shown in Figure 4-26, is a vinyl, 8.5-mil-thick flexible material. It works well in temperature extremes ranging from 0° to 220°F (–18° to 105°C). The flexible vinyl tape resists abrasion, moisture, alkalis, acids, and corrosion. It is also designed to seal against moisture and provide mechanical protection.

Standard electrical tape is rated at up to 600 V. It has an **Underwriter's Listing (UL)** of UL 510 and a Canadian **CSA Standard** listing of C22.2. Thicker tapes are used for higher voltage protection requirements. Standard electrical tape can be used to cover and bind the thick tapes in place.

How Is Electrical Tape Applied?

Electrical tape should be applied in a half-lapped layer with sufficient tension to produce a uniform wind throughout the cover area. Wrap tape from a smaller diameter surface to a larger diameter surface. Reduce the tension on the final wrap to prevent flagging. Some techs will make a ¼-in. tab at the end of the run if the tape is going to be removed soon after installation. The tab allows for quick identification and removal of the tape.

Finally, many electrical tapes have a 5-year shelf life from the date of manufacture when stored in a humidity-controlled environment of 50° to 80°F at 75% relative humidity.

SERVICE TICKET

As a new technician, the service manager asks you to make a list of fasteners you will need to carry in your service van. Use several sources to develop a comprehensive fasteners list, including the information in this unit. Go to an air conditioning supply house and make of list of fasteners you may need. Check the Internet for help. Finally, check with your fellow technicians to determine which fasteners would be best to stock.

SUMMARY

Knowing the names of fasteners and how they work is a valuable tool for all HVACR professionals. Technicians should know as much as they can about everything so that they will be a valuable asset to a company and to themselves. The HVACR professional tends to be a jack of all trades and a master of HVACR. The true professional must know air conditioning, heating, and refrigeration. He or she must also know electrical work, plumbing, and some carpentry. Our career calls for a well-rounded, skilled person or at least for a person who is willing to learn skills outside of the narrow HVACR skill set.

Knowing about the various types of fasteners will help you complete jobs in a quick and efficient manner. For example, quick connectors speed up the processes of checking and changing out components. Knowing what fasteners are available, how to use them, and calling them by their correct name makes the technician a professional. You will find that there are many different terms for the fasteners we discussed in this unit. When discussing fasteners it is important for you to be on the same page as the person with whom you are communicating.

REVIEW QUESTIONS

1. What size wire nut would be used to join two 14-AWG wires?
2. What size wire nut would be used to join four 14-AWG wires?
3. What size wire nut would be used to join two 12-AWG wires?
4. How much wire should be stripped prior to inserting it in a wire nut?
5. You will need a nut driver that will remove 1/4-in. head screws. What color nut driver should you select?
6. You will need a nut driver that will remove 5/16-in. head screws. What color nut driver should you select?
7. What device is used to determine the size of a bolt?
8. How does a technician know what wire size goes into a quick connector?
9. What are the three sizes of nails used in HVACR?
10. What is the difference between a screw and bolt?
11. You are working on a job where a bolt was sheared off. After investigating the problem you determine that a strong bolt is needed. The number on the head of the sheared bolt is #6. What number bolt will you request?
12. Identify the connector shown in Figure 4-27. What wire size will it handle?
13. Identify the connector shown in Figure 4-28. What wire size will it handle?
14. Identify the connector shown in Figure 4-29. What wire size will it handle?
15. Identify the connector shown in Figure 4-30. What wire size will it handle?
16. Identify the connector shown in Figure 4-31. What wire size will it handle?
17. What is the life expectancy of electrical tape? Under what conditions should electrical tape be stored?

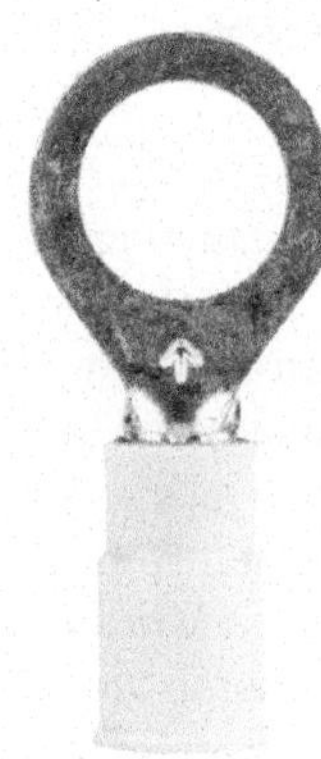

Figure 4-27 Use this connector to answer Review Question 12.

Figure 4-30 Use this connector to answer Review Question 15.

Figure 4-28 Use this connector to answer Review Question 13.

Figure 4-31 Use this connector to answer Review Question 16.

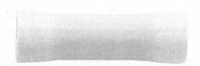

Figure 4-29 Use this connector to answer Review Question 14.

UNIT 5
Power Distribution

WHAT YOU NEED TO KNOW

After studying this unit, you will be able to:

1. Describe the power distribution system.
2. Name the types of transformers used to supply power to HVACR equipment.
3. Explain how single-phase power is supplied to a power user.
4. Explain how three-phase power is supplied to a power user.
5. Explain how electricity is generated.
6. Name the sources of fuel for generating electricity.

The delivery of commercially available electricity is divided into two major parts: **power generation** and **power distribution**. This unit focuses on power distribution or how power gets from the generation site to the end user. At the local level the power distributor is also known as the utility company, electrical provider, electric company, or light company. For troubleshooting purposes it is important to understand how power is supplied to end users. If HVACR equipment does not have the correct supply voltage, it will not function properly. The power distributor is responsible for supplying the correct voltage. In most cases, once the power goes through a building's meter, supply of the correct voltage becomes the building owner's responsibility. As a technician, you need to establish where a problem lies. Is it the fault of the power distributor or is the problem on the owner's side of the distribution system? Wrongfully blaming the power company for a voltage problem is embarrassing for the technician and his or her service company. The technician needs to get this right, even if it requires asking for help from the service manager or supervisor.

This unit provides important information about establishing a troubleshooting diagnosis. You want to get this diagnosis correct before calling out the power company. Determining the right voltage is the first step in the electrical troubleshooting process.

According to the Edison Electric Institute, the U.S. electric transmission grid consists of more than 200,000 miles of high-voltage transmission lines. High-voltage transmission lines are 230 kV and greater. **Transmission lines** carry electricity from power-generating plants to areas where electricity is needed. Electricity travels at nearly the speed of light, arriving at a destination at almost the same moment it is produced. Reliable electric service and regional electricity markets depend on strong transmission systems. Because of our reliance on electricity, reliable electric service is considered a national security issue.

This unit focuses on the power distribution grid and the various voltages supplied to the end user. The technician needs to realize that various voltage levels can be supplied to the customer.

First, we review how power is generated. This includes single-phase and three-phase power generation. Next, we discuss how the power is transformed and provided to the end user. Finally, we investigate how the power is distributed once it is supplied to the building.

5.1 WHEN DID POWER DISTRIBUTION START?

In 1882, Thomas Edison built the first small, but workable electric system at the Pearl Street Station in New York. Electric utilities began to develop in the 1890s. The development began in urban areas because many customers were required to make installation of the distribution system pay off. The electrical industry had a "natural" monopoly. A natural monopoly is used because it is the most efficient way to allow for one provider to deliver expensive goods or services to many customers. Monopolies have been used for services vital to the economic and social fabric of society. Monopolies are operated when it is important to have a large, integrated network that people rely on for service. Finally, generating and transmitting electricity was and still is a capital-intensive project that requires a great amount of start-up money with no guarantee of success.

By the 1920s most urban areas had been electrified. Exclusive utility monopoly or franchise rights came with the obligation to serve all customers in the service area. This included customers who were not conveniently in line with the electrical hook-ups. From this obligation to provide electricity to everyone, the rural electrification program soon followed.

5.2 ENERGY LAW OF CONVERSION

A law of science states that "Energy can neither be created nor destroyed. It can only be transformed or converted from one form to another form." Electricity generators make use of this scientific law.

We live in a world that is surrounded by energy conversion. Here are some technologies that we see working in the HVACR industry and examples:

- *Chemical to thermal:* a furnace that uses oil, natural gas, or wood to heat
- *Chemical to thermal to mechanical:* a vehicle engine
- *Chemical to electrical:* a fuel cell
- *Electrical to mechanical:* an electric motor that converts electrical magnetic force into movement
- *Electrical to radiant:* the use of electricity to make heat or generate light
- *Fuel to energy:* a power-generating plant.

5.3 ELECTRICAL DISTRIBUTION TERMS AND ABBREVIATIONS

The following terms are used when discussing electrical generation, distribution, and use. It is important to understand the language used when communicating about this topic.

Connected load: This is the combined manufacturer's rated capacity of all motors and other electric-powered devices on the customer's premises that may be operated at the will of the customer.

Entrance panel: The name given to the breaker box, which is also known as the fuse box or panel box. The entrance panel may be located inside or outside the building. Some electrical codes do not allow new installations in a clothes closet or general storage room.

Kilowatt (kW): A unit of power equal to 1,000 watts (watts = volts × amps).

Kilowatt-hour (kWh): A unit by which residential and most business customers are billed for monthly electric use. The kilowatt-hour represents the use of one kilowatt of electricity for one hour.

Megawatt (MW): A unit of power equal to one million watts.

Megawatt-hour (MWh): The use of one million watts of electricity for one hour. The term is used most often for large-scale industrial facilities and large population centers. According to the Edison Electric Institute, the average U.S. household uses 11.2 MWh (11,202 kWh) of electricity every year.

Meter loop: The opening in and extension of the customer's service entrance conductors provided for installation of the electric company's meter.

Power: Power equals current times the applied voltage. It is measured in watts and written:

$$\text{Power} = \text{volts} \times \text{amps}$$

Power factor (PF): The power factor is the ratio of power flowing in a circuit versus actual power used. The power factor is used as part of billing considerations in commercial systems. In a resistive circuit this ratio is equal to a PF of 1. In an inductive or capacitive circuit, the amount of power flowing in the circuit is less than the actual power used; therefore, the power factor would be less than 1.

Service drop: The overhead service conductors extending from the electric company's overhead distribution system to the customer's service entrance conductors at the point of electricity delivery.

Watt (W): The basic unit of measure of electric power. The watt is the power dissipated by the current of one ampere flowing through the resistance of one ohm.

5.4 HIGH-VOLTAGE DISTRIBUTION

How does the electrical power distribution system work? Figure 5-1 shows a common electrical distribution system:

- Number 1 is the electricity leaving the power plant.
- Number 2 is the voltage increase at a step-up substation.
- Number 3 is the long-distance transmission line to the area where the electricity will be used.
- Number 4 is the step-down of voltage at a second substation.
- Number 5 is the final distribution point.
- Number 6 shows the step-down transformer that supplies power to the end user. The end users will receive power through overhead lines or underground power lines.

Figure 5-2 is a detailed schematic of the sequence of electrical generation and distribution. Notice that the final power can be delivered underground or overhead.

How is power generated? As shown in Figure 5-3, the most common way power is generated is by the use of a steam turbine turning a generator. Common fuels used to create steam are coal, natural gas, and nuclear energy.

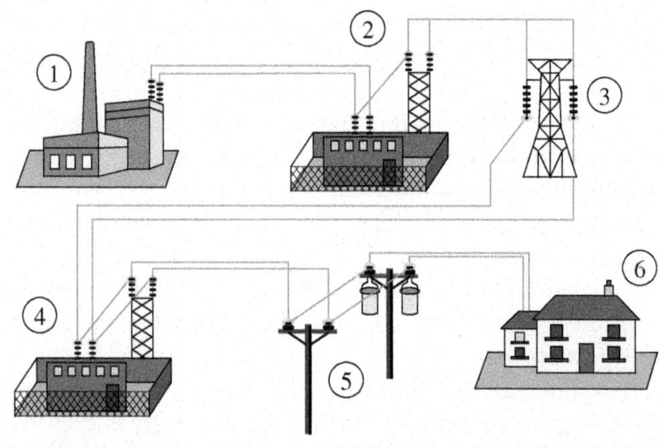

Figure 5-1 This schematic shows a common electrical distribution system.

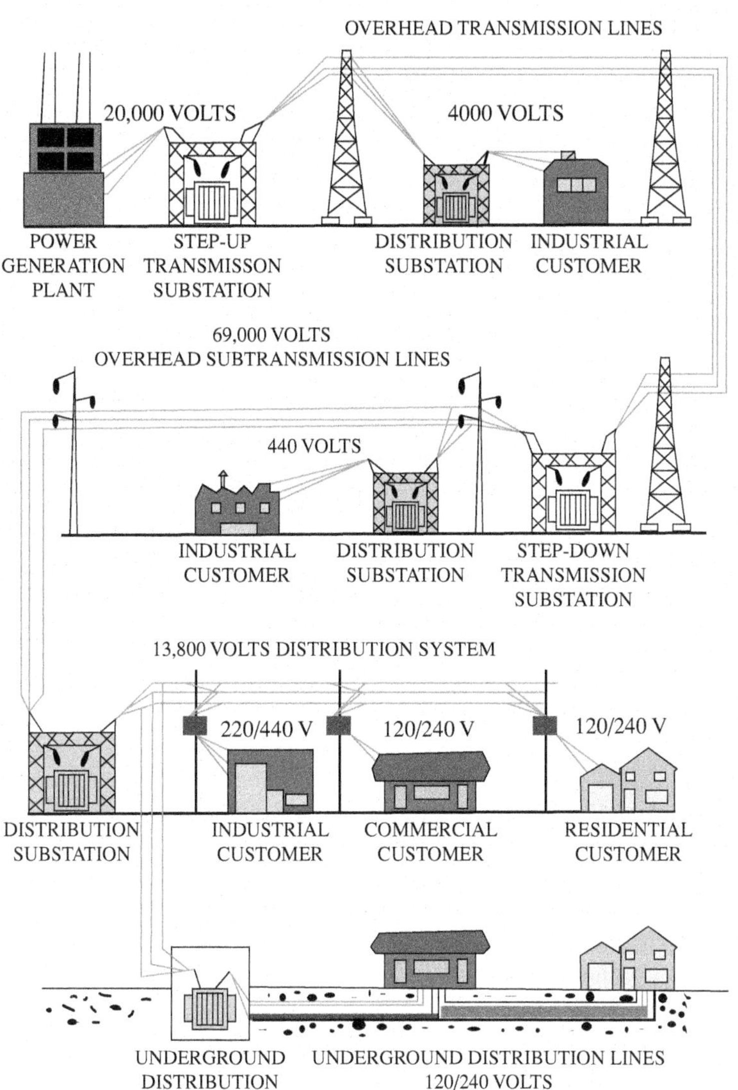

Figure 5-2 This is a detailed schematic of power distribution from the generator to the end user. Notice that several different voltages can be supplied to the end user and that voltage can be supplied overhead or underground.

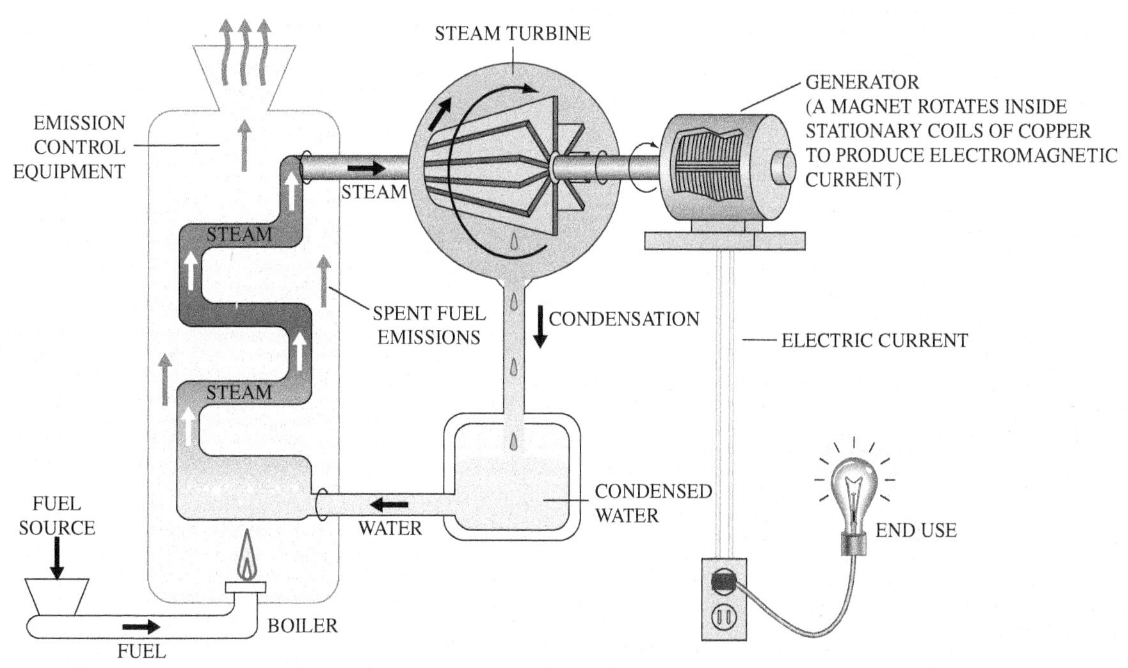

Figure 5-3 This is how power is generated. Fuel is used to boil water and develop steam. The steam is use to rotate a steam turbine, which turns a generator that produces electricity.

A **generator** makes electricity for transmission to its customers. You can think of a generator as a motor in reverse. The rotational force of the turning generator is converted into electricity. Electricity is not easily stored; therefore, it must be generated at the rate at which it is consumed. Generators are shut down or capacity reduced to match the megawatt demand. Most electrical providers are tied together through a network or grid. The grid allows electricity to be shared if one area of the country needs more power than another or when storms damage generation facilities, substations, or transmission lines. The nationwide grid reduces power interruptions, thus providing certain and secure lines of service. Our national security and welfare are very dependent on reliable electric service.

Finally, Figure 5-4 shows the breakdown of the various fuels used to generate electricity in the United States. As you can see most areas of the country use coal, natural gas, or nuclear energy to generate power. The Pacific west coast has a high percentage of hydroelectric electricity generation, which is the use of the force of water to turn generators. Water is not a fuel per se; it is a force used to move a generator to develop electricity.

Figure 5-5 shows, by region, the fuel mix and percentage of each type of fuel used in the United States to generate electricity. Coal is a leading fuel for making steam to create electricity. By 2035 use of coal is still expected to be high, but somewhat reduced at 44% instead of 48% of the national fuel mix. According to the Edison Electric Institute, natural gas is used in 21% of power plants. Almost 95% of the new plants are natural gas based. Natural gas is projected to comprise 21% of the national fuel mix by 2015. At this time green energy renewable sources such as hydro, wind, and solar are a small part of the electricity-generating mix.

Figure 5-6 shows that power generation is 68% of the cost of generating electricity, followed by distribution costs at 24% and transmission costs at 7%. After the initial cost of constructing the power plant, a large part of the cost of power generation is the fuel cost.

On June 25, 2010, the Massachusetts Institute of Technology published a major report stating that natural gas use will increase as older, more polluting coal-powered generating plants are retired. The report concluded that the United States has 92 years' worth of natural gas and gas shale at the present consumption rates.

Once the electricity has been generated, it must be immediately transmitted and distributed to customers. The transmission lines carry high voltage through thick wires on tall towers. This design elevates the wires safely above any activity below them. The wires are separated so that the extremely high voltage does not arc between the power lines. Some of the transmission lines can be as high as 13,800 volts or greater. The voltage on the transmission system will vary, depending on the electrical design. The step-down voltage is around 2,400 volts. From this point the voltage is reduced to levels used by the consumer. Common voltages are:

- Single-phase 208 or 240 volts for residential operation
- Three-phase 440 or 480 volts for commercial and industrial three-phase applications
- Any number of other voltages used in industrial applications.

This book limits the discussion of voltage sources to those found in residential and light commercial applications.

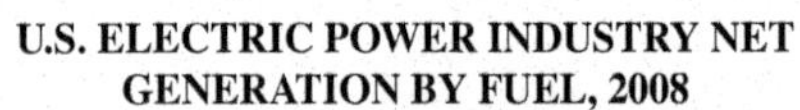

U.S. ELECTRIC POWER INDUSTRY NET
GENERATION BY FUEL, 2008

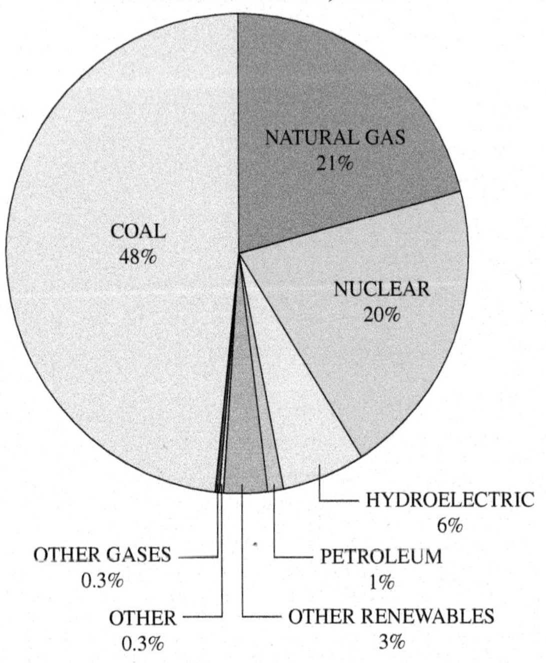

Figure 5-4 Various fuels are used to generate electricity. Coal is used to generate 48% of all U.S. electricity. The United States has large deposits of coal that will be a major fuel for decades.

5.5 END USE DISTRIBUTION BY TRANSFORMERS

This section discusses the various types of transformers that supply voltage to the end user. We discuss both residential and commercial transformers. Before we begin this discussion it is necessary to understand the fundamental operation of a transformer. The fundamentals are the same whether it is a small control transformer or a larger three-phase transformer.

A transformer is a device that is used to change voltage. In this case, it will reduce or "step down" the voltage. When the voltage is stepped down, the current is increased at the same proportional rate. For example, if the voltage to the secondary is stepped down by half, the current-carrying capability of the secondary will be doubled.

Placing transformer windings in series increases the voltage output. Wiring transformers in parallel keeps the voltage

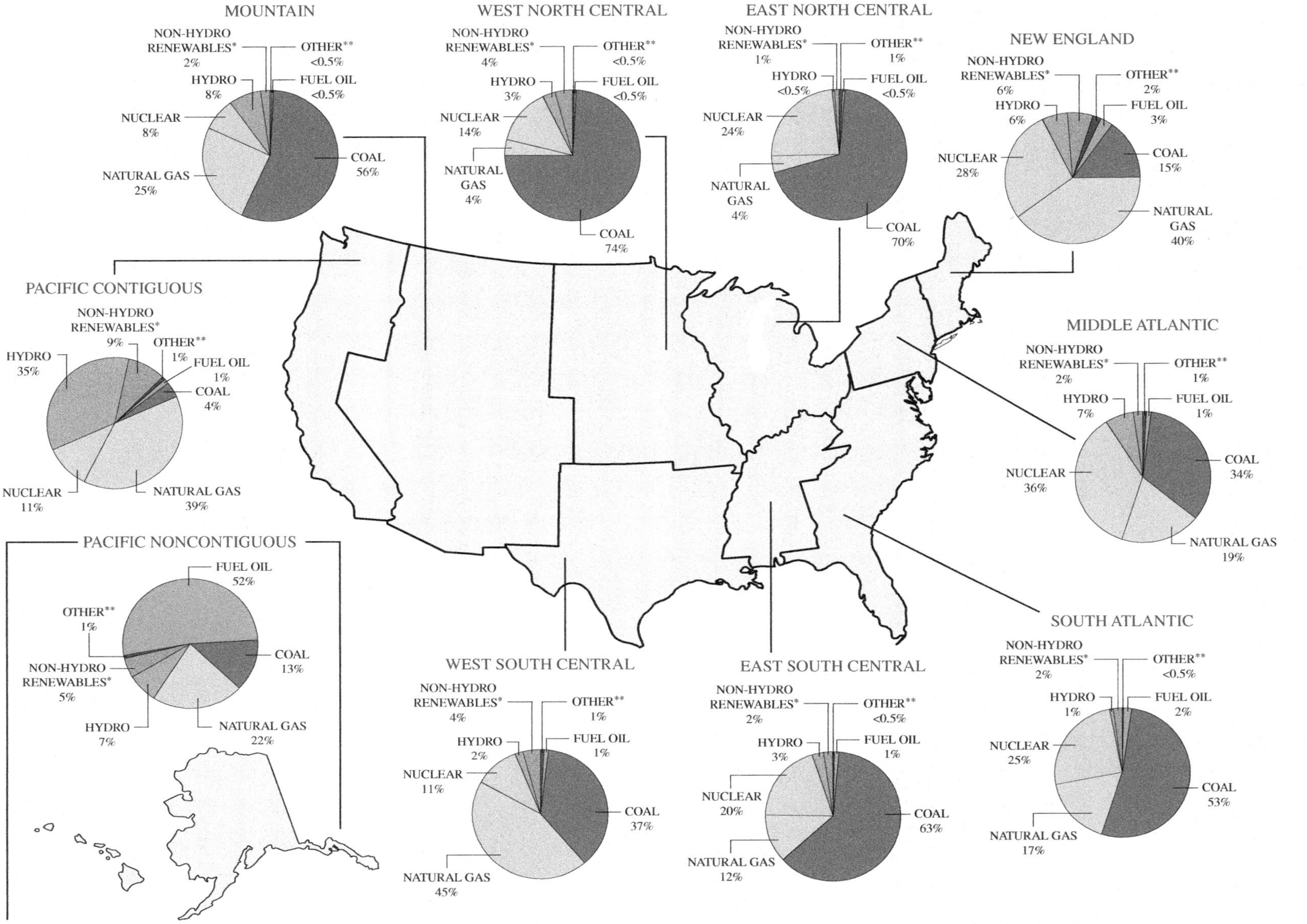

*INCLUDES GENERATION BY AGRICULTURAL WASTE, LANDFILL GAS RECOVERY, MUNICIPAL SOLID WASTE, WOOD, GEOTHERMAL, NON-WOOD WASTE, WIND AND SOLAR.

**INCLUDES GENERATION BY TIRES, BATTERIES, CHEMICALS, HYDROGEN, PITCH, PURCHASED STEAM, SULFUR AND MISCELLANEOUS TECHNOLOGIES.

SUM OF COMPONENTS MAY NOT ADD TO 100% DUE TO INDEPENDENT ROUNDING.

Figure 5-5 Different regions of the country use different fuel mixtures to generate electricity.

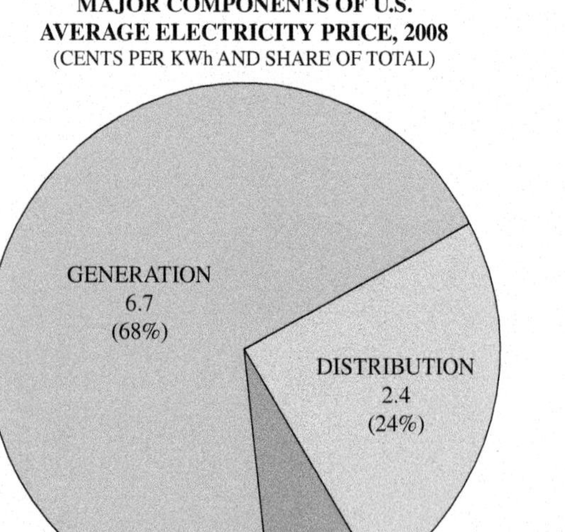

Figure 5-6 Power generation is 68% of the cost of generating electricity, followed by distribution costs at 24% and transmission costs at 7%.

the same, while doubling the current output. A tapped transformer winding will have lower voltage output when compared to a full winding. If you remember these fundamentals, it will be easier to understand the following information on transformers.

Single-Phase and Three-Phase Power from the Electric Company

Single-phase and three-phase power is provided by the electric company to the end user. Figure 5-7 compares the single-phase and three-phase sine waves. Notice that single-phase power has one sine wave (the upper sine wave), whereas three-phase power has three overlapping sine waves (the lower waves). Notice that Phase A is 120 degrees out of phase with Phase B. Phase B is 120 degrees out of phase with Phase C. You can see that the three-phase sine waves fill up more of the space, while single-phase voltage has more than a 50% gap in supplied voltage. Three-phase voltage is more continuous.

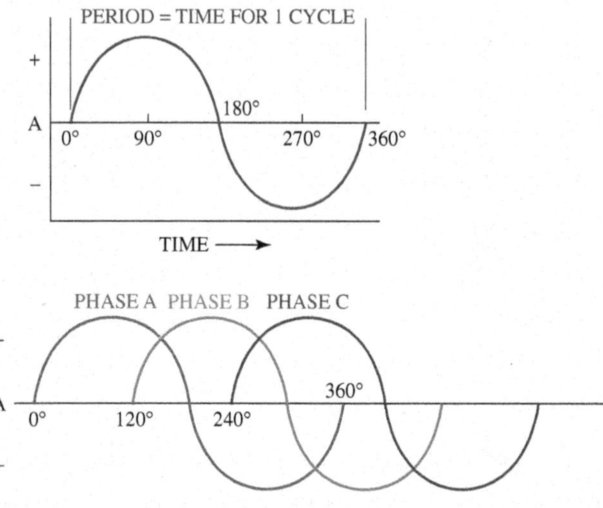

Figure 5-7 The upper sine wave is single phase. The lower sine waves are for three-phase voltage.

If you look at the three-phase power example in Figure 5-7, you will notice that the positive and negative voltages take up more voltage space when compared to single-phase voltage. In other words, with three-phase power more power is applied across a wider range of the cycle when compared to the single-phase sine wave, which has at least a 50% positive and negative gap where no voltage is applied. For this reason, three-phase voltage offers better motor-starting torque and higher running efficiency. Note, however, that equipment must be designed to operate on three-phase power. Hooking up the wrong power source will more than likely burn out the electrical component.

Single-phase power or voltage comes from the secondary side of a three-phase transformer. Three-phase power is generated for commercial power operation and rarely used in residential applications.

Let's look at single-phase power in more detail. Figure 5-8 shows a diagram of the type of high-voltage transformer commonly used in residential applications. The primary voltage from this single-phase utility transformer is very high. It has a range of 2,400 to 4,800 volts, depending on the power company in the service area. The secondary side of the single-phase power supplied to a residence is around 208 to 240 volts. In some parts of the country, only 120 volts is supplied to a residential area.

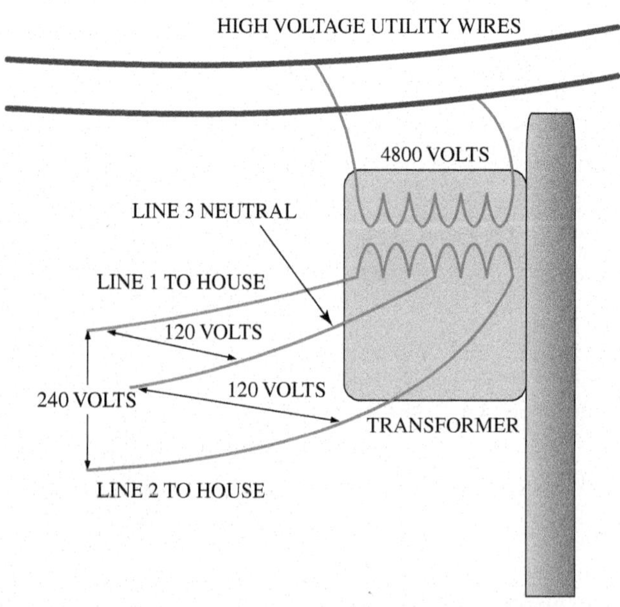

Figure 5-8 This type of transformer is commonly used in residential applications. The single-phase power supplied to a residence is around 208 to 240 V.

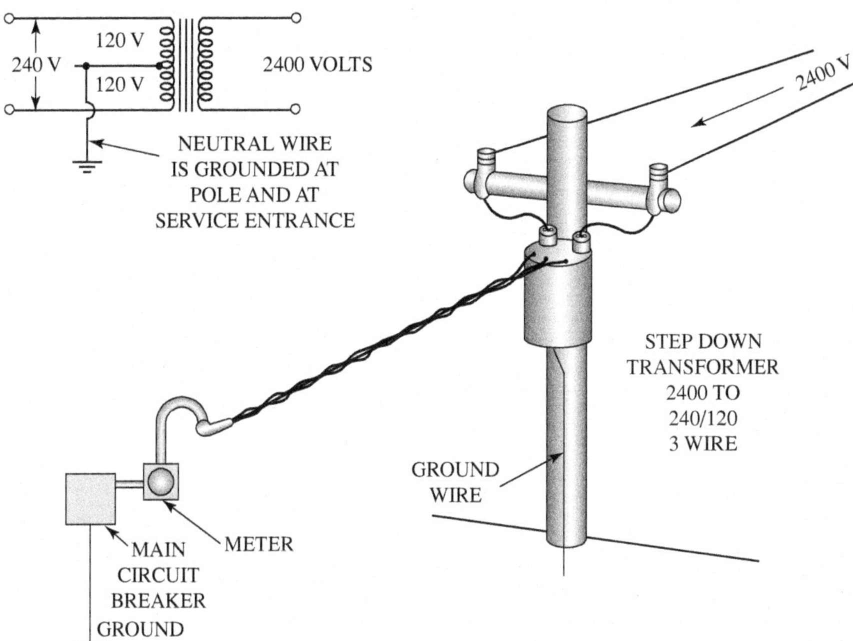

Figure 5-9 This is a diagram of the supply voltage to a residence.

Figure 5-9 is a breakdown of the supply voltage to a residence. One leg of the 240-V transformer can be lost with interesting results in the customer's home. All 240-V circuits would be dead and half of the 120-V circuits would not work. The electric company should be contacted to solve the problem.

The power company is required to supply power at ±10% of the rated voltage. For example, if the power company is to provide 240 V to the customer, ±10% is ± 24 V. In this case, the voltage from the electric provider should range between 216 and 264 V.

TECH TIP

The single-phase power supplied to a residence is around 208 to 240 V. It is rare to find 120 V supplied to a residence, but it is possible in very old structures or communities where the power distribution has not been upgraded. It is important to know the voltage when installing new equipment. Most new residential HVAC systems operate on 240 volts.

Referring back to Figure 5-9, the power supplied to the secondary side of the transformer is 240 volts and from one side of the transformer to neutral is 120 volts. There are three conductors coming from the pole transformer to the meter section. Two conductors are used to provide 240 V and the third, bare conductor is used as the neutral. The neutral wire is a current-carrying conductor that can shock you if touched. The neutral wire is also used to support the other two current-carrying conductors. The 240 V enters the main circuit breaker panel and can be used as 240 V or can be split into 120-V circuits. In Figure 5-9, notice how the 240-V secondary is divided into two 120-V circuits.

TECH TIP

Sometimes terms and words used in our profession are confusing and subject to misunderstanding. This is true for voltage also. For example, some techs use the term 110 volts, 115 volts, or 120 volts. In essence they are close enough to be the same. Decades ago, 110 volts was the common voltage used by appliances and provided by power companies. Now 120 volts is the norm. Other common voltages that are now supplied at a higher level are 240 and 480 volts. Previously they were supplied as 220 and 440 volts and in some cases are still supplied at this lower level. Most voltages used today are 120, 208, 240, or 480 volts. You may hear of a voltage that is a little lower, but more than likely the voltages you will come across are those listed here.

5.6 COMMERCIAL TRANSFORMERS

This section deals with commercial and to some degree industrial transformers that supply power to businesses and industries. The common three-phase transformers use the wye and delta configurations.

The Wye Transformer

Three common types of **wye transformers** are used in the field. They have the following outputs:

- 120- and 208-V outputs
- 120-, 208-, and 240-V outputs
- 277- and 480-V outputs.

The wye winding uses the letter "Y" to relate to its symbol as shown in Figure 5-10, which is a diagram of a wye four-wire,

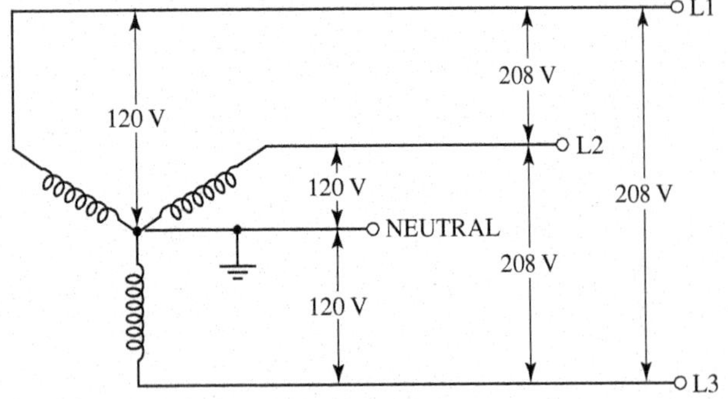

Figure 5-10 This four-wire wye transformer provides 120 single-phase voltage and 208 three-phase voltage to a customer. The word *wye* comes from the way the winding is drawn, like the letter Y.

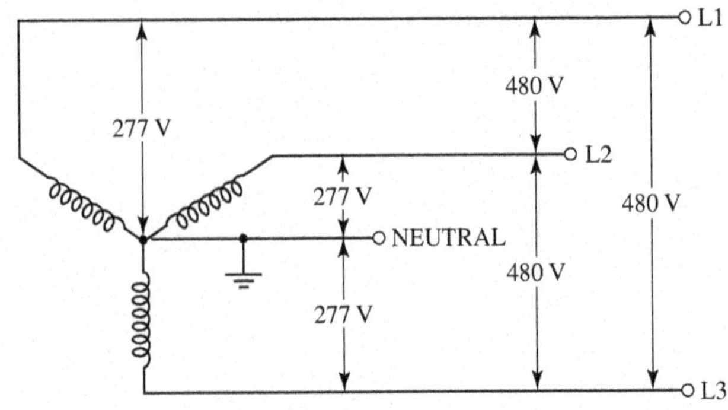

Figure 5-12 This three-phase, 277/480-V transformer is wired to develop 277 volts, which is used for commercial lighting circuits.

208-V, three-phase transformer. This wye transformer can supply 120 and 208 volts to the customer. Reviewing Figure 5-10, note that 208 V can be developed across any two of the L_1, L_2, and L_3 legs. The 120 V will be developed from any leg to ground, which is center tapped on the transformer.

Figure 5-11 shows the voltage between each fuse-protected leg of the transformer shown in Figure 5-10. Neutral is not fused, but it is a current-carrying conductor. A fused disconnect with a 208-V wye transformer will have 208 V between the phases and 120 V to ground.

TECH TIP

Power company distribution transformers are rated by **kilovolt-amps** or **kVA**. Input and output voltage is also an important consideration. The kilovolt-amps determine how much amperage will be supplied to the panel box. The transformer is sized by the power company to provide the voltage and amperage rating of customers' panels. Customers will need to contact their power provider if they increase the load (amperage or voltage) by installing a new main supply panel that is capable of handling higher amperage.

Figure 5-12 is a diagram of a wye four-wire, 227/480-V, three-phase transformer. This wye transformer can supply 277 and 480 V to the user. The 277-V circuit is commonly used for lighting circuits in commercial buildings. The 480-V circuit is used to operate larger HVACR systems. The 480-V circuit can be stepped down to operate 240-V equipment.

Figure 5-13 shows the voltage between each fuse-protected leg of the transformer shown in Figure 5-12. Neutral is not fused, but it is a current-carrying conductor.

The Delta Transformer

Figure 5-14 shows a **delta transformer.** The delta winding uses the Greek letter delta, Δ, as its symbol. As you can see in Figure 5-14, the transformer appears as an inverted delta coil or inverted triangle symbol. This example is a four-wire, 240-V three-phase system. This delta transformer configuration can develop single-phase 120- and 208-V output as well

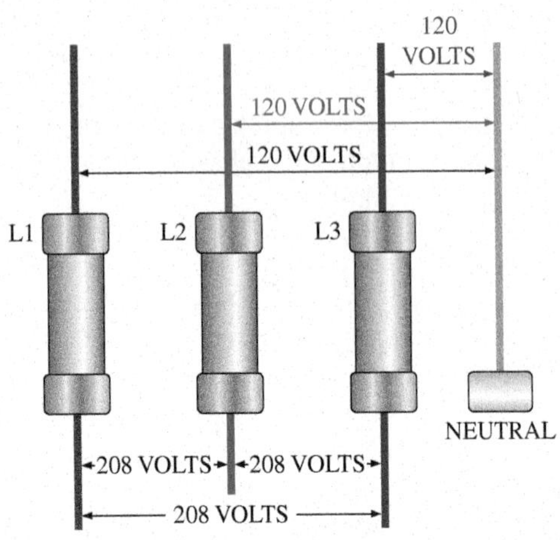

Figure 5-11 A fused disconnect with a 208-V wye transformer will have 208 volts between the phases and 120 volts to ground.

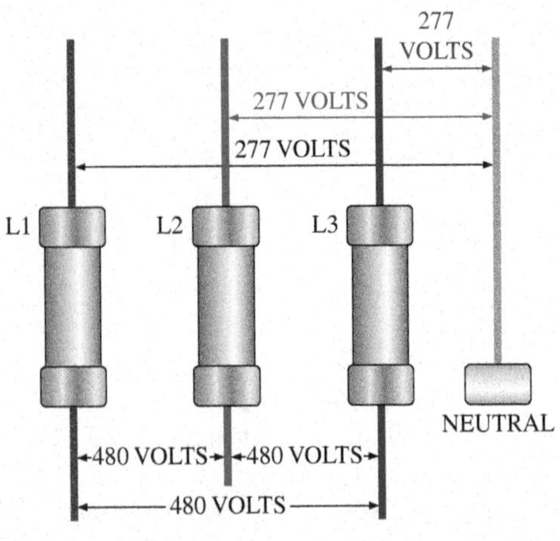

Figure 5-13 This is a wye, three-phase 480-V fuse protector used between each of the phases; 277 V can be wired to lighting circuits.

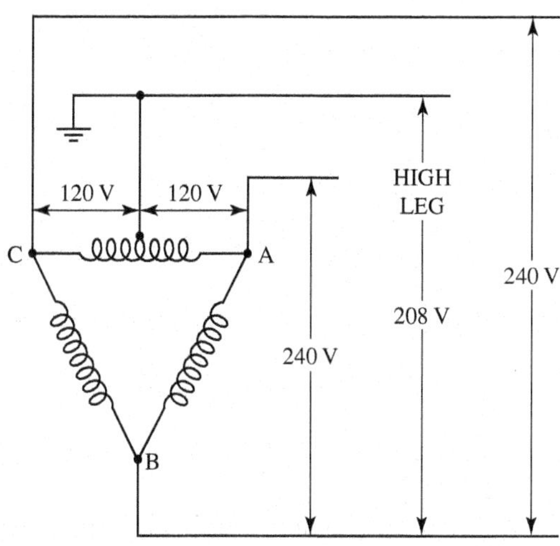

Figure 5-14 This is a three-phase, 120/208/240-V delta wiring configuration. It is also known as a high leg or wild leg power supply. The term *high leg* is used because one leg to ground has 208 V instead of the expected 240 V.

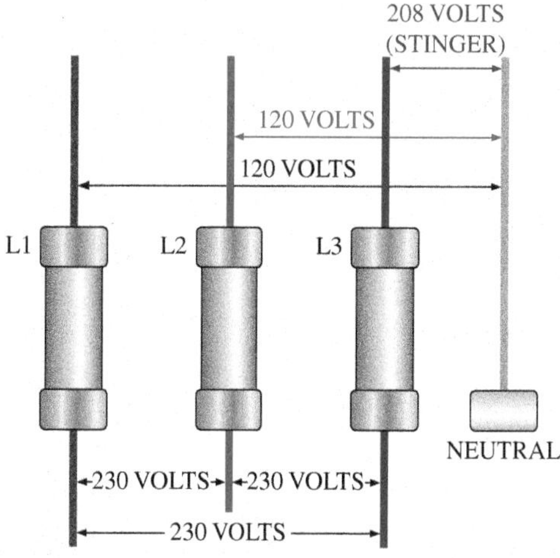

Figure 5-15 This shows the three voltages supplied by a 230-V delta transformer. One leg will measure 208 V to ground, while the other two legs will measure 120 V to ground.

as three-phase 240-V output. Notice how the 208-V output is developed on this delta transformer. The 208-V output is between the lower hot leg and the center tap or ground potential of the upper winding, between C and A. This is also known as the **high leg** or wild leg. The term *high leg* is used since a 208-V circuit is created when going from the center tapped winding to ground. It is important to understand that 120 or 208 volts can be developed to ground on a delta arrangement.

Some technicians use this power supply for temporary voltage for their recovery unit or vacuum pump at the condensing unit or disconnect. As you can see, a tech can get two different voltages. If you need 120 volts to operate a piece of equipment, measure the voltage before plugging in the 120-V equipment because if the 208-V source is present, you could burn up the 120-V piece of equipment when you plug it in.

To better understand this delta arrangement, remember that three transformers coils are used to create the delta design. One of the transformers is larger than the other two. The larger transformer is center tapped to create the 208-V option. The larger transformer must be able to supply power for both three-phase and single-phase loads. It is important to understand the voltage on this transformer arrangement so that it is not considered as a "problem" power supply.

Figure 5-15 shows the three fused voltages supplied by a 230-V delta transformer.

Kilovolt-Amp Rating

For all transformers, the kilovolt amperage (kVA or 1,000 volt-amps) will determine how much amperage will be supplied to the panel box. As mentioned earlier, the transformer is sized by the power company to provide the voltage and amperage rating of the customer's panel served. The customer will need to contact the power provider if they increase the load (amperage or voltage) by installing a new main supply panel.

5.7 RESIDENTIAL SERVICE

Figure 5-16 illustrates two ways power is provided to residences. Older installations use overhead wiring, whereas newer installations use underground wiring. **Overhead wiring** is easier and quicker to repair than underground wiring, but **underground wiring** is protected against storms and falling debris. Many new installations use underground installations to reduce wire damage and reduce visible wiring.

Figure 5-17 is a line diagram of an electricity system serving a residence. Let's follow the sequence of power entering the building and supplying power to the condensing unit:

- At point 1, power leaves the pole transformer.
- Point 2 is where power reaches the mast riser.
- At point 3, power goes through the meter and into the panel box. The panel box distributes power through the breaker to the fused disconnect. A picture of a disconnect is shown in Figure 5-18. The disconnect allows a technician to control the power at the condensing unit without the fear and danger of someone turning it on while it is being serviced.
- At point 4 the power feeds electricity to the air conditioning condenser.

The next paragraphs explain and illustrate what this looks like.

The power to a residence comes from overhead or underground. Figure 5-19 shows three overhead wires on the right side of the transformer that will feed overhead power to a residence. There are two power conductors and one neutral. The neutral is usually bare wire with no insulation. Many of these conductors are aluminum. The bare, neutral conductor is also used to support the two power conductors. The neutral in this case is ground potential and does not carry current.

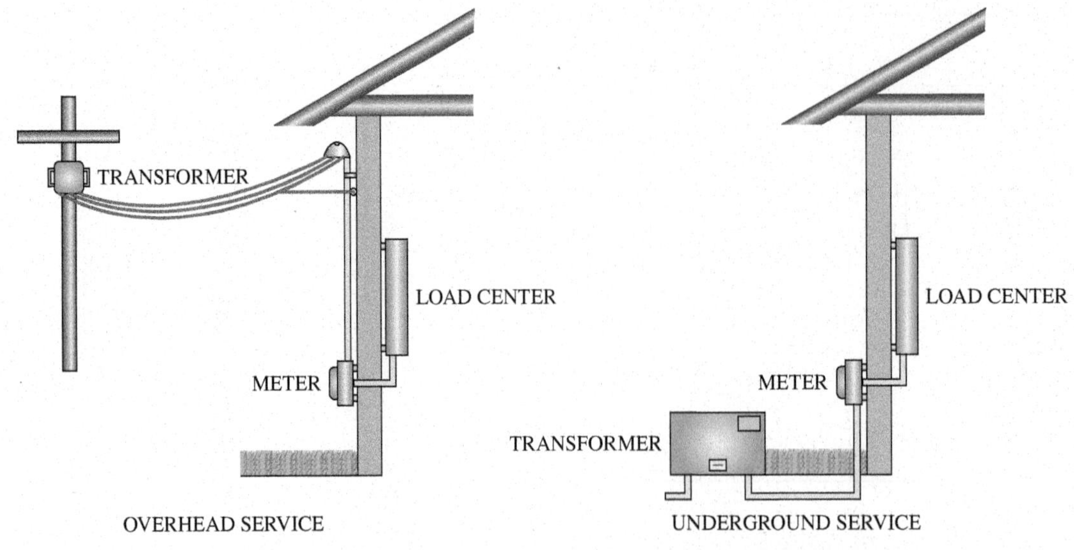

Figure 5-16 Power to residences is provided in two ways: overhead service and underground (lateral) service.

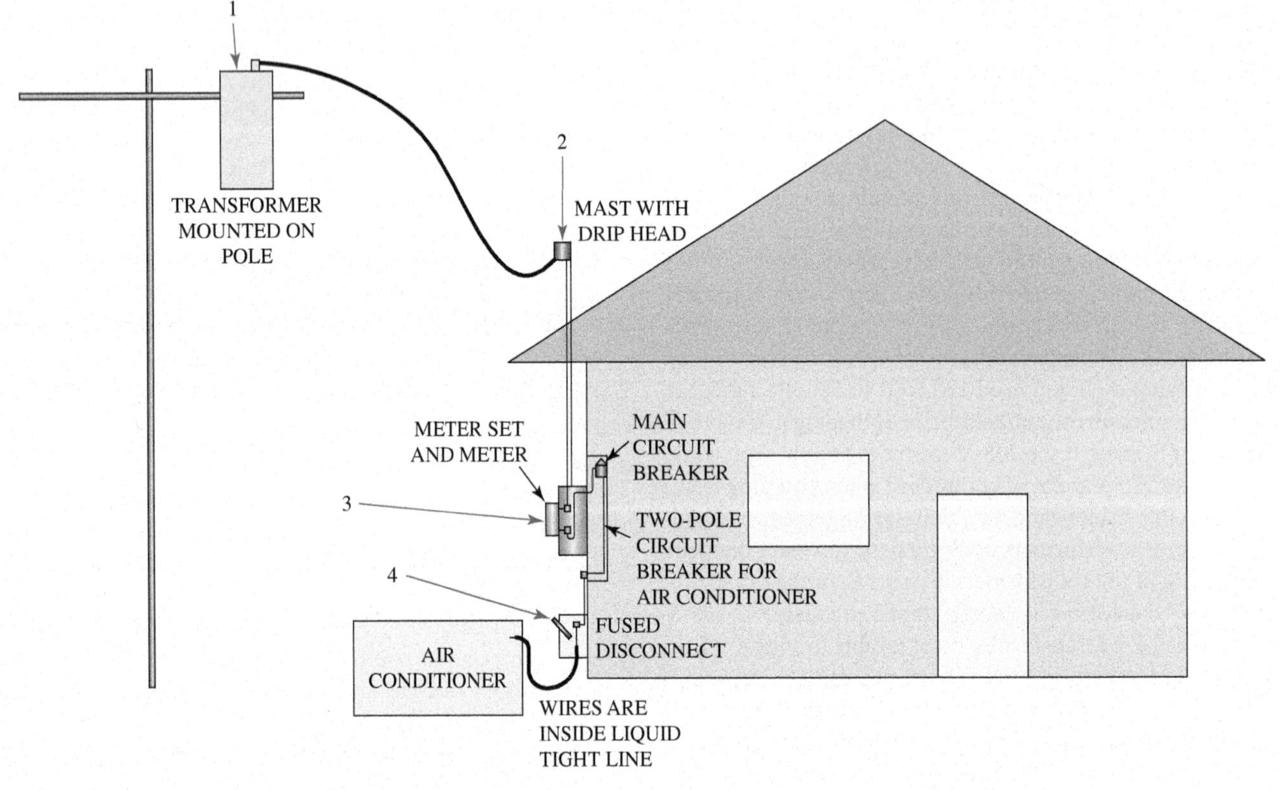

Figure 5-17 This line diagram shows the path electricity takes into a residence.

The transformer shown in Figure 5-19 steps down the high voltage to a usable voltage. This particular transformer supplies power to a residence. It will supply 240 volts to a panel box. The fuse transformer is located on the upper left side of the transformer. If this is open, there will be no power to the buildings that are fed by this transformer.

Figure 5-20 illustrates how the three conductors enter the residence. A drip or sag loop is placed in the conductors so that water will not enter the weather head or meter and panel box. The weather head is designed to separate the wires and allow them to enter the top of the riser conduit. Notice that there are other conductors coming into the structure. These low-voltage conductors are for telephone, cable television, and Internet service.

Figure 5-18 A disconnect provides a safe means of removing power from a condensing unit while a technician is servicing the outdoor unit.

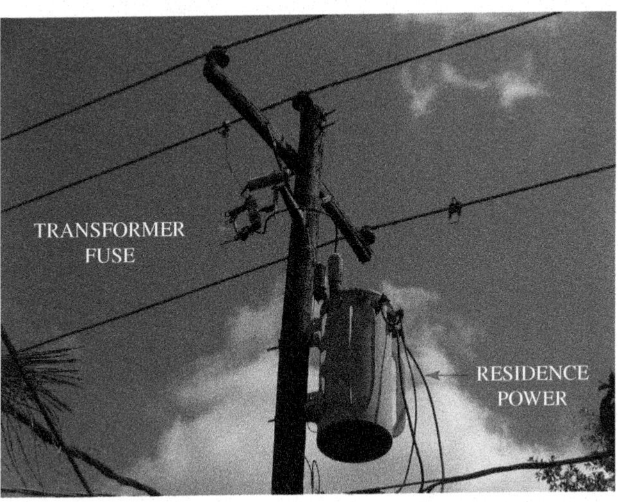

Figure 5-19 This transformer steps down the high voltage to a usable voltage. This particular transformer supplies power to a residence.

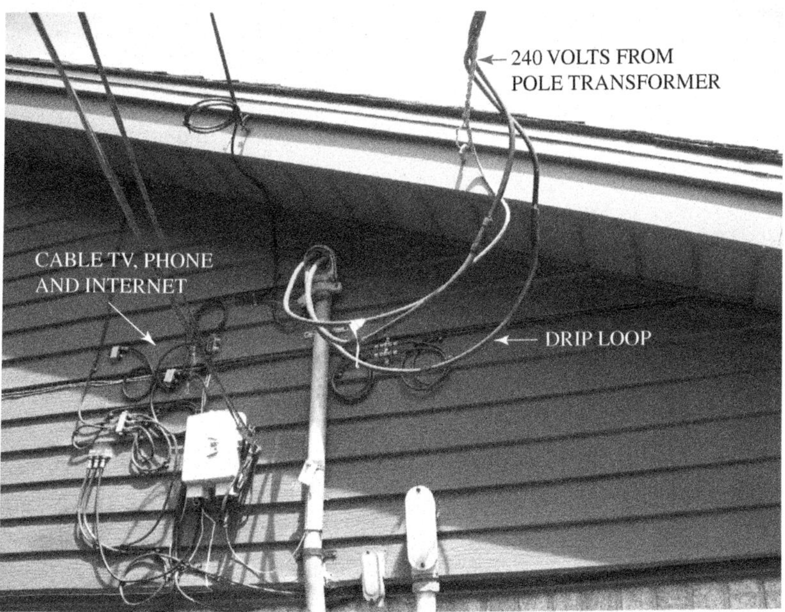

Figure 5-20 The right side of this image shows three conductor wires coming into the weather head.

The residential service shown in Figure 5-21 is divided up into the meter box and the panel box. In most cases the power company is responsible for all wiring up to and including the meter box. The building owner is responsible for the overcurrent protection (breakers or fuses) and wiring once the power leaves the meter can.

A close-up view of a smart meter is shown in Figure 5-22. This residential smart meter sends a signal to the electric company that provides data such as hourly and daily power use. The customer can receive a weekly report comparing temperature and current energy use as well as energy used the previous week.

GREEN TIP

Figure 5-23 is a copy of a report generated by the "smart" or "green" meter shown in Figure 5-22. The report is e-mailed to the customer. The meter collects data and gives the consumer a weekly report of kilowatt use in graphic form. It also provides the cost of the electricity, weather data, and a comparison of energy use from the previous week. This is considered a "green feature" since the consumer is quickly made aware of energy use and can take steps to address high energy bills such as adjusting the thermostat, changing the air filter, or scheduling service to help resolve high energy use issues.

Figure 5-21 Meter box and panel box. Some customers lock their panel box to prevent burglars from turning off power prior to breaking and entering.

Figure 5-22 This is a "smart meter" used to transmit electricity use data.

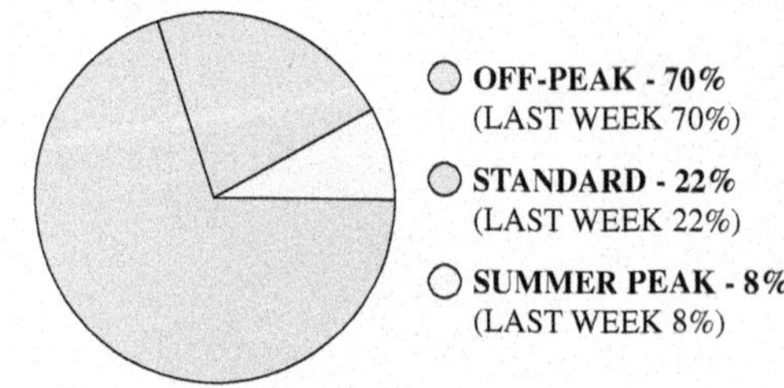

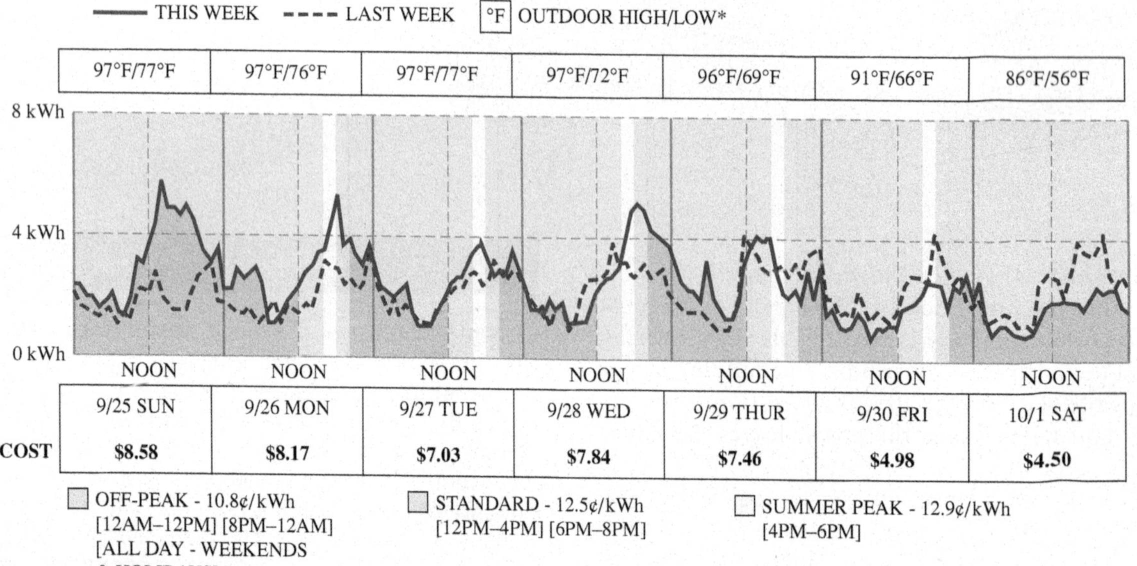

Figure 5-23 The smart or green meter from Figure 5-22 generated this report for the homeowner.

Figure 5-24 This residential panel box has had its cover removed. At the upper part of the panel are the two hot wires that make up the 240-V input. The large wire to the right is the neutral connection. The neutral bus bar is also part of the grounded system.

Figure 5-24 shows an open panel box with the breaker cover removed to expose the breakers and wires. To safely remove the panel cover, the main circuit breaker should be turned off. Before turning off any breakers to the building, the owner should be notified because critical systems may be operating that need to be backed up prior to power loss. This can include dedicated medical equipment, computer systems, or areas of a building that do not have windows. It is always a good idea to notify the customer of a power cut, even if the power is only off for a short period of time. Customers do not like surprises, especially surprises that relate to losing power. Keep them informed.

In Figure 5-24, notice that three wires are coming from the meter can above it, which feeds power to the breaker box. Two of the wires feed 208 or 240 V to a power or bus bar located behind the breakers. Figure 5-25 shows a closer view of the three wires entering the top of the breaker box. The two parallel wires feed power into the main disconnect. Measuring voltage across the breaker entrance lugs will

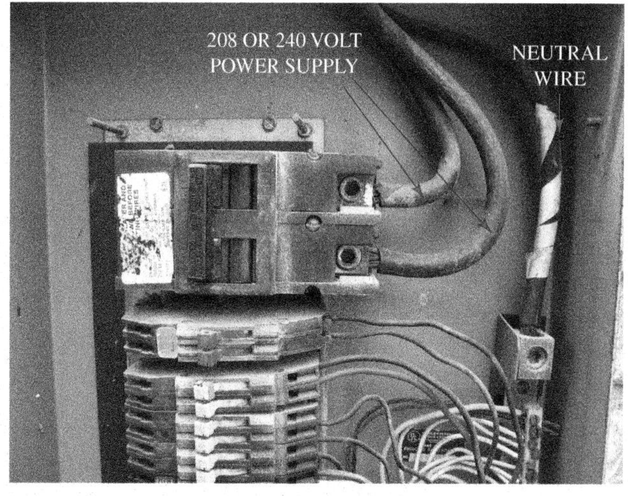

Figure 5-25 Turn the main circuit breaker off prior to removing the panel cover. Two power wires and a neutral wire enter the top right side of this breaker box coming from the electrical meter.

Figure 5-26 The two upper breakers with black wires feed 120 V to a circuit in the building. Next, there is a gap where another breaker can be installed.

indicate how much electricity is being supplied by the power company. The white-taped wire on the right is the neutral conductor. The neutral wire is connected to the neutral bus bar. The ground, not shown, is connected to the bottom part of the neutral bar. Neutral and ground are at the same potential. The difference is that neutral is a current-carrying conductor and ground is not. If the ground is carrying current there is a problem that may require the expertise of an electrician to solve.

Once the power enters the panel box, the following color code is used:

- The white wires are hooked to the neutral bar.
- The black wire is 120 V.
- The red wire is a 240-V circuit.

In a residential panel box, you can check the power being supplied to the various circuits in the building. Residential breakers can be single pole or double pole. Figure 5-26 shows two upper single-pole breakers with black conductors, a breaker gap, and a double-pole breaker with a black and red conductor. Normally a single-pole breaker will have a black wire. The double-pole breaker will have a black and red wire or two black wires. If two black wires are used, one of the black wires will be taped with the same color of tape on both ends. A white wire is used to represent the neutral circuit. These color codes are commonly used, but caution should be taken. Not all wiring is done by knowledgeable electricians or electricians who can identify colors (color-blindness). Never assume that the color code is correct.

5.8 COMMERCIAL SERVICE

The differences between residential and commercial service are that different types of circuit breakers are used and commercial service applications usually have a higher amperage capacity. The commercial panel box may have single- and double-pole breakers, but it will also have three-pole breakers. Single- and double-pole breakers are used to handle single-phase loads. Three-phase breakers are used for some or all of the commercial loads. The capacity of the commercial panel box may be higher than that of a residential panel. Residential panel boxes are normally no higher than 200 to 250 A. Commercial panel boxes can be smaller but many are larger, up to a 1,000 A and higher.

Figure 5-27 shows the distribution of commercial power from the utility transformer to the panel box. Here is a step-by-step explanation of the distribution process:

Step 1: The utility company provides 13.8 kV (13,800 V).

Step 2: Power of 480/277 V comes from the secondary side of the utility transformer.

Step 3: Power goes through the utility company service switchboard. The service switchboard includes a power quality meter and a disconnect.

Step 4: Power is distributed to 240-V single-phase and three-phase circuits along with 120-V single-phase circuits.

Step 5: Power arrives at three distribution switchboards, which include meters and disconnects. Note that the disconnect on the left is single phase. The other two are three phase.

TECH TIP

In the power distribution system, why is alternating current (AC) transmitted instead of direct current (DC)? Alternating current is chosen over direct current because AC has lower transmission losses. Even though the transmission losses are lower with AC, transmission losses from the generator to the end user can be as high as 15%. Wire has resistance. Using Ohm's law, the higher the resistance, the lower the power delivered to the end user. Long-distance transmission has high voltage drops or losses.

Alternating current is also easier to transform to a lower voltage. This is a big advantage when you want to reduce the high voltages that are transmitted over the power grid.

SAFETY TIP

Stay away from the electric power distribution system. The voltage is so great that even coming close to it can kill or injure you. The voltage can jump an air gap to the human body and you will become part of an electrical conductor to ground. It only takes 1/10th of an amp (100 milliamps) to kill a person.

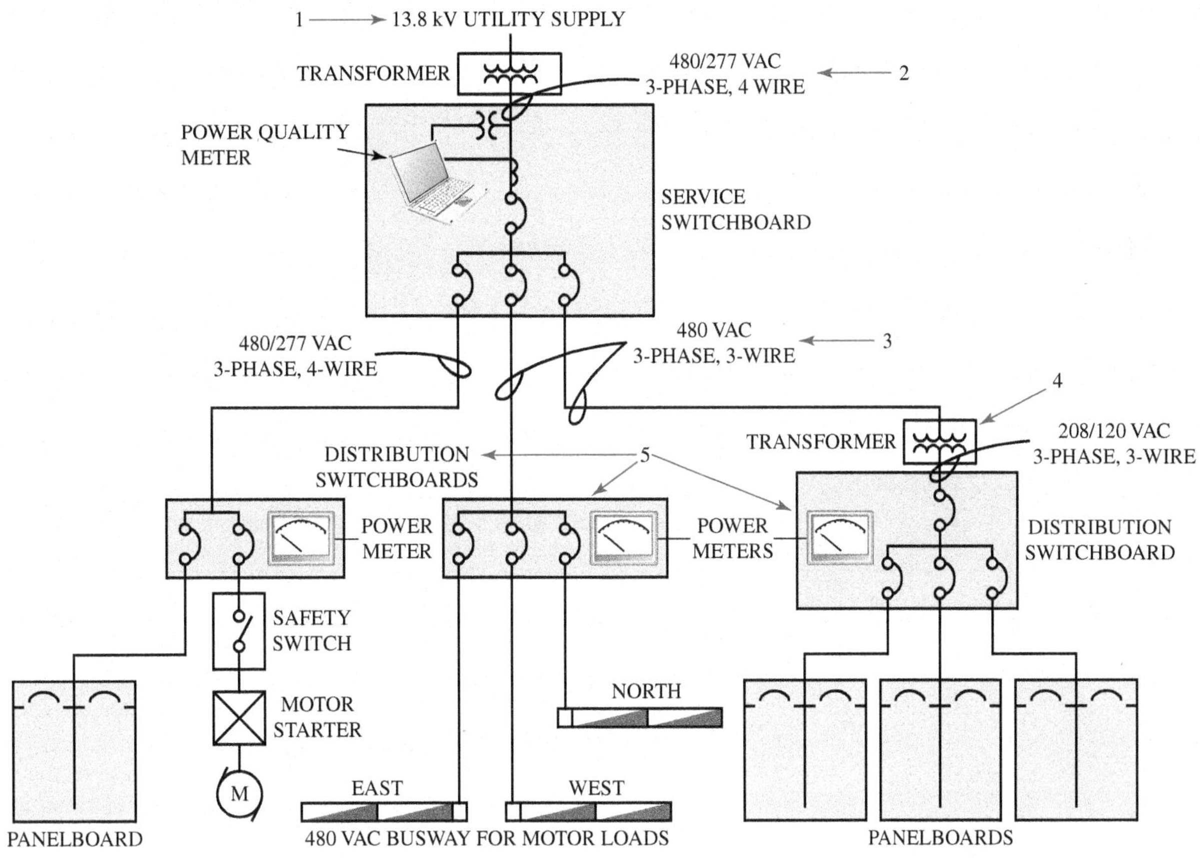

Figure 5-27 Distribution of commercial power from the utility company supply of 13.8 kV to single-phase and three-phase panelboards.

TECH TIP

In summary, the panel box has 240- and 120-V circuits available for use. The single-pole breaker supplies 120 V and the two-pole breaker supplies 208 or 240 single-phase voltage where required.

Three-phase power from the electric company commonly comes from delta (Δ) or wye (Y) transformers. As discussed earlier, use of the words *delta* or *wye* comes from the shape of the transformer windings. The delta winding is shown as a triangle and the wye winding as the letter Y.

When troubleshooting, you will be looking at the secondary side of a three-phase transformer. As a tech you will not directly troubleshoot the power company's transformer. You will need to understand what voltage is being supplied by the transformer to determine if the problem is coming from the power company or the customer's side of the meter. The technician does not want to be embarrassed by blaming the power company when the voltage issue is on the customer's side of the electrical service.

5.9 WHAT IS THE CURRENT STATUS OF ELECTRICAL USE?

The bar graph in Figure 5-28 compares residential energy use patterns in 1980 and 2005. Space heating energy use has dropped. Appliance and lighting energy use has increased almost 50%. Water heating energy use has increased almost 5%, while energy use for air conditioning has doubled.

The pie chart in Figure 5-29 illustrates how electricity is used in homes. This chart is average annual use in a home found in the United States. If the home is in a northern, colder climate the percentage of space heating would swap places with the air conditioning percentage. The refrigerator and water heating percentages are stable and not influenced by the climate.

GREEN TIP

Figure 5-30 shows a more detailed breakdown of residential energy use than Figure 5-29. It is good practice to know where energy is being used in a home before making recommendations to a customer about how to save energy. Yes, you can recommend more efficient heating and air conditioning systems—that is your job. Giving the customer other energy efficiency recommendations, such as using insulation or energy-saving lighting, will bond you to the customer for future HVAC sales and service.

As you can see in the figure, lighting is a rather large percentage of this mix. Since all lighting creates heat, any reduction in lighting use will reduce the amount of time the air conditioning system will operate. Installing high-efficiency equipment can make a building more "green," and the technician can give advice to help homeowners achieve the goal of having the most efficient building possible.

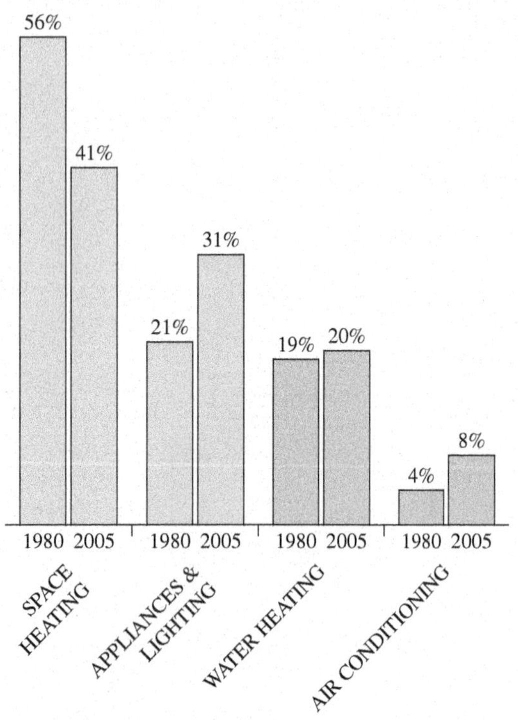

Figure 5-28 This bar graph compares residential energy use patterns in 1980 and 2005.

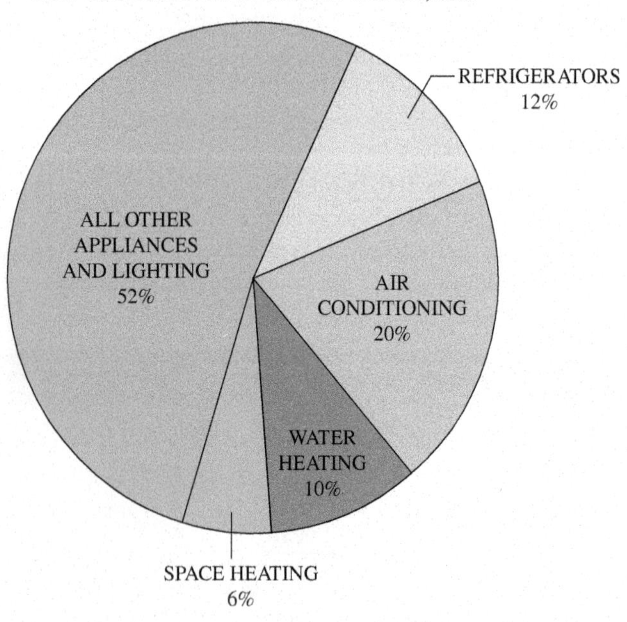

Figure 5-29 Electricity use in homes varies with the climate zone where the statistics are gathered. This pie chart indicates average annual use in a home found in the United States.

Figure 5-31 shows the national fuel mix over a 1-year time period. Most of the fuel used is mined, drilled, or processed in the United States. Coal is the major fuel source followed by natural gas and nuclear energy. A number of other smaller energy sources are used to complete the energy requirements in the United States.

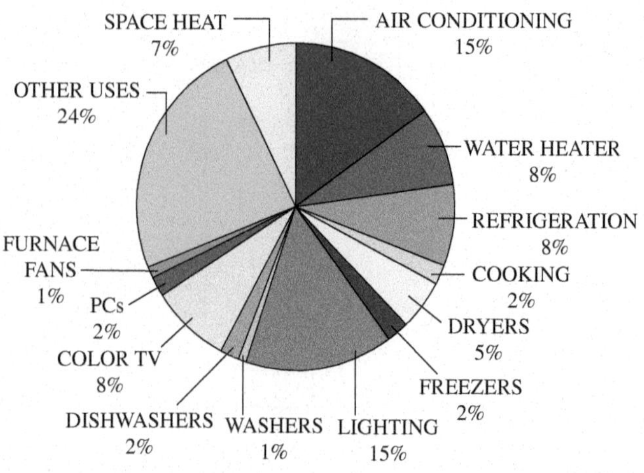

Figure 5-30 This pie charts shows a detailed breakdown of residential energy use. It is good know where energy is being used in a home before making recommendations to a customer about how to save energy.

Figure 5-32 illustrates that natural gas is the fuel of choice for generating electricity during the summer months. The amount of natural gas used during summer increases

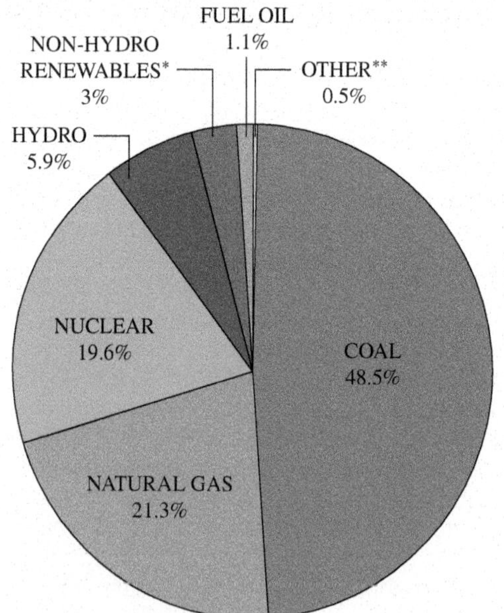

*INCLUDES GENERATION BY AGRICULTURAL WASTE, LANDFILL GAS RECOVERY, MUNICIPAL SOLID WASTE, WOOD, GEOTHERMAL, NON-WOOD WASTE, WIND, AND SOLAR.

**INCLUDES GENERATION BY TIRES, BATTERIES, CHEMICALS, HYDROGEN, PITCH, PURCHASED STEAM, SULFUR, AND MISCELLANEOUS TECHNOLOGIES.

SUM OF COMPONENTS MAY NOT ADD TO 100% DUE TO INDEPENDENT ROUNDING.

Figure 5-31 The national fuel mix over a 1-year time period.

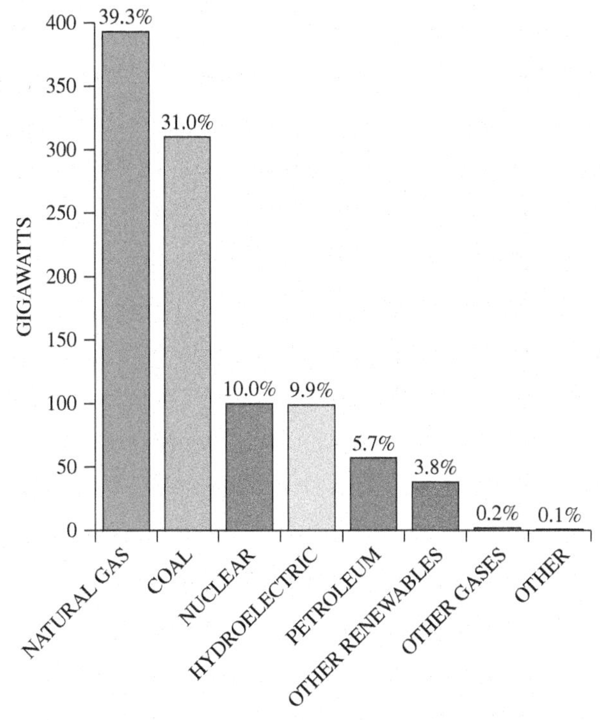

Figure 5-32 Natural gas is the fuel of choice for generating electricity during summer months. Coal is the second choice, followed by others.

compared to its annual energy use. Coal is the second choice, followed by a menu of lesser, but important fuels such as nuclear energy and hydroelectric. Having diverse ways to produce electricity is important so that one fuel does not dominate the mix. An interruption in the natural gas or coal supply would create power production problems. This mix of fuels creates healthy competition for the energy resources used to generate electricity, without total dependence on any one source.

GREEN TIP

What is the future for electricity? Electricity use is expected to increase at the same rate of economic growth. Figure 5-33 shows that electric use followed economic growth through 2007. The economic slowdown of 2008–2010 saw a decline in economic growth and electrical usage. The economic recovery will once again bring electrical growth.

Figure 5-34 shows projections about the demand for electricity. Demand is projected to increase by 28% by 2035. Increased electricity use is expected because (1) the population will increase, (2) demand for electronic products will increase, and (3) electric use increases with normal economic growth. With government-instituted HVAC efficiency requirements, however, the expected energy use per unit will drop as older, less efficient stock is replaced.

As illustrated in Figure 5-35, annual residential energy use is increasing rapidly. Providing high-efficiency heating and cooling options will help reduce this rapid rise in energy use. The government has mandated efficiency improvements in an effort to slow and stabilize the rapid rise in residential energy use.

Figure 5-36 graphs the reductions in pollution that have been experienced, despite increases in power generation. The next challenge for the power industry is to reduce carbon dioxide output, which is linked to global warming. Servicing and installing HVACR equipment properly will also reduce operating costs and utility use because systems operating at peak performance are more efficient. Most techs do not realize that they are at the forefront of "green" energy use in their attempts to reduce the amount of time systems are in use and improve equipment efficiency while it is running.

Many customers complain about their "high" electric bills. The bar graph in Figure 5-37 shows increases in the cost of selected consumer goods between 1988 and 2008. Note that the cost of electricity went up 54%, but that is a lower than most items used in our everyday life. One way

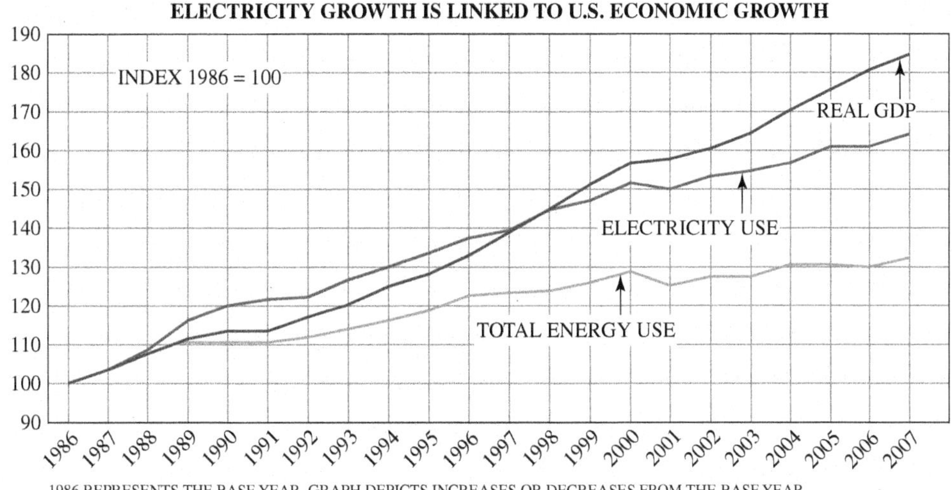

Figure 5-33 This graph shows the growth of the economy and the increase in use of electricity.

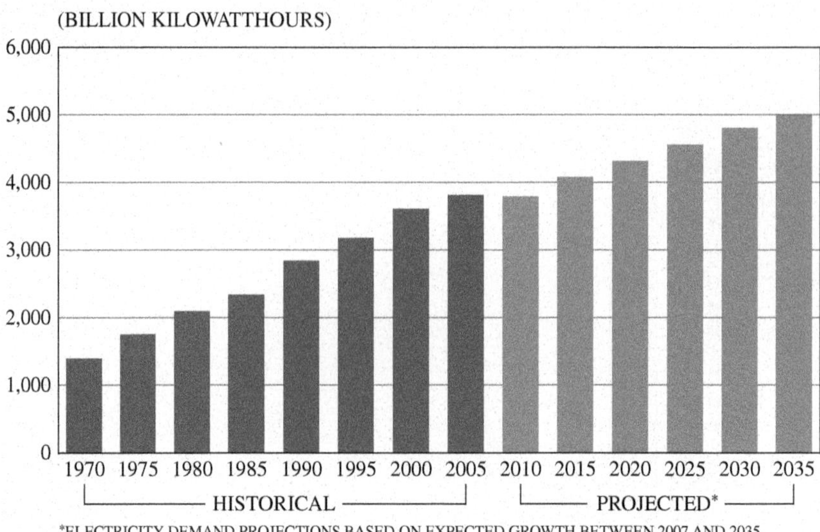

Figure 5-34 Demand for electricity is projected to increase by 28% by 2035.

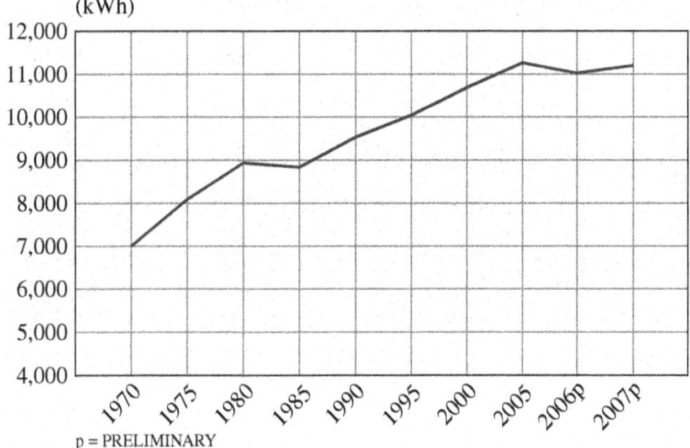

Figure 5-35 Annual residential energy use is increasing rapidly. The government hopes mandated efficiency improvements will help slow the rapid growth.

to help customers save on their electric bill is to tell them to request annual service and purchase high-efficiency equipment when replacing cooling and heating appliances.

Figure 5-38 shows an increase in air conditioning use in all parts of the country. As expected the south leads the country in air conditioning use. This growth is expected to continue because customers are requesting full comfort all year long. This growth in usage will promote higher efficiency installations.

TECH TIP

As shown in Figure 5-39, there is a transformer fuse on the power pole. If there is no power to the building, go outside and look up at the transformer. Spot the fuse transformer. If it is open, contact the power company so they can correct the problem. In some instances it is a matter of resetting the fuse. Other situations require transformer replacement. This is a power company job!

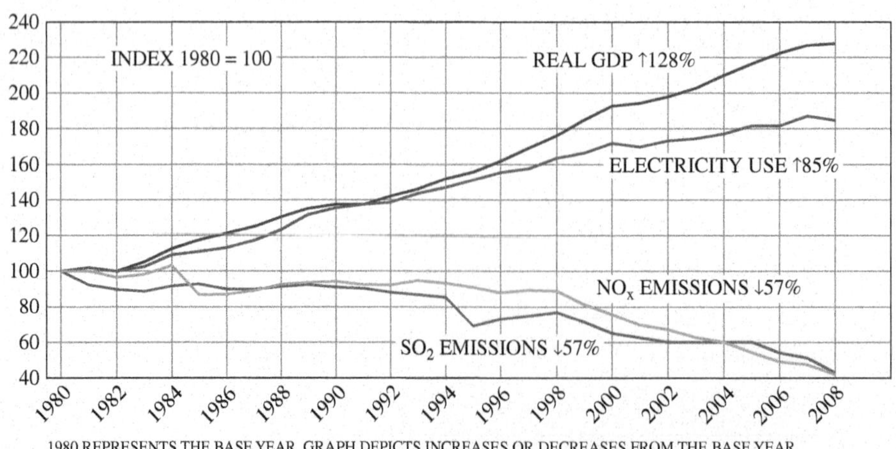

Figure 5-36 Power plants have reduced their emissions despite increasing electricity demand, 1980–2008.

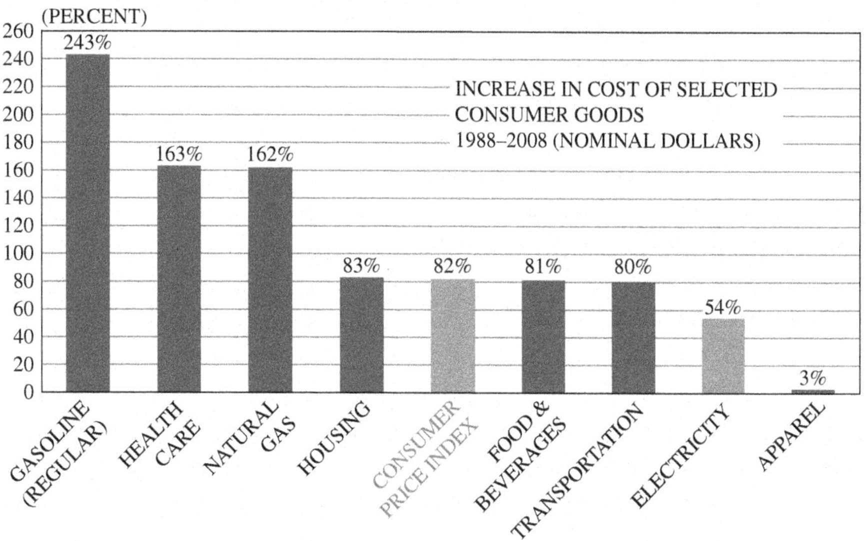

Figure 5-37 Increase in cost of selected consumer goods. Electricity costs vary by region. Electricity costs have not increased as much as they have for many common items found in everyday life.

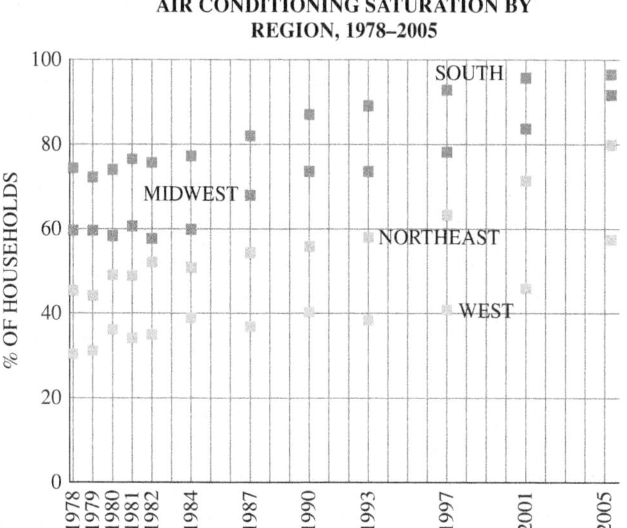

Figure 5-38 This chart from the U.S. Energy Information Administration shows an increase in air conditioning use in all parts of the country.

Figure 5-39 The arrow points to the main transformer fuse. This will be separated from the fuse holder if it is open. When there is no power to a building, look at the power pole to see if the pole fuse is open.

SERVICE TICKET

The second service call of the day is a "no cooling" call from a residence. The customer who called says she is glad you arrived so quickly in the middle of the summer. While doing your ACT (airflow, condensing unit, and thermostat operation) checks, you noticed that the indoor blower is operating. The airflow from the ducts "seems" strong. The house temperature is 81°F. The homeowner states that she keeps the house at about 73°F. She had noticed that the condensing unit was not operating even though the indoor blower section was operating. She understood that something was wrong; the first step of troubleshooting.

You explore the problem at the condensing unit. The gauge pressure seems good. There is no voltage measurement between L_1 and L_2. There is no voltage between L_1 and ground. There is 120 V between L_2 and ground. You check the panel box circuit and notice that it seems okay. You decide to safely reset the breaker by placing a glove on the hand you are going to use to reset the breaker. Standing to the side of the breaker, not in front of it, you turn the breaker to the off position. Next you flip the breaker to the on position. Nothing happens.

You check the voltage again with the same results. You decide that either the breaker is defective or the utility company power circuit is not complete. You remove the panel cover. There is no voltage between L_1 and L_2. There is a 120-V measurement between L_2 and ground. No voltage is measured between L_1 and ground. This is a power provider problem.

You notify the customer that the problem is with the utility company. She asks you to contact them and report the problem. Before leaving you ask the customer if she has noticed that some of the lights and appliances in the house are not working. She tells you the TV in the kitchen, the computer, and the lights in the secondary bathroom are not working. You explain to her that once the power company repairs the problem the air conditioning system as well as the other items in the house will begin to operate. You ask the customer to turn all nonworking items off until the power company addresses the problem. You tell her to watch for the power company employees since her transformer is on a pole in her backyard. They will need to come into her yard to repair the problem.

Finally, since it was not a true air conditioning problem, your company decides not to charge the customer for the service call. That is one way to keep customers happy and to get that return call when there is a real air conditioning problem.

SUMMARY

The information in this unit may not seem relevant to the HVACR profession, but technicians must know how power is supplied to a building. Connecting the wrong voltage source may damage the equipment.

The power supplies discussed in this unit are not the only electrical sources provided to our customers. These are the most common types. Single-phase power is used with residential homes and apartments. Various three-phase power options are used in commercial and industrial installations.

Common voltages in residential buildings are single-phase 208 or 240 volts. Commercial structures use three-phase voltage nominally rated at 208, 240, 440, or 480 volts. The voltage supplied by the power company should be within ±10% of the rated voltage. Low- or high-voltage conditions can create equipment damage.

It is important to know the power distribution output. Understanding the supply voltage is the first step in basic troubleshooting.

Finally, the good news for the HVACR professional is that more homeowners in all areas of the country are using air conditioning, as shown in Figure 5-38.

REVIEW QUESTIONS

1. What is the minimum amperage that will kill a human?
2. Define connected load.
3. What is a kilowatt?
4. What is a kilowatt-hour?
5. Define megawatt.

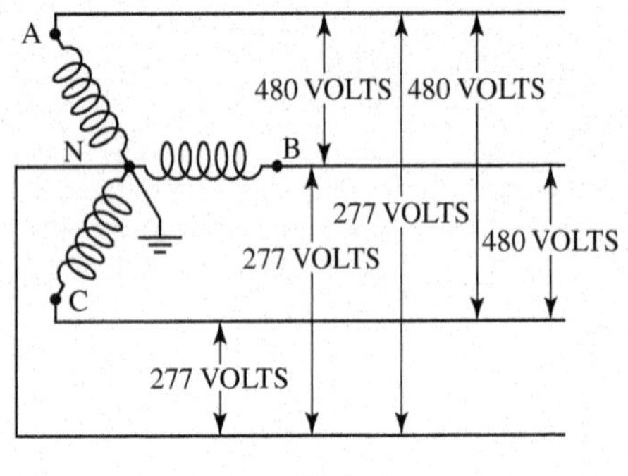

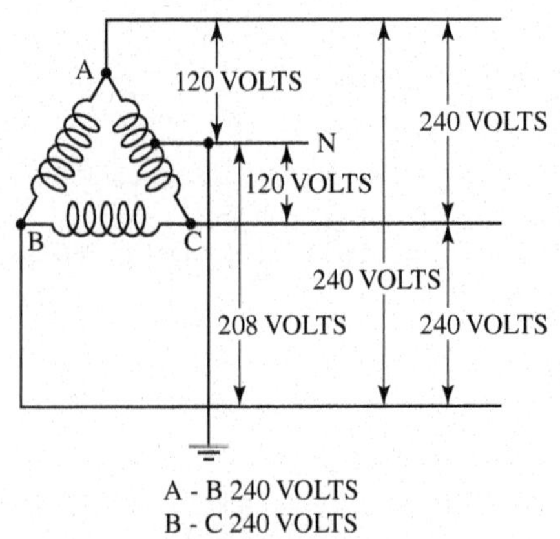

Figure 5-40 Use this image to answer Review Questions 14, 15, and 16.

6. Describe a meter loop.
7. What is a service drop?
8. Define watt.
9. Which power transformer will supply 120, 208, and 240 volts?
10. Which power transformer will supply 277 volts for a lighting circuit?
11. Why is it important to know what voltage is being supplied by the electric company when servicing a piece of equipment?
12. How is 120 V developed from a single-phase transformer that is used to provide power to a residence?
13. What are the two most common sources of fuel for generating electricity?
14. Are the transformers shown in Figure 5-40 single phase or three phase?
15. The transformer on the left side of Figure 5-40 is a ________________ transformer.
16. The transformer on the right side of Figure 5-40 is a ________________transformer.
17. What are the voltages in the boxes in Figure 5-41?

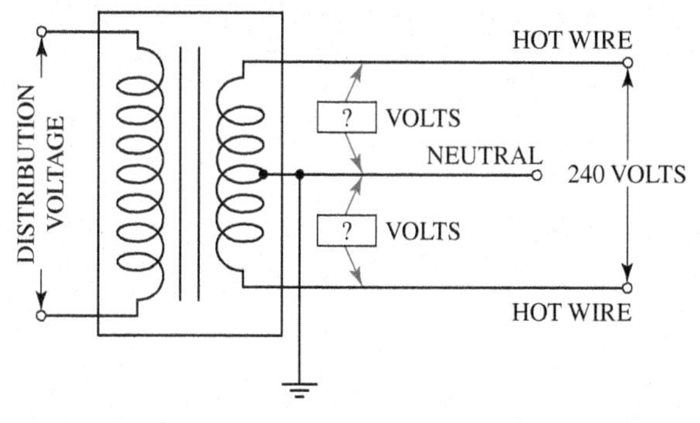

Figure 5-41 Use this image to answer Review Question 17.

UNIT 6

National Electrical Code®

WHAT YOU NEED TO KNOW

After studying this unit, you will be able to:

1. Describe how the National Electrical Code (NEC)® can assist with safe electrical HVACR installations.
2. Define the terms used in the NEC.
3. Use Table 310.15(B)(16) to select wire size based on amperage.
4. Discuss HVACR installation practices that are regulated by the NEC.
5. List fuse characteristics.

The **National Electrical Code (NEC)®**, also known as **NFPA 70®**, is a single, standardized source of minimum electrical code requirements for safe electrical installations in the United States. It has electrical, air conditioning, and insurance industry support. The NEC has been approved as a U.S. national standard by the American National Standards Institute (ANSI). It is formally identified as ANSI/NFPA 70.

Work on the NEC is sponsored by the **National Fire Protection Association (NFPA)**, based in Quincy, Massachusetts. The NEC was developed and is updated by the NFPA's Committee on the National Electrical Code, which consists of 19 code-making panels and a technical correlating committee. Although the NEC is not itself a national law, use of the NEC is commonly mandated by state or local law. This code is adopted and modified by local jurisdictions such as cities, counties, governmental districts, or states. In some cases it is modified and used outside the United States.

First published in 1897, the NEC is updated and published every 3 years. The 2011 code is the most recent edition (see Figure 6-1). Most states adopt the most recent edition within a couple of years of its publication. As with any "uniform" code, a few jurisdictions regularly omit or modify some code sections or add their own requirements (sometimes based on earlier versions of the NEC or on locally accepted practices). However, the NEC is the least amended model code.

In the United States any entity, including the city issuing building permits, may face a civil liability lawsuit (be sued) for negligently creating a situation that results in loss of life or property. Those who fail to adhere to well-known best practices for safety have been held negligent. This means that the city should adopt and enforce building codes that specify standards and practices for electrical systems (as well as other departments such as water and fuel-gas systems). A city can best avoid lawsuits by adopting a single, standard set of building code laws. This has led to the NEC becoming the standard set of minimum electrical requirements. No court has faulted anyone when the NEC standard was correctly followed.

Figure 6-1 Current copy of the National Electrical Code. The code is updated every 3 years; therefore, the next code will be coming out in 2014.

6.1 THE NEC AT WORK

The goal of this unit is to give the technician an overview of the National Electrical Code as it relates to HVACR. In doing so, we will just scratch the surface of the code.

Mechanical, electrical, plumbing, fuel gas, and building codes govern our profession. Many of these regulations relate to manufacturers of our equipment. It is important to understand how to use the codes and how they affect our daily lives. The NEC states "The purpose of the National Electrical Code is the practical safeguarding of persons and property from hazards arising from the use of electricity." This protection is not only against shock hazards, but also fire, which can be started by overloaded wiring and circuits.

Note that there may be exceptions to the information provided in this unit. The NEC has many exceptions to its published code. It is important to look for those exceptions when reading any code book. Consult the NEC or local electrical code officials to determine what is required. Some code enforcement authorities allow the HVACR technician to do a limited amount of electrical work outside of their jurisdiction. Again, check with the local inspector to determine if the work you are doing is within the jurisdiction of the mechanical code and not crossing over into the electrical code.

One of the most important aspects of understanding the NEC is learning how to use Table 310.15(B)(16). This table is used to size wire according to its current-carrying capabilities and the ambient temperature around the conductor (wire). It is also important to understand the nameplate on a piece of HVACR equipment. The nameplate helps us understand the required conductor size and overcurrent protection. We will discuss these important pieces of information in this unit, but first we begin by reviewing some of the terminology used in the NEC and in this unit.

Uniform Mechanical Code

The Uniform Mechanical Code, or UMC as it is commonly known, is the code that directly regulates safety practices when it comes to HVACR equipment. The UMC, however, defers all electrical requirements to the NEC; therefore, it is not discussed in this unit.

6.2 DEFINITIONS

Technicians should understand the following terminology, which is used in the NEC and in this unit:

Ampacity: The current, in amperes, that a conductor (wire) can carry continuously under conditions of use without exceeding its temperature rating. The ampacity rating is found on the condenser nameplate and is used to size wire to the condensing unit (see Figure 6-2). Ampacity is stated as minimum circuit ampacity or MCA. In this example, it is 11 A. This is used to size wire for this circuit.

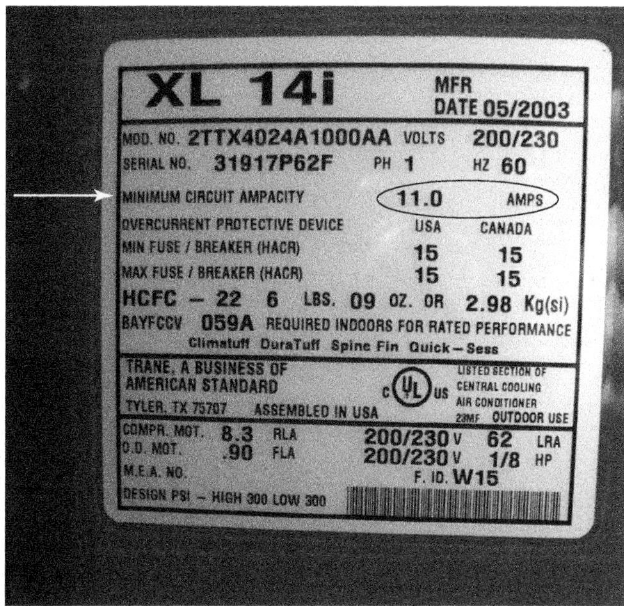

Figure 6-2 Minimum circuit ampacity, or MCA, for this condensing unit is 11 A (note circle). The MCA is used to size the conductors (wires) used with this condensing unit.

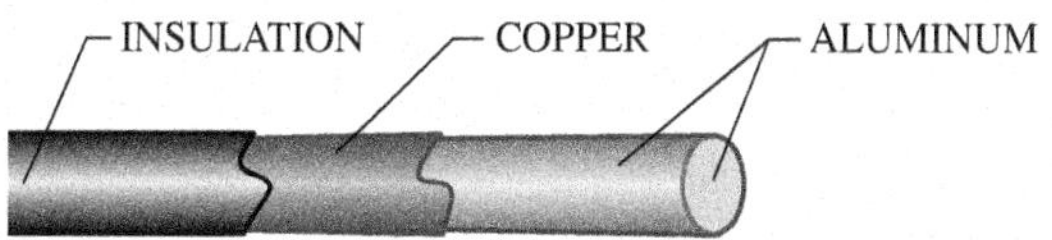

Figure 6-3 A copper-clad aluminum conductor is simply an aluminum wire coated with a thin layer of copper.

Article: The word *article* is used to define a section in the code book. Think of an article as a chapter or unit of the NEC.

Code jurisdiction: A city, town, county, state, or governmental agency that adopts and modifies the NEC is said to have code jurisdiction. Code jurisdictions may require a permit and inspection of any modified or installed electrical equipment, which includes HVACR equipment.

Conductor: Wire used as a path for electrical flow. Normally just called *wire*.

Copper-clad aluminum conductor: Aluminum wire that is coated with copper. An example of a copper-clad aluminum conductor is shown in Figure 6-3. This wire is classified as an aluminum conductor since it is primarily aluminum. If the copper-clad aluminum is not identified on the insulation or spool of wire, scraping the copper coating will reveal the aluminum conductor. Aluminum conductors carry less amperage than comparably sized copper conductors. Also known as CU-AL.

Disconnect: A device used to disconnect the supplied power from a circuit. An equipment disconnect is shown in Figure 6-4.

Grounding: Grounding is a safety mechanism designed to prevent electric shock if a person touches the

Figure 6-4 A disconnect on a package unit that can be seen by the technician. The convenience of the disconnect leads to safe and quick service.

Figure 6-5 Bare ground wire, connector, and 8-foot ground rod connected to a service panel (not shown).

equipment. Grounding is a conductor from the service panel to earth ground. The ground conductor is usually a bare wire strapped onto a copper-clad rod driven into the earth about 8 feet. Figure 6-5 shows an approved ground connection on an 8-foot ground rod. If a piece of equipment does not have a ground conductor, then a ground rod and bare wire connection may be installed at the unit. In most cases the ground wire comes with the current-carrying conductors to the piece of equipment it is servicing.

The grounding conductor should not be carrying current unless there is a problem. If the ground is removed and there is voltage between the ground wire and equipment case or if there is voltage between the removed ground wire and a grounding rod, there is a grounding problem. In other words, there is a current going to ground that should not be going to ground. This is an unsafe condition. Find and correct the shorted or grounded condition.

Locked-rotor amps: The term *locked-rotor amps* (LRA) or *locked-rotor current* (LRC) refers to the amp draw if a compressor is physically locked up and cannot move. The LRA measurement is also the amount of amps drawn momentarily when a motor is started. Figure 6-6 shows a compressor nameplate with the LRA rating circled. The rated load amps (RLA), not listed, will be many times lower than the LRA. RLA is the *maximum* amperage the compressor can draw during normal operation. Check the RLA, but do not charge a system using the RLA rating.

NEC Article 440: This article covers air conditioning and refrigeration equipment. Other code material related to HVACR is found scattered throughout other sections of the NEC.

Overcurrent protection: Overcurrent protection is provided by a breaker or fuse that will trip or open if the current rating is exceeded. A sample of circuit breakers that provide overcurrent protection is shown in Figure 6-7. The circuit must be sized to the wire or equipment it is protecting. Oversizing of the overcurrent protection will cause the wire or equipment to burn out or catch fire before tripping or opening. Figure 6-8 shows examples of common breakers used in the circuit panel box. Figure 6-9 shows fuses for overcurrent protection. Note the two important ratings on the fuse labels: amperage and voltage.

Rated load current: The rated load current (RLC) for a compressor motor is the amperage it will draw at its rated load, rated voltage and rated frequency of its design operation. RLC is also known as rated load amps or RLA and maybe be found on the compressor nameplate or the condensing unit nameplate.

Figure 6-6 The locked-rotor current (LRC) or locked-rotor amps (LRA) listing on this nameplate (circled in red) indicates the very high amperage that the compressor motor will draw when starting or when the compressor motor is mechanically locked up.

Figure 6-7 Circuit breakers provide overcurrent protection.

Figure 6-8 Circuit breakers provide overcurrent protection. The double-pole breaker on the left and single-pole breaker on the right are used in breaker panels.

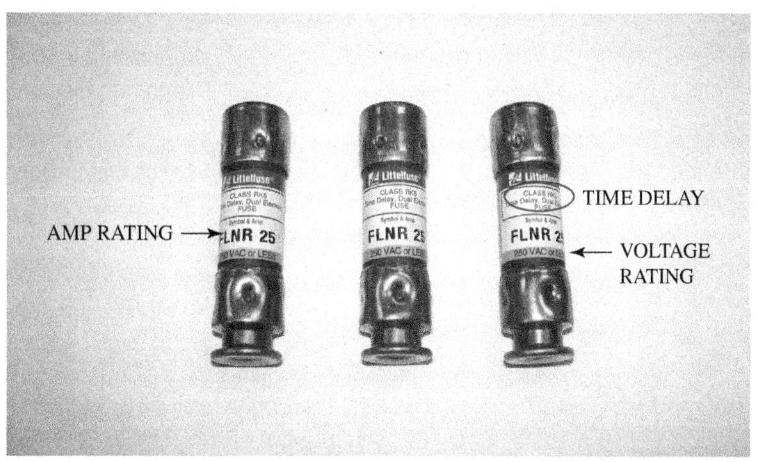

Figure 6-9 Fuses are used as overcurrent devices. These are 25-A, time-delay fuses rated for up to 250 V.

6.3 HOW TO FIND INFORMATION IN THE NEC

The NEC has a table of contents near the front of the code book and an index near the end. Use the table of contents to locate large, general topics. Use the index to find specific topics. The index does not use page numbers. It uses article numbers, which are arranged like chapters and page numbers. For example, if you are searching for electrical definitions, the index refers the reader to Article 100-1. Article 100 is near the front of the code book. See Table 6-1 for an example of the NEC table of contents and Table 6-2 for an example of the index found in the rear of the book.

6.4 ELECTRIC HEATING EQUIPMENT

A positive means of disconnect must be provided to remove electric power from electric heaters, which may include the blower motor, often called an **air handler**. The air handler may include the blower and evaporator sections. Electric heat strips are optional because the HVACR customer may not need heat. The contractor can select the amount of heat output by selecting a number of electric heat strips options. Most single-phase heat strips are rated at 4 to 5 kW (*kW* means kilowatts or 1,000 watts). A common 5-kW heater will produce about 17,000 Btu of heat at 240 V (5,000 W × 3.4 = 17,000). Lower voltage supplied to heat strips produces lower heat output.

We next discuss the safe means of removing power from an electric heating system. This safe removal of power is known as a disconnect. The ground conductor should never be disconnected when the main power supply is connected to a piece of equipment. If more than one power source serves the heating system, the disconnecting source must be grouped and marked. This will prevent the technician from thinking that turning off one disconnect will kill the power to the whole unit. It is a good idea always to check the power in a circuit prior to coming in contact with it.

Table 6-1 Table of Contents from the National Electrical Code

The NEC uses the term *article* to indicate divisions in the code book.

Table 6-2 NEC Index

The numbers in this index are not page numbers. The numbers after the topic are article numbers.

-A-

SAFETY TIP

Remember these steps to ensure that power is disconnected from the voltage source:

- Open (break the circuit) the disconnect or power sources.
- When the disconnect is open, there should be no voltage measured to the piece of equipment or to ground.
- Turning the equipment on at the thermostat should not operate the system.
- Measure voltage at the points where voltage was present. The meter should read no voltage or **OL.**
- At this point no voltage should be present. To be sure the circuit is safe, short the component to ground using a screwdriver or alligator clip.
- "Back hand" the circuit prior to touching or grabbing components. "Back handing" means to quickly touch the assumed dead circuit.

Following these steps will ensure that the equipment is electrically dead and that you will not be injured.

The disconnect device shall be sized to 125% of the combined amperage load of the heat strips, motors, and other current-drawing devices the disconnect serves.

As shown in Figure 6-10, the disconnect shall be within sight of the equipment it controls. This allows the technician to see if someone tries to turn on the equipment while the technician is working on it.

An out-of-sight disconnect must be provided with a means to lock it open so that only the technician can close the circuit to the equipment being serviced. This is known as a **lockout/tagout (LOTO)** procedure. The LOTO procedure prevents someone from turning the equipment on while the technician is working on it. A tag with the technician(s) name on it should be placed on the lock. The tag identifies the person who locked out the disconnect. The technician should have the only key to the lock.

Figure 6-11 The component parts of a LOTO: the lock with only one key, the locking mechanism, and the lockout tag.

Figure 6-11 shows the individual components of a LOTO: the lock with only one key, the locking mechanism, and the lockout tag. Figure 6-12 shows a locked-out disconnect. It only has one key and the key is controlled by the technician working on the equipment. Note that more than one technician may need the equipment secured, so several technicians can lock out the same piece of equipment. Each technician locking out a particular piece of equipment places a tag with his or her name on it.

According to the NEC, the nameplate of an electric heater shall be easily visible and shall contain:

- Model number
- Voltage
- Wattage or amperage rating.

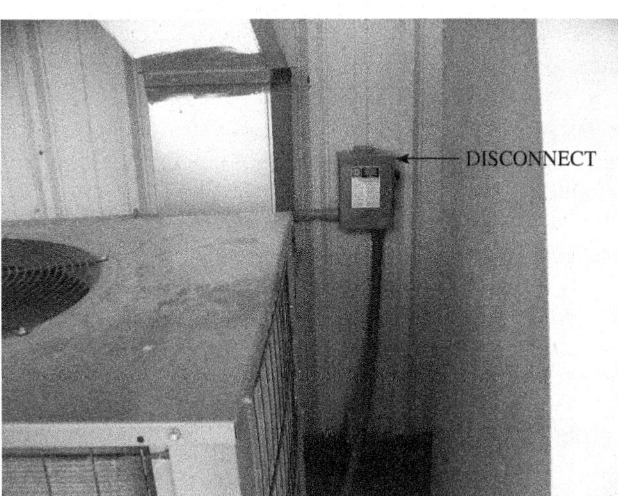

Figure 6-10 Disconnect within sight of the technician working on this package unit.

Figure 6-12 This is a lockout/tagout procedure. The disconnect was locked out after it was opened.

6.5 DUCT HEATERS

Duct heaters are usually electric heaters that are installed to supplement heat to a zone or duct. Duct heaters used in the supply air duct system must be "listed" for air conditioning or heat pump installation. "**Listed**" means that the heater conforms to and has been tested using nationally recognized standards developed by organizations such as Underwriters Laboratory (UL). Condensation or moisture can be an issue when a component, such as a duct heater, is placed in the cold, supply air stream.

A duct heater must be installed so that the airflow is even across the heating elements. Low air flow or uneven airflow causes hot spots on the element. A lack of airflow can burn out an electric heating element. Uneven airflow can occur when the duct heater is within 4 feet of the blower or evaporator. Duct heaters within 4 feet of an elbow may require turning vanes to straighten the airflow and allow good air distribution. Duct elbow turning vanes will improve the efficiency and flow pattern of airflow through any duct system. As you can see, even the NEC requires knowledge of the fundamentals of duct design and airflow properties.

The duct heater and the supply air fan must be interlocked so that the fan operates when the heater is energized. A fan time delay is allowed. A short fan delay is sometimes used to allow the heat strips to warm up prior to blower operation.

6.6 SELF-CONTAINED ELECTRIC HEATING UNITS

Some **self-contained electric heating** units have a built-in circuit breaker that is accessed from a cut-out in the outer panel. Figure 6-13 shows an example of an air handler with a breaker that controls heat strips. The cut-out allows access to the breaker or equipment without having to remove a door or panel; in other words, it is **readily accessible.** Some code jurisdictions allow this design for the disconnect, while other code jurisdictions require an external disconnect. It is important to know the code requirements for the area in which the equipment will be installed and inspected. The unplanned and additional cost of a disconnect, connecting whip, and the labor to install it will reduce a company's profit margin on a job.

Figure 6-13 This air handling unit (AHU) has a blower and electric heat strip. The AHU has a 30-A breaker sized to protect the circuit.

6.7 MOTORS

Per NEC, a common motor shall be marked with the following nameplate information:

- Manufacturer's name
- Rated voltage
- Full load current
- Frequency or hertz
- Single or three phase
- Rated full load speed
- Rated temperature or insulation system class and rated ambient temperatures
- Time rating (The time shall be 5, 15, 30, or 60 minutes or continuous operation.)
- Rated horsepower (hp) if 1/8 hp or more
- Code letter or locked-rotor amps if alternating current is rated at ½ hp or greater [The code letter amperage is found in the NEC, Table 430.7(B), which is reprinted here as Table 6-3.]
- A motor provided with thermal protection shall be marked "Thermally Protected." In some cases this may be abbreviated "T.P." This normally means that amperage protection is provided inside the motor, embedded in the windings.

Some motor manufacturers provide additional nameplate information not listed here.

For single-phase, AC motors, the conductors shall be rated to handle a minimum of 125% of the full load current. For multispeed motors the conductors shall be sized to handle 125% of the highest amperage on the motor. The selection of conductor size for motors will depend on other variables such as the type of motor, AC or DC, and single-phase or three-phase operation. This book does not pretend to give the reader all of the information needed to correctly size conductors for all applications. Refer to the NEC or an electrical contractor when there is a question about selecting wire size.

It is valuable to understand the relationship between wire size and load amperage draw when troubleshooting burned-out or overheated conductors. Undersized conductors will overheat and burn when placed in overrated load amperage conditions.

Disconnects are required for each motor. The disconnect must be sized to handle the rated current of all

Table 6-3 NEC Table 430.7(B): Locked-Rotor Indicating Code Letters

Code Letter	Kilovolt-Amperes per Horsepower with Locked Rotor
A	0–3.14
B	3.15–3.54
C	3.55–3.99
D	4.0–4.49
E	4.5–4.99
F	5.0–5.59
G	5.6–6.29
H	6.3–7.09
J	7.1–7.99
K	8.0–8.99
L	9.0–9.99
M	10.0–11.19
N	11.2–12.49
P	12.5–13.99
R	14.0–15.99
S	16.0–17.99
T	18.0–19.99
U	20.0–22.39
V	22.4 and up

(Reprinted with permission from NFPA 70-2011®, National Electrical Code®, © 2011 National Fire Protection Association, Quincy, MA 02169. This reprint material is not the complete and official position of the NFPA on the referenced subject, which is represented by the standard in its entirety.)

electrical components it serves. The disconnect can be an electricity-removing device as simple as a light switch that is used to disconnect power to a gas furnace. The common light switch disconnect is rated at 15 amps. Here is a list of some standard disconnect sizes:

- 30 amps
- 60 amps
- 100 amps
- 200 amps
- 400 amps
- 600 amps
- 1,000 amps.

Larger size disconnects are used in large motor or power distribution applications. A common 60-A disconnect is

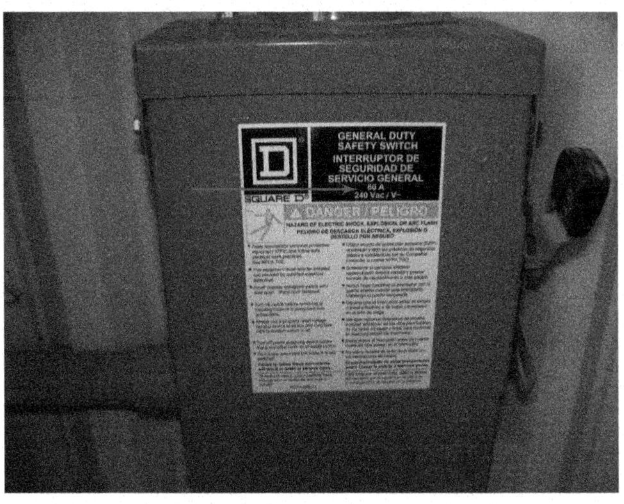

Figure 6-14 The red arrow points to the rating of a common 60-A disconnect. Sending more than 60 A through the disconnect will cause it to overheat and damage it.

shown in Figure 6-14. The red arrow points to the rating of a common 60-A disconnect. Sending more than 60 amps through the disconnect will cause it to overheat and damage it. This disconnect is in the "ON" or in the closed position. Electricity has the potential to flow from the power source to the equipment. When the black-handled lever is in the lower position, it is "OPEN" and has no current flow. Always check the disconnect if you think it is open.

The disconnect can be single phase or three phase. It might or might not have current protection. Current protection can be a fuse or circuit breaker, as shown in Figure 6-15.

TECH TIP

A nonfused disconnect has a movable knife blade that fits tightly between two holding stationary metal clips as shown in Figure 6-16. Notice that the three blades are disconnected from the clips. The disconnect is in the open position. This breaks the flow of electricity from the power source. The green, ground wire is not broken. Ground wires should never be broken.

Two problems could occur with this type of design. The movable blade may not slip completely into the stationary clip or it may not release from the stationary clip. In either case there is danger to the technician. Never assume that the movable blade has slipped into the stationary clip or that it is completely released to disconnect power.

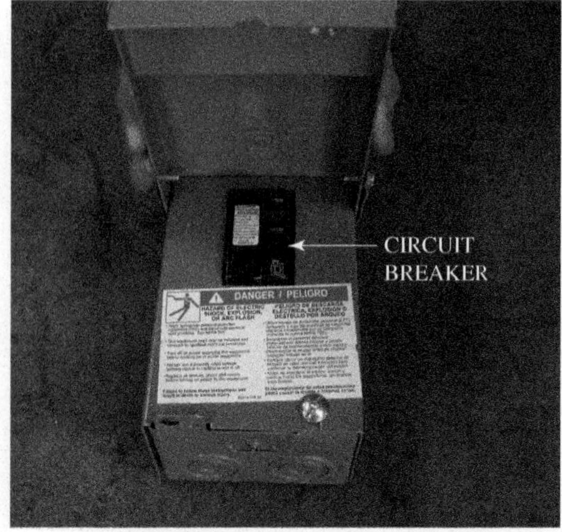

Figure 6-15 Single-phase disconnect with breaker protection.

Figure 6-16 Nonfused, three-phase disconnect.

The NEC addresses the issue of motor thermal overload protection. The selection of thermal overload protection is controlled by NEC requirements and electrical engineering practices. Each motor shall be protected against overheating by a thermal or amperage protective device. The overheated device can protect against both conditions. When replacing a thermal overload of any type, only use the exact replacement taking into account:

- Amperage rating
- Voltage rating
- Design.

For example, there are external thermal loads that are wired in series with the common winding of a single-phase motor. Many of these overloads are clipped to the side of the compressor housing near the compressor terminals. They are clipped to the side of the compressor so that they can be influenced by the compressor temperature as well as the compressor amperage. They actually respond to the compressor amperage in the common winding and the temperature of the compressor shell. High amperage or high compressor shell temperature or a combination of both will cause the thermal overload to the common winding of the compressor to open. This will prevent winding damage. Replacing the thermal overload with one that cannot be attached to the compressor shell will diminish its amperage protection value. It is never a good practice to replace a thermal protection overload with a lower or higher amperage device. This places your action into a liability situation if the system burns up or the motor is damaged in another way that may not be related to what you have done. You do not want to be part of a lawsuit for the design of a piece of equipment. When replacing parts, go back with the original equipment manufacturer's (OEM) part. Do not be pressured into using the wrong safety component in order to get a system operating.

6.8 COMPRESSOR MOTORS

Certain nameplate information is useful when servicing **hermetic compressors**. Per NEC, a hermetic refrigeration motor (sealed motor in the refrigerant environment) nameplate will list:

- Manufacturer's name or trademark
- Phase
- Voltage
- Frequency
- Locked-rotor current (LRC or LRA) (on the nameplate of all but the smallest motors)
- If provided, the words "Thermally Protected" shall be on the nameplate. (In some cases this may be abbreviated "T.P.")

Figure 6-17 shows a nameplate from a three-phase, scroll compressor. Notice that the information on the nameplate follows the requirements of the NEC.

The rated load current will be on the compressor nameplate or the condensing unit nameplate or both. Many manufacturers do not include the rated load current on the compressor, but it is on the condensing unit. It is important to realize this when a compressor has been changed. The replacement compressor may have a higher or lower amperage rating when compared to what is stated on the condenser nameplate. A good practice dictates that the condenser nameplate be changed to reflect the amperage of the replacement compressor.

6.9 CONDENSING UNITS

Condensing units and other HVACR systems shall have a disconnect within line of sight of the system, as shown in Figure 6-18. The disconnect can be installed on the unit if it does not cover the nameplate and does not create restrictions to servicing or removing doors or panels from the system or interfering with any other nearby system. Figure 6-19 shows

Figure 6-17 This nameplate from a three-phase, scroll compressor follows the NEC requirements. Starting at the upper left red circle, the nameplate shows the manufacturer's name, voltage (V), phase (PH), frequency or hertz (HZ), and locked-rotor amps (LRA).

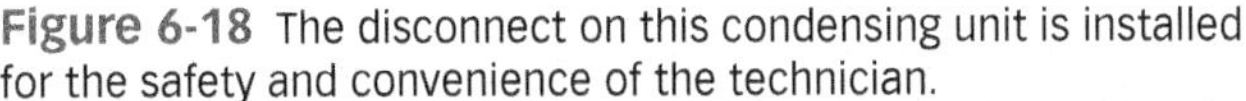

Figure 6-18 The disconnect on this condensing unit is installed for the safety and convenience of the technician.

Figure 6-19 Close-up view of the 60-A, single-phase pull disconnect for the condensing unit shown in Figure 6-18.

a close-up of the residential, 60-A, single-phase disconnect shown in Figure 6-18. The pull disconnect has been removed to kill power to the condenser. The disconnect pull is shown on top of the disconnect box. This design does not use over-current protection such as fuses or breakers.

6.10 CONDENSING UNIT NAMEPLATE

The condensing unit nameplate illustrated in Figure 6-20 contains electrical and refrigeration circuit information for the installing and servicing technician. This section focuses on the electrical information starting at the upper left-hand section or point 1.

Point 1: The condensing unit contains two compressors. The operating voltage is 575 V; three phase; 60 Hz; with a 23.7 rated load amps (RLA) and a 132 locked-rotor amps (LRA) rating. The 575-V voltage is sometimes used in Canada. The amperage ratings are for each individual compressor. To the right side of point 1 refrigeration data is listed, with test pressures listed to the right of that.

Point 2: This is condenser fan data. There are three condenser fans. Two are the same size and one is larger as indicated by higher amp draw and horsepower (HP). The fan data includes the operating voltage (Volts), phase (PH), hertz (Hz), full load amps (FLA), horsepower (HP), and kilowatts (KW).

Point 3: This section is used to size the condensing unit disconnect, wire size, and overcurrent protection such as fuses or breakers. According to this nameplate, the minimum circuit ampacity (MCA) is 61.1 amps, which means you would use 65- or 70-amp fuse breaker protection. The maximum ampacity (MOCP) is 80 amps.

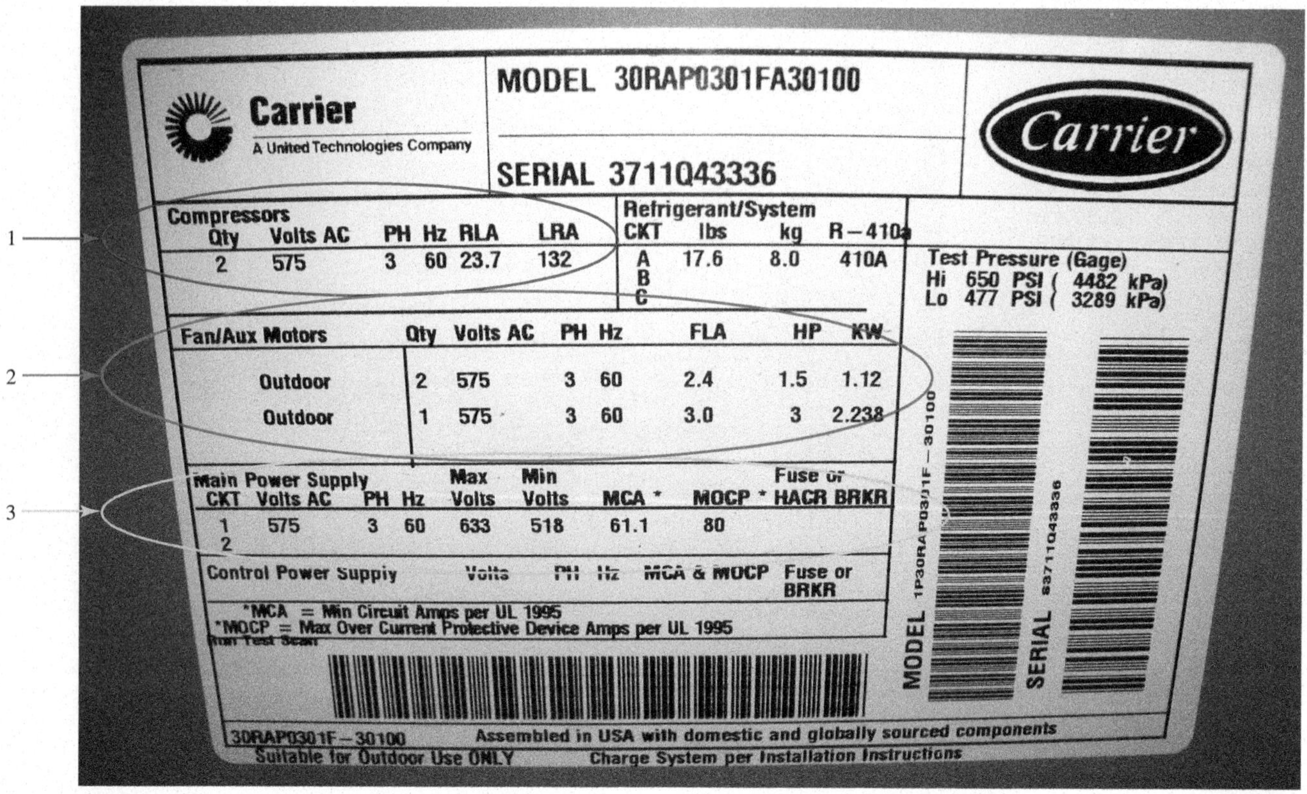

Figure 6-20 Point 1 is information on the two compressors in this condensing unit. Point 2 is information on the condenser fan motors. Point 3 is data for sizing the wire and overcurrent protection.

6.11 ELECTRICAL CONDUCTORS IN AIR DUCTS

Electrical conductors or wires in environment air ducts shall be protected. Article 300.22 of the NEC addresses this protection. The conductors shall be covered and protected in one of several ways:

- Conductors shall be in an impervious metal cover.
- Conductors shall be installed in electrical metallic tubing (EMT).
- Any flexible metal tubing is not to exceed 4 feet in length.
- A corrugated metal sheath is allowed.

Basically, electrical conductors in air ducts should be inside metal tubing. This code applies to general areas that are considered ducts such as common return air spaces found above ceilings and in under-floor duct spaces. In some instances the ceiling space or under-floor space can be pressurized as supply air space. Conductors should be protected in these instances also, since this is a passageway or air duct for airflow.

Nonmetallic conductors and unprotected (insulated, but not in conduit) conductors are generally allowed in residential ducts, but not anywhere in commercial or industrial buildings.

6.12 SIZING CONDUCTORS

The amperage of a conductor or wire will depend on the following:

- Conductor metal type, usually copper or aluminum. Copper is a better conductor than aluminum and will handle more amperage.
- Diameter of the wire or **wire gauge**, known as **American Wire Gauge** or **AWG**. Larger wire carries more current.
- Length of the wire. Short wires carry more current. Conductor amperage tables assume the length of wire is 1,000 feet. Remember that the conductor is measured by a round-trip to and from the load.
- Temperature of the wire. Cooler conductors carry more current because they have less resistance.

All wires have resistance and follow the current-carrying limits stated by Ohm's law.

Table 6-4 shows the NEC table that is used to determine the size of wire needed to handle maximum amperage at a rated temperature. This table gives the correct wire size to use if the user knows the amperage draw, application temperature, and wire type, whether it is copper or aluminum or copper-clad aluminum. A copper-clad aluminum conductor is simply an aluminum wire coated with a thin layer of copper as shown earlier in Figure 6-3.

It is important to understand how to read Table 6-4 when troubleshooting or installing new equipment. When installing new or replacement equipment, the conductor must be sized to handle the correct size amperage. Larger wire can be used, but this will be more costly. Additional cost may prevent your company from getting the bid even if it is better to use oversized wire.

First, let's review the structure of NEC Table 310.15(B)(16) (see Table 6-4), then practice using it to select the correct

Table 6-4 NEC Table 310.15(B)(16) (Formerly Table 310.16): Allowable Ampacities of Insulated Conductors Rated Up to and Including 2000 V, 60°C through 90°C (140°F through 194°F), Not More Than Three Current-Carrying Conductors in Raceway, Cable, or Earth (Directly Buried), Based on Ambient Temperature of 30°C (86°F)

	Temperature Rating of Conductor [See Table 310.104(A).]						
	60°C (140°F)	75°C (167°F)	90°C (194°F)	60°C (140°F)	75°C (167°F)	90°C (194°F)	
Size AWG or kcmil	Types TW, UF	Types RHW, THHW, THW, THWN, XHHW, USE, ZW	Types TBS, SA, SIS, FEP, FEPB, MI, RHH, RHW-2, THHN, THHW, THW-2, THWN-2, USE-2, XHH, XHHW, XHHW-2, ZW-2	Types TW, UF	Types RHW, THHW, THW, THWN, XHHW, USE	Types TBS, SA, SIS, THHN, THHW, THW-2, THWN-2, RHH, RHW-2, USE-2, XHH, XHHW, XHHW-2, ZW-2	
	Copper			Aluminum or Copper-Clad Aluminum			Size AWG or kcmil
18	—	—	14	—	—	—	—
16	—	—	18	—	—	—	—
14**	15	20	25	—	—	—	—
12**	20	25	30	15	20	25	12**
10**	30	35	40	25	30	35	10**
8	40	50	55	35	40	45	8
6	55	65	75	40	50	55	6
4	70	85	95	55	65	75	4
3	85	100	115	65	75	85	3
2	95	115	130	75	90	100	2
1	110	130	145	85	100	115	1
1/0	125	150	170	100	120	135	1/0
2/0	145	175	195	115	135	150	2/0
3/0	165	200	225	130	155	175	3/0
4/0	195	230	260	150	180	205	4/0
250	215	255	290	170	205	230	250
300	240	285	320	195	230	260	300
350	260	310	350	210	250	280	350
400	280	335	380	225	270	305	400
500	320	380	430	260	310	350	500
600	350	420	475	285	340	385	600
700	385	460	520	315	375	425	700
750	400	475	535	320	385	435	750
800	410	490	555	330	395	445	800
900	435	520	585	355	425	480	900
1000	455	545	615	375	445	500	1000
1250	495	590	665	405	485	545	1250
1500	525	625	705	435	520	585	1500
1750	545	650	735	455	545	615	1750
2000	555	665	750	470	560	630	2000

*Refer to 310.15(B)(2) for the ampacity correction factors where the ambient temperature is other 30°C (86°F).

**Refer to 240.4(D) for conductor overcurrent protection limitations.

wire size. The far left vertical column and the far right vertical column are shaded with the same color and represent wire gauge or thickness of wire. Wire gauge is used to determine the wire size required for a specific amperage draw using a specific wire insulation. At the upper left side of the gauge column, the table starts with 18-gauge copper wire or 18 AWG. AWG means American Wire Gauge. You will notice that the far right column starts with 12 AWG since this column relates to the aluminum or copper-clad aluminum conductors. Aluminum conductors below 12 gauge are not normally used.

The top, horizontal shaded row is the maximum temperature and insulation type at which the wire will operate at the rated amperage. The temperature is listed in degrees Celsius (°C) and in parentheses is the Fahrenheit (°F) temperature. Lower temperature wire operating in a higher temperature environment will not carry as much current. The installation can be influenced by the temperature in the conduit or the surrounding environment. Conductors in hot attic spaces need to be sized for the higher temperature wire insulation. Notice the temperature correction chart at the bottom of the table. Higher temperatures reduce the current-carrying capabilities of the wire.

There are three repeated temperature columns. The temperature columns are 60°, 75°, and 90°C. The three temperature columns on the left are for copper conductors. The same three temperature columns on the right are for aluminum or copper-clad conductors.

The third horizontal row on NEC Table 310.15(B)(16) is the type of insulation cover over the wire. In our profession, THHN and THWN are commonly selected insulation types used to hook up HVACR equipment. Here is what these insulation letters mean:

1. The *T* stands for thermoplastic insulated cable.
2. A single *H* means the wire is heat resistant.
3. A double *HH* means that the wire is heat resistant and can withstand a higher temperature. This wire can withstand heat up to 194°F.
4. A *W* means that the wire is approved for damp and wet locations. This wire is also suitable for dry locations.
5. An *X* means the cable is made of a synthetic polymer that is flame retardant.
6. The *N* is for the nylon coating that covers the wire insulation.

The following are examples and applications of THWN, THHN, and THW wiring types:

- **THWN** is flame retardant and heat resistant. Its insulation can be used in dry and damp locations. This wire type is not to be used in wet locations, however.
- **THHN** wire is commonly used in conduit installations in residential and commercial installations. Due to its nylon insulation, it is smaller and more flexible than the more rigid plastic-coated wire.
- **THW** is heat resistant, flame retardant, and moisture resistant. It can be used for dry, damp, and wet locations. It can also be used for underground feeds that are subject to moisture.

Finally, the body of the table provides the conductor amperage ratings. The copper wire and aluminum wire entries are separated by a bold line.

EXAMPLE 6.1 USING NEC TABLE 31.15(B)(16) TO DETERMINE AMPERAGE

Using NEC Table 310.15(B)(16) (see Table 6-4), determine how much amperage a 4-gauge AWG wire will handle.

SOLUTION

Select a number 4 AWG wire from the far left or far right columns. To make it easier to follow, place a straight edge underlining both sides of the AWG 4 wire row. The copper conductors that have a 60°C insulation rating will handle up to 70 amps. The 75°C insulated wire will handle up to 85 amps. The 90°C insulated wire will handle up to 95 amps. Higher temperature insulation on wire can handle more amperage.

The 4-gauge AWG aluminum or copper-clad aluminum conductors that have a 60°C insulation rating will handle up to 55 amps. The 75°C insulated wire will handle up to 65 amps. The 90°C insulated wire will handle up to 75 amps. When compared to the copper wire, aluminum wire will carry less amperage. Higher temperature insulation on aluminum wire will allow the wire to handle more amperage, but never as much as copper wire.

Table 310.15(B)(16) has exceptions that could get you into trouble if you do not pay attention to them. Let's review some of these exceptions. Notice that there is an asterisk (*) on AWG conductor sizes 14, 12, and 10. (Older code additions used an obelisk or cross to note an exception to the amperages published in this table.) This exception derates the wire's ampacity, which refers to reduced current-carrying capabilities. For example, look at the AWG 10 entry that is listed as a copper conductor that has a 60°C insulation rating. It will handle up to 30 amps. The 75°C insulated wire will handle up to 35 amps. The 90°C insulated wire will handle up to 40 amps. For normal HVACR operations the asterisk (*) or exception will lower the amperage draw of each wire size by 5 amps. Now the derated maximum amp draw of these wires is 25, 30, and 35 A, respectively.

The normal voltage rating of a conductor is up to 600 V. Higher voltage wire is available, but not used in standard HVACR equipment. Excessive voltages in conductors will arc through the wire insulation.

Finally, NEC Table 310.15(B)(16) applies to installations where you have three or fewer current-carrying conductors in a single wire way or conduit.

TECH TIP

When measuring wire you must consider the length of the wire going to and returning from the source or load that is using the power. For example, a condensing unit is 120 feet from the electrical panel box. The contractor must select the wire size based on a 240-foot wire.

6.13 ROMEX

Romex® is a trademark for what the NEC refers to as **nonmetallic sheathed cable.** Romex is a registered trademark of the General Cable Corporation. The NEC, Article 336-2, defines nonmetallic sheathed cable as a factory assembly of two or more insulated conductors having an outer sheath of moisture-resistant, flame-retardant, nonmetallic material. Romex is used for home wiring. Romex is not allowed in commercial applications since conduit is required to protect the conductors. Romex in not easy to pull through conduit and in some cases it is against code to install Romex inside conduit.

Romex concealed in a wall requires a nail protective plate when the cable is less than 1 1/4 in. from a stud edge. This protects the wire from nail or screw damage. The nail plates shall be at least 1/16 in. thick per NEC 300-4(a)1.

Figures 6-21, 6-22, and 6-23 show common Romex conductors. Figure 6-22 shows an older Romex 10/2 wire. This is 10-gauge wire with two conductors and a ground. The Romex has a white sheath, which was the common color for all gauges of older Romex. The color of newer Romex sheathing designates its gauge. For example, 10-gauge Romex would be orange.

Romex has two, three, or four conductors with a green insulated or bare conductor used for grounding equipment.

Older nonmetallic sheathed cable or Romex was normally coated with a white, gray, or black insulation. The gauge and number of conductors was printed on the sheathing. At this time the color of the Romex outer insulation jacket indicates the gauge of the wire. The following describes this color coding and gauge relationship.

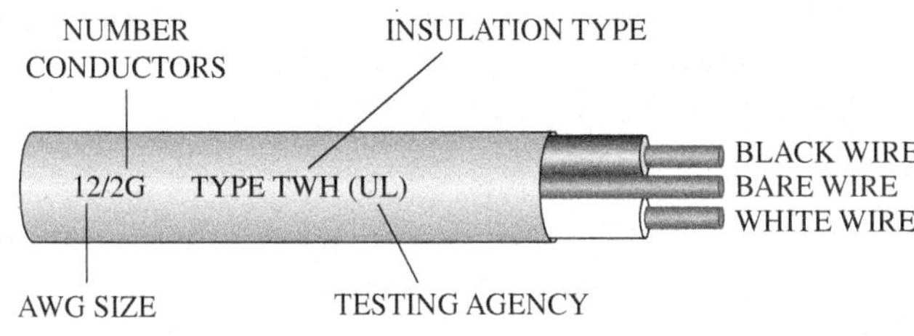

Figure 6-21 This is an example of the information on Romex wire. The gauge is 12 AWG, with two conductors. The *G* indicates a bare ground wire. The insulation type is TWH and it is UL approved.

Figure 6-22 This is an older Romex 10/2 wire. It is 10-gauge wire with two conductors and a ground.

Figure 6-23 Roll of white 14-2G Romex wire. This is 14-gauge wire with two conductors and a ground.

TECH TIP

Individual wire color does not have a recognized standard, but there are norms that you see in the field. For example, in a single-phase, 120-V circuit, you will find the following colors:

- Black wire is hot or 120 V.
- White wire is neutral.
- Bare wire is ground.

On 208/240-V circuits you will find the following colors used most often:

- Black and red colored wires are the power source.
- Bare wire is ground.

Unfortunately, these colors are not always used as listed.

White-Colored Nonmetallic Sheathed Cable

The white wire sheath houses 14-gauge wire, as shown in Figure 6-23. This type of wire is used for 15-A circuits in your home. This type of wire is used primarily in lighting circuits or gas furnaces.

Yellow-Colored Nonmetallic Sheathed Cable

Yellow wire sheath encloses 12-gauge wire that is rated for 20-A circuits. General power for outlets and appliances is the main use for this size of wire feed. Figure 6-24 illustrates an example of yellow-colored, 12-gauge Romex wire.

Figure 6-24 Yellow Romex wire is a 12-gauge wire with a bare ground conductor. Most Romex is rated for at least 600 V.

Figure 6-26 Black Romex wire is 6 gauge and has three conductors.

Orange-Colored Nonmetallic Sheathed Cable

The orange wire sheathing is set aside for 10-gauge wire. It is able to handle 30-A circuit loads. These loads include air conditioner, water heater feeds, and any other 30-A loads. Figure 6-25 illustrates an example of orange-colored, 10-gauge, three-conductor Romex wire with a ground.

Black-Colored Nonmetallic Sheathed Cable

The black-coated wire shown in Figure 6-26 is shared for both 6- and 8-gauge wire. As you may know, 8-gauge wire is good for 45-A circuits and 6-gauge wire is capable of handling 60-A circuits. The 6-gauge wire is better for feeding a subpanel, electric heat strips, an electric range, or a double oven, depending on the amperage rating listed on the appliance.

Figure 6-25 Orange Romex wire is a 10-gauge, three-conductor wire with a ground. This could be used on three-phase circuits since it has three conductors.

Gray-Colored Nonmetallic Sheathed Cable

There is another colored sheathing that has more to do with installation areas than with wire size. This is the gray-colored nonmetallic wire. It is used for underground installations and comes in varying sizes. It has water-resistant qualities and is sometimes resistant to other things like oil and sunlight.

Nonmetallic Sheathed Cable's Outer Jacket Labeling

As with all nonmetallic sheathed cable, the outer jacket is labeled with letters that show how many insulated wires are concealed within the sheath. This wire count does not, however, include the uninsulated wire that is used as a ground wire. For instance, if the cable lists "12-2 WG," then the wire has two insulated 12-gauge wires (a black and a white wire), but it also has a ground wire. If the label says "12-3," this is a three-conductor, 12-gauge cable with a bare copper ground wire included.

TECH TIP

Here is the test for excessive voltage drop between the supply voltage and the load:

Step 1: Check the voltage at the electrical panel with the load off. Let's assume that the load is an electric heater. Record the voltage with the load off.

Step 2: Turn the heater on. The voltage should not change by more than a few volts or a maximum of 2% of the supply voltage. If the voltage changes by more than a few volts or 2%, the conductors feeding the panel are too small or there is a bad connection.

Step 3: Measure the operating voltage at the panel box and then at the load. If there is a voltage drop of 2% or greater, the conductor is too small or the connection is the problem.

Step 4: Connection problems can be caused by a loose screw or lug connection. Tighten all connections. Verify that fuses and breakers are tight. Tighten wire nuts. This step should be routinely done as part of a preventive maintenance program.

6.14 FUSE SIZING

The NEC allows **fuse sizes** of up to 225% of compressor run load amps (RLA) or branch circuit selection amps, whichever is greater, plus 100% of other motor amperage draws provided that all motors are equipped with overload protection. This value is rounded down to the nearest standard fuse size. It is better for a fuse to blow a little early than late.

The fuse size is the smallest standard fuse size that system tests indicate. It can provide system operation over the expected range of conditions without nuisance trips. The recommended or minimum fuse size can be smaller or equal to the maximum fuse size, but not larger. A large overcurrent device would not give the desired equipment protection.

Fuses also have a voltage rating. The voltage rating ensures that the blown fuse has adequate space between the open fuse elements so that voltage does not jump across the open element. For example, do not use a car fuse in your digital multimeter. Most car fuses are rated at 32 volts. Using a car fuse in a DMM may provide a path for voltage to arc across an open element, endangering the technician and the equipment.

6.15 BRANCH CIRCUIT

A **branch circuit** is made up of conductors carrying power from the main or subpanel box to the HVACR equipment. In other words, it is power that branches off from a central power source or breaker box. The minimum branch circuit ampacity is calculated and published by the manufacturer of the listed equipment. This information is located on the equipment nameplate as seen in Table 6-5.

Table 6-5 Nameplate for a 3-Ton Condensing Unit

The nameplate shows the minimum and maximum current required to protect the condensing unit.

MOD. NO. TTX036C100A1		Volts 200/230	
Serial No. J36266476		PH 1 HZ 60	
Minimum Circuit Ampacity		23.0	AMPS
Overcurrent Protective Device		USA	Canada
Recommended Fuse/Breaker (HACR)		40	40
Max Fuse/Breaker (HACR)		40	40
HCFC-22 9 LBS. 08 OZ.		OR 4.31	Kg(al)
BAYFCCV 077A	Required Indoors for Rated Performance		
The Trane Company Tyler, TX 75711-9010 Made in USA		Outdoor USE	
COMPR. Mot. 17.0 RLA		200/230 V	91 LRA
O.D. MOT. 1.50 FLA		200/230 V	1/5 HP
M.E.A. NO. 56-92E			
Design PSI - High 300 Low 300		F.ID. X04	
MOD. NO. TTX036C100A1		The Trane Company	
Serial No. J36266476	X04	Tyler, TX75711-9010	

(Source Trane Application Bulletin APB99-75, December 1999.)

In Table 6-5 the minimum circuit ampacity of 23.0 A is circled. The ampacity is used to size the wire from NEC Table 310.15(B)(16). In this example the technician would use a 25-A rated wire as required by NEC Table 310.15(B)(16). If the maximum overcurrent protection is used, in this case 40 A, you would select a conductor that would handle 40 A. Match the wire gauge with the overcurrent protection used in the circuit.

Figure 6-27 shows a condensing unit nameplate. Let's see what valuable electrical information it contains:

- The voltage rating is 200 to 230 V. The voltage can vary 10% above and below this rating.
- The nameplate lists a minimum circuit ampacity of 11 A.
- The maximum circuit ampacity is 15 A. Since 11-A conductors are not available, you would select the 15-A wire and 15-A overcurrent protection for this installation.
- The compressor draws a maximum of 8.3 RLA and 62 LRA.
- The maximum amperage draw or rated load amps should be avoided, even on hot days. The 62 locked-rotor amps is expected momentarily on start up.
- The outdoor condenser fan motor (O.D. MOT.) uses 0.90 FLA. The 0.90 A will be measured when the fan is moving the correct amount of air through the condenser. A dirty condenser will move less air and draw lower amperage since the fan is moving less air.

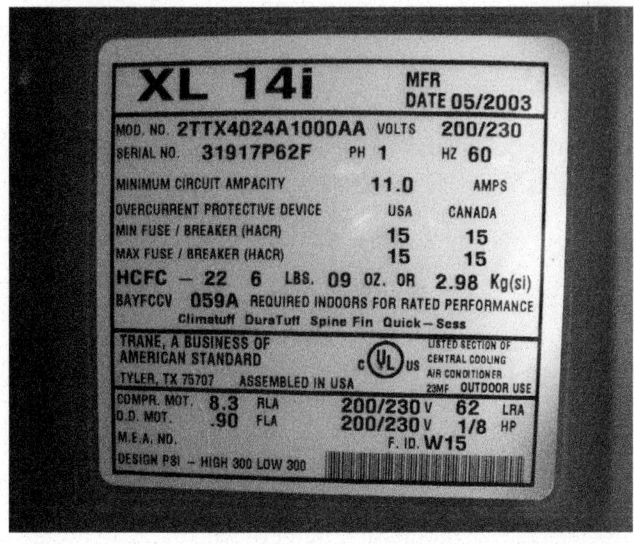

Figure 6-27 This condensing unit's nameplate tells the technician what wire ampacity is required. The ampacity will determine the wire size.

EXAMPLE 6.2 USING WIRE

This sample problem relates to the nameplate from a 3-ton condensing unit. Use the nameplate information given below and Table 6-6 to select the wire size based on condenser ampacity and distance the wire travels.

Here is the nameplate information from Table 6-5 for a 3-ton condensing unit and other information you will need:

- The ampacity on the nameplate is 23 A.
- The minimum wire size would be selected for 25 A.
- There is no 23-A wire.
- The technician should not reduce the wire size below the minimum rating.
- The maximum fuse/breaker size is 40 A.

SOLUTION

In this case, match the wire size with the maximum breaker size that will be used. Selecting the minimum wire size is legal and the material is less expensive. Selecting the minimum wire size and minimum overcurrent protection may result in nuisance trips at condensing unit start-up.

Minimum branch circuit ampacity	23 A
Maximum fuse size	40 A
Recommended fuse size	40 A
Compressor branch circuit selection current	17 A
Outdoor fan motor amps	1.5 A

Calculate minimum branch circuit ampacity:

$$(17\text{ A} \times 1.25) + 1.5\text{ A} = 22.75\text{ A (round to 23 A)}$$

Select the wire size from Table 6-6.

Sizing 3-ton condensing unit for minimum ampacity:

Minimum ampacity = 23 (round up to 25 A)

Sizing the wire for 25 A, up to 45 ft = #10 AWG copper or #8 AWG aluminum wire (75C). See blue oval on Table 6-6 for this point.

Sizing the wire for 25 A, up to 80 ft = #8 AWG copper or #6 AWG wire (75C). See red oval on Table 6-6 for this point.

Sizing 3-ton condensing unit for maximum ampacity:

Maximum fuse size = $(17 \times 2.25) + 1.5 = 39.75$ amps (In this case we are allowed to round up to 40 A.)

Sizing the wire for 40 A, up to 45 ft = #8 AWG copper or #6 AWG wire (75C). See green oval on Table 6-6 for this point.

Sizing the wire for 40 A, up to 80 ft = #6 AWG copper (75C wire). See yellow oval on Table 6-6 for this point.

The recommended maximum breaker/fuse is 40 amps. The copper wire size will be 8 or 6 AWG, depending on the distance of the wire from the breaker to the condensing unit.

Table 6-6 Branch Circuit Wire Sizing Table

This table is used to determine the branch circuit wire size for a condensing unit. To use this table to select a wire size, the technician needs to know the circuit ampacity and round-trip distance the wire travels from the power supply to the load.

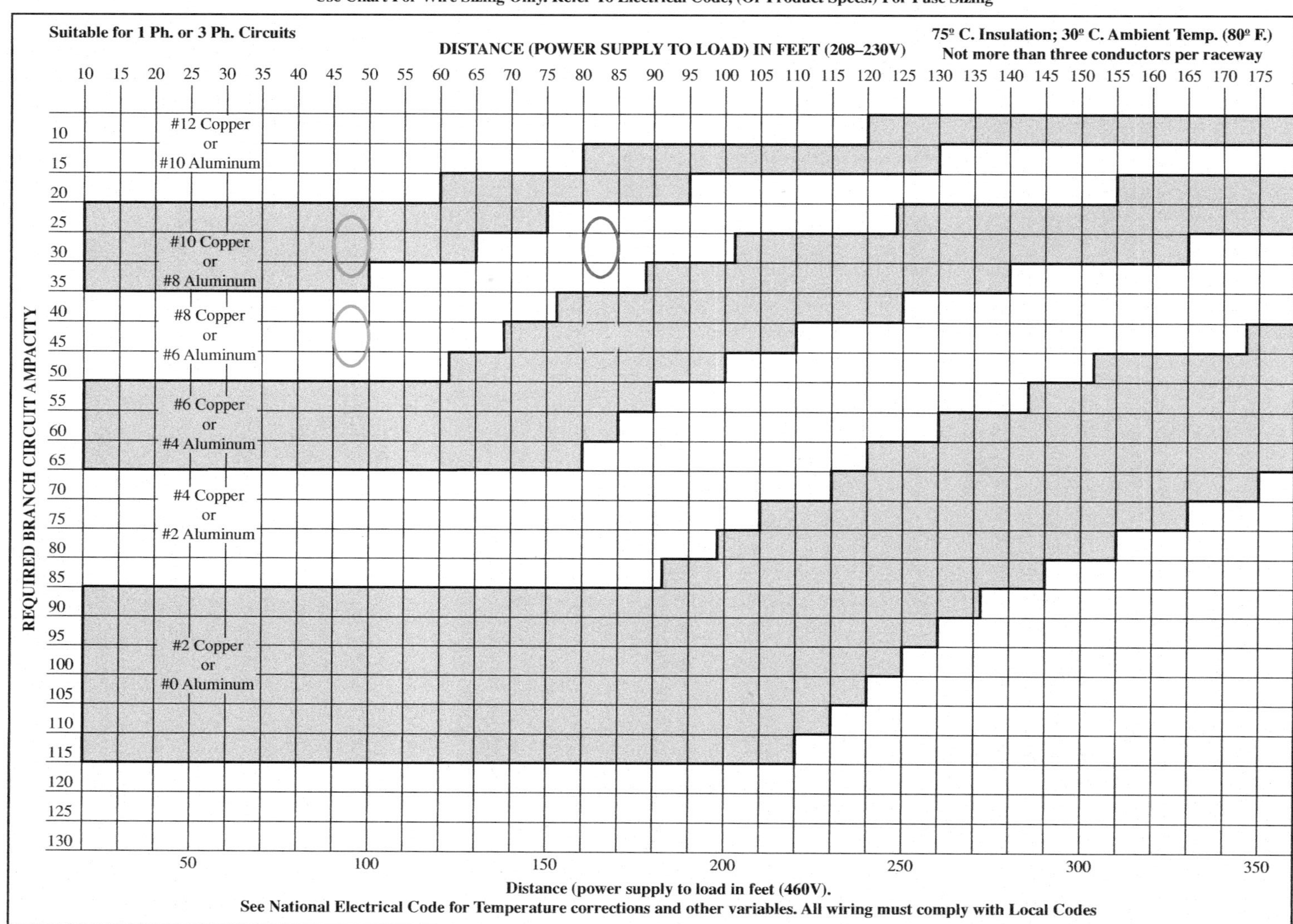

(Courtesy Trane Corporation.)

SERVICE TICKET

The second service call of the day is a poor cooling complaint. You arrive on the job around 10 a.m.; the customer confirms that the cooling system does not cool well when the temperature rises during the day. The customer states that the cooling seems to "come and go." He says, "The air from the ductwork is continuous, but sometimes it is warm and sometimes it is cold. The system cools well later in the evening and in the morning."

You start the logical troubleshooting process (ACT) of checking the airflow from the ductwork. The airflow is cold and seems to be flowing at a good velocity. The air filter is clean. The thermostat is set to cooling and the temperature is low enough to keep the house comfortable. The temperature is lowered so that it does not cut off while troubleshooting. The indoor temperature is 75°F and the outdoor temperature is 85°F. The condenser is rejecting heat as indicated by the warm air felt coming from the top of the condenser. The liquid line is warm and the suction line is cold. The manifold gauge set is hooked up. The pressures are good. The amperage to the common wire on the compressor is 18 amps. The nameplate on the condenser shows the rated load current to be 19 amps. You check the superheat and subcooling and they are surprisingly perfect. What is the problem?

The amperage draw is a little high. You do not expect the amperage to be this close to the nameplate amperage unless the outdoor temperature is much higher. You verify the amp reading with a second clamp-on ammeter. The second ammeter reads 18.2 amps. The input voltage and control voltage are good. The voltage drop from the panel to the unit is around 2%. The equipment has been operating for 25 minutes without shutting off. The indoor temperature is dropping.

You stop the system and check the running components. The contactor contacts and run capacitor are good. All wire connections are fairly tight. One wire on the contactor is loose and you tighten it. When tightening the wiring you notice that the white conductor coming from the disconnect to the contactor is darkened. This could be an indication of overheating. After obtaining permission from the customer, you change the wire. The wire from the disconnect is 12 AWG, which is good for 20 A. The nameplate on the condensing unit requires a 30 ampacity conductor. The 30 ampacity rating will cover 125% of the compressor motor amps plus the condenser fan motor amperage, which is usually a few amps.

You decide to change the conductors to 10 AWG, 90°C, THHN wire. After the condensing unit is restarted, the compressor amperage drops to 16.8 A. Replacing the disconnect conductor and tightening all connections drops the amp draw by about 12%. This might be the solution to the problem.

After showing the customer the overheated wire, you tell him that the wire needed replacing and explain what was done and why. You notify the customer that the equipment did not fail when it was being serviced and tell him to contact you immediately if he has any further problems. The invoice is completed and signed by the customer. You collect the agreed-on fee and proceed to the next service call.

Note: An intermittent problem is one of the big challenges faced by a technician and the customer. Unless the problem occurs while the technician is on the job, an attempt at a solution may not truly solve the problem. If this is the case, the customer may become impatient. Be sure to let the customer know what steps were taken and that the problem may not have been solved. Assure the customer you did the obvious, but to immediately call your office if the problem reoccurs. Show the customer the phone number on the invoice.

SUMMARY

The purpose of this unit was to familiarize you with the National Electrical Code and how it impacts our profession. The NEC is a safety code that covers electrical installations and electrical components. It is important to understand that although HVACR has a code called the Uniform Mechanical Code, we must also follow all building, plumbing, electrical, and life safety codes.

This unit discussed several electrical requirements that must be considered when installing HVACR equipment. For example:

- Required disconnects (unless the breaker can is nearby and can be seen by the technician)
- Wire sizing
- Use of nameplate data to size wire and breakers
- Electrical requirements directed by the NEC.

All HVACR equipment shall have a positive means of removing or disconnecting power. The electrical disconnect is generally sized to handle 125% of the rated load amperage.

General wiring in commercial and industrial buildings shall be protected inside conduit. Open wiring, meaning wiring not inside conduit, is allowed inside residential installations. Even if not required in many residential installations, it is a good idea to have all conductors in conduit when passing through supply or return air ducts. Conductor insulation burning inside a duct can quickly distribute smoke throughout a structure.

NEC Table 310.15(B)(16) is one of the most important sections in the NEC. Table 310.15(B)(16) is used by technicians to determine the correct wire size based on conductor type (copper or aluminum) and application temperature. If a wire burns open or overheats, the wire table should be consulted. Burned or overheated wires are not always caused by undersized conductors. This condition can also be caused by loose connections or a load that is partially shorted. An asterisk (*) on wire sizes in NEC Table 310.15(B)(16) means that the maximum wire size should be reduced by 5 A. This exception affects wire sizes AWG 14, 12, and 10.

We learned that there is important information on wire insulation. The cover or jacket of the wire shows the AWG wire size, and the letters on the insulation indicate the features of the insulation coating. THHN and THWN wire insulation is commonly used when hooking up HVACR equipment. Understanding and using the NEC is extremely important for the advanced technician.

REVIEW QUESTIONS

1. What are two ways to locate information in the NEC?
2. What is the purpose of a disconnect on a piece of HVACR equipment?
3. The rated load power requirement of a condensing unit is 100 A. What size circuit breaker is needed?
4. Where is open wiring or wire not protected by conduit allowed?
5. What three conditions create high resistance in wire conductors?
6. What copper wire size is selected for an air handler that requires 10 ampacity at 9°C?
7. What copper wire size is selected for a condensing unit that requires 35 ampacity at 90°C?
8. What aluminum wire size is selected for a condensing unit that requires 35 ampacity at 75°C ?
9. Most wire discussed in this unit has what maximum voltage rating?
10. What are the two important ratings found on fuses?
11. Use Table 6-7 to answer this question: What wire size is required when installing heater model BAYHTRN105A; unit model WCCO18F1? State wire type, temperature, and gauge.
12. Use Table 6-7 to answer this question: What wire size is required when installing heater model BAYHTRN105A; unit model WCCO42F1? State wire type, temperature, and gauge.
13. Use Table 6-8 to answer this question: What happens to the Btuh heat output when comparing operations at 208 and 240 V?
14. Use Table 6-8 to answer this question: What is the breaker size and wire size for strip heater model #BAYHTR117A? State the wire type, insulation, and temperature rating.
15. Using the nameplate shown in Figure 6-28, what four pieces of information on the nameplate are required by the NEC?
16. Using the nameplate shown in Figure 6-28, what is the LRA?

Table 6-7 Table for Review Questions 11 and 12

Single Circuit Power Ampacity and Over Current Protection

Single Power Entry Kit	Heater Model	Unit Model	Min CKT AMP	Max Over-Current Device	Single Power Entry Kit	Heater Model	Unit Model	Min CKT AMP	Max Over-Current Device
BAYSPEK047A	BAYHTRN105A	WCCO18F1	44	45	BAYSPEK048A*	BAYHTRN117A	WCCO30F 1	113	125
		WCCO24F1	45	50			WCCO35F 1	116	125
		WCCO30F1	52	60			WCCO42F 1	122	125
		WCCO36F1	52	60			WCCO48F 1	126	150
		WCCO42F1	58	60		BAYHTRN123A	WCCO48F 1	156	175
	BAYHTRN108A	WCCO18F1	57	60	BAYSPEK049A*	BAYHTRN108A	WCCO30F 1	63	70
		WCCO24F1	58	60			WCCO36F 1	66	70
	BAYHTRN310A BAYHTRN310F	WCCO36F3	50	50			WCCO42F 1	72	80
		WCCO42F3	54	60		BAYHTRN110A	WCCO24F 1	70	80
		WCCO48F3	59	60			WCCO30F 1	76	80
	BAYHTRN410A BAYHTRN410F	WCCO36F4	26	30			WCCO36F 1	78	80
		WCCO48F4	29	30			WCCO42F1	84	90
		WCCO60F4	32	35			WCCO48F1	89	100
	BAYHTRN415A	WCCO36F4	33	35		BAYHTRN112A	WCCO24F1	78	80
		WCCO48F4	37	40			WCCO30F1	83	90
		WCCO60F4	39	40			WCCO36F1	86	90
	BAYHTRN420A	WCCO48F4	44	45			WCCO42F1	92	100
		WCCO60F4	46	50			WCCO48F1	96	100
	BAYHTRN430A	WCCO60F4	61	60	BAYSPEK050A*	BAYHTRN330A	WCCO60F3	122	125

(Courtesy Trane Corporation.)

Table 6-8 Table for Review Questions 13 and 14

Single-package cooling and heat pump systems without single-power entry kits will normally have two branch circuit power supplies if electric heaters are included: one branch circuit to serve the single-package system and one to serve the electric heater.

Heater Model	Volts	AMP	KW	MCA	Max. Fuse or HACR CKT BKR Size	Canada Only Maximum CKT BKR Size
BAYHTRN105A	208	18	3.74	22	25	30
	240	21	4.98	26	30	30
BAYHTRN108A	208	28	5.76	35	35	40
	240	32	7.68	40	40	40
BAYHTRN110A	208	36	7.47	45	45	50
	240	42	9.96	52	60	60
BAYHTRN112A	208	42	8.64	52	60	60
	240	48	11.52	60	60	60
BAYHTRN115A	208	54	11.21	67	70	70
	240	62	14.94	78	80	100
BAYHTRN117A	208	62	12.97	78	80	100
	240	72	17.28	90	90	100
NONE						

(Courtesy Trane Corporation.)

Figure 6-28 Use this image to answer Review Questions 15 and 16.

UNIT 7

Electrical Installation of HVACR

WHAT YOU NEED TO KNOW

After studying this unit, you will be able to:

1. Describe the electrical requirements of a HVACR installation.
2. Describe what makes an installation "green."
3. Size installation wiring.
4. Size disconnects.
5. Use a punch list to determine if the final installation is prepared for safe operation.
6. State electrical requirements when replacing HVACR equipment.

The most time-consuming part of an installation is setting up the main HVACR components, such as the equipment, refrigerant lines, and ductwork. Making sure the high- and low-voltage wiring is correct is a minor but very important part of the job. Improper high-voltage wiring can lead to fires, electrocution, or the system not working. Incorrect control voltage wiring will not allow the system to operate properly.

This unit discusses both low- and high-voltage wiring and other electrical assemblies that must be considered when installing or replacing HVACR equipment. (Note that use of the term *low-voltage wiring* in this book refers to control voltage wiring or thermostat wiring.) Regarding replacement equipment, it is more than a matter of unwiring and rewiring a piece of equipment with the same control voltage and high-voltage wiring. The wiring or overcurrent protection (fuses or breaker) may need to be upgraded. Some replacement equipment may not have been grounded. All equipment needs to be properly grounded. Additional low-voltage wiring may need to be run when upgrading equipment with more control options.

A word about the mechanical and electrical code mentioned in this unit as well as other sections of this book: These codes are developed as *minimum* safety standards for our industry to follow. Codes reduce the likelihood of fire and electrical damage, making the community safer. Having safer communities translates into fewer injuries and death. This leads to lower insurance rates.

When a town, city, county, or state adopts these codes, they may modify and mold them to the needs of the community they service. Code information stated in this unit is taken directly from the 2011 National Electrical Code (NEC) or 2013 Uniform Mechanical Code (UMC). The codes and the requirements in your community may vary from those discussed in this unit. Enforcement of the code on installation and service work will vary among communities. This creates an interesting situation if you practice HVACR where there are multiple jurisdictions—many jurisdictions modify the codes for their particular needs. The normal installation or equipment change-out that passes inspection in one city may not pass inspection in another. Before starting a job, check with the local jurisdiction and ask about which code they are using and if the code has been modified. Ask the local mechanical building official or inspector about code requirements prior to starting a new installation or equipment change-out.

Here is an example of how various code modifications can affect the profitability or bottom line of a job. The contractor you work for wins the bid on a simple change-out of a residential air handler with a 5-kW heat strip. The homeowner wants the contractor to pull a permit and get the job inspected as required by the local mechanical code. The contract states that your company will be paid in full after the inspector approves the installation. The NEC requires a properly sized disconnect for this change-out. Many of these electric furnaces/air handlers have a set of circuit breakers flush mounted on the panel of the equipment as a disconnect. An example of this is shown in Figure 7-1.

Figure 7-1 This is an air handler circuit breaker that is sized for a 5-kW heat strip and the blower. When this is installed it may or may not be approved as a disconnect. In some localities an external disconnect is also required.

Figure 7-2 This is an external pull disconnect. It will be sized for the total load. This is not an overcurrent device like a fuse or breaker. The pull will be removed to disconnect power from the equipment being serviced.

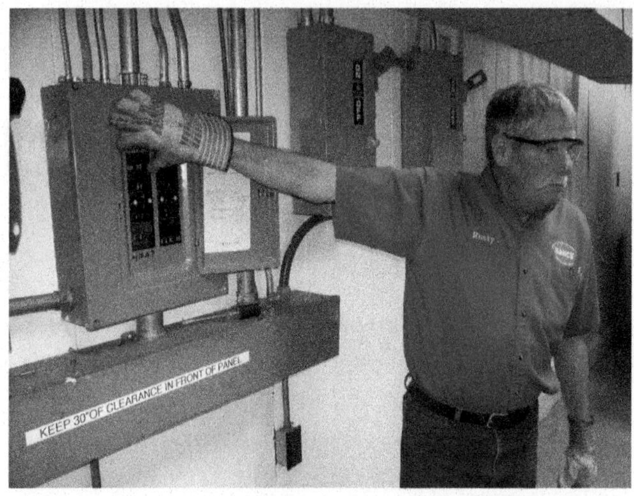

Figure 7-3 This is a safe way to open or close a breaker.

Is this the required disconnect or will the contractor need to install a separate disconnect that is not part of the equipment? This is important to know before the job is bid, because having to install a separate disconnect will add to the cost and increase the installation time for the job. Some code jurisdictions require a separate or external disconnect, while other jurisdictions accept the manufacturer's mounted breakers as a disconnect. On a small job like this one, having to add a separate disconnect would reduce or eliminate the contractor's profit margin. The contractor could not expect the customer to pay the extra cost required to bring the equipment up to the local code if an external disconnect is required. Having to install an unexpected disconnect is one lesson that will not be forgotten.

Finally, some codes require that a separate disconnect *not* be mounted directly on the air handler case. In this example, conduit and/or wire would need to connect the disconnect to the equipment as shown in Figure 7-2.

SAFETY TIP

Closing a breaker on a shorted load can create an explosion at the breaker. If you know that a load is shorted, disconnect from the power source prior to turning on the breaker. Do not stand in front of a panel box when opening or closing the breaker. Turn your face away from the breaker when closing a breaker. Wear gloves and safety glasses. Finally, it is best to have the panel cover on the electrical service when turning breakers off and on. Figure 7-3 shows a safe way to close or open a breaker.

GREEN TIP

What is a **"green installation"**? A green installation is a drastic improvement over the minimum efficiency standards that are mandated by the Department of Energy (DOE) or other energy efficiency standards. Efficiency standards for air conditioning systems are expressed as **SEER** ratings. SEER means **seasonal energy efficiency ratio**. The higher the SEER rating, the less energy that is used. In 2013 the minimum SEER rating on new or replacement residential air conditioning systems was SEER 13.

To determine the SEER rating of a piece of equipment, a complex test is completed by the manufacturer of the equipment. Some models are then retested by a third-party, independent laboratory to verify the results. SEER tests the efficiency of a system over a range of operating outdoor temperatures and is a better indication of efficiency than the old EER measure of energy savings.

Here is a simplified formula that will help you better understand what SEER is:

$$\text{SEER} = \frac{\text{Btu's removed in a cooling season}}{\text{watts used in a cooling season}}$$

A green installation would use an air conditioning system with a SEER rating of 16 or higher. In a few years a SEER 16 rating may be the minimum standard. As the cost of energy increases, the SEER rating will increase. Some types of residential equipment have SEER ratings in the 20s.

Similar to air conditioning systems, the efficiency of a gas furnace is measured as a ratio called **AFUE**. AFUE means **annual fuel utilization efficiency**. AFUE is the ratio of Btu's delivered to the conditioned space divided by the Btu's input into the furnace. Stated another way:

$$\text{AFUE} = \frac{\text{Btu output}}{\text{Btu input}}$$

The minimum efficiency standard for a gas furnace is around 80% AFUE. A furnace that has an input of 100,000

Btuh and a delivered heat output of 80,000 Btuh has an AFUE of

$$\frac{80{,}000}{100{,}000} = 0.80 \text{ or } 80\%$$

An 80% AFUE furnace is better than the older gas furnaces, which had efficiencies of about 55% to 60%. A green gas furnace would have an efficiency of 90% or greater. This means that 90% of the Btu's delivered to the furnace by burning natural gas is used to heat the conditioned space. The other 10% of heat goes into the venting system or is lost by heat transferring through the furnace case.

Replacing an electric heating system with a heat pump is a step toward a green system provided that the SEER rating on the cooling system is 16 or greater. When comparing a heat pump to electric heat, the savings are realized in the heating mode. There will be some cooling savings provided that the SEER rating is higher than that of the system being replaced. Customers who switch from electric heat to a heat pump can expect to reduce their heating electrical costs by half. The higher the SEER rating, the more efficient the heating rating of a heat pump. The design that makes the air conditioning side more efficient transfers to the heating side of the heat pump system. For example, a compressor that draws fewer amps in the cooling mode will draw fewer amps in the heating mode. Even the least efficient heat pump rating will dramatically reduce the heating costs when compared to electric strip heat.

A green installation is about more than efficiency. It includes having the correct refrigerant charge and airflow. More than 67% of all residential air conditioning systems are charged improperly. Many systems have undersized duct systems.

Having a green system includes having the equipment properly sized, not oversized or undersized. Correctly sized and sealed ductwork is important. After incorrect charge, the second biggest problem with air conditioning systems is undersized ductwork. A system with incorrect airflow cannot be charged properly. The airflow should be 350 to 400 CFM per ton of air conditioning. Heat pumps require at least 400 CFM in the heating mode. More airflow in the heating mode would be better.

You can have the most efficient, green equipment, but it will not perform at the rated energy saving standards unless the whole system is installed using "best practices."

TECH TIP

Manufacturers' instructions offer valuable information for technicians. It is important to read this information prior to arriving at the job site. In addition to giving the technician information on installing the equipment correctly, these instructions can help the technician determine if any special components are required to make the installation a success without having to leave the job site to go to pick up parts. As one chief mechanical inspector said, "Some installers think that the paperwork that comes with the equipment is nothing more than junk mail that comes from the post office." These instructions are great assets. Use them!

SAFETY TIP

When replacing like condensing units, one important practice is to determine the amperage requirements of the new unit. In most cases, the replacement condenser will draw less amperage because it is more efficient. Therefore, the new condenser will require lower circuit protection. For example, an old 3-ton condenser may require overcurrent protection of 40 A. The more efficient replacement unit may require only 30 A of overcurrent protection. The fuse or breaker must be a lower rating to protect the equipment properly.

In many localities changing out the breaker will require a licensed electrician and an electrical permit, particularly if the breaker is changed in the main panel box. Some jurisdictions allow the air conditioning contractor the option to install a protected disconnect near the new condenser. The protected disconnect should have an appropriately sized fuse or breaker that will protect the new condensing unit. This measure will prevent having to change the breaker in the main panel box while providing the correct equipment protection at the disconnect. A condensing unit that draws less amperage should not require changing the wire size if it was sized properly for the higher amperage unit that is being replaced. In such a case the wire will be slightly oversized, which is actually beneficial.

7.1 NEW INSTALLATIONS

New installations require a high-voltage connection and in many cases low-voltage (**control voltage**) connections. Let's see what this looks like in terms of an electrical diagram. Figure 7-4 shows a labeled diagram with low- and high-voltage wiring. Notice on the diagram that the voltage source is 208/230 and 24 volts. Notice that the diagram is drawn in two different ways. The upper and lower diagrams are the same yet appear different. **L1** and **L2** are connected to the 208/230-V circuit found on the lower left of the diagram. The **L1, L2** and ground for the upper diagram is on the right side. The 24-V power supply is connected to **C** and **Y1** found on the far right side of the diagram.

The solid lines connecting components are high voltage. The dashed lines represent field-installed wiring that may be low voltage or high voltage. Some diagrams will use a thick dashed line to represent high-voltage field wiring. A thinner dashed line will represent low voltage of which 24 V is very common. Low-voltage wiring is also called control voltage or thermostat voltage. In Figure 7-4 the highlighted yellow dashed lines are high-voltage, field-wired circuit options. The red dashed lines are low-voltage thermostat wiring. Some diagrams use a dashed line around a component such as the thermostat. In this case it does not designate wire, just a separation from the other components.

In this diagram the optional high-voltage circuits are the crankcase heater and hard start kit. The crankcase heater is located in the middle, at the far left. The purpose of the heater is to keep the compressor oil warm so that refrigerant does not condense in the compressor during the off cycle in low-temperature conditions. The hard start kit helps the compressor

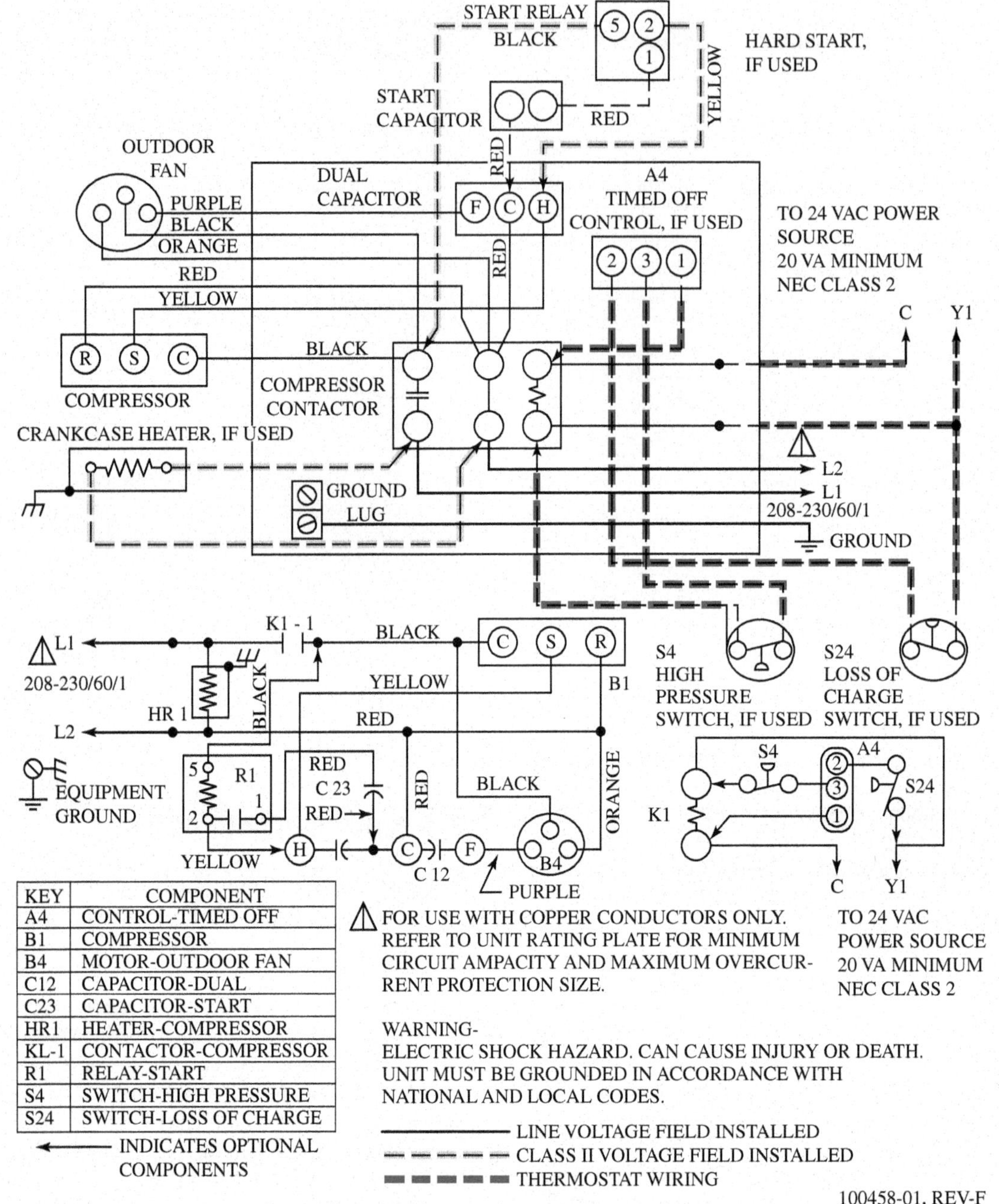

Figure 7-4 Notice in this diagram that the voltage source is 208/230 and 24 V. The dashed lines represent field-installed wiring that may be low voltage or high voltage. The highlighted yellow dashed lines are high-voltage, field-wired circuit options. The red dashed lines are low-voltage thermostat wiring.

start easier. It reduces the current draw on compressor start-up. It has a start capacitor and start relay that is matched to the compressor motor. This will be discussed in a later unit.

Some systems, such as commercial and residential refrigerators and window units, have their own control voltage wired to a line voltage thermostat. These will not require a control voltage thermostat.

The following common electrical items should be checked and sized properly when installing new equipment or replacing equipment:

- Conductors or wires that supply voltage to the equipment
- Disconnect amperage capacity
- Breaker, fuse, or overload protection device size
- Number of thermostat wires.

Conductors are sized by their current-carrying capacity and voltage drop. For wiring shorter than 100 feet, voltage drop is not a consideration. Use the National Electrical Code wiring table, Table 310.15(B)(16), which is reprinted as Table 6-4 in Unit 6 of this text, to select the size of the wire based on amperage draw, insulation type, and type of material (copper or aluminum).

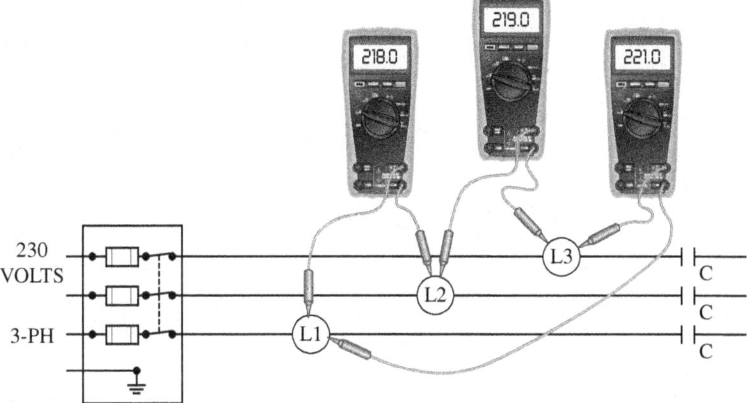

Figure 7-5 The static voltages are within ±10% of the supplied voltage. The de-energized voltages between the three phases are similar enough to be acceptable. The voltage may drop once the equipment is started, but the three voltage readings should remain close to one another.

TECH TIP

In three-phase systems, the three de-energized voltage readings should be similar, as shown in Figure 7-5.

After the equipment is started, the voltage and amperage readings should still be close between the three phases. Figures 7-5 and 7-6 show voltage and current measurements that are not equal, but are within acceptable standards.

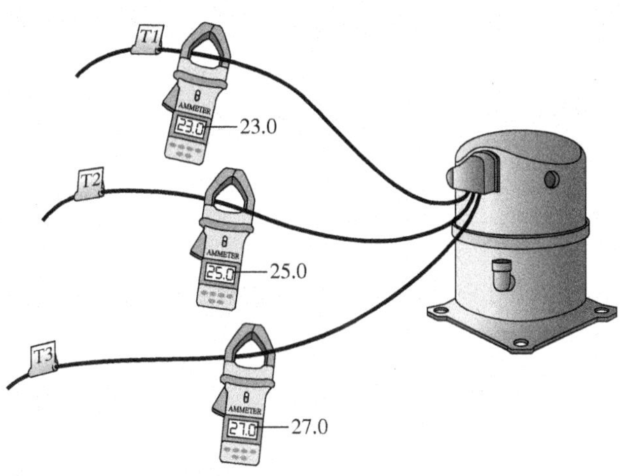

Figure 7-6 The RLA on this compressor is 30 A. The amperage on each leg is close enough to be acceptable.

7.2 HIGH VOLTAGE

Before we start this section, we will define a few terms that will make the learning process a little easier. First we discuss MCA and MOP.

Maximum Circuit Ampacity (MCA)

Maximum circuit ampacity, or **MCA**, is a calculation used to size a wire for a piece of equipment. The MCA is the minimum wire size needed to guarantee that the wiring will not overheat under all operating conditions for the life of the product.

The MCA rating determines the wire size that can be selected from NEC Table 310.15(B)(16). The MCA is found on the condensing unit nameplate as shown in Figure 7-7. The MCA for sizing wire in this example is 11 amps. An 11A conductor is not available, therefore you would select a 15A conductor. Looking at NEC Table 310.15(B)(16) in Unit 6 you would most likely select a 15-gauge (AWG) copper conductor for this installation.

Other important electrical data on the nameplate is the compressor motor RLA, which is listed as 8.3 RLA and 62 LRA. The outdoor fan motor is rated at 0.90-amp FLA. It is a dual-voltage condenser, operating on 200 to 230 V, which is allowed a ±10% deviation.

Maximum Overcurrent Protection (MOP)

Maximum overcurrent protection, or **MOP**, determines the size of the highest fuse or breaker that should be used on a piece of equipment. It is sometimes simply called an *overcurrent protective device* as seen on the nameplate in Figure 7-7. The MOP is the maximum allowable circuit breaker or fuse size that will properly disconnect power to the equipment under any anticipated fault condition. You can install a breaker or HACR-type fuse that is smaller than the listed MOP. A high-current surge

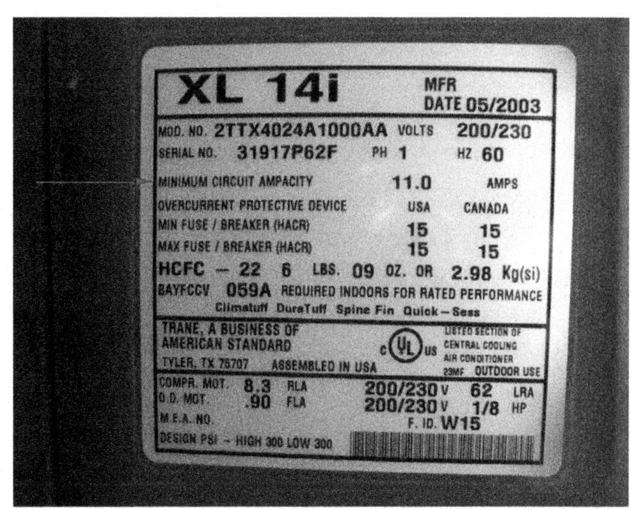

Figure 7-7 This nameplate points to a minimum circuit ampacity of 11 amps. This will be used to size the wire connecting this condensing unit to the power supply. The overcurrent protective device, which is the fuse or breaker, is sized for no more than 15 amps.

is common with motor circuits. An HACR fuse is a time-delay fuse that will not trip immediately because of the high, but temporarily locked, rotor starting amps (LRA). Installing a larger breaker or fuse permits the equipment or wiring to burn up before the overcurrent protection trips.

Installing an undersized overcurrent protective (fuse or breaker) device may cause "nuisance" breaker trips or blown fuses. Nuisance trips happen when the overcurrent device opens periodically, killing the power to the unit. The trip usually occurs when the equipment draws LRA at start-up or in a low-voltage condition. Nuisance trips can also be caused by voltage fluctuations or short power outages. In a common scenario, the tech goes to the nuisance trip job, checks out the equipment, and changes the fuse or resets the breaker. After cycling the equipment a few times, everything works as it should. But sometime later, the overcurrent protection opens again with the same results. Installing the largest recommended overprotection device should eliminate the nuisance callback.

The calculations for MCA and MOP are based on NEC/NFPA 70 requirements and CSA C22.1, the Canadian Electrical Code (CEC). The formulas assume that the motors will have a significant current surge at start-up and that the motor will draw more current as it ages. HACR equipment uses time-delay fuses or a breaker that may be labeled "HACR."

Sometimes a maximum overcurrent protective device is labeled with the abbreviation MOCP instead of MOP.

Calculating MCA and MOP

If the MCA is unknown, the gauge of the high-voltage wiring and circuit protection can be determined using the following formulas:

$$\text{MCA} = (125\% \times \text{current of largest motor}) + \text{current sum of all other loads}$$

Percentages (%) must always be converted to a decimal form in order to use them in a formula. We do this by putting a decimal point two places to the left of the last whole number in the percentage. After converting the percentage to a decimal the formula will look like this:

$$\text{MCA} = (1.25 \times \text{current of largest motor}) + \text{current sum of all other loads}$$

The MCA has been calculated in Figure 7-7. Using the information we have learned, let's see how the nameplate came up with an MCA of 11.0 amps. As seen near the bottom of the nameplate, the compressor's RLA is 8.3 amps and the fan motor FLA is 0.90 amp. Here is the calculation:

$$\text{MCA} = (125\% \times \text{current of largest motor}) + \text{current sum of all other loads}$$

$$\text{MCA} = (1.25 \times 8.3\text{ A}) + 0.90\text{ A}$$

$$\text{MCA} = 10.4\text{ A} + 0.90\text{ A}$$

$$\text{MCA} = 11.3\text{ A}$$

The MCA is rounded down to 11 amps. This helps us select the wire size, known as wire gauge or AWG. Eleven-amp wire is not common, therefore a 12-gauge conductor would be used for this condensing unit. This calculation is limited to a condensing unit that is 50 feet from the panel box. At 50 feet, the circuit has a round-trip of 100 feet (50 + 50 = 100 feet) of conductor. This formula is limited to a total of 100 feet of conductor run. Longer conductor runs will require larger diameter wire that will be a lower gauge; in this case if the length had been more than 100 feet, a 10-gauge conductor would have been used.

EXAMPLE 7.1 DETERMINING WIRE SIZE USING THE MCA OF A CONDENSING UNIT

Let's do an example using a condensing unit with the following amperage conditions:

Compressor RLA: 35 A

Condenser fan motor RLA or FLA: 4 A (FLA is similar to RLA except it is for fan motors)

Control circuit FLA: 1.5 A.

Determine the correct wire size by calculating the MCA.

SOLUTION

$$\text{MCA} = 1.25 \times \text{current of largest motor} + \text{current sum of all other loads}$$

$$\text{MCA} = (1.25 \times 35\text{ A}) + 4\text{ A} + 1.5\text{ A}$$

$$\text{MCA} = 43.75\text{ A} + 4\text{ A} + 1.5\text{ A}$$

$$\text{MCA} = 49.25\text{ A}$$

A 50- or 55-A wire would be selected for this condensing unit. For a safe installation, never go down in wire size—always go up.

EXAMPLE 7.2 DETERMINING WIRE SIZE USING THE MCA OF AN AIR HANDLER

Let's do another example, this time using an air handler unit (AHU) with electric heat strips. Here is the AHU information:

Fan motor FLA: 5 A

Heat strip (3 × 20 A) FLA: 60 A

Control circuit FLA: 2 A.

SOLUTION

The formula we use is a little different from the one we used for a condensing unit:

$$\text{MCA} = (125\% \times \text{FLA of largest motor}) + (125\% \times \text{FLA of all electric heaters}) + \text{sum of other loads}$$

$$\text{MCA} = (1.25 \times 5\text{ A}) + (1.25 \times 60\text{ A}) + 2\text{ A}$$

$$\text{MCA} = 6.25\text{ A} + 75\text{ A} + 2\text{ A}$$

$$\text{MCA} = 83.25\text{ A}$$

You would select wire that could handle 85 amps. A larger current-carrying wire would be more readily available. A 90- or 100-A conductor would most likely be used because it is more common than a wire that can carry 85 amps.

MOCP or MOP is the maximum size of an overcurrent protection device. We are talking about the breaker or HACR fuse size. As previously stated, a smaller overcurrent device may be used, but nuisance trips may become a problem. Maximum overcurrent protection (MOP) is 2.25 times the current of the largest motor plus the FLA of the remaining motors. When the 2.25 is expressed as a percentage, it is 225%. Here, we moved the decimal point two places to the right to create a percentage.

$$\text{MOCP} = (2.25 \times \text{amp draw of the largest motor}) + \text{sum of all other loads}$$

Round down to the next lower standard breaker or fuse rating, but not lower than the MCA value. For example, if the MOP is 26 amps, do not install a 30-A overcurrent protective device. Install a 25-A breaker or HACR fuse. When doing these calculations it is a safe practice to round down for overcurrent protection and round up for wire size.

7.3 CONTROL VOLTAGE

Control voltage or low-voltage wiring is usually operated around 24 VAC or some low DC voltage. Some control voltages use 120 or 240 volts. The most common residential and light commercial control power is 24 V, which can be stated as 24 VAC. Systems with sophisticated direct digital controls (DDC) and installations with communicating systems use low-voltage DC signals to control HVACR systems. We will discuss equipment with the common 24-V circuits because this is the low-voltage control found on most HVACR systems.

The 24-V control wiring is usually an 18-AWG conductor. Smaller conductors such as 20 AWG are also available. When using a 40-VA transformer, the wire needs to handle at least 1.67 RLA. It is good practice to include an extra conductor in the thermostat wiring bundle in case one of the conductors becomes open or has a broken wire. Having extra conductors available may help when it comes to upgrading a system that requires extra control voltage wiring.

Control voltage should not travel in the same conduit as high voltage. High-voltage current flow may induce magnetic energy into the low-voltage wire and create enough current flow to cause a low-voltage component to be energized. In any event, there may be enough induced voltage to create a "ghost voltage" that will be measured by a digital meter. The ghost voltage will register a voltage reading on a digital meter, but the current-carrying capability of this voltage is so low that it will not operate most low-voltage controls. An analog meter will not measure ghost voltage because its low impedance causes the voltage to short out or dissipate and read zero volts.

In most installations, placing control voltage wiring in conduit is not required. Note, however, that best practices require the control voltage to be placed in conduit when it is exposed to the outdoor elements. This reduces the possibility of animal, human, or storm damage. Control voltage wiring will need to be "plenum rated" if it inside a commercial return air or supply air duct or plenum. In the event of a fire, plenum-rated wire produces less flame and smoke development when compared to the insulation on standard thermostat wire.

There is more than one way to solve a problem. The Troubleshooting 7.2 feature explains another way to determine if a thermostat wire is defective.

TROUBLESHOOTING 7.1: OPEN THERMOSTAT WIRE

How do you find an open thermostat wire? The following steps will help you diagnose and find an open wire if there is one:

1. Verify that the control voltage is turned off before you do this check. Expose the thermostat subbase, or all the wires going to the thermostat, and connect pairs of conductors together with an alligator clip jumper (see Figure 7-8). Jumper (tie) wires yellow and red together. Jumper wires green and white together.
2. Go to the other end of the thermostat wire in the air handler or furnace and disconnect the wires (see Figure 7-9).
3. Measure the resistance of each pair of wires (see Figure 7-10). The non-jumpered ends of the yellow and red pairs should have a reading of less than 1 Ω. This low resistance reading means that both the yellow and the red wires are good. Do this with each set of thermostat wires.
4. If you measure infinity (open conductor) in any pair of wires, it will be expressed as a digital display of **OL** on

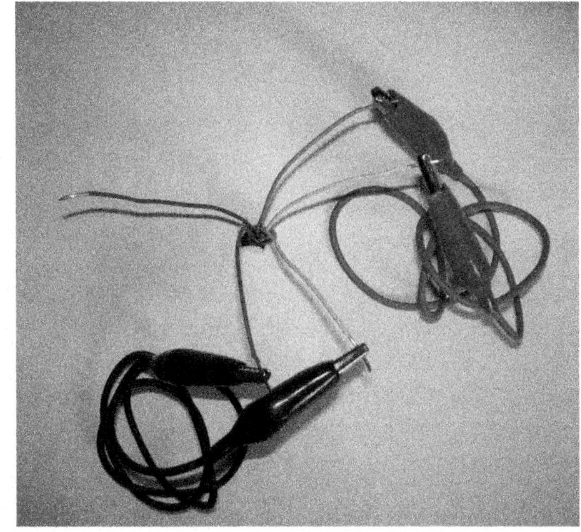

Figure 7-8 Use alligator clip jumpers to connect pairs of conductors.

(continued)

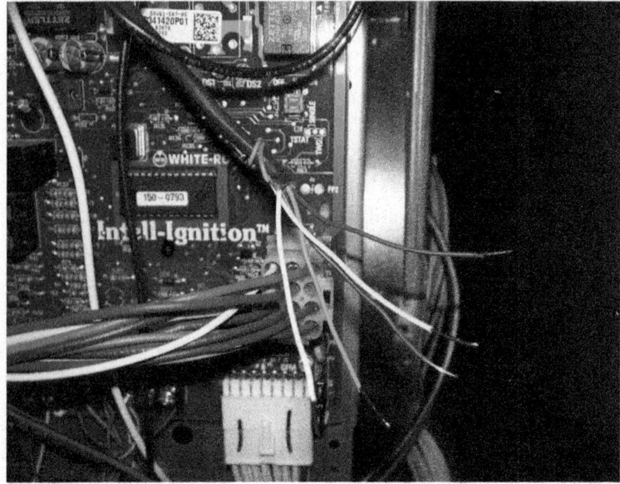
Figure 7-9 Disconnect the other ends of the wires.

most meters. This means that one or more wires are open in that pair and can no longer be used. Even a measurement as low as 5 Ω is not acceptable and will create too much of a voltage drop in the thermostat wire.

5. Determine which wire is open. This can be accomplished by separating the open pair of wires and connecting them to one of the conductors in the good pair of wires.

6. Take one of the suspect open wires and connect it to one of the good thermostat wires. Measure the resistance again. Now the pair with the infinity reading of **OL** has the one open conductor and it is not the conductor that was verified as good.

Figure 7-10 Measure the resistance of each pair of wires.

TROUBLESHOOTING 7.2: ALTERNATE METHODS OF TROUBLESHOOTING THERMOSTAT WIRE

Let's look at two more ways to determine if a thermostat has a defective wire using an ohmmeter. Here is one way:

1. Tie all the thermostat wires together at the air handler or furnace termination as shown in Figure 7-11.

2. From the other end of the thermostat wire, measure the resistance between R, Y, W, G, and the common wire as shown in Figure 7-12. All good wires should have the same resistance. If there is an open conductor, one or more wires will **not** show resistance, which is represented by **OL** on the meter's display.

A third way to check for an open control wire is to do a voltage check provided that the R conductor from the control voltage source is good. Check the voltage from R to the other conductor at the thermostat subbase as set up in Figure 7-13. The thermostat should be in the off position. The wire that does not have voltage is the open conductor.

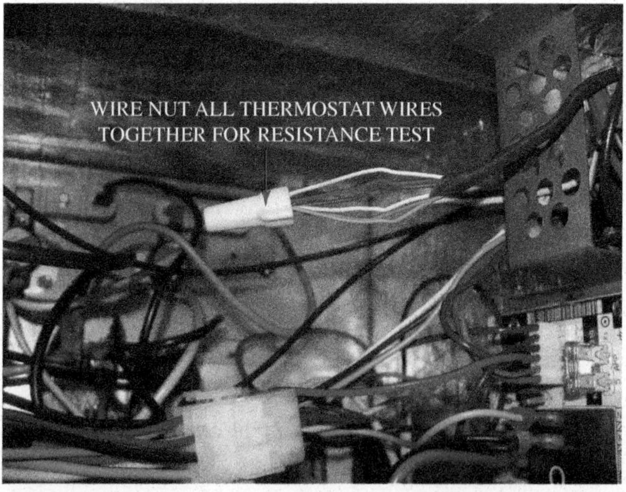

Figure 7-11 Use a wire nut to tie all thermostat wires together.

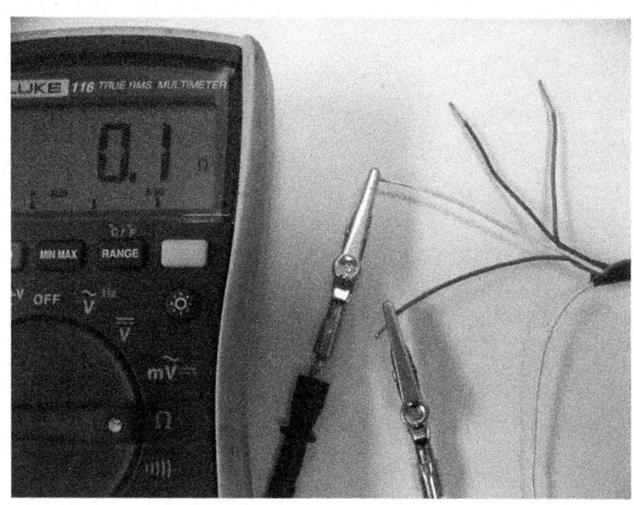

Figure 7-12 Use a multimeter (set on ohms) to measure the resistance between R, Y, W, G, and the common wire.

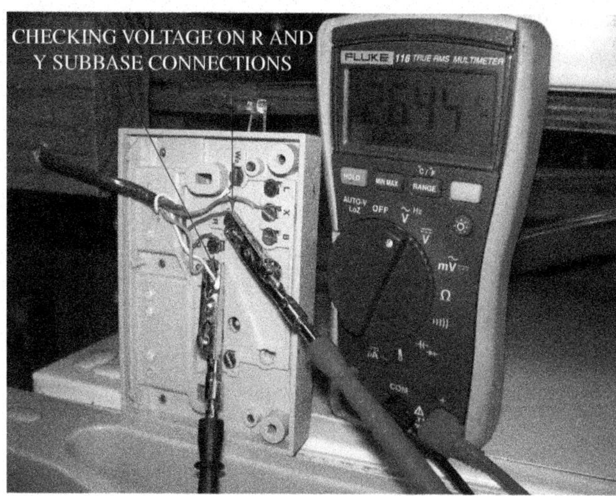

Figure 7-13 Use a multimeter (set on volts) to check for open conductors.

TECH TIP

Long thermostat wire runs of more than 100 feet will create high voltage drops and resistance to current flow. In this case, use lower gauge thermostat wire (bigger wire) and/or a larger VA transformer.

SAFETY TIP

The blades on a throw disconnect, as shown in Figure 7-14, do not always open or close. It is especially dangerous when one or more of the blades stick in the closed position, which means that voltage is still supplied to the equipment it is powering. The technician assumes that the disconnect has removed the power from the equipment. The tech should do a voltage measurement to ground to ensure that the equipment is harmless. A visual check of the disconnect is another way to be certain it is open and safe. Finally, shorting the area to be touched to ground is the final safety check after opening the disconnect circuit. Never assume that the circuit is dead, even if you took all of the correct precautions!

Figure 7-14 The blades on a throw disconnect do not always open or close. A visual check of the disconnect is one way to be certain it is open and safe.

7.4 REPLACING HVACR EQUIPMENT

This section deals with replacing a condensing unit or furnace. Before changing out HVACR equipment, it is important to understand the high- and low-voltage requirements of the replacement equipment. The following questions need to be answered:

- Is the voltage and current the same on the new units?
- Does the replacement equipment require additional control voltage conductors to the condensing unit, furnace, or thermostat?
- Will the thermostat operate with the replacement equipment?
- Does the replacement equipment require a disconnect? Or a disconnect upgrade?

Figure 7-15 Dip switches are used to set the blower speed for heating and cooling. They are found on the circuit board between the two white wiring connectors.

In summary, four components of the power wiring or control voltage to the replacement equipment should be closely checked:

- Conductor or wire size
- Disconnect ampacity
- Breaker or fuse size
- Number of control wires.

These are important aspects of a change-out that must be considered before a bid is given to a customer. The need to add wiring, disconnects, or control circuits must be considered when determining the cost of materials and the labor required to install these items. Often residential, high-efficiency air handlers need to have their blower operating conditions set. This setup is required for a gas furnace or straight air handler with or without electric heat. This is sometimes overlooked by the installers or start-up technician. Many designs use dip switch settings as shown in Figure 7-15.

Table 7-1 shows the guideline used to set the dip switches. It indicates that there is a factory "default" setting, under which the blower will run at the factory-selected speed.

Table 7-1 Dip Switch Settings

This table is used to set the fan operation of an air handler using a variable-speed motor. The dip switches must be set for the desired airflow.

BASED ON 350 CFM/TON (SETUP SWITCH SW1-5 OFF)

MODEL SIZE	SETUP SWITCH SW3 POSITIONS							
	1 2 3 OFF	1 2 3 OFF	1 2 3 OFF	1 2 3 OFF	1 2 3 OFF	1 2 3 OFF	1 2 3 OFF	1 2 3 OFF
060, 3.5T080	DEF.	525_2	700	875	1050_1	1225	1225	1225
5T080, 100	DEF.	700_2	875	1050	1225	1400	1750_1	1750
120	DEF.	700	875_2	1050	1225	1400	1750_1	2100

BASED ON 400 CFM/TON (SETUP SWITCH SW1-5 ON)

MODEL SIZE	SETUP SWITCH SW3 POSITIONS							
	1 2 3 OFF	1 2 3 OFF	1 2 3 OFF	1 2 3 OFF	1 2 3 OFF	1 2 3 OFF	1 2 3 OFF	1 2 3 OFF
060, 3.5T080	DEF.	600_2	800	1000	1200_1	1400	1400	1400
5T080, 100	DEF.	800_2	1000	1200	1400	1600	2000_1	2000
120	DEF.	800	1000_2	1200	1400	1600	2000_1	2100

1. DEFAULT A/C AIRFLOW WHEN A/C SWITCHES ARE IN OFF POSITION
2. DEFAULT CONT. FAN AIRFLOW WHEN CF SWITCHES ARE IN OFF POSITION
3. SWITCH POSITIONS ARE ALSO SHOWN ON FURNACE WIRING DIAGRAM

(Courtesy Carrier Corporation.)

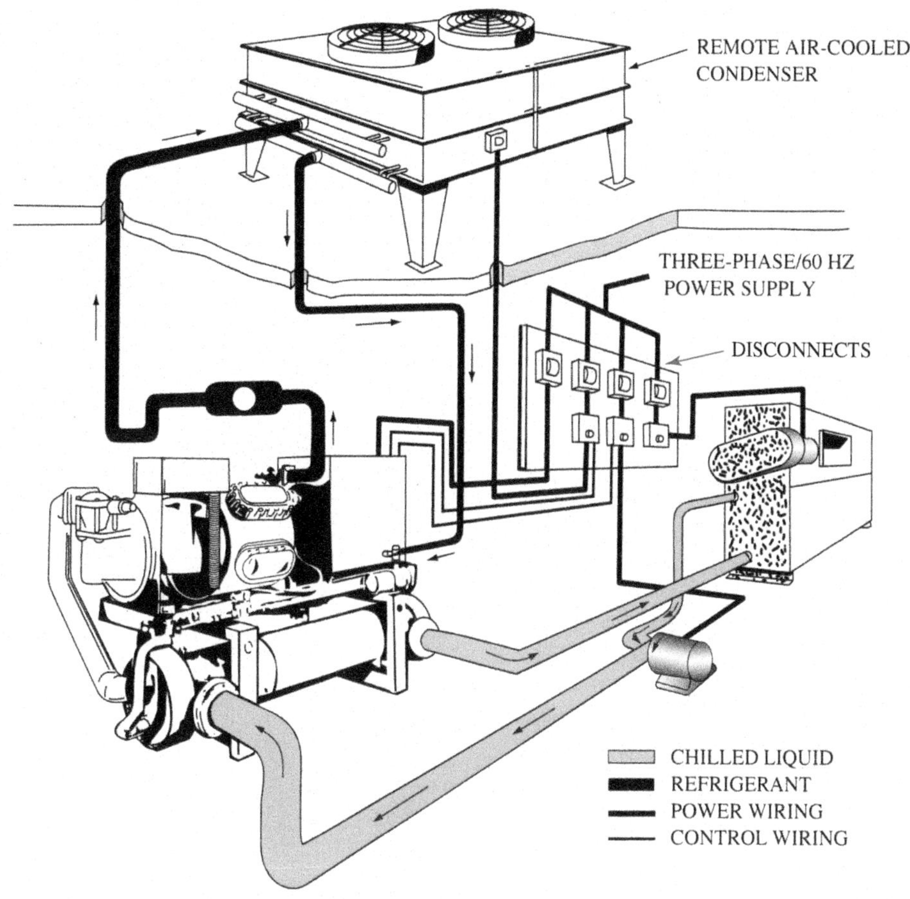

Figure 7-16 This chilled water system has four electrical circuits feeding four different parts of the system.

The blower dip switches should be set for the particular application, which is usually different than the factory default setting. For example, in hot and humid climates, the blower speed should be set for 350 CFM per ton, not 400 CFM per ton. The lower speed will allow more air contact time with the evaporator, thus removing a greater amount of moisture from the air.

Using Table 7-1, you would select the dip switch setting for 350 CFM/ton for model size 120 for a 4-ton condensing unit. This is a good switch selection for warm and humid climates. This will be 350 CFM × 4 tons = 1400 CFM.

In the upper section of Table 7-1 locate model size 120 and set for 350 CFM per ton. Go to the right of the table until 1400 is located. Go up the top of the table to see the selected dip switch settings. The dip switch settings are the white, rectangular boxes. The dip switch is a white plastic slide switch as previously shown in Figure 7-15. Here are the required dip switch settings for this example:

Switch 1 will be in the on position.

Switch 2 will be in the off position.

Switch 3 will be in the on position

These dip switch settings will ensure that the 4-ton blower will develop 1400 CFM or 350 CFM/ton. This is good setting for hot and humid climates.

7.5 LARGE HVACR SYSTEMS

Large HVACR systems have multiple electrical requirements. Each piece of equipment may have a different wire size requirement because they probably use different voltages and have different amperage needs. Each piece of equipment has a different disconnect, as required by the NEC. The disconnect should be within line of sight of the equipment it services. In some cases, a breaker panel can be the disconnect if it is near the equipment it serves. Older installations were not required to have a separate disconnect; therefore, you will find many older systems with no disconnect. A disconnect will be installed when the equipment is replaced, because code updates are required when changing out major components of a system.

Figure 7-16 is an example of a chilled water system that has four circuits feeding four different parts of the system. Four disconnects must be sized for the piece of equipment that is hooked to the power source. The amp draw of the piece of equipment will determine the size of the disconnect. In this figure the disconnects are wired to the following system components:

- Chiller compressor motor
- Remote air-cooled condenser
- Liquid chiller pump
- Air handling unit.

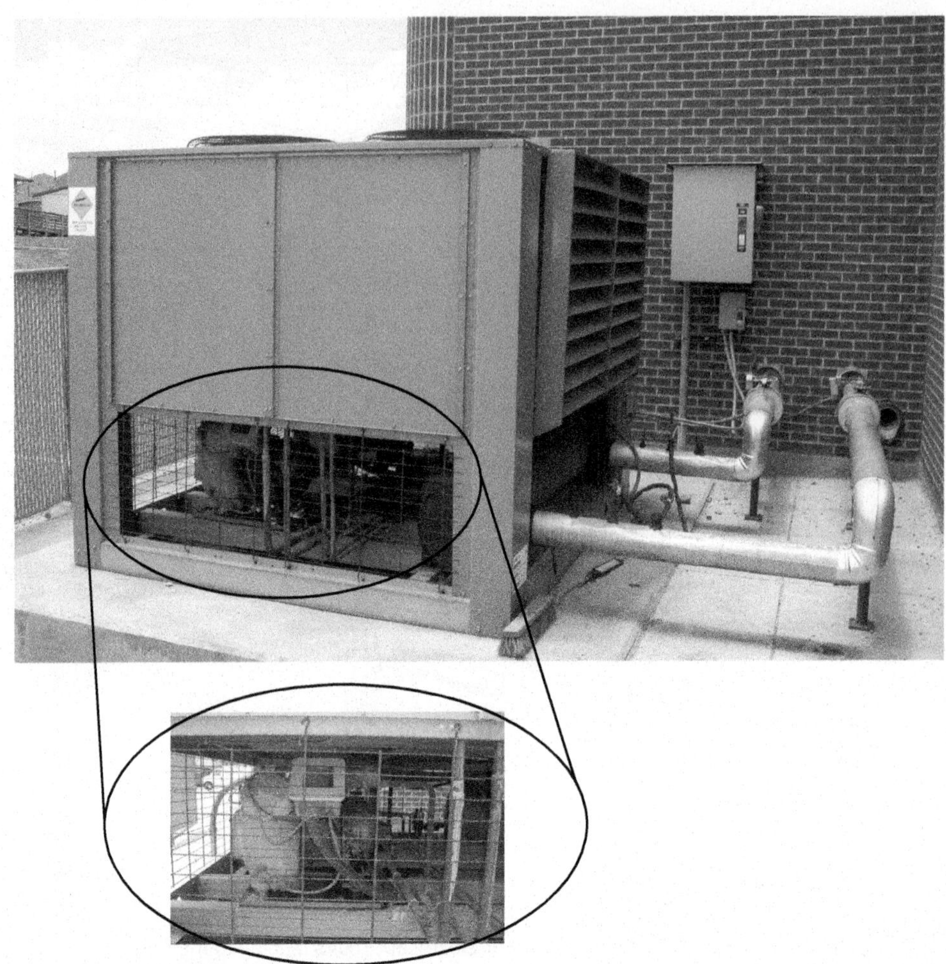

Figure 7-17 A large disconnect is found on high-tonnage equipment. The disconnect is mounted on the brick wall. The smaller disconnect, located below the main disconnect, is used to feed power to a supplemental control in the chiller. The compressor inset is shown to emphasize the largest current drawing device in this large package chiller unit.

Figure 7-17 illustrates the large disconnect found on high-tonnage equipment. The disconnect is located on the brick wall to the right of the chiller. Note that there is another small disconnect below it. The fan grilles, as seen on the top of the unit, indicate that it is an air-cooled system.

7.6 ADDITIONAL ELECTRICAL REQUIREMENTS

In addition to having the correct conductors supplied to the HVACR equipment, there are other electrical requirements per the Uniform Mechanical Code (UMC) and National Electrical Code (NEC). Some of these code requirements make the work environment easier on the technician, while other requirements make it a safer working area. Let's review some of these requirements.

Bonding Jumper

A **bonding jumper** is a wire that connects all of the equipment to a ground rod. According to the NFPA and NEC, the bonding jumper is a reliable conductor to ensure required electrical conductivity between the metal casings of equipment. If the continuity is broken, the equipment can become a dangerous shock hazard if a hot wire touches the case. Figure 7-18 shows a green bonding wire. Green wire indicates a ground.

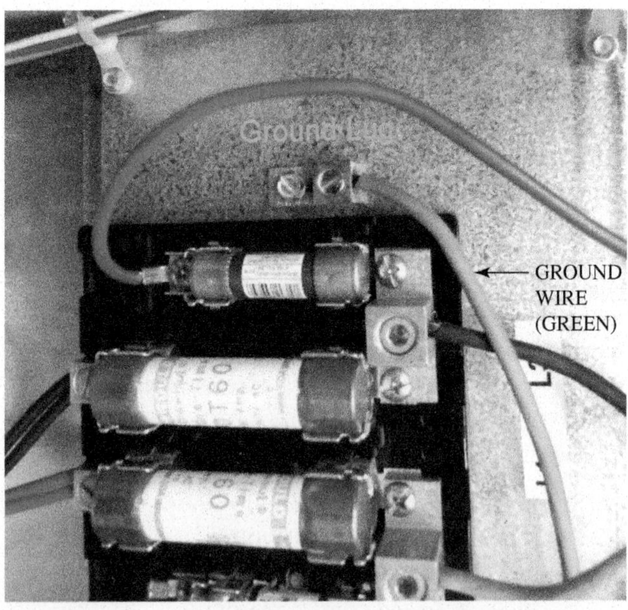

Figure 7-18 A bonding jumper is a wire that connects all of the equipment to a ground rod.

SAFETY TIP

It takes less than an amp to kill you. Figure 7-19 illustrates the effects of current on the human body. A poor or broken ground can kill you. If you ground yourself while touching equipment that has a short, you will become the bonding jumper or ground.

Permanent Lighting

According to the NFPA, permanent lighting shall be provided in attic and basement areas near the equipment. Permanent lighting shall be provided at the roof access for roof-top units. The light switch shall be located inside the building near the attic, basement entrance, or roof access. Figure 7-20 shows an installation's light switch and light arrangement. The switch is in the attic near the entrance.

Protection

The Uniform Mechanical Code requires that HVACR appliances be protected against damage in garages, warehouses, or other areas subject to mechanical damage. The specific type of protection is not stated in the code. The UMC does specify

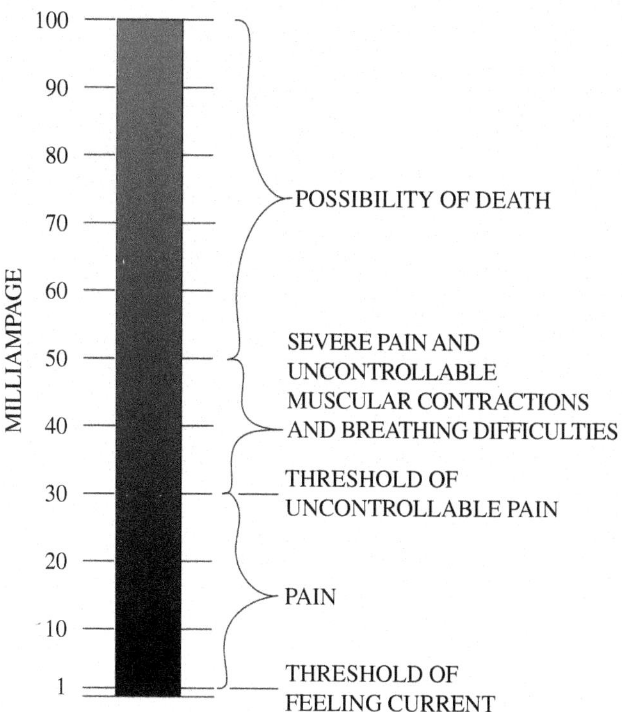

Figure 7-19 Amperage rating of electric current creating various shock effects from 1 to 100 mA. It takes less than an amp to kill you.

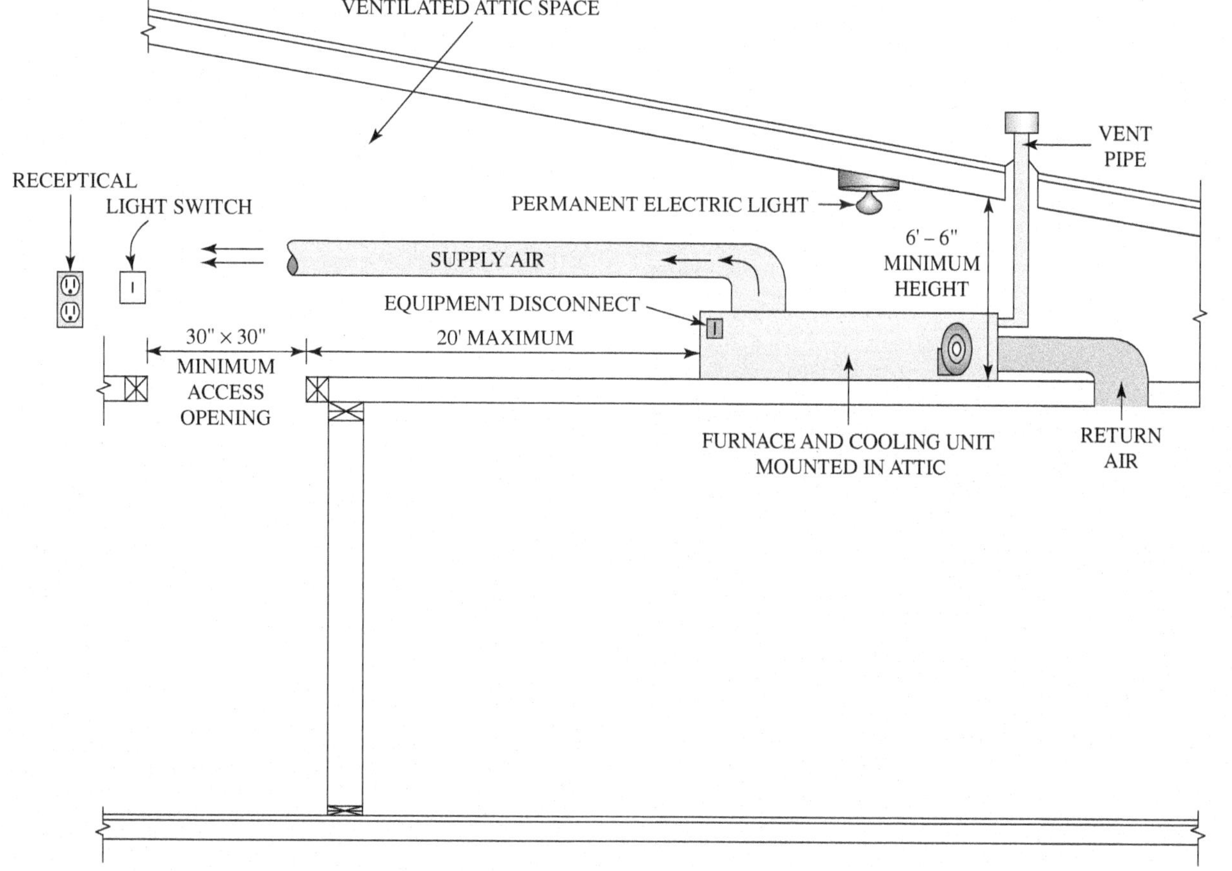

Figure 7-20 Permanent lighting and light switch placed according to NFPA requirements. This drawing also shows other access requirements for an attic installation such as a 30-in. × 30-in. access and a distance of no more than 20 feet from the access point to the equipment.

Figure 7-21 A disconnect is required on a condensing unit unless it is within line of sight of an electric panel box.

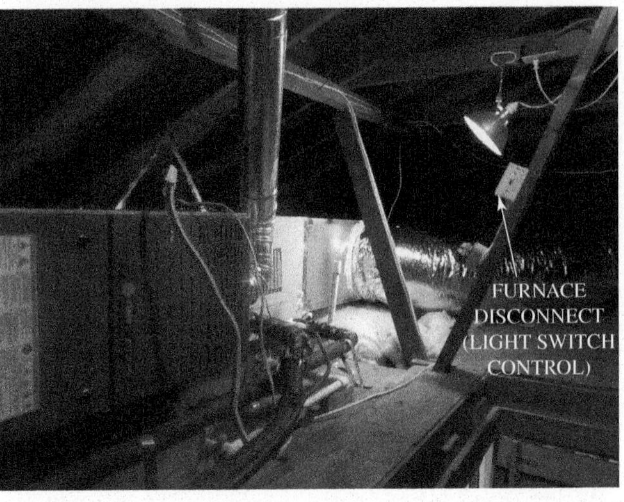

Figure 7-23 This is a clean attic installation of an air conditioning evaporator with a gas furnace. The disconnect for the furnace is a simple light switch.

that heating and cooling equipment located in a garage shall be installed at least 18 in. above the floor level.

Electrical Connections

Electrical disconnects should be within sight of the equipment. Equipment with more than 50 volts shall have a disconnect. A "pull-type" condensing unit disconnect is shown in Figure 7-21. Figure 7-22 is a close-up view of the condensing unit disconnect. When the disconnect is pulled, there is no high-voltage power to the condenser. The 24-V control voltage may still be present at the unit. This disconnect does not remove control voltage. Control voltage power is usually controlled from another, indoor source.

Figure 7-22 This pull disconnect for a condensing unit is rated for 60 A, which will more than handle the total load of the condenser, which is 25 A. The "pull" is sitting on the top of the disconnect, ensuring that no power is supplied to the condensing unit.

The gas furnace in Figure 7-23 has a light switch style of disconnect. A 120-V receptacle shall be located within 25 feet of the equipment for service and maintenance. According to the UMC the receptacle does not need to be located on the same level as the equipment. This statement leads to code interpretation issues. For example, some code jurisdictions allow 25-foot access to a receptacle even if the power for an extension cord has to go through an operable window. Other code interpretations consider the outside or the attic or basement as the same area. They don't consider access through windows, doors, or walls to power as acceptable. In other words they would require a 120-V receptacle near the outdoor equipment. Again, check with the local inspector or code jurisdiction before making any equipment change-outs. Changing out equipment may require a code update, which means lighting and access to 120-V power as well as other code-required upgrades.

Low-voltage wiring (50 volts or less) should be installed to prevent physical damage to the conductors. In most installations this does not require conduit inside or outside the building. Best practices dictate that outside low-voltage wiring be protected in some form of conduit or piping outside the building. Some homeowners' associations do not allow open wiring, which means it must be enclosed in conduit.

SAFETY TIP

Combustibles within commercial supply or return air plenum ducts are not allowed. Combustibles create flame spread and smoke development. Combustibles include standard electrical conductors that are not in conduit, thermostat wiring, and most communication cables. **Plenum-rated cable** or fire-rated cable can be purchased and used in these air ducts. The insulation sheath on the insulation of plenum-rated wire meets the code requirements and is safe for this type of installation. Ducts serving residential structures are exempt from the "no combustibles in the air duct" code requirements. Some residential ducts even have wood or drywall in the return air duct as well as unrated thermostat wiring.

To be clear, commercial and industrial wiring is required to be in conduit inside or outside the ductwork. In most residential installations the wire does not need to be in conduit unless it is exposed outside the structure. For example, Romex can be run in residential attic spaces or between walls and under floors. Outside residential wiring will need conduit protection.

7.7 INSTALLATION INFORMATION

Table 7-2 shows electrical and physical data that is valuable when doing a new installation or a condenser change-out.

Let's review some of the information so that you will understand its value. On the left side of the table, start with Model 2AC14**18P. The voltage, hertz, and phase column is next. This model uses 208–230 volts at 60 hertz and is single phase. The 208–230 volts is allowed a ±10% variance below 208 volts and above 230 volts. The ±10% voltage range is calculated in the next column as 197–253 volts. The minimum circuit ampacity is 9.7 A, which means the wire size should handle 10 amps. The maximum overcurrent device for this model number is listed as 15 amps. The fuse or circuit breaker should be no larger than 15 amps, so the compressor RLA should draw no more than 6.9 amps. The LRA or starting amps is 35 A. The fan motor FLA is 1.1 A, at 1/10 horsepower at 1,075 RPM. The refrigerant charge is 106 ounces, and the condenser weighs 185 pounds. Other important installation notes are listed at the bottom of the table.

7.8 INSTALLATION NOTES

Most installation instructions have a section called *Notes*. Review the notes under Table 7-2. This is additional information that must be followed to ensure the safety and design performance of the installation. This information is not intended to be all encompassing, but gives an idea of the important subjects covered in this often overlooked section. Here is an excerpt from one manufacturer's notes found in a natural gas furnace installation *Notes* box:

1. If any original wire is replaced, use wire rated for 105°C.

Table 7-2 Electrical and Physical Data for Various Condensers

This table gives the installing technician electrical information about various condensing unit model numbers.

				Maximum	Compressor		Fan Motor				Weight (lbs.)
Model	Voltage/Hz/ Phase	Voltage Range	Min. Circuit Amp.	Over Current Device (amps)	Rated Load (Amps)	Locked Rotor (Amps)	Full Load Amps	Rated HP	Nom. RPM	Refrig. Charge (OZ.)+	Wire Guard/ Louvered
2AC14**18P	208–230/60/1	197–253	9.7	15	6.9	35	1.1	1/10	1075	106	185
2AC14**24P	208–230/60/1	197–253	11.6	20	8.4	39	1.1	1/10	1075	129	201
2AC14**30P	208–230/60/1	197–253	15.2	25	11.3	51	1.1	1/5	1075	154	227
2AC14**36P	208–230/60/1	197–253	17,2	30	12.9	61	1.1	1/5	1075	165	226
2AC14*42P	208–230/60/1	197–253	21.5	35	16.2	82	1.2	1/4	1075	204	262
2AC14*48P	208–230/60/1	197–253	22.6	35	17.1	100	1.2	1/4	1075	208	261
2AC14*60P	208–230/60/1	197–253	28.8	50	21.4	137	2.0	1/3	1075	264	270

Do Locate the Unit:

- With proper clearances on sides and top unit (a minimum of 12″ on the three sides, service side should be 24″ and 48″ on top
- On a solid, level foundation or pad
- To minimize refrigerant line lengths

Do not Locate the Unit:

- On brick, concrete blocks or unstable surfaces
- Near clothes dryer exhaust vents
- Near sleeping area or near windows
- Under eaves where water snow or ice can fall directly on the unit
- With clearance less than 2 ft. from a second unit
- With clearance less than 4 ft. on top of unit

(Courtesy Allied Air Enterprises.)

Table 7-3 Required Field-Installed Accessories for Air Conditioners

This table lists various installation requirements.

Accessory	Required for Low-Ambient Cooling Applications (Below 55°F/12.8°C)	Required for Long Line Applications* (Over 80 ft. 24.38 m)
Crankcase heater	Yes	Yes
Compressor start assist capacitor and relay	Yes	Yes
Evaporator freeze thermostat	Yes (For non-Infinity systems only)	No
Liquid line solenoid valve	No	See long-line application guideline
Low-ambient pressure switch	yes (For non-Infinity system only)	No
Support feet	Recommended	No
Thermal expansion valve (TXV) hard shutoff	Yes	Yes
Winter start control	Yes (For non-Infinity systems only)	No

* For tubing line sets between 80 and 200 ft. (24.38 and 60.96 m) and/or 20 ft. (6.09 m) vertical differential, refer to Residential Split-System Longline Application Guideline.

(Courtesy Carrier Corporation.)

2. Only use copper wire between the disconnect box and furnace connections.
3. Symbols are electrical representations only.
4. Replace circuit board fuse with a 3-amp fuse.
5. Blower off delay, gas heating selections are 90, 120, 150 and 180 seconds, cooling or heat pump 90 seconds or 5 seconds when dehumidifying call is active.

Table 7-3 shows information about accessories that will be valuable when evaluating the electrical and mechanical parts of the condenser. A condensing unit has several electrical components.

Figure 7-24 gives the technician more information that relates to the crankcase heater listed in the accessories table, Table 7-3. This crankcase heater is controlled by the discharge line temperature. When the discharge line is warm, the crankcase heater will open the power supply to the heater. The crankcase heater does not need to be operated when the compressor is running. This will save a little energy and extend the life of the heater. This could be considered a "green circuit" since it saves energy.

Figure 7-25 shows an insertion-type crankcase heater, which you might find on a semi-hermetic compressor. The heater keeps the compressor oil warm during the off cycle to prevent refrigerant from condensing and diluting the lubrication.

Figure 7-26 illustrates the location of a **liquid line solenoid valve**, which is an option for the condenser model number found in Table 7-3. When the solenoid closes, the

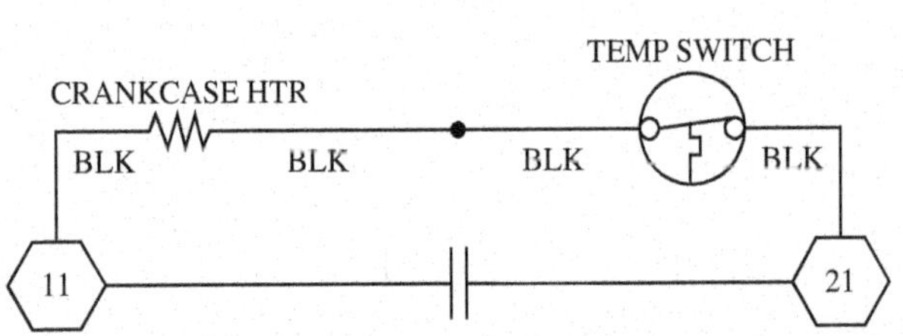

Figure 7-24 The crankcase heater is not on all the time. When the temperature drops the thermostat closes and operates the heater. This could be considered a "green circuit" since it saves energy.

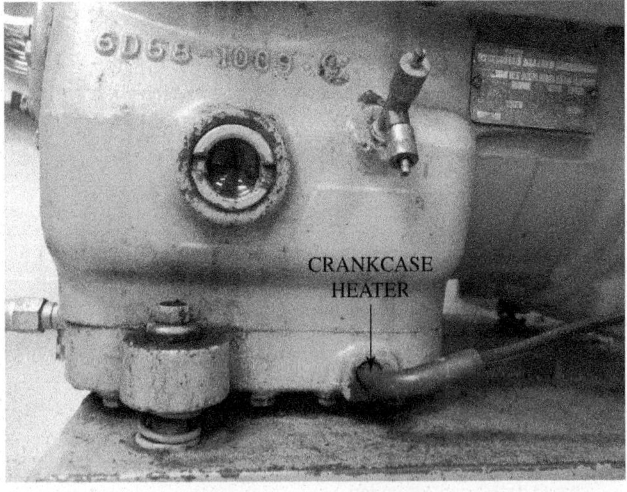

Figure 7-25 This crankcase heater is installed in the crankcase well. It is used to keep the oil warm to prevent refrigerant from condensing in the compressor oil.

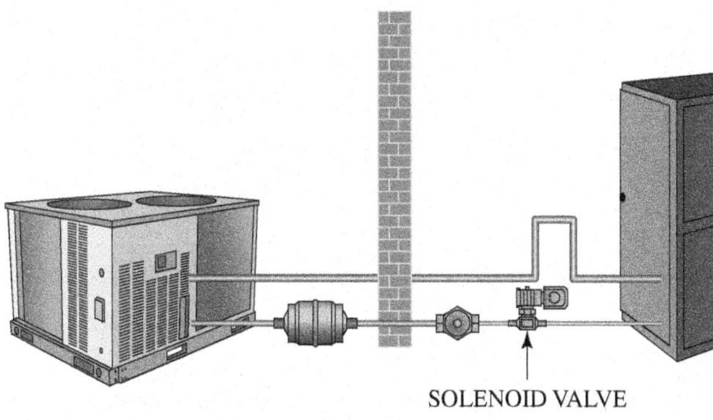

Figure 7-26 A system that includes a liquid line solenoid valve, which is usually located near the air handler or metering device.

compressor continues to operate, pumping down the system until it is shut down by opening the low-pressure switch. This design removes refrigerant from the evaporator and suction line when there is a long refrigerant line set. Removing most of the refrigerant will prevent compressor flooding. The solenoid is usually installed near the air handler or metering device for the best design and operation.

SAFETY TIP

When **stranded wire** (sometimes called braided wire) is used, a single strand of wire may separate and create a hazard. Wire should be inspected to ensure no strand of wire is loose. The loose wire may ground or develop a short circuit to a wire of a different potential. Stranded wire should be hooked around a screw or all conductors slipped into the electrical connector and crimped, then connected. One way to prevent a stray wire is to solder the stranded wires together or place them inside a wire **ferrule**. A wire ferrule is a small metal tube. The stranded wire is slipped into the ferrule and the ferrule is crimped closed. The ferrule is slipped under a connection point and tightened. This will also improve the overall circuit conductivity at that connection.

7.9 ELECTRICAL CHECKLIST

This electrical checklist is used as a final punch list before turning over the operation of the equipment to the owner. You will notice that it does not include any of the other important aspects of an installation such as airflow, ductwork, and charging. A separate checklist should be developed for those critical items.

- Tighten all field and factory connections.
- Check incoming voltage prior to starting equipment.
- Start equipment. Record voltage and current readings while adjusting the charge.
- Check operation of all pressure switches by creating a low- or high-pressure condition to cut off the equipment.
- Operate the thermostat in all modes: heating, cooling, fan auto, fan on, and fan off. Check emergency heat with a heat pump system.
- Program the thermostat per customer's recommendations.
- Show the customer how to operate the thermostat.
- Give the owner the thermostat and equipment owner's manual.
- Record voltage, amperage, pressure, and charging information in a database to be used as a future service record.

7.10 REWIRING EQUIPMENT

In some cases a piece of equipment will need to be rewired. A piece of equipment can be analyzed and repaired if the tech has a wiring diagram and the wiring is still intact from the factory. When equipment wiring is modified without documenting the changes on a wiring diagram, the technician will have trouble. One indication that wiring modifications have occurred is when the tech opens a panel door and a bundle of wire falls out. Another is the sight of spliced wire and additional components inside the cabinet.

The solution is not easy, but it is possible to solve the problem. The answer is to remove all of the wiring and rewire the entire piece of equipment. Use the original wiring diagram or one that is compatible with the unit that is being rewired.

Wiring up the total unit and energizing it will *not* make it easy to find a problem quickly. Instead, do it in stages. Start with the easiest components first. For example, when rewiring a condenser, start with the crankcase heater or control voltage section. Wire up the control voltage and test to see if the contactor will energize. This may include wiring up all pressure switches and safety controls if they are in the control voltage section. Next, hook up the condenser fan motor. After wiring the motor, energize the condenser to determine if the fan motor is operating properly. Finally, hook up the compressor and test the system.

This process allows you to go step by step until the equipment works properly. If there is a problem with one of the steps, it will be easier to solve since you know what wiring was done just before finding the problem. It can be quickly checked, reviewing what was just wired. In summary, this step-by-step wiring approach is best when rewiring a piece of equipment or in any wiring process. Wire up and test the control voltage circuit first. Follow by wiring and testing individual circuits one at a time until the whole unit is working properly.

7.11 OPEN WIRING AND CONDUIT

Open wiring is insulated, high-voltage wiring that is not in conduit or some other protective covering. Open wiring is allowed in most residential installations inside the walls, attic

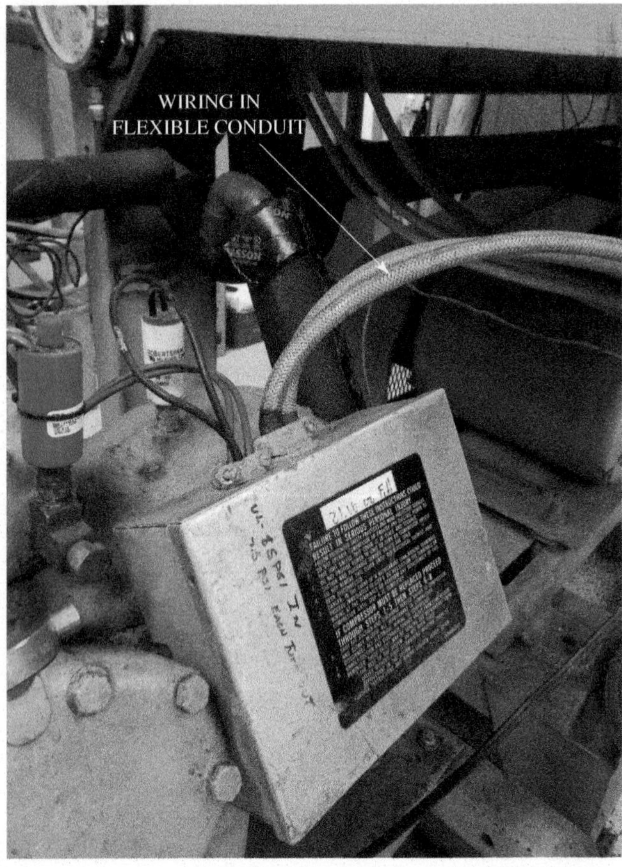

Figure 7-27 The arrow points to flexible conduit that supplies three-phase power to a compressor.

areas, and basement spaces. Overhead open wiring is allowed in all types of electrical distribution systems. Some of the open wiring in a utility electrical distribution system is not even insulated. If a commercial building is wired according to the NEC, it will not have open wiring inside the building. All wiring will be in metal or plastic conduit in all parts of the building. Figure 7-27 illustrates flexible conduit feeding power to a three-phase compressor. Conduit can be rigid or flexible.

SAFETY TIP

Clearance in front of a breaker panel is important for quick access to the breakers. The breaker may need to be opened to prevent equipment or personal damage. It seems as if areas devoted to panel boxes are also magnets for storage. It is difficult to keep stored items away from these serious electrical devices. Figure 7-28 shows a sign warning that 30 inches of clearance should be kept in front of the panel box. Higher voltage will require higher clearance requirements in front of the panel box. The clearance includes a clear pathway to walk up to the panel box. Stored items should be placed so that the path to the panel box stays clear. Clear panel access is required for quick service should a breaker need to be turned off in an emergency.

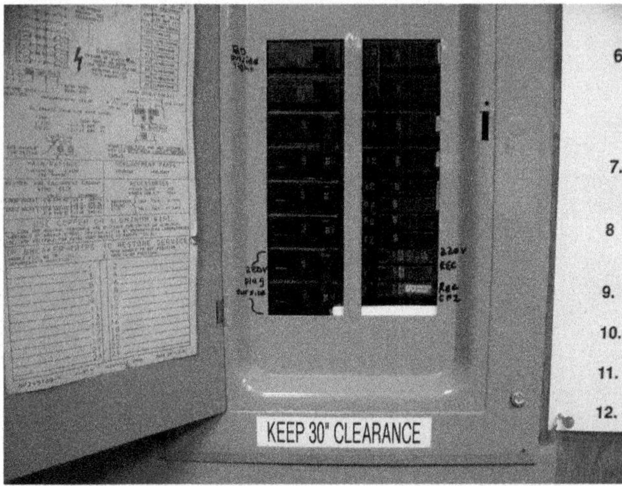

Figure 7-28 Clearance in front of a breaker panel is important for quick access to the breakers. The breaker may need to be opened to prevent equipment or personal damage.

SAFETY TIP

The technician should not hold a meter in one hand along with the two probes, one in each hand. The *improper* way to handle this meter is shown in Figure 7-29. Using the meter on the wrong scale or a high-voltage spike may cause the meter to explode in your hand. Fewer injuries would occur if the meter were set down and not held.

To reduce the risk of shock, put the meter down and measure voltage with one probe attached to one side of the circuit, while using the other probe to skip around measuring voltage. If the components are close together you can place both probes in one hand and measure voltage. That will take a little practice. If this is difficult, learn by practicing on a dead circuit. Measuring voltage with two probes in one hand is a difficult skill, and not practical if the components you are testing are more than a few inches apart.

Figure 7-29 The tech does not have a watch or ring on when measuring voltage. Two other safe practices are *not* to hold the meter and to use one hand to hold two probes.

SERVICE TICKET

For this Service Ticket exercise, use Electrical Diagram ED-1, which appears with the Electrical Diagrams package that accompanies this text.

The system you have been called to service does not work in the heating or cooling mode. The fan does not operate. A tenant has recently rented the lease space and would like the system to be in working order prior to moving into the space. After the initial inspection you notice that the wiring has been heavily modified. Wires have been spliced and some connections seem to be taped together. You have several options. The natural inclination is to troubleshoot the system in its existing state. Maybe the solution will be something simple. Yes, it is a good idea to spend a few minutes measuring voltages, breakers, and system pressures. There may be an easy or obvious fix.

After taking basic voltage and pressure measurements, you find it difficult to determine the solution to this nonfunctioning package unit. The input voltage and transformer output voltage are correct. The refrigeration system has good pressure readings. The wiring mess and modified wiring make it difficult to troubleshoot.

The second option is to draw the existing wiring diagram and use the diagram as a troubleshooting tool. Since the wiring has been heavily modified that is not the best option, because it will take a long time to develop a useful diagram. Also, errors in the drawing will prevent accurate troubleshooting.

You decide to rewire the package unit according to the original electrical diagram. The tenant is made aware of the condition of the unit. She is shown the condition of the wiring. She approves a time and material bid with a limit of 2 hours to get the unit rewired and an estimate of the problem. Let's take a look at the steps that will make this a quick success. The following steps are not law and they can be rearranged for the benefit of the tech.

Start with rewiring and testing a simple circuit using the following steps:

1. Remove all wiring from the package unit. Have additional wire and electrical connectors available. Electrical Diagram ED-1 does not list the color of wire though that would be useful. Label the wire colors on the diagram as the components are hooked up. This will be useful for future troubleshooting. Placing wiring number labels on the wire and the diagram will be helpful. Figure 7-30 shows an example of how to start the numbering sequence that will aid in wiring the unit. The rule for numbering the circuit is to use the same number on all wiring points that are connected together. When the diagram has any type of change, the number will change. Going through a switch, fuse, or load would be a change. For example, in Figure 7-30 start with the number 1 at **L1**. Then **N** (neutral leg) is labeled 2. Once the power goes through the switch, the number changes to 3. When the power crosses the fuse, the number changes to 4. All wire connections on the far left side of the diagram are numbered 4. This process continues until all wire connection points are numbered. It does not matter where you start, but beginning at the power supply with a 1 and 2 is a logical starting point.
2. Between each step turn off the voltage and do the wiring. It is understood that power will be applied after each wiring step. Then test and verify operation.
3. Start with one of the easy circuits first. For example in Electrical Diagram ED-1, the crankcase heater (**CH**) is wired to the high voltage through the normally closed **CR** contacts 4 and 5. The high voltage is applied and the crankcase heater warms the compressor oil. Good start!

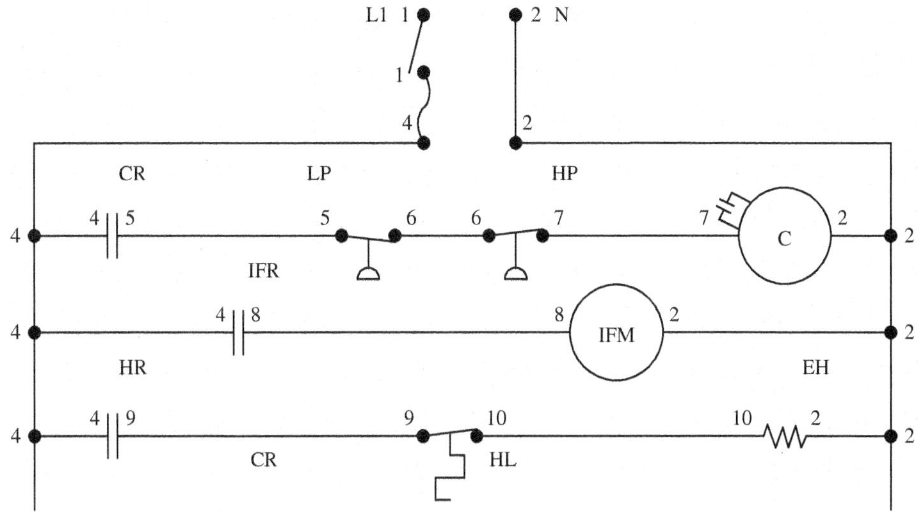

Figure 7-30 This diagram illustrates how to start the numbering sequence that will aid in wiring the unit.

4. Next, wire the primary side of the transformer to the 120-V power supply. The 24-V control voltage is measured on the secondary of the transformer.
5. The indoor blower circuit is another simple circuit that can be hooked up and tested. Wire the thermostat **R** and **G** connections. Connection **R** is hooked to the hot side of the 24-V transformer; **G** is wired to terminal 5 on the normally closed **HR** contacts. **HR** terminal 4 is wired to one side of the **IFR** coil. The other side of the **IFR** coil is hooked to the common side (right side) of the transformer. Sometimes the common side of the transformer is labeled **C**. Close the "ON" fan switch on the thermostat. The contacts on **IFR** terminals 2 and 4 should be closed. Use an ohmmeter to verify that these contacts are now closed. Hook up **IFR** terminal 2 to L1 and terminal 4 to the indoor fan motor **IFM**. Wire the other side of the motor to **N**. Apply the 120-V power supply. With the fan switch in the "ON" position the **IFM** will operate. Finally, turn the thermostat to the "AUTO" and "COOL" setting. Turn the thermostat below the room set point. The blower should now operate in the "AUTO" mode.
6. The heating circuit is a simple circuit. Wire thermostat **W** to one side of the **HR** coil. The other side of the **HR** coil is hooked to the common side of the transformer, which is not labeled. Some wire designs ground the common side of the transformer. If the right side of the transformer is grounded, the right side of the **HR** coil would be grounded with no wire to the transformer. Terminal 1 of the **HR** contacts is wired to **L1**. The terminal 3 connection of **HR** contacts will be wired to the **HL**, the heater overload. The other side of **HL** is wired to the heat strip **EH**. The opposite side of **EH** is hooked to **N**. Finally, in the control voltage section, the **HR** normally open contacts 6 and 4 are wired. Terminal 6 is wired to left side of the thermostat and transformer junction. Terminal 4 is hooked to the left side of the **IFR** coil and terminal 4 of the normally closed **HR**. The other side of the **IFR** coil was already wired to the common side of the transformer. Keep the thermostat in the "OFF" position with the power applied. If nothing happens, proceed by turning the thermostat to the heating mode. Turn the thermostat up a few degrees above the room temperature so that the **HR** coil is energized. This will bring on the heat strip and indoor fan motor. Measure the amps on the heat strip. The measurement should be about 16 to 20 amps. *Note:* The **IFR** is controlled by normally closed **HR** contacts 5 and 4 when the thermostat is in the "AUTO" position.
7. Next, the cooling circuit is wired in two steps. Wire the condenser motor first, then the compressor.
8. Hook the thermostat **Y** terminal to one side of the **CR** coil. Wire the other side of the coil to the common side of the transformer. Wire normally closed **CR** terminal 4 to **L1** and terminal 5 to one side of the crankcase heater **CH**. Apply power and check the crankcase heater for warmth.
9. Wire the outdoor fan motor. Terminal 4 of **CR** is wired to **L1** and terminal 6 to one side of the outdoor fan motor (**OFM**). Hook the other side of the **OFM** to **N**. Turn the thermostat to cooling and test. The condenser fan should operate along with the indoor fan motor.
10. The final connection is the compressor. Wire **CR** normally open terminal 1 to **L1**. Terminal 3 is hooked to one side of the low-pressure (**LP**) switch. The other side of the low-pressure switch is wired to one side of the high-pressure (**HP**) switch. The other side of **HP** is wired to the common on the compressor. **N** is wired to the run winding and the run capacitor. The other side of the run cap (not shown) is wired to the compressor start winding. Hook up the clamp-on ammeter to the common wire of the compressor. Operate and test. Measure the amp draw and check the system charge.

This wiring method will ensure that the system will work according to the manufacturer's design. Rewiring a system without using a step-by-step procedure will lead to more lost troubleshooting time if part or all of the system does not work properly.

SUMMARY

Installing power and control voltage wiring is not the major time-consuming element when it comes to installation or equipment change-out—but it is the most important step in the installation because improper wiring may create a hazardous condition that can cause a fire or electric shock hazard to the technician or the consumer. The wire size, overcurrent protection, and proper grounding are the important safety aspects of good installation or system exchange. When doing calculations for selecting wiring and overcurrent protection, it is a safe practice to round down for overcurrent protection and round up for wire size.

When troubleshooting an electrical problem that involves "butchered" wiring, it may be easier to rewire it rather than spend time trying to figure out the problem. After removing the wiring, start by hooking up the control voltage circuit first. Test each circuit as it is wired. Next, hook up and test components in the high-voltage section. Start with the easiest component first. For example, if you are rewiring a condensing unit, test the control voltage circuit first. In the high-voltage section, wire the crankcase heater or condenser fan motor first. Proceed with more complex components like the compressor.

REVIEW QUESTIONS

1. What term is used when referring to air conditioning efficiency?
2. What term is used when referring to gas furnace efficiency?
3. What efficiency rating number is considered acceptable for a "green" air conditioning unit?
4. What efficiency rating number is considered acceptable for a "green" gas furnace?
5. The UMC and NEC codes are developed as ____________ safety standards for our industry to follow.
6. Describe the safest way to reset a circuit breaker.
7. What is the AFUE of a furnace with a Btuh output of 90,000 and a Btuh input of 100,000?
8. What are two major things that prevent an air conditioning system from operating at its peak efficiency?
9. What measure should be taken regarding the overcurrent protection on a replacement condensing unit?
10. What common electrical items should be checked and sized properly when installing new equipment or replacing equipment?
11. What is the purpose of the maximum circuit ampacity rating on a nameplate?
12. What is the purpose of the maximum overcurrent protection rating on a nameplate?
13. When does "nuisance" tripping usually occur?
14. What is the difference between minimum circuit ampacity (MCA) and the maximum size of an overcurrent protection device (MOCP or MOP)?
15. Using the condenser nameplate shown in Figure 7-31, what is the smallest wire size you would select to wire this unit?
16. Using the condenser nameplate shown in Figure 7-31, what is the largest overcurrent protection device you would select to protect this unit?
17. Determine the MCA of a condensing unit that has a compressor motor with a 20 RLA and two condenser fans motor that draw 2 amps each.
18. Use data from the nameplate shown in Figure 7-31 to answer this question. The condensing unit has been nuisance tripping on its overcurrent protection device, which is rated at 40 amps. After cycling the unit a few times and checking the voltage, amp draw, and pressures, everything seems to be working properly. What would you do to reduce nuisance breaker tripping?
19. What electrical questions need to be asked when changing out a condensing unit?
20. What is the color of a ground wire?

Use Table 7-4 to answer Review Questions 21 through 25.

21. In Table 7-4, what is the operating range of voltage for the 460-V unit?
22. In Table 7-4, what is the only type of wire material allowed according to this manufacturer's table?
23. In Table 7-4, what is the maximum fuse size that is allowed on the Model 50EE-060 unit that uses 460 V and is three phase?
24. In Table 7-4, how much current should a conductor carry that serves the Model 50EE-060 unit that uses 230 V and is three phase?
25. In Table 7-4, how much current should a conductor carry that serves the Model 50EE-060 unit that uses 230 volts and is single phase?
26. Dashed lines are used in some wiring diagrams. What do dashed lines indicate to the technician?
27. What is the supply voltage in Figure 7-30 (or Electrical Diagram ED-1)?
28. What is the control voltage in Figure 7-30 (or Electrical Diagram ED-1)?
29. Redraw the control voltage section in Electrical Diagram ED-1. Using the numbering method of organizing a diagram, number the control voltage circuit on your diagram.

CONTAINS HCFC – 22		DESIGN PRESSURE	
FACTORY CHARGE		278	HI PSIG
12 LBS	8 OZS	144	LO PSIG
ELECTRICAL RATING		NOMINAL 208/230 VOLTS	
1 PH	60 HZ	MIN 197	MAX 253
COMPRESSOR(S):(1)		FAN MOTOR(S): (1)	
PH	1	PH	1
RLA	23.8	FLA	1.7
LRA	129	HP	1/4
MIN. CKT AMPACITY AMPERAGE MINIMUM	31.5	MAX FUSE OR CKT.BKR. FUSIBLE/COUPE CIRCUIT (HACR PER NEC)	50
FOR OUTDOOR USE			

Figure 7-31 Use this condensing unit nameplate to answer Review Questions 15, 16, and 18.

Table 7-4 Table for Review Questions 21 through 25

Model 50EE	V-PH	Oper Voltage*		COMPR		IFM	OFM	Max Fuse† or HACR CKT BKR Amps	MCA
		Max	Min	LRA	RLA	FLA	FLA		
018	208/230-1	253	187	50.0	8.3	4.0	1.1	20	15.5
024				48.0	10.0	4.0	1.1	25	17.6
030				65.0	14.1	4.0	1.1	35	22.7
036				82.0	17.2	4.2	1.5	40	27.2
042				95.4	21.5	4.0	2.1	50	33.0
048				110.0	23.7	4.5	2.4	55	36.5
060				142.0	28.9	6.7	2.8	60	45.6
030	208/230-3	253	187	53.0	8.1	4.0	1.1	20	14.7
036				67.5	11.0	4.2	1.5	25	19.5
042				82.0	13.7	4.0	2.1	30	23.2
048				92.0	15.2	4.5	2.4	40	25.9
060				124.0	19.2	6.7	2.8	45	33.5
036	460-3	506	414	33.8	6.0	2.0	0.7	15	10.2
042				41.0	6.9	2.0	1.2	15	11.8
048				46.0	8.0	2.0	1.2	20	13.2
060				62.0	9.6	3.4	1.4	25	16.8

FLA — Full Load Amps
HACR — Heating, Air Conditioning and Refrigeration
IFM — Indoor Fan Motor
LRA — Locked Rotor Amps
MCA — Minimum Circuit Amps
OFM — Outdoor Fan Motor
RLA — Rated Load Amps
*Permissible limits of the voltage range at which units will operate satisfactorily.
†Maximum dual element fuse.

NOTE: Use copper wire only.

(Courtesy Carrier Corporation.)

UNIT 8
Transformers

WHAT YOU NEED TO KNOW

After studying this unit, you will be able to:

1. Describe how a transformer works.
2. Name the types of transformers.
3. Draw the symbols for two types of transformers.
4. Calculate the VA of a transformer.
5. Describe how to troubleshoot a transformer.
6. Describe how to wire transformers in parallel.
7. Discuss how to find a short in a control circuit.

The purpose of a **transformer** is to "transform" or change input voltage. A transformer has an input voltage and output voltage. A common transformer found in HVACR applications is shown in Figure 8-1. Sometimes the word *transformer* is shortened to *xformer*. Figure 8-2 shows the electrical symbol for a transformer.

A transformer is a device that transfers electrical energy from one alternating current circuit to another with a change in output voltage or current. Most transformers used in HVACR reduce, or step down, the voltage being supplied. The primary winding is hooked to the incoming power source, and the secondary is the reduced voltage normally used in a control voltage circuit.

8.1 HOW DO TRANSFORMERS WORK?

First, let's learn about the construction and operation of a transformer. The foundation of a transformer is the iron or steel core, as shown in Figure 8-3. The core has two sets of windings wrapped around it. One set of windings is called the primary winding. The primary winding is the also known as the input voltage winding. The second set of windings is the secondary winding. This is the transformer output voltage.

The key to understanding how transformers work is to understand magnetism. Remember that magnetism is created by electricity flowing through a wire. In the simplest explanation, magnetism from the primary side induces voltage into the secondary side through the transformer iron core, as shown in Figure 8-3. This is also known as magnetic induction.

In summary, transformers are electrical devices that produce an electrical potential in the secondary circuit through electromagnetic induction. As discussed in the previous paragraph, a transformer has two sets of windings. The primary winding is wrapped around a laminated steel core. When the primary voltage is applied to the transformer, it develops a magnetic field in the steel core.

Next, we will explore the different types of transformers.

Figure 8-1 This type of common step-down transformer is found in air conditioning and heating systems. The output voltage is reduced to 24 volts and used to operate the control circuit.

Figure 8-2 This symbol is used for transformers.

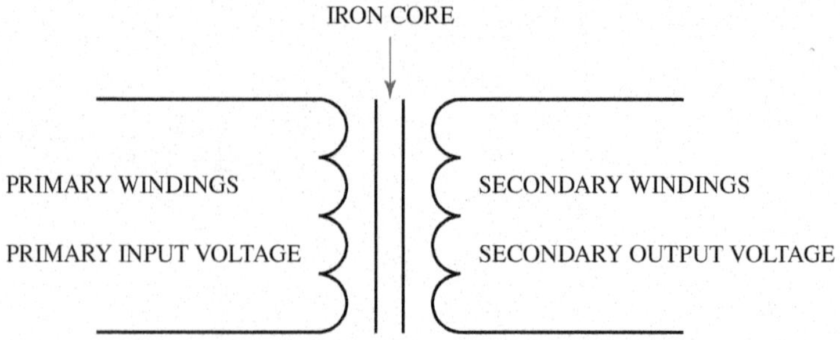

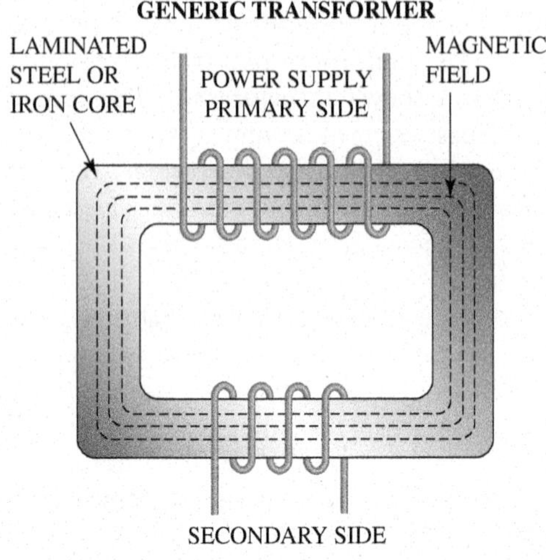

Figure 8-3 Transformers have an iron or steel core and two sets of windings.

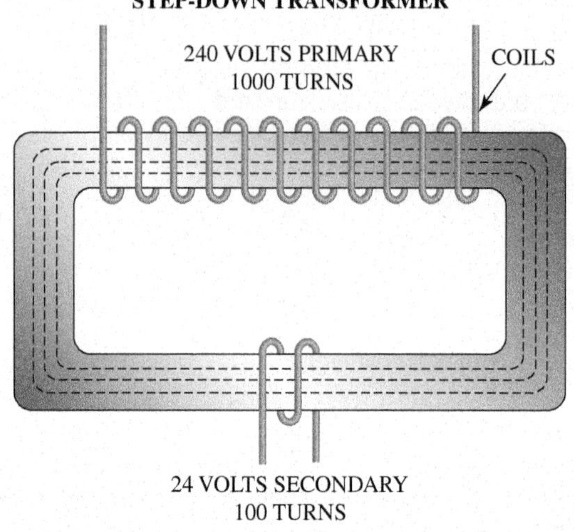

Figure 8-4 This step-down transformer has an input voltage of 240 V and an output voltage of 24 V. Both the turns ratio and resistance ratios will be 10:1.

8.2 TYPES OF TRANSFORMERS

Three types of transformers are used in air conditioning units:

- Step-down transformers
- Step-up transformers
- Isolation transformers.

We discuss these next, along with another type of step-down transformer, the multi-tap transformer.

Step-Down Transformer

The step-down transformer lowers the operating voltage to a reduced and safer output for the technician and the equipment. With lower control voltage the electrical components and wiring can be smaller in size with lower current-carrying requirements. Thermostats operating on low voltage are safer in case a customer decides to investigate this thermal switch. The step-down transformer is the most common type of transformer used in HVACR equipment. However, we do not want to mislead the reader—step-down transformers can still have a dangerous voltage output. For example, some step-down transformers reduce the voltage from 480 to 240 volts.

The primary of the transformer can be 120, 208, 240, or 480 V. The most common secondary voltage for a step-down transformer is 24 V. This is known as a control transformer. A control transformer is one in which the secondary side of the transformer supplies voltage to the control circuit. It is usually a low voltage, low-current operation. Here are some characteristics of a step-down transformer:

1. It has more turns of wire in the primary winding than in the secondary winding.
2. The turns ratio is the same as the voltage ratio. For example, if the turns ratio is 10:1 and the output voltage is 24 volts, then the primary voltage will be 240 volts. In this example, if the resistance of the primary is 120 ohms, the resistance of the secondary will be about 12 ohms. Figure 8-4 shows the turns in a 240-V to 24-V step-down transformer. The "turns ratio" is 1,000 turns on the primary and 100 turns on the secondary, a 10 to 1 ratio. The resistance ratio between the primary and secondary will be approximately 10:1 as well.
3. Twenty-four-volt secondary transformers are used for most control circuits. Figure 8-5 shows the symbol for a step-down transformer. Notice that there are more windings on the primary side of the symbol than on the secondary side. Some symbols show an equal number of windings on both sides of the drawing even if it is a step-down transformer.
4. The amperage on the secondary side of a step-down transformer is greater than the amperage on the primary side. The current flow in the secondary can increase up

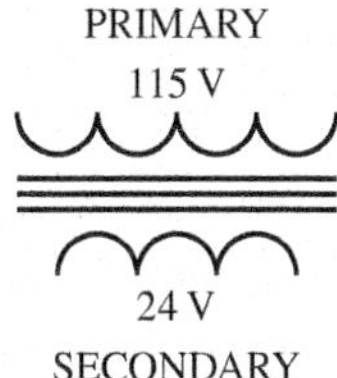

Figure 8-5 This is the symbol for a step-down transformer. Notice that there are more windings in the primary side than the secondary side. The resistance ratio of the winding will correspond to the voltage drop.

to the ratio between the primary and secondary sides. Let's look at an example to understand the current ratio:

- The input voltage and current of a step-down transformer are 120 volts and 1 amp, respectively.
- The output voltage is 24 V. That is a ratio of 5 to 1.
- The 5:1 ratio translates into an increase of the primary amperage from 1 to 5 amps. The maximum voltage the secondary can handle is 5 amps. Secondary amperage greater than 5 amps will overheat and burn out the secondary winding.

In summary, the step-down transformer reduces the input voltage and increases the available output amperage. You will see that the opposite occurs in the step-up transformer.

Step-Up Transformer

The step-up transformer increases the input voltage. The common primary voltages in a step-up transformer are 120, 208, or 240 volts. Other voltages can be stepped up, but the voltages just listed are more often found in our industry. Here are some characteristics of a step-up transformer:

1. It has more turns of wire in the secondary winding than in the primary winding.
2. The input voltage is lower than the output voltage. The turns ratio is the same as the voltage ratio. For example, if the turns ratio is 1:2 and the output voltage is 240 volts, then the primary voltage will be 120 volts. In this example, if the resistance of the primary is 20 ohms, the resistance of the secondary will be about 40 ohms. See Figure 8-6 for an example of a step-up transformer.
3. The amperage on the secondary of a step-up transformer is less than the primary voltage. The current flow in the secondary will go down to the ratio between the primary and secondary. Let's look at an example to understand the current ratio:

   - The input voltage and current of a step-down transformer are 120 volts and 1 amp, respectively.
   - The output voltage is 240 V. That is a ratio of 1 to 2.
   - The 1:2 ratio translates into a decrease of the primary amperage from 1 to 0.5 amp. The maximum voltage the secondary can handle is 0.5 amp. Secondary amperage greater than 0.5 amp will overheat and burn out the secondary winding.

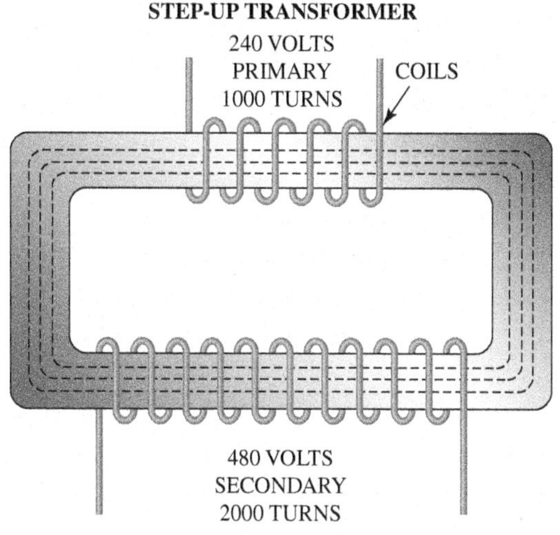

Figure 8-6 This step-up transformer has an input voltage of 120 V and an output voltage of 240 V. The primary has 1,000 winding turns. The secondary has 2,000 winding turns.

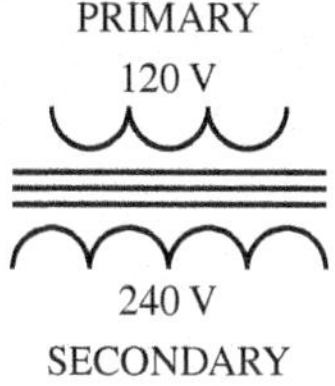

Figure 8-7 This is the symbol for a step-up transformer. Notice that there are more windings in the secondary side than the primary side.

Figure 8-7 shows the symbol for a step-up transformer. Notice that there are fewer windings on the primary side of the symbol.

In summary, the step-up transformer increases the input voltage and decreases the available output amperage. This is the opposite of what occurs in a step-down transformer.

Isolation Transformer

The isolation transformer is the least used transformer in HVACR systems. The isolation transformer does not change the voltage input to output. The purpose of the isolation transformer is to separate the HVACR system voltage from the incoming voltage. Some electrical utilities or electricity generators develop "dirty power." Dirty power means that the supplied voltage is not a clean and smooth sine wave. The voltage varies between each cycle, and voltage spikes can occur. This uneven voltage can affect the operation of some sensitive electronic components, so an isolation transformer is used to reduce the transfer of these voltage imperfections. Figure 8-8 compares an ideal voltage sine wave and a dirty power sine wave.

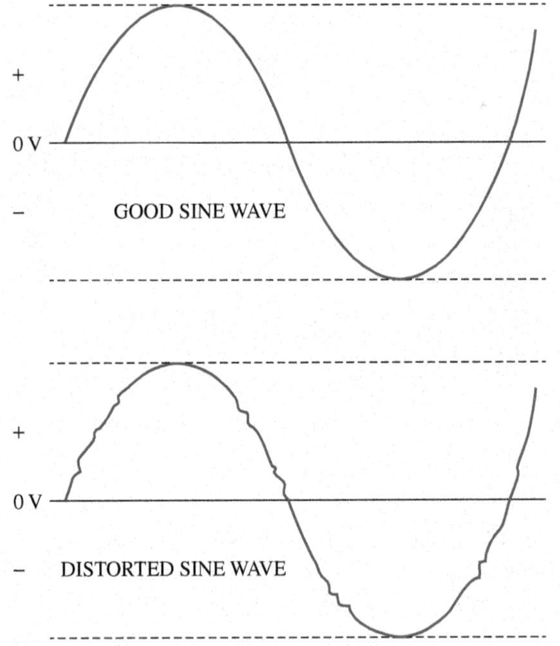

Figure 8-8 The ideal voltage supplied by an electrical utility or generator will have a smooth and proportional sine wave. "Dirty power" will have spikes in the sine wave or uneven positive or negative voltage swings.

Here are some characteristics of an isolation transformer:

1. The primary and secondary windings have the same number of turns of wire.
2. The input voltage equals the output voltage. That is a ratio of 1 to 1.
3. The input and output current are nearly the same.

Multi-Tap Transformer

The **multi-tap transformer** is used in step-down transformer applications. The term *multi-tap* means that the primary side of the transformer has multiple input voltage taps. Figure 8-9 shows several transformer models and the various input options. It also illustrates what input voltages are required for several multi-tap transformer designs. The upper transformer has a single voltage input. The black wire is common. The color of the input wire—white, red, orange, brown, or black/red—determines the correct input voltage requirement. The secondary output of all of these transformer options is 24 volts. The middle transformer has 208- or 240-volt input options. The bottom transformer is a useful one because it has three of the most common input voltage options. This is a good transformer to stock for most applications that need a replacement transformer.

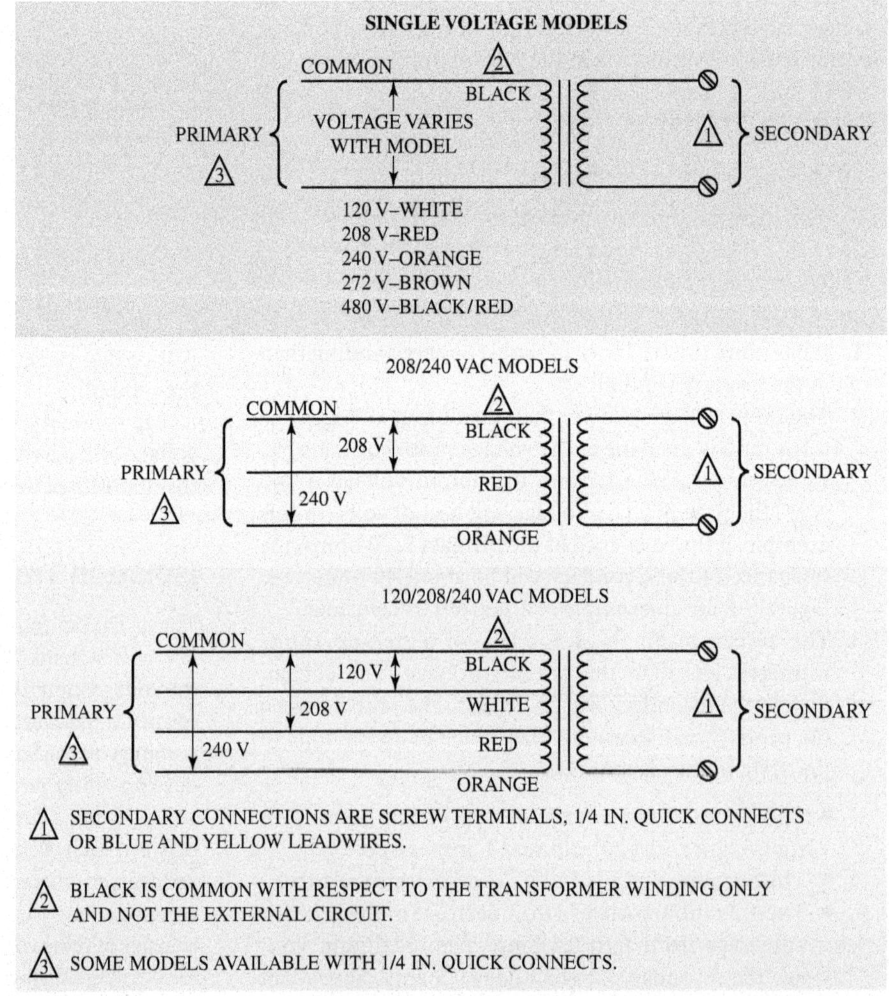

Figure 8-9 Multi-tap transformers have different designs and various input options.

Figure 8-10 This is a nameplate for a multi-tap transformer. Notice the 75 V.A. output rating. The input voltage can be 200, 230, or 460 V. The output voltage is 24 V.

Unfortunately, the HVACR industry does not use a common wire color code. One manufacturer may use black as a common wire, while another manufacturer may use red or white. The only common color code is green for ground and the use of two brown wires to a motor run capacitor.

Figure 8-10 shows the nameplate for a multi-tap transformer. The side shown is the 24-V or secondary side and it has several input options such as 200, 230, and 460 V.

Figure 8-11 illustrates another example of a multi-tap transformer with the black, white, red, and orange wires on the input side and the blue and yellow wires on the 24-V output side. The input, on the right, can be wired to use 120, 208, or 240 volts. The output on the left is 24 volts. Having this transformer as part of inventory stock reduces stock requirements because it offers three input voltage options. This transformer has a common 24-V, 40-VA output. The manufacturer of the transformer shown in Figure 8-11 uses the following color code on the primary side:

Black is common.

White is 120 V.

Red is 208 V.

Orange is 240 V.

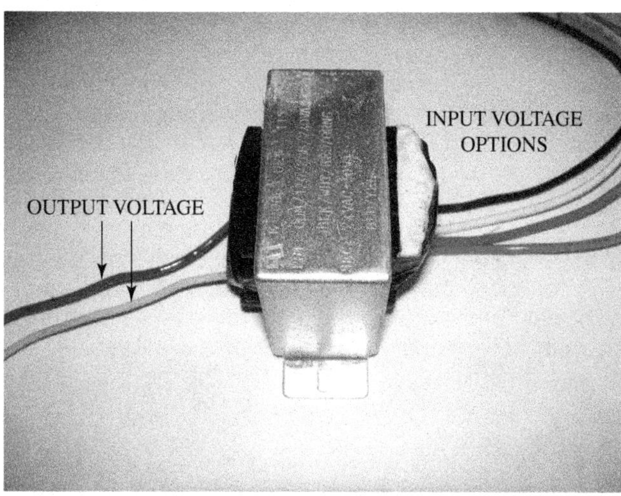

Figure 8-11 This multi-tap control transformer has three input voltage options.

Commercial Transformers

Commercial or **industrial transformers**, illustrated in Figure 8-12, use different options to provide different voltages. Figure 8-12A shows examples of this type of transformer without any specific connections. This is the symbol for a two-input voltage commercial transformer. The input can be 240 or 480 V. The diagram on the left shows the transformer wired in series for a 480-V primary. The diagram on the right connects the primary in parallel for 240-V operation.

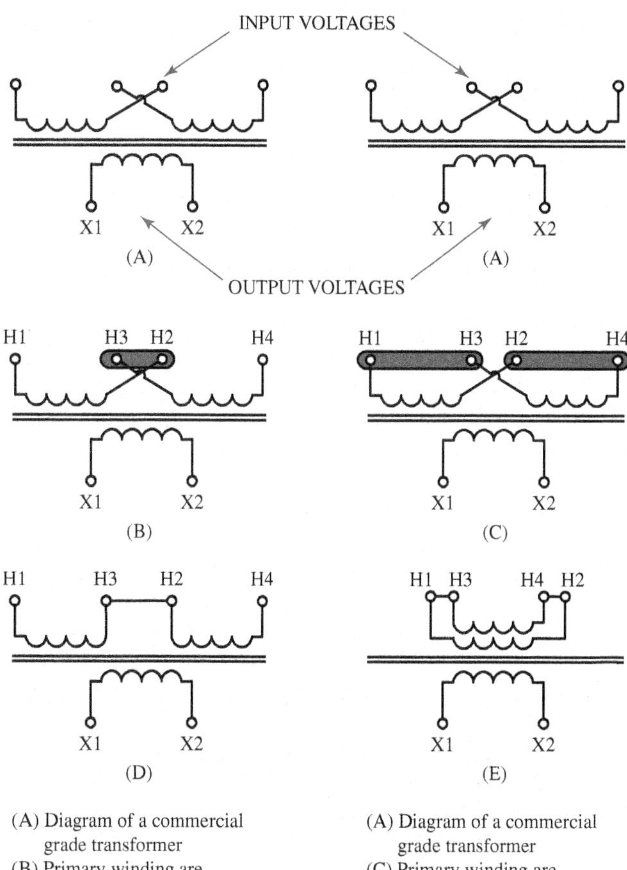

Figure 8-12 These are the symbols for two-input voltage commercial transformers.

Figure 8-13 This is classified as a commercial or industrial transformer.

The primary and secondary identifications are different when compared to noncommercial transformers (see Figures 8-12B and 8-12C). The input connections are identified as **H1, H2, H3,** and **H4.** The output voltage is identified as **X1** and **X2** and can be 120 or 240 V depending on how the primary windings are connected.

The transformer in Figure 8-13 is classified as a commercial or industrial transformer. The input voltage can be wired for 240- or 480-V inputs. The input connections are labeled **H1, H2, H3,** and **H4.** Wire the connections in a series for 480-V input. Wire the connections in parallel for 240-V input. The output will be 240 or 480 volts depending on the wiring arrangement of **X1, X2, X3,** and **X4.**

TECH TIP

Only use two of the input wires when hooking up a multi-tap transformer. Hooking up more than two wires will burn out a multi-tap transformer. Verify the input voltage. Measure the input voltage before hooking up a transformer. Only use two wires on the secondary of a transformer. Tape off the end of any unused wires to prevent them from grounding out the transformer.

8.3 CALCULATING VOLT-AMPS (VA)

The term **volt-amps** or **VA** is used to determine the output rating of a transformer. As you may have noticed, VA is volts multiplied by amps, which is the same formula as watts. The term VA is used to differentiate transformers from other devices that use watts. The VA rating is found on the transformer as shown in Figure 8-10. What does this mean? Let's look at an example of a common 40-VA rated transformer with a 24-V secondary.

EXAMPLE 8.1 DETERMINING AMPERAGE RATING

Determine the amperage rating of the secondary of a 40-VA transformer, which is one of the most common transformers in our industry. The formula is volt-amps = VA.

SOLUTION:

To determine the amperage rating of the secondary side of a 40-VA transformer, convert the formula to:

$$A = \frac{VA}{V}$$

$$A = \frac{40\text{ VA}}{24\text{ V}}$$

$$A = 1.67\text{ amps}$$

What does this mean? It means that the secondary of a transformer can handle no more than 1.67 amps. Amp draw in excess of 1.67 A will overheat the transformer. At some point the transformer will burn out and open the secondary winding.

Short transformer life due to overload should be investigated:

- Check the VA rating of the transformer.
- Calculate the secondary amperage rating.
- Measure the secondary amperage.

TECH TIP

Measuring very low secondary current can be a challenge. The amperage flow is usually 1 or 2 amps, possibly less. If your clamp-on ammeter does not measure low amperage, you can convert your existing ammeter into one that does measure low amperage. Wrap one jaw of the meter with a wire that comes off one side of the transformer as shown in Figure 8-14. If you make 10 wraps around the jaw of the meter, the meter reading will be divided by 10. Wrapping the wire increases the magnetic lines of force being induced by the laminated iron plates in the jaws. The meter sees a higher magnetic field and registers that as a higher current flow. You are fooling the meter into reading a false high-amperage condition. Dividing the reading by the number of wire wraps gives you the correct low-amperage draw.

Some clamp-on ammeters can measure less than 1 amp. Most clamp-on ammeters, however, cannot measure very low amperage, so you must use the jaw wrap method shown in Figure 8-14.

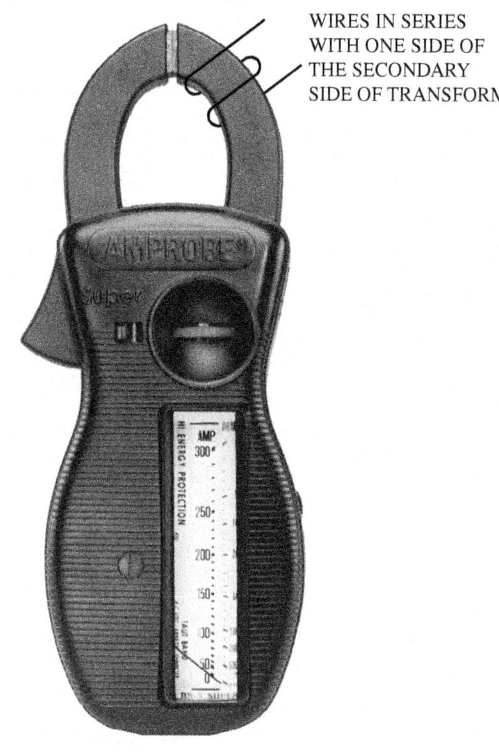

Figure 8-14 Low current can be measured with a clamp-on ammeter by wrapping the jaws and dividing the reading by the number of wraps.

8.4 CHECKING INPUT VOLTAGE ON A TRANSFORMER

The first step in troubleshooting a transformer is to check the input voltage. No input, no output!

Figure 8-15 illustrates this basic troubleshooting step with a transformer wired into a circuit. The primary voltage should be within ±10% of the rated voltage. Therefore, if the input voltage is 120 V, the input voltage should be no less than 108 V and no more than 132 V. How did we come up with those voltages? Let's work through the next example to find out.

EXAMPLE 8.2 DETERMINING TRANSFORMER INPUT VOLTAGE

Determine the primary voltage if the input voltage is 120 V.

SOLUTION:

To find 10% of 120 volts, we need to solve this equation:

$$120\text{ V} \times 10\% = ?$$

First, convert the 10% to a decimal amount by moving the decimal point that would be between the number 0 and the percentage sign (%) two spaces to the left:

$$120\text{ V} \times 0.10 = ?$$

Next, multiply the voltage and the converted decimal amount:

$$120\text{ V} \times 0.10 = 12\text{ V}$$

Now subtract 12 V from 120 V to find the low end of the input voltage range. Then add 12 V to 120 V to get the high end of the input voltage range:

$$120\text{ V} - 12\text{ V} = 108\text{ V (low end of input voltage range)}$$
$$120\text{ V} + 12\text{ V} = 132\text{ V (high end of input voltage range)}$$

A transformer with a primary of 120 V should have input voltage in a range of 108 to 132 V. The voltage is measured at the input voltage to the transformer. The transformer primary should be connected to live power when testing; this includes the secondary side with a connected load.

Low input voltage will create low output or low secondary voltage. High input voltage will create high output voltage. Having a slightly higher input voltage is usually beneficial to a piece of equipment. In a magnetically inductive circuit, like a transformer or motor, the voltage and amperage are opposites of each other. As the voltage goes up, the current draw goes down. As the voltage goes down, the current draw goes up. A somewhat above-average voltage will translate into lower operating amperage. Lower operating amperage will reduce the amount of heat produced by the electrical device. Lower heat levels should translate into longer component life.

Here are voltage ranges for 208-, 240-, and 480-V transformers when assuming an input ±10% rating:

208-V rating = 187 to 229 V

240-V rating = 216 to 264 V

480-V rating = 432 to 528 V

If the correct input voltage is present when the transformer is hooked to the power supply, the correct secondary voltage output should be present. If the secondary is not present, the transformer is defective with most likely an open primary or secondary winding.

8.5 CHECKING OUTPUT VOLTAGE ON A TRANSFORMER

Now let's look at how to check the transformer output voltage under load, as shown in Figure 8-15. "Under load" means that the transformer should be providing power to the coil of a relay, contactor, or other energy-consuming device. The secondary voltage may drop when placed under load. It should not drop to lower than 10% of the rated output.

EXAMPLE 8.3 DETERMINING ALLOWABLE SECONDARY VOLTAGE DROP

Determine the allowable secondary voltage drop of a 24-V transformer.

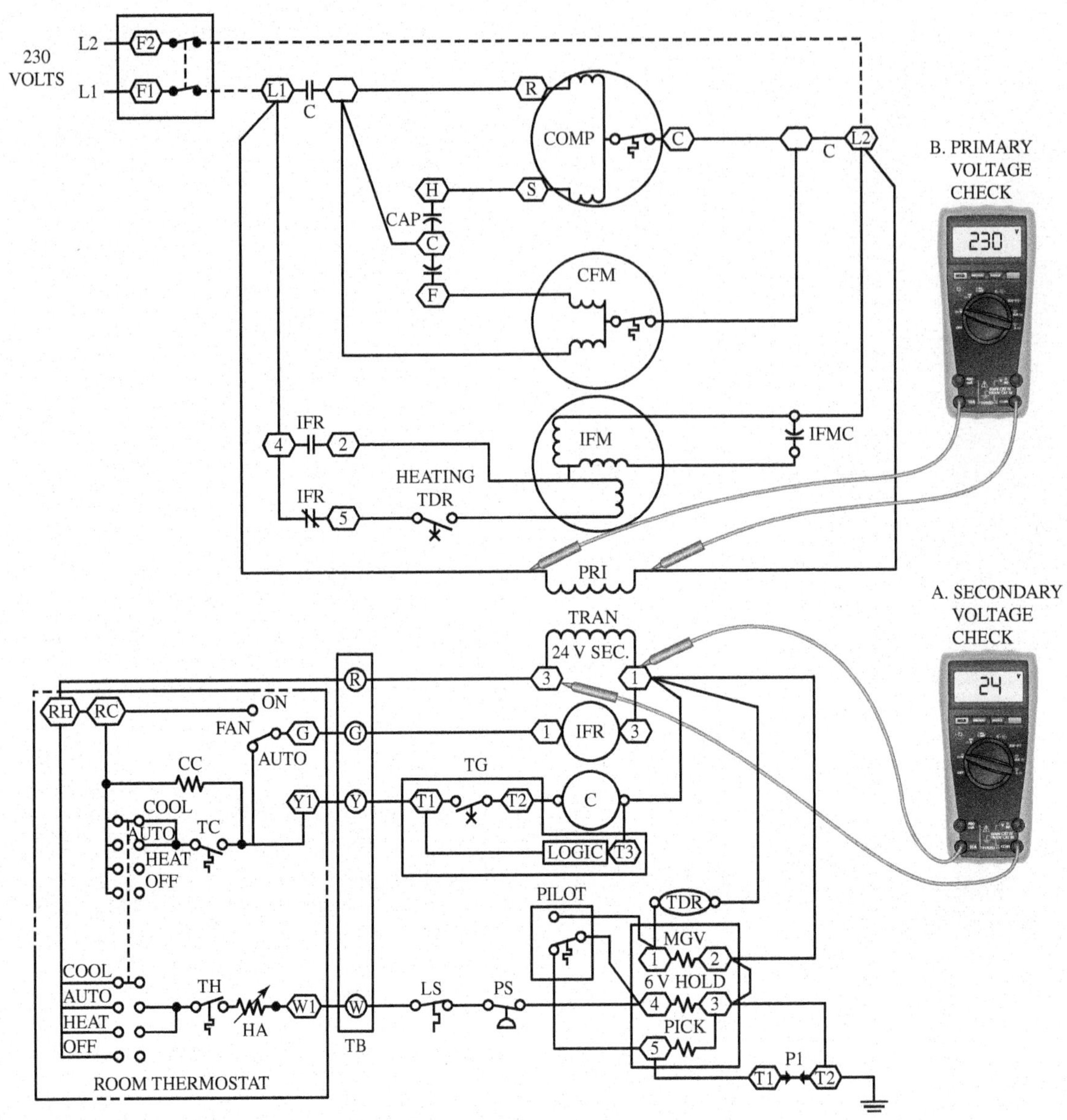

Figure 8-15 This diagram shows a quick way to determine if the system has an incoming voltage problem or control voltage problem. If the primary voltage on the transformer is good, check the voltage on the secondary side.

SOLUTION:

For a 24-V transformer, the voltage should drop no lower than

$$24\text{ V} \times 10\%\ (0.10) = 2.4\text{ V}$$

So the voltage must be at least

$$24\text{ V} - 2.4\text{ V} = 21.6\text{ V}$$

8.6 MORE TRANSFORMER TIPS

A final way to troubleshoot a transformer is to use an ohmmeter. There should be resistance on the primary and secondary of the transformer. The resistance is usually below a couple of hundred ohms. Use the low ohm scale on the meter, R × 1. It is unusual for a transformer to short to ground, but it can happen. Check the primary and secondary windings to the transformer's steel-laminated plates or the transformer housing. Some transformers are coated with a protective coating that looks like a varnish. Scrape the coating off and expose an area of bare steel when checking from the winding to ground.

Use the touch test as a troubleshooting tool. A transformer will be slightly warm when fully loaded. It should never be hot to the touch. Hot transformers will soon burn out.

When checking a transformer, remember the importance of the "turns ratio" we discussed earlier in this unit. For example, a 240-V primary step-down transformer with a 24-V secondary will have an approximately 10:1 resistance ratio. In other words, the resistance of the primary will be about 10 times greater than that of the secondary. If this ratio does not match, there may be a winding-to-winding short. Note that the ratio will not be exact, but it should be close.

You will notice that in Figure 8-9 the higher voltage hook-ups have more windings in series and, therefore, more resistance than the lower voltage windings.

Finally, when wiring in fuse or breaker protection on the secondary side of a transformer, hook it to the "hot side," not the common side of the transformer. The "hot side" goes to the R connection on the thermostat. The other side of the transformer is the common side. The common side may or may not be grounded. The word *common* does not necessarily mean ground or negative. It means that there are several "common" connection points.

TECH TIP

Hook up a temporary fuse in the secondary side of a replacement transformer. The transformer may have burned out because of a short circuit or overloaded condition on the secondary side of the control voltage. Prior to wiring up the secondary of a new transformer, "ohm" the control circuit. This can be accomplished by hooking the ohmmeter to the two wires that go to the control circuit that would normally be connected to the secondary side of the transformer. If there is a short-circuit condition, the resistance will be near zero ohms. The secondary fuse should be sized for short-circuit protection. Size the fuse for 150% of the secondary amperage rating. For the 40-VA transformer, with a maximum amperage draw of 1.67 amps, the fuse size should be 3 A. Many manufacturers use a 5-A fuse to protect against a short-circuit condition. This will not, however, protect against an overload condition, which is slightly above the maximum rated amp condition. Some transformers, like the one shown in Figure 8-16, have built-in circuit protection. This is a resettable circuit protection, preventing damage to the transformer windings in an overload or short-circuit condition.

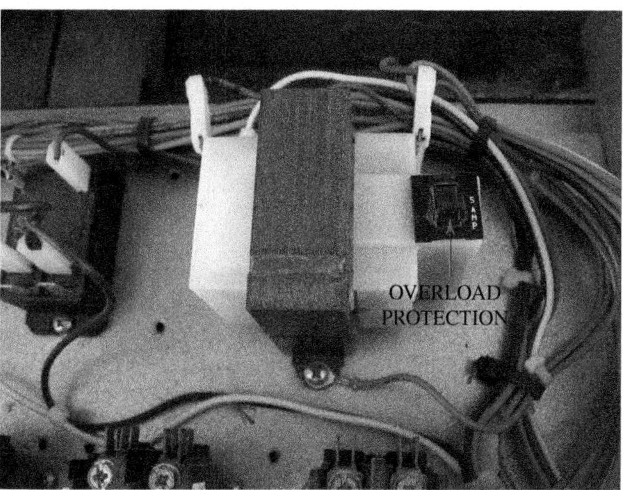

Figure 8-16 This control transformer has built-in overcurrent protection. The 5-amp overcurrent device protects the secondary side of the transformer winding from overloads or short-circuit conditions.

8.7 WIRING TRANSFORMERS IN PARALLEL

Transformers can be wired in parallel to increase the total VA rating. This is also known as "phasing transformers." Two 24-V, 40-VA transformers wired in parallel will have an 80-VA rating. As the load on a transformer increases, the VA rating needs to increase or the transformer will overheat and burn out. You can change out the transformer and install a higher rated VA transformer or wire another transformer in parallel.

It is important to understand how to wire the two transformers in parallel. Wiring the transformer incorrectly will double the secondary output voltage. For example, a 24-V output can become 48 volts. The higher voltage output will burn out coils and other loads that require 24 V. Here are the steps to successfully wire the secondary windings without creating a double voltage condition. Figure 8-17 illustrates how two transformers are wired in parallel.

1. Hook up the appropriate primary voltage to the two transformers.
2. Hook up one secondary wire of transformer 1 to one lead on transformer 2. This can be done on a test basis using a set of jumper cables (alligator clips with connecting wire) prior to the final wiring job.
3. Turn on the power to the transformers and read the voltage between the remaining unconnected leads, as shown in Figure 8-17. When powered up, do not allow the secondary wires to touch each other or ground. This will damage the transformer.
4. If the voltmeter reads "0 V" when connected, the connection is correct. Remove the voltmeter, turn off the power, and connect the final two secondary transformer leads together.
5. If the voltmeter in Figure 8-17 reads double the rated voltage, reverse the secondary leads. You will now read 0 V. Hook the leads together and measure 24 V. Connect the two secondary leads to the control circuit.

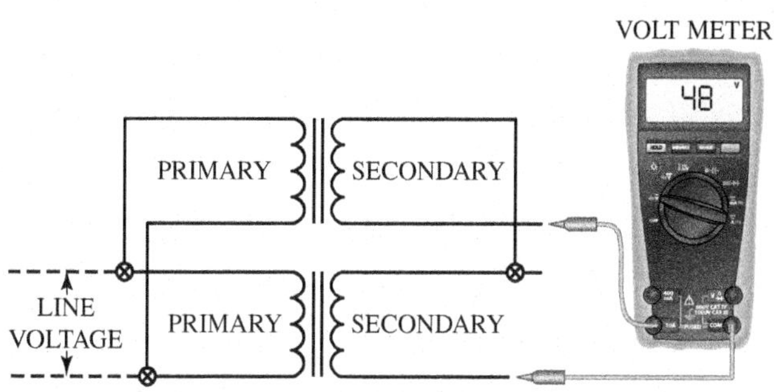

Figure 8-17 Phasing or wiring the secondaries of transformers in parallel is the intermediate step in measuring the final connection.

6. Finally, double check the power to the paralleled secondary winding. If the voltage is 48 V, disconnect the primary power and switch the wiring on secondary.

TECH TIP

Do not "spark" a transformer. Sparking a transformer is the practice of quickly touching the bare wires of the secondary side of a transformer together to see if it creates an electrical spark. Some techs use this as a way of determining that the transformer has secondary voltage. This may be a way to bypass using a voltmeter to measure the secondary voltage, but it is not a good practice. A transformer with an internal or external fuse will blow even with a quick touch or spark. Sparking a transformer is a poor substitute for a voltmeter. A spark can be created at lower, unacceptable voltage. Be a professional and use a professional's instrument for measuring voltage.

TECH TIP

Some transformers have a built-in secondary transformer fuse that will not be obvious to the tech. This type of transformer will be identified as "fused" or "internally fused." In some instances you can remove part of the winding cover, which is paper or plastic, and expose an open fuse link. Wire in an external fuse and continue to use the transformer. This should be considered a temporary fix. Replace the transformer with a new one.

Some protected transformers have a built-in circuit breaker. Reset the breaker with the power off. Power up and try to determine the reason the breaker tripped. A temporary low-voltage situation will create a high-amperage condition that can trip the breaker. By the time you arrive at the job site, the low-voltage condition may have disappeared, but the breaker has been tripped. Reset and test.

SERVICE TICKET

You are called to troubleshoot a package unit (see Figure 8-18) that is not cooling. You quickly determine that the control transformer is burned out. The transformer appears overheated and the secondary winding is open. Replacing the transformer, you install a temporary, 5-amp, inline circuit breaker similar to one shown in Figure 8-19. The purpose of the inline breaker is to protect the secondary of the transformer in case there is a short or overload in the control circuit. You plan to remove the temporary inline breaker after an operational check. As soon as the power is applied, the inline breaker trips. An inspection of the control wiring and control components does not reveal an obvious problem.

Troubleshooting any HVACR system with a control voltage short is always a challenge. You start with these steps:

1. For the package unit shown in Figure 8-18, the first step in determining a short in the control voltage circuit is to disconnect everything from the left side of the transformer, at point 1. Powering up the primary of the transformer without it being connected in the circuit should not short out the secondary unless this transformer windings is already damaged.
2. Using the diagram in Figure 8-18 as our example, connect one control voltage circuit at a time until the inline circuit breaker trips. The inline circuit breaker will be installed, in series, between the left side of the transformer and the **R** connection on the thermostat. On the left side of the diagram, at point 1, hook up the **R** wire to the transformer. The right side of the transformer will also be connected. The circuit breaker does not trip.
3. Next, close the system switch at point 2 that follows the **R** connection. This is the **On** and **Off** switch found on some thermostats. The circuit breaker does not trip.
4. Now switch (close) the fan switch to **Auto,** point 3. The circuit breaker does not trip.
5. The (IFR) coil and contactor coil (C) have been disconnected as part of the troubleshooting process. Next, you turn the thermostat temperature low enough so that the cooling circuit is closed, point 4. The circuit breaker does not trip.
6. When you attach the (IFR) coil, at point 5, the breaker does not trip.
7. Finally, connecting the contactor (C) circuit, at point 6, trips the breaker. You have established that the short circuit is in the contactor circuit. "Ohm out" the contactor coil or look for shorted wiring in this line of the diagram. The ohmmeter indicates that the contactor coil measures 2 Ω, which is very low for a contactor coil. Replacing the contactor solves the problem.

Finding a shorted component or wire can be a real test of your "figuring it out" skills. We just completed an exercise in finding a short-circuit problem. Eliminate as many circuits as possible by disconnecting the circuit from the transformer. Reattach one circuit at a time and apply power after each circuit is applied.

TECH TIP

On older air handlers and furnaces, transformers are combined with a relay. This is called a fan center. A fan center is shown in Figure 8-20. The transformer may be permanently mounted to a plate on the fan center. A burned-out transformer will need to be replaced with a new fan center or by mounting another transformer near the fan center. The relay is usually plugged into the fan center and can easily be replaced if defective. The diagram in Figure 8-21 shows the electrical hook-up for a fan center showing the transformer and fan relay.

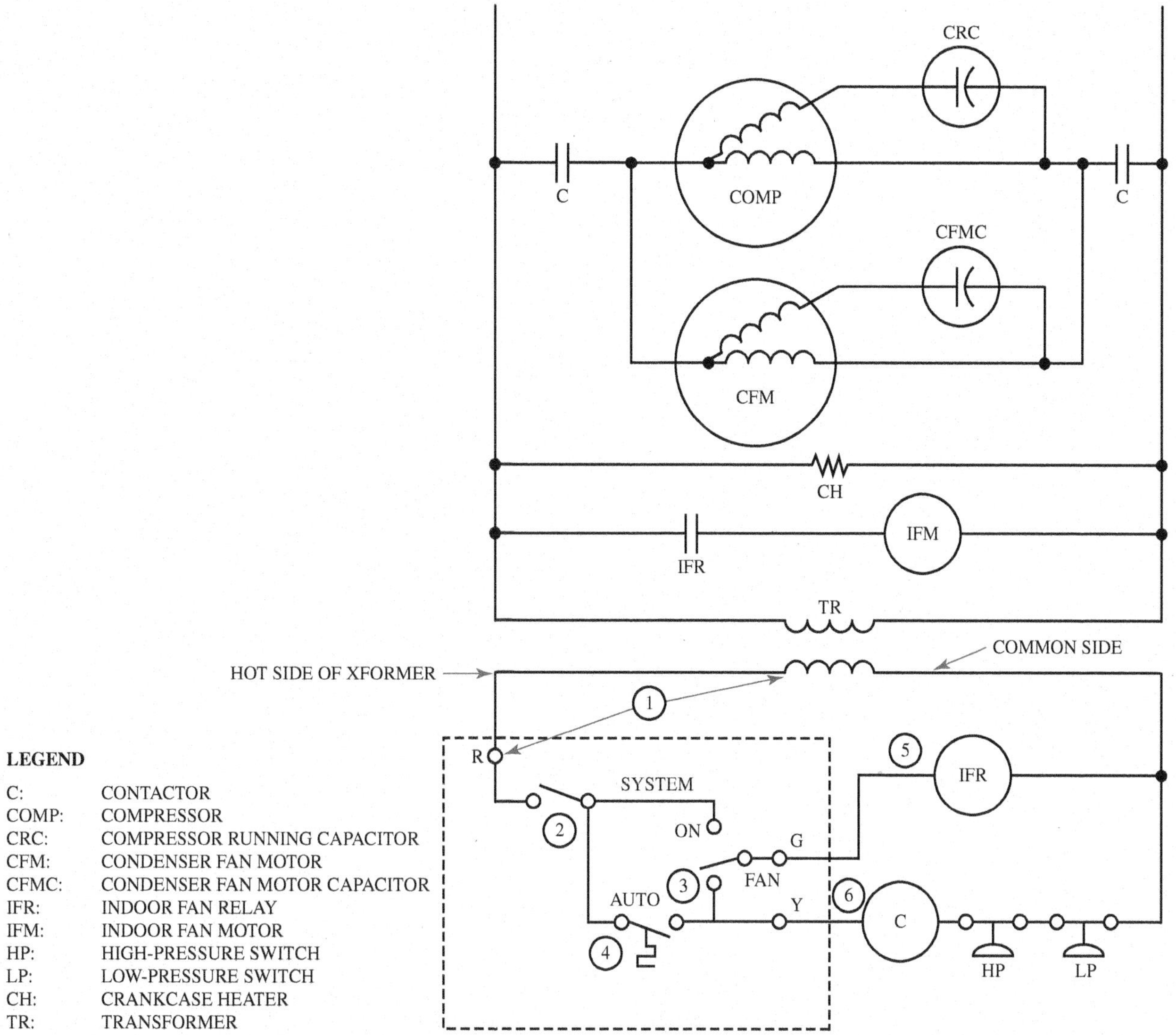

CRC
COMP
C
C
CFMC
CFM
CH
IFR
IFM
TR
HOT SIDE OF XFORMER
COMMON SIDE
1
R
SYSTEM
2
ON
G
3
FAN
AUTO
Y
6
C
4
HP
LP
5
IFR
LEGEND
C: CONTACTOR
COMP: COMPRESSOR
CRC: COMPRESSOR RUNNING CAPACITOR
CFM: CONDENSER FAN MOTOR
CFMC: CONDENSER FAN MOTOR CAPACITOR
IFR: INDOOR FAN RELAY
IFM: INDOOR FAN MOTOR
HP: HIGH-PRESSURE SWITCH
LP: LOW-PRESSURE SWITCH
CH: CRANKCASE HEATER
TR: TRANSFORMER

Figure 8-18 The package unit shown here has a short in the control voltage side of the system. The diagram does not show the short. Troubleshooting steps are necessary.

HOOK UP WIRES
RESET BUTTON

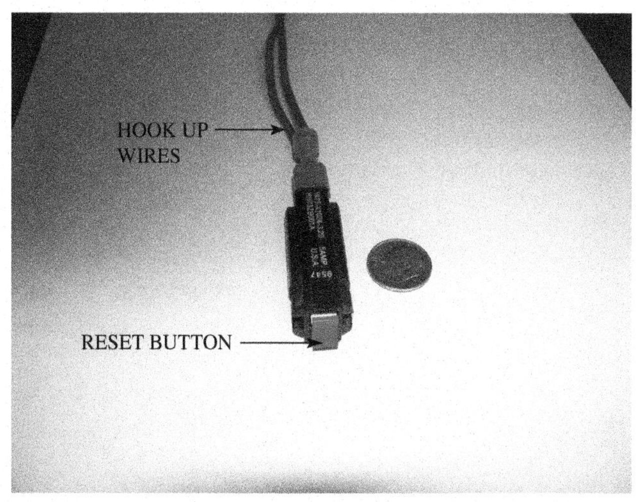

Figure 8-19 This is a temporary, 5-amp, inline circuit breaker. This should be placed on the "hot side" of the secondary of the transformer to protect it against overcurrent or short-circuit conditions.

TRANSFORMER
RELAY

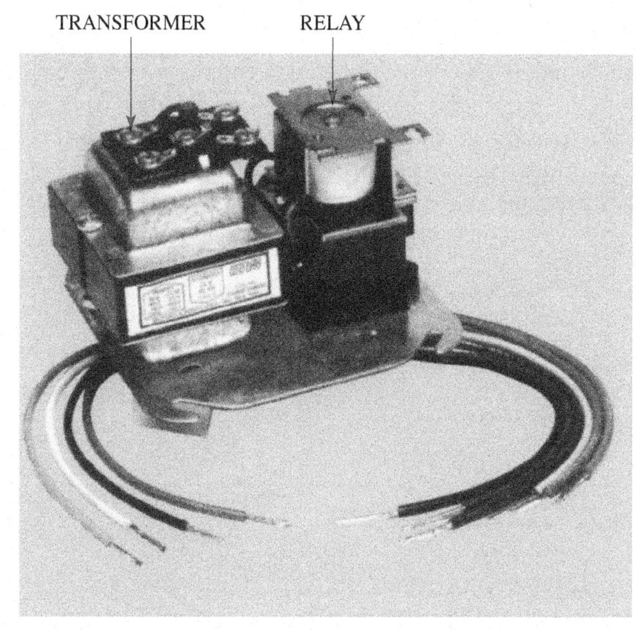

Figure 8-20 The fan center has a combined transformer and fan relay.

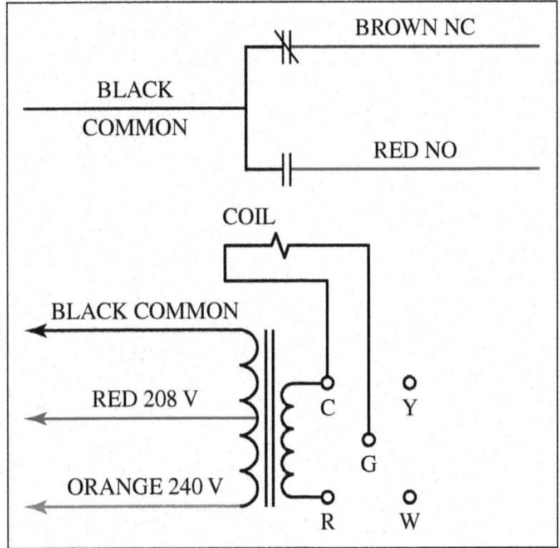

Figure 8-21 This diagram of a fan center shows the symbols for the transformer, relay coil, and relay contacts. The R, C, G, W, and Y screw connects are on the transformer.

SUMMARY

It is important to understand how transformers operate because they are one of the first pieces of equipment that should be checked when troubleshooting. If the input voltage is present, you can determine that part or all of the system has voltage. If the output voltage is present, you can continue the troubleshooting process on the control voltage side. This troubleshooting process will depend on the problem and may not be the most expedient way to proceed. It is a place to start if you do not know anything about the circuit operation.

The first step to checking a transformer is determining if there is input voltage. Next check the output voltage. Some transformers burn out due to unknown reasons. Some burn out due to overloading (too much current draw) on the secondary side. Others burn out due to shorted components in the secondary side of the electrical circuit. Check for circuit overloading by measuring the amp draw on the secondary side of the transformer. Check for short circuits or low-resistance circuits with an ohmmeter prior to attaching the secondary of the transformer. Place a properly sized fuse in the secondary of a replacement transformer. This can be a temporary fuse that you can remove and wire directly or better yet a permanent fuse that will protect the transformer from a short-circuit condition. Size the fuse at 150% of the secondary amperage rating to prevent nuisance blown fuses yet still protect against a short-circuit condition. This will not protect against an overload condition. An overload condition is one in which a higher than normal amp draw is occurring, which will overheat the transformer and eventually burn it out. Use the touch test. A transformer will be slightly warm when fully loaded. It should never be hot to the touch. Hot transformers will soon burn out.

REVIEW QUESTIONS

1. Describe how a transformer works.
2. Name the three common types of transformers.
3. What are four common transformer input voltages?
4. What is the most common step-down transformer voltage output?
5. Does the amperage go up or down in a step-down transformer?
6. What is the maximum amperage output of a 75-VA transformer on a 24-V secondary?
7. What information is required when ordering a replacement transformer?
8. What size fuse would you install in the secondary of a 40-VA, 24-V transformer to protect against a short-circuit condition?
9. Your clamp-on ammeter measures no lower than 5 amps. How would you measure amperage on the secondary of a 24-VA transformer?
10. The technician phases two 24-V, 40-VA transformers to increase the rating to 80 VA. After the technician wires the secondary of the transformer, 48 volts is measured. What does the technician need to do to obtain the desired 24-V output?
11. List three ways to troubleshoot a transformer.
12. What is the advantage of a multi-tap transformer?
13. What is the purpose of an isolation transformer?
14. Which transformer will increase the output voltage and decrease the output amperage?
15. A technician wants to increase the voltage output of a transformer. The tech wires up several input voltage leads with the hope of obtaining increased voltage. Will the tech get increased voltage? If not, what will happen to the transformer?
16. What type of transformer is shown in Figure 8-22? What is its voltage input?

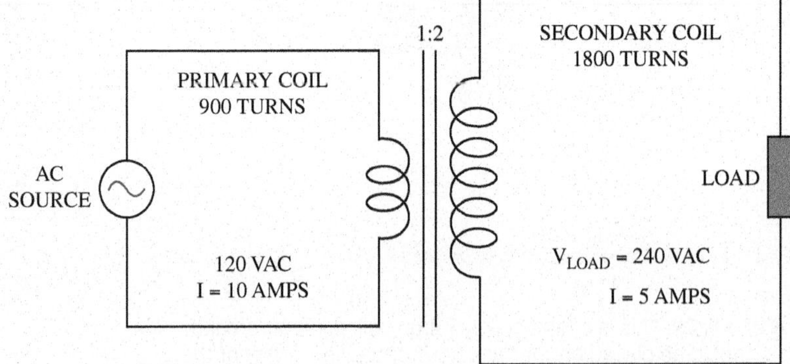

Figure 8-22 Use this diagram to answer Review Questions 16 and 17.

17. In the transformer shown in Figure 8-22, what is the maximum amperage that the secondary can handle?
18. What type of transformer is shown in Figure 8-23? What is its voltage output?
19. In the transformer shown in Figure 8-23, what is the maximum amperage that the secondary can handle?
20. What wire connections are used to hook up a 480-V commercial transformer?

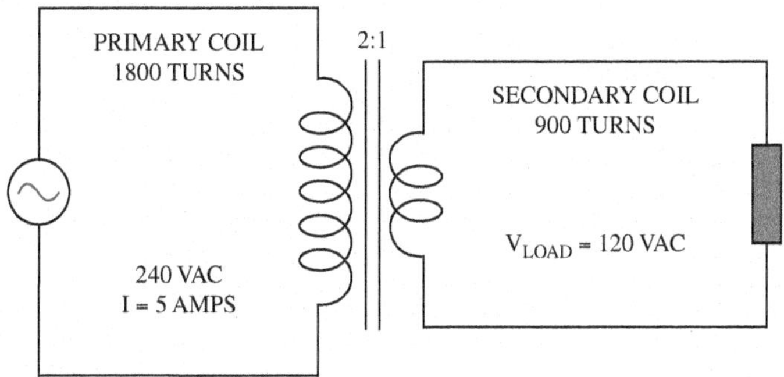

Figure 8-23 Use this diagram to answer Review Questions 18 and 19.

UNIT 9

Relays, Contactors, and Motor Starters

WHAT YOU NEED TO KNOW

After studying this unit, you will be able to:

1. Describe the operation of a relay.
2. Troubleshoot a relay.
3. Describe the operation of a contactor.
4. Troubleshoot a contactor.
5. Describe the operation of a motor starter.
6. Troubleshoot a motor starter.
7. Select a replacement relay, contactor, and motor starter.

The purpose of this unit is to describe the similarities and differences among a relay, a contactor, and a motor starter. **Relays**, **contactors**, and **motor starters** are used to control, that is, turn on and off, an electrical circuit. They are all electric switches. They have a coil of wire that is used to magnetically and mechanically open or close a set of contacts (switches). The main difference between the relay and the contactor or motor starter is the rated amperage that can run through the contacts or switches. Relays are low-amperage devices, whereas contactors and motor starters can handle high amperage through their contacts. The difference between a contactor and motor starter is that motor starters have an overload protection device installed. The **overload protection** device opens up the control circuit when excessive current is measured to the load. One common use of these devices in the HVACR industry is to start fan or pump motors and compressors.

There is another class of relays that help motors start and/or run. These very specific types of relays are associated with specific types of motors. They will be discussed in Unit 15 on motors. They go by the names of centrifugal switch, potential relay, current relay, and solid-state relay.

This unit also describes how to troubleshoot these switching devices and what information is needed to order a replacement part. We start by explaining the standard features found in all of these types of switching devices.

9.1 COMMON FEATURES

The relay, contactor, and motor starter have many common features. This section explores these similarities.

Coil Voltage

All of these control devices have a coil of wire that is energized to magnetically open or close a set of contacts. The coil will energize a magnetic field and the magnetic force will mechanically move the contacts when the proper voltage is applied. A simplified illustration of this is shown in Figure 9-1. In Figure 9-1A, the coil is de-energized and the contacts are held closed by upward spring force. This would be called a normally closed (NC) relay since the contacts are closed when the relay coil is de-energized. The coil in Figure 9-1B is energized and the contacts are opened by the magnetic attraction of the coil. These are called electromechanical devices because they rely on the magnetic field generated by current flow to mechanically move a set of contacts. The spring is a mechanical force that causes a set of contacts to open or close. Some contacts are gravity operated in the de-energized mode. They will close by the force of gravity. These

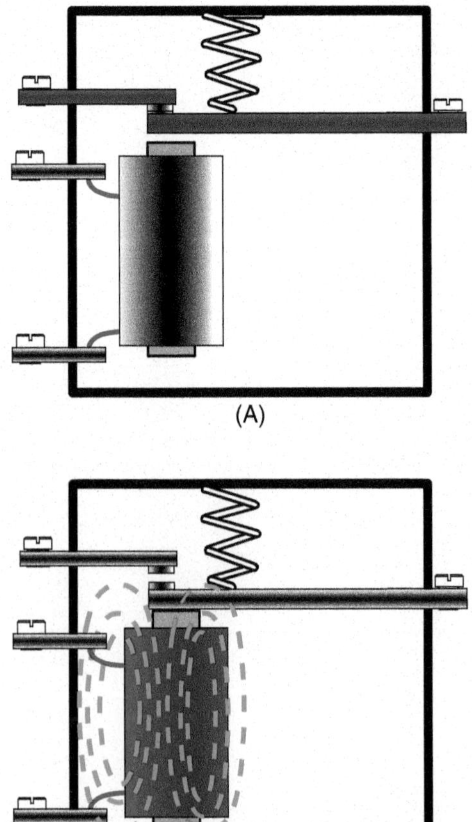

Figure 9-1 (A) This is an example of normally closed contacts. (B) Here the normally closed contacts are shown switching position when the coil is energized.

types of contacts must be mounted in a specific direction to make use of the gravity field.

Common coil control voltages are 24, 120, 208, or 240 volts. If the control voltage is too low, it will not generate enough magnetic force to change the position of the device. For example, if a 24-V coil has 18 volts applied to it, it may not operate the device. Low voltage on the coil could cause the mechanism to "**chatter.**" Chatter occurs when the voltage to the coil is low and the coil creates just enough magnetism to operate the mechanism. The word *chatter* comes from the rapid closing and opening of the contacts, which creates a chattering noise (like the chattering of teeth when a person is cold). Very low voltage will not create chatter since there is not enough magnetic force to operate the contacts. A chatter relay is unnerving and noisy.

High voltage on the coil will cause the coil to overheat and burnout. Sometimes voltages in a system get crossed. For example, a 24-V control circuit accidentally comes in contact with 120 volts. The higher voltage will burn out the 24-V coil quickly. Check the coil voltage before replacing a component that has a burned-out coil.

Number of Contacts

Each device has one or more movable **contacts** or switches. The contacts may be normally open (NO) or normally closed (NC). They are represented in electrical diagrams by the symbols shown in Figure 9-2. Figure 9-2 show the relay symbol for one normally open contact, one normally closed contact, and a relay coil. (Other symbols for a coil are also used, such as a circle or the letter *C* inside a circle.)

Replacement components will need to be selected to meet the number of required contacts. For example, a relay may have a normally open (NO) and a normally closed (NC) set of contacts. One or more contacts may be open, while other contacts may be closed. It is acceptable to replace a component that has one contact with one that has extra contacts; the extra contacts would just not be used.

Rating of Contacts

Contacts are two semi-rounded surfaces that strike together to close and create a complete circuit. The contacts are usually silver plated for good electrical conduction and long life. The contacts are rated by the voltage and amperage they can carry.

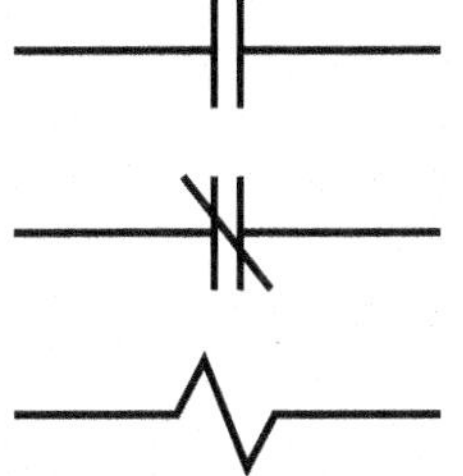

Figure 9-2 These are the symbols for a simple relay with one normally open contact (top), one normally closed contact (middle), and a relay coil (bottom).

Table 9-1 Basic Rating Information for a Relay

The label found on the side of a basic relay includes important information such as coil voltage, contact amperage rating, and maximum contact amperage rating. Manufacturers may arrange this information differently, but the coil voltage, rated load amps, and operating voltage are all important parts of selecting a relay.

Relay #ABC123 24-Volt Coil Low-Current, Magnetic Switch Contact Ratings		
AFL	**ALR**	**Voltage**
11 amps	73 amps	120 volts
5.5 amps	35 amps	277 volts
3 amps	17 amps	480 volts
16 amps resistive	—	277 volts
10 amps resistive	—	480 volts

A typical label, like that shown in Table 9-1 for Relay #ABC123, provides important information about the relay. Using the label we can determine the following information: It has a 24-V coil and is listed as a low-current, magnetic switch. The three columns are contact amperage ratings. These are per pole or per contact set ratings. Starting with the left column, the contact ratings is stated as **AFL** or **amps full load.** This is also known as rated load amps or RLA. This is the maximum amperage operating through the contacts at 120 volts. The 120-V rating is shown in the far right column. A load operating above this amperage level may weld the contacts closed or simply burn out the contacts.

The middle column is listed as ALR or amps locked rotor. This is commonly known as locked rotor amps (LRA). The phrase *locked rotor amps* refers to the condition in which very high amperage draw is occurring. This condition is found when the motor first starts or when the motor is mechanically locked up and cannot rotate. You will notice that the ALR is about six times higher than the AFL. It is normal for the starting amps or ALR to be this high for a brief second on start-up. If the motor stays at the locked rotor amperage draw for much longer than a few seconds, the overload protection device will open the circuit and de-energize the coil or power to the motor windings. A relay like this may have two or more normally open contacts and two or more normally closed contacts. These options allow for many different control options.

Table 9-2 illustrates common information found on a contactor. This is for a double-pole generic contactor, Contactor #XYZ123. As shown on the label in Table 9-2, the coil operates on 24 volts and is listed as a high-current, magnetic switch. The information found on the nameplate is a per pole contact rating. Table 9-2 has three columns that relate

Table 9-2 Basic Rating Information for a Common Two-Pole Contactor

The label found on the side of a common two-pole contactor shows the coil voltage and operating current conditions under various voltage inputs to the load. AFL means amps full load or maximum running amps. ALR means locked rotor amps, also known as starting amps.

Contactor #XYZ123
24-Volt Coil
High-Current, Magnetic Switch
Per Pole Contact Ratings

AFL	ALR	Voltage
28 amps	120 amps	277 volts
12 amps	55 amps	480 volts
8 amps	34 amps	600 volts
48 amps resistive	--------	277 volts
40 amps resistive	--------	480/600 volts

to the voltage and amperage draw allowed on each pole. The far right column shows the range of volts. For example, the first voltage row is 0 to 277 volts. The second row is 278 to 480 volts, the third row is 481 to 600 volts, and so on. AFL refers amps full load or the amps the system operates at when running at normal capacity. This is commonly stated as rated load amps (RLA).

The center column shows the ALR or amps locked rotor. As with the relay, you will notice that the ALR is about five or six times higher than the AFL. As mentioned earlier, it is normal for the starting amps or ALR to be this high for a brief second on start-up. Remember, however, that if the motor stays at the locked rotor amperage condition for much longer than a few seconds, the overload protection device will open the circuit, preventing the motor from starting.

The bottom two rows in Table 9-2 deal with amp draw in resistance amperage (amp res) circuits. A resistance amps circuit is a circuit that handles electric strip heat or incandescent light loads. Most of the loads in HVACR are inductive, not resistive. Inductive loads include motors, relay coils, or transformers. Inductive loads draw high amounts of starting amperage compared to resistive loads.

Selecting the right amperage rating for the contacts is important. You can install a relay or contactor with a higher amperage rating if necessary. A high-amperage contactor will cost more because the contacts are larger and designed to handle higher amperage.

In summary, to select a replacement relay or contactor, you need to know the coil voltage, number of contacts, rated amperage rating on the contacts, and the type of load, whether it is resistive or inductive. Figure 9-3 shows a rating label on the side of a contactor. You will need to know the operating voltage and operating amperage to select the correct replacement. A larger amp rating is acceptable. A replacement contactor with a smaller amp rating will quickly burn out the contacts. The label also usually states the coil voltage. It lists information similar to that given in Table 9-2.

9.2 DIFFERENCES AMONG RELAYS, CONTACTORS, AND MOTOR STARTERS

The major difference between a relay and a contactor is the amount of current a relay contact can handle. A relay is considered to be a low-current switch. Low current in this case means the relay contacts handle less than 15 amps. Contactors and motor starters handle much higher amperage. Relays normally have one or more normally open

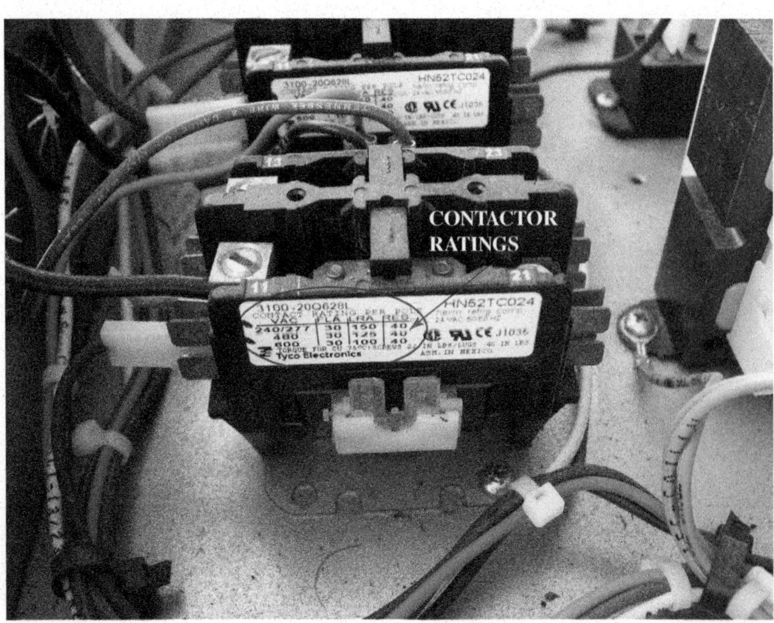

Figure 9-3 Note the rating label on the side of this contactor. A closer view is shown in Table 9-2.

Figure 9-4 Top view of a common relay. Some contacts are NO and some are NC.

(NO) contacts and one or more normally closed (NC) contacts. Contactors and motor starters have normally open contacts.

The coil voltage is normally 24, 120, 208, or 240 volts. Figure 9-4 is a common relay used in the HVACR industry. It has NO and NC contacts. Notice the contact symbols embossed on the relay surface. There are NC contacts between connections 1 and 2, and NO contacts between connections 1 and 3.

In Figure 9-5 we saw that the fan relay and contactor are wired without regard to the location of the component parts. This can be a bit confusing since the component parts of the relay or contactor are one, yet they appear to be separate when looking at an electrical diagram like the one shown in Figure 9-5.

Finally, there are several configurations of relays. A common type was introduced in Figure 9-4. Some relays plug into a socket. Many of the socket-type relays have a clear plastic case over them, so you can see the coil and contacts. These are sometimes called ice cube relays since they have a clear case and are relatively cubical.

Another type is more difficult to identify. Those are relays mounted on a circuit board. They are small and not identified by the manufacturer.

9.3 RELAYS

Several things can go wrong with a relay. You will use an ohmmeter and/or voltmeter to troubleshoot this device. When troubleshooting a relay, it is removed from the circuit so that the technician can safely check the relay condition with an ohmmeter. A voltmeter may be used if the ohmmeter does not reveal a problem.

As discussed in Units 26 and 27, the hopscotch troubleshooting method should be used to isolate a problem component. This section shows you how to conduct simple checks on the suspect component.

Check the Coil

Coil resistance is usually low, less than a 1 kΩ (≤1,000 ohms). Most coil resistance is around 100 Ω or less. As you will learn in other units in this text, some specific-purpose relays have a much higher coil resistance, some as high as 6 kΩ. A coil with a few ohms means the coil wire is shorted and will not develop enough magnetic strength to operate the relay contacts. A coil with a few ohms of resistance may create a high amperage condition and burn out the secondary of the transformer. If the coil appears burned, it will be open or shorted. Low resistance in a circuit creates a high amperage draw.

Check the Contacts

Next check the contacts. Normally closed (NC) contacts should measure 0 Ω. Resistance on closed contacts means the contacts have a carbon buildup, which creates resistance even when the contacts are closed. Resistance on closed contacts could also mean that the closing mechanism (spring, magnetism, or gravity) is not holding the contacts tightly closed. Replacement is required.

Normally open (NO) contacts should read infinity. Any resistance indicates that the contacts are mechanically stuck or welded closed. Replacement is required.

Many relays have extra, unused contacts. If a set of contacts becomes defective, another unused set on the relay may be used.

If the ohmmeter check does not reveal a problem, you will need to troubleshoot the relay "hot," meaning with voltage applied. First, measure the voltage applied to the coil. The proper coil voltage will cause the contacts to switch position. In a moderately quiet environment, you can hear the relay click and change position when the coil is energized. Remove the voltage from the side of the coil and place the wire back on the terminal to hear the relay click and change positions. The clicking sound does not guarantee that the contacts are changing position. Low coil voltage will cause the relay to "chatter" or rapidly open and close positions.

When the coil is in a **de-energized state** and with voltage applied to an open contact or switch, a voltmeter will read the applied voltage in that circuit, as shown in Figure 9-6. When the coil is energized, the contacts or switch will close and the voltmeter reading will drop to 0 V (see Figure 9-7).

Next, test the normally closed contacts. When the normal voltage is applied to a closed set of contacts, the voltmeter reading will be 0 V, since a closed contact is the same as measuring voltage on a single piece of wire at the same potential. When the relay coil is energized, the contacts open and the applied voltage will be measured

SEQUENCE OF OPERATION 9.1: RELAY AND CONTACTOR LOCATION

The electrical diagram in Figure 9-5 shows relay and contactor locations. It is important to understand that the contacts and coil are found in one component but are located in different parts of the diagram. The fan relay coil is located in the lower right, control voltage section. It is labeled **IBR** and highlighted in tan (see point **1** in Figure 9-5). The relay

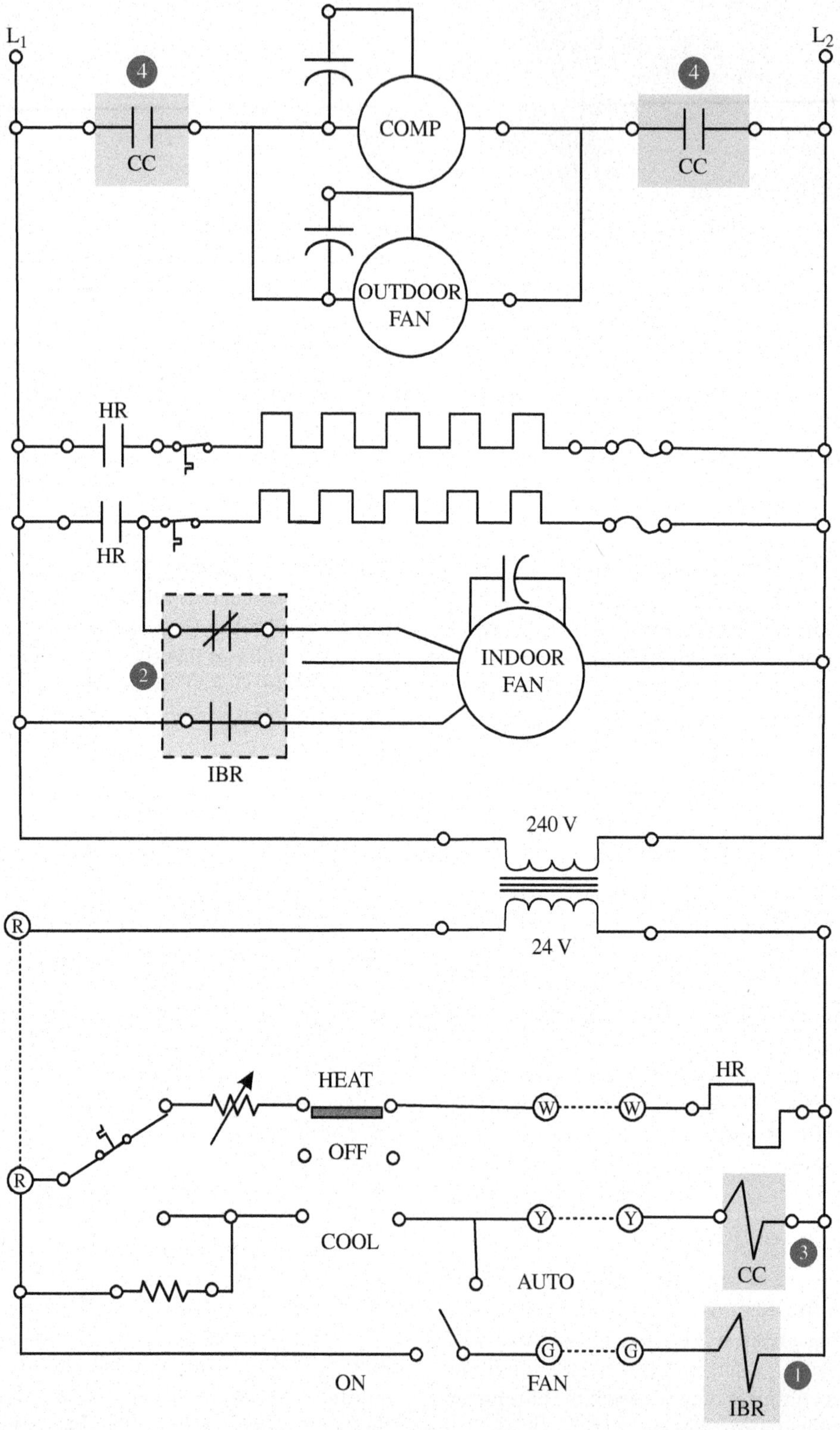

Figure 9-5 Note the relay and contactor locations in this electrical diagram. IBR is the blower relay; CC is the contactor.

(continued)

NO and NC contacts are found in the upper diagram, to the left side of the indoor fan motor, outlined by a dashed rectangle and highlighted in tan (see point **2**). A review of the symbols back in Figure 9-2 indicates that the component is all together in one area. This is true. Figure 9-5 shows how the component is wired without regard to location of the component parts.

1. The fan relay coil is located in the lower part of the diagram, on the right side of the control voltage section. It is labeled **IBR** (see point **1** on Figure 9-5).
2. The relay NO and NC contacts at point **2** are found in the upper part of the diagram, to the left side of the indoor fan motor, outlined by a dashed rectangle and tan shading. When the indoor blower relay **IBR** coil at point **1** is energized, it closes the NO contacts and brings power to operate the indoor fan.
3. Point **3** highlights the contactor coil **CC**, highlighted in blue, found in the lower right side control voltage section.
4. When the coil is energized by 24 volts, it closes the double-pole contacts at the two points labeled **4**.

When the double-pole contacts close the compressor, an outdoor fan motor will begin to operate.

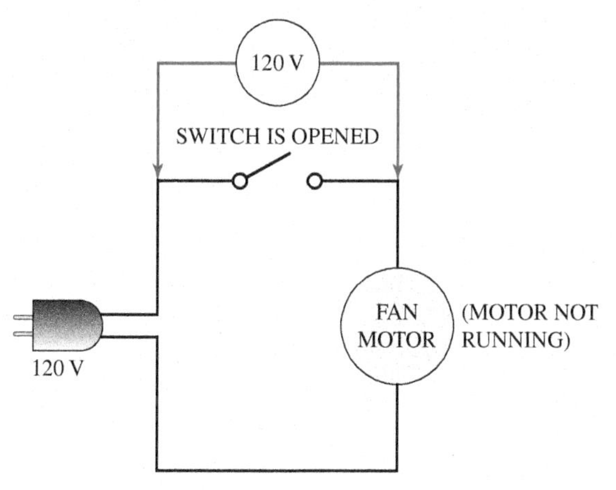

Figure 9-6 A voltmeter will measure the applied voltage across an open wire or open switch.

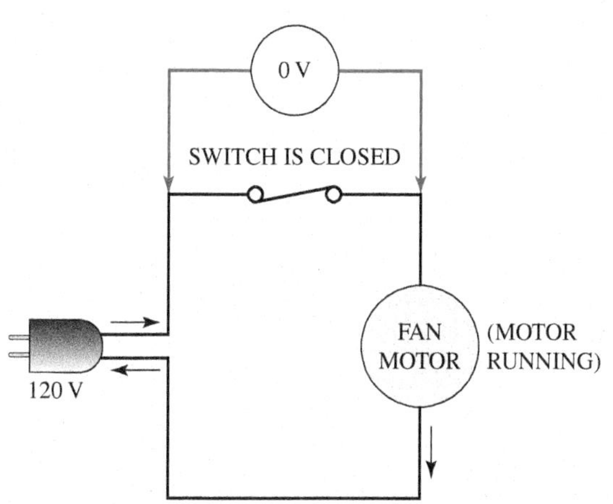

Figure 9-7 There is voltage operating the fan motor and there is voltage in this wire, but the meter measures 0 volts because there is no potential difference across the closed set of contacts.

across the open contacts. Be careful here: Does measuring 0 V across a seemingly closed set of contacts mean the contacts are closed or that there is no voltage to the contacts? The voltmeter may also be defective and not measuring any voltage even though voltage is present and dangerous.

Relay Replacement

When replacing a relay you will need the following information:

- Coil voltage
- Number of normally open contacts required
- Number of normally closed contacts required
- Type of load: inductive or resistive.

9.4 CONTACTORS

Contactors are high-amperage switches. Contactors are used to switch high-amperage loads such as compressors, large motors, or electric heat strips. Figure 9-10 shows two different symbols for a contactor. Contactors have one to four sets of contacts. These are called poles. Figure 9-11 is a **single-pole contactor** used in single-phase circuits. Note that only one of the power circuits is broken. The other power circuit feeds power directly to the other side and is "hot" all the time, even if the contactor is not energized. The unbroken power line has a bar passing voltage from one side to the other without being interrupted. The power bar is near the bottom of this picture. The interrupted power side has two sets of contacts. A flash shield or cover contacts sparks that may develop if the contactor supplies voltage to a short circuit.

TROUBLESHOOTING 9.1: RELAYS

To recap what we just talked about, let's review the major steps.

1. Check the coil with an ohmmeter, as shown in Figure 9-8. The resistance will be below 100 Ω, but not close to 0 Ω.
2. Check the contacts with the coil energized, as illustrated in Figure 9-9. The normally open contacts between 1 and 3 should be closed and measure near 0 Ω.

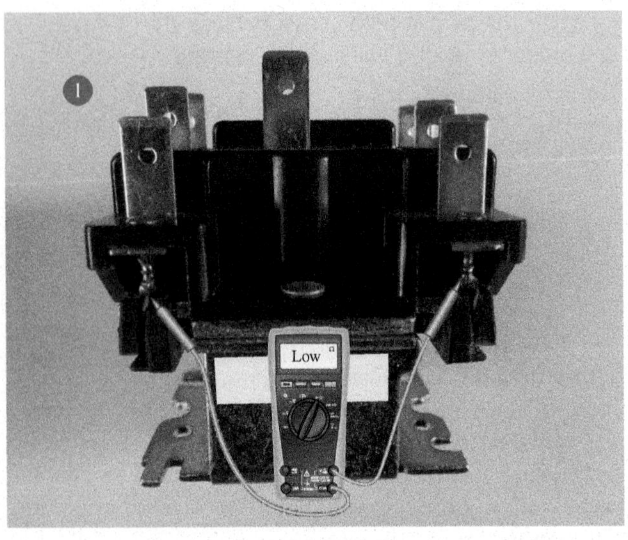

Figure 9-8 Measuring coil resistance. The coil wires are soldered to the points indicated by arrows. Check here or on the connection terminal above the solder joint.

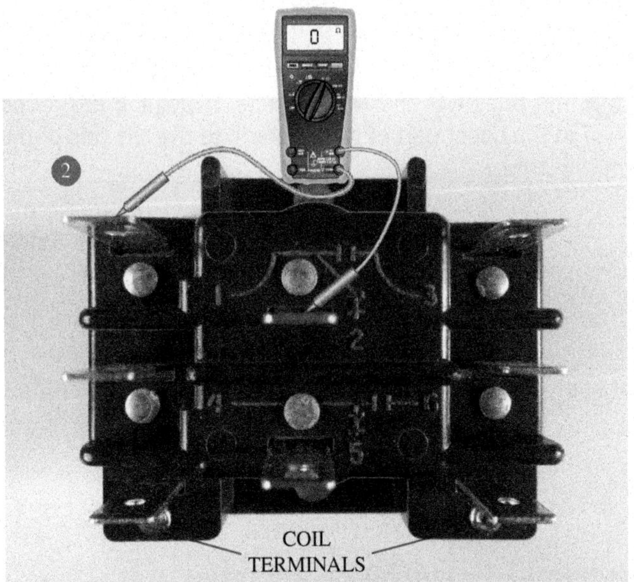

Figure 9-9 Measure the resistance of the contacts. The closed contacts (1–2 or 4–5) should read near zero ohms. No resistance on the open contacts.

SAFETY TIP

The contactor shown in Figure 9-11 has one side powered up at all times. When working around this type of contactor, kill the disconnect so that no power is going to through the contactor. Remember, there can be control voltage going to the coil even when the main power is removed. The control voltage may be coming from a different source. The normal coil voltage for a smaller system is 24 V. Higher voltage coils are sometimes used in commercial and industrial applications. Never assume that a circuit is dead even if the breaker, fuse, or disconnect is pulled. Take other steps to ensure the power is dead before putting your hands into the system.

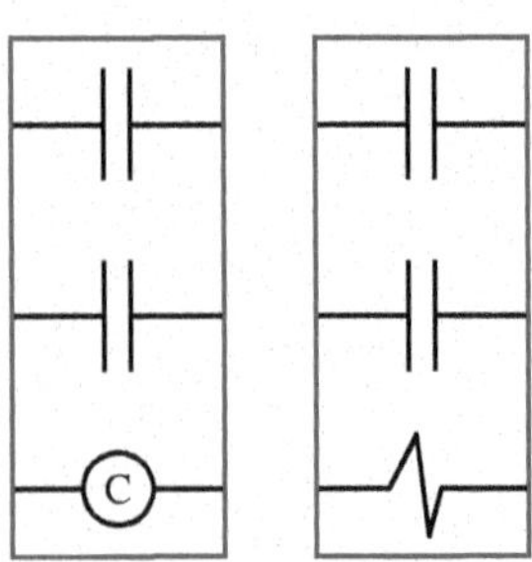

Figure 9-10 Both of these symbols are used to indicate a two-pole contactor. There are two normally open contacts and a coil to close the contacts.

The **two-pole contactor** is the most common single-phase device, and the **three-pole contactor** is used in three-phase circuits. Figure 9-12 shows a double-pole contactor. Figure 9-13 gives a side view of the contactor showing the contactor coil. Some contactors have a hook-up on each side of the contactor as shown here, or have the coil hook-up on the same side of the contactor, as shown in Figure 9-14.

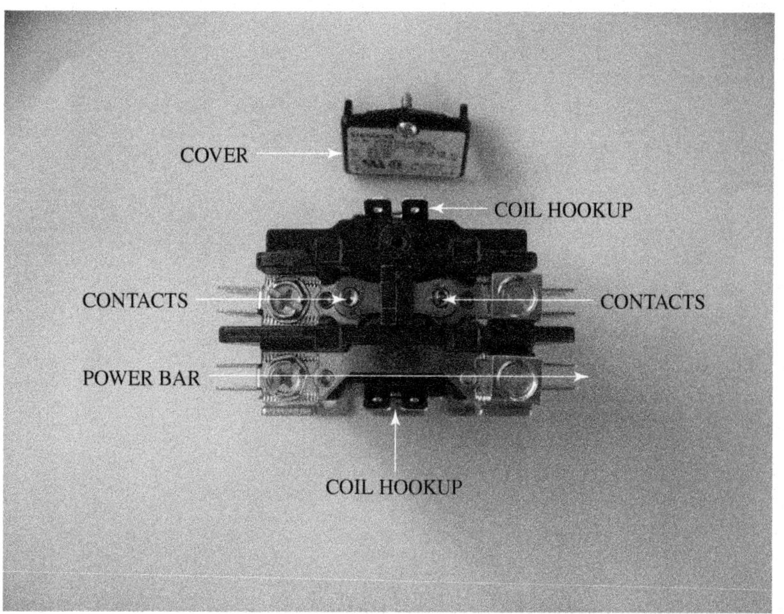

Figure 9-11 The single-pole contactor breaks one power line.

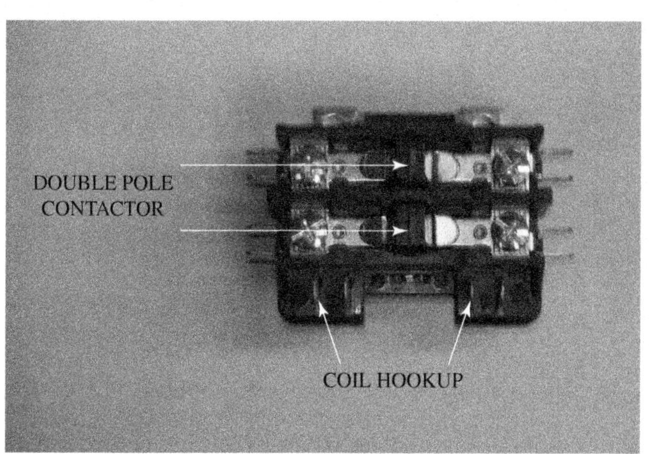

Figure 9-12 The double-pole contactor breaks both voltage sources. This is a common single-phase motor control. The coil voltage hookup is shown at the bottom of this photo.

Figure 9-13 The coil shown is used to energize the contactor. Hook the control wire to this side of the contactor and hook the control wiring to the other side of the contactor. Do not hook up both control wires on the same side of the contactor. This will create a short and burn out the transformer.

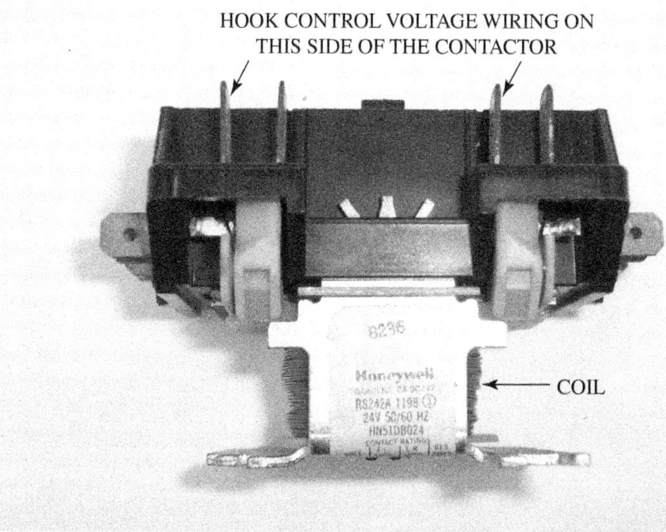

Figure 9-14 This is a side view of a double-pole contactor. The coil is shown on the bottom of the contactor. Both control voltage wires to the coil are hooked up on this side of the contactor.

Figure 9-15 These banks of double-pole contactors are installed in a commercial air conditioning system. The contactors are used to control the compressor and condenser fans.

Figure 9-15 shows a bank of double-pole contactors used to control a compressor and condenser motor fans on a commercial system. Multiple contactors are not normally found on residential units.

Figure 9-16 shows a **four-pole contactor**, which might be used in a three-phase application. The fourth set of contacts can be used for another purpose such as bringing on a single-phase condenser fan motor.

The contacts on a contactor can handle more amperage than a relay. These contacts are larger. The contacts are normally open (NO). The contacts are usually not replaceable; therefore, the whole device must be changed when one or more contacts become defective or the coil burns out.

The coil voltage is normally 24, 120, 208, or 240 volts. The 24-volt coil is very common.

It is important to understand that the contacts and coil make up one component, but are located in different parts of an electrical diagram. Let's look back at Figure 9-5 and break this part of the diagram down:

- The contactor coil is located in the lower part of the diagram, in the control voltage section (point **3**), and labeled "**CC.**"
- The contactor NO contacts are at points **4** in the upper diagram, to the left and right of the compressor motor **COMP**. They are labeled "CC," meaning compressor contacts.

This diagram shows how the contactor is wired without regard to location of the component parts. This is sometimes confusing for new technicians.

Figure 9-16 This is a four-pole contactor with a flash shield covering the contacts. Notice the voltage and current operating table on the flash shield. The 24-V coil connection is below the voltage connections L1–L4.

9.5 TROUBLESHOOTING CONTACTORS

The main way to troubleshoot a contactor is by checking the coil and the contacts and by making sure the contactor has freedom of movement.

Checking the Coil

Troubleshooting a contactor is similar to troubleshooting a relay. Check the resistance of the coil, which will be less than 1 kΩ (≤1,000 ohms), closer to 100 Ω or less (see Figure 9-17). A contactor coil with a few ohms of resistance or less is considered to be shorted. These shorted

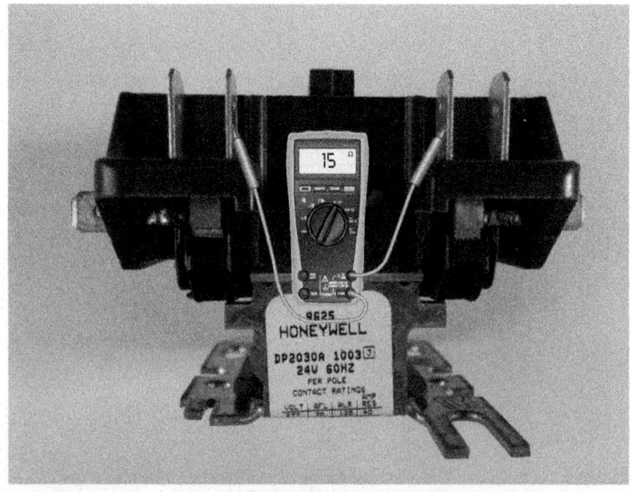

Figure 9-17 When troubleshooting, check the coil on a contactor. The coil has two sets of terminals.

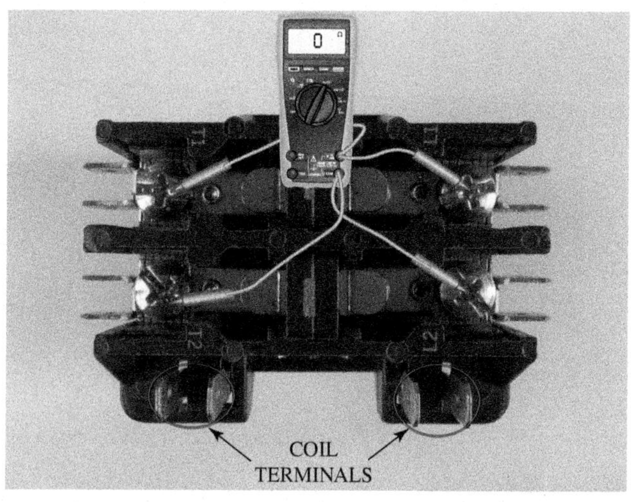

Figure 9-18 Close the contacts and measure resistance across the closed contacts.

conditions will most likely burn out the transformer due to excessive amperage draw. If the coil appears to be burned, it will be open, shorted, or have experienced overheating stress. Low resistance creates a high amperage draw, which can burn out the transformer.

Checking the Contacts

Next, check the contacts (see Figure 9-18). Do the contacts appear to be burned? Burned contacts indicate a buildup of carbon caused by arching when the contacts open and close. This buildup of carbon is normal, but eventually creates contactor failure. The resistance across a set of closed contacts should be 0 Ω. An ohm or two of resistance can be tolerated. The greater the resistance drop across carbonized contacts, the less voltage delivered to the load. There are two ways to test the integrity of the contacts.

First, with the power removed from both sides of the contacts, energize the contactor coil, which will pull in and close the contacts. Measure the resistance across the closed contacts. Resistance should be close to 0 Ω. Resistance of greater than 2 Ω means the contacts have excessive resistance or the mechanical mechanism is not pulling the contacts tightly closed. Low coil voltage could prevent the coil from closing tightly.

Another test procedure is to disconnect the power from both sides of the contacts. Measure the resistance while pushing the contacts closed. Again the resistance should be near 0 Ω. The first method discussed to close contacts is the best method. The disadvantage of the second method is that it forces the contacts closed. Physically forcing the contacts closed may override the problem caused by low coil voltage or mechanical sticking.

Check each contact for resistance. If one set of contacts is damaged, a temporary fix can be completed that bypasses the defective contacts. This will allow this part of the circuit to operate. This jumper should not be left permanently installed since the next technician may come upon the job and not realize that part of the circuit is directly wired to the power supply and get electrocuted. Only use this technique if you are going to pick up the replacement contactor immediately. Place a notice on the unit stating that one side of the contactor is temporarily bypassed while obtaining a replacement contactor. Too many times a technician does a temporary fix that turns out to be permanent and forgotten. In this situation, the next technician's life is in danger if the bypassed contacts are not noticed.

Mechanical Movement

The contactor should move back and forth freely when energized. With no voltage applied, push the contactor in with a high voltage–insulated tool. The contactor should easily go in and spring back once released. If you suspect that the system has a short condition, do not push the contacts because this may cause an explosion at the contactor! In this case, allow the normally operating controls to close the contacts.

Contactor Replacement

When replacing a contactor you will need the following information:

- Coil voltage
- Number of contacts
- Amperage rating of contacts
- Type of load: inductive or resistive.

9.6 MOTOR STARTERS

A motor starter is a contactor with built-in or added overload (OL) protection. It is a combination device used to start, stop, and run a motor by means of commands received from an operator or control circuit. It looks like an oversized contactor because of added overload protection device. The current passes through the contacts and series overload protection. Figure 9-19 illustrates the current or contactor bearing component of a motor starter. There is an OL for each set of contacts. The overload section is shown in Figure 9-20. The OL section is hooked in series with load contacts. The three poles (in the upper section) slip into the bottom of the motor starter and are tightened in place. The three-phase wiring is hooked into the bottom of the OL (not shown). The blue R button is the reset button. The reset button is pushed out when an overload condition exists. This OL is adjustable, with a range of 2.9 to 4 amps. Other devices for disconnecting and short-circuit protection will be needed for a complete safe circuit. These devices will be fuses or breakers.

There are single-phase and three-phase motor starters. Three-phase motor starters are more common. The contacts are rated like the contacts on relays or contactors. The

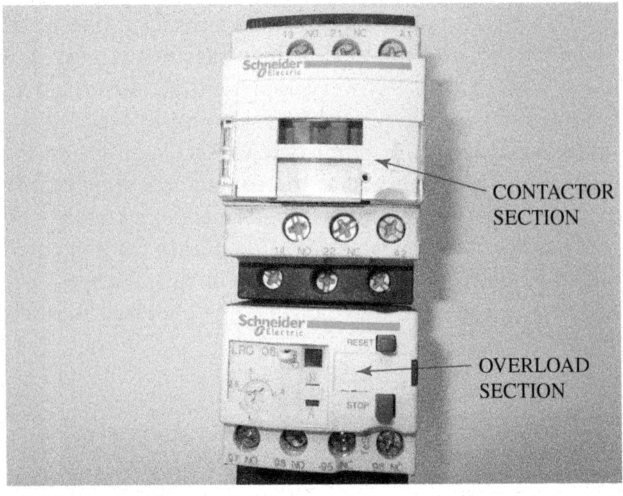

Figure 9-19 This is the contactor and overload section of a motor starter. It has a coil, three sets of load contacts, and a set of auxiliary contacts. Auxiliary contacts are located on the bottom of the overload section.

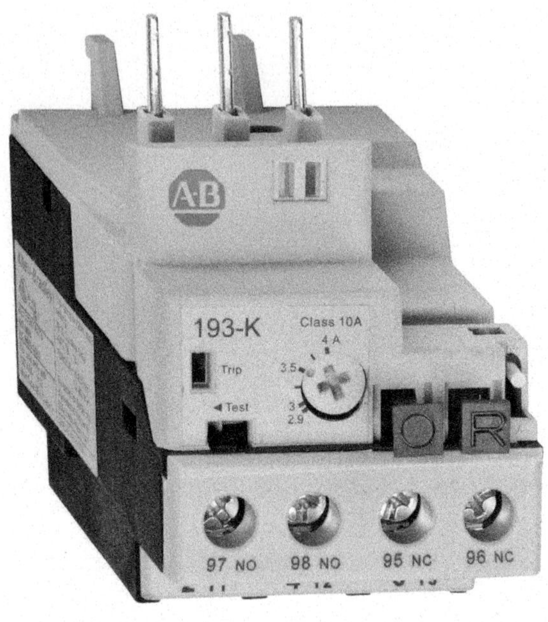

Figure 9-20 This is the overload section of the motor starter. The normally open (NO) and normally closed (NC) auxiliary contacts are located in the lower part of this device.

overloads are sized for the amount of current that is being controlled. The size of the overload can be changed depending on the motor load it is controlling. The main purpose of the overload is to open the control circuit if the load it is controlling draws too much current. Excessive current overheats the load, causing it to become hot and burn out.

Some overloads are set for a specific amperage draw. Some overloads are adjustable. Adjustable overloads can be adjusted about 25% above the rated amperage.

The coil voltage is normally 24, 120, 208, or 240 volts. A 24-VDC coil can also be used.

The information on a motor nameplate will help the technician decide if the overload is sized correctly. Table 9-3 indicates that the running load motor amperage is 7.6 amps. The OL should be sized for this amperage condition. The measured amperage on a motor should be no greater than 8 A. In this instance the applied voltage should be ±10% of 460 V, which is a range of 414 to 506 V. The voltage and amperage should be measured when the motor is operating near full load conditions to determine the correct operation of the motor starter.

The symbol for a single-phase motor starter is shown in Figure 9-21. Notice that the overload sensing device is on one side of the circuit. Current flow through **L1** and **L2** of the single-phase circuit is the same; therefore, only one **OL** sensing device is required. The **T1** and **T2** connections are hooked to the load such as a motor.

The symbol of a three-phase motor starter is illustrated in Figure 9-22. Each power line has overload protection. The dashed line near the upper part of the diagram indicates a mechanical linkage, meaning that the contacts are tied together and move as one unit.

Table 9-3 Basic Information for Selecting a Motor Starter Is Found on the Motor Nameplate

A motor's nameplate provides valuable information about the motor. Understanding this data will aid in troubleshooting. This nameplate is explained in the unit on motors.

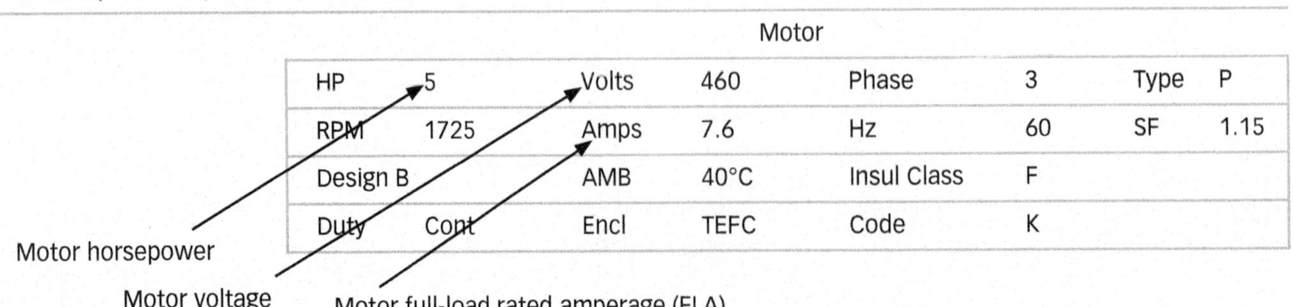
Motor

HP	5	Volts	460	Phase	3	Type	P
RPM	1725	Amps	7.6	Hz	60	SF	1.15
Design B		AMB	40°C	Insul Class	F		
Duty	Cont	Encl	TEFC	Code	K		

(*Source:* www.automationdirect.com/specs.)

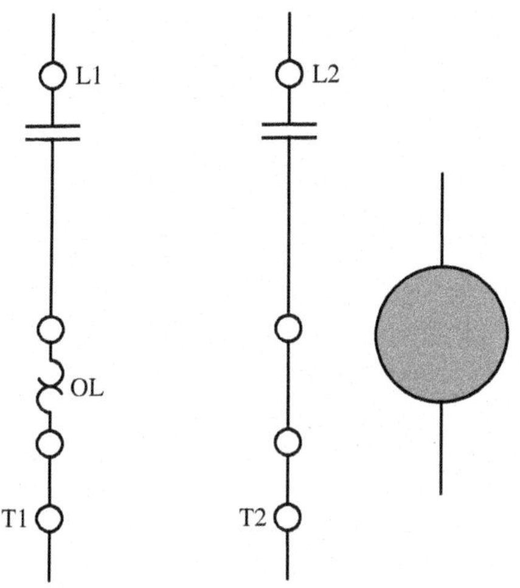

Figure 9-21 This is the symbol for a single-phase motor starter. The starter has two contacts and one overload sensing device labeled "**OL**." Only one sensing element is required since the current flow should be the same in both **L1** and **L2**. The magnetic coil, the symbol for which is a circle, is shown at right.

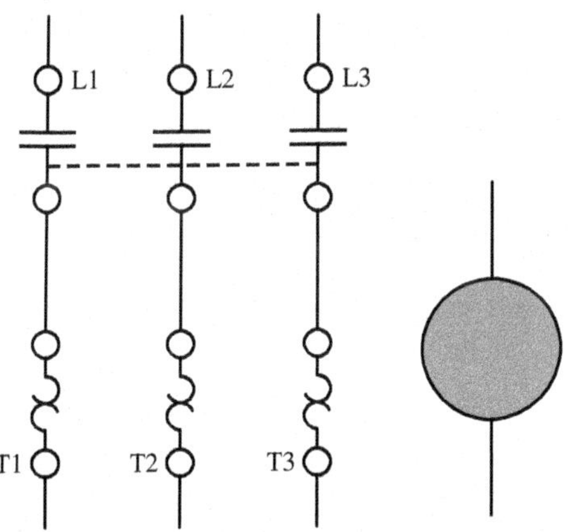

Figure 9-22 This is the symbol for a three-phase motor starter. The starter has three normally open contacts, an overload device (**OL**) in each line, and a magnetic coil (at right). The horizontal dashed line near the upper part of the diagram between the phases means that the three are mechanically linked together.

9.7 ELECTRICAL DIAGRAMS

Figure 9-23 is a simple diagram illustrating the hook-up for a single-phase motor starter that operates a motor. The **START** push button switch completes power to the **M** coil, which closes the **M** contacts under the **START** push button switch and closes the two sets of **M** contacts that starts the motor. The overload (**OL**) protection is a thermal device that opens normally closed contacts if the current flow is excessive. With excessive heat from the current flow, the normally closed **OL** contacts open and kill the voltage to the **M** coil. The motor stops since the de-energized **M** coil opens the three **M** contacts.

Figure 9-24 shows all of the individual components that [illegible] hook up a three-phase motor. The motor starter

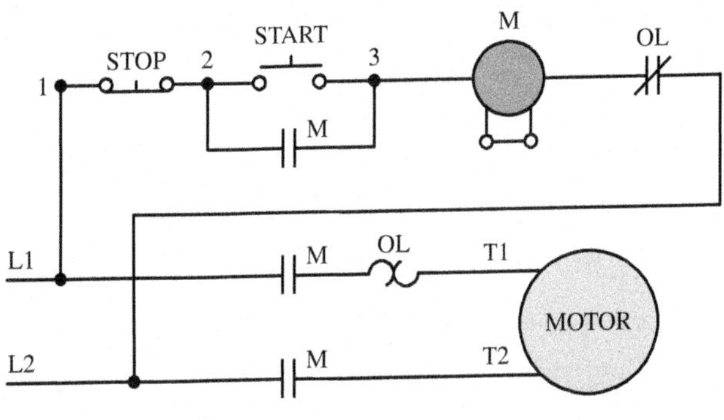

Figure 9-23 This diagram illustrates how to hook up a single-phase motor starter that operates a motor.

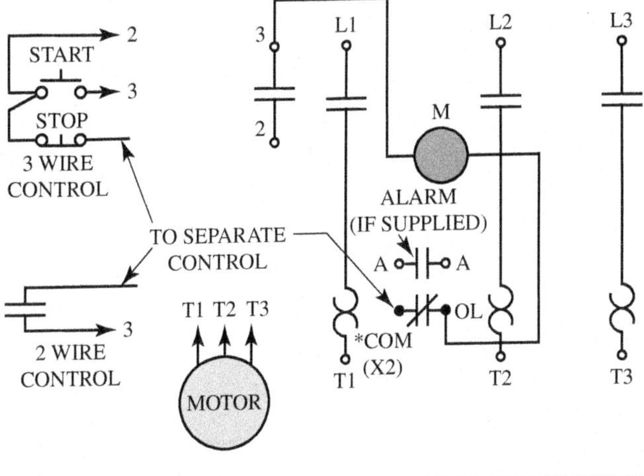

Figure 9-24 These are the individual components of a motor starter circuit.

diagram is shown in the right half of the diagram. Notice that the starter has NO AA auxiliary contacts. The **START** and **STOP** button is at the upper left in this diagram. The three-phase motor is located toward the bottom center of the diagram and shows wire connections **T1**, **T2**, and **T3**.

Figure 9-25 illustrates how the components in Figure 9-24 will be wired to operate a three-phase motor circuit. Let's review the operation of this circuit. First there are two separate voltage circuits: a single-phase voltage source for the upper control section, and a three-phase voltage source to operate the motor.

To start the motor, push the **START** button. The **START** button is a momentary switch that springs open after it is released. It only takes a second to close the **START** switch and start the operation of the circuit. When the switch is closed, it energizes the motor starter coil **M** through the normally closed **OL** switch. The three **M** contacts on the motor circuit close, which operates the motor. If one of the **OL** devices in line with **L1**, **L2**, or **L3** draws excessive amperage, the normally closed **OL** contacts in the control circuit opens. The open **OL** contacts kill voltage to the **M** coil, which opens the M contacts under the **START** button and opens the three **M** contacts that feed power to the motor.

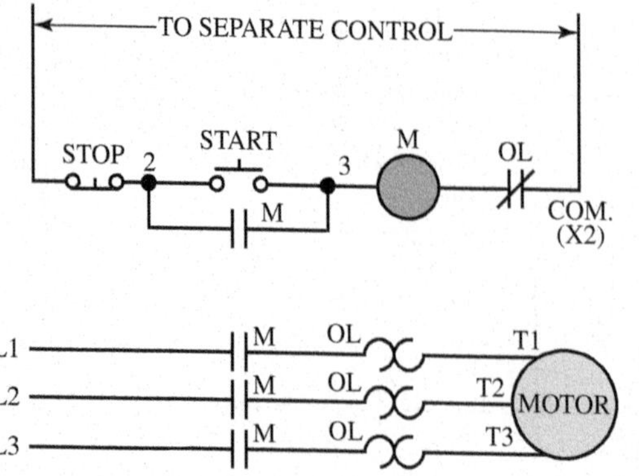

Figure 9-25 This is how the components in Figure 9-24 are wired up as an operational circuit.

Finally, to stop the motor under normal operation, pushing the **STOP** button will temporarily break the circuit and de-energize power to the **M** coil, which will open all four **M** contacts and stop the motor.

Figure 9-26 is a more detailed drawing of a motor starter in a three-phase motor circuit. The diagrams are drawn in two different configurations. The diagram on the left is classified as a wiring diagram. Not all of the components are wired. The diagram on the right is classified as an elementary diagram or ladder diagram. It shows the hook-up of the components in the wiring diagram. Notice that the diagram uses a transformer to reduce the control voltage. The secondary is listed as **X1** and **X2.** The secondary voltage is not stated on this diagram. It is most likely 120 volts. The secondary side of the transformer is fused (**FU1**).

SEQUENCE OF OPERATION 9.2: LADDER DIAGRAMS

We now review the operation of the ladder diagram on the right side of Figure 9-26.

1. At point **1** in blue, the motor can be started by pushing the **START** button or energizing the two wire control contacts shown above the **START** button.
2. Either way, secondary voltage from **X1** is provided to energize the **M** coil (point **2** in blue) through the fuse, through the **STOP** button since it is closed, through the now-closed control contacts **M** or the two-wire control, through the normally closed **OL** contacts, and back to the transformer **X2**.
3. At point **3** in blue, the energized **M** coil will close the three **M** contacts that will start the motor.
4. Any high current condition in any of the three **OL** sensors will cause the normally closed **OL** contacts to open, de-energizing the **M** coil, and opening the three **M** contacts, thus removing power to the motor.

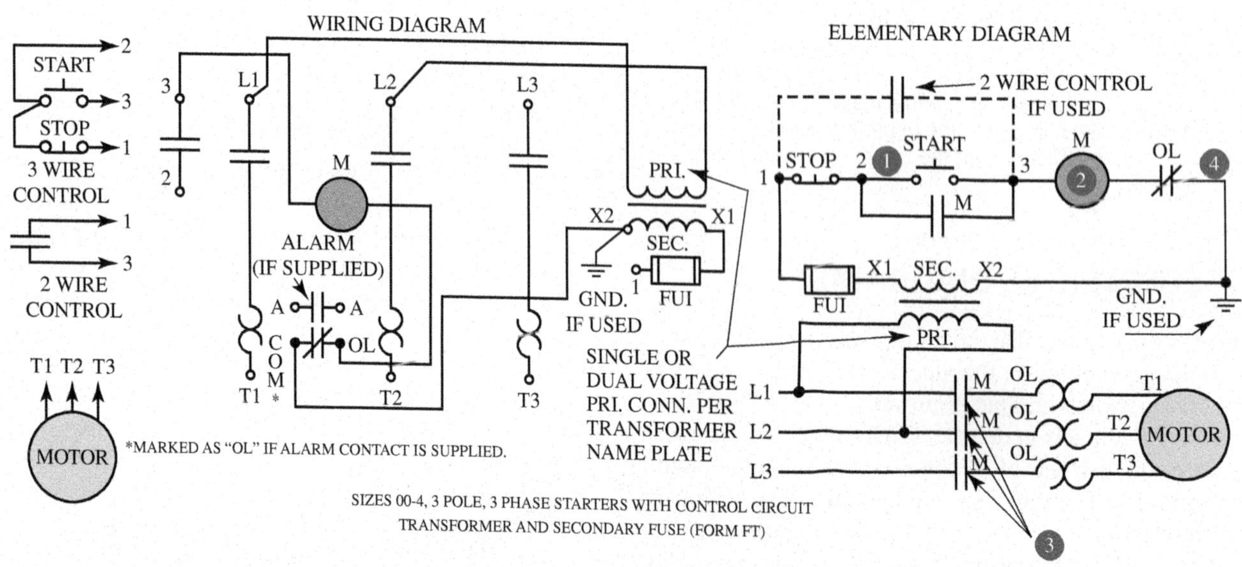

Figure 9-26 This shows the wiring diagram for a three-phase motor starter on the left. The elementary diagram or ladder diagram is shown on the right. These diagrams are the same, but drawn differently.

Overloads

Overloads use a bimetal or heater device to open the normally closed OL contacts. When excess current flows through one or more of the bimetal protection sensors OL in Figures 9-25 or 9-26, all of the three phase **M** contacts are opened. The heater overload device works differently but with the same affect. The heater is a small electric heater. Excess current going through the heater warms lead in a small cylinder. Excessive current and heat cause the lead to soften in a small cylinder that contains a spring-loaded shaft. Once the lead gets hot enough, the shaft will turn and open the contacts to the **M** coil. The lead cools quickly and solidifies around the spring-loaded shaft. The motor starter reset button will need to be pushed to reactivate the circuit.

The overload found in a motor starter is different than what is called a motor relay overload like that shown in Figure 9-27. This is an external overload relay. It is strapped to the outside of the compressor shell. It is in line with the common compressor connection in a single-phase compressor and in series with at least two different windings in a three-phase motor. When excess amp draw or overheating is sensed, the overload will warp and a bimetal switch will open the circuit. The motor will stop. After a cooling period the bimetal switch will cool and the overload will reset and close, providing power to the motor again. If the overload or overheating condition is not corrected, the safety device will open again.

Auxiliary Contacts

Auxiliary contacts are installed on motor starters. These contacts can be NC or NO. These contacts switch positions when the overload device is energized, meaning that a high current condition existed. The contacts switch position until the manual reset button is pushed, allowing the contacts to return to their normal position. The auxiliary contacts are not energized by the motor starter coil.

In Figure 9-26 the NO auxiliary contacts are found under the left-hand section with the WIRING DIAGRAM heading. The auxiliary contacts are located below the **M** coil. In this diagram the normally open contacts are not being used, but could be hooked to an alarm or other monitoring device. The normally closed contacts are wired in series with the **M** coil. If a high current condition is sensed in any of the three phases of the motor, the NC will open and stop the motor operation.

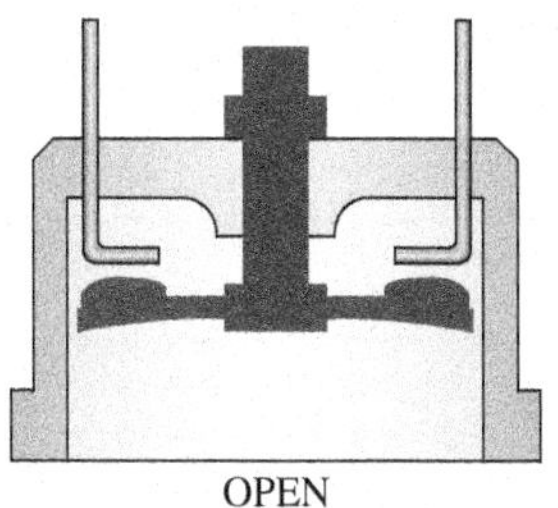

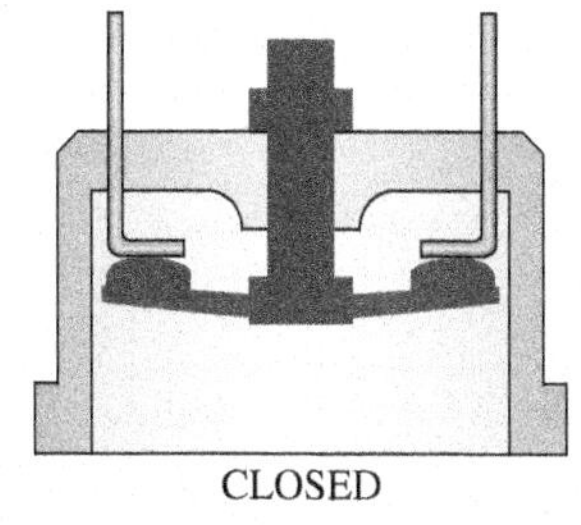

Figure 9-27 This external overload would be found secured to the outside of the compressor shell.

The motor starter uses a manual design, therefore the motor will not operate until someone pushes the reset button. Even then the motor may not operate if it is defective.

SAFETY TIP

Working around motor starters means that you are handling high-amperage equipment. Less than 1/10 of an amp will kill you. Ensure that no voltage is present when changing out components by turning off the power and locking out the power source. Measure the voltage on the component you are going to change. If there is no voltage on the component or no voltage to ground, short the component with a screwdriver or wire to ground. Finally, backhand the area you are going to touch. The component should be safe to touch. The order of these steps is essential so you do not get injured.

9.8 READING ELECTRICAL DIAGRAMS

Electrical diagrams can be drawn in several different ways. Figure 9-29 represents a different way to draw a motor starter diagram than we have not seen thus far. The disconnect, contactor, and overload section are shown isolated inside the rectangular, dashed lines. The contactor section and overload section make up what is known as the motor starter. The additional components shown in the diagram add to the safety and convenience of the circuit. The control power supply is 24 VDC. Even though the diagram is drawn differently, it is essentially the same circuit as the other three-phase motor starter circuits in this unit.

9.9 MOTOR STARTER REPLACEMENT

When replacing a motor starter you will need the following information:

- Coil voltage
- Number of contacts
- Applied voltage
- Amperage rating of contacts
- Overload amperage rating (heaters).

Many parts of a motor starter can be changed. For example, on some models, the tech can change the contacts or coil of the motor starter shown in Figure 9-19. The overload protection section we saw in Figure 9-20 is definitely interchangeable since the OL is sized for each load it needs to control. Motor starters have a long lifetime since replacement parts are available for many years after their purchase. Sometimes, however, it is difficult to obtain replacement parts on older motor starter models.

Troubleshooting the motor starter is similar to checking out a relay or contactor except you are required to check the overload section as well. Check the coil with an ohmmeter

SEQUENCE OF OPERATION 9.3: MOTOR STARTERS

Here is another example of the sequence of operations of the motor starter shown in Figure 9-28:

1. Pushing the **START** button will temporarily close the normally open switch at point **1**. The push button is spring loaded and the contact will reopen when button pressure is removed.
2. Power will come from **L1** and go through the normally closed contact at point **2**.
3. This will energize the starter coil (M) at point **3**.
4. The starter coil will close the **MA** contacts under point **1** and close the three-pole contacts **M** under point **4**.
5. The motor at point **5** will now energize.
6. To manually stop the motor operation, push the **STOP** button at point **2**. This will temporarily open the starter coil circuit and stop the motor operation. The motor will not restart unless the **START** button is pushed again.
7. The motor will also stop if the overload relay senses high amperage or an overheating condition occurs. When this condition occurs, the normally closed overload contact (**OL**) will open and operation will cease.

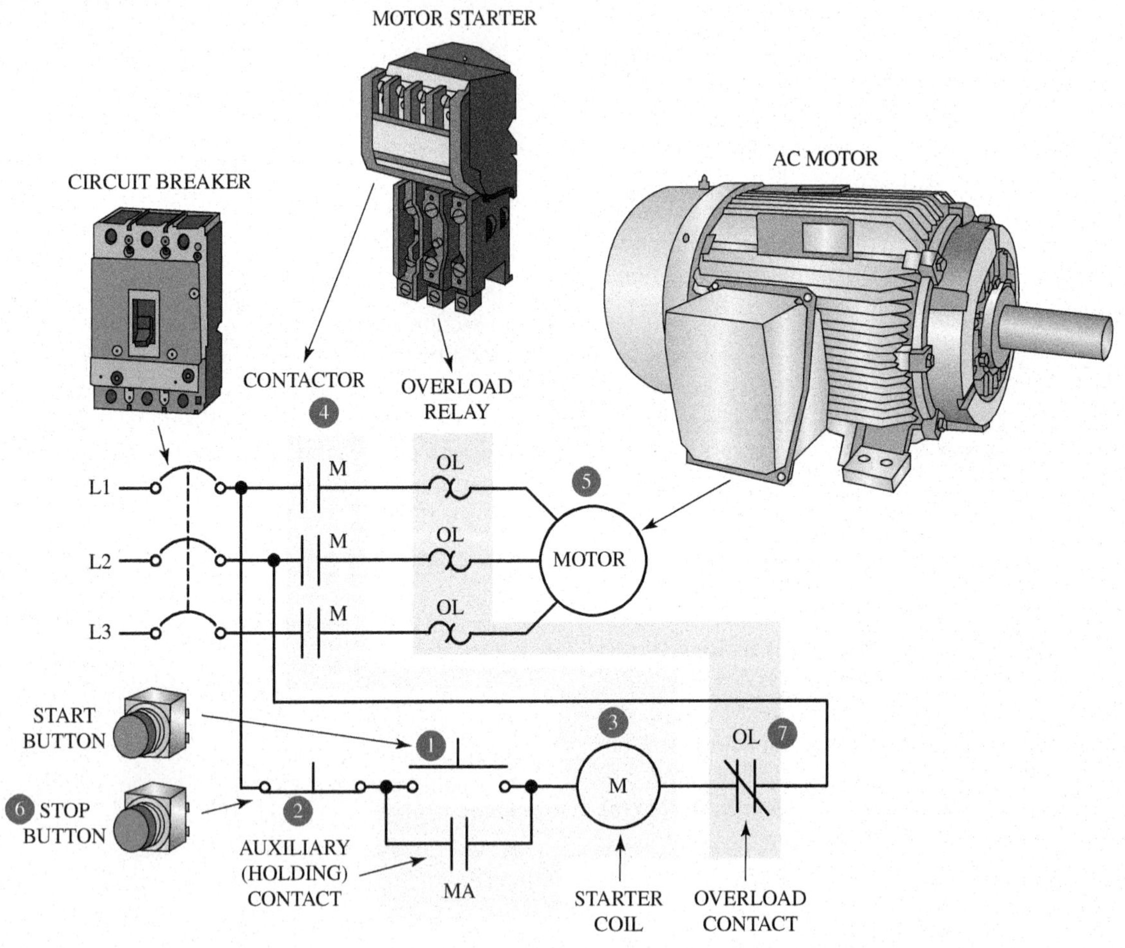

Figure 9-28 This diagram illustrates the sequence of operations for a motor starter.

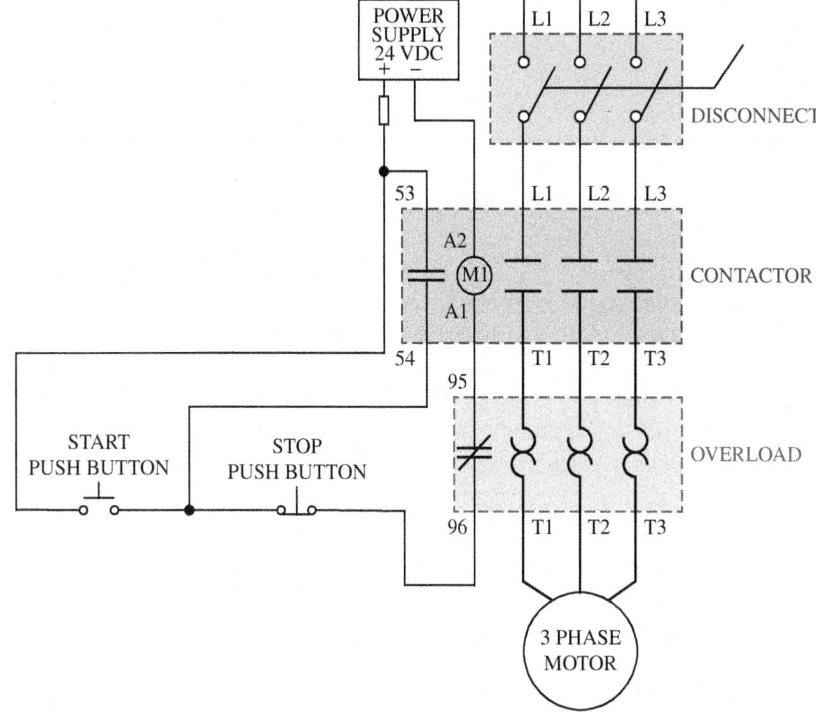

Figure 9-29 This diagram shows the major components of a motor starter circuit. The disconnect, contactor, and overload are outlined with dashed lines. The diagram shows the **START** and **STOP** buttons and a 24-VDC control power supply.

and expect a resistance of less than 100 ohms, but more than a few ohms. Visually inspect the coil for burned or overheated spots. Measure applied voltage to the coil. Do the motor starter contacts change position when energized? Can you hear the contacts click, indicating opening or closing? As mentioned earlier motor starters have replacement parts. For example, a burned-out coil can be replaced. Contacts can be replaced.

Check motor starter contacts the same way you would check contactor contacts. When closed, the contacts should have less than 1 or 2 Ω of resistance. Burned or carbonized contacts should be replaced.

The overload protection should read no resistance (0 Ω). If used, the auxiliary contacts should also be checked. Normally closed and normally open contacts are used as auxiliary contacts.

TROUBLESHOOTING 9.2: MOTOR STARTERS

Using Figure 9-30, troubleshoot the following components with the power removed:

1. At point **1** in blue on the diagram, measure resistance across coil **M**. Higher voltage coils generally have higher resistance coils.
2. At point **2** in blue measure the resistance of the three contacts when you energize the coil or force the contacts closed. Do not have voltage applied to contacts.
3. Check the overload OL closed contacts (point **3** in blue).

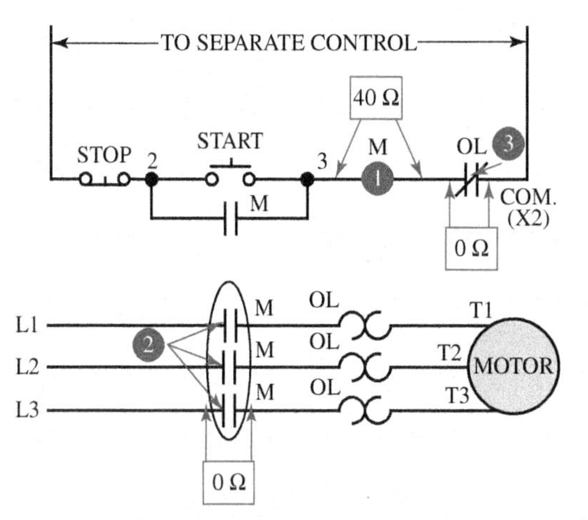

Figure 9-30 Use this motor starter diagram with the Troubleshooting 9.2 feature.

SERVICE TICKET

You are called out on a service call for a 100-ton compressor motor that is not operating. The system was installed in March and now it is late May. This is the first repair call on this new installation and it is covered under a one-year warranty. After spending 10 minutes doing an equipment survey, you notice that the manual reset button on the motor starter is pushed out. You resist the temptation to push the reset button since your gauges and clamp-on ammeter are not hooked up yet.

The first step in troubleshooting is to do an equipment survey to determine what is working and what is not working. Then you hook up test instruments. You notice that the RLA on the compressor nameplate is 24.9 A. After installing your manifold gauge set and clamp-on ammeter on one of the compressor power leads, you push the motor starter reset button. The compressor immediately begins to operate. Pressures seem okay and amperage draw is 21.5 A. After a couple minutes of operation, the compressor stops and the motor starter overload is tripped as indicated by the reset button being pushed out. What is the problem?

This is a challenging problem since everything seems to be okay. The operating pressure and amperage draw are within operating range, but the motor starter overload takes out the compressor. Do you have any ideas, before we turn to the solution?

The equipment was installed and tested during the spring, a cool time of the year that creates lower head pressures and, therefore, lower operating amperage. The warmer weather has raised the operating head pressure, which increases the amperage draw on the compressor. The solution to this problem is to change the overload section of the motor starter. The installed starter is set to open the circuit around 20 A. The RLA on the compressor motor is 25 A. After installing the correct overload device, the system ran without tripping on overload. Installing the correct overload size solved the problem.

Note: Many motor starter manufacturers do not sell their units with a specific overload device because the required amperage protection device varies among equipment. The overload protection is ordered separately.

SUMMARY

This unit covered the selection, operation, and troubleshooting of relays, contactors, and motor starters. These electromechanical devices are the switches used on HVACR fans, pumps, and compressors. It is important to know the fundamental operation of each device in this family of switches. Common selection criteria include knowing the coil voltage, contact amperage, voltage rating, and number of contacts required for a replacement unit.

This unit included troubleshooting tips. Many loads, such as motors, are replaced unnecessarily because the technician does not understand the controlling switches such as the relay, contactor, or motor starter. Understanding these components leads to a big advancement in a technician's ability to troubleshoot HVACR systems.

REVIEW QUESTIONS

1. How are relays, contactors, and motor starters similar?
2. What is meant by *chatter* as it relates to a relay or contactor?
3. What are some common coil control voltages?
4. What will happen to a coil if the wrong voltage is applied?
5. When replacing a relay or contactor, what information do you need in order to obtain the correct replacement?
6. What is the difference between rated load amps (RLA) and locked rotor amps (LRA)?
7. Review Table 9-2. How many running amps will the contactor handle if it is operating a motor load at 480 volts? 240 volts?
8. Review Table 9-2. How many locked rotor amps will the contactor handle if it is operating a motor load at 600 volts? 480 volts?
9. What is the difference between a relay and a contactor?
10. How do you troubleshoot a relay coil?
11. What resistance is found on a set of normally open (NO) contacts when de-energized?
12. What resistance is found on a set of normally closed (NC) contacts when de-energized?
13. A three-pole contactor will have how many sets of contacts?
14. How is a motor starter different from a relay or a contactor?
15. How are motor starter overloads sized?
16. Where can you find the motor voltage and current draw used to select a contactor?
17. What letter designates the coil of a motor starter?
18. Describe the starting sequence of operation for a motor starter.
19. Which overload (OL) will need to open to stop the motor starter circuit?
20. What are the two types of overloads used in motors to protect them from damage?
21. After being energized, what would cause auxiliary contacts to change position?
22. List several steps to troubleshooting a motor starter coil.
23. What is the resistance on a closed set of contacts?
24. Of the devices discussed in this unit, which is the most complex?
25. Identify the component shown in Figure 9-31.
26. Identify the component shown in Figure 9-32.
27. What is the symbol on the left side of Figure 9-33?
28. What are the full load amps and the locked rotor amps in the diagram on the right side of Figure 9-33?
29. Is Figure 9-34 a single-phase or three-phase contactor? How many poles does this contactor have?

Figure 9-31 Use this image to answer Review Question 25.

Figure 9-32 Use this image to answer Review Question 26.

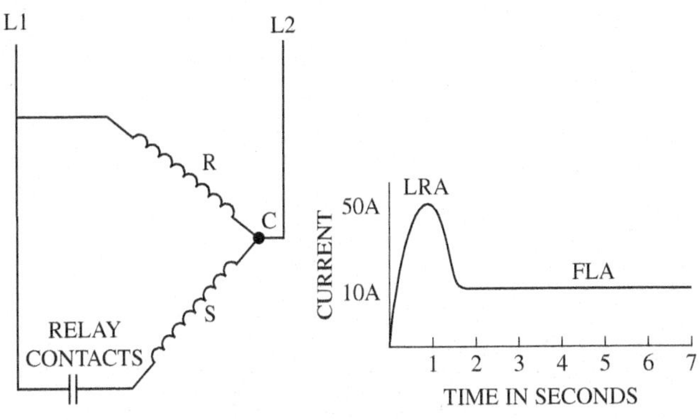

Figure 9-33 Use this diagram to answer Review Questions 27 and 28.

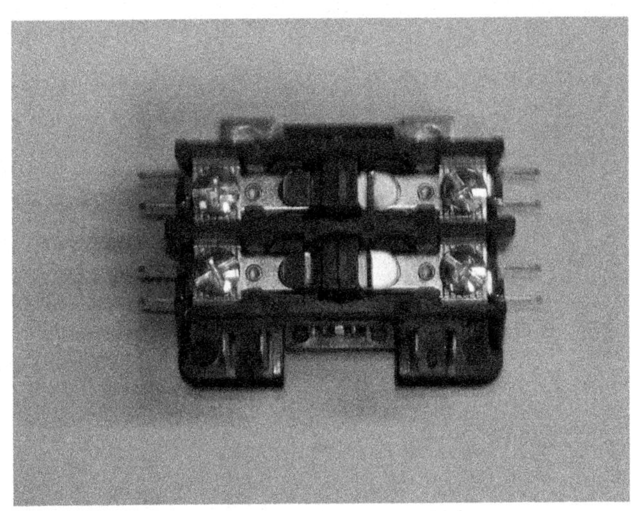

Figure 9-34 Use this image to answer Review Question 29.

UNIT 10
Capacitors

WHAT YOU NEED TO KNOW

After studying this unit, you will be able to:

1. Describe the types of capacitors and their construction.
2. Show how to make a capacitor safe to handle.
3. Determine capacitor ratings.
4. Describe how capacitors are installed.
5. Determine the capacitor rating if wired in series or parallel.
6. Troubleshoot capacitors.

The **capacitor** is one of the most misunderstood components found in an electrical circuit. Some techs think that the purpose of the capacitor is to give the motor a shock, a boost, or an electrical bump. This is really not true. The capacitor is used to start a motor or make a motor run efficiently once it is started. The **start capacitor** is used to assist in starting a motor. A **run capacitor** is used to keep the motor running efficiently. This unit will help you understand the function, operation, installation, and troubleshooting of capacitors. This unit will also discuss the safe handling and installation of capacitors. Correct capacitor installation is important. An improperly installed capacitor may damage the motor it is trying to help.

Finally, this unit covers troubleshooting. A defective capacitor will prevent a motor from starting or running. The tendency is to diagnosis the motor as defective, when a defective capacitor may actually be the problem.

10.1 CAPACITOR SYMBOL

Two symbols are used to represent a capacitor, as shown in Figure 10-1. The preferred symbol is a straight line and a curved line. This symbol is commonly found in electronic circuits. The other capacitor symbol, two parallel lines, is used in many HVAC capacitor diagrams. The two parallel lines, however, can be interpreted as normally open contacts. When first learning to read a diagram, the similarity between the symbols for capacitors and contacts can be confusing. To avoid confusion, note that a motor capacitor is usually closer to the motor diagram.

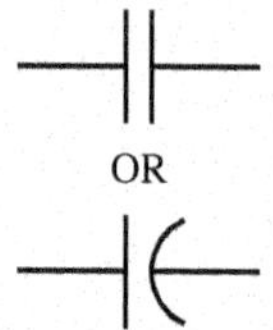

Figure 10-1 Two symbols for a capacitor. The preferred symbol is the bottom one with the curved line.

10.2 CAPACITOR CONSTRUCTION

The construction of a capacitor is simple. A capacitor consists of two foil plates separated by electrical insulation. The insulation is designed to store charged electrons. Figure 10-2 illustrates the internal construction with the metallized foil plates and electrically insulated material. In this illustration the insulation is listed as electrolyte or polypropylene film and shown in Figures 10-2A and 10-2B, respectively. These two types of insulating materials collect electrons but prevent them from crossing onto the metal capacitor plates.

Figures 10-2C and 10-2D show how a capacitor is installed in a motor circuit. The phase shift between the current and voltage improves the starting torque and increases the running efficiency. Improved running efficiency means less amp draw. Figure 10-2D represents a common single-phase motor installation. The capacitor is in series with the start winding and in parallel with the run capacitor. Incorrect wiring will damage the motor. A run capacitor also limits the current flow through the start winding, which prevents this winding from burning out.

In summary, a capacitor is constructed with foil plates that are separated by an insulated material.

TECH TIP

Working with capacitors requires knowing safe handling practices. A capacitor can hold an electrical charge. Discharge the capacitor before handling it. Some techs discharge a capacitor by placing a screwdriver or wire across the terminals. This is not the recommended way to discharge a capacitor. The recommended way is to use a 5,000-Ω, 2-W resistor to safely discharge the capacitor. Remember: A discharged capacitor can become recharged if you test it with an ohmmeter or capacitor checker. These instruments have a battery power supply that is used to test a capacitor, charging it up again. The shock may not be fatal, but it could cause a secondary injury such as a fall.

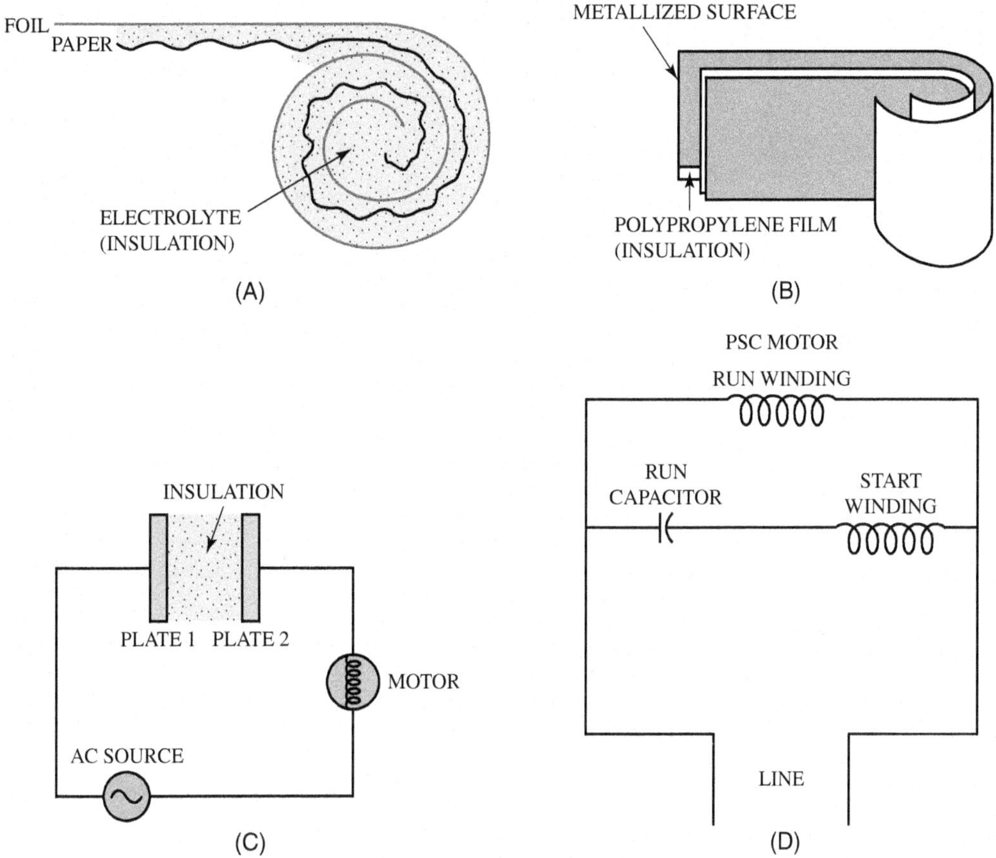

Figure 10-2 A capacitor is constructed of two foil plates separated by electrical insulation. The insulation is designed to store charged electrons, but not to allow electron flow.

10.3 CAPACITOR OPERATION

Without going into a lot of scientific detail, a capacitor is designed to create a phase shift between the current and the voltage. In a resistive load, such as an electric heat strip, the current and voltage are in phase with one another as shown in Figure 10-3A. A capacitor is charged up by current electrons. When this happens the current leads or is ahead of the voltage pressure as indicated in Figure 10-3B. When the current leads the voltage in a motor circuit, the motor's starting torque (turning power) is improved and the amount of running current it draws is reduced. Reduced current draw results in less wattage being used, which in turn lowers the operating costs.

In summary, a capacitor has:

- *Insulation material.* This is used to separate the foil plates.
- *Metal plates.* The surface area of the plates will determine the capacitance rating. The greater the surface area, the greater the capacitance.
- *The ability to improve motor performance.* A capacitor is designed to shift the phase between the voltage and current flow, thus improving starting torque and lowering the running current draw.

10.4 TYPES OF CAPACITORS

Three types of capacitors are frequently used in HVACR equipment:

- Run capacitors
- **Dual capacitors**
- Start capacitors.

Run Capacitor

A run capacitor (see Figure 10-4) is identified by its case. It has a metal or gray plastic case. The start capacitor has a black plastic or Bakelite™ case. Run capacitors can be round, oval, square, or rectangular. The round or oval design is the most common.

If a motor has a start and run winding, the run capacitor is installed in series with the start winding and parallel with the run winding as shown in Figure 10-5. This design increases **starting torque** (twisting motion) and improves running efficiency by reducing running amps. The main purpose of the run capacitor is to reduce the running load amperage. The starting torque is a side benefit. The run capacitor does not provide the maximum starting torque as seen with the start capacitor.

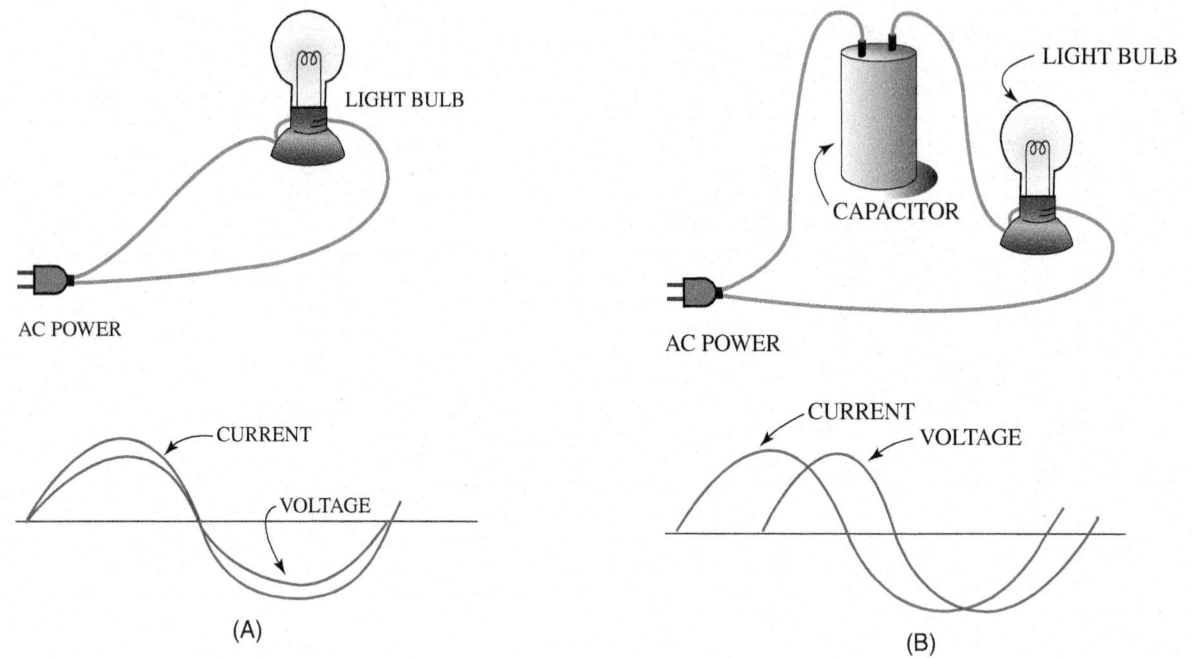

Figure 10-3 (A) In a resistive load, the current and voltage are in phase with one another. (B) The graph shows a resistive and capacitive circuit. The current leads the voltage development. The purpose of a capacitor in series with a start winding is to create a phase shift between the voltage and the current.

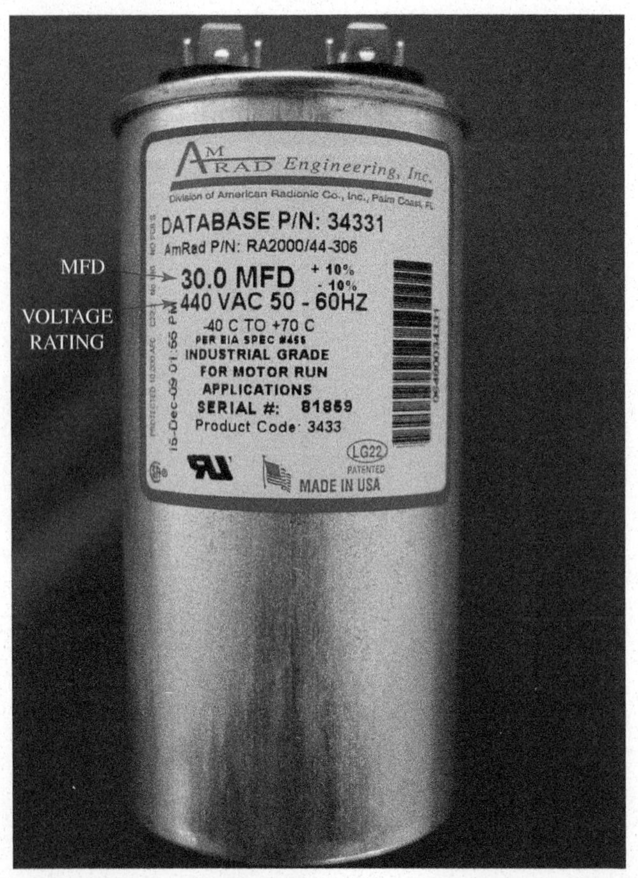

Figure 10-4 This run capacitor has a rating of 30 mfd and a maximum operating voltage of 440 V.

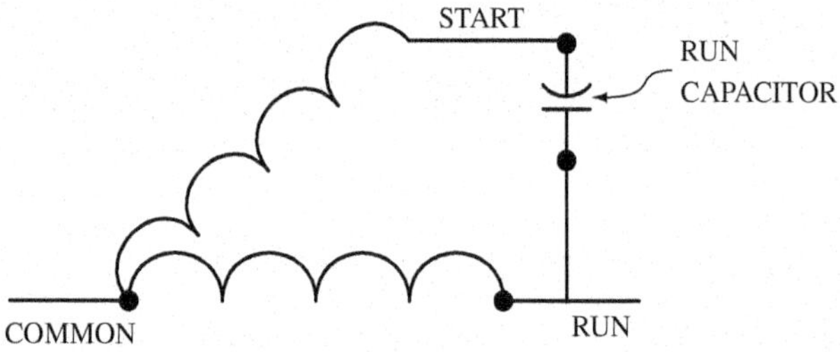

Figure 10-5 This permanent split capacitor motor requires a capacitor to start and run. The run capacitor is wired in series with the start winding and parallel with the run winding.

TECH TIP

Most motors need a capacitor that is designed, tested, and selected to operate a specific motor application. Some motors only have a run winding, while others have a start and run winding. Our industry also has motors that only use start capacitors.

Dual Capacitor

As the name expresses, the dual capacitor is two capacitors in one shell. It serves the purpose of two run capacitors. A common application is in a condensing unit. A dual capacitor is used with the unit's condenser fan motor and compressor motor. It has a common power connection. Figure 10-6 shows a dual capacitor. The top of the dual run capacitor is illustrated in Figure 10-7.

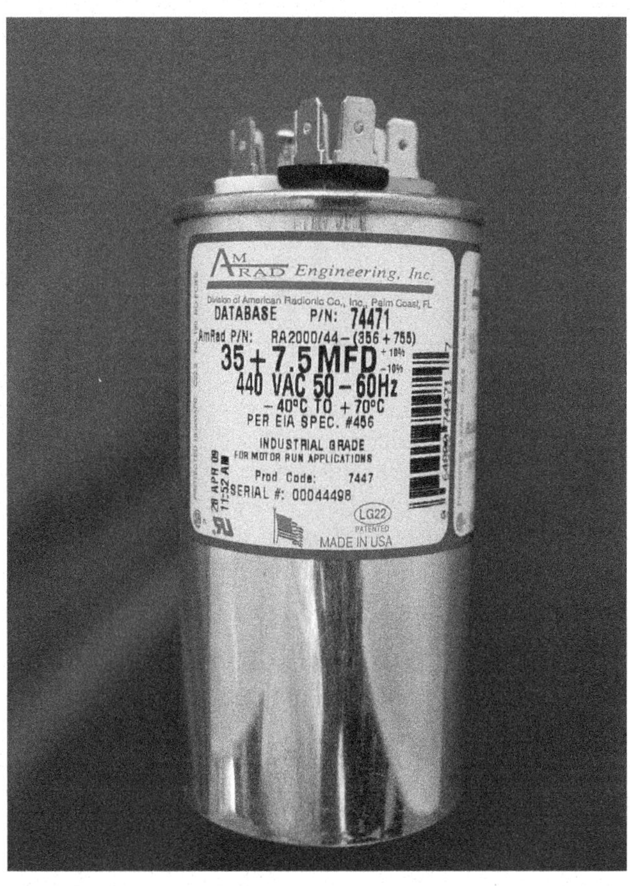

Figure 10-6 Note that this dual capacitor has two microfarad ratings, 35 and 7.5 mfd. The voltage rating is 440 V.

Start Capacitor

The start capacitor is enclosed in a round, black plastic or Bakelite™ case that is not conductive. Unlike the run capacitor it is easily identified, as shown in Figure 10-8. The start capacitor is used to create a 90-degree phase shift between the starting current and the voltage. The 90-degree phase shift will give the motor the maximum starting torque. In selecting the correct capacitor, note that the start capacitor must match the motor design. A capacitor that is too small will not develop enough starting torque. An oversized capacitor will overshoot the 90-degree phase shift and reduce the starting torque. Sizing within 10% must be maintained for the best starting torque.

A start capacitor should not be left in a circuit for more than a second or two. Start capacitors are not designed to remove internal heat like a run capacitor. Some switching device must be used to remove the start capacitor from the circuit very quickly once the motor reaches 75% of its speed. Two types of motors use a start capacitor. One motor type is the capacitor start and induction run motor, abbreviated CSIR. The other motor type is the capacitor start and capacitor run motor, abbreviated CSCR. Motors will be studied in more detail in future units. At that time the importance of capacitors will become more apparent.

SAFETY TIP

Some start and run capacitors have a **bleed resistor** across the terminals, as shown in Figures 10-8 and 10-9. The main purpose of a bleed resistor is motor protection. A bleed resistor discharges the capacitor quickly in case there is a short cycle (on and off quickly) condition. A charged capacitor in a short cycling condition will discharge back into the motor. This can damage the motor. A second benefit of this resistor is to reduce shock risk to the technician. If a bleed resistor is installed, it should be used any time the motor is in operation.

Figure 10-7 The top of the dual capacitor has embossed labels indicating the connection. The letter **C** is for a common power connection. The **HERM** is the connection to the compressor motor start winding. The **FAN** connection goes to the condenser fan motor start winding.

Figure 10-8 Start capacitors are round with a black plastic or Bakelite™ case.

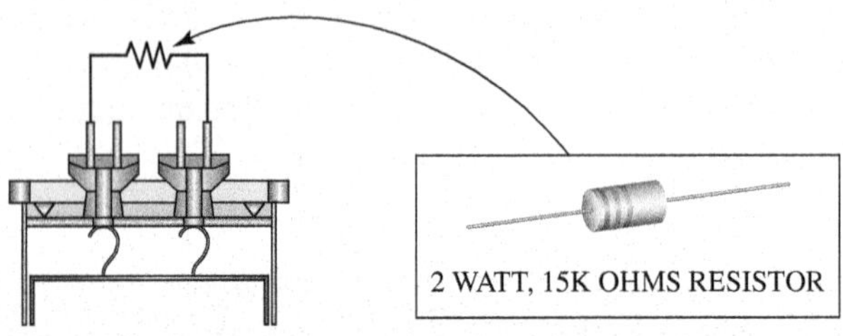

Figure 10-9 A bleed resistor is installed across the terminals of a run or start capacitor. The purpose of the bleed resistor is to remove the charge from a capacitor so that it does not damage the motor in a short cycling condition.

10.5 HOW ARE CAPACITORS RATED?

Capacitors are rated by type, start and run, microfarads, and voltage. We have discussed the type of capacitors. Let's discuss the microfarad rating. A **farad** is a large **capacitance rating**. The higher the farad rating, the greater the capacity to collect and store electrons. The farad is too large a capacitor to be used in our industry. This is why we use a much smaller rating, the **microfarad**. *Micro* means 1/1,000 or 0.001; therefore, we are talking about thousandths of a microfarad. The abbreviations *μfd* and *mfd* are used for microfarads.

Microfarad Rating of Run Capacitors

Run capacitor microfarad ratings start around 5 mfd and increase by fives up to 50 mfd. Larger, single-phase motors will have larger microfarad rated capacitors.

Note: Three-phase motors do not need capacitors to start or run the motor. A capacitor is an identifying feature of a single-phase circuit.

Microfarad Rating of Start Capacitors

Start capacitors have higher microfarad ratings than run capacitors. They start around 40 mfd and extend up to around 250 mfd. The start capacitor usually has a range of capacitance. For example, a start capacitor can have a rating of 88 to 108 mfd. When this capacitor's rating is checked, it should be between 88 and 108 mfd in order to be classified as good. If the start cap is outside this rating, it needs to be replaced.

Voltage Rating

The next rating to be concerned about is the voltage rating. The voltage rating of a run capacitor is about 150% of the applied voltage. The voltage across the run capacitor "sees" the applied voltage plus the voltage generated by the rotation of the motor. The voltage generated by the motor is called *back electromotive force* or, for short, *back EMF*. Measure it when the motor is operating. You will be surprised. A capacitor with an undersized voltage rating will overheat and burn out fairly quickly. Common voltage ratings for run capacitors are:

- 370 V
- 440 V.

Either of these can be used on 120- or 240-V circuits. The higher voltage rated capacitors cost more than lower voltage rated capacitors. The higher volt capacitors are also larger.

SAFETY TIP

A swollen or blown-out capacitor is a bad capacitor. It should be handled with care, especially if the pressure in the capacitor has not been relieved (see Figure 10-10). Some capacitor manufacturers use an interrupter design. When the capacitor bulges it becomes unsafe. The interrupter design separates the terminals from the plates, making the capacitor safe.

10.6 TROUBLESHOOTING CAPACITORS

There are several ways to troubleshoot a capacitor. First, inspect the capacitor for bulges, burns, holes, seepage, and connections. Replace a damaged capacitor or a burned wire connection leading to the capacitor. Another way to determine if a capacitor is defective is to replace it with a like capacitor. This is not always possible, since most service vehicles do not carry a complete assortment of capacitors.

The professional way to check a capacitor is to use a capacitor checker. Many quality DMMs have a built-in capacitor checker option, as shown in Figure 10-11. The capacitor should be within ±10% of its labeled microfarad rating. Some manufacturers have a range as low as ±5%. Look at the capacitor label

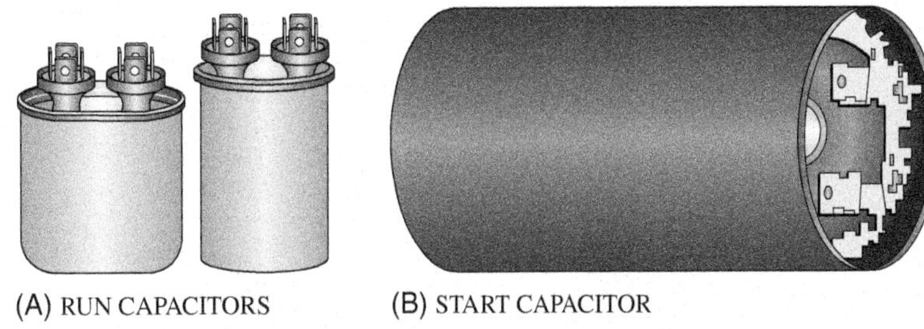

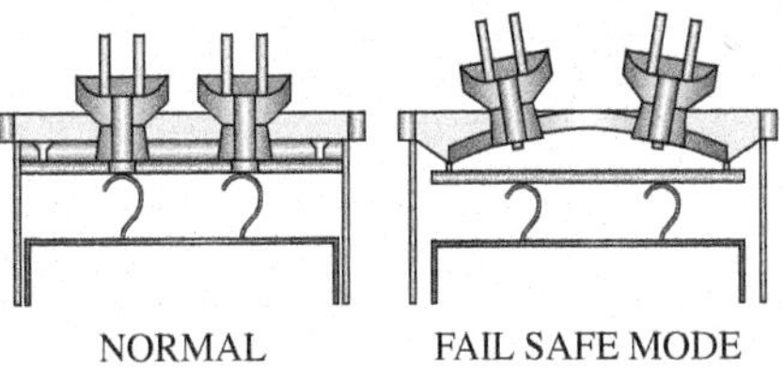

Figure 10-10 A run or start capacitor can become swollen or explode, which makes it unusable.

to determine the acceptable range. Start capacitors have a range printed on the label. For example, you might see "88–108 MFD." The microfarad rating should not fall outside this range. A weak capacitor may prevent the motor from starting or, if started, may draw high amps and cause the motor to cut out on overload.

Some techs use the ohmmeter to check a capacitor. Placement of the ohmmeter probes on the capacitor actually charges the capacitor so that it reads near 0 Ω for a second or two. After a brief moment near the 0-Ω reading, the resistance reading drifts toward infinity (∞). The low resistance reading shows that the capacitor is charging or receiving electrons from the battery in the ohmmeter. After the capacitor is charged, it stops receiving electrons and the ohmmeter reading increases toward infinity. This is best done with an analog (needle movement) ohmmeter. It can be done with a digital meter, but the readout is more confusing. Even if a capacitor shows a charge using an ohmmeter, this does not tell the technician the microfarad rating of the component. The capacitor can be off by more than 10% and show a charge condition. The ohmmeter is good to check for a grounded metal case. Test each terminal to the case. The ohmmeter reading from the capacitor terminal to the metal case should be infinity at all times.

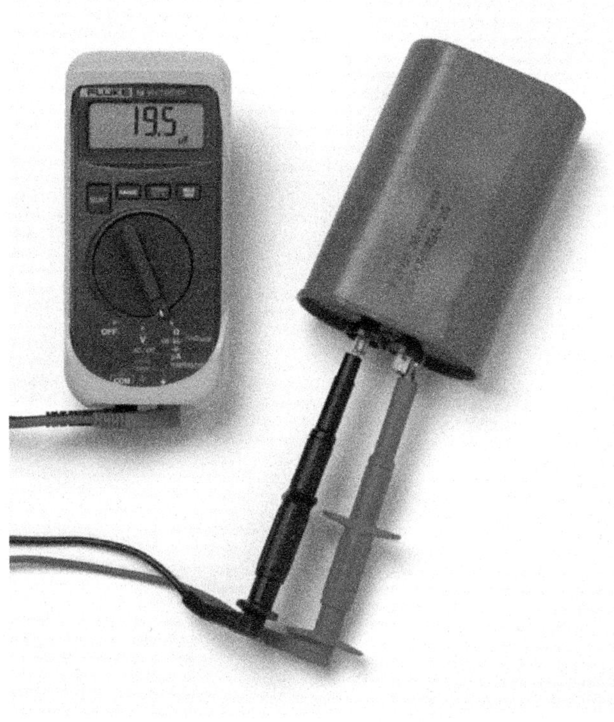

Figure 10-11 A capacitor should be checked with a capacitor checker.

One side of the bleed resistor needs to be removed when checking a capacitor with a capacitor checker or ohmmeter. If the bleed resistor is not removed, the meter reading will indicate a short or the resistance of the bleed resistor. If the capacitor is still usable, the bleed resistor should be installed back on the capacitor prior to operation. Careful planning when removing the resistor will make it easier to reinstall it when the job is finished.

A dual capacitor is checked differently than a single capacitor. Check each side of the dual capacitor for the correct microfarad rating and case ground. One section of the capacitor can be good while the other is defective. The good section can continue to be used, while the defective section can be replaced by a separate, second capacitor. Support the replacement, single-rated capacitor so that it does not dangle by it wire connections.

Another device that is useful when troubleshooting capacitors or motors is the multiple capacitor set, as shown in Figure 10-12. This is available as a start or run capacitor set. The multicapacitor set can be used as a temporary capacitor substitute. The capacitors in this device can be wired in series or parallel to decrease or increase capacitance. (This option is discussed in the next section.) Also, having the option to vary

Figure 10-12 These universal replacement run capacitors can be used as a temporary or permanent capacitor substitute. The universal replacement capacitor can be a run or start capacitor.

capacitance may be useful when troubleshooting a motor. A small improvement in capacitance can start a motor or keep it running. Excessive capacitance is not the answer. More is not better when it comes to sizing capacitors.

TECH TIP

Replace the capacitor and other start components when replacing a defective motor. Some manufacturers require this action in order to fulfill the warranty obligation. Note, though, that it is a good service practice even if the warranty does not require it. Have the motor and replacement components placed on the same receipt to satisfy the warranty requirement.

10.7 CAPACITORS IN SERIES AND PARALLEL

The purpose of wiring a capacitor in series or parallel is to change the total capacitance used by the motor. Hooking capacitors in parallel increases the microfarad rating. Wiring the capacitor in series decreases the microfarad rating.

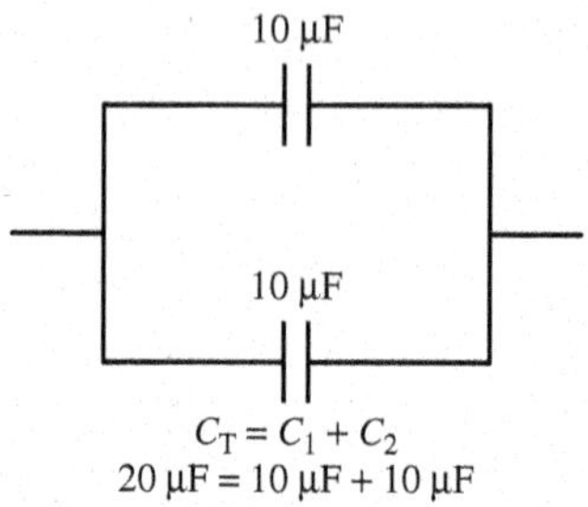

Figure 10-13 These capacitors are wired in parallel. Parallel capacitors increase the microfarad rating because the plate area is increased.

Parallel Capacitors

Capacitors wired in parallel, as shown in Figure 10-13, are used to increase the capacitance. Placing the capacitors in parallel increases the capacitor surface area. The increased surface plate area improves the capacitance or microfarad rating. Therefore, two parallel 10-mfd capacitors increase the microfarad rating to 20 mfd. Hooking additional capacitors in parallel will continue to increase the microfarad rating. After hooking up the parallel capacitors, measure the capacitance with a capacitor checker to verify the wiring and desired microfarad rating. A dual capacitor can be wired in parallel, which will add its microfarad rating together.

The voltage rating of parallel capacitors does not change.

The formula for calculating capacitors in parallel is the same formula that is used to calculate resistors in series:

$$C_T = C_1 + C_2 + C_3 + \text{etc.}$$

EXAMPLE 10.1 FINDING THE TOTAL MICROFARAD RATING

Let's say you have 5-, 10-, and 20-mfd capacitors. What is the total microfarad rating?

SOLUTION

$$C_T = C_1 + C_2 + C_3$$

$$C_T = 5 + 10 + 20$$

$$C_T = 35 \text{ mfd}$$

Series Capacitors

Series capacitors are wired as shown in Figure 10-14. Series capacitors decrease the capacitance because the two middle plates are shared as one plate. The formula for finding the

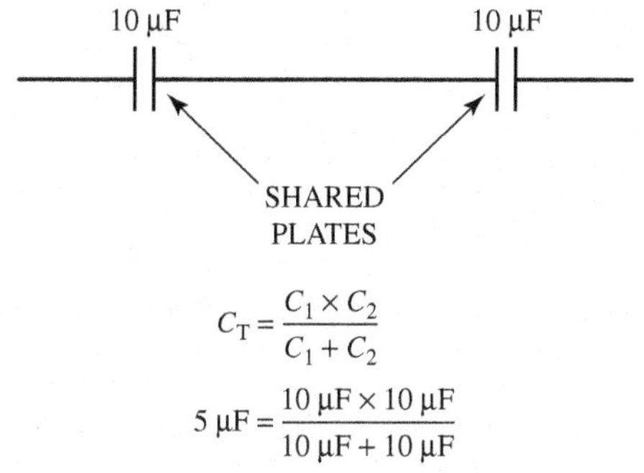

Figure 10-14 These capacitors are wired in series. The two middle plates are shared as one plate.

microfarad rating for series capacitors is the same formula used to find resistors in parallel:

$$C_T = \frac{C_1 \times C_2}{C_1 + C_2}$$

EXAMPLE 10.2 FINDING THE MICROFARAD RATING FOR SERIES CAPACITORS

What is the microfarad rating of two series capacitors that are 10 mfd each?

SOLUTION

$$C_T = \frac{C_1 \times C_2}{C_1 + C_2}$$

$$C_T = \frac{10 \times 10}{10 + 10}$$

$$C_T = \frac{100}{20}$$

$$C_T = 5 \text{ mfd}$$

Capacitors in series follow a couple of general rules:

- The total capacitance in series capacitors will always be less than the smallest rated capacitor in the series circuit. For instance, if you have a 2-mfd capacitor and a 100-mfd capacitor, the total microfarad rating in series will be less than 2 mfd.
- The same rated capacitors in series will be equal to the microfarad rating of one capacitor divided by the number of capacitors in the series circuit. For example, consider three 15-mfd capacitors that are wired in series. Do the calculation: 15/3 = 5 mfd.
- The voltage rating of capacitors in series doubles the voltage that one capacitor can handle. For example, two capacitors in series, each with a 370-V rating, will now increase the voltage rating to 740 V.

TECH TIP

The best way to check a capacitor's integrity and capacitance is to use a quality capacitor analyzer. A capacitor analyzer can measure the capacitor dielectric and determine if the microfarad value is proper or has deteriorated.

If a capacitor analyzer is not available, use a quality analog ohmmeter to check the resistance between the capacitor plates. The needle will deflect toward 0 Ω and slowly rise to a higher resistance. This will not tell you the microfarad rating of the capacitor. Using a DMM on the resistance scale is confusing to the user and not recommended.

10.8 INSTALLING CAPACITORS

Generally, installing a capacitor is not a problem as long as it is placed in the circuit as designated by the wiring diagram. Some run capacitors should be installed in a specific way if one of the terminals is "marked." Capacitors used in the HVACR field are not polarized; however, capacitors used in electronic circuits are polarized. So some HVACR capacitors may be "marked" so that the motor will not be damaged if they short to the metal capacitor case. "Marked" means a terminal can have a negative sign (−), positive sign (+), or white or red paint. Only one terminal will be marked. Capacitor manufacturers vary the mark they use to identify the terminal that should be connected to the power side of the diagram. Even if the capacitor is wired incorrectly, it will work properly unless the capacitor plates short to case ground. In Figure 10-15, the marked side of the capacitor is wired incorrectly. The marked side shows a + sign. If the capacitor shorts to ground, it will burn out the start winding as

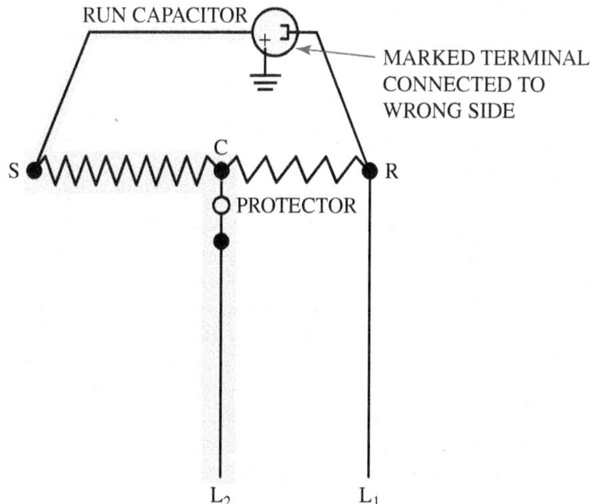

Figure 10-15 This capacitor is hooked up incorrectly. If the capacitor shorts to the metal case, it will keep the start winding in the circuit and burn it out.

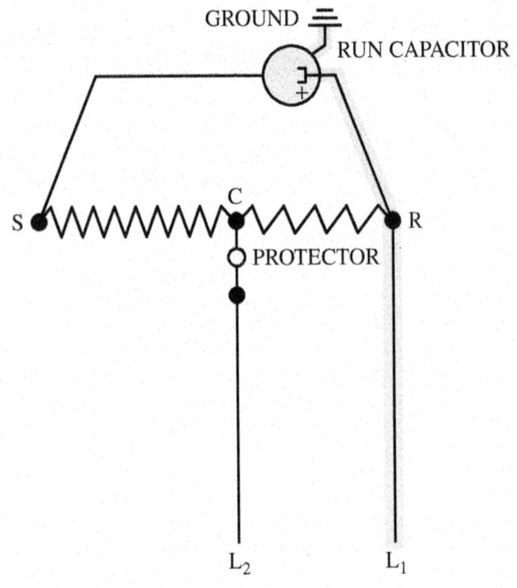

Figure 10-16 This capacitor is correctly wired; therefore, a shorted capacitor will not damage the start winding.

seen by the shaded voltage path from L2, through common, the start winding, and the case of the run capacitor. When this occurs the start winding will burn out. The motor start winding is not designed to have current running through it on a continuous bases. In this instance, a shorted capacitor will take out the start winding, thus damaging the motor. If the capacitor has a marked terminal, it should go to the power supply. When wired this way, the power will go to the shorted case from L1 and not damage any windings. Most likely the overprotection device will blow and stop the motor from operating.

Figure 10-16 shows a shorted capacitor that is wired correctly. Notice that the when the capacitor is shorted to ground it will trip the breaker without damaging the motor start winding. If a capacitor does not have a marked terminal, it can be wired up to any terminal.

SAFETY TIP

Any capacitor showing signs of leaks or bulging must be replaced immediately. Do not operate the compressor with a capacitor showing these signs. Capacitors may contain a charge. Using a voltmeter or capacitor checker may cause them to charge up. Be careful of a shock hazard when handling capacitors.

SERVICE TICKET

You are assigned to a "no cooling" call. Upon a quick inspection using the ACT method of troubleshooting (checking airflow, condenser operation, and thermostat), you find that the compressor is not working. You take control of the condenser by shutting it off at the nearby disconnect. An inspection inside the condensing unit does not reveal an obvious problem. The disconnect is turned on after the manifold gauge set is connected, and the clamp-on ammeter is clamped on the common compressor wire. When energized, the condenser fan begins to operate while the compressor draws high amperage for a couple of seconds but does not start. You suspect that the compressor internal overload prevented the motor from starting. What will prevent a compressor from operating?

You should start by inspecting for obvious signs such as burned wires or damaged capacitors. In this case there was no obvious problem. One easy check is to measure the microfarad rating of the run capacitor. In this unit the run capacitor has a 45-mfd rating at 370 V. A capacitor checker revealed that the capacitance was 6 mfd, which means the capacitor is defective. Checking the service truck inventory, you find 5-, 10-, 20-, and 25-mfd capacitors with 370- or 440-V ratings—no 45-mfd capacitor in stock. It is 4:30 p.m. and the nearest supply house is 30 minutes away and closes promptly at 5:00 p.m. You do not think it is not worth the risk of trying to make it to the supply house. You decide to place a 20-mfd capacitor and a 25-mfd capacitor in parallel to solve the problem.

You explain the options to the customer and she is satisfied with the parallel capacitor arrangement. Even though this solution meant the customer had to pay for two capacitors, she saved the cost of you having to return tomorrow with the exact replacement part and she was also not without air conditioning all night.

SUMMARY

It is important to know the basic operation of a capacitor so that a good motor is not replaced due a lack of knowledge or experience. A capacitor can be used to start or run a motor or both start and keep the motor running. A start capacitor creates a 90-degree current and voltage phase shift to create the maximum starting torque or starting turning power. A run capacitor allows the motor to run at a lower amperage condition plus offers a small amount of starting torque assistance. Single-phase compressors require a run capacitor connected in parallel to the run winding and in series with the start winding. This puts the two windings out of phase from one another and allows the compressor motor to start and run efficiently.

Capacitors are rated in microfarads and voltage. The voltage rating for a start capacitor should be equal to or greater than the starting voltage. The voltage rating for a run capacitor is about 100 volts higher than the supplied voltage. Standard voltage ratings on capacitors are 370 and 440 V. A higher voltage rating can be used, but using lower voltage rated capacitors will cause the capacitor to overheat and burn out.

Capacitors can be wired in series or parallel to, respectively, decrease or increase their microfarad rating. Calculating capacitance in series or parallel is the opposite of

Table 10-1 Limits and Applications for Run and Start Capacitors

Limits for Run Capacitor Ratings

Specific Rating	Maximum Addition
10 to 20 mfd	+ 2½ mfd
20 to 50 mfd	+ 5 mfd
Over 50 mfd	+ 10 mfd

Facts about Capacitors

Capacitor Type	Compressor Motor Type	Characteristics
Start Capacitor	CSIR and CSR	• Designed to operate for only a few seconds during start • Taken out of start winding circuit by relay • Excessive start capacitor mfd increases start winding current, increases start winding temperature, and may reduce start torque • Capacitors in CSR motors should have 15.000 ohm, 2 watt bleed resistor across terminals • Capacitor rated voltage must be equal to or more than that specified • Capacitor mfd should not be more than that specified
Run Capacitor	CSR and PSC	• Permanently connected in series with startwinding • Excessive mfd increases running current and motor temperature • Fused capacitors not recommended for CSR and not required for PSC motors • Capacitor rated voltage must be equal to more than that specified • Capacitor mfd should not exceed rated limits

(Courtesy Tecumseh Service Handbook.)

calculating resistance in series and parallel circuits. Knowing this information will allow you to get a piece of equipment running without having to run to the HVACR supply house.

When troubleshooting a capacitor use a capacitor checker unless the capacitor is obviously damaged. The microfarad rating must be within 10% of the labeled rating to be considered good. Very few new capacitors are defective right out of the box. Test new capacitors for the correct microfarad rating prior to installing.

Finally Table 10-1 provides valuable information on the limits and applications for run and start capacitors. In short, the table summarizes many of the topics that were discussed in this unit.

REVIEW QUESTIONS

1. Name two common types of capacitors.
2. Describe the differences in the physical appearance of the two common types of capacitors.
3. What are the two important capacitor ratings?
4. Briefly describe what a capacitor does in a motor circuit.
5. Which capacitor is used to improve the efficiency of a motor?
6. What is the best way to troubleshoot a defective capacitor that appears to be good?
7. What is the microfarad rating of two 5-mfd capacitors that are connected in series?
8. What is the microfarad rating of two 5-mfd capacitors that are connected in parallel?
9. Why is it important for a run capacitor to have a high voltage rating?
10. A tech is checking a capacitor with an ohmmeter. Reading across the terminals, the capacitor charges to 0 Ω, then drifts toward the infinity scale. A 2-Ω reading is found from each terminal to the metal case. Is there a problem? If so, what is the problem?

UNIT 11
Thermostats

WHAT YOU NEED TO KNOW

After studying this unit, you will be able to:

1. Draw a basic heat-cool thermostat diagram.
2. Recite the four thermostat wire color codes and connection points.
3. Wire a thermostat to a control circuit.
4. Discuss the difference between a mechanical and digital thermostat.
5. List the features and benefits of a programmable thermostat.
6. Describe the purpose of the heat anticipator.
7. Describe the purpose of the cooling anticipator.

The **thermostat**, or *t'stat* as it called in the field, is an automatic switch used to control an HVACR system. The word *thermostat* has Greek origins. *Thermos* means "hot" and *statos* means "standing." The first systems to use thermostats controlled heating systems, and keeping a space at a "hot standing" temperature was desired. Cooling control came later with the advent of air conditioning.

The word *controller* is also used when referring to a thermostat. The controller regulates the heating, cooling, or refrigeration system. Sometimes the word *controller* is used when referring to **digital** or **solid-state thermostat** controls. Most people simply call it a thermostat.

Most thermostats operate several parts of an HVACR system, including the blower and the cooling and heating components. The blower is normally a fan with a duct system that evenly distributes the air to given spaces. The cooling or heating arrangement may control airflow or water flow.

A thermostat is a device that regulates the temperature inside a designated area. The designated area could be a building or a refrigerated box. The purpose of the t'stat is to maintain a desired **setpoint** temperature in the space. The setpoint is the temperature at which the thermostat is set.

The human comfort range for most people is 72° to 80°F and 40% to 60% relative humidity. Sometimes the indoor air in the winter months gets too dry. A dry indoor environment, low humidity, will create a static electricity shock when touching a door knob or someone else. A **humidistat** is used to control the moisture or humidity level in an area. A humidistat can be used in conjunction with a humidifier or dehumidifier. A humidifier adds moisture to an area, whereas a dehumidifier removes moisture from an area.

This unit focuses on the common low-voltage, 24-V thermostat used to control blower operation and maintain desired heating or cooling temperatures. The thermostat can turn on or off the flow of air or water that is used for heating or cooling. The thermostat uses a sensor to cycle the equipment off and on.

The oldest thermostatically controlled heating units in the late 1800s used a mercury switch. Today's modern digital thermostats use a solid-state **thermistor** to sense the air or water temperature. The semiconductor thermistor element changes resistance with the slightest change of temperature. In a digital, solid-state thermostat, the resistance is converted into a signal that will operate the various functions that the t'stat may have available. A thermistor's solid-state sensor is so sensitive that it can maintain a space temperature within ±0.5°F of its setpoint. This sensitivity allows the thermostat with a thermistor sensor to take thermal measurements to within ranges of ±0.02% to ±0.05%, an accuracy that makes them very effective thermometers.

11.1 TYPES OF THERMOSTATS

When you use the word *thermostat,* most people think of the control on the wall that operates a heating or cooling system. Figures 11-1 and 11-2 show common thermostats found in homes or business applications. The **mechanical thermostat** uses moving parts to achieve temperature control. The **digital thermostat** has few if any moving parts. The digital t'stat uses solid-state technology to achieve temperature control.

Figure 11-1 This digital thermostat has a Cool Setting of 74°F. If the inside temperature is 75°F or warmer, the system would be in the cooling mode. The fan is in the Auto selection, which means the blower will cycle with the condensing unit.

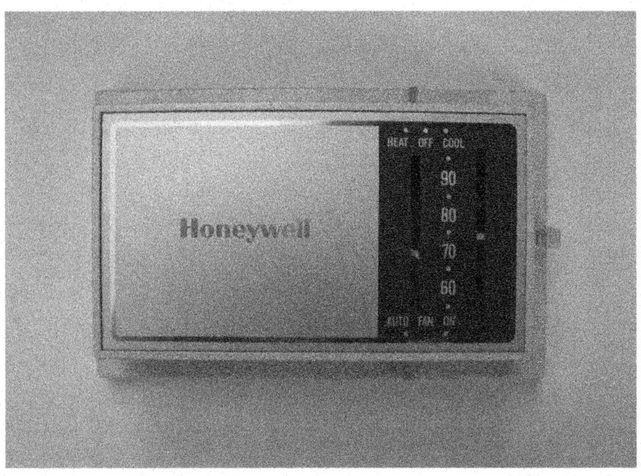

Figure 11-2 This is an example of a mechanical heat-cool thermostat. The room thermometer registers about 67°F.

This unit focuses on the digital or solid-state thermostat because their use is so widespread. Digital thermostats are more reliable and have fewer moving parts than mechanical ones, and they keep the temperature within ±1°F of its setpoint. Their display is easy to read and interpret. We do, however, still need to understand the basic operation of the old-style mechanical thermostat because some are still in use, so let's discuss that type first.

A mechanical thermostats is secured to a **subbase** that is mounted to the wall. With the cover off the thermostat, as shown in Figure 11-3, we can see the control wires attached to the subbase. A close-up of a subbase is shown in Figure 11-4. Each subbase is designed for a specific thermostat.

Next, there is the **line voltage thermostat**, as shown in Figures 11-5 and 11-6. In Figure 11-6 the sensing bulb is placed in the return air stream. When the temperature of the air or water warms the bulb, it causes the pressure diagram or bellows to expand, pushing the movable contacts closed to operate the cooling or refrigeration circuit.

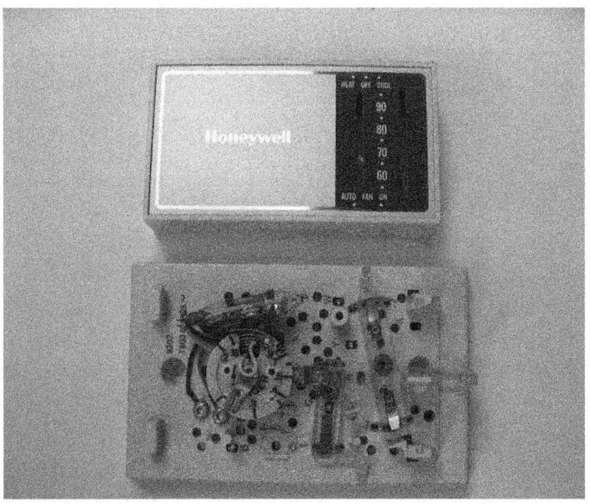

Figure 11-3 The cover has been removed from this mechanical thermostat. This type of thermostat would be mounted on a subbase. The subbase would be mounted to the wall and control wires would be attached to the subbase.

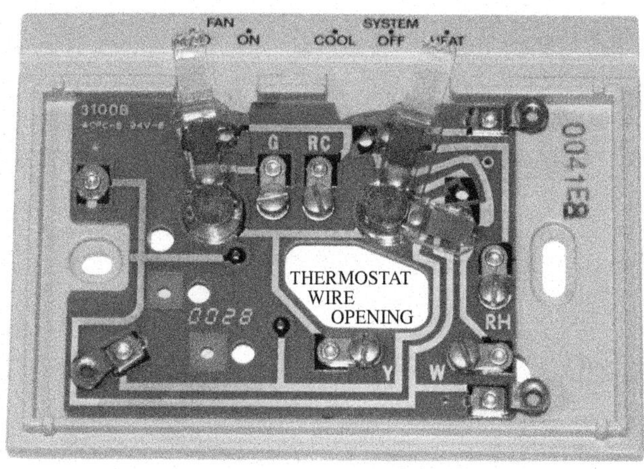

Figure 11-4 This is a thermostat subbase. The subbase is mounted to the wall, and the thermostat is screwed onto the subbase.

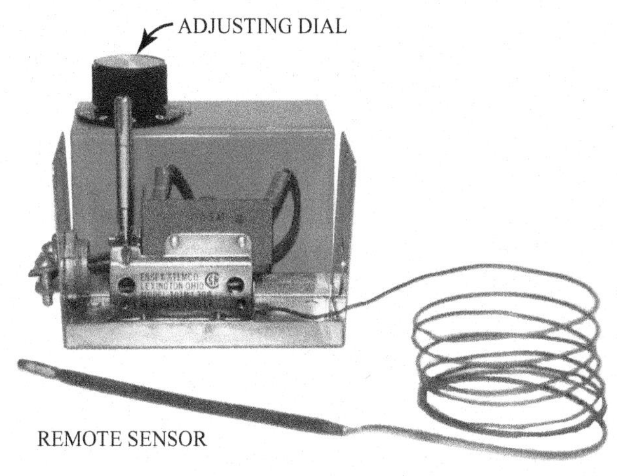

Figure 11-5 This is a line voltage thermostat. The blue cover has been removed to expose the inside parts. The remote sensor is usually refrigerant charged to respond to air temperature changes. The sensor should be in the return air stream.

Line voltage thermostats have a pressure-charged sensing bulb that should be in the return air stream. You might find this type of thermostat in a walk-in cooler, a domestic refrigerator, or on a window unit just behind the air filter. The first thermostats used in heating and refrigeration applications were line voltage thermostats. Even today, line voltage thermostats are used in domestic and commercial refrigeration systems. They are also found in window air conditioners. In many designs the line voltage thermostat must be able to handle the current draw of the compressor since it controls the compressor and a blower motor directly. They are selected to handle the temperature range and the contact's RLA and LRA. The temperature range is:

- High temperature for air conditioning and/or heating applications

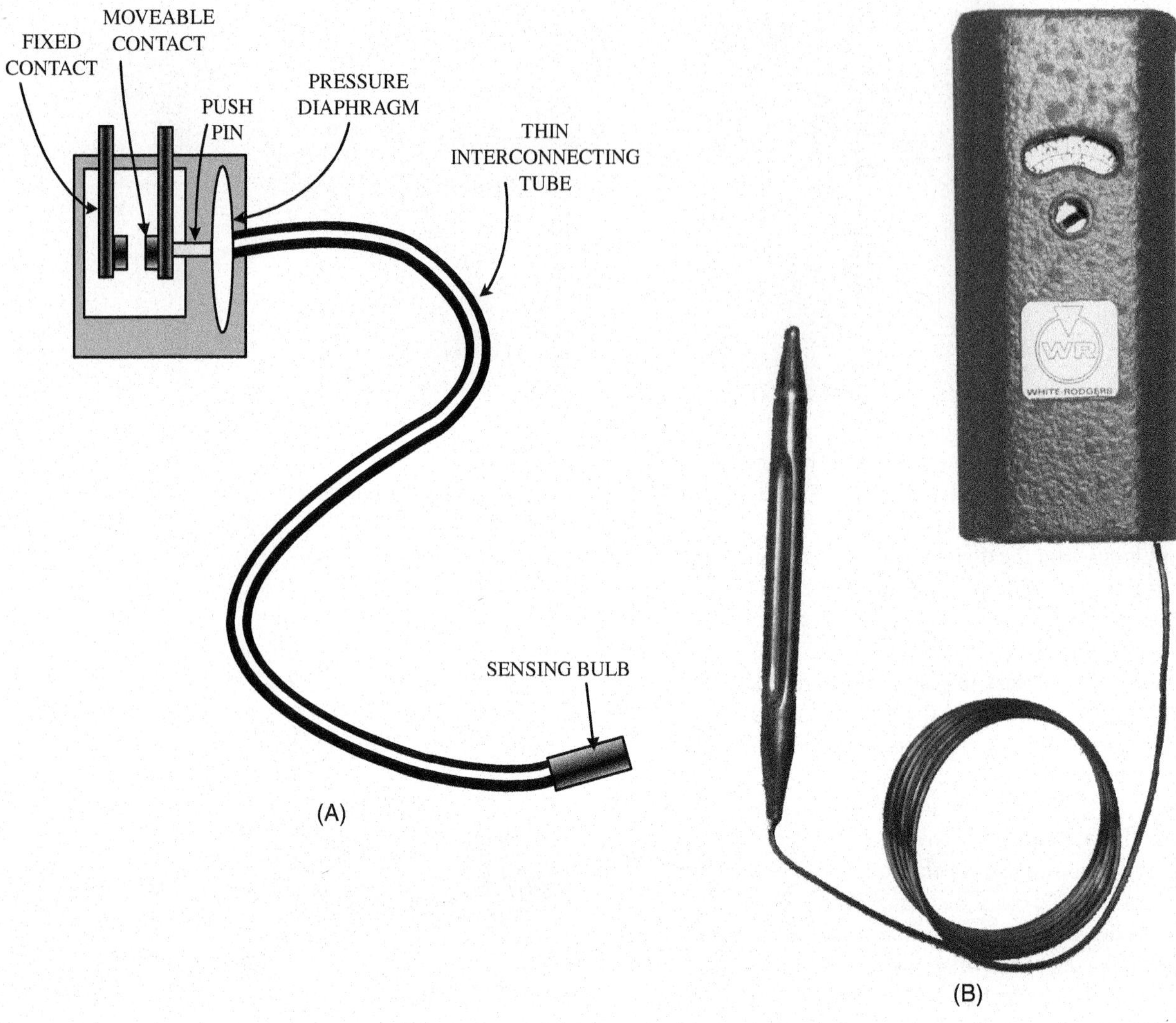

Figure 11-6 (A) Cross-sectional view of a line voltage thermostat. (B) Line voltage thermostat with a sensing bulb.

- Medium temperature for refrigeration (≈40°F) applications
- Low temperature for freezing applications.

Many other types of temperature controls are used in our industry. Not all thermal switches are called thermostats. For example, the internal or external overload in a compressor is a thermostat that opens the power circuit when too much current is drawn or the motor winding becomes too hot. There are discharge line thermostats that open the control circuit to the compressor if the discharge line becomes excessive.

11.2 PARTS OF A THERMOSTAT

The thermostat is a thermal switch that controls heating, cooling, or refrigeration. A common comfort heating and/or cooling thermostat is shown in Figure 11-7 and is composed of two parts:

- *Thermostat:* thermal switch and switch options such as cooling, heating, and fan on/off operation
- *Subbase:* connection for the thermostat wires and wall mounting holes. The thermostat fits over the subbase.

Not all thermostats have a subbase.

11.3 OPERATING VOLTAGE

The thermostat can be selected to operate on low voltage or high line voltage. A low-voltage thermostat can handle around 30 V. High voltages can damage the device. Line voltage thermostats can handle low or high voltages.

Most thermostats used in residential or light commercial air conditioning applications use a 24-V control design. Some residential and commercial systems use low-voltage DC systems. The term *comfort control* is slowly replacing the word *thermostat*. You will hear comfort control used when the system uses a DC voltage or for systems that use communication signals between the thermostat and the rest of the system. Finally, almost every refrigeration system and window unit uses line voltage thermostats.

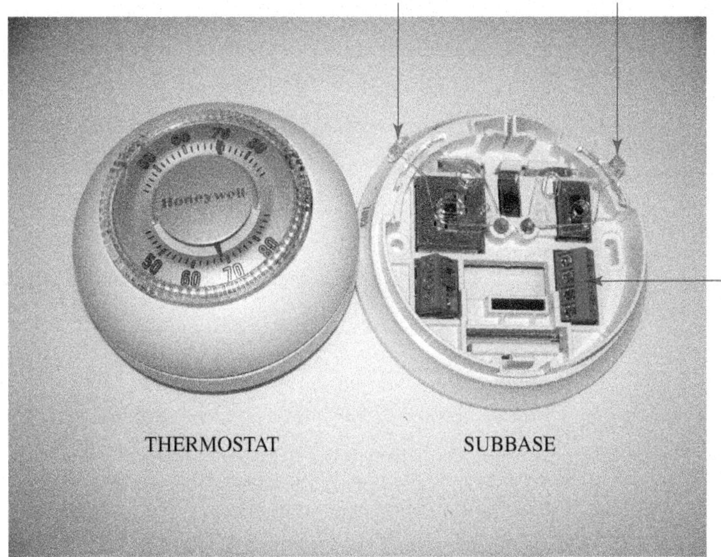

Figure 11-7 The thermostat is composed of two major parts. The thermostat has a thermal switching device, which shows the room temperature. The subbase has the wire connections and switching selections such as heating, cooling, and fan on/off operation.

Here is a breakdown of control voltages found on thermostats or comfort controls:

Low Voltage

- 24 V
- Low DC voltage.

Line Voltage

- 120 V
- 208 to 240 V
- 460 to 480 V.

Figures 11-8, 11-9, and 11-10 show three views of an air conditioning line voltage thermostat. Shown are the front (Figure 11-8), the thermostat with the cover removed (Figure 11-9), and the back (Figure 11-10). These types of thermostats do not have sensing bulbs like the line voltage thermostats shown in the previous figures. To be effective they must be near the return air stream to control air temperature.

Notice in Figure 11-9 that the line voltage thermostat is less complex than a low-voltage thermostat. Figure 11-10 illustrates the factory 14-AWG wire gauge, which is much larger than the 18-AWG wire found in low-voltage temperature control devices.

Figure 11-11 appears to be a line voltage thermostat, but it is a low-voltage device. The amperage rating on the

Figure 11-8 This line voltage thermostat has a built-in sensor; therefore, it should be mounted on the wall near the return air stream. It is an air conditioning line voltage thermostat.

Figure 11-9 This is the same line voltage thermostat as shown in Figure 11-8, but the cover has been removed to reveal its simple design. The amperage rating is located on a tag inside the thermostat cover.

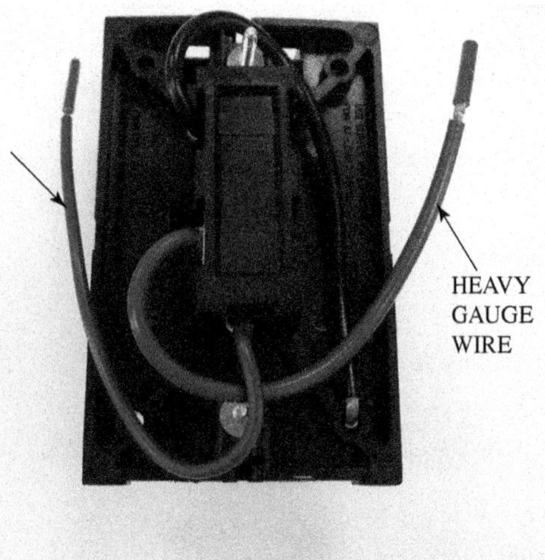

Figure 11-10 This is the back of the line voltage thermostat shown in Figure 11-8. The factory wiring is a heavy-gauge wire that can handle higher amperage conditions than a low-voltage thermostat.

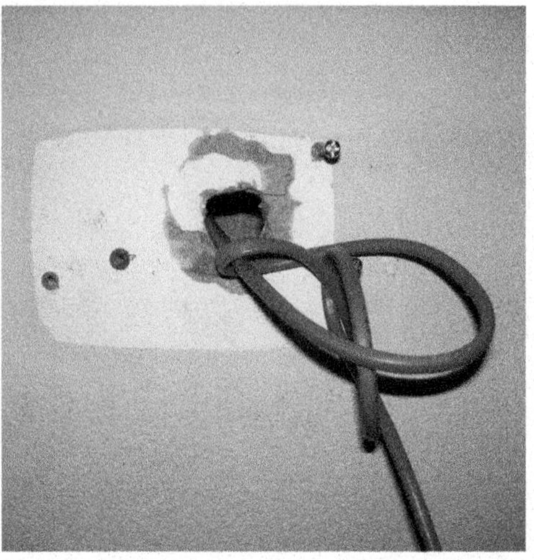

Figure 11-12 The hole behind the thermostat will need to be sealed to prevent air in the wall from affecting the thermostat. Even an inside wall can have outside air influences. The hole can be sealed with expanding foam or by caulking it.

Figure 11-11 This is a low-voltage thermostat that has a similar appearance to a line voltage thermostat. This thermostat is often used in commercial applications, but can also work in a residence.

thermostat will determine if it is a line voltage or low voltage thermostat. This thermostat is often used in commercial applications, but can also work in a residential installation.

11.4 THERMOSTAT INSTALLATION

The installation of a thermostat's control will affect its operation. The installation should be in a location that is free from outdoor temperature influences like the sun or drafts near an outside door. A thermostat mounted on a wall that is directly connected to the outside will be affected by outdoor and radiant temperatures on the wall. The t'stat should be isolated from internal heat sources such as lights, electronics, or people. The best location is on an inside wall that is near the return air grille.

Here are good thermostat installation practices:

- Follow manufacturer's recommendations.
- Mount 52 to 60 inches above the floor.
- It should be level, even if it is a modern electronic thermostat.
- Conditioned air should not blow on it.
- Do not install on an outside wall or behind a door or in corners.
- Do not install where the sun or radiant heat from chimneys or pipes can affect it.
- Do not mount in a kitchen.
- Mount near the return air stream. A central hallway location is a good choice.
- Seal the wall opening around the t'stat wire. The air coming around this opening may affect the thermostat's operation. Even an inside wall can have outside air influences. Figure 11-12 shows a hole that will need to be sealed.
- Normally 18-AWG multiconductor wires are used when hooking up a t'stat. If a run is over 100 feet, larger gauge wire should be used.

11.5 THERMOSTAT WIRING

A basic thermostat uses four wires tied to connections **R**, **Y**, **W**, and **G**. The following color code should be used to make it easier to trace conductors between the thermostat, air handler, and condensing unit:

R = red wire **Y** = yellow wire

W = white wire **G** = green wire

TECH TIP

Here is a great tip from one of the educators who reviewed this book:

Red = Hot, power.
White = Winter snow; needs heating.
Green = The letter G looks like a blower housing from a side view.
Yellow = Color of the sun; needs air conditioning.

Now we will discover what each of these thermostat terminals controls. The back of the thermostat or thermostat subbase has terminal connections with many different letters. The purpose of this section is to explore what these letters control. Figure 11-7 shows what you might expect on the subbase or back of a thermostat.

R Terminal

The R terminal supplies power from one side of the 24-V control transformer. The transformer is usually located in the air handler or heating system or in the control section of a package unit.

RC Terminal

If an RC terminal is present, it is designed for the power for the cooling control. Some HVAC systems have two transformers. One transformer is for cooling and one is for heating. In this case the power from the transformer in the cooling system would go to the thermostat. A jumper will need to be installed between the RC and RH terminals for a system with one transformer.

RH Terminal

The RH connection is for power for the heating circuit. As previously noted RC and RH can be jumped (connected) for combined heating and cooling operation.

Y Terminal

The Y or Y1 terminal controls the air conditioning circuit. The Y wire goes to and makes a connection to the contactor coil. It may pass through the air handler terminal before passing to the condensing unit. Sometimes the Y has a "1" with it: Y1. This means this terminal controls the first stage of a two- or three-stage cooling system.

Y2 Terminal

A Y2 terminal is used for the second stage of cooling if a system has that option. A light blue wire is used to wire up the second stage of cooling, but color coding at this level varies.

W Terminal

The W or W1 terminal controls the heating circuit. The W wire energizes the heating control such as a gas valve or electric heat sequencer. If the terminal is marked "W1," then this control is for the first stage of a possible multiple-stage heating system.

W2 Terminal

A W2 terminal is required for two-stage heating, which is usually found on a heat pump system. The first stage of heating is the heat pump system; the second stage is the auxiliary heat such as an electric heat strip. A brown wire may be used to hook up the control circuit.

G Terminal

The G terminal is used to control the indoor fan blower. The G circuit operates the fan in the cooling mode. Some thermostats do not energize the G circuit in the heating mode when the thermostat is set in the Auto position. In the heating mode, the Auto fan is energized through other wiring designs.

C Terminal

The C, or common connection, is the other power wire coming from the transformer. Use of the C or common does not mean that this circuit is necessarily grounded or negative. Most mechanical thermostats do not have a C terminal. The C terminal is common on digital thermostats because they require a continuous 24 V. In some cases a black conductor is used to designate this connection.

O or B Terminal

The O or B terminal is found on heat pump thermostats. One of these terminals will be wired to the reversing valve solenoid. The O terminal will be wired to the reversing valve solenoid when the solenoid is energized in the cooling mode. The B terminal will be wired to the reversing valve solenoid when it is energized in the heating mode. The orange wire is used to wire the O terminal, and the dark blue wire is used to wire the B terminal. Many heat pump manufacturers use the O connection design. If the heat pump solenoid fails when using the O connection (cooling), the system will be in the heating mode. For human survival, heating is more important than cooling; therefore, many heat pump manufacturers prefer to use the O connection to energize the cooling mode.

E Terminal

The E connection is used to energize emergency heat in a heat pump system. This is usually the second stage of heating. When switching the heat pump t'stat to emergency heat, the thermostat turns off the heat pump and energizes all the auxiliary heat. The E or emergency setting also bypasses all outdoor thermostats and allows 100% heating operation. There is no specific wire color for the E terminal.

Table 11-1 Summary of Terminal Connections for Use with 24-V Control Circuits

Terminal Connections and Common Wire Colors			
Terminal	**Wire Color**	**Controlled Device**	**Explanation**
C	Black	24-VAC common side of the transformer.	From one side of the 24-VAC transformer. This side of the transformer may be grounded.
R or V	Red	24-VAC power to be switched.	The power side of the 24-VAC transformer.
Rh or 4	Red	24-VAC power for the heating control.	Same as R, but the "h" shows dedication to the heating part of the thermostat.
Rc	Red	24-VAC power for the cooling control.	Same as R, but the "c" shows dedication to the cooling part of the thermostat.
G	Green	Fan control.	Fan switch on thermostat. It is connected to R when the On or Auto switch is in the fan position.
W	White	Supplies 24 V to operate the heating and humidifier circuits.	Is connected to R or Rh when thermostat calls for heat. Might be jumpered to Y on a heat pump.
Y	Yellow	Supplies 24 V to operate the cooling circuit.	Is connected to R or Rc when thermostat calls for cooling. Also cooling or first-stage heating on a heat pump. Connected to G when Fan switch is set to auto.

Heat Pumps and Multiple-Stage Systems			
Terminal	**Wire Color**	**Signal**	**Description**
Y2	Blue or orange	Second-stage cooling.	The first stage of cooling will be energized prior to the second stage of cooling.
W2	Varies	Second-stage heating.	First-stage auxiliary heating on a heat pump.
E	Blue, pink, gray, or tan	Emergency heat relay on a heat pump. Active all the time when selected; usually not used.	Disables the heat pump and turns on all stages in the auxiliary heating mode.
O	Orange	Energize the reversing valve in the cooling mode.	Energize to cool and activate the cooling mode.
B	Blue, black, brown, or orange	Sometimes common side of transformer. Needed on some electronic thermostats or if you have indicator lamps. OR B energizes the reversing valve in the heat pump in the heating mode. York and Trane sometimes use B as common.	Can be heating changeover if used in a heat pump or common on connections of older thermostats.
X	Varies	Used as a thermostat power to operate lights or other indicators.	May be common or used to energize the emergency heat relay.
X2	Varies	Second-stage heating or indicator lights on some thermostats.	This function varies, depending on the thermostat design.
T	Tan or gray	Outdoor anticipator reset.	Used on GE/Trane/American Standard and some Carrier products.
L	Varies	Service light	

Aux Terminal

The Aux terminal is the auxiliary terminal. This is used in the heating mode of a heat pump thermostat. Some refer to this as backup heat. There is no universal color for this connection. It should be wired to the backup heating circuit.

S1 or S2 Terminal

The S1 or S2 terminal is the rarest of all terminals. It is used for outdoor air sensors. Many uses require a separate, shielded cable to prevent the electromagnetic force generated by other wires from interfering with the signal inside the wire.

11.6 SUMMARY OF TERMINAL CONNECTIONS

Table 11-1 provides valuable information that you can use when working with most 24-V control circuits. *Caution:* Not all systems will be wired using this color code. Also, wires spliced together may change color. Not all technicians are able to distinguish color differences.

11.7 MECHANICAL THERMOSTATS

Thermostats are nothing more than temperature-operated switches with the addition of a fan control. There are few new 24-V mechanical thermostats available for installation or replacement. Yes, there are tens of thousands of mechanical thermostats still doing a good job controlling temperatures. But digital thermostats are so superior that they are replacing the mechanical thermostats of the past. However, the solid-state design and electrical diagram of the digital thermostat is protected and rarely made available by the manufacturer. And even if we had the electrical diagram for a digital thermostat, it would be difficult to understand without an electronics background.

So this section discusses the mechanical thermostat to give you an understanding of how a thermostat works. Knowing how a mechanical thermostat works will transfer to understanding the digital type.

Simple Control Voltage Thermostat

We will build your knowledge of temperature control by first looking at the simple thermostat shown in Figure 11-13. This diagram is a plain drawing of a control voltage thermostat. The thermostat has dashed lines around it. The dashed lines indicate that the component or wiring is installed in the field. In addition, **R** is power from the transformer, **Y** is power to the air conditioning contactor coil, and **G** is power to the fan relay coil. The "common" connection is not labeled with a letter, but it is on the bottom side of the transformer completing the power to the contactor coil and fan relay.

Let's review the sequence of operations of Figure 11-13:

1. First the switch (point 1), SW, must be closed to supply 24-V power to the thermostat.
2. Reducing the temperature setpoint will cause the thermal switch to close (point 2), sending power to the A/C contactor coil and fan relay coil. The contactor should energize the compressor and condenser fan. The fan relay will energize the blower that will circulate cool air. When the desired temperature is reached, the thermostat will open, killing power to the contactor and fan relay.

Adding a Heating Circuit to the Diagram

The diagram shown in Figure 11-14 is another step toward developing the common thermostat diagram. This thermostat includes a heating circuit. This thermostat is more complex than the one shown in Figure 11-13. The heating circuit **W** is added to this circuit. Let's go through the sequence of operations:

1. The sequence of operations starts with closing SW (point 1). The fan is in continuous operation when the

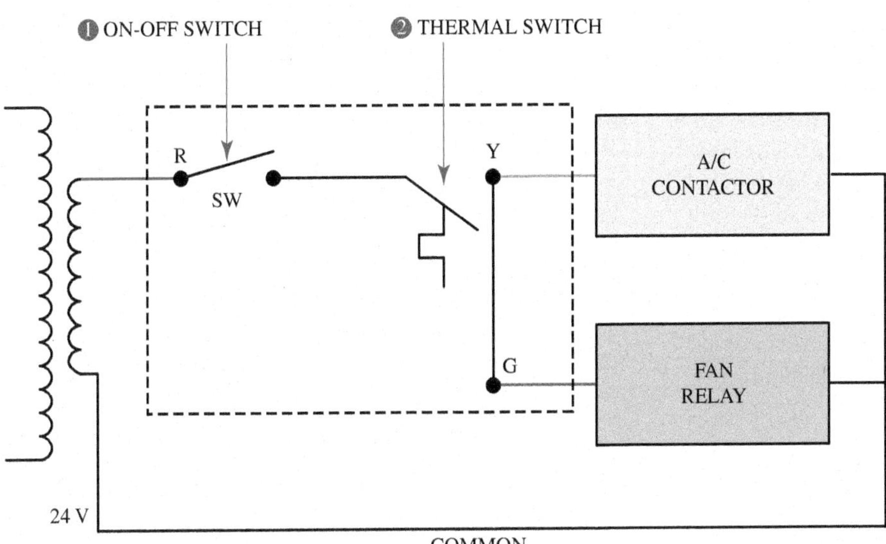

Figure 11-13 This is a simple, cooling-only thermostat.

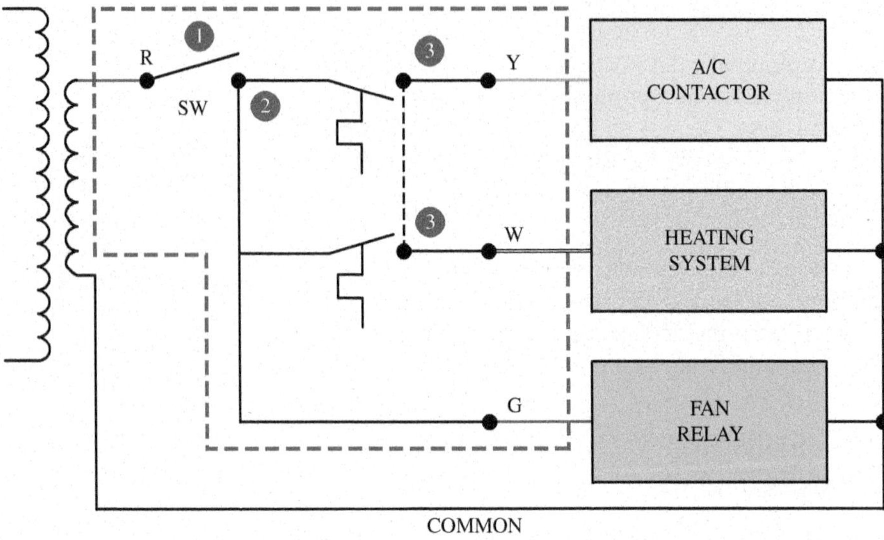

Figure 11-14 This thermostat is more complex than the one shown in Figure 11-13.

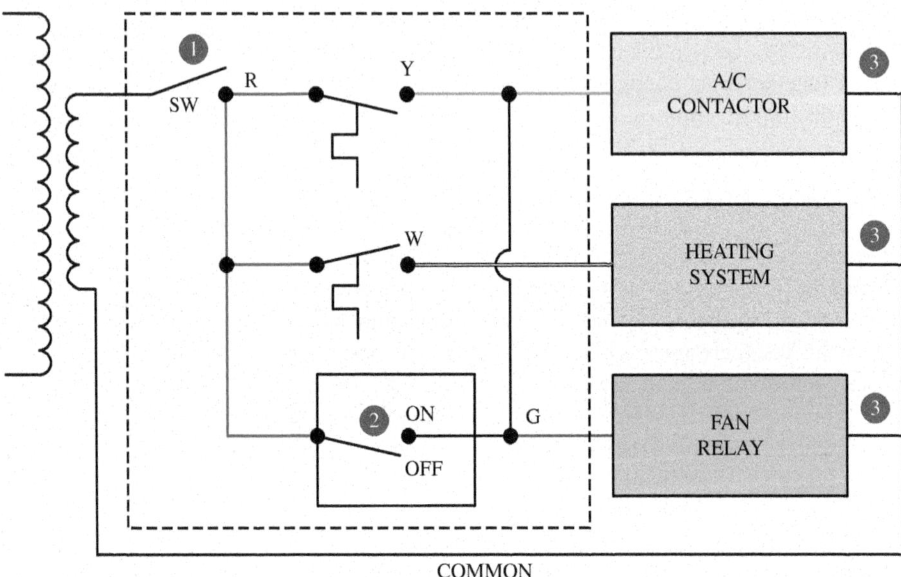

Figure 11-15 The next step in making the thermostat more complex is providing fan control. When the fan switch is in the Off position, the fan will not operate until the heating or cooling circuits are energized. The fan switch can operate the blower continually if it is in the On position.

thermostat switch, SW, is closed. Notice the power supply path from point 2 to **G** and to the fan relay.

2. Power is supplied to the three parallel circuits (point 2): cooling, heating, and fan operation. The fan relay is energized.
3. The cooling or heating circuits operate on their individual thermostat (points 3). The dashed lines between Y and W let the technician know that these switches are mechanically linked together. As one switch is closed, the other is opened. This prevents cooling and heating from operating simultaneously. The difference in temperatures between the two temperature settings is called the **deadband temperature,** which means that the heating and cooling systems do not operate when the temperature is between these two setpoints. The deadband is usually about 5 degrees. The deadband prevents the system from going to cooling then to heating and back to cooling again.

Adding a Fan Control to the Diagram

Figure 11-15 is the final example we will review prior to learning about a common thermostat. This diagram adds a fan control (not in the heating mode). Many heating systems use a different method to control the fan operation. For example, the blower in the heating mode may be operated after the thermostat calls for heat and a time delay of 1 minute. Another way to operate the blower is to use a thermal switch. Once the furnace heat exchanger gets hot enough, a thermal switch closes to operate the blower motor. This allows the heat exchanger to heat up prior to blower operation.

The cooling part of the thermostat in Figure 11-15 simultaneously operates the indoor blower and the condensing unit:

1. At point 1 the thermostat switch will need to be closed to operate the t'stat. This is the system's on/off switch.
2. At point 2 the fan switch On position will select continuous fan operation. The fan Off position will operate the fan in the cooling mode only. The fan in the heating mode is controlled by another switch in the heating circuit, not in the thermostat. There is a connection from the **Y** terminal to the **G** terminal. When the cooling setpoint closes the thermostat contacts, 24

SEQUENCE OF OPERATION 11.1: FAN CIRCUIT OPERATION

Using Figure 11-16, let's look at a step-by-step explanation of how the thermostat operates the fan circuit:

1. Trace the easiest circuit, the Fan Relay, first. At point 1, turn the fan switch to the On position. There will be a control voltage path from the top side of the transformer to the **R** terminal to the closed Fan switch.
2. Next, power goes through the closed Fan contacts to the **G** terminal to the Fan Relay coil (point 2).
3. The other side of the coil, point 3, is wired to the transformer common connection. The Fan Relay will now energize the indoor blower. The following discussion of cooling and heating circuits will have the fan switch in the Auto position.

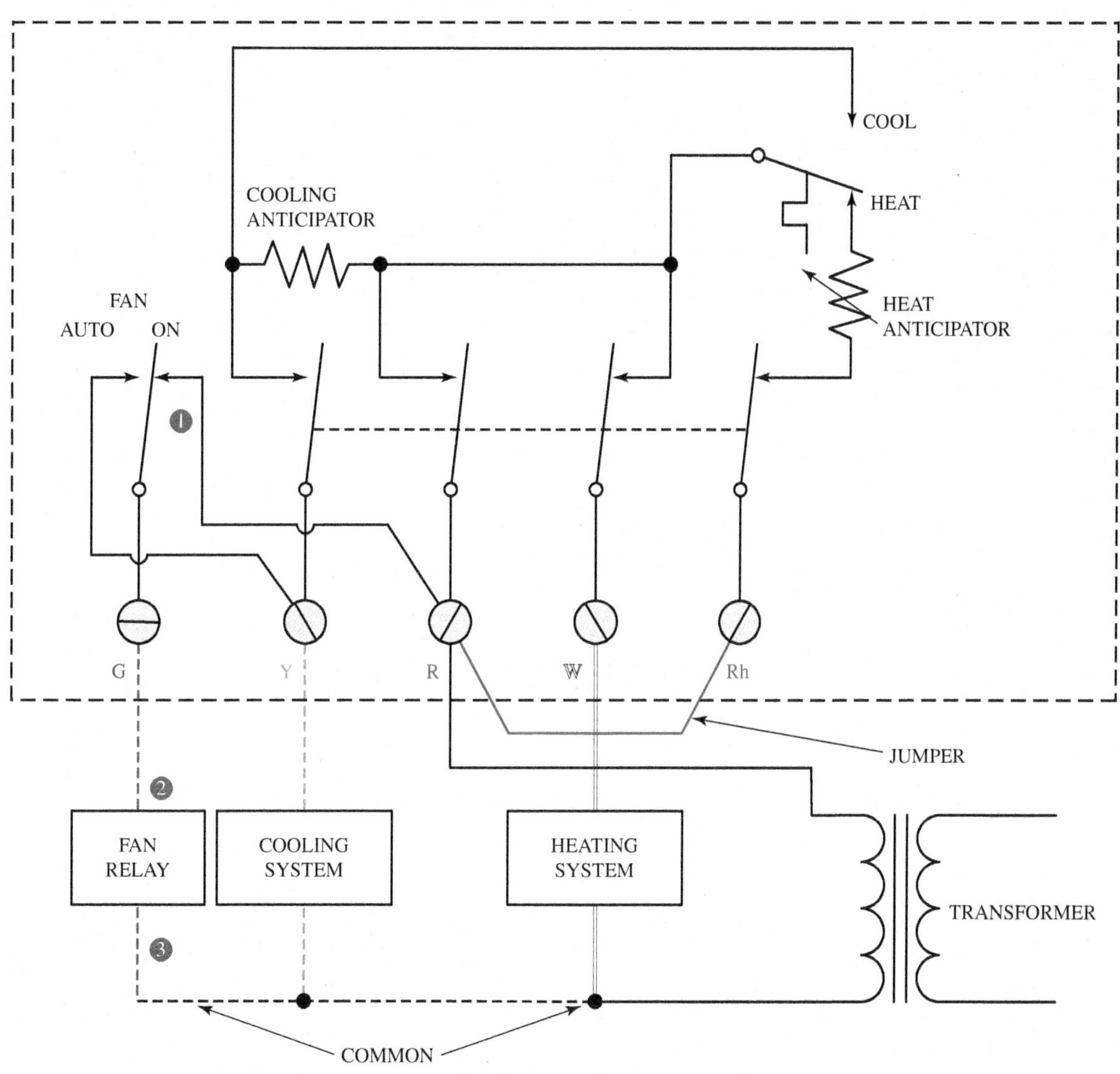

Figure 11-16 This is a common diagram of a mechanical thermostat with fan, cooling, and heating controls. Notice that the thermostat is set in the continuous fan operation and heating position.

volts is applied to the condensing unit and fan relay. The compressor, condenser fan, and indoor blower begin to operate.

3. At point 3, a common connection is provided, completing the circuit to the transformer.

The thermostat diagram has many pathways. In this section we simplify and break down the major pathways for a thermostat. We also explore the sequence of operations of the fan circuit, heating circuit, and cooling circuit.

SEQUENCE OF OPERATION 11.2: HEATING CIRCUIT OPERATION

We will use Figure 11-17 to review the sequence of operations of the heating circuit. With the fan switch in the Auto position and the heat thermostat contacts closed, we can trace the heating circuit. Starting at the top of the Heating System control:

1. Start at the **W** terminal (which is found above the heating system control).
2. Go through the thermostat closed contacts and heat anticipator.
3. Go through the **Rh** and **R** terminals.
4. Now go to the hot side of the transformer.
5. Complete the circuit with power from the Common side of the transformer to the bottom side of the Heating System control.

Note: The Fan Relay is not energized through the thermostat when selecting the Heat-Auto position. There will be another way the fan is energized, but not through the thermostat.

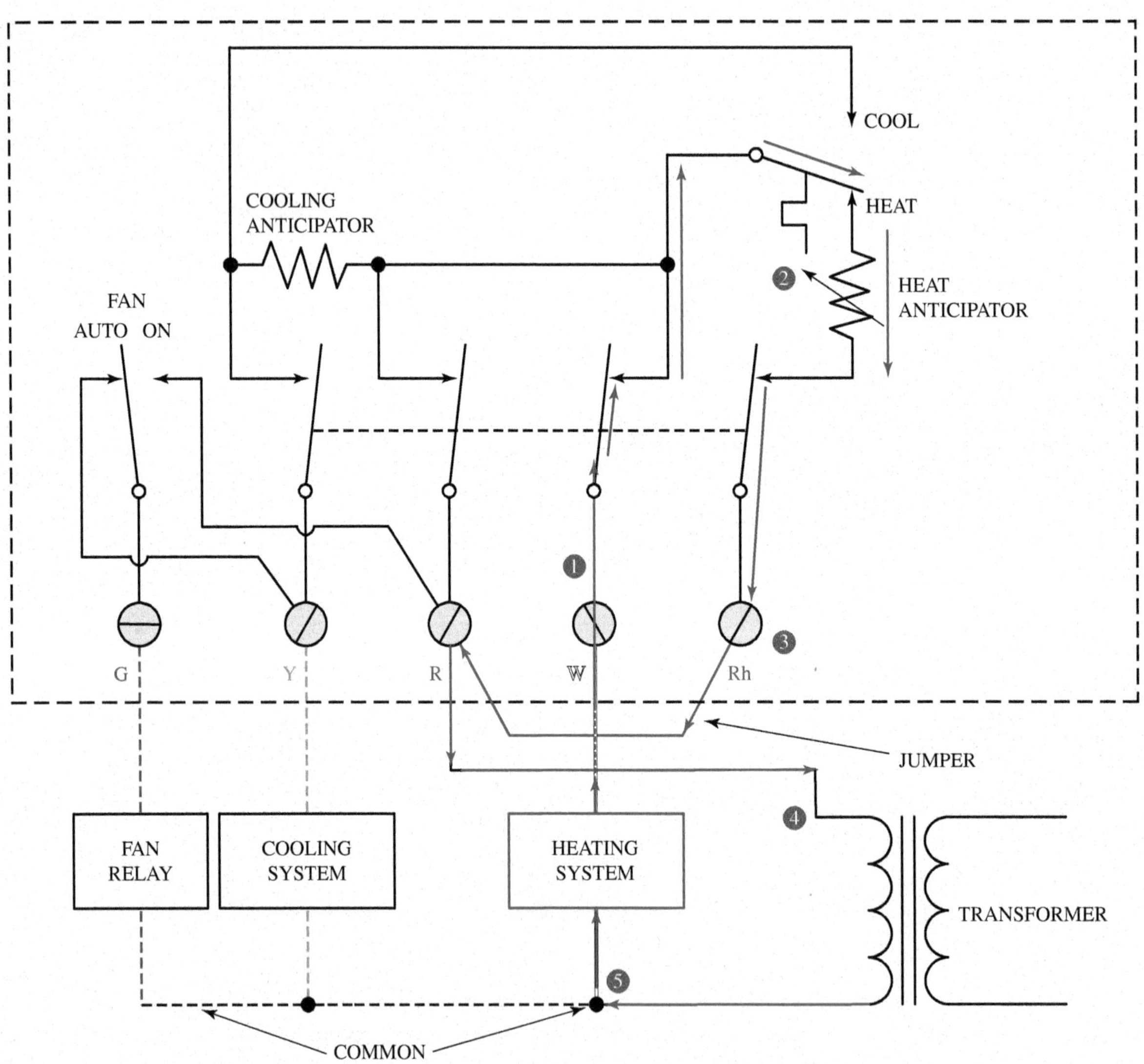

Figure 11-17 This diagram shows a thermostat control circuit through the heating side. The fan is in the Auto mode. In some heating control designs, the Fan Relay does not energize the fan, but it is energized through a fan heating control.

SEQUENCE OF OPERATION 11.3: COOLING CIRCUIT OPERATION

Next, trace the cooling circuit using the t'stat in Figure 11-18. The mechanically linked heat-cool switch will be shifted to the left. This will open the heating switch contacts above **W** and **R**. This will close the cooling contacts above **Y** and **R**. Starting at the top of the Cooling System control, the control voltage path flows as follows:

1. From the top of the Cooling System control to **Y** (point 1).
2. From **Y** go to closed contacts below the Cooling Anticipator.
3. Next, go through the cooling thermostat (point 3).
4. Go to the close switch below and to the right of the Cooling Anticipator (point 4).
5. Now move to the **R** terminal (point 5) and hot side of the transformer.
6. The bottom side of the Cooling System control will be energized by the transformer Common connection.

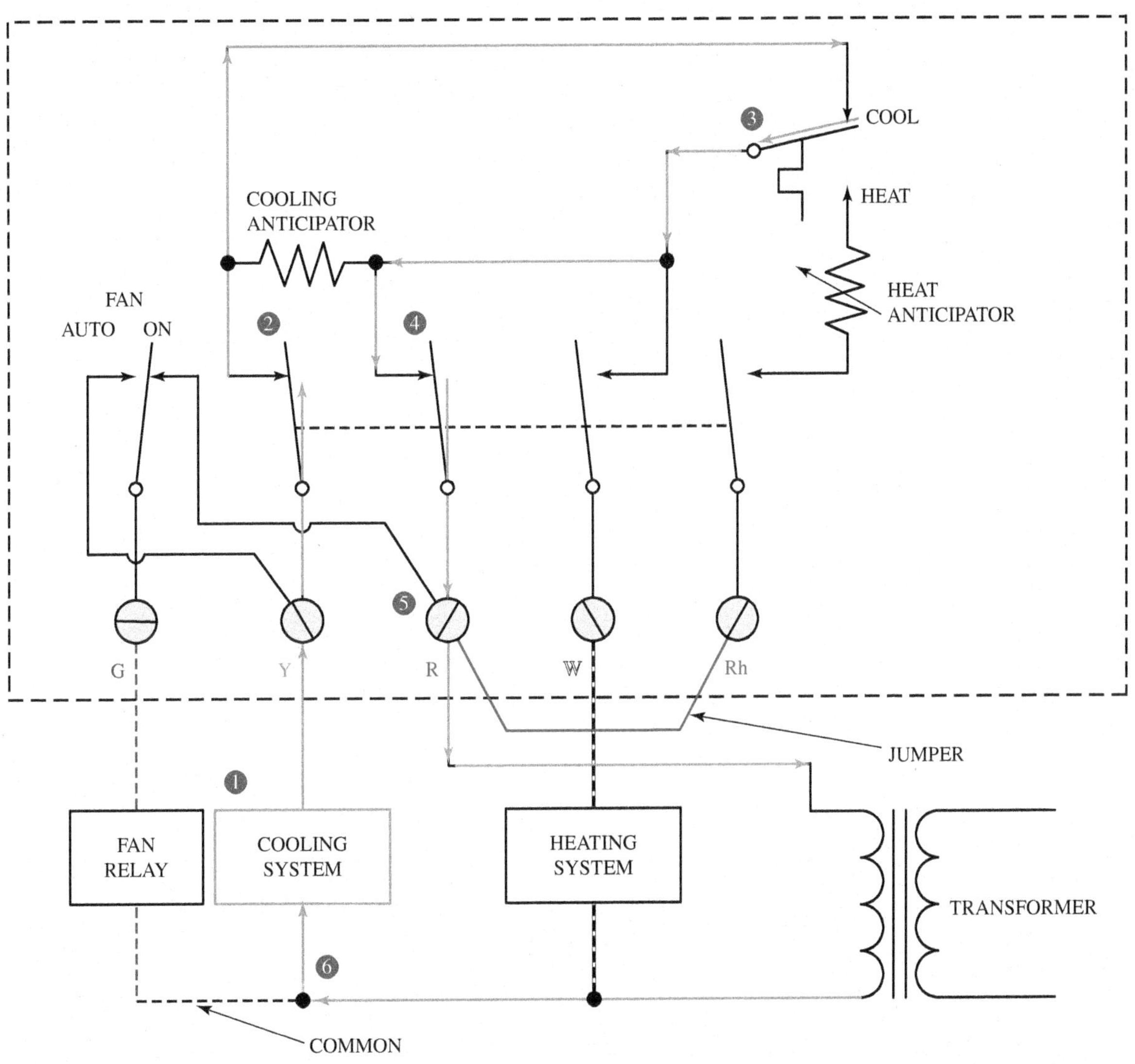

Figure 11-18 Cooling sequence of operations. The fan is set for continuous operation. If the fan is set to the Auto position, the blower will cycle when the cooling system is energized.

SEQUENCE OF OPERATION 11.4: FAN CIRCUIT OPERATION

The fan circuit is also energized in the cooling mode as illustrated in Figure 11-19. Starting at the top of the Cooling System control, the control voltage path flows as follows:

1. From the top of the Cooling System control (point 1) to **Y**.
2. From **Y**, go to closed Fan Auto switch (point 2).
3. Next, go to **G** (point 3).
4. Move from **G** to the top of the fan relay (point 4).
5. The bottom of the fan relay is tied to the common on the transformer (point 5). The indoor blower will now be energized.

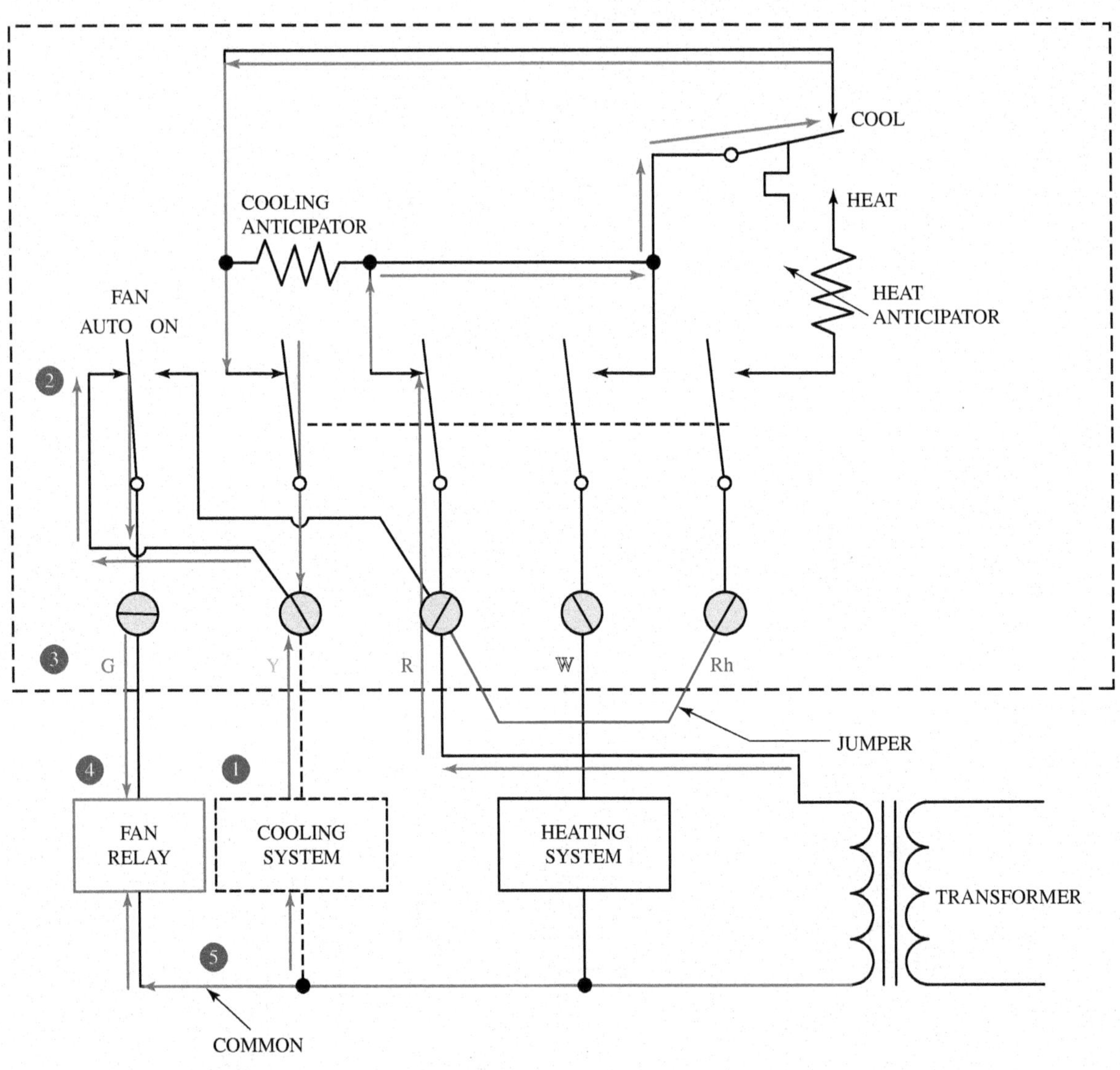

Figure 11-19 Fan sequence operation when the Fan switch is set to the Auto position. The fan is energized when the Cool thermostat is closed.

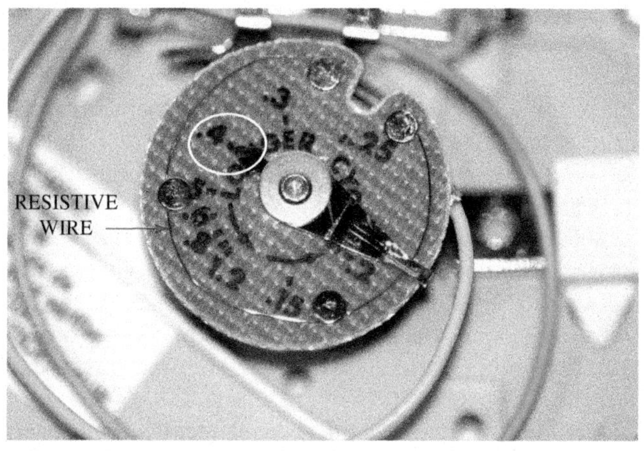

Figure 11-20 This heat anticipator is set for 0.4 amp.

11.8 HEATING ANTICIPATOR

The **heat anticipator** is a wire resistor; it is a little heater found in mechanical thermostats. The purpose of the heat anticipator is to generate heat to open the heating circuit a degree or two early so that the heating system does not overshoot its setpoint temperature. Most heating systems do not abruptly shut down when the setpoint temperature is reached. For example, a gas heating system at setpoint temperature will close the gas valve, thus extinguishing the burner flames. The blower will continue to operate for a minute or two, pushing heat from the heat exchanger and duct system in the conditioned space.

As shown in the thermostat wiring diagram in the upper right side of Figure 11-17, the heat anticipator is located in series with the heating circuit. It is known as an "on cycle" anticipator. This diagram shows the symbol for a variable resistor, which has a diagonal arrow going through the middle of it. The variable resistor has an adjustable resistance range starting at 0 ohms. The heat anticipator is set by measuring the heating circuit current. This can be measured between the R and W circuit. The amperage will be very low, probably less than 1 amp. You will need to set up the clamp-on ammeter to measure this low amperage by wrapping wire around the jaws of the meter about 10 times. The clamp-on amp reading is increased by 10 times so it should be divided by 10 to obtain the correct amperage reading.

For example, the clamp-on ammeter jaw is wrapped 10 times and the reading is 4 amps when the heating system is energized. Divide 4 by 10 (4/10) and the correct anticipator setting is 0.4 amp. Set the heat anticipator to the amperage reading. See Figure 11-20 for a 0.4-amp heat anticipator setting. This correct setting will reduce furnace short cycling or overshooting of the setpoint, which will improve customer comfort and result in fewer trouble calls. This setup should be done in a new or replacement installation or when the thermostat is changed out.

In summary, the anticipator heater, which is a resistive wire, is energized when the furnace is in the heating mode. The heat generated by the heat anticipator warms the thermostat so that the furnace does not overshoot the setpoint temperature.

Most **digital thermostats** use either thermistor devices or integrated logic elements for the anticipation function. This does not need to be set. The thermostat may need to be set up for the specific type of heat (e.g., gas, electric, oil). In some electronic thermostats, the thermistor anticipator may be located outdoors, providing a variable anticipation depending on the outdoor temperature.

Figure 11-21 is another drawing of a heat circuit with the heat anticipator in series with the heating control. Notice that this is less complex than the diagram shown in Figure 11-19. This heat anticipator is set and not adjustable. Here is how the circuit will operate the heating system:

- When the temperature drops, the thermostat switch will close between **W1** and **R1**.
- Power is provided from the top and bottom of the transformer.
- The bottom of the transformer sends power to the left side of the heater control, which may be a gas valve, electric sequencer, or oil heating control.

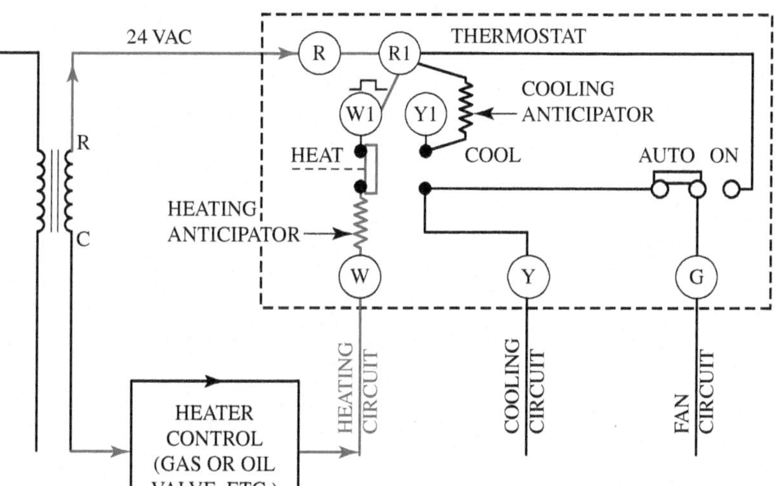

Figure 11-21 Here is another drawing of a heat circuit with a preset heat anticipator in series with the heating control.

- The top of the transformer provides power through the R and R1 connections, the closed t'stat switch, heat anticipator, the **W1** connection, and the right side of the heater control.
- The heating system will begin to operate.

SAFETY TIP

Do not short across a heating control, like a gas valve, when the thermostat is calling for heat. If you short the **W** to the **C** connections on the heating control as shown in Figure 11-17, the 24 volts will burn out the heating anticipator. The heat anticipator is in series with the operating heating circuit. The heating control, such as the gas valve, reduces the current flow through the heat anticipator, protecting it from burning out.

How would this anticipator burn out? Instead of using a multimeter, some techs will quickly touch and release the two control wires together at the heating control to see if there is a spark. The spark indicates that the transformer is working, but this will burn out the heat anticipator. As we saw in Figure 11-20, the heat anticipator is a small-gauge heater wire that can only handle small current flow, thus burning it open if the full voltage from the transformer is applied. The open heat anticipator circuit will prevent the heating portion of the thermostat from operating. In the case of an open anticipator, turn the heat anticipator to its lowest settings. This will operate the heating side of the thermostat without heat anticipation control. The temperature will overshoot the setpoint, but it will allow heating system operation until the t'stat is replaced.

11.9 COOLING ANTICIPATOR

You will notice that the cooling anticipator is wired in parallel with the cooling contacts as illustrated in Figure 11-18. The cooling anticipator is bypassed when the t'stat is closed in cooling mode. The cooling anticipator is a fixed carbon resistor usually found on the subbase of the thermostat. The cooling anticipator heats the carbon resistor element when the thermostat contacts are open. It is called the "off cycle" anticipator since it generates heat when the cooling thermostat is open. Heating the thermostat causes the contacts to close a degree or two before the t'stat reaches the cooling setpoint. Air conditioning systems do not cool immediately. Activating the cooling system a degree or two before the setpoint allows the refrigeration system to start the cooling process early since the evaporator will take a while to reach full cooling capacity. The cooling anticipator projects the need for cooling and starts the A/C system sooner so that the space temperature does not overshoot the thermostat setpoint.

Figure 11-22 is a diagram showing that the cooling anticipator is energized when the thermostat is not calling for cooling. The heat generated by the energized resistor will cause the cooling circuit to begin operation a degree or so early. The resistor generates heat to close the cooling thermostat early so that the cooling process can begin before the cooling setpoint is reached. The cooling system takes a few minutes before maximum cooling develops. This anticipator prevents overheating of the space temperature before the cooling system reaches its full capacity.

TECH TIP

Some comfort controls (thermostats) have many sophisticated options. For instance, some comfort controls use a communication system. The comfort control, condensing unit, and heating system "talk" to one another. The control can be programmed to contact the owner or call a specified contractor to correct any problems that might arise. Settings on the comfort control can be changed or monitored via Internet communications.

The comfort control, condensing unit, and heating system will need to be matched and designed for this type of communication system. Also, wireless technology is becoming popular with thermostat controls. This technology eliminates having to wire the t'stat and gives the installer more options for finding an ideal installation location.

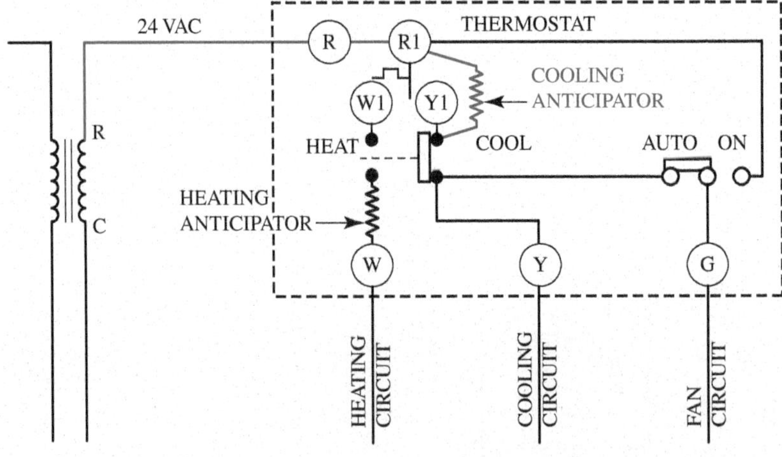

Figure 11-22 The cooling anticipator is only in the circuit when the cooling thermostat contacts are open.

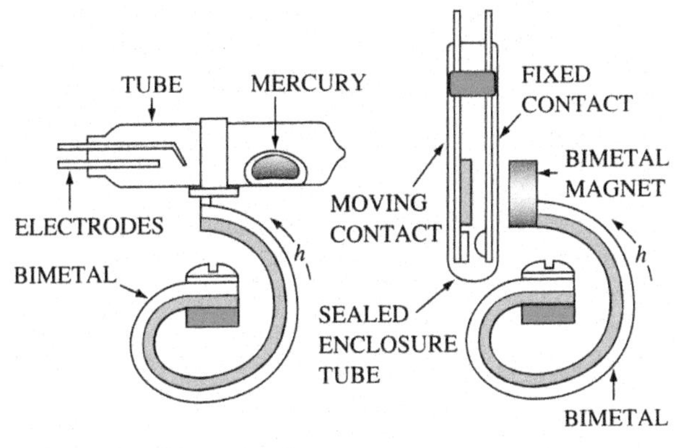

Figure 11-23 The thermostat element on the left is a mercury-filled glass bulb. The snap action thermostat on the right closes when the magnet gets close to the moving contact.

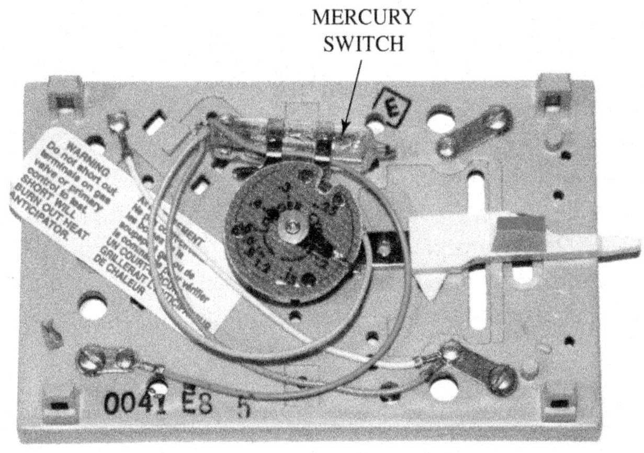

Figure 11-24 This is a mercury-filled, mechanical thermostat. The mercury in the bulb will cover the sealed contacts for the heating or cooling operation. The sealed switch will prevent dirt from collecting on the contacts. The thermostat will be screwed to the subbase.

11.10 TYPES OF MECHANICAL THERMOSTATS

The common types of mechanical thermostats are the mercury-filled bulb and the magnetic switch thermostat. The magnetic switch t'stat is also called a *snap action* thermostat. Figure 11-23 is a drawing of these two types of thermostats. As shown in Figure 11-23, both temperature-controlled switch contacts are placed on the end of a bimetal spring. A *bimetal* consists of two different metals that have different expansion and contraction rates. They are bonded together to respond to temperature changes.

The thermostat element on the left is a mercury-filled glass bulb. When the glass bulb moves downward and to the left, the mercury slides over the contacts and completes the circuit.

The snap action thermostat shown on the right closes when the magnet gets close to the moving contact. The magnet secures a good bond and connection, since the bimetal material will want to open and close at the setpoint temperature. The magnet will create a quick, snap action opening after the setpoint is made.

The **snap action thermostat**, as shown in Figure 11-23, has exposed or sealed terminals that snap closed or open by a strong magnet near the contact points. When the bimetal pushes the contacts close together, but not touching, the magnet pulls the contacts together for a good, positive connection. This prevents the rapid opening and closing of contacts ("chattering") that would cause short cycling of the cooling or heating controls. Think of chattering as your teeth hitting together when you are cold. The rapid short cycling is not good for the equipment or relay switch contacts.

Many mechanical thermostats have a subbase, as we saw in Figures 11-4 and 11-7. The thermostat wires come in through the back of the subbase and are screwed to the appropriate terminals. The thermostat is screwed to the subbase. A cover snaps over the thermostat. The face of the cover usually has a thermometer that can be calibrated with a screwdriver. Figure 11-24 shows a mercury thermostat switch. This model uses three screws to attach it to the subbase.

11.11 BASIC COOLING THERMOSTAT HOOKUP

The basic cooling circuit is hooked up as shown in Figure 11-25. Normally the t'stat wiring on a new or replacement installation would go from the indoor unit, to the thermostat, and then to the outdoor unit, though the direction of wiring does not make a difference.

1. The **C** connection goes from the indoor air handler (or furnace) to the thermostat and then to the outdoor condensing unit. This is one-half of the power from the 24-V control transformer. Not all thermostats will have the **C** connection. The **C** connection will be picked up somewhere in the circuit or there will be no 24-V circuit to operate the system. Remember that the voltage circuit is only complete if the power can return to its opposite side after it passes through a load.

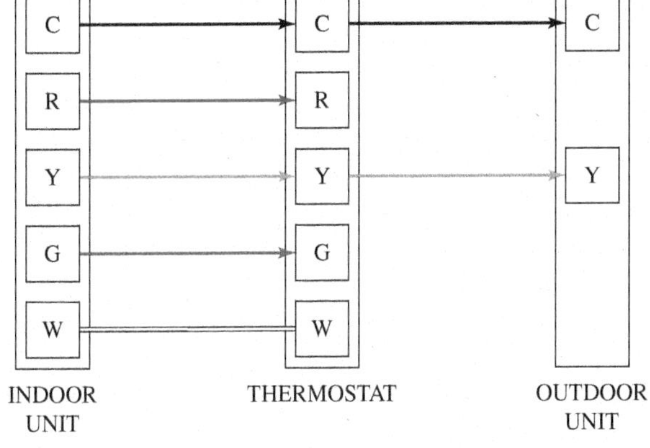

Figure 11-25 This is a basic control wiring hookup.

2. The **R** connection comes from the **R** terminal on the indoor air handler to the thermostat. This is the hot side of the transformer. The transformer is usually located inside the air handler or furnace.
3. The **Y** connection comes from the indoor air handler to the thermostat and then to the outdoor unit.
4. The **G** connection comes from the indoor air handler to the thermostat.
5. The **W** connection comes from the indoor unit to the thermostat. This operates the heating system.

SAFETY TIP

To prevent transformer or component damage, wire up and test one circuit at a time. Start with the easiest circuit first, which would be the fan circuit. Wire and test the fan operation. In this case the fan will work in the On position. It would not work in the Auto position since the cooling circuit is not connected. Next, wire and test the cooling circuit and then, finally, the heating circuit. If you experience a problem after a step, you know where to look—the problem must be with the last item wired since you checked all earlier wiring steps. If you wire everything first and then test and the transformer burns out, finding the problem is more difficult. Was the transformer bad? Are there shorted wires? Was something wired incorrectly? "Wire, test, wire" is the best sequence for any wiring sequence.

11.12 BASIC HEATING OPERATION

The operation of a heating system varies among manufacturers. This section explores some of the common operations found in gas heat, oil heat, electric heat, heat pumps, and coal heating systems to give you a better understanding of the thermostat control required for these systems. The units covering gas, electric, and heat pumps will go into more detail on this operation. Gas heat is covered in Unit 22, electric heat in Unit 23, and heat pumps in Unit 24.

Operation of Gas Heating for Forced Air or Water Systems

1. The thermostat closes, with a call for heat.
2. The inducer draft motor begins to operate, purging the heat exchanger and venting the system for about 1 minute.
3. The gas igniter glows or sparks, creating a hot surface for igniting the gas.
4. The gas valve opens and the gas is lit by the hot surface ignition (HSI) system.
5. The HSI system de-energizes.
6. The heat exchanger heats and the blower or water circulator pump comes on.
7. When the thermostat reaches its setpoint temperature, the contacts open. The gas valve opens and the blower will continue to operate for about 1 minute, forcing heated air from the heat exchanger and duct system. End of cycle.

Operation of Oil Heating Systems

1. Oil heating systems are similar to gas systems.
2. Instead of using a gas valve, the oil furnace or boiler will start an oil pump and pressurize the fuel to be injected into the burner through a high-voltage igniting spark.
3. The spark ignites the fuel oil.
4. The heat exchanger heats and after a short period of time the water pump comes on, forcing heated water through the heating system.
5. When the setpoint temperature is reached, the oil pump stops and the heat in the heat exchanger is purged from the duct system.
6. The water pump or blower turns off. End of cycle.

Operation of Electric Heating Systems

1. The electric heating system energizes the first heat strip and blower immediately after the thermostat contacts close. It is important to have airflow when the heat strip warms up. Overheated heat strips will have a shorter life.
2. If the heating system uses a time-delay device to energize the next stages of heat, they will most likely use a sequencer. The sequencer is a timing device in the electric heating system designed so as to not overload the electric system. Having all the heat strips come on at the same time will overload the wiring. One sign the wiring is overloaded is a dimming of the lights when a load is energized. Loose connections also create the same effect.
3. When the thermostat setpoint temperature is satisfied, the sequencers time out in reverse order. The last sequencer to be energized will be the first one to be de-energized. The blower and the first stage of heat are shut down last. End of cycle.

Operation of Heat Pumps

The heat pump thermostat is different from a conventional single-stage cooling and heating control. It uses two stages of heating control. The first stage of heating is through the refrigeration cycle heat pump, and the second stage is the auxiliary heat. The auxiliary heat is usually electric heat, but it can be a fossil fuel like natural gas. When using fossil fuels, the refrigeration side of the heat pump should not operate. The hot air from the fossil fuel furnace blown into the indoor heating coil (evaporator in the cooling mode) generates high pressure. This high pressure will cause the compressor to overheat or cause the system to cut out on its high-pressure switch or compressor thermal overload.

The heat pump cooling cycle operates though the thermostat like a conventional air conditioning system. The reversing valve can be energized in the heating mode or cooling mode. Most manufacturers have the call for cooling energize the reversing valve. With this design, the reversing valve does not need to be energized in the heating mode. In this case, if the reversing valve coil fails, it will be in the heating mode. Having heat is generally more critical than having cooling.

Remember, the heat pump t'stat has a two-stage operation. This is the basic operation of the heat pump heating circuit:

1. As the temperature drops, the first stage of heating is energized through the t'stat.
2. The heat pump outdoor and indoor sections begin to operate simultaneously. The heat pump uses refrigeration heating since the indoor and outdoor coils switch operation. The indoor coil is now the condenser, heating up the indoor space. The outdoor coil becomes the evaporator, absorbing heat from the cold outdoor air.
3. If the temperature in the space drops a degree or two more, the second stage of heat is energized.
4. The second stage or auxiliary heat is usually electric heat.
5. The second stage of heat will drop out as the temperature rises to the thermostat second-stage setpoint.
6. The first stage of heat will be de-energized as the heat pump provides enough heat to reach the thermostat temperature setpoint.
7. The indoor and outdoor sections turn off. End of cycle.

Figure 11-26 shows a common way of connecting the thermostat, air handler, and heat pump unit. Here is the connection sequence:

1. The **R** from the t'stat is connected to the **R** on the air handler and heat pump. This provides half of the 24-V power. The control voltage comes from a transformer in the air handler.
2. The **G** from the t'stat goes to the air handler **G** connection to operate the fan.
3. The **C** from the t'stat goes to the air handler and heat pump. This provides the other side of the 24-V power, sometimes called the common side.
4. The **W** from the t'stat goes to the second stage or auxiliary heat **W1** and **W2** of the air handler and the heat pump.
5. The **E** connection wires the emergency heat, **E**, to the second stage of heat (auxiliary heat **W**).
6. The **O** t'stat connects to the **O** connection on the heat pump. The **O** energizes the reversing valve in the cooling mode. (The **B** terminal is not shown in this diagram. A wired **B** terminal would energize the reversing valve in the heating mode.)
7. The **Y** t'stat connection goes to the **Y1** on the heat pump. This is the first stage of heat or the reverse cycle heat where the indoor coil is the condenser and the outdoor coil is the evaporator.

Operation of Coal Heating and Alternate Fuels

Coal and alternate forms of fuel, such as corn, wheat, barley, wood pellets, or cardboard, are not as common as they once were, but are gaining in limited popularity. The thermostat will start a screw or auger to bring the fuel into the firebox. A burning bed of these materials must be available at all times since they take a while to generate heat when first ignited.

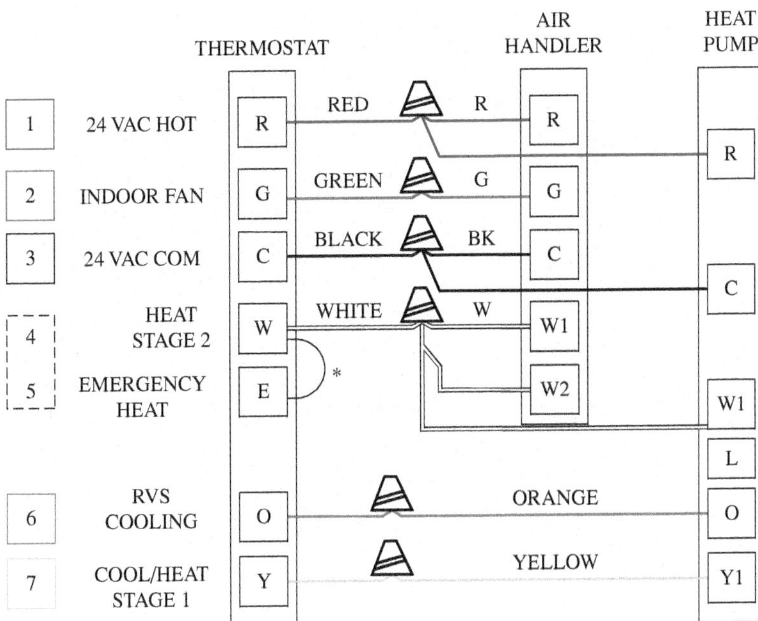

Figure 11-26 This heat pump thermostat is connected to the air handler and heat pump.

GREEN TIP

The **programmable thermostat** can be called a "green" thermostat since it can save energy. An example of a green thermostat is shown in Figure 11-27.

Saving energy reduces utility costs and reduces greenhouse gases. Programmable thermostats are used to adjust the space temperature when a building is unoccupied. The **setpoint temperature** is increased or the thermostat is turned off during unoccupied periods in the summer. This change in temperature is called the **setback temperature**. Conversely, in the winter the setpoint is reduced or the thermostat is turned off in the winter. To derive any savings from this type of thermostat, it should be programmed to a changed setting for at least 8 hours.

This thermostat is designed for those families or businesses that have a set schedule and do not need as much cooling or heating when unoccupied. Even if the energy saving feature is not used, the programmable thermostat will maintain a setpoint temperature within ±0.5°F or less. Less fluctuation in temperature translates into more comfort for the customer. The cooling or heating temperatures do not experience *overshooting*. Overshooting is a condition in which the thermostat cools or heats a couple of degrees or more over its setpoint. Overshooting creates a wider split in the call for cooling or heating. It will get too cold or too warm before the system reaches it setpoint and stops operating. The temperature will need to change several degrees before the cooling or heating system begins to operate again.

A digital thermostat uses a thermistor solid-state component to sense the air temperature. A thermistor, as shown in Figure 11-28, changes resistance with the slightest change in air temperature. The slight change in temperature causes the resistance to change. The change in resistance may cause the selected heating or cooling operation to begin.

The option to change the thermostat a few degrees or simply turn it off will depend on each condition. For example, if the heating or cooling system is undersized, it may take hours for the system to recover to a desired, comfortable temperature. In this case adjusting the setpoint may cause the unit to operate more than expected and create a recovery time longer than 1 hour. If the structure is unoccupied for a period longer than 8 hours, the thermostat may have an "automatic recovery program." This recovery program has the heating or cooling system starting near the setpoint (point at which the building will become occupied again). The recovery program will check itself. If the setpoint was reached prior to the occupancy time, it will start later the next day. If the setpoint is reached quickly, the next day the recovery time will begin closer to the setpoint programmed time.

The conservative rule of thumb for energy savings is 1% to 2% per degree change. No savings are expected in the fall and spring of the year when the outside temperature is in the range of 60° to 70°F.

Some HVAC systems, particularly the ductless systems, have a remote control thermostat as shown in Figure 11-29. Because the unit is mounted near the ceiling, the remote makes it easier to set. The remote comes with a holster that should be mounted near and under the indoor section. This is a wireless t'stat control.

Figure 11-27 A digital thermostat allows the user to program in times to automatically set and reset HVAC. This option can save energy and is considered a "green" thermostat.

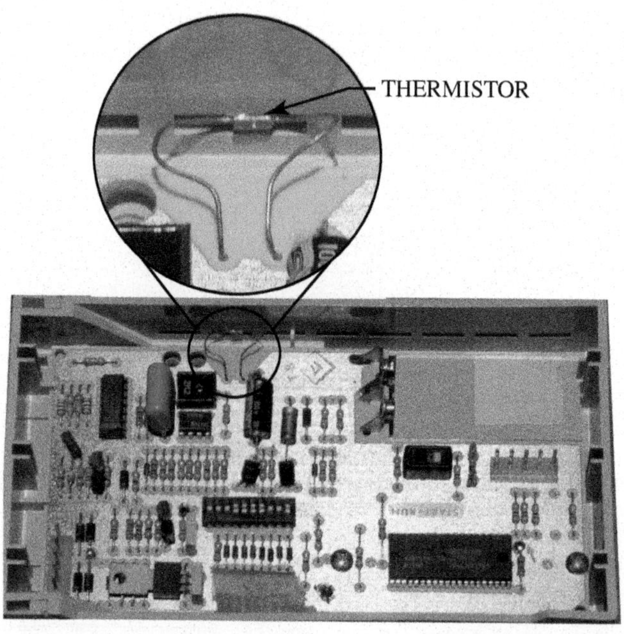

Figure 11-28 This digital thermostat has had its cover removed to expose the thermistor. The thermistor varies resistance at the slightest temperature change. The resistance change is transferred into the operation of the thermostat.

Figure 11-29 Remote wireless thermostat control for a ductless air conditioning system. The remote should be mounted near, and under, the indoor section so that it can easily be accessed. The air sensor is located inside the system.

Figure 11-30 This touch screen thermostat is common technology found on cell phones and other electronic devices. Note that there is no manual switch on this control.

TECH TIP

Many customers have a difficult time with a programmable thermostat, so it is a good idea to offer to program it for them. If they want you to program it, have them write down the desired hours for setback cooling and heating temperatures. Show them the features and benefits of their t'stat. Have the customer push buttons to get familiar with their control. Leave the instruction booklet with them and note the times they have selected.

Most programming instructions are on the cover of the thermostat. Many thermostats can be programmed separately from the subbase. They will need a battery backup installed while doing this method of programming. Some thermostats contain a small lithium battery just as your laptop does. The battery may be rechargeable or it can be implemented using a supercapacitor. These are very high-value capacitors that can run the t'stat for a day or so. The manufacturer may not even tell you about them. A "Remove/Pull this paper" tab is a good indication that the thermostat contains an internal battery.

11.13 TOUCH SCREEN

Many digital thermostats have an LCD touch screen that allows for easy use of the control. A touch screen t'stat is shown in Figure 11-30. The common display options include:

- ✔ Temperature setting
- ✔ Indoor temperature and relative humidity
- ✔ Fan control: auto, on, and off
- ✔ Outdoor temperature
- ✔ Time and date or day
- ✔ Dirty filter alert.

11.14 PROGRAMMABLE THERMOSTAT LANGUAGE

The programmable thermostat has a setting for common applications such as the setpoint temperature during an unoccupied period or changing setpoint temperatures at night. We now discuss common terms you will find on the thermostat face or in the user's manual.

Wake

Wake is the temperature setting you desire when you wake up or, in the case of a business, the time the first person arrives at work. If you awake at 7:00 a.m., you may want to set the wake time for 7:00 a.m. at whatever temperature you desire. Some "smart" thermostats are programmed to obtain your setpoint once you have set the time and temperature. For example, in a heating setting, if you want the temperature to be 70°F by the time you wake up at 7:00 a.m. the t'stat will start the heating system early enough to obtain the desired temperature at the designated time. The thermostat "learns" as it tries to meet the setpoint and programmed time. If the designated temperature is not reached by the programmed time, the heating system will be started earlier the next morning. This adjustment will continue until the target temperature is reached at the programmed time. Explain to the customer that it may take a week or so for the t'stat to fully adjust or learn the homeowner's "temperature requirements."

Leave

Leave is the time the last person leaves the building or living space for the day or, in a business installation, the evening. If a person usually leaves the house for work at 7:00 a.m., then the thermostat can be set to change the Leave temperature to

60°F at 6:30 a.m. The cooling down or "no heating" setpoint will begin before the occupant leaves the area.

For cooling the *Leave* temperature would be higher than the normal desired temperature. For example, when the building is occupied the customer keeps the thermostat at 74°F. The *Leave* temperature may be set to allow the cooling system to drift up to 80°F before operating again. Note, however, that turning off the cooling system completely and allowing a high temperature rise may create moisture problems. In addition, it may be difficult to pull the temperature back down in a reasonable time before being occupied again. To save energy for heating or cooling, the setback or *Leave* temperature should be at least 8 hours. *Exception:* Do not reset the temperature for a heat pump in the heating mode. All savings would be erased since the electric heat strips would come on when heating the space back to the desired temperature.

Return

Return is the time that the first person arrives home for the day. (In the case of a business, it would be the time the first person arrives at work.) If that person arrives home at 5:00 p.m., then the heating time and temperature can be set for, say, 70°F at 4:30 p.m. That way, when the person arrives home, the home is at the desired temperature. Again, some programmable thermostats will start heating or cooling to meet the temperature at the target time, while some will start heating or cooling at the timed setpoint.

Sleep

Sleep is the time when the last person goes to bed for the night. If everyone is in bed by 10:30 p.m., then the heating thermostat can be set to change the temperature to a lower setting for the night. The temperature will not drop until the setback time is reached, which in this example is 10:30 p.m. (The *Sleep* setback point does not apply to a business installation since the *Leave* setpoint will cover those conditions.)

Override

Override allows the user to bypass the programmed settings until the user releases the override button again. Some override functions have time limits. For example, pressing the override function may allow bypass of the program for several hours. After the override time period, the thermostat reverts to the programmed time settings.

11.15 PROGRAMMING OPTIONS

Thermostats offer several programming options, depending on their design:

- The 5-2 programming option
- The 5-1-1 programming option
- The 7-day programming option (see Figure 11-31).

One common option is the 5-2 day. The "5" represents Monday through Friday operation and the "2" the weekend days of Saturday and Sunday. In the 5-2 design, Monday through Friday have programmed times and temperatures that are all the same. The Saturday and Sunday temperature settings are the same. With homes occupied on an irregular schedule on the weekend, the homeowner may not want to change the thermostat setting for these days. In other words, they would leave them unprogrammed. This would leave the temperature at whatever setpoint is selected throughout the whole weekend. If the 5-2 thermostat is used in a business building that is unoccupied on weekends, a setback temperature may be programmed to save energy. Sometimes changing the temperature for 2 days presents a challenge to the heating or cooling system to create a comfortable setting when employees arrive on Monday. In this case, it is best to have the system start up at least an hour before the occupants arrive.

Figure 11-31 This programmable thermostat offers the 7-day option. This is the most versatile time setting thermostat. It gives the user a setback option for each day of the week.

The 5-1-1 option programs Monday through Friday on the same setpoint times, but then Saturday and Sunday can be programmed differently.

A 7-day programmable thermostat will allow the home or building occupant to change the temperature setpoint for each day of the week. The 7-day option can handle the most irregular schedule. This thermostat is the most flexible of the group.

SAFETY TIP

Many thermostats have cooling time delays. Once the cooling system or heat pump cycles off, there will be an approximately 5-minute time delay before it cycles on again. This is for the protection of the compressor. The time delay allows the pressures in the system to equalize for easier starts. The time delay also allows the power fluctuation to level out prior to starting the compressor. This is a good option for most pieces of refrigeration equipment.

SAFETY TIP

Wiring two control voltage transformers together is called *phasing transformers*. A few systems use two transformers to control the heating and cooling circuits. Some transformers are wired in parallel to improve their volt-amp (VA) rating. For example, if you need a 60-VA transformer, phase two common 40-VA transformers in parallel to create an 80-VA rating. The 80-VA rating will more than handle the 60-VA requirements. The primary windings and the secondary windings are hooked up in parallel to develop the 24-V output. Miswiring the secondary can also create 48 volts. A 48-V output will damage most 24-V components. Check the voltage output after wiring transformers in parallel.

11.16 DUMMY THERMOSTAT

"Dummy" thermostats are sometimes used to satisfy the personal need to control a thermostat; they do not control anything. A dummy t'stat is installed, at the customer's request, when the occupants of a home or business "play with" or continually adjust the thermostat. This is especially true if the thermostat is in a common area with easy access for fiddling with the settings. The system will still have a preset (preset for heating and cooling) thermostat mounted in the return air duct or other area that is hidden or unknown to the users in the space. Dummy thermostats are more common in the business environment than the home environment because businesses have more individuals in a given space with a wide variety of comfort zones. No one temperature will satisfy everyone; therefore, each individual wants to change the thermostat to accommodate his or her particular comfort level.

The dummy t'stat works on the "placebo effect." In this case, the placebo effect of adjusting the temperature satisfies the desire to control the area temperature.

SAFETY TIP

In most cases it is not good practice to completely turn off the thermostat. The heating and cooling systems need time to recover to meet the temperature setpoint on the hottest and coldest days. Occasional conditioned air circulation will keep the indoor air from becoming stagnant. In the summer season a periodic cooling cycle will reduce mildew, mold, and moisture buildup in the bathroom, kitchen, and laundry room.

GREEN TIP

Did You Know? 

The Energy Star website states that the average household spends more than $2,200 a year on energy bills—nearly half of which goes to heating and cooling. Homeowners can save about $180 a year by properly setting their programmable thermostats and maintaining those settings. The $180 savings assumes a typical, single-family home with a 10-hour daytime setback of 8°F in winter and setup of 7°F in summer, and an 8-hour nighttime setback of 8°F in winter and a setup of 4°F in summer.

The preprogrammed settings that come with programmable thermostats are intended to deliver savings without sacrificing comfort. Depending on your customer's schedule, they can see significant savings by sticking with those settings or adjusting them as appropriate for their family's particular schedule.

The key is to establish a program that automatically reduces heating and cooling in a home when the occupants do not need as much conditioned air. Table 11-2 shows the Energy Star recommendations for a programmable thermostat.

SAFETY TIP

Working on a thermostat has killed many transformers! Crossing or touching the control wires together too long will create high amperage in the transformer and burn it out. You can avoid this by placing a fuse in line with the secondary of the transformer before you work on the thermostat. For the common 40-VA transformer, use a 5-amp fuse. The maximum amperage load a 40-VA transformer can handle is 1.67 amps. The 5-amp fuse is selected to protect against a direct short and not an overload condition.

Make up a fuse holder with alligator clips on each end. Temporarily wire this series with the hot side of the secondary of the transformer while working on the thermostat. A blown fuse is easier and cheaper to change than a transformer.

Table 11-2 Programmable Thermostat Setpoint Times and Temperatures

Setting	Time	Setpoint Temperature (Heat)	Setpoint Temperature (Cool)
Wake	6:00 a.m.	≤70°F	≥78°F
Day	8:00 a.m.	Set back at least 8°F	Set up at least 7°F
Evening	6:00 p.m.	≤70°F	≥78°F
Sleep	10:00 p.m.	Set back at least 8°F	Set up at least 4°F

Note: The author does not agree with all of these recommendations. Most customers would not tolerate the Sleep and Setpoint temperatures for cooling. For instance, turning up the cooling temperature 4°F when many customers want it cooler at bedtime is difficult to recommend. At best, keep the sleeping temperature the same as the evening temperature. Many customers like it a little cooler when they sleep whether it is summer or winter.

11.17 THERMOSTAT DIAGRAMS

This section takes what you have learned about thermostats and applies it to complete diagrams like those you will see in the field.

Connection and Control Points

Figure 11-32 does not show the thermostat diagram, but it does show all the connection and control points. The thermostat wire connect terminals are found in the rectangle at the bottom of diagram. It is labeled in the rectangle as **R W G Y**. Below the 24-V transformer is an insulated plate that has connection points for **R**, **C**, **G**, **W**, and **Y**. The rectangular **R W G Y** thermostat connection point has wires running to the:

- **R** to the transformer (red)
- **W** to the gas valve (white)
- **G** to the fan relay coil (green)
- **Y** to the air conditioning condensing unit contactor (yellow).

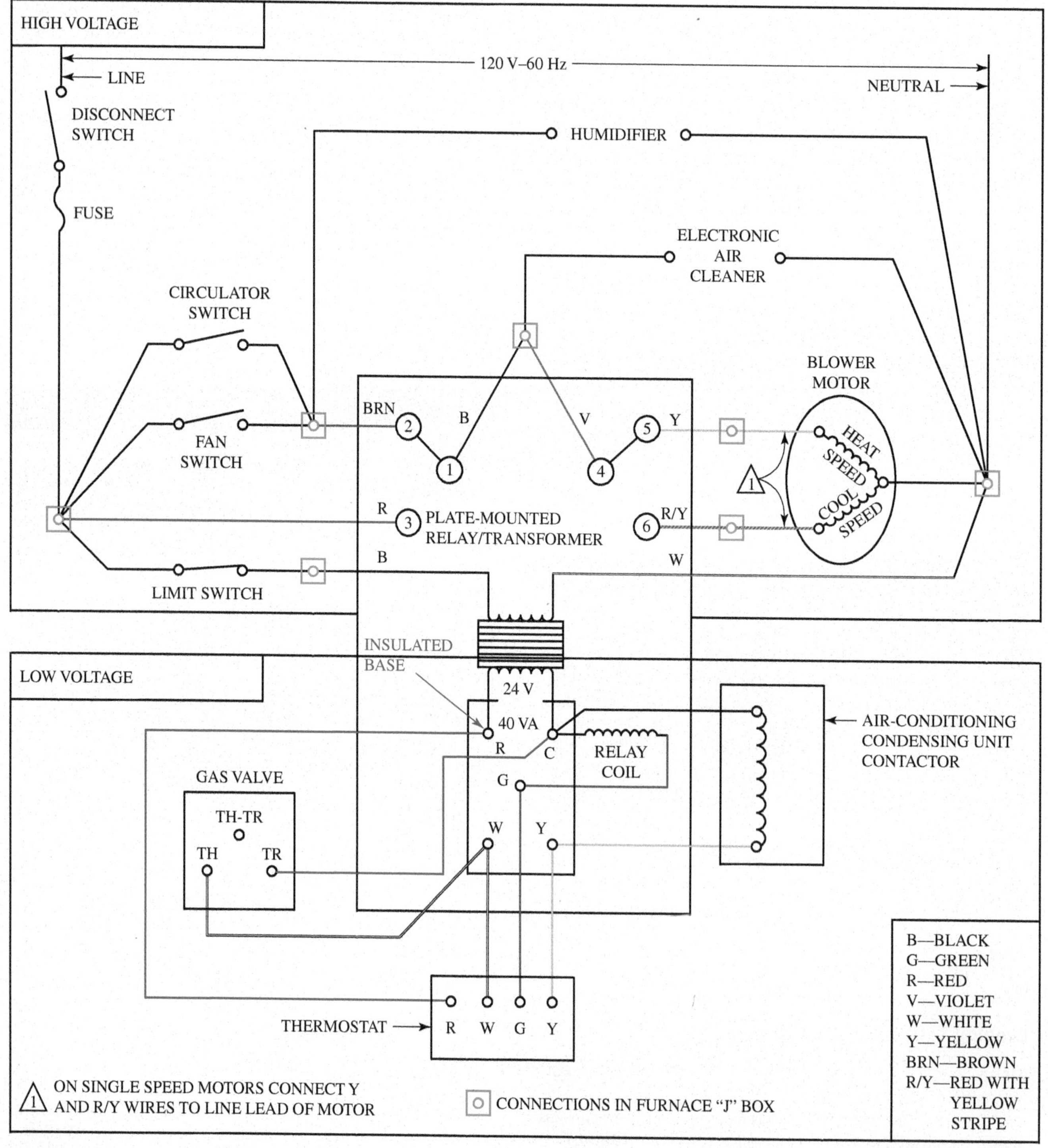

Figure 11-32 The thermostat connections are located at the bottom, center of this diagram. Trace out the control circuit starting with **R** to **W**; then **R** to **G**; and finally **R** to **Y**.

SEQUENCE OF OPERATION 11.5: THERMOSTAT COOLING MODE

Next, let's trace the cooling mode operation using Figure 11-33. (A larger version of this figure appears as Electrical Diagram ED-2 in the Electrical Diagrams package that accompanies this book.) A few notes before we begin:

Notice that the supply voltage in Figure 11-33 is three phase, 480 V, and the control voltage is single phase, 120 V. The **RCSTAT** is a manual hand switch that can be placed in the Fan On, Cool, or Off position. In the Off position with power still supplied, the control transformer and crankcase heater are the only two components that are operating. In the Fan position, the cooling system will not operate. The blower runs continuously by energizing the (EF) coil, which closes the three-pole **EF** contacts feeding power to the evaporator fan motor. The **RCSTAT**'s normally open contacts are found on the left side of the control voltage diagram inside a rectangular box.

1. Manually switch **RCSTAT** to the Cool operation, point 1.
2. This closes the Cool contacts at point 2 and found in the **RCSTAT** box.
3. Sending power to the (R1) coil at point 3. The (R1) relay coil between 7 and 8 energizes when the **RCSTAT** is closed.
4. The (R1) coil closes **R1** (point 4) contacts 3 and 5 (to the right of **RCSTAT**) plus another set of (R1) contacts 4 and 6. To the right of the **RCSTAT** box, contacts 3 and 5 bridge the Fan and Cool circuit so that the evaporator fan cycles with the cooling system.
5. In the Cool position, power goes from the left side of the control voltage through **RCSTAT** and the (R1) coil and up through contacts 5 and 3. Power through contacts 5 and 3 energizes the (EF) coil at point 5. This closes the evaporator fan motor three-pole contactor and starts operating the evaporator fan motor.
6. The **EF** coil also closes (EF) contacts 13 and 14 found below the **RCSTAT** box at point 6.
7. **EF** contacts 13 and 14 are closed and this is in line with the (C) coil, found below the **RCSTAT** box, point 6. The (C) coil at point 7 will be energized when **EF** and **R1** are closed. When the (C) coil energizes, the three-pole Compressor #1 contactor will close, turning on the compressor. The **C** contacts 13 and 14 (near point 6) in the **LSV** line (liquid line solenoid coil) will also be energized when the (C) coil is energized.
8. When the **LSV** is energized it will open the liquid line, allowing refrigerant to flow to the liquid line and evaporator. The evaporator pressure will begin to rise rapidly. The compressor will begin to operate when the low-pressure switch, **LP**, closes.

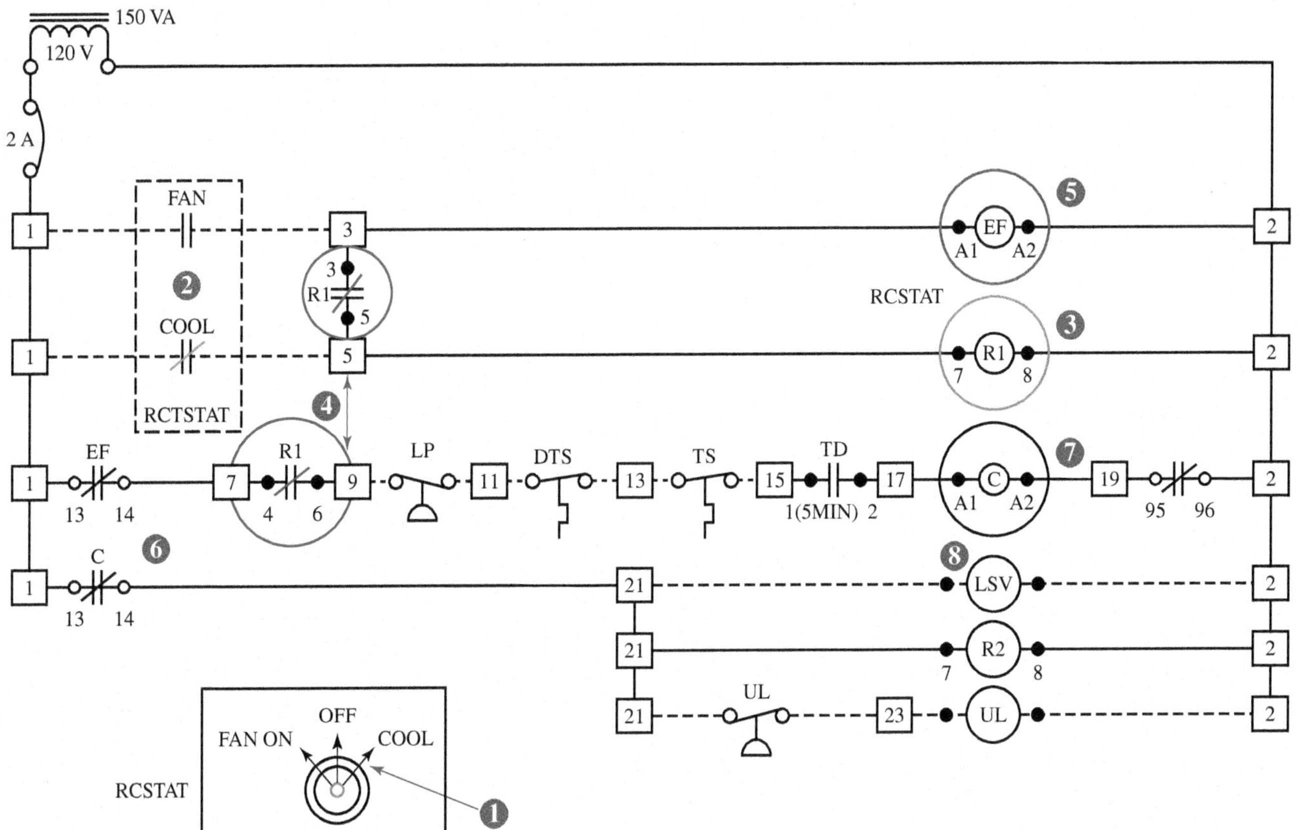

Figure 11-33 This is the control voltage section of enlarged Electrical Diagram ED-2 found in the Electrical Diagrams package that accompanies this book.

Tracing tip: When tracing a wiring diagram, cover it with a clear page protector and use erasable markers to trace the circuit. Change the color of the marker when crossing a load.

Using Figure 11-32, trace the diagram one control circuit at a time. Trace the thermostat being closed to operate the heating circuit. This will create a connection between **R** and **W** on the thermostat connection board at the bottom of the diagram. When tracing this circuit start at **R**, go to the left, and up to left side of the transformer. The power will exit through the right side of the transformer common (**C**) and travel to **TR** on the gas valve. The 24 V will then travel through the gas valve coil. This opens the gas valve and the power leaves through the **TH** connection to complete the circuit to **W** on the insulated base and back to **W** on the thermostat.

Next, trace the **R**-to-**G** connection. This will energize the relay coil and start the fan operation.

Finally, the **R**-to-**Y** connection will energize the air conditioning unit contactor. Of course, the thermostat will not energize the heating and cooling mode at the same time. If this happens you have a defective thermostat or control wires touching each other.

In summary, energize the heating (**W**), fan (**G**), or cooling (**Y**) circuit; power will be provided from the transformer to the thermostat to the **R** connection. Here is how the power through the thermostat will operate each of those separate systems:

Heating operation **R** to **W**
Fan operation **R** to **G**
Cooling operation **R** to **Y**.

Using Figure 11-33, the following safeties and switches must be closed in the Ⓒ coil circuit in order for the compressor to start:

- **LP** switch is closed (low pressure).
- **DTS** switch is closed (discharge temperature switch).
- **TS** is closed (thermostat).
- **TD** is timed out for 5 minutes (time delay).
- Contacts 95 and 96 are normally closed safety contacts that are not controlled by components on this diagram.

Finally, **C** contacts 13 and 14 provide a parallel path to coil (R2). Coil (R2) will control an air-cooled or water-cooled condenser. The condenser section is not shown on this diagram.

A note on **UL**: **UL** is part of an unloader circuit, but not part of the thermostat electrical control circuit. The unloader is a refrigeration flow control device. The unloader reduces the refrigerant in circulation when the cooling demand is reduced. Some unloaders are energized when system capacity is reduced, whereas others are energized at full system refrigeration capacity.

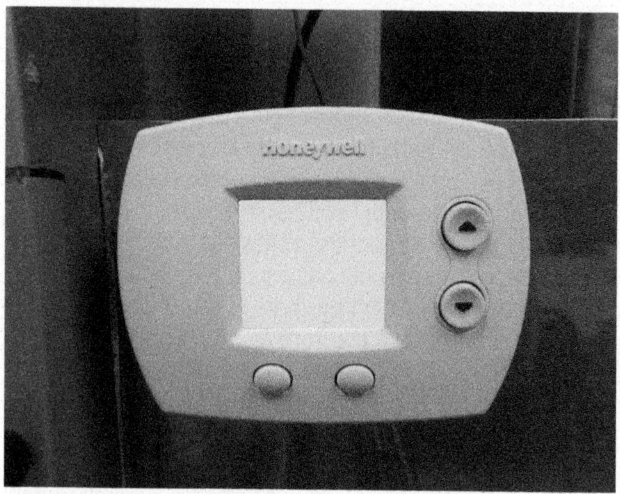

Figure 11-34 No thermostat display. No control voltage.

11.18 GENERAL TROUBLESHOOTING STEPS

Determining if a malfunction is in the thermostat, control wiring, or another component is not difficult. Here are troubleshooting steps that will be helpful:

1. Use the ACT method of troubleshooting, which will be explained in detail in Unit 25, *How to Start Electrical Troubleshooting*. Here is a summary of the ACT method: The ACT method of troubleshooting means you are going to check the duct and condenser airflow (A), the compressor operation (C), and the thermostat setting (T) to help you narrow down the problem.
2. If it is a digital thermostat, does it have a display? If there is no display as shown in Figure 11-34, check the 24-V transformer output. If there is no transformer output, check the input voltage as shown at points in Figures 11-35 and 11-36. If no input voltage is present, trace the primary wire back until you find the reason for the loss of power. The input power is usually 120 or 208/240 volts. To recap:
   No display ⇨ Check transformer secondary ⇨ No voltage, check primary and voltage source
3. Turn the thermostat off and on. Adjust the setpoint several degrees above the room temperature for heating and several degrees below the room temperature for cooling. Many thermostats have a 5-minute time delay after the system is turned off. Inspect the equipment while waiting for the timing sequence to time out. In rare cases of "operator error," the setting or setpoint will be the only problem, in which case, cycle the system several times before you leave the customer. Allow the thermostat to reach a setpoint and shut down. You do not want to be called back because of an intermittent problem.
4. At this point the problem is most likely the thermostat or control wiring. Figure 11-37 shows a way to isolate the problem. Remove the t'stat and expose the control wires or the subbase. With a jumper, go from **R** to **G**. This will

Figure 11-35 No voltage on the secondary of the transformer

Figure 11-36 No voltage on the primary of the transformer.

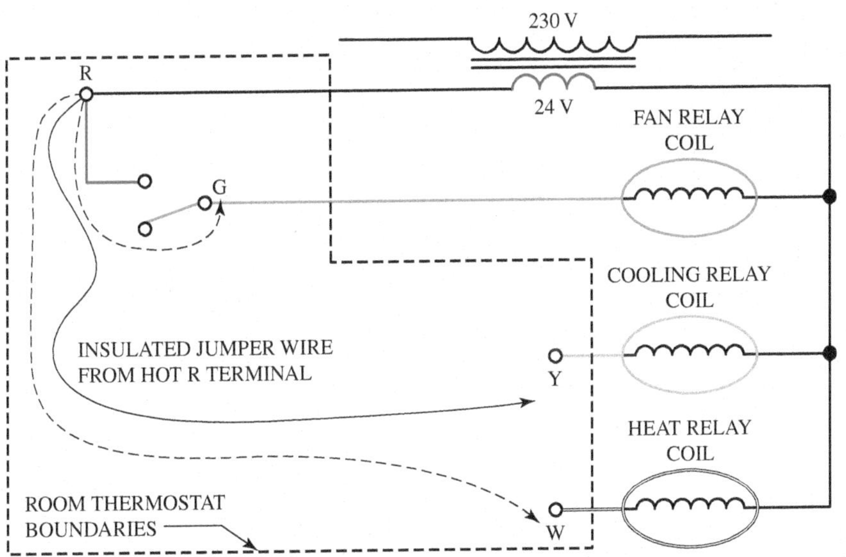

Figure 11-37 This shows a troubleshooting process. Remove the thermostat and use a jumper wire to check each circuit. Jumper from R to G to check the blower section. Jumper from R to Y to check the condensing unit operation. Jumper from R to W to check the heating circuit operation.

operate the blower. It would be best to leave the blower operating throughout the remaining troubleshooting sequence. Next, with a jumper wire, go from **R** to **W**. This will operate the heating system. With a jumper wire, go from **R** to **Y**. This will operate the cooling system.

5. If the jumper activity operates all of these systems, then the thermostat is defective and the connecting wires are all good. If one jumper does not make the system operate, it is an open thermostat wire or some other component issue in that line. Be patient when using this jumper sequence. Some of these circuits may have a 5-minute time delay that is not part of the t'stat. Timed circuits are commonly found in cooling circuits.
6. Finally, you can troubleshoot the thermostat directly. Remove the thermostat (and subbase if it has one) from the wall and disconnect all thermostat wires. With an ohmmeter, check the continuity between **R** and the other connections **G**, **Y**, and **W**. Test one circuit at a time until all the circuits have continuity. Remember to switch the circuit you are testing. For example, if you are testing the cooling circuit, make sure the thermostat is set to cooling and the temperature is turned down several degrees below the setpoint. The resistance should be near zero ohms, except for the **W** connection, which may have some resistance through a heat anticipator. The heat anticipator is a high-resistance heater wire. When checking a mechanical t'stat, it will need to be level. Digital thermostats require batteries or some other power source. You do not want to measure battery voltage when your meter is in the Ohms range. You want to measure the t'stat switches. Measuring voltage will blow the meter fuse or damage the meter.

SAFETY TIP

Many older mechanical thermostats use a mercury switch. Mercury is considered to be a hazardous material. Mercury-containing thermostats must be recycled. Many HVACR supply houses are recycling centers for mercury thermostats as shown in Figure 11-38. Do not throw mercury t'stats into the trash. They can pollute landfills and drinking water.

Figure 11-38 Mercury thermostat recycling bin found in some air conditioning supply houses. Mercury recovery and recycling are important to reduce landfill and water supply contamination by this mineral.

SERVICE TICKET

Your third service call of the day takes you to a customer who purchased a programmable thermostat with the idea of saving money. The customer replaced an old-style mechanical thermostat with this new one. After wiring the thermostat, the system did not work. There is a visual display on the face of the thermostat. The flashing battery icon on the thermostat face indicates that this control is operating solely on battery backup. There are three AA alkaline batteries located in the back of the thermostat. Where do you start? Think about it for a minute before you proceed.

Removing the thermostat from the wall base, you notice that all of the wires are in their correct position. The R was connected to the R terminal, the Y to the Y terminal, etc. There is no voltage measurement between R and C. There should be 24 volts feeding power to the thermostat and measured between R and C. Where is the 24-V control source?

The common control voltage (24 V) is found in the air handler or furnace. You locate the furnace, remove the furnace door, and notice a circuit board. A close look at the circuit board reveals that it has a 3-amp, plug-in type fuse. The fuse appears to be in working order but when checked with an ohmmeter, it is found to be open. Before replacing the fuse you go back to the thermostat and make sure there are no shorted wires at the connection points or nicks in the wire insulation. No shorts are found at the thermostat location or around the circuit board connections.

The fuse is replaced and the thermostat begins to operate properly. To verify the results, you cycle it twice in the heating and cooling modes. You ask the customer for the instructions and proceed to program the thermostat to the customer-desired temperature settings. Once you've finished your work, you review the programming and operation with the homeowner. The owner is presented with the invoice, which is explained to the customer's satisfaction. Unfortunately, many customers think it is easy to change out a thermostat. In this case, the homeowner most likely shorted two wires together and blew the circuit board fuse. However, the blown fuse saved the transformer.

SUMMARY

A thermostat is merely a switch that automatically controls a heating, cooling, or refrigeration system. The modern digital thermostat is more accurate, easier to read, and controls the setpoint temperature more closely than do the old-style mechanical thermostats. Some models are programmable. To save energy, the contractor or homeowner can set times of the day when the equipment can be turned up or down.

Some digital thermostats actually notify the user that there is a system problem. The thermostat can notify a customer or their contractor by phone or via the Internet. The thermostat can transmit a fault indicator so that the technician will have an easier time finding the problem.

The common thermostat terminals are **R, Y, W,** and **G.** The **C** or common terminal is found on some thermostats. The **R** terminal is a red wire and is the power from one side of the transformer. The **Y** terminal is yellow and sends power to the cooling contactor coil. The **W** terminal is white and sends power to the heating control. The **G** terminal is green and controls the indoor blower operation. The **C** terminal is the other side of the 24-V transformer and is required to complete the circuit for a digital thermostat or to operate heating and cooling controls.

Finally, troubleshooting a thermostat is not difficult. First do a system survey using the ACT method or some other logical method of troubleshooting. If the solution to the problem leads back to the thermostat, remove the t'stat and, from the subbase, jumper from **R** to **Y**, **R** to **W**, and **R** to **G.** Think of the jumper sequence as a manual thermostat. The fan, cooling, and heating circuits should work when jumping. The cooling, heating, and blower should all work. If one of the circuits checked does not work, the problem is that specific thermostat wire or that specific control.

REVIEW QUESTIONS

1. What do the R, Y, W, and G terminals control on a thermostat?
2. A heat pump thermostat has a B and O terminal. The O terminal is wired and the B terminal is not wired. Is the reversing valve energized in the cooling mode?
3. What is the purpose of the cooling anticipator?
4. What is the purpose of the heating anticipator?
5. What is the purpose of a humidifier?
6. What is the common control voltage used in most thermostats?
7. What are two advantages of a digital thermostat over a mechanical thermostat?
8. What is a line voltage thermostat?
9. What is the mounting height of a thermostat?
10. Should a thermostat be placed in the supply air or return air stream?
11. What gauge wire should be used for thermostat wire?
12. The **C** terminal is found on what type of thermostat?
13. How is the second stage of heating and cooling designated on a thermostat?
14. How do you measure a heat anticipator's low amperage using a clamp-on ammeter?
15. How is the cooling anticipator adjusted?
16. Is the auxiliary heat in a heat pump thermostat energized with the Y1 or W1?
17. What type of thermostat is considered a "green" thermostat?
18. What is the "override" function of a programmable thermostat?
19. What programmable thermostat would be selected for a business that has a varied schedule every day of the week and on weekends?
20. Why is it important to properly phase a transformer?
21. What improvement is obtained by phasing two transformers together to create a 24-V output?
22. What is the purpose of a dummy thermostat?
23. Refer to Figure 11-39. Select the best answer. The thermostat in this drawing is a:
    a. cooling-only thermostat.
    b. heating-only thermostat.
    c. heat pump thermostat.
    d. cooling and heating thermostat.
24. Refer to Figure 11-39. Select the best answer. The thermostat in this drawing has:
    a. a heating anticipator only.
    b. a cooling anticipator only.
    c. heating and cooling anticipators.
    d. no heating and cooling anticipators.
25. Refer to Figure 11-39. Select the best answer. The thermostat in this drawing has:
    a. two thermostat wires.
    b. three thermostat wires.
    c. four thermostat wires.
    d. five thermostat wires.
    e. six thermostat wires.

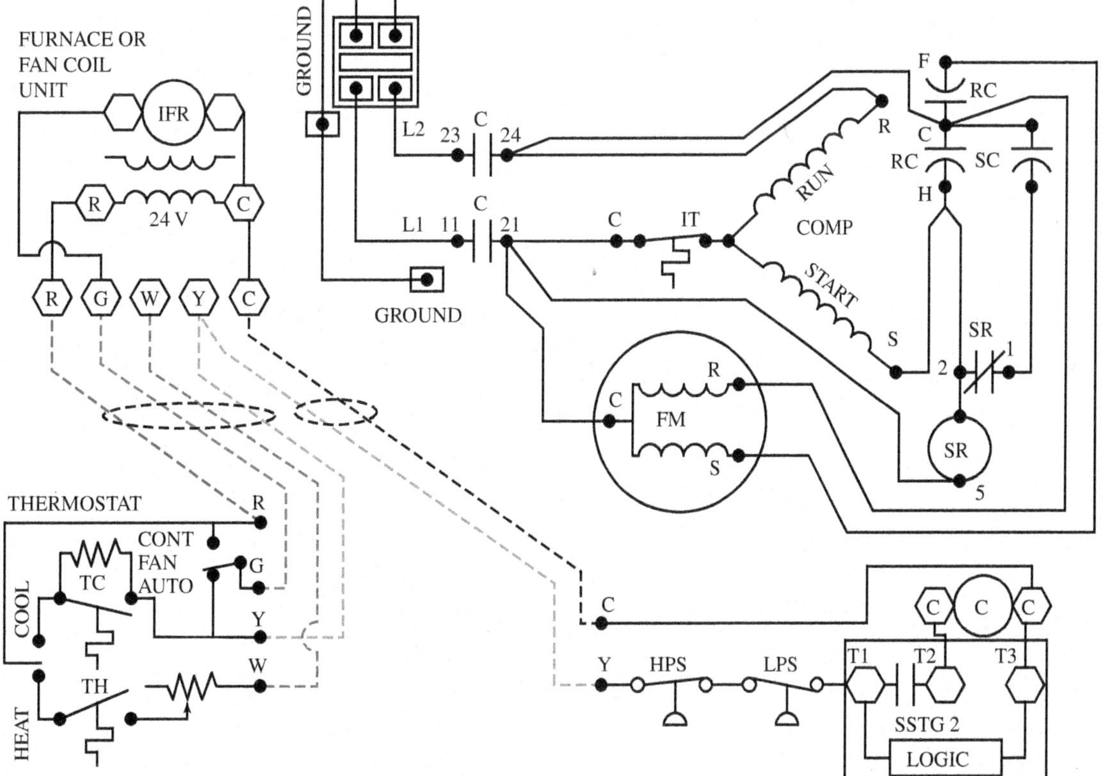

Figure 11-39 Use this drawing to answer Review Questions 23, 24, and 25.

UNIT 12

Pressure Switches

WHAT YOU NEED TO KNOW

After studying this unit, you will be able to:

1. Install a low-pressure switch.
2. Install a high-pressure switch.
3. Install an oil pressure switch.
4. Install a fan cycling switch.
5. Troubleshoot a low-pressure switch.
6. Troubleshoot a high-pressure switch.
7. Troubleshoot an oil pressure switch.
8. Troubleshoot a fan cycling switch.

Various types of pressure switches are used in air conditioning and refrigeration equipment. They are considered safety devices that will temporarily or permanently shut down a piece of equipment should a pressure switch open.

This unit discusses the most common types of pressure switches found in air conditioning and refrigeration equipment. They are the:

- Low-pressure switch
- High-pressure switch
- **Fan cycling switch** (not a safety device)
- **Oil safety switch.**

This unit discusses the installation, operation, and troubleshooting of these control or safety devices. We show how these devices fit in a wiring diagram and also give tips on how to test the pressure switch after installation. It is not only important to know how to troubleshoot these devices, but it necessary to understand how to test and replace pressure switches.

Generally, the two types of pressure switches are the mechanical type or transducer type. A mechanical pressure switch relies on a flexible diaphragm that causes a set of contacts to open or close when pressure is increased or decreased.

The **pressure transducer** converts pressure into an analog electrical signal. The word *transducer* means "to transform" from one type of energy to another. Pressure applied to a pressure transducer produces a deflection of the diaphragm, which introduces a strain that can be detected by diaphragm movement. The strain produces an electrical resistance change that is proportional to the pressure. Pressure transducers are similar to mechanical-type pressure switches. The major difference is that the control wiring leads back to a solid-state circuit board or other advanced control circuit.

At this time, most pressure transducer switches are of the less expensive mechanical design type. You will, however, find pressure transducers in more sophisticated HVACR equipment. For example, the **electronic expansion valve (EXV)** will most likely have pressure transducers to control superheat via a data feed to a microprocessor.

12.1 LOW-PRESSURE SWITCHES

The purpose of the low-pressure switch (**LPS**) is to stop the compressor when the refrigerant pressure is too low. Low refrigerant pressure reduces system capacity and reduces the amount of gas returning to cool the motor windings. The motor can overheat and trip out on its overload. Low charge also freezes over the coil, which reduces capacity and can cause liquid flooding since little heat is being transferred into the refrigerant. A frozen coil will block airflow through the air handler and duct system.

The symbol for a low-pressure switch is illustrated in Figure 12-1. The semicircle on the lower part of the symbol represents a refrigerant pressure dome. When the pressure falls below its setpoint, the pressure in the dome drops, which opens the switch contacts.

A low-pressure switch can be preset to a specific pressure or the pressure setting can be adjustable. Figure 12-2 shows a

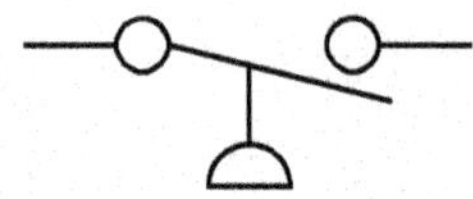

Figure 12-1 This is the symbol for a low-pressure switch. The semicircle at the bottom of the symbol represents a pressure dome. As the pressure drops the linkage connected to the top of the dome drops, opening the circuit.

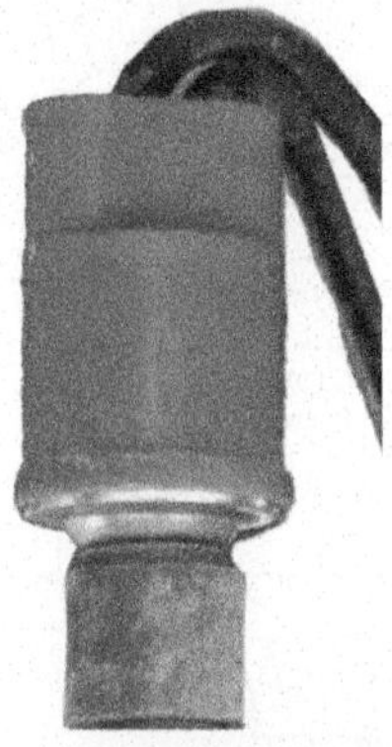

Figure 12-2 This preset low-pressure switch is used in air conditioning applications. The switch is screwed on to a Schrader valve fitting that is located on the compressor or suction line.

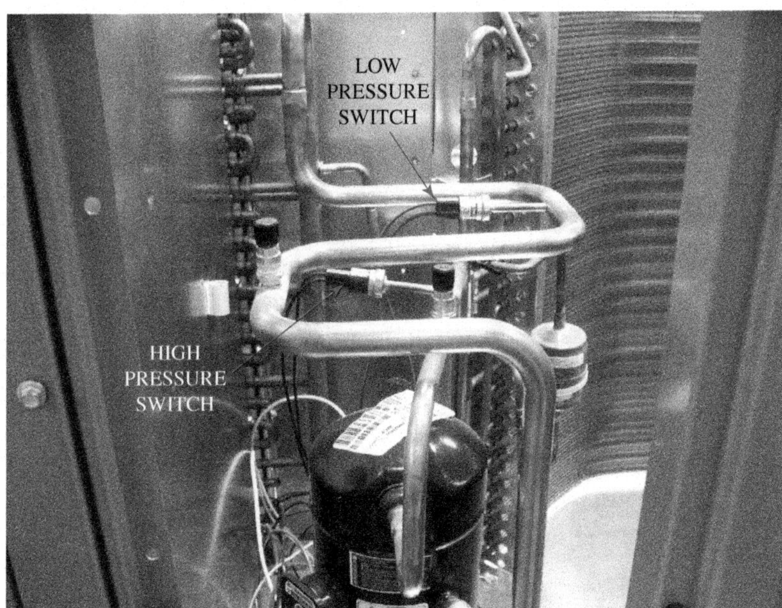

Figure 12-3 A preset low-pressure switch is brazed onto the suction line of this condensing unit. A high-pressure switch is below it, brazed to the discharge line.

preset pressure switch. Figure 12-3 shows a low-pressure switch brazed onto a suction line. A low-pressure switch can also be installed on the low side fitting of a semi-hermetic compressor.

When the suction pressure drops below a preset level, the low-pressure switch opens the control circuit, stopping the operation of the compressor or condensing unit. Here are some common low-pressure cut-out pressures for air conditioning applications:

- R-22 = 20 to 25 psi
- R-410A = 60 psi.

The **cut-in** or closing pressure is 20 psi higher than the **cut-out** pressure. The closing pressure for R-22 would be 40 to 45 psi. The closing pressure for R-410A would be 80 psi. The cut-in and cut-out pressures vary by equipment manufacturer, but these pressures are reasonable for air conditioning systems. Refrigeration applications operate at lower pressures than air conditioning units. The low-pressure cut-out setting for refrigeration applications is 0 psi or in a vacuum measured with inches of mercury (Hg).

Figure 12-4 shows an adjustable low-pressure switch. The advantage of this switch is that it can be used with a range of different refrigerants and applications. Another advantage of the adjustable low-pressure switch is the operating amperage. The adjustable low-pressure switch handles higher starting and running amps when compared to the preset switches. In some smaller condensing units, the compressor amperage runs through the low-pressure switch.

Adjusting this low-pressure switch is easy. Set the desired cut-out and cut-in pressures. Some other models are more difficult to set up. Let's look at an example using this type of low-pressure switch.

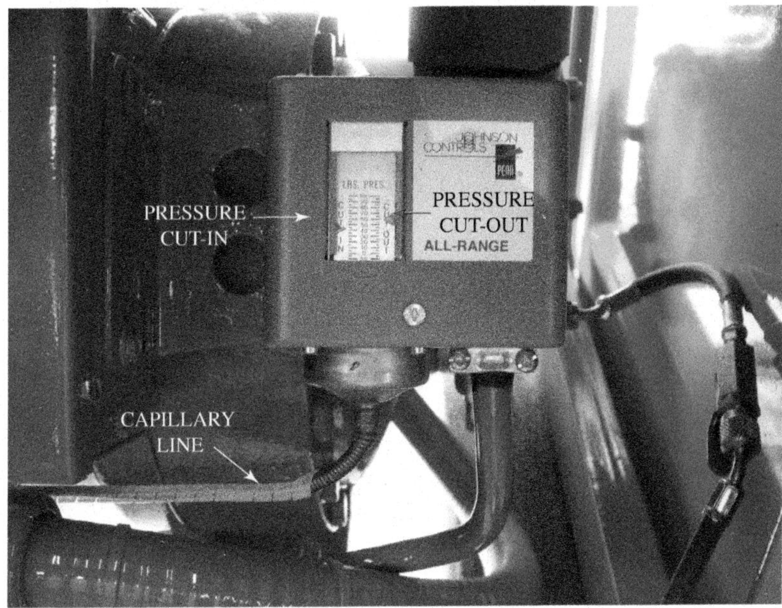

Figure 12-4 This adjustable low-pressure switch has cut-out and cut-in settings that should be adjusted to the refrigerant type and application. The low-pressure capillary line comes from the bottom of the pressure switch and is attached to the suction side of the compressor. The cap tube is protected with a plastic cover to shield it from possible damage and vibration wear, which could create a leak.

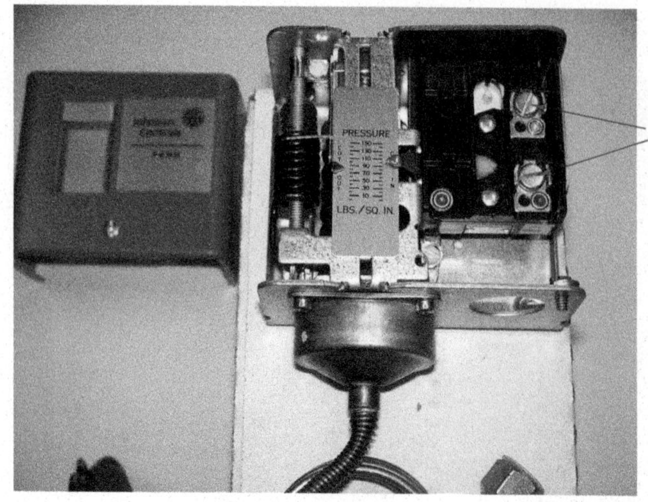

Figure 12-5 View of a low-pressure switch with the cover removed. The switch is set to open at 80 psi and close at 100 psi. These are approximate settings and must be checked prior to installing the switch on the system. The pressure cut-out and cut-in may be off ±5 psi or more. The wire connections are the two screws on the right side of the switch.

EXAMPLE 12.1 SETTING A LOW-PRESSURE SWITCH

Use Figure 12-5 for this example. What are the low-pressure cut-out and cut-in pressures?

SOLUTION

Let's review the settings so that you can use this information to set an adjustable low-pressure switch. Setting this low-pressure switch is easy. Subtract the cut-out pressure from the cut-in pressure. Some low-pressure switches have a different way to set up the cut-out and cut-in pressures. In this case the cut-out is about 80 psi and the cut-in is 100 ± 5 psi. If the operating low side pressure drops below 80 psi, the control circuit will open. When the pressure rises to 100 psi, the low-pressure switch will close and the compressor will begin to operate. Depending on how the system is wired, the compressor will cut off and the condenser fan may or may not cut out.

EXAMPLE 12.2 SETTING A MORE COMPLEX LOW-PRESSURE SWITCH

Here is a different example of a low-pressure setup that requires more thought. Figure 12-6 shows one type of low-pressure switch that you may encounter. Adjusting this type of low-pressure switch can be a little confusing.

The low-pressure switch will state something like the following information on the cover:

$$\text{Cut-out} = \text{cut-in} - \text{differential}$$

This formula can be stated in another way. For example, you could say:

Pressure switch opening pressure = Closing pressure minus the pressure differential between opening and closing the contacts

For this example we notice that the low-pressure switch settings in Figure 12-6 are set as follows:

- Cut-in pressure of 100 psi (right side)
- Differential of 75 psi (left side).

SOLUTION

Let's plug this information into the formula. Adjust the pressure cut-in to read 100 psi. Adjust the differential pressure to read 75 psi.

$$\text{Cut-out} = \text{Cut-in} - \text{differential}$$
$$\text{Cut-out} = 100\text{ psi} - 75\text{ psi}$$
$$\text{Cut-out} = 25\text{ psi}$$

The low-pressure switch will open the circuit at 25 psi. The low-pressure switch will close when the pressure rises to 100 psi. This pressure switch can be used for a variety of refrigerant pressures. You will need to know the recommended cut-out.

Figure 12-6 This type of low-pressure switch requires a minor calculation to find the pressure cut-out.

This is just an example—it is not intended for any specific refrigerant type or operating system. Looking at the range of pressure settings, this pressure switch could also be used as a fan cycling switch (discussed later in this unit). The low-pressure switch in Figure 12-6 will need to be brazed or flared into the suction side of the system. When brazing, best practices require a suction line with a nitrogen purge to prevent carbon contamination in the refrigerant lines.

12.2 LOW-PRESSURE SWITCH INSTALLATION

Most replacement low-pressure switches are good from the factory. However, if desired, a switch can be quickly checked prior to installation. Here are two steps that will save you extra work and confusion if a switch does happen to be defective:

1. The pressure switch will be open or have an infinite resistance reading before attaching it to a test pressure. Check it for an infinite resistance reading.
2. Next, hook the ohmmeter to the low-pressure switch leads and pressurize the switch with nitrogen or some other inert gas. The low-pressure switch should close near the pressure setpoint for the refrigerant type and application.
3. When testing with nitrogen, it is best to control the test pressure through the low side of the manifold gauge set for the most accurate pressure reading.
4. Finally, you do not want to hook up a pressure switch that will not open or close. The pressure switch should also be tested after it is hooked up to the suction line and wired to the control circuit.

Most low-pressure switch installations use either a flare fitting or a brazed fitting. Figure 12-2 shows a flare installation. Figure 12-3 shows a brazed installation. The low-pressure flare installation may have a depressor pin that pushes open a Schrader valve when it is screwed onto the suction line. If the low-pressure switch does not have a depressor pin, the Schrader valve core will need to be removed before the switch is installed. Without a depressor pin the Schrader valve will not be pushed open to monitor the suction pressure. The low-pressure switch will not "see" pressure and, therefore, will be open.

The low-pressure switch will need to be tested by creating a low-pressure condition on the suction side of the system. There are several ways to safely create a low-pressure condition. One easy way is to do the following steps:

Step 1: Hook the **low side gauge** to the suction side.

Step 2: Start the equipment and allow the pressures to stabilize for a few minutes.

Step 3: Close the **liquid line service valve** or receiver service valve and watch the suction pressure drop.

Step 4: Notice the suction pressure at which the pressure switch opens and the compressor turns off. The pressure should be within ±10 psi of the pressure setting. You may be able to hear a click when the pressure switch opens or closes. Depending on the noise in the area, the click may or not be heard.

Step 5: Open the liquid line service valve and notice the cut-in pressure when the compressor starts to operate again. The cut-in or closing pressure should be within ±10 psi of the pressure setting.

Figure 12-7 shows the wiring diagram with the low-pressure switch in a blue circle. (This diagram is also found as enlarged Electrical Diagram ED-1, which appears in the Electrical Diagrams package that accompanies this book.) In this example, the pressure switch is in series with the compressor. Another option is to wire the low-pressure switch in series with the (CR) coil in the low voltage control section. This low-pressure switch will need to have contacts capable of handling the compressor RLA and LRA. This requires a heavy-duty and more expensive pressure switch since the contacts will be much larger than a switch installed in the control voltage circuit, which has lower amperage requirements.

12.3 LOSS OF CHARGE SWITCH

The diagram in Figure 12-8 shows another type of low-pressure switch. It is called a "loss of charge" pressure switch instead of a low-pressure switch. The loss of charge pressure switch will open when the suction pressure gets near 0 psi, which is when most of the charge is gone.

You may find a loss of charge pressure switch in a heat pump system or commercial refrigeration system. These types of units operate at a very low suction pressure; therefore, they need a loss of charge low-pressure switch, not the conventional low-pressure switch. Some manufacturers place the loss of charge pressure switch on the liquid line, especially in a heat pump application. In this case, it would open at 5 psi and reset or close around 20 psi.

12.4 DETERMINING IF A LOW-PRESSURE SWITCH IS DEFECTIVE

Troubleshoot the low-pressure switch by hooking up the manifold gauge set, attaching a clamp-on ammeter to the compressor, and starting the **hopscotch troubleshooting** procedure. The hopscotch troubleshooting procedure is discussed in detail in Unit 25. Basically, with the hopscotch method, you measure voltage using a ladder diagram from the left side of a diagram toward the right side or load side of the diagram. When the voltage drops to zero (0 V), it means

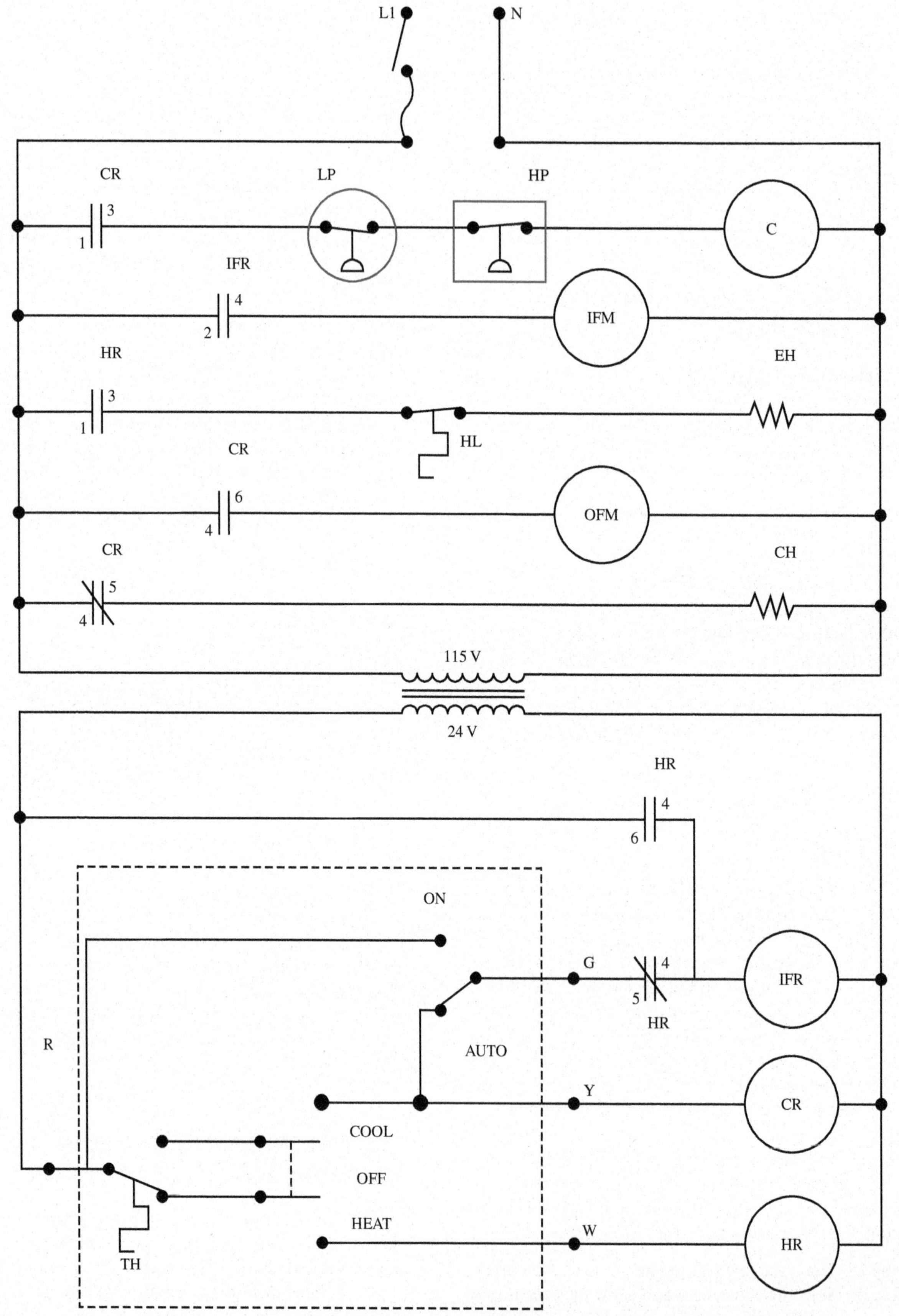

Figure 12-7 The low-pressure switch is circled in blue. The high-pressure switch is outlined with a red square. These pressure switches are in line with the compressor **C**; therefore, they need to be able to handle the high amperage conditions required by compressor operation.

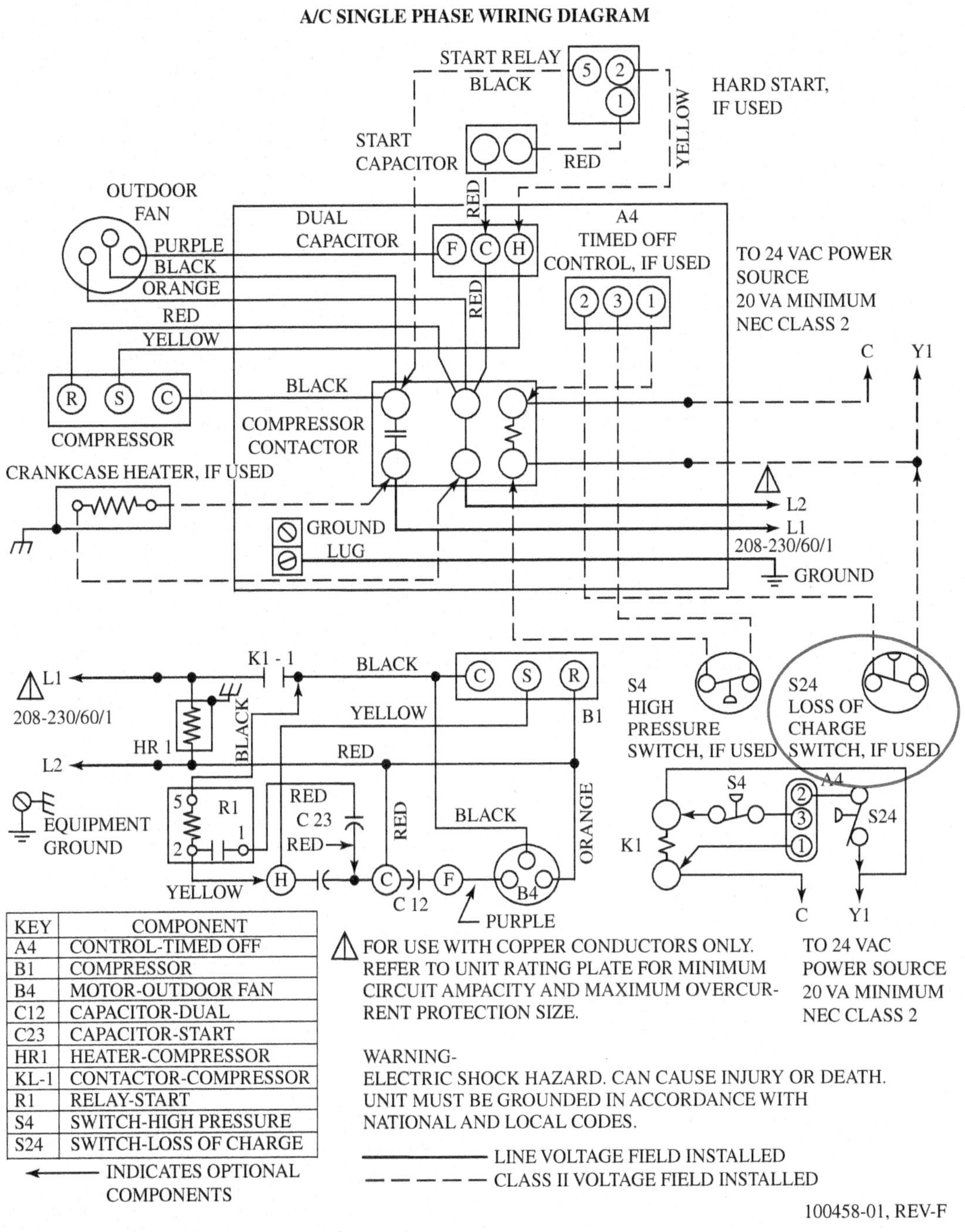
A/C SINGLE PHASE WIRING DIAGRAM
START RELAY
BLACK
HARD START, IF USED
YELLOW
START CAPACITOR
RED
OUTDOOR FAN
PURPLE
BLACK
ORANGE
RED
YELLOW
DUAL CAPACITOR
A4 TIMED OFF CONTROL, IF USED
TO 24 VAC POWER SOURCE 20 VA MINIMUM NEC CLASS 2
C
Y1
BLACK
COMPRESSOR
COMPRESSOR CONTACTOR
CRANKCASE HEATER, IF USED
GROUND LUG
L2
L1
208-230/60/1
GROUND
L1
208-230/60/1
K1 - 1
HR 1
L2
EQUIPMENT GROUND
R1
RED
C 23
YELLOW
C 12
PURPLE
B1
B4
ORANGE
S4 HIGH PRESSURE SWITCH, IF USED
S24 LOSS OF CHARGE SWITCH, IF USED
A4
S4
S24
K1
C
Y1
KEY
COMPONENT
A4 CONTROL-TIMED OFF
B1 COMPRESSOR
B4 MOTOR-OUTDOOR FAN
C12 CAPACITOR-DUAL
C23 CAPACITOR-START
HR1 HEATER-COMPRESSOR
KL-1 CONTACTOR-COMPRESSOR
R1 RELAY-START
S4 SWITCH-HIGH PRESSURE
S24 SWITCH-LOSS OF CHARGE
INDICATES OPTIONAL COMPONENTS
FOR USE WITH COPPER CONDUCTORS ONLY. REFER TO UNIT RATING PLATE FOR MINIMUM CIRCUIT AMPACITY AND MAXIMUM OVERCURRENT PROTECTION SIZE.
TO 24 VAC POWER SOURCE 20 VA MINIMUM NEC CLASS 2
WARNING-
ELECTRIC SHOCK HAZARD. CAN CAUSE INJURY OR DEATH. UNIT MUST BE GROUNDED IN ACCORDANCE WITH NATIONAL AND LOCAL CODES.
LINE VOLTAGE FIELD INSTALLED
CLASS II VOLTAGE FIELD INSTALLED
100458-01, REV-F

Figure 12-8 The loss of charge switch is circled in blue. It will open near 0 psig. The high-pressure switch is shown to the left of the loss of charge switch.

that specific switch is open. Use this step-by-step sequence to determine if the low-pressure switch is defective:

Step 1: After hooking up the gauges and clamp-on ammeter, turn the system on.

Step 2: The suction pressure and high side pressure should be within range of their air conditioning or refrigeration application. If the system pressure is below the low-pressure setpoint, the refrigeration system will not operate. In this case, the system will need to be charged up after the leak is repaired. Prior to charging, change the filter-drier if the refrigeration system is exposed to the air and evacuated to 500 micron. The system will cycle on and off on the low-pressure switch if the system has a low charge. This on and off condition is known as **short cycling**. The suction pressure may be high enough to operate for a short period of time before dropping low enough to open the low-pressure switch.

Step 3: If the compressor or condensing unit is off and the pressures are adequate, start the hopscotch sequence of troubleshooting. As you jump or hopscotch through the circuit, the 24 volts will be lost on the right side of the open low-pressure switch. Referring back to Figure 12-7 or enlarged Electrical Diagram ED-1, 120 volts will be measured on the left side of the low-pressure switch and 0 volts on the right side. Here is another example: When referring to Figure 12-8, 24 volts will be measured on the left side of the loss of charge pressure switch (in the blue oval) and 0 volts on the right side of the open pressure switch.

Step 4: The low-pressure switch or loss of charge switch is suspected to be open. Measuring 24 volts across the low-pressure switch in Figure 12-7 indicates that the switch is open. Measuring 24 volts across the loss of charge pressure switch in Figure 12-8 indicates that the switch is open.

Step 5: Turn off the power supply to the unit. Remove at least one wire from the pressure switch control circuit. An open low-pressure switch will measure infinity. At this point be certain that the system has enough refrigerant pressure to close the pressure switch.

Step 6: When you have determined that the pressure switch is defective, notify the customer in writing of the total charge to replace the component. It is good customer service practice to give a firm price quote in advance of servicing a piece of equipment. This prevents a misunderstanding and allows the customer to get a second opinion if desired.

Step 7: Test the replacement pressure switch before installing.

Step 8: Install the loss of charge switch to the system and test it, by creating a low-pressure condition.

Step 9: Finally, check the system amperage and refrigerant pressures to ensure correct operation.

Step 10: Review results with the customer and present an invoice. Describe the materials, parts, and labor charges on the invoice. The invoice should correspond with the price quoted.

This section discussed the operation, installation, and troubleshooting of low-pressure switches. The LPS opens when the suction pressure drops too low for normal system operation. Operating the system at low pressure reduces system capacity and prevents a good flow of refrigerant gas to cool the compressor motor. The compressor windings overheat. Under low-pressure conditions in the air conditioning mode, the evaporator and suction line will begin to frost over since the coil is operating below freezing temperatures.

Most low-pressure switches are automatically reset. This means that when a low-pressure switch opens, it will close once the pressure increases. This cycling on and off will drastically reduce system capacity, and the customer will report a problem.

TROUBLESHOOTING 12.1: OPEN LOW-PRESSURE SWITCHES

Use the following steps to pinpoint the problem of an "open" low-pressure switch (LPS).

1. As shown in Figure 12-9, hook up the refrigeration gauges and clamp-on ammeter. The pressure on the low side gauge should be above the setpoint of the LPS to ensure the switch is closed.
2. As shown in Figure 12-10, measure voltage across the low-pressure switch. An open switch will measure the control voltage.
3. Disconnect the power and measure resistance across the suspect low-pressure switch. As shown in Figure 12-11, an open pressure switch will read infinity or a very high resistance.

Figure 12-9 Hook up the gauges and clamp-on ammeter.

(continued)

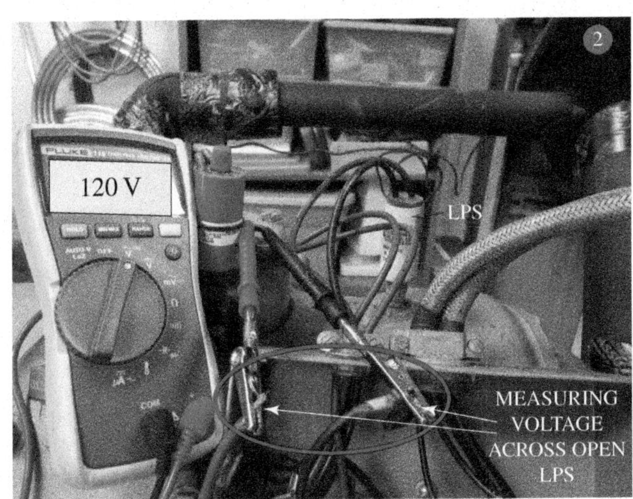

Figure 12-10 Measure the voltage across an open LPS.

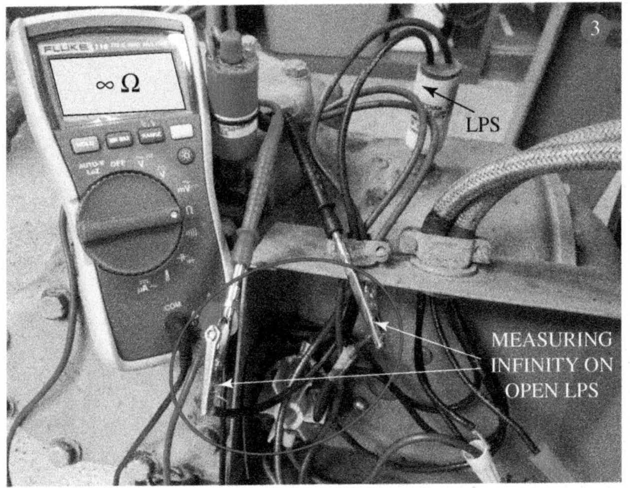

Figure 12-11 No resistance is measured on an open LPS.

SAFETY TIP

A low-pressure condition in an air conditioning system will result in ice on the evaporator coil that can become several inches thick. The coil can be thawed out in one of several ways. The blower fan can be set to continuous fan operation until the ice clears. This works well if there is light frosting. Thick ice will take hours to thaw using this technique. It is too dangerous to use an open flame to remove the ice. A refrigerant leak in the coil will add to this danger. Using a heat gun plugged into a portable **ground fault circuit interrupter (GFCI)** outlet is a safe way to quickly remove ice. If it is a heat pump system, place it in the heating mode for a few minutes to clear the ice. Note, however, that a heat pump should not be operated for an extended period of time when the outdoor temperature is above 65°F.

12.5 HIGH-PRESSURE SWITCHES

The purpose of a high-pressure switch is to protect a compressor from high-pressure conditions. A high-pressure condition translates into higher temperature operation, which shortens compressor life. Overheating causes the motor windings to deteriorate. Additionally, the overheated condition will cause compressor lubrication breakdown.

A high-pressure switch is more important than a low-pressure switch. Both safety devices will be used in a premium piece of equipment, although they are more common in commercial systems than in residential equipment. Top-of-the-line residential systems, however, will have pressure switches to protect the equipment. The customer pays for a better product and receives a piece of equipment that has more protection and, thus, a longer life.

Figure 12-12 is the symbol for a high-pressure switch. The semicircle on the bottom of the symbol represents a pressure dome. As the pressure under the dome increases, the connecting rod is pushed up and breaks the circuit. This will stop the compressor or condensing unit.

Figure 12-13 illustrates the operation of a high-pressure switch. The bellows inside the switch will open the switch contacts if the pressure setpoint is exceeded. The switch is in series with the control voltage and compressor contactor that controls the compressor's power supply. The switch will open when the pressure exceeds the setpoint, thereby stopping the compressor operation. These pressure switch contacts will not carry high current; therefore, they must be wired into a low current-carrying control circuit. Low current in this case will be an amp or two. There are high-current pressure switches that can carry enough current to be placed in series with a small compressor.

Like the low-pressure switch, the high-pressure switch can have a fixed or adjustable opening pressure. Here are four options:

- Fixed high-pressure switch with automatic reset
- Fixed high-pressure switch with manual reset

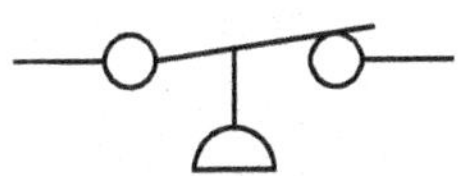

Figure 12-12 This is the symbol for a high-pressure switch. Symbolically, increased pressure under the dome will push the connecting rod up, thus opening the contacts on the right side and breaking the circuit.

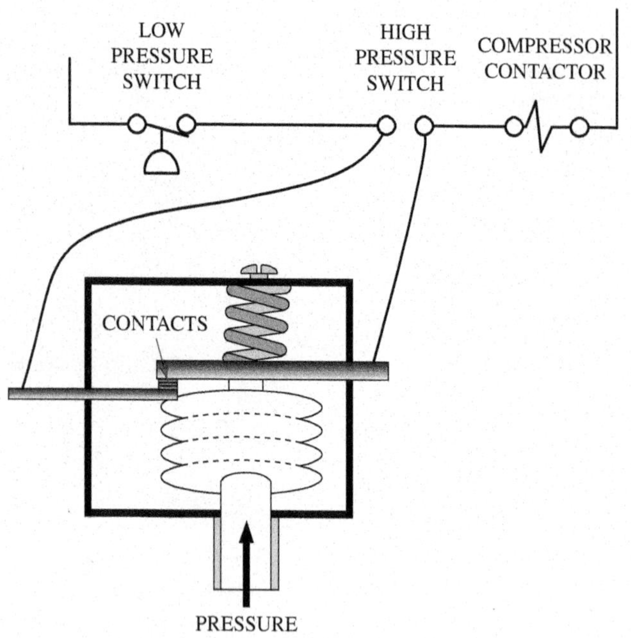

Figure 12-13 Pressure under the bellows pushes up on the high-pressure switch. Excessive bellows pressure will open the high-pressure switch.

- Adjustable high-pressure switch with automatic reset
- Adjustable high-pressure switch with manual reset.

Figure 12-14 shows a fixed high-pressure switch that has an automatic reset. If the pressure opens due to an excessive pressure condition, it will automatically close when the pressure drops by a prescribed amount.

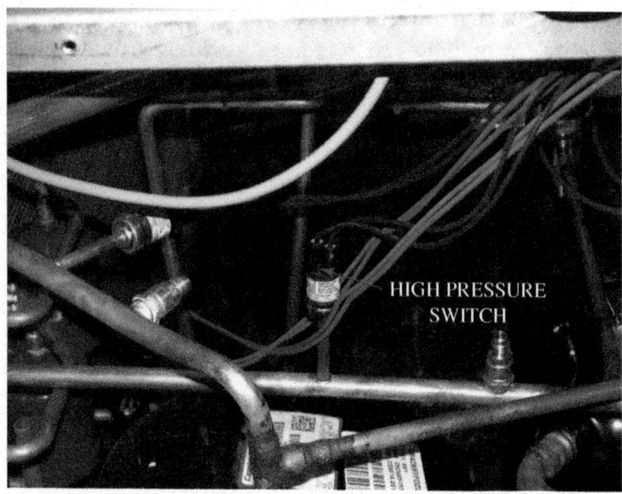

Figure 12-14 High-pressure switch with an automatic reset. This pressure switch is brazed into the refrigeration system. The high-pressure switch will open when the high side pressure becomes excessive. The pressure switch will close when the high-pressure condition drops. This style of pressure switch will only carry control voltage current.

Figure 12-15 This is an R-22 high-pressure switch set to cut out at 410 psi. The high-pressure switch has a flare fitting that will screw onto a ¼-in. flare fitting. The right side shows the red reset button. The current safety code does not allow a Schrader valve core to be installed when connecting a high-pressure switch. The valve could be blocked and not allow high pressure to be sensed.

Figure 12-15 shows a fixed high-pressure switch with a manual reset. The red button on the right of the switch is pushed if the high-pressure switch is tripped.

Figure 12-16 shows a high-pressure switch installed on a small water-cooled condenser. This switch is set for a lower pressure when compared to an air-cooled condenser with refrigerant, because water-cooled condensers operate at lower head pressures.

Figure 12-17 shows an adjustable high-pressure switch. Because it is adjustable, it can be used on a variety of air- and water-cooled condensing units.

Figure 12-18 shows the installation of a fixed cut-out high-pressure switch and a fixed cut-out low-pressure switch.

Figure 12-16 This high-pressure switch is installed on a tube-in-tube water-cooled condenser. The outer tube of this condenser holds condenser pressure. The inside tube carries the cooling water. The high-pressure switch must be pushed (red button) to reset it if open.

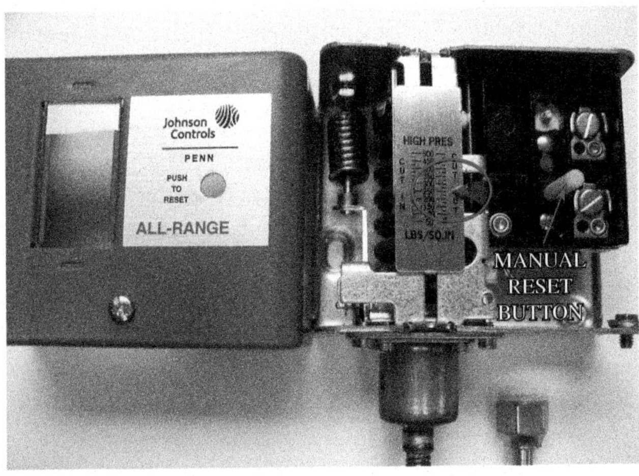

Figure 12-17 This is an adjustable high-pressure switch that has a flare connection. This pressure switch is set to open at 250 psi. The differential pressure is preset for 75 psi lower than the cut-out pressure; therefore, the switch will close at 175 psi. This is a manual reset device—once opened, it will not allow a complete circuit until the reset button has been pushed.

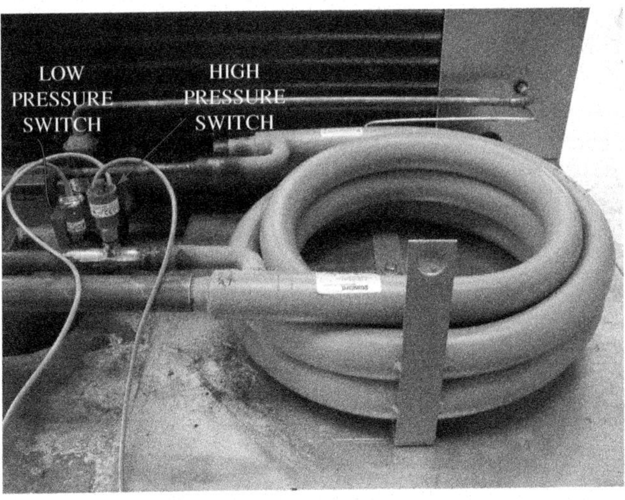

Figure 12-18 This is a small, water-cooled condenser that is in the process of being installed. The high-pressure switch has a manual reset. The low-pressure switch is to the left of the high-pressure switch. The low-pressure switch is an automatic reset device.

The high-pressure switch is installed on a water-cooled condenser. The high-pressure switch is a manual reset device as indicated by the red button. The low-pressure switch is an automatic reset device.

Here are two common high-pressure cut-out switch settings for an air-cooled condenser:

- R-22 = 420 to 450 psi
- R-410A = 610 to 650 psi.

The differential or cut-in pressure is around 50 ± 10 psi lower than the cut-out pressure. If the R-22 high-pressure switch opens at 420 psi, it will close at approximately 370 psi. If the R-410A high-pressure switch opens at 620 psi, it will close at approximately 570 psi.

A high-pressure reset may be designed to automatically close the pressure switch when the pressure drops or a manual reset may be required to close the pressure switch. A manual reset requires a technician to go to the job and push a button on the pressure switch to get it to close.

In summary, high-pressure and low-pressure conditions can do damage to equipment. A high-pressure condition tends to do more damage than a low-pressure condition. Many high-pressure switch devices have the manual reset design. If a piece of equipment is left to cycle on and off for a period of time, the manual reset high-pressure switch is preferred by many manufacturers. It is hoped that the tech will find the reason for the high-pressure cut-out before leaving the job. High-pressure nuisance trips are not common.

TECH TIP

Short cycling on any pressure switch will cause oil to log up in the evaporator. Short cycling with automatic reset does not allow enough time for the refrigerant oil to be returned to the compressor. If you notice a low oil level in a compressor sight glass and the system has been short cycling and starting up again, the low oil charge is most likely caused by the trapped oil in the evaporator section. Find the cause and correct the reason for the short cycling. Allow the compressor enough operating time to return the oil. Unless you notice a large oil leak, the oil is trapped in the evaporator or suction line. Large oil leaks are most likely to be in the compressor area.

12.6 HIGH-PRESSURE SWITCH INSTALLATION

You will find that installing a high-pressure switch is similar to installing a low-pressure switch. Most replacement high-pressure switches are good from the factory, but if desired you can quickly check a switch prior to installation. Here are two steps that will save you extra work and confusion if the switch is defective:

Step 1: The pressure switch will be closed and have a zero resistance (0-Ω) reading before attaching it to a test pressure.

Step 2: Hook the ohmmeter to the high-pressure switch wire leads and pressurize the switch with nitrogen or some other inert gas. It should open near the pressure rating. It is best to control the test pressure though the high side of the manifold gauge set for the most accurate pressure reading.

SAFETY TIP

When testing a high-pressure switch, you will be working with test pressures in the range of 400 to over 600 psi. It is important that you take safety precautions when working with any pressures. Control the test pressure through the high side of the manifold gauge set. Hook the center hose to the source of the test pressure such as a nitrogen regulator. Use safety glasses and gloves when handling these high pressures. Remember that pressure will be in the hoses even if the gas valve pressure is closed. When finished, crack the hose to relieve high pressures.

You do not want to hook up a pressure switch that will not open or close. The pressure switch should be tested after it is hooked up to the discharge or liquid line and wired to the control circuit.

Most high-pressure switch installations are either a flare fitting or a brazed fitting. Figure 12-17 shows a flared installation. The flare installation may have a depressor pin that pushes open a Schrader valve when it is screwed on the discharge or liquid line. As required by Underwriters Laboratory (UL), a new high-pressure switch will *not* have a Schrader valve. The rule basically states that "no valve shall be installed between a high-pressure safety device and the pressure it is monitoring." If the Schrader valve were to stick closed, it would take the high-pressure switch out of the circuit and eliminate the protection provided by this safety device.

The high-pressure switch will need to be tested by creating a high-pressure condition on the high side of the system. There are several ways to safely create a high-pressure condition. One easy way is to follow these steps:

Step 1: Hook the high side gauge to the discharge or liquid line.

Step 2: Start the equipment and allow the pressure to stabilize for a few minutes.

Step 3: Next, create a high-pressure condition in the condenser. This can be accomplished by covering the condenser coil with plastic to reduce airflow through the air-cooled condenser. In the case of a water-cooled condenser, reduce the water flow through the condenser. Watch the high side pressure rapidly rise.

Step 4: Notice the high-pressure reading causing the pressure switch to open and the compressor to turn off. The pressure should be within ±30 psi of the pressure setting. You may be able to hear a click when the pressure switch opens or closes.

Step 5: Remove the plastic or restore the water flow. Notice the reset pressure. The manual reset high-pressure switch may click, but it will not close until the manual reset button is pushed.

Refer back to Figure 12-7, which shows the wiring diagram with the high-pressure switch in a red square. Figure 12-8 shows the high-pressure switch to the left side of the loss of charge, low-pressure switch. In the Figure 12-7 example, the pressure switch is in series with the compressor. Another option is to wire the high-pressure switch in series with the (CR) coil in the low voltage control section. The high-pressure switch in Figure 12-7 will need to have contacts capable of handling the compressor RLA and LRA. This will be a more expensive pressure switch since the contacts will be much larger than a switch installed in the low voltage control circuit.

TECH TIP

It is difficult to replace a pressure switch when the refrigeration suction line is under full pressure. However, the suction pressure will be lower when the compressor is operating. You can also reduce the suction pressure by pumping the system down to 20 psi. Quickly remove and replace the low-pressure switch while the operating pressure is low. Do not allow the system to go into a vacuum. Many low-pressure switches connect to a Schrader valve fitting; therefore, the refrigerant release will be *de minimus*. Recovering most of the refrigerant will also do the trick. Leave some refrigerant pressure in the system so that you will not need to pull a vacuum.

12.7 DETERMINING IF A HIGH-PRESSURE SWITCH IS DEFECTIVE

Troubleshoot the high-pressure switch (**HPS**) by hooking up the manifold gauge set, attaching the clamp-on ammeter to the compressor and starting the hopscotch troubleshooting procedure. Use the following sequence to determine if the high-pressure switch is defective:

Step 1: Install manifold gauges and clamp-on ammeter then turn the system on.

Step 2: The suction pressure and high side pressure should be within range of their air conditioning or refrigeration application. If the system pressure is above the high-pressure setpoint, it will not operate. In this case, the system charge will need to be reduced by recovering the refrigerant or cooling the condenser. The high-pressure condition could be an overcharge, a dirty coil, or in the case of a water-cooled condenser, it could be low water flow or fouled condenser tubes. A high side restriction such as a closed solenoid valve can also cause high pressures. If the high pressure is below the pressure switch setpoint, proceed to the next step. If the pressure is about the setpoint, correct the high pressure problem.

Step 3: Make sure that the manual reset button has been reset. If the compressor or condensing unit is off and the pressure is not excessive, start the hopscotch sequence of troubleshooting. If the pressure switch is open, the voltage will be lost on the right side of an open high-pressure switch. Referring back to the diagram in Figure 12-7, 115 volts will be measured on the left side of the high-pressure switch and 0 volts on the right side.

Step 4: With these voltage readings the high-pressure switch is suspected to be open. In this case, the measurement across the open pressure switch is 115 volts. In other systems, the measured voltage will be the applied voltage through that pressure switch, which may be 24 or 240 V.

Step 5: Jumper the high-pressure switch. The system should operate properly.

Step 6: Turn off the power supply to the unit. Remove at least one wire from the pressure switch circuit. An open high-pressure switch will measure infinity.

Step 7: Notify the customer in writing of the total repair cost before changing out the pressure switch.

Step 8: Field test the replacement pressure switch prior to installing.

Step 9: Install the high-pressure switch to the high side and test by creating a high-pressure condition.

Step 10: Finally, check the system amperage and refrigerant pressures to ensure correct operation.

Step 11: Review results with the customer and present them with an invoice.

TROUBLESHOOTING 12.2: OPEN HIGH-PRESSURE SWITCH

Use the following steps to pinpoint the problem of an "open" high-pressure switch (**HPS**).

1. Figure 12-19 illustrates how to connect the manifold gauge set and the clamp-on ammeter.
2. Figure 12-20 shows the **HPS** with the cover removed. Voltage will be measured across an open **HPS**.
3. Figure 12-21 shows the ohmmeter measuring infinity (open) across an open set of contacts. Power is disconnected and the wiring from one or both sides of the **HPS** is disconnected.

Figure 12-20 When power is applied and control voltage is measured, the HPS is open.

Figure 12-19 Hook up the gauges and clamp-on ammeter. The high-pressure gauge reading should be lower than the HPS set pressure.

Figure 12-21 When infinity is measured across a HPS, the switch is open.

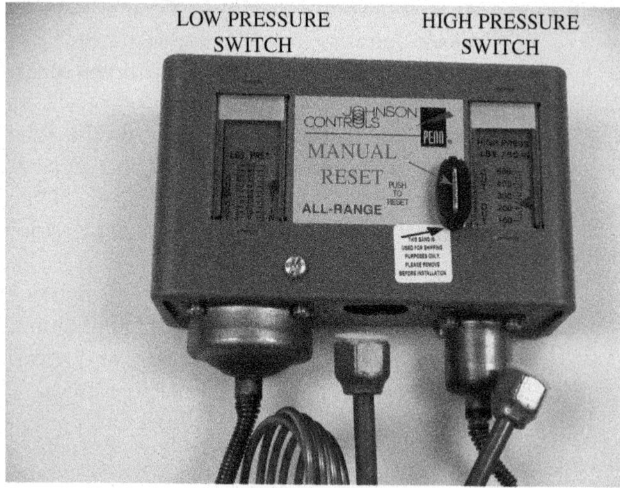

Figure 12-22 On this dual-pressure safety switch, the low-pressure connection is on the left and the high-pressure connection is on the right. The high-pressure switch has a manual reset button. The switch will open if the suction pressure drops too low or if the head pressure increases beyond its cut-out pressure. The low- and high-pressure differentials are preset on this dual-pressure switch.

This section discussed the operation, installation, and troubleshooting of high-pressure switches. The pressure switch opens when the high side pressure increases above normal system operation. Operating the system at excessive pressure reduces system capacity and overheats the compressor motor. This leads to shorter compressor life and possibly compressor damage.

Many high-pressure switches are of the manual reset design. This means that when a high-pressure condition opens the switch, it does not close once the pressure drops.

12.8 DUAL-PRESSURE SWITCH

Figures 12-22 and 12-23 show a dual-pressure switch. It will open the control circuit or compressor circuit if the low side pressure becomes too low or the high side pressure becomes excessive. There is only one set of contacts in a dual-pressure switch. The dual switch can have preset pressures or it can be adjustable as shown in these images.

TECH TIP

The excess capillary line on a pressure switch should be wound in a circle and secured so that vibrations will not cause rubbing between the bare copper. Vibration and rubbing action will eventually cause a refrigerant leak. Looping and securing the excess cap tube will create a professional look on your installation or component change-out.

12.9 CONDENSER FAN CYCLING SWITCH

The condenser fan cycling switch (see Figure 12-24) is used to turn off one or more condenser fans when the condensing pressure drops. The pressure at which the switch opens to stop the condenser fan is found by using the following formula:

$$\text{Cut-out} = \text{cut-in} - \text{differential}$$

The symbol for a fan cycling switch is the same as the low-pressure switch symbol. This is also known as head pressure

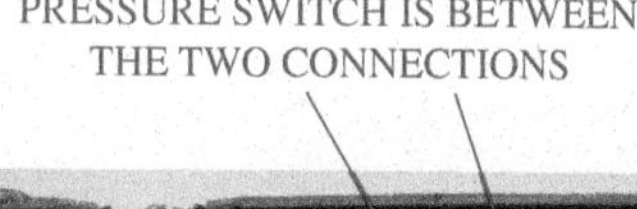

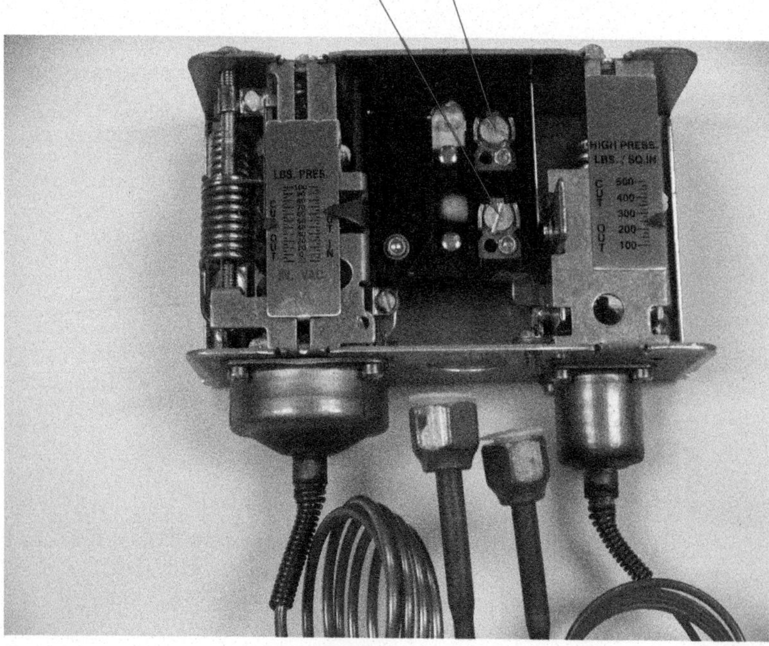

Figure 12-23 The cover is removed from the dual-pressure switch shown in Figure 12-22. The left side has the adjustments for the low-pressure cut-out. It has a cut-out and cut-in pressure adjustment. The right side has the adjustment for the high-pressure cut-out. The cut-in pressure is preset and not adjustable. There is one set of contacts for a dual-pressure switch.

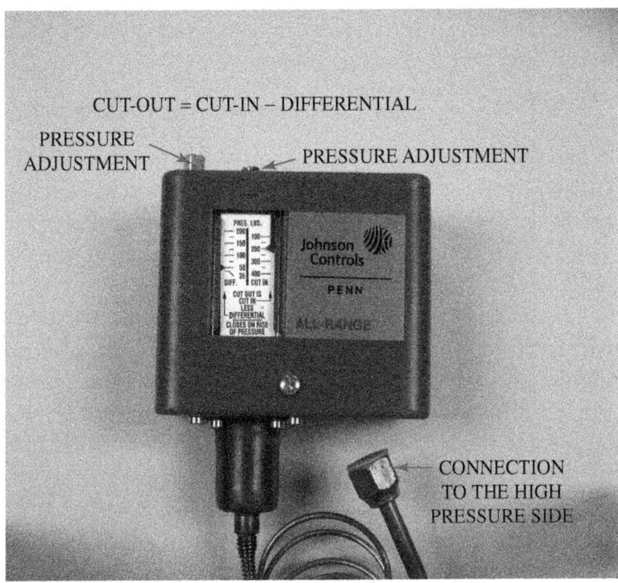

Figure 12-24 This is an adjustable fan cycling switch. Some fan cycling switches look like a low- or high-pressure switch. The fan cycling switch must match the refrigerant being used in the system. The fan switch must be able to handle the amp draw of the fan motor(s).

control. Most metering devices need a 100-psi differential between the high side and low side. The high pressure is used to push the refrigerant through the metering device during normal operating conditions. The outdoor temperature and condensing temperature track each other. As the outdoor temperature drops, so does the high side pressure. If a condenser has a fan cycling switch, it will open the fan circuit at a set pressure. Think of the fan cycling switch as a low-pressure switch for the condenser fan.

TECH TIP

The fan cycling switch opening and closing pressures vary somewhat among manufacturers. Here are some guidelines to help you determine what these pressures should be:

- Set the fan cycling switch to open at an approximately 90°F corresponding saturation condensing temperature.
- Set the fan cycling switch to reclose at approximately 120°F.

For example, for a condenser using R-22 refrigerant, the fan cycling control will cut off (open) at 168 psi (90°F) and cut in (close) at 260 psi (120°F). This will give you a 92-psi condensing pressure differential.

If there are two condenser fans, set the second fan to open at a 60°F outdoor ambient temperature and close at 70°F. Again, switches from different manufacturers vary somewhat on these settings.

There are ways to keep the head pressure high other than cutting on the fans. Some designs use a liquid line or discharge line thermostat that opens the fan circuit when the refrigerant line temperature drops. Other designs include closing off dampers to the condenser coil. Even with the condenser fan off, cold prevailing winds will create enough airflow to drop the condensing pressure. Hot gas bypasses and flooded condensers are refrigerant designs that keep the head pressure high enough to push refrigerant through the coil.

TECH TIP

Testing a fan cycling switch in warm weather is a challenge. The condensing temperature can be lowered by spraying water on the coil. Be careful to stay away from any open wiring. This may drop the high side pressure low enough to cause the fan switch to open the circuit to the condenser motor. Have the high gauge installed to see at what pressure the fan stops and cycles back on.

12.10 OIL PRESSURE SAFETY SWITCH

We now turn to a discussion of the operation, installation, and troubleshooting of oil pressure safety switches. An oil pressure safety switch is designed to stop a semi-hermetic compressor from operating when the oil pressure is low. Figure 12-25 shows the symbol for an oil pressure safety switch. You will notice that the symbols appear as two pressure switches in one. Figure 12-26 shows a complete component diagram with the heater and normally closed (NC) contacts.

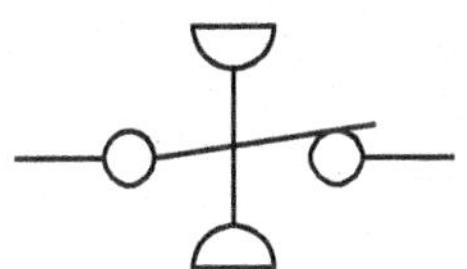

Figure 12-25 Symbol for an oil pressure switch. The oil pressure will be monitored by the lower semicircle or pressure dome. The suction pressure is monitored by the upper pressure dome.

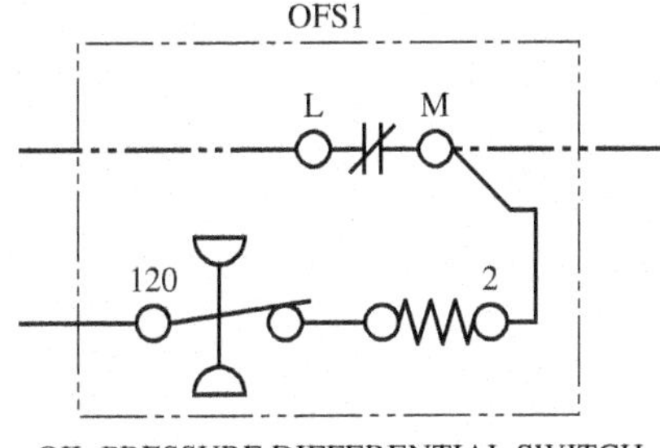

Figure 12-26 This drawing shows an oil pressure safety switch with its internal components such as the heater and normally closed (NO) contacts. OFS1 is the acronym for "oil failure safety switch #1." The switch can be wired to handle a 120- or 240-V control circuit.

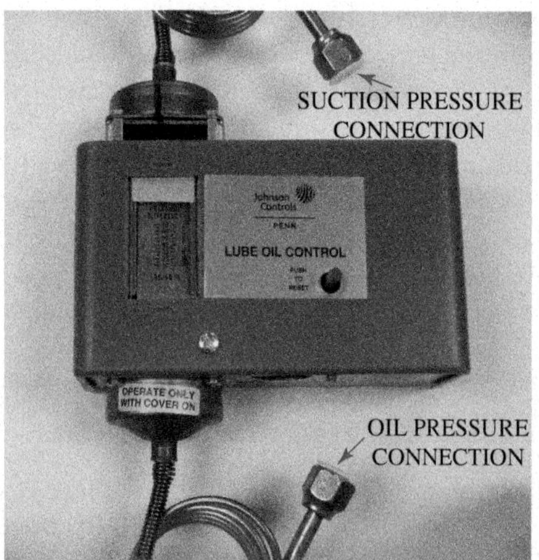

Figure 12-27 The oil pressure safety switch measures the net oil pressure by sensing the oil pump pressure (bottom connection) and the suction pressure (top connection). The net oil pressure is adjustable on this safety switch. Some pressure switches have preset pressure settings.

The oil pressure safety switch can trip due to one or more of the following factors:

- Low oil capacity in the compressor crankcase
- Refrigerant flooding or excess liquid in the compressor crankcase
- Restricted oil pump inlet strainer
- Oil pump wear or failure
- Defective oil safety switch
- "Brown-out" conditions (low voltage).

The oil pressure safety switch operates on a pressure differential. The pressure differential is the oil pump pressure minus the suction pressure. The differential is called **net oil pressure.** It can be expressed as follows:

Net oil pressure = oil pump pressure − suction pressure

When the compressor starts, the oil pump pressure and the suction pressure are equal. For this example we will use 100 psi as the starting suction pressure inside the compressor. The suction pressure will begin to drop as the compressor pulls gas into the compressor's cylinder. The oil pressure will increase since it is pumping oil. The oil pressure switch will allow the compressor to operate for 120 seconds, allowing the pressure differential to develop. In this time period the minimum oil pressure difference of 9 psi must develop to prevent the safety switch from opening. Some oil safety switches have a higher pressure differential requirement.

If the pressure differential is not reached in 120 seconds, the heater in Figure 12-26 will open the normally closed contacts and stop the compressor. The oil safety switch must be manually pushed closed in order to reset the circuit.

Pictured in Figure 12-27 is an oil pressure safety switch. The oil pressure switch receives two pressure inputs. One input is from the oil pump pressure and the other from the suction side of the compressor. There are no pressure adjustments. The pressure switch is preset at the factory with a specific oil pressure differential. For example, the manufacturer may set the pressure switch differential at 9 psi. Oil safety switches are reset manually should the oil pressure differential fail. It is extremely important to have good oil pressure. A lack of oil pressure will quickly damage a compressor's moving parts.

The control has two diaphragms. One is attached so that it can measure the suction pressure in the crankcase. The other diaphragm measures the pump outlet pressure. Many manufacturers recommend a 9- or 30-psi net oil pressure for proper lubrication. Figure 12-28 shows the capillary tube connections to the side of the compressor. One line senses

Figure 12-28 Connections to the oil safety switch. One connection goes from the oil pump. A second connection comes from the suction side of the compressor.

the compressor oil pressure. The other line senses the suction pressure.

Enlarged Electric Diagram ED-3, which appears with the Electrical Diagrams packaged with this book, is a two-chiller diagram in which each circuit uses an oil safety pressure switch. These switches are labeled as oil pressure differential switches and abbreviated **OFS1** for chiller circuit number 1 and **OFS2** for chiller circuit number 2. On ED-3, the **OFS1** is located to the far right of the diagram, about halfway down. **OFS2** is located at below it, at the bottom right-hand corner of the diagram.

12.11 TROUBLESHOOTING THE OIL SAFETY PRESSURE SWITCH

Few oil safety switches fail. It is a challenge to find the reason why an oil safety switch opens due to a low-pressure differential. Emerson Climate Technologies (which makes compressors under the Copeland name) publishes the following information, which applies to most semi-hermetic compressor manufacturers.

Reasons for low oil pressure failure are many. Check the following:

- Is the oil level in the compressor sight glass correct when it is in operation? You will need to know the manufacturer's sight glass recommended level. There is no standard across all manufacturers. One may require a half a sight glass of oil. Another may require a 3/8 oil level or 7/8 oil level.
- The pressure in the compressor crankcase should be slightly lower than the suction pressure for good oil return. Pressure flows from high pressure to lower pressure. Even a few psi difference will help with oil return to the compressor crankcase.
- Poor oil return can be caused by low refrigerant velocity. This could be caused by operating a compressor unloaded for long periods of time, oil traps in the suction line, an oversized suction line, or an incorrect oil and refrigerant match.
- Short cycling causes poor oil return. The small amount of oil that is discharged in the refrigerant piping does not have time to return due to the short operating cycle.
- Oil foaming on start-up is caused by excess refrigerant in the compressor oil. This is known as refrigerant migration or liquid floodback on start-up. Check the crankcase heater operation to reduce refrigerant migration during the off cycle. Investigate piping conditions that will create flooded starts. For example, check suction piping for traps or refrigerant draining from the evaporator directly into the compressor.
- The wrong oil and refrigerant match will also cause oil return problems. Some lubricants and refrigerants are not compatible; therefore, the oil will pool in the evaporator and suction line and not return to the compressor.

Reasons for oil safety tripping when the oil level is good in the compressor:

- The oil temperature could be excessive. Discharge line temperatures in excess of 250°F means that the oil temperature in the compressor cylinder is 300°F, which is the temperature at which oil will break down. The viscosity and pumping pressure of hot oil are lower than those of cooler oil.
- Low superheat will cause liquid flooding. Superheat as low as 5°F can entrain liquid refrigerant in the middle of the returning vapor and dilute the compressor oil.
- The crankcase has liquid refrigerant settling under the oil. Refrigerant is heavier than oil. The refrigerant will be sucked into the oil dip tube on start-up. The refrigerant will boil off and prevent the oil pressure from developing.
- The oil screen pick-up may be restricted.

Failure of the compressor will occur when:

- Oil circulation becomes blocked in the refrigeration circuit. This can occur in an oil separator when the trapped oil does not return oil to the compressor.
- The compressor oil pump fails.
- The oil safety switch is reset without finding the reason for oil pressure trips.
- Some three-phase compressors have a specific motor rotation. Running the oil pump in the reverse direction prevents the oil pumping action.
- Off-cycle refrigerant migration causes flooded starts or at a minimum the oil is diluted with refrigerant. Check the crankcase heater and piping design to combat these problems.

If the oil pressure differential and oil level are good, but the oil safety switch opens, the oil pressure safety switch may be faulty. Also, if the oil safety switch is energized during a "no compressor" operation, there will be no oil pressure differential, which will cause the safety switch to open. After this occurs the compressor will not start, because it is locked out by the oil safety switch.

TECH TIP

Oil safety switches can experience nuisance or mystery trips. Before you diagnose the problem as an unknown cause, review the operation as we discussed earlier. One mystery trip that is difficult to repeat is a "brown-out" condition. This is a low-voltage condition that causes the amp draw to increase, opening the compressor overload. The oil safety device will still have voltage even if it is low. The safety opens because there is no pressure differential, and there is no oil pressure because the compressor is off. The compressor will not restart after the overload cools and closes until the oil safety switch has been reset. If you suspect that low voltage is the problem, install a current-sensing relay to control the safety switch. When the compressor is not drawing amps, it will shut off the voltage to the oil safety switch.

SEQUENCE OF OPERATION 12.1: COMPRESSOR CIRCUIT #1

Here is the sequence of operation of the compressor #1 (**COMP-1**) circuit using enlarged Electrical Diagram ED-3 and image outtakes. This explanation includes the controls in that circuit and the oil pressure differential switch (**OFS1**).

1. Locate the **START** button as pointed out by the red arrow in Figure 12-29, which is located between points 5 and 7, left side of Electrical Diagram ED-3, midway. The **START** button is pushed.

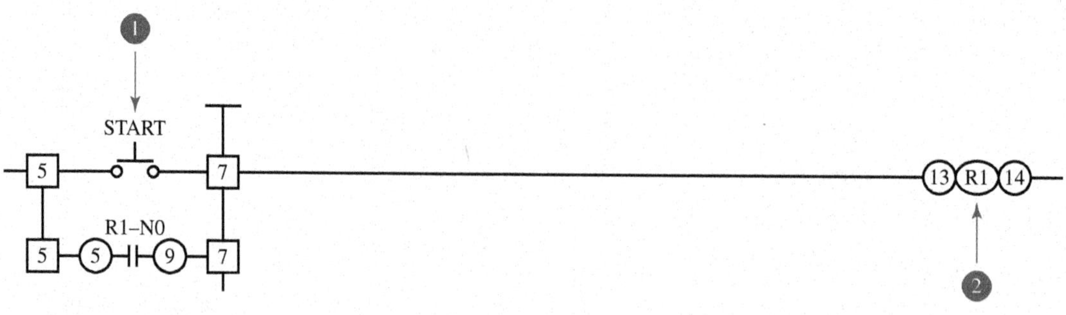

Figure 12-29 Starting (red arrow) and stopping switches for Electrical Diagram ED-3. **R1** (blue arrow) will be energized when the **START** button is pushed.

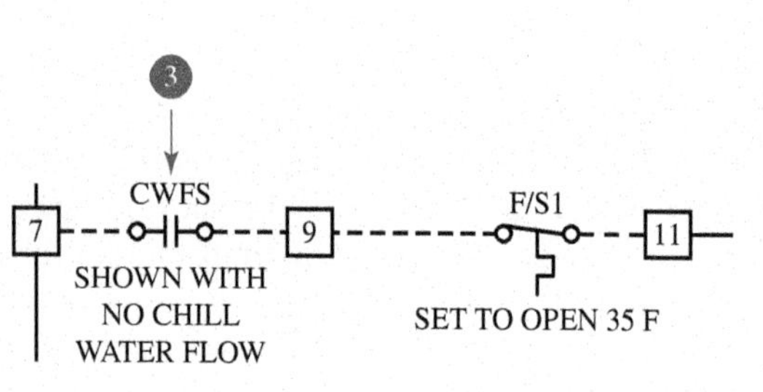

Figure 12-30 The red arrow points out the chilled water flow switch (**CWFS**).

2. In Figure 12-29, the **R1** coil is now energized (see blue arrow), closing the **R1-NO** contacts. These contacts are located on the diagram below the **START** button between terminals 5 and 9.
3. The chilled water flow switch (**CWFS**), as shown in Figure 12-30, must be closed to continue the circuit operation. This indicates chilled water flow.
4. In Figure 12-31 or Electrical Diagram ED-3, follow line 7 through freezestat **F/S1**, thermostat **T/S1**, low-pressure switch **LPS1**, high-pressure switch **HPS1**, and timer solid-state contacts **TMR1**.
5. This will energize the **CR1** coil and liquid line solenoid **LSV1** coil through the **OFS1** closed contacts **L** and **M**, as shown in Figure 12-32.

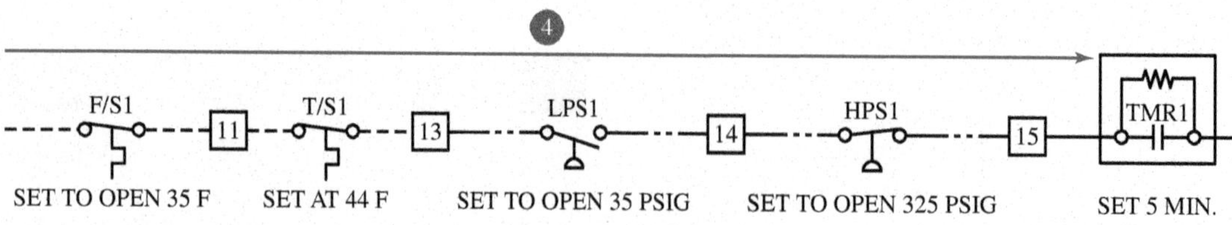

Figure 12-31 Follow line 7 through freezestat **F/S1**, thermostat **T/S1**, low-pressure switch **LPS1**, high-pressure switch **HPS1**, and timer solid-state contacts **TMR1**.

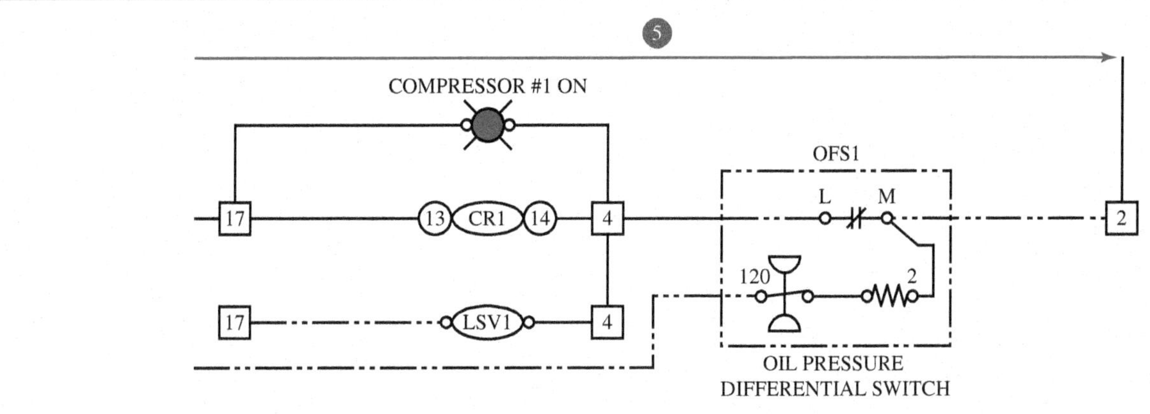

Figure 12-32 Power continues from timer **TMR1** to **CR1**, **L,** and **M** and on to the right side of the diagram.

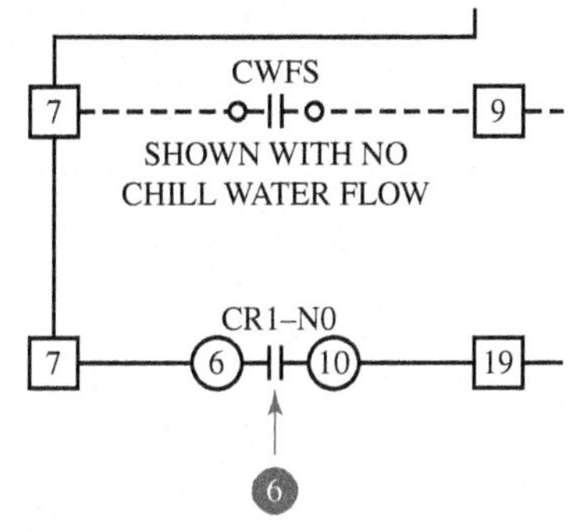

Figure 12-33 The **CR1-NO** contacts will close when the **CR1** coil is energized.

6. The **CR1** coil controls two sets of contacts. It will close the **CR1-NO** contacts at the bottom of line 7, under the **CWFS**, as shown in Figure 12-33. The **LSV1** coil will open the liquid line solenoid valve.
7. The second pair of **CR1-NO** contacts between terminals 7 and 11 is also closed. See Figure 12-34. These closed contacts will energize coil **SC1**.
8. The **SC1** contacts in the red oval will close, operating compressor #1 or **COMP-1**, as shown in Figure 12-35.

Note: If the oil pressure differential is not adequate, the **OFS1** heater will cause the normally closed **L** and **M** contacts to open, thus stopping the control voltage to chiller compressor #1 (**COMP-1**).

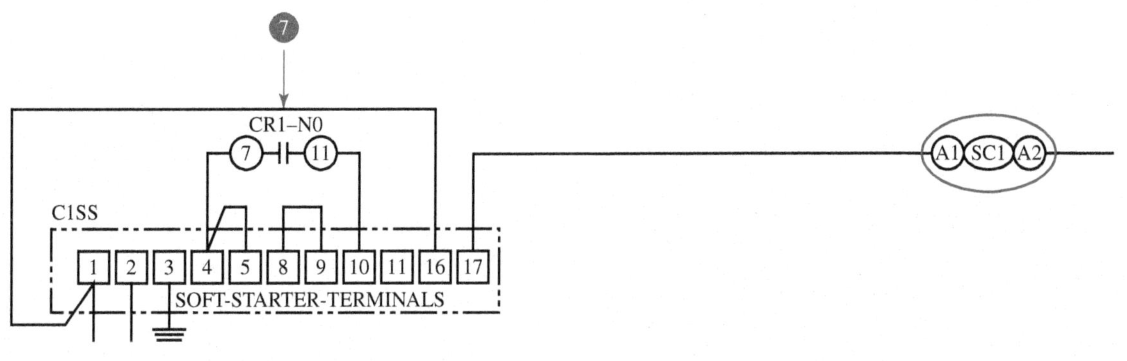

Figure 12-34 The **CR1-NO** contacts will close, energizing the **SC1** coil.

(*continued*)

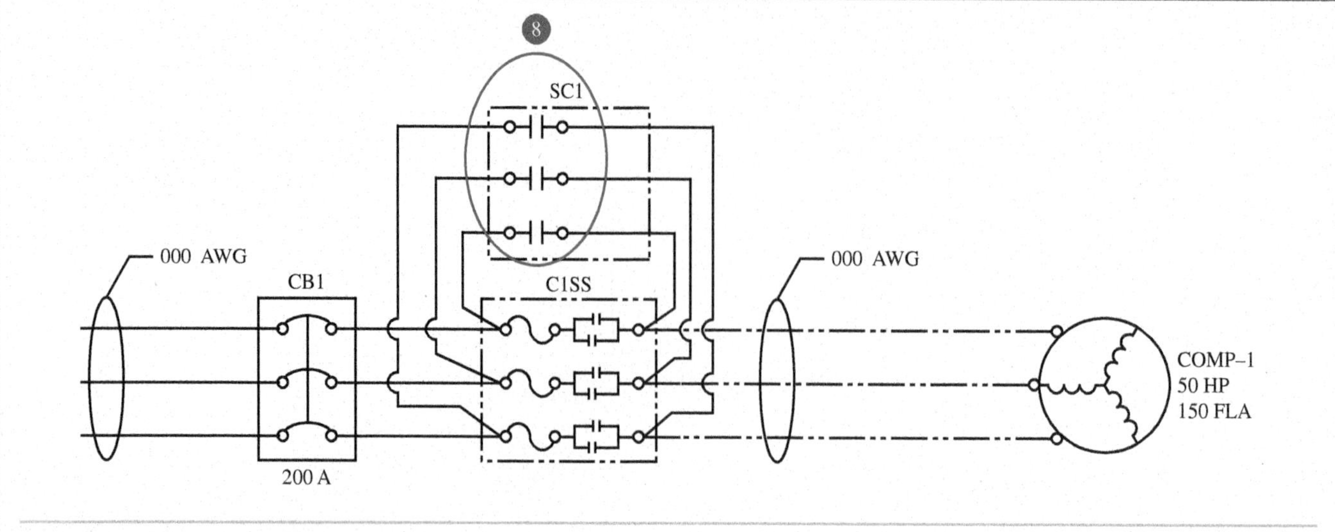

Figure 12-35 The three-pole contactor **SC1** closes, providing power to the compressor.

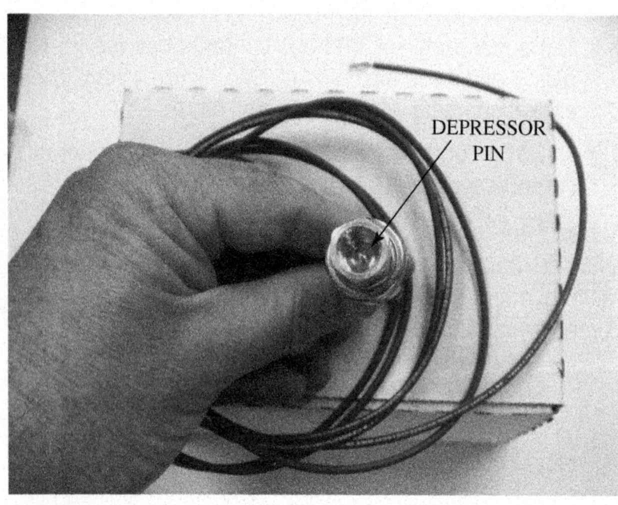

Figure 12-36 A depressor pin is found in the head of some flare fittings.

TECH TIP

Figure 12-36 shows the depressor pin found in the head of some flare fittings. The pin is important if a Schrader valve fitting is used. Not all refrigerant fittings have a valve and not all safety switches have a depressor pin.

SERVICE TICKET

Day 1: In the spring you are asked to respond to a "no cooling" call at a recreation center. When you arrive at the job site, the manager tells you that this is the first time air conditioning has been needed since early November. The building is about 80°F and uncomfortable because of the high humidity condition. The manager states that she turned on the air conditioning about 3 hours earlier and the temperature has continued to rise. It is especially warm in the office, which has three employees, three computers, a large copy machine, a coffee bar, and an operating television set.

Beginning the **ACT** method of troubleshooting, you notice that the 10-ton semi-hermetic compressor is not operating. The manifold gauge and clamp-on ammeter are hooked up to the condensing unit. The pressures on the high and low sides of the R-410A system are equalized close to 240 psi for each side. After more inspection you notice that the oil safety pressure reset button is pushed out. You push the reset button and the compressor begins to operate. Next, you notice the following:

- The 12-amp draw on the compressor is below the 15 RLA amps.
- The head and suction pressures are falling in line for an R-410A system.
- After 10 minutes of run time, the supply air is 60°F.
- The air filter, evaporator coil, and condenser coil look clean.
- The net oil pressure is 25 psi and, after running, the oil level looks good for that model of compressor. The oil level was checked prior to starting the compressor.

Everything seems to be okay except for the oil safety trip. You ask the manager if she wants a spring clean and check. She tells you that service is scheduled for next week. Since you were not able to find the problem, the dispatcher told you to suspend any charges, but write up a service ticket documenting the reading and problems. It is difficult to charge a regular customer when a problem has not been solved. You explain the "no charge" invoice to the manager and she signs and returns it.

Is there anything else that should have been checked?

Day 2: The service manager asks you to come to the office prior to your first service call, which you notice is the recreation center again. It is another "no cooling" call. After a discussion with the service manager, you are planning a different approach to solving the problem.

When you arrive on the job, the manager says it is not cooling again. After the routing inspection, you notice that the oil safety switch is tripped again. This time you hook up a suction pressure gauge to the oil pump's Schrader valve, the suction side, and discharge side before hitting the reset button. This requires two sets of manifold gauges: two compound gauges to measure oil pressure and suction pressure. Only one gauge is required to measure the high side pressure. The clamp-on ammeter is hooked up. The oil level in the compressor sight glass is higher than yesterday. The compressor is not yet running.

The reset button is pushed and the compressor begins to operate immediately. The oil level in the sight glass begins to drop. The suction pressure drops and stabilizes at 188 psi (40°F). The oil pressure fluctuates between 190 and 196 psi. The compressor sight glass has oil foaming and seems to drop to a normal level after a minute or two of operation. After 120 seconds of operation, the compressor shuts down. The oil safety switch is pushed out again. It is definitely a refrigerant dilution problem since the net oil pressure differential was not developed. At least 9-psi differential is required to keep the system running after 120 seconds. What do you do next? If the reset button is pushed again it may work as it did yesterday.

In the meeting with the service manager, he asked you to check the following:

1. *Inspect the crankcase heater*. The crankcase heater is not warm. Taking an amperage measurement reveals no amp draw. The heater will need to be replaced.
2. *Inspect the suction line for refrigerant traps*. The suction line does a loop over the evaporator and drains directly into the suction side of the compressor. No trap in this design. Looping the suction line higher than the evaporator coils prevents liquid refrigerant from draining the evaporator into the compressor during the off cycle. Good design.
3. *Check the oil level*. The high oil level noticed prior to starting the compressor could be a result of liquid refrigerant condensing in the crankcase when it gets cool at night. During the night, the unit is off until the manager arrives in the morning. Foaming is a sign of liquid refrigerant boiling in the crankcase.

The solution is to replace the open crankcase heater. The compressor may come back on if the oil safety reset button is pushed a second time. Instead of taking the risk of a second oil failure, you turn your heat gun on and direct it toward the crankcase of the compressor. The next step is to go to the supply house and pick up the replacement crankcase while the refrigerant is warming and boiling to a vapor. You return and install the crankcase heater. The crankcase heater is energized and draws a couple of amps. It begins to feel hot to the touch. The reset button is pushed and the oil pressure differential after 120 seconds of operation is now 40 psi. This will keep the compressor operating. You complete the invoice and discuss the work with the recreation center manager.

Your service manager calls the customer in the morning and she says that the system is cooling. Sometimes oil safety problems take a couple of trips to solve, especially if the reset button is pushed without having your gauges hooked up to measure oil pump and suction pressure differentials. This problem could have been solved on the first trip with a little more observation and by checking the common causes of oil pressure problems. Lesson learned.

SUMMARY

This unit discussed some of the common pressure switches used in air conditioning and refrigeration equipment. Other types of pressure switches are also used in various heating, air conditioning, and refrigeration applications.

The low-pressure switch is used to temporarily break power to a circuit when the suction drops below a preset level. Low refrigerant pressure will cause a compressor motor to overheat, since the volume of the cool gases is reduced. Low refrigerant can also cause the evaporator coil to freeze up, thus reducing heat transfer and airflow. The ice layer can become several inches thick and will need to be thawed prior to operating the system again.

The high-pressure switch is used to temporarily or permanently stop the operation of a compressor. The temporary open circuit is called an automatic reset, which means that the high-pressure switch will close when the pressure drops to a preset level. The high-pressure switch with a manual reset will require the tech to push a button on the switch that closes the contacts and restarts the compressor. It is important for the compressor to be protected from high-pressure and overheating conditions. These conditions will severely damage the compressor if it is left to operate in this high-pressure condition.

The fan cycling switch is used to turn off the condenser fan when the head pressure drops too low to effectively push liquid refrigerant through the metering device. Most metering devices require a 100-psi pressure differential between the high side and low side in order to achieve the required refrigerant flow rate to cool properly.

The oil pressure safety switch keeps the compressor from being damaged due to a lack of lubrication. The oil safety switch operates on a pressure differential. This is called *net oil pressure*. The oil safety switch operates on the pressure difference between the suction pressure and oil pump pressure. Before the compressor starts, the suction pressure and oil pump pressure are the same. As the compressor starts, the suction pressure drops and the oil pump pressure increases. The oil safety switch must develop at least a 9-psi differential in 120 seconds of running time. If the differential is not developed in the allotted time, the oil safety switch will open the control circuit and stop the compressor.

This unit also discussed the installation and troubleshooting process required of these pressure switches. It is important to check the "failure" mode operation of the switch when it is replaced. The replacement installation will require that the switch be checked to ensure it has been wired correctly.

It is a "best practice" to check any replacement part under the conditions in which it will be operating. If it is a safety device, it should be tested under the conditions it was designed to protect against; for example, you could create a low-pressure condition to see if the low pressure stops the compressor when the suction pressure drops below its setpoint.

REVIEW QUESTIONS

1. What is the purpose of a low-pressure switch?
2. What is the purpose of a high-pressure switch?
3. What is the purpose of a fan cycling switch?
4. What is the purpose of an oil pressure safety switch?
5. What would cause a low-pressure switch to trip?
6. What would cause a high-pressure switch to trip?
7. When would you expect to see a fan cycling switch stopping a fan?
8. What are two reasons an oil safety switch will open?
9. How do you test a replaced low-pressure switch?
10. How do you test a replaced high-pressure switch?
11. What are the two refrigerant connections for an oil pressure safety switch?
12. How do you test a fan cycling control?
13. What are the low-pressure settings for a system with an R-22? An R-410A?
14. What are the high-pressure settings for a system with an R-22? An R-410A?
15. Here are readings from gauges connected to a compressor that is using an R-410A:
    - Suction side: 90 psi
    - Head pressure: 385 psi
    - Oil pump pressure: 9 psi.

    What is the net oil pressure?
16. Why is it important to have a working crankcase heater?
17. List six reasons why an oil safety switch will open the control circuit to the compressor.
18. Refer to Electrical Diagram ED-3. If **OFS1** opens due to inadequate oil pressure, will **COMP-2** continue to operate?
19. What is the safety device shown in Figure 12-37?
20. Does the safety device in Figure 12-37 have a fixed or adjustable setting?
21. Refer to Figure 12-38. Where are the capillary lines connected in the refrigeration system?
22. Refer to Figure 12-39 or Electrical Diagram ED-2. Does this diagram have a low-pressure switch? High-pressure switch? Oil safety switch?
23. Refer to Figure 12-39. If a safety device opens, will the evaporator fan continue to operate?

Figure 12-37 Use this image to answer Review Questions 19 and 20.

Figure 12-38 Use this image to answer Review Question 21.

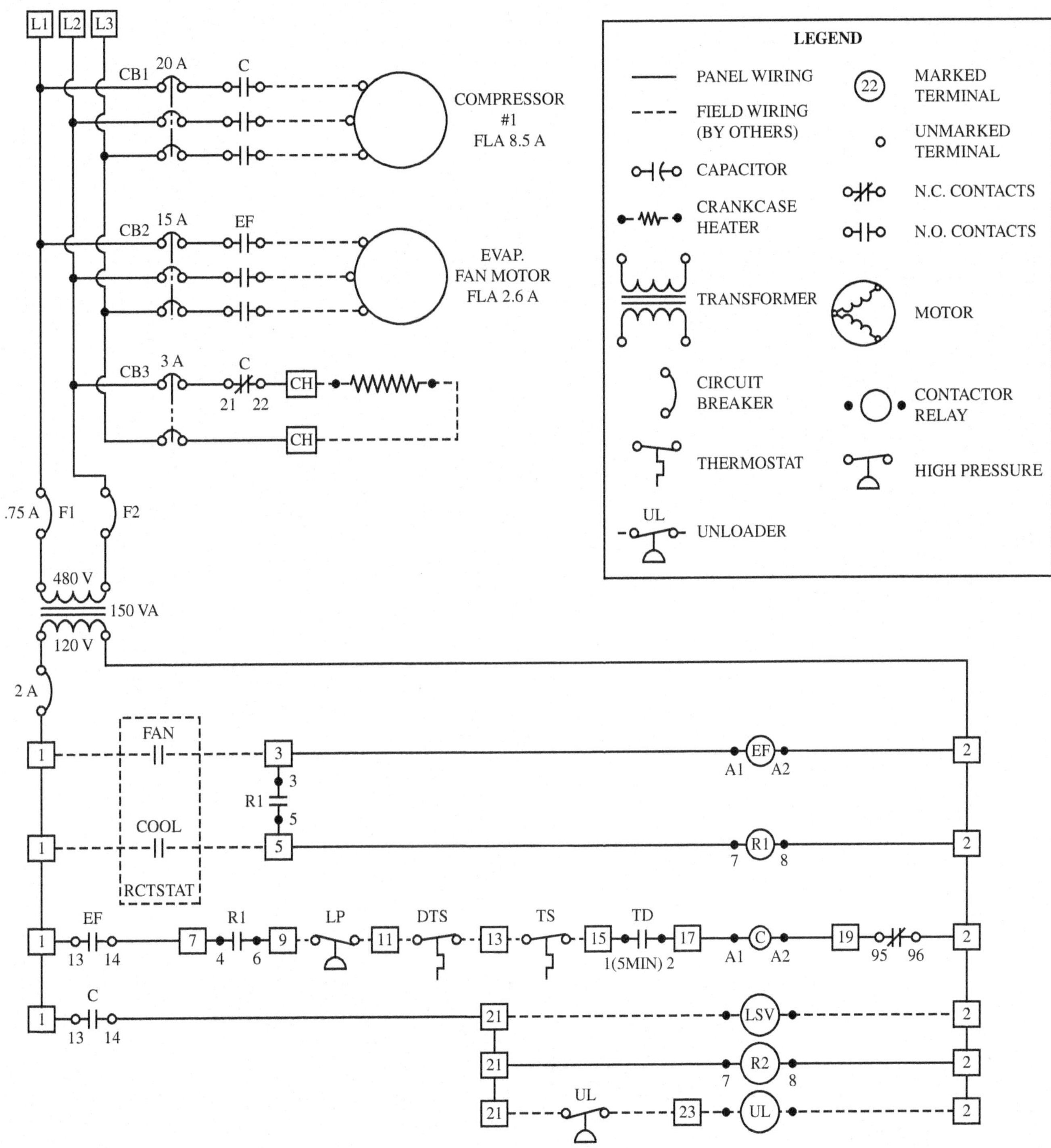

Figure 12-39 Use this image to answer Review Questions 22 and 23.

UNIT 13

Miscellaneous Electrical Components

WHAT YOU SHOULD KNOW

After studying this unit, you will be able to:

1. Describe and troubleshoot a crankcase heater.
2. Describe and troubleshoot a solenoid valve.
3. Describe and troubleshoot an electric unloader.
4. Describe and troubleshoot a hot gas sensor.
5. Discuss the purpose of a time-delay relay.
6. Explain the operation of a lockout relay circuit.
7. Discuss situations in which an explosion-proof system is required.
8. Explain the operation of a line voltage monitoring module.
9. Describe and troubleshoot an overload relay.

This unit is designed to explain important electrical components that were not covered in other parts of this book. In some cases the component information was quickly mentioned, but this unit expands on the information to provide a better understanding of each component. Many of the components covered in this unit may not be found in the basic HVACR system. However, these components are accessories that add to the safe operation of HVACR equipment.

Each component's operation is discussed by showing its installation in a wiring diagram. Then troubleshooting of each component is outlined.

13.1 CRANKCASE HEATER

The crankcase heater is an electric resistance heater. The abbreviation for a crankcase heater is CCH and its symbol is shown in Figure 13-1. The same symbol is used for a resistor or electric heater and these three components share the same electrical diagram.

Figure 13-2 shows an immersion-type crankcase heater. Figure 13-3 illustrates the installation of this type of crankcase heater, which is designed to keep oil warm. The heater fits into a well located below the oil crankcase. Figure 13-4 is a belly band type crankcase heater. This type of heater must be installed low enough to keep the oil warm.

Figure 13-1 Electrical symbol for a crankcase heater. This same symbol is used for a resistor or electric heat strip.

In Figure 13-5 the red oval highlights where the crankcase heater is installed in an electrical diagram. (Figure 13-5 is also supplied as part of the Electrical Diagrams package that accompanies this text and is labeled ED-2.) The crankcase heater reduces the amount of refrigerant that will condense in the crankcase during the off cycle. Condensed refrigerant can occur when the ambient temperature is cool around the compressor. These heaters are high voltage and normally operate on the same incoming voltage. Many crankcase heaters are wired so that they are energized and heating any time the power is applied to the condensing unit. The crankcase heater does not, however, need to be energized when the compressor is operating. The heat from the motor, the heat of compression, and movement of the oil will keep the refrigerant from condensing in the compressor crankcase.

Some manufacturers will design a crankcase heater to cut off when the compressor is operating or when the outdoor temperature is above 70°F. Such a design will save a little energy, particularly if the compressor has a long operating cycle. Referring back to Figure 13-5, we can see that the crankcase heater is de-energized when the contactor is energized. Notice that the normally closed (NC) contacts between 21 and 22, to the left of crankcase heater, are closed when the compressor is off. The contacts open when the compressor circuit is energized.

13.2 TROUBLESHOOTING THE CRANKCASE HEATER

The crankcase should be warm or hot to the touch. To prevent a skin burn, it would best to test it with an infrared thermal gun or a contact temperature probe. At any rate, do not touch it directly. The crankcase heater draws little current. In some cases an amp or two is the maximum amp draw. You will need to wrap the jaws of your clamp-on ammeter to calculate the low amperage reading. Divide the ammeter reading by the number of jaw wraps to obtain the true amp draw. You can also check the heater with an ohmmeter. Do this by disconnecting the power to the heater and measuring the heater resistance. The resistance will usually be less than a few hundred ohms. Use the low scale, R × 1 resistance setting of the ohmmeter.

The heater can be shorted. This presents an interesting problem since the fuse or breaker will open when a crankcase heater is shorted. You will need to eliminate several components

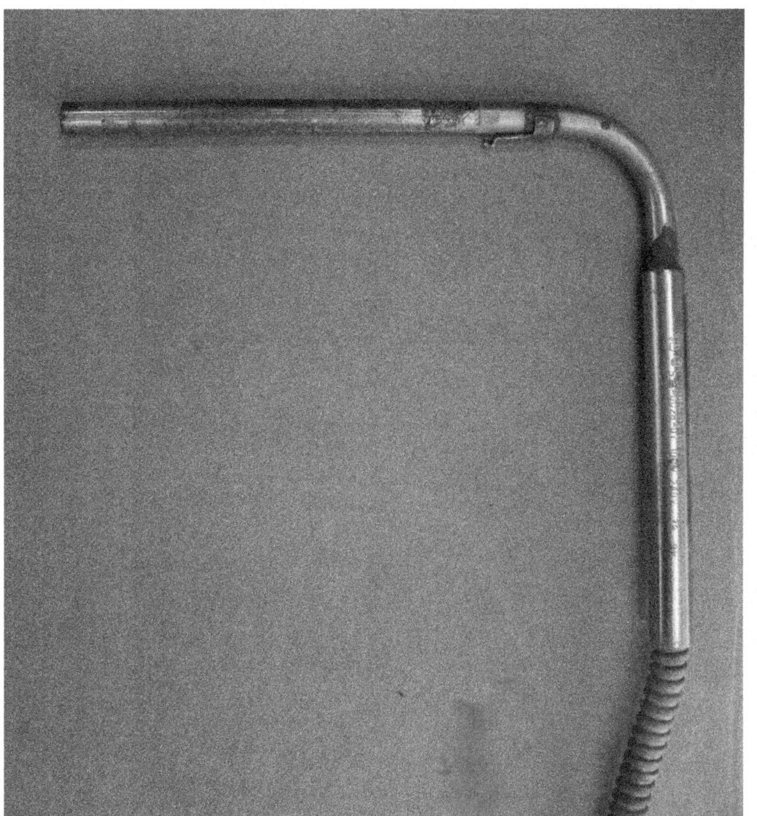

Figure 13-2 This immersion-type crankcase heater fits into a compressor crankcase well. The resistance will usually be less than a few hundred ohms.

Figure 13-3 This is an installed immersion-type crankcase heater. The heater probe is placed into a well that is below the oil sump.

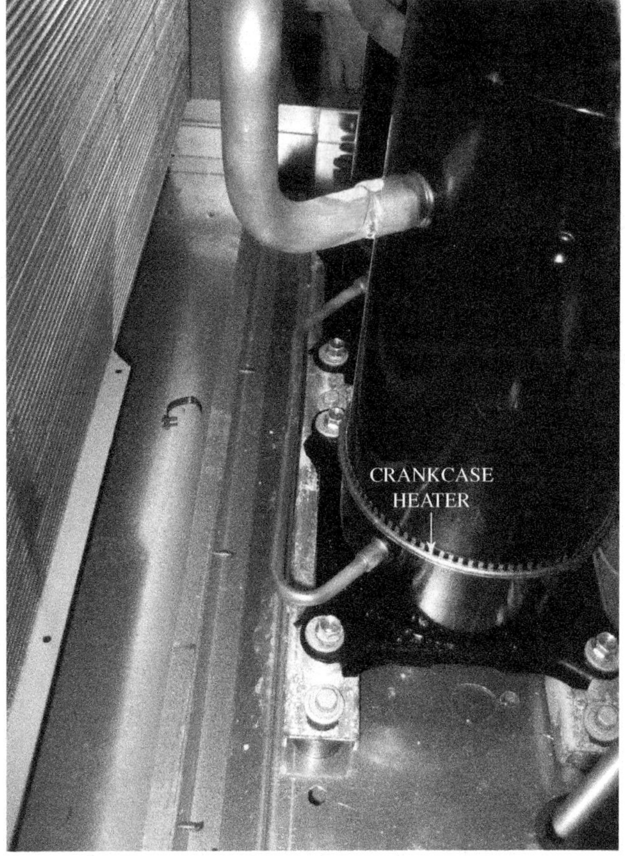

Figure 13-4 This is a belly band type crankcase heater. It is installed low on the compressor to heat the oil that is found near the bottom of the compressor shell.

to determine that the heater is the exact cause of the problem. The process of tracking down a short circuit will be discussed in the troubleshooting units later in the text.

13.3 SOLENOID VALVE

The **solenoid valve** is nothing more than an electromechanical opened or closed valve. The solenoid valve is fully open or fully closed. There is no "in-between" position. If it is

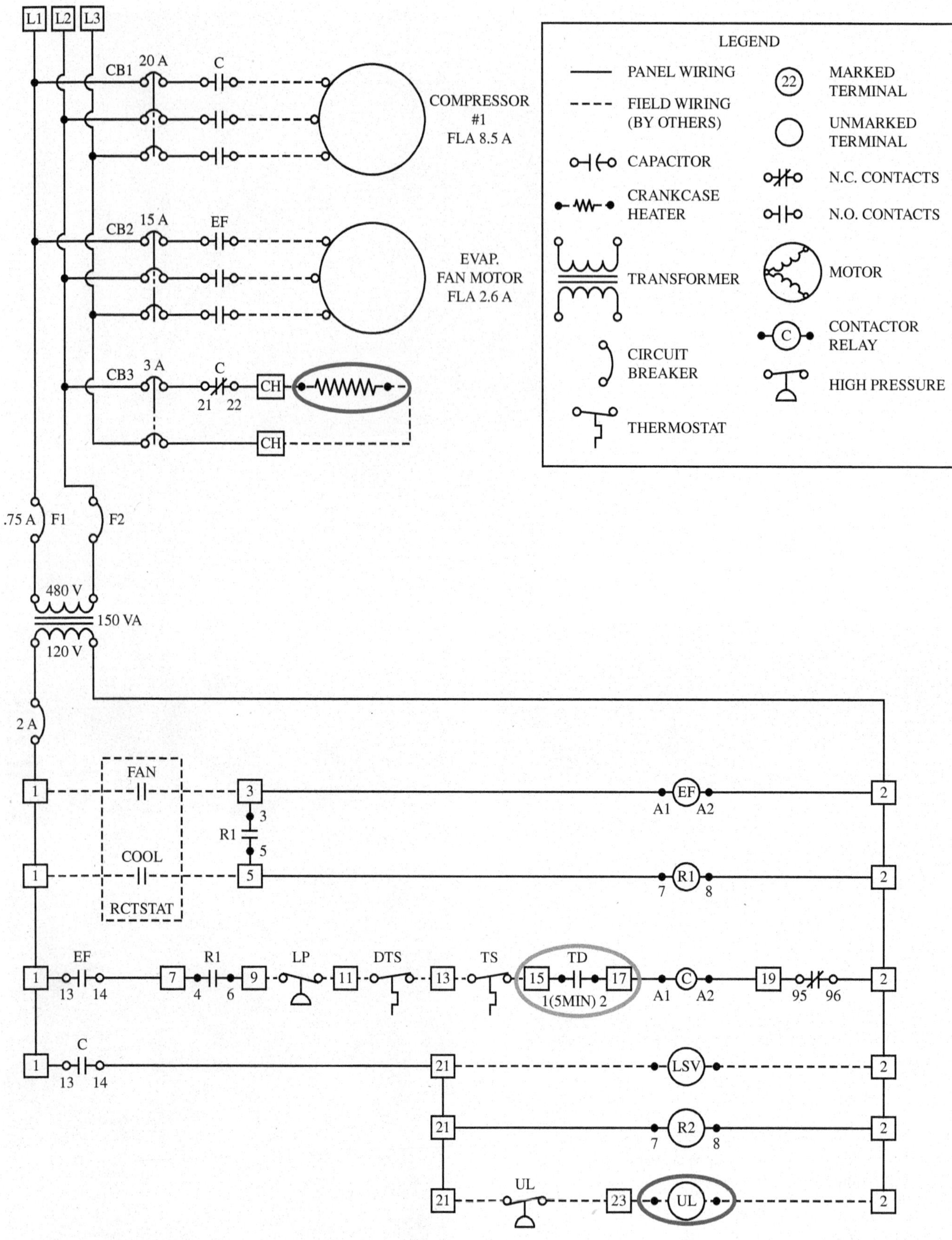

Figure 13-5 The crankcase heater in this diagram is indicated by the red oval. The unloader solenoid coil (UL) is indicated by the blue oval, the 5-minute time-delay relay by the gold oval.

partially open or closed, it is said to be leaking and is considered defective. The solenoid has a coil of wire surrounding an iron core, much like a transformer or relay. The coil is a cylindrical coil of wire surrounding a movable iron coil that moves along the length of the coil when current flow is passed through it. The solenoid relies on magnetism to move a plunger up or down. The plunger opens or closes the valve. Most solenoid valves used in HVACR are normally closed (NC), which means they will block fluid or gas flow when they are de-energized. The normally open solenoid valve is less common, but is used in some applications.

The symbols for a solenoid valve are shown in Figure 13-6. The symbol represents a coil of wire. Pictured in Figure 13-7 are the body and coil of a solenoid valve. It is installed in a liquid line. The coil is mounted on top of the valve. The coil can be removed and replaced if it burns out. The valve body can be disassembled. The orientation of the valve is not important, but the direction of flow is critical. Some valves have a directional arrow indicating the direction of flow.

In summary, the solenoid is constructed of two major components:

- The magnetic coil, which, when energized, is used to open or close the valve
- The valve and the valve body, which has liquid or gas passing through it.

Figure 13-6 These two symbols are used to represent a solenoid valve coil. These symbols are also used to designate the coil on a relay or contactor coil, which can also be considered a solenoid action.

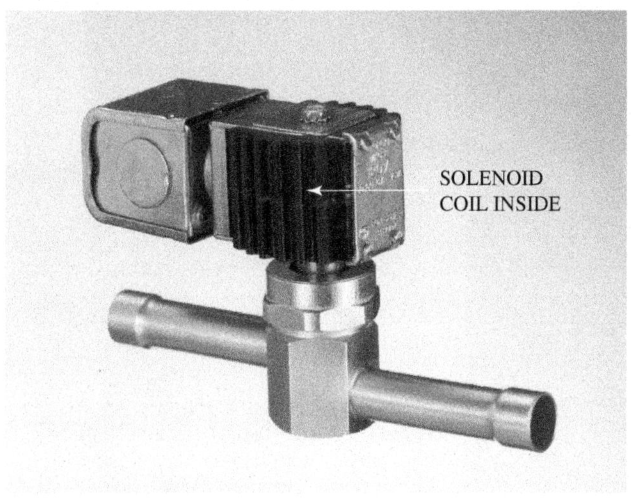

Figure 13-7 This is the body and coil of a solenoid valve.

Figure 13-8 This a normally closed solenoid valve in the liquid line of a walk-in cooler. When the valve is energized, it will open the liquid line for refrigerant flow to the metering device and evaporator. This coil is operated on 120 volts.

The coil needs to be matched with voltage. Common coil voltages are 24, 120, and 240 volts.

Figure 13-8 is a solenoid valve assembly installed in the liquid line of a walk-in cooler. The valve must be kept cool when replaced. This can be done by purging it with nitrogen and wrapping the valve body with a wet rag or using a heat-absorbing gel. The valve can be also be disassembled and purged with nitrogen to prevent internal damage. Figure 13-9 illustrates an exploded view of a solenoid that can be disassembled.

13.4 TROUBLESHOOTING THE SOLENOID VALVE

A quick check is to touch a steel screwdriver to the coil. The magnetic force of the coil will attract the steel shaft. An energized coil does not mean that the valve has switched positions from closed to open or vice versa. It is difficult to determine if the magnetic field is strong enough to change the valve position. A low-voltage condition can create a weak, yet attractive magnetic field. If the valve is not changing position, check the voltage **under load**. What we mean by "under load" is to measure the voltage going to the coil to see if the voltage does not drop below 10% of its required voltage.

A sign of a defective coil is overheating. The coil should not be excessively hot to the touch. If the paint or label is burned, do not touch it—it will burn out soon. An overheated coil results from too little or too much supply voltage.

Most solenoids will make a clicking sound when they are energized. Again this is not a foolproof way of determining that the valve is completely opening or closing. It is merely a sign that the coil is probably good enough to cause a change in valve position.

Sometimes the valve gets stuck in an open, partially open, or closed position. Tapping the valve may jar it loose.

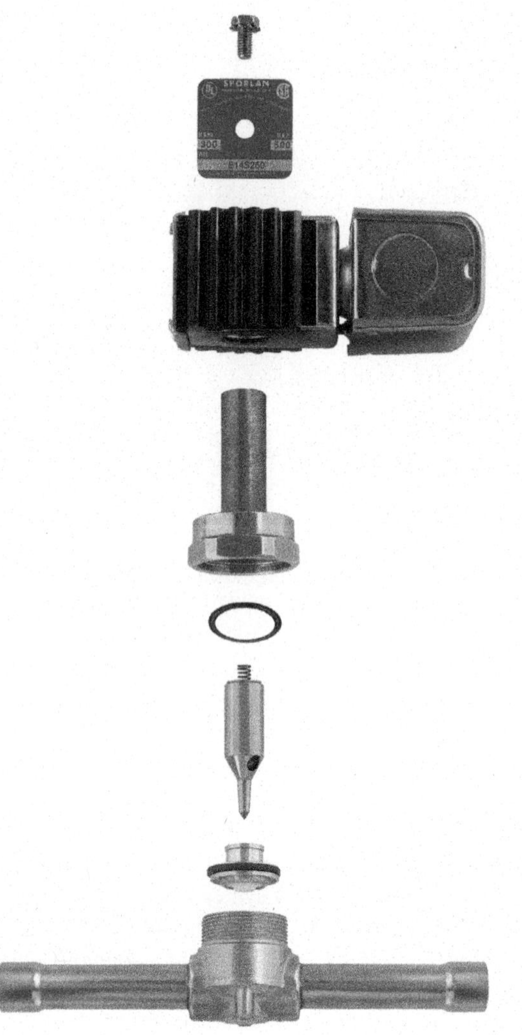

Figure 13-9 This is an exploded view of a liquid line solenoid valve.

There is no guarantee the valve will continue to open or close once it has been tapped and changes position.

In some cases you can replace just the defective part of a valve. For example, if the coil is open, it can be replaced if one is available. Older solenoid valves have old designs and replacement parts may not be available. In such a case, the whole valve would need to be replaced. Be sure to test a replacement valve before installing it.

SAFETY TIP

Do not energize a solenoid coil unless it is installed over the valve shaft. The coil will burn out if there is no load or no magnetic field absorbing metal in the middle of the magnetic field.

13.5 ELECTRIC UNLOADERS

Unloaders are used to reduce compressor capacity so that it runs at a lower Btuh output. Running a compressor at lower output has the following advantages:

- Reduced starting and stopping capacity losses
- Fewer starts and, therefore, less stress on the motor windings since there are fewer LRA starts
- Better humidity control since the system has longer run times.

An unloader can be a **mechanical unloader** or **electric unloader**. This section addresses the electric unloader. An electric unloader, as pictured in Figure 13-10, is used to reduce the capacity of a compressor. That cylinder or bank of

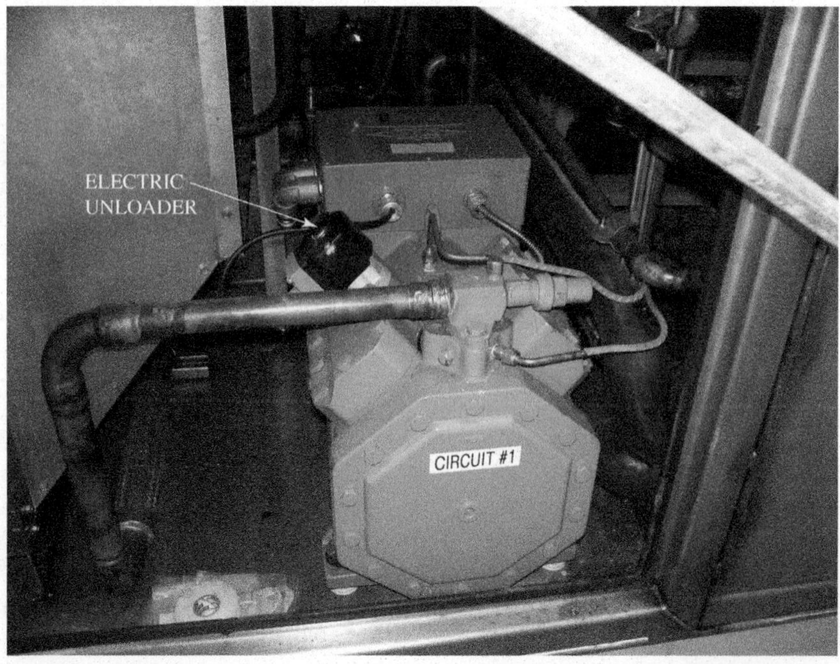

Figure 13-10 This is a semi-hermetic compressor with an electric unloader in one of the cylinder heads. The unloader is a solenoid valve that reduces the capacity of the compressor when less capacity is required.

cylinders is prevented from pumping refrigerant. This electric unloader is similar to a solenoid valve.

The electric unloader can be designed to unload in the energized or de-energized mode. Look back at Figure 13-5 to locate the unloader coil (UL). The (UL) solenoid has a blue oval around it and is found at the bottom of the diagram. This unloader works when the **UL** low-pressure switch, located between points 21 and 23 on the diagram, drops to a low operating pressure, thus providing power to the unloader (UL) coil solenoid.

In the air conditioning mode the unloader will open at a suction pressure that is equal to a temperature of about 32°F. For example, an R-410A system would unload around 101 psi (32°F) and load back up to full capacity at 118 psi (40°F). This is a 17-psi differential. An unloader's energizing pressure and difference settings vary among compressor manufacturers and system applications. In refrigeration systems, unloading a compressor will be done at a much lower pressure since they operate below freezing. When the unloader is energized, it reduces refrigerant pumping capacity. The suction pressure will rise and the amp draw will drop. In many cases you will notice a difference in the compressor sound. When the suction pressure rises to its setpoint, the unloader will de-energize and the compressor will be at full capacity again.

13.6 TROUBLESHOOTING THE ELECTRIC UNLOADER

Here are some troubleshooting tools you can use to ensure that an electric unloader is working properly:

- For an air conditioning system, the unloader should be energized when the suction saturation temperature is around 32°F. Check for amp draw on the unloader around 32°F. The compressor amp draw will be lower when compared to full load capacity. The operating sound of the compressor may change as it unloads and loads.
- You can also check the amp draw on an electric unloader. The amp draw on the solenoid will be low, possibly an amp or two. Use the low-amperage setting on your clamp-on ammeter to see if the unloader is energized. The unloader can be energized and not change position or move the solenoid plunger. In other words, the unloader is stuck in the unloaded or loaded position.
- Check the voltage supply to the unloader when it is in the unloaded and energized condition. Place a steel screwdriver blade up to the solenoid coil to check for magnetic pull of the coil. A magnetic pull does not mean that the solenoid is working, but does indicate that there is voltage at the coil. The voltage may not be high enough, however, to create the magnetic strength needed to open or close the unloader.
- Next, remove the power and check the unloader coil with an ohmmeter. If the voltage is applied and coil resistance is good, the electric unloader should be working. The unloader solenoid port could be struck open or closed.

13.7 REPLACING THE ELECTRIC UNLOADER

Here are the steps for replacing an electric unloader:

Safety first: Verify that power is removed from the compressor and the unloader.

1. Tag out and lock out the circuit(s) feeding the compressor and unloader.
2. The refrigerant from the compressor may need to be recovered or pumped out to 0 psig.
3. Change out the unloader. Have a new gasket available for the change-out.
4. Pressure test with nitrogen.
5. Purge nitrogen. Pull vacuum on the compressor.
6. Charge the compressor with the operating refrigerant.
7. Open (back seat) the service valves and put the system back on line.
8. Test unloader operation by reducing the suction pressure and measuring the amperage drop of the compressor.

13.8 SOLID-STATE TIME DELAYS

Solid-state timers are used to prevent the compressor or condensing unit from short cycling. Short cycling could be caused by someone turning the cooling thermostat on and off in rapid succession. Short cycling could be caused by a power fluctuation or by a pressure switch opening and closing in less than a minute. Timers protect the compressor by:

- *Allowing the pressures to equalize during the off cycle.* Compressors starting under unequal pressures between the high side and low side draw excessive amps, thus overheating the motor windings. *Note:* Start caps improve starting torque and reduce the initial amp draw for systems that must start under unequal pressures.
- *Allowing the power to stabilize.* When power fails or fluctuates, it may not come on at the rated voltage. The voltage may be higher or lower than the ±10% rating. Out-of-range voltage will cause high amp draw, thus tripping the motor out on overload protection, which is not good for the motor.
- *Allowing scroll compressors to stop turning in the reverse direction.* When a scroll compressor stops, the moving scroll rotates in the reverse direction while the pressure equalizes across the high and low side. If the compressor starts up when it is rotating backward, it will run backward. The scroll running backward will make an unusual noise, overheat, and trip the motor overload. The customer may call about the noise, triggering an unnecessary service call. Overheating the compressor shortens motor life. The compressor does not cool in the reverse mode of operation.

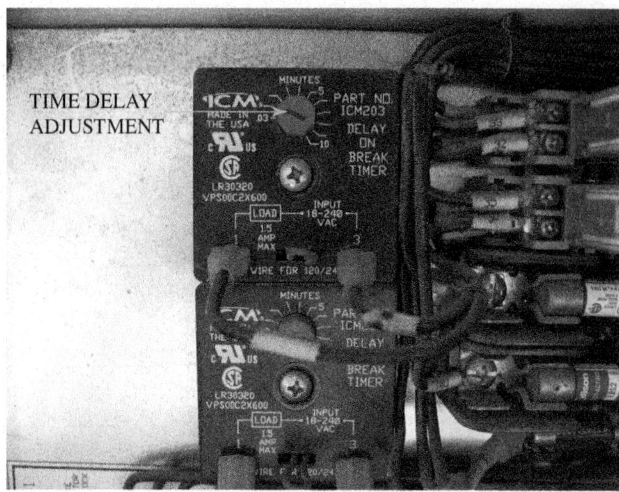

Figure 13-11 Solid-state adjustable time delay. The time delay will not allow the control circuit to operate the compressor for the set time selected on the timer. This allows for the system pressure to equalize and the voltage to stabilize.

Figure 13-11 shows two adjustable "delay on break" timers. One timer would be used on each cooling system, therefore we recognize that there are two compressors.

Timers can be "delay on break" or "delay on make":

- *Delay on breaker timer:* In most installations the delay on break timer is the best option since the timer starts its timing sequence once power is removed from the timer, which is usually at the end of the cooling cycle or if there is a power interruption.
- *Delay on make timer:* A delay on make timer starts it timing sequence when power is applied. With this timer the cooling system will not operate until it goes through its set timeout period. Most customers and techs expect quick responding cooling. This will delay the start of cooling and possibly cause concern for the customer.

This timer is installed in the control circuit as shown in the gold circle in Figure 13-5. Most timers are preset or manually set for a 5-minute time delay. This is usually enough time for the pressures to equalize and the power quality to be at the required voltage level when energized.

13.9 TROUBLESHOOTING THE SOLID-STATE TIMER

Troubleshooting a solid-state timer is not difficult. If it is an adjustable timer, turn the time to zero and start the equipment. The system should come on in a few seconds. If the system does not come on, measure the voltage across the two wires going to the timer. If the timer is open, the control voltage will be read on the voltmeter. If the control voltage is 0 V, then the timer is not the problem. A 0-V reading means that the switch is closed or it can indicate that there is no voltage in that circuit. If the control voltage is read across the timer and it is not allowing the system to operate, jumper the timer and the system should start. Replace the defective timer and test the replacement device by cycling the control circuit off and on at least one time.

13.10 LOCKOUT RELAY

The **lockout relay** is similar to a conventional relay, as shown in Figure 13-12. Another name for a lockout relay is a **holding relay**, a **reset relay**, or an **impedance relay**. The purpose for wiring a lockout relay into a holdout circuit is to prevent the compressor or condensing unit from operating when a safety device opens. It is a built-in manual reset design. In other words, the power to the circuit will need to be removed and reapplied for the lockout circuit to be reset. If a safety device opens again, the system will return to the lockout condition.

The major differences between a lockout relay and a conventional or multipurpose relay are the resistance of the coil and the way it is wired into the circuit. A lockout coil's resistance is very high when compared to a standard relay. We will review the diagram shown in Figure 13-13 so that you will understand how the lockout relay differs from a standard relay. (Figure 13-13 can also be found in the enlarged Electrical Diagrams package that accompanies this text; it is labeled ED-4.)

The lockout relay in Figure 13-13 is called a reset relay in the diagram. The **RESET RELAY** coil is located in the control voltage section, near the bottom right side of the diagram; it is circled in red. The coil will be matched to the correct control voltage, which is commonly 24, 120, or 240 volts. The normally closed reset relay contacts in the diagram are in the blue rectangle.

Figure 13-12 This lockout relay looks similar to a standard relay. The normally closed (NC) contacts between terminals 4 and 5 are used in the lockout circuit. The coil resistance is very high compared to that of a conventional relay or contactor.

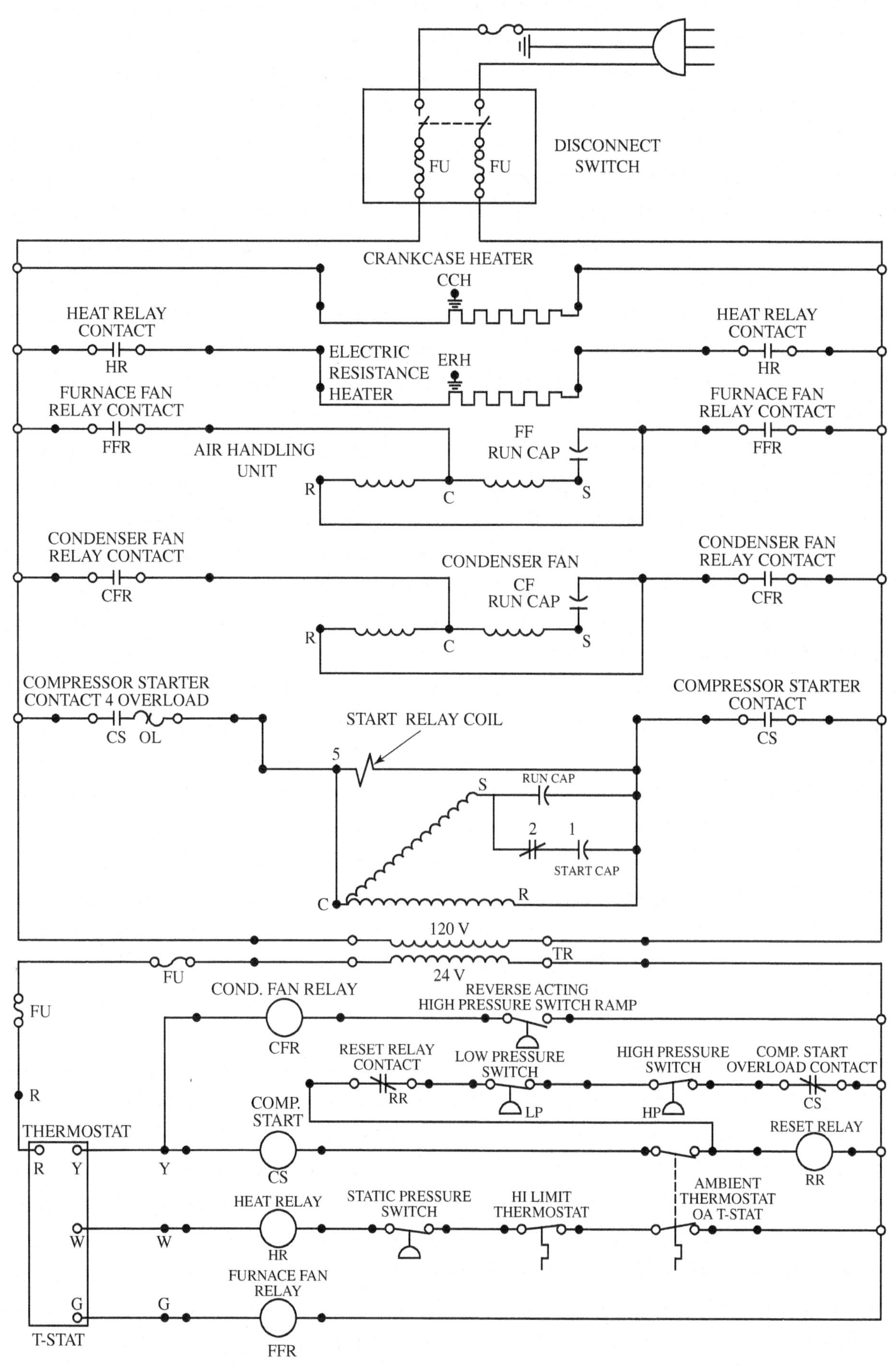

Figure 13-13 This control circuit diagram illustrates placement of a lockout relay circuit, which is referred to as a **RESET RELAY** in this diagram.

SEQUENCE OF OPERATION 13.1: NORMAL OPERATION OF THE LOCKOUT RELAY CIRCUIT

Follow the highlighted circuit in Figure 13-14 for a closer look at the lockout circuit.

Here is the sequence of operations when there are no problems with the system:

1. Starting at the left side of the transformer, point 1, power goes through fuses to the **R** terminal of the thermostat.
2. Next, power goes through the t'stat contacts **R** and **Y** at point 2.
3. Power continues through the compressor starter **CS** at point 3.
4. The voltage continues through the closed ambient thermostat at point 4.
5. Since the resistance of the reset relay coil is extremely high, the current will travel through the zero resistance switches of the closed reset relay contacts at point 5.
6. Power continues through the closed low-pressure switch at point 6.
7. At point 7 the circuit is complete through the closed high-pressure switch.
8. The power goes through the closed compressor starter (CS) overload contact at point 8.
9. Finally, at point 9, the power returns to the right side of the transformer to complete the power loop in the low voltage control circuit.

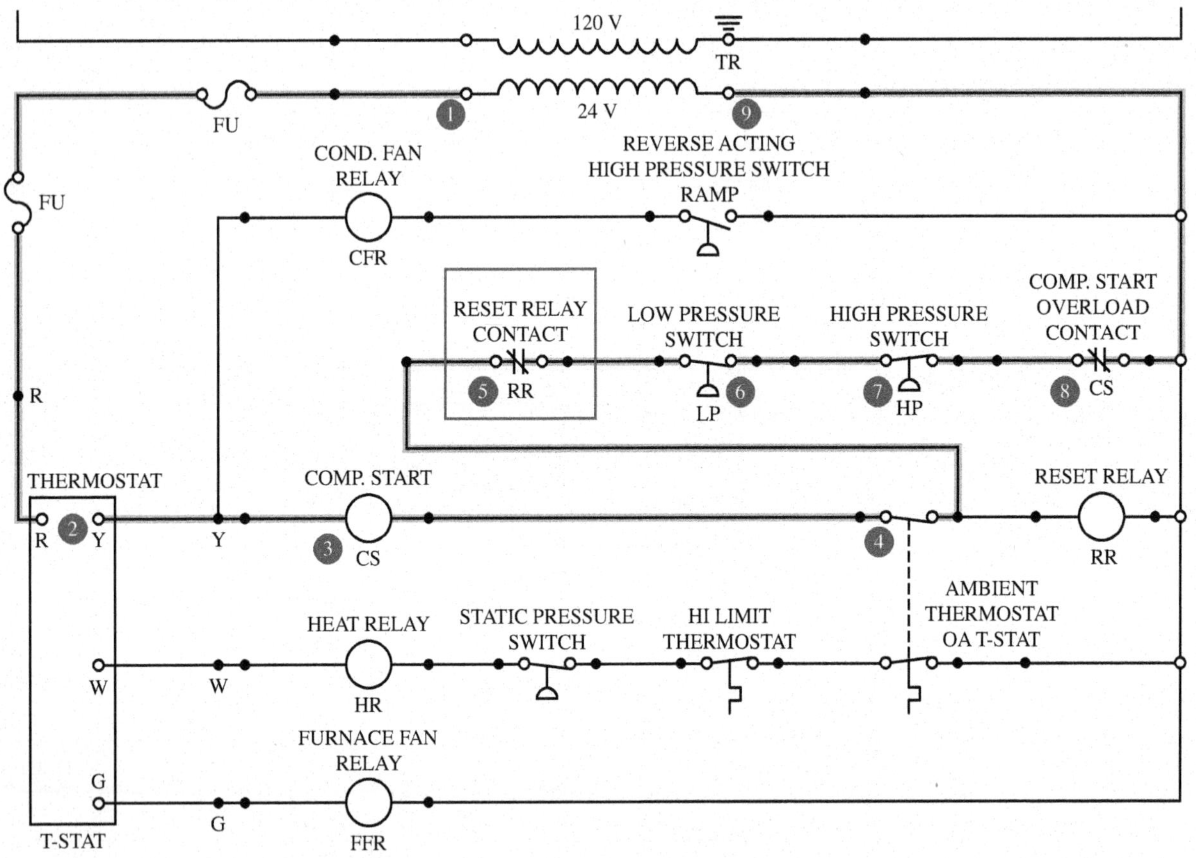

Figure 13-14 This control circuit diagram shows a lockout relay circuit in place. The reset relay coil is in line with the **Y** thermostat and compressor starter (motor starter or contactor). Since the resistance of the reset relay is so high, the current will travel through the zero resistance switches of the closed reset relay contacts, closed low-pressure switch, closed high-pressure switch, and closed compressor starter overload contact.

13.11 RESET RELAY

Let's look at Figure 13-14 again. When the low-pressure, high-pressure, or compressor starter overload contacts open, the current will be redirected through the COMP START coil and the RESET RELAY coil. The reset relay coil has many times the resistance of the compressor starter coil. Remembering Ohm's law, the higher the resistance, the greater the voltage drop in a series circuit. In this case the voltage drop across the reset relay coil will be near 24 volts

SEQUENCE OF OPERATION 13.2: SYSTEM PROBLEM WITH THE LOCKOUT RELAY CIRCUIT

The path for the control voltage in the lockout condition is highlighted in Figure 13-15. The reset relay will de-energize when all of the safeties are closed and when the thermostat circuit, the **Y** circuit, is broken momentarily. It can be reset by turning the thermostat off, by turning the thermostat above the cooling temperature setpoint, or by turning off the power to the control voltage.

Let's see how the current will flow in the circuit when there is an open safety. If any of the following safety devices listed below are open, the system will stop and go into lockout:

- Low-pressure switch (LP)
- High-pressure switch (HP)
- Compressor starter overload contact (CS).

1. Starting at point 1, left side of transformer, the power will go through the fuses to the **R** connection on the thermostat.
2. Power will pass from **R** to **Y** at point 2 in the thermostat.
3. Next, the power will flow through the compressor starter coil (contactor), point 3.
4. Since the circuit above the compressor starter coil circuit is open, the power will continue through point 4 on the ambient thermostat and energize the reset relay coil, point 5.
5. The reset relay at point 5 is a high resistance (high impedance) coil. This will create a large voltage drop across the reset relay coil, leaving inadequate voltage available to energize the compressor starter coil.
6. At point 6 the reset relay coil is energized and opens the reset relay contacts.
7. Point 7 shows that power returns to the right side of the transformer to complete the circuit and lockout the compressor operation.

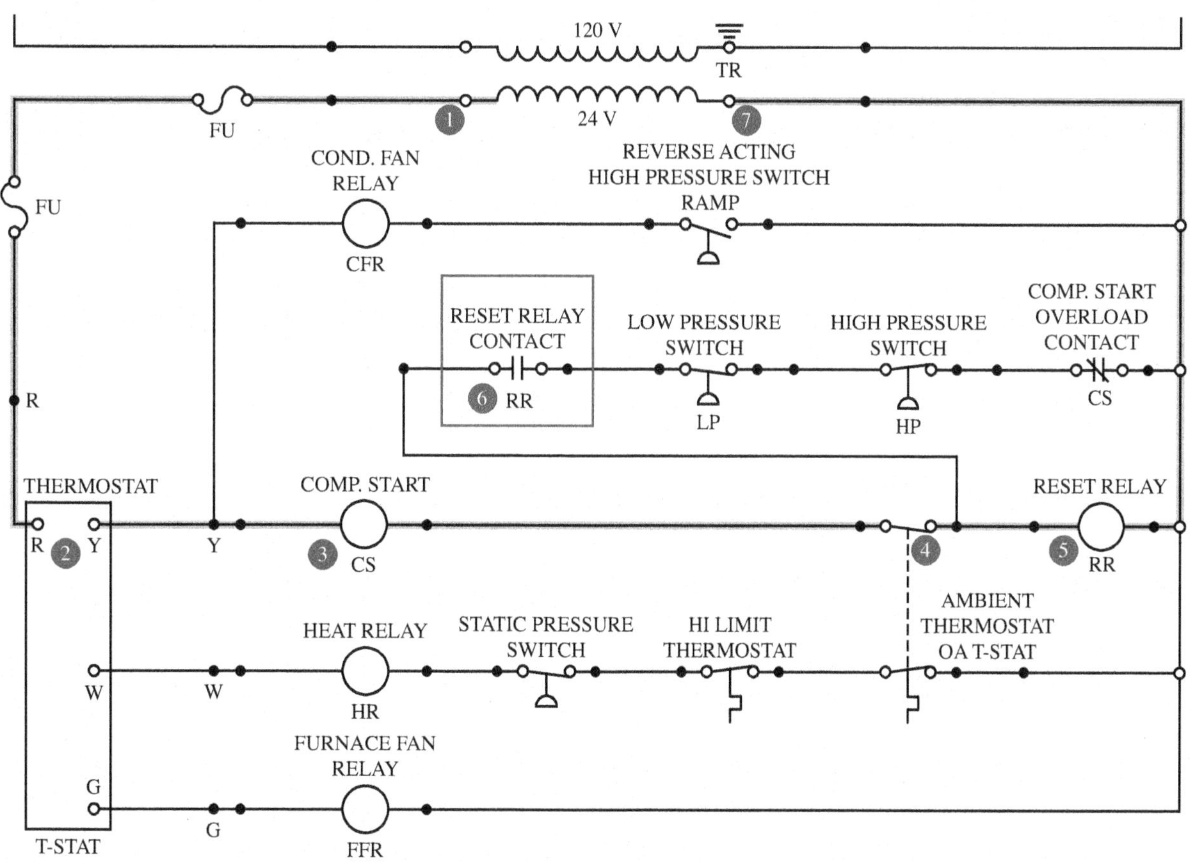

Figure 13-15 This control circuit diagram shows a lockout relay circuit in the lockout position. One of the safety devices had opened, which caused the reset or lockout circuit to energize. The control voltage now goes through the compressor starter coil, (CS); the ambient thermostat; and the reset relay coil, (RR). Reset relay coil (RR) has high resistance and, therefore, a high voltage drop. This high voltage drop prevents the (CS) coil from being energized. When the (RR) coil is energized, it opens the normally closed **RR** contacts.

and the voltage drop across the comp start coil will be a volt or two. The low voltage across the compressor start coil will not be enough to energize the coil; therefore, the compressor circuit will not be energized. The reset relay contacts in the blue rectangle in Figure 13-15 (above the COMP START coil) will be open since the reset relay coil is energized. Even when the low- and high-pressure switches reset or the comp starter overload contact is manually reset, the control circuit will remain open as long as the reset relay coil is energized and the reset contacts open.

13.12 TROUBLESHOOTING THE LOCKOUT RELAY

Troubleshooting of a lockout relay is similar to troubleshooting of a standard relay except that the lockout relay has a very high resistance coil. Sometimes the resistance could be as much as 10,000 Ω. It is important to have an ohmmeter or resistance scale that will measure at least 10 kΩ in order to check this coil. The closed contacts will have no resistance or near 0 Ω when de-energized. The contacts should read infinity when the reset relay is energized.

In the rare case when a reset relay needs to be replaced, it should be checked for correct operation. This can be done while the system is operating. Remove power to one of the safety controls. This will create a lockout condition that will stop the compressor. Measure the control voltage across the lockout relay coil and the normally closed contacts will open. Read the control voltage when the reset relay contacts are open.

13.13 HOT GAS THERMOSTAT

A hot gas thermostat is a device that senses the discharge line temperature and shuts the compressor down if it goes beyond a high-temperature setpoint. The symbol for a hot gas sensor is the same as that for a furnace thermostat (see Figure 13-16). The hot gas t'stat will be strapped to the discharge line, near the compressor. The purpose of the hot gas t'stat is to prevent the compressor from overheating due to a low charge, low oil level, or anything that will cause abnormally high gas temperature.

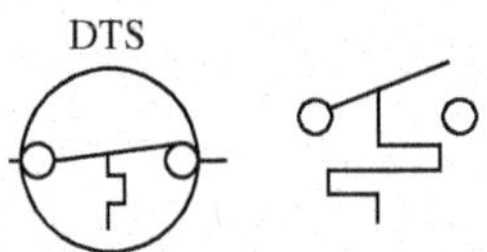

Figure 13-16 These symbols represent a discharge line thermostat. They are the same symbols used by a heating thermostat since they open in response to an overheated condition, in this case excessive discharge line temperature. Some open around 225°F, others at higher temperatures.

Hot gas temperatures are accompanied by motor overheating and lubrication breakdown. The temperature on the discharge line is about 50° to 75°F cooler than the internal compression chamber. If the discharge line is 250°F, the compressor cylinder can easily be over 300°F. Many compressor oils break down at 275° to 300°F. If you need to check the operation of this device, you should measure this temperature no more than 6 inches from the compressor discharge line.

13.14 TROUBLESHOOTING THE HOT GAS THERMOSTAT

The resistance of the hot gas t'stat should be near 0 Ω. If this safety device is cutting the compressor out, measure the temperature of the discharge line. You will need a measuring device that will record at least 250°F or greater. Use a thermocouple clamp-on temperature probe. Most infrared (IR) thermometers are not accurate enough for this reading. Also, check the refrigerant charge. An undercharged system will generate high discharge temperatures since there is not enough returning cool gas to keep the motor from overheating.

SAFETY TIP

When working around the discharge line, take precautions against burning your skin. Your skin will detect a hot surface around 120°F. As the temperature rises from this point, it will damage your skin. If you need to determine the temperature of a discharge line, use a temperature-measuring device such as a clamp-on thermistor device capable of measuring over 250°F. Infrared (IR) guns do not have the accuracy required for most pipe temperature measurements since an IR device requires a black, dull surface to be accurate.

13.15 LINE VOLTAGE MONITOR

Line voltage monitors are designed to protect compressor motors and other three-phase loads. There is also a protector that protects single-phase motors. A line voltage monitor (see Figure 13-17) protects motors from premature failure or damage due to common voltage faults. The voltage monitor protects against:

- *Voltage imbalances:* A voltage imbalance can cause high current draw or damage to one of the three phases.
- *Low- and high-voltage conditions:* These conditions can cause high current flow and winding damage.
- *Phase reversal:* Phase reversal prevents reverse rotation of the motor. Screw and scroll compressors cannot tolerate reverse rotation.
- *Phase loss:* Losing a phase will cause the other two phases to draw high amperage, overheat, and burn out one or more windings. Phase loss may prevent the motor from starting.

Figure 13-17 This three-phase, line voltage monitor has several important functions. It has an adjustable time delay and it protects against voltage imbalances and low- and high-voltage conditions. When wired correctly it will protect a compressor; it will sound an alarm if so wired.

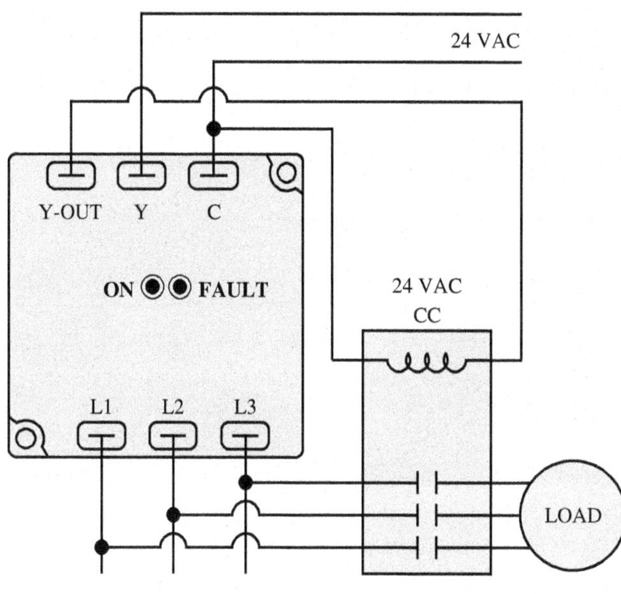

Figure 13-18 This is a wiring diagram for a line voltage monitor. The line voltage monitor reduces motor failure caused by a voltage imbalance, low- and high-voltage conditions, phase loss, and short cycling.

- *Faulty power:* Power variations cause amperage increases and overheating conditions.
- *Short cycling:* A time delay allows pressures to equalize and voltage to stabilize.

Here are some common specifications for a line voltage monitor:

Inputs

- High-voltage range: 190 to 600 VAC
- Frequency: 50 and 60 cycle options
- Control voltage range: 18 to 30 VAC.

Outputs

- Relay type
- Single-pole, single-throw (SPST)
- Contact rating: 10- and 20-amp options.

Figure 13-18 is a wiring diagram of a phase monitor. You will notice that the wiring is fairly simple. Here are the directions:

Step 1: With all power off, wire the voltage monitor to the input voltages L1, L2, and L3.
Step 2: Wire 24 volts to the Y and C terminals.
Step 3: The contactor coil (CC) will be wired to terminals Y-OUT and C.

This module has an ON and FAULT light. The ON light will give the technician a quick indication that the system has power. The FAULT light will alert the tech to failures using various sequences of blinking lights.

13.16 EXPLOSION-PROOF SYSTEMS

This section briefly discusses **explosion-proof systems.** Explosion-proof systems are required on oil drilling rigs, grain elevators, and gas stations. *Explosion proof* means that electrical devices such as compressors, motors, and switches have been designed to contain explosions or flames produced within their shell or terminal box. These devices may generate arcs, sparks, flashes, or hot surfaces that must be isolated. To be classified as explosion proof, these devices must be isolated and contain the igniting gases inside the component. Conduit connections will be poured and sealed to prevent the hazardous environment from entering these components. The National Electric Code (NEC) has a section that addresses explosion-proof environments.

An explosion-proof containment is required in an atmosphere of flammable or combustible gases, vapors, or dust. Explosions or fires resulting from arching and sparking devices or hot surfaces need to be contained to the electrical device such as a motor or contactor. Therefore, this device will be installed in what is known as a NEMA 7 enclosure. Figure 13-19 shows an explosion-proof window unit. Notice the oversized seal on the compressor. This seal was added to a conventional compressor.

Figure 13-19 Explosion-proof window unit. Notice the seal on the compressor and the fan motor at the rear of the unit. Explosion-proof components are large because they require large, sealed connection boxes.

Figure 13-20 shows an explosion-proof condenser fan motor. The seal is important to keep any explosion contained in the component.

Motors have a rating called the **minimum ignition temperature (MIT)**. This temperature relates not only to the outer temperature of the motor, but also to the inner motor winding and laminated plates or core.

Nonsparking tools must be used in an explosion-proof environment. Also, moving fan blades and blower wheels must be spark free. For example, a fan blade cannot be made of steel. It must be made of aluminum or plastic, thus eliminating sparks when touching a similar surface.

Grounding is also important. Static electricity or a loose ground spark can create an ignition source that may cause an explosion. Figure 13-21 is an additional ground not required in a normal installation.

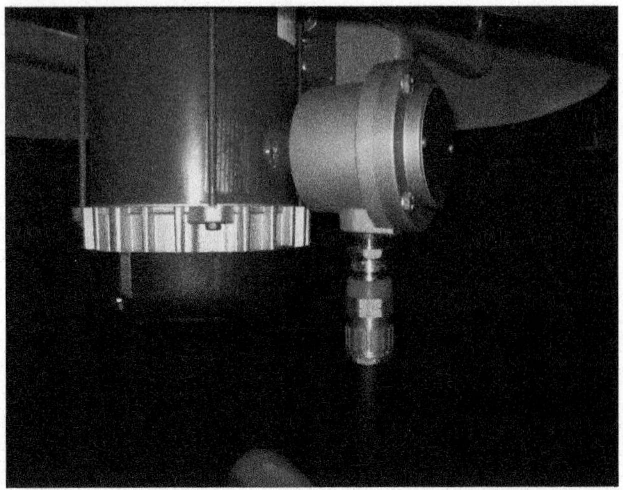

Figure 13-20 This is an explosion-proof condenser fan motor. It has a spark-free, aluminum fan blade (not shown).

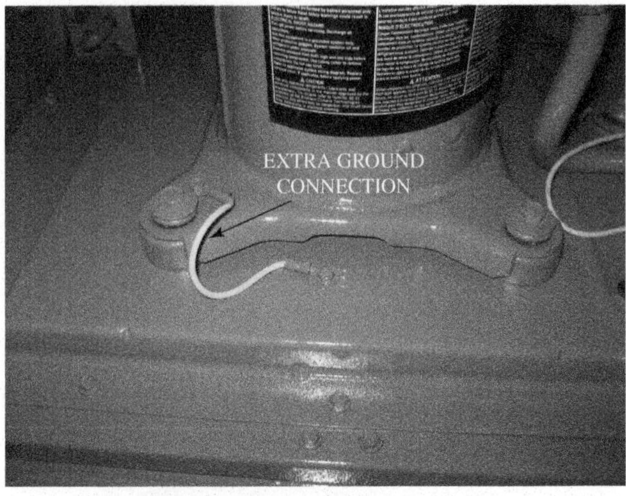

Figure 13-21 The additional ground wire ensures that the compressor is truly grounded, since this compressor is mounted using bolts that have a rubber insulating gasket.

SAFETY TIP

There is a difference between *explosion proof* and *water resistant*. We have seen some examples of explosion-proof installations. Figure 13-22 is an example of a weather-tight or water-resistant enclosure, not an explosion-proof enclosure. When explosion-proof enclosures are open for service, they must be carefully resealed to maintain their integrity.

13.17 OVERLOAD RELAY

Most of the time it is called simply an *overload*, but the full, correct term is *overload relay*. Overloads have been briefly mentioned in previous units and you will see more discussion

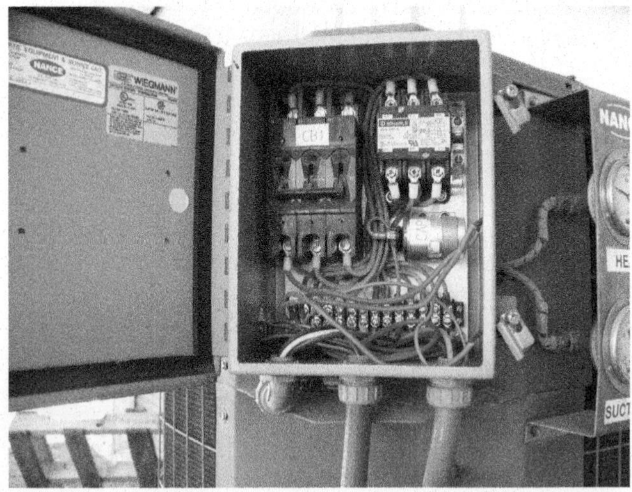

Figure 13-22 This is not an explosion-proof enclosure, but it is classified as water tight or water resistant.

of them as we progress through this technical material. At this point we will discuss them in more detail so that you will understand their safety function.

An overload prevents a component from overheating or getting so hot that open windings are burned or a fire started. Fuses and breakers are considered to be system overload devices. This section covers an overload that will protect an individual component part such as a motor or electric heat strip. Overloads are sized to handle about 125% of the normal rated amperage, but this will vary among manufacturers. The best practice is to use the required manufacturer overload device. Do not substitute overloads.

An overload can be external or internal. It will open if the current through it becomes excessive or it overheats due to high temperatures. The overload is connected in series with the section of the component that draws the most current. Figure 13-23A shows a drawing of an external overload that you would find on a hermetic compressor. Figure 13-23B illustrates the wiring of the overload in series with the common terminal on the compressor. This external overload uses a bimetal strip that snaps open if the current or heat becomes too high. As it cools the snap action closes and allows current to flow again.

Figure 13-24 shows a diagram with an external and internal overload on the compressor. It is not standard to have two overloads in series with the common connection. The internal thermal overload is optional and installed as a manufacturer's option. Notice that the overloads are wired in series with the common connection that carries the most amps. The overload is connected to the common motor lead since that is the point where the most motor amperage is drawn.

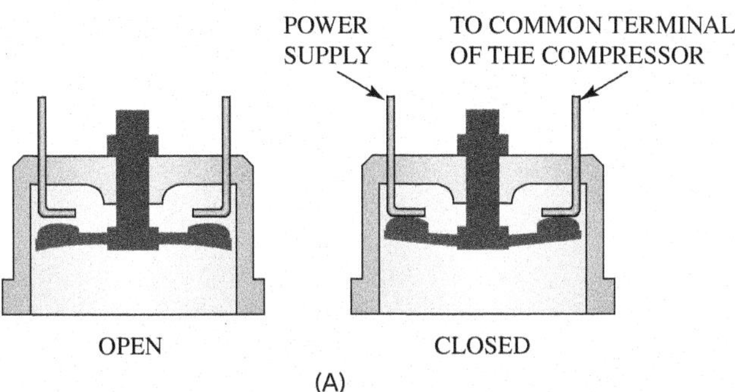

(A)

(B)

Figure 13-23 (A) This is a drawing of an external overload. This uses a bimetal strip that snaps open if the current or heat becomes too high. As it cools, the snap action closes and allows current to flow again. One side is connected to the power supply, and the other side is connected to the common terminal of the compressor. *(Courtesy Tecumseh Products Company.)* (B) Here an external overload is hooked up to a common terminal. The overload, on the right, will be secured to the side of the compressor.

13.18 TROUBLESHOOTING THE OVERLOAD RELAY

It is fairly simple to troubleshoot an overload. The overload should have no resistance. When it is open, it will measure infinity. Allow the overload to cool before checking it. This is especially true of an internal overload that may take a considerable time to cool and reset. If an internal overload is defective, the motor will have to be replaced. Regarding replacing external overloads, always use the manufacturer's recommendations. Avoid the temptation to use a substitute. You would become responsible for any damage that occurs if the device were to burn or explode. This is the best policy for replacing any safety device. Use the recommended replacement.

SERVICE TICKET

Use Figure 13-13 to troubleshoot this problem. Your third service call of the day presents an interesting challenge. The customer states that the system stopped cooling about 4 hours ago. A quick check using the ACT method of troubleshooting reveals the following:

- The compressor is not operating. The high and low pressures are equalized and the compressor is not drawing amperage.
- The thermostat is set to cool and is several degrees below the inside temperature.
- The indoor blower is operating.
- The condenser fan is operating.

You turn off the thermostat. When the t'stat was turned back on, the compressor immediately started. With the compressor now operating, you take the system's vital signs. The charge is correct and the compressor amp draw is an amp or two below RLA. The coils and filters are clean. A clean and check was done on this unit a couple of weeks ago.

You wait another 15 minutes, after which everything seems to be running correctly. You tell the customer that you could not find a problem and to contact your

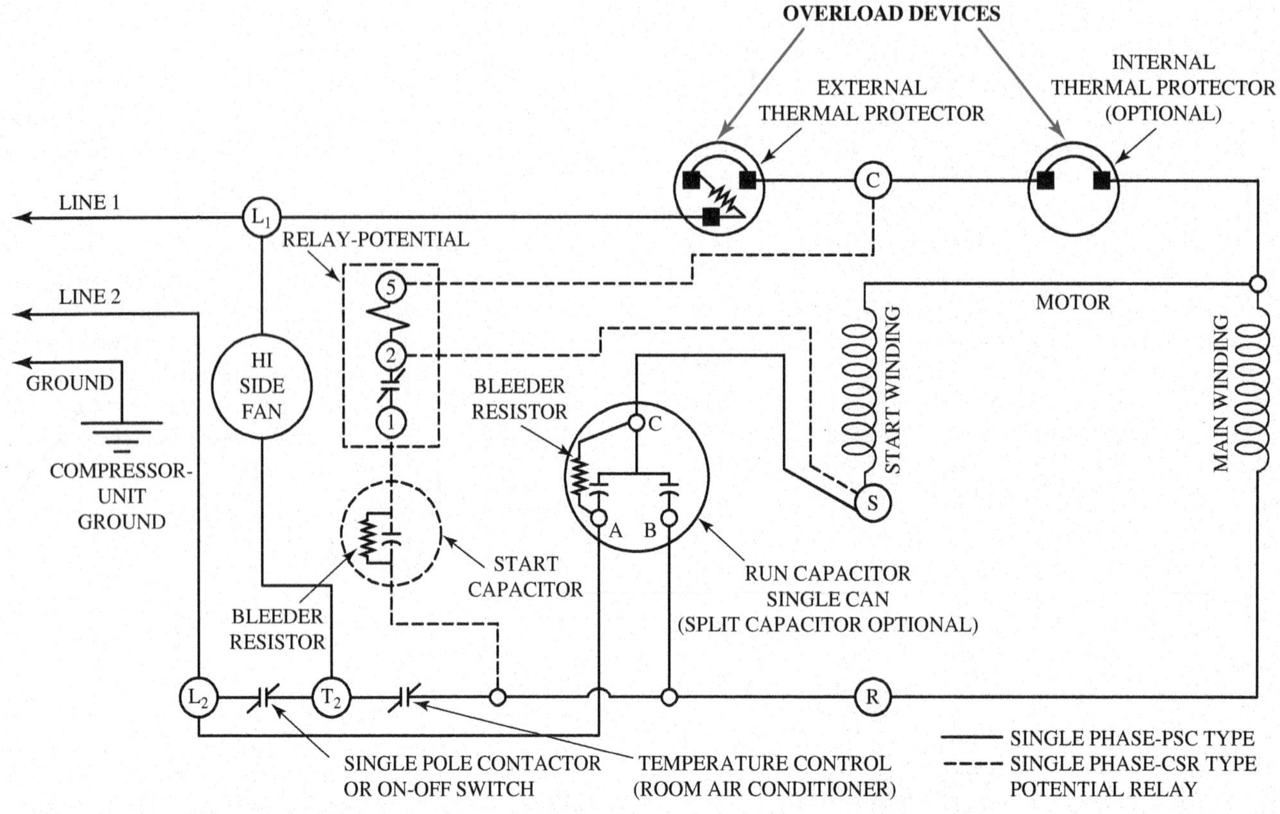

Figure 13-24 This compressor has two compressor overloads. One is an external overload protector and the other is an internal overload protector. These are identified in this diagram as "external thermal protector" and "internal thermal protector."

company if the problem occurs again. You explain that nuisance problems happen sometimes, but to call if the equipment "acts up" again.

What was the problem? What caused the compressor to stop operating? Was it an undetermined nuisance trip?

Three hours later you are heading back to the shop and are told by the dispatcher to return to this customer. The building is getting warm again. You notice from the wiring diagram that this system has a lockout relay design like that shown in Figure 13-13. This time you realize that turning the thermostat off and on would reset the lockout circuit and you would lose the ability to verify that the unit was indeed "locked out."

This time you find the same thing. The compressor is not working. You decide to focus on the lockout circuit. The reset relay (lockout relay) coil has 23 volts across it. The reset relay contacts are open since the coil was energized. You read 24 volts across the reset relay open contacts. This voltage reading means that the three safety devices are closed. The next question is "Which safety device opened and caused the reset relay to energize, opening the reset relay contacts?"

Was it the low-pressure switch, high-pressure switch, or compressor starter overload contacts? What do you think opened? Why?

All three of these safety devices are now closed since 24 volts was measured across the open reset relay contacts. You start the compressor by resetting the thermostat, meaning that you turn it off and on. Everything checks out okay as on the previous visit. You wait around for 30 minutes for something to create an open safety switch. The system operates normally; nothing happens. You call the service manager with the findings. You cannot wait around until the problem surfaces again. The service manager tells you to place a low-amp fuse across each of the safety switches. You install a 1/10-amp fuse across each of the three safety switches. You notify the owner of the problem and promise to come back in the morning to check the condition of the unit.

When you arrive the next day, the compressor is off again. The fuse across the high-pressure switch is open. When the high-pressure switch opened, the fuse carried the load momentarily and then opened. The problem is a high-pressure issue.

How do you find out what is causing the high-pressure switch to open? What do you think caused it to open?

You check the amp draw on the single-phase condenser fan motor. It measures 4 amps. The fan motor RLA is 3.2 amps. This is a problem. You check the capacitor and finds that it is weak. The 5-mfd cap measures 3 mfd.

This could be the cause of the high motor amps. You check the replacement cap before installing it. It measures 4.9 mfd, which is within the ±5% range of the label rating of 5 mfd. The motor is started and the measured amp draw is 3 amps. You think that the problem was excessive condenser fan motor amps causing it to overheat and trip on the motor overload after it ran for a while. When the fan motor cycles off on its overload, the condensing pressure will rise, causing the high-pressure switch to open and activating the reset relay circuit.

You monitor the system for 15 minutes while writing up the invoice and cleaning the area. You explain the problem, its solution, and the charges to the customer. You ask the customer to call the dispatcher immediately if another problem arises. This solved the challenging problem.

SUMMARY

This unit discussed common electrical components used in HVACR equipment. Understanding these common components will help in the troubleshooting process. There will always be electrical components and refrigeration systems that you will not understand. When this occurs, review the component's features and its position in the electrical system. This may help you determine the component's function. If the undetermined component is in the working part of the circuit, ignore its operation. If the undetermined component is in the nonworking part of the circuit, you may need to call for help. Calling for help is not a sign of weakness, since no one technician can possibly be familiar with all electrical components, especially with all of the new components in use today.

The first component covered was the crankcase heater. The heater is important for keeping the refrigerant warm. The heat keeps the oil warm and prevents the refrigerant from condensing in the compressor crankcase.

A solenoid valve is an electromechanical valve that opens or closes by energizing a magnetic coil. The solenoid valve is commonly found in a liquid line, including water and other fluid lines.

Electric unloaders are used on semi-hermetic compressors to reduce the capacity of a compressor when lower capacity operation is required. This allows the compressor to continue running without cycling off.

A solid-state timer will delay the operation of a compressor for about 5 minutes. This time delay allows the pressures to equalize and the power quality to stabilize.

The lockout relay is a type of device that is usually found in a control circuit. As the name implies, the lockout relay prevents or locks the compressor circuit from operating when a safety device opens. The lockout circuit is reset when the power is turned off and back on.

A hot gas sensor is designed to keep compressors from overheating and burning out. The discharge line will become excessively hot when there is a low refrigerant charge or inadequate compressor lubrication.

The line voltage monitor provides many protective features. The monitor protects against short cycling, voltage imbalances, phase reversals, and low-or high-voltage conditions.

The unit discussed explosion-proof designs. It is important to be able identify explosion-proof designs and be able to access the enclosed wiring and connections. Equally important is making sure you seal the enclosure back to its original condition so that it does not create a danger to the surrounding environment.

The final section talked about internal and external overload devices. These safety devices are used to protect the motor from high amperage conditions that could burn out a motor winding or cause the motor shell to get hot enough to ignite a fire.

REVIEW QUESTIONS

1. What is the operation voltage of the unloader shown in Figure 13-5?
2. What is the operation voltage of the crankcase heater shown in Figure 13-5?
3. Draw the symbols for a crankcase heater and a solenoid coil.
4. What is the purpose of a crankcase heater?
5. What are three reasons to use a solid-state timer?
6. What is the voltage across a timer's contacts that will not close?
7. The cooling circuit shown in Figure 13-13 is operating normally. What is the voltage across the reset relay coil?
8. The cooling circuit shown in Figure 13-13 is not operating because one of the safety switches opened and reclosed. What is the voltage across the reset relay coil?
9. Refer to the electrical diagram in Figure 13-13. What safety device(s) will cause the reset relay to be energized?
10. Refer to the electrical diagram in Figure 13-13. Why does the reset relay coil have higher resistance when compared to the comp start coil?
11. When is an explosion-proof enclosure required?
12. A line voltage monitor protects a motor from what six conditions?
13. Draw a simple line voltage monitor hooked to a three-phase compressor circuit.
14. How can technicians determine which safety switch is opening as a result of a problem if all switches are closed when they arrive on the job?
15. What is the purpose of a discharge gas thermostat?

UNIT 14
How Motors Work

WHAT YOU NEED TO KNOW

After studying this unit, you will be able to:

1. Describe how a motor operates.
2. Identify the basic parts of a motor.
3. Determine the difference between a single-phase and a three-phase motor.
4. Describe the voltage and current requirements for a motor.
5. Explain the difference between RLA and LRA.
6. Determine motor speed from the number of poles.
7. Define power factor.
8. List the types of motor bearings.
9. Interpret the information on a motor nameplate.
10. Find information on the "unknown soldier."

GREEN TIP

Installing efficient air conditioning components and equipment makes you, as a technician, part of the green energy movement. Motors are the number one energy-consuming device in our profession. Ensuring that the correct motor and running components are installed is the gateway to energy efficiency. Upgrades are best done when a motor needs to be replaced. Any time you can replace a damaged motor with a more efficient one, you should do so.

Simply retrofitting a good motor is not a good practice, even if a replacement motor is more efficient. The replacement motor may save energy, but will it save enough energy to offset the energy used to manufacture it?

A new type of green motor will be discussed in a future unit. It is called the electronically commutated motor, or ECM. Its operation and features are different from the motors discussed in this unit; therefore, a unit of this book, Unit 16, has been dedicated to the ECM.

It is important to understand motor operation so that you do not misdiagnose a motor with embarrassing results. No one likes to be wrong. Misdiagnosis reflects poorly on the technician and the company providing the service. Manufacturers report that 25% of the motors returned under warranty are found to have "no problem." In the case of a compressor, the misdiagnosis can be electrical or mechanical. Plain and simple, the tech returning the compressor did not know how to troubleshoot the motor. Returned, general-purpose motors, such as fan or pump motors, also have an unacceptable return rate with "no problem found." In this case it would be a misdiagnosed electrical problem. General-purpose motors, unlike compressor motors, have only a slight chance of coming up with a mechanical misdiagnosis. If a general-purpose motor spins freely, it is mechanically good. You will learn in Unit 16 that the electronically commutated motor (ECM) will not spin freely when the trigger voltage is removed.

This unit begins by discussing how motors work. The unit is divided into common single-phase and three-phase motor types. Within each of these categories, we explore the various types of motors. A section on troubleshooting explores problems with each type of motor. The troubleshooting section may be the most important part of this unit because it gives you the vital information you will need to determine a true solution to a problem. Remember, when in doubt ask for help.

14.1 BEFORE WE START!

Before we start our discussion of motors, we should review the terms that are used in this unit. Some of these terms are found in previous units. A review will be useful to the understanding of this unit's information and will help make you a proficient troubleshooter.

RLA and LRA

RLA is the abbreviation for rated load amps. This information is found on condensing unit nameplates and sometimes on **motor nameplates**, as shown in Figure 14-1. The RLA entry in Figure 14-1 is located near the middle of this nameplate. This is the maximum amperage the compressor should draw at its design rating, which includes operating pressures, operating voltage, and compressor Btu capacity at high ambient temperatures. For a single-phase compressor motor, this amperage will be measured on the common connection **C** of the compressor. For a three-phase compressor motor, this amperage will be very close to being the same on any of the three legs. All three legs should be checked for amp draw.

MODEL ABCD48
SERIAL NO. 1234
A.C. VOLTS 240 PHASE 1Φ HERTZ 60
VOLTAGE RANGE MIN. 186 MAX. 264
MIN. CIRCUIT AMPS 30
MAX. FUSE AMPS OR HACR TYPE CIRCUIT BREAKER 40
(HACR CIRCUIT BREAKER FOR U.S. ONLY)
COMPRESSOR RLA 18 LRA 56
MAX. WORKING PRESSURE 450
FAN MOTOR FLA 1.5 H.P. 3/4
FACTORY CHARGE OZ. R22 135
FACTORY TEST PRESSURE PSIG LOW 150 HIGH 300

Figure 14-1 A condensing unit nameplate shows a compressor rated load amps (RLA) of 18 A and locked-rotor amps (LRA) of 56 A. The full load amps (FLA) of the condenser fan motor is 1.5 A.

Another compressor manufacturer defines rated load amps as the amp draw that is within 71% of the maximum continuous current rating of the overload protection. This does not mean that the motor can safely operate at a higher amp draw above the 71% level.

LRA is the abbreviation for locked-rotor amps. The LRA condition is created in two different ways. The most common way is on a motor start-up. When a motor begins to rotate from a standstill position, it draws a large amount of current as it tries to develop the magnetic field that will start motor movement. The LRA is about four to six times higher than the rated load amps. The LRA condition only exists for a split second at motor start-up.

The LRA condition will also be seen when a motor is mechanically locked up or if the starting components are defective.

TECH TIP

Check running current. The readings should not exceed the manufacturer's full load rated amps, also expressed as rated load amps. Low amps are normal during low load conditions. Excessive current may be due to:

- Shorted or grounded windings
- Bad capacitor
- Faulty start relay
- Excessive bearing wear
- Operation in an overloaded condition.

14.2 USE OF METERS

To electrically troubleshoot motors, the technician will need to be able to use a multimeter and ammeter. As a review, the tech will need to check voltage, measure resistance, and determine amperage draw (see Figure 14-2). It is important for a motor to have the correct voltage input. The measured voltage should be within ±10% of the nameplate voltage when the motor is operating. The torque drops rapidly as the applied voltage is reduced. With reduced voltage the motor will run slower and hotter due to the reduced airflow over the motor frame. As the motor voltage goes down, current draw goes up.

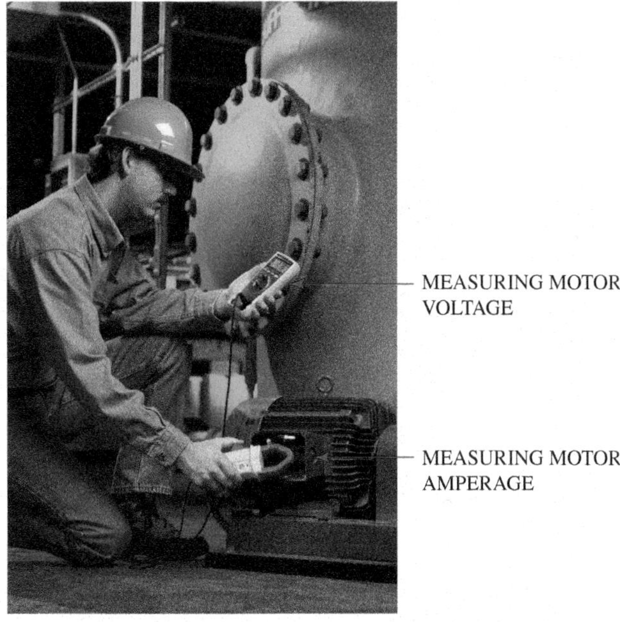

Figure 14-2 Technician measuring the voltage and amperage of an operating motor. Measure the voltage before and after the motor starts.

Amperage should be measured on one wire, as shown in Figure 14-3. Notice that a glove is used for extra protection. The clamp-on ammeter is measuring the magnetic energy in the wire. Higher amperage creates more magnetism in the wire. The ammeter translates this magnetism into an amp reading on the meter. As stated earlier, measure a single-phase compressor on the common connection of the compressor. For a good three-phase compressor, this amperage will be the same on any power leg. Measuring two wires at the same time with different potential will cancel or modify the amperage reading. It is important to measure only one wire for an accurate amp reading.

Two scales are used when measuring motor winding resistance. Use the R×1 (i.e., the lowest) meter scale when measuring the motor winding. Most motor windings are below 100 Ω. Some of the high-speed windings on a fractional horsepower motor may be a little higher. Large three-phase motors will have a resistance of less than 1 Ω, since these windings are large in diameter and, thus, have lower resistance. Use the highest resistance scale when measuring the motor windings to the motor case ground. For a good motor, the resistance to ground is usually in the megohm range (1,000,000 Ω) and greater.

Before condemning any motor, take your readings again and when possible have a coworker verify the results. If you are unsure of your diagnosis, in private, call your supervisor and discuss the action.

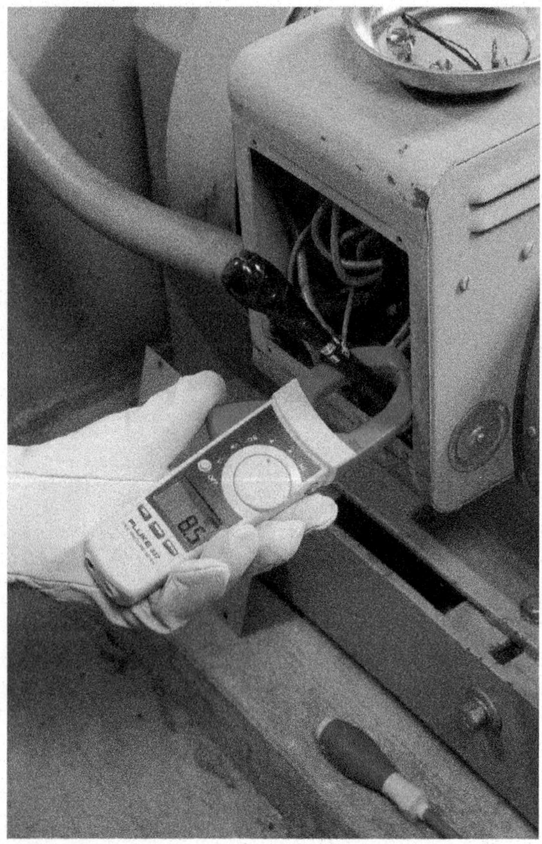

Figure 14-3 Measuring amperage on one motor wire. The tech uses a glove as protection against accidental shock from a nicked wire or exposed connection.

SAFETY TIP

Wear safety glasses when measuring voltage or amperage. Even the best multimeter may explode if placed on the wrong setting. For example, the multimeter may blow up if it is set on the amperage scale and voltage is measured. Wearing gloves affords protection against the lower voltages. But standard, dry gloves should not be expected to prevent shock—gloves are just one measure of protection. In the case of failure or use of the wrong meter setting, do not hold a multimeter when measuring voltage.

14.3 STARTING TORQUE

Starting torque is one of the most misunderstand motor concepts. Starting torque is the force or twisting motion used to get a motor moving. A single-phase motor with a properly sized start capacitor has the best starting torque. A three-phase motor also has high starting torque. Capacitors are not used with three-phase motors.

Basically, high starting torque is created when there is a 90-degree phase shift between the current and voltage. For the best starting torque, the starting current should lead the voltage by 90 degrees. When using a start capacitor, the motor manufacturer tests the motor for that 90-degree phase shift to determine optimal starting torque. An undersized capacitor or oversized capacitor will reduce the motor starting torque. The start cap is not used to bump, boost, jolt, or kick the motor but to shift the current and voltage phase so that it will have the greatest starting torque or twist under a load. In HVACR a compressor load would be starting under a pressure difference between the high side and low side. Simply getting the mass of a fan motor moving is considered starting under a load, even though this would be a light load.

14.4 RUNNING OPERATION

Many single-phase motors have run capacitors. The run capacitor provides some starting torque, but the main purpose of the run capacitor is to produce low running amps. Having a weak or oversized run capacitor will increase running amps and may reduce starting torque. The run capacitor should be within ±5% of the microfarad rating. Some capacitor manufacturers, however, may accept a ±6%, ±7%, or ±10% range. The capacitor should be checked with a capacitor checker. If a capacitor checker is not available, a new cap can be installed to determine if the original capacitor was the problem. Three-phase motors do not have run capacitors.

The run capacitor also limits the current flow through the start winding once the capacitor has been charged. This electron charge of a capacitor occurs in less than a second. The start winding should have no or limited current flow after it is used to start the motor. A shorted run cap will burn out the start winding. Some run caps have marked terminals. The mark may be an equal (=) or a minus (−) sign. It could be a dab of paint near the marked terminal. The marked terminal should be connected to the power side, not the start winding. The start winding is less likely to be damaged if the marked or designated terminal of the run cap is connected to the line side of the power source.

14.5 VOLTAGE

The motor voltage should be within ±10% of the listed or **rated voltage** when the motor is in operation. Voltage that is lower or higher will cause the motor to overheat. Check the motor voltage prior to operating the motor and when the motor is energized. The voltage may change after it is energized.

TECH TIP

Check line voltage at the breaker panel first with the motor off and then with the motor on. The voltage should be within ±10% of the motor rating.

Low line voltage will cause the motor to draw more current than normal. Severe undervoltage leads to overheating and premature failure.

High line voltage causes excessive inrush current at motor start. This eventually breaks down the motor winding insulation and leads to motor failure.

14.6 HOW DO MOTORS WORK?

The goal of this book is keep the theory as simple as possible. We want give you enough information to understand how an electrical device works and how to troubleshoot it. How a motor works is complex. This section will explain motor operation in basic terms and by using examples. The discussion is aimed at field technicians, not electrical engineers.

Motors work on the principle of **magnetism.** As shown in Figure 14-4A, magnetism is created in a wire when voltage is applied. If you apply direct current (DC), like a battery, the magnetic field is developed in the wire and does not change. When you apply alternating current (AC) in a wire, the magnetic field builds and collapses, then builds and collapses in the opposite direction. Coiling the wire increases the magnetic strength, as indicated in Figure 14-4B. Magnetism is created in the stationary winding known as the **stator.** Take a look at the stator windings in the two different size motors in Figures 14-5 and 14-6.

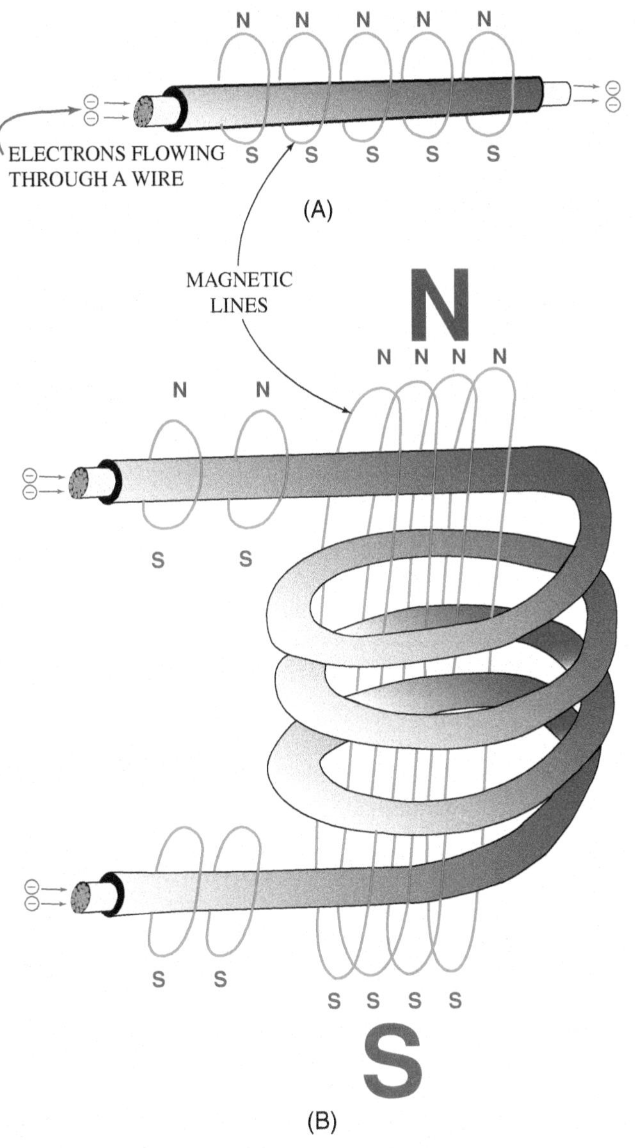

Figure 14-4 (A) The electrons flowing through a straight wire form a weak magnetic field around the wire. (B) With the same wire coiled, the weak magnetic field around each wire combines to create a stronger magnetic field around the coil.

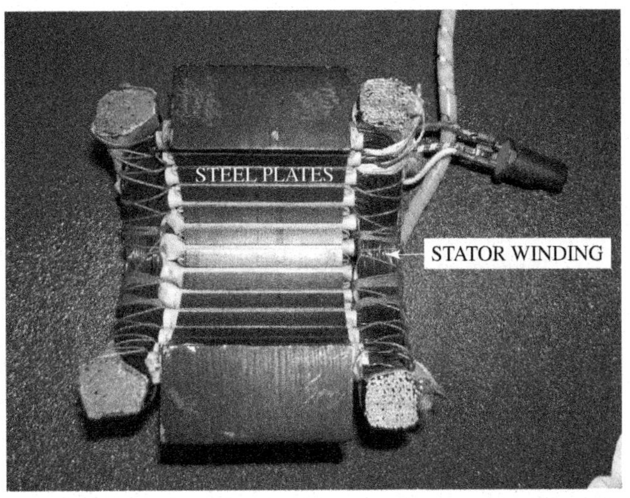

Figure 14-5 This is the stator or stationary winds embedded in a stack of steel-laminated plates. Voltage is applied to the stator motor windings. The black cylinder on the right is an internal overload used to protect the motor from high amperage and overheating conditions.

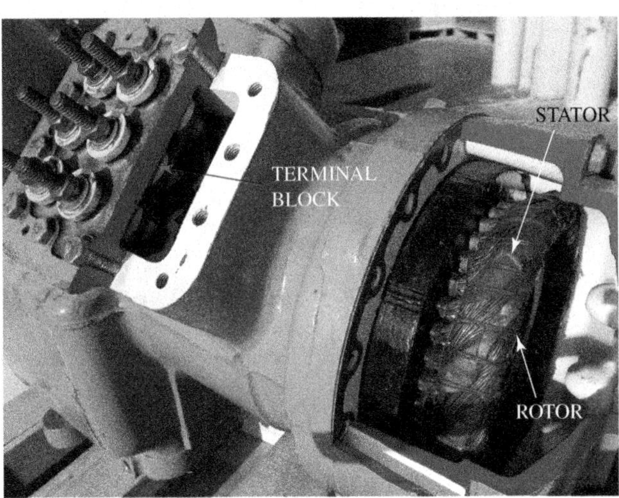

Figure 14-6 Cutaway of the stator winding on a semi-hermetic compressor. The upper left side is a cutaway view of the terminal block that ties the motor winding to the external power connections.

The stator windings are wrapped around an iron core to further intensify and concentrate the magnetic field, as illustrated in Figure 14-7. The changing of magnetic fields pushes and pulls the rotating part, the rotor, creating the movement necessary to do the work it was designed to do. Figure 14-8 shows a **rotor** and stator for a fractional horsepower motor, sometimes abbreviated FHP.

Figures 14-9A and 14-9B and Figure 14-10 illustrate the various component parts of a motor. Figure 14-9A is the rotor or moving part of the motor. Figure 14-9B is the stator winding that is energized to create the movement of the magnetic

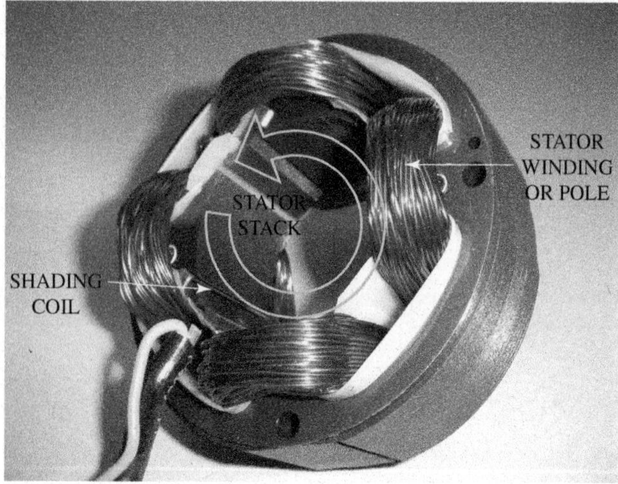

Figure 14-7 The stator windings are woven through the stack of steel-laminated plates. This is a four-pole, shaded pole motor.

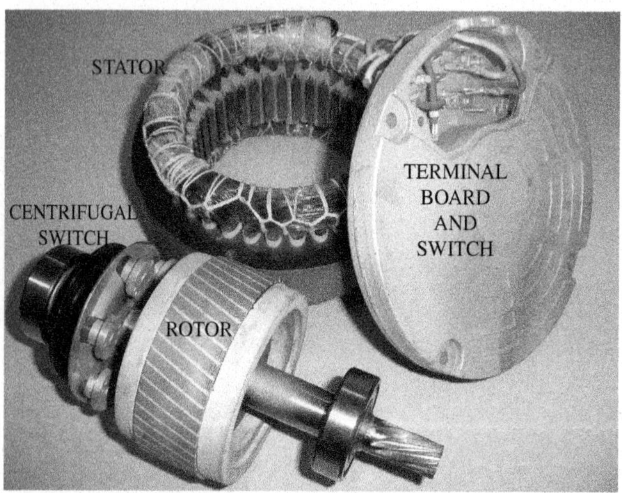

Figure 14-8 The rotor or moving component of the motor is shown in the lower part of this photograph. The upper part shows the inside. The copper-colored windings are the start windings. The outer or green windings are the run windings. This is a split-phase motor that uses a centrifugal switch to remove the start winding.

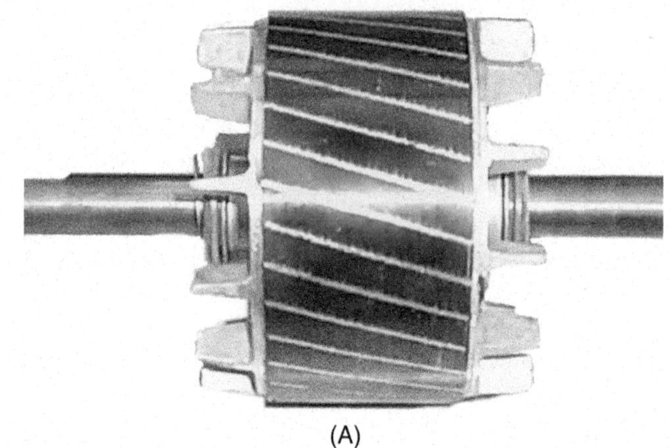

(A)

(B)

Figure 14-9 (A) The rotor is the moving part of a motor. (B) This is the stator part of a motor. Voltage is applied to the stator winding to create a moving magnetic field in the stator. The rotating magnetic field causes the rotor to move.

field that causes the rotor to move. Figure 14-10 focuses on the design of the rotor. This is called a squirrel cage rotor because its design resembles a squirrel cage. The iron core and conduction bars are designed to be create the rotor movement in a clockwise (CW) or counterclockwise (CCW) direction.

You have played with magnets and realize that magnets can attract or repel each other. This attraction/repulsion phenomenon is created in the stationary or stator winding and is used to move the rotor. As you know, simply placing two magnets together will cause the attraction or repulsion action, but the magnets will not move unless you forcibly change the direction and create another attraction or repulsion action. Of course, this would not work for motors. Motors need a changing magnetic field to cause the rotor to move by attraction and repulsion. The attraction or repulsion action of the stator is created by the positive and negative current fields of the 60 Hz/second power supply. Since direct current (DC) is stable and does not modulate up and down to change the magnetic field, DC motors use commutators to change and develop the magnetic field to create rotor motion. DC motors are not commonly found in HVACR, but the author believes it is important to mention this since we are learning about how a motor moves.

Finally, the magnetic field in the rotor must be shifted to create rotor motion in the clockwise or counterclockwise direction. Without a shift in the magnetic field, the motor will not rotate. It is like putting two magnets together—they will not move unless you push them in a different direction. In the following sections on motors, we discuss various ways to shift the phase so that the motor rotates in the desired direction. Some

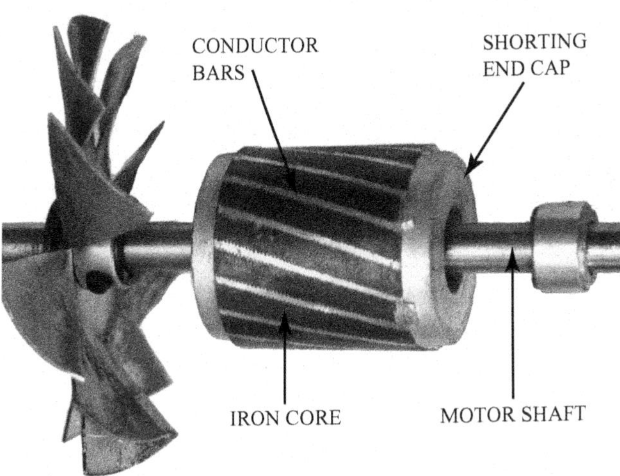

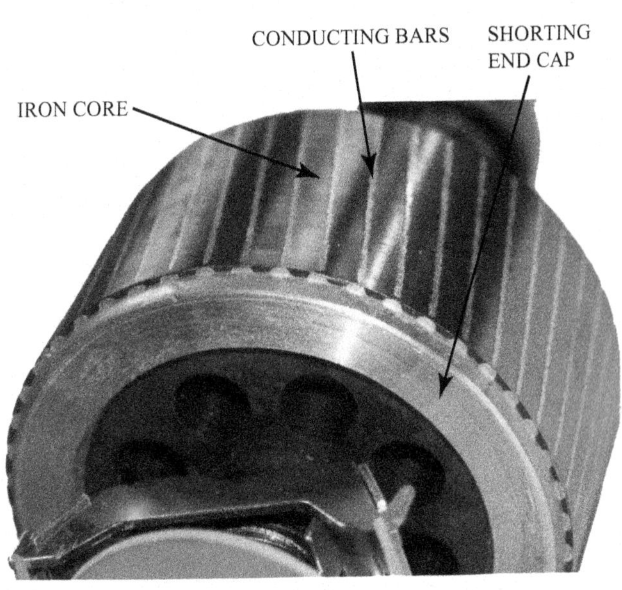

Figure 14-10 This is a squirrel cage motor. The upper section shows the rotor and connecting motor shaft. The lower section shows a close-up of the rotor construction. It is composed of conducting bars and an iron core.

motors concentrate the magnetic energy to force the direction of the motor. Other designs involve splitting the phase windings, the use of run capacitors, and three-phase operation.

TECH TIP

What is NEMA? **NEMA** is the abbreviation for **National Electrical Manufacturers Association**. A NEMA motor rating refers to the physical dimensions, electrical characteristics, and class of insulation used in constructing motors. When a motor has "NEMA" on its nameplate, it means the manufacturer followed NEMA standards in the design and development of its product. The NEMA standard is followed by most motor manufacturers to provide consistency and quality. The use of NEMA standards by manufacturers makes it easier for us to find the correct replacement motor.

14./ OTECTION

Most moto... overload or thermal protective device. Th... tection is to keep the motor from getting s... ys itself or causes a fire. Motor protectors ha... d contacts mounted on bimetal strips that act ... f safety device inside the motor windings. Whe... verload protection device senses excessive heat, ... trip expands and opens the protector, removing ... nding from the supply voltage.

Some thermal protectors are e... windings. Other overload protectors are ... the motor outside the motor shell. The symbol for an ... externally, overload device is shown in Figure 14-11.

Thermal protection can be automatically o... reset. The auto reset device automatically reconne... ually to a motor after the motor winding has cooled. The m... wer reset overload has a reset button that is used to manually close the protective device once it has cooled. A third, rare type of protection is called a "one-shot" protective device. The one-shot device opens a winding permanently, after which the motor cannot operate.

Always allow the motor time to cool off. Some larger motors may require a couple of hours to cool prior to resetting on open overload device.

Trip and reset cycling of the thermal protection device is a common application problem for fractional horsepower (FHP) motors. Overheating is an indication that the motor has been incorrectly applied. Nuisance tripping is usually caused by one of the following conditions:

- The motor is overloaded. If the amp draw is more than 10% greater than the nameplate amperage, it is overloaded.
- The motor is underloaded. If the amp draw is less than 75% of the nameplate rating, it is underloaded.

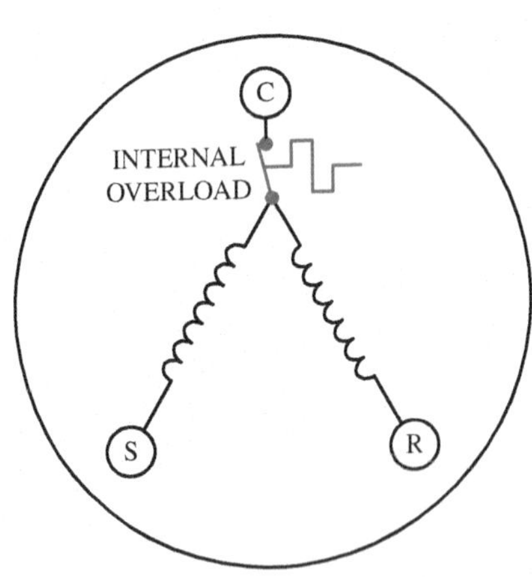

Figure 14-11 This is the symbol for a single-phase, split-phase motor. It has a start winding and run winding. The thermal or internal overload is in series with the start and run windings.

- The capacitor rati[...] capacitor could be too large or too [...]or microfarad rating could be lower [...] on the capacitor.
- Low airflow [...] housing. A reduced ventilation ope[...]ow a motor to cool. A hot motor will [...]e.
- Defective [...]ectors are rare.

TECH TIP

Table 14-1 shows a motor information worksheet that will help you select the correct motor replacement if you do not use an exact replacement. This information will help you match up the correct motor replacement. Complete the information prior to contacting your supplier or searching the Internet for a motor replacement.

Table 14-1 Motor Information Worksheet

Use this table to organize the information you will need to change out a motor.
(Courtesy of Fasco Motors.)

FASCO	**Motor Info Worksheet**
Motor Type	__________
Motor Diameter	__________
Voltage	__________
Horsepower	__________
Amps	__________
RPM or Poles	__________
Number of Speeds	__________
Iron Stack	__________
Motor Length	__________
Cap Rating	__________
Rotation	__________
Mounting	__________
Ventilation	__________
Bearing Type	__________
Shaft Diameter	__________
Shaft Length	__________
Number of Shafts	__________
Special Features	

14.8 NAMEPLATE INFORMATION

Most manufacturers include their name and address on the motor nameplate. Some have good motor information on their website. The National Electrical Code (NEC) states that the motor nameplate must show most of the following information:

- Manufacturer's name and address
- Model and serial number
- Rated voltage
- Rated load amps
- Locked-rotor amps
- Frequency
- Phase
- Horsepower
- RPM speed
- Multispeed single-phase motors
- Insulation class and rated ambient temperature
- Service factor
- Frame size
- Power factor
- Time rating or duty.

In addition to this required information, motor nameplates may also include the NEMA design letter, full-load efficiency, and power factor. Finally, some nameplates may even include identification numbers, a certification code, the manufacturer's serial number, and symbols and logos. See the sample nameplates shown in Figures 14-12 and 14-13 and compare their similarities and differences.

The nameplate in Figure 14-13 provides many valuable pieces of information. The most important pieces of information are identified as listed here:

1. The model and serial number
2. The horsepower or ability to do work
3. The RPMs or revolutions per minute, which indicate the speed of the motor
4. The frame size, which is the physical size of the motor; this information is used when selecting a similar replacement.

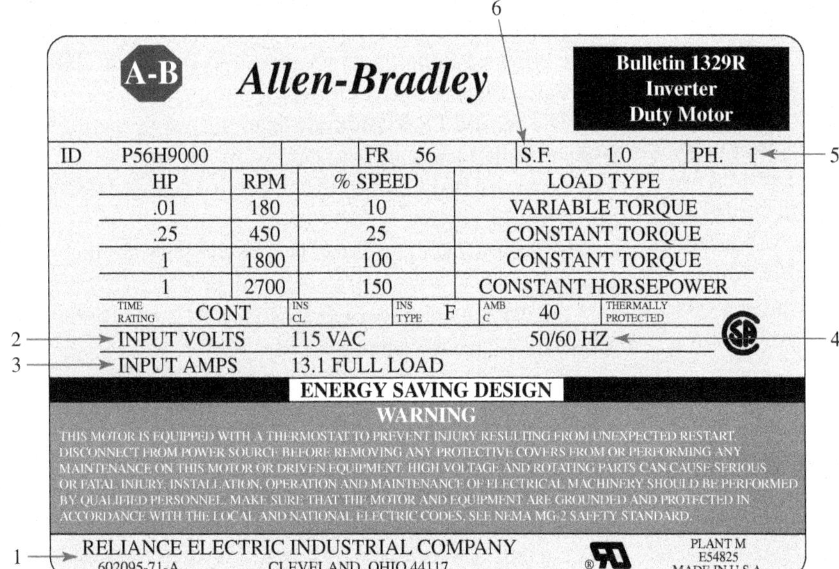

Figure 14-12 This is the nameplate for a variable-speed motor or variable-horsepower motor.

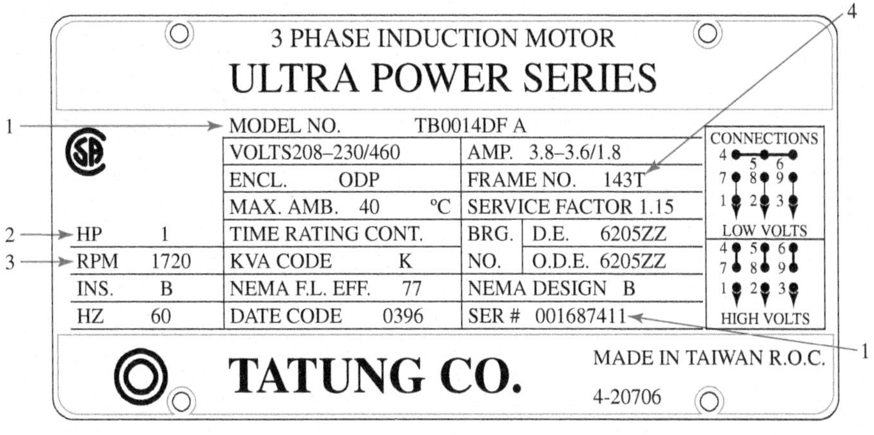

Figure 14-13 This nameplate has information on how to wire a low-voltage or high-voltage connection. The horsepower and speed are set.

Model and Serial Number

Each motor will have a model number and may have a serial number. This information will be useful if the motor needs to be replaced. This information will allow you to choose an exact replacement or select a substitute motor. A sample of a model number is indicated by the number 1 in Figure 14-13.

TECH TIP

Will a 230-volt motor run on 208 volts? The motor is allowed to operate at ±10% of the nameplate voltage. In this case, the performance will be reduced by 10% to 12%. The motor will run at a higher temperature. In this case, there will be no undervoltage reserve.

14.9 RATED VOLTAGE

The rated voltage is the voltage that is applied to the motor. The rated voltage is ±10% of the nameplate voltage. For example, a voltage rated at 480 V will have a ±48-V range above and below the 480 volts. Hence, in this case the operating voltage should be 432 to 528 volts. Operating outside this range will cause the motor to overheat and "trip out" its overload protection.

If it is a dual-voltage motor, the ±10% will apply to the lowest voltage and highest voltage on the dual-voltage motor. So, for a 208/240-V motor, acceptable voltages are found by subtracting 20.8 volts from the 208-V side and adding 24 volts to the 240-V side. Therefore, the range will be 187.2 volts to 264 volts. For long motor life, it is best to operate the motor near the voltage rating. A slightly higher voltage is best. It is almost impossible to operate on the exact design voltage such as 240 or 480 volts. Measure motor voltage under a normal operating load.

TECH TIP

Some common colors for single-phase motor wires are:

- White or yellow = common power wire
- Brown or brown with a white tracer = capacitor connections
- Black = high speed
- Blue = medium speed
- Red = low speed
- Green = ground.

Note that this color code is not used by all motor manufacturers and color codes from overseas manufacturers vary widely.

14.10 RATED LOAD AMPS (RLA)

As the torque load on a motor increases, the amperage required to power the motor also increases. When the rated load torque and horsepower are reached, the corresponding amperage is known as the rated load amperage (RLA). On a compressor motor the RLA condition may be reached when a fully charged air conditioning system is operating at its full Btuh rating on a very hot day. When possible it is best to run the compressor below the rated load amps. The lower the amp draw, the cooler the temperature at which the motor will operate. Cooler operation equals longer and more efficient motor life.

A blower motor or pump will operate at RLA when they are moving the designed airflow or liquid flow. A lack of airflow around a fan or pump motor will cause the motor to overheat. This is noted by the number 3 in Figure 14-12.

Fan motors commonly use the term *full load amps* (FLA) instead of rated load amps.

14.11 LOCKED-ROTOR AMPS (LRA)

Locked-rotor amps or locked-rotor current is found when a motor starts up in four conditions, as discussed next.

When alternating current motors are started with full voltage applied, they create an inrush current that is usually many times greater than the value of the running full-load current. This very high amp condition occurs momentarily. The value of this high current can be important on some installations because it can cause a voltage drop that might affect other equipment. High inrush current can also dim lights and affect sensitive electronic equipment.

The second LRA condition is found when the motor is mechanically locked up and cannot move. In this case high amperage will be drawn until the overload device opens the circuit to the motor.

The third LRA condition is created when the incoming voltage is low.

Finally, a LRA may occur when a start capacitor or relay is defective or wired up incorrectly.

There are two ways to find the LRA value of the motor:

1. Look it up in the motor performance data sheets as provided by the manufacturer. It will be noted as the locked-rotor current. LRA may also be found on the motor nameplate.
2. Use the locked-rotor code letter that defines the inrush current of a motor that is required when starting it.

Sometimes a locked-rotor letter code is used instead of an amperage number. The higher the alphabet letter, the greater the starting amperage. The code letter is found as an amperage in the NEC.

TECH TIP

One of the reviewers for this book has a good Tech Tip to share with you: The direction of the shaft of a replacement condenser motor is an important factor in choosing a replacement. If you are using the original equipment manufacturer's (OEM) recommended motor, this will not be a problem. If using a generic substitute, however, be aware that not all motor bearings are designed to handle a downshaft load.

14.12 FREQUENCY

To operate successfully, a motor's **frequency** must match the power system (supply) frequency, now called hertz (Hz). In North America, this frequency is 60 Hz (cycles). In other parts of the world, the frequency may be 50 or 60 Hz. The frequency has to do with the number of times the magnetic field changes in the power supply, which causes a change in the motor windings per second. For example, a 60-Hz motor will change to create a positive and negative cycle 60 times a second. When two motors are equal in design, the lower hertz motor will turn slower. Frequency is a major factor in the speed or RPMs of the motor. This is noted by the number 4 in Figure 14-12, which indicates a 50- or 60-Hz motor.

14.13 PHASE

The motor voltage phase will be single phase or three phase. Two-phase voltage is highly unusual in the world. Some people get confused and think that 208/240 volts is two phase. It is normally single-phase or three-phase power. The single-phase characteristic is indicated by the number 5 in Figure 14-12.

14.14 HORSEPOWER

Horsepower is the measure of how much work a motor can be expected to do. This value is based on the motor's full-load torque at full-load speed ratings. The standardized NEMA table of motor horsepower ratings runs from 1 to 450 hp. If a motor's horsepower is below 1 hp, the motor is known as a fractional horsepower motor (FHP). If a load's actual horsepower requirement falls between two standard horsepower ratings, you should generally select the larger size motor for your application. The horsepower is noted by the number 2 in Figure 14-13.

Electric motors are typically rated in horsepower or output watts in some European countries. The nameplate horsepower on a general-purpose motor is the horsepower the motor is rated to develop when connected at the rated voltage and frequency. NEMA defines general-purpose motor horsepower on the basis of breakdown torque.

Special-purpose or FHP motors are not typically designed to NEMA standards but rather for a specific machine. They may carry ratings specified by the finished equipment manufacturer. Unless requested to do otherwise, the motor manufacturer will generally rate a motor's horsepower using NEMA standards.

The motor may be marked with the actual horsepower developed at the rated RPMs at the OEM buyer's request, which may be considerably less than the NEMA rating. In short, unlike the consistent horsepower ratings of large, general-purpose motors, the ratings of special-purpose motors are tied to a single unit and may vary considerably for similar output motors.

These inconsistent horsepower rating methods tend to complicate amperage ratings. Whereas in the United States, the amp ratings typically indicate the current a motor will draw at normal load, Canada requires the ampere rating to reflect the actual current when the motor is loaded at the rated output, regardless of actual load speed.

14.15 RPM SPEED

Revolutions per minute (RPM) is the rated full-load speed. The RPMs are indicated by the number 3 in Figure 14-13. This is the motor's approximate speed under full-load conditions, when voltage and frequency are at the rated values. A somewhat lower value than the actual laboratory test result figure is usually stamped on the nameplate because this value can change slightly due to factors such as manufacturing tolerances, motor temperature, and voltage variations. This is known as slip. **Slip** is the difference between the rotating magnetic field and the speed at which the rotor is turning. The rotor is trying to keep up with the motor's rotation magnetic field in the stator. On standard induction motors, the full-load speed is typically 96% to 99% of the no-load speed. Table 14-2 shows the actual speed of various motors. The **synchronous speed** is the RPMs listed on the motor nameplate. In summary, the difference between the synchronous speed and actual speed is called *slip*.

TECH TIP

You should consider a couple of things before replacing a defective condenser or blower motor. If the motor can still rotate, mark the rotation direction on the fan or blower wheel. Measure the distance between the fan or blower and the venturi or fan housing. This will assist in lining up the fan or blower when changing out the motor. Finally, mark the shaft of the motor where the fan hub or blower wheel touches the shaft. If you use an exact replacement motor, you can mark the new shaft prior to installing the fan or blower wheel and finish the installation with a perfect alignment.

Table 14-2 Synchronous Speed versus Actual Speed

The actual speed of a motor is slightly lower than the speed listed on the nameplate. The greater the number of poles, the slower the motor speed.

Number of Poles	Synchronous Speed	Actual Speed
2	3,600	3,450
4	1,800	1,725
6	1,200	1,140
8	900	850

14.16 MULTISPEED SINGLE-PHASE MOTORS

Multispeed motors are actually multi-horsepower motors. The motor speed windings are in series with the high-speed winding, as shown in Figure 14-14. The speed depends on the winding resistance. As you can see, the higher the motor winding resistance, the slower the motor speed. The windings are tapped so the motor is weaker, running slower under load. For example, for a three-speed 1/3-hp motor:

- High speed is 1/3 hp.
- Medium speed is 1/4 hp.
- Low speed is 1/6 hp.

Lower speeds have more turns per coil, thus adding resistance to current flow, which reduces torque. This can be compared to letting up on the gas in your car. Less gas reduces the engine output and the car runs slower. The additional resistance of the low-speed windings weakens the motor and, as a result, it runs slower. At idle or no load, a multispeed motor runs the same on any speed connection just as your car will accelerate rolling downhill with the key turned off.

Figure 14-15 is the electrical diagram for a multispeed permanent split capacitor (PSC) motor. The run winding is the speed winding. Tapping the run winding changes the motor speed. Do not wire up two speed windings in parallel. This will burn up one or more run windings. Some common PSC wire color codes are:

- Black = high speed
- Blue = medium speed
- Red = low speed
- Brown or brown/white = capacitor wires
- Green = ground.

The main voltage color may be white, blue, or another color. Not all motor manufacturers use this color code. Also there may be two lead rotating wires. This changes the rotation of a motor. The motor can go clockwise (CC) or counterclockwise (CCW). This reduces the number of motors that a supplier needs stock in inventory.

Multispeed motors must be under a load to change speeds. A multispeed blower removed from the blower compartment will run at high speed, no matter which speed tap is used.

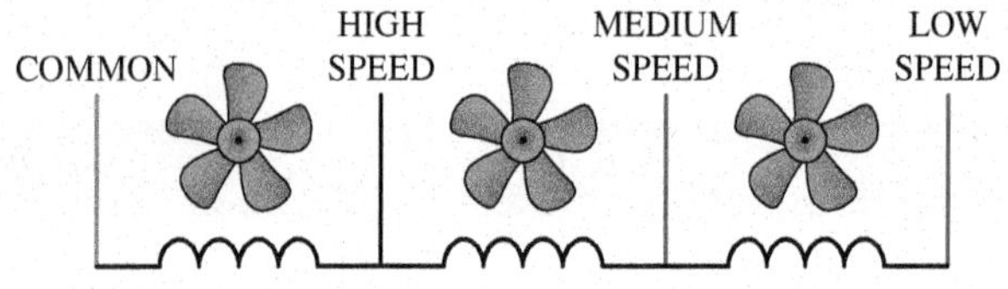

Figure 14-14 The multispeed motor has series motor windings. The slower speeds have more windings in series with the high-speed winding.

TECH TIP

The low-speed winding can be open and the higher speed winding can be good and operate the motor at high speed. The motor can be wired to a higher speed until it can be replaced if the lower speed is required.

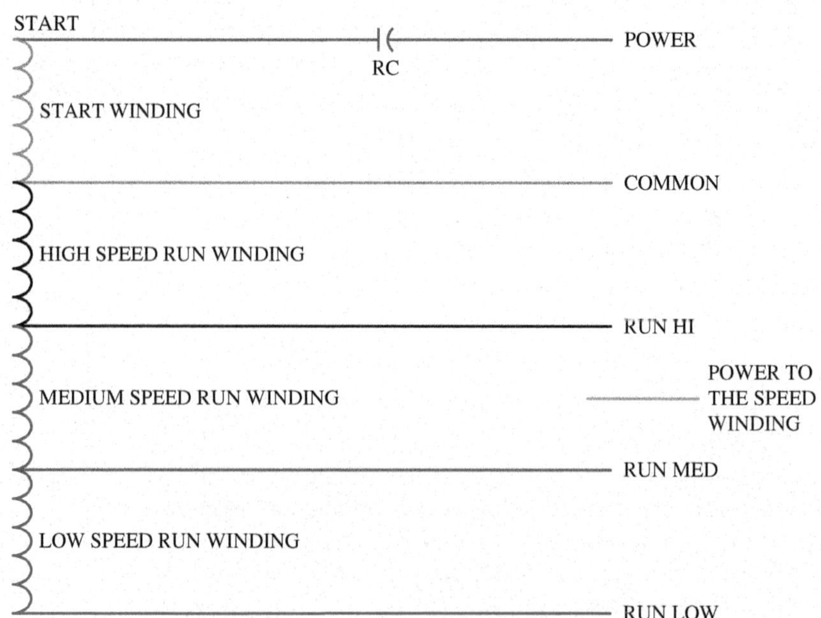

Figure 14-15 In this multispeed PSC motor, the run winding is the speed winding. The lower speed windings have higher resistance. One side of the power supply is connected to the right side of the run capacitor and one of the speed taps. The second power wire goes to the common connect. Applying power to two speeds will damage the motor winding.

TECH TIP

The abbreviations CC and CCW can be confusing. To reduce the mistakes using these abbreviations some motor manufacturers use CCWSE, CWSE, CCWLE, or CWLE. Here are the translations:

CCWSE = counterclockwise rotation as seen from the shaft end of the motor
CWSE = clockwise rotation as seen from the shaft end of the motor
CCWLE = counterclockwise rotation as seen from the lead end of the motor
CWLE = clockwise rotation as seen from the lead end of the motor.

Many condenser fan motors are easily reversed by switching two sets of wires. The manufacturer designs a quick connector on the motor leads to change motor rotation in a matter of seconds. It is important to observe the rotation of any fan or pump when changing out the motor.

14.17 INSULATION CLASS AND RATED AMBIENT TEMPERATURE

A critical element in motor life is the maximum temperature that occurs at the hottest spot in the motor. The temperature that occurs at that spot is a combination of motor design (temperature rise) and the ambient (surrounding) temperature. The standard way of indicating these components is by listing on the nameplate the allowable maximum ambient temperature, usually 40°C (104°F), and the class of insulation used in the design of the motor. Available insulation classes are B, F, and H. Classes B and F are the most common.

The **insulation class** letter designates the amount of allowable temperature rise based on the insulation system and the motor service factor. Table 14-3 shows the insulation group, maximum operating ambient temperature, and allowable temperature rise before a motor is considered to be overheated. An overheated motor will break down insulation and shorten the life of a motor. It is common to see temperatures given in Celsius.

Table 14-3 Motor Insulation Class

Insulation Class	Ambient Temp.	Temp. Rise	Total Temp.
A	40°C	65°C	105°C
B	40°C	90°C	130°C
F	40°C	115°C	155°C
H	40°C	140°C	180°C

The insulation class is divided into four different insulation groups: A, B, F, and H. The temperature that occurs at that spot is a combination of motor design (temperature rise) and the ambient (surrounding) temperature. The standard way of indicating these components is by showing the allowable maximum ambient temperature expressed as total temperature, usually 40°C (104°F), and the class of insulation used in the design of the motor. The higher lettered motor will withstand a higher operating temperature.

Finally, the nameplate lists the maximum ambient temperature at which the motor can still operate and be within tolerance of the insulation class at the maximum temperature rise. Ambient temperature may be abbreviated as **AMB** on a nameplate.

14.18 SERVICE FACTOR (SF)

The **service factor** is the number by which the horsepower rating is multiplied to determine the maximum safe load that a motor may be expected to carry continuously without exceeding the allowable temperature rise of its insulation class. For example, a 10-hp motor with a service factor of 1.15 will deliver 11.5 horsepower continuously without exceeding the allowable temperature rise of its insulation class.

14.19 FRAME

The frame designation refers to the physical size of a motor as well as certain construction features such as the shaft and mounting dimensions. The frame number is pointed out by number 4 in Figure 14-13.

Under the NEMA system, most motor dimensions are standardized and categorized by a **frame size** number and letter designation as shown in Table 14-4. In fractional horsepower motors the frame sizes are two digits and represent the shaft height of the motor from the bottom of the base to the middle of the shaft in sixteenths of an inch. For example, looking at Table 14-4, a 56-frame motor would have a shaft height ("D" dimension) of 56/16 of an inch. That would be 56 divided by 16, or 3 1/2 inches.

On larger three-digit frame size motors, 143T through 449T, a slightly different system is used where the first two digits represent the shaft height in quarters of an inch. For example, again looking at Table 14-4, a 326T frame would have a "D" dimension of 32 one-quarter inches, or 8 inches. Although no direct inch measurement relates to it, the third digit listed in a three-digit frame size (e.g., in the 326T case, the "6") is an indication of the length of the motor's body. The longer the motor body, the longer the distance between mounting bolt holes in the base (i.e., a greater "2F" dimension). For example, a 145T frame has a larger 2F dimension than does a 143T frame.

When working with metric motors (IEC types), the concept is the same as previously noted with one exception: The shaft height above the base is now noted in millimeters rather than inches. The frame size is the shaft height in millimeters.

Table 14-4 Frame Size Chart

This table shows how the NEMA frame size chart is used. For example, dimension "D" is the height of the motor from the base to the middle of the motor shaft. The standardized measurements will assist the technician in selecting the correct replacement motor.

Frame Size Chart

Standardized motor dimensions as established by the National Electrical Manufactures Association (NEMA)

Frame	D	2E	2F	BA	H	N	U	V	Keyway	Frame	D	2E	2F	BA	H	N	U	V	Keyway
48	3	4-1/4	2-3/4	2-1/2	11/32	1-1/2	1/2	--	29/64 Flat	**284T**	7	11	9-1/2	4-3/4	17/32	4-7/8	1-7/8	4-5/8	1/2 × 1/4
56	3-1/2	4-7/8	3	2-3/4	11/32	1-7/8	5/8	--	3/16 × 3/16	**284TS**	7	11	9-1/2	4-3/4	17/32	4-7/8	1-7/8	4-5/8	3/8 × 3/16
56H	3-1/2	4-7/8	3	2-3/4	11/32	1 -7/8	5/8	--	3/16 × 3/16	**286T**	7	11	11	4-3/4	17/32	4-7/8	1-7/8	4-5/8	1/2 × 1/4
56HZ	3-1/2	4-7/8	3	2-3/4	11/32	1-7/8	7/8	--	3/16 × 3/16	**286TS**	7	11	11	4-3/4	17/32	4-7/8	1-7/8	4-5/8	3/8 × 3/16
143T	3-1/2	5-1/2	4	2-1/4	11/32	2-1/4	7/8	2	3/16 × 3/32	**324T**	8	12-1/2	10-1/2	5-1/4	21/32	5-1/2	2-1/8	5-1/4	1/2 × 1/4
145T	3-1/2	5-1/2	5	2-1/4	11/32	2-1/4	7/8	2	3/16 × 3/32	**324TS**	8	12-1/2	10-1/2	5-1/4	21/32	5-1/2	2-1/8	5-1/4	3/8 × 3/16
182T	4-1/2	7-1/2	4-1/2	2-3/4	13/32	2-3/ 4	1-1/8	2-1/2	1/4 × 1/8	**326T**	8	12-1/2	12	5-1/4	21/32	5-1/2	2-1/8	5-1/4	1/2 × 1/4
184T	4-1/2	7-1/2	5-1/2	2-3/4	13/32	2-3/4	1-1/8	2-1/2	1/4 × 1/8	**326TS**	S	12-1/2	12	5-1/4	21/32	5-1/2	2-1/8	5-1/4	3/8 × 3/16
213T	5-1/4	8-1/2	5-1/2	3-1/2	13/32	3-3/8	1-3/8	3-1/8	5/16 × 5/32	**364T**	9	14	11-1/4	5-7/8	21/32	6-1/4	2-3/8	5-7/8	5/8 × 5/16
215T	5-1/4	8-1/4	7	3-1/2	13/32	5-3/8	1-3/8	3-1/8	5/16 × 5-32	**364TS**	9	14	11-1/4	5-7/8	21/32	6-1/4	2-3/8	5-7/8	1/2 × 1/4
254T	6-1/4	10	8-1/4	4-1/4	17/32	4	1-5/8	3-3/4	3/8 × 3/16	**365T**	9	14	12-1/4	5-7/8	21/32	6-1/4	2-3/8	5-7/8	5/8 × 5/16
256T	6-1/4	10	10	4-1/4	17/32	4	1-5/8	3-3/4	3/8 × 3/16	**365TS**	9	14	12-1/4	5-7/8	21/32	6-1/4	2-3/8	5-7/8	1/2 × 1/4

(Source Reznor, Thomas and Betts Technical Form RGM 608-A, page 1, 1-800-695-1901; www.ReznorOnLine.com.)

14.20 POWER FACTOR (PF)

A motor's **power factor** is the ratio of motor load watts divided by volt-amps at the full-load condition. The power factor for a motor changes with its load. The power factor is at its minimum at no load and increases as additional loads are applied to the motor. The power factor usually reaches a peak at or near full load on the motor. Common power factor numbers are 0.8 or 0.9. A perfect power factor would be 1.0, which is not possible with the pure inductive load that is found in motors. Capacitors help create a power factor closer to 1.0.

14.21 TIME RATING OR DUTY

Standard motors are rated for continuous duty (24/7) at their rated load and maximum ambient temperature. The term for this is *time rating* or *duty* and it is found on the nameplate. Motors designed to be operated all the time are referred to as *continuous* and abbreviated **CONT**. If the running time is less than 60 minutes per hour, it will be listed on the motor nameplate. A noncontinuous motor must be turned off at the prescribed times or the motor will overheat and burn out prematurely.

14.22 BEARING TYPES

The bearing type may not be listed on the motor nameplate. Choosing the correct bearing type is important when selecting a replacement motor. When the load is heavy, the wrong bearing type will reduce the life of the motor. When noise is a consideration, ball bearings should not be selected.

Most of today's small, air-moving motors use a sleeve bearing system. The bearings may be either iron or bronze with self-aligning bearings or rigidly mounted steel bearings. Belt drive applications that apply a heavy radial load on the motor require ball bearings.

The following general types of bearings are commonly used in HVACR:

- **Self-aligning sleeve bearings**
- **Babbitt lined bearings**
- **Unit bearings**
- **Ball bearings**.

Self-aligning sleeve bearings are quiet, inexpensive, and have long life capability. They are typically found in motors that are 5.6 inches in diameter or less. They are designed for light, balanced loads and are made of either sintered iron or bronze.

Babbitt lined, rigid steel bearings are quiet, have a long life, and are capable of carrying heavy direct-drive air-moving loads. The bearings are pressed into a cast metal end shield and therefore dissipate heat better than self-aligning bearings can.

A unit bearing is a single sleeve extending into and supporting the rotor and shaft assembly. They are limited to use in low FHP motors with light load applications. Unit bearing motors are commonly used in commercial refrigeration evaporators.

Table 14-5 Sleeve Bearing Temperatures

This table shows how cooler bearing surfaces have longer sleeve bearing lives. The table is for a sleeve bearing that had a single oiling when it was manufactured.

Bearing Temperature		Expected Motor Life (single oiling)
104°F	40°C	100,000 hrs.
120°F	49°C	50,000 hrs.
140°F	60°C	40,000 hrs.
160°F	71°C	30,000 hrs.
180°F	82°C	20,000 hrs.

(Source Fasco Motors.)

Ball bearings are used in applications where heavy radial loads are encountered. Typical applications for small motors are belt drive units or those with unbalanced loads.

According to Fasco, a leading manufacturer of motors, *permanently lubricated* simply means the bearing has no provision for easy replenishment of the lubricant. In other words the bearings do not need external lubrication.

- Sleeve bearing motors may be replaced with ball bearing models if the additional noise of ball bearings is not a concern. However, ball bearing motors must be replaced with a similar ball bearing model. Table 14-5 shows the expected life of a sleeve bearing under various operating temperatures. Sleeve bearings are very common in fan motors.
- Under normal conditions, sleeve bearings will outlast ball bearings. In a well-designed and properly loaded sleeve bearing system, there is minimal if any metal-to-metal contact. As with many lubricated surfaces, the shaft is supported by a film of oil between the shaft and bearing.
- Ball bearings are maintenance free and beneficial in harsh environments. They are also useful in fan units where windmilling is a problem. In Table 14-6 you will notice that the life expectancy of a ball bearing is shorter than that of a sleeve bearing because they are placed in heavy load conditions. A ball bearing can handle a heavier load than can a sleeve bearing.

Finally, bearing wear on noncompressor motors can be detected by spinning the motor by hand (the exception to this is the ECM). If the shaft turns rough or the motor bearing makes a noise, there may be a problem. Make sure the motor bearings are greased or oiled per manufacturer's recommendations. Excess lubrication can damage the motor windings. The only way to detect internal compressor motor

Table 14-6 Ball Bearing Temperatures

Do not substitute ball bearing motors for sleeve bearing motors if noise will be a problem. The life expectancy of a ball bearing is shorter than that of a sleeve bearing. A ball bearing can handle a heavy load when compared to a sleeve bearing.

Bearing Temperature		Expected Motor Life
104°F	40°C	80,000 hrs.
120°F	49°C	40,000 hrs.
140°F	60°C	20,000 hrs.
160°F	71°C	10,000 hrs.
180°F	82°C	6,000 hrs.
200°F	94°C	4,000 hrs.
212°F	100°C	3,000 hrs.

(Courtesy of Fasco Motors.)

bearing wear is by doing an autopsy, which means opening the compressor. This may be done with semi-hermetic compressors after failure. This is difficult to do on hermetic compressors.

TECH TIP

Multispeed motors must be under a load to change speeds. For example, a multispeed blower removed from a blower compartment will run at high speed, no matter which speed tap is used.

14.23 SERVICE CHECKLIST

Take the following steps for more accurate motor replacement:

- Before removing the motor, scratch an arrow on the case of the motor indicating the direction of rotation.
- Record the voltage. In some cases it is not listed on the nameplate. Measure the voltage if you are uncertain.
- Record the number of speeds.
- Note the type of motor. Is it a PSC motor? Shade pole motor?
- Measure the shaft diameter. Does the motor have two shafts?
- What are the capacitor size(s)? What type(s) of capacitor are used?
- Provide all the nameplate data. Include all data.
- Provide special application considerations. The name of the OEM unit is important. The model number and serial number of the unit may help you find the exact motor replacement.
- Use the worksheet in Table 14-1 to organize your information.

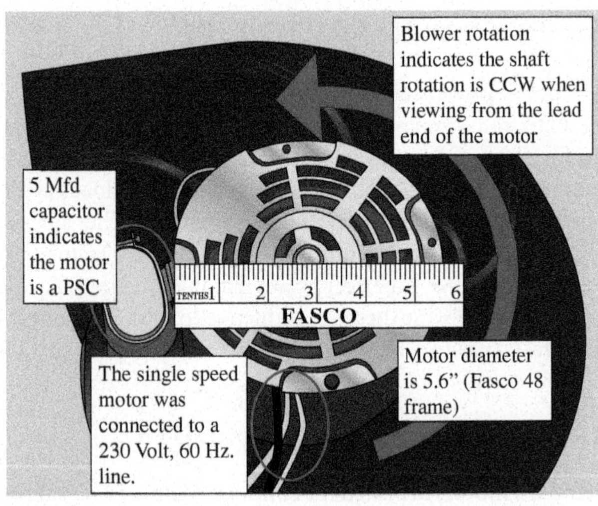

Figure 14-16 Some of the information that can be collected on a motor can be viewed in this common air-handling motor assembly. The motor is a PSC type since it has a run capacitor. Measure the incoming voltage. The motor diameter is 5.6 inches. The motor rotates CCW from the lead in. The motor leads will determine how many speeds are required on the replacement motor.

Figure 14-16 shows some of the information that can be derived by inspecting and measuring up an "**unknown soldier.**" Just a little information will help you select the correct motor when the nameplate is unreadable.

TECH TIP

An overloaded motor (i.e., one that is too small for the job) will have lower speed, an amperage draw above 10% of RLA, and will overheat.

An underloaded motor (i.e., one that is too big for the job) will have little change in speed and an amperage draw that is 25% below RLA, and it will overheat.

SERVICE TICKET

You have determined that the blower motor shown in Figure 14-17 is defective and needs to be replaced. Your job is to call the dispatcher so that a replacement motor can be ordered and delivered to your location as soon as possible. What information are you going to give the dispatcher to ensure that you receive the correct motor and get the air handler back on line quickly?

It is important to give the dispatcher as much information as you can in case an exact replacement is not available. With adequate motor information, it may be possible to select a substitute motor. Before you continue with this exercise, test your skills. Inspect the motor nameplate in Figure 14-17. Write down all the information you think you will need in order to get the correct motor, then read the solution.

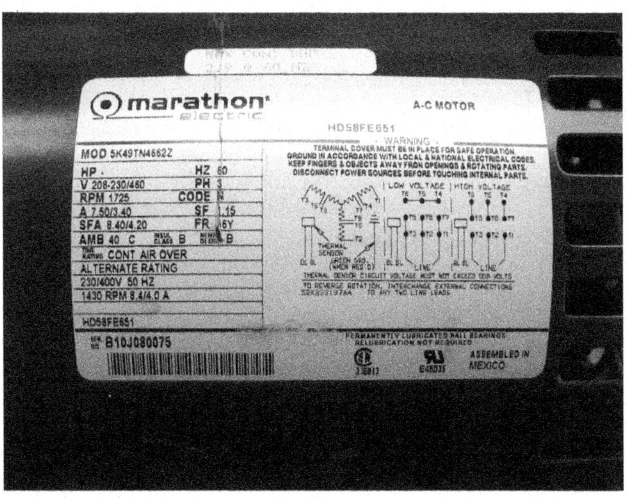

Figure 14-17 This type of motor might be found in a commercial application.

Here is the information you give the dispatcher:

1. Model #5K49TN4662Z
2. Hertz 60
3. Phase 3
4. Voltage 208/230 or 460
5. RPM 1725
6. Service Factor 1.15
7. Frame # 66Y.

With this information the dispatcher should be able to find you a direct replacement or an acceptable substitute.

TECH TIP

Regarding the Service Ticket discussed here, it is a best practice to test the motor amp draw after it is installed. The motor should be operated under normal conditions such as with the V-belts or blower installed on the motor. In the case of an air handler, all of the doors and covers should be installed on the unit. Not testing the motor installed under normal operating conditions will give you a false amp reading. For example, with air handler doors or panels missing, the blower will move more air and thus draw a higher amperage. You may think that the motor is defective if it is pulling excess amps. This condition may also overload the motor and cause it to overheat.

SUMMARY

The focus of this unit was to help you understand how motors work. We discussed how magnetism is used to get a motor moving in the right direction. Single- and three-phase motors are used in our industry. The nameplate will tell you the voltage type and many other pieces of information to determine motor operation and motor capabilities. It is important that you understand all of the information provided on the nameplate, but especially the voltage source, maximum amperage capabilities, RPM, shaft size and rotation, and frame number.

Learning the many features of a motor is beneficial when it comes to troubleshooting and replacing motors. Most techs do not realize what goes into a motor's design until it fails and they are not able to obtain an exact replacement. The air conditioning supply houses or motor suppliers can help you find the correct replacement. In some cases a motor is so specialized that only the original equipment manufacturer's (OEM) motor will do. Defective motors can be rewound and bearings replaced, but that usually takes a few days.

Twenty-five percent of motors returned under warranty are not defective. What are technicians thinking when they install a new motor and it does not work? Do they think they have a new defective motor? It could happen, but the chances are very slim. Try to check out the motor before it is totally installed. At least check the resistance of the windings. Damage can occur during shipping and it is best to know this before the motor is installed. Actually, checking the resistance at the supply house is the best practice, but is rarely done.

Replacing starting components will ensure long motor life. A weak capacitor or erratic starting relay may not be apparent when a new motor is replaced. The short life of the new motor will have you double checking the starting components. Yes, there will be a motor replacement warranty but it may not cover travel and labor costs. Some warranties require starting component replacements also.

REVIEW QUESTIONS

1. What is the purpose of NEMA?
2. When checking motor windings, do you select the high-resistance or low-resistance range on the ohmmeter? Explain.
3. What does the term *locked-rotor amps* (LRA) refer to?
4. A motor with an unknown horsepower rating needs to be replaced. A motor of the same type, current rating, voltage rating, and stack length will have nearly the same output. If the motor amperage is known, what is the maximum amperage rating that should be selected?
5. What is starting torque?
6. A motor has a nameplate voltage rating of 480 volts. How much can the voltage vary above and below the nameplate voltage?
7. A motor has a nameplate voltage rating of 208/240 volts. List the voltage range that this motor can operate at without major problems.
8. What causes the rotor of a motor to move?
9. What is the purpose of motor thermal protection?
10. What are four reasons a motor will create a nuisance trip on its overload?

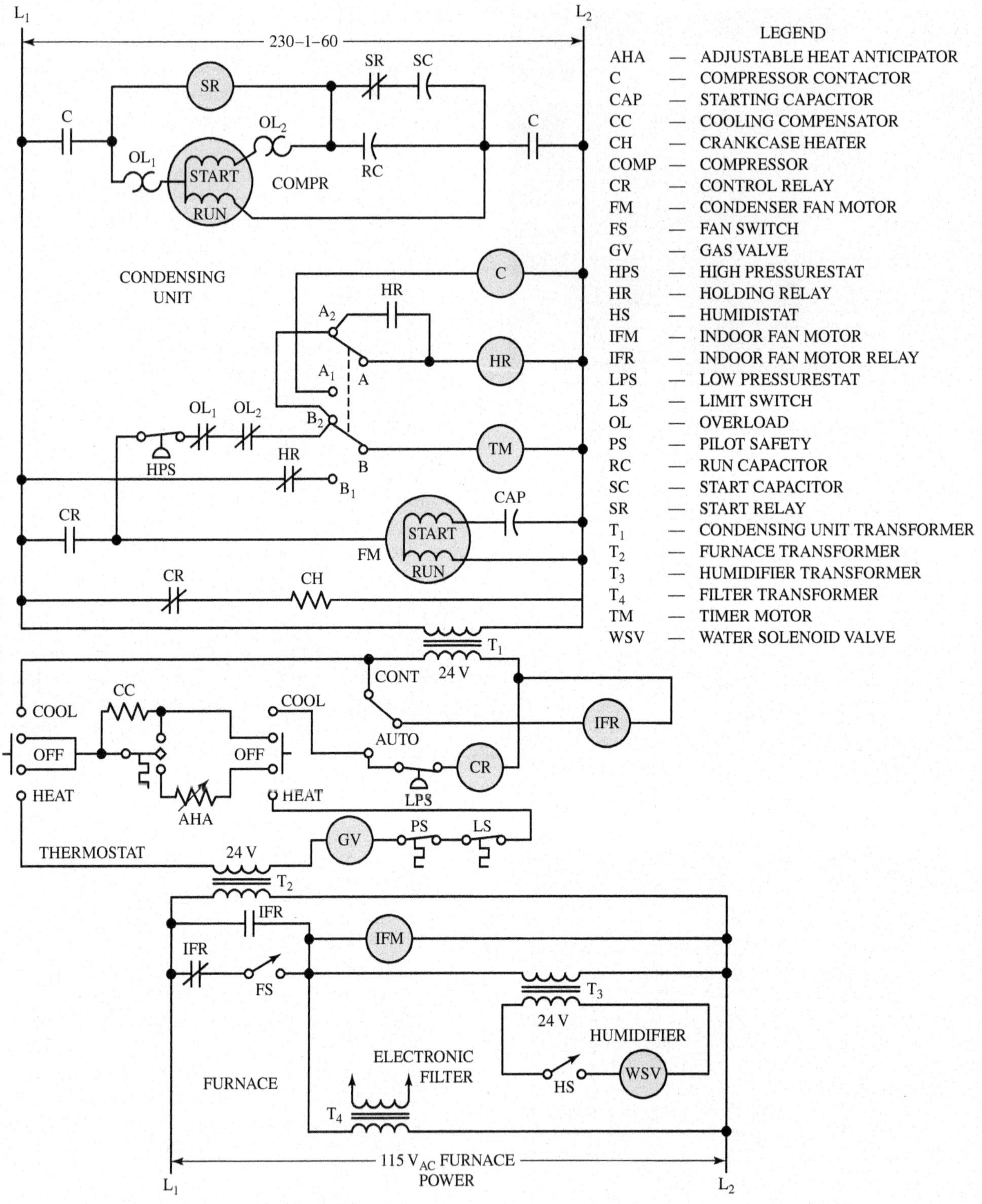
L1
L2
230–1–60
SR
SC
C
OL2
C
OL1
START
RUN
COMPR
RC
CONDENSING UNIT
HR
A2
A1
A
OL1 OL2
B2
HPS
HR
B
B1
CR
CAP
START
FM
RUN
CR
CH
T1
CONT
24 V
CC
COOL
COOL
OFF
OFF
AUTO
HEAT
HEAT
AHA
LPS
PS
LS
THERMOSTAT
24 V
T2
IFR
IFM
IFR
FS
T3
24 V
HUMIDIFIER
ELECTRONIC FILTER
FURNACE
HS
T4
115 VAC FURNACE POWER
LEGEND
AHA — ADJUSTABLE HEAT ANTICIPATOR
C — COMPRESSOR CONTACTOR
CAP — STARTING CAPACITOR
CC — COOLING COMPENSATOR
CH — CRANKCASE HEATER
COMP — COMPRESSOR
CR — CONTROL RELAY
FM — CONDENSER FAN MOTOR
FS — FAN SWITCH
GV — GAS VALVE
HPS — HIGH PRESSURESTAT
HR — HOLDING RELAY
HS — HUMIDISTAT
IFM — INDOOR FAN MOTOR
IFR — INDOOR FAN MOTOR RELAY
LPS — LOW PRESSURESTAT
LS — LIMIT SWITCH
OL — OVERLOAD
PS — PILOT SAFETY
RC — RUN CAPACITOR
SC — START CAPACITOR
SR — START RELAY
T1 — CONDENSING UNIT TRANSFORMER
T2 — FURNACE TRANSFORMER
T3 — HUMIDIFIER TRANSFORMER
T4 — FILTER TRANSFORMER
TM — TIMER MOTOR
WSV — WATER SOLENOID VALVE

Figure 14-18 Use this diagram to answer Review Question 19.

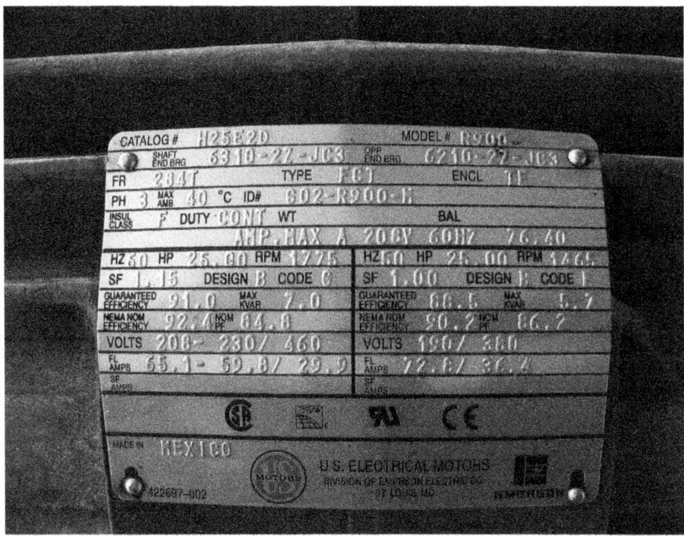

Figure 14-19 Use this nameplate to answer Review Question 20.

11. You are doing scheduled maintenance on an air conditioning system. One of the requirements is to record the amperage draw of the blower motor. The nameplate has the FLA and LRA. Which amperage do you use to indicate that the motor is operating properly?
12. On a multispeed single-phase motor, compare the high-speed and low-speed winding resistance. Which speed has the higher resistance?
13. A single-phase motor's nameplate is unreadable. How do you determine the motor speed?
14. To purchase a replacement motor, you need to determine the frame size or frame number. What is the frame size for a motor that measures 3.5 inches from the middle of the motor shaft to the motor base?
15. You are troubleshooting a motor and suspect a short to ground. What ohmmeter range do you use to test it?
16. Why is knowing a motor's insulation class and rated ambient temperature important?
17. Which type of motor bearing will have a longer life under a heavy load?
18. What is the advantage of a motor that has a higher service factor?
19. How many motors are shown in Figure 14-18? (You may also use enlarged Electrical Diagram ED-5 to answer this question.)
20. What is the phase of the motor described by the nameplate shown in Figure 14-19? What is its voltage?
21. What will cause excessive motor amp draw?

UNIT 15
Motor Types

WHAT YOU NEED TO KNOW

After studying this unit, you will be able to:

1. List five types of single-phase motors used in HVACR.
2. Describe how a shade pole motor operates.
3. Describe how a capacitor start/induction run motor operates.
4. Describe how a permanent split capacitor motor operates.
5. Describe how a capacitor start/capacitor run motor operates.
6. Describe how a three-phase motor operates.
7. Explain how to troubleshoot single-phase motors.
8. Explain how to troubleshoot three-phase motors.

This unit discusses the common motors you will find in the HVACR industry. It is important to gain an understanding of each type of common motor so that you will be able to recognize whether it is working properly. Some of the motors have start or run components—in other words, the motor requires external components to operate.

We will discuss common motors used to operate condenser fans, blower fans, and water pumps. It is important to thoroughly understand how to check single-phase and three-phase compressor motors. Many single-phase motors need external components such as capacitors and/or relays. Three-phase motors do not require special components. They can operate when power is applied through a contactor or motor starter.

This unit discusses troubleshooting of each type of motor and its associated components. In many instances the motor is not defective, but the associated start or run component is the problem. A defective or weak capacitor may keep the motor from starting or running correctly. A defective start relay may prevent the motor from starting. Wiring these components incorrectly may damage the motor or even prevent the motor from starting. Low- or high-voltage conditions can shorten motor life.

15.1 TYPES OF MOTORS

This section introduces the various types of motors used in our industry. We will start with the simplest and advance to motors that are more complex. Here is the order in which we'll make our motor discoveries:

Single-Phase Motors

- **Shaded pole motor**
- **Split-phase motor**
- **Capacitor start/induction run motor**
- **Permanent split capacitor motor**
- **Capacitor start/capacitor run motor.**

Three-Phase Motors

- **Delta three-phase motor**
- **Wye three-phase motor.**

An electronically commutated motor (ECM) is considered a "green" motor because of its lower energy use and variable-speed capabilities. Because ECMs are one of the most complex types of motors, they have been given a unit of their own, Unit 16.

15.2 SHADED POLE MOTORS

The shaded pole motor is the most basic motor used in HVACR equipment. Shaded pole motors are used in small direct-drive fans and blowers in heating and ventilating equipment where low efficiency and low start torque are acceptable. They do not have capacitors and they are normally fractional horsepower (FHP) motors, which means they are less than 1 hp in size. Here are some examples of fractional horsepower motors: 1/10, 1/4, 1/3, 1/2, and 3/4 hp.

Figure 15-1 illustrates that the shaded pole motor has coils of varnished copper wire wound through a stack of

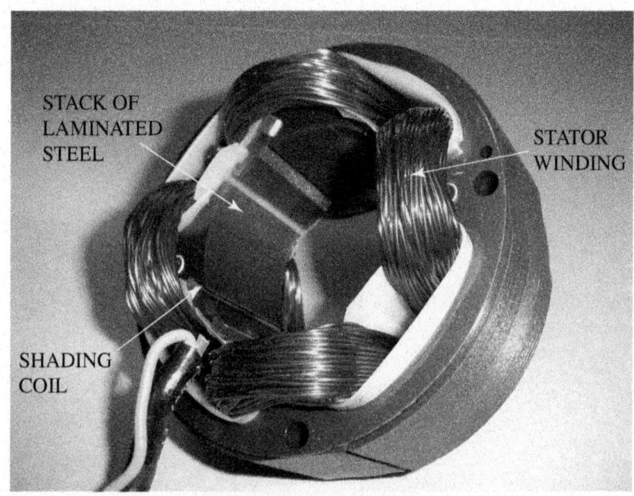

Figure 15-1 The stator windings are woven through the stack of laminated steel. This is a four-pole, shaded pole motor.

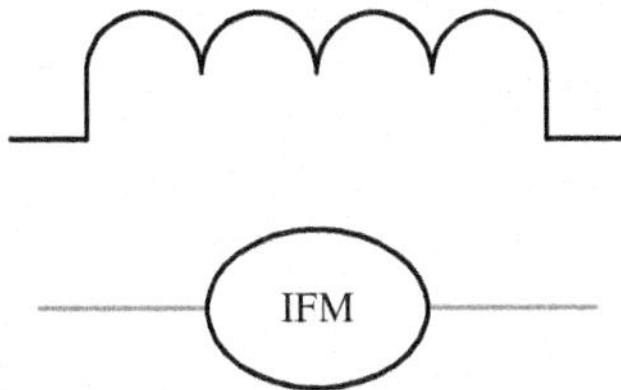

Figure 15-2 These are electrical diagram symbols for a shaded pole motor. These symbols are also used for a coil in a relay, contactor, or solenoid. If the circular symbol is used for something other than a shaded pole motor, there will be no letters inside the circle or they will have different letters such as R or C.

steel laminated plates, which make up the stator winding. The motor shown here is a four-pole motor. The coils are wired in pairs. The greater the number of paired coils, the lower the speed of the motor, which is stated in revolutions per minute (RPMs). The RPM measurement indicates how many complete motor shaft turns are made per minute. The electrical diagram for a shaded pole motor is simply a coil or circle as shown in Figure 15-2. When the circular symbol is used, sometimes it is labeled with initials like "IFM" for indoor fan motor or "CFM" for condenser fan motor.

The shaded pole motor has a shading coil as illustrated in Figure 15-1. The purpose of the shading coil is to concentrate and focus the magnetism in the coil and steel plates to setpoints in the stator winding. Let's explain how that works.

The 60-Hz voltage produces a positive and negative sine wave, which attracts (pulls) and repels (pushes) the motor rotor 60 times per second. Like most motors, this is how the motor creates its motion and direction of rotation. How does the shaded pole motor know which direction to turn? The magnetic field builds up in the stator when the voltage sine wave moves toward a positive peak or when it moves toward a negative peak. Once the magnetic field reaches a positive or negative peak, the magnetic field collapses since the supplied voltage is dropping. The collapsing magnetic field concentrates the stator magnetic field into one spot (the shading coils) on the rotor winding, as shown in Figure 15-3. The motor turns in the direction of the shading coil or in the direction of the concentrated magnetic field. Even if the motor is not operating, you can determine the direction of rotation by removing the end bell of the motor and inspecting the shading coil.

Figure 15-4 illustrates a simple shaded pole motor. The motor windings are in series and wrapped around each of the four poles. The slit holding the shading coil concentrates the magnetic lines of flux and causes the motor to turn counterclockwise.

Figure 15-5 is a diagram of an air conditioning package unit that has two shaded pole motors. The (CFM) and (IFM) circles indicate shaded pole motors. The (CFM) is the condenser fan motor and the (IFM) is the indoor fan motor. They are drawn with the initials inside a circle. A circle with or without initials could also represent a coil. One way to determine that this is a motor is by looking at the legend, as seen on the right side of diagram. The motors do not have a capacitor, therefore they have to be shaded pole motors.

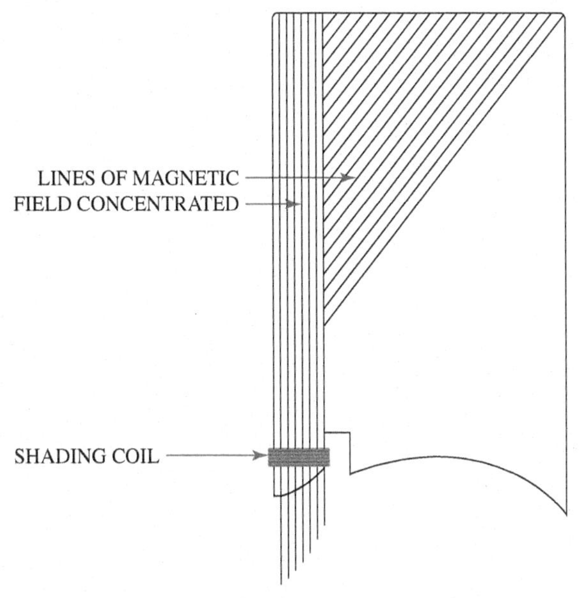

Figure 15-3 The magnetic field is concentrated toward the shading coil to force the motor in a counterclockwise direction. The rotor will turn in the direction of the shading coil.

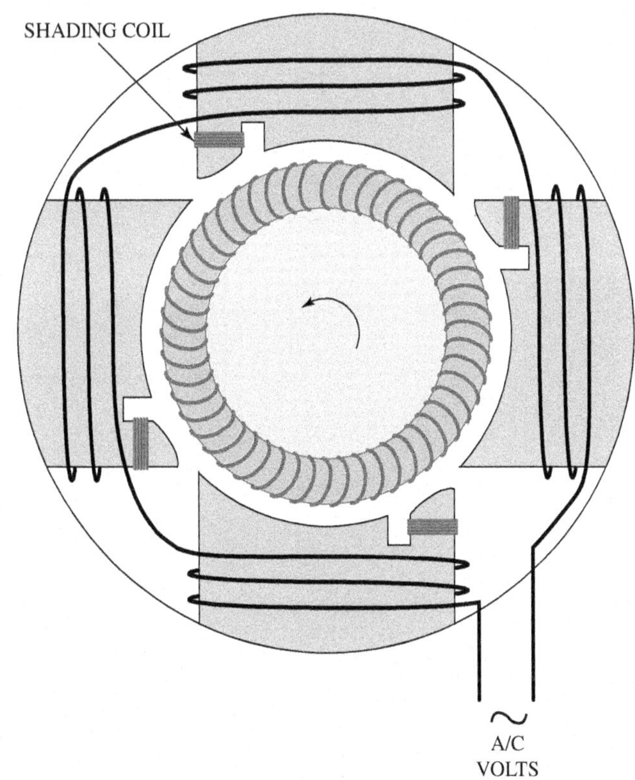

Figure 15-4 The rotor will turn in the direction of the shading coils. The shading coils concentrate the magnetic energy into one area of the stator poles and create the rotation of the motor, in this example counterclockwise.

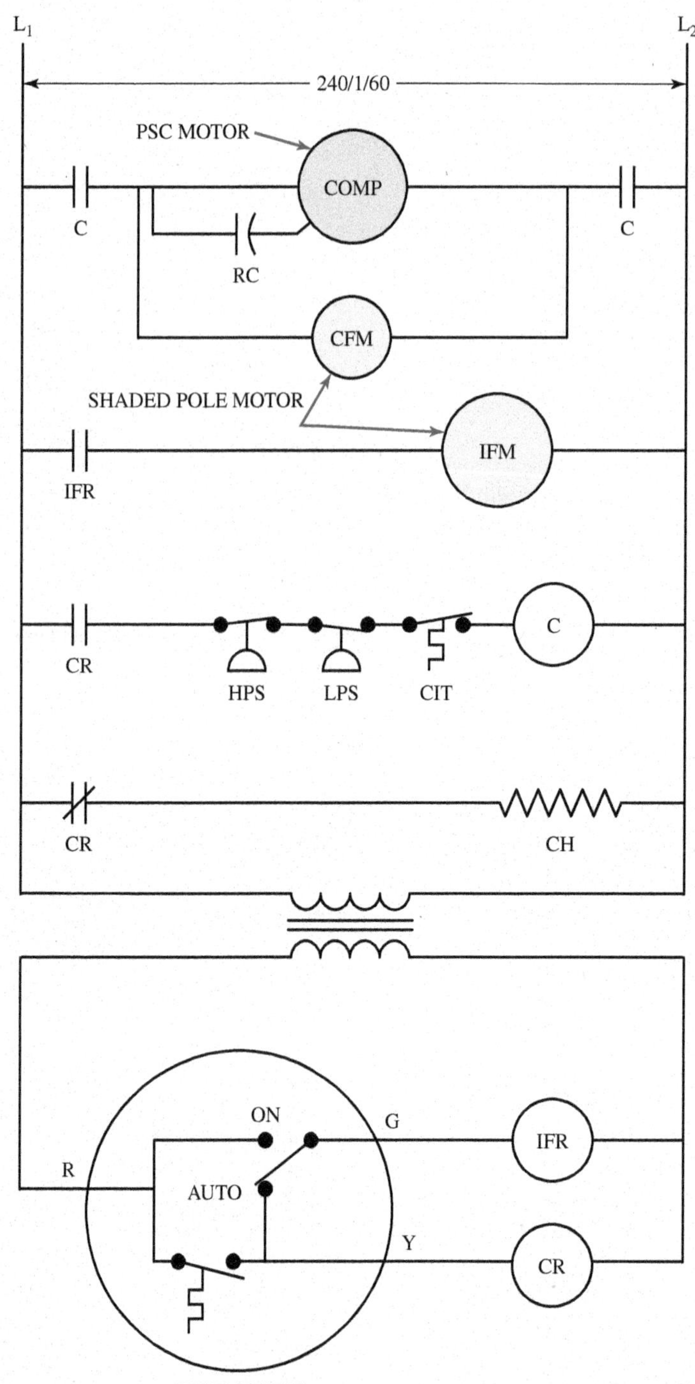

Figure 15-5 This single-phase air conditioning diagram is used to point out the shaded pole motor and PSC motor diagrams.

Let's compare the advantages and disadvantages of the shaded pole motor:

Advantages

- Designed for constant operation
- Well suited for multispeed applications
- Low starting current
- Very high reliability
- Very low cost

Disadvantages

- Low efficiency (25% to 40%)
- Not easily reversed
- Low starting torque
- Only useful for small loads

15.3 TROUBLESHOOTING THE SHADED POLE MOTOR

Because it is so simple, without any external components, the shaded pole motor is the easiest motor to troubleshoot. It has one set of windings, therefore a continuity check is in order. Resistance is usually less than 100 Ω through the motor winding. An open winding would read infinity. The resistance reading from the motor lead to ground should be infinity or a very high resistance reading.

Note: Your ohmmeter should be able to read at least a million ohms. Some multipurpose clamp-on ammeters with a resistance feature have a limited resistance range. Some of these multipurpose meters can only read as high as 1,000 ohms. This would not be a good meter for checking for shorts to ground or for checking the coil resistance of some gas valves or relays. Know the limits of your instruments or you will misdiagnosis a perfectly good motor.

Finally, if the bearings are good, the rotor should turn freely. Give it a spin.

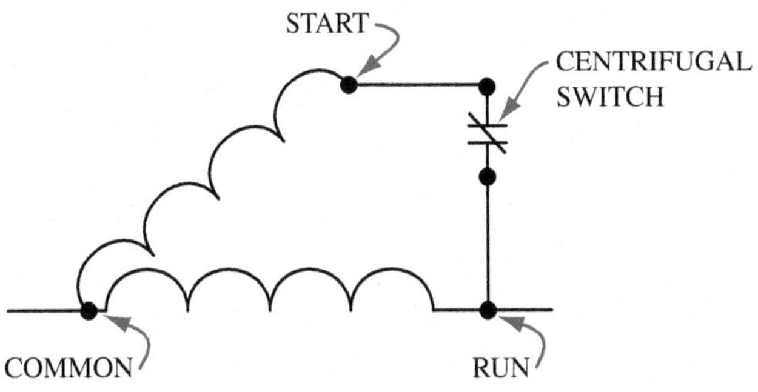

Figure 15-7 This schematic is for a split-phase motor with a centrifugal switch. The centrifugal switch is used to remove the start winding when the motor reaches 75% of full motor speed.

15.4 SPLIT-PHASE MOTORS

The split-phase motor has two sets of windings as shown in the electrical symbol in Figure 15-6. One winding is used to start the motor, and the other winding is used to keep it running. A split-phase motor has a start winding and a run winding designed electrically to cause a phase-splitting effect on the incoming current and voltage; thus the name *split phase*. The phase shift allows these motors to develop higher starting torque when compared to a shaded pole motor. As the motor reaches load speed, a switch cuts power to the start winding, allowing the motor to operate continuously on the run winding only. Load speed is about 75% of the motor RPMs; at that speed the start winding is taken out of the circuit. The start winding is only in the circuit for a second or less. Figure 15-7 is a diagram of a centrifugal switch in series with the start winding. The centrifugal force of the turning motor opens the switch and removes the start winding from the circuit.

You might hear the centrifugal switch "click" when the motor is shut down. This is the sound of the centrifugal switch closing. When this happens the start winding becomes available should the motor start up immediately.

Split-phase motors are used in mechanical duty equipment that requires improved starting torque and moderate running efficiency when compared to the shaded pole motor. Typical applications include laundry equipment, conveyor systems, machine tools, belt-drive blowers, and water pumps.

Figure 15-8 is an electrical diagram of a split pole motor that uses a current relay to remove the start winding. Figure 15-9 is a picture of a current relay. Let's discuss how this relay operates to energize the start winding, and then quickly remove the start winding from the circuit.

Current relays are used on small air conditioning and refrigeration appliances. The current relay has a heavy-duty coil and a set of normally open contacts. The contacts are fairly loose, therefore the current relay must be mounted so that the contacts are held open by gravity. The current relay is installed upward, or pointed up in the direction of the arrow shown on the side of the relay. The relay is directional. Installing the current relay upside down will close the relay contacts, which is not what is wanted for this type of motor.

In summary, the current relay coil is in series with the main (run) winding. As the run winding is energized,

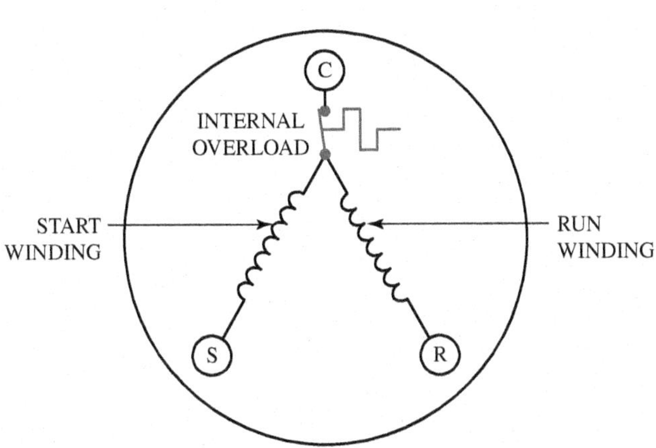

Figure 15-6 A split-phase motor has start and run windings. Some motors have internal overloads to protect the winding from high-amperage or overheating conditions.

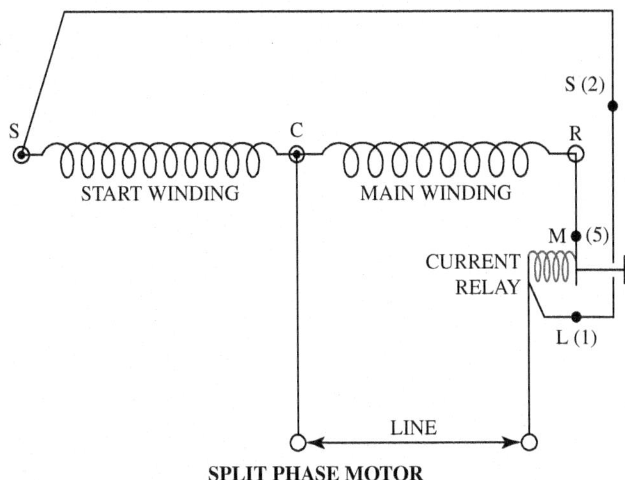

Figure 15-8 The current relay coil is in series with the main (run) winding.

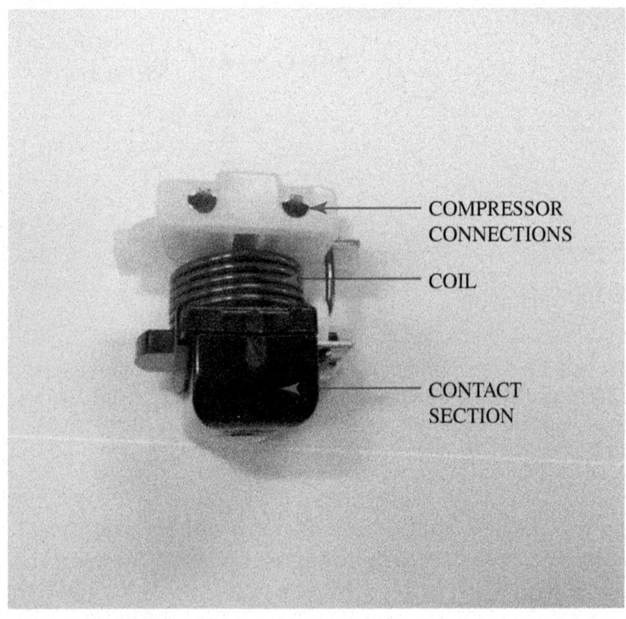

Figure 15-9 The current relay coil is constructed of heavy-gauge wire that is in series with the run winding. LRA creates a short burst of magnetism that closes a set of contacts for a moment, thus energizing the start winding. The contacts are located directly above the coil.

it draws locked rotor current (RLA). The temporary high-current condition causes the magnetic energy in the current relay coil to close the relay contacts for a moment. The closing of the contacts between points M(5) and L(1) for a second causes the start winding to be energized long enough to start the split-phase motor. The contacts quickly open as the magnetic field created by LRA is reduced.

You will find this motor design on small appliances such as domestic refrigerators, freezers, and under-the-counter ice makers. In many cases the current relay slips over the run and start terminal of the compressor. Power is then run to the common and run terminal on the current relay.

Here is a comparison of the advantages and disadvantages of the split-phase motor:

Advantages	**Disadvantages**
■ Medium motor efficiency	■ Possible damage to start windings caused by frequent starting or high inertial loads
■ Starting torque up to 200% load torque	■ Very high starting current
■ High reliability	■ Limited to multiple
■ Low cost	speeds or low speeds

SEQUENCE OF OPERATION 15.1: SPLIT-PHASE MOTOR

1. In Figure 15-10 the current relay coil is in series with the main (run) winding. The coil is pointed out by the blue arrow. The current relay coil is between points (5) and L(1).
2. In Figure 15-11, as the main winding is energized, it draws locked rotor current (RLA).
3. In this case, the relay contacts between points L(1)(5) and S(5) close for a second, causing the start winding to be energized long enough to start the split-phase motor. The LRA condition lasts for a split second, allowing the contacts to snap closed then back open. The closed relay contacts are pointed out by a red arrow.
4. This split second condition is long enough to energize the start winding and get the motor rotating. This particular design has a matched current relay for the motor. The run winding will keep the motor running once it is started. There is no run or start capacitor in this design.

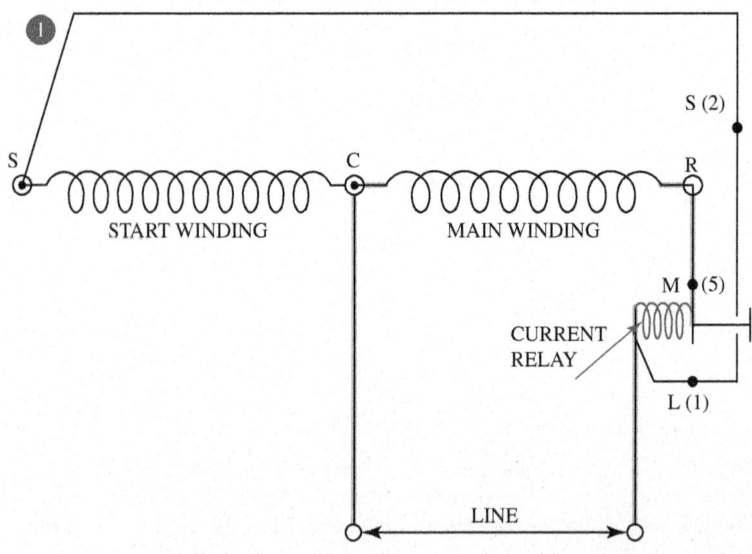

Figure 15-10 Tracing the yellow path shows the sequence of operation when the compressor is first energized. During the first split second the system is energized, the compressor will develop locked rotor amps in the main (run) winding and in the current relay coil.

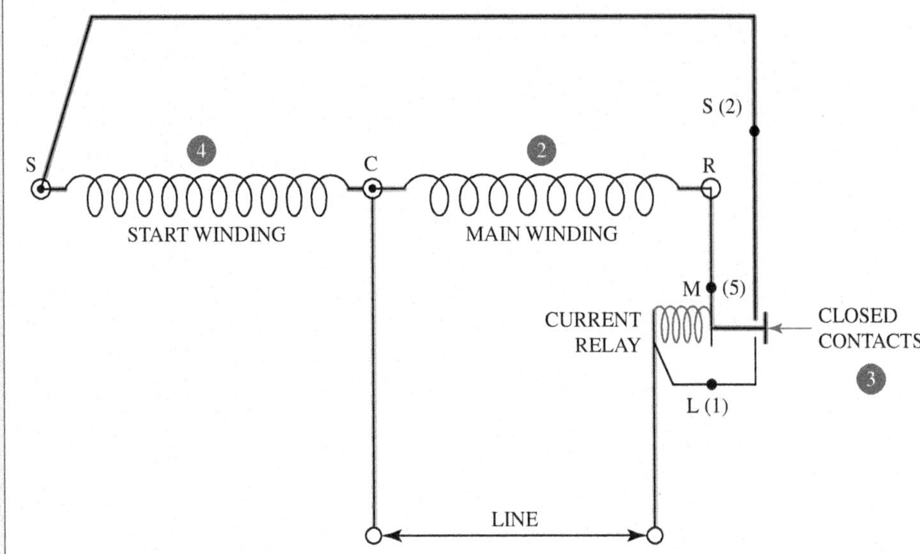

Figure 15-11 Drawing LRA for about a second or less will cause the current relay contacts to close momentarily to energize the start winding. The start winding will be taken out of the circuit to prevent damage to it. The relay contacts will reset and reopen after this brief operation. The current relay coil stays in series with the main winding throughout the compressor's operating period.

15.5 TROUBLESHOOTING THE SPLIT-PHASE MOTOR

The split-phase motor has two sets of windings. The resistance of each set of windings will not be equal. The resistance of the start winding will be several times greater than that of the run winding. Resistance to case ground should be infinity or a very high resistance. Finally, the rotor should turn freely and smoothly unless it is carrying a large load such as a pump or large blower wheel.

Check the current relay. The coil resistance will be near 0 Ω since it is a large coil that is very short in length. The contacts should be open when positioned correctly. Check the closed contacts' resistance by turning the relay upside down. It should have very low resistance. Shake the current relay. The contacts should rattle, meaning that contacts are loose and not stuck open or welded closed.

Replace the current relay with the correct match when changing out the motor. This is one of the "good practices" and in many cases a manufacturer's requirement.

15.6 CAPACITOR START/INDUCTION RUN MOTORS

Capacitor start/induction run (CSIR) motors are the next step up from the split-phase motor. The capacitor start motor is a split-phase motor with a properly sized start capacitor in series with the start winding. The motor must have a device that can quickly remove the start capacitor (or *start cap*) and start winding after motor start-up. A current relay, potential relay, or centrifugal switch is commonly used to do this.

The CSIR design improves the starting torque, or the ability to start a motor with a load on it, such as a heavy fan blade or water pump. The start capacitor also helps in reducing starting current, known as locked rotor amps or LRA. Like a split-phase motor, a "Cap Start" is designed with a run winding and a start winding displaced electrically to cause a phase-splitting effect. As the motor reaches load speed, a switch or relay cuts power to the start winding, allowing the motor to operate continuously on the run winding only.

The split-phase motor shown previously in Figure 15-6 can easily become a CSIR motor. This is accomplished by installing the properly sized start capacitor in series with the start winding. A switching device is required to remove the start cap.

Figure 15-12 shows the electrical diagram for a CSIR motor with a centrifugal switch. The starting switch that removes the start capacitor and winding may be a centrifugal switch, potential relay, or current relay. Figure 15-13 shows a CSIR motor that has a centrifugal switch for removing the start winding. The motor has been disassembled to show its internal construction.

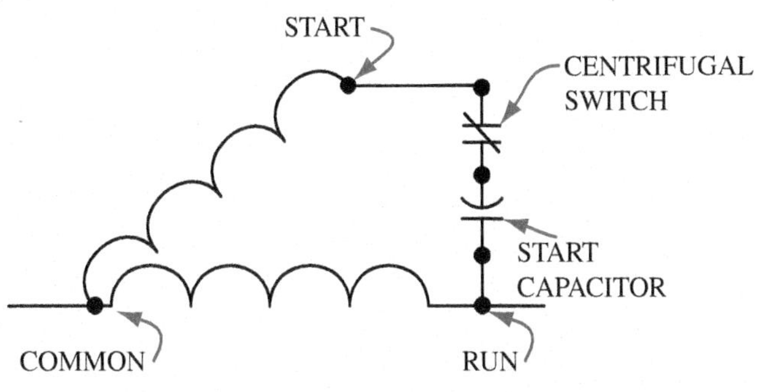

Figure 15-12 In this CSIR schematic diagram, the centrifugal switch is used to remove the start capacitor and start winding when the motor reaches 75% of full motor speed.

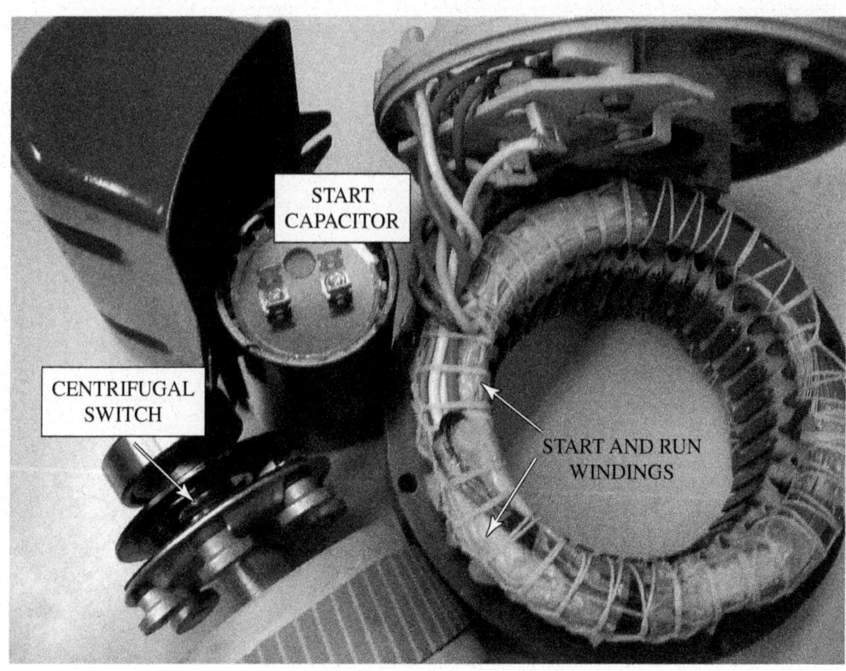

Figure 15-13 This split-phase motor uses a centrifugal switch and start capacitor. The start capacitor gives the motor better starting torque. The centrifugal switch is used to remove the start capacitor from the start winding a split second after the motor starts. The centrifugal switch will open at 75% of motor speed.

SEQUENCE OF OPERATION 15.2: CSIR MOTOR

1. A common current relay used to remove the start circuit is shown in Figure 15-14. The current relay coil is in series with the run winding as pointed out by the blue arrow. As the run winding is energized, it draws locked rotor current (LRA).
2. The red arrow points indicates where the temporary high-current condition causes the magnetic energy in the current relay coil to close the normally open contacts for a moment. The closing of the contacts between points M(5) and L(1) for a second causes the start winding to be energized long enough to start the split-phase motor. Notice in the starting sequence that the start capacitor, as pointed out by the black arrow, is in series with the start winding. This creates good starting torque for the motor. The contacts quickly open as the magnetic field is reduced.
3. The contacts are closed long enough to energize the motor and capacitor without damaging the start motor winding or start cap. This usually takes place in less than a second.

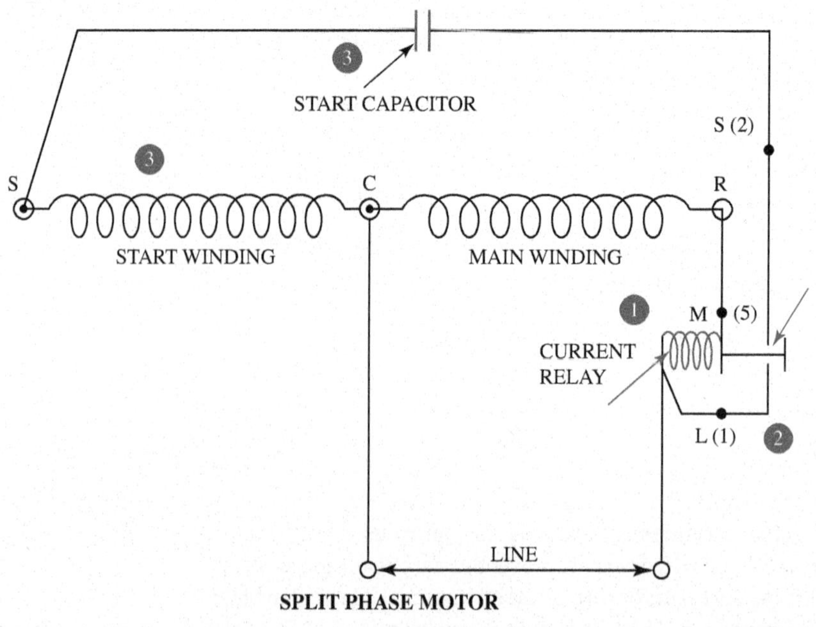

Figure 15-14 A current relay coil and start capacitor are used with the CSIR motor.

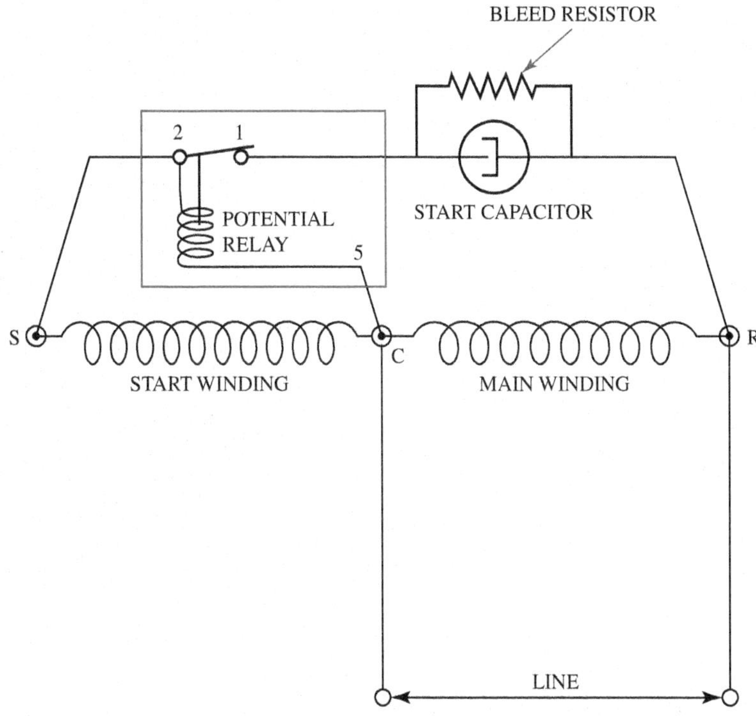

Figure 15-15 This CSIR diagram uses a potential relay with the start capacitor. The potential relay is outlined in red. This is not as common as the current relay or centrifugal switch design. The potential relay and start capacitor must be sized for maximum starting torque.

The current relay and start capacitor are sized for a specific motor. Start components have a limited life and should always be replaced when replacing any motor.

In summary, the current relay coil is in series with the run winding. As the run winding is energized it draws locked rotor current (RLA). The temporary high-current condition causes the magnetic energy in the current relay coil to close the open contacts for a moment. The closing of the contacts between points M(5) and L(1) for a second causes the start winding and start capacitor to be energized long enough to start the split-phase motor. The contacts quickly open as the magnetic field is reduced.

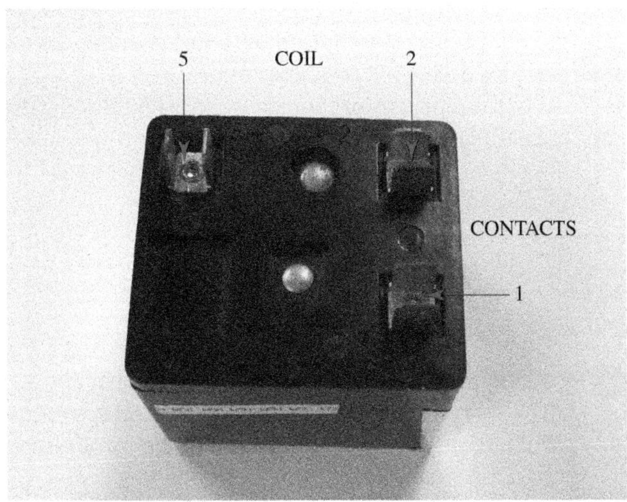

Figure 15-16 This potential relay is used to remove a start capacitor from the motor's start winding circuit. The relay uses connections 5, 2, and 1 as labeled on the relay. Sometimes the relay has extra terminals that are not used except for the placement of a wire connection.

Figure 15-15 shows the diagram for a CSIR motor that uses a potential relay to remove the start cap and start winding from the circuit. As shown in Figure 15-16, the potential relay has connection numbers 5, 2, and 1. Between connections 5 and 2 is a high-resistance coil, usually thousands of ohms of resistance. Connections 2 and 1 make up a set of normally closed contacts. Figure 15-17 is a potential relay with the cover removed so that you can see the contacts and coil. The coil is energized by high voltage, called *back electromotive force* or *back EMF*, created by the motor windings when they are in motion. On motor start-up the high voltage on the potential relay coil will energize and open the normally closed contacts. This will remove the start cap and start winding. We will see this circuit again when the capacitor start/capacitor run motor is discussed.

Here is a comparison of the advantages and disadvantages of the CSIR motor:

Advantages	Disadvantages
■ Medium motor efficiency	■ Possible damage to start windings caused by frequent starting or high inertial loads
■ Starting torque up to 200% load with moderate starting amps	■ Not easily reversed (standstill only)
■ Limited to multiple speeds or low speeds	■ Reliability reduced by additional components

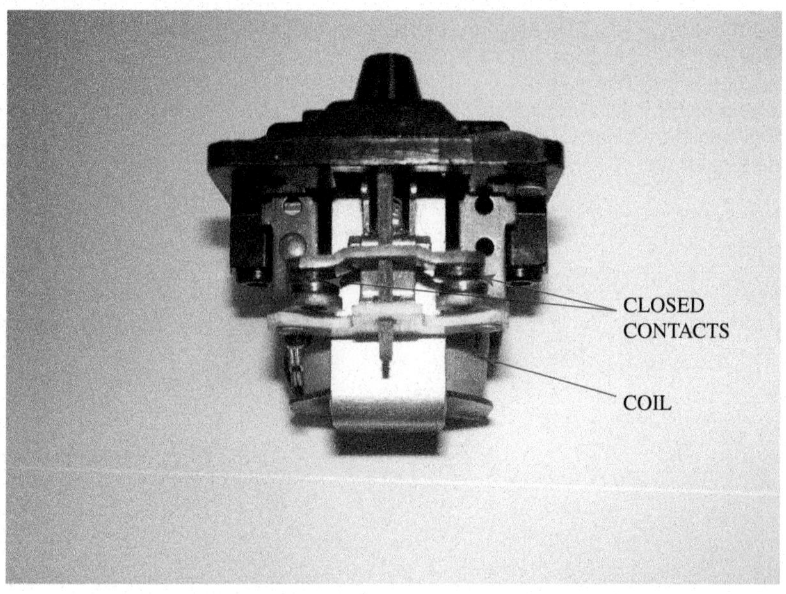

Figure 15-17 The cover has been removed from this potential relay to expose the contacts and coil. The coil is between connections 5 and 2. The closed contacts are between 2 and 1.

15.7 TROUBLESHOOTING THE CAPACITOR START MOTOR

The capacitor start motor is also called a split-phase motor. The windings should have resistance. The start winding will have a greater resistance than the run winding. Check the winding to the motor case ground. Resistance should be infinity or extremely high.

Next, check the start capacitor. It should be in the range as listed on the capacitor shell. Most start capacitors have a range of capacitance. For example, the capacitor label may read 88 to 108 mfd. Use a capacitor checker to determine if it is within the listed range. An oversized or undersized capacitor may prevent the motor from starting. These motors are designed to have a specific capacitor. Use it.

The centrifugal switch or current relay should also be checked. Before checking the centrifugal switch, check the rating of the start capacitor. Attach a clamp-on amp meter to the start winding circuit. When the motor starts the amp meter will register an amp spike for a split second. If the meter does not show an amp spike, the centrifugal switch or current relay is defective. Some low-end clamp-on ammeters cannot register a short amp spike. It pays to have good instruments.

Check the current relay. The current relay coil will have less than an ohm of resistance. The contacts should close and open when it is turned upside down and inverted back to its normal position.

If a potential relay is used, check the coil and normally closed contacts. The coil resistance will be 3 to 7 kΩ and the closed contacts will be near 0 Ω of resistance. Check the operation of the potential relay contacts by measuring the amperage between terminal 1 and the start capacitor. The amp draw should quickly jump up and drop out after the motor is started. This occurs in about a second.

Finally, the motor rotor should turn freely, indicating no rotor or bearing damage.

TROUBLESHOOTING 15.1: CAPACITOR START MOTOR

We will troubleshoot a CSIR compressor motor similar to the one shown in Figure 15-18.

1. Check the run winding as shown in Figure 15-19.
2. Check the start winding as shown in Figure 15-20. The start winding is good as indicted by the resistance, which is many times higher than that of the run winding.
3. Check for a grounded compressor as shown in Figure 15-21. No resistance is measured.
4. Measure the capacitance of the start capacitor as shown in Figure 15-22 to see if it is equals the correct microfarad rating. Remove one side of the bleed resistor and reconnect after testing the cap.
5. Check the current relay coil as shown in Figure 15-23. Coil resistance should be very low. Also check the contacts by turning the relay upside down and right side up.

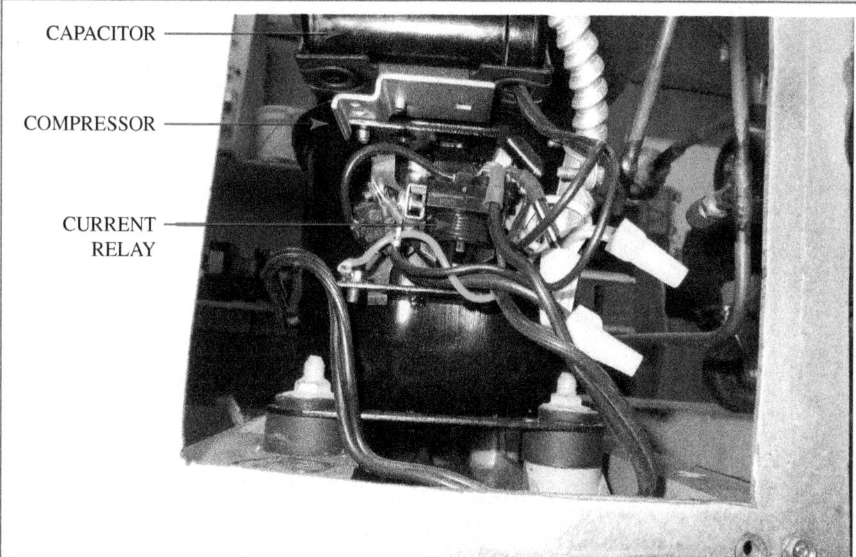

Figure 15-18 Capacitor start/induction run compressor motor.

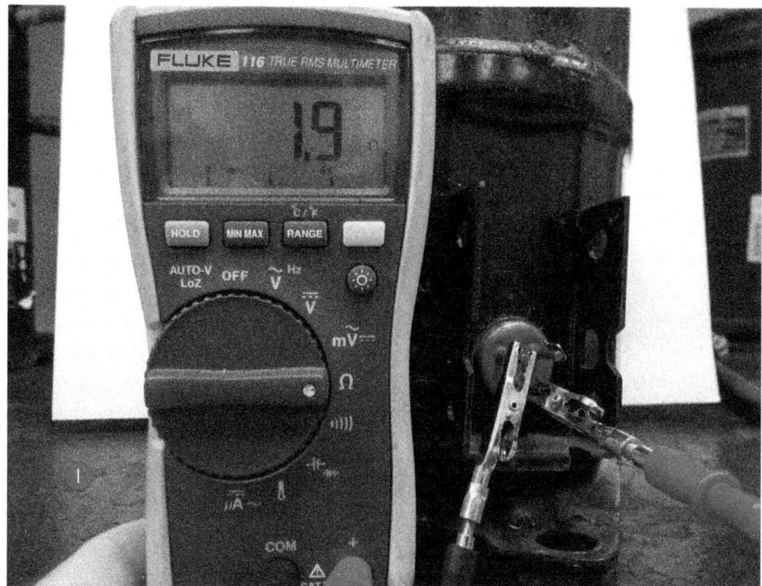

Figure 15-19 Checking the compressor motor run winding.

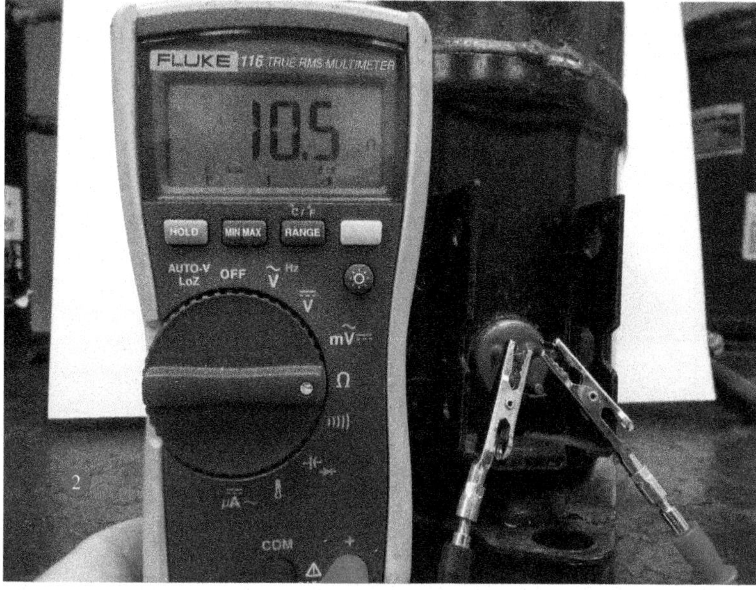

Figure 15-20 Checking the compressor motor start winding.

(*continued*)

Figure 15-21 Checking the motor winding to ground.

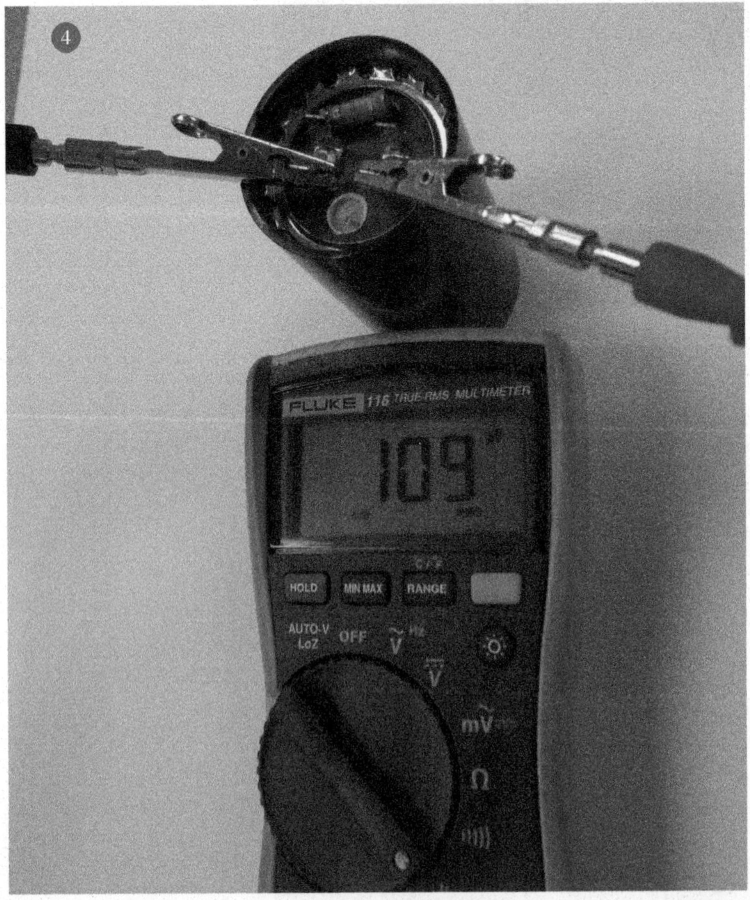

Figure 15-22 Checking the start capacitor with the bleed resistor disconnected.

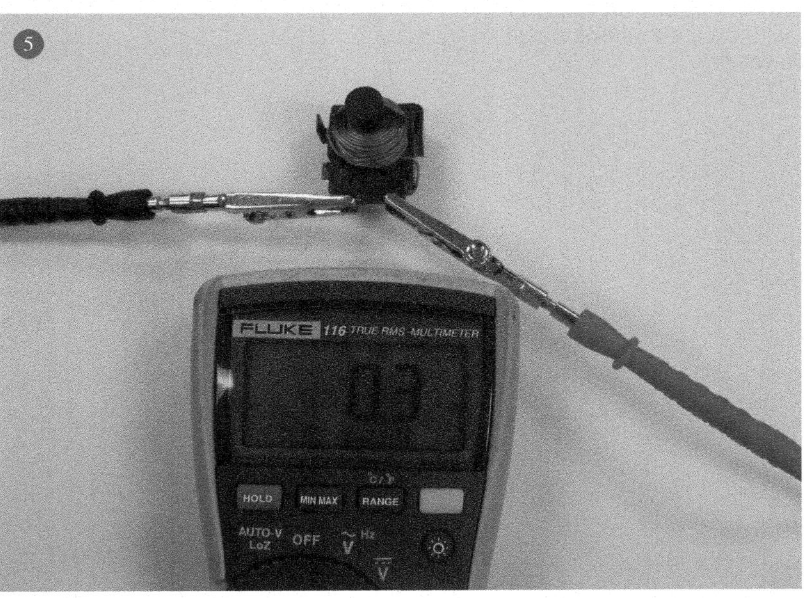

Figure 15-23 Checking the coil resistance of the current relay.

15.8 PERMANENT SPLIT CAPACITOR MOTOR

The permanent split capacitor (PSC) motor is the motor choice for direct-drive fans and blowers in heating, air conditioning, and ventilating equipment. PSC motors have also gained popularity in small pump and compressor applications. Figure 15-24 shows a common PSC motor. Most single-phase compressor motors are PSCs.

A PSC motor includes a main (run) winding and a start winding. A run capacitor is connected in series with the start winding. The PSC has good starting torque and good running efficiency. The PSC motor run capacitor improves the starting torque over a noncapacitor-type motor. The main purpose of the PSC motor is to improve running efficiency, which means lower running amperage draw. A defective run capacitor may or may not allow a PSC to start. A defective capacitor will definitely cause high running amperage.

Figure 15-24 These are the windings of a PSC fan motor. The start winding is the copper-colored wired located closest to the rotor. The run winding is green and located outside the start winding. The two brown wires go to the run capacitor as seen on the right.

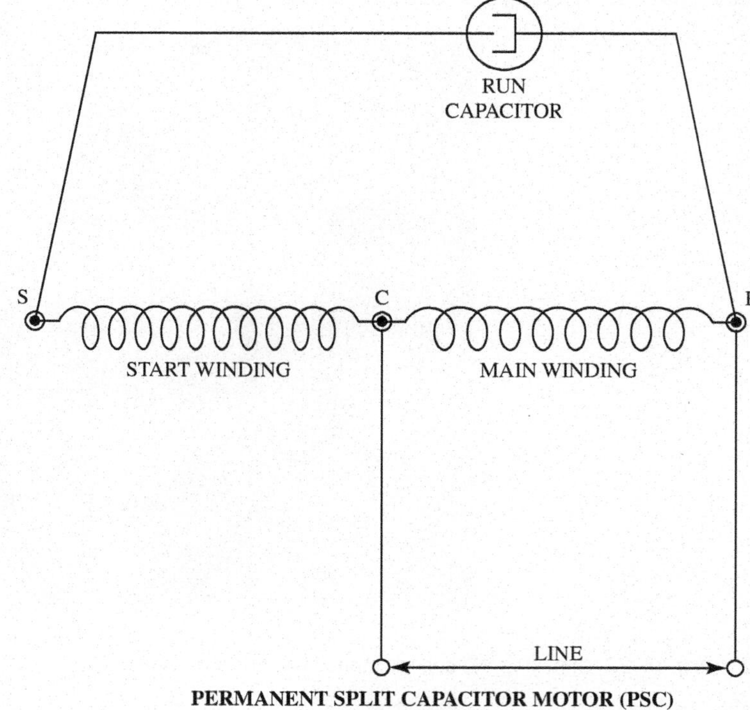

Figure 15-25 In this PSC circuit diagram, the run capacitor is wired in series with the start winding and in parallel with the run or main winding.

Figure 15-25 shows a diagram of a PSC circuit. Refer back to Figure 15-5 to see the compressor PSC motor. The run capacitor is wired in series with the starting winding and is in parallel with the run winding. In addition to creating lower running amperage, the run capacitor reduces the current flow through the start winding. At the beginning of the cycle, the motor is energized through the start winding and run capacitor. Current does not travel through the run capacitor. Current does charge up on the run capacitor's plates. Once the capacitor plates are charged with electrons, current flow stops or is greatly reduced. The run cap is in series with the start winding, therefore current is also dramatically reduced in the start winding. Even though the run cap is not a switch, the high resistance reduces the current flow enough to prevent the start winding from damage.

Check the resistance of the run capacitor by measuring the resistance across it. You will notice that the run capacitor will charge toward 0 Ω, then increase to a very high resistance. The run cap is charging through the voltage in the ohmmeter, which transfers into a resistance reading. The lower the electron flow, the higher the resistance.

PTC Solid State Resistor

Finally, there is another type of a PSC motor as shown in the electrical diagram in Figure 15-26. This diagram has a PTC solid-state resistor. The resistance of the PTC device varies with its temperature. This is a starting device. PTC mean **positive thermal coefficient.** At normal ambient temperature the resistance of a PTC device is very low. This allows full current flow through the start winding that starts the motor. Current flow through the PTC device quickly causes the resistance to increase and safely reduces the current flow through the start winding. Think of the PTC as an electronic start winding switch. The rest of the circuit operates as a PSC motor once the PTC resistance increases to many thousands of ohms.

Here is a comparison of the advantages and disadvantages of a PSC motor:

Advantages

- Highly efficient (up to 70%)
- Designed for constant operation
- Good for multispeed applications
- Instantly reversible
- Low starting current
- Very high reliability
- Low cost for its advantages

Disadvantages

- Moderate starting torque
- Cannot start a motor under a heavy load

PSC Motor Checklist

Since this is a split-phase motor, the start and run windings should be checked. The resistance of the start winding is several times greater than that of the run winding.

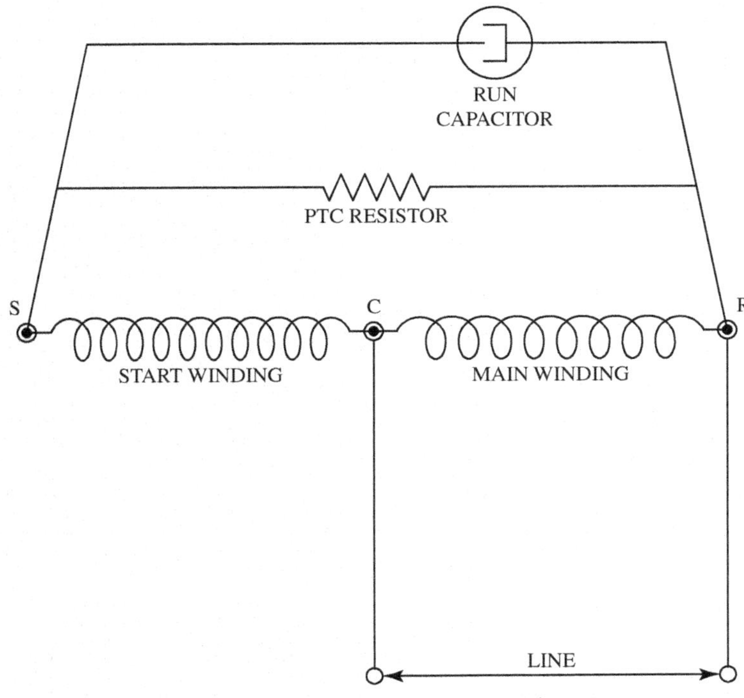

Figure 15-26 This PTC solid-state resistor changes resistance as it heats from current flow.

The resistance to motor case ground should be infinity or a very high resistance reading. The microfarad rating of the run capacitor should be within ±5% of its nameplate rating. Some capacitor manufacturers accept a slightly high tolerance of being acceptable. An oversized or undersized run cap will cause high starting and running amps. Always check the running amperage when you change out a run cap. On a compressor motor, the RLA should be lower than the nameplate rating unless it is an extremely hot day or unless there is another problem such as the system being overcharged. If a fan motor does not start, hand-spin the fan blade in the opposite direction of normal operation. If the motor continues to rotate, the problem is most likely the run capacitor. The run winding will draw high amperage and the motor will overheat, likely tripping on the motor overload.

Running a PSC motor with a defective or weak capacitor will overheat the motor. The internal overload will overheat due to excessive amperage and cause the common connection in the motor to open until it cools. The motor will operate when the overload closes and the cycle will occur all over again. Eventually this will burn out a motor winding.

The PTC should have low resistance once it has cooled for 5 minutes. The cool resistance can be as low as 5 Ω or high as 50 Ω. Check the PTC's operation by monitoring the starting amperage on the start winding. The amperage should spike and return to a low level.

The motor will spin freely when bearing surfaces are good. Some PSC motors require annual lubrication. This will require a nondetergent oil or turbine oil. Most PSC motors have sealed bearings, so they do not require and, in fact, cannot be lubricated.

TROUBLESHOOTING 15.2: PSC MOTOR

Here is a summary of how to troubleshoot a permanent split capacitor motor:

1. Check the run winding as illustrated in Figure 15-27.
2. Check the start winding as shown in Figure 15-28.
3. Check for a grounded winding as shown in Figure 15-29.
4. Check the run capacitor, which is good, as shown in Figure 15-30. Remove the power and wiring from at least one side of the capacitor's terminals to obtain a safe and accurate reading. Check grounding.

(continued)

Figure 15-27 Checking the resistance of the run winding.

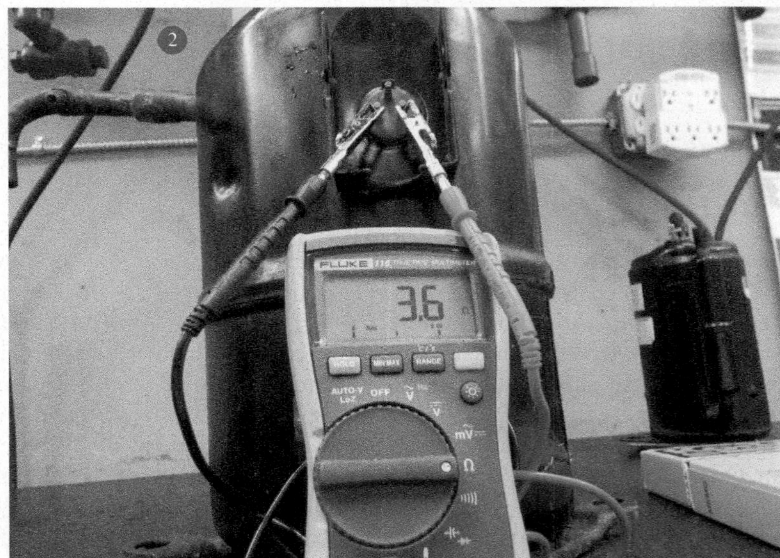

Figure 15-28 Measuring the resistance of a compressor motor's start winding.

Figure 15-29 Checking the resistance for grounded motor windings.

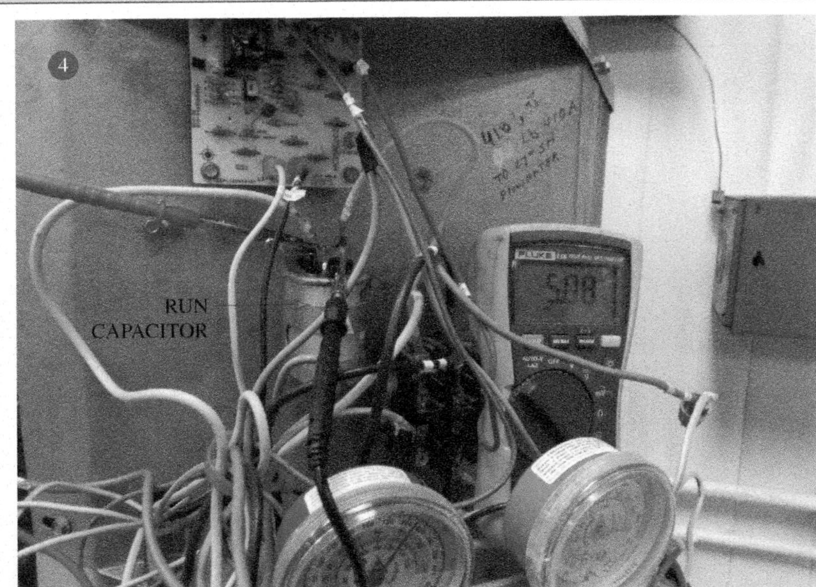

Figure 15-30 Checking the run capacitor.

TECH TIP

When checking a blower motor, you must have all the ductwork connected and panel doors in place to get an accurate amperage reading. Amp measurements may be higher when panels are off and ductwork removed because the motor moves more air and, thus, is pulling more amps.

15.9 CAPACITOR START/CAPACITOR RUN MOTOR

The capacitor start/capacitor run motor is abbreviated **CSCR** or **CSR.** The capacitor start/capacitor run motor has the best features of the "old-style motors" that we have discussed thus far. Of the single-phase motors, it has good starting torque and good running efficiency. The reason we say "old-style motor" is because, as discussed in Unit 16, a fairly new motor, the electronically commutated motor, or ECM, is starting to be used instead. The ECM does not require start components such as capacitors or special relays. The ECM has good starting torque and high efficiency. The ECM represents a big advance in motor technology.

The CSCR motor uses a start capacitor, run capacitor, and potential relay. The start capacitor and potential relay are used to generate good starting torque. These two start components are seen on a condensing unit in Figure 15-31. The potential relay has connection numbers 5, 2, and 1. Connections 5 and 2 represent a high-resistance coil, usually thousands of ohms of resistance. Connections 2 and 1 make up a set of normally closed contacts. The coil is energized by high voltage, called *back EMF,* created by the motor winding. On motor start-up the high voltage on the potential relay coil will energize and open the normally closed contacts.

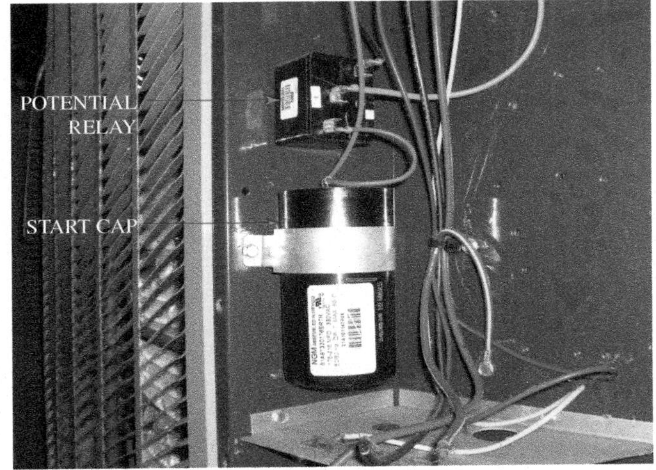

Figure 15-31 Here is a start capacitor and potential relay found on a condensing unit.

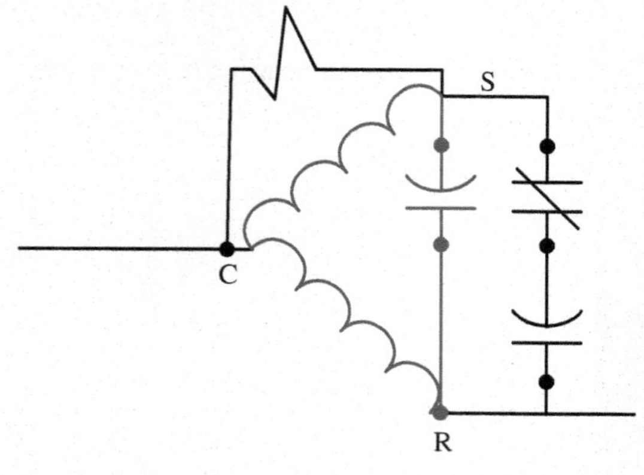

Figure 15-32 This is a schematic diagram for a CSCR motor.

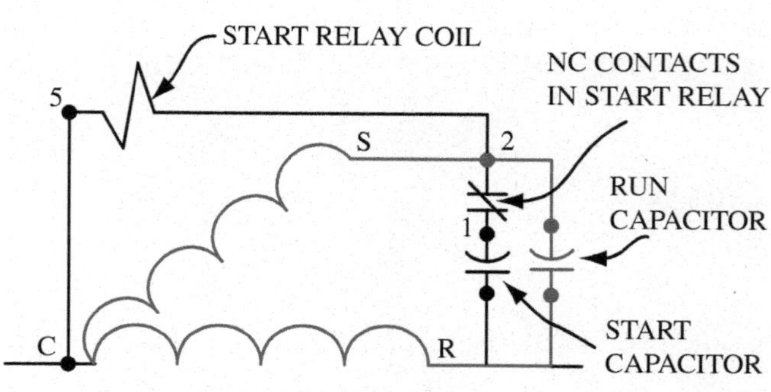

Figure 15-33 This diagram shows a CSCR motor after a start cap and potential relay have been wired to a PSC motor.

This removes the start capacitor and start winding from the circuit.

The potential relay is designed to remove the start capacitor from the circuit a moment after the motor is started. This is at 75% of the full motor speed. The start capacitor and start winding need to be removed from the circuit or one—and possibly both of them—will burn out. The run capacitor makes the motor run at peak efficiency, thus lowering operating costs. If you evaluate the CSCR motor circuit shown in Figure 15-32, you will notice that it is a PSC design (blue lines) with added start capacitor and potential relay components. Figure 15-33 is the diagram for the start capacitor and potential relay after they have been wired to a PSC motor. The PSC motor circuit is shown in blue. Adding the start components to a PSC motor is done with a *hard start kit* (discussed in a later section). The start capacitor and potential relay must be matched to the motor. Unmatched start components will not provide the maximum starting torque required to turn and start the motor.

In summary, Figure 15-34 shows an electrical diagram for a capacitor start/capacitor run motor. The diagram is basically for a PSC motor with the addition of a start capacitor and start relay. The start relay is normally a potential relay with connections 5, 2, and 1. There may be other nonconnected terminals on the potential relay because some manufacturers use the same relay housing for several different relay designs. Power is supplied to this motor from the common **C** connection and the run winding connection.

SEQUENCE OF OPERATION 15.3: CSCR MOTOR

Figure 15-34 shows what an actual CSCR motor will look like as part of a wiring diagram. Notice how the CSCR components are not isolated, but woven into the diagram. The start components are:

- SR coil (part of potential relay)
- SR closed contacts (part of potential relay)
- SC start capacitor.

The run components are:

- Cap 1 (run capacitor)
- Cap 2 (run capacitor).

1. In Figure 15-34 note the two points labeled in the blue circle 1, which are the contactor **C** contacts between 21 and 11, and the contacts between 13 and 23, which are closed when the Ⓒ coil is energized in the control voltage circuit.
2. At point 2 the (SR) coil (potential relay) is energized by the counter-EMF (sometimes called *back EMF voltage*) generated by the start winding. Note that the (SR) coil (between points 5 and 2) is wired in parallel with the start winding. The start winding is like the secondary winding of a transformer with the primary winding being the run winding. The counter EMF is almost 100 volts higher than the input voltage.
3. When the (SR) coil is energized by the back EMF, it opens the normally closed **SR** contacts, at point 3.
4. This action takes the start cap, **SC** at point 4, out of the circuit.
5. The run capacitors **CAP1** and **CAP-2** at point 5 are left in the circuit to reduce current flow through the run winding and improve running efficiency.

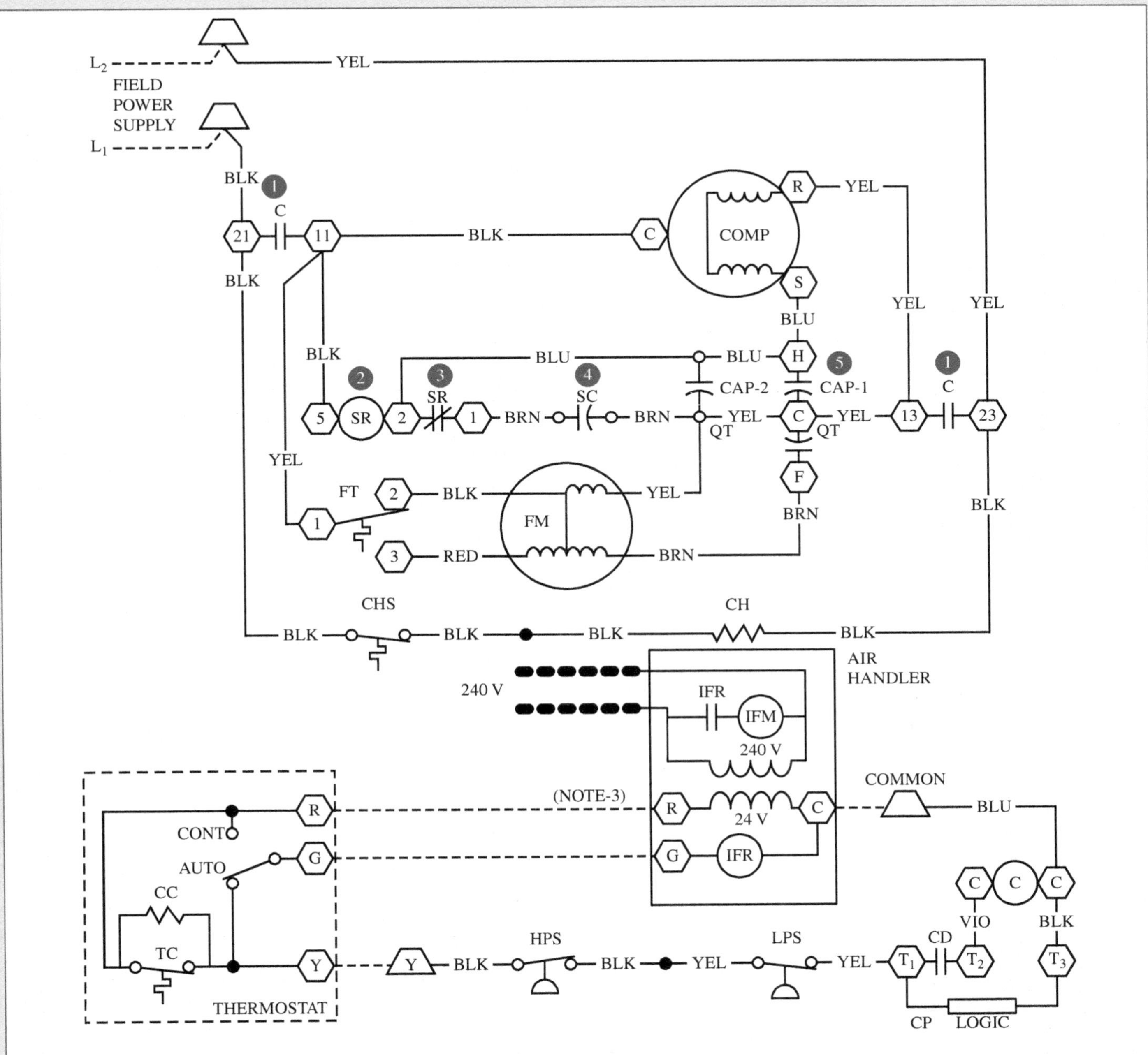

Figure 15-34 This is an electrical diagram for a CSCR motor. Locate the potential relay, which is abbreviated SR. The SR has a coil and normally closed contacts. The start capacitor, SC, is found in series with the CAP-1 and CAP-2 and the start winding. The condenser fan motor is a two-speed PSC motor.

15.10 HARD START KITS

Table 15-1 shows the start capacitor and potential relay that must match if they are to be used to make a capacitor start motor. Adding these start components to the correct existing PSC motor is referred to as a *hard start kit*. A true hard start kit must use the motor manufacturer's recommended start capacitor and potential relay. Let's look at an example of how to select a hard start kit from Copeland Compressor Company for compressor model #ZPS20K4E-PFV (see Table 15-1).

The start capacitor will be an 88- to 108-mfd capacitor with a minimum operating voltage of 330 V. This chart

Table 15-1 Recommended Hard Start Kits for Selected Compressor Models

This table shows the matching start capacitor and potential relay required to make a motor a true capacitor start motor. This is what the manufacturer recommends to be certain the motor will have high starting torque. Use of other capacitors or start relays would be just a guess in terms of how well they would work with the selected motor. The wrong size start capacitor will not provide the required starting torque. The wrong potential relay may even damage the capacitor or motor.

	Start Components							
	Start Capacitors			Relays				
Model	**MFD**	**Volts**	**Part Number**	**G.E. p/n**	**Copeland p/n**	**Pick-up Volts**	**Drop-out Volts**	**Coil Voltage**
ZPS20K4E-PFV	88-108	330	014-0036-03	3ARR3KC3LP	040-0140-08	140-150	40-90	332
ZPS26K4E-PFV	88-108	330	014-0036-03	3ARR3KC3P5	040-0001-79	170-180	40-90	332
ZPS30K4E-PFV	88-108	330	014-0036-03	3ARR3KC5M5	040-0001-62	150-160	35-77	253
ZPS35K4E-PFV	88-108	330	014-0036-03	3ARR3KC3P5	040-0001-79	170-180	40-90	332
ZPS40K4E-PFV	88-108	330	014-0036-03	3ARR3KC3P5	040-0001-79	170-180	40-90	332

Courtesy of Emerson Climate Technologies, Inc.

provides a recommended part number for the capacitor. However, a generic start capacitor with the stated rating can be used, or a cap with a higher voltage rating can be used. A lower voltage rated cap will quickly overheat and burn out.

Next, select the potential relay. Copeland gives the technician two compatible options. When selecting the potential relay, you should consider the pick-up volts, the drop-out volts, and the coil voltage. For this compressor model, the pick-up voltage is between 140 and 150 volts. This means that the normally closed contacts on the potential relay will open once 140 to 150 volts is applied across the relay coil. This voltage removes the start capacitor from the circuit in a split second.

The drop-out voltage is 40 to 90 volts. When the coil volts drops below this voltage range, the potential relay will reset, which closes its contacts. This will allow the start capacitor to be in the circuit should the motor quickly start up again.

The potential relay coil can handle up to 332 volts, which should be below its operating voltage. The relay coil does have a high voltage on its winding. The coil has the back EMF or voltage on its windings. The back EMF across the relay coil comes from the compressor winding, which creates a higher voltage when it rotates. A single-phase 240-V motor generates over 300 volts of back EMF. Measure the voltage across a run capacitor. That is the EMF generated by a motor. Now measure the voltage across a potential relay coil. It will also be higher than the applied voltage.

Here are the advantages and disadvantages of CSCR motors:

Advantages

- Highly efficient (up to 70%)
- Designed for constant operation
- Low starting current
- Very high reliability
- Low cost for its advantages

Disadvantages

- More starting components
- Additional cost for start capacitor and potential relay

15.11 TROUBLESHOOTING THE CSCR MOTOR

Troubleshooting a CSCR motor is like troubleshooting a PSC motor with two added components: the correctly matched start capacitor and potential relay. You can check the start capacitor with a capacitor checker. Unlike a run capacitor, the start capacitor has a microfarad range. A common start capacitor might be expressed as 88 to 108 mfd. If the capacitor has a lower or higher rating, it needs to be changed. The voltage rating is the supply voltage. It does not need to be higher since the start capacitor is only in the circuit for a split second. A defective start cap will be manifested as high starting amperage. The motor may not start with a defective start cap.

You can troubleshoot a potential relay with an ohmmeter or a clamp-on ammeter. Use an ohmmeter to check the potential relay's normally closed contacts and relay coil. The closed contacts should register near 0 Ω. The closed contacts on the relay are found between terminals 2 and 1. The potential relay coil resistance will be high. The resistance range will be above 4,000 Ω. The high-resistance coil is between terminals 5 and 2 on the relay.

Check the potential relay with a clamp-on ammeter. Refer back to Figure 15-34; clamp the ammeter on either side of the start capacitor **SC** or the **SR** contacts. When the potential relay is operating normally, there will be a high current draw for a split second when the motor starts. If there is no amp draw, the start capacitor could be open. If the amp draw continues for several seconds, the potential relay contacts may be welded together, preventing the break in the starting circuit. The continuous current flow will damage the start capacitor.

Although this may not be a common practice, the starting components on a CSCR motor should be changed out whenever the motor is changed for electrical problems. This will include changing the:

- Run capacitor
- Start capacitor
- Potential relay.

SAFETY TIP

When checking compressor motor winding resistance, do not measure the resistance on the compressor terminals. The compressor terminals may be the weakest part of a sealed compressor. High-pressure refrigerant and oil could blow out on you.

Measure the resistance of the motor winding from the connections on the contactor. Keep in mind that the conductors from the contactor to the compressor terminal may be shorted. This will lead you to believe that the compressor windings are shorted. Before condemning a compressor for defective windings, recover the refrigerant to a low level and measure the resistance on the compressor terminals. Be certain the compressor is the problem! Remember: 25% of motors are returned with no problem found.

15.12 HOW TO DETERMINE COMPRESSOR WINDINGS

A single-phase compressor has three connections and they are called *common*, *start*, and *run* (CSR). Remember that there are only two windings and they are called the *start winding* and the *run winding*. There is no common winding. The common terminal is a point where the start and run windings are tied together. Figure 15-35 shows the symbol for this connection.

Here are some guidelines for determining the common, start, and run connections on a single-phase compressor. These guidelines are *not* 100% foolproof. As a responsible technician you must take the time to verify the CSR connections when changing out a compressor or after you have removed the compressor wires for troubleshooting. Even though the circuit has overcurrent protection and an internal overload, the compressor is not protected against miswiring. A miswired compressor can burn up in a second or two without releasing any smoke or sparks, tripping a circuit breaker, or giving any other indication that there is a problem.

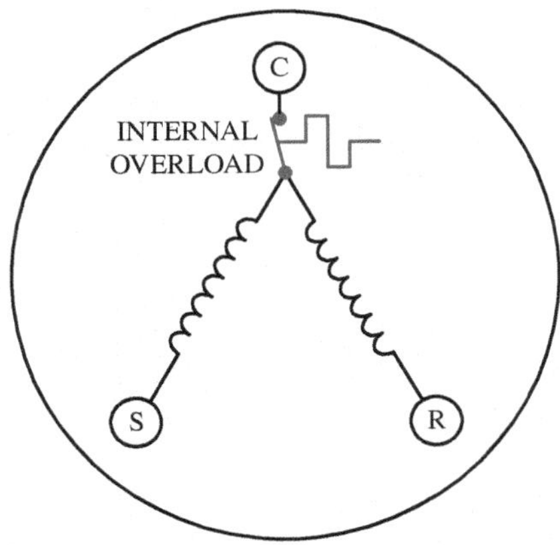

Figure 15-35 This is the symbol for a single phase, split-phase motor. It has a start winding and run winding. The internal overload is in series with the start and run winding.

Helpful guidelines to determine CSR connections (does not work 100% of the time; check the terminals or wires going to the terminals):

- Determining compressor terminals is like reading a book.
- The reading or connections will be in the following order: **C**ommon, **S**tart, and **R**un.
- Using the "reading the book" method, read from top to bottom and from left to right.
- In most cases the top winding will be common. Always check and verify with an ohmmeter.
- Common does not mean a ground connection. Common means that there is a shared connection point for the start and run windings. One side of the power supply will be supplied to this terminal.

SAFETY TIP

Always check CSR connections prior to installing any new or replacement motor. Why? First, compressor or motor damage could have occurred during shipment. Second the wire connections inside the motor could have been changed or miswired at the motor manufacturer. Finally, check the motor winding to ground. The resistance should be infinity or very high resistance. The technician would not want to install a compressor or other motor only to find out after installation that it is defective.

Figures 15-36 and 15-37 shows two different examples of compressor wiring configurations that follow the reading the book method. You will, however, see other configurations in the field. Never assume that the hook-up method expressed here is used. The next section will describe how to verify the CSR connections.

15.13 VERIFYING CSR TERMINALS

The single-phase compressor motor is an example of a split-phase motor. The split-phase motor has a start winding with much higher resistance than the run winding. In summary, the highest resistance is between the start and run winding connections because the resistance of the two windings is additive. The lowest resistance reading is the run winding, which is measured between the common and run connection. With this information in mind, here is a logical sequence for determining common, start, and run terminals:

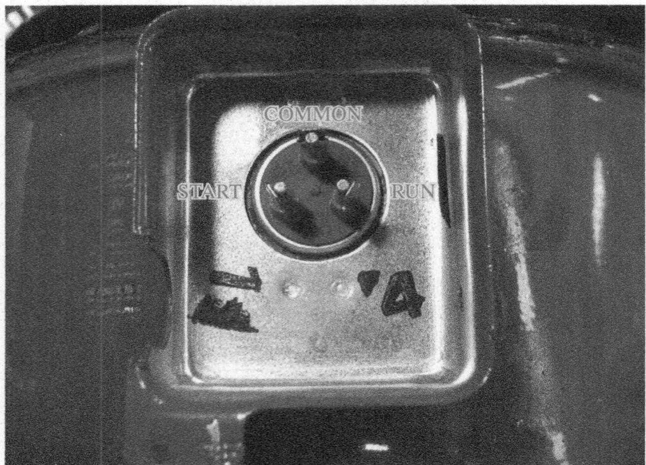

Figure 15-36 Remember *common*, *start*, and *run* (CSR) to help you check the compressor connections. Using the "reading a book" rules, common will be at the top. The "second line" of the page, so to speak, would be the start winding, followed by the run winding on the "last line" of the page. Always verify this guideline because there are exceptions. Wiring a compressor incorrectly will burn out the motor quickly.

Figure 15-37 Using the "reading a book" rules, common will be at the top. The "second line" would be the start winding. The "last line" is the run winding. Always verify this guideline because there are exceptions. Wiring a compressor incorrectly will burn out the motor quickly.

Label the three terminals T_1 (C), T_2, and T_3 (R) as illustrated in Figure 15-38. The letters C and R stand for the common and run winding connections, respectively. The letter S is hidden from view and is associated with the T_2 connection.

- Use the R × 1 resistance scale.
- Measure the resistance between each of the terminals. There will be resistance between terminals T_1 and T_2, T_1 and T_3, and T_2 and T_3.
- The highest resistance reading will include the start and run windings. These two windings are in series with each other.

Figure 15-38 When determining the common, start, and run terminals, label the compressor with numbers or letters. The highest resistance will include the start and run windings. The lowest resistance will be the run winding. The winding with the middle resistance reading will be the start winding.

EXAMPLE 15.1 DETERMINING COMMON, START, AND RUN TERMINALS

Let's apply the information we've just learned to determine the common, start, and run terminals if we are given the following resistance readings:

Terminals $T_1 - T_2 = 8\ \Omega$
Terminals $T_1 - T_3 = 2\ \Omega$
Terminals $T_2 - T_3 = 10\ \Omega$.

SOLUTION

The highest resistance is between terminals T_2 and T_3. This ohm reading is the combined resistance of the start and run windings. The highest resistance reading helps us determine the common connection. The terminal that is *not* part of the highest reading is terminal T_1. This is the common connection terminal. Once the common terminal is determined, finding the start and run winding is easy. The highest resistance from the common connection is the start winding. The lowest resistance from the common connection is the run winding. Therefore, based on our resistance readings, terminal T_2 is the start winding, and terminal T_3 is the run winding.

So we have T_1 = common; T_2 = start; and T_3 = run.

Notes

- If the windings seem good, check at least one terminal to ground. This can easily be done by measuring from any terminal to the suction or discharge line piping. You do not need to scratch paint off of the compressor housing to check to ground.
- Compressor wire color code is not strictly followed. Do not use wire color as a standard for determining CSR terminals. Sometimes the electrical diagram color may not be correct. For example, running out of wire on a

condensing unit assembly line does not necessarily mean production will stop. Wires burn open and techs replace the wire with whatever color they have on hand. Do not trust the color code. It is a good practice to mark wires that are removed when doing any HVACR work.

- Finally, many common PSC fan or pump motors follow the same guidelines for determining the common, start, and run connections. Measure the resistance to determine the winding type. High-voltage motors (460 V) do not follow these guidelines. Some high-voltage, split-phase motors actually have the same resistance for the start and run windings. Measure the resistance of at least one winding to the case ground. Resistance to ground should be infinity or a very high resistance.

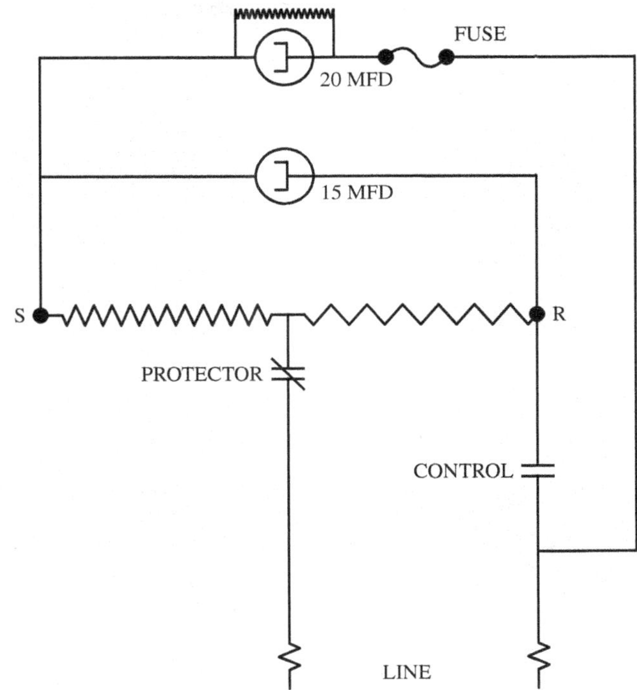

Figure 15-39 The 220,000-Ω bleed resistor drops the voltage to the start winding, but allows a trickle voltage that keeps the start winding warm during the off cycle. This start winding acts as a crankcase heater.

TECH TIP

Figure 15-39 is a diagram of a compressor circuit that uses a bleed resistor to keep the start winding warm, thus acting as a crankcase heater. Although not common anymore, a number of circuits are still wired this way. When replacing the capacitor, reinstall the bleed resistor. The parallel run capacitors will create a 35-mfd capacitance. Also, reconnect the bleed resistor after it has been removed to check the capacitor. It is important to understand the purpose of the resistor. Many techs think it is to protect them from shock. It is not a safety device even though it will discharge the capacitor and prevent discharge shock.

SAFETY TIP

Figure 15-40 shows how miswiring the compressor will damage the start motor winding. The diagram on the left is correct. The wiring on the right has continuous voltage applied to the start winding. The motor is not protected against miswiring conditions. The start winding will burn out quickly.

INTERCHANGING "COMMON" AND "START" CONNECTIONS

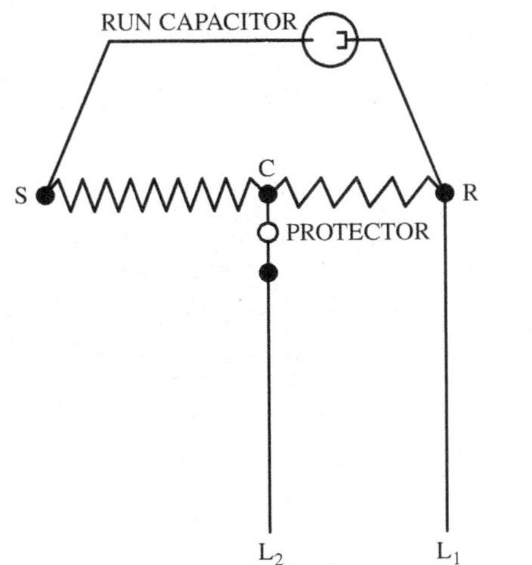

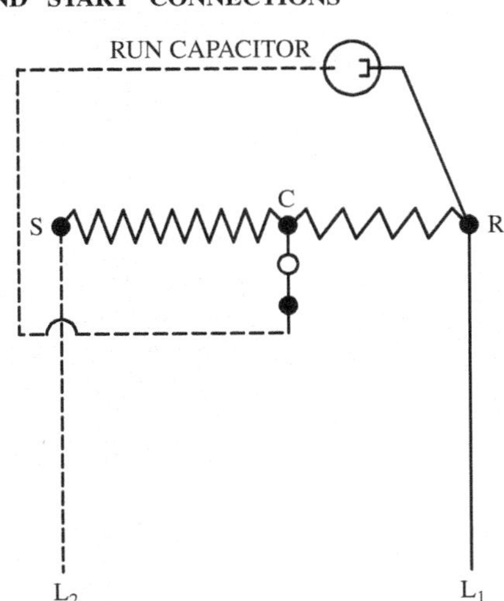

Figure 15-40 Miswiring the compressor will damage the start motor winding.

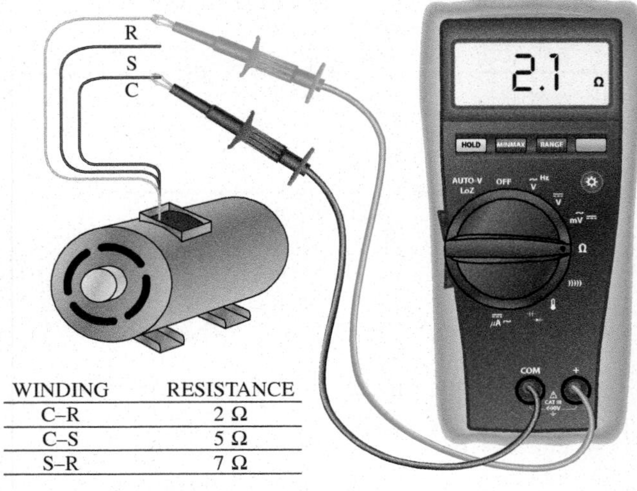

WINDING	RESISTANCE
C–R	2 Ω
C–S	5 Ω
S–R	7 Ω

Figure 15-41 Many common split-phase motors follow the same guidelines as those used to determine CSR terminals. Measure the resistance to determine the winding type.

Finally, the CSR terminals of other noncompressor PSC motors can be checked by measuring the resistance of the motor windings. The lowest motor winding will be the run winding as shown in Figure 15-41. The exception to this rule is a 460-V PSC motor. In this type of motor, the start and run windings may be close to having the same resistance.

15.14 THREE-PHASE MOTORS

Three-phase motors have three sets of windings. The power to the three run sets of windings is 120 electrical degrees apart, as shown in Figure 15-42. These windings (see Figure 15-43), when excited by a three-phase power supply, set up a revolving field that causes the rotor to develop good starting torque. No additional starting devices, capacitors, or switches are required for operation.

There are two common types of three-phase motors: the delta motor and the wye motor. There are other three-phase motor configurations, but these are the most common. In

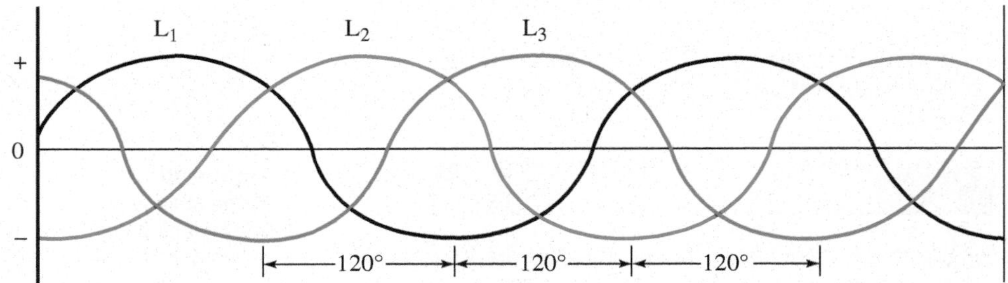

Figure 15-42 This is a three-phase power supply to a three-phase motor. Each of the phase voltages is 120 degrees out of phase with each other. A three-phase motor will have three sets of winding. Each winding is energized by a different voltage phase.

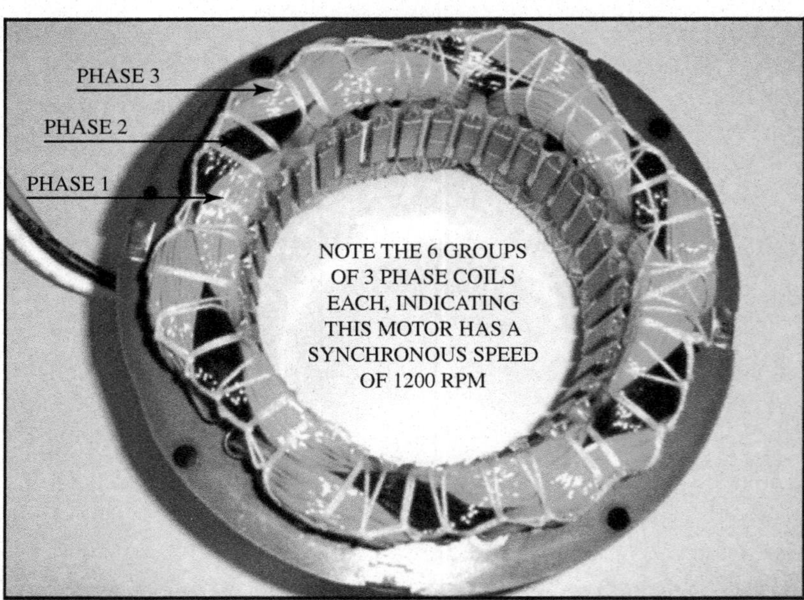

Figure 15-43 This is the stator of a three-phase motor. Three power wires (upper left) feed voltage to two sets of windings.

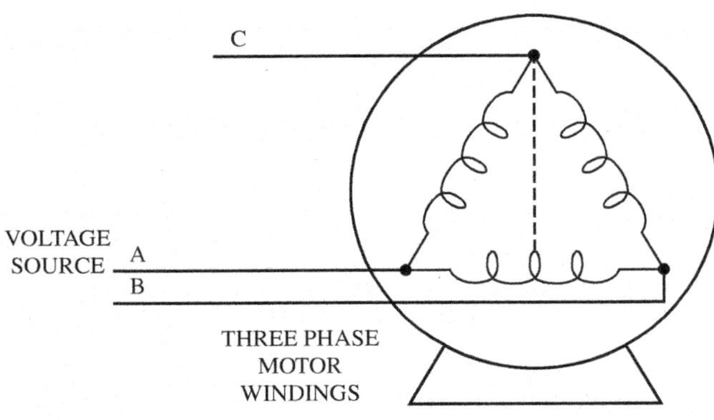

Figure 15-44 This is the electrical symbol for a delta three-phase motor.

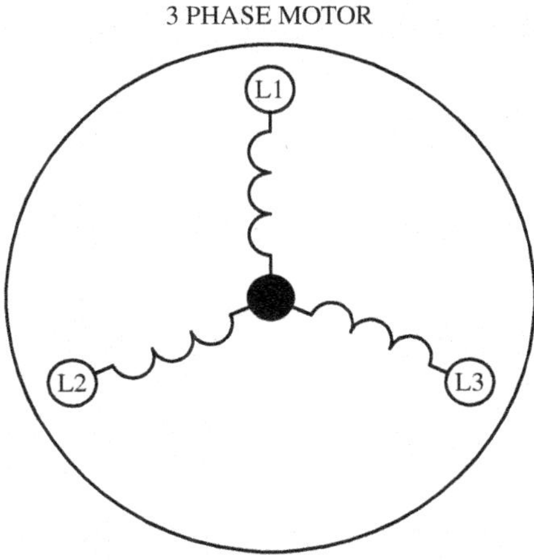

Figure 15-45 This is the electrical symbol for a wye three-phase motor.

some designs they have 6, 9, or 12 windings. Since these are three-phase motors, the motor windings will be in sets of three.

Delta Motors

Three-phase motors have three or six windings. Figure 15-44 shows the electrical symbol for a delta motor with three windings. It is called *delta* because the winding is drawn like the Greek letter delta Δ, which looks like a triangle. The resistance of the windings is the same.

TECH TIP

Generally, the larger the three-phase motor, the lower the resistance of the windings. Fractional horsepower three-phase motors may have 10 to 20 Ω of resistance. Large horsepower motors will have very small resistance, sometimes less than 1 Ω. The larger three-phase motors have lower gauge (larger diameter) wires, therefore the resistance will be low. Some techs think that the motor windings are shorted since the resistance is less than an ohm. Not so!

Wye Motors

Figure 15-45 shows the symbol for a wye motor (pronounced like the letter "Y"). Its windings look like a Y. In this three-phase motor, two windings are in series with each other. Like the delta motor, the resistance is the same across each winding.

It does not matter whether a motor is a delta or wye configuration. The replacement motor, delta or wye, must match the horsepower, amperage, and voltage requirements, however, since these three-phase motors have different wiring characteristics.

TECH TIP

Three-phase motors may be used in any application where shaded pole, PSC, or split-phase motors are used provided three-phase power is available.

Let's look at the advantages and disadvantages of three-phase motors:

Advantages	Disadvantages
■ Very high efficiency	■ Application limited by three-phase power availability
■ Very high reliability	■ Limited multispeed capabilities
■ High starting torque	
■ Moderate amp draw	
■ Moderate cost	
■ Instantly reversible	

15.15 TROUBLESHOOTING THREE-PHASE MOTORS

Troubleshooting three-phase motors is simple. Each winding should have nearly the same resistance within. In most cases the resistance from the winding to ground should be infinity or at least hundreds of thousands of ohms. For compressor motors, the resistance to ground should be at least 1,000 Ω per volt. For example, a 480-V compressor motor should have at least 480,000 Ω to ground. Most good motors have a reading of millions of ohms to ground or infinity ∞. Many digital multimeters display "**OL**" as indication of infinity. OL can mean outer limit or overlimit, but it is an indication of a high resistance.

TECH TIP

Reversing the direction of a three-phase motor is as simple as reversing two of the three incoming power leads. Reversing all three power leads will not change the motor rotation. The direction of three-phase motor rotation is important on some compressors. Many compressors can operate in either rotation. Check with the manufacturer if in doubt. Three-phase scroll compressors are rotation sensitive. A scroll in reverse rotation will make excess noise and trip out on overheating or overload condition. Another indication of rotation reversal is that the suction and discharge pressure will not develop normally.

When starting a new or replacement three-phase compressor, you should be at the compressor site to shut down the unit should reverse rotation be experienced. Take control of the equipment at the condenser site, not at the thermostat. Have the manifold gauges and clamp-on ammeter hooked up. This is a recommended troubleshooting sequence.

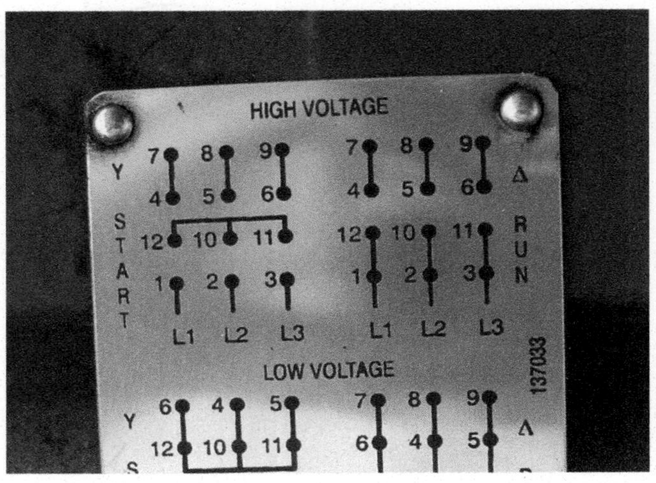

Figure 15-47 This is the nameplate for a dual-voltage, three-phase motor.

15.16 DUAL-VOLTAGE MOTORS

Some three-phase motors can be operated on 230 or 460 volts. Figure 15-46 shows a wiring diagram for a dual-voltage motor. The motor can be hooked up to operate on either voltage. Let's examine how this is accomplished. Looking at the diagram, the term "Low Voltage" refers to the 230-V hook-up, whereas the term "High Voltage" refers to the 460-V hook-up.

When wiring a delta motor for 230 volts using the diagram on the left side of Figure 15-46, you would:

- Connect motor wires 1, 7, and 6 together.
- Next, connect wires 2, 8, and 4 together.
- Finally, connect wires 3, 9, and 5 together.

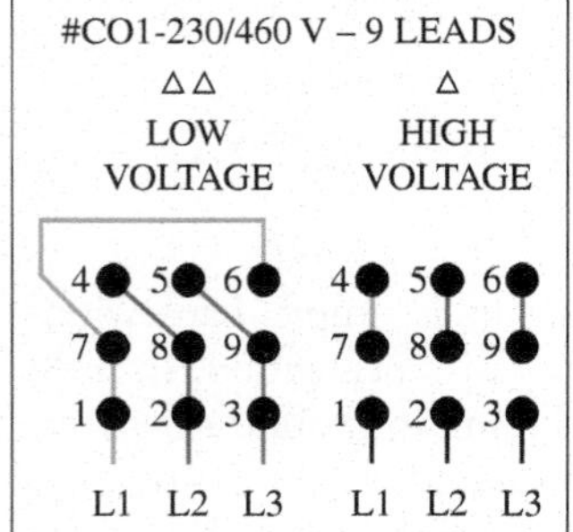

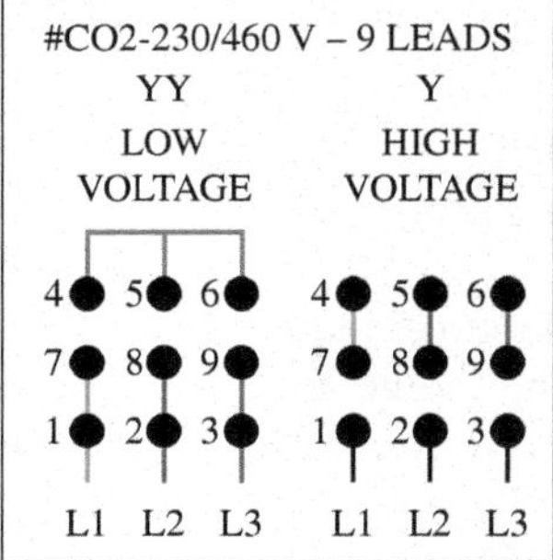

Figure 15-46 This is a hook-up diagram for a nine-lead dual-voltage motor that can be wired in a delta or wye configuration at 230 or 460 volts. The diagram on the left is for a delta motor configuration. The diagram on the right is for a wye configuration. The term "Low Voltage" means it is wired for 230 V. "High Voltage" refers to the 460-V hook-up.

To wire a delta motor for 460 volts using the diagram shown in Figure 15-46, you would

- Hook wires 1, 2, and 3 to L1, L2, and L3 of the power supply.
- Next, connect wires 4 and 7 together.
- Finally, connect wires 5 and 8 together and wires 6 and 9 together.

Measure the incoming voltage before starting the motor. Have the clamp-on ammeter hooked up on one leg and quickly check each leg for amps when starting. Each leg should have approximately the same amperage. If they do not, quickly turn off the motor and verify the wiring. Note, however, that an amperage imbalance can have other causes, as discussed in the troubleshooting units.

Figure 15-47 is a nameplate for a dual-voltage motor. The upper half of the nameplate is the high-voltage or 480-V connection. The lower half is the low-voltage or 240-V connection.

Dual-Voltage Wiring Diagram

Figure 15-48 shows a three-phase, dual-voltage motor diagram. This motor can also be wired as a delta or wye motor. Again, it is important to immediately check the voltage and amp draw on motor start-up. This may prevent motor damage in case of miswiring or a bad wiring connection. This motor can be wired as 240 volts, which is considered to be low voltage (note the upper wiring diagram in the red box). It can also be wired as 480 volts, which is considered to be high voltage (note the lower wiring diagram in the red box).

In the red box, you will notice on the left "Y START" and on the right "Δ RUN." The Y START windings are used to start the motor and the Δ RUN windings are used

MOTOR WIRING DIAGRAM
12 LEAD, DUAL VOLTAGE, WYE START/DELTA RUN, BOTH VOLTAGES
OR
6 LEAD, SINGLE VOLTAGE, WYE START/DELTA RUN

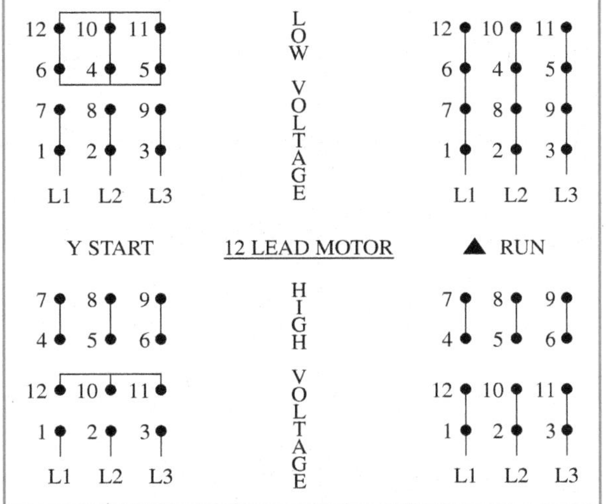

CONNECTION DIAGRAM
P1 P2 G
THERMALLY PROTECTED
GROUND (GREEN)

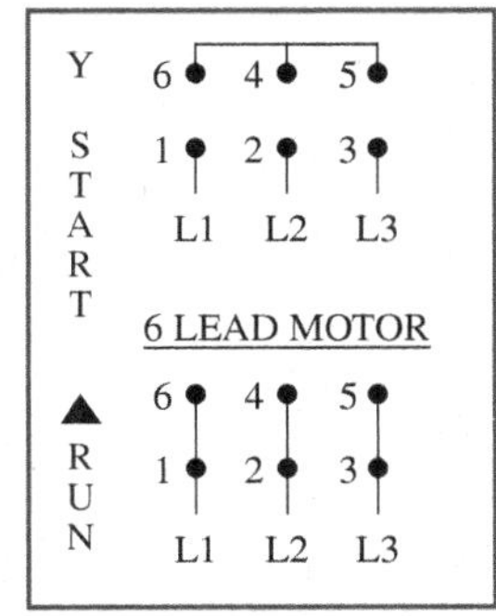

Figure 15-48 This dual-voltage motor wiring diagram can be used for a wye or delta motor. This motor can be wired as 240 volts (low voltage) or 480 volts (high voltage).

to keep the motor running. Actually the Y START windings are wired in series to reduce the starting amperage and develop maximum starting torque. For example, if you draw out the Y START winding and label the leads L1, L2, and L3 you will notice that the L1 and L2 are in series. L1 and L3 are in series. L2 and L3 are in series. Reviewing Ohm's law, series circuits have more resistance than parallel windings and reduced amp draw; in this case, less amp draw on start-up or less LRA when starting.

The blue box shows a six-lead connection. The yellow box illustrates the internal thermal motor connection and ground connection.

15.17 THREE-PHASE DIAGRAM

Figure 15-49 is a diagram for a commercial air conditioning condenser. Let's review some of the basic components to get an understanding of this three-phase diagram. The bottom of the diagram shows a component arrangement box within the red rectangle. When removing the cover to the condensing unit, you would find this list of components laid out in this fashion. The components outside the component arrangement box are scattered throughout the condensing unit.

At the top of the diagram you will notice that the supply voltage is 460-3-60 or stated another way, 460 volts, three phase, and 60 cycles.

The only component operated on three-phase power is the wye wound compressor motor. The other components are operated by single-phase 460-V or single-phase 230-V motors. Three-phase power is supplied to the compressor through a double-pole contactor. The L2 power is continuous and is not broken by the contactor. This can be a dangerous condition if the tech does not realize that power is applied even when the compressor is not operating.

In Figure 15-49, below the compressor is a single-phase condenser fan motor. The motor is a 240-V PSC design. From your knowledge you should determine that the lower winding of **FM** is the start winding. The start winding is in series with the run cap. The upper winding is the run winding.

A 460- to 230-V transformer is used to supply a lower voltage to the timing delay circuit. A field-supplied 24-V transformer is shown in dashed lines. The 24-V circuit is a safe way to provide control voltage. In addition, the use of 24 volts means that smaller conductors can be used and the size of components reduced.

TECH TIP

Air flowing through an idle ventilator or blower will cause it to rotate in the direction opposite of its normal rotation. This is referred to as "windmilling." Windmilling can cause problems such as these:

- The fan may windmill so fast in reverse on start-up that the motor cannot reverse the blade so that it turns in the wrong direction. A stronger motor with ball bearings may be necessary to correct this problem.
- At speeds below 500 RPM, oil may not be returned to the oil reservoir, rapidly depleting the motor's lubricating capacity. This will significantly reduce the life of the motor.

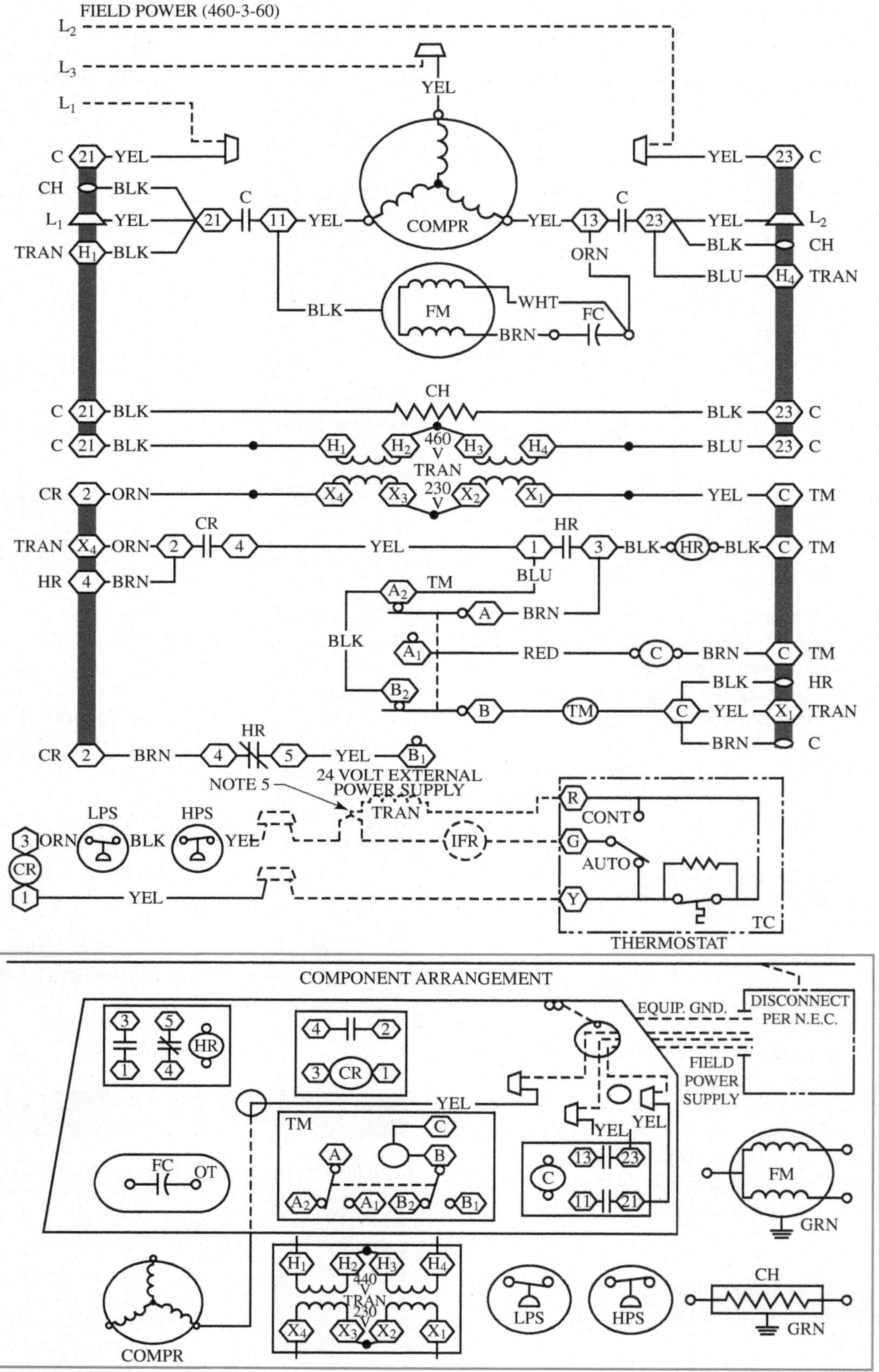

Figure 15-49 This is a diagram for a three-phase, 460-V condensing unit. The compressor uses a three-phase, 460-V motor. The remaining components use a single-phase 460- or 230-V motor.

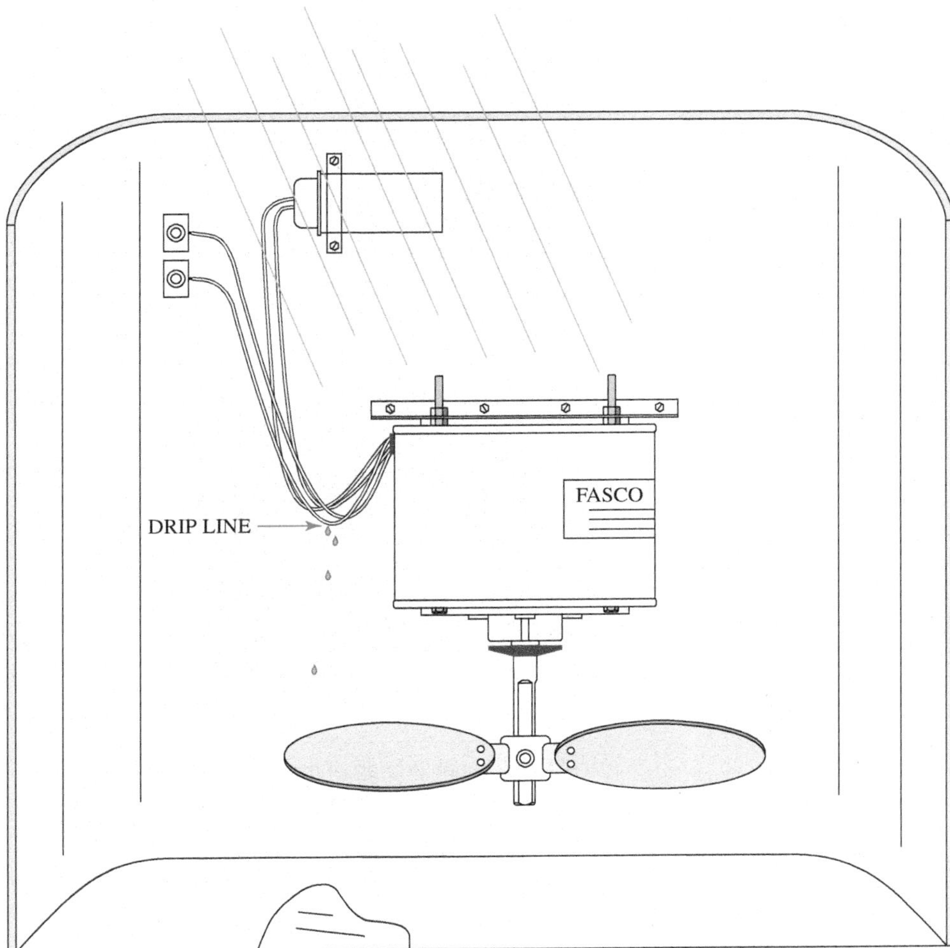

Figure 15-50 To protect a motor from rain or moisture, it is always a good practice to leave a drip loop in the lead wires so water will not be channeled directly into the motor housing. Be careful to keep the loops short enough to prevent the wires from making contact with the load.

15.18 MOTOR MOUNTS AND ACCESSORIES

Some of the common motor mounts are:

- Rigid
- Cradle
- Belly band
- Stud
- C-frame.

Motors have several mounting accessories including resilient bases, bracket assemblies, band mounting kits, and adapter plates.

TECH TIP

When installing a condenser fan it must be protected against rain and moisture. The wiring must also be placed so that the wire does not become a path for water flow into the motor's electrical compartment. Figure 15-50 shows an example of a drip loop used to prevent water infiltration.

15.19 MOTOR VENTILATION

When selecting a replacement motor, the ventilation pattern of the replacement motor must match the original motor as closely as possible for safe and maximum motor life. Moisture and vapor exposure must be considered. Here are the common motor ventilation styles:

- Open, fully vented
- Open drip-proof
- Totally enclosed, nonvented
- Totally enclosed, fan cooled.

Open Motors or Fully Ventilated Motors

Open or fully ventilated motors that are protected from the weather and are not built into a combustible structure benefit from having a fully or partially ventilated frame. Figure 15-51 shows ventilation slots in the motor shell and end shield that cool the motor's windings and bearings. Heat from the coils and bearings is allowed to dissipate into the ambient air, lowering the temperature and maximizing motor life.

Figure 15-51 Totally open ventilated motors have vent slots in the shell and end shield. These motors may be smaller than enclosed motors.

Figure 15-52 Open drip-proof motors are used in equipment where open ventilation is needed for cooling, but where the motor may be exposed to moisture drips onto the windings.

Open, fully vented motors are used in indoor applications that are protected from moisture and vapor. When an open vented motor is required in equipment that is exposed to rain, snow, or other type of moisture, a partially open motor is available with closed end shields with the shaft up or down to protect the windings and bearings from liquids.

Open Drip-Proof Motors

The open drip-proof design shown in Figure 15-52 is used where motors are exposed to the weather or where there may be a fire hazard from molten or flaming materials. The frames are typically totally enclosed or drip-proof to allow for lower operating temperatures. The upper half of the shell is closed to prevent moisture from dripping into the motor. These are used in shaft horizontal installations only.

Totally Enclosed Nonvented (TENV) Motor

The totally enclosed nonvented motor has no vent openings (see Figure 15-53). It is designed for maximum protection from weather and contaminants. The TENV motor may have the shaft installed in any direction. These motors run hotter than open-type motors, therefore they should never be used to replace an open vented motor.

Totally Enclosed Fan-Cooled (TEFC) Motor

Totally enclosed fan-cooled motors have no opening in the shell or end shields. Cooling is accomplished by an external fan system to move the air over the motor shell and end shields. These motors are used in equipment where no cooling air is provided by the load. Figure 15-54 shows a TEFC motor.

15.20 MOTOR REPLACEMENT AND AMP RATINGS

One way to ensure that a replacement motor with similar output is chosen (when the horsepower is unknown) is to select a motor with a current rating that is very similar to that of the failed motor. Motors of the same type, current, and voltage ratings, and stack length will likely have very similar output. Table 15-2 will help the technician choose a replacement

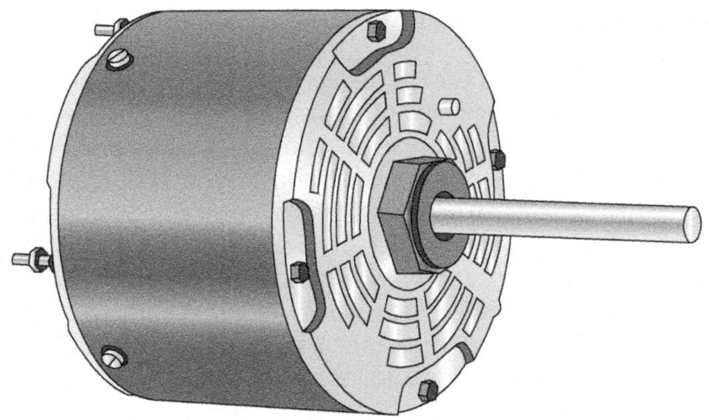

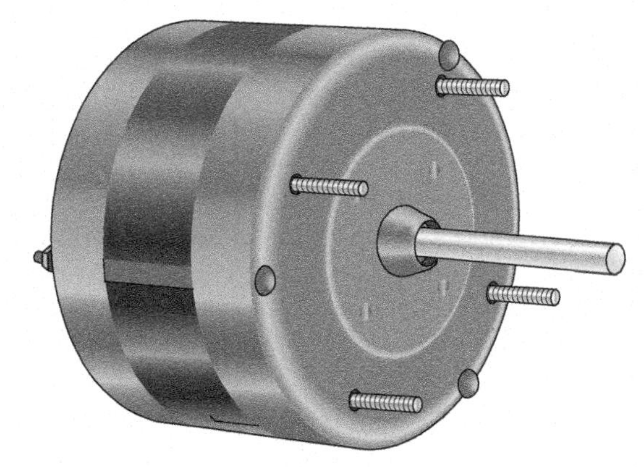

Figure 15-53 The totally enclosed, nonvented (TENV) motor has no shell ventilation openings.

Figure 15-54 The totally enclosed fan-cooled (TEFC) motor does not have any vent openings in the shell or end brackets. The motor has an internal fan that moves air over the motor shell and end shields.

Table 15-2 Amp Ratings for Use When Choosing a Replacement Motor

Nameplate Amps of the Defective Motor	Nameplate Amp Range of an Acceptable Replacement
1.0	1.0–1.25
1.1	1.1–1.37
1.2	1.2–1.52
1.3	1.3–1.62
1.4	1.4–1.75
1.5	1.5–1.87
1.6	1.6–2.0
1.7	1.7–2.13
1.9	1.9–2.37
2.0	2.0–2.5
2.2	2.2–2.75
2.4	2.4–3.0
2.6	2.6–3.25
2.8	2.8–3.5
3.0	3.0–3.75
3.3	3.3–4.12
3.6	3.6–4.5
4.0	4.0–5.0
4.4	4.4–5.5
4.8	4.8–6.0
5.0	5.0–6.25
5.5	5.5–6.87
6.0	6.0–7.5
6.5	6.5–8.12
7.0	7.0–8.75
7.5	7.5–9.37
8.0	8.0–10.0
8.5	8.5–10.62
9.0	9.0–11.25
9.5	9.5–11.87
10.0	10.0–12.5
10.5	10.5–13.12
11.0	11.0–13.75
11.5	11.5–14.4
12.0	12.0–15.0

(Courtesy Regal Beloit.)

motor if the amperage of the defective motor is known. The current rating of the replacement motor should be the same or no more than 25% greater than that of the motor being replaced.

One of the steps to insure a replacement motor with similar output is selected (when the HP is unknown) is selecting a motor with a current rating very similar to that of the failed motor. Motors of the same type, current and voltage ratings and stack length, will likely have very similar output.

A replacement motor should be selected with a current rating the same and not more than 25% greater than the defective motor.

TECH TIP

Some motors bearings are sealed, while others require periodic lubrication. The lubrication may be in the form of a light, nondetergent oil or grease. Use the manufacturer's recommendations for type and frequency of lubrication. Overlubrication may damage the motor.

15.21 MOTOR ROTATION

Motor rotation in most cases is important. For many compressors, the direction of rotation does not matter. This means that if the compressor is wired correctly it will turn in the right direction to create compression and pumping action toward the discharge line. Some three-phase compressors require a specific rotation or the oil pump will not create the needed oil pressure.

Many replacement single-phase fractional horsepower motors used in condensing units or air handlers have directional wiring options. Single-phase motors that do not have this option may have one of the following abbreviations on the nameplate:

- **CCLE** (counterclockwise lead end)
- **CCSE** (counterclockwise shaft end)
- **CWLE** (clockwise lead end)
- **CWSE** (clockwise shaft end)

This will tell you the shaft rotation and the rotation orientation. For example, if the nameplate has the abbreviations **CWSE**, it means that when looking at the shaft end (SE), the motor will turn in a clockwise direction (CW). **CWLE** refers to viewing the motor from the lead end (LE), or looking at the motor where the wires enter it. In this view, the motor will turn in a clockwise direction (CW).

SERVICE TICKET

You are troubleshooting a residential 4-ton air conditioning unit. At the beginning of the troubleshooting process, you do all the quick checks using the ACT method. Air is flowing from the indoor registers and the condensing unit. The thermostat has been switched to cooling and the setting is low enough to energize the condensing unit. After hooking up the clamp-on ammeter and manifold gauge set, you notice there is no amperage on the compressor and the high and low pressures are equalized. What does this mean? The compressor is not operating! A minute later the compressor groans, draws locked rotor amps, and shuts off again. What is the problem?

The problem seems to be the compressor or the compressor circuit. You check the start and run windings from the contactor wire and run capacitor terminals. They measure 15 and 5 ohms, respectively, which seems to be good. The resistance reading from the compressor to ground was infinity or **OL** on the digital meter reading. The electrical features of the compressor are good. The compressor could be mechanically locked.

Next, check the electrical components. The compressor motor is CSCR. A visual inspection reveals that the capacitors and potential appear to be in good shape. The run and start capacitors check out within their microfarad rating. The potential relay coil measures 5.1 kΩ. Potential relay coils measure high resistance, starting around 4 kΩ and going as high as 8 kΩ. The potential relay contacts measure infinity. The potential relay contacts, terminals 2 and 1, should measure near 0 Ω. You have found the problem. Without the normally closed (NC) potential relay contacts, the start capacitor was not in the starting circuit. The compressor would draw LRA on start-up and the compressor would open the compressor internal overload.

This can also be checked with a clamp-on ammeter. Place the clamp-on ammeter over the wire that is in line with the potential relay 2 and 1 terminals. When the motor starts there will be an amperage spike for a split second until the NC contacts open. If the contacts are open or the start capacitor is defective, there will be no amp spike.

The solution to the problem is to replace the potential relay with an exact replacement. Test the system by starting the compressor while monitoring the amperage on the common winding of the compressor and gauge pressures.

SUMMARY

This unit has provided you with much information about motors. Table 15-3 provides a summary of the motors discussed. It includes the motor type, starting torque, running efficiency, operating cost, and usual applications. This motor comparison table will help you review and contrast the motors discussed in this unit.

Finally, use Table 15-4 to aid in troubleshooting motor problems. Some of the conditions listed in the table were covered in this unit. This table will give you more ideas on what needs to be checked to solve problems. You will notice in the table that some of the problems are electrical and mechanical.

Table 15-3 Motor Comparison Chart

Motor Type	Starting Torque	Running Efficiency	Operating Cost	Applications
Shaded pole	Low	Poor	Low	Small fans and pumps
Split-phase	Moderate	Poor	Moderate	Pumps and fans
Capacitor start/induction run (CSIR)	High	Poor	Moderate	Pumps and fans
Permanent split capacitor (PSC)	Medium	Good	Moderate	Single-phase motors, compressors, and fans
Capacitor start/capacitor run (CSCR)	High	Good	Moderate	Single-phase motors, compressors
Three-phase	High	High	High	Commercial and industrial uses

(Source Motor Comparison Table 05-03-06.)

Table 15-4 Motor Troubleshooting Chart

This table offers electrical and mechanical suggestions for solving motor problems.

Your motor service and any troubleshooting must be handled by qualified persons who have proper tools and equipment.

Trouble	Cause	What to Do
Motor fails to start	Blow fuses	Replace fuses with proper type and rating
	Overload trips	Check and reset overload in starter.
	Improper power supply	Check to see that power supplied agrees with motor nameplate and load factor.
	Improper line connections	Check connections with diagram supplied with motor.
	Open circuit in winding or control switch	Indicated by humming sound when switch is closed. Check for loose wiring connections. Also see that all control contacts are closing.
	Mechanical failure	Check to see if motor and drive turn freely. Check bearings and lubrication.
	Short-circuited stator	Indicated by blown fuses. Motor must be rewound.
	Poor stator coil connection	Remove end bells, locate with test lamp.
	Rotor defective	Look for broken bars or end rings.
	Motor may be overloaded	Reduce load.

(continued)

Table 15-4 *(continued)*

Trouble	Cause	What to Do
Motor stalls	One phase may be open	Check lines for open phase.
	Wrong application	Change type or size. Consult manufacturer.
	Overload	Reduce load.
	Low voltage	See that nameplate voltage is maintained. Check connection.
	Open circuit	Fuses blown, check overload relay, stator and pushbuttons.
Motor runs and then dies down	Power failure	Check for loose connections to line, to fuses and to control.
Motor does not come up to speed	Not applied properly	Consult supplier for proper type.
	Voltage too low at motor terminals because of line drop	Use higher voltage on transformer terminals or reduce load. Check connections. Check conductors for proper size.
	Starting load too high	Check load motor is supposed to carry at start.
	Broken rotor bars or loose rotor	Look for cracks near the rings. A new rotor may be required as repairs are usually temporary.
	Open primary circuit	Locate fault with testing device and repair.
Motor takes too long to accelerate and/or draws high amp	Excessive load	Reduce load.
	Low voltage during start	Check for high resistance. Adequate wire size.
	Defective squirrel cage rotor	Replace with new rotor.
	Applied voltage too low	Get power company to increase power tap.
Wrong rotation	Wrong sequence of phases	Reverse connections at motor or at switchboard.
Motor overheats while running under load	Overload	Reduce load.
	Frame or bracket vents may be clogged with dirt and prevent proper ventilation of motor.	Open vent holes and check for a continuous stream of air from the motor.
	Motor may have one phase open	Check to make sure that all leads are well connected.
	Grounded coil	Locate and repair.
	Unbalanced terminal voltage	Check for faulty leads, connections and transformers.
Motor vibrates	Motor misaligned	Realign.
	Weak support	Strengthen base
	Coupling out of balance	Balance coupling.
	Driven equipment unbalanced	Rebalance driven equipment.
	Defective bearings	Replace bearing.
	Bearings not in line	Line up properly.
	Balancing weights shifted	Rebalance motor.
	Polyphase motor running single phase	Check for open circuit.
	Excessive end play	Adjust bearing or add shim.

Table 15-4 (*continued*)

Trouble	Cause	What to Do
Unbalanced line current on polyphase motors during normal operation	Unequal terminal volts	Check leads and connections.
	Single phase operation	Check for open contacts.
	Unbalanced voltage	Correct unbalanced power supply.
Scraping noise	Fan rubbing air shield	Remove interference.
	Fan striking insulation	Clear fan.
	Loose on bedplate	Tighten holding bolts.
Noisy operation	Airgap not uniform	Check and correct bracket fits or bearing.
	Rotor unbalance	Rebalance.
Hot bearings general	Bent or sprung shaft	Straighten or replace shaft.
	Excessive belt pull	Decrease belt tension.
	Pulleys too far away	Move pulley closer to motor bearing.
	Pulley diameter too small	Use larger pulleys.
	Misalignment	Correct by realignment of drive.
Hot bearings ball	Insufficient grease	Maintain proper quantity of grease in bearing.
	Deterioration of grease or lubricant contaminated	Remove old grease, wash bearings thoroughly in kerosene and replace with new grease.
	Excess lubricant	Reduce quantity of grease, bearing should not be more than 1/2 filled.
	Overloaded bearing	Check alignment, side and end thrust.
	Broken ball or rough races	Replace bearing, first clean housing thoroughly.

(Courtesy Regal Beloit.)

REVIEW QUESTIONS

1. What components should be changed out when replacing a CSCR motor with an electrical burnout?
2. What are the advantages of a three-phase motor over a single-phase motor?
3. What is a split-phase motor?
4. Which has more resistance, the start winding or the run winding?
5. Which single-phase motor has the best starting torque and running efficiency?
6. Of all motor types discussed in this unit, which has the best starting torque and best running efficiency?
7. What are three ways to troubleshoot a shaded pole motor?
8. What are four ways to troubleshoot a permanent split capacitor motor?
9. What are six ways to troubleshoot a CSCR motor?
10. What components make up a hard start kit?
11. You find a defective 55-mfd run capacitor on a compressor. You have a 40- and a 75-mfd capacitor in the service van. Is one of these capacitors an adequate replacement for the defective cap? Explain.
12. You are going to replace a CSCR compressor. Are there any electrical components that must also be replaced? If so, what are they?
13. Summarize the 11 points that you will need to know when selecting a replacement for a motor that does not have a nameplate.
14. What resistance scale is used when checking motor windings?
15. What resistance scale is used when checking the motor to ground?
16. What single-phase motor would you select if starting torque or running efficiency is not a concern?
17. What single-phase motor would you select if starting torque is not a concern, but running efficiency is?
18. Which single-phase motor(s) have good starting torque?
19. When a motor is operating, is the coil voltage on a potential relay lower, the same as, or higher than the applied voltage?

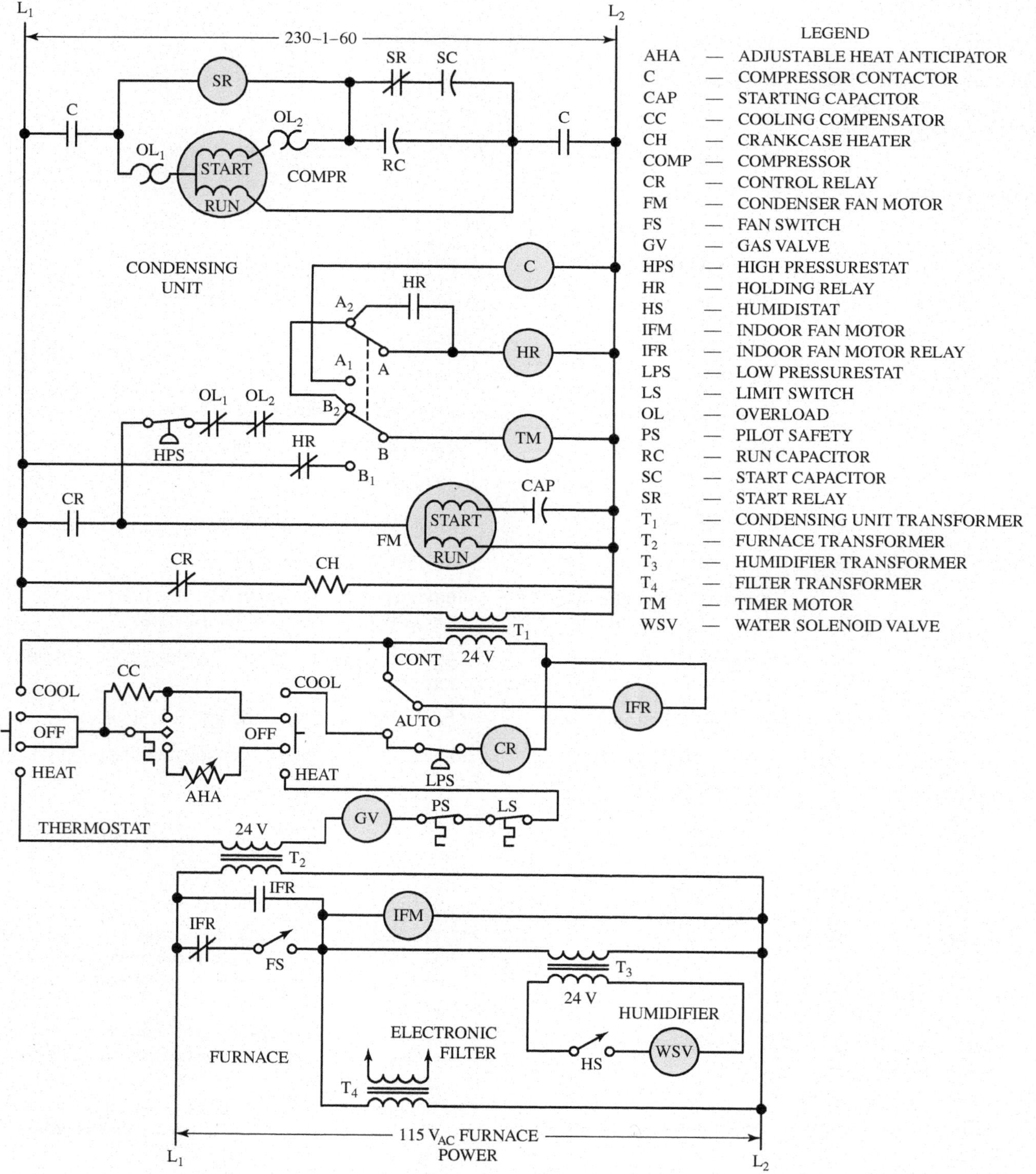

Figure 15-55 Use this diagram to answer Review Questions 27 through 31.

Figure 15-56 Use this image to answer Review Question 32.

Figure 15-57 Use this image to answer Review Question 33.

Figure 15-58 Use this image to answer Review Question 34.

20. How long is the start capacitor in a compressor circuit?
21. When troubleshooting a potential relay coil your reading is 5,440 Ω. Is there a problem?
22. The reading between connections 2 and 1 on the potential relay is 0.5 Ω. Is there a problem? If so, what is the problem?
23. When troubleshooting potential relay contacts, your reading is near 0 Ω. Is there a problem?
24. Name four common motor mounts.
25. Name the five types of single-phase motors discussed in this unit.
26. Why is it important to change the start components when changing out a motor?
27. How is the potential relay identified in Figure 15-55 (or enlarged Electrical Diagram ED-5, which appears with the Electrical Diagrams package that accompanies this book)? What is it called?
28. What type of motor is the compressor in Figure 15-55?
29. What type of motor is the fan motor in Figure 15-55?
30. What type of motor is the indoor fan motor in Figure 15-55?
31. How many motors are shown in Figure 15-55?
32. What type of motor enclosure is shown in Figure 15-56?
33. What type of motor enclosure is shown in Figure 15-57?
34. What type of motor enclosure is shown in Figure 15-58?
35. What type of motor is shown in Figure 15-59?
36. Is the diagram of the potential relay on the right side in Figure 15-60 correct? If not, draw the correct relay on the left diagram. What is the highest coil voltage rating on this relay?

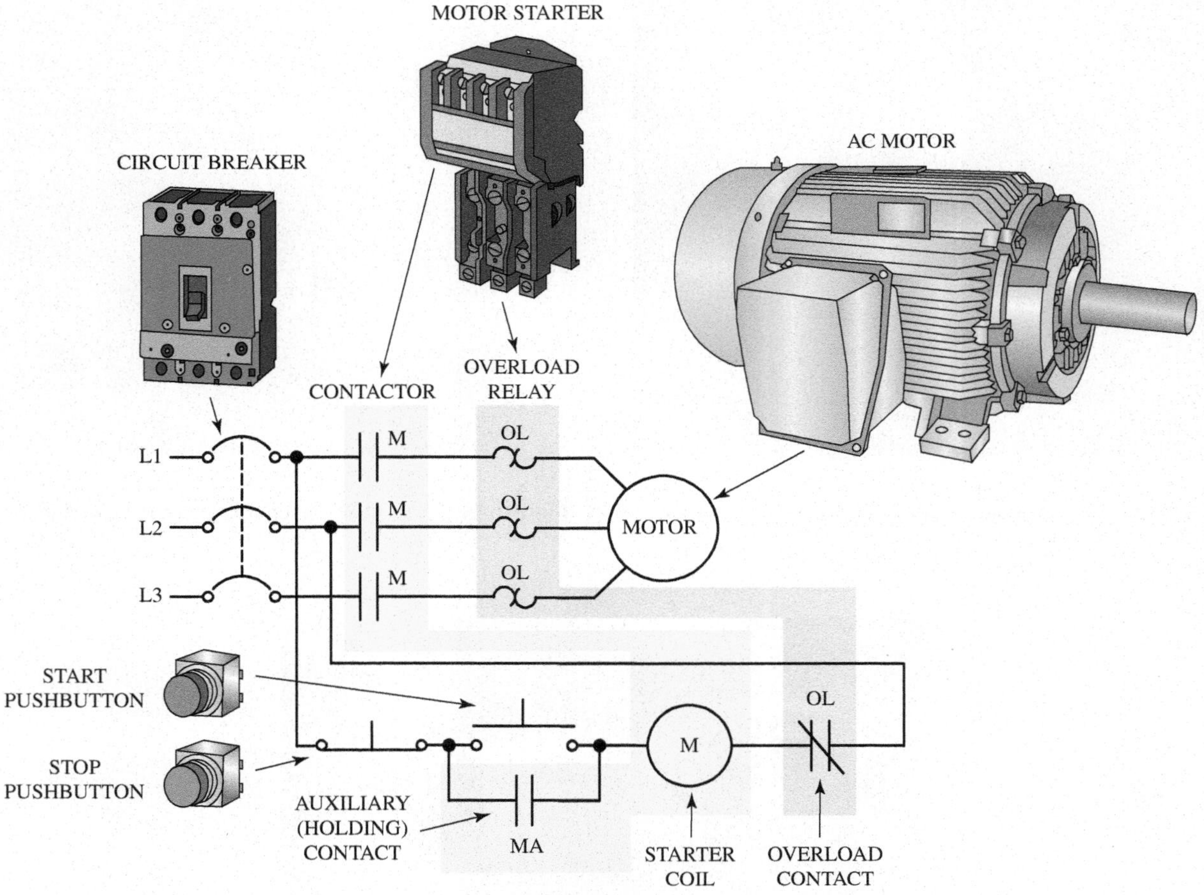

MOTOR STARTER
CIRCUIT BREAKER
AC MOTOR
CONTACTOR
OVERLOAD RELAY
L1
L2
L3
M
OL
MOTOR
START PUSHBUTTON
STOP PUSHBUTTON
AUXILIARY (HOLDING) CONTACT
MA
STARTER COIL
OVERLOAD CONTACT

Figure 15-59 Use this image to answer Review Question 35.

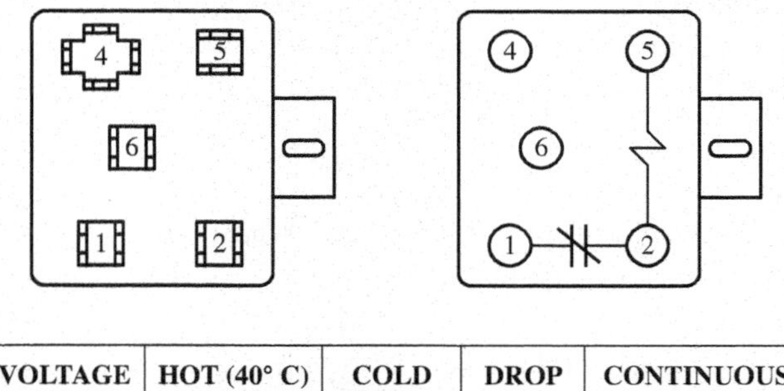

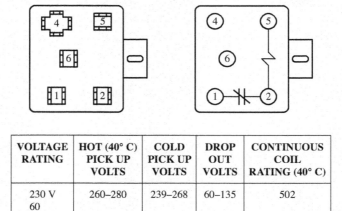

VOLTAGE RATING	HOT (40° C) PICK UP VOLTS	COLD PICK UP VOLTS	DROP OUT VOLTS	CONTINUOUS COIL RATING (40° C)
230 V 60 1 ϕ	260–280	239–268	60–135	502

Figure 15-60 Use this image to answer Review Question 36.

UNIT 16

ECM: The Green Motor

WHAT YOU NEED TO KNOW

After studying this unit, you will be able to:

1. Describe how an electronically commutated motor (ECM) operates.
2. List the basic components of an ECM.
3. Discuss the applications of the ECM.
4. Describe basic ECM troubleshooting.
5. Compare efficiencies of the PSC motor and the ECM.
6. Describe how to select a replacement ECM.

The purpose of this unit is to acquaint you with the **electronically commutated motor (ECM)**, which is simply a high-efficiency motor. A basic understanding of the ECM motor is important to conduct fundamental troubleshooting. This unit is not intended to replace the excellent training material developed by Regal Beloit. More information on the ECM can be found at the Regal Beloit website.

The ECM is so efficient it is sometimes called a "green motor." The word *green* refers to an environmentally friendly method of getting a job done. Why is the ECM a friend of the environment? It uses much less electricity than a permanent split capacitor (PSC) motor. According to Genteq®, the average PSC is 12% to 45% efficient, while the ECM is 65% to 80% efficient. Most energy sources used to generate electricity use a pollution-generating fuel like coal, natural gas, or oil. Because the ECM uses less electricity, less pollution is generated. Also, if properly set up, an ECM helps reduce moisture in an air-conditioned spaced. Lower moisture or lower relative humidity means more comfort, which normally translates into a customer raising the thermostat temperature by a degree or two. The higher temperature setting results in less running time for the equipment—therefore, less energy is used.

This green motor is found in over 10 million high-end products, including variable-speed products. With proper installation, setup, and periodic maintenance, these motors provide reliability in today's high-end HVACR systems. The ECM is used in energy-efficient residential, commercial, and refrigeration products. *Note:* When ECMs were first developed, they were called **integrated control modules**, abbreviated **ICM**. You may run across older sources that use the ICM term.

Most ECMs are used in variable-speed blower applications. When designed and properly setup in the field, the ECM in an air conditioning system air handler can vary the motor speed depending on the load and duct design. For example, if a room is warm and the equipment is in the air conditioning mode, the ECM will run at full rated capacity or maximum airflow rating. When the temperature in the room falls close to the thermostat setpoint, the programmed circuit board will reduce the speed of the air handler. The reduction of the motor speed will use less energy and the air going through the evaporator coil will make more contact with the cold coil surfaces. The increased coil contact time will improve the amount of moisture removed by the coil. This will result in less moisture in the space, which reduces the relative humidity (RH) in the space. Lower RH means drier air. Drier air in hot and humid climates is more comfortable; therefore, customers tend to increase the thermostat setting by a degree or two, thereby reducing air conditioning system run time. Again, less run time equals less energy use and less pollution.

Each original equipment manufacturer (OEM) designates a special program unique to the ECM for each piece of its equipment. This program provides multiple airflow and comfort options for each demand of the system in which the motor is installed. For this reason, any replacement ECM must come from the OEM.

Additionally, because an ECM uses less energy, it runs cooler. Less heat is added to the air stream, and cooler motors tend to last longer.

In summary, the ECM reduces the energy used by the blower motor at lower motor speeds. The ECM can be programmed to reduce the airflow through an evaporator, thus reducing the moisture content of the air. If the air is drier, the customer is more likely to increase the thermostat setting an energy-saving degree or two. ECMs are also used on condenser fans and compressor motors. They are also found moving air through the evaporator of a walk-in cooler or freezer.

TECH TIP

The air handler blower on a commercial system usually operates continuously. The continuous operation provides air circulation in the space and brings in fresh, outside air to reduce indoor air pollutants. On a residential air conditioning system, the blower can operate continuously or it can cycle with the condensing unit. Customers need to know the advantages and disadvantages of each option so that they can decide which type of operation to use. Here is information that will help them make a decision about whether to use continuous fan circulation or cycle the fan with the condensing unit:

- Continuous fan operation will improve air quality, because the air is always being filtered. Note, though, that the air filters may need to be changed more often in this case since more air is being filtered. Continuous operation in the cooling mode will also increase the relative humidity in the space. The off cycle air movement will evaporate some of the moisture on the coil and in the drain pan. This moisture will be re-evaporated into the air. Finally, air duct leaks and air leaks around the air handler will either pull in outside air or exhaust conditioned air from the space.
- Cycling the fan with the condensing unit will reduce the fan motor operating costs and increase the life of the motor. There will be less infiltration dust and unconditioned air due to air duct leaks. The fan cycling operation usually allows the evaporator coil to chill for a minute prior to starting the blower. The fan continues to operate about a minute after the condensing unit cycles off to push the cooled air from the duct system into the space.

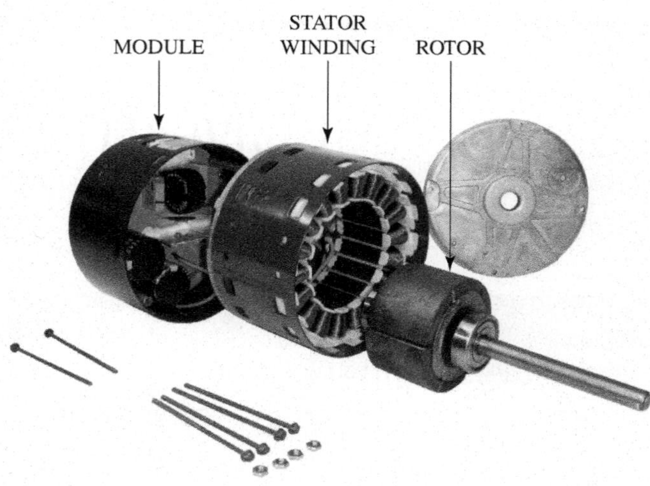

Figure 16-1 Basic components of an ECM. At left is the control module, in the middle the three-phase stator, and on the right the rotor, which is pulled out of the motor. Not shown is the circuit board.

16.1 INDUSTRY STANDARDS

The Extended Industry Standard Architecture (EISA) took aim at motors because from mining to air movement, motors account for nearly 50% of total U.S. energy used in those areas and two-thirds of all electrical energy used in industrial settings. Even small increases in motor efficiency could lead to large energy savings over the lifetime of a motor.

EISA is aimed at manufacturers. Since December 19, 2010, manufacturers have not been allowed to make general-purpose motors of 1 to 200 hp at less than NEMA Premium efficiency for use in the United States. (Motors made prior to that date can still be sold and installed.) EISA is applicable to motors constructed and shipped individually or as components within equipment.

To comply with EISA, engineers need to check their specifications to ensure that the new minimum motor efficiencies are brought up to date for fans, pumps, cooling towers, chillers, etc. ECMs exceed these new energy efficiency requirements.

16.2 WHAT IS AN ECM?

An electronically commutated motor is a three-phase, brushless DC motor. It is comprised of three major components: a circuit board, an electronic control module, and a three-phase motor with a permanent magnet rotor, as shown in Figure 16-1. The module and motor make up the actual ECM. The circuit board is used to set up cooling and heating applications.

The circuit board is usually shared by other switches and controls in an HVAC system. There are dip switches and jumper pins on the HVAC system circuit board that allow changes to airflow and comfort options related to ECM operation. The control module is controlled through the circuit board, which has other functions such as controlling the cooling and heating system. Figure 16-2 illustrates a common circuit board used to control an ECM as well as heating and cooling operations. Figure 16-3 is a close-up of the circuit showing dip switches that must be set in the on or off position per manufacturer's recommendations.

The control module in Figure 16-4 converts 120- or 240-V single-phase power to essentially three-phase DC power to operate the motor. The control module is a power inverter, which means it converts AC to DC. As shown in Figures 16-5 and 16-6, the motor appears as if it were a regular motor. The additional length is the electronic module that converts the single-phase alternating current to three-phase direct current power. Figure 16-7 shows the module and motor connected to the high- and low-voltage wiring.

In summary, the ECM control converts AC single-phase power to DC power and then pulses that DC voltage reversing polarity to the three-phase motor windings.

An ECM can change the 60-Hz frequency that controls the speed of the motor. An ECM can control the frequency of the cycle rate it delivers to the motor; that is, it is not a slave to the 60-Hz power supplied by the utility company like PSC motors are. Remember that frequency is an important part of the RPM formula. The lower the frequency, the lower the RPM or motor speed. As the motor speed decreases, the current draw decreases, making the motor more efficient. An ECM not only operates at lower RPMs than most PSCs, but it is also more efficient at those speeds.

Figure 16-2 The circuit board directs the thermostat, which in turn controls the ECM. This thermostat has dip switches or settings that must be set up to operate the blower based on heating operation and cooling tonnage.

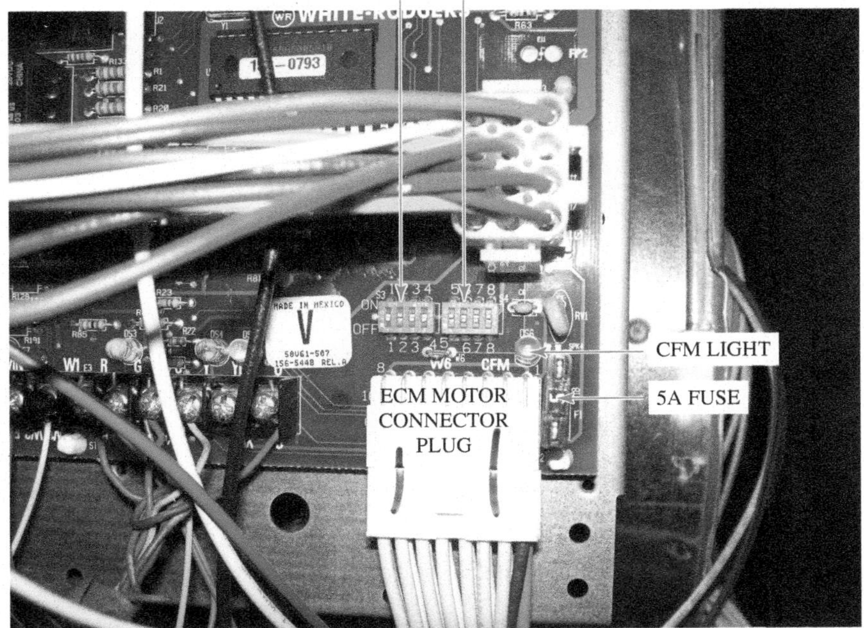

Figure 16-3 On this circuit board eight dip switches must be set to properly cycle the heating and cooling equipment. Also shown are a green CFM light and a 5-amp circuit board fuse.

Here are the reasons ECMs are gaining so much popularity among equipment manufacturers:

- Has high-efficiency operation. ECMs are about 80% efficient when compared to the best PSC motor at 60%. This improves the efficiency rating of a condenser, air handler, or furnace.
- Adds little heat to the air stream. ECMs run cooler, thus adding less heat to the air circulating through them.

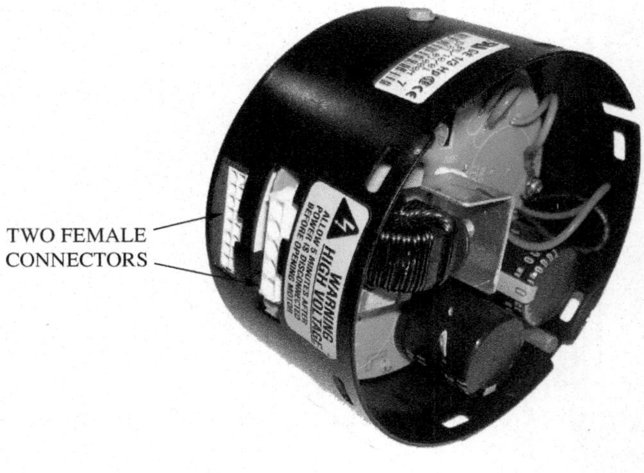

Figure 16-4 This is a control module that has been removed from a motor. Two sets of wires plug into the side of the module.

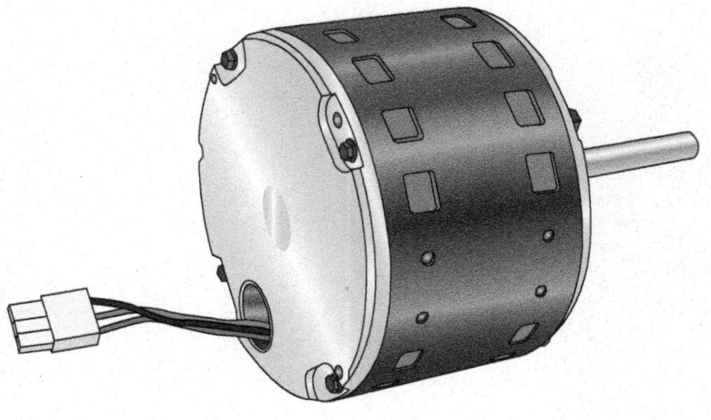

Figure 16-5 This is a three-phase DC motor. The motor resistance is approximately equal. The resistance is less than 20 ohms between the windings.

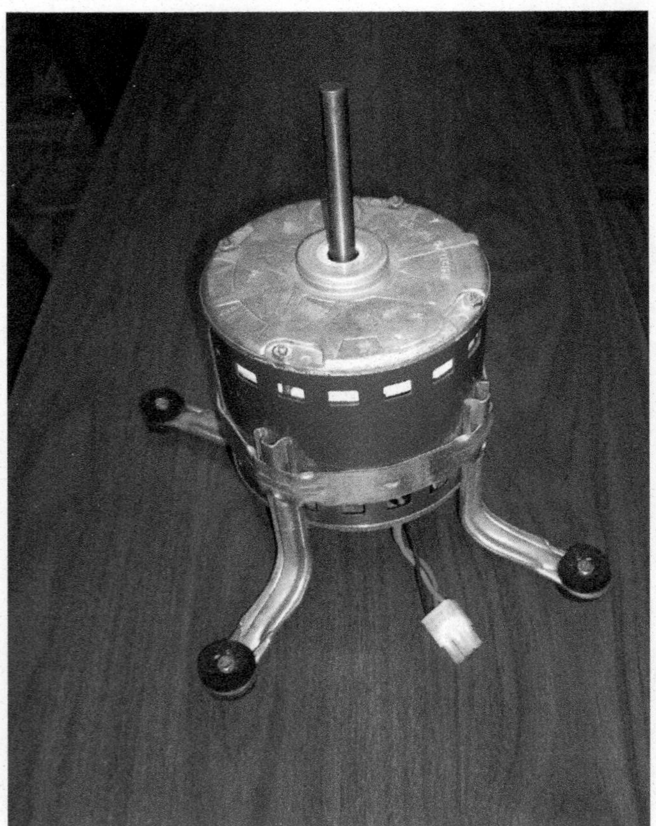

Figure 16-6 Another view of an ECM without the control module. The belly band is placed around the motor so as not to cover the ventilation openings. The connector is plugged into the control module.

Figure 16-7 The lower black module has two connectors. The upper connector is the high-voltage source. The lower connector is the low-voltage control source from the module. The gray motor is in the upper part of the picture. The circuit board is not shown, but will be connected to the other end of the two connectors.

- Reduces relative humidity in the cooling mode when programmed for lower air speeds and ramping up and down air speeds during the cooling cycle.
- Starts unloaded. The ECM starts up slowly, which reduces locked rotor amps and places less stress on the blower wheel, fan blade, and blower housing. This benefit eliminates sudden gusts of air from the duct as well as the abrupt shutdown of airflow at the end of the

air cycle. The customer is less likely to notice that the HVAC equipment is cycling on and off. This is a selling point for this type of equipment.

- Uses standard ball bearings. This makes for long bearing life and the ability to carry heavy loads.
- Is able to set the motor speed across a wide range of settings for heating, cooling, and condenser pressure.
- ECMs used as air handlers can be set to operate at a specific CFM. For example, on the circuit board most ECMs are set up to operate at 400 CFM per ton. A setting of 350 CFM per ton can easily be selected to slow the blower down and allow for more moisture removal when compared to 400-CFM operation. Speeds for condenser fan motors will have a higher CFM rating.
- Lower air speeds or ramping the airflow up and down traps more dust in the air filter. This makes the air filter more efficient, but it also means that the filter will need to be changed more often. Higher air speeds creates more contamination blowby in air filters.
- Overcomes minor undersized duct designs with greater airflow. The ECM is not designed for this purpose, but it does offer this benefit.

SAFETY TIP

Do not trust the blower door switch for power disconnection or reconnection. Frequent arcing across the door switch may lead to door switch failure. Use the main disconnect to remove power on any system with an ECM motor before removing or replacing panels with a door switch. Always confirm with a voltmeter that power has been de-energized. Check voltage line to line and each line to ground when you think the power is off. Finally, short the "dead" power to ground before touching it with your body.

In summary, Figure 16-8 provides a cutaway view of an ECM motor and module sections. The control module is on the left side and the motor is on the right side. The cutaway exposes the stator and rotor motor section.

Figure 16-9 is a flowchart explaining what we discussed in this section. It gives an example of how an ECM motor will operate in an air handler or furnace application. The ECM is comprised of the ECM control module and three-phase motor. Starting at the left side of the flowchart, on a call for heating or cooling the thermostat closes and completes the circuit to the circuit board in a HVAC system. In most cases the t'stat is the switch that closes and provides a path for 24 volts to the circuit board. The circuit board is programmed to operate the ECM at a specific CFM or at full speed. Many circuit boards can be set to start the ECM at a low speed and slowly ramp up to the programmed speed over a period of 5 to 8 minutes. The motor will slow down and stop when the

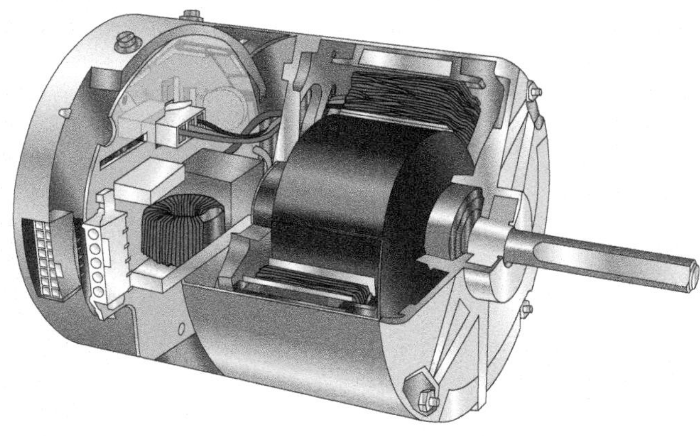

Figure 16-8 A cutaway view of an ECM motor and module sections. The control module is on the left, the motor on the right.

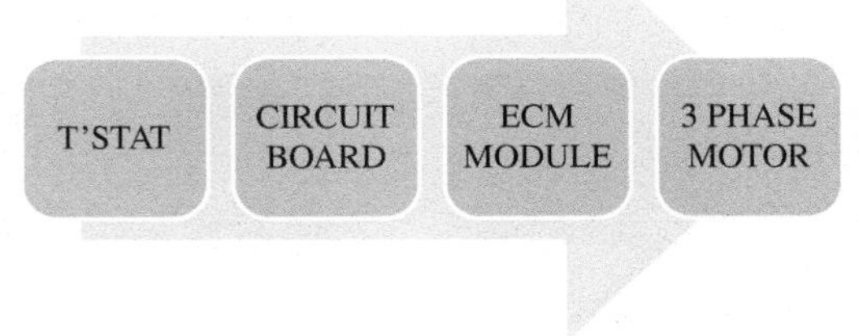

Figure 16-9 Flowchart illustrating ECM motor operation in an air handler or furnace application.

thermostat is satisfied and the fan switch is set to the automatic position.

16.3 ECM WIRING

Genteq ECMs are built in the following applications: 115, 120/240 dual voltage, 230, or 460 volts. Check this prior to applying any voltage. The module has two electrical connections, as shown in Figure 16-10. The plug connect close to the motor is for continuous AC power and the other connection is low voltage, which comes from the circuit board. As shown in Figure 16-11, the module does not need to be attached to the ECM body.

Figure 16-12 is a diagram showing the high-voltage electrical connections. The connector on the left has a black, white, and ground wire. If it is 120 volts, the black wire will be hot and the white wire will be neutral. The green wire is ground. The connector on the right is a 240-V motor. The black and white wires will both be hot. There will be no neutral. Remove the jumper wire that goes between terminals 1 and 2.

Figure 16-13 is the high-voltage connector that would plug in to the motor. It has a black, red, and green ground wire. If it is 120 volts, the black wire will be hot and the white

Figure 16-10 This is the module end of an ECM. The lower connector is the low-voltage single input. The upper connector with the black, white, and green wire is the continuous voltage input.

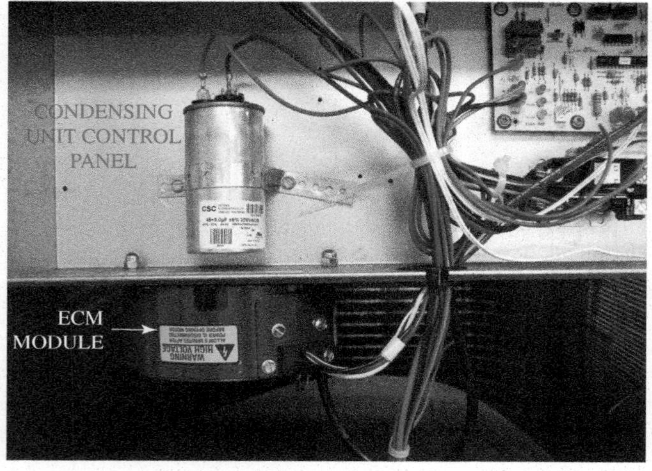

Figure 16-11 The module does not need to be attached to the ECM motor body. Here a condenser fan ECM module is installed under the condensing unit control panel since there is not enough room for the full assembly if installed in one piece.

wire will be neutral. The green wire is ground. If it is a 240-V motor, the black and red wires will both be hot. There will be no neutral. The designs of the high-voltage and low-voltage connectors are different to prevent mistakes during hook-up. These connectors are designed to come together in one way, eliminating a connection error on the module.

Electrical power is supplied to the module at all times even when the motor is not running. Power to the motor should be disconnected before removing the connect plugs. Plugging or unplugging the module with power applied may cause arching in the module. This arching may damage the module. Also, before opening the module and separating it from the motor, allow 5 minutes for the large capacitors to discharge.

TECH TIP

The shaft of a de-energized ECM will not turn freely without being powered up. When power is removed the shaft will have a drag or "cogging feel" from the permanent magnets that are used to make up the rotor. It will seem as if the ball bearings are worn and dragging.

16.4 INSTALLATION SETUP

An air handler or furnace with an ECM must be set up prior to initial operation. If the dip switches on the circuit are not set up for the specific application, the motor will operate at factory setting speeds that may or may not be what is best for the application. An ECM condenser fan does not normally need to be set up. It is programmed and ready to operate per OEM specifications.

For an air handler or furnace, it is important to first check the polarity of the 120-V power supply. The hot power leg and neutral must be connected to the right places. The hot wire is "usually" black and the neutral white. The ground wire is green. To check the hot wire, 120 volts should be measured from the hot wire to ground. If 120 volts is not measured, then you are not measuring the hot wire or there is no ground or no supplied power. Next, measure the neutral to ground. It should measure no voltage or 0 V. Without the proper 120-V polarity to the circuit board, the equipment will not work at all or it will operate erratically. What happens with reversed power polarity varies from unit to unit; therefore, we cannot go over all the possible problems that can arise when this condition occurs. There will, however, be some really unusual

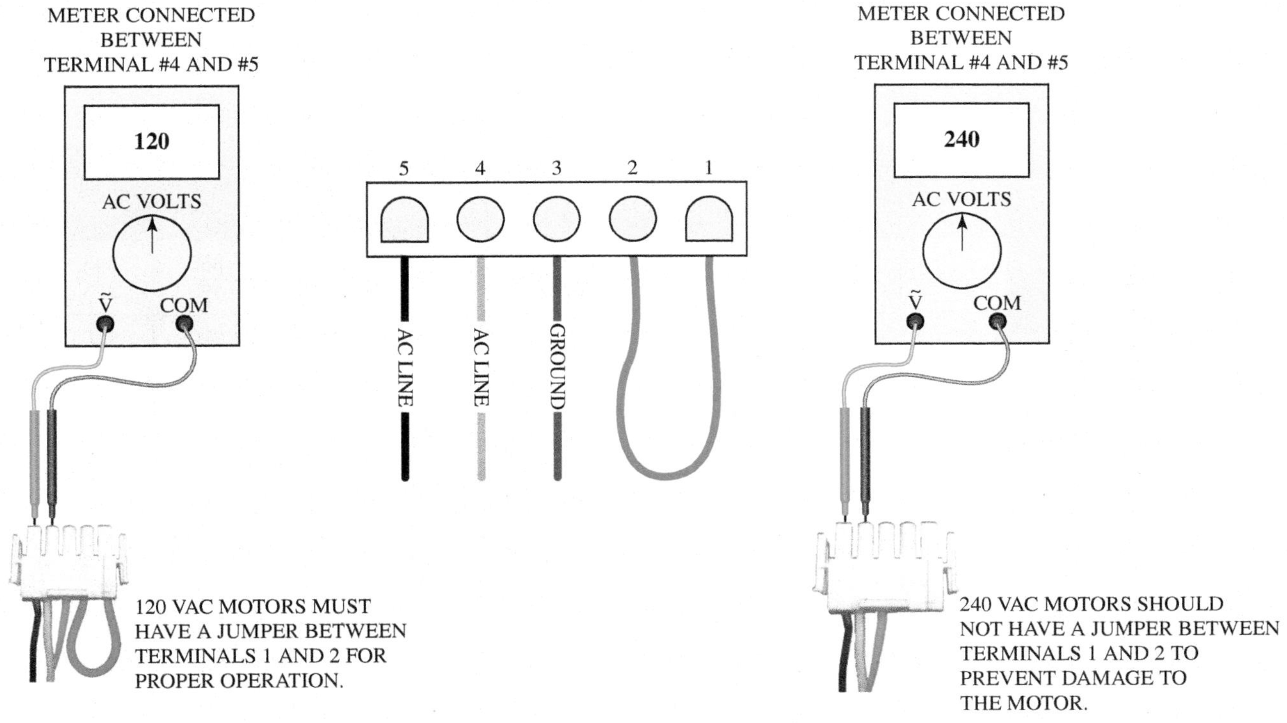

Figure 16-12 This is the high-voltage connector closest to the motor.

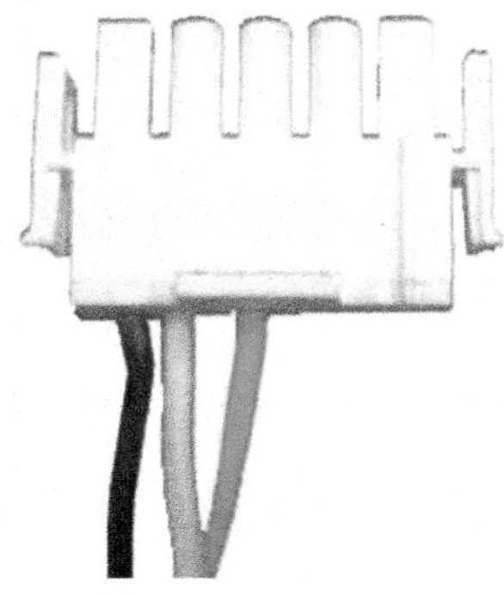

Figure 16-13 This connector goes into the module input nearest the motor. This is a 240-V connector. The wires are color coded. Black and white or red is high voltage and green is ground.

problems. A good ground is important for all pieces of equipment, with or without circuit boards. In many instances it is a matter of safety to protect the technician from getting electrocuted.

With 240 volts, there are two hot wires and a ground. There is no polarity issue here, but again the proper ground is extremely important.

After the correct voltage and polarity have been established, the next step is to follow the OEM setup procedures. This is often overlooked in an installation and start-up. The tech doing the install thinks that the equipment is already set up from the factory for her specific application. For example, the focus is usually the circuit board. A jumper wire may need to be added or a jumper wire may need to be clipped under specific operating conditions. As seen back in Figure 16-2, many circuit boards have small dip switches or rotary switches that need to be set for cooling and heating operations. The user interface or thermostat may need programming.

Table 16-1 is a chart from the inside door of a gas furnace that lists the dip switch settings required for a specific ECM operation. For example, let's assume the system installed is a 3-ton unit. The tech wants to set the blower for enhanced moisture removal. He therefore selects the 350 CFM/ton airflow row for a 3-ton unit. Here is the setup for the dip switches:

SW 1 is on.

SW 2 is off.

SW 3 is off.

SW 4 is on.

This setting will give the 3-ton setting of 1,050 CFM when the ECM is working at 100% speed.

The CFM and watts of the motor can be determined if the **external static pressure (ESP)** is measured. ESP is a measurement of ductwork resistance to airflow. In the previous example, an external static pressure (ESP) at 0.5 for the setting described here will tell the tech that the airflow is 1,000 CFM and the power consumed is 260 watts.

Table 16-1 Dip Switch Settings for a Gas Furnace

This chart is found inside the door of a gas furnace. The settings must be set to operate the furnace and cooling system at the designed conditions.

*UD080R9V3K Furnace Cooling Airflow (CFM) and Power (Watts) vs. External Static Pressure with Filter

Outdoor Unit Size (Tons)	Airflow Setting	Dip Switch Setting					External Static Pressure				
		SW 1	SW2	SW3	SW4		0.1	0.3	0.5	0.7	0.9
2.5	LOW (350 CFM/TON)	OFF	ON	OFF	ON	CFM WATTS	880 120	875 155	860 190	845 225	840 245
	NORMAL (400 CFM/TON)	OFF	ON	OFF	OFF	CFM WATTS	1020 170	1000 205	990 240	980 280	960 320
	HIGH (450 CFM/TON)	OFF	ON	ON	OFF	CFM WATTS	1110 210	110 260	1110 320	1100 350	1100 385
3.0	LOW (350 CFM/TON)	ON	OFF	OFF	ON	CFM WATTS	1040 190	1010 220	1000 260	1000 310	990 340
	NORMAL (400 CFM/TON)	ON	OFF	OFF	OFF	CFM WATTS	1200 250	1200 320	1190 370	1190 415	1175 450
	HIGH (450 CFM/TON)	ON	OFF	ON	OFF	CFM WATTS	1340 355	1340 425	1330 475	1320 530	1300 570
3.5**	LOW (350 CFM/TON)	OFF	OFF	OFF	ON	CFM WATTS	1215 265	1210 330	1210 375	1200 430	1185 465
	NORMAL** (400 CFM/TON)	OFF	OFF	OFF	OFF	CFM WATTS	1430 415	1415 457	1410 520	1385 575	1330 580
	HIGH (450 CFM/TON)	OFF	OFF	ON	OFF	CFM WATTS	1430 415	1415 475	1410 520	1385 575	1330 580

NOTES:

1. *First Letter may be "A" or "T"
2. **Factory setting
3. Continuous Fan Setting: Heating or Cooling airflow is approximately 50% of selected Cooling value.
4. For Variable Speed: low speed airflows are approximately 30% of listed values.
5. LOW 350 CFM/TON is recommended for Variable Speed application for Comfort & Humid Climate setting; Normal is 400 CFM/TON; High 450 CFM/TON is for Dry Climate setting

(Courtesy Trane Corporation.)

You will notice that with an ECM, the CFM stays about the same even as the airflow becomes restricted (increased ESP). The number of watts used by the motor increases as the ESP increases. Increased watts means the motor is going to run hotter. This prolonged heat can shorten the module's life.

Again it is important to measure the external static pressure (ESP) across the ECM. The device used to measure static pressure is called a *pressure differential gauge*. The common brand name is made by Dwyer Instruments and it goes by the trademarked name of Magnehelic. You are more likely to hear the word *Magnehelic* rather than the correct name of pressure differential gauge. The Magnehelic measures very low pressure drop, which is expressed as inches of water column and abbreviated ″WC. Here is an example of the low pressure differential we are discussing here. One-tenth of an inch of water column (0.1) is equal to 0.04 psi.

ESP in excess of 0.90″WC is usually damaging to the module due to excess heat. Check the manufacturer's specs to determine the maximum ESP. High static (restricted airflow in the duct system) will cause the motor to operate on the high end of its design conditions, which leads to high current draw, an overheated motor module, and shortened component life.

Figure 16-14 shows a circuit board with a green CFM light. For this model, one blink equals 100 CFM; therefore, a 10-blink sequence will mean the air handler is moving 1,000 CFM (10 blinks × 100 CFM). The air handler or furnace panels and ductwork for the unit need to be installed to obtain an accurate measurement since removing the door may allow

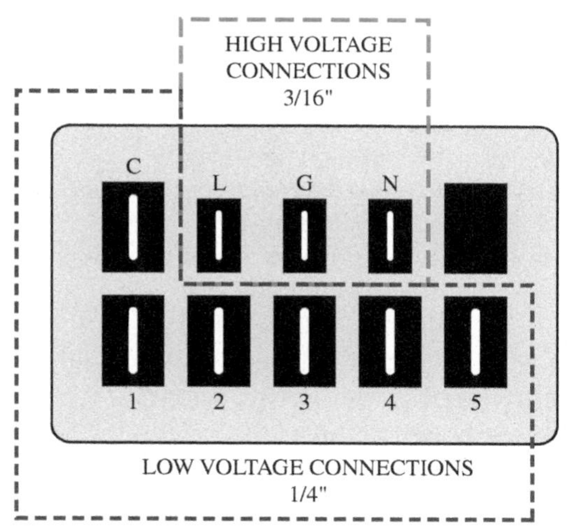

Figure 16-19 Terminals C, 1, 2, 3, 4, and 5 are used for the low control voltage connection. Some of these control connections may not be hooked up.

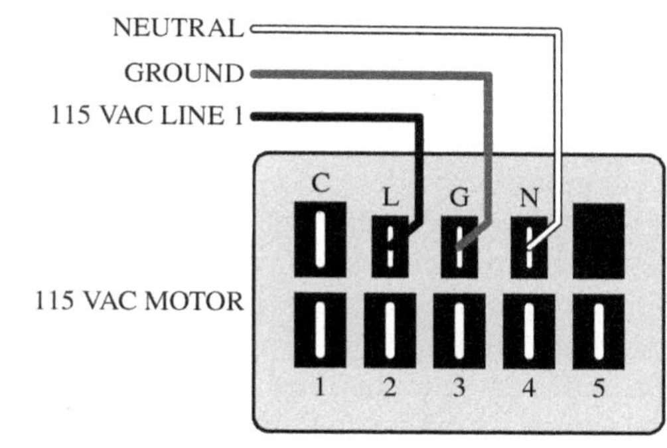

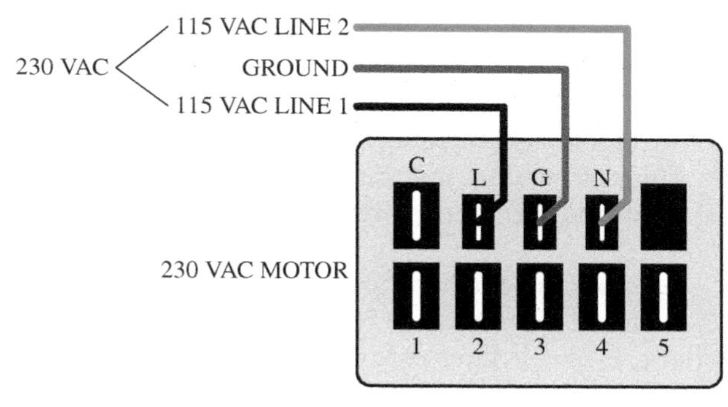

Figure 16-20 The high-voltage section can be designed for 115, 230, or 460 volts.

The motor is operated by two other programs. One program is the HVAC OEM torque value per demand and the other is the constant-torque program.

Installation and Setup of a Constant-Torque Motor

As previously shown in Figure 16-17, a constant-torque motor has one connection block for both the high-voltage and low control voltage connections. Figure 16-19 shows the connection diagram. Terminals **C**, **1**, **2**, **3**, **4**, and **5** are used for the low control voltage connection. The low-voltage hook-up uses a 1/4-in. female connector. Some of these control connections may not be used or wired. The **L**, **G**, and **N** connectors are used for 115/120-, 230/240-, or 460-V supply voltage. The motor voltage needs to be stipulated when changing the motor or changing the module.

The high-voltage hook-up uses a 3/16-in. female connector. The different connector size reduces the chance of cross connecting the high- and low-voltage motor circuits. This high voltage will be energized at all times, even when the motor is off. It is used to control the ECM electronics and drive the motor.

The high-voltage section can be wired for 115, 230, or 460 volts. The X13 is not a dual-voltage motor. The motor must be used as built with the proper line voltage. In Figure 16-20, note that the polarity must be observed when wiring 115 volts. The **L** connection is hot. The **G** connection is ground. The **N** connection is neutral. Normally the black wire is hot and white is neutral, but check it to be sure. *Note:* In Canada the hot wire must be marked or flagged if a white wire is used.

The 230- and 460-V connections require the power to be connected to **L** and **N**, and ground to be connected to **G**. There is no polarity with this voltage.

The **C** on the motor terminal in Figure 16-20 is 24 volts AC common. A 9- to 23-V DC signal can also be used on this connection. The **1** through **5** are communication inputs, which determine what torque is programmed into the motor. Only one of these terminals is energized to give the desired operating characteristics. Figure 16-21 shows a simple heat-cool setup, where **C** is the common 24-V

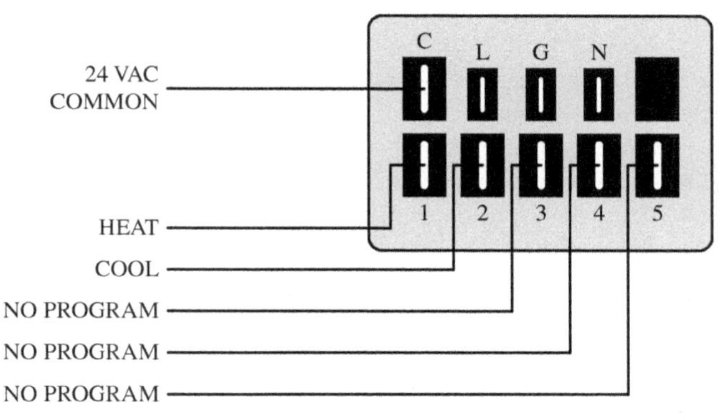

Figure 16-21 This shows a simple constant-torque heat-cool setup.

connection. Terminal **1** is for heat blower operation and terminal **2** is for air conditioning blower operation. The remaining terminals are not programmed on this motor. The high voltage will be connected to **L** and **N**, and **G** will be the motor ground.

Taps have programmed torque values, not speeds. OEMs can program taps in any order with each tap having a unique torque value for its intended purpose.

Off delays can only be programmed by the design of the OEM. Off delays programmed into the motor are not adjustable. On delays can be created by delaying the communication to the motor.

OEM manuals are required for proper setup:

- Never change taps to adjust airflow without the manufacturer's manual.
- Settings determine performance and capacity.
- Improper settings may cause performance and capacity issues and premature parts failures.

In summary, one benefit of the constant-torque ECM is that torque is maintained if static pressure changes, for example, if the air filter or coil is dirty. A constant-torque motor also has better airflow, better system performance, and is more efficient than a PSC motor.

16.7 TROUBLESHOOTING THE VARIABLE-SPEED ECM

Knowing how to troubleshoot the ECM is important. Regal Beloit (Genteq), one of the manufacturers of these motors, reports that 40% of their motors returned under warranty have no fault detected. In other words, the technician misdiagnosed the problem. The goal of this section is to give you a few tools to prevent you from wrongfully condemning a good ECM. If the ECM is good, what are you going to say when you replace it and the system still does not work? These are hardy, yet expensive motors. You can ill afford to make a mistake.

An ECM replacement is much more expensive than a PSC motor. Motors out of warranty will cost the customer more than the customer wants to hear about. If there is doubt about the motor being the problem, ask for help.

As with most troubleshooting procedures, start off with a visual inspection. If it is an ECM condenser fan problem, check the following:

- Incoming voltage, which is usually 240 volts
- Connectors, to determine if they are firmly in place on the module.

If it is an air handler or furnace, check the following:

- Incoming voltage
- Correct voltage polarity
- Air handler ground
- Circuit board fuse
- The two connectors (to determine if they are firmly in place on the module)
- The high-voltage and communication connections on the circuit board
- 120, 240, or 460 volts on pins 4 and 5
- 120-V motors must have a jumper between terminals 1 and 2 for proper operation
- 240-V motors will not have a jumper between terminals 1 and 2 to prevent motor damage.

What Information Will I Need to Replace an ECM?

The easiest way to replace an ECM is to contact the local supplier of the equipment in which it is installed and give them the model number and serial number of the piece of equipment. For example, if the replacement module or motor is in a furnace, contact the local brand supplier (OEM) with the equipment model and serial number to see if the components you need are in stock. If the furnace, air handler or condensing unit model or serial number is not available, you can use the motor model number if it is available. You may also need the supply voltage. It is a good practice to count the number of wires in the connector. Not all ECMs use a 16-wire connector. This might be important when selecting a replacement motor. Remember that not all ECMs have the same pin number or pin configuration.

The five-pin module connector is the high-voltage connector. Have as much information available as you can prior to contacting a supplier. This should be a common practice when replacing any component. It is best to call ahead to be certain the replacement component is available. Place the item on "will call" to ensure it will be there when you arrive to pick it up. You will find that many suppliers may reduce inventory at certain times of the year, therefore you want to be certain your trip is not wasted. Your employer will establish guidelines that relate to supply house procedures.

Each motor and module is designed for a specific piece of equipment. There are no universal ECM replacement control modules or motors. Always use the exact replacement part from the OEM or a reliable source.

In summary, step 1 requires you to know the model number and serial number of the equipment using the ECM. List the number of wires going to each connector on the module. Note the voltage and application. Finally, call ahead to the equipment manufacturer supply outlet to see if they have an exact replacement ECM. Place the component replacement(s) on will call to ensure that the parts will be ready for pickup. Take the damaged component(s) with you and verify the replacement part by inspecting it before leaving the supply outlet. Additionally, the supply outlet may require you to swap the damaged component for the replacement part if a warranty is involved.

TROUBLESHOOTING 16.1: VOLTAGE INPUT TO A HIGH-VOLTAGE CONNECTOR

Here is the correct procedure for checking the voltage input to a high-voltage connector:

1. Figure 16-22 illustrates how to safely remove power from the motor by opening (off position) the disconnect or breaker that supplies power to the motor.
2. Remove the high-voltage connector to the module by disconnecting the white connector, *not* by pulling on the wires. The high-voltage connector in Figure 16-23 has been removed or disconnected from the motor. Voltage can also be checked with the motor connected and the voltage connector in place. Use a long, thin probe extension inserted into the back of the connector. It is sometimes more difficult to check with the motor connected. The blower wheel or fan blade needs to be on the motor shaft when running the ECM.
3. Check the input power (see Figure 16-24). Input power can be in the ±15% range of the nominal 120 or 240 VAC supplied to the motor.
4. Check polarity when the input power is 120 VAC. The black wire is hot for 120-VAC operation. The white wire is neutral.

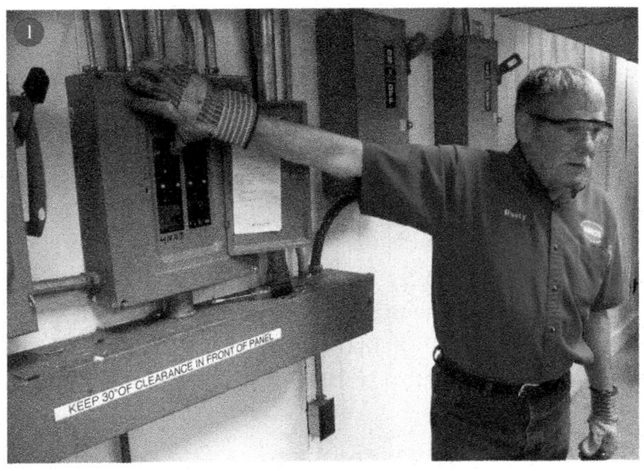

Figure 16-22 Turning off the power to the air handler and ECM motor.

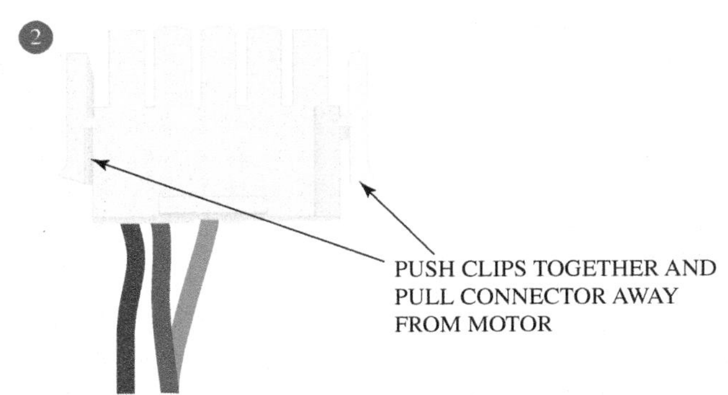

Figure 16-23 Remove the high-voltage plug from the ECM motor.

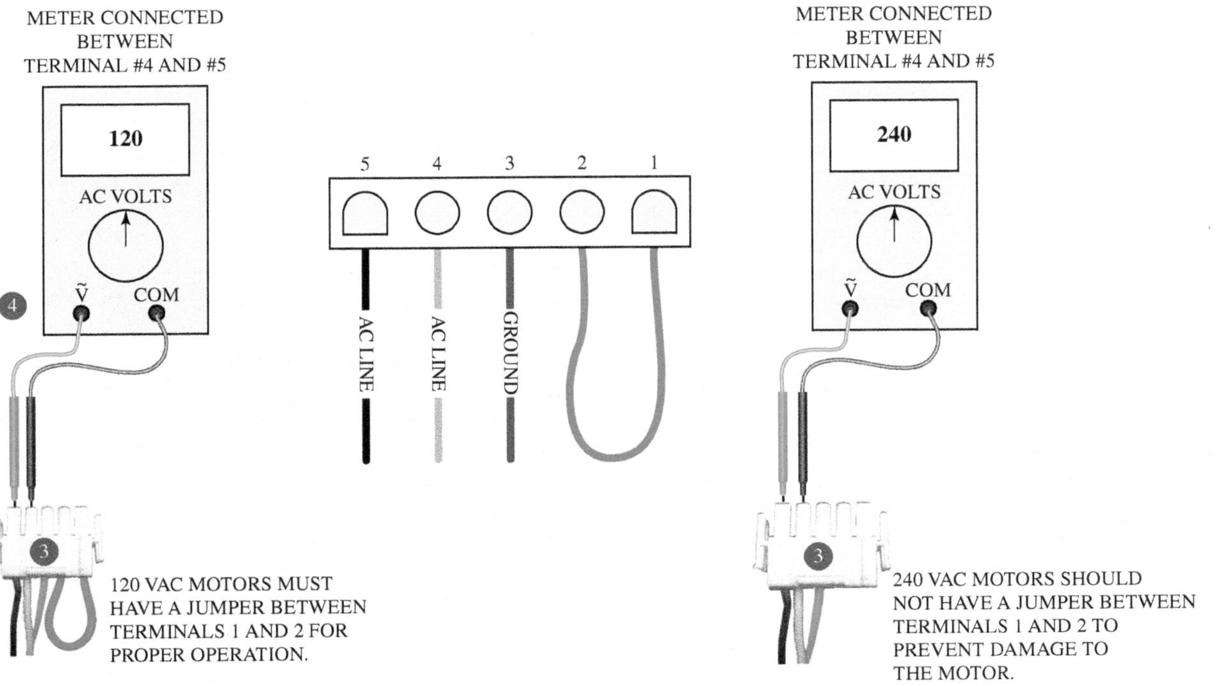

Figure 16-24 Measuring the high-voltage connector.

16.8 AIRFLOW PROBLEMS

In some cases the airflow in an air handler or furnace on a new or replacement unit is too low or too high. Table 16-1 states in one of the footnotes at the bottom that the factory settings for the unit in the last row are for a 3.5-ton system. This furnace can also be used on a 2.5- and 3-ton system. As you can see, the airflow setting may be incorrect on anything other than a 3.5-ton system. Even if the system is 3.5 tons, you may require higher or lower airflow than the factory settings.

The circuit board needs the correct CFM setting. Normal airflow is 400 CFM per ton, but 350 CFM per ton can be used for better dehumidification. High CFM settings (e.g., 450 CFM) will increase humidity levels in a cooled space.

Some circuit boards have a CFM flashing light indicator. For example, one flash will equal 100 CFMs. Count the number of flashing lights in a sequence and multiple by 100. If you count 12 flashes, the CFM would be 1,200 (12 flashes × 100) or 3 tons. The flashing light indicator will tell the technician if the dip switch settings are correct.

Finally, an undersized return or supply duct system will make the ECM increase the air velocity, which will make the duct system noisy. The airflow through a very restricted duct system will not be corrected by the ECM. Low airflow will also reduce heat transfer in the evaporator and cause the coil to freeze up.

An incorrect airflow setting in the heating mode is generally not noticed unless the system is a heat pump. Low airflow in a heat pump in the heating mode will cause excessive head pressure and overheat the compressor. The heating system will cut out on the high-pressure switch or the compressor internal overload.

TECH TIP

Manufacturers report few failures of the ECM motor. When it does fail, it is usually the electronic module. In most cases the module can be replaced without replacing the motor component. The biggest reason for module failure is water or moisture contamination of the module and prolonged operation at high total ESP. Water can drip onto a module from an overflowing drain pan or drain line from the evaporator, humidifier, or condensing furnace. Install a drip loop as previously shown in Figure 16-15 to prevent the water from tracking down the wire and into the module.

16.9 MOTOR RESISTANCE

The motor windings are wound as a three-phase motor, therefore the winding has close to the same resistance. The resistance is low, usually less than 20 ohms each. The winding resistances should be within ±10% of each other. If the motor winding has good continuity measurements, do a ground check. Measure at least one winding to ground. The resistance should be infinity or at least 100,000 Ω to ground. The motor should be replaced if the ground resistance is lower than 100,000 Ω.

16.10 HELPFUL INSTRUMENTS

The TECMate Pro is an instrument that can be used to assist in determining problems (see Figure 16-25). It is used to simulate a communication signal to the motor. This tool

Figure 16-25 The TECMate Pro is a troubleshooting instrument that makes ECM diagnostics easier and more certain.

does not work on the X13 or outdoor fan motors. Here is information on the TECMate Pro from Regal Beloit:

> Begin set-up with the TECMate Pro switch in the off position. The TECMate Pro has two wires with alligator clips for connection to a 24-volt power supply. The power supply is not polarized. The green led on the TECMate Pro will turn on when properly connected to 24 volts. The 16-pin connector from the TECMate Pro is connected to the motor in place of the HVAC OEM 16-pin harness or 4-pin connector on the model 3.0. The 5-pin high-voltage connector must be connected to the motor with its power confirmed and the system turned on after the connections are made for this test.
>
> If the motor runs when the TECMate Pro is turned on, the problem is before the motor in the HVAC system. If the motor does not run with the TECMate Pro, then the motor control module has failed and will need to be replaced. The motor, however, may still be good.

Another instrument that can be used with the ECM is the variable-speed Zebra (see Figure 16-26). This device will help when working with the ECM. Here are some features of this ECM test instrument:

- Accurately diagnoses exactly which component has failed in a variable-speed system that is not working correctly.
- Assists in system setup by reporting CFM airflow (and RPM) as room registers and other airflow controls are adjusted.
- Allows easy experimentation with all of the adjustable settings and options available on the variable-speed motor to provide optimum customer comfort.

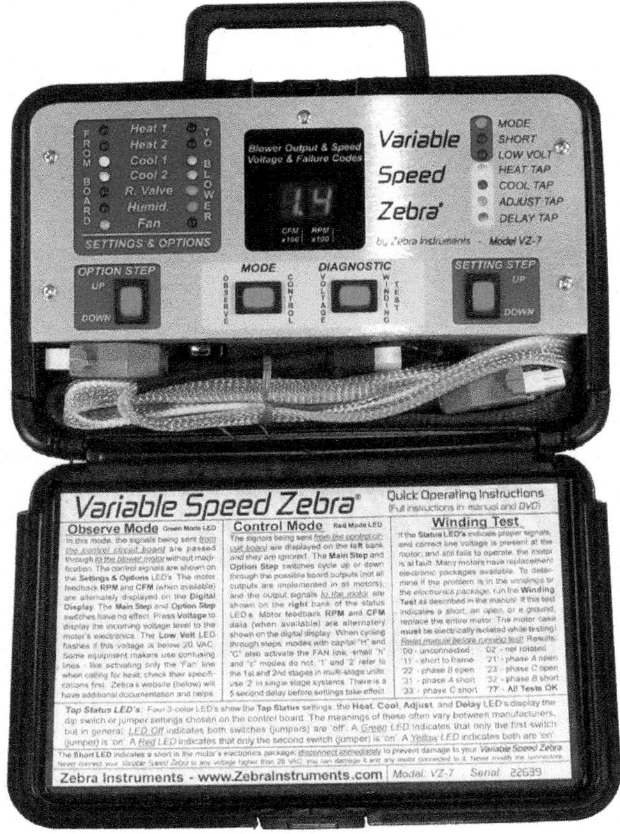

Figure 16-26 This is another ECM motor tester that is useful when troubleshooting ECM components.

16.11 THE CONSTANT-TORQUE ECM CHECKLIST

The constant-torque ECM has several troubleshooting steps that are similar to those used with its variable-speed brother.

- Does the motor turn? Can you move the blower with your hand? It may not move smoothly, but it should move.
- Check the high-input voltage circuit at the motor terminal. Is the 120-V supply the correct polarity? Is 230 or 460 volts measured on the 4 and 5 terminals? Is the motor properly grounded?
- Is the motor running? Is the correct speed tap selected? Check the tap selection using the manufacturer's guidelines. If it is correct, is the duct system noisy or is the coil freezing up? These are indications of a lack of airflow. Check the air filter. Is the evaporator coil clean? Check the size of the return and supply ducts. Are any registers closed or are there other duct restrictions? If the motor is running, the problem is not the motor.
- Measure the total external static pressure (ESP). If it above the manufacturer's rating, usually 0.90″WC or greater, there is an airflow problem. Most well-designed return and supply ducts systems have an ESP of around 0.500 = WC.

If the correct high voltage and low voltage are present, then the motor has failed. Like the variable-speed ECM, the constant-torque motor requires an exact OEM replacement. There are no universal replacement ECMs. The replacement motor belly band should not cover the ventilation openings on the motor.

SAFETY TIP

Is the motor properly grounded? Is the piece of equipment you are working on properly grounded? How do you check for a proper ground? Knowing the answers to these questions may save your life. Proper grounding can be verified using a couple of different methods. For example, let's check the ground on an ECM device. Using the voltage scale, you should measure voltage from the line side of the motor to ground. It would about 120 volts for both a 120- or 240-V power supply. No voltage reading means that the ground is broken or there is no power supply output.

You could use the ohm scale to measure a very low resistance from the ground on the motor to a ground outside the equipment housing. Do not get across a power supply with your ohmmeter. This will damage the meter or at least blow the meter fuse. The resistance should be zero ohms or no more than a few ohms. A good spot to make this measurement outside the equipment would be the ground of the disconnect or the ground on the switch providing power to the equipment. This would assure you that the ground is connected to the power supply.

TROUBLESHOOTING 16.2: CONSTANT-TORQUE ECM

The troubleshooting steps for a constant-torque ECM motor will depend on the motor's input voltage. Here are troubleshooting the steps for each of the common voltage sources:

115/120 Voltage Source

1. As shown in Figure 16-27, check the input voltage. Check the polarity of the 115/120 ±10% volt power supply. A 115/120-V reading should be measured from constant-torque ECM motor line **1** to ground. The same reading should also be measured from line **1** to neutral.

230/240 Voltage Source

2. As shown in Figure 16-28, check the input voltage. Check the 230/240 ±10% volt power supply. A nominal 230 volts should be measured between line **1** and line **2**. A 115-V reading should be measured between line **1** and ground and also from line **2** to ground.

460/480 Voltage Source

3. As shown in Figure 16-29, check the input voltage. Check the 460/480 ±10% volt power supply between line **1** and line **2**. Also, 277 volts should be measured between line **1** and ground and also between line **2** and ground.

Low-Voltage Source

4. Is the low-voltage communication voltage available? The low-voltage connect will use **C** and one of the terminals **1** through **5** that are programmed to operate at a specific torque, as shown earlier in Figure 16-19 and here in Figure 16-30. Some of the terminals may not be programmed. If no low voltage is measured between **C** and one of the wire taps, check the system wiring, controls, and thermostat setting.

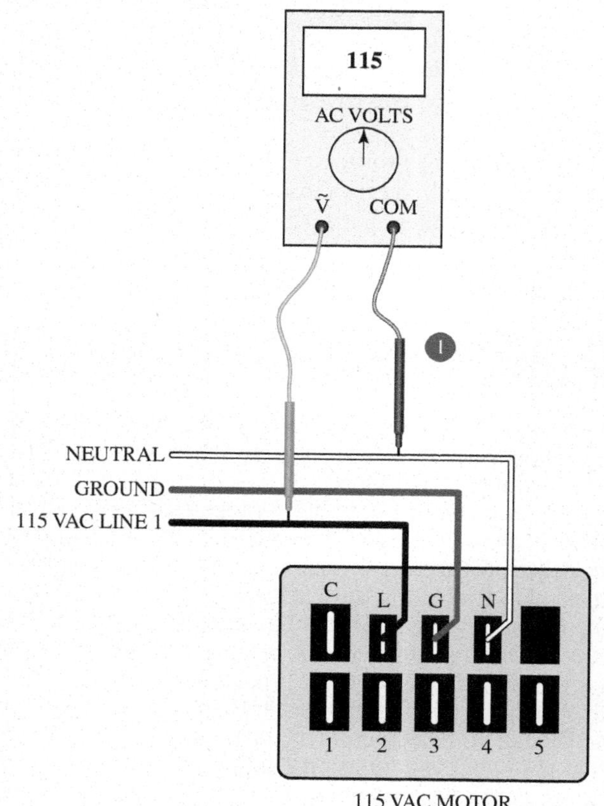

Figure 16-27 Check the correct polarity of the 115-V power supply. A 115-V reading should be measured from line **1** to neutral and also from line **1** to ground.

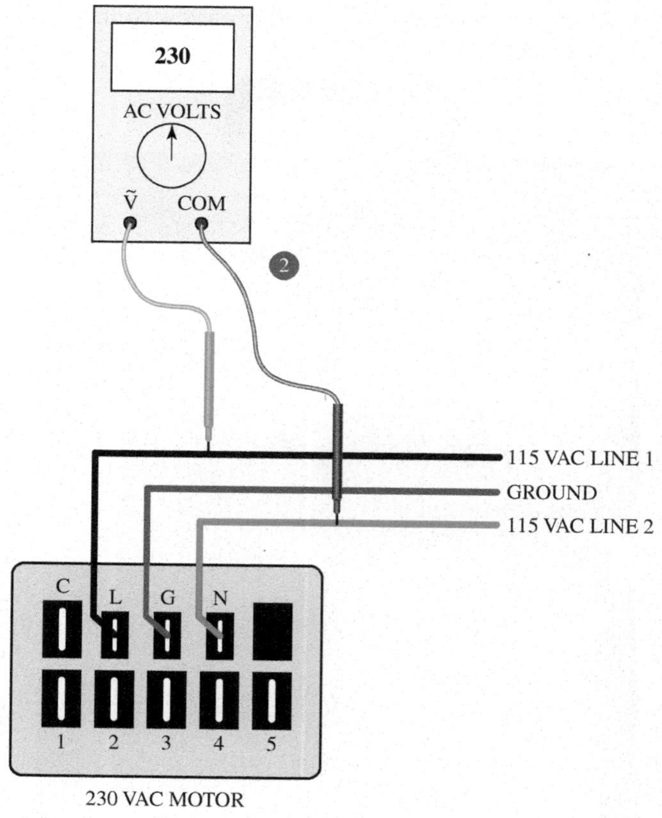

Figure 16-28 Check the 230/240-V power supply between line **1** and line **2**. A 115-V reading should be measured between line **1** and ground and also from line **2** to ground.

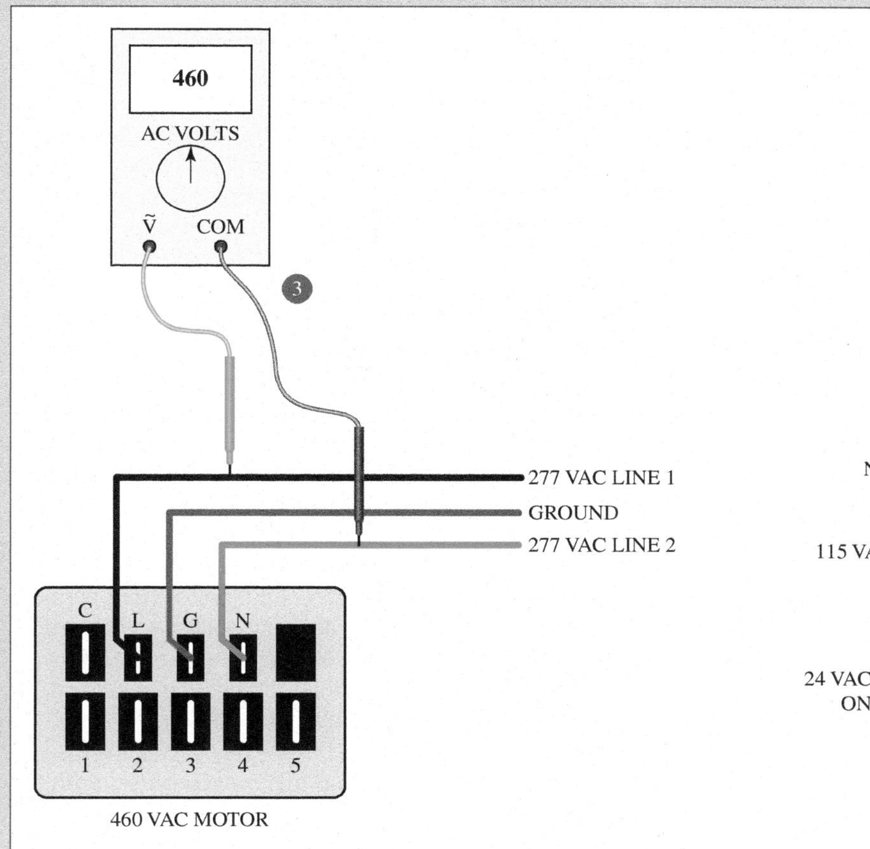

Figure 16-29 Check the 460/480-V power supply. The voltage between line **1** and ground should be 277 volts. You should also measure 277 volts from line **2** to ground.

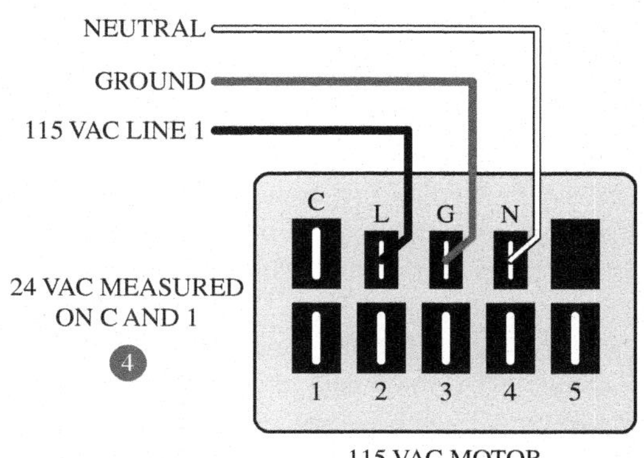

Figure 16-30 Checking the high and low voltage on a constant-torque motor. The correct polarity voltage is measured on **L** and **N**. The 24-V communication signal is correct.

TECH TIP

An ECM air handler or furnace installed in an undersized duct system will move more air than a conventional PSC or shaded pole motor. The ECM is not designed to overcome poor duct design, which is usually an undersized supply or return duct system or both. It will try to compensate for the undersized duct by increasing airflow, which increases total static pressure in the duct system. Running an ECM at high static pressure will increase energy consumption also. The motor will run hotter than normal with an expected shortened control module life. An ECM installed in an undersized duct system will increase airflow and increase duct noise especially at the return or supply grilles. If a customer states that the airflow is better when a replacement ECM air handler is installed, it probably means that the duct system is undersized or the previous blower was too small.

TECH TIP

The ECM may rock back and forth before turning in the correct direction. Also, X13 motors stop much faster than other motors. This is a normal programmed braking function.

SAFETY TIP

This caution is from Genteq, the leading manufacturer of ECMs:

> Full load amps (FLA) is published on the label of the motor as required by UL (Underwriters Laboratories). UL is an electrical testing agency that examines and certifies that electrical appliances are safe to use. However, due to the efficiency of the motor and its programming, it will

most likely never run at or near the FLA unless the external static pressure (ESP) is very high. High ESP normally means that the airflow is restricted. Even then the speed limit will limit the motor operation for protection (in Genteq products). Checking the amperage can be misleading depending on the type of motor being used and what the purpose is. Using amperage to gauge motor operation can be misleading because it may be low even when operating properly. If using amperage to gauge efficiency, it can again be misleading because the total power (watts) is a formula that contains voltage, amperage, and power factor, not just amps.

SERVICE TICKET

The first two scheduled service calls of the day involved working with electronically commutated motors.

On the first callout, a new installation required a system setup and checkout. The heat pump system had a variable-speed air handler. What steps are required to ensure quality and a long life?

You first turn off the power to the air handler and condensing unit. Condensing units with crankcase heaters should have been powered up about 24 hours before starting. The crankcase heater drives the liquid refrigerant away from the compressor.

Several items can be checked first. You start with the ECM settings as follows: Establish the system capacity in tons. Determine the manufacturer's instructions for the dip switch or speed settings for the rated capacity. Set the dip switches or speed settings for the rated tonnage. Some circuit boards have start delay settings and stop delay settings. Table 16-2 is a sample chart showing the cooling-off delay settings for a specific manufacturer's air handler.

Set the switches according to contractor recommendations or customer preferences. Next, you turn the power on and check the power polarity and equipment ground. You also control the thermostat so that the system does not energize until it is ready for operation.

Next you check the outdoor unit. The outdoor unit should be turned off at the disconnect. Remove the cover on this section. Check the input voltage and ground. Check the setting for the defrost circuit. Common defrost options are 45, 60, or 90 minutes. Some heat pumps do not have a defrost timer setting, but rely on sensing cold refrigerant temperatures to activate the defrost circuit. The outdoor unit fan ECM is usually factory set. You install the manifold gauge set, paying attending to the gauge location so that the gauges are not damaged by high pressure when the system is changed from cooling to heating. Turn the thermostat setting to the cooling or heating mode. Hook up the clamp-on ammeter to the common of the compressor. Close the disconnect and watch the pressure and amperage readings. Check the indoor blower to ensure good airflow. Use the manufacturer's recommendation to check pressures, amperage, superheat, and subcooling. You record all of these readings to develop a baseline for future reference. Log the information in the address database, which can then be used on the next service call.

The second service ticket of the day is diagnosing a gas furnace problem. The customer woke up this morning to a cold house. Upon investigating the basement upflow furnace, you notice that the burners are fired up. After about 1 minute of operation, the burners shut down. The blower does not operate. The burners seem to be cycling on the high limit control. What are some quick checks you could do to narrow down the problem?

The furnace power supply seems to be good since the burners are operating. The 24-V control voltage is required for the gas valve and burner operation. The power is removed and the squirrel cage blower rotated. The motor is not restricted or locked up. The motor movement is not smooth, but this is typical for a constant-torque ECM. Next, the power polarity and ground connections are verified as correct. The 120-V power on the motor is correct as checked on the motor connections (see Figure 16-30). The 24-V communication signal measures a strong single, as shown in Figure 16-31. Since the input voltage and communication voltage are correct, what is the problem?

This is a motor problem. Is the problem the module or motor? Without a motor tester instrument the actual diagnosis is not clear. Further inspection reveals that when the module is opened, some condensation pours from a seam between the module and motor. The module is water damaged, but is the motor okay? How is the motor checked? You inspect the motor and don't see any apparent water inside the motor housing or around the winding.

The ECM is a three-phase DC motor, therefore all of the windings will have the same resistance, usually lower than 20 ohms. The winding readings are 6.8, 6.9,

Table 16-2 Cooling-Off Delay Options

This is a cooling-off delay chart for a specific air handler that uses an ECM. The specific manufacturer's chart should be used for each air handler or furnace.

Switch Settings		Selection	Nominal Airflow
5-OFF	6-OFF	NONE	SAME
5-ON	6-OFF	1.5 MINUTES	100%*
5-OFF	6-ON	3 MINUTES	50%
5-ON	6-ON	**	50–100%

(Courtesy Trane Corporation.)

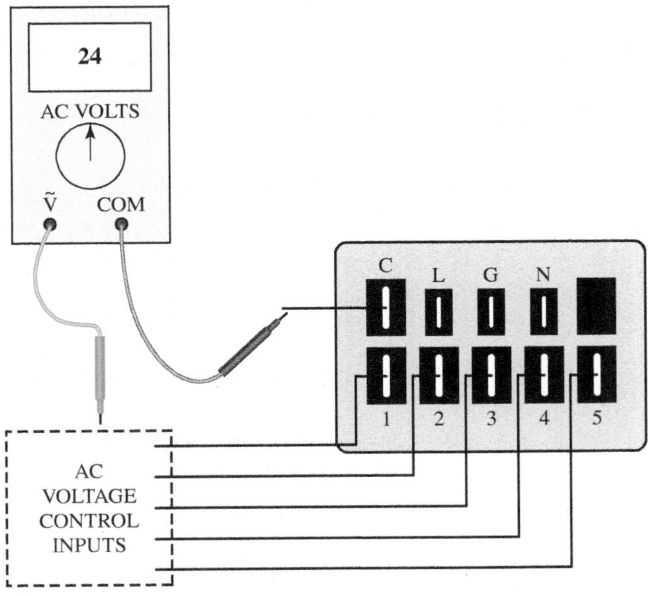

Figure 16-31 A 24-V communication signal will energize the motor when **C** and **1** are powered. The other connections may have different airflow outputs or the other connections may not be used by the constant-torque motor.

and 7.1 Ω. These windings are good. The resistance from the motor housing to the winding is 10 MΩ. This is good.

You call ahead to the supply house to be certain that a module is available for purchase. The supply house asks for the model and serial number of the furnace. It is in stock and available for pickup. The module is placed on "will call." When the motor is picked up, the model number of the module is verified before leaving the supply outlet. When you go to pick up the new module, you take the older ECM so you can compare similarities. It looks like a match.

The ECM is reassembled and installed in the furnace. You check the dip switches for the correct setting prior to starting. The system is cycled a couple times in the heating and cooling mode to establish correct operation before leaving the job.

The evaporator drain pan seems to be running slow, so you clean it and blow out the drain line. You pour a gallon of water into the drain and it now drains quickly. The cause of the motor damage has been solved. The invoice is completed and explained to the customer. You recommend annual system cleaning and maintenance to keep this problem from occurring again.

SUMMARY

The ECM motor is found in high-end equipment. The OEM has many options for choosing the amount of airflow, on and off delays, and comfort profiles for each system. It is important for the installing contractor to set up the ECM operation based on the many motor options. Without a correct ECM setup, the system may not operate properly, meaning the customer will not be satisfied with the system's performance.

Current generation ECMs are much improved over earlier ones. Reliability improvements over the generations include the following:

1. Fully encapsulated module electronics to protect against moisture damage, the number one reason an ECM fails
2. EMI filters to provide protection against line transient and voltage spikes
3. Speed limiting to prevent overcurrent operation due to extremely high static pressure operation
4. Durable ball bearings on all models.

The ECM has more troubleshooting steps when compared to a conventional PSC motor. Even so, a high percentage of these motors are returned under warranty with no problem found. It is important for the ECM setup to be complete before beginning the troubleshooting process. This is especially true of a new installation. Check the input voltages. The motor is a DC, three-phase motor, therefore the motor windings should all have the same resistance. The resistances should be equal and be less than 20 Ω. Finally, an ECM test instrument is a valuable asset when diagnosing ECMs.

Work safely around an ECM. Disconnect the power to the ECM before removing the two connectors. Also wait about 5 minutes before separating the control module from the motor. Figure 16-32 shows the inside of a module. Note that the large charged capacitors can hold a dangerous charge.

Table 16-3 is a comparison chart that will help you identify the vast benefits of constant-torque and variable-speed

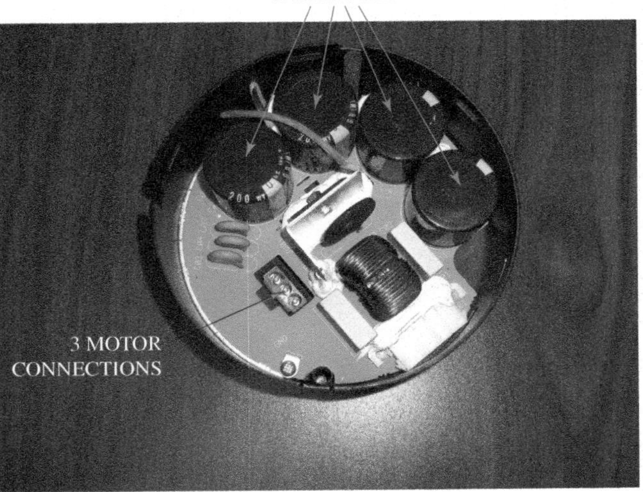

Figure 16-32 The four large capacitors inside the module can be slow to discharge. Allow the motor to set disconnected from the power supply at least 5 minutes prior to disassembly. This will allow the caps to safely discharge. The three motor connections can be easily identified.

Table 16-3 Comparison between PSC Motors and Standard Constant-Torque ECMs and Premium ECMs

MOTOR COMPARISON

genteq™

		PSC MOTOR	STANDARD ECM (X13)	PREMIUM ECM (ThinkTank™ 2.3 & 3.0)
PRODUCT SPECS	TECHNOLOGY	Induction Motor	Brushless DC	Brushless DC
	TYPICAL OUTPUT RANGE	1/3, 1/2, 3/4, 1 HP	1/3, 1/2, 3/4, 1 HP	1/3, 1/2, 3/4, 1 HP
	INPUT POWER	120V model or 240V model	120V model or 240V model	120V/240V in one model
	HIGH VOLTAGE POWER	On or off with speed taps by relay or circuit board	Constantly powered	Constantly powered
	MOTOR EFFICIENCY	59%	80%	80%
	SPEED RANGE	800-1100 RPM	600-1100 RPM	200-1300 RPM
	SETTINGS/SPEEDS	2 to 5 speeds—set by motor design	2 to 5 torques—pre-programmed	Programmed/adjustable variable airflow or torque
	INPUT COMMANDS	High voltage power switching with speed taps	Low Voltage torque taps: AC or DC	Low voltage AC or DC airflow/torque taps or fully variable (PWM)
	BEARINGS	Sleeve or Ball	Ball	Ball
	ELECTRONICS	None	Encapsulated	Encapsulated
	RUN CAPACITOR	Required	None	None
PRODUCT FEATURES	AIRFLOW CONTROL	By speed taps only—limited by external static pressure (ESP)	Constant torque with fewer ESP limitations	Constant airflow or constant torque
	OFF-DELAY/ON-DELAY/ SPEED RAMPING	Delay is external timer controlled/no ramping	External or programmed off-delay/External timer controlled on-delay/no ramping	Adjustable off-delay & on-delay/wide ramping range
	OUTPUT INFORMATION	None	None	Programmable: Speed, CFM, or BlaKBox™ Diagnostics
	TROUBLESHOOTING	Multimeter, confirm proper voltages	Multimeter, confirm proper voltages	TECHMate Pro™ (excluding 2.5)
	MOTOR SERVICE	Motor as one piece	Motor as one piece	Replaceable control module or motor module
HOMEOWNER BENEFITS	CUSTOMIZED COMFORT	NO	NO	YES. Select your geographic climate and the system automatically delivers ultimate comfort year-round.
	CONSTANT AIRFLOW	NO	NO	YES. Maintains constant airflow to ensure comfort with static changes related to blocked vents, dirty filters, or other obstructions.
	MOST QUIET OPERATION	NO	Quieter than a PSC	YES. The most quiet of all motors. Slower ramp-up and ramp-down so that there is less noise.
	EFFICIENT DEHUMIDIFICATION	NO	SOME	OPTIMAL. Widest speed range, efficient constant fan, and intelligent climate adjustments allow Think-Tank to provide ultimate comfort.
	ELIGIBLE FOR UTILITY REBATES & TAX CREDITS	NO	YES	YES. Higher-efficiency systems that meet federal or local guidelines are eligible for rebates. With an ECM, these systems can receive an extra rebate.
	LOWER OPERATING COSTS	NO	SOME. System saves more over a PSC but not as much as the premium ECM.	YES. The ultra-low-speed constant-fan setting and highest efficiency saves the most on energy bills.

(Continued)

MOTOR COMPARISON

	PSC MOTOR	STANDARD ECM (X13)**	PREMIUM ECM (ThinkTank™ 2.3 & 3.0)
TECHNOLOGY	Induction Motor	Brushless DC	Brushless DC
AVAILABLE HORSEPOWER	1/3, 1/2, 3/4, 1 HP	1/3, 1/2, 3/4, 1 HP	1/3, 1/2, 3/4, 1 HP
SPEEDS	Select 1-5 taps	Select 1-5 taps	Variable-speed
SPEED RANGE	800-1100 RPM	600-1100 RPM	200-1300 RPM
CLIMATE PROFILES	NO	NO	YES
INTELLIGENT CONTROLS	NO	Constant Torque	Constant Airflow
ADVANCED FEATURES	NO	NO	Communicating Motor, BLAKBox™ Technology
PEAK MOTOR EFFICIENCY	59%	80%	80%
AVERAGE POWER * (COOLING CFM)	552 W	413 W	413 W
AVERAGE POWER * (CONTINUOUS FAN CFM)	515W	200 W	83 W
SUMMARY	No Features, Baseline Product	Limited Features, High Efficiency	Full Features, Maximum Efficiency

* Average power consumption based on specified setting in a typical three-ton system.

www.theDealerToolbox.com

**

(Courtesy Genteq®.)

ECMs when compared to PSC motors. It is good to know the benefits of the ECM so that you can talk intelligently with your customer about the features and benefits of this high-end product.

REVIEW QUESTIONS

1. What does the abbreviation ECM mean?
2. Which is more efficient: the shade pole motor, the PSC motor, or the ECM?
3. What are two reasons that manufacturers select the ECM for an air handler or furnace?
4. What components make up an ECM?
5. Which motor would have a higher starting amp draw: an ECM or a PSC?
6. What bearing are used by an ECM? What is the advantage of this bearing type?
7. What precaution should you take before removing an ECM's connectors?
8. There are two electrical plugs on an ECM. Describe where to find the high-voltage connection.
9. You have purchased a new ECM replacement motor. Before wiring it you notice that the motor shaft turns unevenly and roughly. Is this normal or an indication of defective motor bearings?
10. True or false: The air handler ECM comes wired and set up from the factory. No change of settings is required.
11. Explain how the ECM is set up to dehumidify and save energy.
12. What is a drip loop on an ECM?
13. What is the main difference between a constant-torque ECM and a variable-speed ECM?
14. What are the high voltages that can be used to power a constant-torque motor?
15. How do you determine the hot leg or hot side of a 120-V power supply?
16. How many CFMs are required per ton of air conditioning?
17. What problem does increased external static pressure (ESP) indicate?
18. How many CFMs would you expect from a 5-ton air handler?
19. What happens to the CFM on an ECM when the airflow becomes restricted as a result of, for example, undersized duct-work, a dirty air filter, or a dirty evaporator coil?
20. What happens to wattage used on an ECM when the airflow becomes restricted?
21. What are four ECM improvements that have helped to extend the life of the device?
22. An ECM motor has a resistance of 20, 10, and 30 Ω. The resistance to ground reads 3.1 MΩ. Is there a problem? If so, what is the problem?

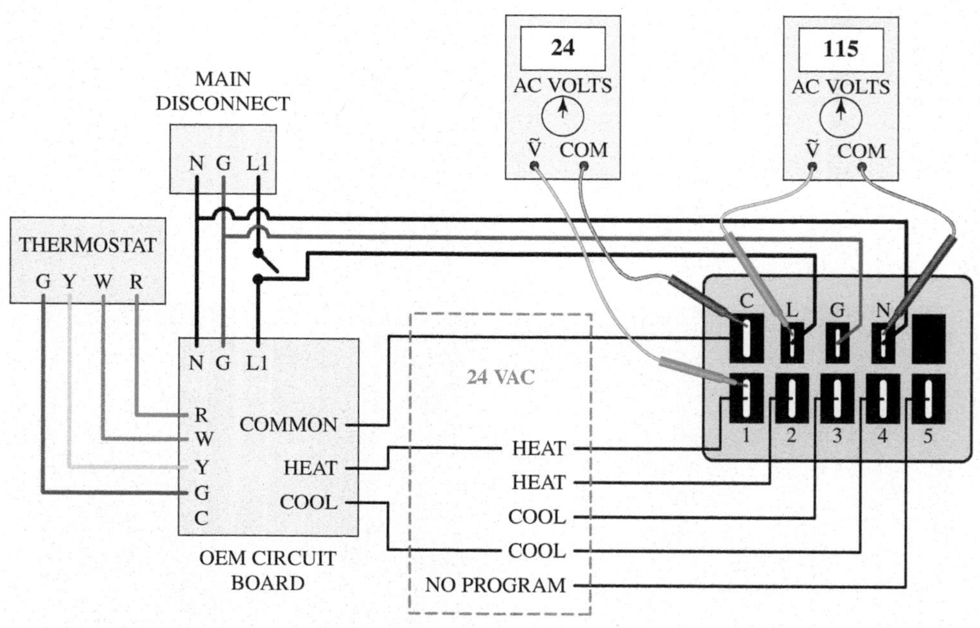

Figure 16-33 Use this drawing to answer Review Question 24.

Figure 16-34 Use this photo to answer Review Question 25.

23. An ECM motor has a resistance of 8.1, 7.9, and 8.2 Ω. The resistance to ground reads 2.1 MΩ. Is there a problem? If so, what is the problem?

24. Refer to Figure 16-33. If you are measuring these voltages on the terminals and the motor does not move, what is the problem and the solution?

25. Refer to Figure 16-34. This is a correctly wired ECM constant-torque motor. Which wire color(s) is the high-voltage section?

UNIT 17

Understanding Electrical Diagrams

WHAT YOU NEED TO KNOW

After studying this unit, you will be able to:

1. Draw and identify symbols used in HVACR schematics.
2. Describe the different types of electrical diagrams.
3. Discuss two reasons for using electrical diagrams.
4. Describe two reasons for developing an electrical schematic.
5. Describe the steps in developing an electrical diagram.

An **electrical diagram** is an organized group of symbols with interconnecting wires. Electrical diagrams help the technician make sense of what is happening in an HVACR system. They are the road maps through an HVACR system. Think of the wire connections as roads through which the electrical circuits travel. The symbols, like landmarks on a road map, are used to identify the components of an electrical system.

You are learning a new language that uses electrical symbols instead of words. This text is designed to help you practice this new language so that you can become proficient in the art of understanding electrical circuits. The more you use the symbols, the more familiar you will become with them. Even an experienced technician has to be able to identify new and different electrical diagrams.

Next, you will learn how these symbols are used to develop circuits that operate air conditioning, heating, or refrigeration circuits. You will also see that identical circuits can be drawn in several different ways. You will find out how to develop an electrical diagram if one does not exist. If the original electrical diagram is not available when troubleshooting, a field-developed diagram can be helpful. In summary, this unit covers the basic types of circuits, electrical symbols, and electrical wiring diagrams and how to develop these diagrams.

17.1 SYMBOLS

Symbols represent a component in an HVACR circuit. Some of the symbols make sense in that they are intuitive and do not need to be memorized. Others will need to be learned. In some cases several different symbols are used for the same component. This unit covers only the most common symbols found in our industry. Figure 17-1 lists the most common symbols used in this textbook and in our industry.

Common Symbols

Let's explore some of the common symbols you will find on air conditioning equipment. We will look at what they mean and how to identify them. Figure 17-2 shows a symbol that represents a heating thermostat switch.

The word *thermostat* is abbreviated *t'stat* or *tstat*. As the air around the thermostat gets cooler, the arm drops and completes the circuit, which starts the heating system. The zigzag line below the arm represents the old style of thermostat control, which uses a bimetal (two dissimilar metals) element to open or close the arm. As the air around the thermostat gets to the temperature setpoint, the bimetal element expands and opens the circuit to the heating system. The heat source, such as gas, stops. Most heating systems will operate the blower for a minute or two at the end of the heating cycle when the thermostat opens. This operation allows the heating system to blow out the heat trapped in the heating system and ductwork, improving its efficiency. Because bimetal temperature sensing is not commonly used anymore, we will simply use the word *temperature sensor* when referring to this type of device. Most temperature sensors on modern t'stat controls are solid-state devices.

In summary, the bimetal element attached to the arm responds to temperature changes. It expands as the air temperature rises and contracts as the air temperature drops. When the bimetal contracts, the arm drops to close the contacts, which starts the heating system.

Figure 17-3 is a symbol for a cooling thermostat switch. Notice how the arm is below the contact point. A temperature rise in the room causes the bimetal element (temperature sensing) to expand, which closes the contacts and starts the cooling system. As the air around the thermostat gets cooler, the arm drops and opens the circuit to stop the operation of the cooling circuit. As the room air heats, the bimetal element again expands, closing the cooling thermostat contacts, and once again starting the cooling system.

The left side of Figure 17-4 shows a symbol for a relay or contactor switch. This example would be identified as a double-pole relay or contactor since it has two sets of contacts. Several different symbols are used for the coils, as seen on the right side of Figure 17-4.

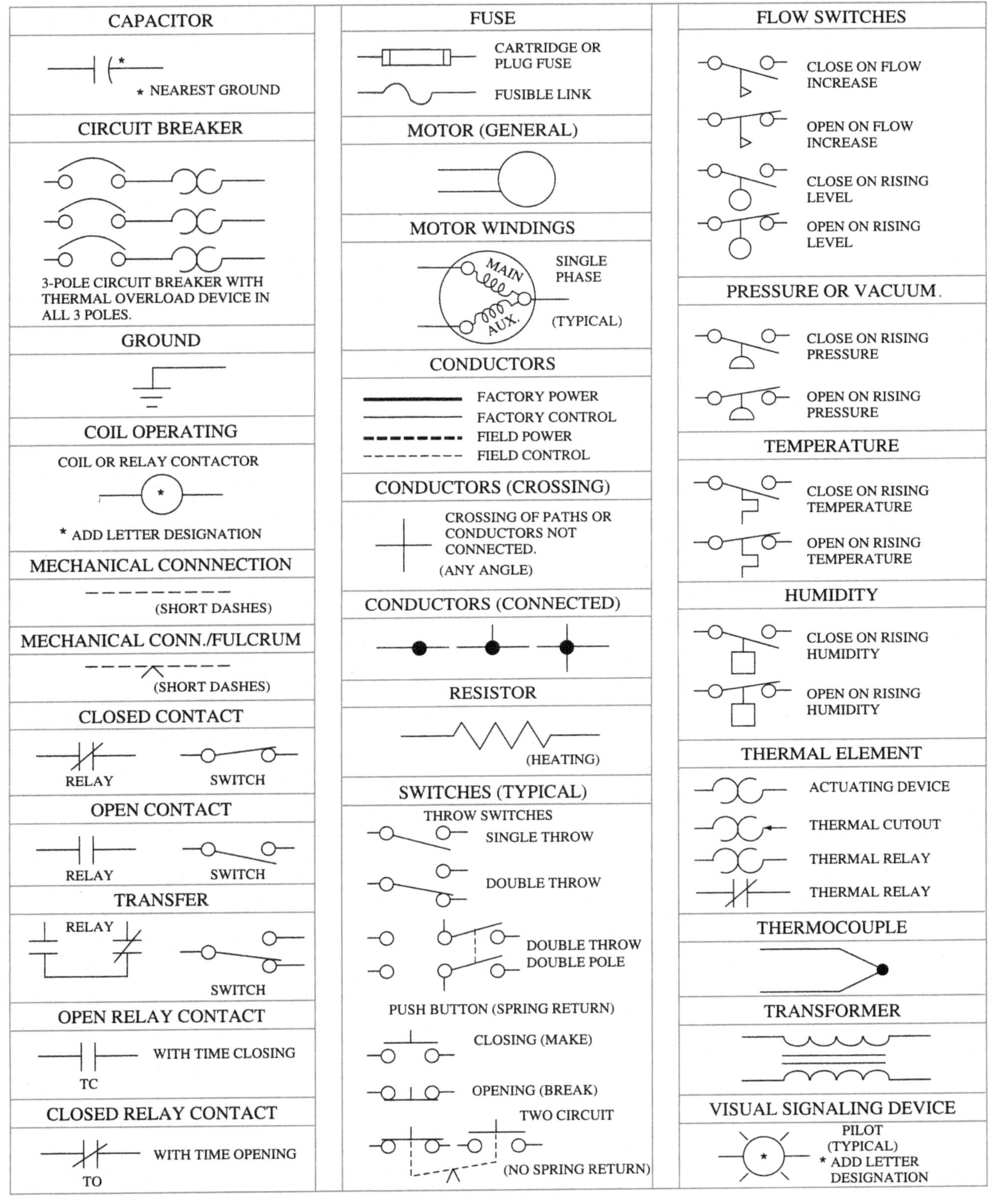

Figure 17-1 Even though these electrical symbols are common, they are not universal in the United States or worldwide.

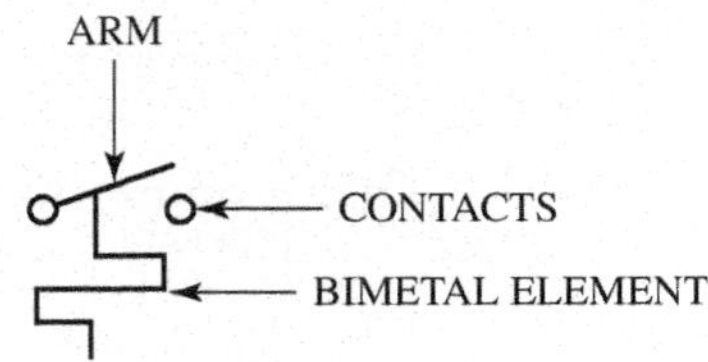

Figure 17-2 This symbol represents a heating thermostat switch.

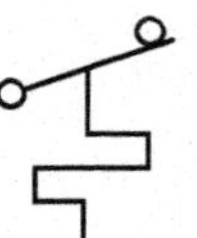

Figure 17-3 This symbol represents a cooling thermostat switch.

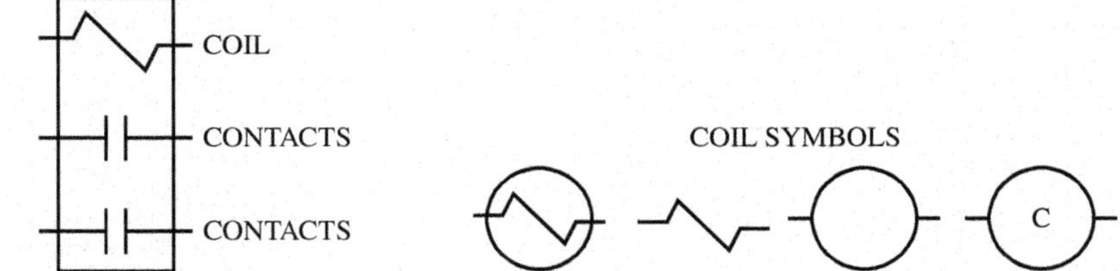

Figure 17-4 The symbol on the left represents a relay or contactor switch. Different symbols are used for coils, as shown on the right.

The switch in Figure 17-4 is a magnetically controlled switch that closes normally open (NO) contacts when energized. The difference between a relay and contactor is the operating amperage of the contacts. Relays operate with lower amperage loads. Contactors operate at amperage loads above 15 amps. They do have a maximum amperage rating. Relays and contactors were discussed in detail in Unit 9 on electrical components.

The high-pressure switch is another common symbol (see Figure 17-5). The hemisphere at the bottom of the symbol represents the system pressure acting on a set of contacts. When the high-side pressure exceeds its design pressure, the rod pushes up on the arm and breaks the circuit. When the system pressure drops, the contacts reset and close to complete the circuit and operate the cooling system. High-pressure conditions can damage the compressor.

Low-pressure switches are not as common as high-pressure switches. Figure 17-6 shows the symbol used for a low-pressure switch. The hemisphere at the bottom of the symbol represents the system's low pressure acting on a set of contacts. When the low-side pressure drops below its design pressure, the rod drops down and the arm breaks the circuit contacts. When the system pressure rises, the contacts reset and close to complete the circuit and start the cooling process.

Figure 17-7 shows a transformer symbol. A transformer is a device that transfers electrical energy from one alternating circuit to another with a change in voltage and current. Most transformers used in HVACR reduce or step down the voltage being supplied. There are also step-up and isolation transformers, which were discussed in Unit 8.

The primary winding of a step-down transformer is hooked to the incoming power source, and the secondary winding is the reduced voltage normally used in a control voltage circuit. The common transformer used in air conditioning is a transformer that uses 24 volts on the output or secondary side. The transformer is constructed by wrapping a set of primary windings around one end of a laminated iron core. The secondary windings are wrapped on the other end of the laminated iron core. The transformer works by inducing magnetism from the primary side to the secondary side. These windings are magnetically but not actually electrically connected together as shown in Figure 17-7.

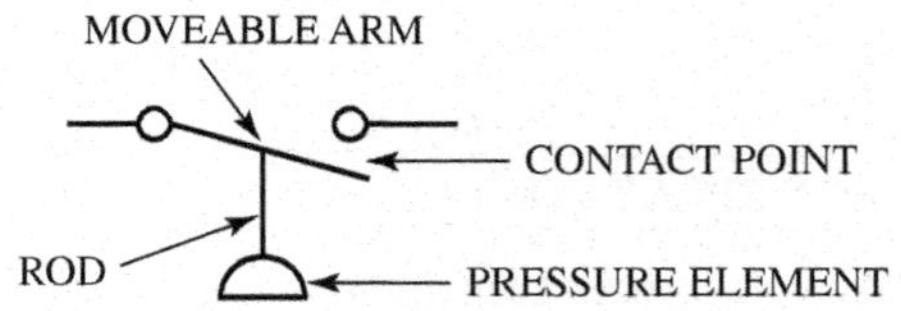

Figure 17-6 This is the symbol for a low-pressure switch.

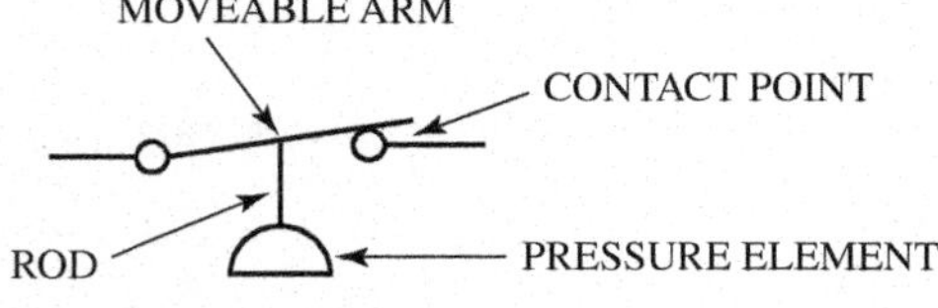

Figure 17-5 This is the symbol for a high-pressure switch.

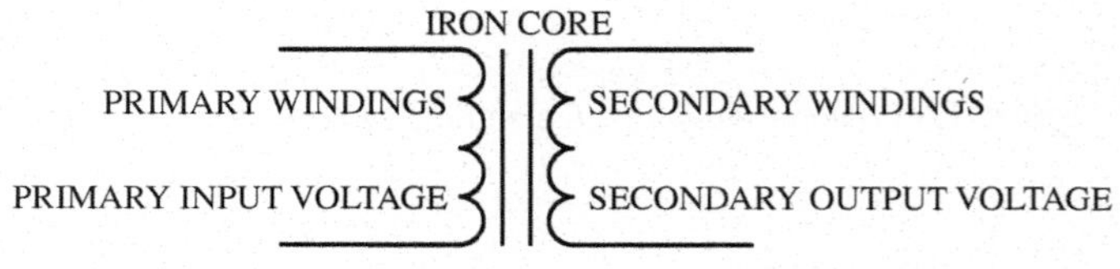

Figure 17-7 This is the symbol for a transformer.

As you can see it is not difficult to learn the symbols if they have some meaning. You will learn these symbols as you work with the diagrams in this textbook and become familiar with them during lab or field work.

17.2 CIRCUIT TYPES

Complete Circuit

A complete circuit is a path that electricity can flow through. It is also known as a closed circuit. The simplest complete circuit includes a power supply, a load or resistance, and interconnecting wires. Figure 17-8 illustrates a complete circuit. It includes a power source, a load (the electric heater in the upper figure), a motor load (the lower figure) and interconnecting wires. Switches, thermostats, and pressure control devices are placed in the circuit as optional devices for convenience and safety during operation of the system. This is not a common circuit; it is used merely to illustrate a complete circuit.

An adequate power source that will push (voltage) electrons (current flow) through the circuit is required. A load is required to consume and dissipate some of the electron flow in the circuit. The load performs useful functions such as operating a compressor motor (to pump refrigerant) or a fan motor (to move air). The path for electron flow or wires interconnects the load and power source to create a complete circuit.

A complete or closed circuit can be a simple series circuit, parallel circuit, or series-parallel circuit. Series-parallel circuits are also known as combination circuits. There are also open circuits, short circuits, and grounded circuits. Next we discuss the various types of circuits.

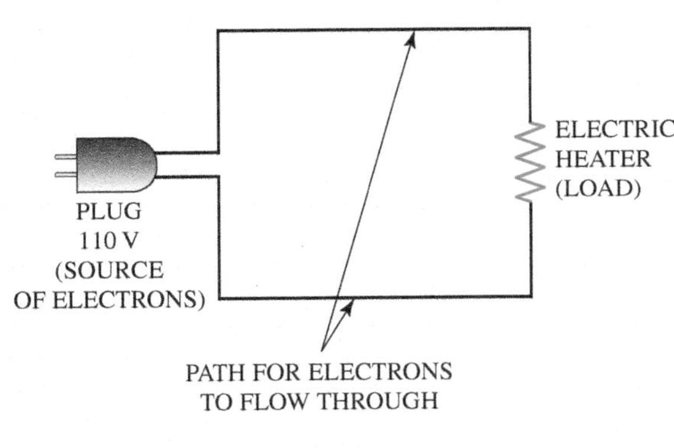

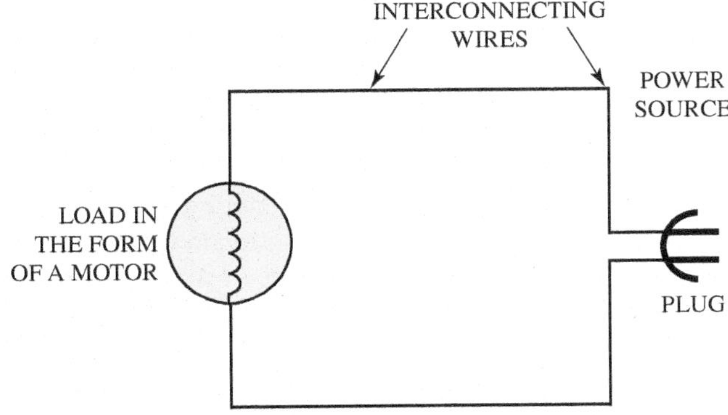

Figure 17-8 A complete circuit is a path that electricity can flow through. It is also known as a closed circuit.

Series Circuit

A series circuit is shown in Figure 17-9. A series circuit has the same current flow at every point in the circuit. If there are two loads in a series circuit, the current flow will be the same everywhere in the circuit, but the voltage will be divided across the loads. If there are two equal loads in a series circuit, as shown in Figure 17-9, the voltage drop (measured voltage) across each heater will be the same. This voltage drop can be calculated by dividing the supply voltage by 2. This is just an example to explain series circuits. By the way, most heat strips and most HVACR components have only one load in series with the power supply, not two as shown in this example.

TECH TIP

Understanding how the current and voltage flow in series, parallel, and combination circuits will help you troubleshoot a circuit. It is also important to know how total resistance changes as these circuit types change.

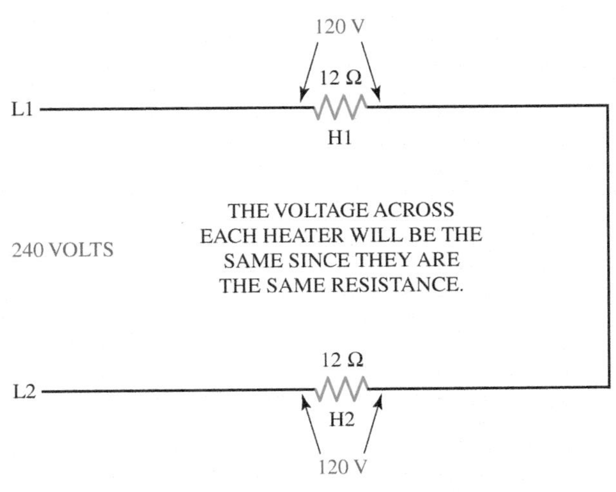

Figure 17-9 This series circuit has electric heat strip loads H1 and H2, which have the same resistance. The current flow is the same everywhere in the circuit. The voltage will be split across each heat strip since they have the same resistance.

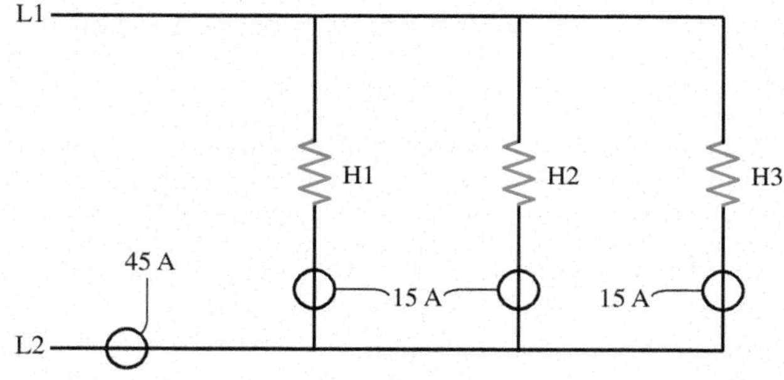

Figure 17-10 In a parallel circuit, the same voltage is applied across each of the heat strips. The total current is equal to the sum of the current in each of the parallel circuits.

Parallel Circuits

Parallel circuits are arranged so that the power source is the same across all the loads. In Figure 17-10, the power source is in parallel with three electric heating elements. The voltage supplied across each element is the same. The current flow through each element is the same because the resistance of each heater is the same. The total current flow equals the sum of the amperage draw of each branch or heater. This is represented by the formula

$$I_T = I_1 + I_2 + I_3 + \cdots$$

If the resistance of the heating elements is different, the current flow through each branch will be different. Using Ohm's law, you can calculate that the heating element with the lower resistance will develop higher current flow because there is less opposition in the heating element. Heating elements with higher resistance develop lower current flow because they oppose the flow of electrons. With this in mind, a shorted heating element will draw extremely high amperage and cause a fuse or breaker to trip.

Series-Parallel Circuits

Combination circuits have series and parallel components in the same circuit, as shown in Figure 17-11. This is the most common type of HVACR diagram. Let's trace through part of the series-parallel circuit shown in this figure.

Point 1: Figure 17-11 shows the contactor circuit, highlighted in yellow, in series with the series load of the compressor and run capacitor.

Point 2: Below and in parallel with the compressor line, highlighted in red, is the a switching relay, which is in series with an indoor fan and condenser fan. These fans are in parallel with each other.

Point 3: The primary of the control transformer is in parallel with the components above it.

Point 4: The primary side of the transformer, highlighted in blue, is also in series with the thermostat, limit switch, and gas valve.

Point 5: This is the secondary or 24-V side of the transformer. It will be used to energize the contactor coil and the relay coil.

Some components are in series and some are in parallel, which is why the word *combination* is applied to this circuit: It combines series and parallel circuits together in the same circuit.

In combination circuits, the voltage is the same across each parallel branch. The current flow depends on the amount of resistance or impedance in each branch. The total current flow is the sum of current flow in each of the branches. Impedance is the resistance to current flow in inductive components such as motors, transformers, or relay coils. An inductive device is any component that uses magnetism to do work. For example, motors require magnetism to create rotating motion.

Simply measuring the resistance of an induction component and using the Ohm's law formula will not give a technician the current flow requirement in a circuit. A special impedance formula is needed to calculate this information. Knowing the exact impedance is not important to the HVACR tech.

A complete circuit can be found in many different configurations. It is important to identify the type of complete circuit in order to understand the voltage and current flow and requirements. Knowing what to expect in the circuit makes the troubleshooting process easier to master.

We have discussed complete or closed circuits. We now turn to a discussion of open, short, and grounded circuits.

Open Circuit

An open circuit means that there is a break or "opening" in the wire or load that stops the flow of electrons. Figure 17-12 has one open switch that creates an open circuit. Figure 17-13 shows an open contactor in a series circuit. A common way to open a circuit is with the use of a thermostat to open cooling, heating, or refrigeration circuits when they are not required.

Short Circuit

A short circuit has the voltage going to ground or around the load. The word *short* means that voltage is taking a shortcut around the load. A short circuit does not need to be a direct short to ground or a routing of the voltage around the load. For example, a condition in which the load has a very low resistance is a short-circuit condition. A load that is supposed to be 10 Ω but is actually 2 Ω is a short-circuit condition.

Figure 17-14 is a diagram of a short-circuit condition. The short circuit is not a circuit by design, but a problem. A short circuit is usually a wiring mistake or a defect that causes the load to be bypassed. A short-circuit condition causes

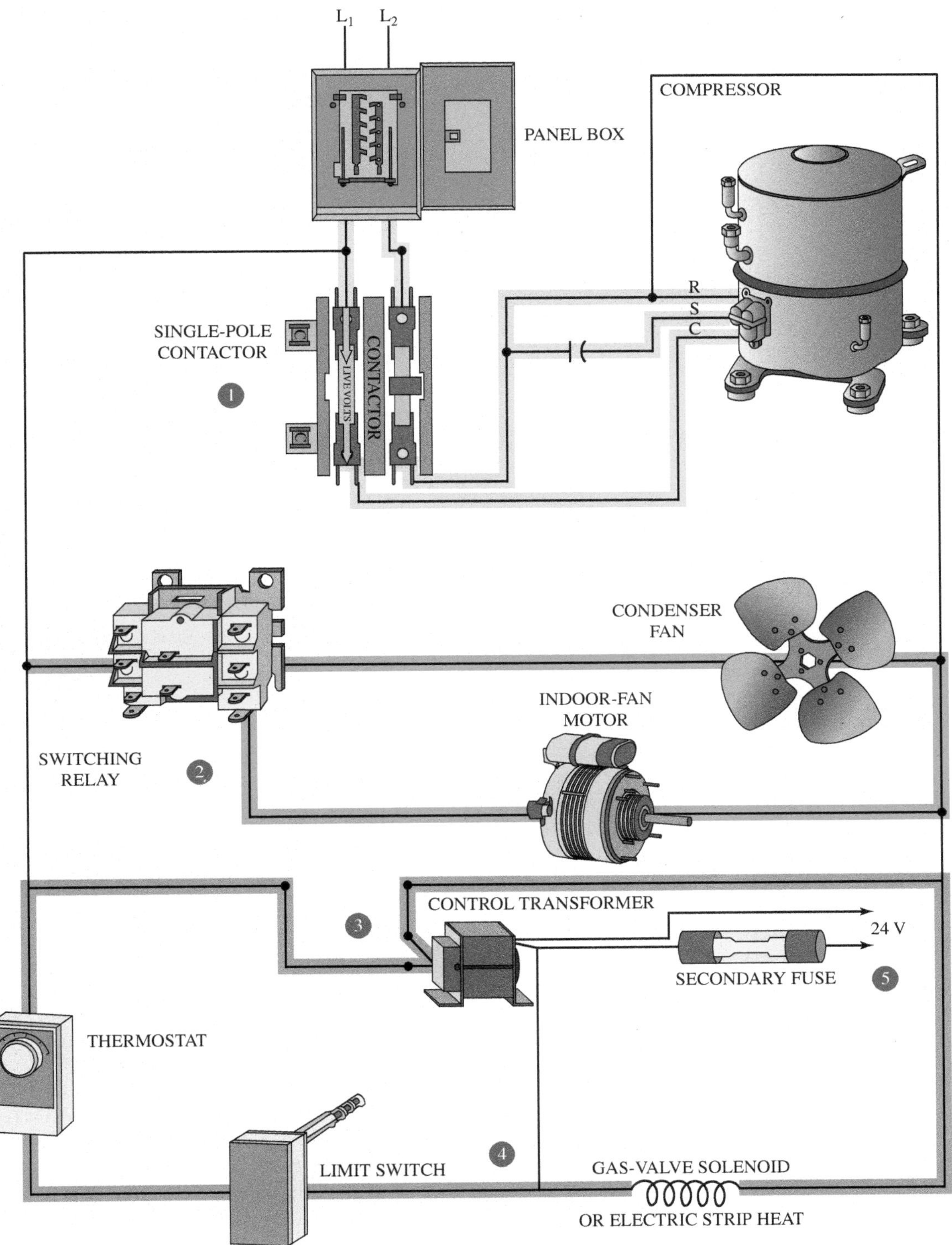

Figure 17-11 This simple ladder diagram of a series-parallel circuit shows parallel loads in series with switches that control their operation.

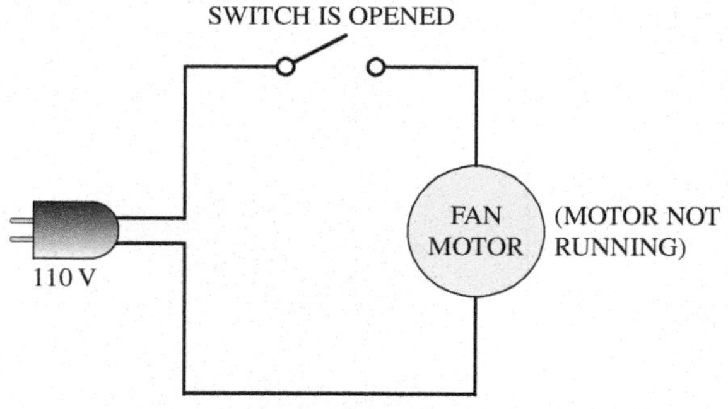

Figure 17-12 This is an open circuit with no current flow. The switch is open, blocking the current flow.

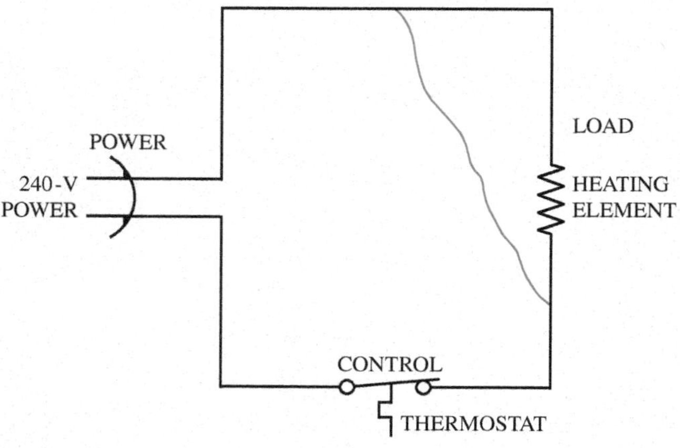

Figure 17-14 This is a short circuit or a power supply that is bypassing the load. A red wire is causing the power supplied to be shorted out. The power supply or wire will be damaged in this shorted condition.

excessive amperage draw, which may open the fuse or circuit breaker or even burn a wire open.

As stated earlier, a short circuit is generally thought to have no resistance. A circuit can be considered as shorted even if it has a resistance reading. Another example: Consider what will happen if the resistance in a parallel branch is 5 Ω and a relay coil in that branch shorts out and reduces the resistance from 5 to 1 Ω. This low resistance will create a high current flow in that part of the branch. If this part of the circuit is served by a transformer, it will cause the transformer to overheat or the fuse to blow. A short circuit can be difficult to troubleshoot because there is resistance in the circuit. Each component needs to be inspected for abnormal appearance, such as a burned coil or overheating. If this does not reap results, each of the parallel branches needs to be checked individually for resistance. Also, remember that a wire or terminal can be touching the case ground and cause a shorted condition.

TECH TIP

Do not test transformer voltage using the spark method. The spark method involves quickly touching a 24-V transformer's leads to each other. Some techs do this to see if a spark is created. A spark indicates that there is some voltage at the output of the transformer. But some transformers have internal or external fuse protection that will blow if this procedure is used. So it is recommended that a voltmeter be used to check the secondary voltage instead of the spark method.

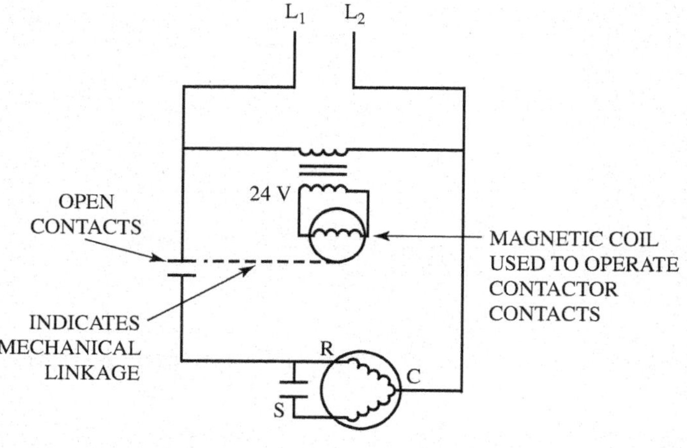

Figure 17-13 The contactor is a magnetic switch that opens and closes a contact or set of contacts to operate a compressor or fan load. A thermostat is used to control the 24 volts going to the contactor coil. This is an open circuit, since the contacts on the left side are open.

Ground Circuit

A ground circuit is similar to a short circuit. A ground circuit is one where the component has less resistance than normal to ground, as shown in Figure 17-15. Normal resistance to ground is called "infinity" or at least has a resistance of millions of ohms to ground. A good motor will have a reading of millions of ohms to ground. If the motor winding to ground is 100 ohms, this would be considered a grounded condition.

Consider another example: A good motor winding may have a resistance of 10 Ω. If some of the windings are shorted to the case, the resistance will be less than 10 Ω, maybe even 0 Ω. The lower resistance means that the component is grounded or some of the windings have a voltage circuit to motor case. This lower resistance to ground stops the motor operation and excess current draw is observed. The excess current will open a fuse or trip a circuit breaker. Sometimes the terms *short circuit* and *ground circuit* are incorrectly used

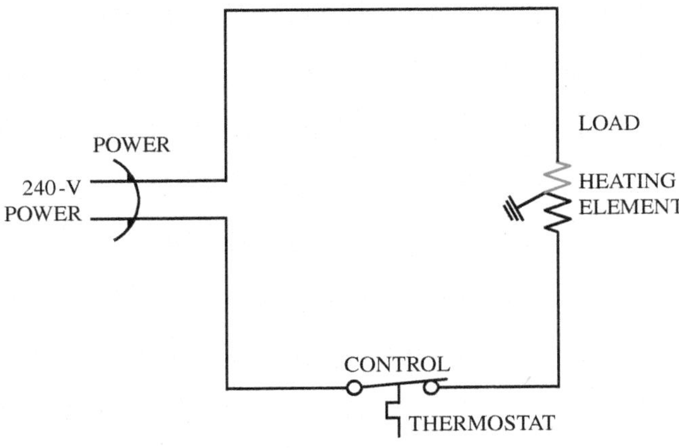

Figure 17-15 This heating element is shorted to ground. This will result in high amp draw, which will possibly trip the circuit breaker. The shorted heat strip will reduce heat output.

interchangeably, but the results are still the same: A defective component is created by a ground or short.

TECH TIP

Most diagrams for HVACR equipment use similar symbols. If a system does not display a symbol legend, the technician can refer to the symbol list shown in Figure 17-1 or locate another system nearby that may have an intact electrical diagram.

17.3 TYPES OF ELECTRICAL DIAGRAMS

There are many names for electrical diagrams, which can cause some confusion. For instance, electrical diagrams may be called wiring diagrams, schematic diagrams, ladder diagrams, pictorial diagrams, external diagrams, connection diagrams, panel diagrams, or hook-up diagrams. HVACR manufacturers and circuit designers have created these names to make it easier for the user to understand a difficult part of our profession. It is important to realize that the meaning of these terms may differ among technicians. When reviewing industry publications, you will find that the terminology varies between textbooks and in manufacturers' literature. But we will attempt to create a working vocabulary that you can use in the field. In the following sections we will define, use, and show examples of:

- **Wiring diagrams**
- **Schematic diagrams.**

The other terms mentioned above fall into one of these two categories.

17.4 WIRING DIAGRAMS

As the name implies, wiring diagrams show how the HVACR system components are wired together. Wiring diagrams are also known as:

- Pictorial diagrams
- External diagrams
- Hook-up diagrams
- Connection diagrams.

A wiring diagram is a snapshot or pictorial view of what a technician can expect to find when the equipment cover is removed from a HVACR system. A wiring diagram is arranged so that the physical location of the components in the diagram is similar to the arrangement of the actual equipment. Figure 17-11 showed a simplified pictorial diagram of a basic packaged air conditioning and heating system. This view is helpful in understanding the basic operation of the circuit but needs more detail if it is going to be valuable to an experienced technician.

Wiring diagrams and some component symbols are illustrated in Figure 17-16. You will notice that not all component symbols are represented in this diagram. For example, the compressor shows **C**, **S**, and **R** but not the windings, while most of the relays show the coil and the contacts. The features of the wiring diagram shown in Figure 17-16 are as follows:

- Connection diagram
- List of component names
- Color code for most wiring
- Dashed lines (-------), which indicate field wiring, both high and low voltage.

Figure 17-16 has enough detail for a technician to understand circuit operation for troubleshooting, though the fan and blower motor part of the diagram (located near the upper right side) does not show the internal motor windings or symbols. As stated above, you will notice that this diagram uses some electrical symbols inside boxes to represent components.

Figure 17-17 shows the electrical symbol for the compressor and condenser motors shown in Figure 17-16. This figure shows the start (**S**) and run (**R**) motor windings and common connection (**C**). The block representation of the transformer, near the middle of Figure 17-16, is shown in symbol form in Figure 17-18.

A wiring diagram is not as detailed as a schematic diagram. Wiring diagrams can be used to wire and troubleshoot a HVACR system. As we will see in the next section, schematic diagrams are more useful for performing troubleshooting diagnostics.

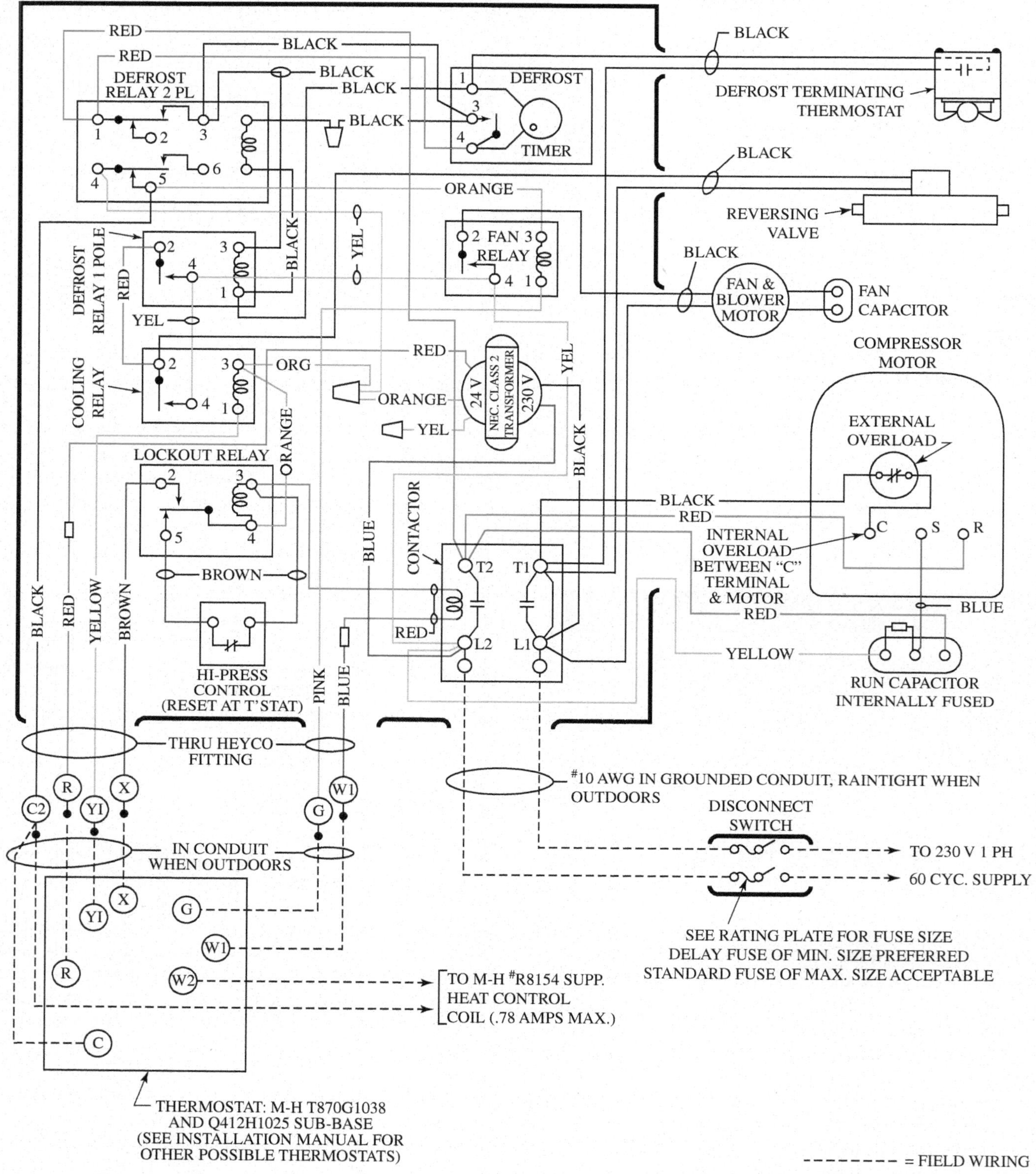

Figure 17-16 This is an example of a wiring diagram for a heat pump outdoor section and thermostat. Notice how the components are "pictured" with connected wiring. This is a snapshot of the components and interconnecting wires. Wires are color coded (black, red, yellow, brown, etc.) and components are identified to make it easier to trace for troubleshooting.

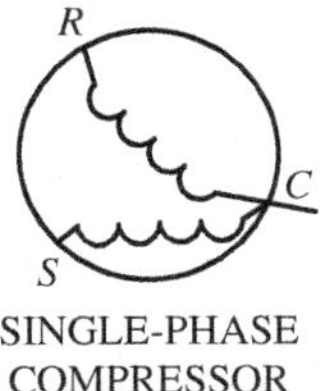

Figure 17-17 This is a symbol for a single-phase blower motor and compressor motor. The windings are not illustrated in the wiring diagram shown in Figure 17-16.

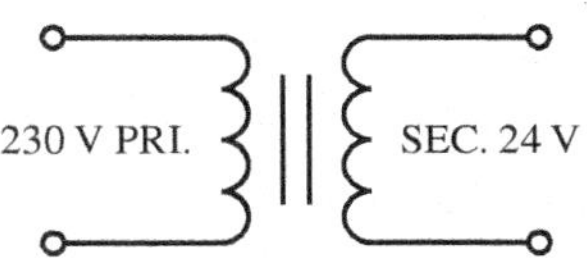

Figure 17-18 This is a symbol for a transformer. The transformer windings are not illustrated in the wiring diagram shown in Figure 17-16.

SEQUENCE OF OPERATION 17.1: AIR CONDITIONING CONDENSING UNIT AND THERMOSTAT CIRCUIT

The wiring diagram shown in Figure 17-19 is for a condensing unit and thermostat control. Notice that the condenser fan motor (**FM**), shaded in green, has a circle around the symbols for the motor windings. The compressor (**COMP**) motor winding, circled in blue and shaded in gray, does have a diagram circle around the symbol. Not all symbols are

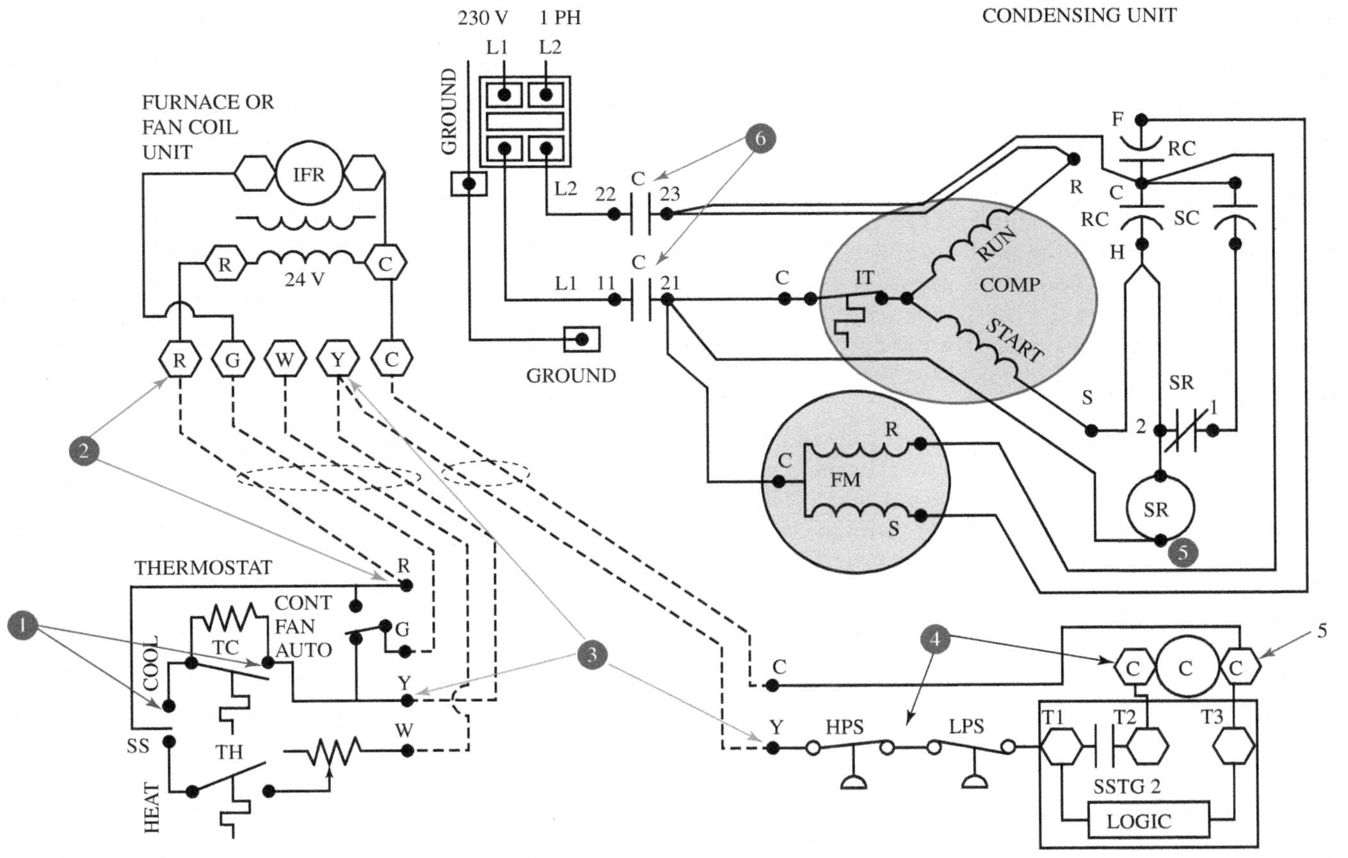

Figure 17-19 This is a wiring diagram for an air conditioning condensing unit and thermostat circuit. The air handler fan motor is not included in this diagram. The dashed lines are low-voltage field wiring that is hooked up by the installation crew. The R, G, W, and Y wires should follow the wiring color code.

(*continued*)

contained within a boundary. The dashed lines represent field-wired, low-voltage control wiring. The control wiring hook-up on the thermostat should use the following color code:

R = red wire

G = green wire

Y = yellow wire

W = white wire.

Here is the basic sequence of operations for the unit shown in Figure 17-19:

1. Close the thermostat's Cool switch (SS) and adjust the thermostat so that the cooling thermostat closes (TC).
2. The 24 volts from the **R** side of the transformer travels to the t'stat.
3. The 24 volts goes through the t'stat to the **Y** connections.
4. The 24 volts goes through the high- and low-pressure switches; control logic to the left side of the contactor coil ©.
5. The 24-V circuit is completed by going through the right side of the coil through the **C** connections and to the transformer.
6. When the © contactor coil is energized, it will close the double-pole contactor, thus starting the compressor (**COMP**) and condenser fan motors (**FM**).

17.5 SCHEMATIC DIAGRAMS

A schematic is another representation of a HVACR electrical diagram. It is also called a **ladder diagram** or **line diagram**. The schematic makes it easier to understand the electrical sequence of operations. When troubleshooting, a schematic diagram is the most helpful type of electrical diagram. A ladder diagram is drawn as a ladder with rungs or steps between the two power wires as shown in Figure 17-20 or Electrical Diagram ED-5, which appears with the Electrical Diagrams package that accompanies this book.

In a ladder diagram, you have power rails and control rungs. The rails provide the lines of power to the circuit. The circuit rungs are stacked like the steps of a ladder. The rung circuits have a load and switches that control the load. The switches open and close, stopping and starting the circuit controlled by the load. The ladder diagram can be drawn in a vertical or horizontal manner. It is easier to understand the vertical diagram when used for troubleshooting purposes. Ladder diagrams can be as simple as that shown in Figure 17-21 or more complex as shown in Figures 17-22 and 17-23.

The schematic diagram divides the system up into a series of individual circuits. Each circuit has an expressed function in the correct operation of the overall system. Figure 17-24 (page 322) is a classic example of the ladder schematic diagram. Notice how this ladder diagram is divided into sections. The upper half is the high-voltage air conditioning section. The lower section is called the low-voltage or control voltage section. The way the schematic diagram is laid out is easier to understand than a wiring diagram.

TECH TIP

In the hopscotch troubleshooting method, we mentioned that the "load" is on the right side of the diagram and the "line" is on the left side. In reality, with an **L1** and **L2** power supply, commonly found in 208/240-V circuits, both sides are the line side. To explain the hopscotch procedure, we need to identify one side as the "load side" or voltage dropping source and the opposite side as the "line side," the voltage supply. In some electrical designs, the load is on the left side of the ladder diagram. In this case, you will reverse the hopscotch procedure by fastening the voltmeter probe to the left side of the circuit and start hopscotching from right to left. The hopscotch procedure works well in both directions as we will see in later troubleshooting units.

17.6 IDENTIFYING THE PARTS OF A DIAGRAM

The basic components of a good diagram include the following:

- The **pictorial diagram** shows the placement of components and their wire color (see the upper part of Figure 17-27, page 324). This diagram also indicates the number connection on each component. This is also known as a wiring diagram or connection diagram.

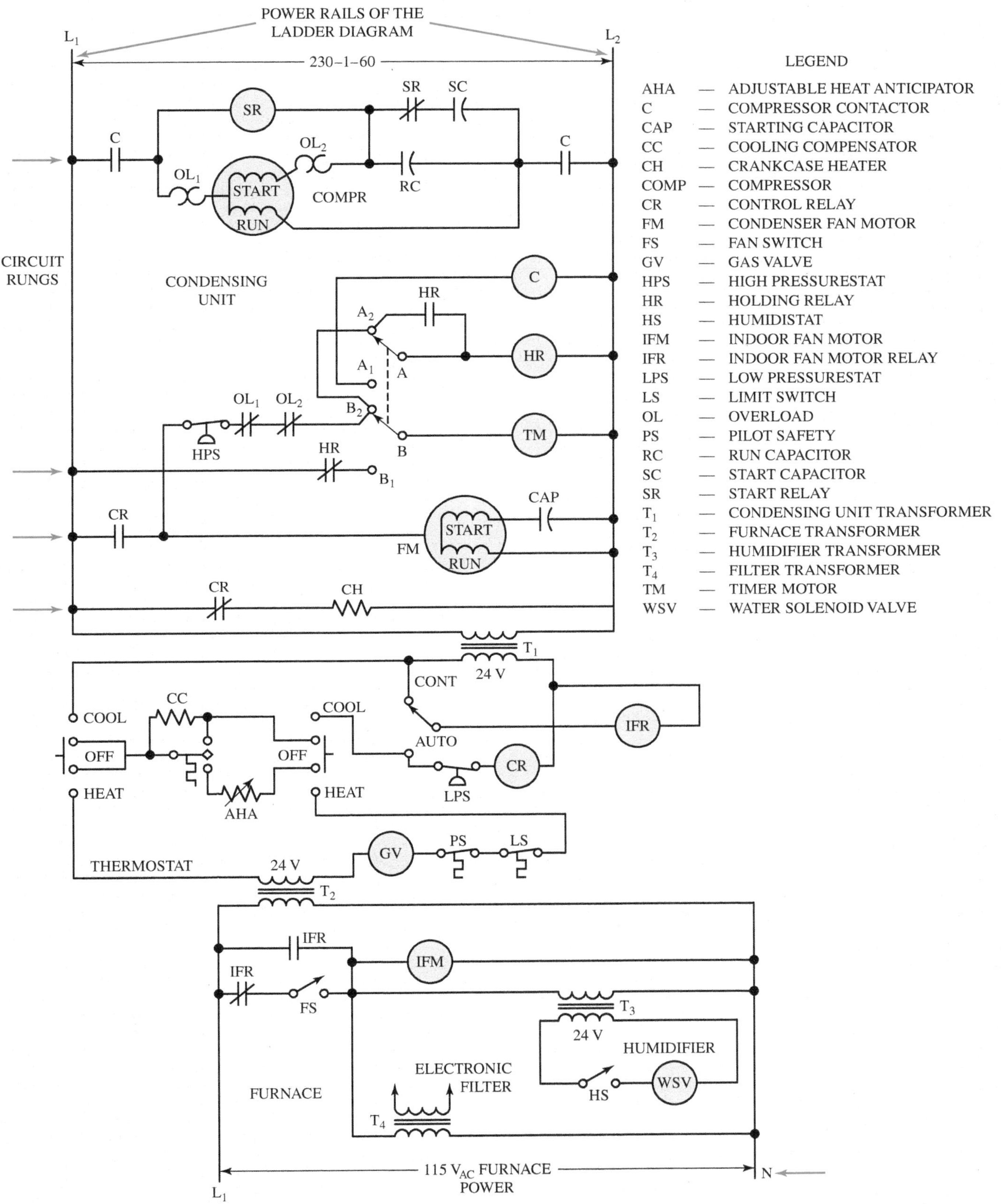

Figure 17-20 This specific type of schematic is a ladder diagram. It is the most useful type of wiring diagram for troubleshooting. The legend located at the upper right side identifies the various component parts found in the ladder diagram.

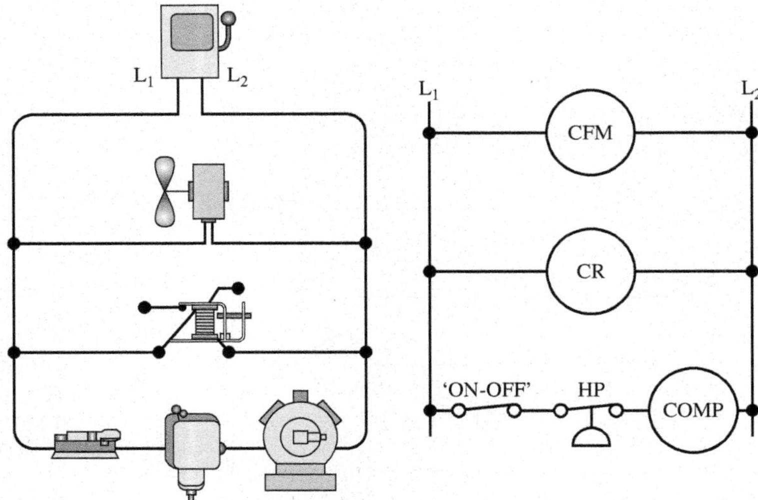

Figure 17-21 On the left side is a pictorial diagram. The schematic on the right shows the same components drawn as a ladder diagram.

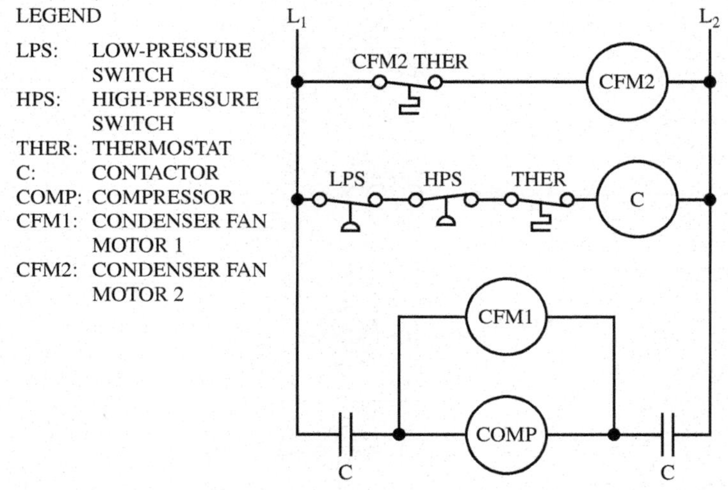

Figure 17-22 This ladder diagram with its legend of components is more complex than the one shown in Figure 17-21.

- The schematic diagram is used to display the current path to various components. This is also known as a ladder diagram (see the lower right side of Figure 17-27, page 324). The ladder diagram is used for troubleshooting and installation purposes. As you see, the ladder diagram is easy to read since it is drawn in a line-by-line format.
- The *legend* can be thought of as a dictionary of the symbol language used in an electrical diagram. The legend in Figure 17-27 translates the letters and symbols used in the diagram to the component name or type of wiring. (For a larger view of this figure, see Electrical Diagram ED-7, which appears with the Electrical Diagrams package that accompanies this book.) In Figure 17-27, the field and factory wiring is included in the legend. The factory wiring is shown as solid, black lines. Field wiring and control wiring are shown as dashed lines.

Figure 17-28 is an example of a comprehensive legend that can be used with most HVACR applications. The designer of the electrical diagram will develop the legend, therefore the form and content will vary between wiring diagrams. There is no standard for a legend or the letters used to represent an electrical device.

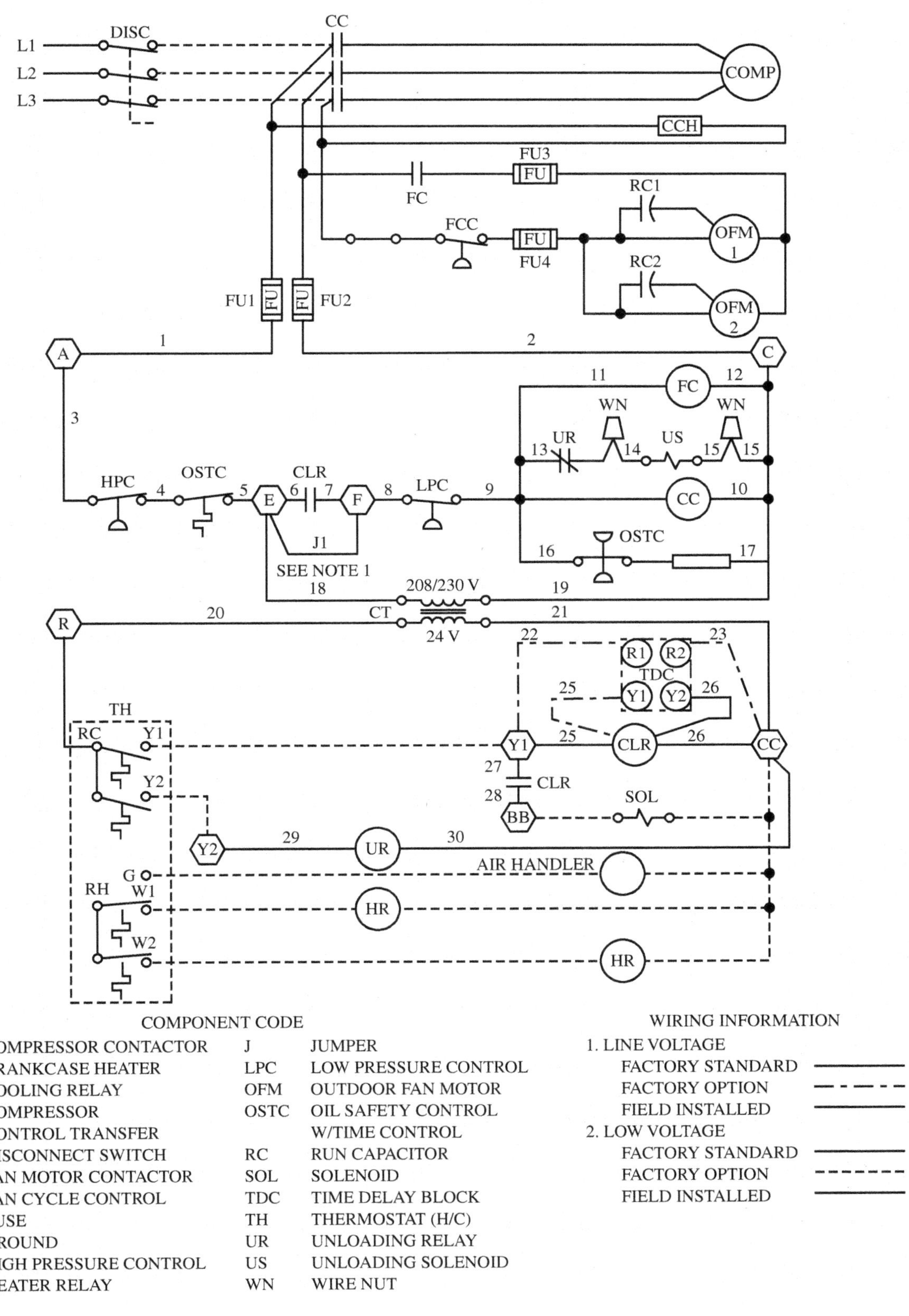

NOTES:

1. WHEN LIQUID LINE SOLENOID VALVE IS NOT USED REMOVE JUMPER J1

Figure 17-23 This schematic ladder diagram is more complex than those shown in Figures 17-21 and 17-22. It is more complete and, therefore, more useful—note the component code, notes, and wiring information at the bottom of the diagram.

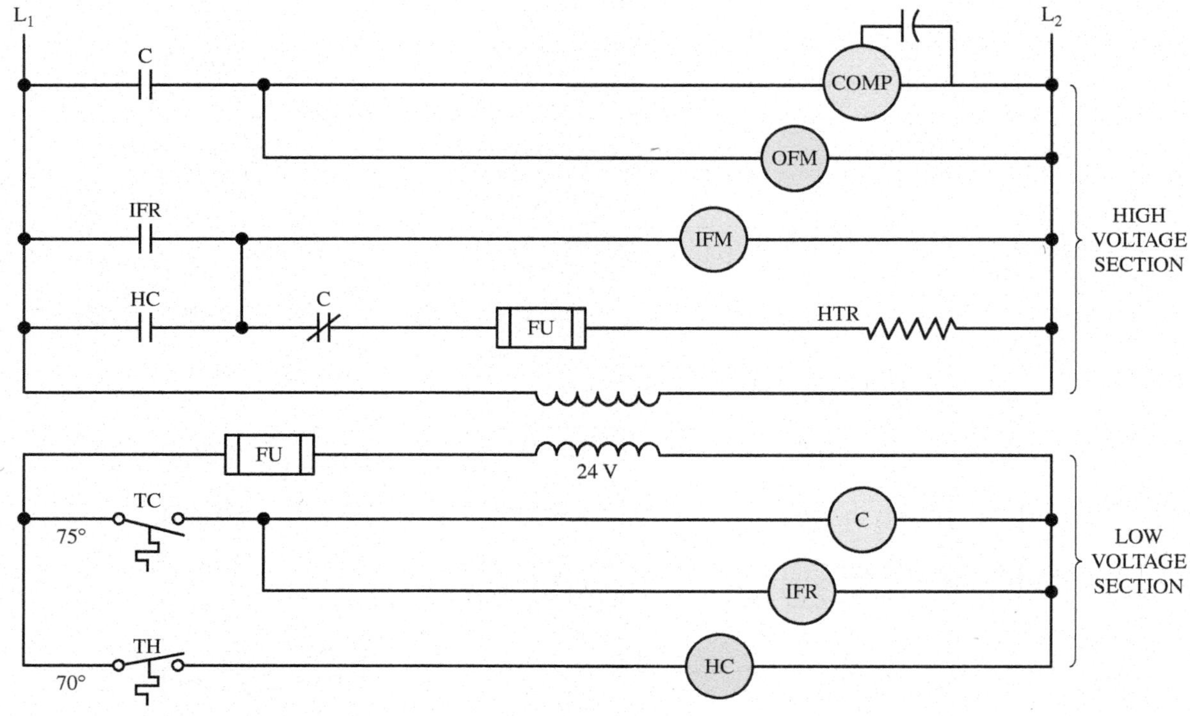

Figure 17-24 In this typical ladder diagram, the upper section is the high-voltage section. The lower section is the low-voltage or control voltage section.

TROUBLESHOOTING 17.1: USING THE HOPSCOTCH TROUBLESHOOTING PROCEDURE

The ladder diagram is used to conduct the hopscotch troubleshooting procedure. Here are the steps in the hopscotch troubleshooting technique. Let's use Figure 17-25 for this example:

1. Determine the load side of the circuit. On the equipment being diagnosed, find the power on the right side of the diagram; in this case it is marked as **L2** and found below the number 1 blue dot.
2. Start by attaching one of the voltmeter probe's to the right side, actually the power side, of any component found in the system as seen on the ladder diagram at the point 2 blue dot in Figure 17-25. The load side is the side of the circuit that has the component with the greatest resistance, in this case the Ⓒ or coil. The line side will have switches such as pressure switches, contacts, and overloads. Connect the one voltmeter probe to the right side or load side of the circuit.
3. Starting at the left side of the ladder diagram, point 3 in Figure 17-25, the second probe is used to measure voltage or jump over switches. You should measure 230 volts up to the coil. Loss of voltage before getting to the coil indicates an open switch or wire.

Note: Let's review how a voltmeter works. As discussed in Unit 3, a voltmeter measures potential difference. In this example, it would be the difference between **L1** and **L2**. Once the voltmeter probes are on the same voltage or potential line, the meter will read 0 V even if the voltage is present.

Note: This is how the **hopscotch troubleshooting** or jumping across components troubleshooting method works. You measure voltage across switches until the voltage is lost or the meter reads zero volts. Having voltage on the left side of a switch, but no voltage on the right indicates that the circuit is open at that switch.

4. What to expect? You should read voltage through all closed switches up the load, which in Figure 17-25 is the coil Ⓒ or point 4. Once you jump over, or to the right, of the coil, the voltage reading on the meter will be **0 V**, since the potential voltage at that point is the same.
5. If voltage is measured on the left side of a switch and zero volts measured when the probe is moved to the right side of the switch, the switch is open. Figure 17-26 shows an example of the voltage dropping to zero volts when the **CR 1** switch is open.

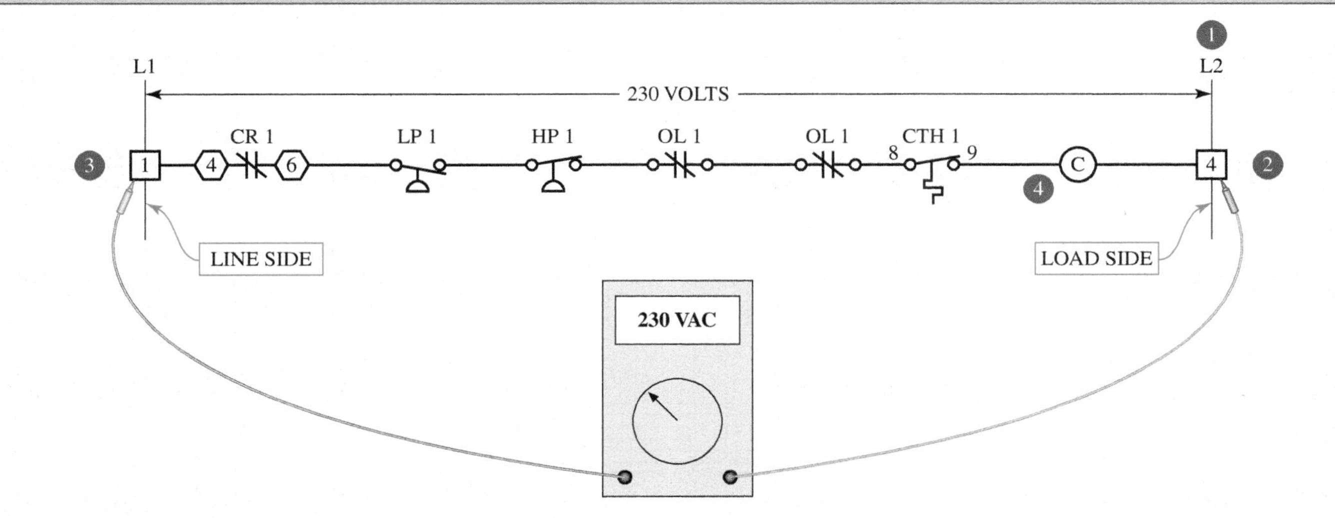

Figure 17-25 Ladder diagrams are useful when using the hopscotch troubleshooting method.

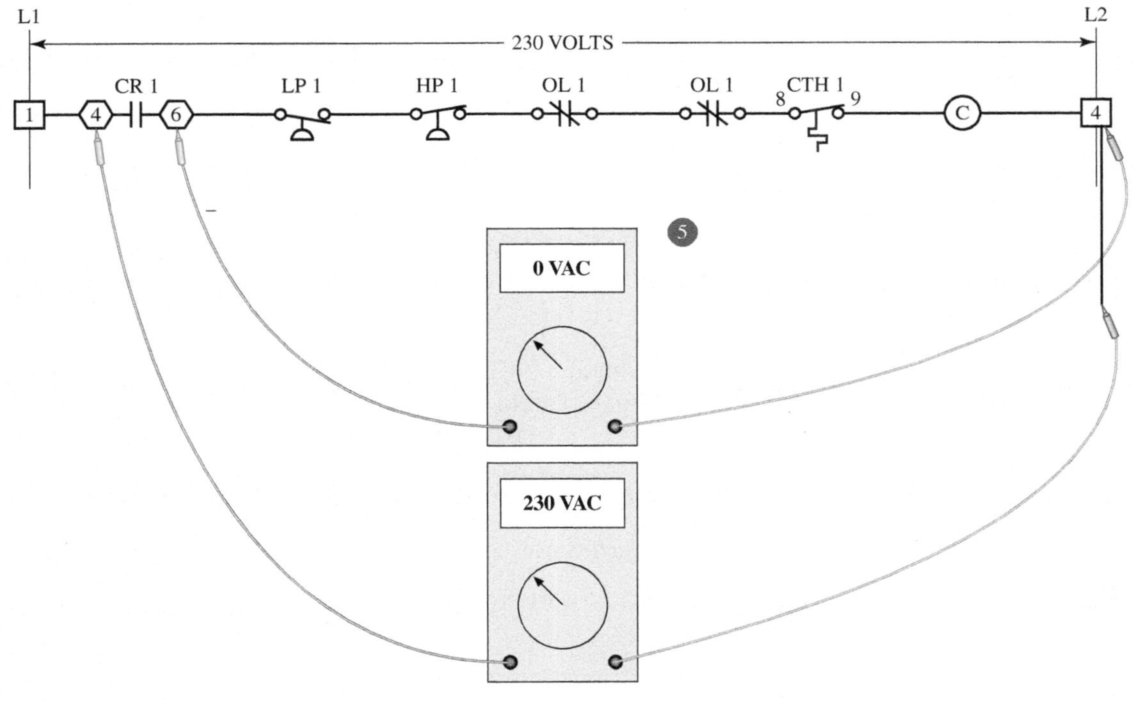

Figure 17-26 When using the hopscotch method of troubleshooting, the voltage will be lost when jumping over an open switch, as shown in this example. **CR 1** is open. There is voltage on the left side or point 4 on the **CR 1** contacts. There is 0 VAC on the right side since it is open.

- Notes are used to clarify or provide additional useful information. An example of notes is found at the bottom left side of Figure 17-27 or Electrical Diagram ED-7.
- Color labels on the diagram identify the color of the wire used in the component. Be cautious when using wire color for tracing circuits using the diagram because, for example, the wiring from the factory or the field may have been changed. The factory assembly line may not stop the wiring process when they run out of one color of wire. In addition, the field tech does

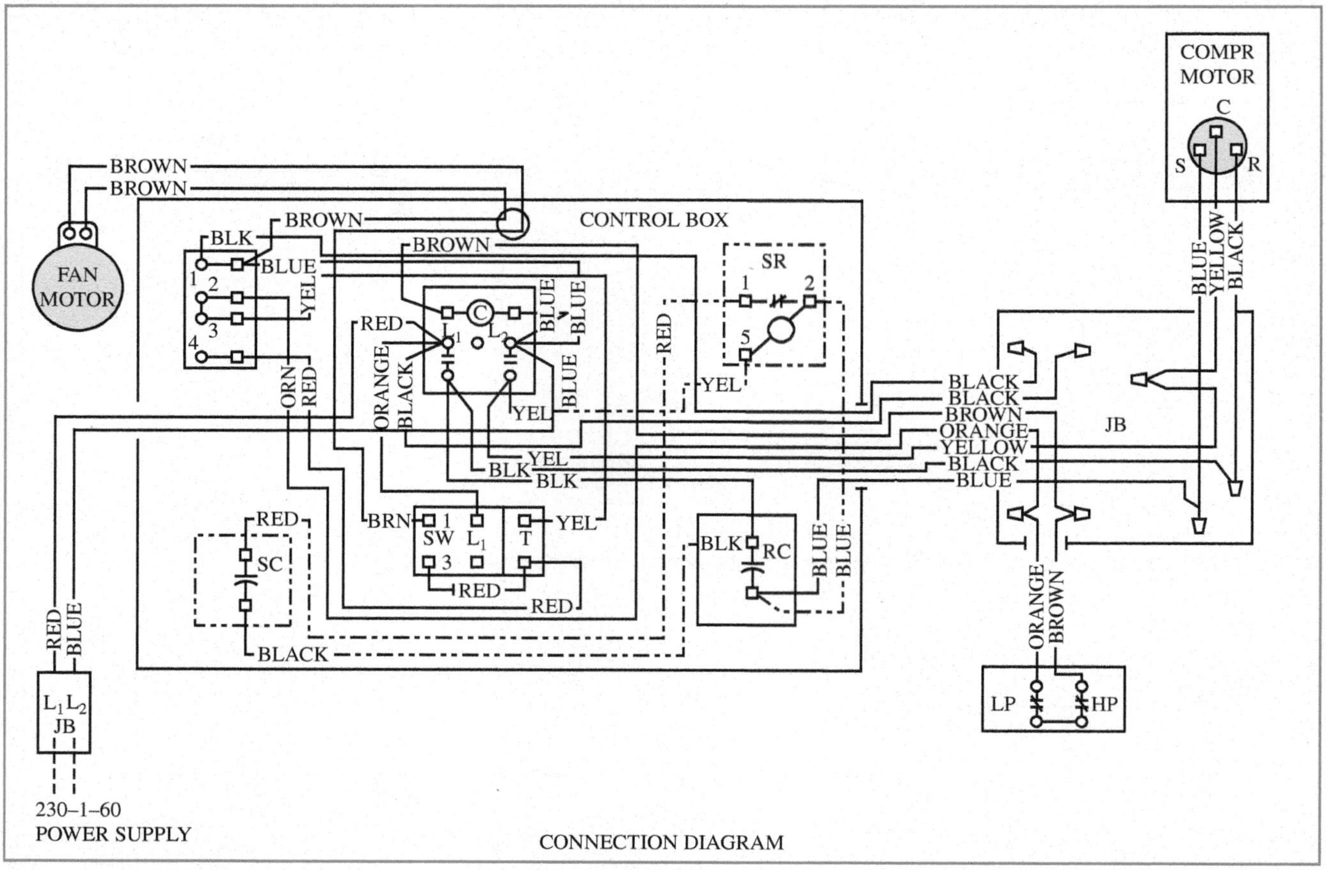

COMPR MOTOR
C
S R
BLUE YELLOW BLACK
BROWN
BROWN
FAN MOTOR
BROWN
CONTROL BOX
BLK
BLUE
BROWN
SR
1 2
3
4
YEL
RED
C
L1 L2
BLUE
BLUE
5
RED
ORANGE
BLACK
ORN
RED
YEL
BLUE
YEL
BLACK
BLACK
BROWN
ORANGE
YELLOW
BLACK
BLUE
JB
BLK
YEL
BLK
RED
BRN 1 SW L1 3
T
YEL
SC
RED
RED
BLK RC
BLUE
BLUE
ORANGE
BROWN
RED
BLUE
BLACK
L1 L2 JB
LP HP
230–1–60
POWER SUPPLY

CONNECTION DIAGRAM

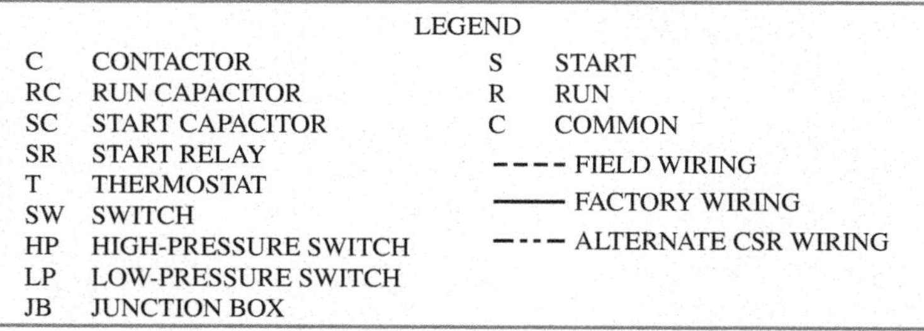

LEGEND

C	CONTACTOR	S	START
RC	RUN CAPACITOR	R	RUN
SC	START CAPACITOR	C	COMMON
SR	START RELAY	----	FIELD WIRING
T	THERMOSTAT	——	FACTORY WIRING
SW	SWITCH	---·-	ALTERNATE CSR WIRING
HP	HIGH-PRESSURE SWITCH		
LP	LOW-PRESSURE SWITCH		
JB	JUNCTION BOX		

NOTES

1. FAN MOTOR PROVIDED WITH INHERENT THERMAL PROTECTOR.
2. COMPR. MOTOR PROVIDED WITH INHERENT OVERLOAD PROTECTOR.
3. MAX. FUSE SIZE 30-AMP DUAL ELEMENT.

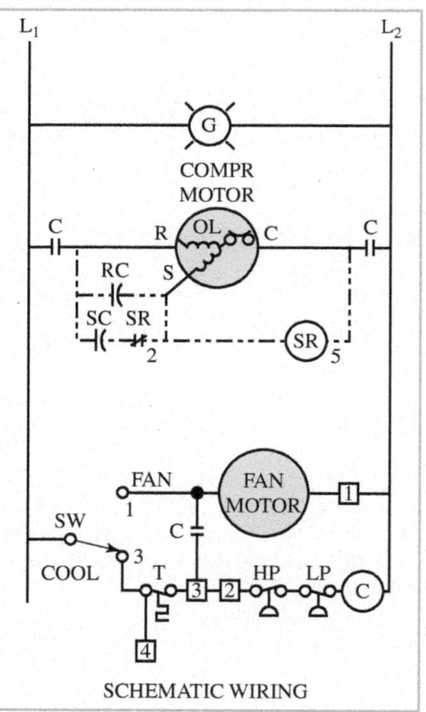

Figure 17-27 This pictorial wiring diagram has labeled components, a connection diagram, a legend, and notes. This is both a connection and a wiring diagram. The diagram in the lower right is a ladder diagram of the connection diagram shown at top.

Relays	
R	Relay, General
CR	Cooling Relay
DR	Defrost Relay
FR	Fan Relay
IFR	Indoor Fan Relay
OFR	Outdoor Fan Relay
GR	Guardistor Relay
HR	Heating Relay
LR	Locking Relay (Lock-in or Lockout)
PR	Protection Relay (Relay in series with protective devices)
VR	Voltage Relay
TD	Time Delay Device
THR	Thermal Relay (type)
M	Contactor
MA	Auxiliary Contact

Solenoids	
S	Solenoid, General
CS	Capacity Solenoid
GS	Gas Solenoid
RS	Reversing Solenoid

Switches	
DI	Defrost Initiation
DT	Defrost Termination
DIT	Defrost Initiation–Defrost Termination (dual function device)
GP	Gas Pressure
HP	High Pressure
LP	Low Pressure
HLP	Combination High-Low Pressure
OP	Oil Pressure
RM	Reset, Manual
FS	Fan Switch
SS	System Switch
HS	Humidity Switch (Humidistat)
TA	Thermostat, Ambient
TC	Thermostat, Cooling
TH	Thermostat, Heating
TMA	Thermostat, Mixed Air
CT	Thermostat, Compressor Motor
HT	High Temperature
LT	Low Temperature
RT	Refrigerant Temperature
WT	Water Temperature

Miscellaneous	
C-HTR	Crankcase Heater
RES	Resistor
HTR	Heater
PC	Program Control
OL	Overload
L	Indicating Lamp
⊕	Manual Reset Device
+	Automatic Reset Device
CAP	Capacitor

Compressors	
C	Common
S	Start
R	Run

Figure 17-28 This is a sample of a common electrical legend.

not carry a full complement of wire colors, therefore a replacement wire may be a different color when compared to the wiring on the diagram. Also consider the fact that a colorblind technician may not be able to distinguish the difference in wire color.

- Reference or margin numbers, as seen on the left side of Figure 17-29, identify a line or component in a horizontal plane in line with the number. If you are calling a tech support service, having reference numbers is helpful. They can be used by a technician for troubleshooting over the Internet or by phone. For example, the technician could explain that he does not understand the safety devices on lines 18 and 25. Tech support can then zero in on this line and explain the operation of each component.
- Some designers place a letter designator along the bottom of the diagram to reference the vertical components in a diagram. This can be used like a road map to locate a specific component. For example, component **C 3** may be a specific compressor control. The letter **C** would be used to locate the vertical axis. The number **3** would be used to locate the horizontal axis. So **C3** would identify the location of the part under discussion. This design is also valuable when reading an explanation of the sequence of operation or step-by-step operating instructions.

17.7 GUIDELINES FOR READING ELECTRICAL DIAGRAMS

Most electrical diagrams follow the guidelines below. Knowing these guidelines will help the technician troubleshoot a system.

1. Reading a schematic diagram is similar to reading a book. Diagrams are generally read from left to right and top to bottom.
2. Electrical symbols are shown in their de-energized positions.
3. Contactor or relay contact symbols are shown with the same numbers or letters used to designate the contactor or relay coil. All contact symbols that have the same number or letter as a coil are controlled by that coil regardless of where in the circuit they are located. An example of this is shown in Figure 17-29, circled in red. The (1M) coil controls the three-pole contactor that

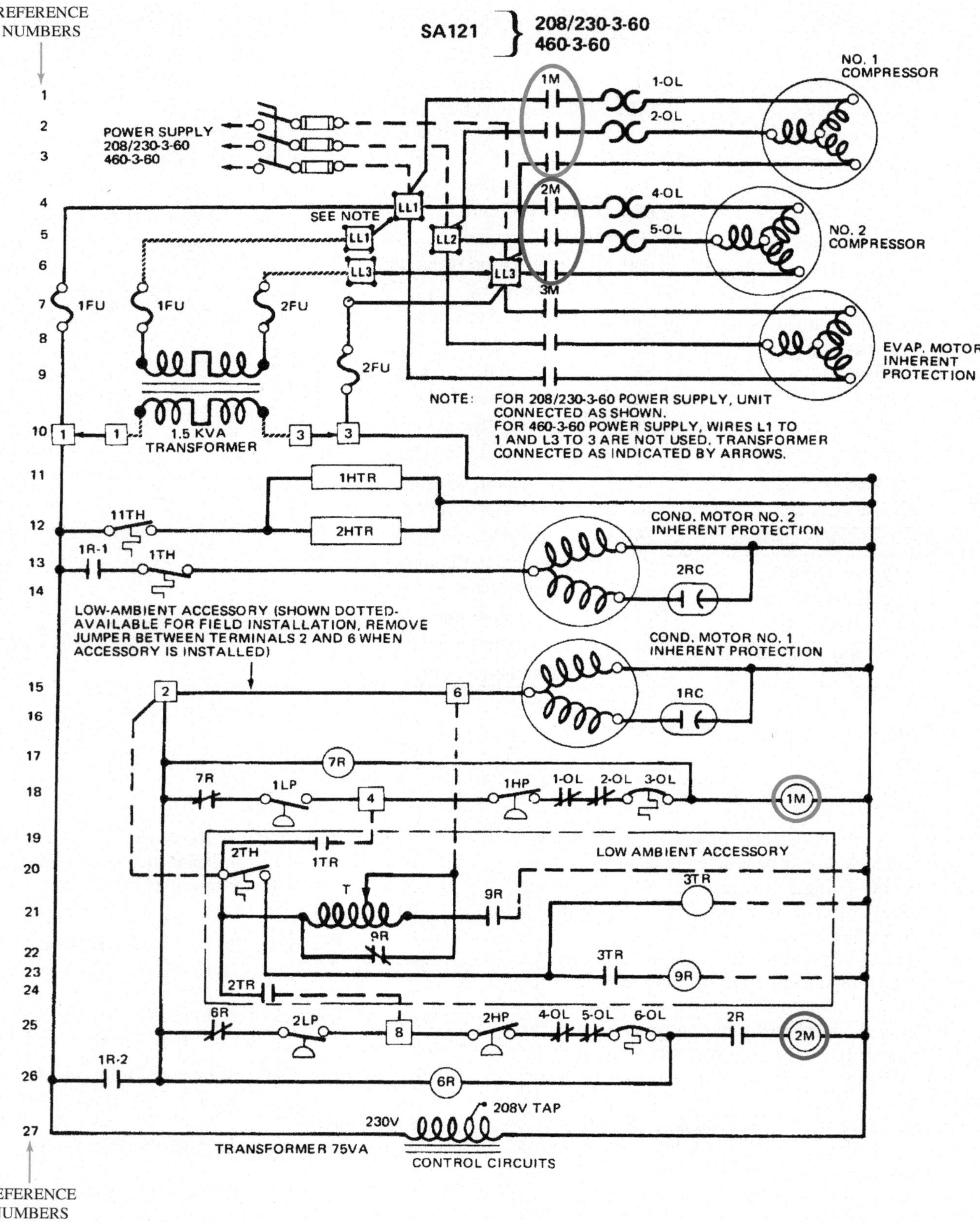

Figure 17-29 Reference numbers or margin numbers, as seen on the left side of this figure, identify a line or component in a horizontal plane in line with the number.

operates the No. 1 compressor. Another example is the (2M) coil circled in blue. The coil closes the three-pole contactor that operates the No. 2 compressor. Other examples like this can also be shown in this diagram.

4. When a relay is energized, or turned on, all of its contacts change position. If a contact is shown as normally open, it will close when the coil is energized. If the contact is shown as normally closed, it will open when the coil is energized.
5. There must be a complete circuit before current can flow through a component. Current leaves one side of the power supply and returns on the opposite side of the power supply.
6. Components used to provide a stop function are generally wired normally closed and connected in series with the loads they control.
7. Components used to provide a start function are generally wired normally open and connected in parallel. If either switch is closed, a current path will be provided to the motor and the motor will be turned on.
8. Switches, like thermostats or contacts, are in series with the loads they control.
9. Most loads are drawn on the right half of the diagram. Switches are drawn to the left side of the diagram.

Reviewing these guidelines before you use a diagram the first few times will help you more fully understand the full operation of a wiring diagram.

17.8 DESIGNING AN ELECTRICAL DIAGRAM

This section discusses the common way an electrical diagram is produced. The procedure is not difficult and can be used to design a diagram in the office or in the field. By using this design method, you should be able to number all of the wires prior to connecting the components in the system. There are wiring number labels that can be used to identify the wire and the circuit connections. The wiring numbers will be useful when using the diagram for troubleshooting. At the end of this process the wire color can also be labeled on the diagram.

Step 1: The first step is to draw a component layout. This can be done freehand, with computer drawing programs, or by use of computer-aided drafting (CAD) software. Whether done by hand or software, allow plenty of room to draw and label wiring. It is difficult and confusing to use an undersized drawing. It is easy to mix up wiring lines when they are too close together. Drawing a diagram with various color markers will also aid in wiring and troubleshooting the system.

Step 2: The next step is to start numbering the wiring on the diagram, starting with the number **1**. Start with the power supply **L1** and **N** (neutral) or **L2** for single-phase and **L1**, **L2**, and **L3** for three-phase supply voltage. Do not skip numbers. Use the same number on the same wire or interconnecting wires.

Figure 17-30 shows an example of how to start the numbering system. Write down the number 1 on the yellow wire connections; number 2 on the green wire connections; and number 3 on the blue wire connection. Figure 17-31 shows a more complex diagram and the way to use the numbering system to develop a diagram.

Step 3: Starting with one leg of the power supply, use the same number for all wiring, wire connections, and the side of any component(s) connected to the power supply. Change the number when you jump over any switch (opened or closed), load, or electrical component.

Let us review the circuit in Figure 17-31 to practice this method of diagram design:

1. You can start with **L1** or **L2**. In this case we start by numbering **L1** with the number "**1**." Color coding the diagram using the same colors used for the actual wiring will help in this process also. In this example, we have selected a blue wire for all number **1** connections. Number **1** is placed everywhere that **L1** is directly connected to that side of the power source. Notice all of the number **1** connections. Number **1** is placed on all wire junctions, the left side of **CC** contacts, the left side of **FR** contacts, and the left side of the thermostat switch. Numbered wire markers can be used if different color wire insulation is not available. Tape the wire number on both ends of the wire for easy identification and verification with

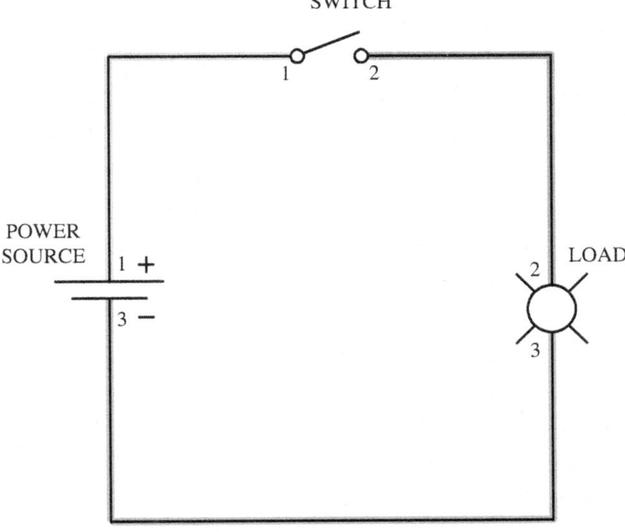

Figure 17-30 This shows how to number a diagram so that it can be used for wiring and troubleshooting purposes. Notice that you use the same number on the same wire. When you jump across a component, the number changes.

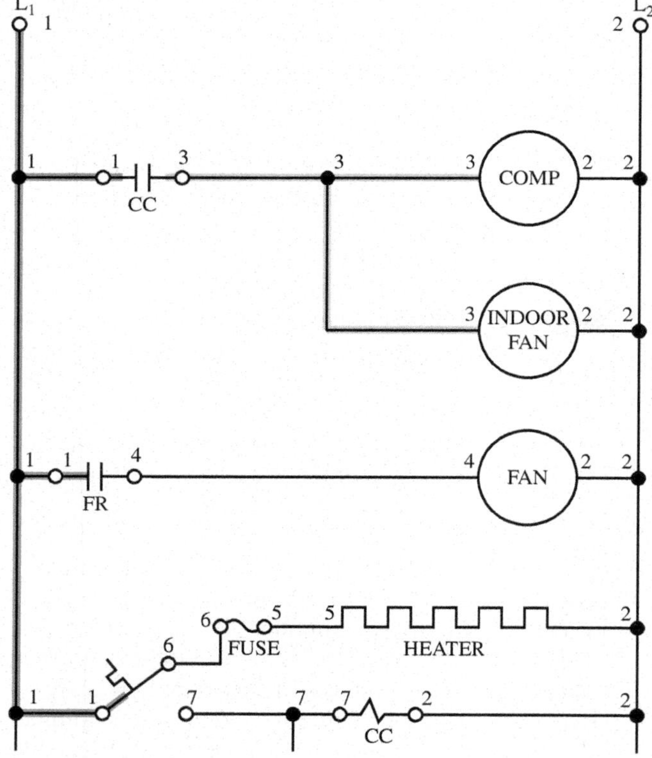

Figure 17-31 This sample circuit is used to demonstrate how a diagram is numbered so that it can be used for wiring and troubleshooting purposes. Notice that the number is the same on a wire or interconnecting wire. Once the wire is broken by a component, the number changes.

similarly numbered wires. All wires at the same connection point will have the same number.

2. The number "2" is used to identify the **L2** circuit connections. The number "2" is used on all right-side wire junctions, the right side of the compressor motor, the right side of the indoor fan motor, the right side of the fan, the right side of the heater, and the right side of the Ⓒ coil.
3. The number "3" or orange wire is place on the right side of **CC** contacts, the left side of the compressor motor, the left side of the indoor fan motor, and one wiring junction.
4. The number "4" is used to identify the right side of the **FR** and the left side of the fan.
5. The number "5" is used to identify the point on the right side of the fuse and left side of the heater.
6. The number "6" is used to identify the right side of the thermostat connection and the left side of the fuse.
7. Number **7** has two connection points in the control circuit (not shown) and the left side of Ⓒ coil.

In summary, designing a useful electrical diagram is not difficult if you use the following steps:

1. Develop a diagram from a pictorial or component layout plan. You can also draw a diagram if one is not available. Space out components on the drawing, allowing plenty of room to draw interconnecting wiring. If this is a hand-drafted diagram, use a pencil for ease of changes and corrections. Colored pencils work well if you are going to color code the wires.
2. Start numbering the wiring at the high-voltage power supply, which is usually **L1** and **L2**, sometimes **L3**, if it is a three-phase power supply. Start with the number **1**. Work from left to right, top to bottom.
3. All components and wire junctions connecting the same line will have the same number.
4. The other side of a component will start with the next number. Do not duplicate a set of numbers on a different circuit.

17.9 DRAWING A FIELD DIAGRAM

Sometimes it is necessary to draw a field diagram in order to troubleshoot a HVACR problem. Drawing a field diagram may be required when the technician cannot figure out what is happening in an electrical problem. Generally, if a pile of wiring rolls out when a service panel is removed, there is a wiring problem. The field drawing is also vital if there is no diagram available or if the circuit has been heavily modified, making the original diagram useless.

Step 1: Decide if a full or partial diagram is required. For instance, if the problem is a condenser fan issue and the compressor circuit is functional, then only draw the condenser fan circuit. It would be best to have a diagram of the complete circuit, but in most cases the technician does not have time to do extra work on the job. It is usually "on to the next service call."

Step 2: Draw out all the components in their general arrangement as shown in Figure 17-32. The wiring diagram can be developed and then converted to a schematic (ladder) diagram. We will describe how to take an air conditioning and heating system and convert it into a diagram that can be used for troubleshooting. The same process can be used to develop furnace, refrigeration, or other HVACR diagrams. If possible, divide the drawing into high-voltage and control voltage components.

Step 3: Draw the interconnecting wires the way they appear in the system. Do not draw them the way they should be, but the way they are actually wired. Miswiring may be the reason the circuit(s) is not working properly.

Step 4: Label the colors of the wire. Remember the "road map": The components can be considered to be landmarks and the wires roads. This step is shown in Figure 17-33, with the high-voltage section being wired up first. Either the high voltage or control voltage can be wired first—the choice is determined in the field, and each job is a little different. Always start with the easiest circuit first.

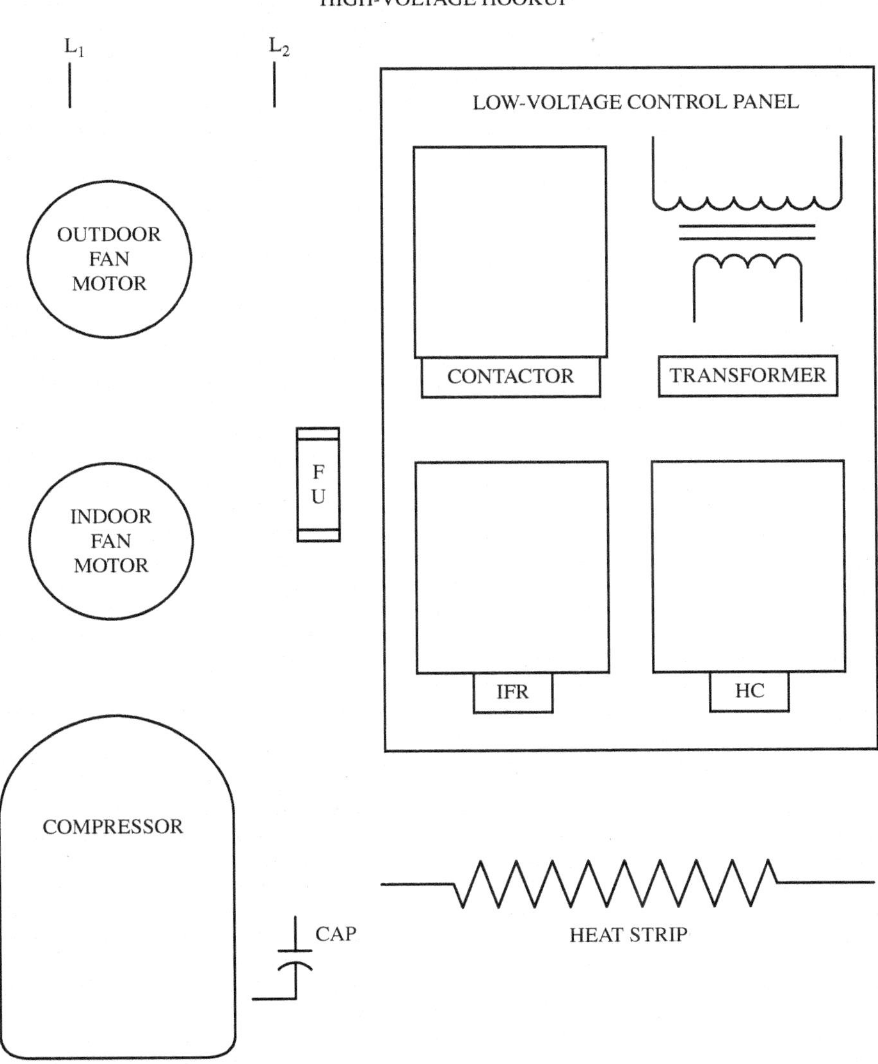

Figure 17-32 This figure illustrates the layout of the components on a cooling and heating system. Draw the components with ample room to draw wires connecting the components.

Step 5: In the high-voltage diagram (see Figure 17-33), draw the wires from the power source **L1** and **L2** to some of the loads. The diagram highlighted in blue shows that the black wire (labeled **BLK** for black) goes from power source **L1** to the input or the line side of the contactor. (*Note:* The outline colors used in the example do not represent the color of the wire.) The blue (**BL**) wire from the contactors connects one side of power to the outdoor fan motor and compressor. If this were a situation in which the technician was rewiring a system, the technician could stop here and apply voltage. Nothing would operate since there is not a complete circuit to close the contactor, but a miswired circuit may create a short, causing an electrical spark and some unwanted smoke. If this happens at this point in the wiring process, a minimal amount of circuitry would need to be investigated to find the problem.

Step 6: Wire a circuit or two, and then test them by applying power. This will let you know where the problem is. If you wire the entire circuit and it trips a breaker when you test it, you may not know where to start looking for a short circuit or miswired condition.

Figure 17-34 shows the additional wiring that completes the heating circuit.

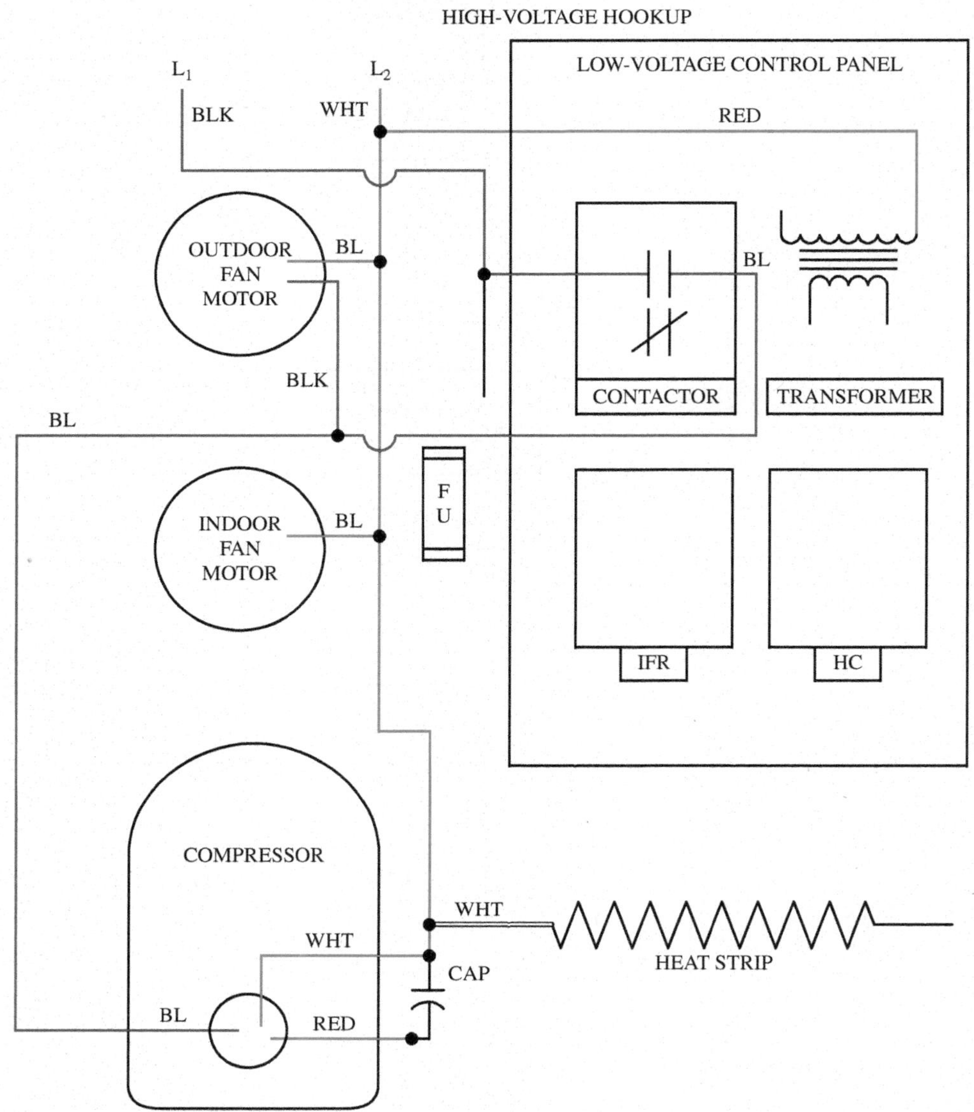

Figure 17-33 This diagram shows the wiring for the high-voltage section of the cooling system from Figure 17-32. The wires colors have been labeled to make it easier to trace and troubleshoot.

This circuit does not show a thermostat or indoor fan relay (**IFR**) and heating contactor (**HC**) coils. These control voltage components are shown in Figure 17-35, and their low-voltage hook-up is shown in Figure 17-36.

Finally, the field diagram can be transformed into a ladder diagram to aid in troubleshooting. The diagrams we have been working with (Figures 17-32 through 17-36) have been transformed into the ladder diagram shown in Figure 17-37.

It is important that the wiring diagram be double-checked prior to using it in troubleshooting. Using an improperly drawn diagram will create confusion and add unnecessary time to the job. So, spending a little extra time up-front may save a lot of time troubleshooting. Trace each line of the diagram, starting with the control voltage circuit. A miswired circuit is a clue to the problem or it may be the problem. Wiring diagrams are not always 100% correct from the manufacturer. Check the diagram for accuracy, plus to see if it makes sense.

In conclusion, drawing an electrical diagram in the field aids the troubleshooting process when a valid diagram in not available or the diagram has been heavily modified beyond use and understanding. Start with a sketch of the basic components. Next, draw and label the colors of the interconnecting wires, which will aid in tracing the various circuits. Convert the drawing into a ladder diagram for hopscotch troubleshooting. Make a copy of the diagram and leave it with the unit. Keep one diagram on file for future reference.

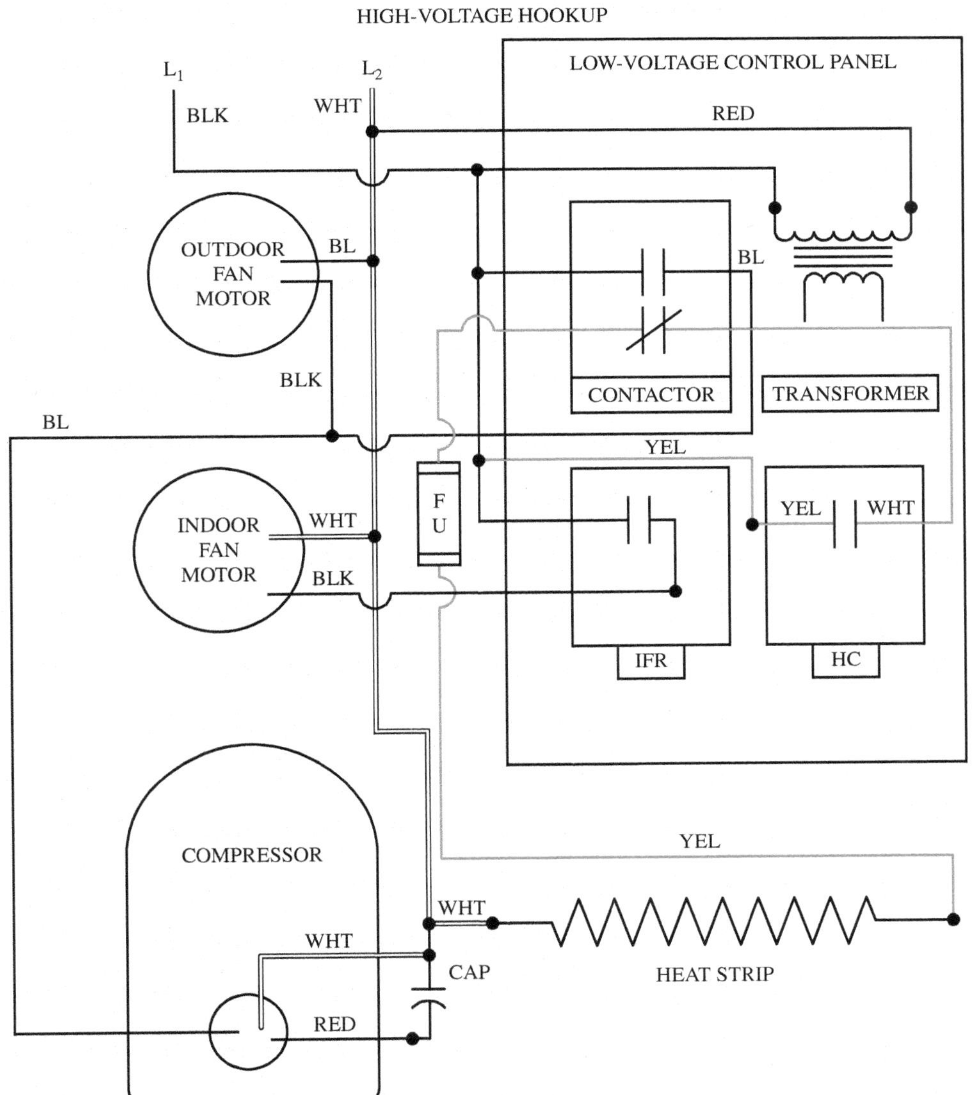

Figure 17-34 The voltage to the other high-voltage circuits for our sample system have been wired. Wiring a circuit or two and then testing them will help you understand where a wiring mistake has been made.

17.10 REWIRING A SYSTEM: THE WIRE AND TEST METHOD

When should a technician remove all the wiring and rewire a system that has a problem? The system should be rewired if the wiring has been modified so much that it would take too long to figure it out. As mentioned earlier, a sure sign that rewiring is in order is if a pile of wiring falls out when the panel is opened. To rewire a system, the technician must have a good wiring diagram to follow or have the knowledge to wire a circuit from experience.

Do not rush through this process. Rewire and test one circuit at a time to determine which circuit is having a problem. Here is a good process to aid in solving this problem:

1. Wire and test a simple circuit first. For example, the technician is rewiring a commercial package unit. The package unit includes heating, cooling, and a blower

TECH TIP

When using an electrical diagram for troubleshooting, cover the diagram with a clear piece of page protector or other washable clear plastic sheet. Tape the clear sheet to the diagram. Use an erasable marker to trace out the diagram. Using different colors for the various circuits will aid in your troubleshooting efforts.

Various types of clear sheets are available. Some will allow you to easily remove the marker tracings and some will not. Try the marker on the corner of the clear sheet. Let it dry and set for about 5 minutes before trying to erase it. This will test the compatibility of the marker on the clear sheet. Some so-called "erasable" markers permanently stain the clear sheet.

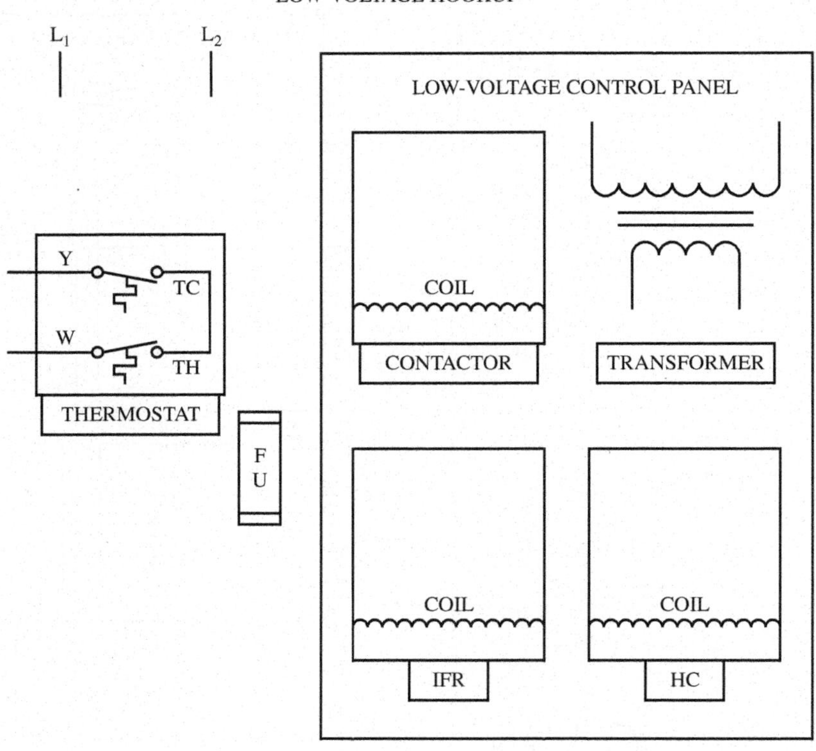

Figure 17-35 Here is the control voltage component layout for our sample system. Leave ample space to draw wiring.

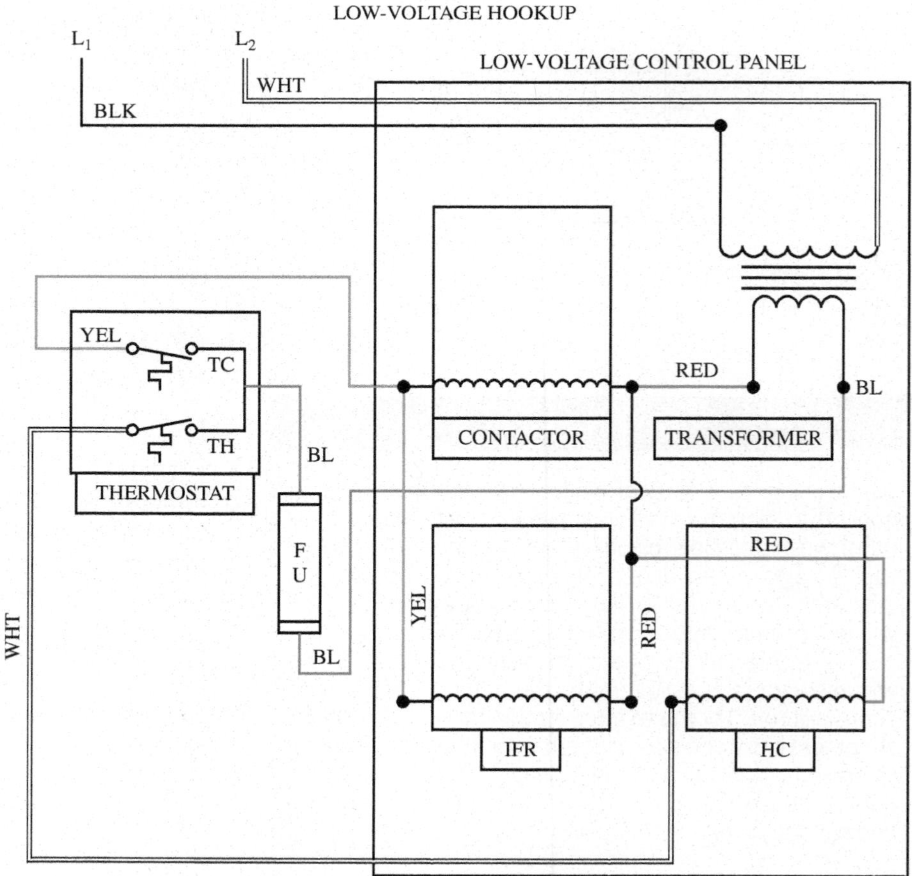

Figure 17-36 This is the wiring diagram for the control voltage components of Figure 17-35. Wire colors identified to aid in tracing wires and troubleshooting.

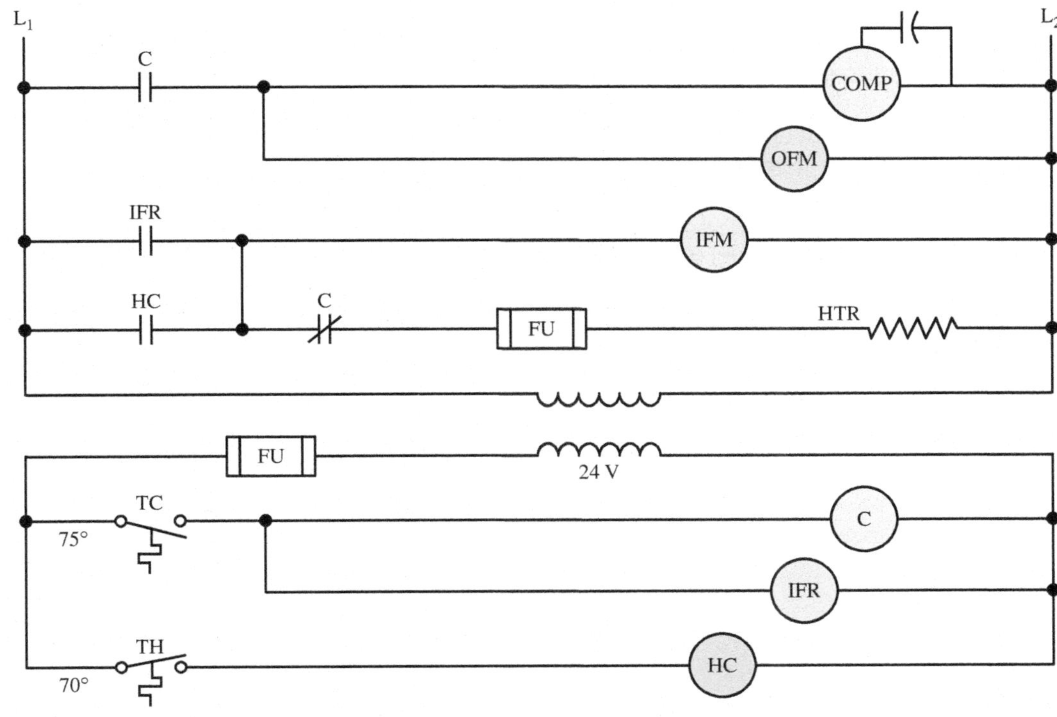

Figure 17-37 This is a ladder diagram designed from Figures 17-34 and 17-36. It is useful when using the hopscotch troubleshooting method. The term *ladder diagram* gets its name from its "ladder-like" configuration. This is a good diagram to use for troubleshooting.

section. Start by wiring the blower section. Wire and test. Go to the cooling circuit.

2. Wire the condenser fan motor. Test the circuit.
3. Wire and test the compressor circuit. Next, wire the heating circuit in gradual steps. It is best to use this sequence of "wire and test" in case there is a miswiring problem or a defective component that may blow a fuse or breaker or let the "smoke out" of a component.
4. The staged wire and test method will identify the problem circuit, since it will be the last circuit hooked up. Remember that the problem represented by a blown fuse or breaker can be a wiring mistake or a defective component in that circuit.

SERVICE TICKET

You've been a technician for 2 years. You are sent on a commercial air conditioning system service call. The air conditioning design is a 10-ton split system. It is about 15 years old and by the look of things it has been serviced by many different technicians over the years. After doing a quick check, you find that the condensing unit is not operating. When you open the large condensing unit panel door, a wad of loose wire falls out almost touching the roof where the unit is mounted. Some of the wires are not connected. The worn, paper wiring diagram and unit information located on the panel door are not readable. The unit nameplate has deteriorated beyond recognition, therefore no make or model number is available. You inspect the major components and do not find any apparent burned or damaged parts.

After thinking about what to do, you decide to strip all wiring from the condensing unit and rewire it. You check the resources in your service van and find a diagram that shows the general components found in a commercial condensing unit.

Diagram in hand, you proceed with the following steps for rewiring the system:

Step 1: First, you wire up one circuit at a time, starting with the simplest circuit—the control voltage to the contactor relay. The 24-V control voltage is hooked up, energized, and the contactor contacts closed.

Step 2: Next, you hook up the condenser fan motor and check its operation. The compressor control circuit is wired and tested without the compressor being hooked up.

Step 3: The compressor is wired and the whole system is ready for one more test. The system's electrical components function properly. The charge is checked and the system problem corrected. You were not sure what the problem was, but the best solution was to strip and rewire. Start and check the sequence of operation.

Step 4: Finally, it is a good idea to make a field wiring diagram in case the unit does not operate in the future. This will be the troubleshooting road map for you or another technician.

In summary, sometimes it is best to remove the wiring from a system and start from scratch to rewire it. In this service call, you did not have a choice. Rewiring the condensing unit was the only option.

SUMMARY

It is essential for technicians to understand electrical symbols and how to use a wiring diagram to troubleshoot. Compare the importance of this information to a truck driver who needs to know where to make a delivery. The driver must understand the symbols and roads shown on a map to get to the destination. A technician needs the same aids.

Understanding this information will help the reader in the upcoming troubleshooting units. You need to know the terminology in order to get help from the office or manufacturer's tech support hotlines. Unfortunately, the terminology is not always the same among technicians, so be aware of different terms in use in the field. The first part of this unit spent time defining terms such as circuit types: complete circuit, open circuit, short circuit, and grounded circuit. We also defined series, parallel, and series-parallel (combination) circuits. These circuit types are universally understood. The misunderstanding comes with the use of terms that relate to electrical diagrams. This unit provided what are considered to be the industry standard names for electrical circuits.

Being able to draft a field diagram is a practical troubleshooting skill. There are many instances when the system wiring has been so badly butchered that it is a wonder the equipment is even operating. This can also be an unsafe condition. Electrical, pressure, and temperature safety devices are sometimes bypassed. Undocumented alternations make the wiring a rat's nest. This is observed when the tech opens the unit panel and a bundle of wire falls out. In such cases it is best to strip out all of the wires and rewire the system. A system wiring diagram will be required for this step. Leave the new wiring diagram on the job and make a copy for future reference.

Finally, practice drawing a wiring diagram and then converting it into a useful ladder diagram. When troubleshooting a diagram, use a clear page protector and erasable marker to trace out the circuit. Use different color markers to trace out different circuits. These steps will help you understand the various paths found in a HVACR circuit.

REVIEW QUESTIONS

1. What is an electrical diagram?
2. Are symbols shown in their energized or de-energized state?
3. What is the difference between a relay and contactor?
4. What is a complete circuit?
5. The amperage in a parallel circuit is 1 amp in branch 1, 2 amps in branch 2, and 3 amps in branch 3. What is the total amperage in the circuit?
6. The amperage in a series circuit is 1 amp in component 1, 1 amp in component 2, and 1 amp in component 3. What is the total amperage in the circuit?
7. What is the current flow in an open circuit?
8. What is a short circuit?
9. What is a wiring diagram?
10. What type of diagram is required for "hopscotch" troubleshooting?
11. What is a pictorial diagram?
12. You are looking at a heating system's wiring information, which you found on the unit panel. You notice an abbreviation on the schematic that is not familiar to you. Where do you look on the wiring information to translate the abbreviation to the name of the electrical component?
13. What is the purpose of the reference numbers or margin numbers seen on a wiring diagram?
14. What are the nine electrical diagram reading guidelines?
15. When would a technician develop a field drawing?
16. What is the purpose of placing a clear page protector over an electrical diagram?
17. Sometimes a HVACR system needs to be rewired. What is the first step in this process?
18. In Figure 17-38, what component is shown on lines 12 and 13?
19. In Figure 17-38, draw the symbol for the major component in line 7.
20. In Figure 17-38, what components are not shown in the symbol list?
21. In Figure 17-39, what do CTD, DTS, and ST mean?
22. In Figure 17-39, which diagram would be best for the hopscotch troubleshooting procedure?
23. List the 15 components shown in Figure 17-40. Use legend abbreviations.
24. In Figure 17-40, the asterisk (*) is used to show optional components that can be field installed or ordered from the manufacturer. List the optional components.
25. Does Figure 17-40 show a connection or option for heating?
26. Compare Figures 17-40 and 17-41. Are they essentially the same piece of equipment? If not, how are they different?
27. In Figure 17-41, what do the dashed lines mean?
28. From Figure 17-42, draw the electrical symbol for 10 components that may be factory options.
29. Figure 17-43 lists cautions and notes for a condensing unit. Study this information to answer the following questions:
    a. What type of wire is required?
    b. What safety code is used to wire the equipment?

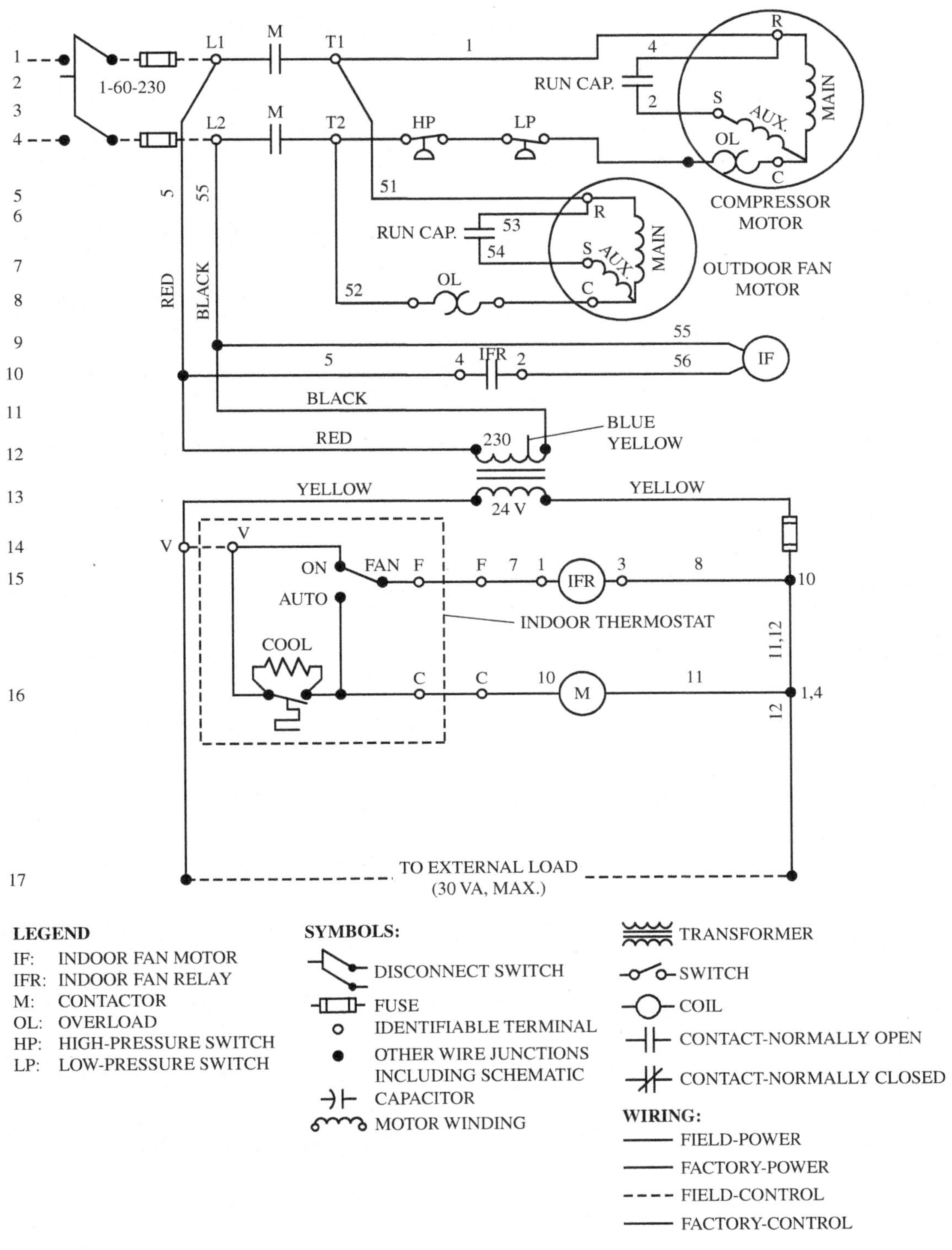

Figure 17-38 Use this diagram to answer Review Questions 18, 19, and 20.

c. It states that CH is not used on all units. What does this mean?

30. Is the compressor shown in Figure 17-44 single phase or three phase? Is the condenser fan motor single phase or three phase?

31. In Figure 17-44, draw the symbol for ground. Draw the symbol for a fan capacitor.

32. Use Figure 17-45 to answer the following questions:
   a. On what line is the condensing unit coil located?
   b. What parts of the system does the thermostat control?
   c. How is the compressor represented in this diagram?

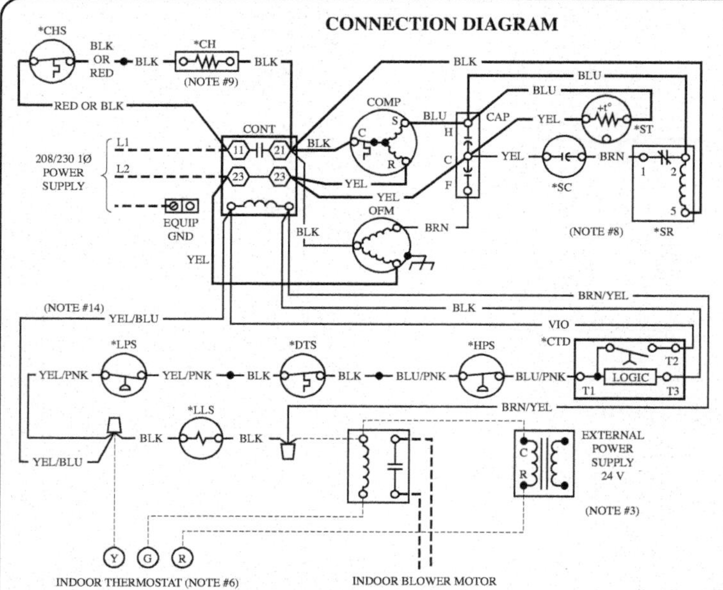

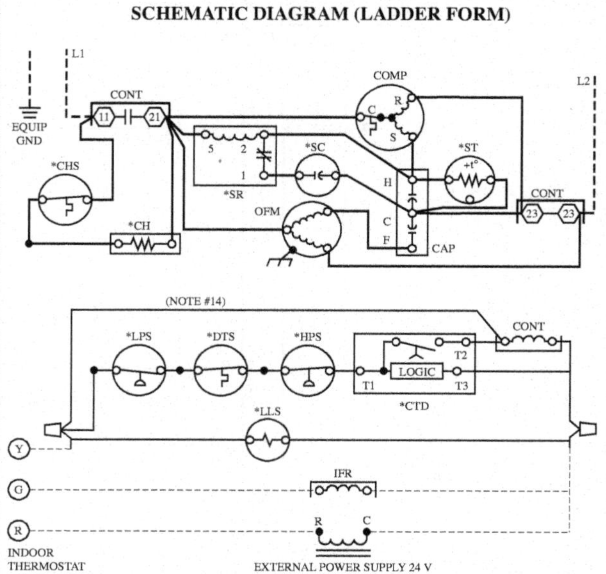

CONDENSING UNIT CHARGING INSTRUCTIONS
For use with units using R-410A refrigerant

REQUIRED LIQUID LINE TEMPERATURE

Liquid Pressure at Service Valve (psig)	Required Subcooling Temperature (°F)					
	6	8	10	12	14	16
251	78	76	74	72	70	68
259	80	88	76	74	72	70
266	82	80	78	76	74	72
274	84	82	80	78	76	74
283	86	84	82	80	78	76
291	88	86	84	82	80	78
299	90	88	86	84	82	80
308	92	90	88	86	84	82
317	94	92	90	88	86	84
326	96	94	92	90	88	86
335	98	96	94	92	90	88
345	100	98	96	94	92	90
354	102	100	98	96	94	92
364	104	102	100	98	96	94
374	106	104	102	100	98	96
384	108	106	104	102	100	98
395	110	108	106	104	102	100
406	112	110	108	106	104	102
416	114	112	110	108	106	104
427	116	114	112	110	108	106
439	118	116	114	112	110	108
450	120	118	116	114	112	110
462	122	120	118	116	114	112
474	124	122	120	118	116	114

COOLING ONLY CHARGING PROCEDURE

1. Only use subcooling charging method When OD ambient is greater than 70°F and less than 100°F, indoor temp is greater than 70°F and less than 80°F, and line set is less than 80 ft.
2. Operate unit a minimum of 15 minutes before checking the charge.
3. Measure liquid service valve pressure by attaching an accurate gauge to the service port.
4. Measure the liquid line temperature by attaching an accurate thermistor type or electronic thermometer to the liquid line near the outdoor coil.
5. Refer to unit rating plate for required subcooling temperature.
6. Find the point where the required subcooling temperature intersects the measured liquid service valve pressure.
7. To obtain the required subcooling temperature at specific liquid line pressure, add refrigerant if liquid line temperature is higher than indicated. When adding refrigerant, charge in liquid form using a flow restricting device into suction service port. Recover refrigerant if temperature is lower. Allow a tolerance of +/- 3°F.

LEGEND

Symbol	Meaning
——— (heavy solid)	FACTORY POWER WIRING
——— (light solid)	FACTORY CONTROL WIRING
- - - - (light dashed)	FIELD CONTROL WIRING
- - - - (heavy dashed)	FIELD POWER WIRING
o	COMPONENT CONNECTION
(field splice symbol)	FIELD SPLICE
—•—	JUNCTION
CONT	CONTACTOR
CAP	CAPACITOR (DUAL RUN)
*CH	CRANKCASE HEATER
*CHS	CRANKCASE HEATER SWITCH
COMP	COMPRESSOR
*CTD	COMPRESSOR TIME DELAY
*DTS	DISCHARGE TEMP SWITCH
*HPS	HIGH PRESSURE SWITCH
IFR	INDOOR FAN RELAY
*LLS	LIQ LINE SOLENOID VALVE
*LPS	LOW PRESSURE SWITCH
OFM	OUTDOOR FAN MOTOR
*SC	START CAPACITOR
*SR	START RELAY
*ST	START THERMISTOR

*** MAY BE FACTORY INSTALLED**

NOTES:

1. Symbols are electrical representation only.
2. Compressor and fan motor furnished with inherent thermal protection.
3. To be wired in accordance with National Electric N.E.C. and local codes.
4. N.E.C. class 2, 24 V circuit, min. 40 VA required, 60 VA on units installed with LLS.
5. Use copper conductors only. Use conductors suitable for at least 75°C (167°F).
6. Connection for typical cooling only thermostat. For other arrangements see installation instructions.
7. If indoor section has a transformer with a grounded secondary, connect the grounded side to the BRN/YEL lead.
8. When start relay and start capacitor are installed start thermistor is not used.
9. CH not used on all units.
10. If any of the original wire, as supplied must be replaced, use the same or equivalent wire.
11. Check all electrical connections inside control box for tightness.
12. Do not attempt to operate unit until service valves have been opened.
13. Do not rapid cycle compressor. Compressor must be off 3 minutes to allow pressures to equalize between high and low side before starting.
14. Wire not present if HPS, LPS or CTD are used.

⚠ CAUTION

1. Compressor damage may occur if system is over charged.
2. This unit is factory charged with R-410A in accordance with the amount shown on the rating plate. The charge is adequate for most systems using matched coils and tubing not over 15 feet long. Check refrigerant charge for maximum efficiency. See Product Data Literature for required Indoor air Flow Rates and for use of line lengths over 15 feet.
3. Relieve pressure and recover all refrigerant before system repair or final disposal. Use all service ports and open all flow-control devices, including solenoid valves.
4. Never vent refrigerant to atmosphere. Use approved recovery equipment.

330440-101 REV.F

Figure 17-39 Use this diagram to answer Review Questions 21 and 22.

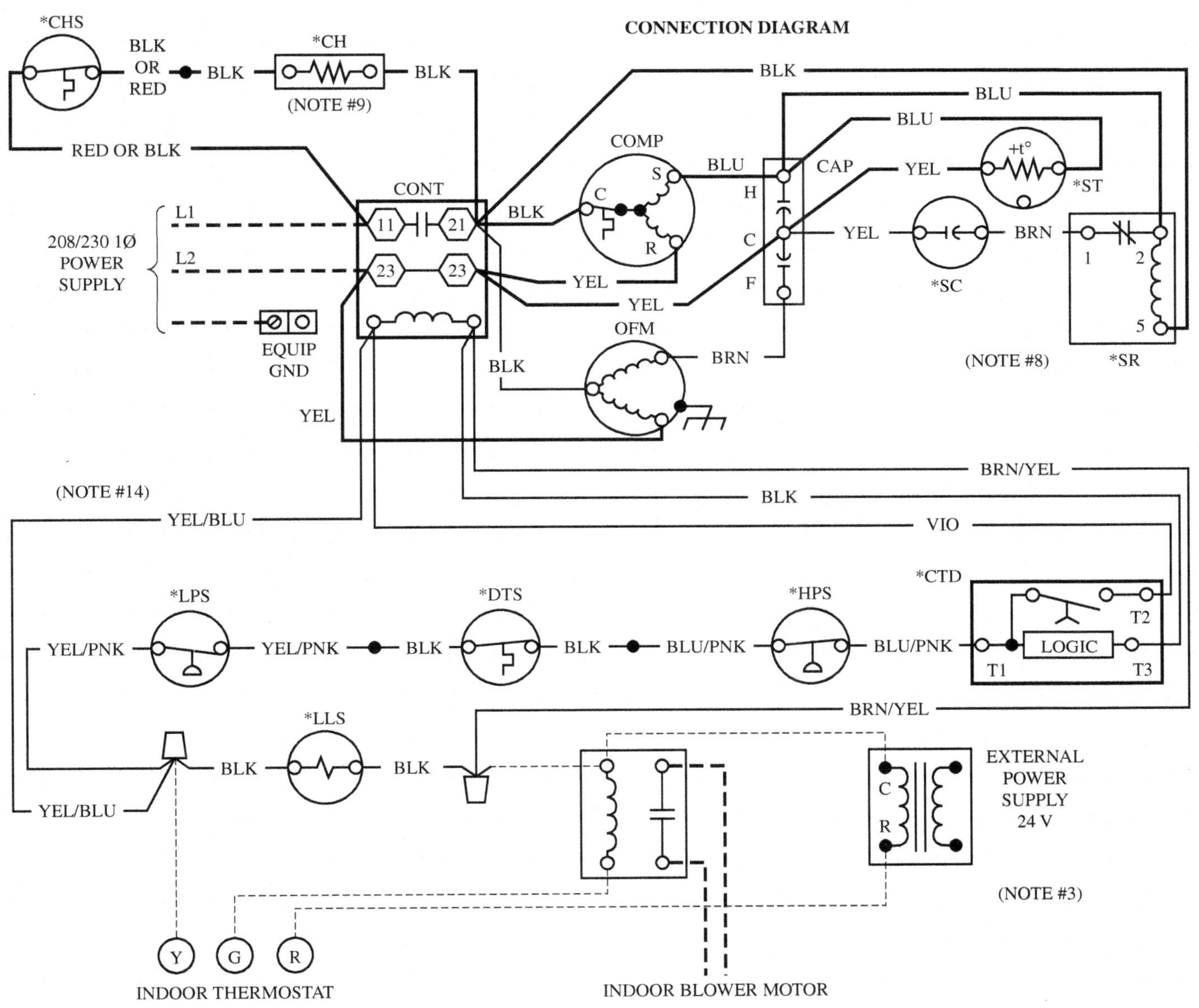

Figure 17-40 Use this diagram to answer Review Questions 23, 24, and 25.

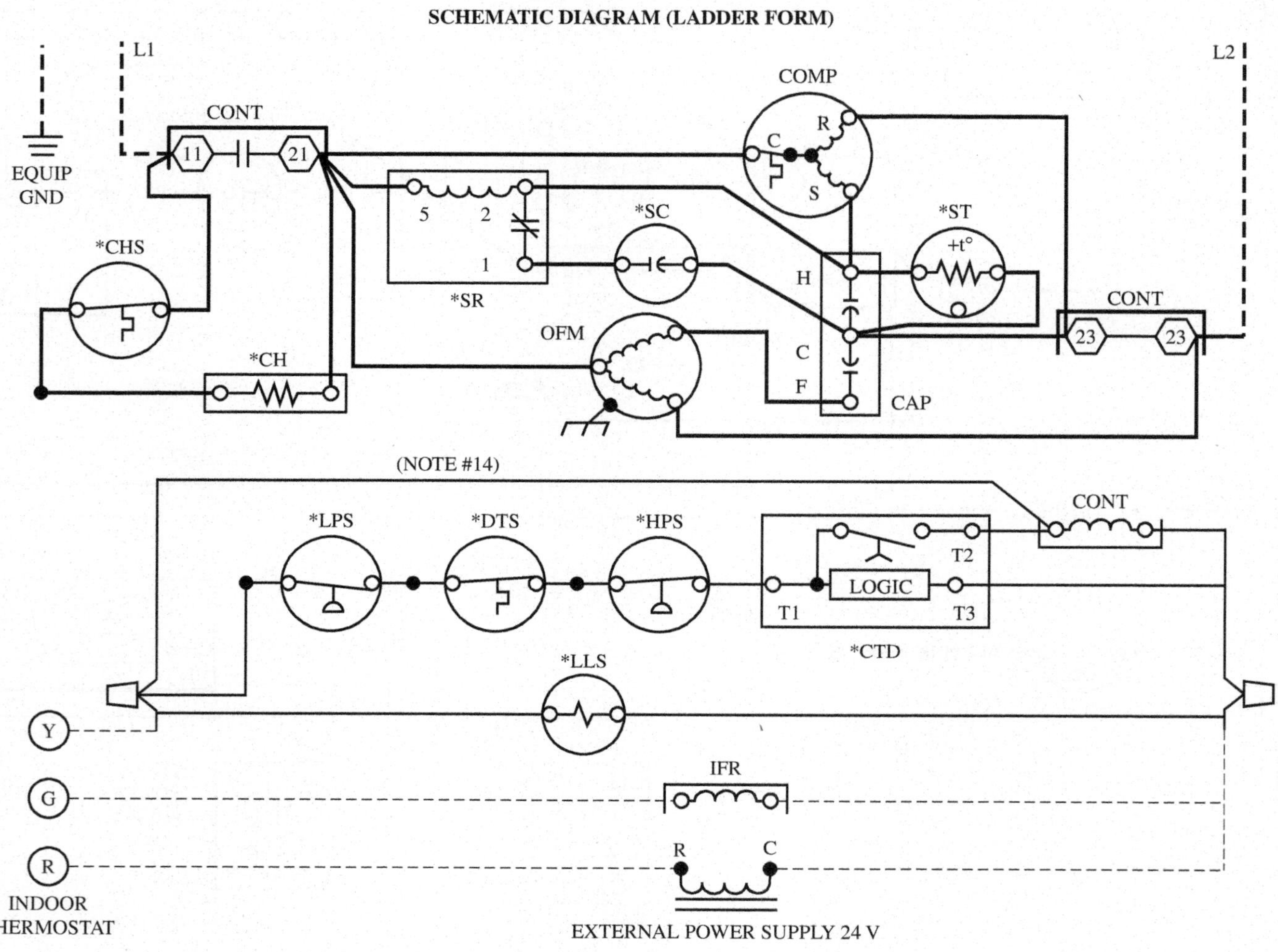

Figure 17-41 Use this diagram to answer Review Questions 26 and 27.

LEGEND

———	FACTORY POWER WIRING
———	FACTORY CONTROL WIRING
- - - -	FIELD CONTROL WIRING
- - - -	FIELD POWER WIRING
o	COMPONENT CONNECTION
	FIELD SPLICE
—●—	JUNCTION
CONT	CONTACTOR
CAP	CAPACITOR (DUAL RUN)
*CH	CRANKCASE HEATER
*CHS	CRANKCASE HEATER SWITCH
COMP	COMPRESSOR
*CTD	COMPRESSOR TIME DELAY
*DTS	DISCHARGE TEMP SWITCH
*HPS	HIGH PRESSURE SWITCH
IFR	INDOOR FAN RELAY
*LLS	LIQ LINE SOLENOID VALVE
*LPS	LOW PRESSURE SWITCH
OFM	OUTDOOR FAN MOTOR
*SC	START CAPACITOR
*SR	START RELAY
*ST	START THERMISTOR

*** MAY BE FACTORY INSTALLED**

Figure 17-42 Use this diagram to answer Review Question 28.

CAUTION

1. Compressor damage may occur if system is over charged.
2. This unit is factory charged with R-410A in accordance with the amount shown on the rating plate. The charge is adequate for most systems using matched coils and tubing not over 15 feet long. Check refrigerant charge for maximum efficiency. See Product Data Literature for required Indoor air Flow Rates and for use of line lengths over 15 feet.
3. Relieve pressure and recover all refrigerant before system repair or final disposal. Use all service ports and open all flow-control devices, including solenoid valves.
4. Never vent refrigerant to atmosphere. Use approved recovery equipment.

330440-101 REV.F

(Carrier m#241bc6 Base16, catalog #24abc6-1w)

NOTES:

1. Symbols are electrical representation only.
2. Compressor and fan motor furnished with inherent thermal protection.
3. To be wired in accordance with National Electric N.E.C. and local codes.
4. N.E.C. class 2, 24 V circuit, min. 40 VA required, 60 VA on units installed with LLS.
5. Use copper conductors only. Use conductors suitable for at least 75°C (167°F).
6. Connection for typical cooling only thermostat. for other arrangements see installation instructions.
7. If indoor section has a transformer with a grounded secondary, connect the grounded side to the BRN/YEL lead.
8. When start relay and start capacitor are installed start thermistor is not used.
9. CH not used on all units.
10. If any of the original wire, as supplied must be replaced, use the same or equivalent wire.
11. Check all electrical connections inside control box for tightness.
12. Do not attempt to operate unit until service valves have been opened.
13. Do not rapid cycle compressor. Compressor must be off 3 minutes to allow pressures to equalize between high and low side before starting.
14. Wire not present if HPS, LPS or CTD are used.

Figure 17-43 Use this diagram to answer Review Question 29.

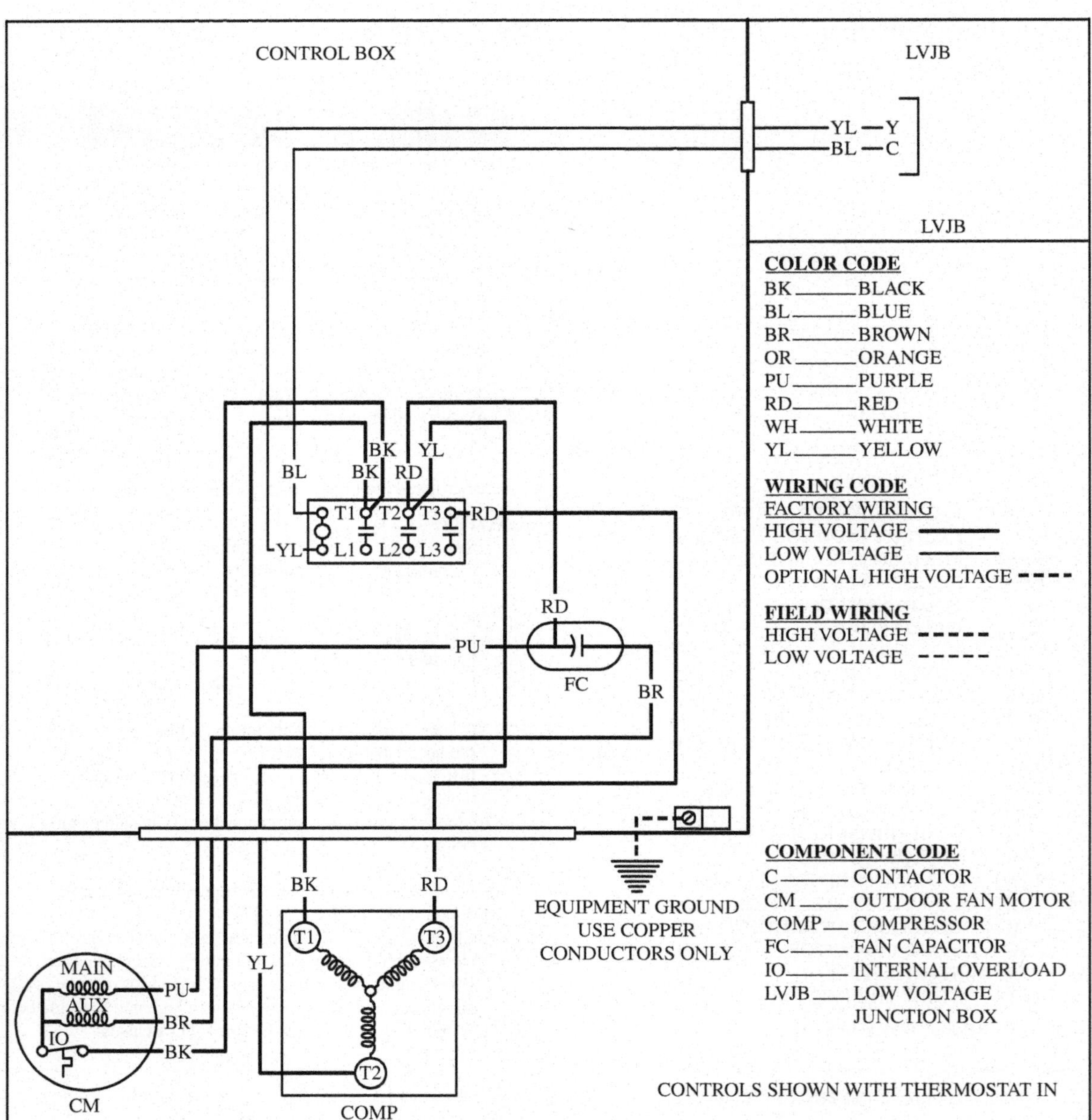

Figure 17-44 Use this diagram to answer Review Questions 30 and 31.

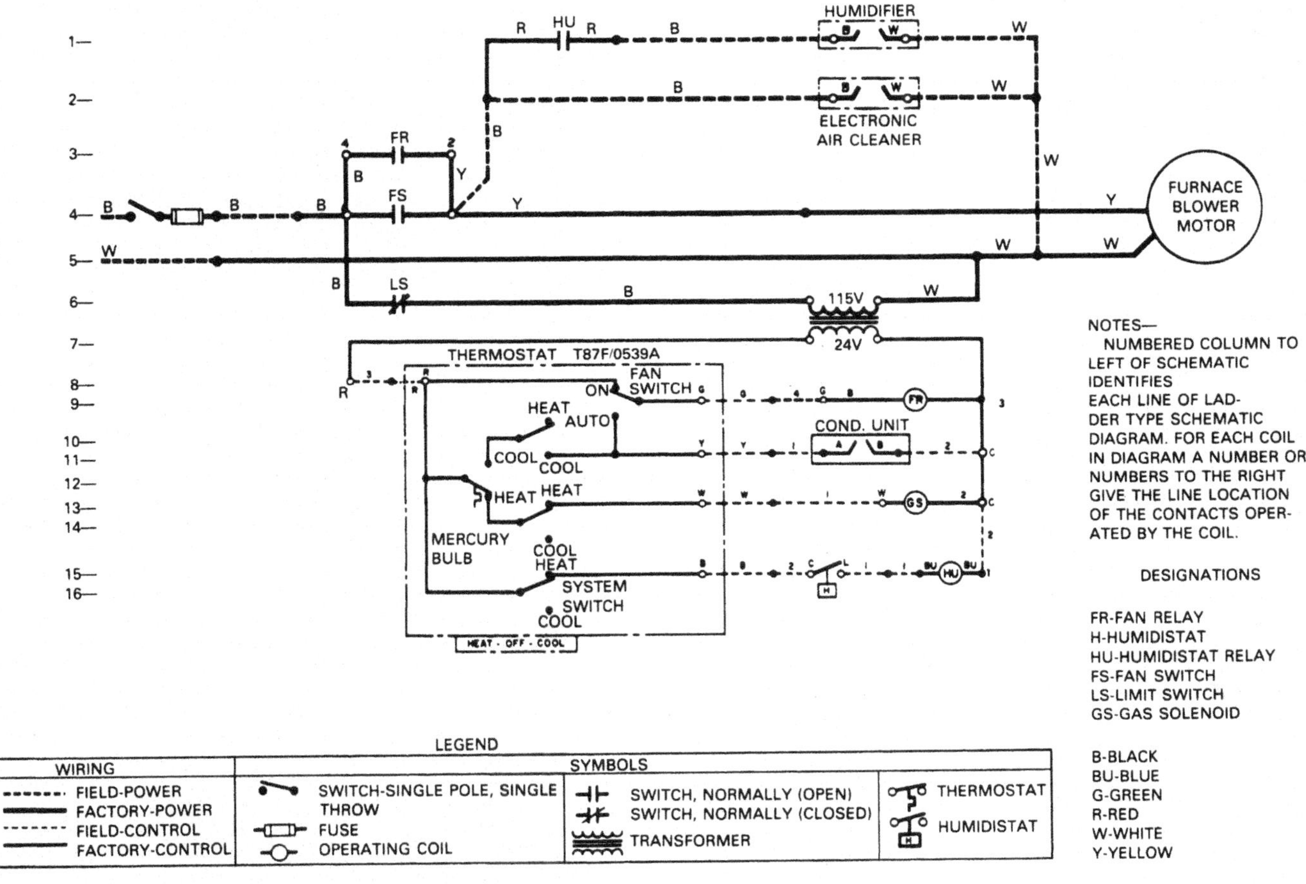

Figure 17-45 Use this diagram to answer Review Question 32.

UNIT 18
Resistors

WHAT YOU SHOULD KNOW

After studying this unit, you will be able to:

1. List types of resistors.
2. Describe resistor ratings.
3. Use the resistance color code to determine resistance.
4. Calculate the resistance of a color-coded resistor.
5. Determine resistor tolerance.

A **resistor** is used to drop the voltage in an electrical circuit. In HVACR it is used on circuit boards and to discharge start capacitors. Resistors are also found in oil safety switches as a way to reduce incoming voltage. It is valuable to learn something about these components because they may come into play during your troubleshooting activities. For example, you find a burned-out resistor on a circuit board. The board should be replaced, but what if it is not available or the order time is too long? Replacing the resistor is a repair long shot, but it might work. You should investigate the reason for the burn-out. Maybe wires touching each other caused the resistor to burn open. It is worth the chance of trying to repair it—but don't get too excited, because in many cases the resistor burned out because of some other defective component that may not be so obvious.

This unit provides the essentials of how to select a resistor for a particular job. It is always best to go with an exact replacement, but what do all those color bands mean? Let's take a look at the answer to that question.

18.1 THE RESISTOR

A resistor is designed to slow down or oppose current flow. As mentioned earlier, resistors are commonly used in modern HVACR equipment and on circuit boards, across capacitors, and in oil safety switches. Like many components in our profession, the resistor is basically ignored because it is a reliable component that rarely causes a problem.

Electric heat strips and crankcase heaters are also resistors. This unit, however, focuses on the resistor component that is used to reduce current flow and not heat air or a fluid.

TECH TIP

Two-watt, 470,000-Ω (470-kΩ) resistors are found across the terminals of start capacitors. The resistor may be soldered or connected using a quick-connect slip-on connector. The purpose of the resistor is to discharge the capacitor immediately after power is removed from this component.

Remove the resistor when checking the capacitor. Replace the resistor across the capacitor terminals prior to placing the capacitor in service. The resistor is used to discharge the capacitor, thus preventing the charge from damaging the motor in a short cycling condition.

18.2 TYPES OF RESISTORS

The two common types of resistors are the **carbon resistor** and **wire-wound resistor**. Carbon resistors use various grades and quantities of carbon wrapped in a small insulated cylinder (see Figure 18-1).

A wire-wound resistor is made of resistive metal alloy such as nickel-chrome wire wrapped in an insulated shell (see Figure 18-2). Some wire-wound resistors have a hollow core that helps dissipate heat from the device. Hollow-core wire-wound resistors should be installed in a vertical position so that heat can rise through the chimney of the resistor.

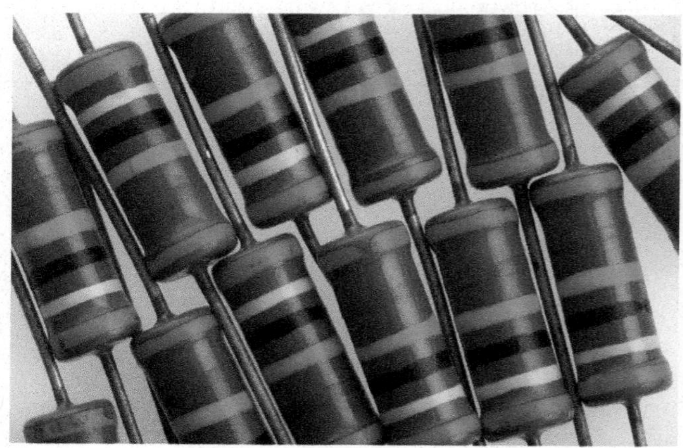

Figure 18-1 On these carbon resistors, the color of the band represents the resistance and the tolerance.

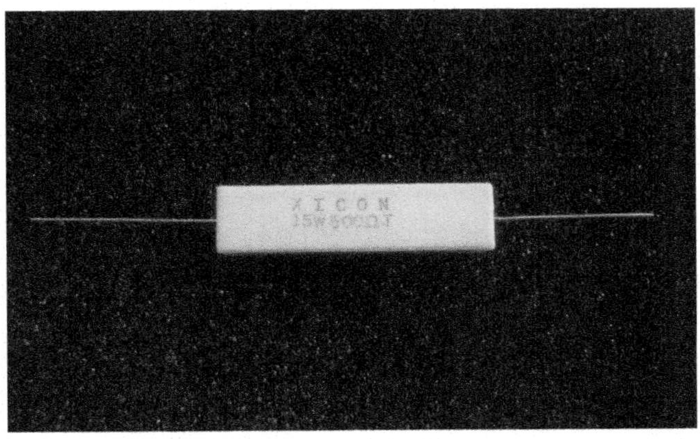

Figure 18-2 Wire-wound resistors use nickel-chrome wire, which offers resistance to electron flow.

18.3 POWER RATING

The common power rating for resistors is in the range of 0.25 to 5 watts and higher. The power or wattage rating is important. Selecting an undersized resistor will overheat the component and it will burn open. Try the touch test. A resistor should not be so hot that you cannot touch it. A hot resistor means that it is undersized or it is carrying more load than it should be. It will soon burn out.

Larger resistors are rated at higher wattage ratings. The resistor on the circuit board in Figure 18-3 is a 1-watt resistor.

18.4 OPERATING TEMPERATURE AND VOLTAGE

Expect most resistors to operate as a neutral temperature component. In other words, normal hot and cold temperatures do not significantly change the resistance rating. The temperature should be near the ambient temperature around the component. Resistors can operate below 0°F and above 200°F with little impact on their performance. Some resistive materials vary with temperature, but they are not covered in this unit.

Most resistors have a maximum voltage range of 200 to 600 volts. The temperature or voltage range is usually not a concern since the resistor is designed and tested for the application.

Figure 18-3 The lower right-hand corner of the circuit board illustrates a common carbon resistor. The color code is yellow, violet, and black. The tolerance color is gold.

18.5 TOLERANCE

The term **resistor tolerance** relates to the percentage by which a resistor varies above or below its resistive rating. Precision resistors can have a resistance of less than ±1%. Most resistors have a tolerance of 5%, 10%, or 20%. For example, a 100-Ω resistor with a ±10% tolerance can have a resistance of 90 to 110 ohms. You can always use a replacement resistor with a lower tolerance rating, but never a higher tolerance rating.

SAFETY TIP

Occasionally you may need to replace a resistor that has burned out on a circuit board or in one of the electrical instruments you use in your profession. In most cases damaged circuit boards are replaced, but it is worth a try to replace an obvious burned-out component such as a resistor, fuse, or open circuit connection. A replacement circuit board may take a day or two to get and it may instead be possible to repair it with a resistor from an electronic parts outlet. You should use an exact replacement resistor, but the tolerance can be better than the one you change out. For example, a damaged resistor with a ±10% (silver band) tolerance can be replaced with a ±5% (gold band) or lower. Remember, you can also wire resistors in series or parallel to obtain the desired resistance. Wiring the resistors in parallel also increases the watt rating.

18.6 CALCULATING RESISTANCE

Resistors are rated in ohms. Some resistors have their value printed on the component body. Most do not. As shown earlier in Figure 18-1, most carbon resistors have color bands that represent resistance and tolerance. Each color band represents a number, and the order of the color represents a number value. You will notice that the color bands are grouped to one end of the resistor. This is the end where you start calculating the total resistance.

Use Figure 18-4 to calculate the value of a banded resistor. The first two color bands indicate a number. The third color band indicates a multiplier or the number of zeros added to the numbers on the first two bands. The fourth

The standard resistor color code table:

Color	1st digit	2nd digit	3rd digit*	Multiplier	Tolerance
Black	0	0	0	$\times 10^0$	
Brown	1	1	1	$\times 10^1$	±1%
Red	2	2	2	$\times 10^2$	±2%
Orange	3	3	3	$\times 10^3$	
Yellow	4	4	4	$\times 10^4$	
Green	5	5	5	$\times 10^5$	±0.5%
Blue	6	6	6	$\times 10^6$	±0.25%
Violet	7	7	7	$\times 10^7$	±0.1%
Gray	8	8	8	$\times 10^8$	±0.05%
White	9	9	9	$\times 10^9$	
Gold				×0.1	±5%
Silver				×0.01	±10%
None					±20%

** 3rd digit - only for 5-band resistors*

Figure 18-4 This table explains the color code that will help you identify the value of a resistor based on its color band. The "1st digit" and "2nd digit" columns are the first two numbers of the resistance. The "3rd digit" column, shaded in gray, is an additional number for use on five-band resistors. The "Multiplier" column is the number used to multiply the first two digits. The "Tolerance" column is the percentage range expected of the resistor. Common tolerance indicators are gold, silver, and no band at all.

band is the tolerance of the resistor. The normal tolerance is ±5%, ±10%, or ±20%.

Some precision resistors have a fifth color band of brown. The brown band, as shown at right in Figure 18-5, represents a ±1% tolerance rating. There is no standard for the fifth band. The brown band on the upper part of a resistor indicates that it is a precision resistor.

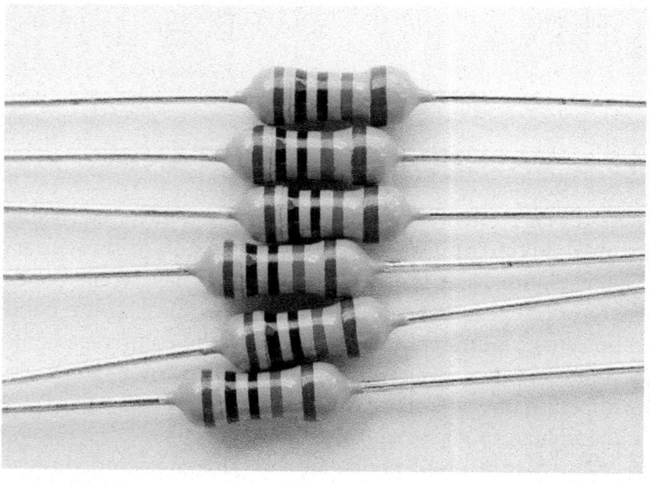

Figure 18-5 These are carbon resistors. The color band count starts on the left side with brown, black, black, and red. Brown is the ±1% tolerance band.

A five-band resistor uses a different formula to calculate resistance. The five-band resistor uses the first three bands as numbers. The fourth band is the multiplier, and the fifth band is the tolerance band. Using Figure 18-4 and the sample resistor shown in Figure 18-5, we can determine that a resistor with brown, black, black, red, and brown color bands will be 1,000 Ω (1 kΩ) (because the red multiplier band adds two zeros). So Figure 18-5 shows a 1,000-Ω resistor with a ±1% tolerance. The tolerance will vary from one manufacturing company to another; the fifth band may indicate 2%, 1%, 0.5%, or an even closer tolerance, according to each manufacturer's own standards.

Let's work a couple of examples to determine resistance.

EXAMPLE 18.1 CALCULATING RESISTANCE: RESISTOR WITH FOUR BANDS

Refer to Figure 18-6, which shows the following color bands: yellow, violet, red, and gold. We said earlier that the color bands of a resistor are stacked toward one end and we start counting the colors nearest the stacked end. For example, in Figure 18-5 you would start on the left side. In Figure 18-6, however, the colors are spaced out such that it might be hard to determine on which end to start the color band count. Here are a couple of tips if the color bands are spaced evenly or there is doubt about which end to start from:

- The color tolerance bands are always at the end. In Figure 18-6, gold is the tolerance indicator and it is at the end of the color band. Gold, silver, or no color is used as a tolerance counter—never a resistance code number.

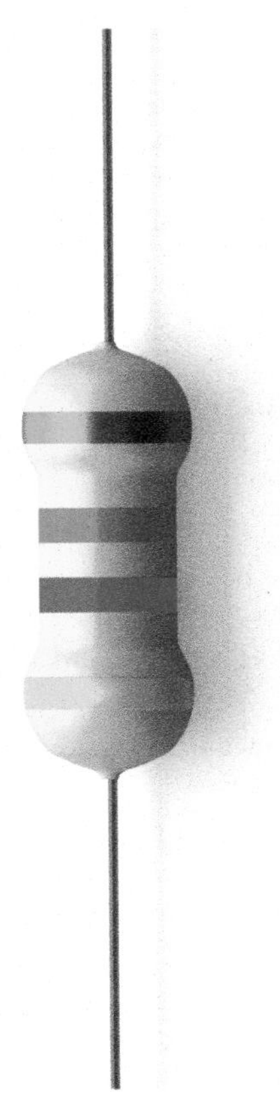

Figure 18-6 This resistor has four color bands. Count the color bands starting with yellow, then violet, and then red. The gold band is the ±5% tolerance band.

- Some colors do not have tolerances such as black, orange, yellow, and white, therefore they will be at the end where you should start counting the color bands.

Let's calculate the value of the banded resistor shown in Figure 18-6.

SOLUTION

Start with the color band that is closest to an end. Use the table in Figure 18-4 to determine the resistor's value. The yellow band is the first band and is equal to 4. The violet band is 7. The red band is the multiplier of 2. The red means add two zeros to the numbers from the first two bands or multiply the first two numbers, 47, by 10^2. That calculation is:

$$47 \times (10 \times 10) = ?$$
$$47 \times 100 = 4{,}700\ \Omega$$

At this point we have a 4,700-Ω or 4.7-kΩ resistor. The fourth band is gold, which represents a ±5% tolerance range. Five percent of 4,700 Ω is:

$$4{,}700 \times 0.05 = 235\ \Omega$$

The resistor is considered to be good if it is within plus or minus 235 Ω of the 4,700-Ω resistor value. This will be a range of 4,465 to 4,935 Ω.

EXAMPLE 18.2 CALCULATING RESISTANCE: RESISTOR WITH THREE BANDS

A resistor has the following color bands: brown, red, red. There is no fourth color band, therefore the tolerance will be ±20%. Calculate the value of this resistor.

SOLUTION

The resistor will be 1,200 Ω, with a ±20% tolerance. The tolerance is ±240 Ω. The tolerance is calculated by multiplying $1{,}200 \times 0.20 = 240\ \Omega$.

TECH TIP

Memorizing the color code is not necessary since it is readily available. If you choose to memorize the color code, here is a sentence that will assist you:

Better Be Right Or Your Great Big Venture Goes West

The first letter of each word helps you to remember the color, as listed in Figure 18-4. For example, **Better** is black, **Be** is brown, etc.

TECH TIP

The fastest way to identify the resistance of a replacement resistor is to measure it with an ohmmeter. The resistor must first be removed from the circuit. Measuring the resistance of a resistor that has been removed from a circuit is an accurate way of determining its true resistance. The resistor may have changed resistance due to overheating or some other abnormal condition. Use the color code method to determine resistance and verify with an ohmmeter.

Resistors have a range or tolerance. Ohmmeters have a tolerance or a range also. Be careful when condemning a precision resistor using an ohmmeter as the only indicator of resistance. If the resistor is close to the correct resistance, try a better quality meter to verify the resistance.

SERVICE TICKET

This service call presents a real challenge. It is a "no heat" call. After completing a checkup, you notice a burned mark on the circuit board. The furnace is over 15 years old and has a first-generation solid-state control device. You call the service manager to order a replacement board. After waiting 15 minutes the service manager calls back and states that the circuit board is no longer available.

One option is to replace the furnace. Prior to talking with the owner, you notice that the resistor is the burned-out component. The option of replacing the furnace or trying to repair the circuit board is discussed. The owner is told that there is an outside chance that changing the resistor will fix the problem. The owner decides to take the chance that changing the resistor will work, realizing, however, that she might be throwing good money out the window.

The next challenge is to determine the resistance. The bands on the resistor are red, violet, orange, and gold. What resistor does the technician need to purchase? What tolerance?

In this case the homeowner and technician were lucky that a replacement resistor did the trick to repair the problem. In many instances, one burned-out resistor is a cover for many other damaged components. Always check the back of a circuit board to investigate carbon burns, known as carbon tracking. The damaged board may indicate that the repair is impossible.

SUMMARY

This unit discussed how to select a resistor based on a color code. Knowing a resistor's tolerance and power capabilities is important when purchasing a replacement component.

Even though troubleshooting a resistor is not a common problem, it is important to be able to understand how to select a resistor. More and more systems are going to solid-state circuits, which use resistors, so it is important to know a little bit about a component we may see every day. There may be a time when you need to refer to technical material to determine resistance, especially if the resistor is open. Keep in mind that an electronic supply house will help you select an appropriate replacement.

REVIEW QUESTIONS

1. What are two common types of resistors?
2. What are two applications of resistors in HVACR equipment?
3. A 1,000-Ω resistor has a silver tolerance band. What is the specific range of resistance?
4. What is the resistance of a resistor that has color bands of red, blue, and yellow?
5. What is the resistance of a resistor that has color bands of orange, violet, and green?
6. What is the resistance of a resistor has color bands of brown, green, and brown?
7. How is the power rating or heat dissipation rate of a resistor expressed?
8. What is a quick way of determining resistance when a resistor is not in a circuit?
9. What is the resistance of a carbon resistor that has color bands of brown, black, blue, and gold? What is the percentage of resistance tolerance?
10. You are out on a job and notice two resistors with the exact same color code. One resistor is larger than the other. Why is there a difference in size?

UNIT 19

Fundamentals of Solid-State Circuits

WHAT YOU NEED TO KNOW

After studying this unit, you will be able to:

1. Describe what is meant by solid state.
2. Discuss the advantages of solid-state systems.
3. Discuss three applications of solid-state components.
4. Understand capacitors used in solid-state circuits.
5. Describe the purpose of a diode.
6. Describe the purpose of a transistor.
7. Describe the purpose of an integrated circuit.
8. Describe the purpose of a rectifier.
9. Describe the purpose of a varistor.
10. Discuss the purpose of a circuit board.

This unit follows the basic guideline set up at the beginning of this book: Keep it simple! This short unit introduces the basic components used in **solid-state circuits** or as we know of them in our industry as the **circuit boards.** It is not necessary to have an in-depth knowledge of electronics to understand solid-state circuits. It is helpful to have a basic understanding of electronics when working with solid-state circuits, but the basic operation can be described in simple terms. As you will learn in Unit 20, *Taking the Mystery Out of Circuit Boards*, the most important thing to understand is that you need specific inputs, like a voltage source or a closed circuit, and a specific output, like a different voltage source to operate a motor, compressor, or other circuit. Simply put, the circuit board components replace the relays, switches, and other controls that we see in older HVACR equipment.

This unit does not dive into the theory of how electrons and protons travel in components to make things work. We instead cover the major components found in solid-state circuits in a way that should help you understand them, but not confuse you with theory. Theory will not fix an air conditioning unit, but having a basic comprehension of how a system operates will. It is important to have some basic information about how solid-state circuits works.

What does *solid state* mean? Solid state, as the name expresses, refers to the use of a solid material to control electrons or the movement of electrons. You have to know what came before "solid state" to understand why the term *solid state* is used. Many electrical components that preceded solid-state devices used glass vacuum tubes or glass tubes filled with special gases to control electron flow. Yes, these were solid devices, but they had vacuum voids or special electron-transferring gases. The advent of the name *solid state* occurred when the electron flow components became totally solid; that is, they had no glass, open areas, or voids.

Solid-state components are current technology. Some solid-state components have been miniaturized compared to the first-generation solid-state components and the older vacuum tube technology. As a comparison, the first vacuum tube computers filled a large room and did not offer as many functions as today's much smaller cell phones.

Here are the advantages of solid-state technology:

- Reduced equipment size
- Increased computing power
- Less heat and energy usage
- Durability.

All of these advantages benefit the growth of electronic technology, which includes the control of HVACR equipment.

TECH TIP

Be careful with systems that use solid-state or circuit board-based technology. Do not use jumpers or other tools to short out components or bypass components unless the manufacturer recommends this troubleshooting technique. Any short to ground from a solid-state device may destroy the electronic module or circuit board.

SAFETY TIP

Having a good equipment ground is important for the safety of technicians and for the operation of solid-state circuits. Hooking up the ground does not guarantee that the equipment is grounded. The ground connection that terminates at the power supply may be loose or develop a corroded connection. In some instances the ground rod or anode has deteriorated or was never installed to National Electrical Code (NEC) specifications. Tips for checking a good ground: Check the ground by measuring full voltage between the hot leg and ground. There should be no voltage between neutral and ground.

19.1 CAPACITORS

Capacitors used on HVACR motors and compressors have a different function than capacitors used in solid-state circuits. Capacitors used on single-phase motors or compressors help

Figure 19-1 Two capacitors that may be found in solid-state circuits. The capacitor on the left is about the size of a penny. The capacity ratings of these devices are very low when compared to the capacitors used in motor circuits.

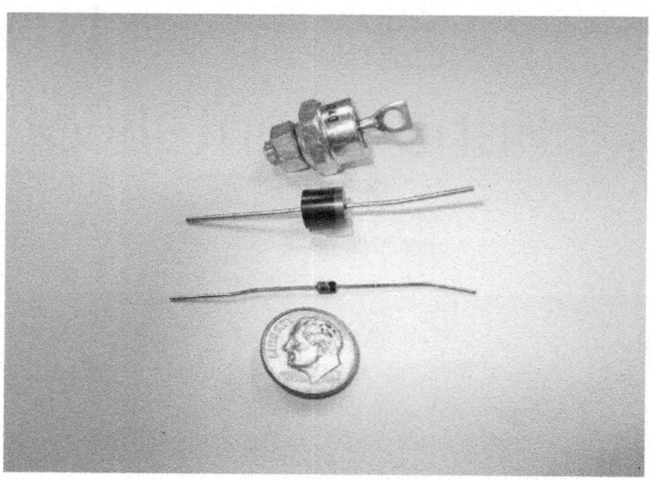

Figure 19-2 The small, glass-enclosed diodes have limited current handling capabilities. Diodes are used to modify the AC sine wave to create a type of voltage that is closer to DC voltage.

start the motor rotation (start caps) or reduce running amps (run caps). The capacitors in solid-state circuits are used to create a smooth direct current voltage from a rectified alternating sine wave. Capacitors in solid-state devices may also be designed to block or control current flow and direction. Figure 19-1 shows two different types of capacitors that you might see in solid-state circuits. Capacitors used in solid-state circuits usually have a very low microfarad and voltage rating when compared to motor caps. Like most solid-state devices, you will not be replacing capacitors, but whole circuit boards.

Capacitors are sometimes compared to batteries because they store electrons and because they both have two terminals. But capacitors are completely different from batteries. The battery creates a chemical reaction that develops electrons on one terminal and absorbs electrons on the other terminal. A capacitor, however, cannot create electrons; it merely stores them on its plates. Electrons do not travel through a capacitor. The capacitor terminals are connected to two plates separated by an electrically insulating material. The insulating material is called a *dielectric*. One way to think of a capacitor is to imagine two pieces of aluminum separated by a piece of insulating paper. Electrons are stored on the capacitor plates.

TECH TIP

The polarity feeding the power supply to a circuit board is important, especially if the power supply is 120 volts. A 120-V supply has a hot leg, a neutral leg, and a ground. Reversing the hot leg and neutral leg will create unusual operating characteristics when used in conjunction with a circuit board. The circuit board may not work or it may operate erratically. The incorrect polarity could also lead you to misdiagnose the problem with the circuit board.

19.2 DIODES

A **diode**, as shown in Figures 19-2 and 19-3, is a two-terminal electronic component that conducts electric current in only one direction. Think of it as an electron check valve. The term usually refers to a **semiconductor** diode, the most common type today. A semiconductor is a crystalline piece of semiconductor material connected to two electrical terminals.

The most common function of a diode is to allow an electric current to pass in one direction, while blocking current in the opposite direction. Thus, the diode can be thought of as an electronic version of a check valve. This unidirectional behavior is called **rectification** or changing. This is used to convert alternating current to direct current and is discussed in more detail in the next section.

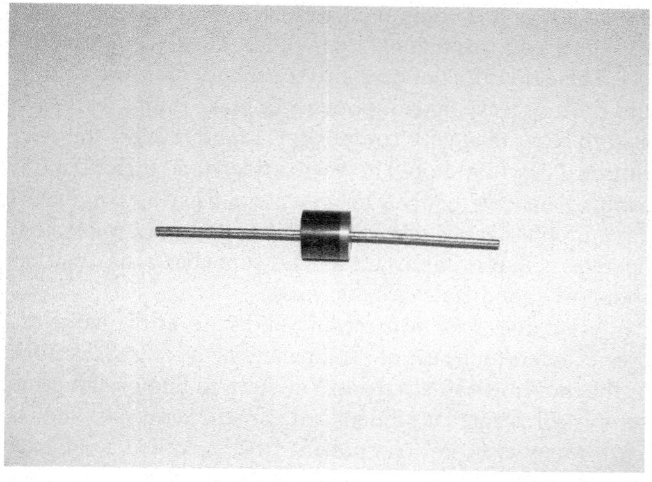

Figure 19-3 On this diode, the cathode side is indicated by the gray circle on the right side of the diode body.

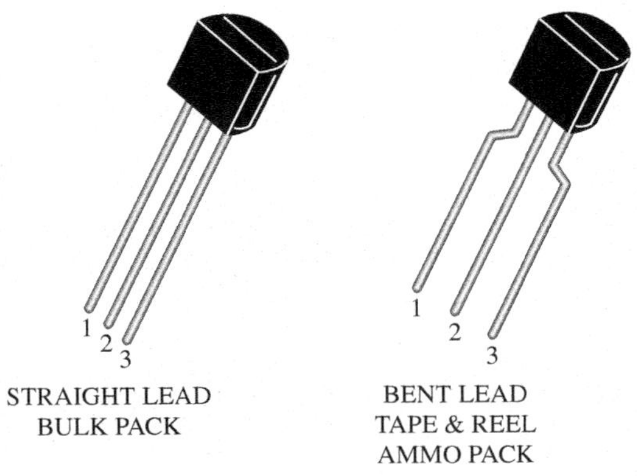

Figure 19-4 Transistors are found in solid-state devices. Some transistors are so small that they require a magnifying glass to see.

Figure 19-6 The three ICs pictured here are miniaturized circuits designed to fit into one shell or case. Miniaturization is the key to this component. The IC can have many transistors and other components inside the IC body. The integrated circuits are mounted on a circuit board and can control several operations.

19.3 TRANSISTORS

The **transistor** is compared to its predecessor, the glass vacuum tube. A transistor, as shown in Figure 19-4, is a semiconductor device used to increase or amplify an electrical signal. It can also be used as an electronic switch. It is made of a solid piece of semiconductor material that normally has three terminal connections as labeled in Figure 19-5. Some transistors are used individually but most are miniaturized and embedded in an integrated circuit (IC). The IC will be discussed later in this unit.

The transistor is the fundamental building block of modern electronic devices. The transistor was first used in its simplest form in the early 1950s. Its lower power requirements and compact size allow for a reduction in equipment size. The earliest example of commercial transistor use is the small portable radio, now known as the larger "boom box." The first computers had few solid-state components and were as big as a room and had less calculating power than today's average laptop and other portable devices. The advent of the transistor and other solid-state devices allowed the computer size to shrink and the power requirements to drop significantly. This paved the way for personal computers (PC), then laptops and the other computing devices that you see every day and probably own.

The small size and weight advantages of the transistor helped start the rush to miniaturize electronic devices. Other advantages of transistors are that they do not require a warm-up period and they use less power; therefore, they have less heat to dissipate. The lower power requirement translates into higher efficiency operation. Transistors have a long life, are highly reliable, and are physically rugged.

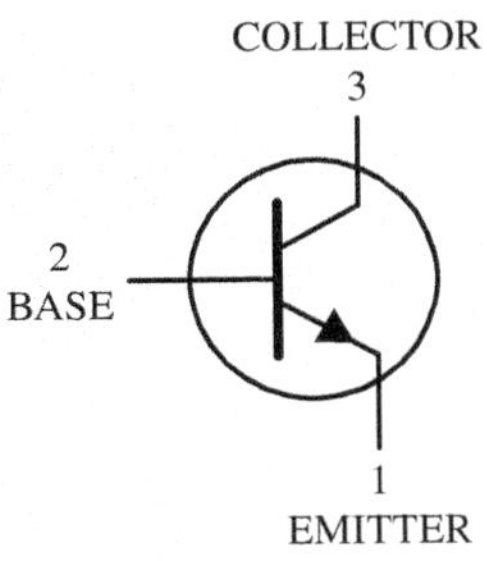

Figure 19-5 This is the common symbol for a transistor.

19.4 INTEGRATED CIRCUITS

An **integrated circuit** or **IC** contains many transistors and other solid-state devices installed in one component and mounted on a circuit board. Sometimes called a *chip*, the IC is connected to the circuit board by metal pins, as shown in Figure 19-6.

The IC is truly a miniaturization of most of the components discussed in this unit. The IC is used in virtually all electronic equipment found in HVACR circuit boards.

19.5 RECTIFIERS

A **rectifier** is an electrical device that converts alternating current (AC) to direct current (DC). Many solid-state devices use DC power. The rectifier blocks part of the AC wave to modify it so that it is closer to a DC voltage. For example, the rectifier can be designed to chop off the bottom half of an AC voltage sine wave. The rectifier by itself does not create a straight-line DC voltage. It may need additional rectifiers, diodes, and capacitors if a straight-line DC voltage is required.

Figure 19-7 shows four rectifiers wired into a **bridge rectifier** circuit. The bridge rectifier is also called a full-wave rectifier. The upper part of the diagram shows a normal AC sine wave, 60 hertz per second. The middle diagram shows how the four rectifiers are wired to create the bridge rectifier circuit. The lower diagram is the modified output voltage. The modified DC is not a flat-line voltage as expected from

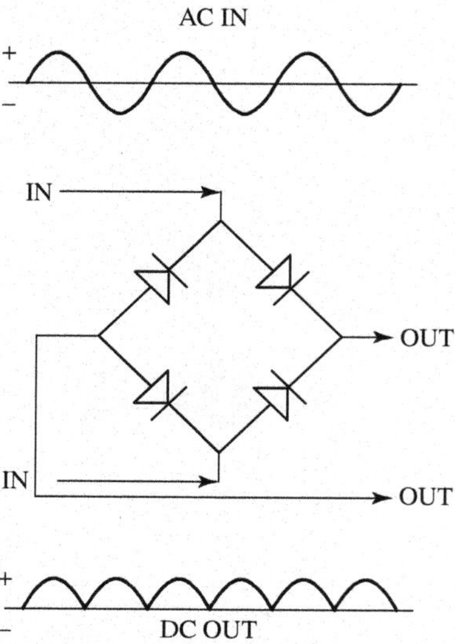

Figure 19-7 The upper part of the diagram represents a standard AC sine wave. The middle section illustrates a bridge rectifier diagram. The lower diagram shows the alternating current that has been modified by the bridge rectifier. The partial DC sine wave will need to be modified further if a pure, straight-line DC signal is required.

a battery voltage, but a choppy positive voltage that may be adequate for some DC circuits. The choppy DC voltage can be cleaned up with capacitors and other solid-state components if required by the DC operation.

Figure 19-8 shows two different types of bridge rectifiers. The identifying feature of a bridge rectifier is the four wire connections. The size of the rectifier will determine its current carrying capabilities.

Rectification may occasionally serve in roles other than to generate DC voltage. For example, in gas heating systems, pilot flame rectification is used to detect the presence of a small flame that will be used to light the burners. Two metal electrodes placed in the pilot flame provide a current path through the flame. The flame is actually a conductor that modifies the AC voltage. The rectification of an applied AC voltage will happen in the pilot flame plasma, but only while the flame is present to generate it. The partial DC voltage is used to signal a safety control that the pilot flame is present and the gas valve can open for burner ignition. This starts the ignition process.

Figure 19-8 These are examples of two different bridge rectifiers. The current carrying capability of these rectifiers varies. Large rectifiers will have greater current carrying capabilities.

TECH TIP

You may have heard of the term **inverter**. An inverter performs the opposite function of a rectifier. The inverter changes DC to AC. This can be handy in a service truck. The inverter can be plugged into the cigarette lighter and be used to charge up small AC devices such as refrigerant leak detectors, flashlights, laptops, or devices that require low power levels.

19.6 VARISTORS

A **varistor** is a two-element semiconductor with reverse resistance in which the resistance drops as the applied voltage increases. They are also called **metal oxide varistors (MOVs)**. The symbols for a varistor are shown in Figure 19-9. A MOV is illustrated in Figure 19-10. At first glance, you may note

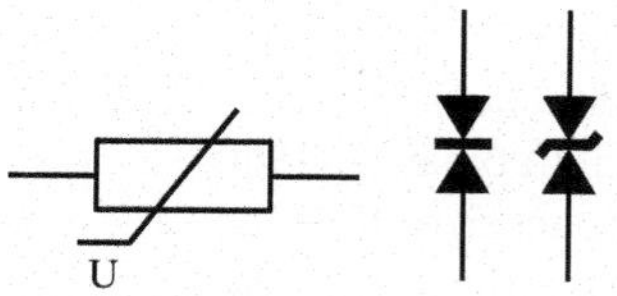

Figure 19-9 These are symbols for varistors.

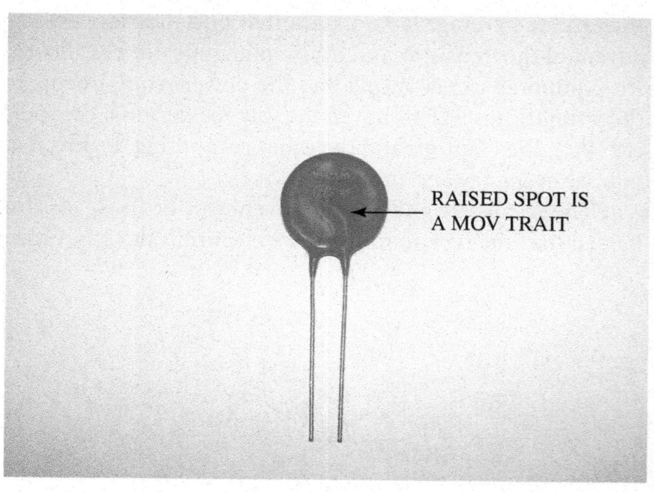

Figure 19-10 The metal oxide varistor (MOV) is used to protect electrical equipment by directing high voltage or voltage spikes away from the component. Some think of the MOV as a lightning rod to divert high voltages (although it is not capable of diverting an actual lightning strike).

that the MOV has the same appearance as a capacitor used in solid-state circuits. But the MOV has a raised line that is not commonly found on capacitors. Think of the MOV as a lightning rod for high voltages, directing dangerous voltage away from the component it is protecting.

Varistors are often used as a safety device to short circuit transient high voltage in electron circuits. A transient voltage is a high-voltage spike that can last long enough to cause component damage. This can occur, for example, when applying a short-term, transient 200 volts to a 120-V circuit. The longer the higher voltage is applied, the more likely it is that the components will be damaged. Even a short-term high-voltage spike can open a motor or coil winding or do damage to sensitive solid-state devices. The MOVs are wired in a circuit so that when they are triggered at a higher than normal voltage they will direct the current created by the high voltage away from the component they are protecting. A varistor is also known as a voltage-dependent resistor (VDR).

The common varistor is the metal oxide varistor (MOV). A typical surge protector power strip is built using MOVs. The MOV is a ceramic mass of zinc oxide grains sandwiched between two metal plates. The result of this design is a highly nonlinear current-voltage characteristic in which the MOV has a high resistance at low voltages and a low resistance at high voltages. The MOV works well in shorting line voltage transients from the component it is protecting. Figure 19-11 shows a MOV on a motor starter. In this example, the MOV is wired across two of the three phases. The MOV causes a high-voltage spike to bypass the component, thus protecting it.

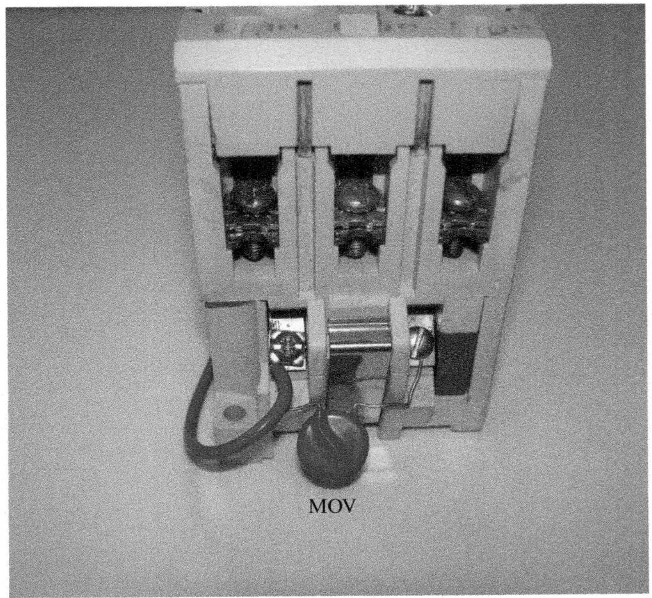

Figure 19-11 This metal oxide varistor (MOV) is protecting a motor starter. The MOV is used to protect electrical equipment by directing high voltage or voltage spikes away from components. It looks like a capacitor, but it has a dimple line; this is one identifying feature of the MOV that capacitors do not have.

What MOVs Do Not Do

A MOV will not control a surge of lighting. Lighting is generally at a higher magnitude than a voltage spike and will destroy the MOV and possibly the equipment it is trying to protect. Use of a series-connected thermal fuse is one way to protect against a catastrophic MOV failure. Varistors with internal thermal protection are also available.

A varistor provides no equipment protection against:

- Inrush current surges (LRA during equipment startup)
- Overcurrent created by a short circuit
- Voltage sags (also known as brownouts); it neither senses nor affects such events.

19.7 MICROCONTROLLERS AND MICROPROCESSORS

The **microcontroller** is basically a dedicated computer that operates one major component. A **microprocessor**, also known as a central processing unit or CPU, is designed to operate complex systems. The CPU can be designed to perform multiple functions in an HVACR system. The electronic expansion valve (EXV) can be controlled by a simple CPU or microcontroller. The electronic control attached to the EXV measures various refrigerant line temperatures, motor temperatures, and converted pressures to control an accurate EXV superheat. Accuracy is within ±1°F of the set superheat temperature.

19.8 CIRCUIT BOARDS

Solid-state components are mounted on *circuit boards*, sometimes called *printed circuits*. The top and bottom of a raw circuit board are shown in Figures 19-12 and 19-13, respectively. Figure 19-12 shows the circuit board without the installed components. The component installations are identified by letters on the board. For example:

- R = resistor
- IC = integrated circuit
- D = diode
- K = relay
- P = plug-in connection.

The bottom side of the board, as shown in Figure 19-13, shows the component connection lines and solder points. The connection lines replace the use of wires. They are usually copper strips, but can be silver or gold in special applications.

Figure 19-14 shows a circuit board used in a modern gas furnace. Several of the solid-state components have been identified. The IC on the board contains many miniaturized transistors and other components. The MOV protects the circuit board from voltage transients. Other identifiable parts on the board include resistors, thermostat connection point, capacitors, resistors, and other connection points.

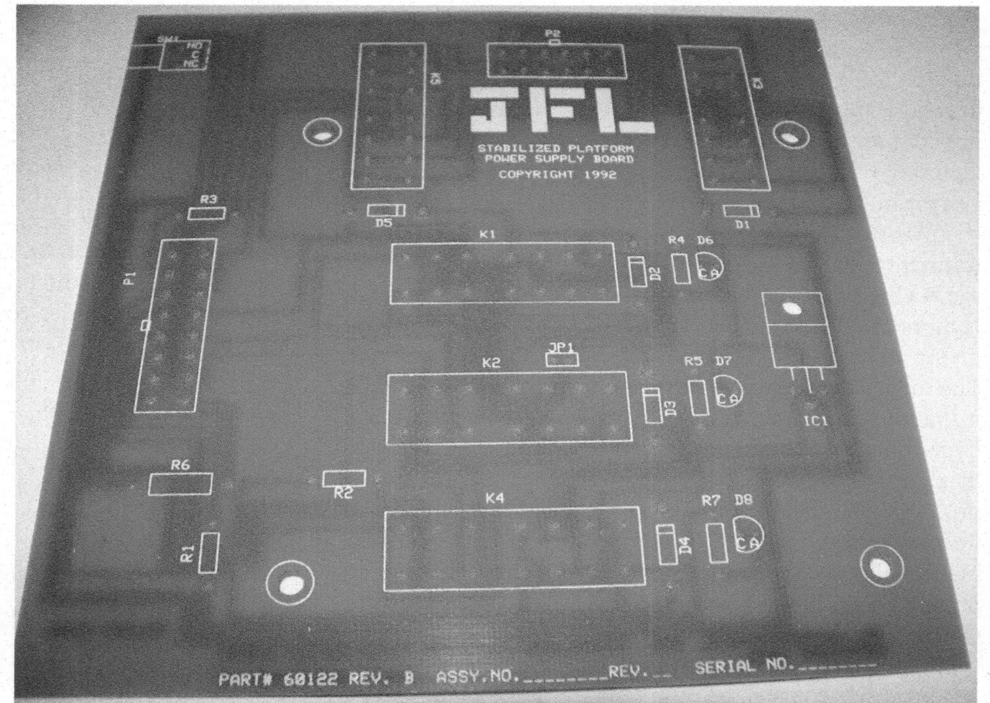

Figure 19-12 In the top part of a circuit board, note that the components are identified by letter. For example, R1 identifies a resistor; D4 is a diode; IC1 is an integrated circuit; K4 is a relay; and P1 is a plug. Other than knowing the names of these components, there is little value in this identification because the manufacturers of these boards do not share the diagram and parts identifications.

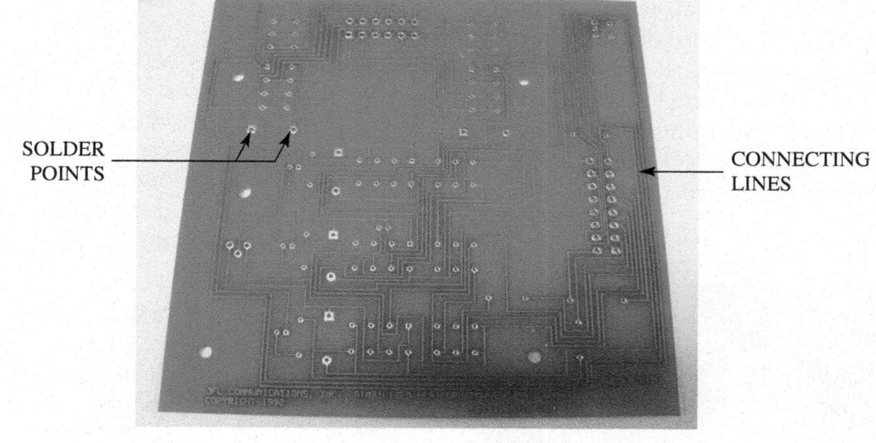

Figure 19-13 This is the bottom side of a circuit board. A component's connection wire will come through a hole in the board that is tinned with solder. Additional solder is added to secure the connection. The connection lines replace the use of wires. They are usually copper, but can be silver or gold.

Figure 19-14 This circuit board is for a gas furnace. Some of the parts discussed in the unit are labeled so that you can apply what you have learned.

SERVICE TICKET

You are sent out on a job where you find that the circuit board is burned out. There is a burned and, thus, open resistor on the top section of the board. When the board is removed, you see that there are many open connection lines on the back of the board. The circuit board will need to be replaced. What do you need to do before changing the board? What information will you need to obtain the correct replacement? What do you think?

Before replacing the board you will need to check the input voltage. Is the correct high voltage or control voltage present? The input high voltage is usually 120, 208, or 240 volts. The control voltage is normally 24 volts. Was the fuse blown on the circuit board? Some common type of fuses used on circuit boards are shown in Figure 19-15. The fuse and fuse holder (see Figure 19-16) are used to protect the secondary side of a replaced transformer. A short or overload on the secondary will damage the replacement transformer. It is easier and less expensive to change a fuse rather than a transformer.

The easiest way to order a replacement circuit board is to get the model and serial number of the system that needs repair. For example, the circuit board in Figure 19-14 came from a high-efficiency natural gas furnace. The board itself has numbers and identification information, but those are not as useful as the model and serial number of the furnace being repaired. Some systems will be able to use improved or modified circuit boards; therefore, it is important to get the latest upgrade.

When picking any replacement part, it is a good practice to take the defective part with you to match up the replacement component. In many cases the replacement component will appear different than the original part, but it should have the same functions and connections or it will not work. Look for instructions to see if the replacement component needs any special modifications.

Finally, in rare instances repairing the circuit board will fix the problem. Resoldering an open connection on the back of the board or replacing a burned-out resistor may work, but is it worth a callback to the customer? A callback would be a "no charge" service call. Your supervisor should be left to make this decision: "Do I fix it, Boss, or replace it?"

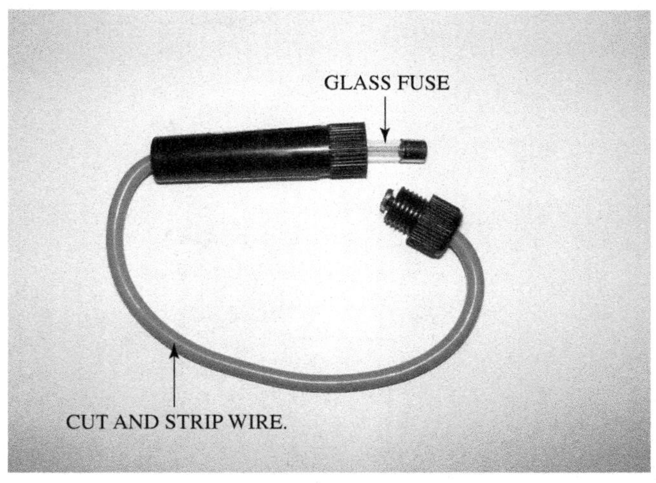

Figure 19-16 This fuse and fuse holder are used to protect the secondary side of a replaced transformer. Cut the middle of the red wire and place it in series with the load you desire to protect. For example, when changing a transformer, fuse the secondary. Check the control circuit for shorts or grounds. It is easier to change a fuse than to replace a transformer.

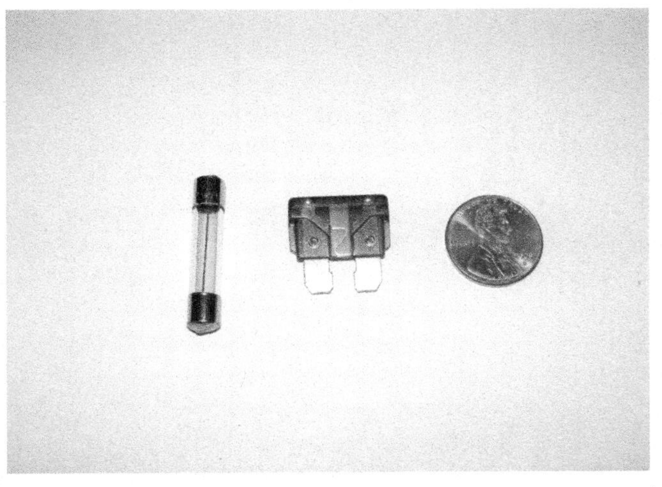

Figure 19-15 Glass and automotive type fuses are found on circuit boards. Check the fuses before condemning the board. A fuse may appear good even if it is blown. Check it with an ohmmeter if it appears to be good.

SUMMARY

This unit was designed to give you a brief overview of electronics without overwhelming you with electron theory. We briefly discussed the operation of diodes, transistors, rectifiers, varistors, and circuit boards.

Electron theory is important if you are working directly with electronic circuit boards. In our industry, however, circuit boards are good or bad. We do not need to know which component is defective. It is rare to attempt any field repairs on circuit boards, other than changing a blown fuse.

This unit has prepared you for the next unit, which is titled *Taking the Mystery Out of Circuit Boards.* That unit goes into general troubleshooting of circuit boards and when to replace them.

Figure 19-17 Use this image to answer Review Question 11.

Figure 19-18 Use this image to answer Review Question 12.

REVIEW QUESTIONS

1. What are four advantages of solid-state technology?
2. What is meant by the term *solid state*?
3. What are three common purposes of a capacitor found in a solid-state circuit?
4. True or false: A capacitor does not allow electrons to flow through it.
5. List four advantages of the transistor over the glass vacuum tube it replaced.
6. What is a diode?
7. What is the purpose of a bridge rectifier?
8. What is in an integrated circuit?
9. What is the purpose of a varistor?
10. What are the limits of a MOV?
11. Identify the device shown in Figure 19-17.
12. Identify the device shown in Figure 19-18.

UNIT 20

Taking the Mystery Out of Circuit Boards

WHAT YOU NEED TO KNOW

After studying this unit, you will be able to:

1. Understand the basics of a circuit board so that troubleshooting will be easier.
2. State the reason why circuit boards have become popular with manufacturers.
3. Describe how to hook up a circuit board.
4. Troubleshoot a circuit board.

Seeing a circuit board or solid-state device seems to bring fear to the HVACR technician. There is no reason for this concern. It is the fear of the unknown that creates this apprehension. This unit will show you that the circuit board should bring a sense of relief to the technician, not a sense of dread. One of the things you will realize is that the purpose of the circuit board may not be completely clear by looking at the diagram. You almost have to know the sequence of operation of the circuit to understand what the circuit is supposed to do. You don't have to know what every component (resistors, relays, capacitors, transistors, etc.) on the circuit board does. All you need to know is what controls the board, given specific inputs, which are transformed into specific outputs.

Circuit boards were developed to reduce the cost of equipment and make troubleshooting easier. Circuit boards allow a large number of components to fit into a small space. Prior to the circuit board, there were many large components and many wires to connect these components. Some of the initial responses to the newer systems with circuit boards were "There is something missing!" or "Where did all the components go?" The equipment has more space now that components have been miniaturized and placed on a circuit board. Instead of using wires, the circuit board uses thin strips of flat copper or copper foil connectors to complete the path for electricity on the circuit board. Let's look at the construction of the circuit board.

20.1 CONSTRUCTION OF CIRCUIT BOARDS

The electrical pathways on circuit boards are created using the following process: The base circuit board is a rigid glass-epoxy-based material. One side of the board is covered with a thin layer of copper foil. The copper foil will later become the path for electrical flow. The board's copper side is printed with a circuit pattern either photographically or by silk screening. Its patterns are printed with a chemically resistant ink called *resist*. The board is then dipped in a corrosive solution, dissolving the parts of the copper film that are exposed and unprotected. Any copper protected by the resist pattern remains. The patterns left after the etching are called the circuit board traces; they are like wire paths that connect the electronic components. The resist is washed or scrubbed off, leaving clean, shiny copper. The name *printed circuit* comes from the exposed "printed" pathways that remain on the board after it has been chemically washed.

The remaining copper paths act as the conductors between the components on the circuit board. The use of low current in the circuits allows the use of the thin foil metal. Figure 20-1 is an example of a circuit board prior to component connection. The components are soldered at a hole at the end of the copper strips. The basic board has flat metal raceways designed on the surface of the board. The connection raceways are commonly copper, but can be other highly conductive metals such as silver or gold. Much of this assembly operation is high volume, thus automated. Finally, the circuit boards are tested before shipment.

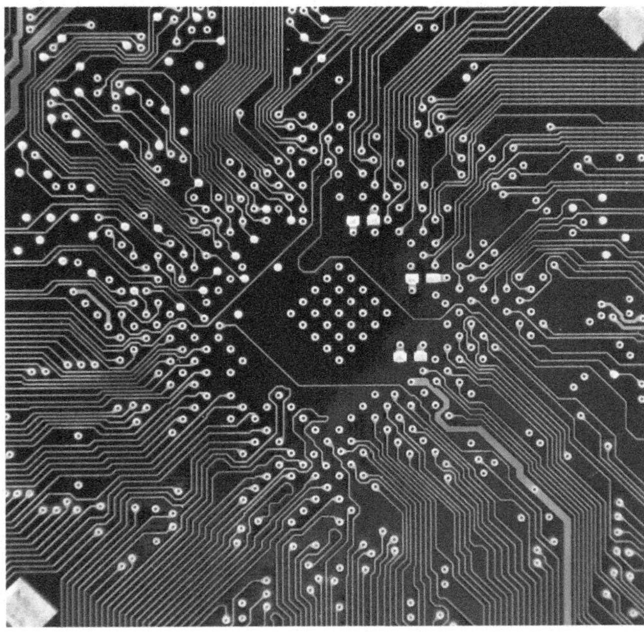

Figure 20-1 This is a circuit board prior to component connection. It is usually soldered wherever components are installed.

When circuit boards were first introduced in HVACR equipment over a couple of decades ago, the learning curve was steep. Because some technicians did not understand the circuit board, many of these high-technology components were wrongfully condemned. With continuing education and advanced industry learning, the condemnation rate of good circuit boards has dropped. It is easy to bad mouth and condemn something we do not understand. Learning and understanding has turned that attitude around. Circuit boards are considered to be an industry advancement.

In summary, circuit boards were developed:

- To reduce manufacturing costs
- To be more reliable when compared to all the components and wires they replaced
- So that fewer and smaller components can be used
- To use lower DC voltage and current, making it safer
- To be more precise and accurate.

Parts on a circuit board are usually smaller than the parts they replace. Some of the common components found on a circuit board are integrated circuits, resistors, capacitors, relays, transformers, dip switches, fuse holders, chips, transistors, LEDs, diodes, and connection points.

SAFETY TIP

Circuit boards are fairly safe, but the technician can damage components accidentally. Dropping the circuit board can create unnoticed hairline cracks that may open and close when heated or cooled. Static electricity can damage the board. Ground yourself out prior to touching a circuit board by touching the bare metal of a cabinet or water pipe. Store new and unused circuit boards in special factory packaging that reduces static electricity buildup.

20.2 UNDERSTANDING CIRCUIT BOARDS

There may be a lot of "magic" that occurs in a circuit board, but we do not need to understand the technical aspects in order to comprehend what a circuit board does. I use the word *magic* very loosely since what happens in a circuit board is engineered, designed, and tested to work the way it does. Engineers know what outputs are required and what limited inputs are available. The circuit board is designed around these conditions. This is not an electronics unit; it is an attempt to show you the best way to easily understand the circuit board device. The best way to grasp the circuit board is to think of it as a "**black box.**" A black box can simply be described as a device having the correct inputs that will produce the desired outputs without you having to know what happens in between the inputs and outputs.

Let's look at an example of a simple black box that can operate a fan when the air temperature warms. A simple black box can be designed to turn on a fan motor when the air temperature rises above 75°F. The black box will have a 120-V power supply, an adjustable thermostat, and an output 120-V receptacle that can operate a fan motor. Even though the black box is not a circuit board, it can be viewed in the same light.

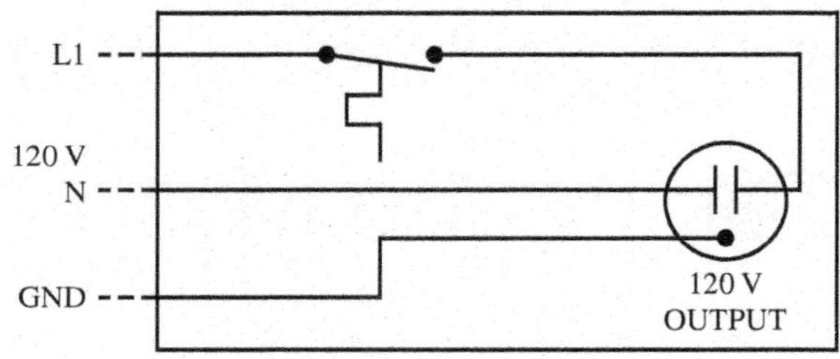

Figure 20-2 This is a diagram for a simple black box. The diagram could represent a circuit board or components wired on a table top.

Figure 20-2 represents a diagram of how a black box should be wired. The simple black box is a mystery until you look at the diagram. The diagram simplifies what is inside the box. Even if you did not have the diagram and did not understand how it worked, you could make the fan operate provided you followed the manufacturer's instructions. The instructions would state that you need a 120-V input and a 120-V fan motor plugged into the receptacle on the box. The receptacle is represented by the round circle, two parallel vertical lines, and a dot. The instructions would recommend a temperature adjustment to determine when the fan would be energized. The diagram could be a circuit board or components wired on a table top or enclosed in a box. The point is that circuit boards are like black boxes. You do not need to know the function of all of the components on the circuit board or in the black box to understand what the device is supposed to be doing, in this case operating a fan when the temperature rises.

Circuit boards have inputs and outputs. The inputs are usually voltage sources such as 24 volts and 120 or 240 volts. Some circuit boards have a 24-V transformer mounted on them, therefore they do not need an external 24-V source. Circuit boards in air handlers or furnaces have connections for the thermostat labeled **R**, **W**, **Y**, and **G**. Heat pump air handlers may have additional terminals such as **O**, **B**, **X**, and $\mathbf{W_2}$.

Circuit board outputs include power to operate a blower, contactor coil, heating system, or electronic air cleaner. Another output is a single LED light to show that the circuit board is powered up or a flashing LED light to indicate a problem The number of flashes will indicate a specific problem. The flashing light signal will vary from manufacturer to manufacturer. This is called the error code for that unit.

20.3 FIVE-MINUTE TIME-DELAY CIRCUIT BOARD

The way to understand a circuit board is to recognize its inputs and outputs. One way to comprehend the circuit board is to review an electrical diagram that includes the circuit board. For our first example we will review a simple circuit board. The circuit board shown in the bottom right-hand

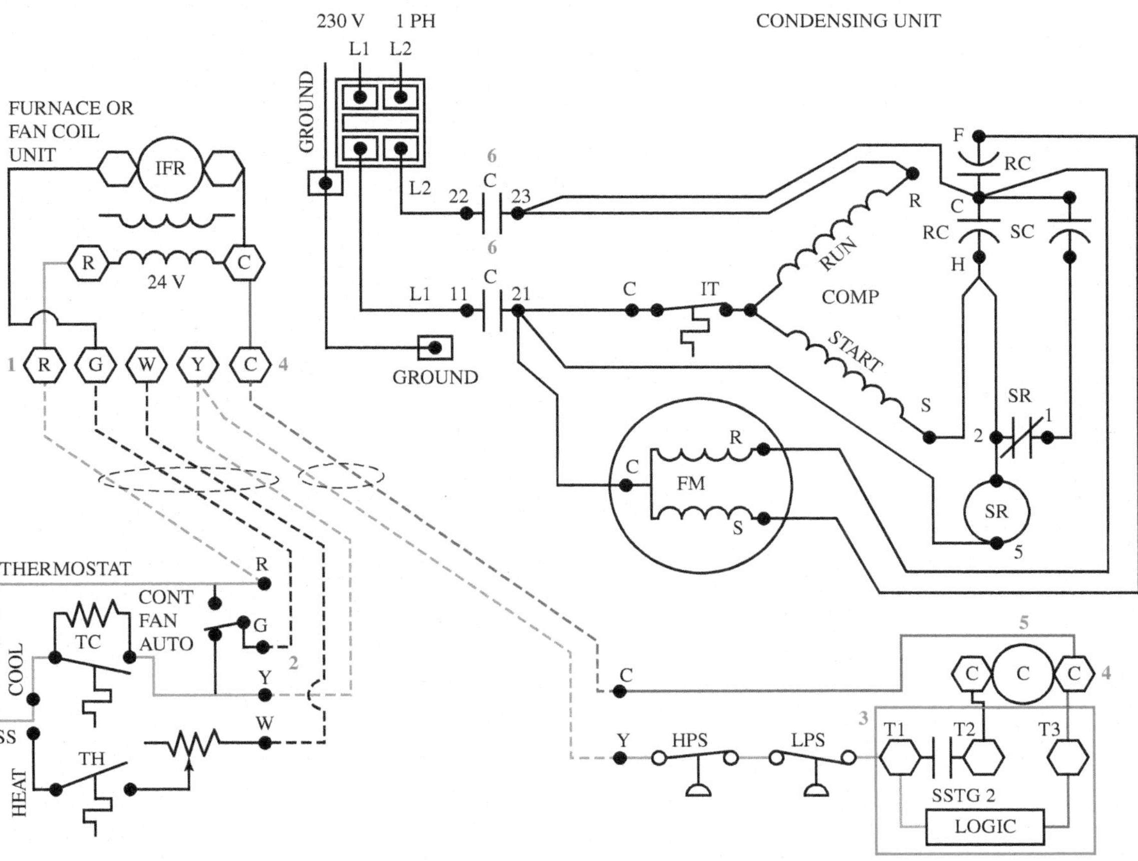

Figure 20-3 The circuit board in the bottom right corner of the diagram (outlined in red) is a time-delay circuit. When the 24-V power to the circuit board is interrupted, it will not energize the contactor coil © for a period of time, usually about 5 minutes.

side of Figure 20-3 shows one input and one output. This is a 5-minute time-delay device. It prevents the condensing unit from short cycling by creating a time delay between off and on starts. Let's review its basic sequence of operations:

1. The **time-delay circuit** board in Figure 20-3 is located on the lower, right-hand side of the diagram, outlined in red. It is labeled in the box as **SSTG 2**. This is a small time-delay circuit board. As with many components on an electrical diagram, the size of the circuit board does not represent the actual size of the device. It is not drawn proportional to the size of the other components on the diagram. Nonproportional drawings are common when electrical designers develop a diagram.
2. This simple circuit board has three connections: **T1**, **T2**, and **T3**. Look carefully at Figure 20-3 and determine the input and output connections.
3. **T1** and **T3** are the input connections. This input provides 24-V power to the circuit board. The diagram in Figure 20-3 traces the input voltage to **T1** and **T3**. You will notice that the 24 volts comes from the **R** terminal at point **1**. Following the dashed yellow lines will aid in tracing this circuit. Next, power goes through the thermostat and out of the **Y** terminal (point **2**) on the thermostat to **T1** (point **3**). The **T3** connection (blue dashes) receives voltage from the common side of the transformer **C**. The **C** connections are listed as point **4**.
4. **T2** and **T3** make up the 24-V source that will energize the contactor coil © at point **5**. At the end of a cooling cycle, when the circuit board is de-energized, it goes through a time-delay sequence of about 5 minutes. The time delay is designed to allow the refrigeration circuit time to equalize pressure between the high side and low side. Starting up with equal pressures is easier on the compressor. It reduces motor starting torque requirements and lowers starting amperage.
5. When the 5-minute time delay after the 24-V circuit has been interrupted has passed, the **T1** and **T2** contacts on the circuit board will close. This will allow for the immediate flow of 24 volts to the contactor coil ©.
6. When the cooling thermostat closes, 24 volts is provided to the circuit board. The **T1** and **T2** contacts will be closed, allowing the contactor coil **C**, at point **6**, to be energized immediately. The two-pole contactor will close, providing power to the compressor and condenser fan.

The time delay has another advantage when it comes to power fluctuations or brownouts. When the voltage drops out and quickly comes back on, it may not be at the correct voltage level required to operate equipment properly. For example, if the voltage fluctuates and comes back on much lower than the rated voltage, it will cause a high current draw on the load. The temporary high current draw event will cause the compressor or other motors to overheat. The motor(s) may trip on their

SEQUENCE OF OPERATION 20.1: CONDENSATE OVERFLOW CIRCUIT BOARD

1. See Figure 20-4. The lower right side of the diagram, at point **1**, shows the circuit board connections to the condensate overflow sensor. The sensor connects to **COND** on the board.
2. The left side of the diagram, at point **2**, has 24 volts going to two terminals labeled **24 VAC** and **GND**. The ground connection is the common connection on the transformer. The 24-VAC connection is the "hot side" or **Y** of the thermostat. To determine which of the two wires is hot and which common, measure each wire voltage to ground. The wire that measures 24 volts to ground is the hot wire and is connected to 24 VAC. The common wire **C** to the GND connection will measure 0 volts to ground.
3. Next, as seen at point **3**, wire one side of the contactor coil to the **C** and **GND** connection on the circuit board.
4. Wire the right side of the contactor coil to the normally open (**NO**) connection on the circuit board, point **4**.
5. The **COM** connection on the circuit board is wired to **Y** as seen at point **5**. Power to the **Y** connection to this device will occur when the thermostat is energized through the **Y** connection.
6. The **NO** open contacts on the circuit board will be closed until the sensor's element detects an overflow water condition in the condensate pan. Water in the pan will create a path for electricity between the sensors, completing a shutoff function. The circuit board will open between the **NO** on the left side of the circuit board. This will stop the condensing unit since there will **not** be a complete circuit to provide 24 volts to the contactor. Test the safety switch by blocking the emergency drain pan water outlet and pouring water into the pan until it stops the condensing unit. You can also test it by placing the sensor in a cup of water. It will not take much water to stop the condensing unit. The pan will need to drain and may need to be dried around the sensor prior to placing the condensate sensor back in the bottom of the pan. The pan sensor can be very sensitive to moisture.

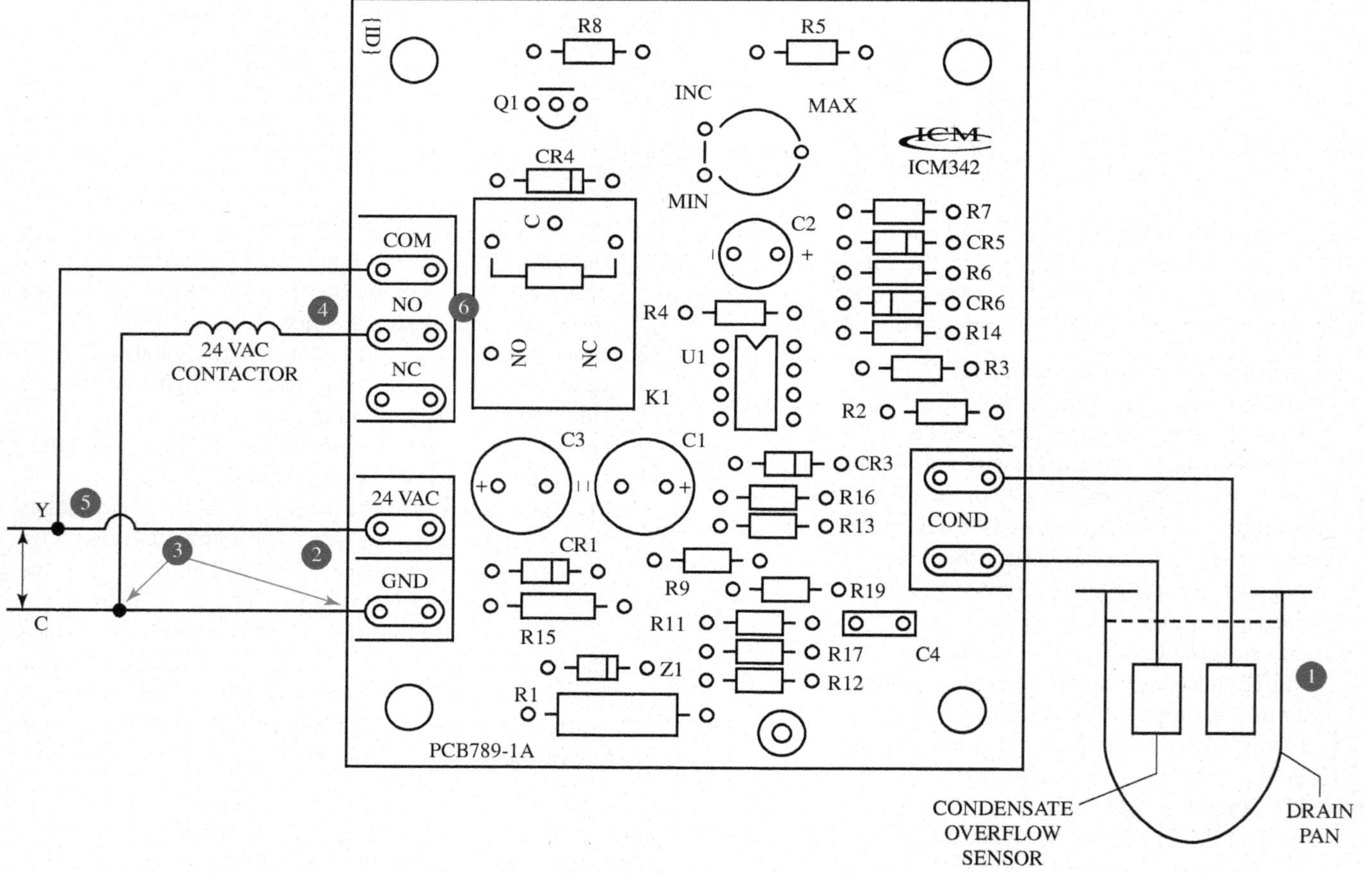

Figure 20-4 This circuit board is for condensate pan overflow sensor control. This control will turn off the condensing unit when the auxiliary drain pan has water in it. The blower fan will also be turned off by this control if the thermostat is in the automatic position.

overload protective device. The 5-minute time delay will allow the voltage variations and the pressures to stabilize in the system. This is an inexpensive way to protect the equipment.

TECH TIP

CPU and **PCB** are common abbreviations you will see on electrical diagrams with solid-state components. What do they mean? The CPU is a central processing unit. Think of it as a computer with a logical process for doing a designed job such as turning on an air conditioner or heating system. The PCB is a printed circuit board. The PCB is the thin, rigid board with wire pathways that are etched for component connections.

Figure 20-5 This circuit board is used to shut down a condensing unit when the main drain pan overflows into the emergency drain pan.

Figure 20-5 is the actual circuit board illustrated in Figure 20-4. This is a condensate sensing device. The sensing device is placed in an emergency overflow drain pan and stops the condensing unit when a water overflow condition occurs.

SEQUENCE OF OPERATION 20.2: HEAT PUMP CIRCUIT BOARD

The next circuit board diagram is more complex than those in the previous examples. The circuit board shown in Figure 20-6 is for a heat pump. First, we examine the inputs.

1. Starting at the upper left side of the board, note the field-selected timing pins, shown as **P1** and labeled as point **1**. This select pin allows choices between different run times before the defrost cycles. Normal run-time selections are 30, 45, or 90 minutes. Humid outdoor winter conditions will require more frequent defrost cycles. In humid winter conditions, you would select the 30-minute run-time defrost cycle. This means that every 30 minutes of run time the circuit board checks to see if the conditions are right (frost buildup) for the defrost operation. The defrost operation cycle reduces a system's efficiency, therefore the defrost cycle should only be energized when necessary. Colder air has less moisture in it because the moisture is changed to snow or ice. Colder climates may be able to use defrost times set for a 45- or 90-minute run-time cycle. This will provide longer heating times and more efficient operation.
2. Below the timing pins are the **test pins**, labeled **2**. These are used to create an accelerated heating and defrost cycle operation. Shorting across the **test pins** with a coin will run the system through the defrost cycle very quickly. The voltage at the test pins is low voltage and, therefore, not a shock hazard.
3. **P5** indicates the compressor delay pins, labeled **3**. Once the thermostat has been de-energized, the timing circuit prevents the contactor coil from being energized for about 5 minutes. The same conditioning will occur if power is removed from the circuit board.
4. The **reversing valve** connections, at point **4**, energize the solenoid valve on the reversing valve in the cooling mode. Some manufacturers energize the reversing valve in the heating mode.
5. The **low-pressure switch** (point **5**) is factory installed and wired to the circuit board.
6. The **defrost thermostat** (point **6**) is wired to the circuit board and is used to terminate the defrost cycle when the refrigerant line warms. The defrost thermostat will be open above a set temperature, around 32°F. This thermostat closes upon a drop in temperature. It can be checked with an ohmmeter above freezing. This will be an infinity reading. When the defrost thermostat is below freezing, it will be closed and measure near zero ohms. The freezing setpoint will vary among manufacturers. One way to test the defrost thermostat is to place it in the freezer with the leads sticking out of the box so that you can measure resistance. It should read near 0 Ω when below freezing.
7. Below the defrost thermostat is the optional **high-pressure switch** connection, labeled **7**. This safety device is a normally closed switch that will be field installed. When the pressure increases the high-pressure switch will open and shut down the condensing unit.

(continued)

8 Looking at the upper part of the board, there is a connection for the **fan**, labeled **8**. This is direct power to operate the blower. Power should be measured at this terminal when the thermostat is set to the fan **ON** mode and when the temperature is adjusted to operate in the heating or cooling mode.

9 Scanning down the right side of the diagram, there are two LED diagnostic lights (point **9**). The use of troubleshooting lights varies among designers. We will use this as an example of some optional LED use. One LED could be a power light indicating the supply voltage to the circuit board. The LED may be a flashing or a solid light. The second LED can be used as a flash code. Another way to use two LEDs is to use them to indicate double-digit numbers such as 10, 20, 30, 33, 35, or any other two-digit number. One LED light is used to indicate the first number in the double-digit sequence. The second flashing light will indicate the second flashing digit. The technician will need to know what the LEDs mean and the corresponding troubleshooting table.

10 Next, is **P2**, the **24-V terminal strip connections** (labeled **10**). This terminal strip is used to connect all of the thermostat wires.

11 To the left of the terminal strip is the 24-V power connection used to power the circuit board (point **11**). This power is also used by the digital thermostat. Power will be supplied by a 24-V transformer, in this case external to the circuit board. Some transformers are mounted on the circuit board.

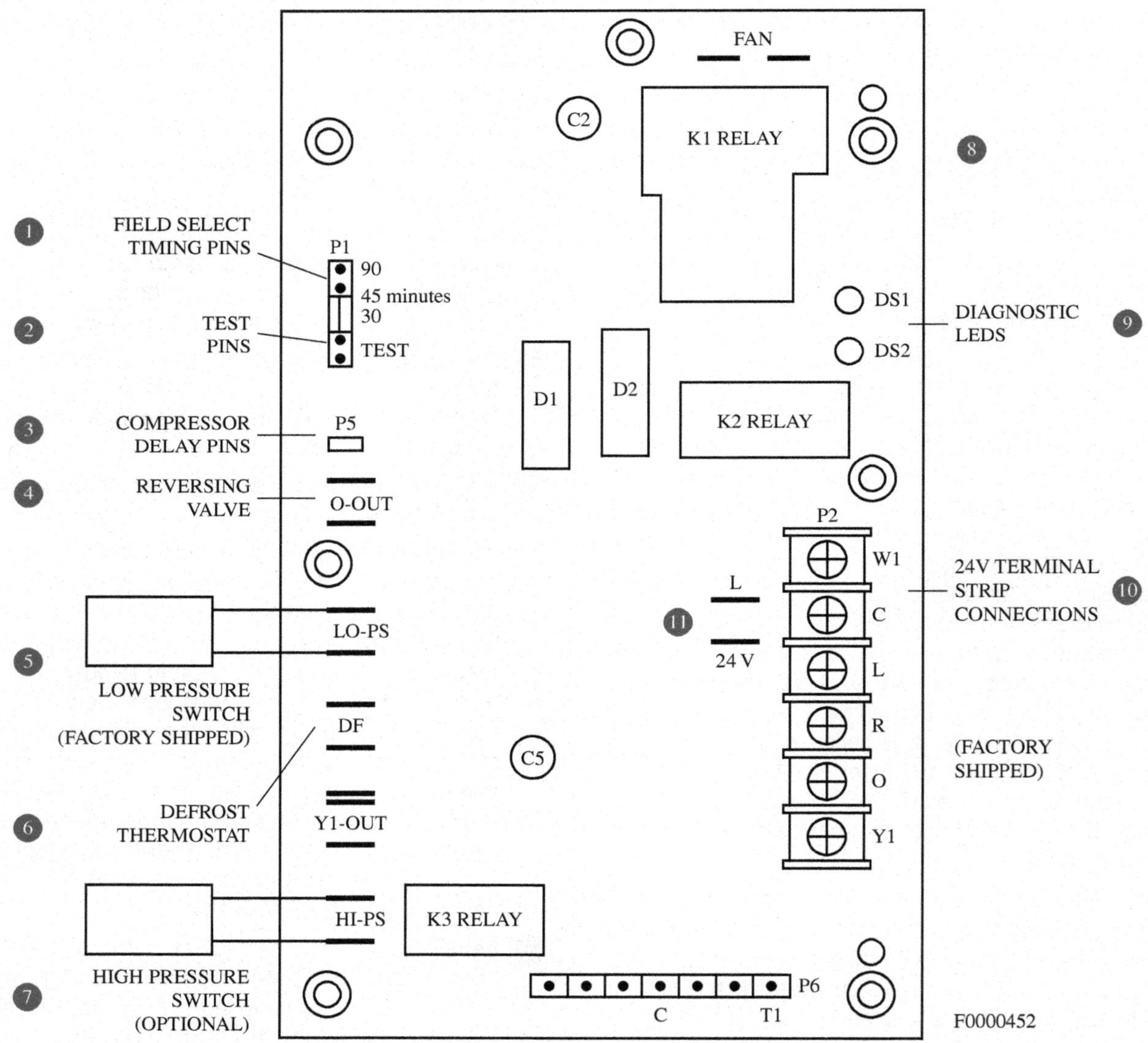

Figure 20-6 On this heat pump circuit board, the inputs are the low-pressure switch, defrost thermostat, high-pressure switch, and thermostat connects found on the 24-V terminal strip connections. The board uses diagnostic LEDs to lead the technician who is repairing the heat pump system.

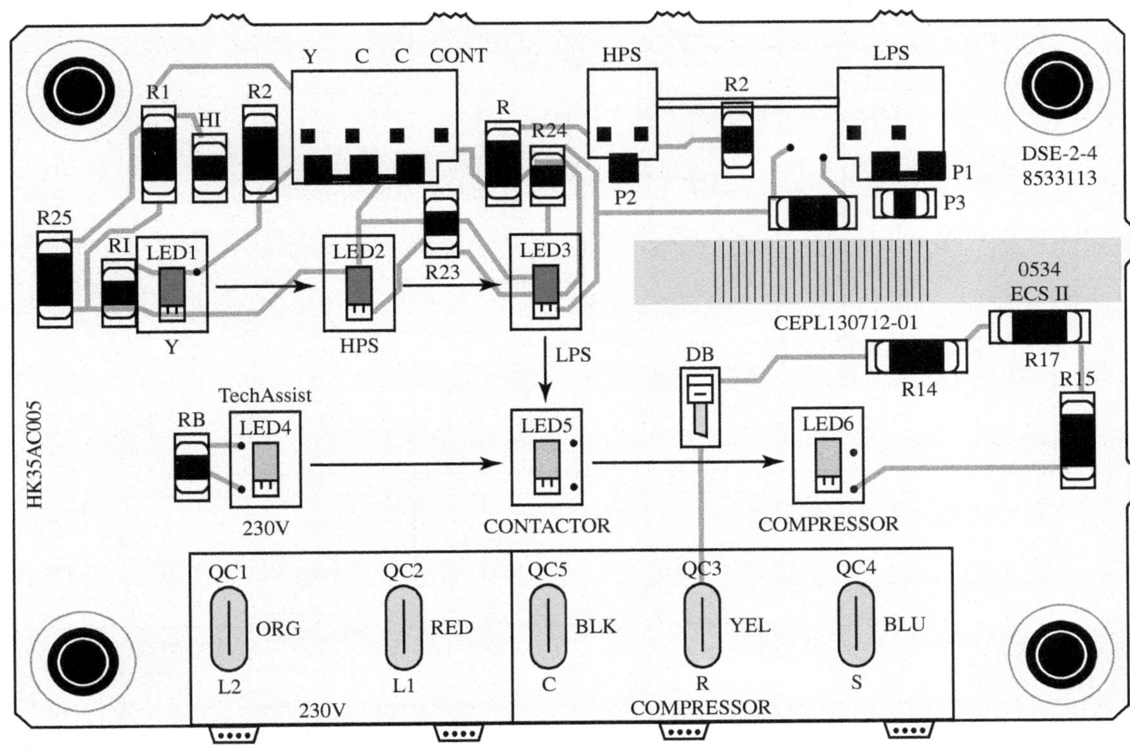

Figure 20-7 The TechAssist® board is used as a visual indication of what is working in a condensing unit. When the system is functioning normally, the LED lights will indicate the supply voltage of 230 volts and the operation of the contactor and compressor.

Understanding the circuit board connections will help the technician establish a logical path for troubleshooting. Measure the inputs and determine if the outputs are what are required to make the system operate properly.

20.4 TECHASSIST® CIRCUIT BOARD

The TechAssist® circuit board is used to give technicians a quick visual display, using LED lights, of what is not operating in a condensing unit. Figure 20-7 shows the board that is used for this specific troubleshooting process.

Let's examine the TechAssist inputs. Starting at the upper left side there is a **YCC CONT** plug connection. Going to the right on the circuit board is the high-pressure switch connection (**HPS**) and low-pressure switch (**LPS**). The pressure switch connections should measure 0 Ω when the refrigerant pressures are correct. In this case the **HPS** would be 0 Ω if the high-side pressure is not excessive. The **LPS** is closed or 0 Ω when there is adequate pressure in the system. These are normally closed safety switches. The bottom left side of the board has 230-V inputs **L1** and **L2.** To the right are the connections to the compressor's **C**, **R**, and **S** common, run, and start circuits.

Six LEDs are seen on the circuit board in Figure 20-7 and on the electrical diagram in Figure 20-8.

The following LEDs are seen on the circuit board in Figure 20-7:

- **LED 1 Y:** Y circuit (center, left-hand side of circuit board)
- **LED 2 HPS**: high-pressure switch
- **LED 3 LPS**: low-pressure switch
- **LED 4:** 230-volt power supply
- **LED 5:** contactor operation
- **LED 6:** compressor operation.

The LEDs in Figure 20-8 are represented by the light symbol, which is the circle with six rays around the circumference. In the diagram, the letter **A** in the light symbol represents high voltage, and the letter **G** represents control voltage.

In Figures 20-7 and 20-8, the TechAssist board includes three green, low-voltage 24-V components under LEDs 1, 2, and 3. The three amber high-voltage LEDs 4, 5, and 6 include the 230-V power light, contactor, and compressor monitoring. Figure 20-9 further isolates the LEDs into a more understandable diagram.

As with most circuit boards, the TechAssist device should reduce service time since it identifies circuits that are not operating. This circuit board does not lock out any components and has no effect on system operation. It simply provides an indication of power with closed contacts or completed circuits through particular components.

Finally, Figure 20-10 shows an operating table that relates to the TechAssist device. For example, when the system is working properly, all green and amber lights will be lit. It

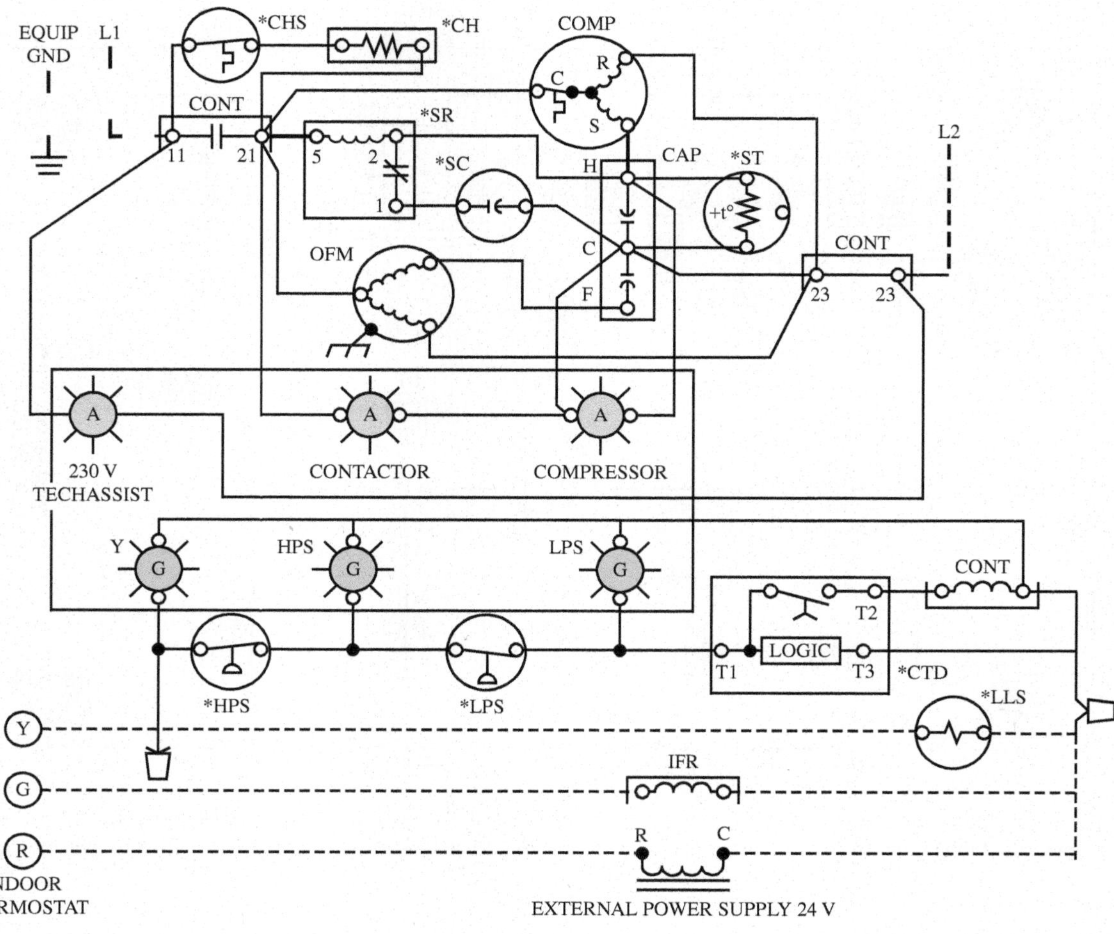

Figure 20-8 This is the full condensing unit diagram for use with the TechAssist troubleshooting and monitoring system. This diagram helps the technician integrate the TechAssist board into the condensing unit operation.

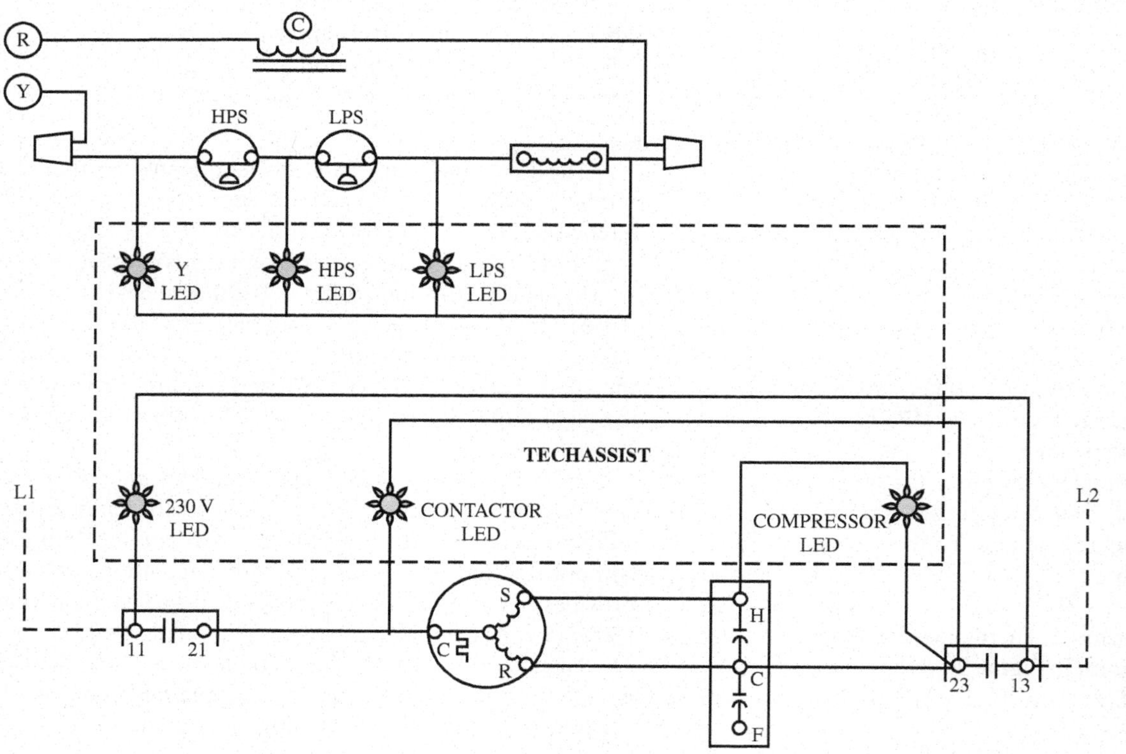

Figure 20-9 This is a scaled-down version of the TechAssist troubleshooting device diagram shown in Figure 20-7. This diagram simplifies the TechAssist operation and its role in the condensing unit wiring circuit.

LED	Description	ON	OFF
Y	Thermostat cooling call	Yes	No
HPS	High pressure switch	Closed	Open
LPS	Low pressure switch	Closed	Open
230 V	AC line power available	Yes	No
Contactor	Contactor energized	Yes	No
Compressor	Compressor running	Yes	No

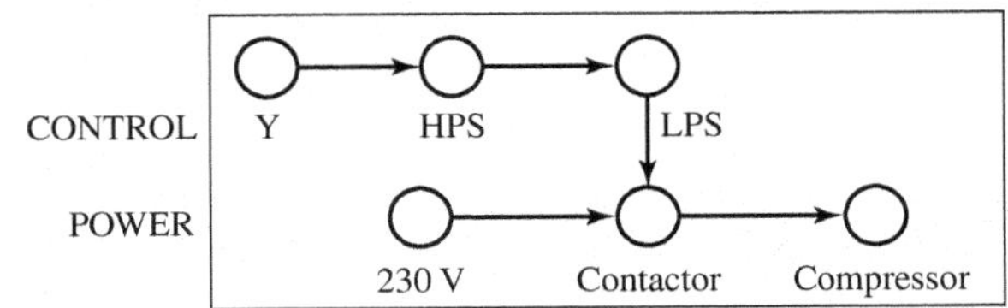

NOTE: Follow arrows to first unlit LED in control (green) ladder or power (amber) ladder to determine problem area. For contactor LED to be ON, all control LEDs (green) as well as 230 V LED (amber) must already be on, and contactor must operate properly

Control ladder			Power ladder			Indication	Possible cause & quick Trouble–Shooting
Green LEDs			Amber LEDs				
Y	HPS	LPS	230 V	Contactor	Compressor		
Off	Off	Off	Off	Off	Off	– No 230 V power to unit – No thermostat call	– Check 230 V disconnect/circuit breaker – Check 230 V L1 & L2 wire connections
On	Off	Off	– –	Off	Off	– HPS switch open	– Check pressures – Check HPS wire connections
On	On	Off	– –	Off	Off	– LPS switch open	– Check pressures – Check LPS wire connections
On	On	On	Off	Off	Off	– No 230 V power to unit	– Check 230 V disconnect/circuit breaker – Check 230 V L1 & L2 wire connections
On	On	On	On	Off	Off	– Contactor open***	– Check contactor and replace if necessary – Check contactor wire connections
On	On	On	On	On	Off	– Compressor not running*	– Check for compressor protector trip, capacitor fault, faulty connections or compressor fault. SEE BELOW.
On	On	On	On	On	On	– Compressor running	– OK
Off	Off	Off	On	Off	Off	– No thermostat call, standby	– OK
Off	Off	Off	On	On	Off	– Contactor stuck closed, compressor tripped	– Check contactor and replace if necessary – Check wire connections for shorts
Off	Off	Off	On	On	On	– Contactor stuck closed, compressor running	– Check contactor and replace if necessary – Check wire connections for shorts

*Compressor not running: If all LEDs are on except the compressor, this indicates that power is applied to the compressor but it is not running. Check the following sequence:

(1) Check all compressor wire connections
(2) Check run capacitor and replace if necessary
(3) Try starting compressor
(4) If compressor LED does not come on, then leave unit standby for up to four hours to allow thermal protector to reset, and all pressures to equalize
(5) Try starting the compressor again
(6) If compressor LED comes on, then turns off quickly, investigate starting issues such as low line voltage
(7) If compressor LED never comes on, compressor may be faulty.

Figure 20-10 This is an LED troubleshooting chart used in conjunction with the TechAssist device. This chart makes it easier for the technician to determine the condensing unit problem.

is important to have a code breaker table like that shown in Figure 20-10.

Using Figure 20-10, let's look at an example of how the TechAssist circuit board is used to find a problem. Just looking at the LEDs may confuse the technician. Consider a TechAssist board where the only lights that are lit are the **Y** circuit LED and the **HPS** LED. Using these lights as the only indicator of a problem, what is wrong? Can you figure it out without looking at the troubleshooting table? It may not be obvious. That's why it is important to have the troubleshooting table shown in Figure 20-10. Using the table we see that the problem is an open low-pressure switch. This table provides a large amount of troubleshooting information. Study it carefully so that you can realize the benefit of this thorough information.

TECH TIP

Check the equipment ground. Circuit boards and other operating components may operate erratically if the ground is inadequate. A grounded piece of equipment is also a safety measure to protect the technician should the power line touch the equipment case. Check the equipment ground by making a measurement from the hot leg to the case. There should be a 120-V reading. Check the voltage between neutral and ground. The voltage should be zero. Sometimes an additional ground is added directly to the equipment.

20.5 TROUBLESHOOTING A CIRCUIT BOARD

There are several ways to troubleshoot a circuit board. For our first example, let's assume the worst and that would be that you do not know anything about the circuit board. You do not have a diagram to help you understand its operation and you do not have the unit wiring diagram. The only thing you do know is that the system you are troubleshooting is not operating properly. Prior to troubleshooting a circuit board, use the ACT method to determine if there is an obvious problem. The ACT method reminds you to check the airflow, condensing unit operation (or furnace), and the thermostat setting. If it is a cooling or heat pump problem, hook up the gauges and clamp-on ammeter to the compressor common wire.

If it is a heating problem, check the fuel source, which is usually gas, electricity, or oil. Assuming that everything you checked was fine, you would now focus on the circuit board. Examine the board for inputs and outputs.

TECH TIP

Figure 20-11 shows a circuit board mounted in front of a blower. Notice the wiring harness going to the blower. The small dip switches numbered 1 through 8 are found between the two wiring harnesses. Airflow is the second major problem found in air conditioning systems. The first problem is incorrect refrigerant charge. Normally the airflow problem is a duct system problem. In some instances the dip switch settings are not correct. The installer and start-up tech assume that the dip switches are factory set for their application. Check the dip switch settings to ensure that they are correct for the installation. It is recommended to check the dip switch settings during the annual clean and check maintenance service. You will find that many dip switches are still set to the factory settings. Check those settings.

Does the circuit board have a fuse as shown in Figure 20-11? Inspect the fuse to see if it is blown. Also check its resistance because some fuses will appear good, but are blown. You should remove a fuse before checking its resistance.

Check the input voltages. This may be marked on the board as 24V, 120V, 240V, or possibly two of these voltages. See the arrow pointing to the input voltage on the left side of the PCB in Figure 20-12. The transformer should supply 24 volts to the points labeled **SEC-1** and **SEC-2.** If 24 volts is not at this point on the board, check the input and output of the transformer. If this cannot be determined, follow the power from the input power supply and output of the transformer to the circuit board. If the board has a mounted transformer, what is the voltage output?

This circuit board in Figure 20-12 also has a 120-V input as pointed out by the arrows. The 120-V input is identified as **PR1** (hot side) and **NEUTRAL.** The standard wire insulation

Figure 20-11 The 5-A fuse (see lower arrow) is used to protect the circuit board from overcurrent damage. The dip switches (see upper arrow) to set the blower speed for heating and cooling are numbered 1 through 8. The tiny dip switches are found between the two white wiring connectors and can be switched with a paper clip or pen point.

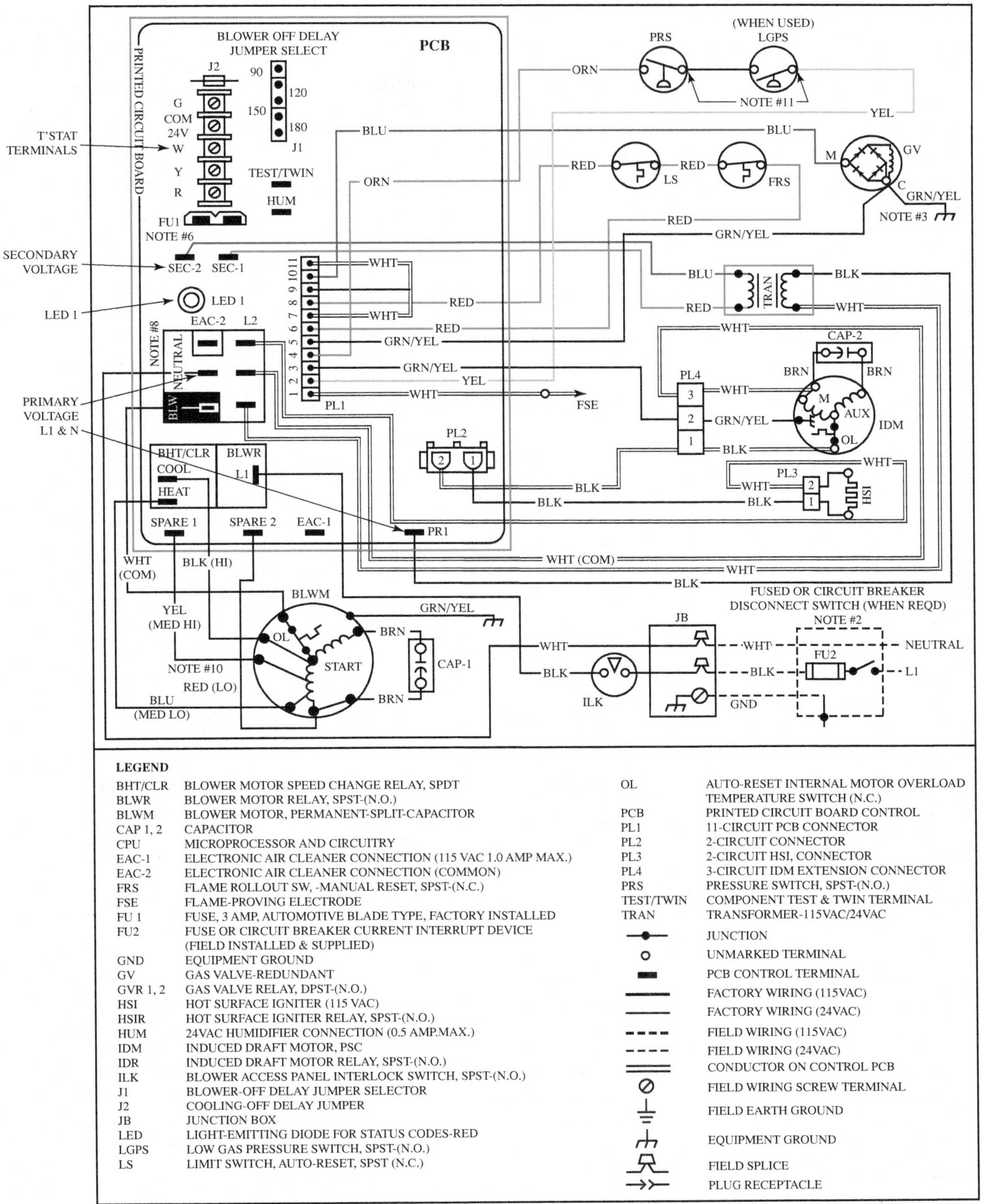
PCB
PRINTED CIRCUIT BOARD
BLOWER OFF DELAY
JUMPER SELECT
J2
90
120
150
180
J1
G
COM
24V
W
Y
R
T'STAT
TERMINALS
TEST/TWIN
HUM
FU1
NOTE #6
SECONDARY
VOLTAGE
SEC-2
SEC-1
LED 1
EAC-2
L2
NOTE #8
NEUTRAL
BLW
PRIMARY
VOLTAGE
L1 & N
PL1
PL2
BHT/CLR
BLWR
COOL
HEAT
L1
SPARE 1
SPARE 2
EAC-1
PR1
WHT
(COM)
BLK (HI)
YEL
(MED HI)
NOTE #10
RED (LO)
BLU
(MED LO)
BLWM
OL
START
BRN
CAP-1
GRN/YEL
WHT (COM)
BLK
WHT
ILK
JB
GND
FUSED OR CIRCUIT BREAKER
DISCONNECT SWITCH (WHEN REQD)
NOTE #2
FU2
NEUTRAL
L1
(WHEN USED)
PRS
LGPS
ORN
NOTE #11
YEL
BLU
RED
LS
FRS
M
GV
C
NOTE #3
TRAN
CAP-2
PL4
IDM
AUX
PL3
HSI
FSE
LEGEND
BHT/CLR BLOWER MOTOR SPEED CHANGE RELAY, SPDT
BLWR BLOWER MOTOR RELAY, SPST-(N.O.)
BLWM BLOWER MOTOR, PERMANENT-SPLIT-CAPACITOR
CAP 1, 2 CAPACITOR
CPU MICROPROCESSOR AND CIRCUITRY
EAC-1 ELECTRONIC AIR CLEANER CONNECTION (115 VAC 1.0 AMP MAX.)
EAC-2 ELECTRONIC AIR CLEANER CONNECTION (COMMON)
FRS FLAME ROLLOUT SW, -MANUAL RESET, SPST-(N.C.)
FSE FLAME-PROVING ELECTRODE
FU 1 FUSE, 3 AMP, AUTOMOTIVE BLADE TYPE, FACTORY INSTALLED
FU2 FUSE OR CIRCUIT BREAKER CURRENT INTERRUPT DEVICE (FIELD INSTALLED & SUPPLIED)
GND EQUIPMENT GROUND
GV GAS VALVE-REDUNDANT
GVR 1, 2 GAS VALVE RELAY, DPST-(N.O.)
HSI HOT SURFACE IGNITER (115 VAC)
HSIR HOT SURFACE IGNITER RELAY, SPST-(N.O.)
HUM 24VAC HUMIDIFIER CONNECTION (0.5 AMP.MAX.)
IDM INDUCED DRAFT MOTOR, PSC
IDR INDUCED DRAFT MOTOR RELAY, SPST-(N.O.)
ILK BLOWER ACCESS PANEL INTERLOCK SWITCH, SPST-(N.O.)
J1 BLOWER-OFF DELAY JUMPER SELECTOR
J2 COOLING-OFF DELAY JUMPER
JB JUNCTION BOX
LED LIGHT-EMITTING DIODE FOR STATUS CODES-RED
LGPS LOW GAS PRESSURE SWITCH, SPST-(N.O.)
LS LIMIT SWITCH, AUTO-RESET, SPST (N.C.)
OL AUTO-RESET INTERNAL MOTOR OVERLOAD TEMPERATURE SWITCH (N.C.)
PCB PRINTED CIRCUIT BOARD CONTROL
PL1 11-CIRCUIT PCB CONNECTOR
PL2 2-CIRCUIT CONNECTOR
PL3 2-CIRCUIT HSI, CONNECTOR
PL4 3-CIRCUIT IDM EXTENSION CONNECTOR
PRS PRESSURE SWITCH, SPST-(N.O.)
TEST/TWIN COMPONENT TEST & TWIN TERMINAL
TRAN TRANSFORMER-115VAC/24VAC
JUNCTION
UNMARKED TERMINAL
PCB CONTROL TERMINAL
FACTORY WIRING (115VAC)
FACTORY WIRING (24VAC)
FIELD WIRING (115VAC)
FIELD WIRING (24VAC)
CONDUCTOR ON CONTROL PCB
FIELD WIRING SCREW TERMINAL
FIELD EARTH GROUND
EQUIPMENT GROUND
FIELD SPLICE
PLUG RECEPTACLE

Figure 20-12 This diagram includes important information such as the circuit board connections and the diagram legend. The printed circuit board, or PCB, is outlined in red.

color is black for the hot side and white for neutral. Correct voltage polarity is important with a circuit board. Reversing the polarity will create unusual circuit operation if it works at all.

Draw a sketch of the circuit board and wire connections. Label the general position of the wire and color of the wires. This is a starting point for troubleshooting the circuit board. No one likes drawing and labeling a diagram, but it will be useful in the troubleshooting process if you do not know much about the circuit.

Next, determine the voltage outputs. You will need to trace the output wires to see what they control. Here are some tips:

- Is there a wire that controls the contactor?
- If so, is there 24 volts or some other control voltage leaving the circuit board?
- If there is 24 volts at the output of the circuit board, is there 24 volts at the contactor coil?
- If there is 24 volts at the contactor coil, will it be energized? The contacts will be closed unless the contactor coil is open or if the contactor is mechanically bound, thus stuck closed.

There is no guarantee that these steps will uncover the problem. But using the information in this section will locate many of the common problems if a circuit diagram is not available.

TECH TIP

The first step in circuit board troubleshooting is checking the input voltage polarity. Circuit boards that use 120 volts have one hot leg labeled **L1** and a neutral leg that has the same potential as ground. Most circuit boards require that this power be hooked up properly. To verify the correct voltage hook-up, use the voltmeter to measure this voltage. The **L1** connection to ground should measure 120 volts. The neutral connection to ground should measure zero volts. A good, tight system ground is also required. A reverse voltage hook-up or a poor ground will result in erratic or no circuit operation.

20.6 TROUBLESHOOTING A CIRCUIT BOARD USING A DIAGRAM

Having a diagram will make the troubleshooting process much easier. This section goes through the basic steps of troubleshooting when a diagram is available.

When troubleshooting, run through all of the pretests (ACT) because many problems are not circuit board related. Using much of the same procedure discussed next is recommended when you have the circuit board diagram and wiring hook-up diagram as shown in Figure 20-12. Use Figure 20-12 for the following troubleshooting example.

This is a diagram for a gas furnace that also controls the condensing unit. You can use it to figure out the basic operation of this circuit board design. Let's break it down to help you understand the major features and operation of the diagram:

- Study the legend at the bottom of the diagram. This will explain the many abbreviations and some of the symbols used in the diagram. The legend will also give you some tips on the various components found in the circuit. It may lead to troubleshooting strategies.
- Highlight the printed circuit board (PCB) outline, as shown in Figure 20-12 in red, so that you do not it confuse it with the other components in the diagram. It is not easy to differentiate the circuit board from the rest of the electrical diagram.
- The circuit board in this example has an **LED** problem code light. **LED 1** is found midway, on the far left side of the board. Now we will look at inputs and outputs that make this printed circuit board work.
- The primary input voltage, as pointed out by an arrow on the board, is **L1** (hot) and **NEUTRAL**. The use of **L1** and **NEUTRAL** (and not **L1** and **L2**) can be determined by looking at the lower right side of the diagram. **L1** and **NEUTRAL** make up a 120-V power source. (If the board had **L1** and **L2** it would be a 240-V power supply.)
- Continuing, **L1** is connected to the circuit board at **BLWR**, which is the abbreviation for the blower relay. **NEUTRAL** is connected on the diagram near the center, left side. On the circuit board measure 120 volts across **L1** and **NEUTRAL**. As discussed in the Tech Tip, verify that the hot and neutral are connected at the correct terminals. The color code is not always followed.
- Next, check the secondary voltage, 24 volts across board terminals **SEC-1** and **SEC-2**. These are the red and blue wires. If 24 volts is not present at **SEC-1** and **SEC-2**, check the voltage to the primary of the transformer. The primary voltage will be the black and white wires to the transformer.
- Assuming that all voltages are present, 120 and 24 volts, another quick check is the t'stat terminal strip, located at the upper left-hand side of the PCB. Using a jumper wire, go from **R** to **G**. The indoor blower should operate.
- Jumper from **R** to **W**; the heating system should operate. There will be a delay for the heating system since the **hot surface igniter (HSI)** will need to be energized prior to the gas valve opening. The burners should ignite in about 1 minute or less.
- Finally, jumper from **R** to **Y**; this will operate the condensing unit.

For a no-heat problem and an **HSI** that is not energizing, take the following steps:

1. Turn off the thermostat and connect the voltmeter across **HSI** at the black wire on **PL2** and the white wire connected to **L2**.
2. If you have the proper voltage (120 V) across the **HSI**, at the correct time in the ignition process, the igniter will start to glow.
3. If you do not have the correct voltage to the igniter, then you have to look closer at the control board or inputs to it.

In summary, these are major points that can be quickly checked. Review the circuit board outputs and determine other points than you can check.

The troubleshooting flowchart in Figure 20-13 is used in conjunction with the circuit board flashing lights. The

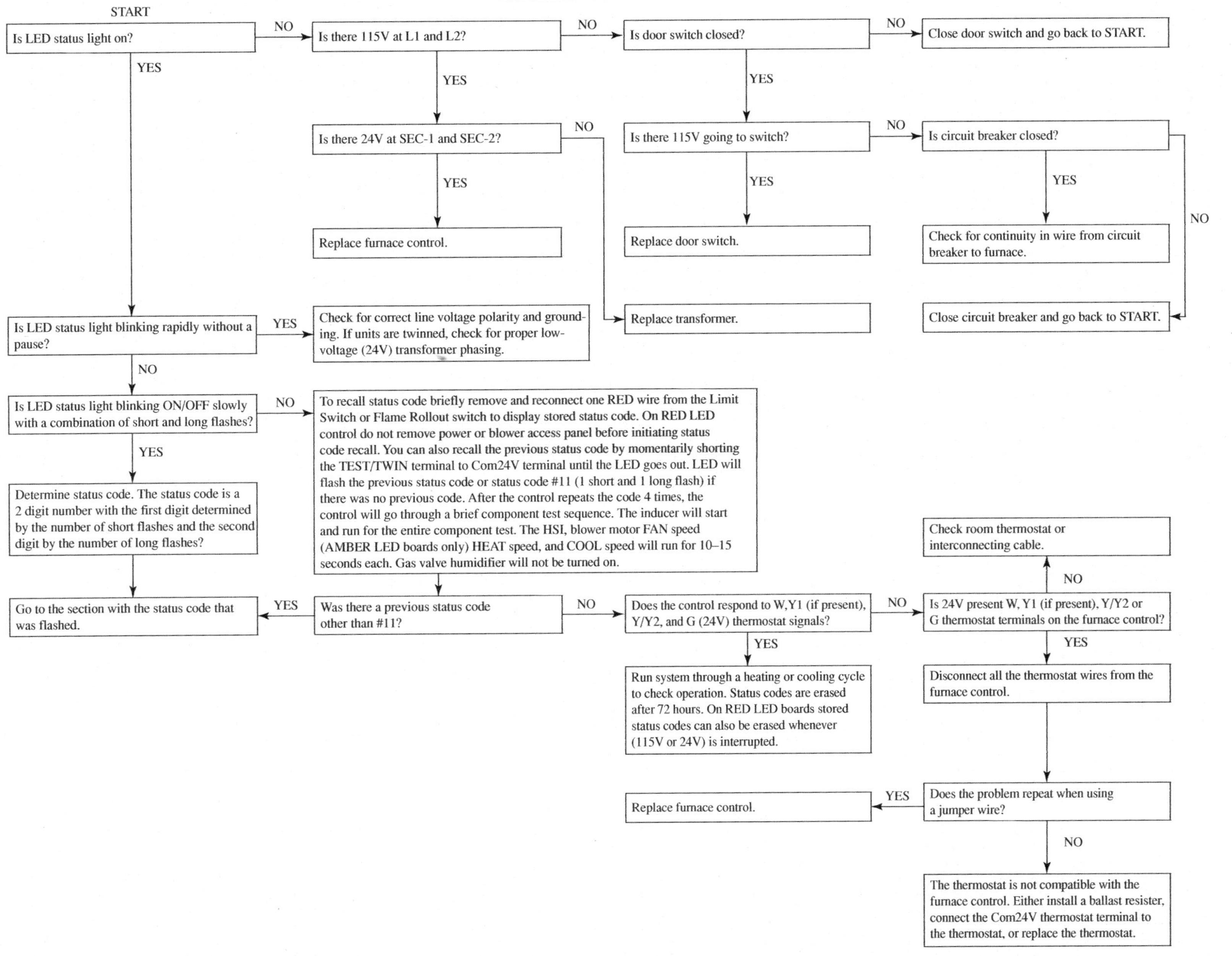

Figure 20-13 This troubleshooting flowchart relates to the electrical diagram shown in Figure 20-12. The flowchart starts at the upper left side, when viewed as a landscape (horizontal) image. Locate the word START and follow the troubleshooting sequence. This allows the technician to follow a logical sequence to track down the problem.

flowchart helps the technician follow a logical sequence to find a solution to a circuit problem. The chart can be used with or without flashing light indicators.

TECH TIP

Mark circuit board connections, plugs, and wires when you remove them from the circuit board. Use wire marker tape or black marker on quick-connect plugs. Most connection plugs can only be installed in one direction. This makes the installation of a replacement circuit board easier if you know the correct direction for reinstalling the connections. Mark other wires that are removed from the circuit board. This makes reinstallation of the same or a new board accurate and easy. Hooking up the wrong connections will cause the board not to operate properly and possibly damage the circuit board or other components.

20.7 MORE PRACTICE WITH CIRCUIT BOARDS

Figure 20-14 shows a gas furnace circuit board. Let's review this board since many other circuit boards have similar connections, although they may be arranged in a different configuration. We will go in a counterclockwise direction, starting at the upper left-hand side at the 24-V thermostat terminals. The 24-V volt terminals are the thermostat wires **G**, **Com**, **W**, **Y**, and **R**. Below this terminal block is a 3-A fuse holder. Below the fuse holder are the two 24-V transformer wires.

Next are the LED operation and diagnostic lights. The diagnostic light legend for this circuit board is found in Figure 20-15. Underneath the light is a power block with connections for the 120-V neutral wire, the blower and electric air cleaner connection **EAC-2**. To the right of the power block is a low-voltage harness connection.

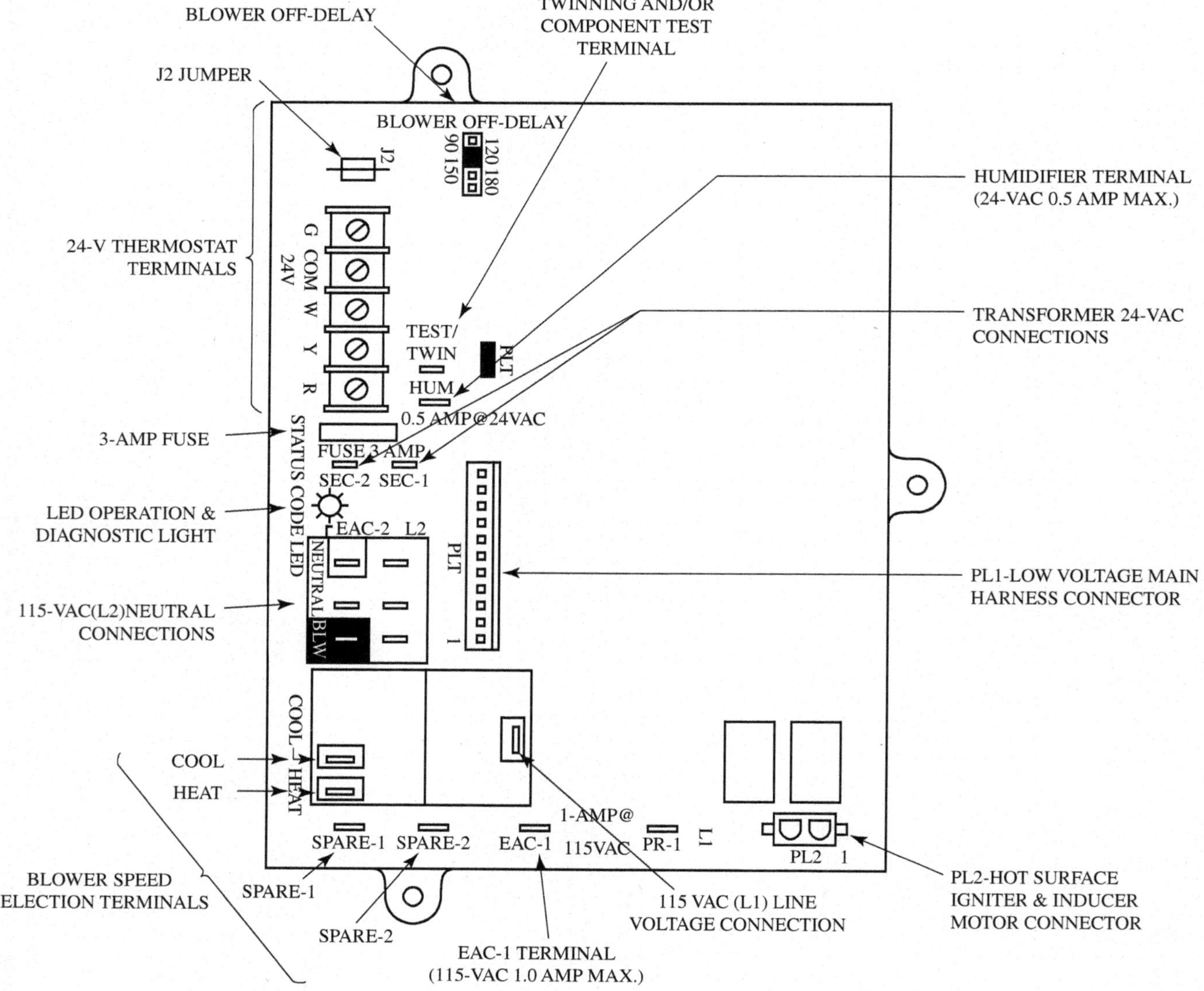

Figure 20-14 This is a circuit board for a gas furnace. Understanding a circuit board like this one will help you understand similar circuit boards manufactured by other furnace companies.

SERVICE

If status code recall is needed, briefly remove then reconnect one main limit wire to display stored status code. On RED LED boards do not remove power or blower door before initiating status code recall. After one status code recall is completed component test will occur.

LED CODE **STATUS**

CONTINUOUS OFF - Check for 115VAC at L1 & L2, & 24VAC at SEC-1 & SEC-2.

CONTINUOUS ON - Control has 24VAC power.

RAPID FLASHING - Line voltage (115VAC) polarity reversed. If twinned, refer to twinning kit instructions.

EACH OF THE FOLLOWING STATUS CODES IS A TWO DIGIT NUMBER WITH THE FIRST DIGIT DETERMINED BY THE NUMBER OF SHORT FLASHES AND THE SECOND DIGIT BY THE NUMBER OF LONG FLASHES.

11 NO PREVIOUS CODE - Stored status code is erased automatically after 72 hours. On RED LED boards stored status codes can also be erased when power (115VAC or 24VAC) to control is interrupted.

12 BLOWER ON AFTER POWER UP (115VAC or 24VAC) -Blower runs for 90 seconds, if unit is powered up during a call for heat (R-W closed) or R-W opens during blower on-delay.

13 LIMIT CIRCUIT LOCKOUT - Lockout occurs if the limit or flame rollout switch is open longer than 3 minutes.
- Control will auto reset after three hours. - Refer to #33.

14 IGNITION LOCKOUT - Control will auto-reset after three hours. Refer to #34.

21 GAS HEATING LOCKOUT - Control will NOT auto reset.
Check for: - Mis-wired gas valve -Defective control (valve relay)

22 ABNORMAL FLAME-PROVING SIGNAL - Flame is proved while gas valve is de-energized. Inducer will run until fault is cleared. Check for:
- Leaky gas valve - Stuck-open gas valve

23 PRESSURE SWITCH DID NOT OPEN Check for:
- Obstructed pressure tubing. - Pressure switch stuck closed.

24 SECONDARY VOLTAGE FUSE IS OPEN Check for:
- Short circuit in secondary voltage (24VAC) wiring.

31 PRESSURE SWITCH DID NOT CLOSE OR REOPENED - If open longer than five minutes, inducer shuts off for 15 minutes before retry. Check for:
- Excessive wind - Proper vent sizing - Defective inducer motor
- Low inducer voltage (115VAC) - Defective pressure switch
- Inadequate combustion air supply - Restricted vent
- Disconnected or obstructed pressure tubing
- Low inlet gas pressure (if LGPS used)
If it opens during blower on-delay period, blower will come on for the selected blower off-delay.

33 LIMIT CIRCUIT FAULT - Indicates a limit, or flame rollout is open. Blower will run for 4 minutes or until open switch remakes whichever is longer. If open longer than 3 minutes, code changes to lockout #13. If open less than 3 minutes status code #33 continues to flash until blower shuts off. Flame rollout switch requires manual reset. Check for: - Restricted vent
- Proper vent sizing - Loose blower wheel - Excessive wind
- Dirty filter or restricted duct system.
- Defective blower motor or capacitor. - Defective switch or connections.
- Inadequate combustion air supply (Flame Roll-out Switch open).

34 IGNITION PROVING FAILURE - Control will try three more times before lockout #14 occurs. If flame signal lost during blower on-delay period, blower will come on for the selected blower off-delay. Check for: - Control ground continuity
- Flame sensor must not be grounded
- Oxide buildup on flame sensor (clean with fine steel wool).
- Proper flame sense microamps (.5 microamps D.C. min., 40.0 - 6.0 nominal).
- Gas valve defective or gas valve turned off - Manual valve shut-off
- Defective Hot Surface Ignitor
- Low inlet gas pressure
- Inadequate flame carryover or rough ignition
- Green/Yellow wire MUST be connected to furnace sheet metal

45 CONTROL CIRCUITRY LOCKOUT Auto-reset after one hour lockout due to;
- Gas valve relay stuck open - Flame sense circuit failure
- Software check error
Reset power to clear lockout. Replace control if status code repeats.

COMPONENT TEST

To initiate the component test sequence, shut OFF room thermostat or disconnect the "R" thermostat lead. Briefly short the TEST/TWIN terminal to the "Com 24V" terminal. Status LED will flash code and then turn ON the inducer motor. The inducer motor will run for the entire component test. The hot surface ignitor, blower motor FAN speed (AMBER LED boards only) blower motor HEAT speed, and blower motor COOL speed will be turned ON for 10-15 seconds each. Gas Valve and Humidifier will not be turned on. **327884-101 REV. B**

Figure 20-15 This is the diagnostic legend for the circuit board diagram shown in Figure 20-14. This will help the tech zero in on the problem. Not all diagnostic legends are the same, even those from the same manufacturer.

Under the power block are the blower speed terminals for heating and cooling operations. The operation of the blower in the heating mode is slower than in the cooling mode. The blower in the heating mode will have about a 1-minute delay on start-up, allowing the furnace to warm up the heat exchanger. This is a programmed delay on shutdown that allows the blower to remove heat from the heat exchanger after it has completed its heating cycle. To the right of the cool-heat hook-up is the 115-V hot connection. Beneath this is the **EAC-1** connection.

At the bottom right corner of the board is the hot surface igniter and inducer motor connector plug. At the top center portion of the board is the blower off-delay switch. This gives the tech the option of setting the blower off delay to 90, 120, 150, or 180 seconds. The blower off delay will allow the heated or cooled air to be blown out of the heat exchanger and duct system.

In Figure 20-15 the LED code or flash legend assists the technician in troubleshooting the gas furnace, which includes the heating system and blower system. If the LED light is off, check the high voltage or control voltage. If the light is continuously on, the system is operating correctly. A rapidly flashing light means the voltage polarity voltage is reversed.

As Figure 20-15 states the first digit of the trouble code will determine the number of short flashes. The second digit is determined by the number of long flashes. Use of a circuit board with diagnostics and an LED code legend will speed up the troubleshooting process. The LED is not always correct, but in most cases it will lead you in the correct direction to solve the problem.

SERVICE TICKET

As an entry-level technician, you are sent out to do a "clean and check" service on a gas furnace. This includes checking the filter, ductwork, furnace safeties, blower motor, vent pipe, and heat exchanger. The furnace will be cycled a couple of times while observing ignition and the flame level. You do this because you cannot do a clean and check on a unit that is not working correctly. As a new technician you were trained to do these procedures.

After greeting the customer, the first thing you were taught to do was to turn the furnace on to make sure there are no operating problems. When doing preventive maintenance it is a good idea to make sure the equipment is not broken. If you do your clean and check without knowing whether the equipment is operating correctly, you will be blamed for any problems that arise. Start off with a correctly functioning piece of equipment.

In this service call you find out that the furnace is not operating. The furnace is a newer model heater with a circuit board. The LED troubleshooting legend located on the panel door has been damaged by water. It seems the evaporator condensate had leaked down through the furnace cabinet and washed most of the print off the legend.

The board has two LEDs and numerous inputs and outputs. The power LED is glowing, but the troubleshooting LED is not. You are not familiar with this brand of board, but decide to do some quick checks prior to calling the dispatcher for help. After a 5-minute inspection everything appears to be okay. Even the fuse on the board appears good.

Not wanting to look dumb and not wanting to call the dispatcher, you check the apparently good fuse. You pull out the ohmmeter and zero it to make sure that the meter and probes are operational. The fuse reads infinity, which means that it is open. It is a common 5-A fuse circuit board that you have on the truck. Prior to replacing the fuse, you contact the customer and dispatcher about your plan. The customer needs to approve this procedure since he will be billed for the fuse. If the fuse does not solve the problem, the customer is billed for a troubleshooting call. Fortunately, after changing the fuse the system operates okay and you proceed with the scheduled clean and check appointment.

The point of this Service Ticket discussion is that some problems are simple and can be quickly fixed. The customer needs to know what is going on so that the technician does not take the blame for something that is not her fault. Communication with the customer and office is paramount and always a good customer service.

SUMMARY

This unit provided an introduction to understanding circuit boards. You learned that circuit boards may seem complicated but with a little investigation they can be understood. Even without the wiring diagram, several quick checks can help you determine if the problem is circuit board related or if the problem is with the inputs to the circuit board. For example, most circuit boards have low-voltage inputs and high-voltage inputs.

Measure the low-voltage input, which is usually 24 volts. The 24 volts can be supplied from a transformer that is located on the circuit board or supplied from an external transformer. The high-voltage inputs can be 120 or 240 volts. The 120-V circuit is polarized, meaning that it is important to hook up the hot and neutral wires to the proper connections. A good ground is required for any voltage. Improperly wired hot/neutral voltages or no or poor ground connections will make the circuit board operate erratically and sometimes not at all.

Check the circuit board's fuse. Other inputs you can check are pressure switches and other safety devices. Check labeled outputs on the circuit board. Again, these are checks you can do without the aid of a diagram or the flashing light error code table.

Having the wiring diagram and the flash code legend makes the troubleshooting process useful and logical. Use these resources before calling for help. Your supervisor or the manufacturer's tech support person will want to know what you have done. This is a professional practice. Record your troubleshooting findings on paper prior to making the call for assistance. This will help you understand what is happening and also assist your support person. With this documented information, tech support can make helpful suggestions.

REVIEW QUESTIONS

1. What are the advantages of a circuit board?
2. What components might you find on a circuit board?
3. Name the safe handling practices for working with circuit boards.
4. What is a quick way to check a circuit board?
5. Looking back at Figure 20-3, what contacts on the circuit will close when the timing circuit is complete (5 minutes)?
6. Again looking back at Figure 20-3, the technician is troubleshooting the time-delay circuit board. After 5 minutes, 24 volts is measured across **T1** and **T2**. Is there a problem? If so, what is it?
7. Use Figure 20-15 to answer this question: The furnace is not operating and is flashing a trouble code of one short flash and three long flashes. What is the problem? What caused the problem?
8. Use Figure 20-15 to answer this question: The furnace is not operating and is flashing a trouble code of two short flashes and four long flashes. What is the problem? What caused the problem?
9. What is the purpose of the dip switches shown in Figure 20-11?
10. What can happen if 120-V power is reversed?
11. Why is it important to have a good equipment ground?
12. Review the diagram shown in Figure 20-16. List five features of the board that were discussed in this unit.
13. Use Figure 20-16 to answer this question: You wish to start the furnace without going to the thermostat to set it to heat and adjust the temperature to make the furnace operate. How would you do this?

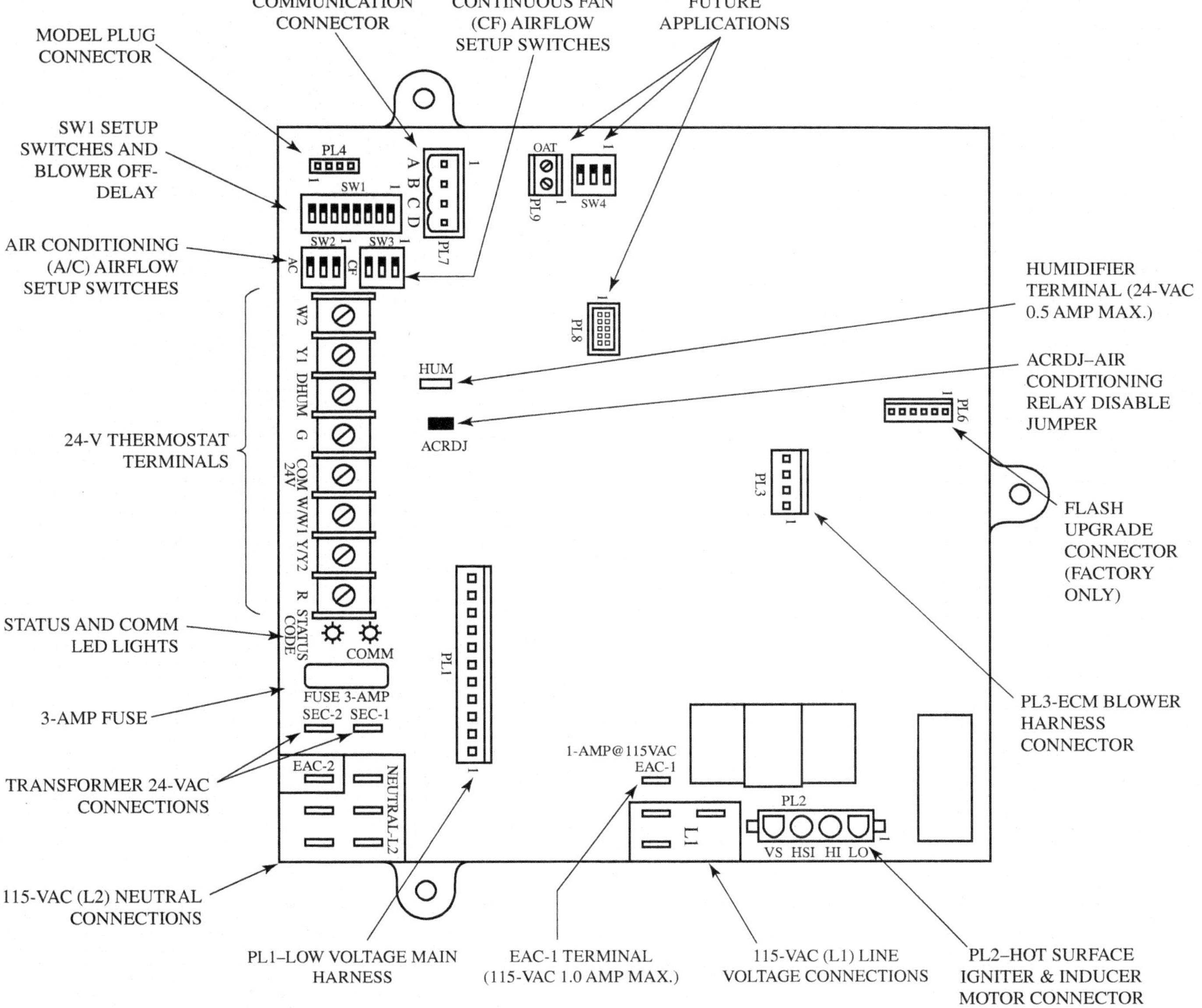

Figure 20-16 Use this diagram to answer Review Questions 12 and 13.

UNIT 21

Air Conditioning Systems

WHAT YOU NEED TO KNOW

After studying this unit, you will be able to:

1. Interpret a window unit air conditioning diagram.
2. Interpret a basic residential air conditioning diagram.
3. Interpret a basic commercial unit air conditioning diagram.
4. Trace the sequence of operation of an air conditioning diagram.
5. Draw an electrical diagram from various air conditioning electrical components.

The purpose of this unit is to pull together and apply many facets of the material learned in the previous units. This is a "**systems**" unit. In other words, we have been learning about parts, pieces, and subsections, but not the total system operation. This unit discusses basic residential and commercial air conditioning and refrigeration operation. We will be tracing the sequence of operation of several electrical diagrams. We first discuss a very fundamental air conditioning system. Then we discuss a window unit air conditioner and finish off with discussions of residential and commercial systems. Let's get started!

21.1 BASIC AIR CONDITIONING SYSTEMS

Before we go through the sequence of operation of a basic air conditioning system, let's briefly identify the main components, power supply features, and their abbreviations. *Note:* Here and throughout this unit, component and coil abbreviations are circled on diagrams; abbreviations for switches are not circled.

Power supplies, **L1** and **L2**

Contactor contacts, **CC**

Compressor, **COMP**

Condenser fan motor, **COND**

Blower relay contacts, **BR**

Blower motor, **BM**

Transformer, primary, **230 V**

Transformer, secondary, **24 V**

Room thermostats, **R**, **Y**, and **G**

Contactor coil, **CC**

Blower relay coil, **BR**

Note: The starting and stopping sequence occurs with the three major components (**COMP**, **COND**, and **BM**) operating in unison. Some designs have the condenser section coming on first, then the blower motor a little later. This design allows for chilling of the evaporator prior to starting the airflow in the ducts. Another design has the compressor operating first, followed by a delay where the condenser fan and blower operate after a 1-minute time delay.

TECH TIP

A good way to trace an electrical diagram is to use a page protector or overhead transparency film taped over the diagram. Use an erasable marker to trace out the diagram. Using different color markers will help differentiate between the control and high-voltage sections or between the numerous circuits in the diagram.

21.2 WINDOW AIR CONDITIONERS

Window air conditioners or **window units** are used to cool a room or two rather than a whole home or area. As the name suggestions, window units are usually installed in a window. In some cases a section of wall is cut out and the window unit is installed in that manner. Window units are primarily used for cooling, but they may be purchased with heating options such as electric heat strips or a heat pump. This section will talk about the air conditioning side of a window unit.

Figure 21-2 shows the basic components of a split air conditioning system. It will have high and low voltage sections. A window unit will have only high voltage, 120V or 240V. The window unit is a very simple air conditioning system with the following basic components as illustrated in Figure 21-2:

- Compressor
- Overload
- Fan motor (combined condenser fan and indoor blower, double shaft)
- Two run capacitors
- Thermostat
- Switch
- Power cord.

In this example the power switch has four terminals. They are 7, 2, 3, and 4. Power is supplied to terminal 7 from the power cord. The switch has three sets of contacts. Depending on the switch position, the contacts will be open or closed. The switch legend is found to the right side of

SEQUENCE OF OPERATION 21.1: BASIC SYSTEM

Figure 21-1 will be used to trace the sequence of operations for a basic air conditioning system. It is a good practice to start tracing a circuit at the control voltage section, which in this case is the lower part of the diagram and is operated by 24 volts.

1. At point **1**, the thermostat is closed, creating a path between **R** and **Y** and between **R** and **G**.
2. The 24 volts will energize the contactor coil (CC) and blower relay coil (BR) located at point **2**.
3. When contactor coil (CC) is energized it will close the single-pole contactor contacts, **CC**, found at point **3**, to the right of **L1**.
4. The compressor and condenser fan motor (COMP) and (COND) at point **4** will begin to operate.
5. When blower relay coil (BR) is energized, it will close the **BR** relay contacts at point **5**. Blower motor (BM) will begin to operate. When the temperature around the room thermostat is satisfied, it will open the connection between **R** and **Y** and between **R** and **G**, stopping the cooling operation.

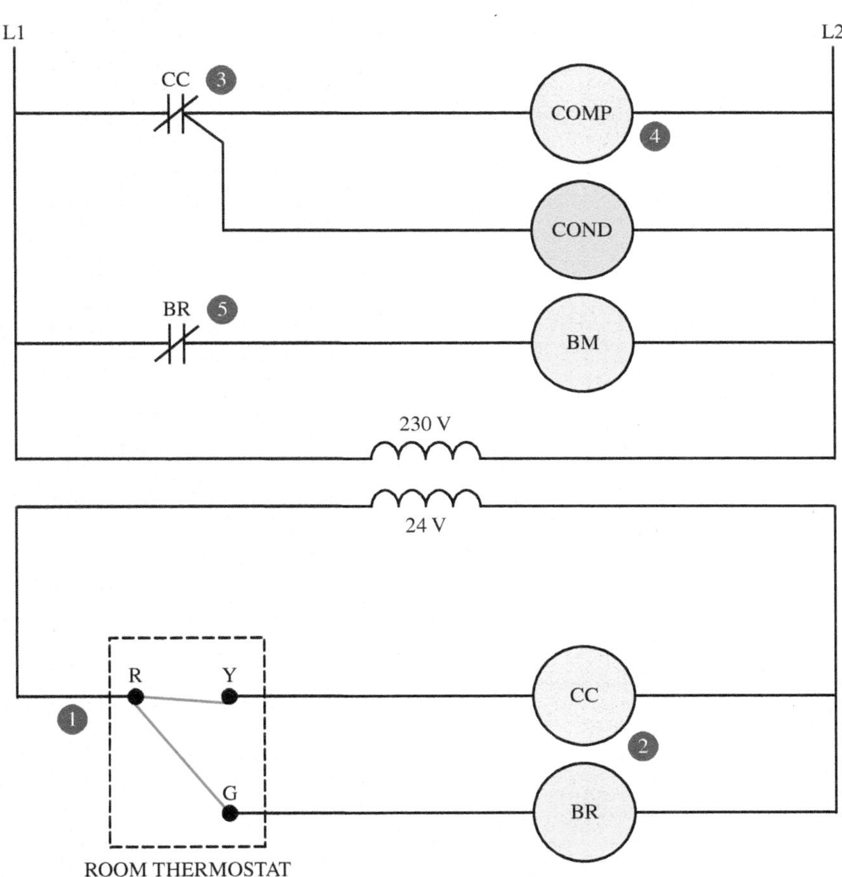

Figure 21-1 This is a basic electrical diagram of an air conditioning system.

Figure 21-2. In the legend the letter "O" means open and the letter "C" means closed. If you select the *Normal Cool* operation contacts from 7 to 2 and from 7 to 3 will be closed. If you look at the switch options and what they control, you will notice that contact 2 goes to the thermostat switch and controls the compressor operation. Switch terminals 3 and 4 go the normal speed and high speed of the motor. At no time will terminals 3 and 4 be closed at the same time. If this should occur because of miswiring or a defective switch, both normal and high-speed motor windings will be energized. Energizing both speeds simultaneously will burn out the fan motor.

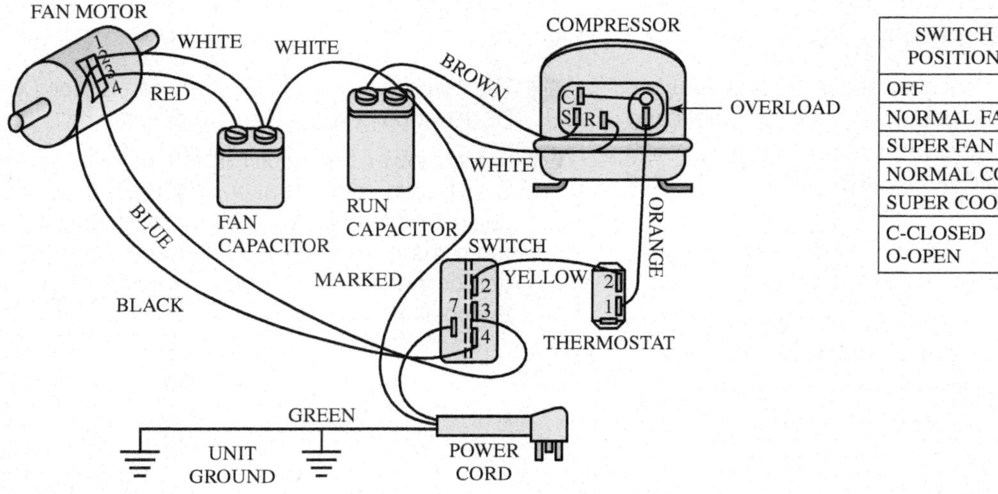

SWITCH POSITION	CONTACTS		
	2	3	4
OFF	O	O	O
NORMAL FAN	O	C	O
SUPER FAN	O	O	C
NORMAL COOL	C	C	O
SUPER COOL	C	O	C
C-CLOSED O-OPEN			

Figure 21-2 The common components of a window unit are illustrated here. Manufacturer's diagrams are not usually shown in this manner. This diagram helps you understand the wiring and connection sequence of the actual components.

SEQUENCE OF OPERATION 21.2: WINDOW UNIT COMPRESSOR OPERATION

We will use Figure 21-3 to track the sequence of operation of the window unit discussed in the previous section.

Select the *Super Cool* position, which will operate the compressor when the thermostat calls for cooling and the fan at high speed. The path highlighted in yellow covers the compressor operation.

1. Start at point **1** where the power enters at the switch at terminal **7** and goes to the switch at terminal **2**.

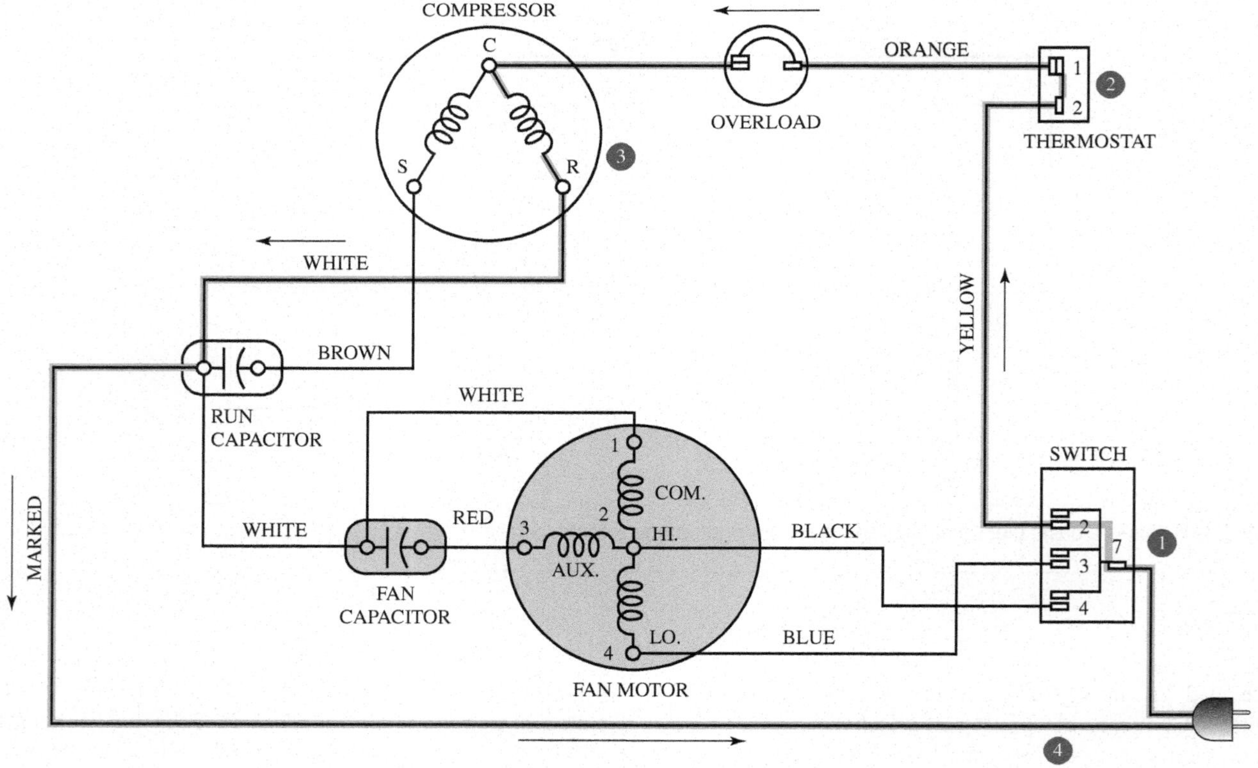

Figure 21-3 This window unit diagram shows the compressor operating sequence briefly after the compressor is started. The start winding is not highlighted because there is little current flow through the start winding after the motor starts.

2. At point **2** on the diagram, terminal **2** on the thermostat, the power goes through the closed thermostat at terminal **1** to the compressor overload.
3. Next, power goes through the external overload and to the common connection **C** on the compressor (point **3**).
4. To complete the power circuit, electrons flow through the run and start windings, returning to the other side of the power supply (point **4**).

Note: The start winding is not shown energized or highlighted because it will have little electron flow after the initial charge of the run capacitor. The start winding did its job by creating a phase shift on start-up, which generates starting torque to get the motor moving. It is not needed after starting. The run cap charges and discharges as the 60-Hz voltage sine wave rises and falls, but the electron flow is minimal once the first electron charge is complete.

SEQUENCE OF OPERATION 21.3: COMBINATION CONDENSER FAN AND BLOWER MOTOR

Now we explore the sequence of operation of the combination condenser fan and blower motor. Figure 21-4 highlights the window unit's fan operation.

1. As in Sequence of Operation 21.2, power is provided from the power cord through terminals **7** and **4** of the switch (point **1**).
2. Electrons flow through the run winding (point **2**), which is labeled **2** and **1** inside the fan motor. The motor will operate on high speed.
3. From run winding **1** inside the fan motor, the power completes the cycle around the left side of both caps and back to the other side of the power cord (point **3**), creating a complete circuit.

Note: As in Sequence of Operation 21.2, the start winding (**AUX**) is not shown as energized because it is not needed once the fan motor starts.

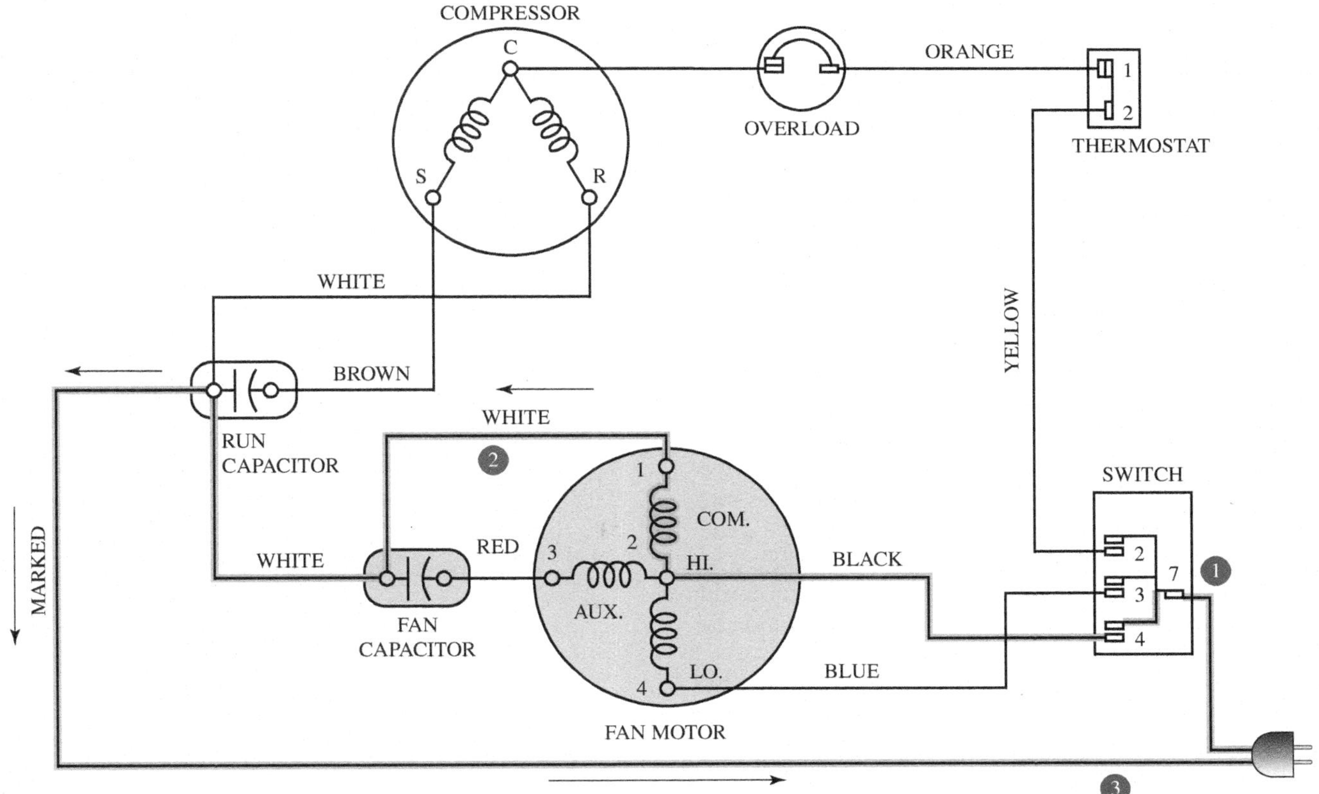

Figure 21-4 This window unit diagram shows the high-speed operation of the fan motor. The **AUX** winding is the start winding. The **AUX** winding is essentially out of the circuit once it helps start the motor and charge up the fan cap.

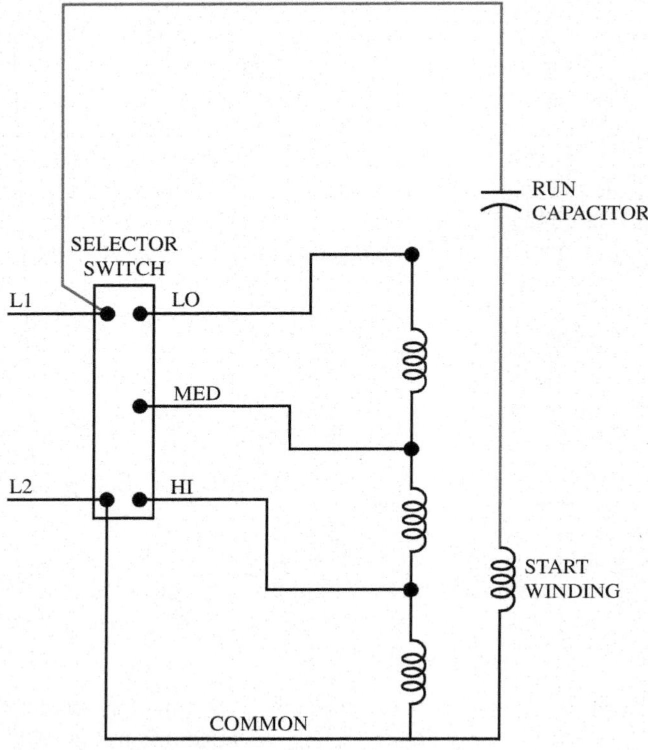

Figure 21-5 This multispeed motor diagram has been extracted from a window unit diagram. Notice that the run winding changes resistance as the speed changes. The high-speed winding has the lowest resistance.

21.3 MULTISPEED FAN MOTORS

In the previous examples we looked at a two-speed fan motor. **Multispeed fan motors** are commonly found on window air conditioning units. This section examines a three-speed fan motor that you might find on a window unit.

Figure 21-5 is a diagram of a multispeed motor and selector switch. You will notice that changing the number of run winding in series with each other causes the motor speed to change. The greater the run winding resistance, the slower the motor speed. The highest speed has the lowest run winding resistance. If the low- or medium-speed winding burns open, the high-speed winding will continue to work. Let's investigate this operation some more.

L1 is the switching power leg. The **L2** power leg stays connected to the common tie point between the high-speed start winding and the start winding. The run winding is attached in parallel with the start winding and in parallel with the run cap. This is the normal run winding installation for any single-phase motor that uses a run cap.

L1 will be switched to either the **LO**, **MED**, or **HI** speed connection. There is no universal motor wiring color code. The wires going to the run cap are usually brown; sometimes, however, one wire is solid brown and the other is brown with a white strip running its length. If you look at the diagram, the brown wires (not labeled) will be coming from the top side of the start winding and from **L1**.

The colors of the speed wires are left up to the motor manufacturer. There is a consistent color code among model numbers from the same manufacturer.

21.4 POWER FOR WINDOW UNITS

The receptacle and plug for a window unit will determine the amperage and voltage requirements for those components. Figure 21-6 shows a chart that will help you select the wall receptacle required for the factory-installed plug on the window unit.

You must use NEC Table 310.15(B)(16) (see Table 6-4 in Unit 6, *National Electrical Code*) to select the wire size for the required amperage. For example, if the window unit you are installing has a 20-A requirement at 125 volts, you will need to install the receptacle that is shown second from the top in Figure 21-6. NEC Table 310.15(B)(16) requires that a 12-gauge wire be installed from the breaker or

RECEPTACLE CONFIGURATION	RATING
	15 A 125 V
	20 A 125 V
	30 A 125 V
	50 A 125 V
	15 A 250 V
	20 A 250 V
	30 A 250 V
	50 A 250 V
	15 A 277 V
	20 A 277 V
	30 A 277 V
	50 A 277 V

Figure 21-6 This chart shows the current and voltage requirements for the plug and receptacle configurations found on window units.

SEQUENCE OF OPERATION 21.4: PACKAGE AIR CONDITIONING UNIT

We will trace the sequence of operation for a package air conditioning unit using Figure 21-7. (For a larger view of this figure, please see Electrical Diagram ED-8, which appears with the Electrical Diagrams package that accompanies this book.)

1. Power (240 volts) is supplied through **L1** and **L2** to the left and right side of the ladder diagram near the top (points **1**).
2. The primary of the transformer (point **2**) is energized, providing 24 volts to the control voltage section located at the bottom of the diagram.
3. When the thermostat switch is closed (point **3**), it will provide power to the **G** and **Y** circuits.
4. Power to **Y** completes the circuit to the (CR) coil (point **4**) and to the right side of the transformer (point **2**).

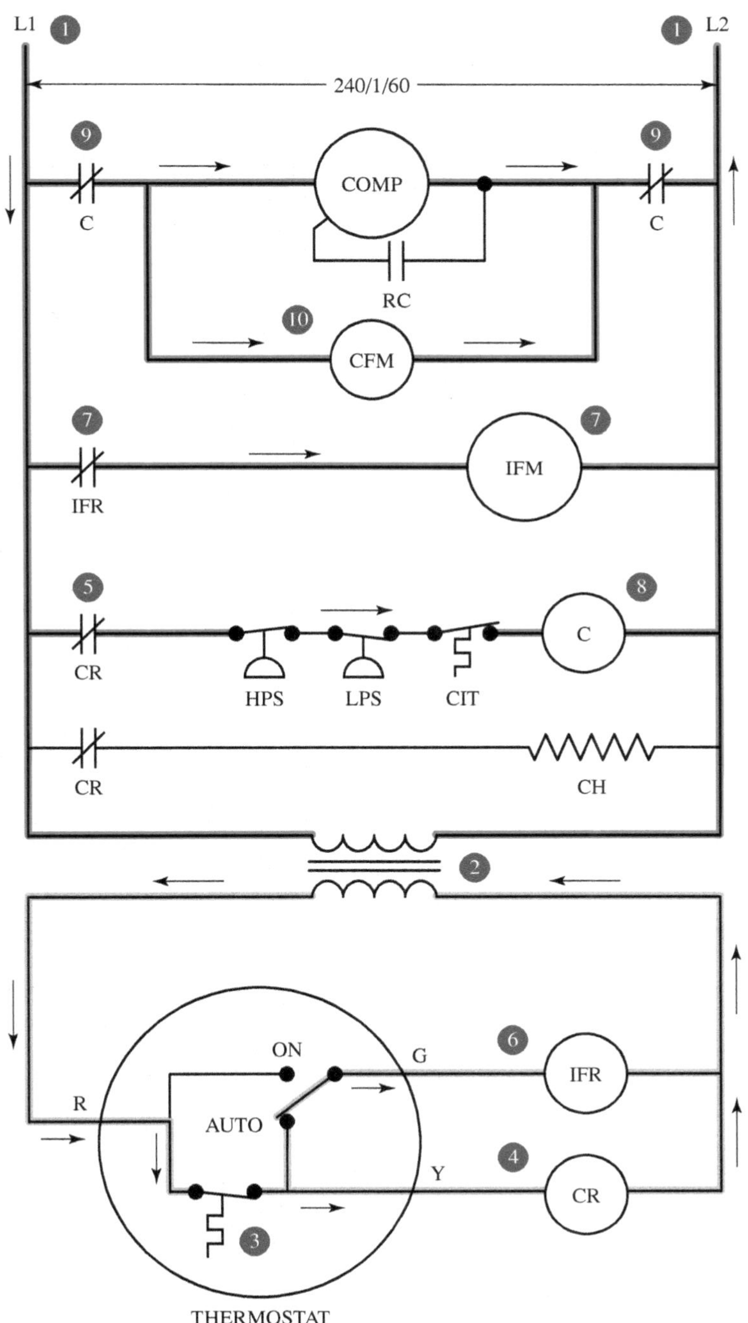

Figure 21-7 This package unit shows the control voltage circuit highlighted in yellow and the high-voltage circuit in red.

(continued)

5 This will close the **CR** contacts at point **5**. The lower normally closed **CR** contacts will open. This will de-energize the crankcase heater **CH**.

6 At point **6**, power from **G** completes the circuit to the (IFR) coil and to the right side of the transformer.

7 This will close the **IFR** contacts and energize the indoor fan motor, (IFM), at points **7**.

8 When **CR** contacts close, power from **L1** to **L2** is available to the contactor coil Ⓒ. Power goes through the closed **CR** contacts at point **5**, the high-pressure switch **HPS**, the low-pressure switch **LPS**, the compressor internal thermostat **CIT**, and to the contactor coil Ⓒ, as shown in point **8**.

Continuing, the red highlighting shows the start of the compressor and condenser fan motor.

9 The double-pole contactor Ⓒ (points **9**) closes when the contactor coil Ⓒ is energized. This switching action supplies power from **L1** and **L2** to the upper circuits.

10 Both the compressor and condenser fan motor begin to operate (see point **10**).

Notice that the compressor has a run capacitor. You learned in the motor unit that this is a permanent split capacitor (PSC) motor. The condenser fan motor does not show a capacitor, therefore it should be a simple shade pole motor.

fuse box to that outlet. If larger gauge wire (AWG), which is smaller wire, is installed to supply this outlet, it may overheat and cause a fire. Even if this does not occur, the voltage drop in an undersized conductor will reduce the amount of voltage supplied to the receptacle. Lower supply voltage equals higher current draw and overheating of the appliance itself.

Even though we are not classified as electricians, we are held responsible for wiring safety. There is an old saying in our trade that says, "The good technician knows the HVACR profession as well as customer service, carpentry, plumbing, and electrical." All of these trades are part of a well-rounded tech's skills—we are expected to wear many hats!

21.5 PACKAGE AIR CONDITIONING SYSTEMS

Next, we trace the electrical diagram for a package air conditioning unit. A **package unit** is an air conditioning system that has all of the major components in one "box." A packaged air conditioning system includes all of the major refrigeration components such as the compressor, condenser, meter device, and evaporator. It includes all of the major electrical components, motors, and controls except the thermostat. To get a package unit operating, you need the power supply connections, thermostat, and ductwork. Ductwork is optional, but is used in most installations.

At this time all major components are in operation. We have broken down the sequence of operation to help you better understand the operation of this package unit. In reality, this diagram has the compressor, condenser fan, and indoor fan coming on simultaneously. There are some equipment designs that delay the indoor fan operation to allow the coil to chill for about a minute. The condenser fan may be designed to energize before or after the compressor begins operation.

21.6 SPLIT AIR CONDITIONING SYSTEMS

A **split system** is an air conditioning system that has two major parts: the condensing unit and the blower/evaporator section. The condensing unit contains the following major components:

- Compressor
- Condenser fan motor
- Condenser coil
- Various electrical components such as a contactor, capacitors, and circuit board.

The blower and evaporator section is comprised of these major components:

- Blower motor
- Evaporator coil
- Metering device
- Various electrical components such as a relay, capacitor, and circuit board.

The thermostat is used to control and operate the major components.

This section discusses the operation of the condensing unit and control voltage section. We will use Figure 21-8 for this example. (For a larger view of this figure, please see Electrical Diagram ED-9, which appears with the Electrical Diagrams package that accompanies this book.) Notice the connection diagram on the left-hand side and the schematic diagram (ladder diagram) on the right half. Figure 21-9 shows the legend for these diagrams.

Before starting the sequence of operations, let's look at what is energized prior to closing the thermostat. The external 24-V power supply (at the bottom of the schematic diagram) will be energized. This transformer is usually installed in the air handler or furnace section of the system.

The crankcase heater **CH**, in the upper left side of the connection diagram, is energized during the off cycle provided the

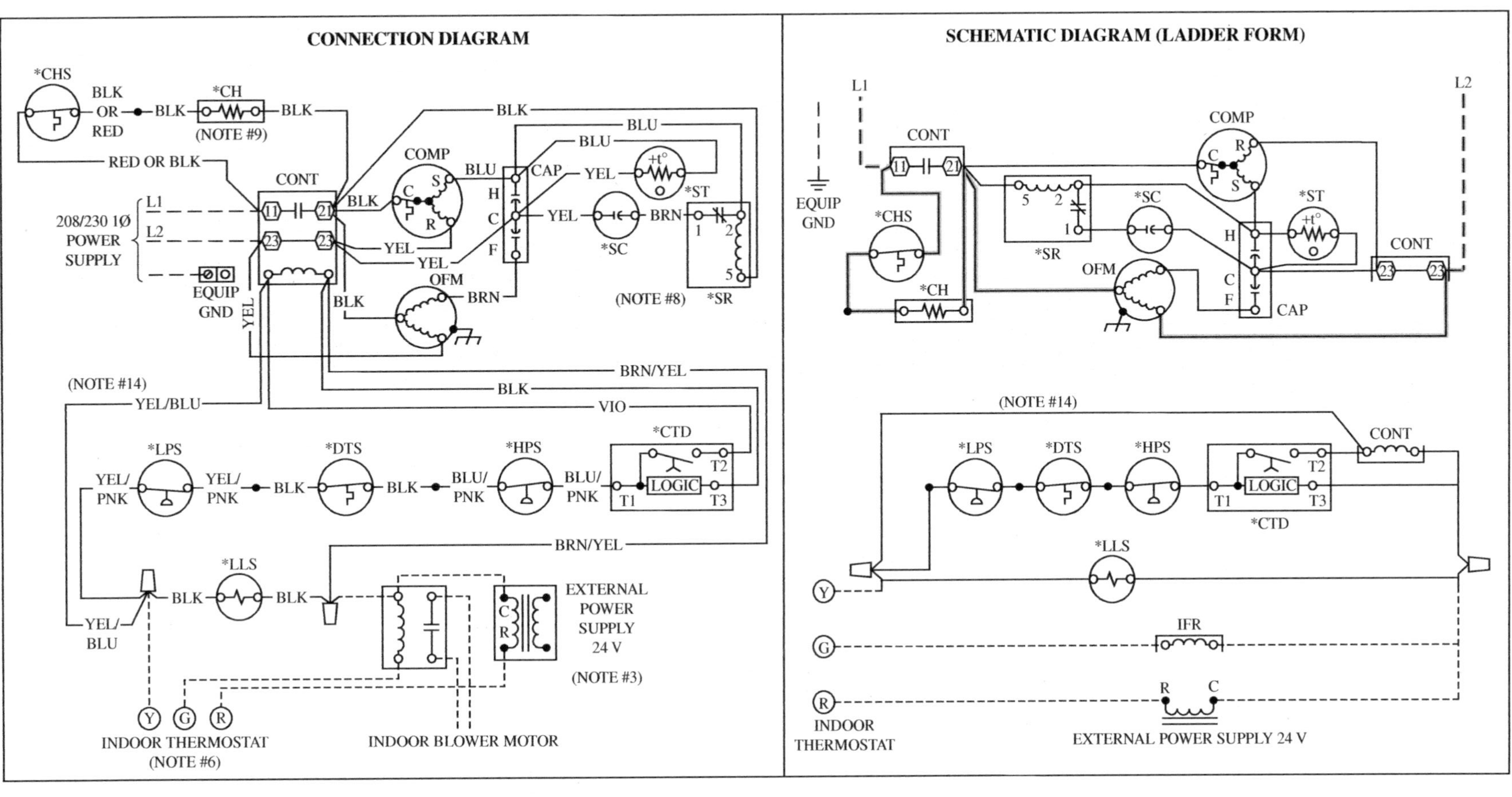

Figure 21-8 These are the connection and schematic (ladder) diagrams for a split system. Both diagrams can be useful when tracing out a diagram. You will notice that the connection diagram has more specific information like the color of wiring. The schematic diagram is designed to be used in troubleshooting.

LEGEND

———	FACTORY POWER WIRING
———	FACTORY CONTROL WIRING
- - - - ·	FIELD CONTROL WIRING
- - - - ·	FIELD POWER WIRING
o	COMPONENT CONNECTION
	FIELD SPLICE
—●—	JUNCTION
CONT	CONTACTOR
CAP	CAPACITOR (DUAL RUN)
*CH	CRANKCASE HEATER
*CHS	CRANKCASE HEATER SWITCH
COMP	COMPRESSOR
*CTD	COMPRESSOR TIME DELAY
*DTS	DISCHARGE TEMP SWITCH
*HPS	HIGH PRESSURE SWITCH
IFR	INDOOR FAN RELAY
*LLS	LIQ LINE SOLENOID VALVE
*LPS	LOW PRESSURE SWITCH
OFM	OUTDOOR FAN MOTOR
*SC	START CAPACITOR
*SR	START RELAY
*ST	START THERMISTOR

*** MAY BE FACTORY INSTALLED**

Figure 21-9 This legend is used to help the tech navigate through the electrical diagram shown in Figure 21-8.

power is on and the crankcase heater switch **CHS** is closed. The crankcase heater path is highlighted in yellow:

- **L1** through a closed **CHS**
- Through the crankcase heater
- To the right side of open contacts at connection point **21**
- Through the run winding of **OFM**
- Completing the path to **L2**.

Finally, you may have noticed the optional ***ST** or start thermistor. The start thermistor is not usually used in conjunction with a start relay. It is a solid-state relay that may be used without the **SR** and **SC** start components. It would be unnecessary to have both starting components, but you will find all kinds of variations in HVACR controls.

21.7 COMMERCIAL AIR CONDITIONING SYSTEMS

The commercial air conditioning system we explore now is a basic three-phase air conditioning system (see Electrical Diagram ED-2, which appears with the Electrical Diagrams package that accompanies this book). This system could be as small as a couple of tons or as large as 10 tons or more. The system we consider has an 8.5 FLA compressor, so it is a low-tonnage system. This system does not have a condenser fan, so it will have a water-cooled condenser or a remote air-cooled condenser. A water-cooled condenser, not shown, will have a supplemental wiring diagram showing the condenser water pump operation. We now explore the sequence of operation for our basic three-phase unit starting with the operation of the evaporator fan motor.

SEQUENCE OF OPERATION 21.5: CONTROL VOLTAGE

First we will trace the control voltage in the schematic diagram shown in Figure 21-10.

Start at the **R** or hot connection on the 24-V transformer and follow the highlighted circuit.

1. At point **1** on the diagram, power goes from the **R** connection on the transformer through the **R** connection on the t'stat, which goes through the closed thermostat to the **Y** connection. This will feed 24-V power to the control components.
2. At point **2**, power continues through the **R** to **Y** circuit in the thermostat to two parallel control circuits.
3. At point **3**, the lower parallel circuit from the **Y** connection supplies 24 volts to open the liquid solenoid valve (**LLS**). When the solenoid valve opens, liquid refrigerant is supplied to the metering device. The low-pressure switch will close after the refrigerant pressure rises above the **LPS** setpoint.
4. At the same time, at point **4**, the upper parallel control circuit from the **Y** terminal has 24 volts going through optional components such as the low-pressure switch (**LPS**), discharge thermostat switch (**DTS**), and high-pressure switch (**HPS**). An asterisk (*) before a component means that these are optional components and will not be found in many models. They can be field installed. If these optional components are not installed, the line above that will feed 24 volts directly to the contactor coil, CONT.
5. Next, at point **5**, power goes through the logic timer. The logic timer is a 5-minute time delay that prevents compressor short cycling. The contacts between **T1** and **T2** close after the time delay is over.
6. After voltage passes through the logic timer, the contactor will be energized as it completes the circuit to the common (**C**) side of the transformer at point **6**.

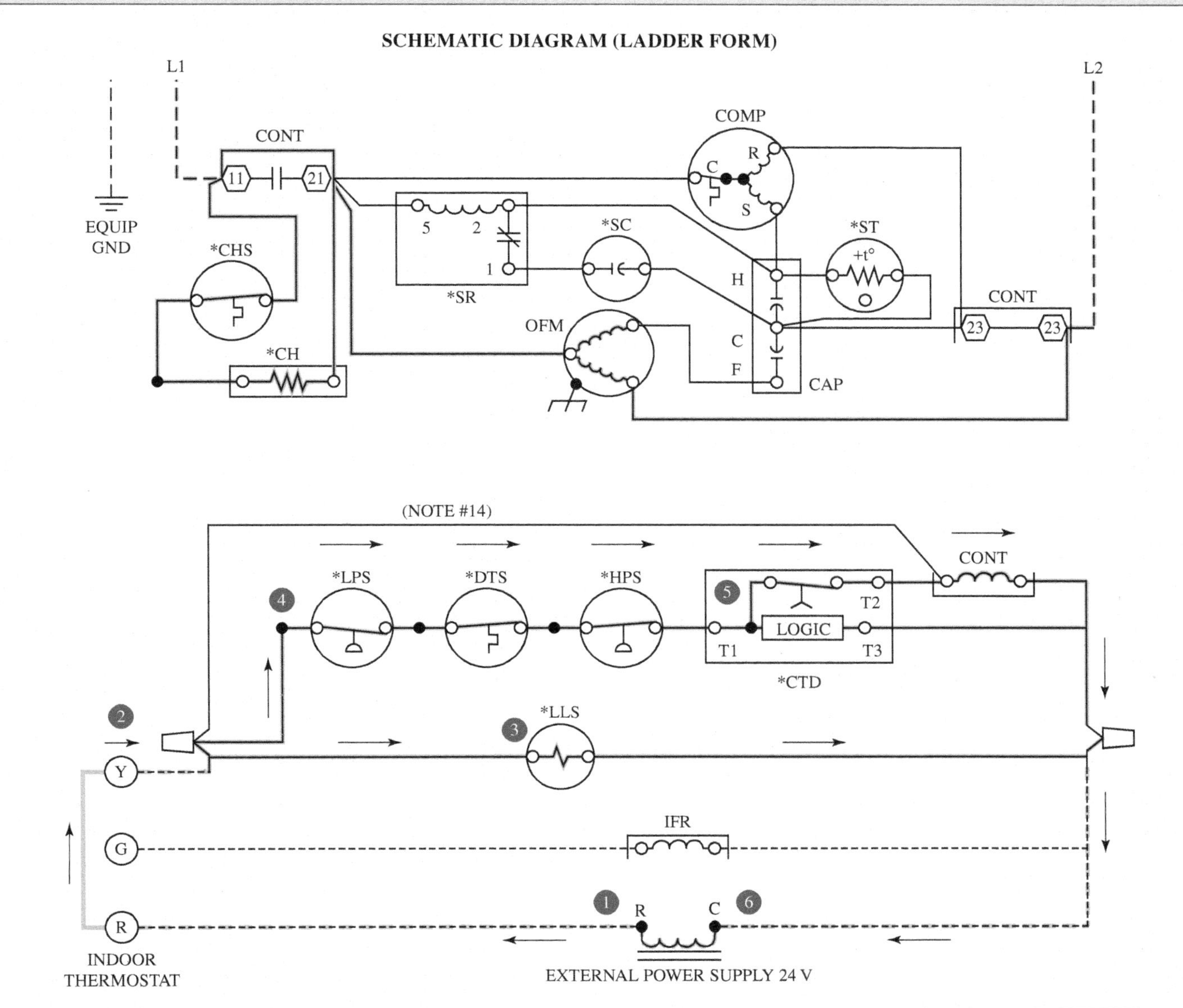

Figure 21-10 This diagram illustrates what is energized in the high-voltage section of the split system before the control voltage is powered up. For example, the crankcase heater **CH** will be hot. The control voltage section is shown in the lower part of the diagram.

SEQUENCE OF OPERATION 21.6: STARTING THE COMPRESSOR AND CONDENSER FAN

This sequence of operation, where the compressor and condenser fans are started, occurs in less than a second. The second part of this sequence is discussed in Sequence of Operation 21.7.

Using Figure 21-11, here is the sequence of operation of the high-voltage circuit immediately after the contactor coil (CONT) is energized:

1. When the contactor coil (CONT) is energized in Figure 21-10, the single-pole contactor between ⟨11⟩ and ⟨21⟩ closes as shown in Figure 21-11, point **1**. The other contactor pole is a straight-through connection and is designated as ⟨23⟩ and ⟨23⟩ with a line between the two numbers. The line means they are connected to power at all times.

(continued)

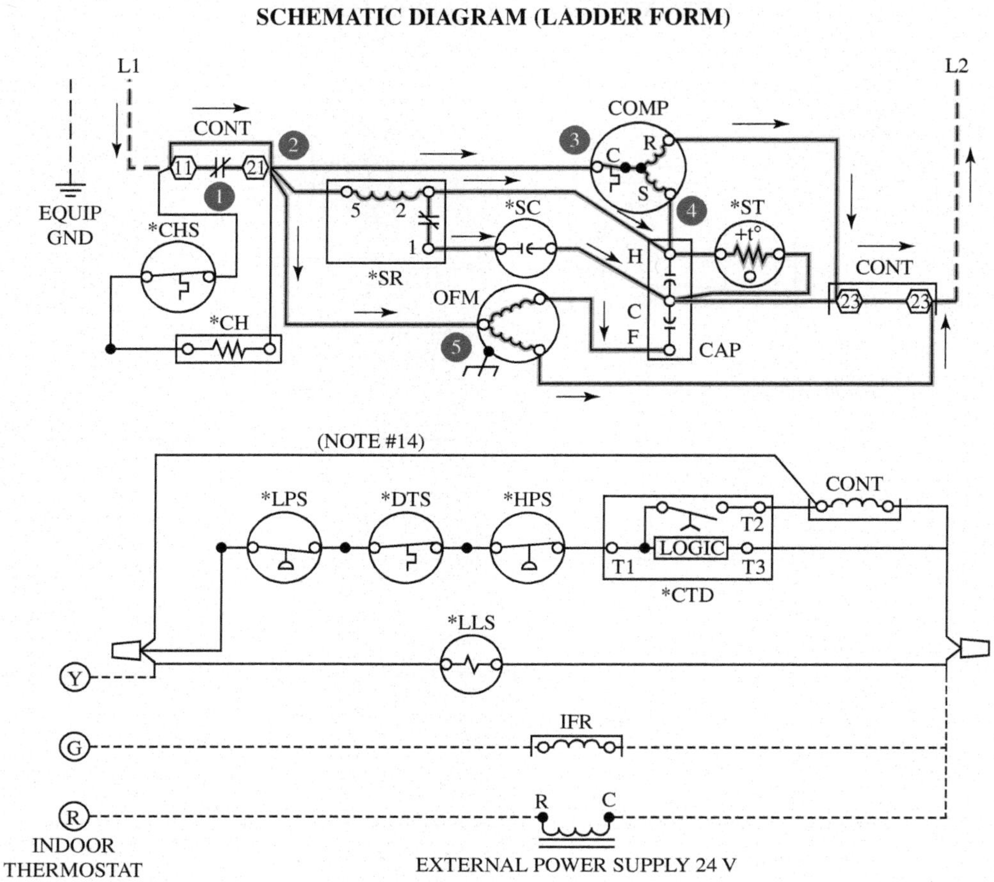

SCHEMATIC DIAGRAM (LADDER FORM)

L1
L2
CONT
COMP
EQUIP GND
*CHS
*CH
*SR
*SC
*ST
OFM
CAP
CONT
(NOTE #14)
*LPS
*DTS
*HPS
LOGIC
T1
T2
T3
*CTD
CONT
*LLS
IFR
INDOOR THERMOSTAT
EXTERNAL POWER SUPPLY 24 V

Figure 21-11 This shows the operation of the compressor and condenser fan.

2 At point **2**, the power is split between three parallel circuits: the compressor circuit (**COMP**), optional compressor starting components (**SR** and **SC**), and the outdoor fan motor circuit (**OFM**).

3 At point **3** power enters the common connection of the compressor run winding circuit. Power to the compressor goes through the common terminal **C** and then to run winding **R** before returning to **L2**.

4 Point **4** shows the compressor start winding circuit: Power goes to the optional start relay ***SR** (potential relay) terminal **5**, through the relay coil terminal **2**, and on to the **H** (herm) connection on the run capacitor. There is another parallel electrical path in the **SR** relay. Power goes through (SR) coil at terminal **5** to terminal **2**, through closed contacts **2** to **1**, and to the start cap. The start cap is charged through the start relay and **C** (common) connection on the run cap. If you carefully trace out the start cap and run cap, you will notice that they are in parallel when the motor starts. This parallel cap arrangement adds the capacitance together. In the next drawing and Sequence of Operation 21.7, we will see that the **SR** contacts open shortly after compressor start-up.

5 At point **5**, the third parallel path is the operation of the **OFM**. The outdoor fan motor receives power through the common terminal and through the run winding. The start winding receives a short-term current flow through the fan (**F**) terminal on the run cap to get the motor started (**CAP**).

SEQUENCE OF OPERATION 21.7: RUNNING THE COMPRESSOR AND CONDENSER FAN

Sequence of Operation 21.6 discussed the starting circuit operation. There are a couple things that happen after the first split second of operation. Use Figure 21-12 for the following discussion:

1. At point **1**, the start relay (***SR**) or potential relay contacts between **2** and **1** will open due to the back electromotive force (EMF) developed by the moving compressor rotor. The back EMF will energize the (SR) coil between terminals **5** and **2**, opening the normally closed contacts **2** and **1**. This voltage back EMF is higher than the applied voltage. For example, if the applied voltage is 240 volts, the back EMF developed by the rotation of the motor will be around 300 volts or greater.

2. When the contacts between **2** and **1** are open, the start cap is removed from the circuit at point **2**. The start cap will burn out quickly if left in the circuit for long periods of time. The **SR** must be sized properly or the start cap will be left in a second or two longer than needed. This will overheat the capacitor and create premature failure. Not leaving the start relay contacts closed long enough may prevent the compressor from starting.

3. After the **OFM** is started, power to the start winding of the **OFM,** at point **3**, is minimal since the run cap blocks current flow once it is energized with an initial charge of electrons.

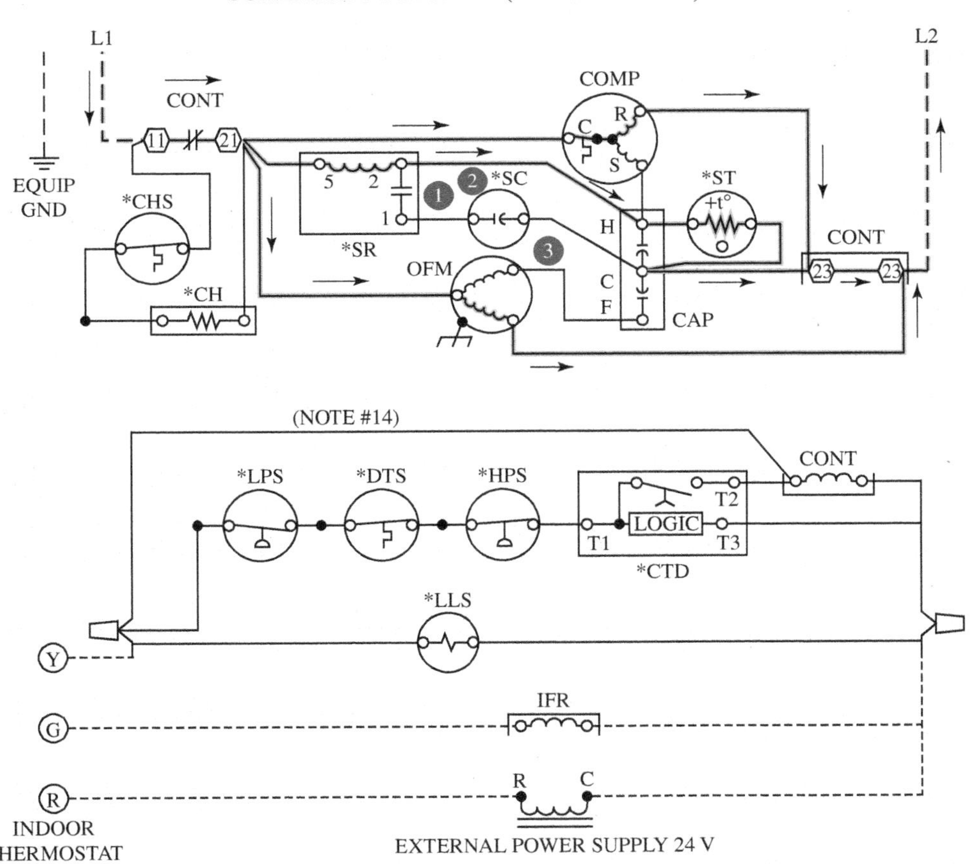

Figure 21-12 This diagram highlights the operation of the split system after it has been energized. Notice that **SR** contacts **2** and **1** are open. This takes the start cap, **SR**, out of the circuit. The compressor and fan motor start windings have little current flow after start-up, therefore they are not highlighted.

SEQUENCE OF OPERATION 21.8: EVAPORATOR FAN OPERATION

Before starting the sequence of operation explanation, let's see what is energized when the power is supplied (see Figure 21-13). When **L1**, **L2**, and **L3** are powered up, the 480-V transformer steps down the voltage to 120 volts. This is the control voltage protected by a 2-A fuse. Notice that the transformer has a 150-VA rating. The crankcase heater **CH** will be deenergized until the contactor coil Ⓒ is energized. The **C** contacts in series with the crankcase heater are normally closed (**NC**) and open when energized. The heat from **CH** is not required when the compressor is in operation.

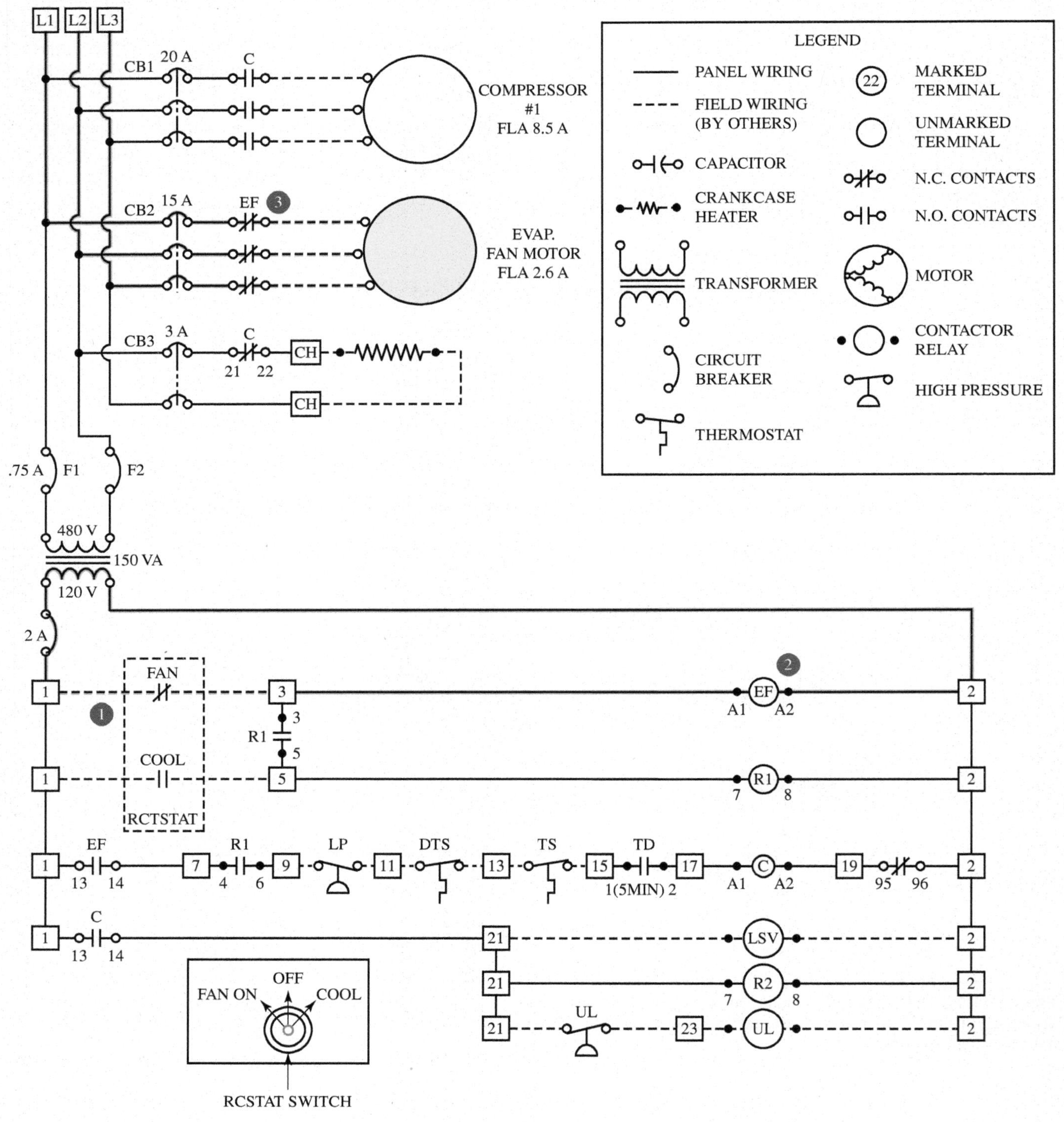

Figure 21-13 This diagram shows power supplied to a basic three-phase unit. The fan switch is selected to operate the blower motor. The crankcase heater is on since the compressor is not operating.

Here is the sequence of operation of the evaporator fan circuit by itself:

1. Turn the **RCSTAT** to the (Fan) ON position. The **RCSTAT** switch is seen on the lower left side of the diagram (see both points **1**). This will manually close the fan contacts and start the blower fan operation only.
2. This will provide a 120-V path to the evaporator fan contactor coil, (EF), point **2**.
3. The energized coil will close the three-pole contactor **EF** at point **3**, supplying 480 volts to the evaporator fan motor. The blower motor comes on.

SEQUENCE OF OPERATION 21.9: CONTROL VOLTAGE OPERATION

Figure 21-14 is used in this sequence of operations explanation:

1. Manually set the **RCSTAT** switch to Cool. This closes the Cool contacts on the switch, point **1**.
2. At point **2**, the (R1) coil will be energized by 120 volts.
3. At point **3**, the (EF) coil is energized through **R1** closed contacts **5** and **3**.

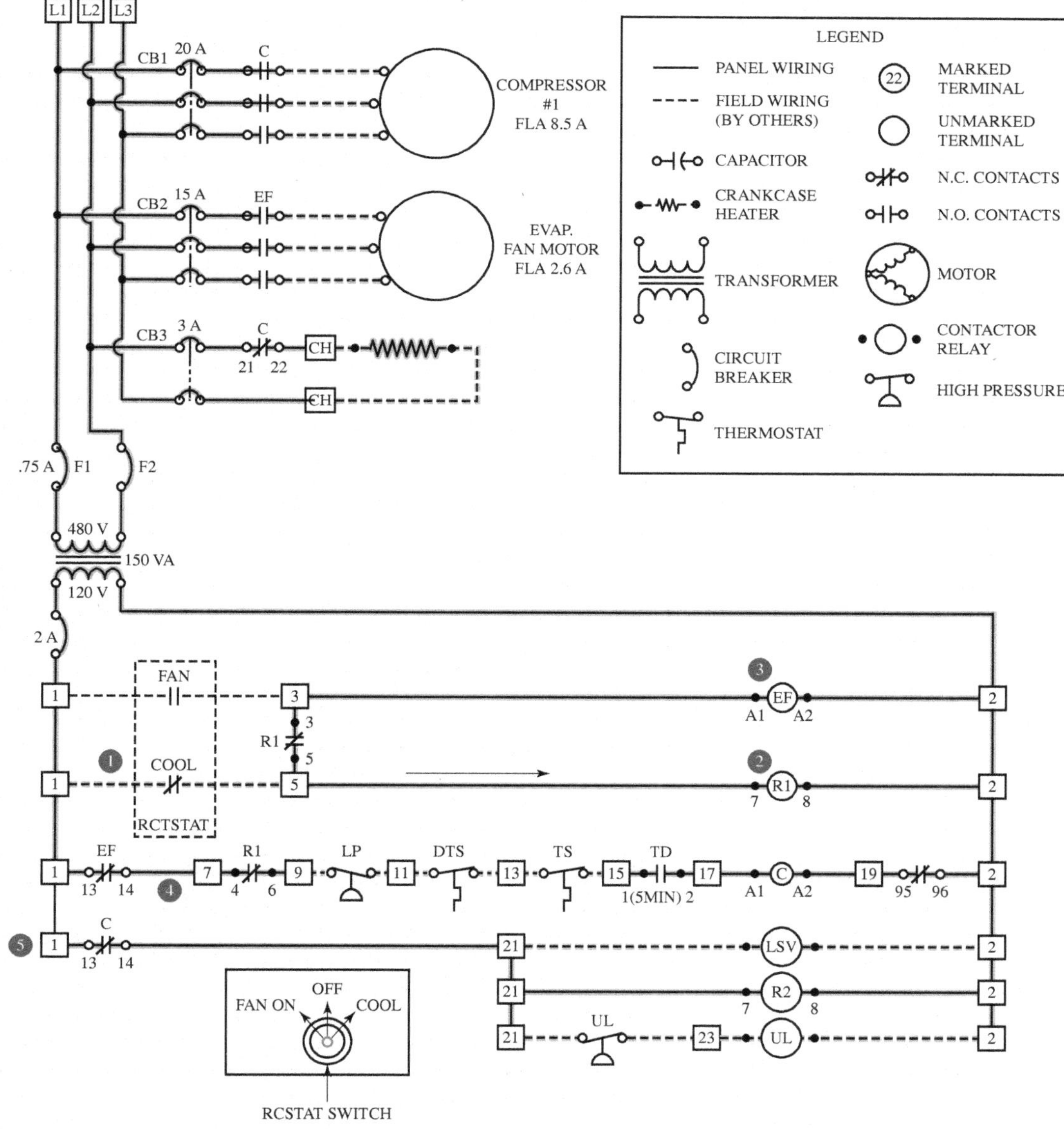

Figure 21-14 In this control voltage diagram, note that there are four parallel control circuits. One line is indicated by point **3**. The other three parallel branches are found to the right of points **1**, **4**, and **5**.

(*continued*)

4 At point **4**, the **EF** and **R1** contacts are closed, completing control voltage to the following closed switches, safety devices, thermostats, or time delays: **LP**, **DTS**, **TS**, and **TD**. *Note:* The closed contacts between terminals **95** and **96** (right of coil Ⓒ) represent a remote control device used to shut down the system from another location such as a control room.

5 With all the switches closed in line 4, contactor coil Ⓒ will be energized. The energized contactor coil will close the C contacts at point **5**.

Compressor and Evaporation Motors

Before we discuss compressor and evaporation motors, let's locate the closed contacts after the (EF) coil and Ⓒ coil have been energized with 120 volts. When the (EF) coil is energized, it will close the three-pole contactor feeding power to the evaporator fan. When the Ⓒ coil is energized, it will close the three-pole contactor feeding power to the compressor. The Ⓒ coil will also open the **NC** contacts in the crankcase heater circuit. The **NO** contacts between **13** and **14** in the lower control voltage circuit will close. In total, four contacts will change position, three closing and one opening.

Immediately, the evaporator fan motor control circuit will be energized as shown in Figure 21-14. There may be a short delay before the compressor begins operation. Point **1** in Figure 21-15 shows evaporator fan **EF** energized. Next, at point **2**, go to the lowest closed contacts **13** and **14**, **C**, found on the bottom left side of the bottom diagram. When **C** closes, power is supplied to three parallel branch circuits labeled [21]. The next three points will describe what happens.

At point **3**, power goes through the closed contact to the liquid line solenoid valve, (LSV). This opens the solenoid valve supplying liquid refrigerant to the metering device and evaporator. This is not a pumpdown system. The (LSV) is simply used to isolate the refrigerant in the liquid line and condenser, preventing refrigerant from entering the evaporator during the off cycle. Think of it as an electrical check valve. You will see this design in marine and commercial refrigeration circuits. Now go the next parallel branch. *Note:* The parallel path for power to the (R2) coil is used to control a remote air-cooled condenser or a water-cooled condenser pump. This condenser is not shown on the diagram.

At point **4**, the bottom line shows the unloader circuit, **UL**. The unloader is used to reduce the capacity of a system when less cooling is needed. This allows the compressor to continue operating at a lower Btuh capacity. The (UL) circuit operates on pressure. As the load drops (returning air or water gets cooler), the suction pressure drops. At a predetermined low-pressure setting, the (UL) low-pressure switch will close, thus energizing the (UL) coil. The electric unloader will reduce the pumping capacity of the compressor. If the suction pressure rises, the pressure switch **UL** opens, bringing on full compressor capacity. Finally, as we saw in Sequence of Operation 21.9, the **C** coil is energized at point **5**, and the three-pole contactor will close, bringing on the compressor motor, at point **6**.

SERVICE TICKET

You are called to service a package unit on the roof of a convenience store. When you arrive a couple of quick checks reveal that the system is not working. The indoor blower will not operate and the digital thermostat face is blank. You climb on the roof to do a visual inspection and notice that the door has been removed from the package unit and—surprise!—most of the wires are missing. The paper diagram on the unit panel is weathered and unreadable.

You report this to the store owner and she becomes visibly upset. She said that a tech came out earlier in the day and said the unit needed to be rewired for $500. The owner said she was going to need to get a second opinion on the diagnosis. After you reported the missing wiring, she suspected that the previous tech did the damage. The blame could not be determined.

After surveying the site you contact the dispatcher for an estimate for the job to rewire the XYZ package unit, Model #3T-36000. The dispatcher thought you could do the job in a couple of hours for $399. The price includes labor, wire, and wire connectors. The convenience store was getting warm and the owner agreed to the price. Figure 21-16 provides an example of what you are dealing with.

Using the information in Figure 21-16, draw a separate wiring diagram. *Hint:* First draw wiring between all of the components, power supply, and controls; then draw the wiring diagram. Compare your diagram with others in the class. There is more than one way to draw a diagram, therefore comparing other diagrams will be a learning experience. Finally, if this were a real service call, you would leave the drawing on the job or give a copy to the equipment owner.

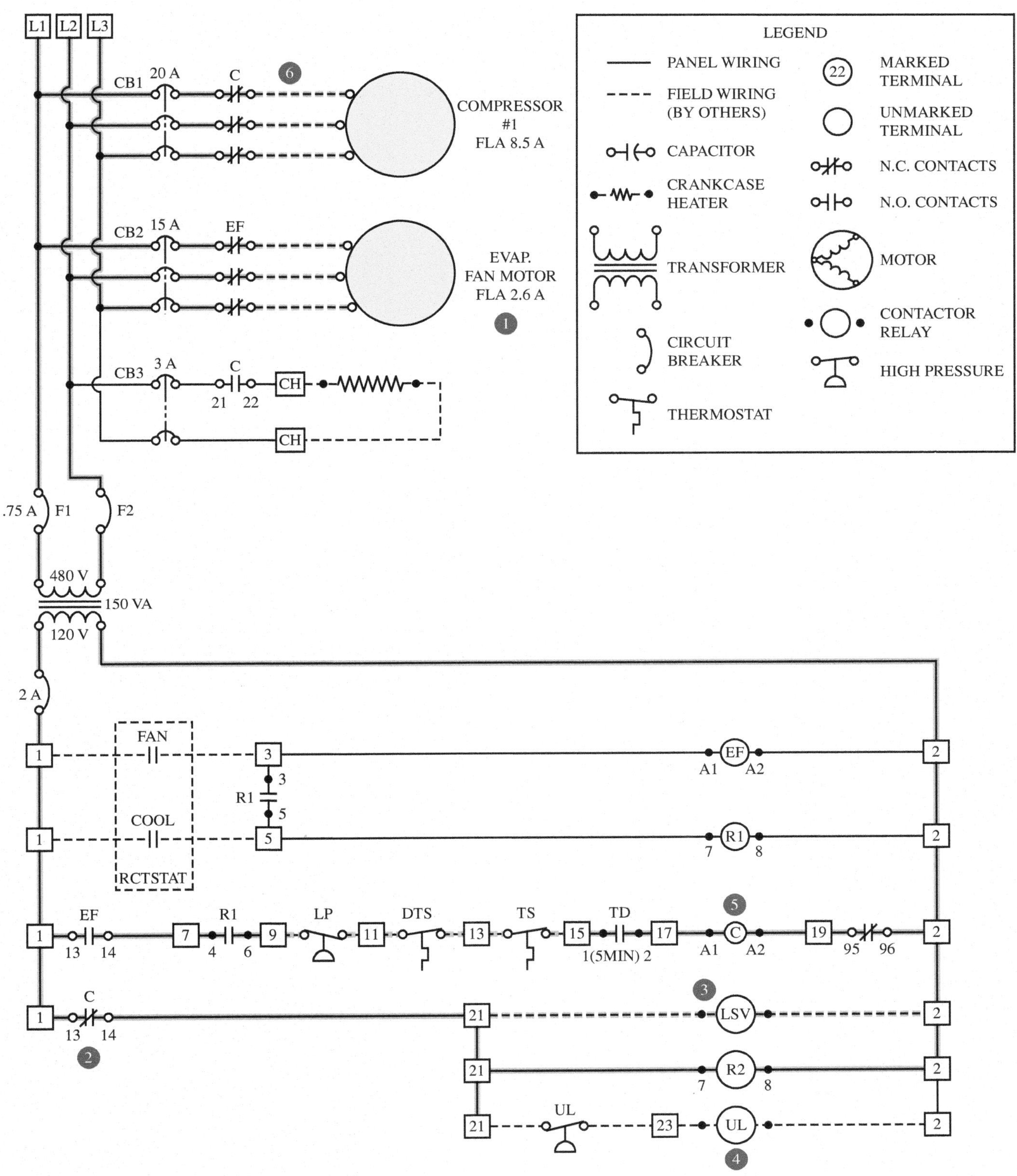

Figure 21-15 This diagram shows the operation of the compressor and evaporator fan motor when the compressor and evaporator fan contactors are energized.

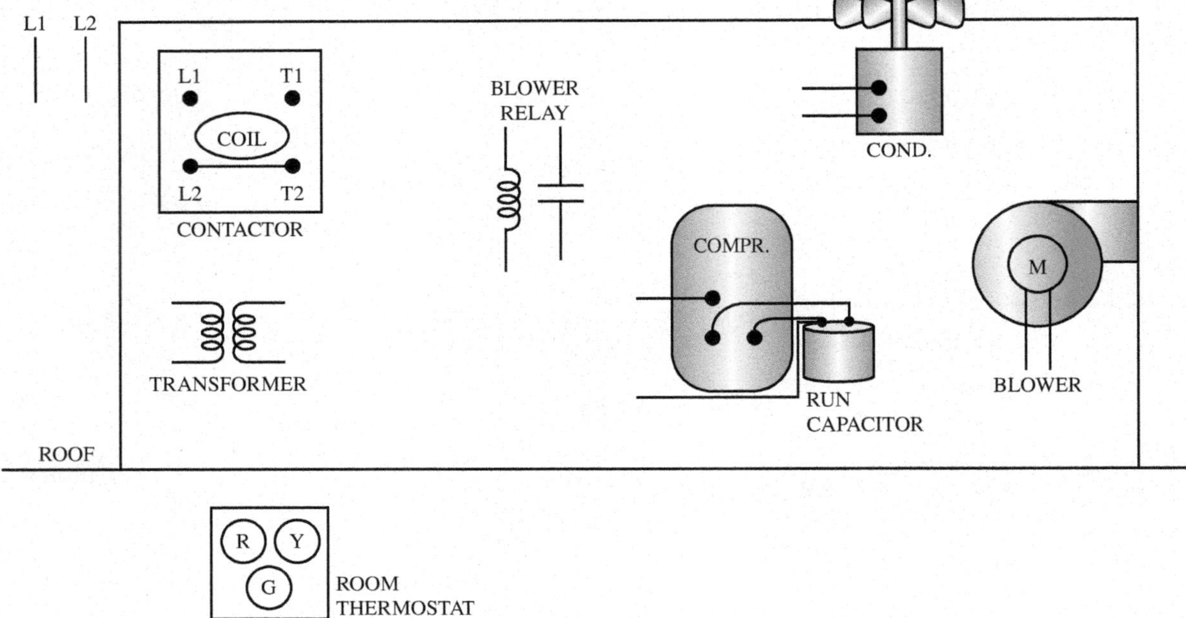

Figure 21-16 This is what the convenience store's package unit looks like.

SUMMARY

This unit was about air conditioning systems. A system is composed of many different parts and pieces that make up an operating air conditioning unit. This unit brought together the various bits and pieces that were presented in many of the preceding 20 units. We discussed single-phase and three-phase systems, package units, and split systems. We also covered a window unit diagram that had not been discussed up to this time. With the knowledge obtained in the prior units you were able to understand an operating system.

This unit traced the sequence of operations of numerous air conditioning systems. The three-phase system discussed could also have been a refrigeration system used to cool products above freezing. It could not have been a freezer since the diagram did not have a defrost option.

REVIEW QUESTIONS

1. The diagram in Figure 21-17 is for what type of air conditioning system?
2. How many speeds does the fan in Figure 21-17 have?
3. Is the compressor in Figure 21-17 wired correctly? If not what is the problem?
4. Convert the components in Figure 21-16 to make a functional air conditioning electrical diagram.
5. Which control(s) are used to pump the system down in Figure 21-18?
6. How many pole contactors are used in Figure 21-18?
7. The purpose of this exercise is to draw a capacitor start/capacitor run compressor. Using Figure 21-18 and using the symbols for the compressor, **SR**, **SC**, and run capacitor, redraw the diagram using only these four components. Use **L1** and **L2** for a direct hook-up.
8. What type of electrical diagram is shown in Figure 21-7?
9. In Figure 21-10, when is the crankcase heater energized?
10. Referring to Figure 21-14 or enlarged Electrical Diagram ED-2, list the pressure controls found in the diagram.
11. Referring to Figure 21-14 or enlarged Electrical Diagram ED-2, list the temperature controls found in the diagram.
12. Referring to Figure 21-14 or enlarged Electrical Diagram ED-2, when the system is switched to Cool, what controls the on and off cycling of the compressor?
13. Referring to Figure 21-14 or enlarged Electrical Diagram ED-2, does the evaporator fan motor cycle on and off with the compressor?
14. Referring to Figure 21-14 or enlarged Electrical Diagram ED-2, what voltage will you measure across the **R1** coil when the system is cooling? What voltage will you measure across the **TD** contacts when the system is cooling?

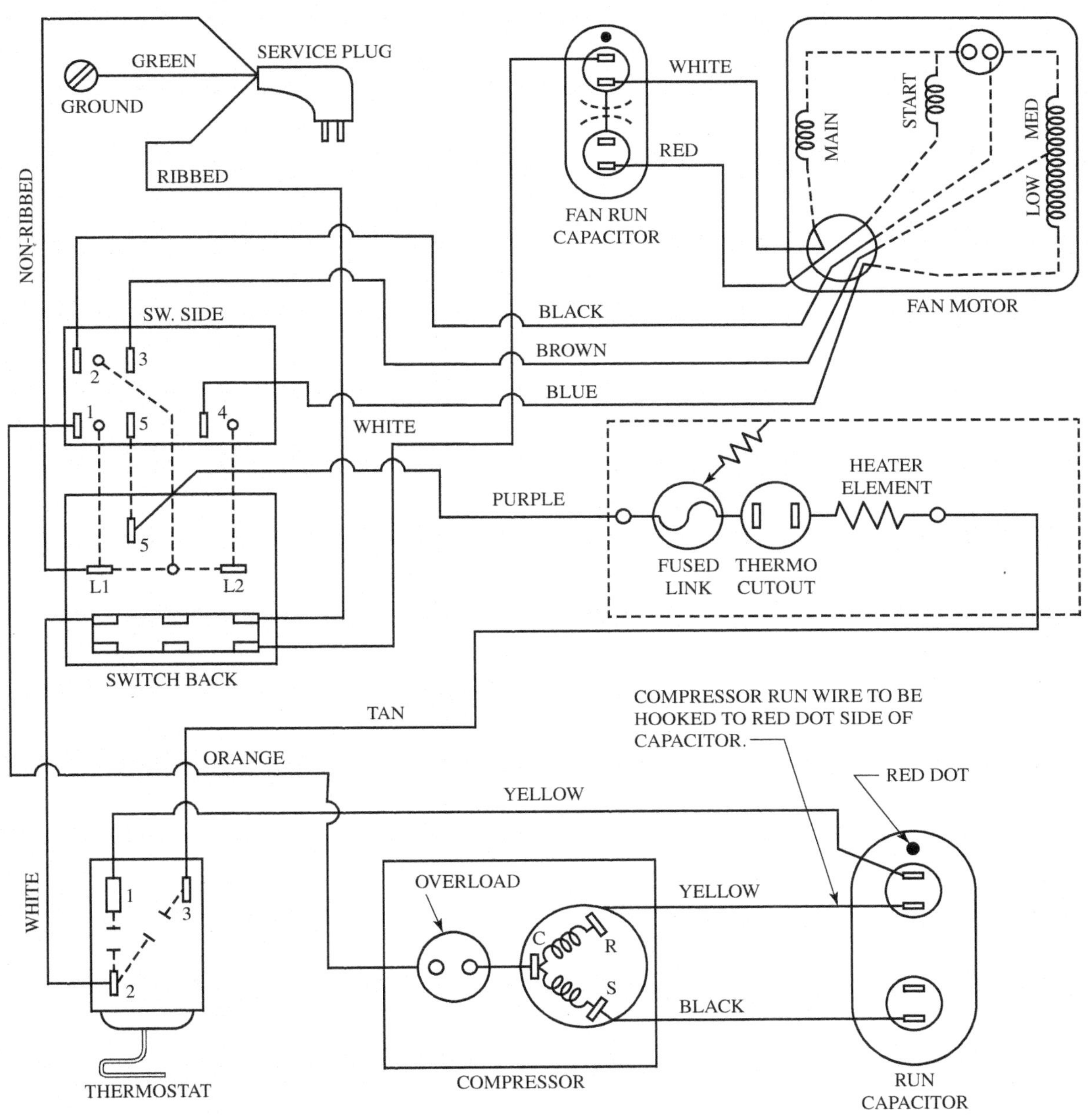

Figure 21-17 Use this diagram to answer Review Questions 1, 2, and 3.

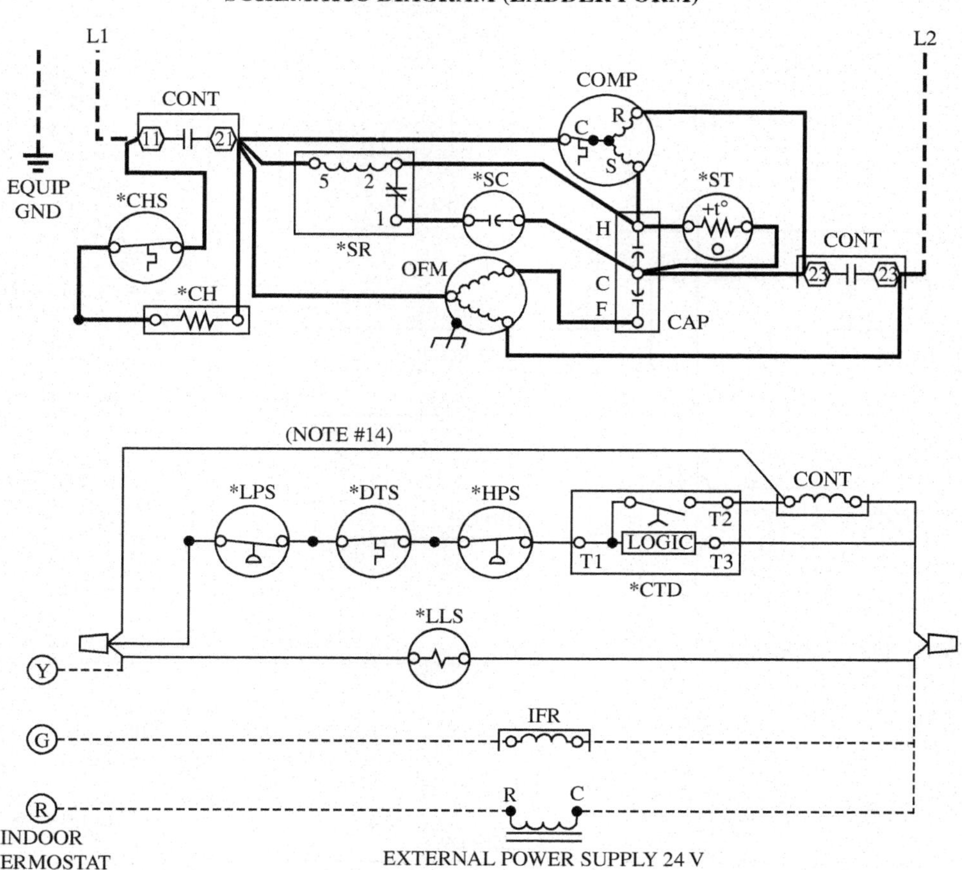

Figure 21-18 Use this diagram to answer Review Questions 5, 6, and 7.

UNIT 22

Gas Heating Systems

WHAT YOU NEED TO KNOW

After studying this unit, you will be able to:

1. Explain how gas heating systems are categorized by efficiency.
2. Describe a gas-burning cycle.
3. Identify basic gas-burning controls.
4. Describe the operation of three components found on a low-efficiency gas furnace.
5. Using an electrical diagram, identify which components of a gas furnace use line voltage and which ones use control voltage.
6. Identify safety controls in a gas furnace wiring diagram.
7. Describe two types of troubleshooting techniques used when working on gas furnaces.

Gas heating systems can be very simple or very complex. The simplest gas-burning systems are those that have the fewest electrical controls. Some gas heating systems have only a gas valve and a temperature control to turn the gas valve on when the temperature drops in a room and turn it off when the set temperature is reached. These systems are typically gas space heaters that do not have a blower to move air through the unit and into the room.

In this unit, a simple gas heating system with a blower is discussed. For contrast, we also present a complex and highly efficient heating system. The differences between these two systems will illustrate the differences in operation required to efficiently control and monitor a gas-burning system. More complex systems require additional electrical control and monitoring devices. As the complexity increases, the technician is required to know and understand more about electrical and **electronic controls**.

Also, as a system's complexity increases, manufacturers provide additional tools that the technician can use to troubleshoot systems. Circuit boards provide basic troubleshooting codes that can be interpreted by the technician. Troubleshooting codes usually take the form of flashing lights on the circuit board. The number of flashes is an indication of the problem. For example, two flashes repeated every 10 seconds may mean that the voltage polarity is incorrect. Some equipment is so sophisticated that it will register a problem on the thermostat or at a connected laptop at the customer's location.

Troubleshooting flowcharts are provided to help guide the technician in conducting the troubleshooting procedure. While both of these things are helpful, the technician must rely on practicing the troubleshooting sequence in a certain way. Just like "hopscotch," the troubleshooting process relies on the technician taking the correct readings in a sequential way to determine if a component is receiving the right voltage or if a switch is open rather than closed.

By the end of this unit, you will have a better idea of:

- The similarities and differences between gas heating systems
- The operation from start-up, through the heating process, to shutdown
- Troubleshooting techniques, processes, and tools available for systems at each end of the efficiency spectrum.

22.1 CLASSIFYING GAS HEATING SYSTEM TYPES

Gas heating systems can be classified in several ways. Natural and **LP** (**liquefied petroleum**, also known as propane or butane) are the two commercial gases that are distributed for general consumption. Units are typically designated and sold as either one or the other type of gas-burning unit. More often, however, gas-burning units are set up as natural gas units, and if a unit needs to be used with LP, an "LP kit" is purchased or provided by the manufacturer to convert the unit from natural gas to LP.

SAFETY TIP

Liquefied petroleum gas is heavier than air, so it can pool in low areas or be confined in a small room. Natural gas is lighter than air and rises. Small gas leaks can result in large explosions that can kill. Before operating any electrical component (gas furnace, electric light switch, etc.), smell for the presence of gas. In their natural state these gases are odorless, so a strong odorant is added to give them a distinct smell. Some gas companies have stated that it smells like rotten eggs. If you can smell gas, ventilate the area before you begin working. If you smell gas before entering the furnace area, it would be advisable to call the fire department or gas company for assistance and do not turn on the light or any other spark-creating device.

22.2 VENTING SYSTEMS AND EFFICIENCY

Gas heating systems are also classified by their efficiency, which relates to how they are vented. **Venting** is the process of removing the combustion by-products of burned gas fuel. The relationship of efficiency to venting has to do with the temperature of the exhaust or chimney gas. Low-efficiency units do not extract enough heat from the burning gas to reduce exhaust vent gas temperatures. Temperatures of up to 300°F can be measured at the chimney connection for low-efficiency units that are rated from 60% to 78% efficient. Mid-efficiency units (between 80% and 85% efficient) extract more heat and reduce the exhaust gas temperature to around 210° to 250°F.

The venting required for low- and mid-efficiency gas furnaces is classified as a **Type B venting** system and typically consists of a double-walled vent pipe. In some cases, a stainless pipe that can resist corrosion from acid is used. Acid is created in the exhaust or flue gas, because when the venting temperature drops below 212°F, moisture that contains acid can condense out of the flue gas and collect in the vent pipe.

When flue gas temperatures drop to as low as 120°F, the exhaust temperature is low enough to allow the use of plastic pipe as a vent material instead of metal pipe. Plastic resists corrosion from acid and can be easily sealed so that both condensing moisture and exhaust gases are completely isolated from the living space. By extracting the maximum amount of heat and exhausting flue gas at the lowest possible temperature, gas furnace efficiencies of more than 90% can be achieved. Some modern furnaces have even been rated 98% efficient. Units with this type of efficiency are classified as high-efficiency gas heating systems. Both natural gas and LP heating systems can qualify for these high-efficiency classifications. The more efficient the gas furnace, the higher its cost, as shown in Table 22-1.

Table 22-1 Furnace Efficiency vs. Cost

Efficiency Type	Vent Type	Efficiency Rating (AFUE)	Furnace Cost
High-efficiency	Plastic vent	90% and higher	High
Mid-efficiency	Double wall (or possibly stainless vent)	80% to 85%	Midrange
Low-efficiency	Double wall vent	Below 80%	Low

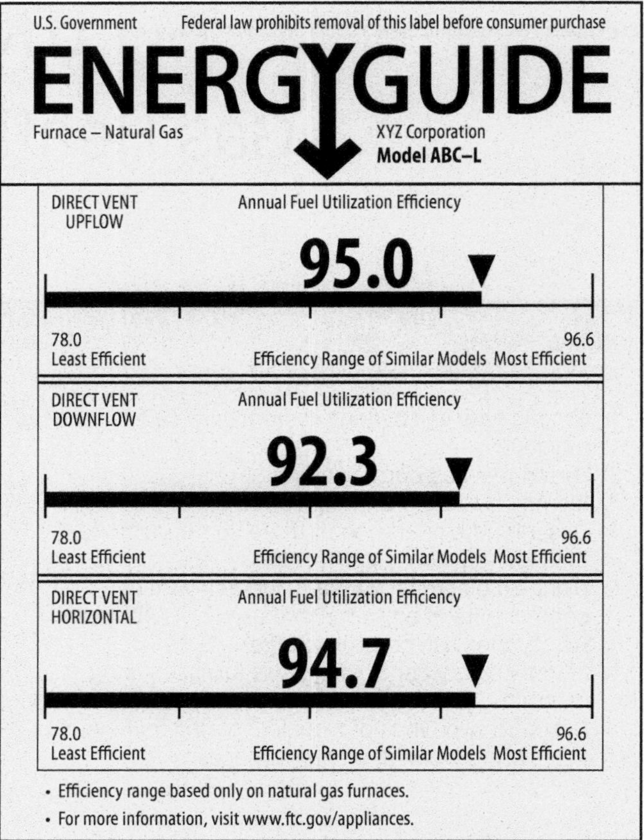

Figure 22-1 An EnergyGuide label, as mandated by the U.S. Department of Energy.

GREEN TIP

Most modern gas furnaces have an AFUE (annual fuel utilization efficiency) rating. Figure 22-1 shows the AFUE label that you will find on a new gas furnace. AFUE is the amount of heat that goes into conditioned space versus the amount of heat developed by burning the gas fuel. A furnace manufacturer completes a complex set of testing standards to determine a furnace's AFUE rating. A simple, but not totally correct way of viewing AFUE is expressed in the following formula:

$$\text{AFUE} = \frac{\text{Btuh output}}{\text{Btuh input}}$$

The U.S. Department of Energy (DOE) has mandated that furnaces have a minimum AFUE rating of 80%. Warmer climates tend to use lower AFUE systems, while colder climates use higher AFUE systems. The DOE mandates use of the EnergyGuide label, as shown in Figure 22-1, which estimates the actual energy consumption of an appliance and provides some information about whether that consumption is above or below the average for that type of product. The dollar amount at the bottom of the EnergyGuide label is the estimated yearly

Figure 22-2 The ENERGY STAR label is the government's symbol for energy efficiency.

operating cost based on the national average cost of electricity or other fuel usage.

The ENERGY STAR label shown in Figure 22-2 is the government's symbol for energy efficiency. It helps consumers easily recognize highly efficient products, homes, and buildings that save consumers energy and money and help protect the environment. The ENERGY STAR label often appears at the bottom of the EnergyGuide label for qualified products. The most efficient gas furnaces may have the ENERGY STAR label. They will have an AFUE of 85% and higher and have a high-efficiency blower (ECM).

TECH TIP

The **National Fuel Gas Code** and other national standards (ANSI Z223.1, Z21.47, Z21.10.3, and Z21.13) categorize gas appliances according to their type of venting, as listed in Table 22-2.

In addition to fuel, efficiency, and venting classifications, gas heating units can be further classified by such things as:

- Type of **ignition system**
- Type of **gas control**
- Type of fan/blower control
- The direction in which air is moved through the unit (upflow, counter flow, and horizontal).

Each type of system will use different electrical and electronic control devices, which makes each type of gas heating system distinct in terms of its electrical components and configurations.

Table 22-2 Gas Appliance Categories

Category	Venting	Condensing	Venting Material
I	Gravity	No	B vent
II	Gravity	Yes	Per manufacturer
III	Fan-assisted	No	Stainless
IV	Powered	Yes	Plastic

GREEN TIP

More on Green Heating Systems

Gas heating systems can be classified as "green systems." The term *green* has many definitions, but if a unit is referred to as green, then it is being compared to another unit that uses energy less efficiently. The AFUE efficiency rating of a gas-burning appliance is listed by the manufacturer and is displayed as an EnergyGuide yellow sticker on new systems. As mentioned earlier, the AFUE rating refers to the percentage of fuel that is extracted as usable heat to warm a space. The higher the AFUE number, the less energy a system wastes and the more it is considered to be "green."

Standard heating systems with standing pilots (pilots that burn continuously) are considered less green than fan-assisted 78% (78 AFUE) systems. The 90 AFUE systems are greener than the 80 AFUE models. The closer a system is to a 100% AFUE rating, the more green the system is, because a high AFUE rating means that most of the heat goes into the space being heated, rather than being vented to the outside. Some gas furnaces have AFUE ratings as high as 98%, meaning that a very minimal amount of heat is vented outside.

As gas heating systems become greener, they become electrically complicated. There are more components to help it become more efficient and more controls to monitor the operation. Very simply, a gas-burning appliance can be monitored by a human being: Gas is turned on, the flame is lit, and the flame is extinguished at the end of use. When basic controls are applied, the gas heating system can be turned on and off automatically with a thermostat. While the system is burning gas, the operation can be monitored and turned off if a safety issue occurs.

In a later section on furnace operation, a simple gas heating system will be described as it moves from one sequence to the next. The basic operation of a gas-burning appliance will be described. This is the same sequence of operation that more sophisticated furnaces follow with the exception that more sophisticated furnaces require additional steps. The additional steps have to do with the extra electrical components that are used to increase efficiency to become "greener."

TECH TIP

Power—nothing works unless line voltage is received. Always ensure that the line voltage and control voltage are present by using the voltmeter. When the blower door is removed, the door switch will turn off power to the unit. This is a safety switch. If at all possible, service electrical systems with the power off.

SAFETY TIP

If the system needs to be powered with the door off, manually close the switch rather than use a jumper. Be sure to check that this switch operates before concluding a service call. While checking with the power on, be sure to use good electrical safety procedures.

22.3 GAS FURNACE COMPONENTS

This section explores the common electrical components found in a forced-air gas furnace. After this explanation we will explore how these components work together as a gas furnace. Here is a list of common electrical items that you may find on a gas furnace:

- **Door switch**
- **Gas valve**
- **Inducer fan assembly (air proving switch)**
- **Hot surface igniter**
- **Flame detector (flame rod)**
- **High limit switch**
- **Flame rollout switch**
- **Circuit board**
- **Blower motor assembly.**

We will be using estimates on the timing sequence of operation. For example, one manufacturer may activate the system blower automatically after 60 seconds, while another may design it to come on 90 seconds after the beginning of the heating cycle. Some circuit boards have options to start the blower at different time intervals. Others are preset to energize at a prescribed time. When a time, resistance, or something specific is mentioned, remember that it will vary among manufacturers.

22.4 DOOR SWITCHES

The door switch shown in Figure 22-3 is used to prevent a furnace from operating without the blower door panel. The door switch is a rocker switch that opens if the panel is removed. This door switch has been bypassed and forced closed with a nylon tie while it is being serviced. The nylon tie or other bypass measure must be removed to engage the safety feature of the door switch. If the furnace were allowed to operate without the blower door cover, the burned gases could be pulled into the return air stream and distributed to the conditioned space. This could be dangerous if these vent gases included carbon monoxide. Carbon monoxide is not always present in burned fuel, but why take a chance? Always remove the bypass before reinstalling the blower door.

SAFETY TIP

A carbon monoxide (CO) detector should be installed in every building that has a fuel-burning appliance within its structure. The best carbon monoxide detector will be battery operated and installed on or near the ceiling. A power failure will not affect the operation of a battery-operated detector. CO is lighter than air, therefore the detector should be installed high on a wall or on the ceiling.

Even a building without fuel-burning appliances can have high CO levels. For example, parking a running car in a garage or below a building will create a poison danger to people in the building. CO is odorless, tasteless, and cannot be seen in the air. Have everyone vacate the area if the CO detector sounds. Call the fire department.

Figure 22-3 The door switch is a rocker switch that opens if the panel is removed.

Table 22-3 Carbon Monoxide Poisoning Levels (in parts per million)

Carbon Monoxide: (CO) Product of Incomplete Combustion
100 PPM - Safe for continuous exposure
200 PPM - Slight effect after six (6) hours
400 PPM - Headache after three (3) hours
900 PPM - Headache and nausea after one (1) hour
1,000 PPM - Death on long exposure
1,500 PPM - Death after one (1) hour
Most codes specify that CO concentrations shall not exceed 50 PPM

Figure 22-4 This is an older style gas valve with a pilot light solenoid and a main solenoid valve. The light blue knob is the ON/OFF selector. It is in the ON position.

SAFETY TIP

Table 22-3 shows various carbon monoxide exposure levels and their consequences. When natural gas (**CH_4**) with the help of two oxygen molecules (**O_2**) burns, it creates heat, carbon dioxide (**CO_2**), and water (**H_2O**). Stated as a formula:

$$CH_4 + 2\,O_2 = CO_2 + 2\,H_2O$$

This is perfect combustion. Carbon monoxide is formed if there is not enough air for the combustion process or if the flame is not at a high enough temperature, which can occur if the flame is touching the heat exchanger or burner tubes. Anything that interferes with complete combustion may cause **CO** to form.

22.5 GAS VALVES

Gas valves, like those shown in Figures 22-4 and 22-5, are control valves used to meter the flow of gas to the burners. A natural gas valve also reduces the pressure from the gas supply, whereas a propane gas valve does little to control the pressure of the gas supply.

The gas valve is a solenoid valve, therefore it has a coil that is energized to open by 24 volts in most cases. The solenoid coil may be a low- or high-resistance coil. The resistance of some coils is as low as 100 Ω, whereas others may have as much as a million ohms of resistance. Use a DMM with a high resistance (more than 1 MΩ) to determine if the gas valve coil is good. If the control voltage is measured at the gas

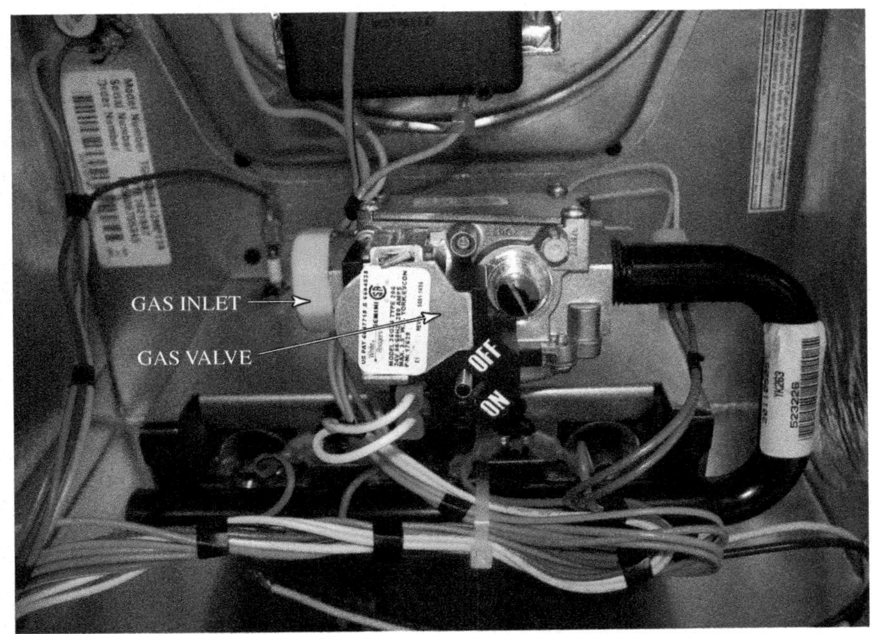

Figure 22-5 Common gas valve on a gas furnace. The gas line will be installed on the left side of the valve when the yellow protective cap is removed. The gas line going through the furnace wall should be hard pipe. The valve is in the OFF position.

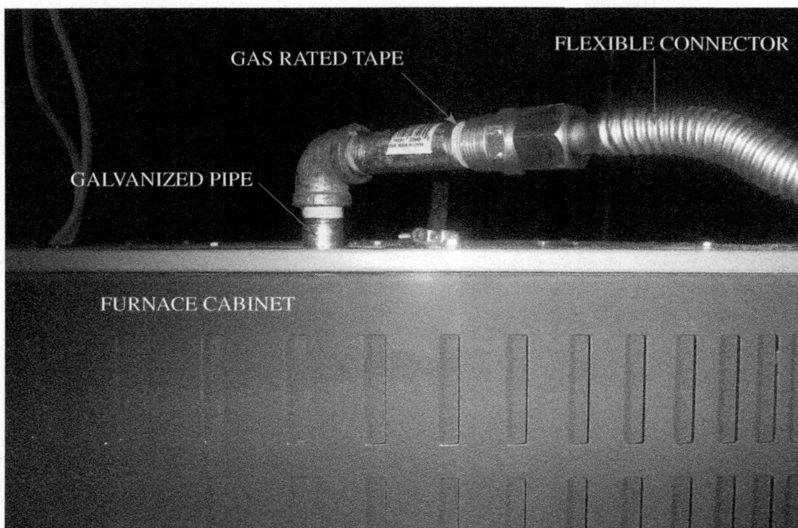

Figure 22-6 The gas connection should be hard pipe to the point where it penetrates the furnace cabinet.

valve, it should click open. If the control voltage is present on the valve and it does not open, it needs to be replaced.

The gas piping should be a piece of hard pipe from the gas valve until the point where it penetrates the furnace wall, as shown in Figure 22-6. A flexible connector going through the furnace wall may develop a leak if it rubs the penetration point. Many local codes require a hard pipe installation inside the furnace housing. A flexible connector has a thin wall compared to hard pipe. The use of flexible tubing penetrating the cabinet could possibly lead to a gas leak if the furnace vibrations cut into the thin flexible connector surface. Yellow or blue, gas-approved Teflon tape is used to seal the joints.

22.6 INDUCER FAN ASSEMBLY

An inducer fan motor is a simple shaded pole motor (see Figure 22-7). Some of these motors are three-phase ECM motors, which therefore must be checked as three-phase motors. This is the first device to be energized in the heating cycle. You can hear it operate at the beginning of the cycle. The inducer fan motor will come on to purge the heat exchanger prior to firing the burners. It will operate for about 15 to 20 seconds before the igniter is energized.

Modern heat exchanger designs are more efficient than older designs, but also more restrictive to the airflow through it. That is why an inducer fan is needed that can pull the flue gas from the heat exchanger and push it into the venting system. It is important for the vent pipe to have good flow for the flue gases. The inducer fan assembly has an air proving switch that closes when there is adequate venting power through the heat exchanger and into the vent. When the switch closes it allows the heating sequence to continue to the next step. In Figure 22-7, notice the gray hose attached to the air proving switch on the inducer housing.

As shown in Figure 22-8, the airflow can be checked by placing a tee in the hose and measuring the very low, negative air pressure. In this case, a negative pressure will be measured since the manometer fluid is being pulled toward the hose connection.

A manometer or Magnehelic® is tied into the hose to measure pressure in inches of water column, which is abbreviated **WC**. The water manometer or a pressure differential gauge can also be used to measure this low air pressure. For example, a reading of 0.42″WC or greater might be acceptable. If the airflow is restricted, the inducer will develop less than the required pressure to close the air proving switch. A reading near 0.00″WC would mean no vacuum or negative air pressure. This condition is most likely the result of an air restriction in the venting system or the motor may not be turning at the correct RPMs to vent properly.

22.7 HOT SURFACE IGNITERS

Generally, after the inducer fan purges the heat exchanger and there is positive airflow, a hot surface igniter (HSI) is activated. The HSI is a hot glowing element used to ignite the gas. The HSI replaces the older style, continuously burning pilot light, which improves energy efficiency. It operates on 120 volts. There are two types of HSIs:

- Silicon carbide
- Silicon nitride.

The silicon carbide igniter is used on older gas furnaces. It has a resistance of about 11 to 20 Ω. The carbide is very delicate and breaks easily. It is designed in the shape of a squeezed **M.** When replacing this type of burned-out HSI, do not touch the replacement igniter. Dirt or oil from your fingers will affect performance. This type of HSI may have a hairline crack that is visually undetectable but will not get hot

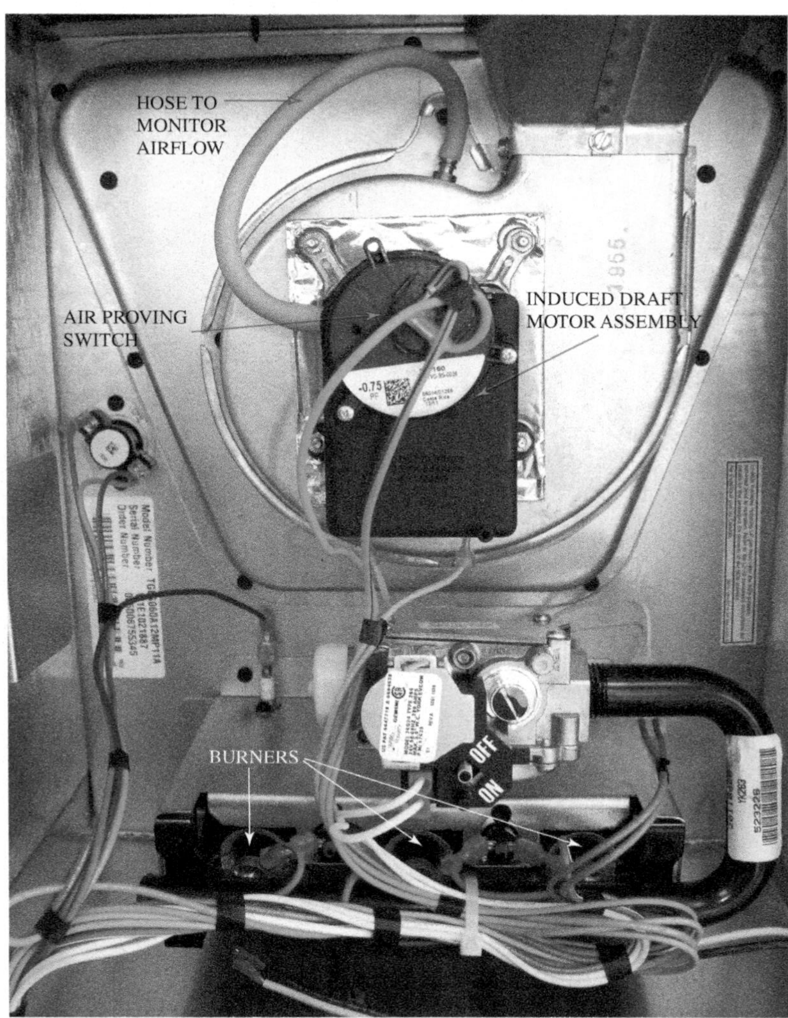

Figure 22-7 An inducer fan motor assembly is used to purge the heat exchanger and develop exhausting of the flue gases.

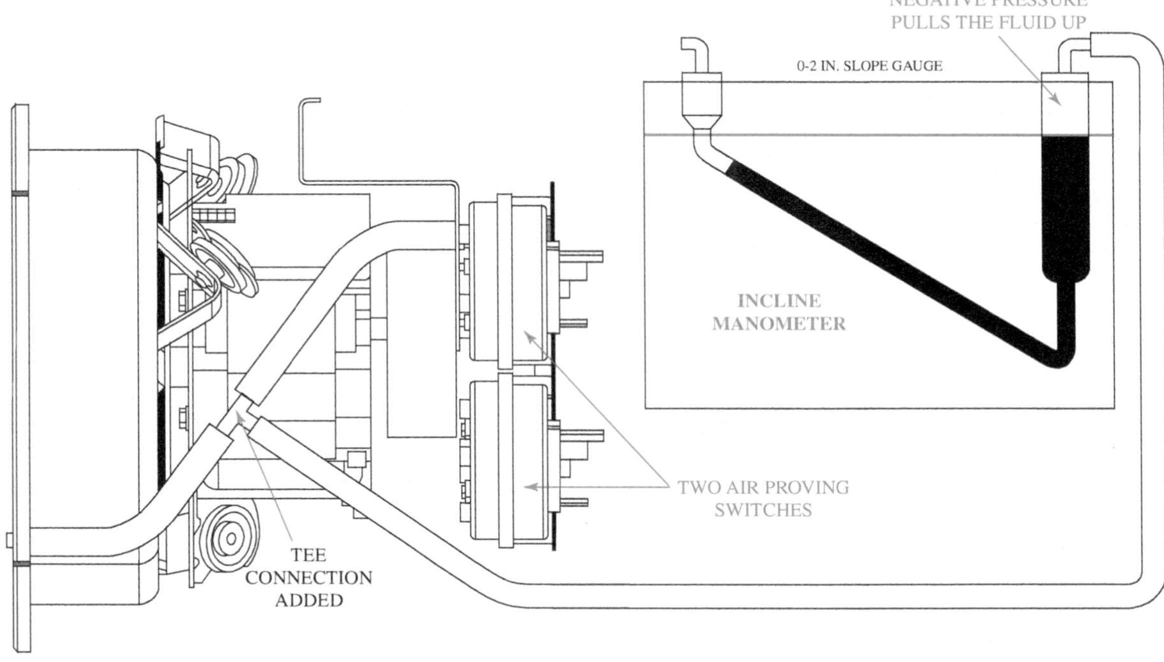

Figure 22-8 This furnace is a two-stage gas furnace. It has two air proving switches and two stages of forced venting. One switch is for first or low stage heat, and the other is for high or second stage heat. Low heat has less power venting. In this example, an incline manometer is used to test the negative pressure developed by the operation of the inducer fan assembly.

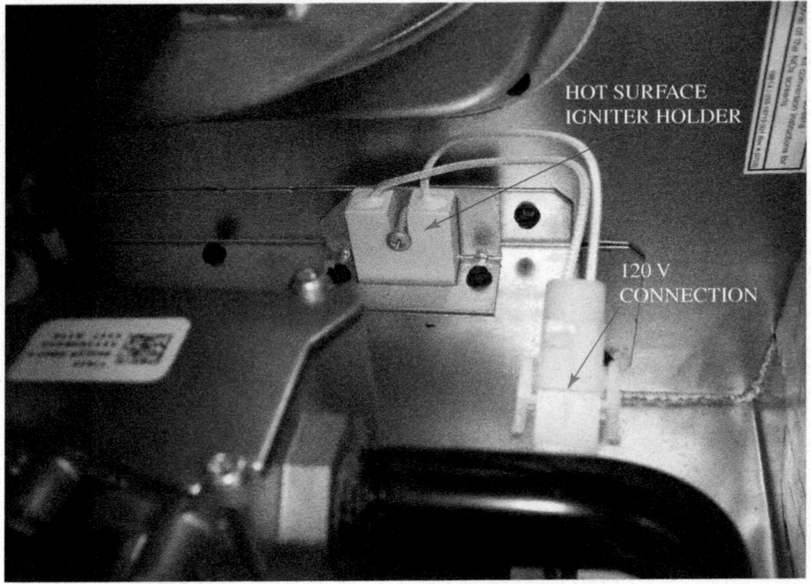

Figure 22-9 In this silicone carbide installation, you are looking at the back side of the igniter as found in the controls area. The other end of the igniter is in the burner gas stream. Note that 120 volts is needed to heat up the silicone carbide and cause the gas to ignite.

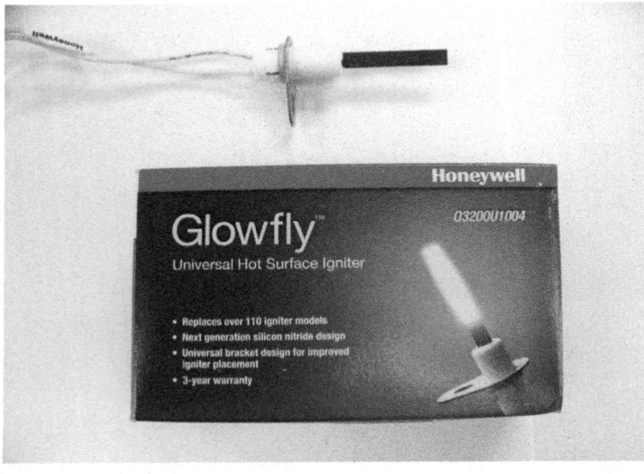

Figure 22-10 This is a silicon nitride hot surface igniter. It is tough and longer lasting compared to a silicon carbine HSI. It is energized for a short period of time to ignite the gas by a 120-V circuit.

enough to ignite the gas. The back side of a carbide igniter is shown in Figure 22-9. The carbide igniter is placed in the gas burner field.

Figure 22-10 shows a silicon nitride igniter. It is an upgrade in technology in that this igniter is very tough. It looks somewhat like a flat nail. The resistance of a good silicon nitride igniter is in the range of 40 to 75 Ω.

For both of these HSIs, high resistance is an indication that its life is about over. Higher resistance will also reduce its heat output, impairing its ability to light off the gas.

22.8 FLAME SENSORS

Modern furnaces use a flame detector or flame sensor to determine if a flame is present. It is also called a flame rod. If a flame is not detected in a few seconds or less, the circuit board will close the gas valve. The heating cycle may start again after a delay. Some designs will retry ignition a few times, then lock out. **Lockout** is the term used for not allowing the gas to cycle additional times until the power has been interrupted. The lockout is a safety device that can shut down the system to prevent raw gas from building up in the furnace.

A flame detector or flame rod is used to determine if a flame is present after the gas burner ignites. A flame detector is installed in the burner flame path and connected to ground. The other end is usually connected to the circuit board, and if a small microamp current flow to ground is not established in seconds, the control board will shut down the heating cycle. If a flame is present across the sensor, a low-current DC signal will be developed in the rod and the circuit board will allow the heating operation to continue.

Figure 22-11 shows the end of the sensor that connects to the circuit board. The other end of the flame sensor (not seen) will sense the burner flame and allow the heating cycle to continue. Figure 22-12 is a flame sensor with a part of the porcelain missing. This may affect the performance of the sensor. The flame sensor adjustment is critical and will vary among designs. Figure 22-13 illustrates one manufacturer's installation recommendations to ensure that the sensor will do its job.

Figure 22-14 shows how to check the function of the flame sensor. When the sensor "feels" the flame heat, a low current will be generated between the flame sensor and the circuit board. Depending on the manufacturer and design, a current flow signal measured in microamps (as low as 3 to 8 μA DC) may be adequate to allow the heating cycle to continue. The DC microamp meter must be placed in series with the flame sensor when taking this reading. The placement of the sensor in the flame is important to develop the proper low-current signal. The sensor cannot be grounded or the

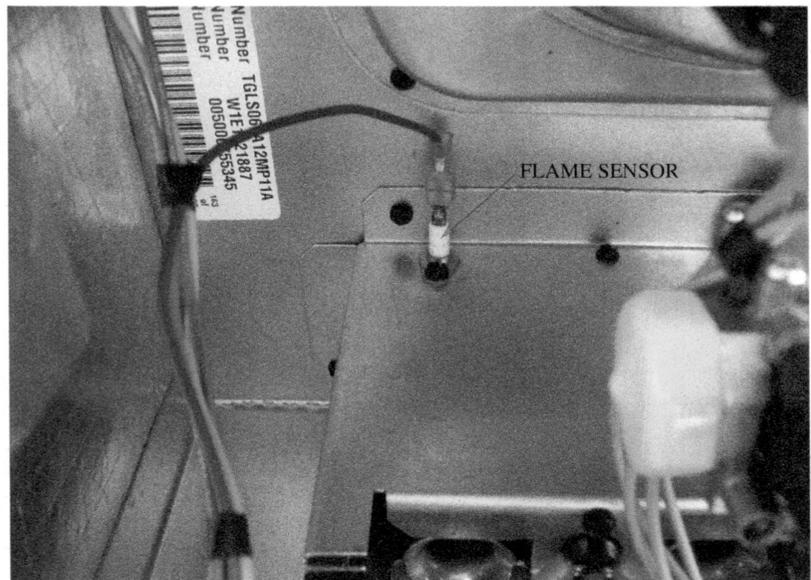

Figure 22-11 This is the back side of the flame sensor. The purpose of the flame sensor is to verify that a flame is indeed present, allowing the gas valve to continue to release gas.

low current will not reach the circuit board. This would stop the heating operation. The flame sensor should not be cleaned with sandpaper. Use steel wool or a scrubbing pad to remove buildup on the rod.

Figure 22-15 shows another type of flame sensor used in older style furnaces that have a continuously burning pilot light. This is a thermocouple used to sense the pilot light. The other end of the thermocouple is screwed into the gas valve. When the pilot burns across the upper half of the thermocouple, it generates a 20- to 30-millivolt DC signal that will energize a solenoid in the gas valve, allowing gas to flow on demand. The gas valve will open when there is a call for heat.

Figure 22-16 shows how to measure DC millivolt output. It will take the thermocouple a minute or two to develop full voltage when exposed to the initial pilot flame.

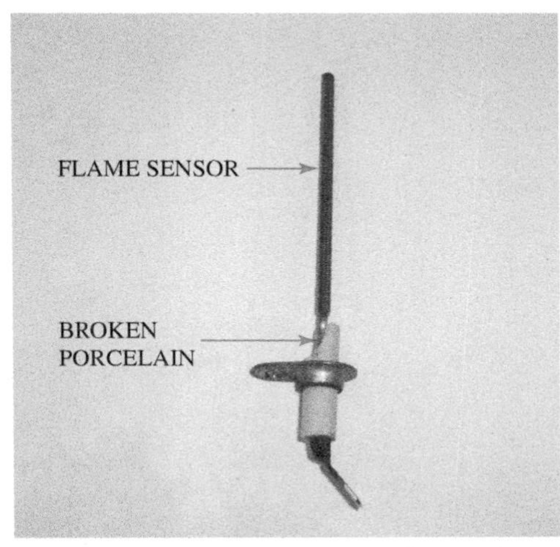

Figure 22-12 This is a flame sensor, also call a flame detector or flame rod. Notice how the porcelain insulator is broken. This might create a problem for flame sensing if it touches the furnace chassis or ground.

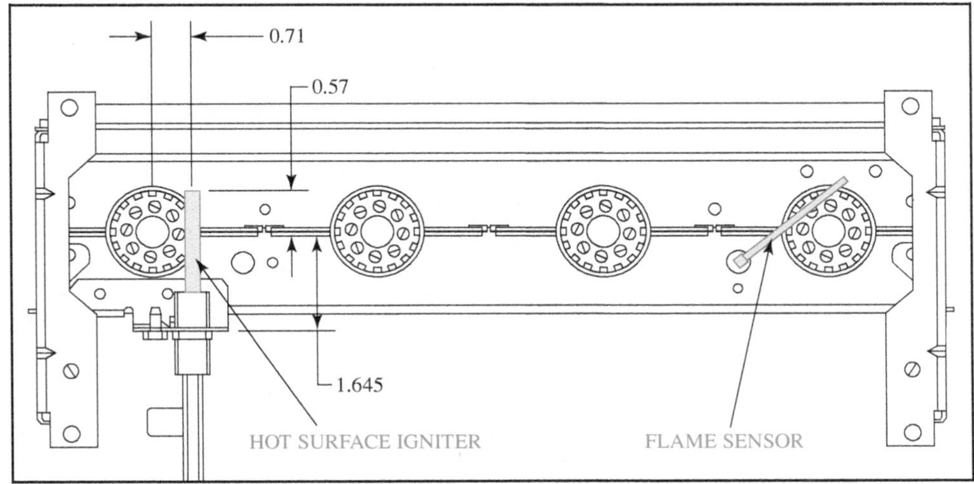

Figure 22-13 This is a manufacturer's recommendation for the installation of a hot surface igniter and the placement of the flame sensor.

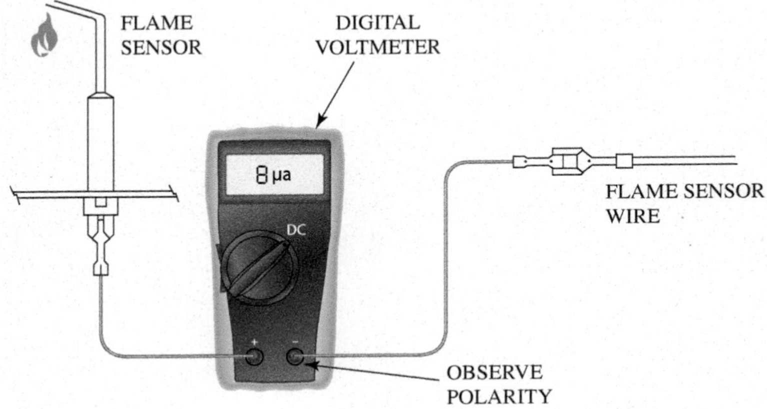

Figure 22-14 This is the setup to check the microamp current flow through a flame sensor. The DC voltmeter is placed in series with the flame sensor and circuit board. Depending on the design conditions, a range of 3 to 8 microamps (μa) DC may be adequate to allow heating to continue.

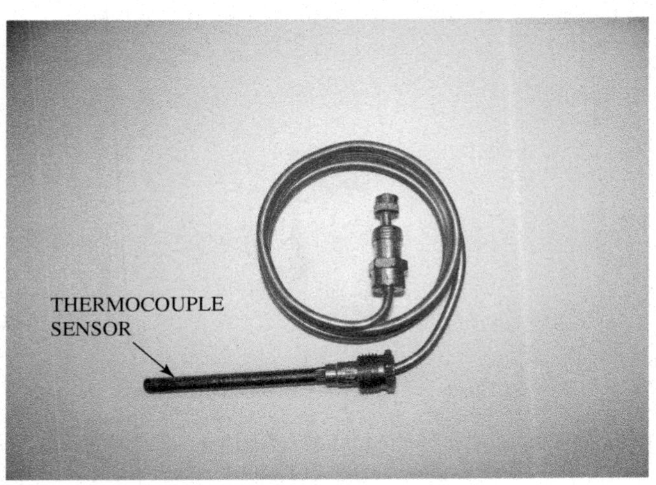

Figure 22-15 This thermocouple is used to sense the pilot light on the older style standing pilot furnaces. The other end of the thermocouple is screwed into the gas valve.

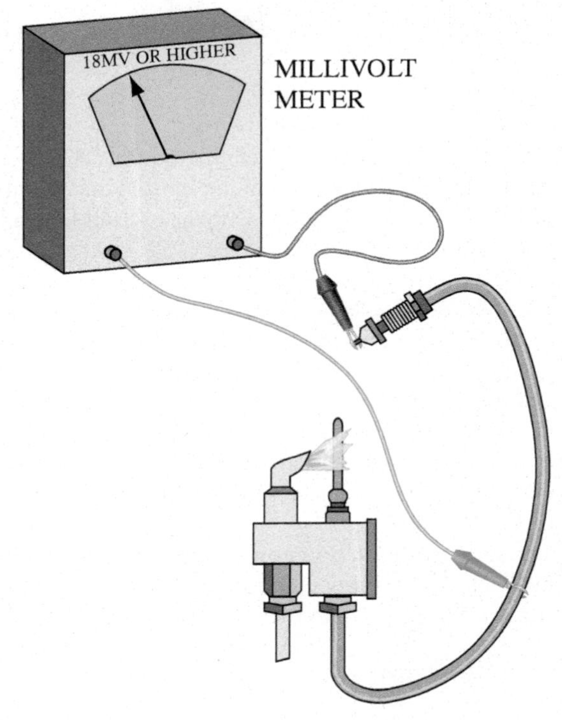

Figure 22-16 This shows how to measure the DC millivolt output of a thermocouple.

22.9 HIGH LIMIT SWITCHES

The high limit switch, as shown in Figure 22-17, is like an eyeball that "sees" excessive heat in the furnace heat exchanger section. Figure 22-18 shows a closer view of the same limit switch. It is a bimetal switch that opens when the temperature around it rises above a set temperature. Some high limits are set to open at 180°F; others are set to higher temperatures. Most of these devices are automatically reset after they open, meaning that when they cool they will snap closed again, allowing the heating cycle to start again.

If the high limit switch is defective, it will have to be replaced with the exact same part. Substituting a limit switch at a different rating will leave you open to a serious lawsuit if a furnace fire occurs—even if the fire is not related to the high limit switch.

Figure 22-19 shows the same limit switch again, but this is the back side of it, which is in the control section of the gas furnace. You can remove the wire from one side of the limit switch and check for continuity. Resistance should be near 0 Ω. Removing the louvered furnace panel will reveal the high limit switch and other critical controls.

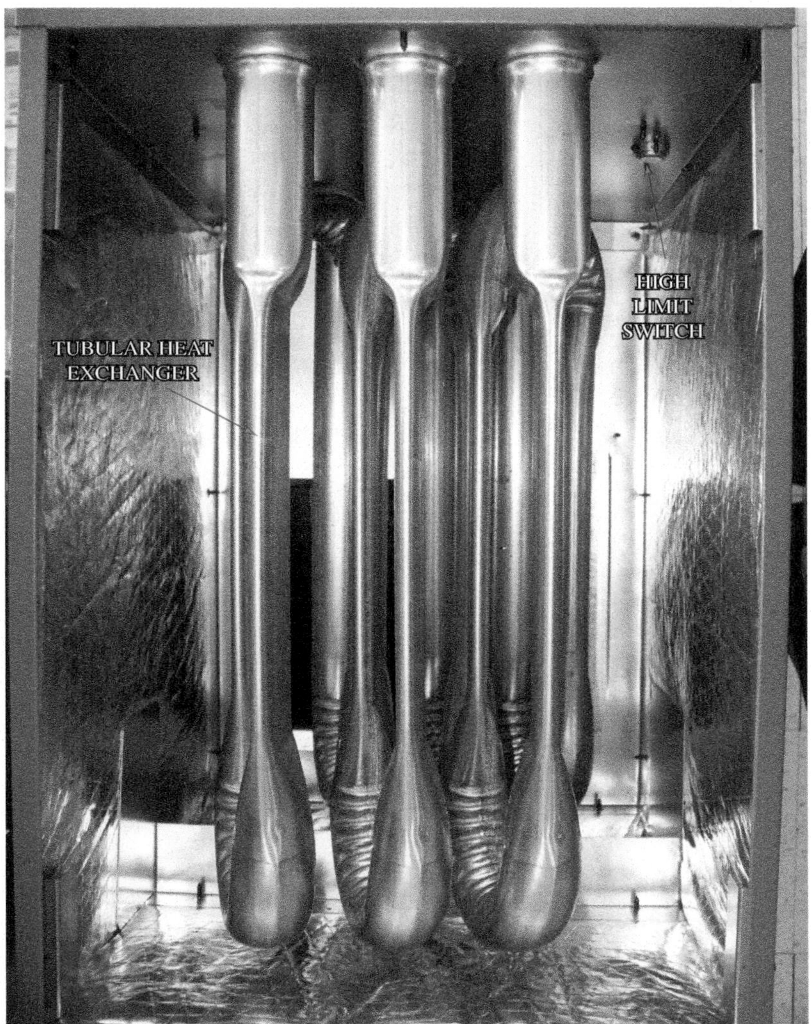

Figure 22-17 This is a view of a gas furnace before it is installed. This furnace has a tubular heat exchanger. The high limit switch is seen on the right. This area will get excessively hot if there is a lack of airflow.

Tripping or opening of the high limit switch is usually caused by low or no airflow. For example, the air filter or evaporator may be dirty. The blower motor may be off or the blower squirrel cage dirty. Anything that reduces the airflow significantly will cause the high limit switch to open.

Finally, Figure 22-20 shows an extended high limit switch found in the air stream of some gas furnaces. The extended device "sees" or "senses" the airflow better since it extends into the air stream. Most of these high limit devices are of the automatic reset design. Most high limit switches fail because they were toggling on high temperatures.

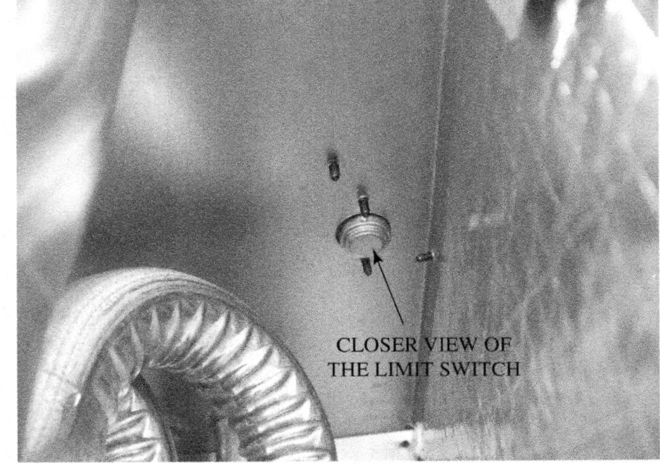

Figure 22-18 This is the same furnace shown in Figure 22-17, but in a closer view. Notice how the high limit switch resembles an eyeball.

22.10 FLAME ROLLOUT SWITCHES

The flame rollout switch is really a high limit switch that must be reset manually. Figure 22-21 shows the red reset button on a flame rollout switch. A flame rollout is a very dangerous event because it means the flames have reached into the

Figure 22-19 This shows the back side of the high limit switch from Figure 22-17 in the control section of the gas furnace. Removing the louvered furnace panel will reveal the high limit switch and other critical controls.

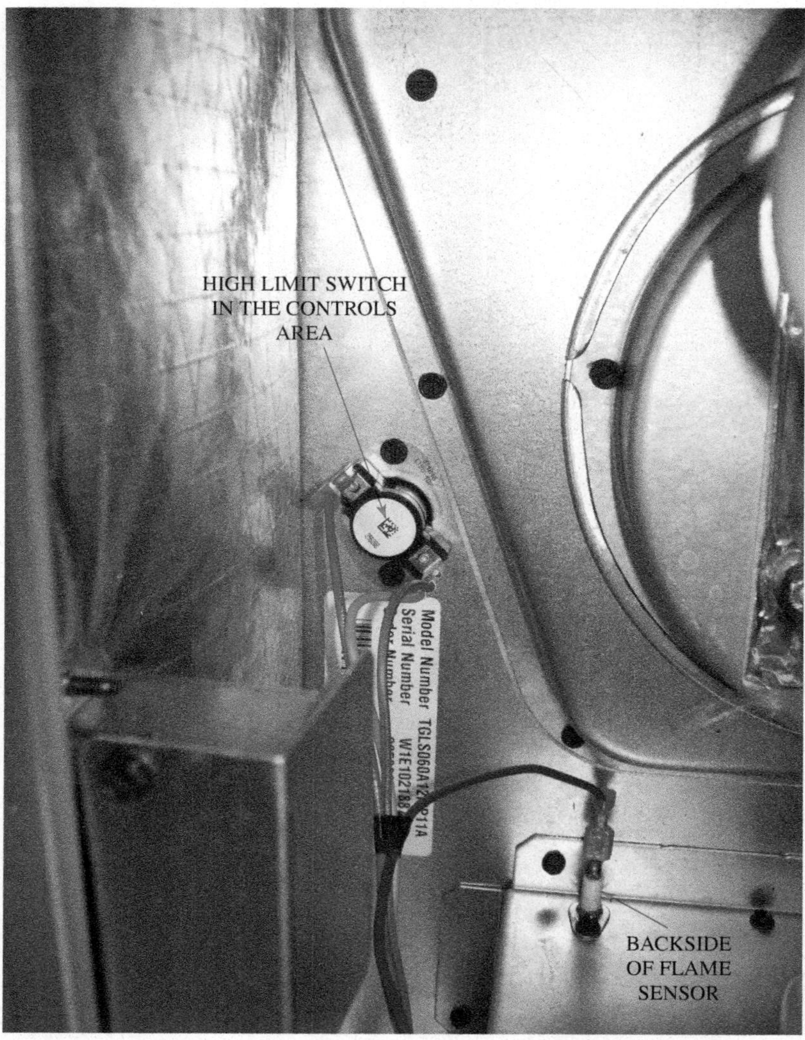

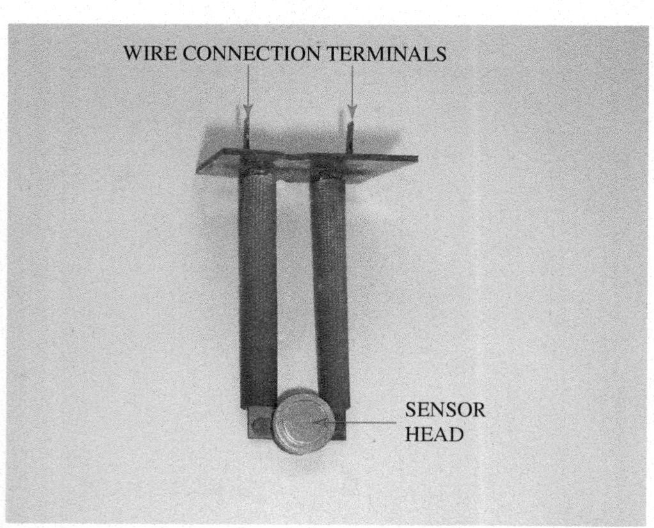

Figure 22-20 Depending on the design this high limit switch extends 3 to 4 inches into the furnace airflow stream. The bimetal sensor head is about the size of a dime.

furnace control area from the burner area. This can burn wires or start a furnace fire. For this reason, a rollout switch must always be of the manual reset design. You will need to push a reset button to close the switch. A tech needs to check out this condition. Flame rollout may be caused by inadequate venting, a lack of combustion air, or very high gas pressures.

To test this safety switch, remove voltage and a wire from one side to test for continuity. Push the red button. Expect 0 Ω if the switch is good. As with any safety device, replace the switch with a like device.

TECH TIP

A high number of the circuit boards returned to distributors do not have any problems. One distributor reports over 50% returns that have no problem. Use routine checks before condemning a board: Check for a good equipment ground, 120-V polarity, connections, manufacturer's troubleshooting sequence, etc.

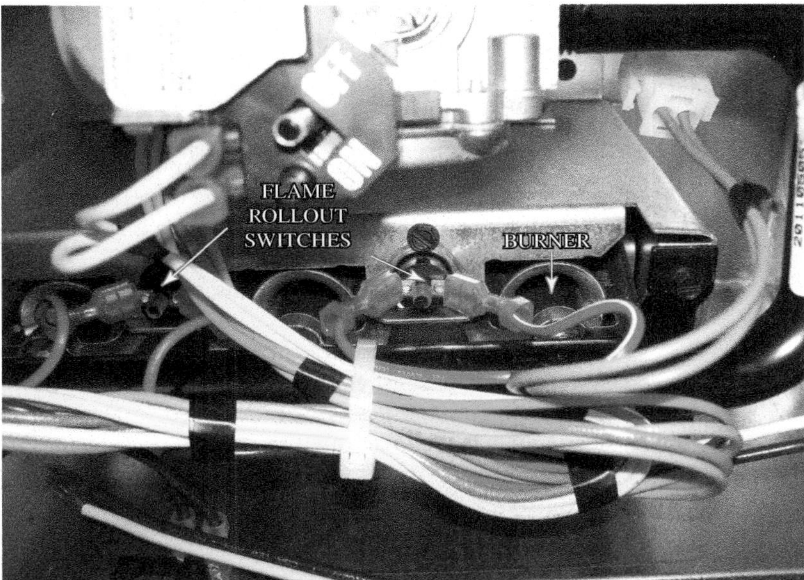

Figure 22-21 Flame rollout switches are installed near the burners in case flames back up out of the burner compartments. Flame rollout switches are manual reset safety devices. Remove voltage and a wire from one side to test for continuity.

22.11 CIRCUIT BOARDS

The circuit or control board, as shown in Figure 22-22, is the brain of a furnace system. It controls signals to the gas valve and blower operation. It keeps the furnace operation safe if one of the external safety devices opens or sends a problem signal. Many circuit boards provide a technician with LED flashing fault codes, airflow status, and CFMs. Table 22-4 is a sample fault code table similar to those found on the inside of the panel door of the blower section of a gas furnace.

A circuit board contains high- and low-voltage connections. It also has a control voltage fuse in the range of 3 to 5 amps. Always use the same size or a smaller replacement fuse. The board can usually handle up to 10 amps for blower motor operation. A transformer may be mounted on the board or simply attached as shown in the lower right side of Figure 22-22.

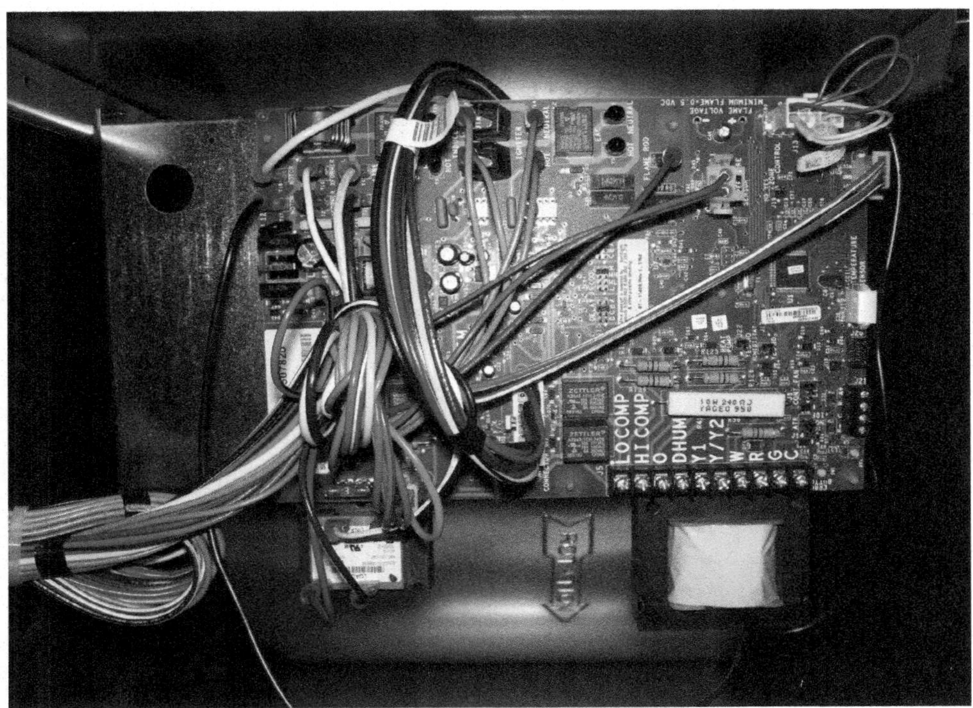

Figure 22-22 This is a circuit board used in a residential gas furnace. This control board is the brain of the system.

Table 22-4 Sample Fault Code Table

Error Code	LED Indication
Normal Operation	On
Hardware Failure	Off
Fan On/Off Modified	1 Flash
Limit Switch Fault	2 Flashes
Flame Sensing Fault	3 Flashes
4 Consecutive Limit Switch Trips	4 Flashes
Ignition Lockout Fault	5 Flashes
Induced Draft Motor Fault	6 Flashes
Rollout Switch Fault	7 Flashes
Internal Control Fault	8 Flashes

TECH TIP

A dusty or dirty circuit board is an indication of a dirty air filter or leaks in the return air ducts. Check out these problems. Use a spray duster, like that used on computers, to gently clean the board. Have your vacuum turned on to remove loose dirt as it breaks loose.

22.12 BLOWER MOTOR ASSEMBLY

The blower motor assembly is behind the circuit board (refer to Figure 22-22). This assembly includes the motor, possibly a run capacitor, the squirrel cage blower, and the blower housing. New blowers are controlled by a circuit board time-delay option. With this option, a blower can be turned on 60, 90, or 120 seconds after the furnace has been energized. This gives the furnace heat exchanger a chance to warm up prior to blower start-up. The blower may come on with a call for heat even if the burner did not fire off for some reason.

The blower cycle is set by the manufacturer and in some cases the tech is given the option of selecting a 60- or 90-second time off delay after the thermostat is satisfied. This gives the blower a chance to purge the residual heat out of the heat exchanger and ductwork. If the blower is a PSC motor, it will most likely run on medium speed in the heating mode. If it is an ECM motor, it will ramp up and down around the call for heat. The ECM will provide smooth starting and stopping and more even airflow for the conditioned space.

22.13 LOW-EFFICIENCY GAS FURNACE OPERATION

Every gas furnace must be able to do the following generic sequence of operation in the following order:

1. The thermostat closes because the room temperature has dropped or the temperature has been adjusted to a higher temperature.
2. The control voltage is sent through the thermostat **W** terminal to start the furnace operating.
3. The inducer draft motor is energized and the air proving switch closes, allowing the next stage to start.
4. The hot surface ignition or standing pilot is ready for the gas supply to begin.
5. The gas valve is energized. It opens, allowing gas to start the ignition process.
6. The flame sensor creates a current flow from the burner flame. This establishes that the burners are functioning.
7. The burners and gas rate are adjusted to regulate the supply of fuel and air mixture.
8. The furnace monitors combustion through safety controls.
9. When the heat exchanger warms, the blower motor moves heat to the living space though a duct system.
10. The thermostat opens the **W** circuit because the air has warmed to its setpoint.
11. The gas valve closes and the burners extinguish.
12. The blower shuts down after the heat exchanger cools to around 100°F. This takes about 1 or 2 minutes after the burners shut down.

How each gas heating system performs these operations depends on the design of the unit.

The signal received from the thermostat in step 1 of the sequence of operation is typically 24-V power sent through a small set of contacts in the thermostat. Contacts close in response to a drop in temperature. When the contacts close and if the circuit is functional, the control circuit to the gas heating system is complete; the signal is received and the gas unit starts the ignition process.

TECH TIP

When a thermostatic switch closes a set of contacts as the result of a drop in temperature, the action is referred to as "close on temperature drop." In this case we are talking about a heating thermostat. Sometimes, the expression most often heard is "close on drop" and refers to the way the switch works. This can be confusing when describing the operation of a switch that works in the opposite way. If the switch opens on temperature drop (cooling t'stat), instead of closing as described above, it could be identified as an "open on drop" or "close on rise" switch.

The ignition process mentioned in step 5 of the sequence of operations can be very different with each type of gas heating system. Low-efficiency units have a "standing pilot." A standing pilot is a small flame that is allowed to burn 100% of the time. The standing pilot (see Figure 22-23) is used to

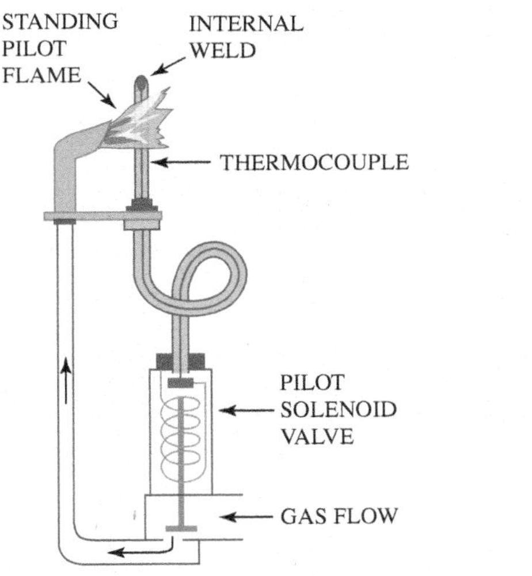

Figure 22-23 The thermocouple develops 20 to 30 millivolts DC to energize the pilot solenoid valve and open gas flow to the standing pilot flame. This allows the main gas valve (not shown) to send gas to the burners and be ignited by the pilot flame.

ignite the main burner when a signal is received from the thermostat.

In the case of a standing pilot, the flame is sensed (step 5) through a thermocouple circuit. The thermocouple is a device that is made of two different metals that are internally welded at the tip of the thermocouple. When heated, the two different metals cause a weak electrical voltage of 20 to 30 mV DC to produce a current of electricity. The current flow is connected to a coil or solenoid that is wrapped around a movable plunger. When the thermocouple is heated, the low current flow is enough to hold the plunger in the open position and gas flows to the pilot. The current is not strong enough to lift the plunger. The pilot flame is lit while manually holding the plunger up, until the thermocouple is heated, which could take 30 to 90 seconds. If the flame is lost, the plunger drops and stops the flow of gas to the pilot. This acts as one of the safety devices (step 4) that monitors the flame.

Figure 22-24 is a simplified diagram for a gas furnace that uses a standing pilot light. The yellow highlighting in

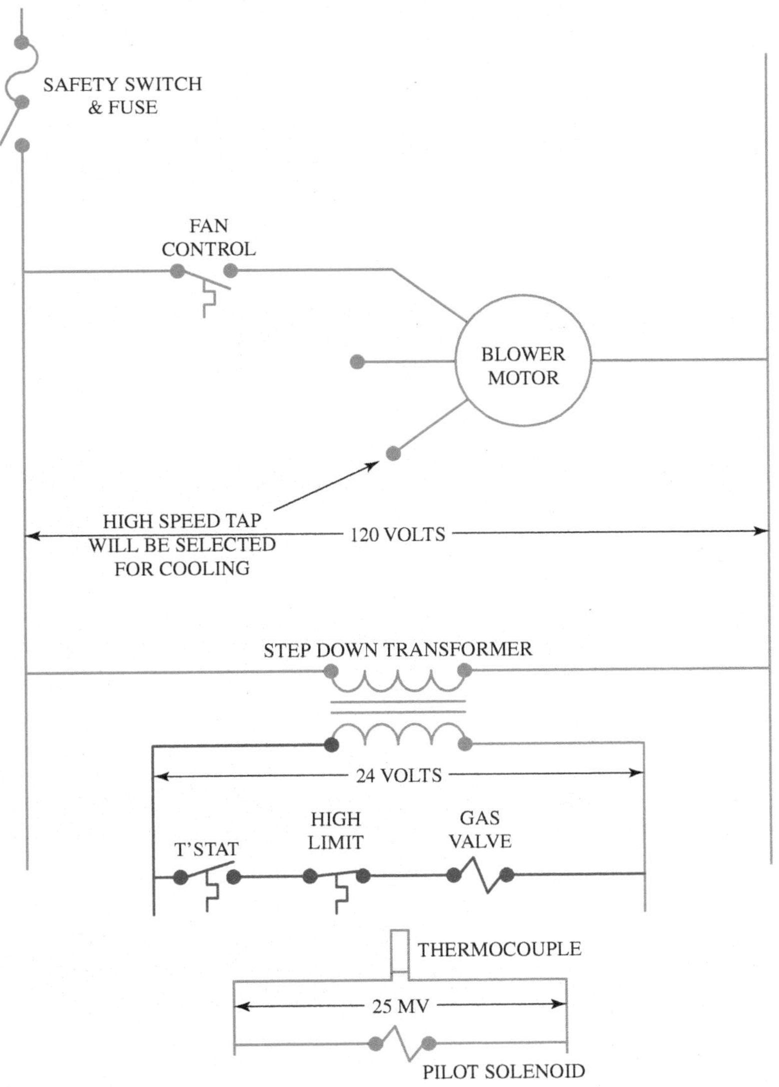

Figure 22-24 This is a simplified diagram of a gas furnace with a standing pilot light. Figure 22-25 traces the circuit's operation.

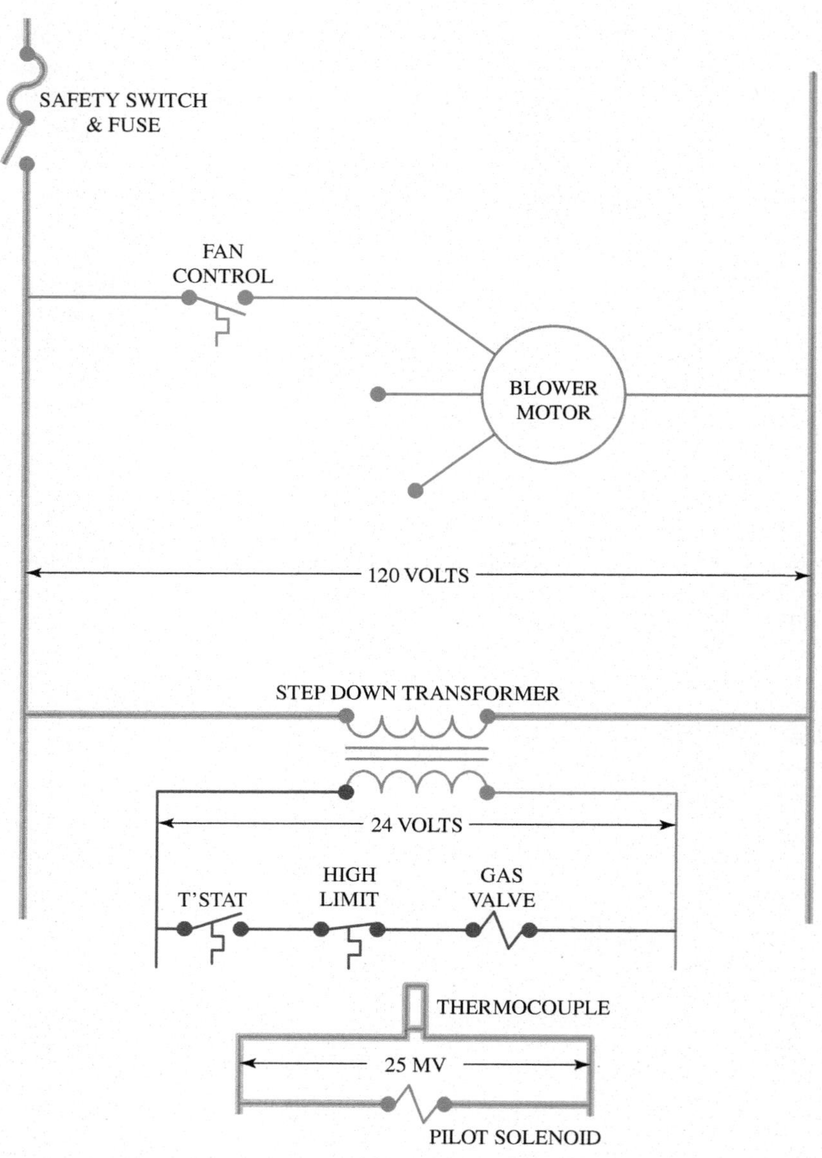

Figure 22-25 The yellow highlighting indicates what is operational prior to a call for heat.

Figure 22-25 shows what occurs when the pilot is lit and prior to a call for heat. This part of the system is electrically active, supplying 120-V power to the control transformer. The safety switch is closed and the power moves down to the step-down transformer, also active and supplying 24 volts to the control circuit. The thermocouple generates 20 to 30 millivolts DC to energize the pilot solenoid when exposed for a pilot flame for at least 60 seconds.

SEQUENCE OF OPERATION 22.1: MAIN GAS VALVE

The regulation of gas (step 5 of the generic sequence of operations listed earlier) to the main burner is accomplished with the main gas valve. Use Figure 22-26 as you study the following sequence of operations:

1. At point **1** the thermocouple is heated by the pilot light. The flame generates a DC millivolt signal that energizes the pilot valve solenoid. The gas valve can now open when energized by the 24-V signal.
2. At point **2**, the thermostat closes, sending the 24-V signal (see yellow highlighting) to the gas valve.
3. When the thermostat closes on temperature drop, the gas valve at point **3** is energized through the 24-V control circuit. In this circuit is another switch, the high limit switch, which needs to be closed. The gas continues to flow through the main gas solenoid valve as long as current is allowed to flow.

4 When the thermostat is satisfied at point **2** and opens at point 4, the electrical flow through this circuit is interrupted and the main gas solenoid closes, cutting off the flow of gas and extinguishing the main burner flame. The pilot continues to burn as long as the gas is provided to the furnace.

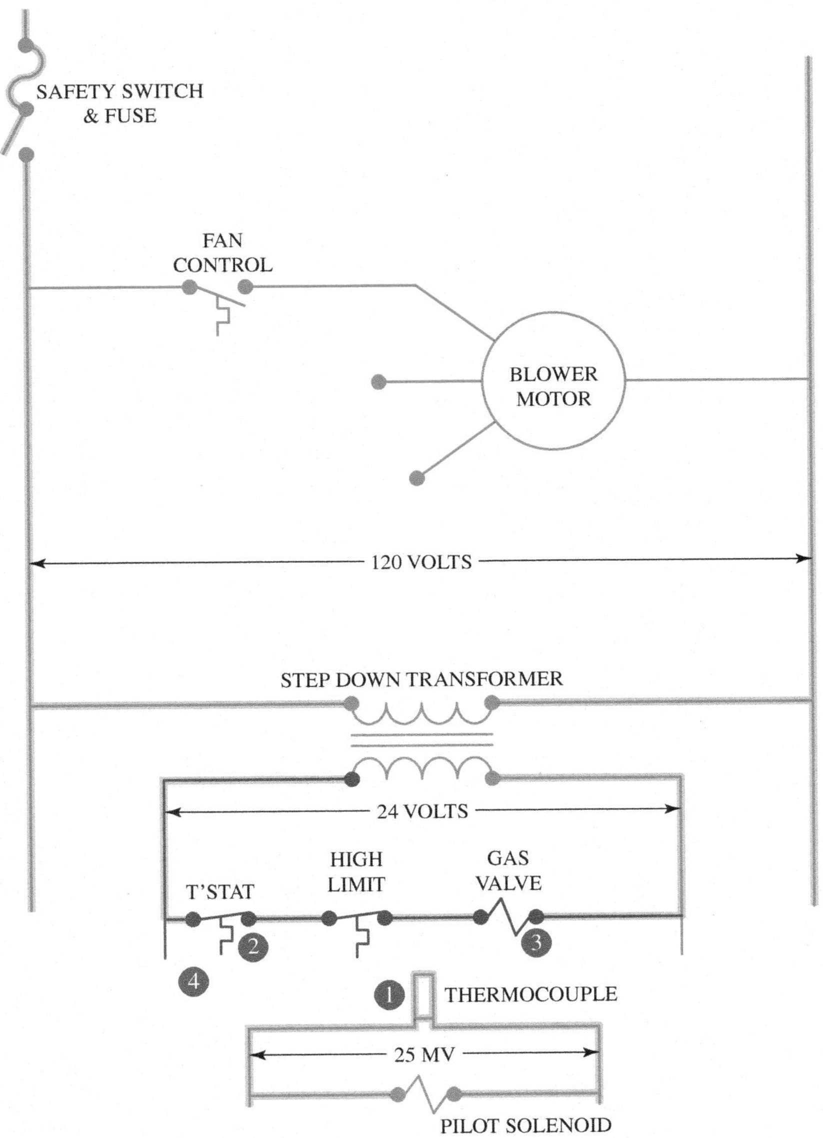

Figure 22-26 When the t'stat in the 24-V control circuit closes, it energizes the gas valve and allows gas to flow to the burners. The fan switch will close in about a minute of operation as the heat exchanger heats. This starts the blower.

Another switch that monitors the flame, indirectly, is the high limit control (step 8). This control senses heat in the heat exchanger and opens on rise to cut the flow of electricity to stop the flow of gas, only if the exchanger temperature is too high. It acts as a safety control to monitor the main gas valve and main burner flame. If, for instance, the flow of air over the heat exchanger is not enough to keep this switch closed, the high limit will open on temperature rise and shut down the main burner. The high limit may also be combined with another switch and act as one control, called the "Fan & Limit." The combo switch will control the operation of the fan when the heat exchanger gets hot and interrupt the gas flow if the heat exchanger gets too hot.

SEQUENCE OF OPERATION 22.2: BLOWER OPERATION

In combination, the fan or blower is operated (step 9 of the generic sequence of operations) with another set of contacts to turn the blower on when enough heat builds up in the heat exchanger. It also turns off the blower when the heat exchange temperature drops. The blower operation is highlighted in Figure 22-27 and explained here:

1. When the gas valve is energized at point **1**, it begins to warm the heat exchanger.
2. The fan switch closes at point **2** when the exchanger reaches its temperature setpoint.
3. The closed fan switch will operate the blower at point **3**. The fan switch at point **2** opens when the gas valve has closed and most of the heat has been purged from the heat exchanger.

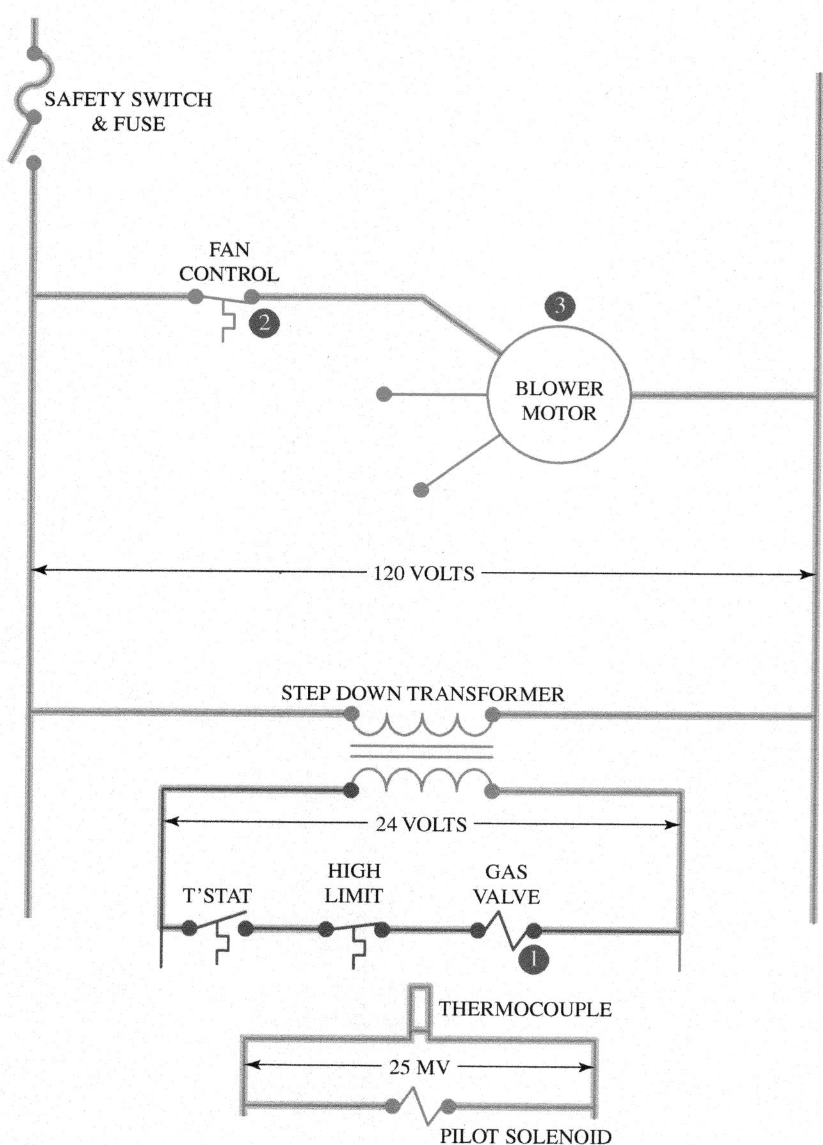

Figure 22-27 This diagram highlights the final stage in the heating operation. When the fan control switch gets warm enough to close, the blower begins to operate on medium or low speed. The high-speed connection (one of the two unwired connections) will be used for cooling operation if needed.

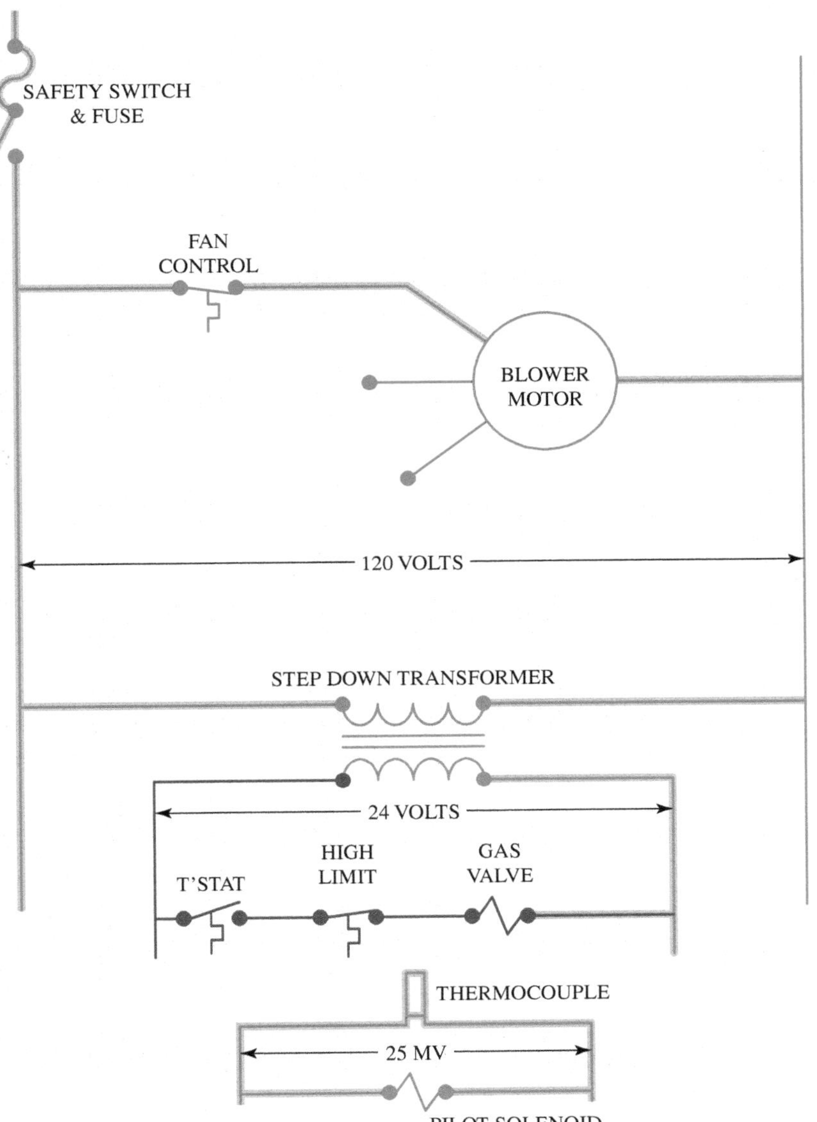

Figure 22-28 When the thermostat is satisfied, the control circuit is de-energized and becomes inactive. The gas valve closes and the gas to the main burner is shut off. The pilot continues to burn. The blower is still removing heat from the heat exchanger.

As shown in Figure 22-28, when the thermostat is satisfied, the control circuit is de-energized and becomes inactive. The gas valve closes and the gas to the main burner is shut off. The same thing would happen in this circuit if the thermostat remained closed (calling for heat) and the high limit switch opened. This is because both switches are in series in this circuit. If either switch opens, the gas valve is de-energized and there is no gas flow.

The rise in space temperature satisfies the t'stat and the main burner flame is turned off. When the thermostat has been satisfied, the fan control ensures that the residual heat in the heat exchanger has been moved to the living space. But, there is still a small flame; the pilot is shown in Figure 22-29 and the blower shuts off with the fan control open. The pilot should not go out. Instead, it should remain burning, waiting for the thermostat to call for the main gas valve and light the main burner once again. If, by chance or by fault, the pilot goes out, the pilot solenoid valve will not have enough current to hold the valve open. At that point the valve will close, sensing that the flame has been extinguished.

22.14 HIGH-EFFICIENCY OPERATION

As gas heating systems evolve from low-efficiency, standard gas-burning appliances to more efficient models that transfer heat to the air more effectively, more electrical devices have been added. Since this is an electrical book our emphasis will be on the electrical components.

Systems that are 80% efficient will have an ignition system, vent fans, and integrated circuits to monitor the various pressure switches and sensors. Each of these systems and individual devices helps the unit burn gas more efficiently. More usable heat is able to be generated from the raw gas and transferred into the living space. There is a trade-off of efficiency

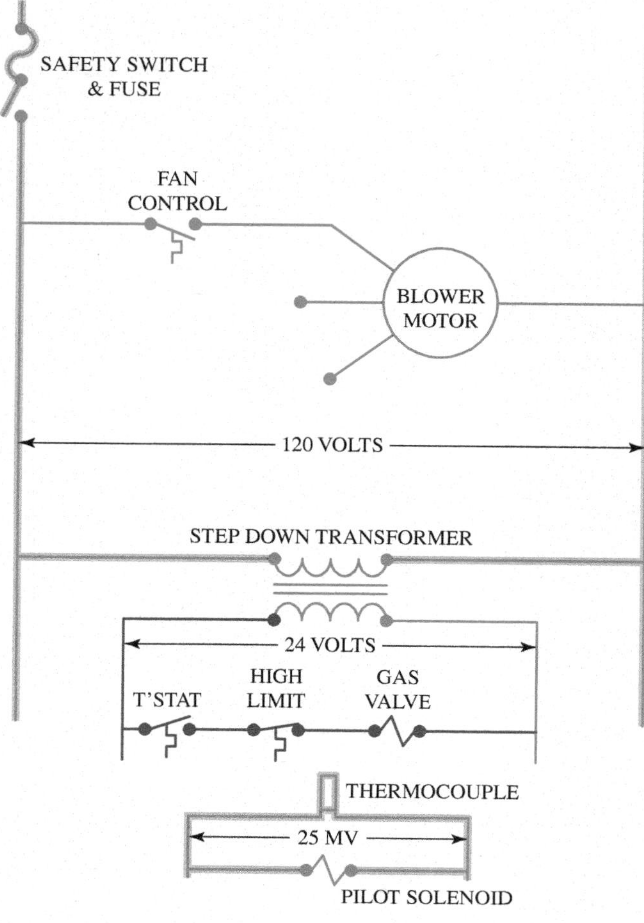

Figure 22-29 When the heat exchanger cools down, the fan control opens, turning off the blower motor. The unit is ready to start again when the thermostat senses a drop in temperature and begins the cycle again.

and price. Gas heating systems that are more efficient generally cost more than lower efficiency units. It is expected that the higher cost is offset by lower long-term gas bills.

The wiring diagrams for 80 AFUE and 90 AFUE systems appear very similar. Both have heat exchangers that have been lengthened, which causes the hot exhaust gas to travel longer distances. This design allows for longer heat exchanger contact with the air and, thus, more heat transfer. What is also interesting is that the exhaust gas is forced to travel downward in the heat exchanger so that the part of the exchanger that is the coolest (near the point where the vent motor is attached) is where the air returned from the living space is first blown. This air is the coolest and tends to cool the exhaust gas to its lowest temperature, thus extracting the last little bit of heat for the living space.

A similar general operating sequence listed in the previous section is found in high-efficiency heating systems:

1. The thermostat closes because the room temperature has dropped or the temperature has been adjusted to a lower temperature.
2. The control voltage is sent through the thermostat to start the furnace operating.
3. The gas valve is energized. It opens, allowing gas to start the ignition process.
4. Sense that a flame is established.
5. Regulate the supply of fuel and air mixture.
6. Monitor combustion through safety controls.
7. When the heat exchanger warms, the blower motor moves heat to the living space.
8. The thermostat opens because the air has warmed to its setpoint.
9. The gas valve closes and the burners extinguish.
10. The blower shuts down after the heat exchanger cools to around 100°F. This takes about 1 or 2 minutes after the burners shut down.

The signal from the thermostat is received from **R** to **W,** as highlighted in blue in Figure 22-30A (found at at the bottom of the diagram). At point **1**, the **IFC** (integrated furnace control) begins the heating cycle. The **IFC** is a circuit board. The inducer vent motor at point **2** is started through the closed limit switch **TCO-B** at point **3** to pre-purge the combustion chamber to remove any unburned fuel or vapor. After the vent motor has operated, the pressure switch closes when it senses negative exhaust pressure in the combustion area. The **IFC** also checks to see if there is enough pressure drop to safely exhaust combustion gases by ensuring the **PRESSURE SWITCH** at point **4** has closed. When pressure drop is sensed, the pressure switch closes, sending the signal to the **IFC**. The pressure switch is connected in series with the limit switch **TCO-A** and the **MANUAL RESET FLAME ROLLOUT SWITCH** at point **5**. These switches would open if the temperature of either switch became higher than their rating. At the beginning of the cycle, these two temperature switches should be closed because they are cool.

TECH TIP

The rollout switch is sometimes a fuse-type safety switch. This means that if the switch is open, it needs to be replaced. Fuse-type safeties cannot be reset. Some, on the other hand, can be manually reset. Rollout switch temperature limits are set by the manufacturer. Always replace the rollout switch with the manufacturer's direct replacement.

At the end of the pre-purge cycle, all three switches must be closed (limit switch, rollout switch, and pressure switch) before the **IFC** (integrated furnace control) turns on the **HSI** (hot surface igniter) as seen at point **1** in Figure 22-30B. The gas valve at point **2** opens to begin the ignition process. At this time gas is flowing to the combustion or burner area. When the gas hits the hot surface igniter (**HSI**), the gas will ignite to produce a flame. The **FLAME SENSOR**, at point **3**, must sense the presence of a flame by allowing

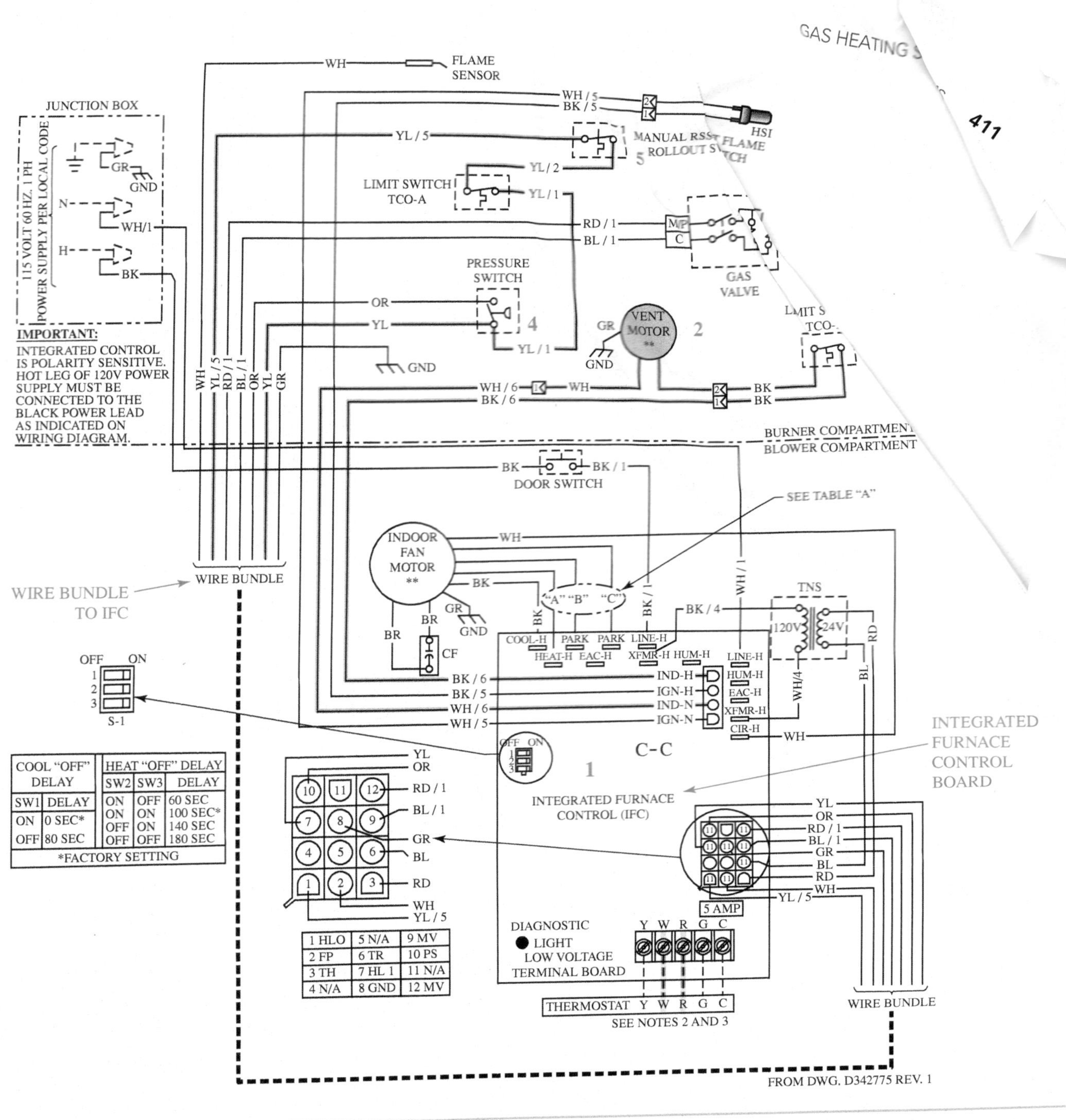

COOL "OFF" DELAY		HEAT "OFF" DELAY		
SW1	DELAY	SW2	SW3	DELAY
ON	0 SEC*	ON	OFF	60 SEC
OFF	80 SEC	ON	ON	100 SEC*
		OFF	ON	140 SEC
		OFF	OFF	180 SEC

*FACTORY SETTING

1 HLO	5 N/A	9 MV
2 FP	6 TR	10 PS
3 TH	7 HL 1	11 N/A
4 N/A	8 GND	12 MV

Figure 22-30A The 24-V signal is received from the thermostat at the **R** and **W** terminals as highlighted in blue (bottom right side). The vent motor is operated to pre-purge unburned gases and vapor from the combustion chamber. The limit switch **TCO-A** and flame rollout switch must be closed in order to "prove" the vent motor operation. The pressure switch will close next to finally prove vent motor operation.

the conduction of a small amount of DC power through the flame and back to the **IFC**, as highlighted in Figure 22-30B. As seen here, if the **IFC** can detect the presence of this small power source, the flame is proven and the combustion cycle proceeds. If not, the **IFC** turns off the gas. Depending on the manufacturer and model, the **IFC** may be programmed to attempt to relight the burner or it may go into the safety lockout mode (always check the manufacturer's literature for operation).

The **IFC** is preprogrammed (see switch setting). The blower in Figure 22-30B, point **4**, is timed to turn on when the heat exchanger is warmed. The blower moves warmed air from the heat exchanger to the living space. At the end of the cycle, the thermostat opens and the heating system starts the shutdown cycle. The gas valve is shut and the blower continues for a predetermined amount of time to ensure that the heat exchanger is cool and the warmed air is driven in the conditioned space.

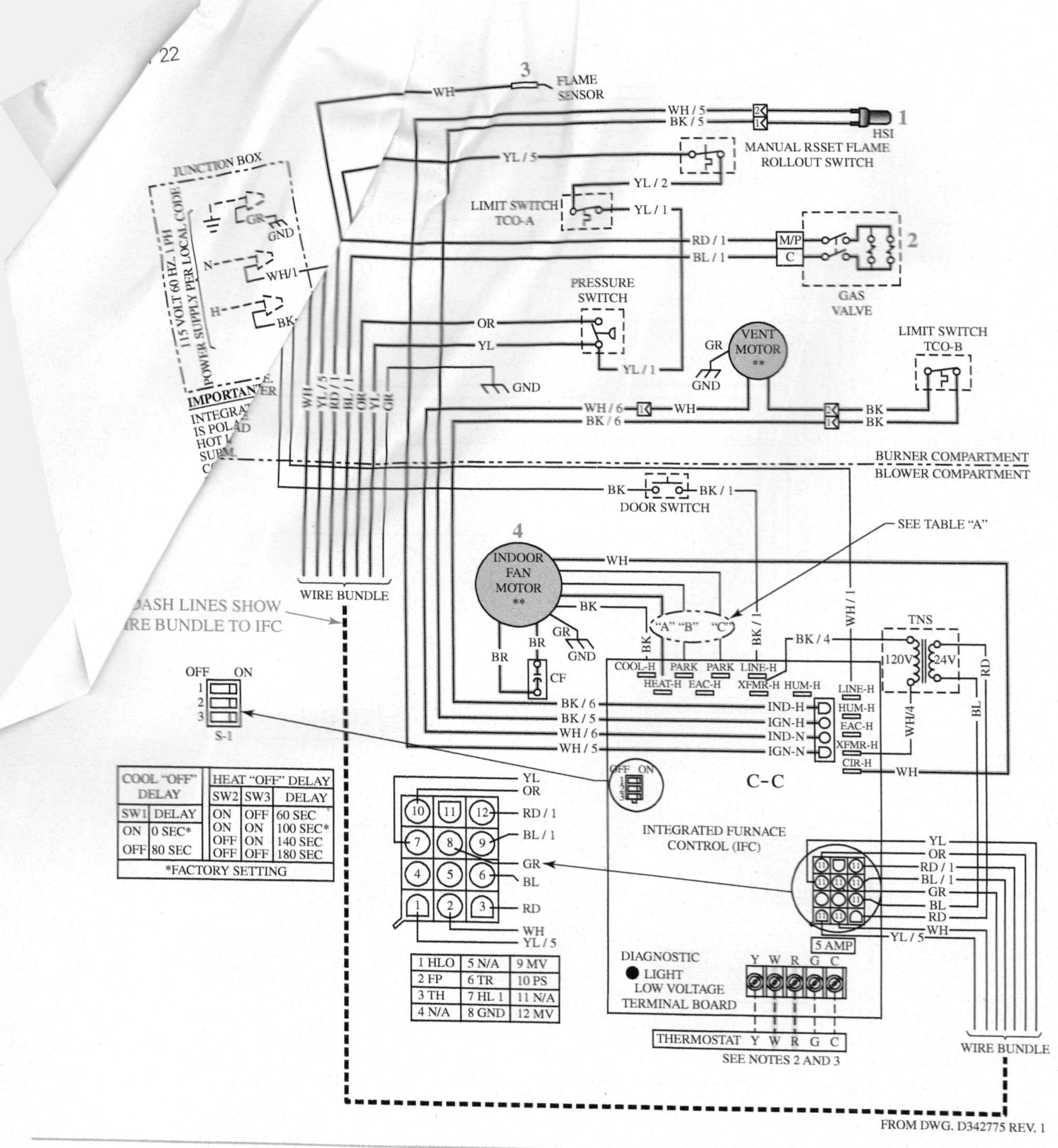

FROM DWG. D342775 REV. 1

Figure 22-30B The flame sensor detects the flame, allowing a small amount of DC power to flow to the integrated furnace control (**IFC**).

22.15 BASIC TROUBLESHOOTING PROCEDURES

Follow these basic procedures when checking gas heating systems:

1. Check for power. This basic check needs to be made routinely before anything else is investigated. Always ensure that the correct line voltage is present. Figure 22-31 gives an example of how to find the correct 120-V polarity.
2. Check for power in the control circuit. Sometimes this basic check and the first basic check can be done at the same time. If 24 volts is measured at the thermostat, then it can be assumed that the correct line voltage is present. If any deviation from 24 volts is measured, the line voltage needs to be checked.
3. Check that the thermostat is set to call for heat. If not, set it a few degrees above the room temperature.
4. Check that all power switches are on and all fuses are good.

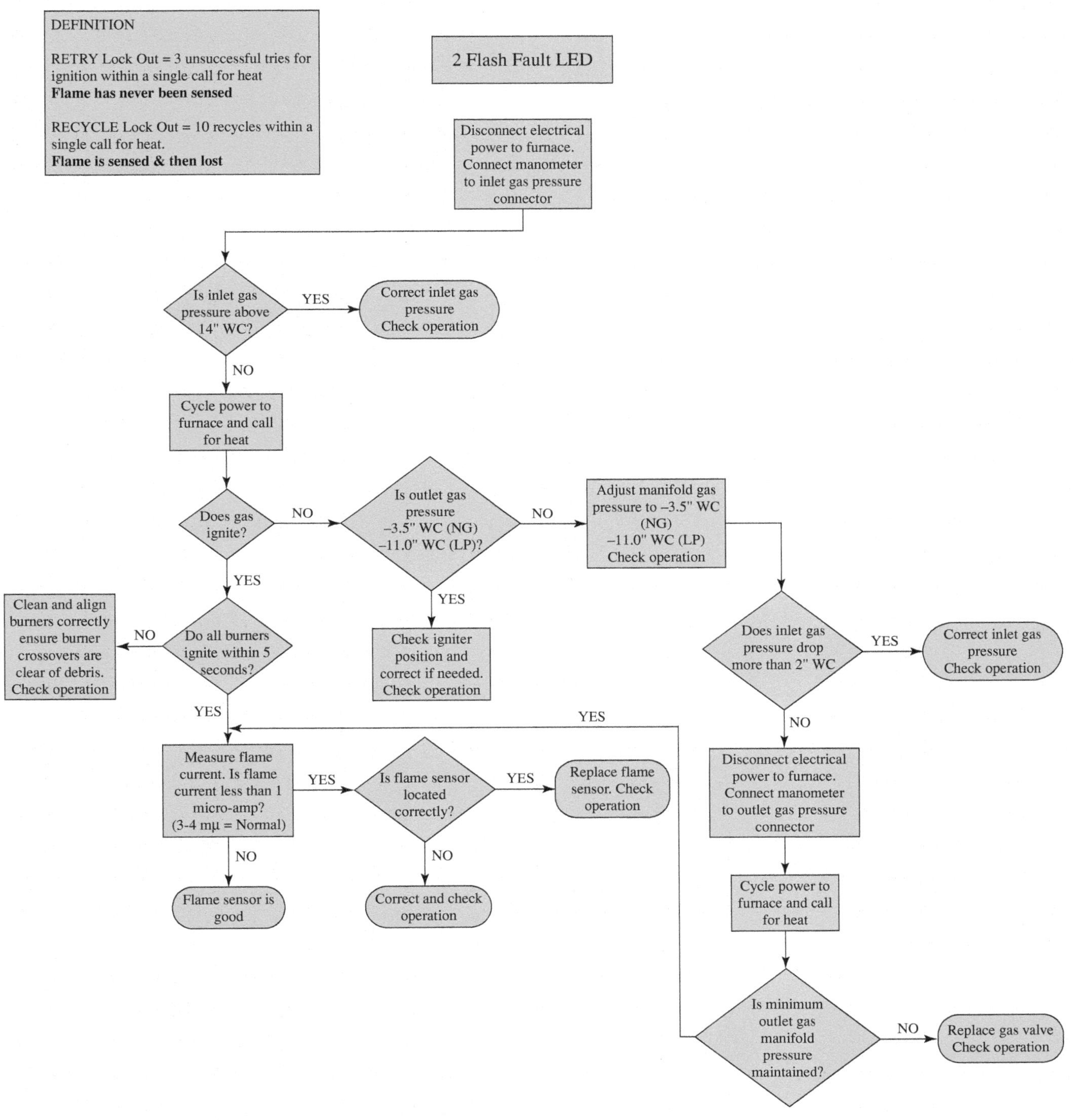
DEFINITION
RETRY Lock Out = 3 unsuccessful tries for ignition within a single call for heat
Flame has never been sensed
RECYCLE Lock Out = 10 recycles within a single call for heat.
Flame is sensed & then lost
2 Flash Fault LED
Disconnect electrical power to furnace. Connect manometer to inlet gas pressure connector
Is inlet gas pressure above 14" WC?
YES
Correct inlet gas pressure Check operation
NO
Cycle power to furnace and call for heat
Does gas ignite?
NO
Is outlet gas pressure –3.5" WC (NG) –11.0" WC (LP)?
NO
Adjust manifold gas pressure to –3.5" WC (NG) –11.0" WC (LP) Check operation
YES
YES
Check igniter position and correct if needed. Check operation
Clean and align burners correctly ensure burner crossovers are clear of debris. Check operation
NO
Do all burners ignite within 5 seconds?
Does inlet gas pressure drop more than 2" WC
YES
Correct inlet gas pressure Check operation
YES
YES
NO
Measure flame current. Is flame current less than 1 micro-amp? (3-4 mμ = Normal)
YES
Is flame sensor located correctly?
YES
Replace flame sensor. Check operation
Disconnect electrical power to furnace. Connect manometer to outlet gas pressure connector
NO
NO
Flame sensor is good
Correct and check operation
Cycle power to furnace and call for heat
Is minimum outlet gas manifold pressure maintained?
NO
Replace gas valve Check operation

Figure 22-32B

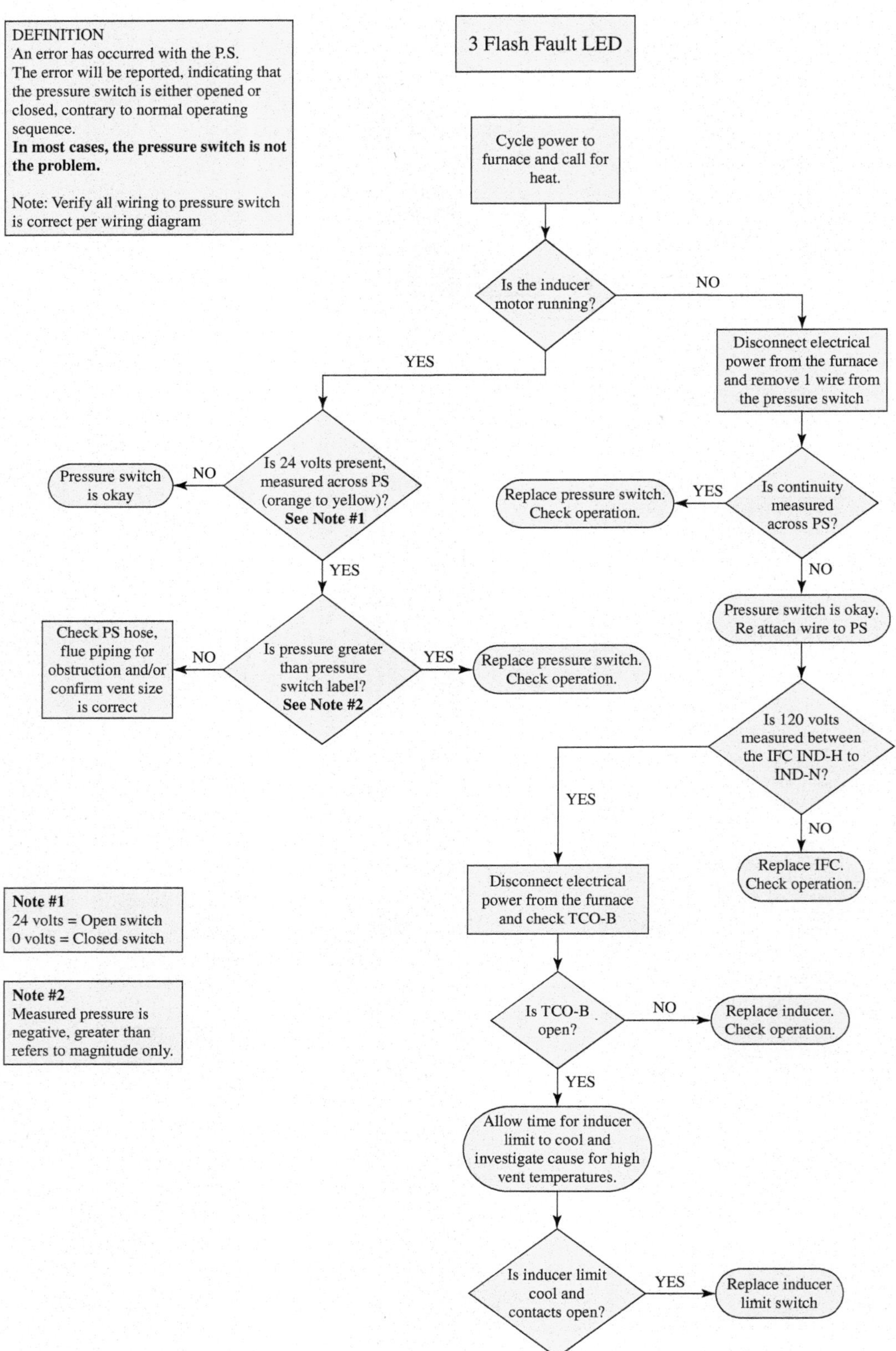
DEFINITION
An error has occurred with the P.S.
The error will be reported, indicating that the pressure switch is either opened or closed, contrary to normal operating sequence.
In most cases, the pressure switch is not the problem.
Note: Verify all wiring to pressure switch is correct per wiring diagram
3 Flash Fault LED
Cycle power to furnace and call for heat.
Is the inducer motor running?
NO
YES
Disconnect electrical power from the furnace and remove 1 wire from the pressure switch
Is 24 volts present, measured across PS (orange to yellow)? See Note #1
NO
Pressure switch is okay
Is continuity measured across PS?
YES
Replace pressure switch. Check operation.
NO
Pressure switch is okay. Re attach wire to PS
YES
Is pressure greater than pressure switch label? See Note #2
NO
Check PS hose, flue piping for obstruction and/or confirm vent size is correct
YES
Replace pressure switch. Check operation.
Is 120 volts measured between the IFC IND-H to IND-N?
YES
NO
Replace IFC. Check operation.
Disconnect electrical power from the furnace and check TCO-B
Note #1
24 volts = Open switch
0 volts = Closed switch
Note #2
Measured pressure is negative, greater than refers to magnitude only.
Is TCO-B open?
NO
Replace inducer. Check operation.
YES
Allow time for inducer limit to cool and investigate cause for high vent temperatures.
Is inducer limit cool and contacts open?
YES
Replace inducer limit switch

Figure 22-32C

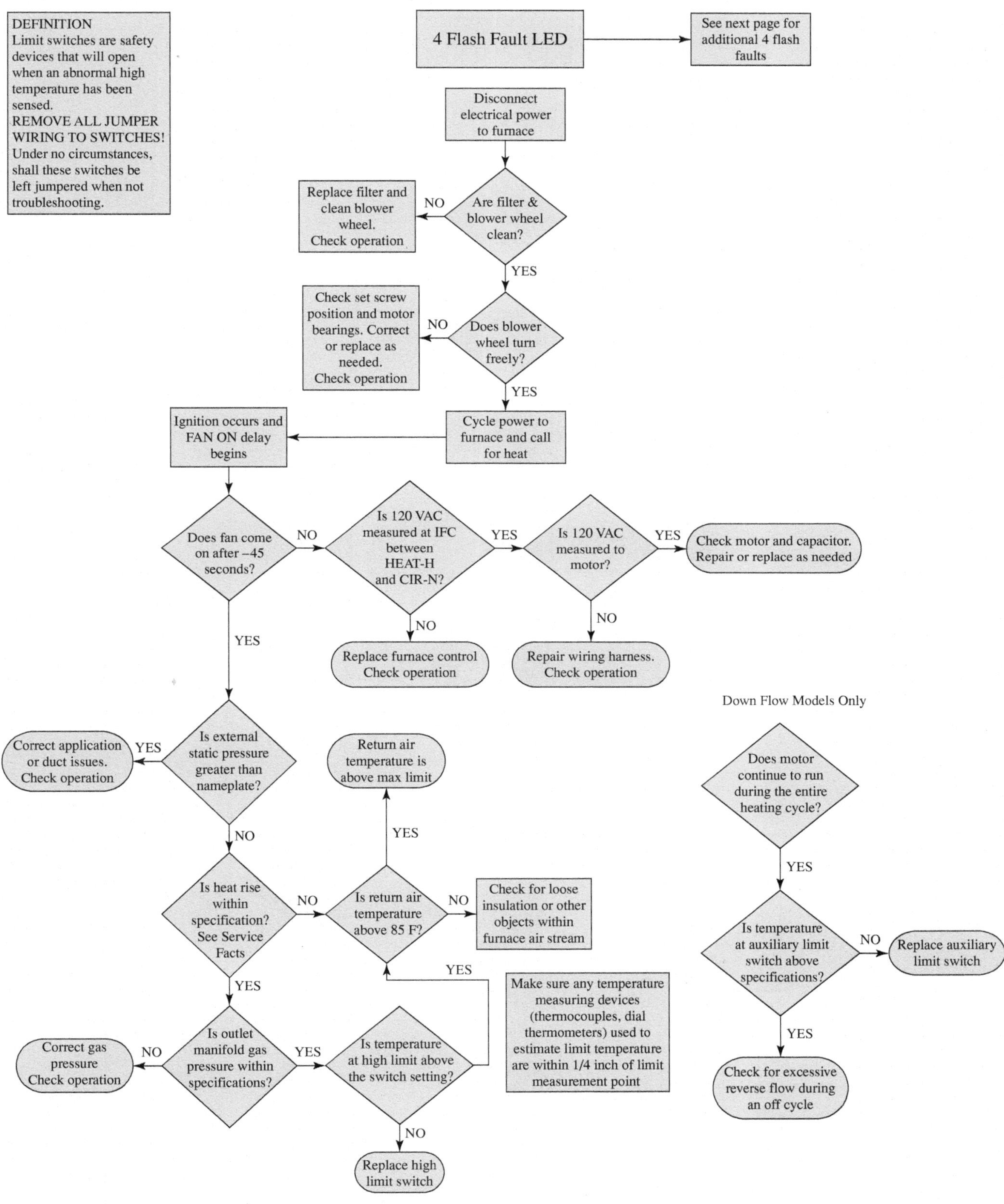
DEFINITION
Limit switches are safety devices that will open when an abnormal high temperature has been sensed.
REMOVE ALL JUMPER WIRING TO SWITCHES!
Under no circumstances, shall these switches be left jumpered when not troubleshooting.
4 Flash Fault LED
See next page for additional 4 flash faults
Disconnect electrical power to furnace
Are filter & blower wheel clean?
NO
Replace filter and clean blower wheel. Check operation
YES
Does blower wheel turn freely?
NO
Check set screw position and motor bearings. Correct or replace as needed. Check operation
YES
Cycle power to furnace and call for heat
Ignition occurs and FAN ON delay begins
Does fan come on after –45 seconds?
NO
Is 120 VAC measured at IFC between HEAT-H and CIR-N?
YES
Is 120 VAC measured to motor?
YES
Check motor and capacitor. Repair or replace as needed
NO
Replace furnace control Check operation
NO
Repair wiring harness. Check operation
YES
Is external static pressure greater than nameplate?
YES
Correct application or duct issues. Check operation
NO
Is heat rise within specification? See Service Facts
NO
Is return air temperature above 85 F?
YES
Return air temperature is above max limit
NO
Check for loose insulation or other objects within furnace air stream
YES
Is outlet manifold gas pressure within specifications?
NO
Correct gas pressure Check operation
YES
Is temperature at high limit above the switch setting?
YES
NO
Replace high limit switch
Make sure any temperature measuring devices (thermocouples, dial thermometers) used to estimate limit temperature are within 1/4 inch of limit measurement point
Down Flow Models Only
Does motor continue to run during the entire heating cycle?
YES
Is temperature at auxiliary limit switch above specifications?
NO
Replace auxiliary limit switch
YES
Check for excessive reverse flow during an off cycle

Figure 22-32D

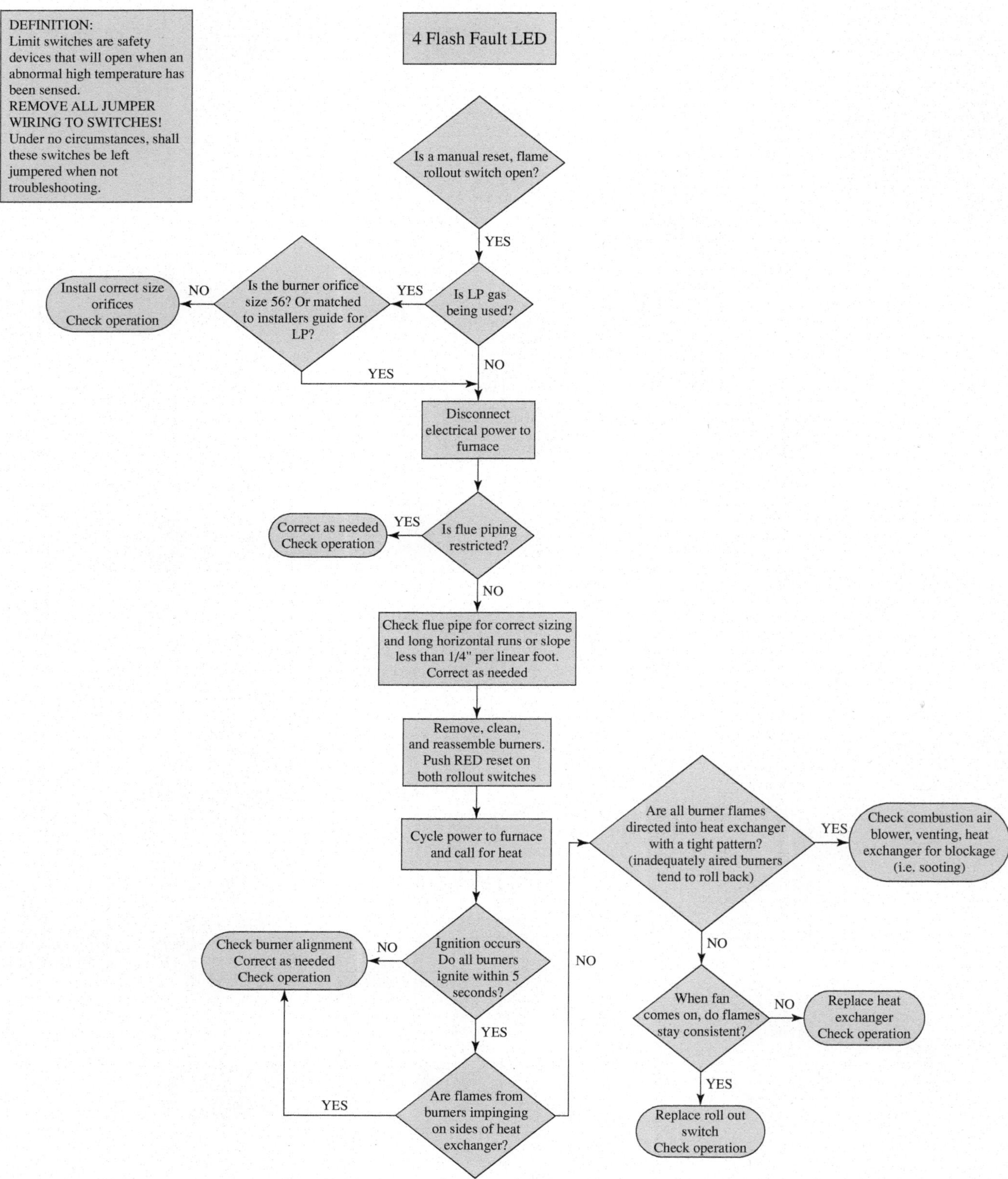

Figure 22-32E

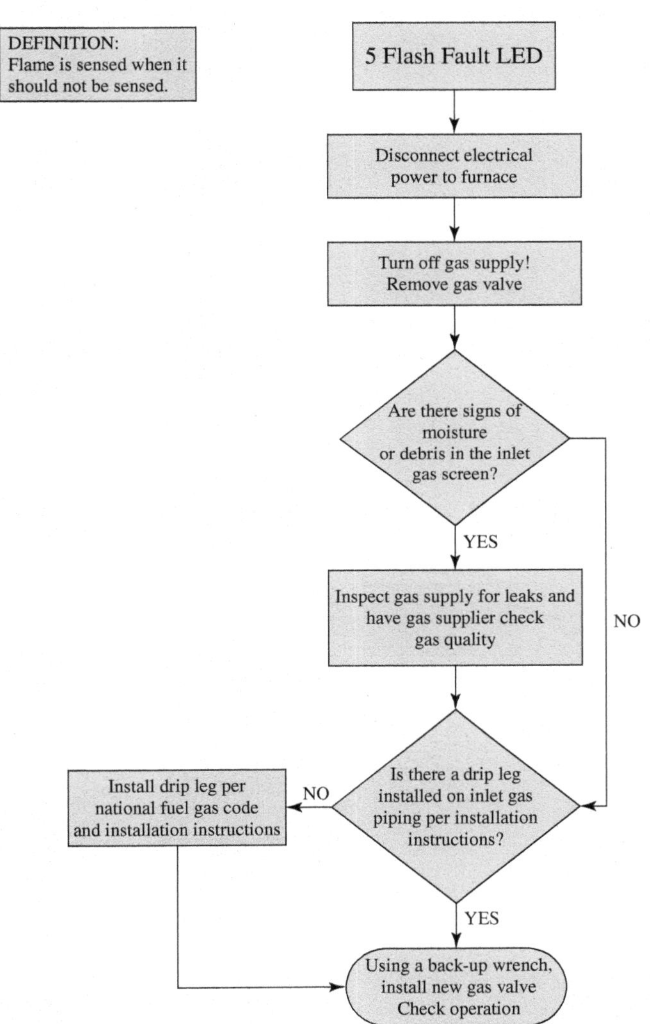
DEFINITION:
Flame is sensed when it should not be sensed.
5 Flash Fault LED
Disconnect electrical power to furnace
Turn off gas supply! Remove gas valve
Are there signs of moisture or debris in the inlet gas screen?
YES
NO
Inspect gas supply for leaks and have gas supplier check gas quality
Is there a drip leg installed on inlet gas piping per installation instructions?
NO
Install drip leg per national fuel gas code and installation instructions
YES
Using a back-up wrench, install new gas valve Check operation

Figure 22-32F

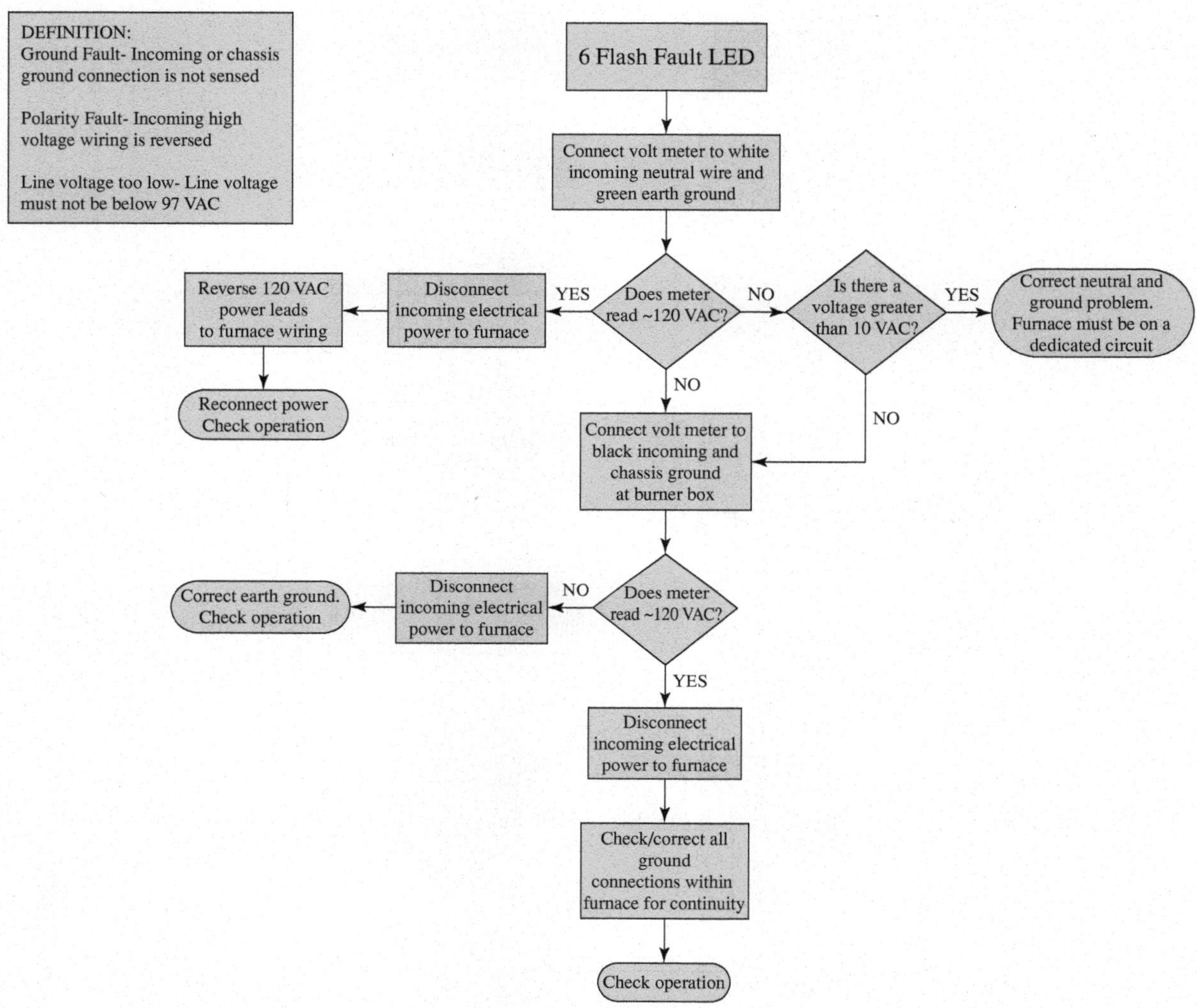
DEFINITION:
Ground Fault- Incoming or chassis ground connection is not sensed
Polarity Fault- Incoming high voltage wiring is reversed
Line voltage too low- Line voltage must not be below 97 VAC
6 Flash Fault LED
Connect volt meter to white incoming neutral wire and green earth ground
Does meter read ~120 VAC?
YES
Disconnect incoming electrical power to furnace
Reverse 120 VAC power leads to furnace wiring
Reconnect power Check operation
NO
Is there a voltage greater than 10 VAC?
YES
Correct neutral and ground problem. Furnace must be on a dedicated circuit
NO
NO
Connect volt meter to black incoming and chassis ground at burner box
Does meter read ~120 VAC?
NO
Disconnect incoming electrical power to furnace
Correct earth ground. Check operation
YES
Disconnect incoming electrical power to furnace
Check/correct all ground connections within furnace for continuity
Check operation

Figure 22-32G

Figure 22-32H

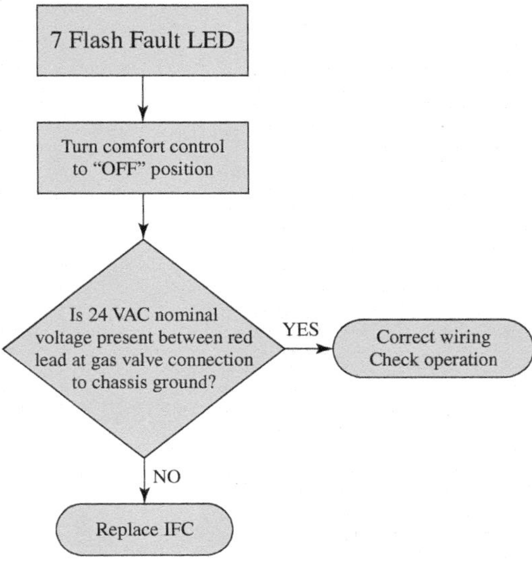
DIFINITION:
External gas valve circuit error (24 volts is present when it should not be present)
7 Flash Fault LED
Turn comfort control to "OFF" position
Is 24 VAC nominal voltage present between red lead at gas valve connection to chassis ground?
YES
Correct wiring Check operation
NO
Replace IFC

Figure 22-32I

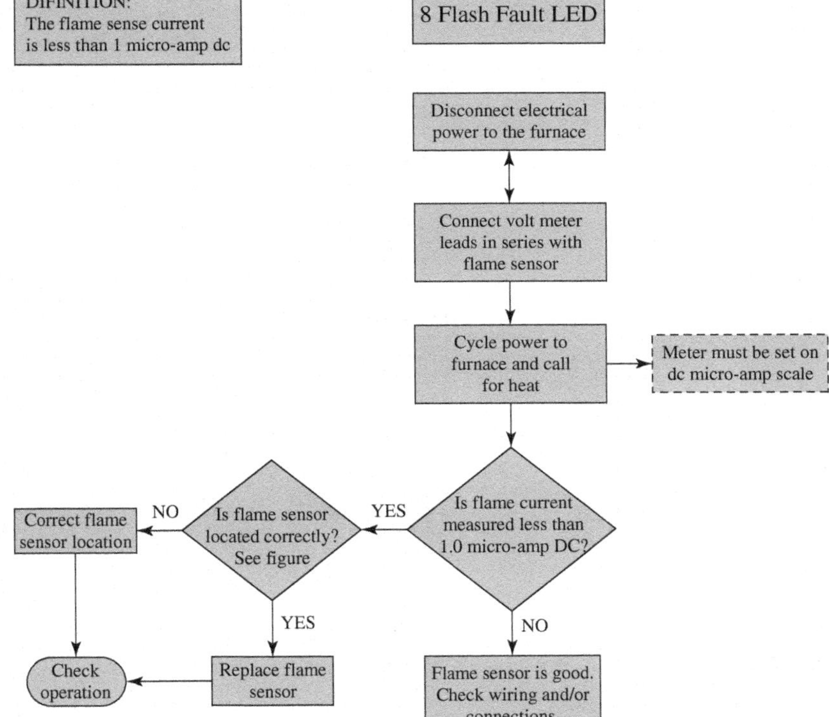
DIFINITION:
The flame sense current is less than 1 micro-amp dc
8 Flash Fault LED
Disconnect electrical power to the furnace
Connect volt meter leads in series with flame sensor
Cycle power to furnace and call for heat
Meter must be set on dc micro-amp scale
Is flame current measured less than 1.0 micro-amp DC?
YES
Is flame sensor located correctly? See figure
NO
Correct flame sensor location
YES
Replace flame sensor
Check operation
NO
Flame sensor is good. Check wiring and/or connections.

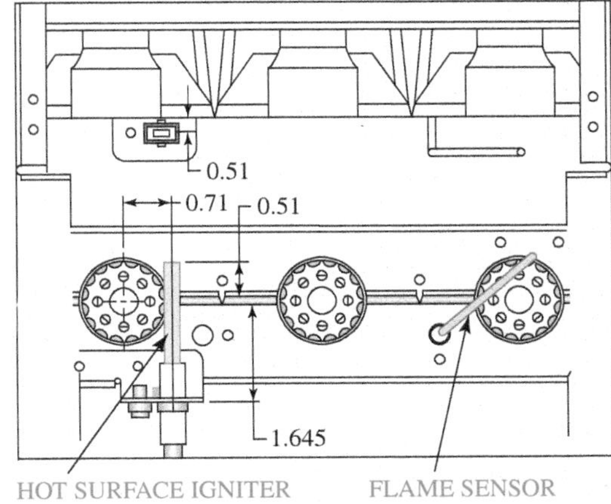
0.51
0.71
0.51
1.645
HOT SURFACE IGNITER
FLAME SENSOR

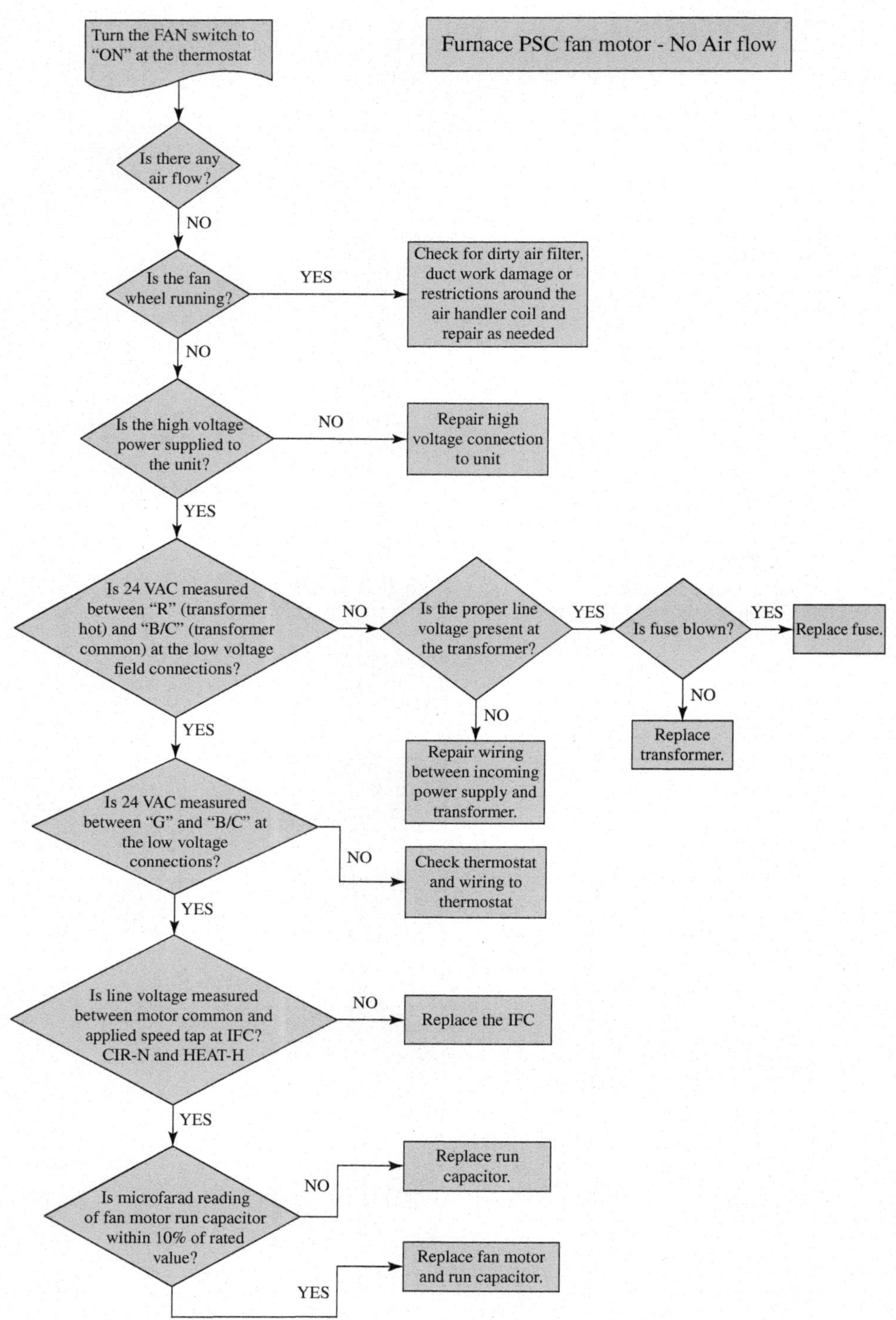

Figure 22-33 This troubleshooting flowchart is designed to guide a service technician through the check sequence for a PSC blower motor. This chart can be used for both low- and high-efficiency gas heating systems.

There is no flash code for "No Air." If the basic checks are used before the troubleshooting process, blower and air trouble will be found. Both low- and high-efficiency systems experience air problems. High-efficiency systems may have an electrically commutated motor or a standard PSC (permanent split capacitor) motor. The Figure 22-33 flowchart could be used for both low- and high-efficiency gas heating systems that use PSC motors.

22.17 ADDITIONAL REVIEW OF GAS FURNACE OPERATION

Now we compare the differences and similarities of a mid-efficiency gas furnace using Figures 22-34, 22-35, and 22-36. (For a complete, enlarged view of these figures, see Electrical Diagram ED-10, which appears with the Electrical Diagrams

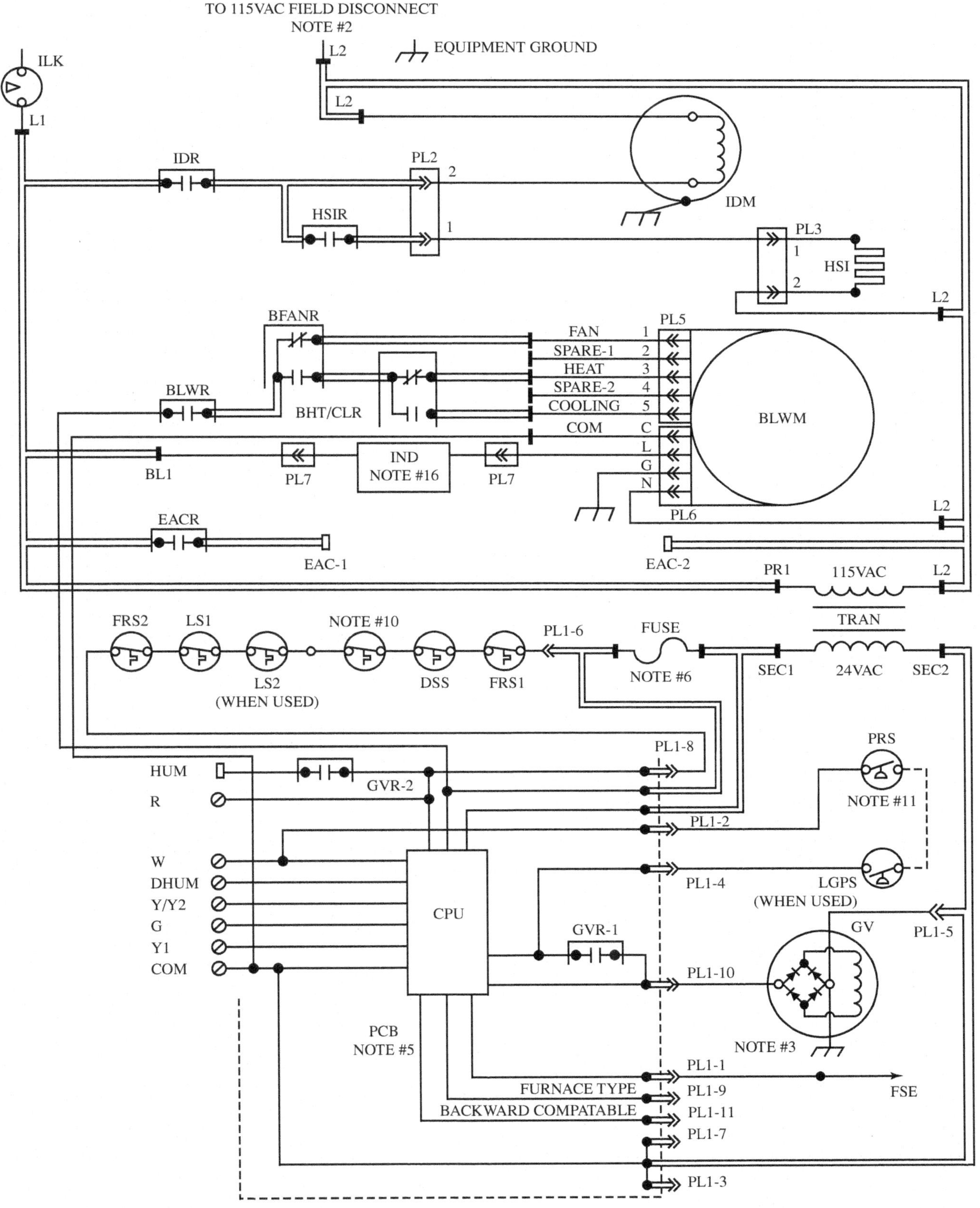

Figure 22-34A This is the ladder diagram embedded in Electrical Diagram ED-10. It is extracted from the large diagram so that it will be easier to trace. It is useful to have a ladder diagram and wiring diagram when tracing or troubleshooting a circuit.

NOTES:

1. IF ANY OF THE ORIGINAL EQUIPMENT WIRE IS REPLACED USE WIRE RATED FOR 105°C.
2. USE ONLY COPPER WIRE BETWEEN THE DISCONNECT SWITCH AND THE FURNACE JUNCTION BOX (JB).
3. THIS WIRE MUST BE CONNECTED TO FURNACE SHEET METAL FOR CONTROL TO PROVE FLAME.
4. SYMBOLS ARE ELECTRICAL REPRESENTATION ONLY.
5. SOLID LINES INSIDE PCB ARE PRINTED CIRCUIT BOARD CONDUCTORS AND ARE NOT INCLUDED IN LEGEND.
6. REPLACE ONLY WITH A 3 AMP FUSE.
7. INDUCER (IDM) MOTOR CONTAINS INTERNAL AUTO-RESET THERMAL OVERLOAD SWITCHES (OL).
8. L2 CONNECTIONS ARE INTERCHANGEABLE WITHIN THE L2 CONNECTOR BLOCK.
9. BLOWER MOTOR SPEED SELECTIONS ARE FOR AVERAGE CONDITIONS, SEE INSTALLATION INSTRUCTIONS FOR DETAILS ON OPTIMUM SPEED SELECTION.
10. FACTORY CONNECTED WHEN BVSS (CHIMNEY ADAPTER ACCESSORY KIT) IS NOT INSTALLED.
11. FACTORY CONNECTED WHEN LPGS IS NOT USED.
12. IGNITION-LOCKOUT WILL OCCUR AFTER FOUR CONSECUTIVE UNSUCCESSFUL TRIALS-FOR-IGNITION. CONTROL WILL AUTO-RESET AFTER THREE HOURS.
13. BLOWER-ON DELAY: GAS HEATING 25 SECONDS, COOLING OR HEAT PUMP 2 SECONDS.
14. BLOWER-OFF DELAY: GAS HEATING SELECTIONS ARE 90, 120, 150 OR 180 SECONDS, COOLING OR HEAT PUMP 90 SECONDS OR 5 SECONDS WHEN DHUM IS ACTIVE.
15. BLWM IS LOCKED-ROTOR OVERLOAD PROTECTED BY REDUNDANT ELECTRONIC CONTROL CIRCUITS.
16. INDUCTOR USED WITH 3/4 HP, 1 HP BLWM.

Figure 22-34B

package that accompanies this book.) Here is a breakdown of the figures:

- Figure 22-34 shows a view of the ladder diagram in Electrical Diagram ED-10.
- Figure 22-35 shows a view of the connection diagram in Electrical Diagram ED-10.
- Figure 22-36 is the legend for these diagrams. The legend will be important in identifying the components between the two diagrams.

The high-efficiency gas furnace can be challenging to a student as well as a technician. Before reviewing the following sequence of operation for a gas furnace, a few important points need to be clarified. You must understand that the blower door must be installed or the door switch depressed for power to be conducted through the blower door interlock switch. The door switch is sometimes bypassed to allow system observation without the door and troubleshooting activities. To maintain the safe integrity of a furnace, its bypass must be removed before installing the blower door. Having the door switch operational is an important safety practice. Power through the door switch will go to the following components:

- 120 V to furnace control CPU
- 120 V to transformer TRAN
- 120 V to inducer motor IDM
- 120 V to blower motor BLWM
- 120 V to hot surface igniter HSI
- 24 V to gas valve GV via the transformer.

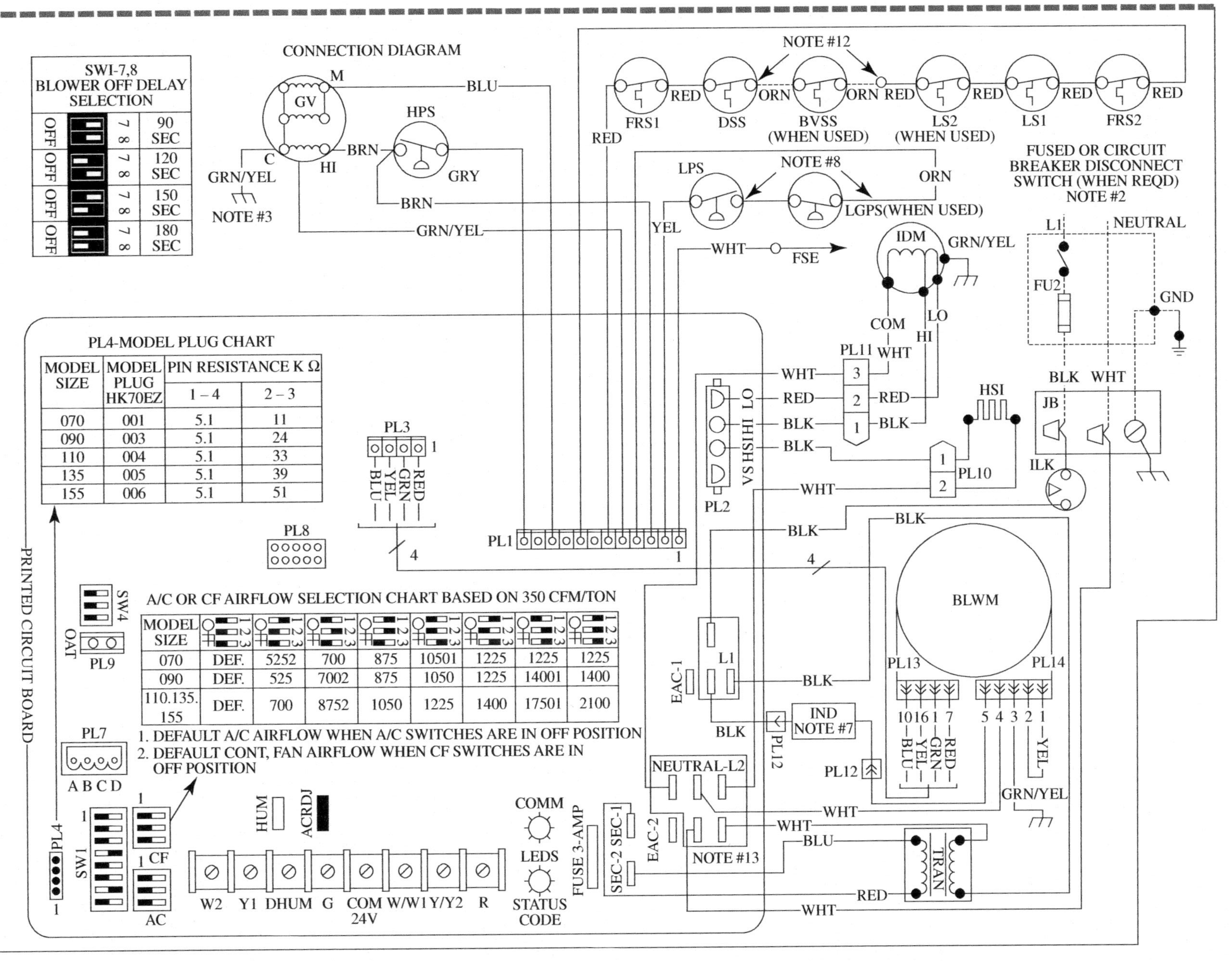

Figure 22-35A This is the connection diagram extracted from Electrical Diagram ED-10.

LEGEND

A/C	AIR CONDITIONING (ADJUSTABLE AIRFLOW -CFM)
ACR	AIR CONDITIONING RELAY, SPST (N.O.)
ACRDJ	AIR CONDITIONING RELAY DEFEAT JUMPER
BLWM	BLOWER MOTOR (ECM)
BVSS	BLOCKED VENT SAFETY SWITCH, MANUAL RESET, SPST (N.C.)
CF	CONTINUOUS FAN (ADJUSTABLE AIRFLOW-CFM)
COMMR	COMMUNICATION RELAY, SPDT
CPU	MICROPROCESSOR 7 CIRCUITRY
DHUM	DHUM CONNECTION (24VAC)
DSS	DRAFT SAFEGUARD SW., AUTO-RESET, SPST (N.C.)
EAC-1	ELECTRONIC AIR CLEANER CONNECTION (115VAC 1.0 AMP MAX.)
EAC-2	ELECTRONIC AIR CLEANER CONNECTION (COMMON)
FRS	FLAME ROLLOUT SWITCH, MAN. RESET, SPST (N.C.)
FSE	FLAME-PROVING SENSOR ELECTRODE
FUSE	FUSE, 3 AMP, AUTOMOTIVE BLADE TYPE, FACTORY INSTALLED
GV	GAS VALVE
GVR	GAS VALVE RELAY, DPST (N.O.)
HPS	HIGH-HEAT PRESSURE SWITCH, SPST (N.O)
HPSR	HIGH-HEAT PRESSURE SWITCH RELAY, SPST (N.C.)
HSI	HOT SURFACE IGNITER (115VAC)
HSIR	HOT SURFACE IGNITER RELAY, SPST (N.O.)
HUM	24VAC HUMIDIFIER CONNECTION (0.5 AMP MAX.)
HUMR	HUMIDIFIER RELAY, SPST (N.O.)
IDM	INDUCER DRAFT MOTOR 2-SPEED, SHADED POLE
IDR	INDUCER MOTOR RELAY, SPST (N.O.)
IH/LOR	INDUCER MOTOR SPEED CHANGE RELAY, SPDT
ILK	BLOWER DOOR INTERLOCK SWITCH, SPST (N.O.)
IND	INDUCTOR (NOTE #7)
LED	LIGHT EMITTING DIODE FOR STATUS CODES
LGPS	LOW GAS PRESSURE SWITCH, SPST (N.O.)
LPS	LOW-HEAT PRESSURE SWITCH, SPST (N.O.)
LS 1,2	LIMIT SWITCH, AUTO-RESET, SPST (N.C.)
PCB	PRINTED CIRCUIT BOARD
PL1	12-CIRCUIT CONNECTOR
PL2	4-CIRCUIT HSI & IDM CONNECTOR
PL3	4-CIRCUIT ECM BLWM CONNECTOR
PL4	4-CIRCUIT MODEL PLUG CONNECTOR
PL7	4-CIRCUIT COMMUNICATION CONNECTOR
PL9	2-CIRCUIT OAT CONNECTOR
PL10	2-CIRCUIT HSI CONNECTOR
PL11	IDM CONNECTOR (3-CIRCUIT)
PL12	1-CIRCUIT INDUCTOR SPLICE CONNECTOR
PL13	16-CIRCUIT ECM BLOWER CTRL. CONNECTOR
PL14	5-CIRCUIT ECM BLOWER POWER CONNECTOR
SW1-1	MANUAL SWITCH, STATUS CODE RECALL, SPST (N.O.)
SW1-2	MANUAL SWITCH, LOW-HEAT ONLY, SPST(N.O.)
SW1-3	MANUAL SWITCH, LOW-HEAT RISE ADJ. SPST(N.O.)
SW1-4	MANUAL SWITCH, COMFORT/EFFICIENCY ADJUSTMENT, SPST (N.O.)
SW1-5	MANUAL SWITCH, COOLING CFM/TON, SPST (N.O.)
SW1-6	MANUAL SWITCH, COMPONENT TEST, SPST (N.O.)
SW1-7,8	MANUAL SWITCHES, BLOWER OFF-DELAY, SPST(N.O.)
SW4-1	MANUAL SWITCH, TWINNING MAIN (OFF)/SEC. (ON)
SW4-2&3	FOR FUTURE USE
TRAN	TRANSFORMER, 115VAC/24VAC

- JUNCTION
- TERMINAL
- CONTROL TERMINAL
- FACTORY POWER WIRING (115VAC)
- FACTORY CONTROL WIRING (24VAC)
- FIELD CONTROL WIRING (24VAC)
- CONDUCTOR ON CONTROL
- FIELD WIRING SCREW TERMINAL
- EQUIPMENT GROUND
- PLUG RECEPTACLE

Figure 22-35B

A/C	AIR CONDITIONING (ADJUSTABLE AIRFLOW -CFM)
ACR	AIR CONDITIONING RELAY, SPST (N.O.)
ACRDJ	AIR CONDITIONING RELAY DEFEAT JUMPER
BLWM	BLOWER MOTOR (ECM)
BVSS	BLOCKED VENT SAFETY SWITCH, MANUAL RESET, SPST (N.C.)
CF	CONTINUOUS FAN (ADJUSTABLE AIRFLOW-CFM)
COMMR	COMMUNICATION RELAY, SPDT
CPU	MICROPROCESSOR 7 CIRCUITRY
DHUM	DHUM CONNECTION (24VAC)
DSS	DRAFT SAFEGUARD SW., AUTO-RESET, SPST (N.C.)
EAC-1	ELECTRONIC AIR CLEANER CONNECTION (115VAC 1.0 AMP MAX.)
EAC-2	ELECTRONIC AIR CLEANER CONNECTION (COMMON)
FRS	FLAME ROLLOUT SWITCH, MAN. RESET, SPST (N.C.)
FSE	FLAME-PROVING SENSOR ELECTRODE
FUSE	FUSE, 3 AMP, AUTOMOTIVE BLADE TYPE, FACTORY INSTALLED
GV	GAS VALVE
GVR	GAS VALVE RELAY, DPST (N.O.)
HPS	HIGH-HEAT PRESSURE SWITCH, SPST (N.O)
HPSR	HIGH-HEAT PRESSURE SWITCH RELAY, SPST (N.C.)
HSI	HOT SURFACE IGNITER (115VAC)
HSIR	HOT SURFACE IGNITER RELAY, SPST (N.O.)
HUM	24VAC HUMIDIFIER CONNECTION (0.5 AMP MAX.)
HUMR	HUMIDIFIER RELAY, SPST (N.O.)
IDM	INDUCER DRAFT MOTOR 2-SPEED, SHADED POLE
IDR	INDUCER MOTOR RELAY, SPST (N.O.)
IH/LOR	INDUCER MOTOR SPEED CHANGE RELAY, SPDT
ILK	BLOWER DOOR INTERLOCK SWITCH, SPST (N.O.)
IND	INDUCTOR (NOTE #7)
LED	LIGHT EMITTING DIODE FOR STATUS CODES
LGPS	LOW GAS PRESSURE SWITCH, SPST (N.O.)
LPS	LOW-HEAT PRESSURE SWITCH, SPST (N.O.)
LS 1,2	LIMIT SWITCH, AUTO-RESET, SPST (N.C.)
PCB	PRINTED CIRCUIT BOARD
PL1	12-CIRCUIT CONNECTOR
PL2	4-CIRCUIT HSI & IDM CONNECTOR
PL3	4-CIRCUIT ECM BLWM CONNECTOR
PL4	4-CIRCUIT MODEL PLUG CONNECTOR
PL7	4-CIRCUIT COMMUNICATION CONNECTOR
PL9	2-CIRCUIT OAT CONNECTOR
PL10	2-CIRCUIT HSI CONNECTOR
PL11	IDM CONNECTOR (3-CIRCUIT)
PL12	1-CIRCUIT INDUCTOR SPLICE CONNECTOR
PL13	16-CIRCUIT ECM BLOWER CTRL. CONNECTOR
PL14	5-CIRCUIT ECM BLOWER POWER CONNECTOR
SW1-1	MANUAL SWITCH, STATUS CODE RECALL, SPST (N.O.)
SW1-2	MANUAL SWITCH, LOW-HEAT ONLY, SPST(N.O.)
SW1-3	MANUAL SWITCH, LOW-HEAT RISE ADJ. SPST(N.O.)
SW1-4	MANUAL SWITCH, COMFORT/EFFICIENCY ADJUSTMENT, SPST (N.O.)
SW1-5	MANUAL SWITCH, COOLING CFM/TON, SPST (N.O.)
SW1-6	MANUAL SWITCH, COMPONENT TEST, SPST (N.O.)
SW1-7,8	MANUAL SWITCHES, BLOWER OFF-DELAY, SPST(N.O.)
SW4-1	MANUAL SWITCH, TWINNING MAIN (OFF)/SEC. (ON)
SW4-2&3	FOR FUTURE USE
TRAN	TRANSFORMER, 115VAC/24VAC

- JUNCTION
- TERMINAL
- CONTROL TERMINAL
- FACTORY POWER WIRING (115VAC)
- FACTORY CONTROL WIRING (24VAC)
- FIELD CONTROL WIRING (24VAC)
- CONDUCTOR ON CONTROL
- FIELD WIRING SCREW TERMINAL
- EQUIPMENT GROUND
- PLUG RECEPTACLE

Figure 22-36 This is the legend for Figures 22-34 and 22-35. The legend will help you understand what the components on the diagram mean. The legend is the abbreviated language of a wiring diagram.

SEQUENCE OF OPERATION 22.3: GAS FURNACE

You will notice in the wiring diagram of Figure 22-37 that all of the 120-V components have a black (**BLK**) and white (**WHT**) wire going to them. Black for the "hot side" and white for neutral are commonly used colors for identifying 120-V circuits. Let's review the sequence of operation for this gas furnace diagram:

1. *Power:* If the door switch is closed, 120 volts will be provided to the furnace. As stated before, power is

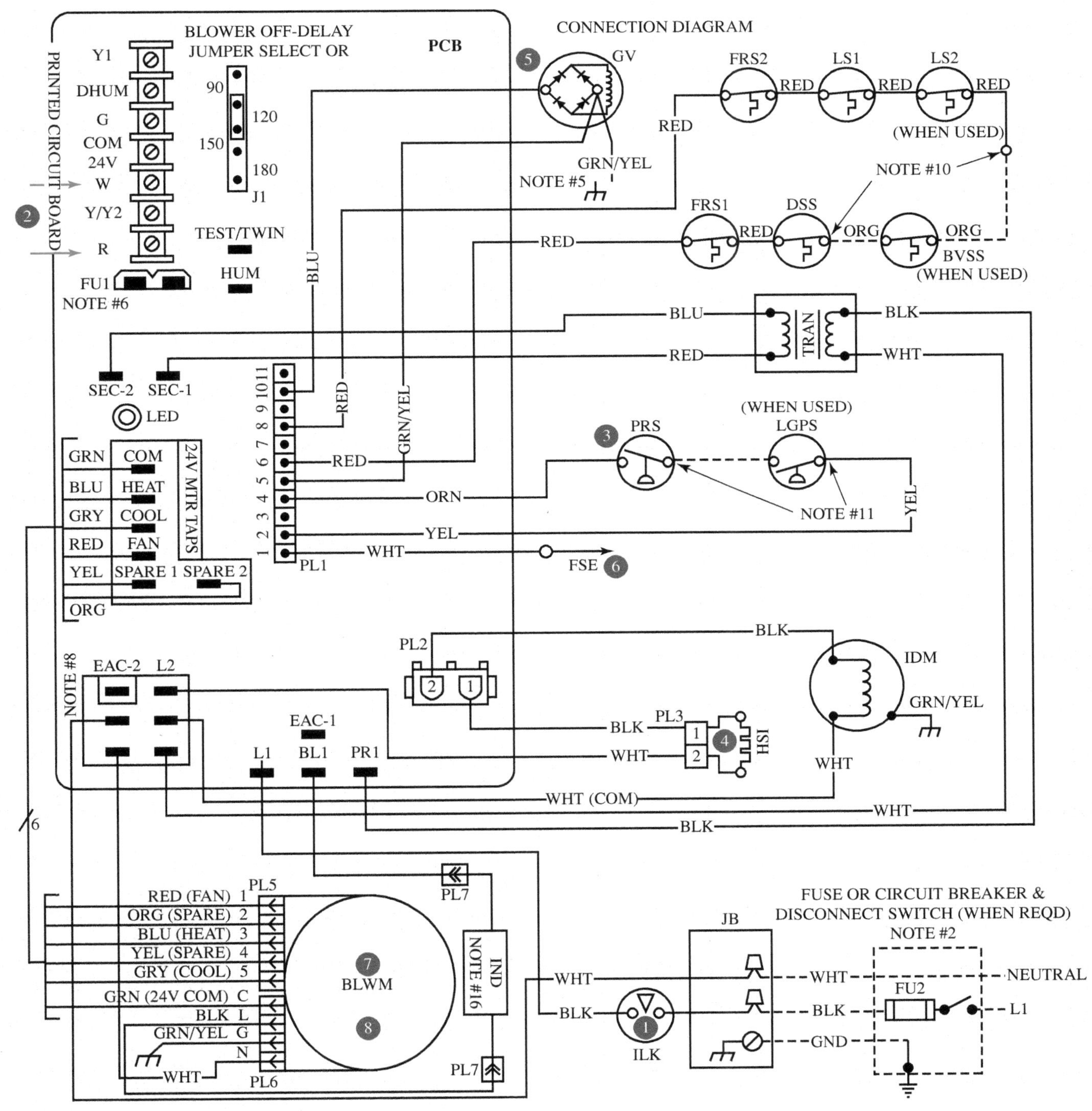

Figure 22-37 This is the wiring diagram for Sequence of Operation 22.3.

(continued)

conducted through the blower door interlock switch, **ILK**, at point **1**.

2 *Call for heat:* The thermostat calls for heat by closing the connection between the **R** and **W** circuit. As seen at point **2** on the diagram, the 24-V control voltage is fed from the thermostat to **W** on the upper left-hand side of the circuit board. This will start the heating sequence.

3 *Pre-purge period:* The furnace circuit board control performs a self-check, verifying that the pressure switch contacts **PRS** are open at point **3**, thus starting the inducer motor **IDM**. Note that **PRS** is drawn as a high-pressure switch. This pressure switch has a special feature in that it will shut down the furnace operations if bypassed by the technician.

When the inducer motor **IDM** comes up to speed, the pressure switch contact **PRS** opens to begin a 15-second pre-purge cycle. This pre-purge cycle flushes any unburned gas and flue gases from the heat exchanger. Pre-purge also provides adequate combustion air to start the light-off and burning process.

4 *Hot surface ignition energized:* At the end of the pre-purge cycle the **HSI**, at point **4**, is energized with 120 volts for 17 seconds.

5 *Gas valve opens:* When the warm-up period is complete, the main gas valve **GV**, at point **5**, opens. **GV** permits gas flow to the burners where it is ignited by the **HSI**.

6 *Flame proving period:* Five seconds after the **GV** opens, a 2-second flame proving period begins. The **HSI** will remain energized until the flame is sensed or until the 2-second flame proving period begins.

When the burner flame is proved at the flame proving sensor electrode **FSE** at point **6**, the furnace control **CPU** begins the blower delay period, allowing the heat exchanger to warm prior to starting the blower. If the burner flame is not proved within 2 seconds, the control **CPU** will open the gas valve **GV**. Depending on the system design, the **CPU** will repeat the ignition sequence several times before going into a lockout condition, which will prevent further flame proving events. In lockout, the gas valve will be closed and the furnace will not operate until the control power is reset.

Note: With some designs the lockout will reset automatically after 3 hours. Most designs will reset the lockout by momentarily interrupting 120 volts to the furnace or by interrupting 24 volts at **SEC1** or **SEC2** on the furnace control **CPU**. As an additional safety design, if a flame is proved when a flame should not be present, the **CPU** will lock out the heating mode and operate the inducer motor **IDM** until the flame is no longer present.

7 *Blower-on delay:* If the burner flame is proven, the blower motor at point **7** is energized on the "Heat" speed 25 seconds after the **GV** is energized. This blower time delay allows the heat exchanger time to warm up before energizing the blower **BLWM**.

8 *Blower-off delay:* When the room temperature satisfies the thermostat, the **R** to **W** circuit is opened. This stops the 24 volts to the **W** terminal on the circuit board (point **2**). This de-energizes the **GV**, stopping the gas flow to the burner. The inducer motor **IDM** will remain energized for a 5-second post-purge of the heat exchanger. The blower motor **BLWM** will remain energized for the time selected. For this model the off-cycle time delays can be selected for 90, 120, 150, or 180 seconds. This time delay clears the heat exchanger of most heat.

22.18 TRACING MORE DIAGRAMS

Next, we have one more gas heating diagram to trace and understand. Figure 22-38 shows the full diagram that we will be discussing. In Figures 22-39 and 22-40 we divide the diagram into the high-voltage and low-voltage sections, respectively. This will make each section larger and easier to follow. Refer back to Figure 22-38 when tracing the larger diagrams.

Let's start with the high-voltage control section as drawn in Figure 22-39. This diagram is illustrated with the applied power but no call for heat. Follow the 115-V source. You notice that the only component energized is the transformer (**TRAN**).

Next we look at the low-voltage section in Figure 22-40. This diagram shows the voltage path prior to a call for heat. Notice that all of the thermal safeties are closed, thus feeding power to the **R** terminal on the printed circuit board. Figure 22-41 illustrates the t'stat closing contacts between the **R** and **W** terminals. This energizes the low-voltage section. The gas valve will open and the HSI will be energized by the high-voltage section in Figure 22-42. The gas valve will remain open if **FSE**, the flame sensor, feels the gas burning.

Finally, Figure 22-43 shows the operation of the blower motor. The blower will begin to operate after a time delay whether the furnace fires off or not. At the end of the heating cycle, the burners will extinguish because the gas valve closes. The blower will continue to operate for about a minute after the thermostat **W** circuit opens.

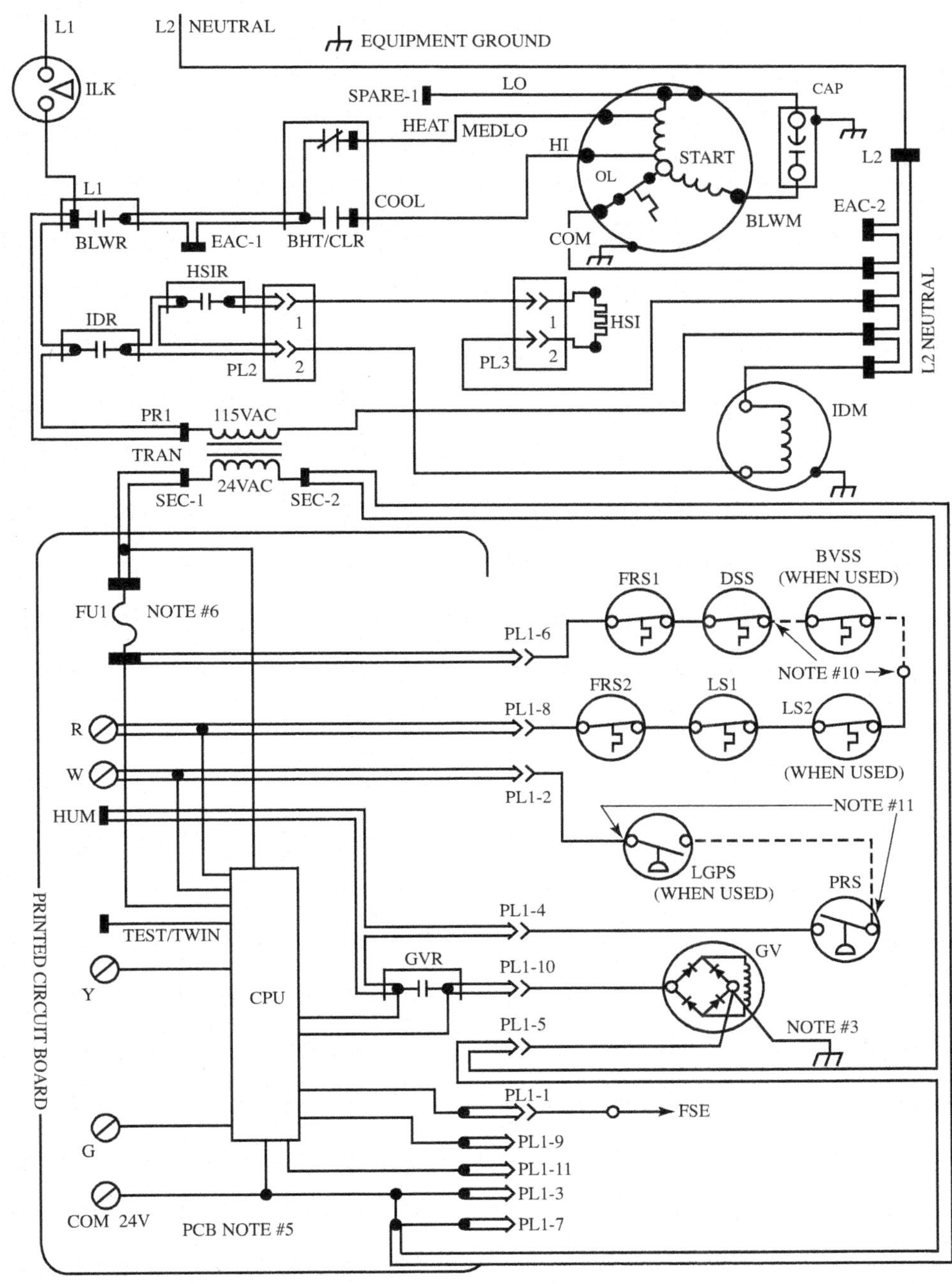

Figure 22-38 In this diagram, the upper section is the high-voltage section. The part below the transformer is the low-voltage section.

HEAT MODE HIGH VOLTAGE SECTION
115V
L1
NEUTRAL
EQUIPMENT GROUND
ILK
SPARE-1
LO
CAP
HEAT
MEDLO
HI
START
L2
OL
L1
COOL
EAC-2
BLWM
BLWR
EAC-1
BHT/CLR
COM
HSIR
IDR
HSI
PL2
PL3
L2 NEUTRAL
PR1
115VAC
IDM
TRAN
24VAC
SEC-1
SEC-2

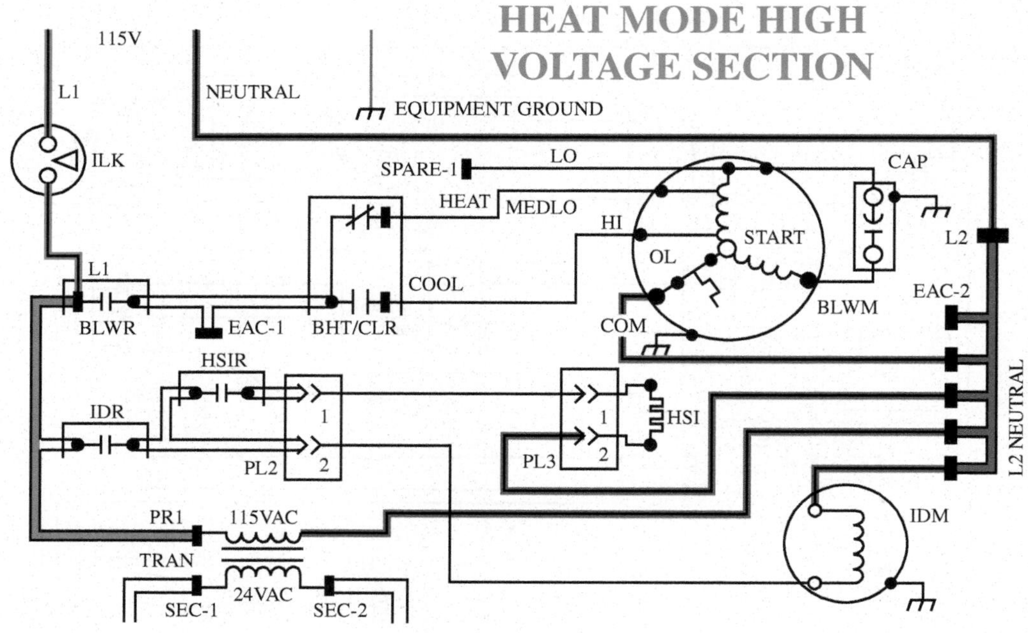

Figure 22-39 This traces the high-voltage section in red and blue before the thermostat calls for heat.

HEAT MODE, LOW VOLTAGE SECTION
TRAN
24VAC
SEC-1
SEC-2
FRS1
DSS
BVSS (WHEN USED)
FU1
NOTE #6
PL1-6
NOTE #10
FRS2
LS1
LS2
PL1-8
R
W
(WHEN USED)
PL1-2
HUM
NOTE #11
FROM T'STAT
LGPS (WHEN USED)
PRS
"W" INPUT TO CPU
PL1-4
TEST/TWIN
GVR
PL1-10
GV
PRINTED CIRCUIT BOARD
Y
PL1-5
NOTE #3
PL1-1
FSE
G
PL1-9
PL1-11
PL1-3
TO C.U.
COM 24V
PCB NOTE #5
PL1-7

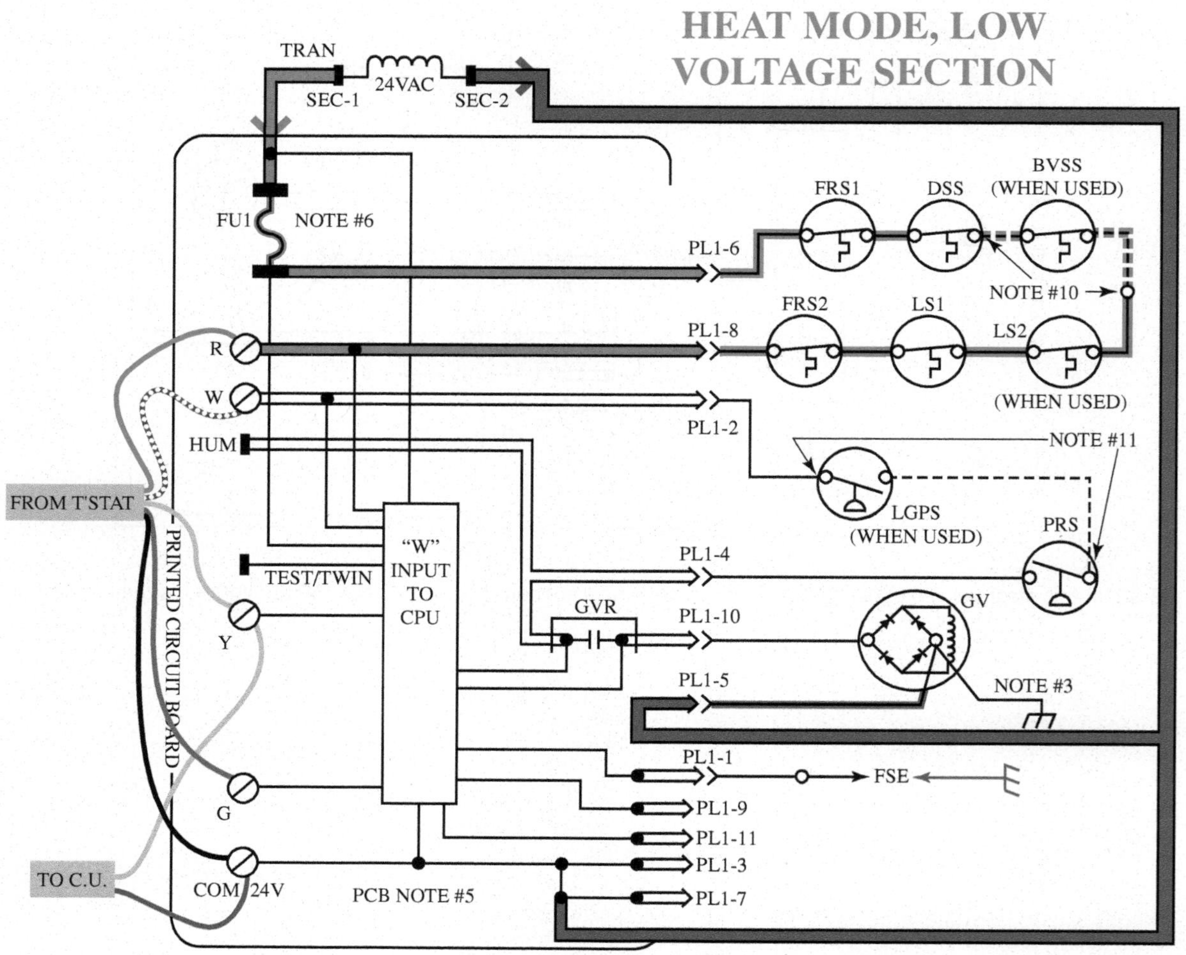

Figure 22-40 This traces the low-voltage section in red and blue before the thermostat calls for heat.

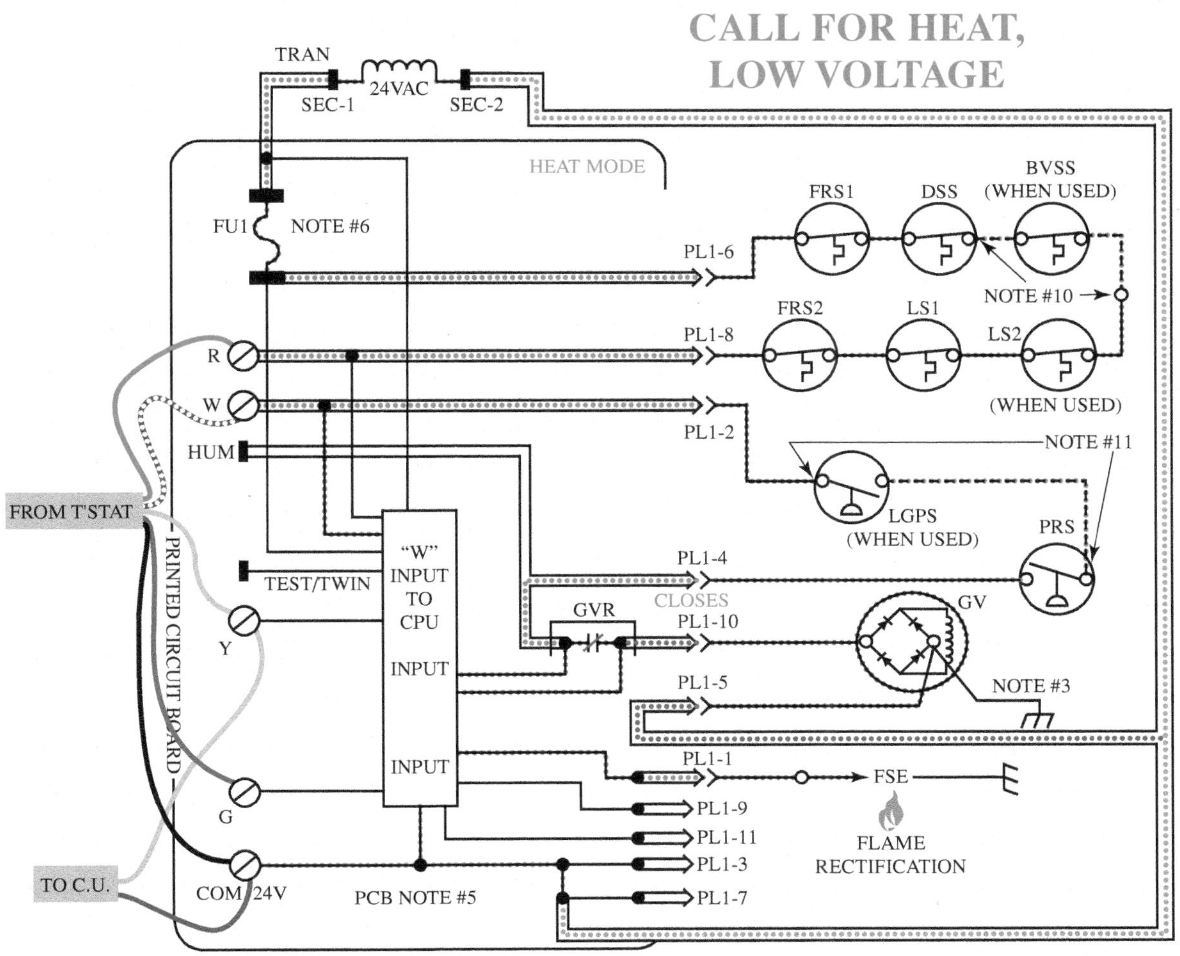

CALL FOR HEAT,
LOW VOLTAGE
TRAN
24VAC
SEC-1
SEC-2
HEAT MODE
FU1
NOTE #6
PL1-6
FRS1
DSS
BVSS
(WHEN USED)
NOTE #10
PL1-8
FRS2
LS1
LS2
(WHEN USED)
R
W
HUM
PL1-2
NOTE #11
LGPS
(WHEN USED)
PRS
FROM T'STAT
PRINTED CIRCUIT BOARD
"W"
INPUT
TO
CPU
TEST/TWIN
PL1-4
CLOSES
GVR
PL1-10
GV
Y
INPUT
PL1-5
NOTE #3
INPUT
PL1-1
FSE
PL1-9
G
PL1-11
FLAME
RECTIFICATION
PL1-3
TO C.U.
COM 24V
PCB NOTE #5
PL1-7

Figure 22-41 For a low-voltage call for heat, the t'stat closes the contacts between the **R** and **W** terminals. Notice that the gas valve **GV** will be energized.

CALL FOR HEAT,
T'STAT CLOSES
115V
L1
NEUTRAL
EQUIPMENT GROUND
ILK
SPARE-1
LO
CAP
HEAT
MEDLO
HI
START
L2
OL
L1
COOL
EAC-2
BLWR
EAC-1
BHT/CLR
COM
BLWM
CLOSES
HSIR
1
HSI
IDR
1
CLOSES
PL2
2
PL3
2
NEUTRAL
PR1
115VAC
IDM
TRAN
24VAC
SEC-1
SEC-2

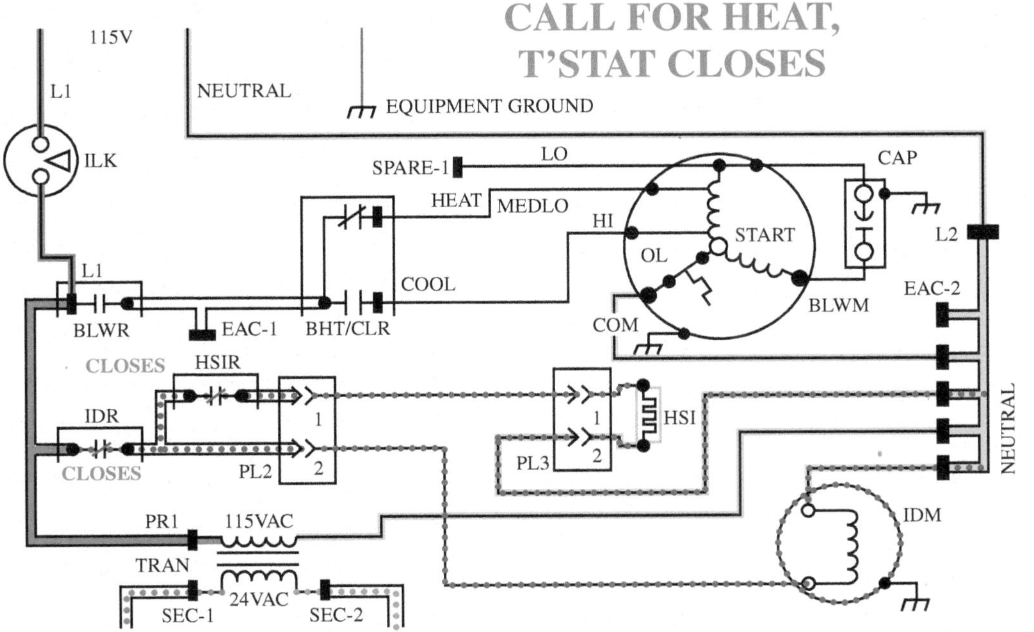

Figure 22-42 The gas valve will open. The HSI will be energized in the high-voltage section to ignite the gas.

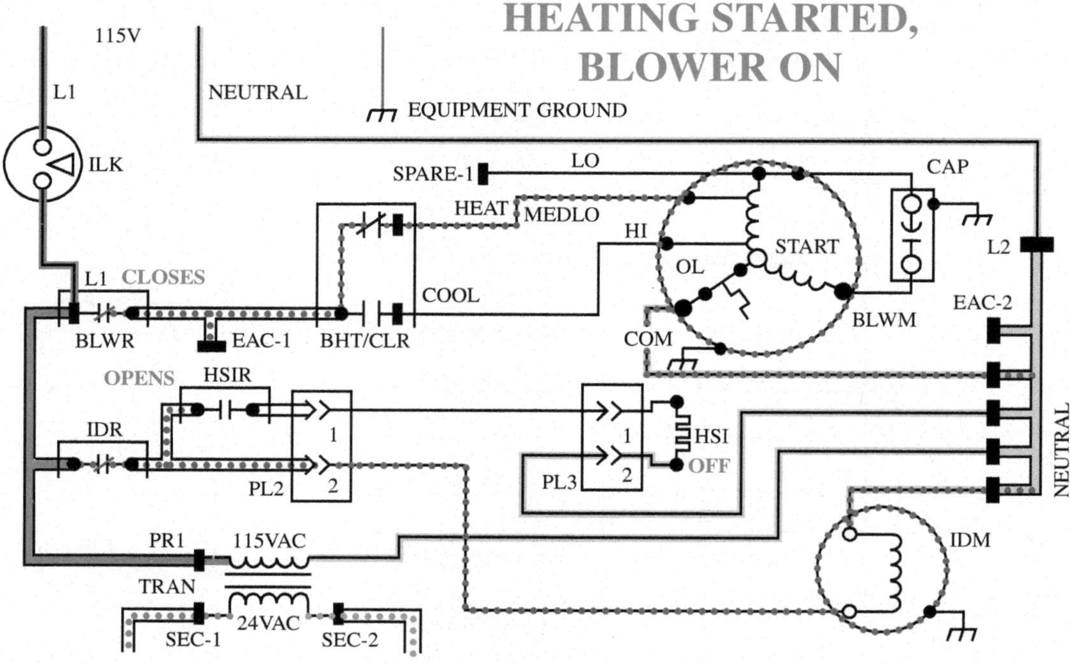

Figure 22-43 This diagram shows the operation of the blower motor.

SERVICE TICKET

You are sent out on a "no heat" service call. A low-efficiency heating system can be separated by the voltages. Line voltage operates the blower motor and supplies power to the transformer. The step-down transformer supplies 24 volts to the control circuit and operates the gas valve, and the thermocouple supplies millivolts to hold open the pilot solenoid—there are three voltages at work. Each of these voltages must exist for the furnace to operate. The loss of any one of these could be the cause of your customer's heating problem. You must find the fault by, first, determining if the voltages are present.

The first step in troubleshooting is to listen to and communicate with the customer. You listen and ask leading questions to determine where to start. The customer may tell you that he saw a standing flame. This simple statement tells the service technician that the DC millivolt thermocouple system is working; one voltage is ruled out. The customer may also tell you that the blower comes on when the fan switch at the thermostat is moved from "Auto" to "On." This verifies that all voltages are present: thermocouple millivolts, line, and low or control voltage (if the system is equipped with a fan relay and a thermostat with a fan switch).

In other cases, you may need to take a voltage reading to determine that voltage is present. If a voltage reading is taken at the thermostat or transformer, for instance, you may read 24 volts, which indicates that line voltage is applied and the step-down transformer is producing control voltage. (*Note:* Control voltage is used to remotely control a gas heating system. This usually means that the thermostat can be placed at a distance from the heating system, in a location that monitors the temperature accurately.) This single check allows you to take one reading to confirm two voltages.

If voltage is present, but the system is not heating, you will need to determine the next course of action, depending on what the customer provides for history. Table 22-6 provides some insight into what you might check first, based on the customer's complaint. The possible problems may be more extensive, but the first place to check leads you from one possible problem to the next. As you move through the troubleshooting procedure, each possible problem can be either confirmed or ruled out.

Using the diagram of the low-efficiency furnace (standing pilot unit) in Figure 22-44, trace the diagram for a customer complaint of "no heat." You would communicate with the customer upon arrival and determine if this condition happened after the customer noticed any other nontypical operation. The customer might relate that there were squealing noises when the furnace was operating and now the burner only turns on and off automatically. This gives you a clue as to where to start. Looking at the table, you should check the blower.

Table 22-6 Troubleshooting Guidelines

Customer Complaint	First Check	Possible Problem
Not enough heat; furnace turns on and off	Filter	Airflow problem: filter; blower; limit control
No heat; furnace turns on and off	Blower motor	Airflow problem: bad blower motor; bad blower bearings; blocked filter; blockage in ductwork
No heat; pilot will not stay lit	Thermocouple	No millivolts: thermocouple is bad; pilot orifice blocked; pilot flame not impinging thermocouple; bad thermocouple connection at gas valve
No heat; pilot is on; no burner flame	Transformer	No control voltage: transformer burned out; broken control wiring; bad gas valve; faulty thermostat; open high limit switch; no line voltage
Adequate heat; cold draft from floor diffuser	Fan control	Wrong temperature setting or switch action in the fan control: bad fan control; wrong setting
Adequate heat; blower runs all the time	Fan control	Wrong temperature setting or switch action in the fan control: bad fan control; wrong setting
Fuel bills are higher than usual; burner turns off and then back on before the blower turns off	Blocked or dirty heat exchanger	Exhaust gas problem: heat exchanger problem; airflow problem; fan control not working properly or set correctly

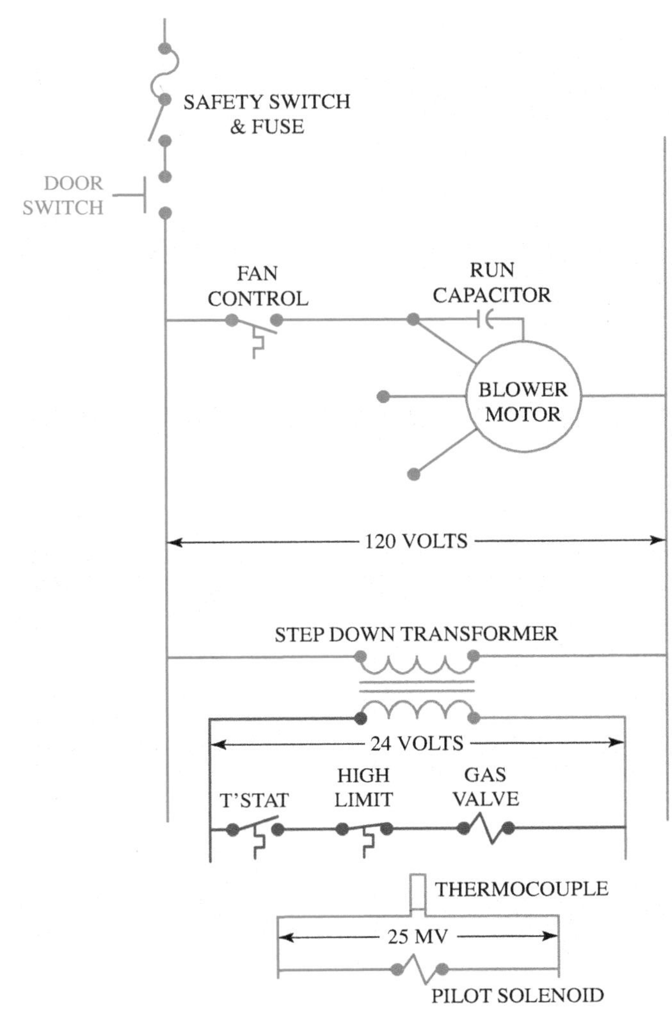

Figure 22-44 This diagram incorporates a blower door switch that turns off the power when the door is removed. Note the "push-button" switch symbol. When the door is in place, the switch is held closed.

Before checking the blower, you must do the routine checks so that nothing will be overlooked (refer to the eight basic checks earlier in this unit). When the basic checks have been completed, you remove the blower door (note there is a door switch that turns off the line voltage in the wiring, just after the safety switch and fuse), verify that the power is off, and visually examine the blower motor.

Let's say that the PSC blower motor is cold, but has a burned smell. When the blower wheel is turned by hand, there is a great deal of resistance. You suspect that the bearings are bad, but the problem is that the motor will not turn. Is power being applied to the blower? Applying voltage (closing the blower door switch) and checking for voltage at the blower terminals, you read line voltage. Turning off the power and disconnecting the motor, you verify that the motor is bad by checking for continuity (resistance); you find none, which means a circuit is open. What could have happened? Here is one possibility: The bearings became worse over time, making the motor operate at slower speeds. The squeaking noise was a telltale sign of bearings getting worse. The motor could not stay cool enough and got hot by trying to turn with bad bearings, which burned out the motor windings.

SUMMARY

This unit featured the types of gas heating systems and their operation. It started with a discussion about classifications of gas heating systems. It was noted that gas-burning systems could either be natural gas or liquefied petroleum (LP) systems. They can be further classified into efficiency types: low, medium, and high efficiency. These are known by their AFUE (annual fuel utilization efficiency) rating or percentage: 70% and below, 80%, and 90% and above.

The unit then reviewed the operation of low- and high-efficiency systems. Low-efficiency systems have a simple design without many additional controls and components. This type of system has three electrical circuits: millivolts, low voltage or control voltage, and line voltage. The operation of the loads found in each of these circuits is controlled by related switches. The fan control operates the blower when the temperature of the heat exchanger is sensed, for instance. A safety control senses when too much heat is in the heat exchanger (high limit) and the thermocouple is used to determine if the standing pilot is lit. This provides safety protection. We do not want gas in the furnace if it cannot be safely ignited.

A high-efficiency system provides for flame and high-temperature safety, in addition to determining if the system is operating at its maximum efficiency by monitoring the venting of the exhaust gases. Exhaust gas pressure switches and electronic ignition help to conserve gas through more efficient gas burning. Because high-efficiency systems require more electrical and electronic devices, the technician has additional troubleshooting tools to use. The circuit board usually has an LED flash code that helps to verify observed problems in the operation. The technician may also have a troubleshooting flowchart to help guide the troubleshooting process.

Whether it is a low- or high-efficiency system, the technician must make eight basic checks. These checks involve using test instruments to determine if the correct voltages are being applied. Some checks are visual or, in the case of temperature, conducted by a careful touch. As with any work conducted on gas heating systems, the work must be done using safe work practices consistent with industry policies and procedures. Always read and understand the manufacturer's installation and service manuals. Follow all safety procedures listed. Above all, use common sense. If you smell gas, do not operate light switches or create conditions that would cause ignition of those gases. A gas smell may be coming from a sewer system, but until it is investigated, all precautions must be taken to protect life and property.

REVIEW QUESTIONS

1. Explain why gas heating systems are categorized by efficiency.
2. Describe a gas-burning cycle.
3. Identify two basic gas-burning controls.
4. List four electrical components not found on low-efficiency systems.
5. List one electrical component found only on a low-efficiency system and what it is used for.
6. Describe what control voltage is used for.
7. Describe the term *line voltage*.
8. Identify the motor used for venting in the wiring diagram shown in Figure 22-34.
9. Identify five safety controls in the wiring diagram shown in Figure 22-35.
10. Describe three types of troubleshooting techniques.
11. What is the device shown in Figure 22-45?
12. Name the three numbered components in Figure 22-46.
13. Wire the heating and cooling circuit in Figure 22-47. Redraw on your answer sheet. Label all components.
14. Use the flowcharts shown in Figure 22-32 to answer this question: If the input power is present, what is the next step?
15. Use the flowcharts shown in Figure 22-32 to answer this question: Five flashes are indicating a problem. There is a good gas supply. What is the problem?
16. Use the flowcharts shown in Figure 22-32 to answer this question: Seven flashes are indicating a problem. You measure 24 volts from the **R** lead on the gas valve to ground. What is the problem?

Figure 22-45 Use this image to answer Review Question 11.

Figure 22-46 Use this image to answer Review Question 12.

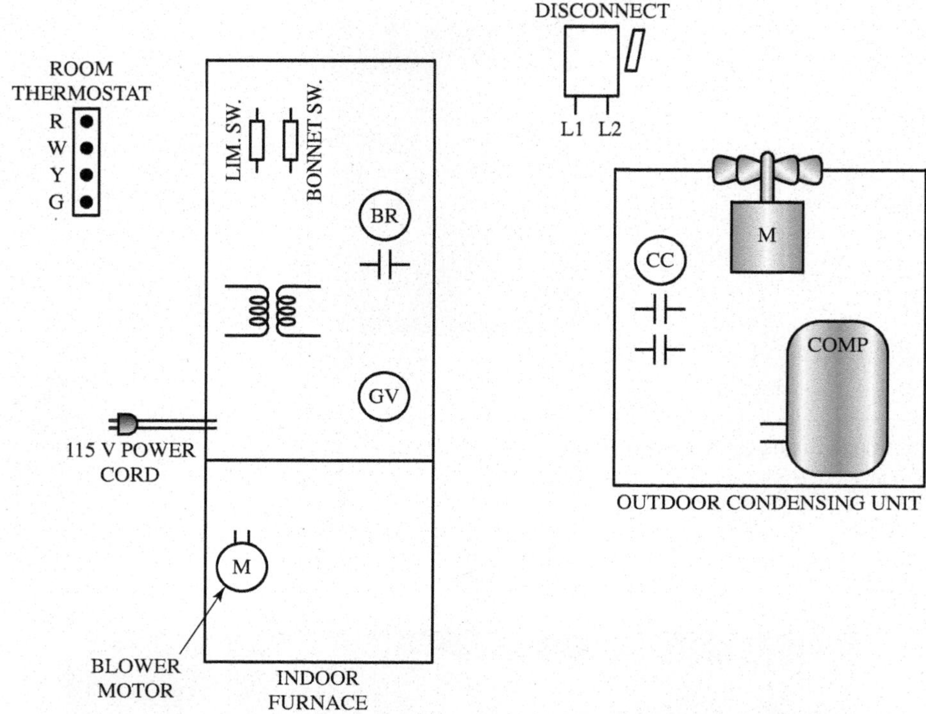

Figure 22-47 Use this diagram to answer Review Question 13.

UNIT 23

Electric Heating Systems

WHAT YOU NEED TO KNOW

After studying this unit, you will be able to:

1. Identify the different types of electric heating systems.
2. Identify the components of electric heating systems.
3. Explain the advantages and disadvantages of electric heat.
4. Determine the CFM of an electric heating system.
5. Trace the sequence of operation of a simple electric heating furnace.
6. Perform basic troubleshooting checks on an electrical heating system.

Electric heating systems are generally less complex than fuel-burning systems. They are also smaller in size when compared to other units that deliver the same number of **Btu's (British thermal units)**. Because they do not need to convert (burn) fuel to heat, the entire fuel-burning portion of the system is not necessary. Not needing to convert fuel to heat eliminates that portion of the system, allowing the electric system to take up less room.

Electric heat units simply apply electricity to a resistive material to create heat to warm air or water. An example of this can be seen from the heat generated by the element in an incandescent light bulb or the heat of a toaster. Electric water heaters or electric boilers are used to warm water for domestic water or space heating purposes. Figure 23-1 illustrates what these resistive heating elements may look like when used for air or water heating.

Electric systems do not use combustion gases, so no venting is required. Not having to exhaust products of combustion and not needing air for combustion are valuable advantages of electric heating systems. Here are more advantages to consider when selecting electric heating systems:

1. Electric heating units can be placed almost anywhere.
2. Installation costs are lower, because there is no need to run vents or fuel piping. That is why electric units can be placed in locations that cannot be used by conventional fuel-burning furnaces.

Because of their flexibility, electric heating systems are produced in many configurations. Popular configurations are discussed in this unit. An electric boiler is mentioned in this unit, but it will not be discussed. Electric boilers are not as popular as electric furnaces. An electric boiler is similar to a domestic hot water heater and operates, generally, in the same fashion.

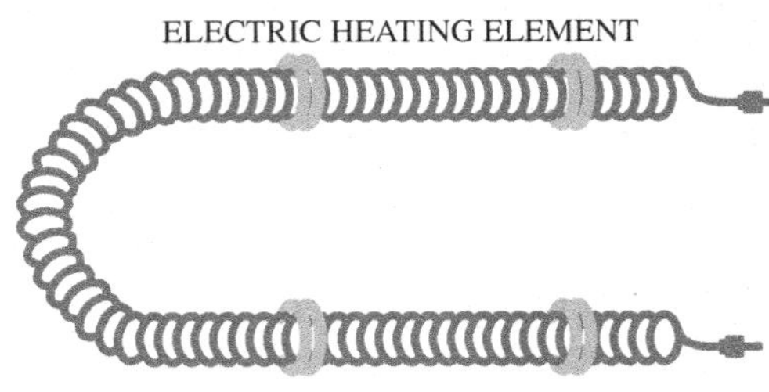

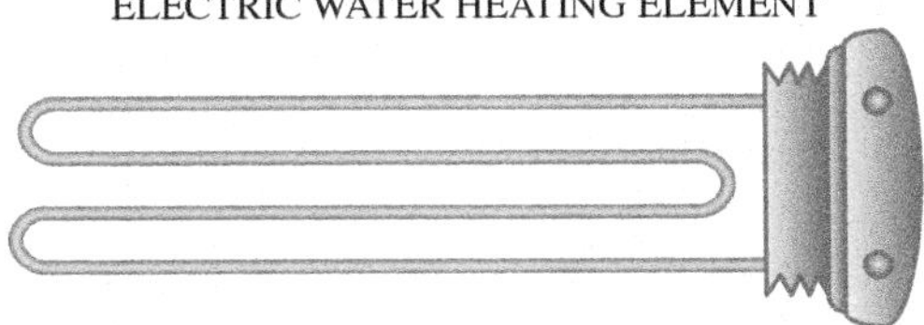

Figure 23-1 Two types of heating elements are shown. The top element is an air-style heating element. The lower element is immersible and used in water-based systems.

Electric heating systems can be primary or secondary sources of heat. They can be installed as an individual room heating system or as an electric furnace to heat an entire building as a central heating system. Most often an electrical heating system is coupled to a heat pump to provide backup heat during defrost periods or as supplemental heat during periods of low outside temperature.

In some locations of the country, utility companies provide special rates for "all-electric" buildings. Utility companies use these incentives to even out the amount of electrical usage throughout the year. Where electrical power consumption is controlled, utility companies offer central control of heating, cooling, and water heating operations. In this situation, the power company may have the ability to turn off or reduce the use of electricity during periods of high demand. They may also offer lower utility rates during low-use periods.

GREEN TIP

Later in this unit we discuss the various ideas of what is considered "green" for direct electric heating systems. Some thoughts are that direct electric heating is not a green technology. Some countries in the world regulate or prohibit the use of electricity for direct electrical heating. But if electricity can be produced by renewable means, does that still mean that direct electrical heat cannot be a "green" technology? For example, there is a farmer in Pennsylvania who is planning on installing solar panels to generate electricity since his oil heating costs are high. The plan is to generate electricity and use small electric heaters throughout the house to supply heating and not rely totally on oil heat. This could be considered a green use of electric heat.

This unit concludes with a short section on troubleshooting. Many of the troubleshooting techniques used for transformers, relays, and safety devices are also used for electric heating systems. When compared to fuel-burning furnaces, electric systems tend to be simpler in both operation and service. There are fewer components to check during a service call. Some things that are checked on other types of furnaces become simpler to check on electric systems. Airflow, for instance, is a simple check. With four measurements and a few calculations using a formula, a technician can easily determine the airflow and adjust the fan speed.

23.1 ELECTRIC HEATING SYSTEM TYPES

Electric heating systems can be stand-alone heating systems or backup heating systems. Stand-alone systems are known as electric baseboard, electric furnace, or electric boiler systems. Electric backup systems include electric heating strips, coils, elements, sections, and duct heaters. Whether a unit is a stand-alone system or a backup, its active elements have the same function and general appearance. These types of units are characterized by having electric heating elements made of tungsten that resist the flow of electricity and become hot. Figure 23-2 shows a common residential electric heat strip found in an air handler. White, round porcelain insulators are used to hold the heat strips in place and prevent shorting to ground. Figure 23-3 is a large commercial or industrial-grade heat strip. The element has a round band or fins used to increase the heat exchange surface. Notice how the heating elements are evenly spaced across the face of the duct system to reduce airflow restrictions.

The configurations of heating elements can vary. Some elements are fashioned in a horizontal row, while others are in a circular pattern. The configuration is dependent on the location and use of the electrical element. If placed in round ductwork, a circular configuration may work the best. As air or water (also known as a "medium") moves across the element, the element is cooled as the air or water is heated.

Figure 23-4 is a cutaway view of the inside of a residential electric furnace with a blower. The components of this furnace are discussed later in this unit.

In an electric boiler element, the element is encased so that water cannot touch the electrical element. The element may have the same appearance as that of a domestic electric water heater element, as shown in Figure 23-1. These elements are sometimes called *immersion elements*, because they are meant to be used in liquids like water.

Another form of electric heating system is called an electric radiant panel. Radiant panels can be small, spot heating units, or they can be entire floors, walls, or ceilings. In these applications, electric resistance wire is placed within a floor, wall, or ceiling. When the wire is powered, the electric resistance heater warms the large surface. When used in a floor, the entire floor is warmed. When cool air comes in contact with the warm floor, heat is transferred by conduction to the air, in turn, warming the air.

Figure 23-2 This is a set of common residential heat strips, also known as a heating element.

Figure 23-3 This is a commercial or industrial-size heat strip. This is only a portion of the heat strip, which is placed in an area of airflow that is 4 ft by 5 ft. The spacing around the heat strip permits good airflow through this section.

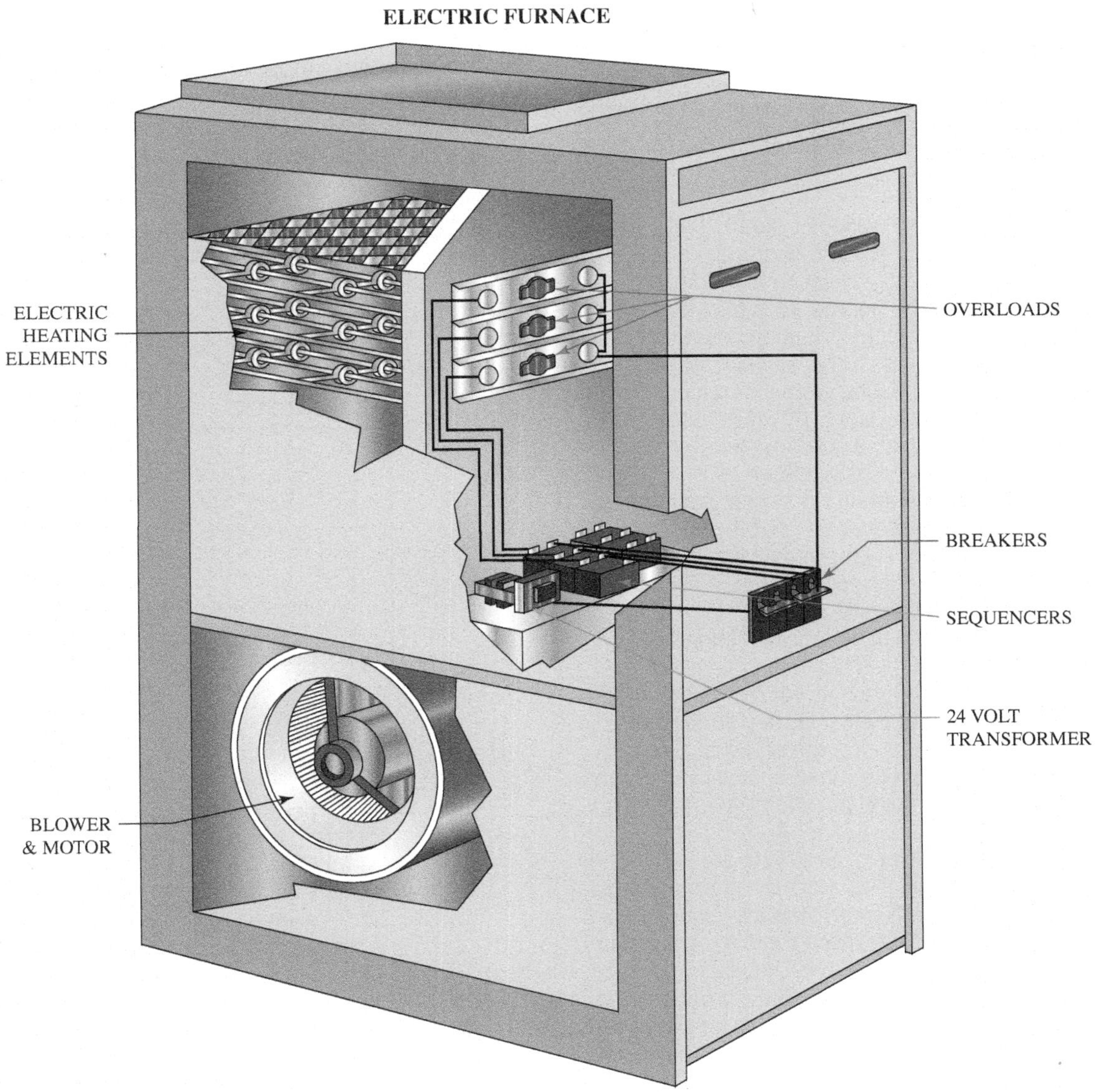

ELECTRIC FURNACE

ELECTRIC HEATING ELEMENTS

OVERLOADS

BREAKERS

SEQUENCERS

24 VOLT TRANSFORMER

BLOWER & MOTOR

Figure 23-4 This electric furnace houses several strips of electrical heating elements. A blower moves air across the elements to keep them cool while powered.

SAFETY TIP

Electric elements may not appear to be hot. When airflow over an electric element is slowed or stopped, the element may glow because heat is not being taken away. Without airflow the element may get too hot and burn out. The same is true in reverse. If airflow is increased, the element will cool, which means that it will not appear to be hot. Never touch an electric heating element.

TECH TIP

Before work on an unknown system is started, be sure of the type and configuration of the system. Electrical heating systems come in many different configurations. The control configuration will determine how the system operates and under what conditions. The customer or owner is the best source of information on past performance.

GREEN TIP

Green Heating Systems?

There are several different opinions about direct electric heating systems. Direct electric heat means that we are using a heating element, similar to those found in a toaster, to heat the air. The heated air is blown into the space needing warmth. Electric heating systems have always been referred to as 100% efficient. What this means is that an electric heating element converts all of its electrical power to heat. Note, however, that 100% efficient does not mean that electric heating costs are low. As a matter of fact, electric heating is one of the most expensive forms of heat. In reference, gas heating systems convert fuel to heat energy. The efficiency of the burning gas is based on how well the conversion is made from fuel to heat and the amount of waste produced; electrical systems have no waste.

The greatest advantage of electrical heating systems is that no chimney or venting is needed. Because of this, there is no need to bring in fresh air for combustion and there is no wasting of heated air that may be exhausted through a chimney or vent during the off cycle. Because electric heating systems do not need an opening to the outside for venting, they have an advantage over other forms of heating systems.

Electrically heated homes, however, are not considered to be economical. The cost to operate direct electrical heating elements is more than the cost of other types of heating systems. The reason for this relationship of efficiency and cost has to do with how the electricity is made. Power companies can only produce and provide electricity using natural gas at an efficiency of less than 50%. This makes the cost of electricity much higher than when natural gas is converted to heat with a direct gas-burning heating system. If, on the other hand, the electricity were produced by photovoltaic panels or wind, electricity would come from a renewable source. Electricity produced in this way would cause opinion to shift. Presently, the installation of direct electric heating systems in some European countries has been suspended because of their nonefficiency. Electric heating is not considered to be "green" in the United States either; however, it is viewed as a convenient backup or emergency heating system.

23.2 ELECTRIC BASEBOARD HEAT

The simplest of electric heating systems is the electric baseboard system. An illustration of a **baseboard heater** and a wiring diagram is shown in Figure 23-5. With this type of system, the home owner can easily adjust the temperature in each individual room.

An electric baseboard system is comprised of electric heating elements encased in a cover and mounted at floor level on the "exposed" or outside wall of each room (see Figure 23-6). Power is supplied to the baseboard through a line-voltage thermostat mounted on an inside wall. The

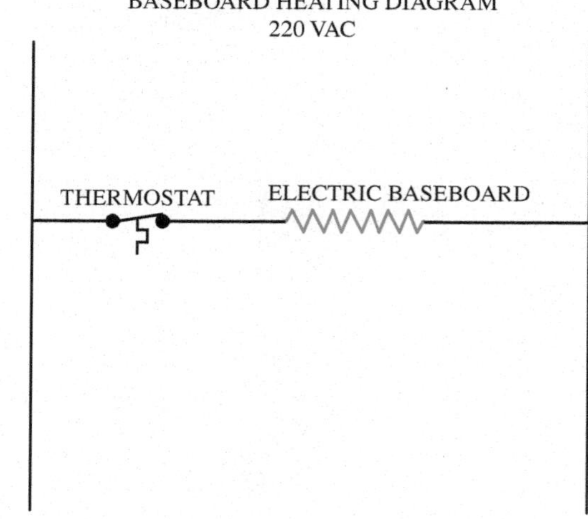

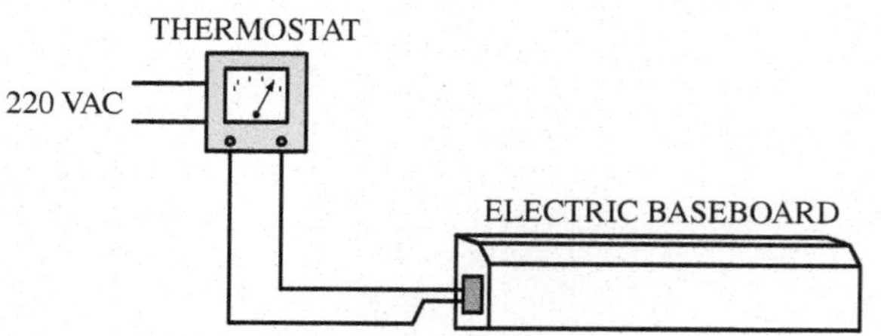

Figure 23-5 Power is supplied to the baseboard heater through a line-voltage thermostat mounted on an inside wall. The thermostat operates to close a set of contacts when the temperature of the room drops, thus supplying power to the baseboard. When the baseboard warms the room, the thermostat opens and turns off the power to the baseboard.

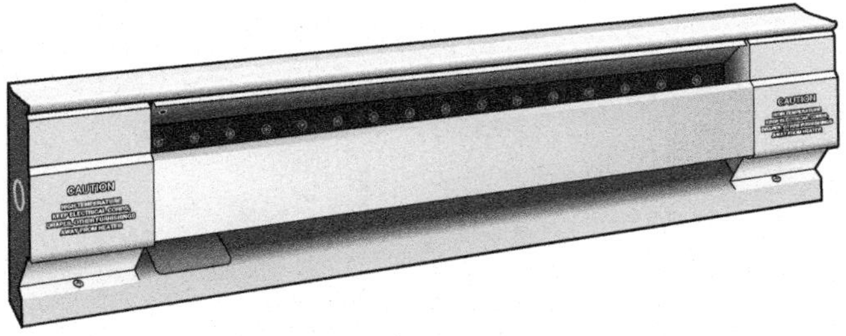

Figure 23-6 Baseboard electric heaters can be used in residential or commercial installations.

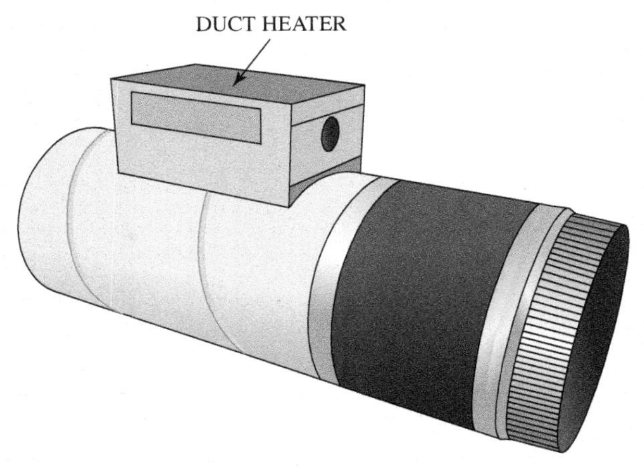

Figure 23-7 This device is an electric heating unit that fits into a round duct. It is used to both heat a space or to boost the heat to a space.

thermostat operates to close a set of contacts when the temperature of the room drops, thus supplying power to the baseboard. When the baseboard warms the room, the thermostat opens and turns off the power to the baseboard.

The baseboard heating system for each room of a structure is designed in the same way. Each room has a heating element and a thermostat. Individual temperatures for each room can be set. If a room is not being used, the heat for that room can be turned off or set to the lowest thermostatic setting.

You may see electric baseboard heating in the entryway of commercial buildings. Baseboard heating is used as a supplemental form of heat in the wintertime in heavy traffic areas like an entryway. The air requires additional warming because of the opening of the door as people enter and leave the building.

Note that electric radiant panels are very similar to electric baseboard configurations. Electrically, they are wired and operate in the same way. The difference is in how heat is moved into the occupied space. Electric baseboard heat conducts heat to air, heating the air. Radiant panel systems radiate heat to objects, including occupants, in a manner similar to how the sun heats objects. The air temperature can be much lower, but occupants feel comfortable because of the radiant energy hitting them.

23.3 ELECTRIC BACKUP HEAT

Electric backup heat comes in many different configurations. The flexibility of the heating element allows a manufacturer to create backup heat in any shape, allowing it to be made to fit any application. It can be configured as an electric heating element installed in a section of ductwork to directly heat or to boost the heat to an individual room. It can also be configured as a bank of heating elements that can take over the heating of an entire home or building if the primary heating system fails. Figure 23-7 is a sample of duct heater that can easily be installed in a duct system.

Electric backup heating is typically staged to turn on only when it is needed. The primary system may be a heat pump or other fuel-burning piece of equipment. The primary system is given the first stage of a multistage thermostat and the backup heat the second; and sometimes the third. In this way, the electric backup only supplies the required amount of heat to stay comfortable. Electrical usage bills can be kept to a minimum by staging electrical usage. Staging also prevents overheating of a heated area. Figure 23-8 is another example of an electric strip heater. This could be used as a primary or secondary heating system.

23.4 ELECTRIC FURNACE

An **electric furnace** can be smaller than a conventional gas-burning system. Because it does not convert fuel to electricity, it does not require heat exchangers or a venting system. However, large electrical wiring and larger breakers will be required. Electric heating elements are housed along with their respective control and safety devices in a metal enclosure that can be connected to supply and return ducting. Manufacturers generally design the furnace in a way that allows the unit to be installed in an upflow, counterflow, or horizontal configuration. Installation clearances are typically "zero." This means that the furnace case can touch combustible surfaces such as wood or drywall. The door panels must be removable for service activities, but they can touch a closed door. Both of these things mean that the furnace can be mounted in any position and placed anywhere. Some of the unusual places in which this type of furnace can be placed are above cupboards in a kitchen or under a counter as well as in a closet, crawlspace, or attic space. Service access is still required.

In Figure 23-9 the upper part of the electric furnace cabinet houses the electric elements (strips) and the blower. The lower portion has an evaporator coil that will need to be installed if the system provides air conditioning. Notice the double-pole breaker found in the upper left side of this electric furnace. This is a positive means of disconnecting power for service. Some jurisdictions may require a separate external disconnect on this unit.

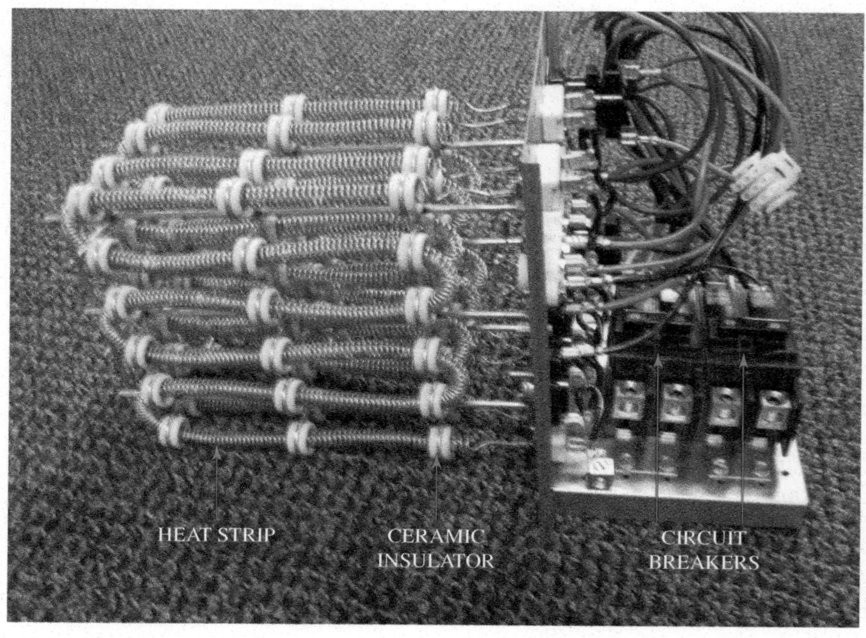

Figure 23-8 This bank of four heating strips might be used in an electric furnace or as a backup to a heat pump system.

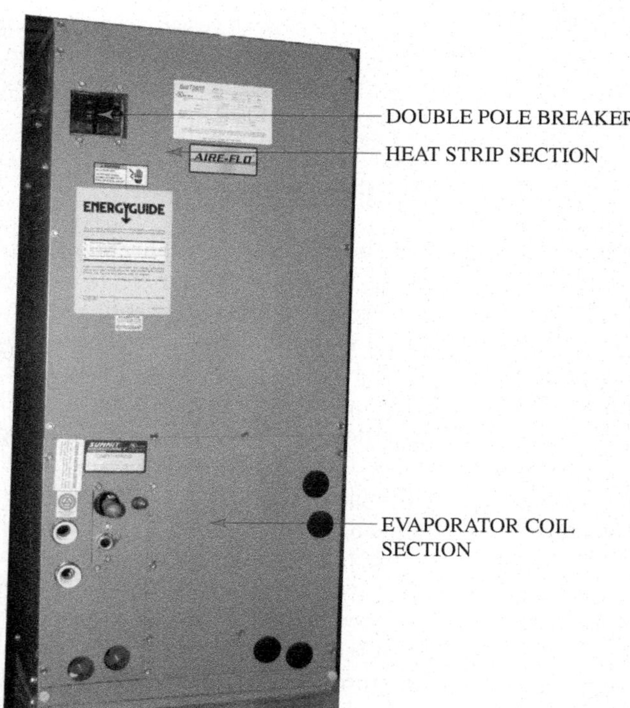

Figure 23-9 The upper part of the electric furnace cabinet houses the electric elements (strips) and the blower. The lower portion has an evaporator coil.

Figure 23-10 is a common air handler without the heat strip option installed. A panel in the air handler is removed and the appropriately sized heat strip assembly is inserted in the slot. This design allows one air handler to be used for several different heat strip options. For example, the installation may only require a 5-kW heat strip. Other options may include 10-, 15-, or 20-kW heat strips. Commercial electric heat will include higher-kilowatt heat strip options.

In summary, the electric furnace costs less to install; there are no pipes, vents, chimneys, or drains to install. Customers like the fact that no open flame is associated with an electrical furnace. With no fuel being burned, there is no concern about carbon monoxide, soot, or smell.

TECH TIP

Heat strips will smell for a few minutes when they are first turned on after not having been used for an extended period of time (e.g., over the summer). A coating of dust collects on the surface of the heat strips. This dust quickly burns off and is not a fire hazard or a danger to the customer. You should warn customers about the smell because they may want to go outside for a short time to avoid the unpleasant odor. Some new heat strips have to burn off a residue when first energized.

The blower compartment is integral to the electric furnace and operates in the same way as other warm-air furnaces. The blower is designed to turn on a few minutes after heating elements are powered on and turns off a few minutes after they are powered off. This action is accomplished with a time-delay relay, also called a sequencer. Next, we discuss some of the common components of an electric heat strip.

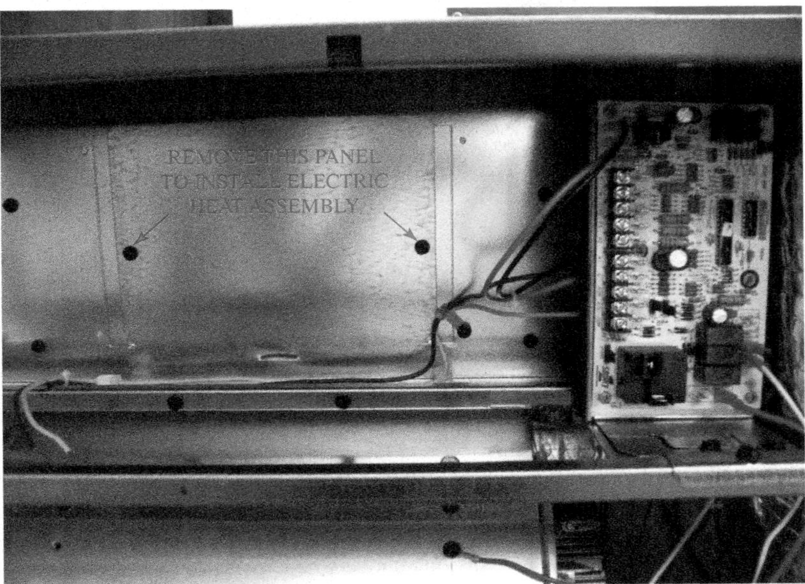

Figure 23-10 This is an air handler without an electric heat option. The lower portion is the blower section. The control circuit board is seen to the right of the removal panel.

23.5 COMMON COMPONENTS OF AN ELECTRIC HEATER

The next sections discuss the common components found in an electric heating system. They are:

- Heat strips or heating elements
- Sequencers or relays
- Overcurrent protection
- Fusible links
- Thermal overloads
- Miscellaneous components.

23.6 HEAT STRIPS

We looked at residential and commercial heat strips, also known as heating elements, earlier in this unit. The heat strip must have continuity or a complete circuit and have some resistance to current flow. The resistance is required to prevent a short circuit as well as generate heat. One usual type of heating element is a ¼-inch-diameter spring. A close-up of this type of heater is shown in Figure 23-11 as part of a replacement restringing element kit. These kits are available if the heat strip opens. The kit includes a new heat strip, ceramic insulators, a fusible link, and miscellaneous hardware.

The heating capacity of heat strips is measured in kilowatts and abbreviated "kW." These are just some of the heat capacities available for single-phase heating elements:

- 4 kW (4,000 watts)
- 4.5 kW (4,500 watts)
- 5 kW (5,000 watts).

One watt will generate approximately 3.4 Btu of heat; therefore, 1 kW or 1,000 watts will develop 3,400 Btu of heat per hour of operation.

Figure 23-11 This is an electric heating element replacement kit. The kit includes a new heat strip, ceramic insulators, a fusible link, and miscellaneous hardware. When 240 volts is applied, the element will develop 3,500 watts of heat.

A 5-kW heat strip will develop 17,000 Btu of heat:

$$5{,}000 \times 3.4 = 17{,}000 \text{ Btu/hour of operation}$$

Here are some rules of thumb regarding common residential or commercial single-phase electric heat strips. A 5-kW heat strip operating at 240 volts will:

- Draw about 20.8 amps.
- Measure about 11.5 Ω.

As you can see this information can be used for troubleshooting heat strips.

One final point on heat strips is the voltage rating. Most single-phase heat strips are rated to operate at around

200 to 250 volts. Using the watts formula of $W = E \times I$, you notice that the higher the voltage, the higher the wattage of the heat strip. A heat strip operating at 240 V will create about 20% more heat than a heat strip operating at 200 V. Measure the voltage when all heat strips are energized since a full-load circuit may significantly drop the supplied voltage. A large drop in voltage should be investigated. It may be caused by:

- Poor connections
- Undersized conductors
- Power supply problems.

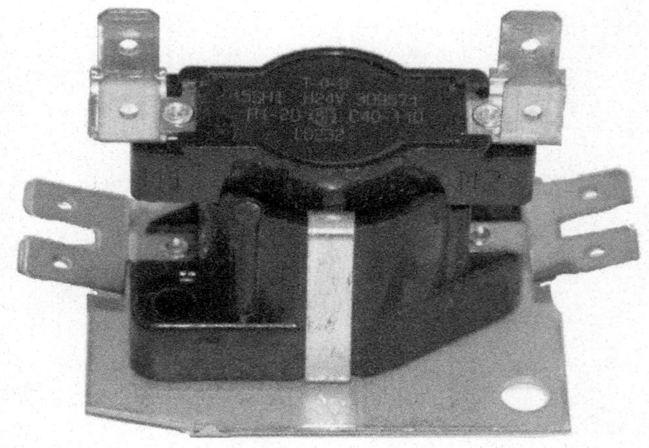

Figure 23-12 This is a single-stage heat sequencer. When energized by 24 volts on the lower, brass connections, the upper part of the switch will close after a short delay. The upper contacts will close and energize the blower and one heat strip. The upper contacts can handle about 25 amps.

23.7 SEQUENCERS

Electrical **sequencers** are used to switch electrical heating elements or strips. Some manufacturers use a contactor to energize heat strips. This section discusses the use of the time-delay sequencer. A single-stage sequencer that is designed to close the upper contacts and operate a blower and one heat strip is illustrated in Figure 23-12. The lower connections (brass colored) are for the control voltage, which is usually 24 volts. Between these connections is a heater coil that warms up and causes a bimetal switch to close after a specific time delay. Between the upper connections (silver colored) are the contacts. The contacts will handle the high-amperage load of a heat strip and blower motor. The contacts are rated around 25 amps.

When electric heating elements are first powered, the electric draw starts out high when the elements are cool, but as the elements heat the amperage drops. Sequencers have large contacts that can dissipate the amp draw of the electrical elements, but more importantly, a bank of several heating elements can be powered one after another. Bringing on heating elements separately is preferable to bringing on several heating elements at the same time. Electrical safety code generally recognizes this high-amperage starting condition and limits electric heater starting loads at 10 kW, which is about 40 starting amps. High-amp starts place stress on an electrical system and may even cause lights to momentarily dim. Staging the timing of when heating strips start up reduces this concern.

Sequencers are a form of time-delay relay. When powered by 24 volts, the sequencer will turn on one or two heating elements. Figure 23-13 illustrates two sequencers tied

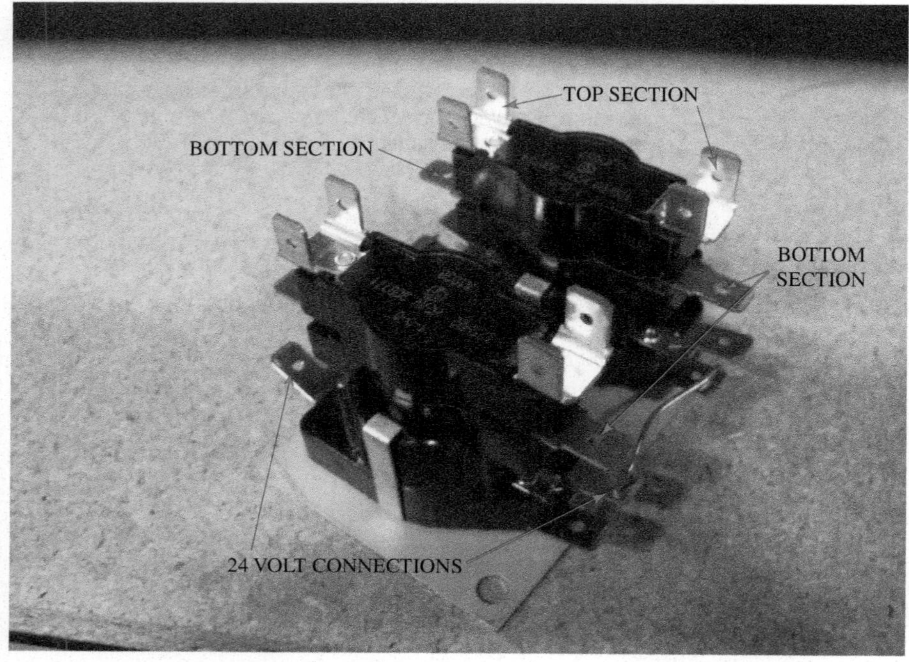

Figure 23-13 This is a dual sequencer. The brass connections are used to wire the control voltage. The upper connections are used to switch the heat strips and blower motor. Each sequencer has a bottom section and top section. The bottom section can switch a heat strip. The top section can switch two heat strips or one heat strip and a blower.

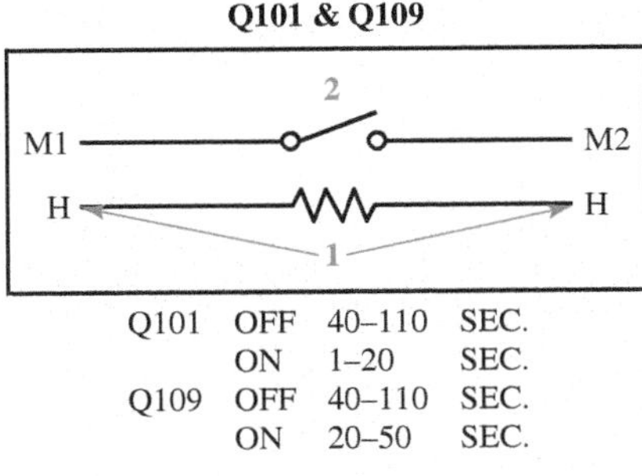

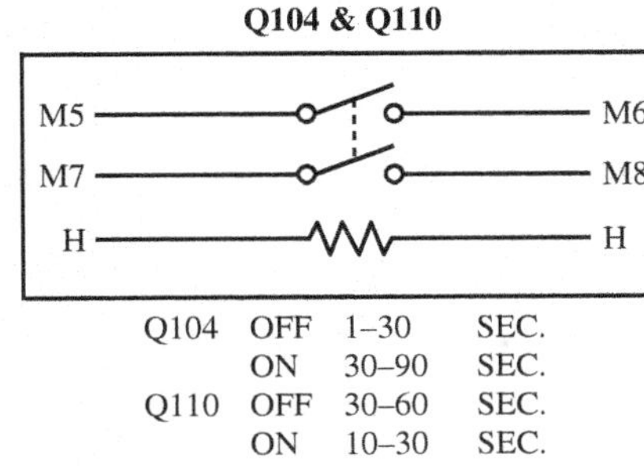

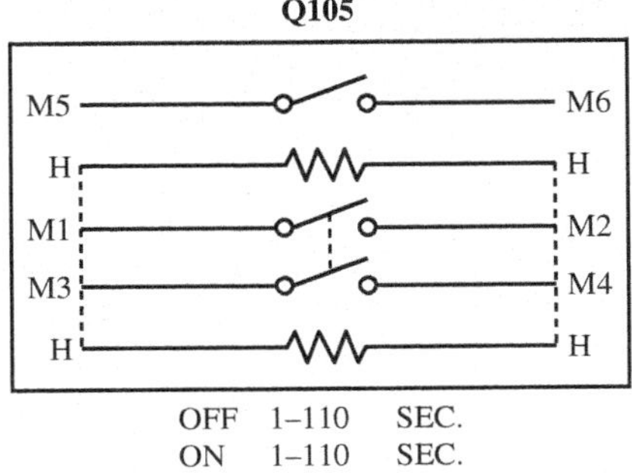

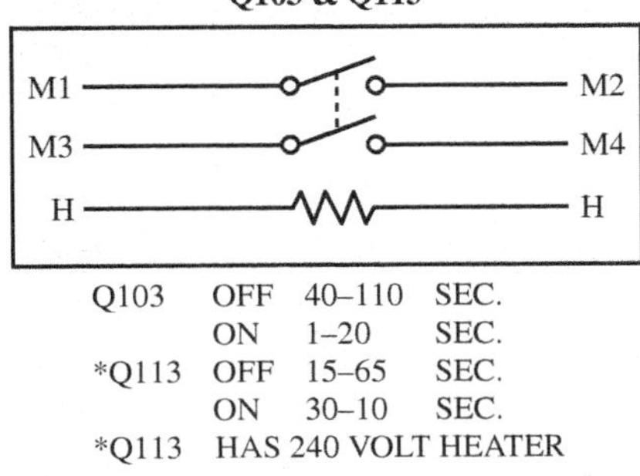

Figure 23-14 The **H** or heater connections are for the control voltage. Most sequencer control voltage is 24 volts. Notice that model Q113 requires a 240-V control. The **M** or main connections will supply high voltage to the blower or heat strips.

together in parallel with the 24-V control wired to the bottom connections. The top two sections will operate two sets of heat strips. The bottom set of heat strips will operate two more sets of heat strips. The brass connection near the bottom mounting place will be energized by 24 volts to close the contacts in a staged manner so that not all the heat strips operate at the same time.

Figure 23-14 shows the various electrical symbols used for a sequencer. The capital letter "**H**" is used to represent the control voltage connections, which are usually 24 volts. The letter "**M**" and a number following **M** represent the sequencer contacts' identification. As stated earlier, the sequencer uses a time-delay circuit, which is accomplished by heating a bimetal strip. When the bimetal is heated, it expands, extending a rod, closing contacts between the **M** connections. When the sequencer **H** coil is de-energized, it cools and the bimetal contracts and opens the **M** contacts. The design of the time-delay bimetal heater creates more time or less time for the contacts to close and open.

The information in Figure 23-14 shows a sequencer model number Q101. When the **H** coil is energized (point 1), it closes contacts between **M1** and **M2** (point 2) after a 1- to 20-second delay. The sequence contacts open, or turn off, after 40 to 110 seconds.

Notice that model Q104 closes the **M** contacts later, at 30 to 90 seconds, while opening the **M** contacts sooner, at 1 to 30 seconds. A common sequencer operation is that the first sequencer will bring on the first heat strip and blower. The first sequence will be the last one to open or go off at the end of the heating cycle. In summary, the first sequence will be the first one on—operating the blower—and the last one off. This ensures continuing blower operation until most heat strips have turned back off.

Using Figure 23-13 and the Q105 diagram in Figure 23-14, let's see how this will work:

1. When the thermostat calls for heat, both 24-V sequencer coils **H** are energized simultaneously.
2. When energized, the sequencer on the left Figure 23-13 will be energized first by the 24-V control voltage, which closes after a short time delay.
3. Refer to Figure 23-14. The blower and two sets of heat strips will begin to operate. This is caused by the contacts between M1-M2 and M3-M4 closing.
4. The second sequencer will close after a longer time delay and close contacts M5-M6.
5. All heat three heat strips and the blower will be operating.
6. When the thermostat is satisfied, it will de-energize all the **H** coils.
7. The second sequencer (M5-M6) will open after a short delay.
8. The last heat strip that was energized will turn off first.
9. The first sequencer **H** coil will de-energize last.
10. The blower and first heat strip will now turn off.

As shown in Figure 23-15, some sequencers close several sets of contacts to turn on multiple strips of electric heat. The sequencers are designed to operate in a timed sequence. For example, the sequencer on the left will close very quickly once energized. It will operate the first heat strip and blower. The middle sequencer will come on after a time delay. The sequencer on the right closes minutes after it is energized.

Sequencers are made in many different styles. Some are solid state and work similarly to PTC (positive temperature coefficient) relays. Others are designed and look similar to an overload or snap disk. It doesn't matter how they are constructed, each type of relay does the same thing.

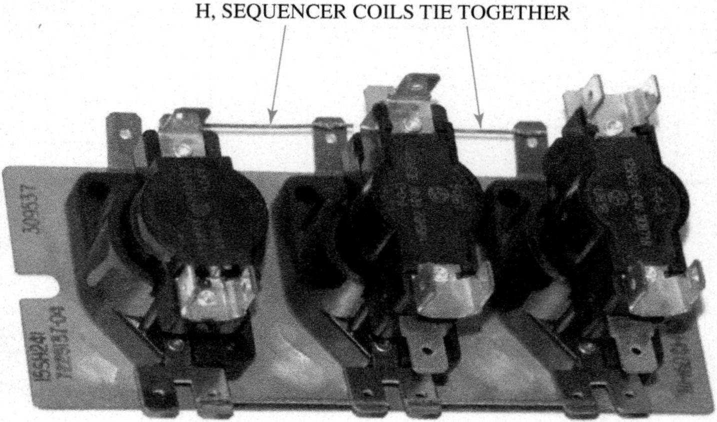

Figure 23-15 Note that there are three sequencers with the control tied together.

Other Types of Controls

Silicone controlled rectifiers (SCRs), also called *silicone gate controlled rectifiers*, are another way of controlling power delivered to electric strip heaters. The SCR is a solid-state device that can turn on and off very quickly. Because it is a "rectifier" (a device that only allows electrical flow to occur in one direction), two SCRs are used to control one electric heating element. The signal to the SCR comes from a circuit board. The way it generally works is as follows:

1. The thermostat signals the need for heat.
2. The circuit board receives the signal and interprets it for the level of heat needed.
3. The circuit board sends both SCRs a pulse signal to turn on part of the AC sine wave.
4. As the sine wave moves through zero volts, the SCRs turn off.

By turning on only part of the AC sine wave, only a portion of the applied voltage reaches the electric heating element. The element only heats to the temperature that meets the heating need of the structure and satisfies the thermostat setpoint.

Working on electric heating systems with SCR controls requires some knowledge of solid-state controls and attention to the manufacturer's recommendations. Always read and understand the manufacturer's installation and service manuals.

23.8 OVERCURRENT PROTECTION

In this case overcurrent protection simply means a fuse, breaker, or fusible link or overload device used to protect the equipment, a circuit in the equipment, or wiring. These terms are often used interchangeably because they have some similarities. In the case of breakers or fuses, they are normally used to protect the whole unit from excessive current, but they can be sized to protect one component in the unit. You will have overcurrent protection for the unit and you may have optional protection for components like the transformer or circuit board.

Figure 23-16 shows two common fuses, the plug-in fuse and the glass (Buss) fuse, used in a control circuit board. These types of fuses can also be found in the secondary side of a transformer. Figure 23-17 shows a circuit board with the plug-in U-type fuse.

Breakers or fuses of the correct amperage and voltage rating should be within easy access of the heating system. Typically, the breaker is the same rating as the maximum amperage listed on the nameplate of the electrical heating unit. The installing contractor may need to interpret the amperage values of an installation to apply the correct size breaker. In some instances, a breaker of 115% of the unit's "minimum" amperage may be specified. A larger breaker should never be used. A breaker is designed to protect the equipment and wire. A breaker of greater amperage will not turn off the electrical supply in the event of overcurrent draw. A breaker that is too small will turn off the power before the maximum current is drawn by the unit. Next, we will discuss the fusible link and thermal overload.

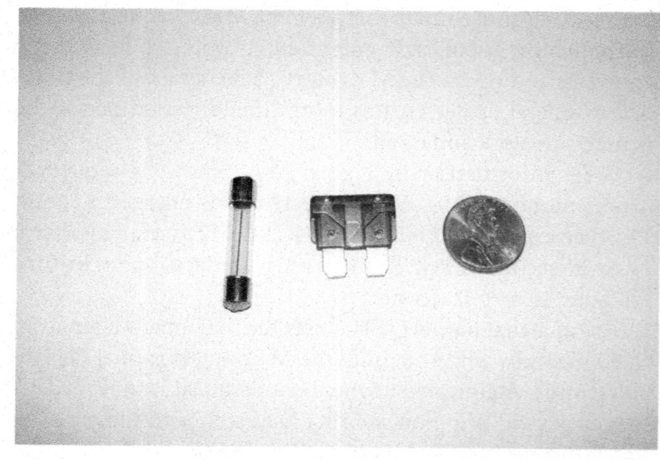

Figure 23-16 Plug-in fuses are used to protect a circuit board from overcurrent conditions. A glass fuse can be used as a plug-in fuse or in a fuse holder. (Penny shown for size reference.)

Figure 23-17 This is a circuit board for an air handler with an option for electric heat strips. Notice the 3-amp plug-in fuse located at the upper left side of the circuit board.

23.9 FUSIBLE LINKS

A **fusible link** (see Figure 23-18) is often wired in series with an electrical heating element. The purpose of the link is to open when either high amperage or high heat is encountered. The fusible link cannot be reset and must be replaced if open. The cylinder is silver and has manufacturer information printed on its surface. The information may include temperature and amperage ratings. The link is a small cylindrical device that has one square end and one tapered end. The taper may be black or red, depending on the color of the material used in its manufacture. The link can be checked for continuity to determine if it is open. Resistance should be zero ohms.

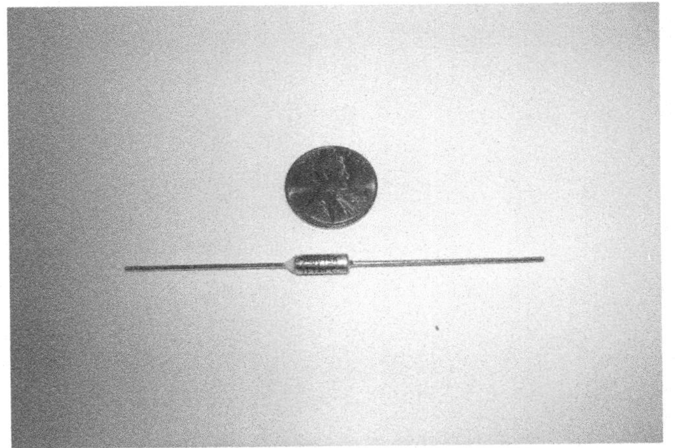

Figure 23-18 This common fusible link is found in series with a heating circuit. It is used to protect the heating circuit from damage. If an overheating or overcurrent condition occurs, it will open and not reset. When this happens, the fusible link will need to be replaced.

23.10 THERMAL OVERLOADS

A thermal overload, where *thermal* refers to heat, is a safety device used to open a heating circuit in case of excessive heat. For example, excessive heat will be created by a lack of

Figure 23-19 Thermal overloads on a commercial electric strip heater. These heating elements are protected by two thermal overload devices. The overload is like an eye, watching over the heating element.

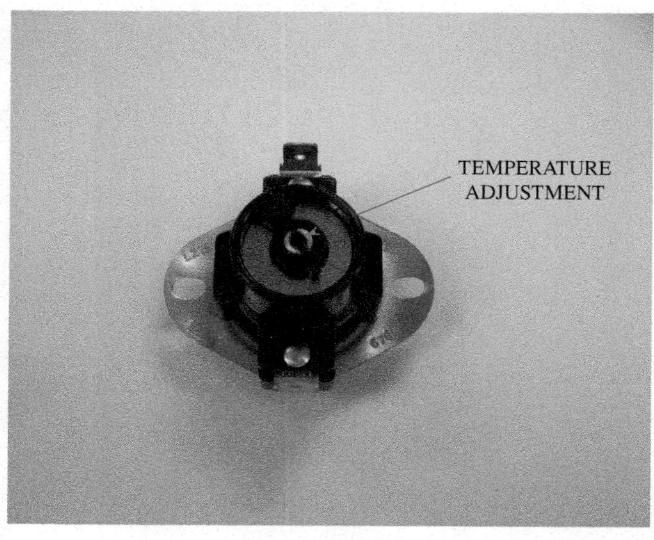

Figure 23-21 This is an adjustable thermal overload. The temperature can be adjusted to the appropriate safety temperature.

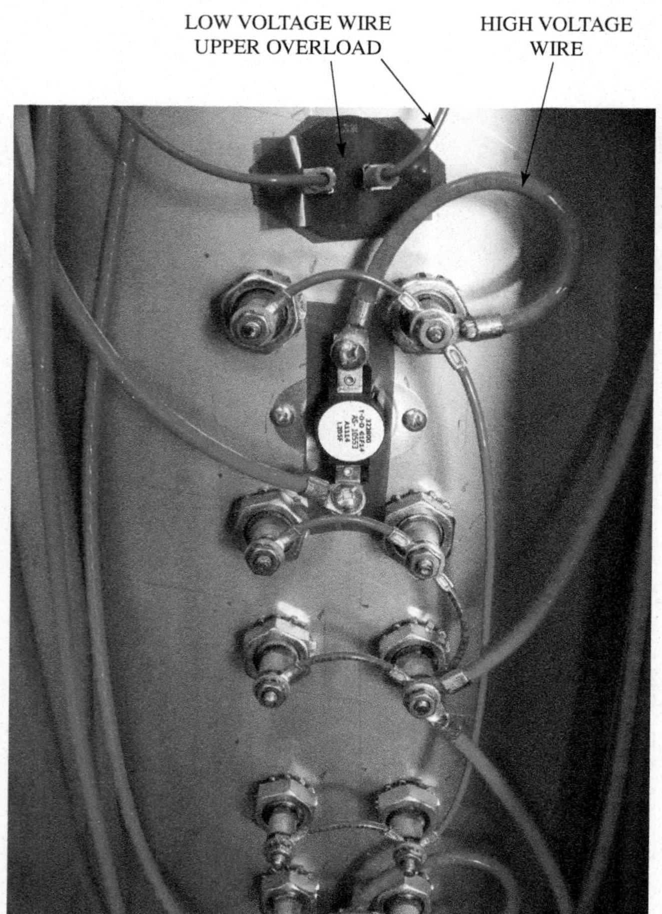

Figure 23-20 This is the backside of the thermal overload. Notice that two gauges of wire are used to hook up the overloads.

airflow if the blower motor stops or if the air filter or evaporator coil is dirty.

The thermal overload is usually a bimetal device that opens when the air temperature rises. These devices are normally round and look like an eye. Figure 23-19 shows two thermal overloads monitoring a large commercial electric heater. Figure 23-20 shows the backside of the two thermal overloads shown in Figure 23-19. Notice that two gauges of wire are used to hook up the overloads. The upper overload is wired in the lower amperage control circuit. The bottom overload is connected in series with the heat strips; therefore, it must be capable of handling the higher amps that run through these heat strips. It has the larger, lower gauge red wire.

Most thermal overloads are the automatic-reset type. This means that when the overload opens, then cools, it will automatically close its contacts and allow the heat strips to operate again. A thermal overload can also be designed with a manual reset option. The manual reset means that the technician will need to push a button on the back of the overload to get it to reset or close.

Finally, some thermal overloads are adjustable. Figure 23-21 shows a picture of an **adjustable thermal overload.** This device allows a technician to stock one thermal overload on the service truck or at the office that can cover various temperature applications.

23.11 MISCELLANEOUS COMPONENTS

The electric heating system has other components necessary for its operation. The furnace will have a blower assembly and possibly a circuit board. It may also have a filter rack, and a filter grille may be placed in the return air duct system. It is also common to find an evaporator coil in the electric furnace cabinet.

23.12 WIRING REQUIREMENTS

Electrical supply wiring for electric heating systems is very important. Wire sizing must conform to the National Electrical Code as well as local code requirements. Be familiar with all applicable codes.

Wiring is based on amp draw and, to a lesser degree, supply voltage. Electric heating units can draw significantly high amperage. The nameplate of all electric heating units will list the amp draw. In some cases the amp draw is part of the unit model number (check with the manufacturer for model number identification). Amp draw charts are based on copper wire gauge. Some installations may use aluminum wire if it is permissible. Some manufacturers may prohibit the use of aluminum wire. If aluminum is used, connections may require the use of a special connection coating or anti-oxidation material. Copper wire does not require a special coating to be applied at connection points.

SAFETY TIP

If electrical supply wire for an electric heating system is noticeably hot, there may be a problem with the size of the wire or the wiring connections or there may be a break in the wire insulation that is allowing oxidation breakdown of the wire (in the case of aluminum). If the technician suspects that the temperature of the wire is too high, a qualified electrician should be called to determine if there is a problem.

SAFETY TIP

Electric wire used to connect heating elements is rated for temperature and amperage conditions. Be sure to use wire that is rated for the condition. Wire not rated for this condition may melt, causing short circuits and an electrical fire.

23.13 OPERATION

A single strip of electrical heating element is easily operated, as in the case of baseboard or radiant panel heating. When multiple strips of heat are used, control of the system becomes more complicated. For purposes of explanation, an electric furnace will be used. Keep in mind that other installations that use multiple stages of electric heat will generally operate in the same way.

An electric furnace with three strips of electrical heating elements and a two-stage thermostat is installed in a residential application. The first two strips of electric heat are sized to handle the residential heating load under normal conditions. If the outside temperature should drop to a point where the first two heat strips are not creating enough heat, the temperature in the residence would drop, causing the second stage of the thermostat to call for additional heat.

Here is a sequence of operation when a thermostat calls for heat:

1. **FIRST-STAGE HEATING OPERATION:** The signal is received by the first sequencer and the blower relay.
2. The sequencer delays for approximately 30 seconds and turns on the first electrical heating element and blower. If the fusible link and the overload are closed, the electric heating element begins to warm.
3. The first sequencer and second sequencer heaters are in parallel; therefore, they are both energized at the same time. The second sequencer will begin its 60- to 90-second time delay. At the end of the second sequencer's time delay, the second heating element is powered and begins to warm.
4. The operating blower moves air across the heated electrical elements, cooling the elements and warming the air to be distributed to the occupied space.
5. When the temperature of the room meets the setpoint of the thermostat, the thermostat opens the **W** circuit.
6. The control voltage signal is lost to the sequencers. The second sequencer turns off the second heat strip after a short time delay. Then the first sequencer turns off the first strip of heat after a longer delay.
7. When the first sequencer turns off, the fan control will turn off the blower.
8. **SECOND-STAGE HEATING OPERATION:** If the first stage of heat (first two heating elements and blower) did not satisfy the thermostat setpoint and the temperature of the occupied space drops, the second stage of heating begins. The second stage of the thermostat would close. This is normally labeled as $\mathbf{W_2}$ on the diagram or thermostat.
9. When the second stage of the thermostat closes, both the first and second strips of heating will already be operating. The third sequencer is then energized, delaying power to the third strip of heat for approximately 60 seconds. The third heating strip operates until the room temperature rises to turn off the second stage of the thermostat $\mathbf{W_2}$. At this point, the third sequencer turns off power to the third heating element. When the room temperature is met, the other heat strips are de-energized as described in steps 5, 6, and 7.

23.14 EFFICIENCY CHECK

An electric heating system has the advantage of being easily checked for efficiency and proper airflow. These systems do not have to factor in the efficiency of fuel conversion or the efficiency of the fuel burned. To check for proper airflow, the following procedure is used:

1. Measure the amp draw of the operating heating element (or elements), including the blower amperage. The system should be fully energized for this measurement. This can be done at the furnace disconnect.
2. Measure the operating voltage.
3. Measure the return air temperature.
4. Measure the supply air temperature. This air temperature must be measured at three diameters of the plenum (at least), downstream of the heating elements. The temperature probe must also be shielded from the radiant energy given off from the elements (if there is direct line of sight). The airflow must be mixed to get an accurate temperature measurement. One way to ensure the accuracy of the temperature measurement is to place the temperature probe past the first supply air elbow so the heat coming directly off the heat strips will be mixed.
5. The following formula is used:

$$\text{Cubic feet per minute (CFM)} = \frac{\text{voltage (V)} \times \text{amperage (A)} \times 3.4\ \text{(Btu's per watt)}}{\text{supply temperature} - \text{return temperature} \times 1.08}$$

The difference between the supply air temperature and return air temperature is also known as delta-T and is abbreviated ΔT. Some formulas will use ΔT in place of the bottom part of the formula that states "supply temperature – return temperature"; thus, it would appear as $\Delta T \times$ **1.08.**

EXAMPLE 23.1 CALCULATING CFM

Using the formula from step 5 above, calculate the CFM given the following information:

Amp draw (with blower)	85.9 A
Voltage	230 V
Return air temperature	68°F
Supply air temperature	118°F

SOLUTION

$$\text{CFM} = \frac{230\text{ V} \times 85.9\text{ A} \times 3.4}{50° \times 1.08}$$

$$\text{CFM} = \frac{67173.8}{54}$$

$$\text{CFM} = 1{,}244\ \text{(rounded off from 1243.95)}$$

The resulting airflow of 1,244 CFM closely matches the amount of air necessary for a 3-ton air conditioning unit. The recommended airflow for air conditioning is 400 CFM per ton. If this were a 3-ton unit, the recommendation would be close to 1,200 CFM. If this were an adjustable belt-drive blower, the motor pulley could be opened a little more to make the blower deliver a little less air. Actually, 1,244 CFM is good for a 3-ton system, which needs 1,200 CFM. Measurement inaccuracy and general instrument tolerance will create an error of around ±5%. If this were a three-speed blower, you would select the motor speed with the closest match possible. A little extra airflow is better than low airflow.

SEQUENCE OF OPERATION 23.1: ELECTRIC HEATING FURNACE

Next, we cover the sequence of operations of a simple heating furnace using Figure 23-22. For a larger view of this figure, see Electrical Diagram ED-1, which appears with the Electrical Diagrams package that accompanies this book and shows the diagram for a heat-cool packaged unit. Figure 23-22 shows just the heating portion of Electrical Diagram ED-1. Refer to both figures to follow this sequence of operations:

1. The first step to energize the heating circuit is to close the (TH) thermostat and close the **Cool Off Heat** switch.
2. This will energize the coil (HR). **HR** is a contactor-type relay that will activate the heating circuit immediately.
3. When the (HR) coil energizes, it will close **HR** contacts 1 and 3.
4. **HR** contacts 6 and 4 will also close.
5. Closing the **HR** contacts will supply voltage to electric heat strip **ER.** The heat strip begins to provide heat.
6. The final step in the heating process is to activate the indoor fan motor (IFM). Closing **HR** contacts 6 and 4 will energize the (IFR) coil.
7. This closes **IFR** contacts 2 and 4.
8. The indoor fan motor (IFM) starts. The blower and heat strip are actually activated simultaneously. The heat strip needs immediate airflow to prevent damage from overheating.

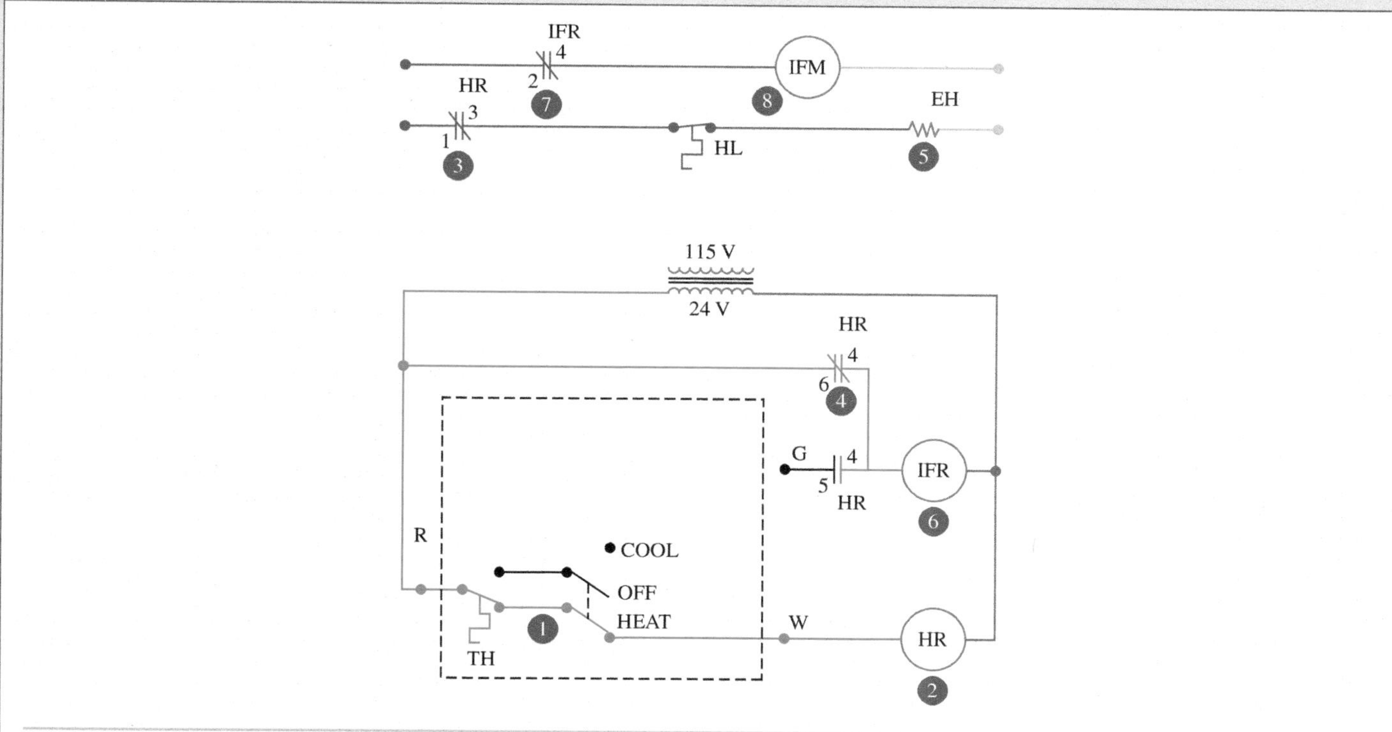

Figure 23-22 Heating diagram used in Sequence of Operations 23.1. This diagram was extracted from enlarged Electrical Diagram ED-1.

23.15 THREE-PHASE HEAT STRIPS

Figure 23-23 is a three-phase electric heating diagram that can be used in commercial or industrial applications. Notice that there are three heat strips fed by the three-phase power supply. The blower and heat strips are designed to come on simultaneously. These large-kilowatt heaters use contactors to activate the heating circuit. Some designs will keep the blower in operation at all times since continuous air circulation is required in commercial occupancies where people are present. With several of these of heat strips wired in parallel, you may easily find heat outputs in the range of 100 kW and greater.

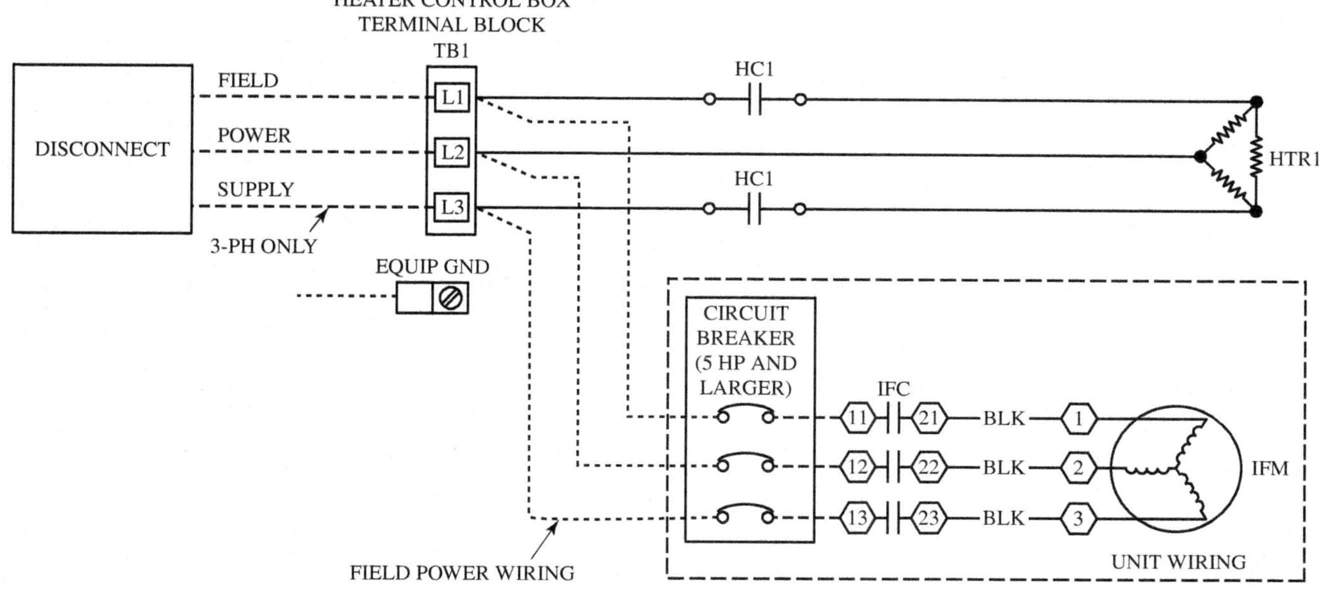

Figure 23-23 This diagram is for a commercial three-phase electric strip heater. Notice that there are three heat strips labeled HTR1, which are fed by the three-phase power supply.

Table 23-1 Basic Electric Heating Troubleshooting Checklist

Customer Complaint	First Check	Possible Cause
No heat	Thermostat and voltage	Check thermostat settings
		Incoming voltage
		Open breaker
		Open fusible link
		Burned-out element
		Open wiring
		Blower motor not functioning
Not enough heat	Amp draw	One strip of heat burned out
		Open fusible link
		Dirty filters
		Blower motor problems
Too much heat	Thermostat setting	Thermostat set incorrectly
		Sequencer stuck closed

23.16 BASIC TROUBLESHOOTING CHECKLIST

Electric heating troubleshooting is generally simple. If a heating element is not working, it will show a loss in amperage. Voltage checking using the hopscotch approach will determine if a sequencer is passing voltage (zero volts across the sequencer contacts **M**) or not (voltage across the sequencer contacts **M**). Table 23-1 can be used to guide the troubleshooting process.

SAFETY TIP

Most electric heating systems use line voltage for most of the relays and sequencers. Line voltage ranges from 208 to 240 volts. Always check supply voltage before replacing any electrical components, especially the control transformer to ensure that the right voltage tap is being used.

TECH TIP

Airflow for electric heating systems is very important. Low airflow is responsible for heating element burn-out. Airflow is necessary to keep the electric elements cool. Low airflow will allow the elements to glow red or white-hot and shorten the life of the element. Likewise, do not block airflow while an electrical element is operating or turn off the blower while troubleshooting.

SERVICE TICKET

With every troubleshooting procedure it is important for you to verify the diagnosis. An example would be that you are called for a "not enough heat" problem. Checking with an ammeter, you determine that one of the electrical heating strips is not pulling amperage. You suspect that the heating element is burned out and open. The next check would be to verify that the heating element has no continuity. To do this, you turn off and lock out power to the heating unit. Remove the power connection from one side of the electric heating element. Set the multimeter to the ohms scale and test from one end of the element to the other. If there is no resistance reading, the problem has been verified as an open (burned-out) heating element. The heat strip should have some resistance, but not zero ohms or infinity (open).

SUMMARY

Electric heating systems are smaller and less complex than fuel-burning systems that deliver the same number of Btu's as larger fuel-burning systems. Because they do not convert fuel to electricity, they do not need venting systems. With no need for venting systems, they can be placed anywhere and cost less to install. Because of their flexibility, electric heat systems can be produced to fit nearly any heating application.

Electric heating systems can be primary sources of heat or secondary sources. They can be installed as an individual room heating system or as an electric furnace to heat an entire building as a central heating system. Often an electrical heating system is coupled to a heat pump to provide backup heat during defrost periods or as supplemental heat during periods of low outside temperature.

How "green" are electric heat systems? Several ways of looking at electrical heating systems were discussed in regard to being energy efficient or using renewable energy. Electric heat is clean, convenient, and flexible. But if electricity can be produced by renewable sources, would direct electrical heat be considered a "green" technology?

The unit concluded with a small section on troubleshooting. Many of the troubleshooting techniques used for transformers, relays, and safety devices are also used for electric heating systems. Electric systems tend to be simpler in both operation and service than fuel-burning units. There are also fewer components to check during a service call. Some things that are checked on other types of furnaces become simpler to check on electric systems. Airflow, for instance, is a simple check. With four measurements and a few basic calculations using a formula, a technician can easily determine the airflow and adjust the fan speed.

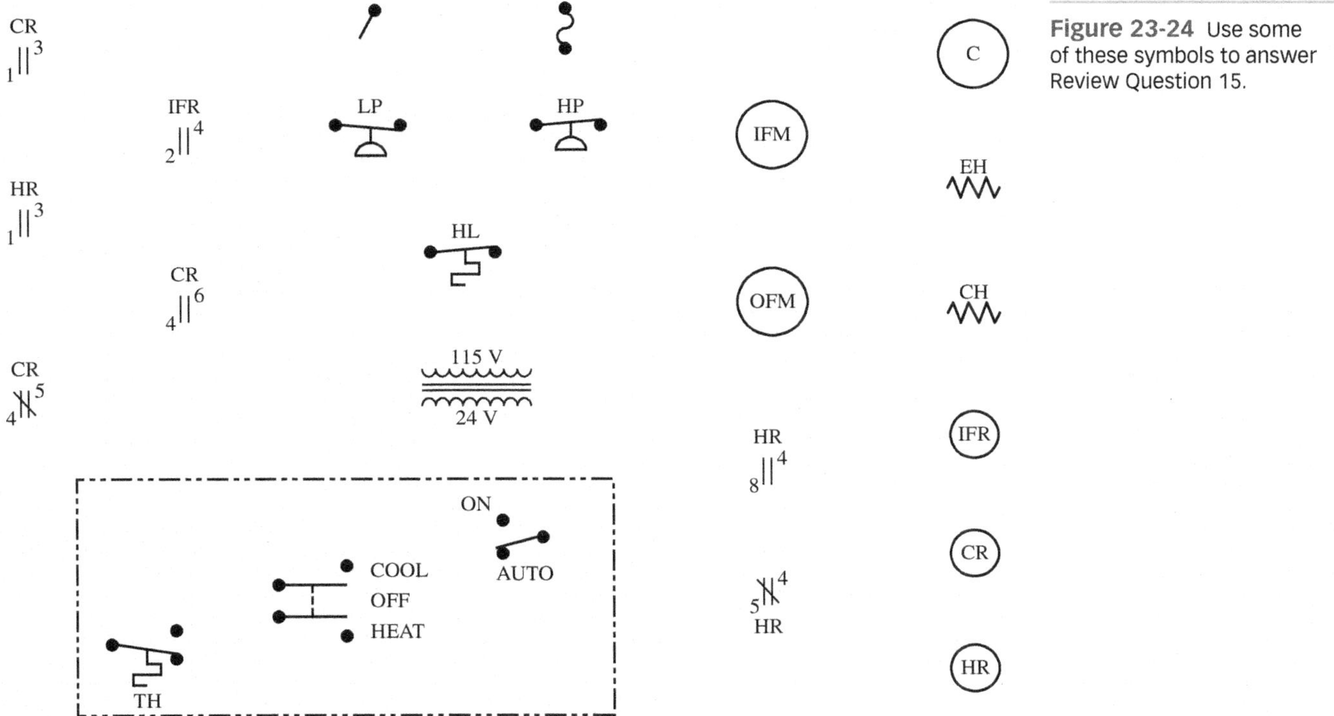

Figure 23-24 Use some of these symbols to answer Review Question 15.

REVIEW QUESTIONS

1. Explain why electric systems are smaller for the same Btu output.
2. How could direct electric heating systems be considered "green"?
3. What function does a sequencer perform?
4. Why are multiple electric elements turned on separately?
5. What measurements are required to determine the CFM of an electric heating system?
6. Why is supply wire size important and how is it determined?
7. What is the difference between backup and primary heat?
8. What is staging and why is it used?
9. What is a fusible link and what is it for?
10. What is the difference between an immersion element and one that is used in air?
11. An electric furnace has a 10-kW heat strip. At 240 volts what is the estimated amp draw?
12. What is the CFM of an electric furnace that has the following readings?:
    - 240 volts
    - 43 amps
    - 68°F return air
    - 110°F supply air
13. When troubleshooting an electric furnace, what is the first thing to check on a "no heat" service call?
14. When troubleshooting an electric furnace, you suspect an open heat strip. The power is safely disconnected and one end of each heat strip is disconnected from the circuit. You measure the resistance of three heat strips. Here are your results:

    Heat strip 1 measures infinity.

    Heat strip 2 measures 11 Ω.

    Heat strip 3 measures 12 Ω.

    What is the problem?
15. Using the symbols shown in Figure 23-24, draw a simple electric strip heat furnace with a blower and control voltage.
16. Name the device in Figure 23-25. What does it do?

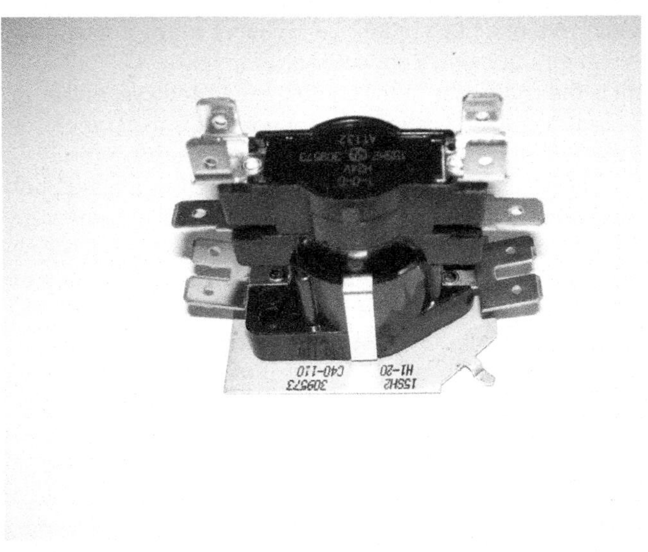

Figure 23-25 Use this image to answer Review Question 16.

UNIT 24

Heat Pump Heating Systems

WHAT YOU SHOULD KNOW

After studying this unit, you will be able to:

1. Explain what is meant by the term *electromechanical* and name components that are considered to be electromechanical.
2. Describe the operation of a reversing valve.
3. Name five common components found only in heat pump systems.
4. Explain how inside electrical components and outside electrical components are connected to function as one heat pump unit.
5. Describe the start-up and shutdown of the heating mode of operation.
6. Describe the heat pump defrost cycle.
7. Recognize when a hard start kit might be used and describe its operation.
8. Discuss the benefits derived from having a microprocessor as part of a heat pump system.

This unit describes the operation of electrical components found in a generic **heat pump**. The most common heat pump looks like a regular split-system air conditioning system. Heat pumps can also be window units and package units. Since the outdoor heat exchanger is used for cooling or heating it is called the outdoor coil, not the condenser. The normal inside heat exchanger is called the indoor coil since it is also used for cooling or heating. Identifying a heat pump's heat exchangers as an outdoor coil or indoor coil reduces confusion because either coil can be the condenser or evaporator. The outdoor coil is the condenser in the cooling mode and evaporator in the heating mode. Likewise the indoor coil is the evaporator in the cooling mode and condenser in the heating mode.

The heat pump refrigeration cycle is briefly discussed, but we do not focus on the heat pump refrigeration cycle because it is not electrical based. Both the descriptions and the wiring diagrams in this unit are designed to convey the operation of any type of heat pump, whether it uses an air source or geothermal source. With small differences, the generic heat pump operates in a similar way to most heat pumps. Understanding the operation and the sequence of events as described here will help your understanding of the basic operation of the heat pump and electrical components. It is suggested that other sources of information about the mechanical operation of a heat pump be used to fully understand heat pump operation.

The principle of a heat pump is to move heat from a place of abundant heat to a place of lower value heat. In the heating mode, a heat pump absorbs heat from a large reservoir of low-temperature heat (usually cold outdoor air) and compresses it to a high temperature, delivering it to the occupied space. In the cooling mode, the heat pump works like a standard air conditioner, removing heat from the occupied space and rejecting it outside.

GREEN TIP

Green Heat Pumps

Heat pumps are considered to be "green" because they use half the energy when compared to heat provided by electric heat. A number rating is given to all heat pumps. A Seasonal Energy Efficiency Ratio (SEER) rating denotes how efficient a heat pump is when operating in the cooling mode. The higher the rating, the less energy that is used to produce the same amount of energy transfer.

The government mandate for all split-system heat pumps as of 2013 is SEER 13. Window unit heat pumps do not need to meet these requirements. A number of manufacturers have three levels of efficiency that start at SEER 13 and rise in increments to levels as high as SEER 20 or more. Heat pumps used for both heating and cooling are sized for the cooling load, not the heating load. The heating rating for heat pumps is called the COP (coefficiency of performance). The COP of a standard heat pump is around 3. This means the heat pump will provide three times as much heat as the equivalent heat energy in electricity. For example, if the electrical equivalent input (watts $\times$ 3.4 = Btu/hour) is 10,000 Btu/hour, the heat output will be 30,000 Btu/hour.

The drawback to heat pump efficiency is the initial cost of the system and its installation. Generally, heat pumps cost more than other types of heating systems and geothermal-source systems are more expensive to install. When compared to electric heat, air-source heat pumps will save you money provided you are not located in an extremely cold climate. Generally, even zero-degree weather will save money when compared to all-electric heat. The savings in the cooling mode are the same. Air-source heat pumps are sold to save heating energy not cooling.

24.1 HEAT PUMP TYPES

Heat pump systems have been classified as **air-source heat pumps (ASHPs)** or **geothermal-source heat pumps (GSHPs)** (or geo-source heat pumps). Air-source heat pumps pull heat from outdoor air to heat occupied spaces.

Figure 24-1 This is a heat pump outdoor unit. It is installed on snow legs to the height of the average snowfall for the area. This elevation reduces defrost problems.

As shown in Figure 24-1, the heat pump outdoor unit looks like a regular condensing unit. Geothermal heat pumps remove heat from the ground or water. Both of these systems move heat in similar ways: by means of a refrigeration system. Both take advantage of the heat of compression to add heat to the air.

GSHPs are further classified as either **groundwater** or **ground-loop** systems. The difference is that the groundwater system uses a well or water source like a lake to pull water for heat exchange, as shown in Figure 24-2A. This system may also use a second well to dispose of the water that has picked up heat or rejected heat. If water is not injected back into the aquifer, the water is released to a pond, lake, or stream. This is an open-loop system since the water may be exposed to the air from the water source.

The ground-loop system uses a water-based fluid that is moved around a horizontal closed piping loop, buried in the ground at a depth of approximately 6 feet. The loop may also be a vertical loop design will depths as much as 200 to 300 feet deep below the earth's surface as shown in Figure 24-2B. The water in the loop is conditioned and either absorbs heat from the ground (heating mode) or releases heat to the ground (air conditioning mode). This is a closed-loop system since the fluid is circulated inside piping and not exposed to air as in the open-loop design. The closed-loop design will have fewer contaminants since it is sealed. The circulating fluid is usually treated to reduce corrosion and prevent freezing.

Under the ground-loop designation, there is another system that is called a *direct-exchange* or *refrigerant loop*. A copper heat exchanger is buried in the ground. Instead of water being pumped in a loop, this system uses a refrigerant loop to absorb the ground heat or release the heat to the ground directly. Within the designation of "ground loop" there are also many different configurations of loops. More information on heat pump configurations, loop types, etc., can be found by conducting a simple online search. A good source of independent information would be the U.S. Department of Energy's Energy Efficiency and Renewable Energy program.

Before we learn about the operating sequence of the heat pump, we will discuss the basic electrical components found in air-source heat pumps. We will then apply these electrical components in the cooling, heating, and defrost operations.

24.2 ELECTRICAL COMPONENTS

The following electrical components are found in heat pumps and not necessarily in standard air conditioning systems. In the next sections, we cover the following components:

- **Reversing valve** with solenoid
- Heat pump thermostat
- Printed circuit board for heat pump applications
- Lack of charge pressure switch
- Defrost controls.

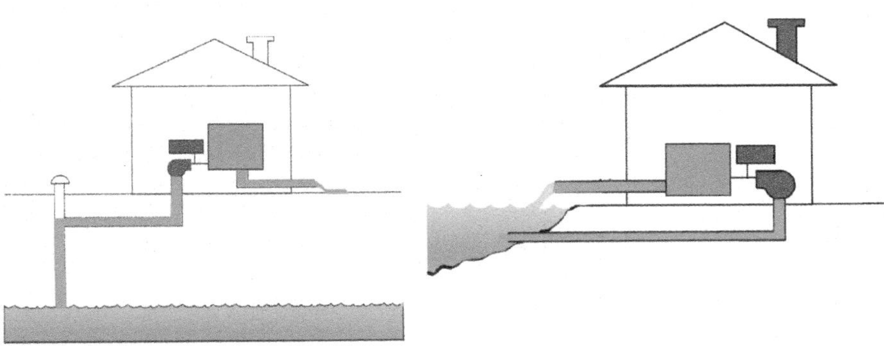

(A)

Figure 24-2 (A) This is an open-loop geothermal system. It uses a well water or groundwater source to absorb heat for the heating mode and reject heat for the cooling mode. (*Continued on next page*)

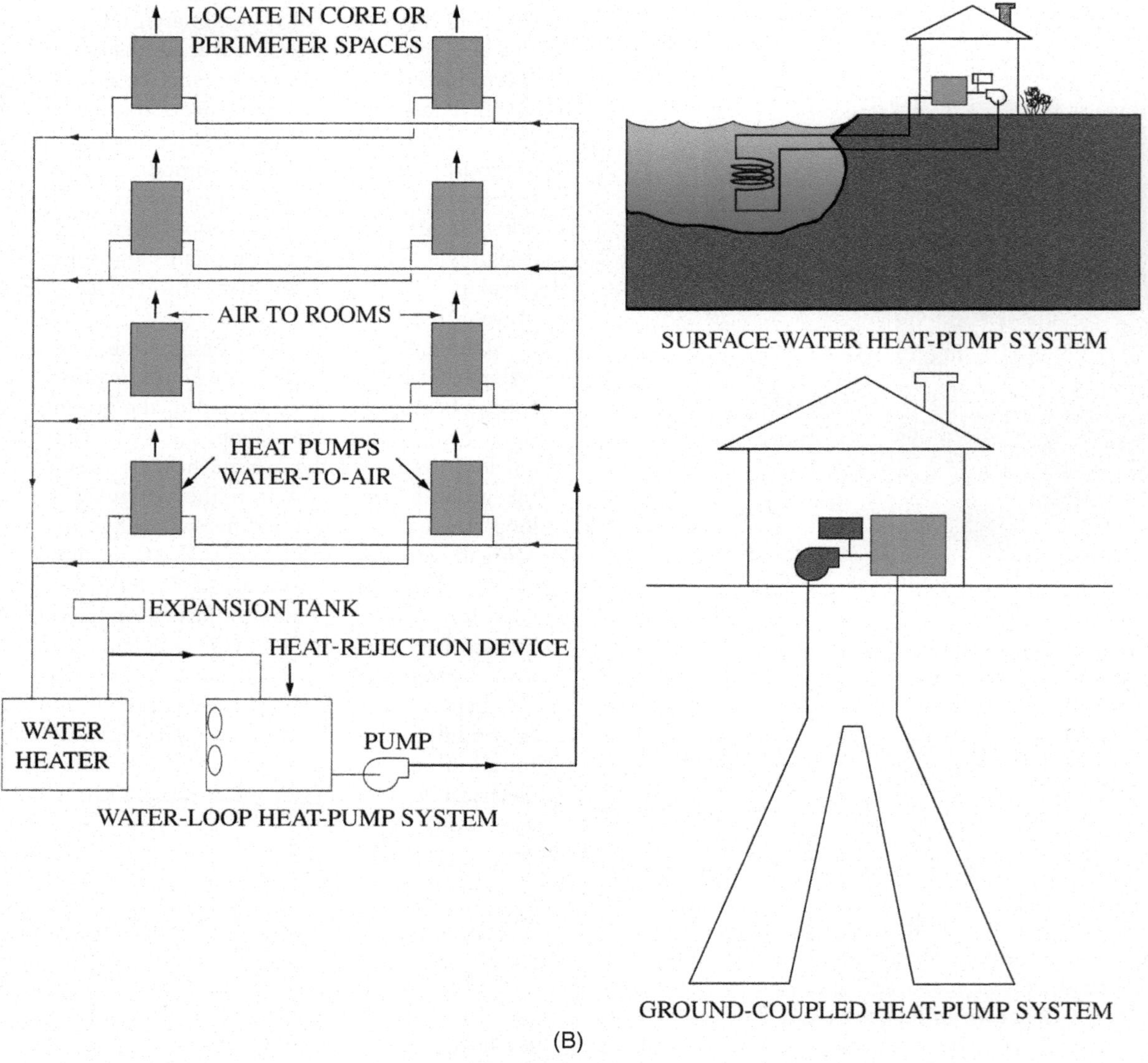

Figure 24-2 (B, left) This shows a commercial or multizone application. It uses supplemental water heater for heating. (B, right) This illustrates a ground-coupled or geothermal heat pump.

24.3 REVERSING VALVE

The reversing valve itself is mechanical, but we are going to emphasize its electrical operation. Our profession uses other names for the reversing valve. You may hear it called or identified as one of the following terms:

- **Four-way valve**
- **Switchover valve** (**SOV**)
- **Heat pump valve**
- **Reversing valve solenoid** (**RVS**).

The electric solenoid valve controls the flow of refrigerant. Figure 24-3 shows the piping arrangement. The fitting labeled with the number 1 goes to the discharge of the compressor. The fitting labeled 3 goes to the suction accumulator or suction side of the compressor. The fittings labeled 2 and 4 will be piped to the indoor or outdoor coils. The other side of the reversing valve has the solenoid valve.

Figure 24-4 is the other side of the reversing valve shown in Figure 24-3. This side has the solenoid coil that changes the pressure in the reversing valve cylinder, which in turn is used to shift the heat pump mode between cooling and heating. The coil can be powered with 24, 120, or 208/240 volts. Do not energize the coil when it is removed from the solenoid shaft. This will burn out the coil if left suspended in the air.

SAFETY TIP

Do not energize the solenoid coil if it has been removed from the solenoid shaft. This will burn out the coil if left suspended in the air. One way to test a solenoid is with a steel tool. The magnet force will attract steel. Be careful because this action may magnetize your tool. Magnetized tools should not be used around electronics like circuit boards or other digital components.

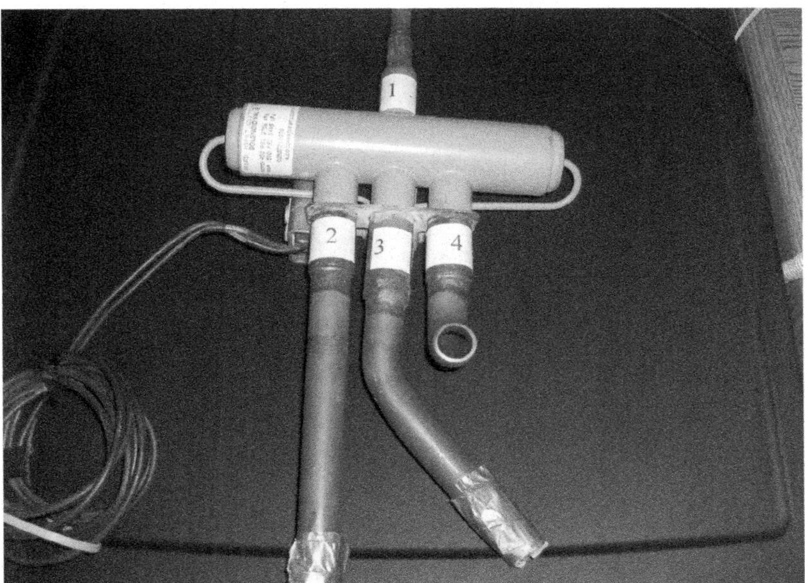

Figure 24-3 This reversing valve has been removed from a heat pump. The reversing valve is normally located in the outdoor section.

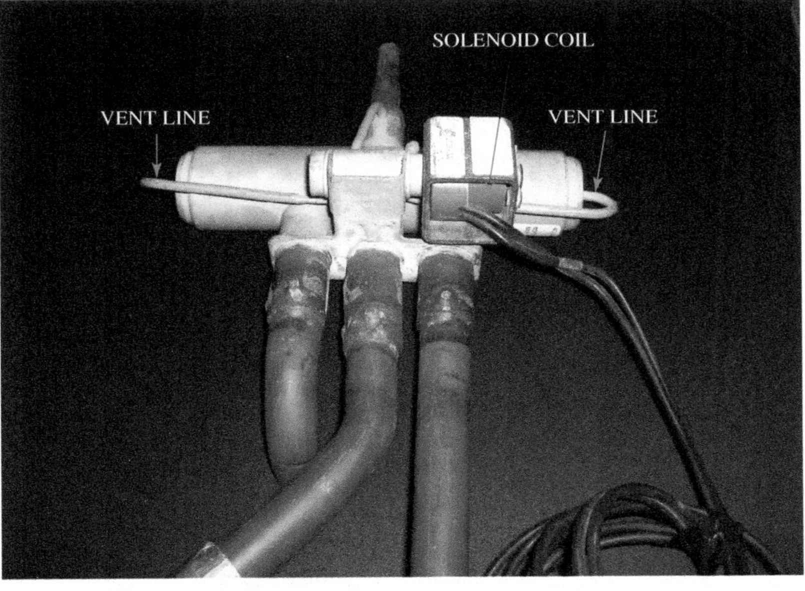

Figure 24-4 This is the other side of the reversing valve shown in Figure 24-3.

Figures 24-5 and 24-6 show a cutaway of a reversing valve. These cutaway images feature the slider part of the valve that moves from side to side to redirect the refrigerant flow to the outdoor coil in cooling mode and indoor coil in heating mode.

Figures 24-7 and 24-8 show a reversing valve that has been installed in an outdoor unit. Some designs have the valve outside the coil and accessible as seen here. Other designs have the reversing valve inside the outdoor coil. If you are having a difficult time find the reversing valve, follow the discharge line and it will lead you to the side of the valve that has one fitting. The middle fitting on the three-port side is the suction pressure and goes to the accumulator or directly to the compressor.

Figures 24-9 and 24-10 illustrate the reversing valve in the cooling and heating modes, respectively. Figure 24-9 shows the reversing valve in the cooling mode, in which the solenoid is not energized. The high pressure in the lower venting line keeps the slider valve pushed to the left side of the discharge line, directing the hot, high-pressure refrigerant to the outdoor coil.

Figure 24-10 shows the same reversing valve in the heating mode with the solenoid valve energized. When the solenoid valve is energized, it pulls up the plunger. The upper vent line is now blocked. The lower vent line is open below the plunger and the high pressure trapped on the right side of the slider drops because the equalizing line is connected to the low suction pressure. At this time there is low pressure

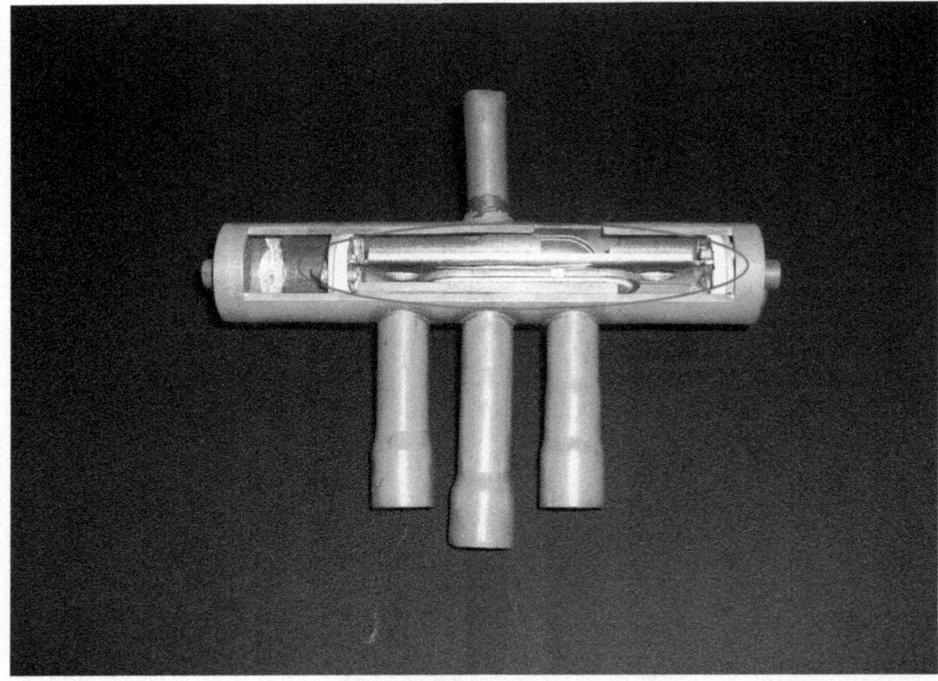

Figure 24-5 This is a cross-sectional view of a reversing valve (inside the red oval). The slider valve has been manually pushed to the right of the reversing valve body.

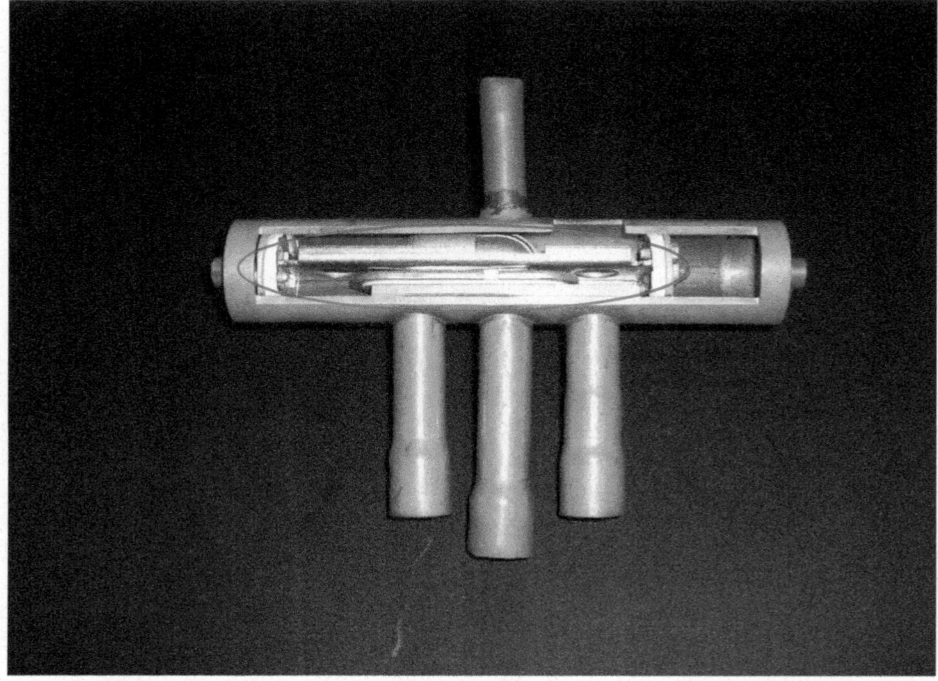

Figure 24-6 In this view of the Figure 24-5 reversing valve, the slider valve has been manually pushed to the left of the valve body.

Figure 24-7 This is a reversing valve installed on an outdoor unit.

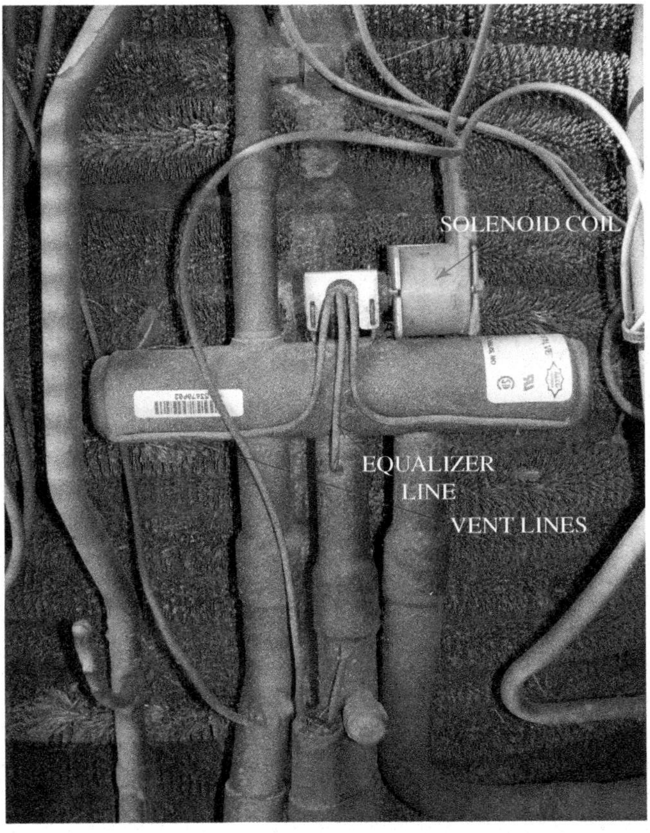

Figure 24-8 Here is a closer view of the reversing valve shown in Figure 24-7. The top fitting is for the discharge line. The middle and bottom fittings go to the suction side.

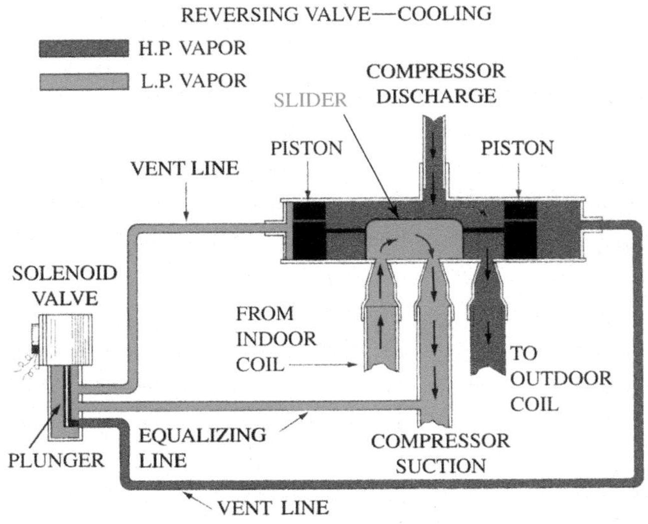

Figure 24-9 This is the cooling mode of operation. The solenoid is not energized.

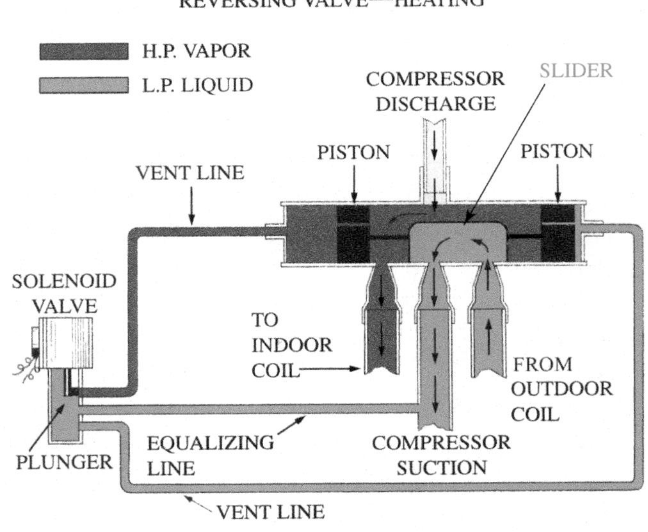

Figure 24-10 This is the heating mode of operation.

on the right side of the slider and the high pressure is building up on the left side. This forces the slider to the right and the discharge gas is diverted to the indoor coil. This starts the heating cycle.

Energizing the solenoid coil in the heating or cooling cycle is the choice of the manufacturer. You will see it either way.

TECH TIP

A 100-psi pressure difference between the high side and low side is required to shift the valve when the heat pump compressor is running. Many reversing valves are replaced because of low charge. Hook up the manifold gauge set to ensure that the amount of pressure difference is enough to move the slider valve when the compressor is operating.

TECH TIP

Sometimes it is difficult to troubleshoot a reversing valve. One question you will have is whether the slider valve moves or shifts positions. One way to detect movement is by placing a rare earth (strong) magnet on the valve body until it grabs steel inside the valve body. The magnet will move as the slider shifts position. See Figures 24-11A and 24-11B as an example. This will not work on valve bodies that are all steel.

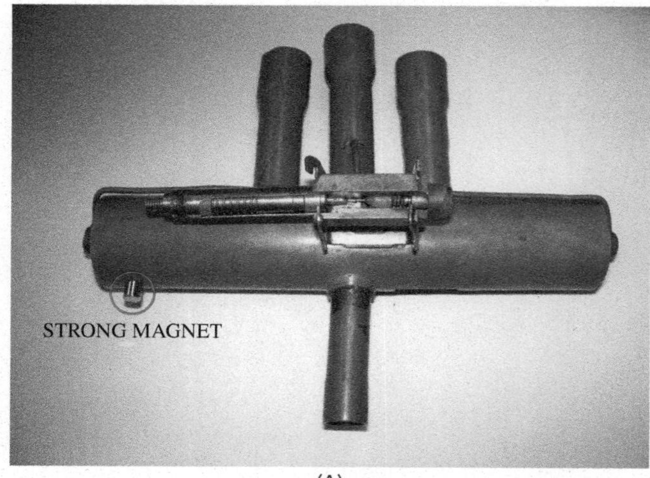

(A)

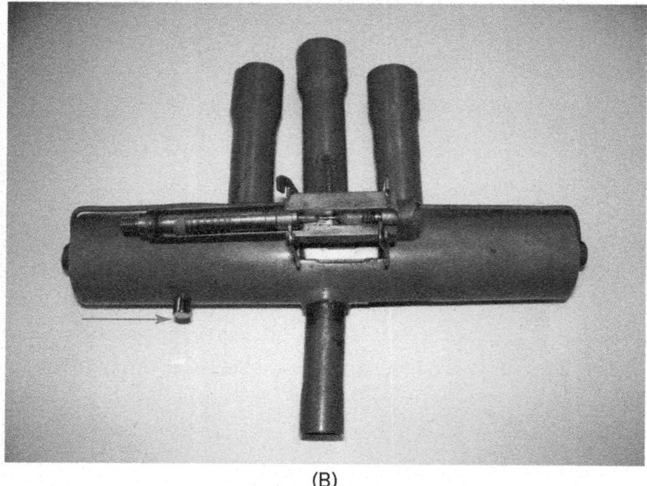

(B)

Figure 24-11 (A) One way to check the shifting action of a reversing valve is to place a strong magnet on the valve body. When the valve shifts position, the magnet will move. This troubleshooting tip will not work on steel valve body, only brass or copper bodies like the one shown here. (B) The magnet shifts position as it follows the slider valve.

24.4 HEAT PUMP THERMOSTAT

The heat pump t'stat is similar to a regular heat-cool t'stat except that it has added features. For example the heat pump t'stat will have two stages of heat known as **W1** and **W2**. The **W1** stage is used to control the heat pump heating mode. The **W2** is the second stage of heat, which is usually electric heat. An electromechanical heat pump thermostat is shown in Figure 24-12.

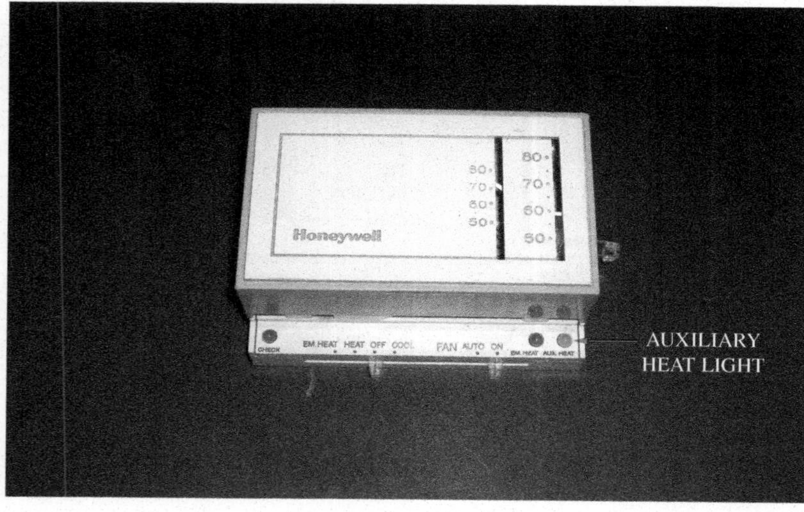

Figure 24-12 This is a mechanical heat pump thermostat. It has features such as two stages of heat and emergency heat. The auxiliary heat is the second stage of heat. When the green light is lit, the second stage of heating and the heat pump are operating.

Figure 24-13 When a thermostat is in the emergency heating mode, it disables the heat pump and brings on the second-stage heat, which is usually electric heat.

Another difference you will notice in a heat pump is the emergency heat option. Figure 24-13 shows the emergency light on a heat pump t'stat. When the thermostat is switched to the emergency heating mode, the heat pump is switched off and the second-stage heat is started up. In essence, you are operating on electric heat or whatever type of second-stage heat is used.

Let's discuss another term you may find on a heat pump t'stat: *supplemental heat.* Supplemental heat means that the second stage of heat is on. You will notice a green light labeled "AUX HEAT" in Figure 24-12 that indicates supplemental heat. When this light is illuminated, it means that the heat pump and the electric heat are operating. Having said this, there is no guarantee that the heat pump or supplemental heat is actually operating. The green light indicator does mean that there is a call for extra heat even if there is a problem and the heat pump and/or supplemental heat are not functioning.

The diagram in Figure 24-14 shows the wiring diagram for a heat pump thermostat, indoor fan coil unit, and outdoor coil. This design has the **O** terminal energized, which means that the reversing valve is energized in the **COOL**ing mode. If a thermostat uses the **B** terminal on a heat pump, the reversing valve will be energized in the **HEAT**ing mode.

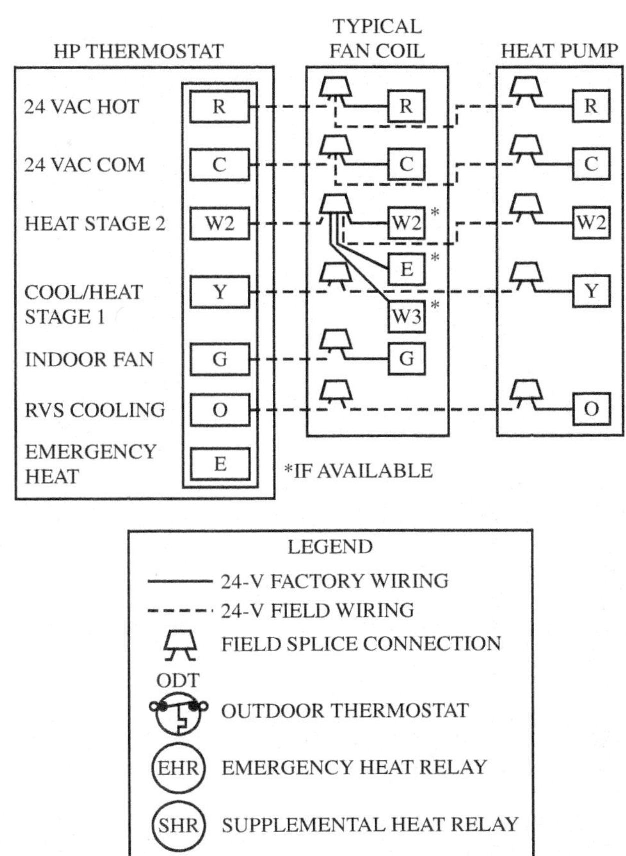

Figure 24-14 This wiring diagram is for a thermostat, fan coil, and outdoor unit. This design has the reversing valve energized in the cooling mode.

24.5 PRINTED CIRCUIT BOARD FOR HEAT PUMP APPLICATIONS

The installed printed circuit board, as shown in Figure 24-15 and used in a heat pump, is similar to the boards used in conventional air conditioning systems except they have a few additional features. The heat pump board will have a defrost cycle circuit to control frost on the outdoor unit in the heating mode. Frost occurs on the outdoor coil when the coil temperature drops below freezing and there is moisture in the outdoor air. This circuit will control how often the heat pump goes into defrost and the way defrost is terminated.

Figure 24-16 shows that the defrost timer on a heat pump circuit board is set for 90 minutes. After 90 minutes of run time, the heat pump will determine if it needs to go into the defrost cycle. The board will receive a signal to terminate the defrost cycle and resume heating. The length of the defrost time (i.e., a 30-, 60-, or 90-minute defrost setting) will depend on the outdoor humidity level. Dry climates will use the 90-minute setting since there is not much moisture in the air to collect on the coil. High moisture climates or climates that have a high winter rain will require a defrost check more often. The **test pins** to the left of the time selections are used to run an accelerated defrost circuit test. Short across the pins in the heat mode and the heat pump will go into an abbreviated test run of the defrost cycle.

Figure 24-17 is a defrost circuit board removed from a heat pump. You can see the various connection points that are required in today's heat pump controls. This circuit board

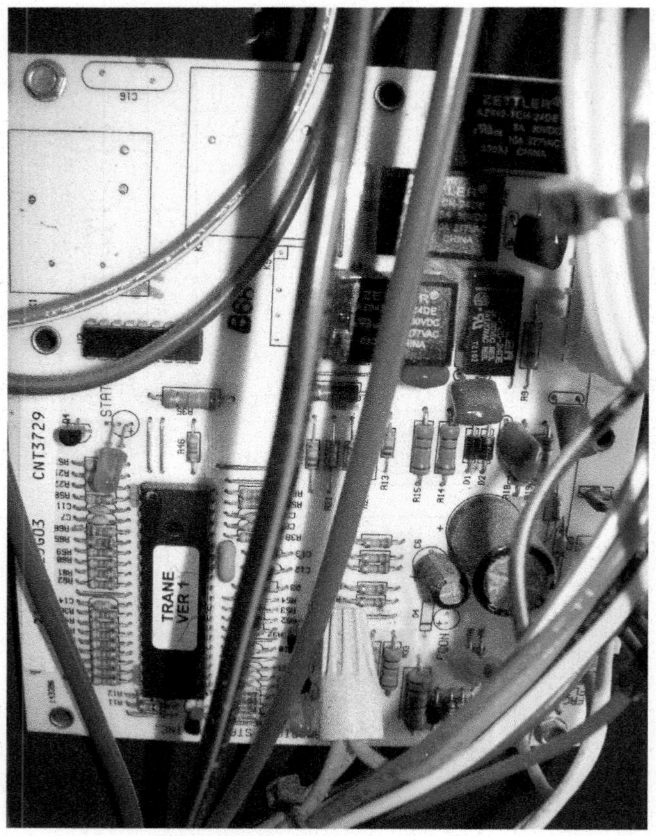

Figure 24-15 This is an installed heat pump circuit board. It controls the cooling, heating, and defrost cycles.

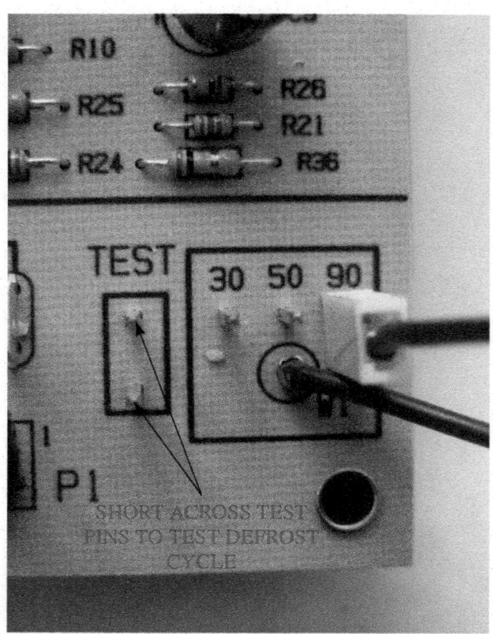

Figure 24-16 This is part of a heat pump circuit board. The defrost timer is set for 90 minutes. The test pins to the left are used to run an accelerated defrost circuit test.

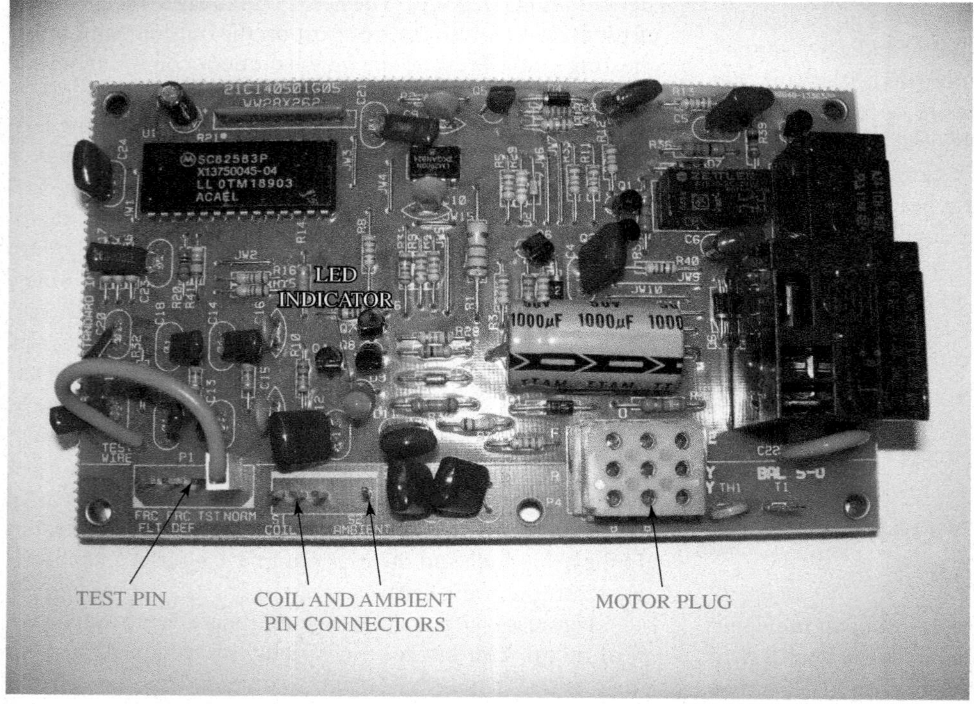

Figure 24-17 This is a defrost circuit board. The thermistor pin connectors are labeled. The board may be placed in the accelerated defrost test mode by moving the red wire (lower left side) to the pin to the left that is labeled TST. The test mode will commence and quickly go through the defrost operation, bringing on the cooling cycle and stopping the outdoor fan.

has actually decreased the number of parts and wires required in a heat pump.

24.6 LOSS OF CHARGE PRESSURE SWITCH

The heat pump's low-pressure switch (LPS) may be called a **loss of charge pressure switch.** In the heating mode when it is very cold outside the low side pressure gets very low. The normal LPS will not work well in the heating side of a heat pump application. A regular LPS will open near the following pressures:

- 23 psig ± 5 for R-22
- 55 psig ± 5 for R-410A.

The heat pump loss of charge switch opens near:

- 7 psig ± 5 for R-22
- 22 psig ± 5 for R-410A.

The loss of charge switch may be installed on the suction line or liquid line. The loss of charge switch is a fixed pressure switch. A common device is shown in Figure 24-18.

24.7 DEFROST CONTROLS

The **defrost control** system is used to remove frost from the outdoor coil in the heating mode. The current defrost system normally includes:

- A microprocessor (also known as a circuit board or defrost control board)
- A defrost thermostat
- A defrost thermistor.

Figure 24-19 This heat pump has a defective defrost component. The frost buildup will prevent heat transfer from the air into the refrigerant.

The buildup of frost is insulation on the outdoor coil and reduces heat transfer from the air into the refrigerant. The defrost cycle temporarily places the heat pump system in the cooling mode to remove the frost. The outdoor fan is also cut off to reduce cold air from blowing through the frosty coil. Figures 24-19 and 24-20 show what can happen if part of the defrost system is defective.

The first defrost controls were electromechanical. The controls were timers that moved and switches that changed position. Many of the mechanical defrost circuits are still in the field and they are quickly being replaced by solid-state devices that use a thermistor to sense temperatures and a circuit board to control the defrost operation. The thermistor shell

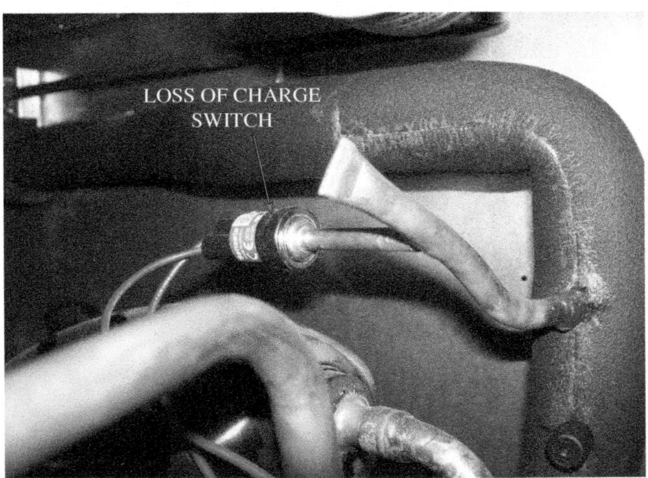

Figure 24-18 This is a loss of charge pressure switch used on a heat pump. The low-pressure switch has been brazed into a process tube that is parallel with the suction line near the compressor. Some loss of charge switches are installed on the liquid line.

Figure 24-20 This is the same heat pump as in Figure 24-19 with the top grille removed. Notice that the compressor, accumulator, and piping are frosted over. The ice can melt in the cool air, above freezing, but it will take hours. You can use water to thaw the unit or a heat gun. Placing the system in the cooling mode will help but it will drop the indoor temperature.

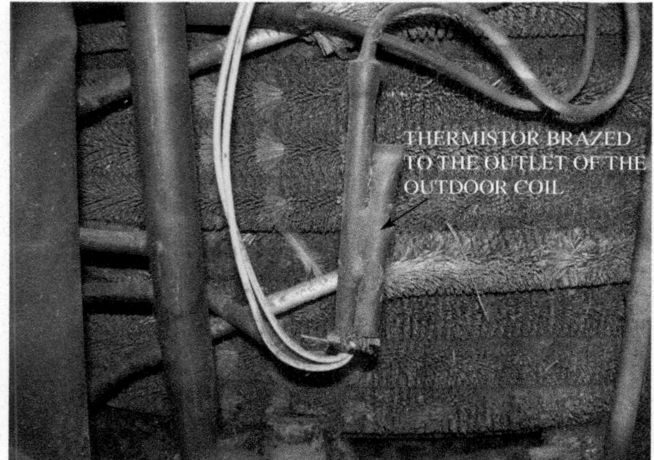

Figure 24-21 This is a thermistor brazed to the outlet of the outdoor coil. The thermistor changes resistance as the coil temperature changes. A temperature chart is consulted to determine if the thermistor is good. The thermistor is hooked to the circuit board.

shown in Figure 24-21 is brazed to the outlet of the outdoor coil. The thermistor is a variable resistor and its resistance varies with the coil temperature.

Figure 24-22 is a thermistor chart. These tables are found in the installation manual and should be used to determine if a sensor has failed. An accurate, contact thermometer is required. Infrared (IR) temperature measuring instruments will not give accurate measurements on refrigerant lines. IR instruments are accurate for measuring the temperature of black, dull surfaces. Notice that the resistance goes up as the temperature exposure of the thermistor goes down. Some thermistors work in the reverse direction. Use the manufacturers' recommended thermistor chart.

Let's review the heat pump thermistor chart found in Figure 24-22. As the temperature drops, the resistance increases. The resistance on the vertical axis is measured in thousands of ohms, which is labeled "KOHMS" in the figure.

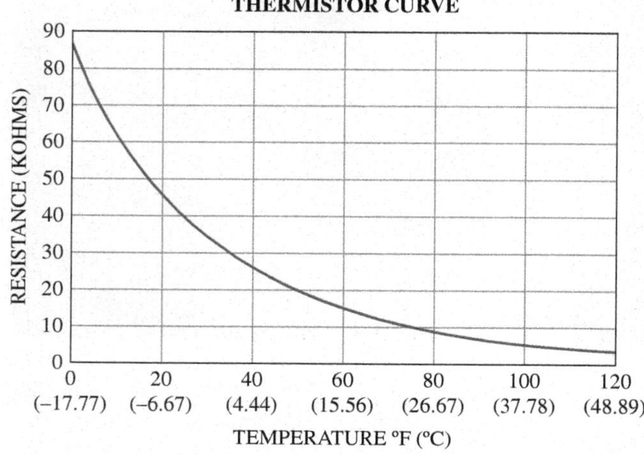

Figure 24-22 Thermistor chart used in the heat pump defrost circuit.

Defrost cycle initiation (start) and termination (end) is done by one of numerous methods:

1. Temperature initiation; temperature termination
2. Time initiation; temperature termination
3. Static pressure initiation; temperature termination
4. Static pressure and time initiation; temperature or time termination
5. Time and temperature initiation; time or temperature termination.

Temperature and Temperature Defrost

The temperature start and temperature stop method of defrost is not very common, but you need to be exposed to all methods of defrost in case you see an old system in the field. This method of defrost will initiate or start the defrost cycle when the difference between the outdoor temperature and the temperature of the coil reaches a designed setpoint, for example, 20°F. The electromechanical control will consist of two temperature measuring devices. One thermostat-like device will measure the outdoor air entering the coil and the other thermostat the outdoor air temperature leaving the coil temperature. Figures 24-23A and 24-23B show an example of a coil line temperature thermostat, also called a defrost thermostat. The defrost t'stat will open and terminate the defrost cycle.

Let's see how this would work. Normally the temperature difference between the outdoor coil and air is about 20°F. The coil will be colder so that heat will transfer from the air into the refrigerant. If the outdoor air is 40°F and the coil is 25°F, you will begin to accumulate frost on the coil since it is below freezing. This is a 15°F delta-T or temperature difference. As the coil loads up with frost, the outdoor coil temperature will drop. When it drops to 20°F delta-T, the defrost circuit will initiate. It will terminate or end the defrost cycle when the coil temperature reaches a designed setpoint, for example, 60°F.

Note: We use general temperatures in this section. The stated temperatures are not to be used in the field. Check with the heat pump manufacturer for specific temperatures, pressure, time, and ways of obtaining defrost.

Time-Only and Temperature-Only Defrost

This is a simple defrosting method used on older systems. When the timer closes a set of contacts after a designed run time, the system goes into the defrost cycle. When the coil temperature rises to a specific temperature, the defrost operation terminates. As you can imagine, the time-only method is very crude. It would go into defrost mode whether the heat pump coil needed it or not.

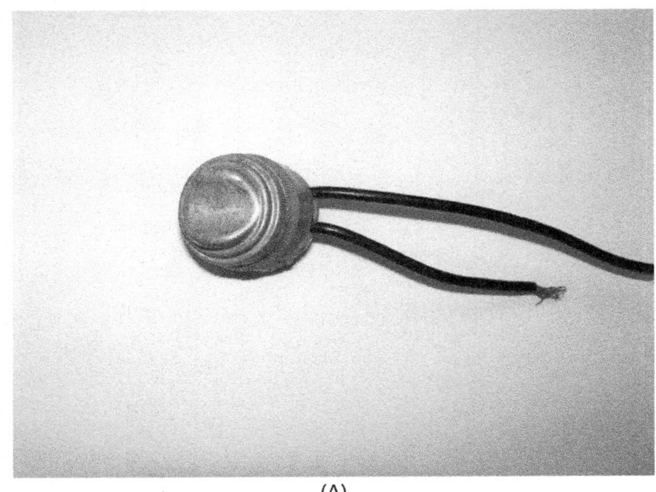

(A)

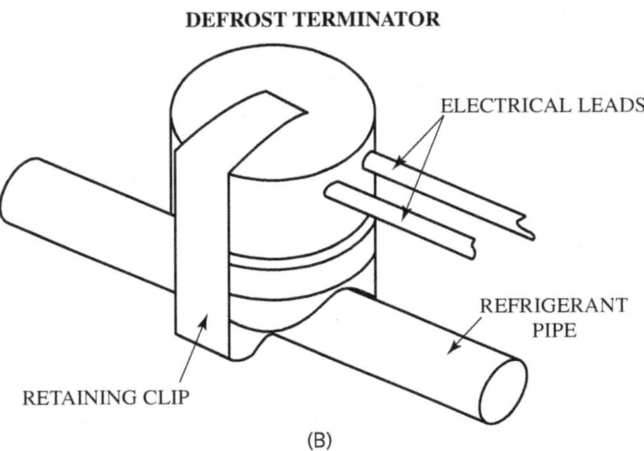

(B)

Figure 24-23 (A) This defrost termination thermostat is used on the outlet of an outdoor coil. When the refrigerant line gets warm, it will open and terminate the defrost circuit. (B) The defrost termination thermostat is strapped to the refrigerant line.

Static Pressure and Temperature

Activating the defrost cycle when there is little frost on the outdoor unit is a waste of energy and unnecessary. The static pressure method is not commonly used anymore but was popular because it only allowed the defrost cycle to run when the coil was frosted. The static pressure method has a sensor that checks the airflow across the coil. Depending on the coil design, a clean coil may have a static pressure drop of about 0.20 inch of water column, stated as 0.20″WC. When the coil gets loaded with frost, the pressure switch would close in a range of 0.50 to 0.65″WC. At this pressure drop the heat pump would go into the defrost cycle. The cycle would be terminated by warming of the coil temperature.

One problem with this defrost method is that a dirty coil will cause more defrost cycles than necessary. Dirt or debris on the coil will create a pressure drop before frost is formed. Over years of use coils tend to collect and hold dirt even with the best cleaning techniques. With this in mind, the technician should clean the coil twice a year. The heat pump maintenance schedule should call for a coil cleaning during the spring checkup for cooling and again during the fall checkup for heating.

Static Pressure, Time, and Temperature

This method is similar to the static pressure and temperature defrost method just discussed. The difference is the addition of time. To overcome false defrost activation due to wind effects, the cycle is designed to initiate defrosting with a certain amount of run time and a high pressure drop across the coil. The defrost cycle will terminate if the time expires (times out), usually after 10 to 15 minutes of defrost run time.

Time and Temperature Defrost

The initiation of time and temperature is required to start the defrost cycle. What this means that after a specific compressor run time of 30, 60, or 90 minutes and a below freezing temperature on the outdoor coil, the defrost cycle with initiate.

Time or temperature is also required to terminate the cycle. Defrost terminates when the defrost timer opens or the outdoor coil temperature rises above freezing. The timer usually opens after around 15 minutes of defrost activity even if the coil is below freezing. This design is commonly found on older mechanical defrost systems and on solid-state designs in use at this time. We will show an older version of the time and temperature defrost method because the solid-state design is similar. Understanding these diagrams will help you comprehend the current method of defrost control. This is how it works: Figure 24-24 is a defrost diagram of a heat pump that uses electromechanical components. We will use this diagram as the basis of our explanation.

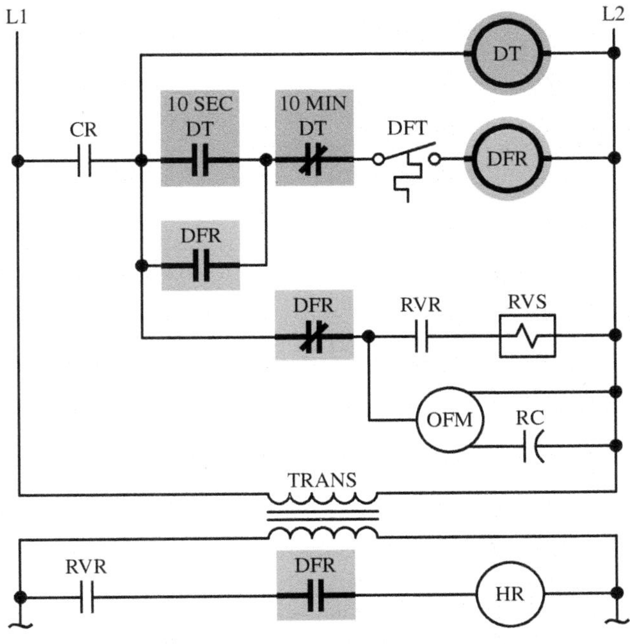

Figure 24-24 This is the defrost part of the heat pump circuit. This uses the time-temperature defrost method.

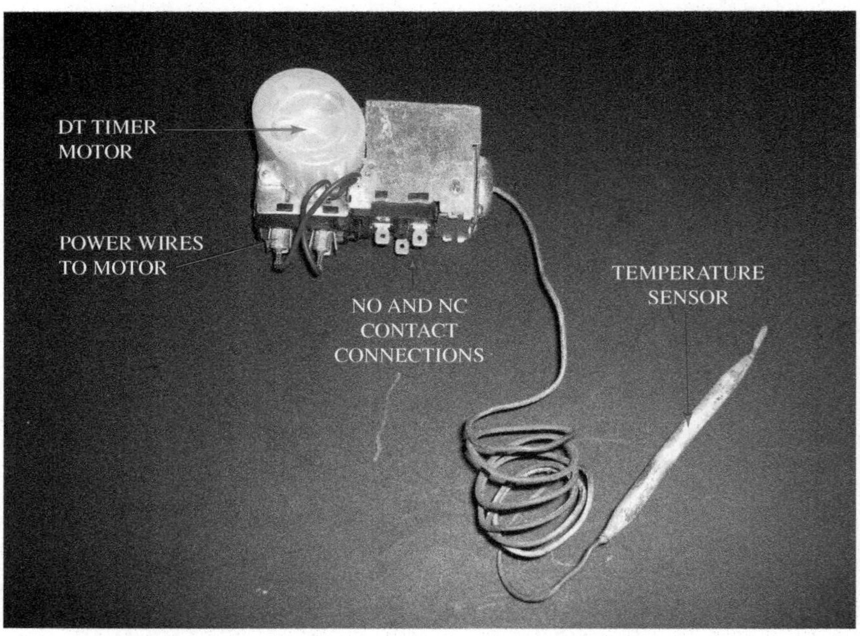

Figure 24-25 This is a mechanical timer and temperature sensor used to initiate and terminate the defrost cycle. The sensor is attached to the outlet of the outdoor coil.

Here is a legend that we will use in discussing the sequence of operation for the following diagrams:

DT defrost timer	**RVS** reversing valve solenoid
CR control relay	**OFM** outdoor fan motor
DFR defrost relay	**RC** run capacitor
DFT defrost relay thermostat	**HR** heating relay
RVR reversing valve relay	**TRANS** transformer

The coils and motors are circled to make it easier to identify these components. The reversing valve coil is marked with a rectangle.

Figure 24-25 is an actual timer and temperature sensor used for this method of defrost. The sensor is attached to the outlet of the outdoor coil to measure temperature in the heating mode. The outdoor coil will be cold from the low-pressure refrigerant and winter air.

We will use Figure 24-24 to explain a time and temperature defrost design. The defrost timer **DT** motor operates when the heat pump is operating. The heating circuit contacts **CR** must be closed for the **DT** coil to be energized. The **CR** contacts are closed when the heat pump is energized in the cooling or heating mode.

Figure 24-24 is the defrost part of the heat pump circuit. It is shown in its normally de-energized position. It is energized by 240 volts through **L1** and **L2**. This is a mechanical time and temperature defrost method. At this time the cooling and heating system is off. The only component energized is the transformer. The green-circled coil **DT** switches the two **DT** contacts. The blue circled defrost relay coil **DFR** switches three **DFR** contacts. There are two contacts in the high-voltage section and one set of contacts in the control voltage section. The **CR** coil (not shown) is energized when there is a call for heat or cooling and it closes the **CR** contacts. The reversing valve solenoid **RVS** is energized in the heating mode. When the reversing valve solenoid is energized, it will shift the discharge refrigerant flow to the indoor coil, thus warming the indoor air as it passes through the heated indoor coil.

The defrost operation will start when the following contacts are closed:

- Control relay contacts **CR**
- Defrost timer contacts **DT** (the 10 SEC and 10 MIN contacts are closed)
- Defrost thermostat contacts **DFT**.

The timer **DT**, as shown in Figure 24-25, operates during the run time of the heat pump in the heating mode. If the timer is set for 60 minutes, it will close as the 10 SEC contacts **DT** for a short period of time after 60 minutes of cumulative operating time. If the temperature of the outdoor coil is below the temperature setpoint, around 32°F, it will close the defrost thermostat **DFT**. Four events need to occur in Figure 24-24 prior to initiating defrost:

- The **CR** contacts close as a result of a call for heat.
- The 10 SEC **DT** contacts close.
- The 10 MIN **DT** contacts close.
- The **DFT** closes.

The (DFR) coil will energize and open the normally closed **DFR** contacts, switching the heat pump to the cooling cycle and stopping the outdoor fan. This temporary cooling operation will heat up the outdoor coil and remove built-up frost faster.

The defrost cycle is terminated when either the timer contact **DT** opens because it has timed out after 10 minutes or the defrost temperature **DFT** rises above its setpoint. It is assumed that the frost is cleared when the time or temperature point is reached. The next sequence of operation will help you understand the step-by-step operation of the defrost cycle.

SEQUENCE OF OPERATION 24.1: START OF THE DEFROST OPERATION

Now we will go through a detailed explanation of this defrost circuit as shown in Figures 24-26 through 24-29.

1. Begin with Figure 24-26. To initiate defrost, the **DT** timer at point 1 will close the **DT** contacts on the left for 10 seconds and the **DT** contacts near the middle of the diagram for 10 minutes.
2. The outlet of the coil temperature as sensed by **DFT**, point 2, needs to be closed if the heat pump is to go into defrost. A frosty coil will cause the **DFT** contacts to close.
3. This will energize the (DFR) coil at point 3.
4. This will close the normally open **DFR** holding contacts at point 4, keeping the heat pump in defrost. This known as a holding circuit.
5. This heat pump is designed to energize the reversing valve solenoid [RVS] in the heating mode. The reversing valve will be de-energized in the defrost mode. Figure 24-27 illustrates that when the (DFR) coil at point 5 is energized.

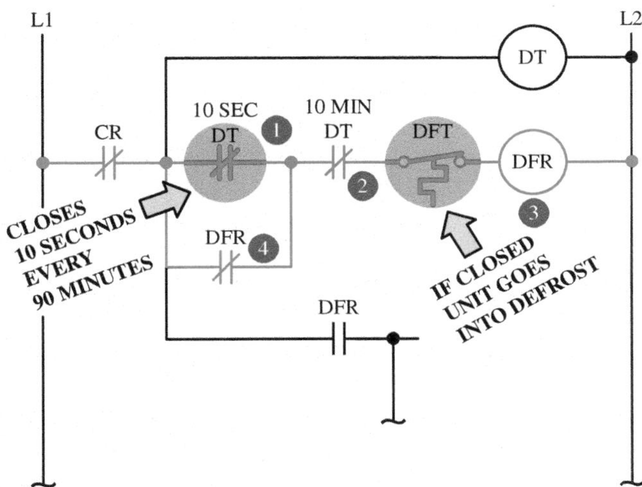

Figure 24-26 Initial defrost cycle.

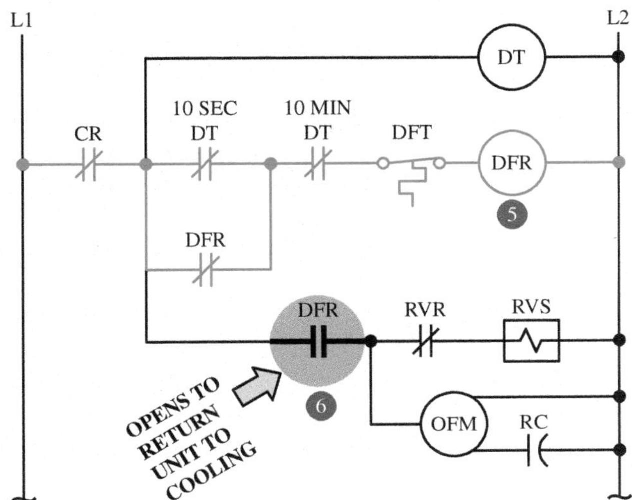

Figure 24-27 When the (DFR) coil is energized it will open the DFR contacts as pointed out here. The [RVS] and (OFM) will now be de-energized. The will put the circuit in the defrost mode.

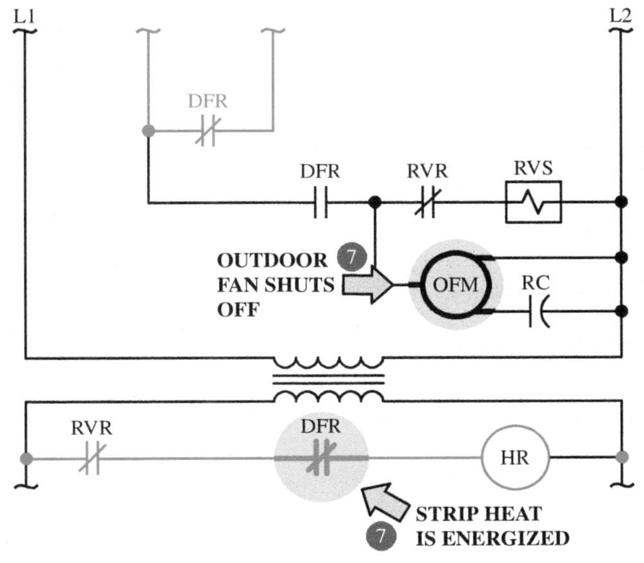

Figure 24-28 This diagram continues the defrost operation. It will de-energize the reversing valve solenoid [RVS] and the condenser fan motor (OFM). The reversing valve relay (RVR) coil (not shown) closes and energizes the **RVS** contacts to power the (HR) coil, which brings on the electric heat.

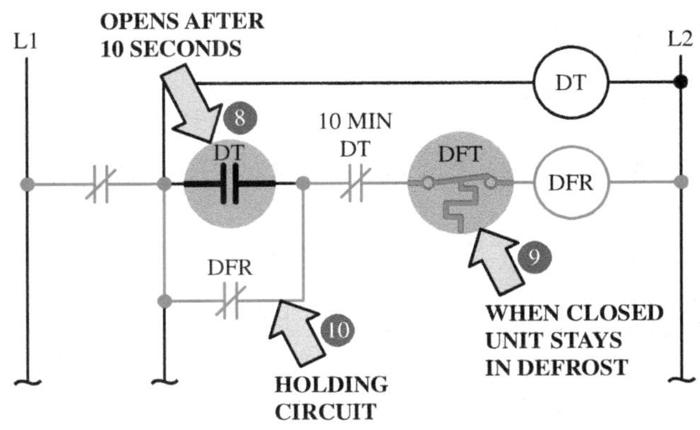

Figure 24-29 The 10-second defrost timer contacts open at the end of a 10-second time period. With the defrost thermostat still closed, the unit will remain in defrost mode because the circuit is completed through the holding circuit.

(continued)

6 When the DFR coil is energized it opens the **DFR** normally closed contacts at point 6, thus de-energizing the reversing solenoid RVS and outdoor fan motor OFM.

7 Point 7 on Figure 24-28 indicates that the outdoor fan shut off and the electric strip heat is energized. Electric strip heating is required because the indoor coil now becomes cold since it is in the cooling mode. The heat strips temper the cold air to reduce cold blow complaints.

8 In Figure 24-29 the 10-second defrost timer contacts open at the end of a 10-second time period as seen at point 8.

9 With the defrost thermostat still closed at point 9, the unit will remain in defrost mode.

10 When the unit is in defrost mode, the defrost circuit is completed through the holding circuit **DFR** (point 10).

SEQUENCE OF OPERATION 24.2: TERMINATION OF THE DEFROST OPERATION

We now discuss how this defrost circuit is terminated or ended. Use Figure 24-30 for the following discussion:

1 To terminate the defrost cycle, the 10-minute timer contacts DT (point 1).

2 Or, the defrost thermostat contacts **DFT** (point 2) must open. When the **DFT** opens on the temperature rise of the outdoor coil, it breaks the power to the defrost relay coil DFR.

3 The normally open **DFR** contacts at point 3 open again.

4 The normally closed **DFR** contacts at point 4 close again.

5 Use Figure 24-31 for the last of our defrost termination discussion. After 10 minutes of run time, the defrost timer **DT** contacts at point 5 will open the defrost termination circuit, placing the heat pump back in the heating mode. The RVS and OFM circuits (highlighted in red in the diagram) will be energized.

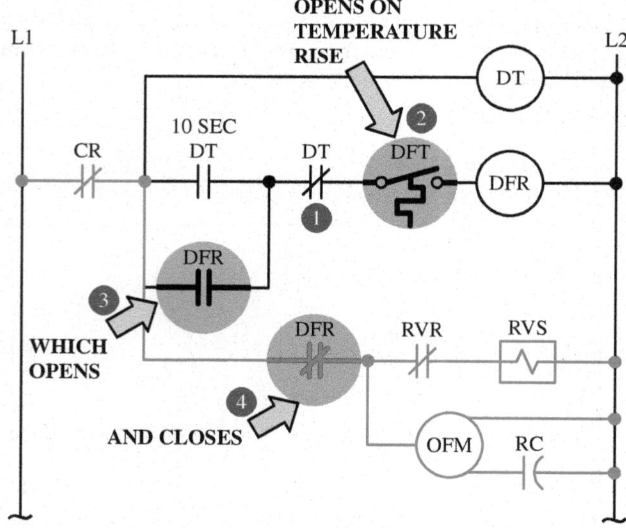

Figure 24-30 The defrost cycle is terminated by time or temperature restraints. This can be done by opening the 10-minute timing contacts **DT** or if the defrost thermostat **DFT** opens because the coil temperature has risen.

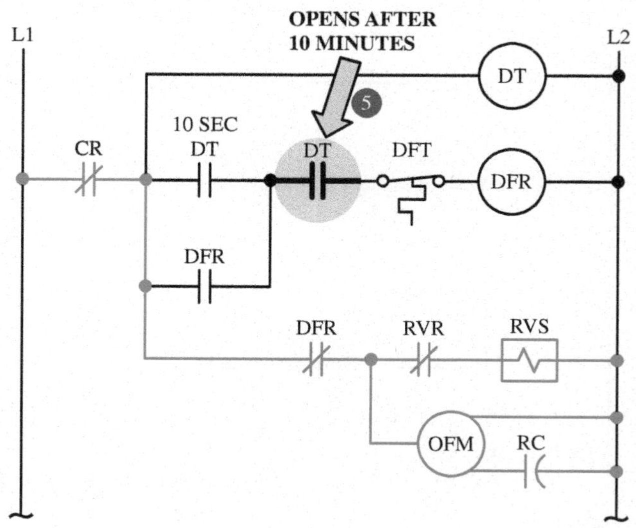

Figure 24-31 The defrost timer **DT**, after 10 minutes of defrost run time, will open the defrost termination circuit, placing the heat pump back in the heating mode.

24.8 CAUSES OF INCOMPLETE DEFROSTING

Incomplete defrosting can have several causes. Here are some suggested causes of this problem:

- Low change
- Reversing valve not changing over to cooling mode in defrost
- Electrical defect in the circuit board, defrost thermostat, thermistor, or defrost timer
- Prevailing winds preventing complete defrost
- Nearby water mist like a cooling tower, sea spray, or landscape sprinklers
- Blocked outdoor unit drainage slots
- Outdoor unit installed below the snow line
- Dirty indoor coil, dirty air filter, or anything that causes a reduction in airflow
- Minimum return air temperature below 60°F
- Too much electric heat during defrost, causing the defrost cycle to terminate early.

24.9 CRANKCASE HEATER

Some air conditioning and refrigeration systems have a crankcase heater (see Figure 24-32) to keep the compressor oil warm so that refrigerant does not condense in the oil when the outdoor temperature drops. Liquid refrigerant is a good cleaning agent and not a lubricant. Having liquid refrigerant in the compressor crankcase will dilute lubrication properties and cause slow but continuous damage to compressor bearing surfaces and moving parts.

The crankcase heater is a resistive heater. It can be installed as a belly band or installed in an insert well located in the compressor crankcase. To check the heater, feel the area around the heater. It will feel warm to hot. You can also use an ohmmeter to check resistance. There should be no resistance to ground. If you use a clamp-on ammeter to check the heater, the amp draw will be low.

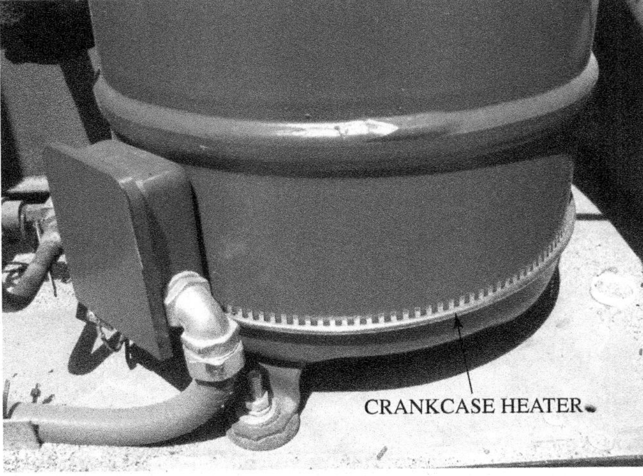

Figure 24-32 Crankcase heater strapped around the lower portion of a hermetic compressor.

24.10 HARD START KIT

A hard start kit is required to provide the torque to start most heat pump systems. The hard start kit consists of a specifically designed potential relay and start capacitor matched to the compressor motor. The compressor will need to start in unequal pressure differences, especially in the heating mode. The hard start components will provide the extra starting torque required for difficult starting conditions. Remember that the hard start kit is matched for that motor, so other relays, capacitors, and generic "hard start kits" will most likely not make the match. Faulty start components or the wrong size start components will prevent the compressor from starting and you may diagnosis the compressor as being defective. Check the relay and capacitor before condemning the compressor.

24.11 AUXILIARY AND EMERGENCY HEAT

All air-source heat pumps need to have an **auxiliary heat** or **emergency heat** system in case the refrigerant system should fail. This is also known as backup heat. Auxiliary heat is the second stage of heat or **W2**. It comes on automatically if the space temperature drops a degree or two lower than the thermostat setpoint. The backup heat source could be a fuel other than electricity, such as propane, natural gas, or oil. These backup systems, if used, would be considered dual-fuel systems. They would use the same integral blower, combustion, and control systems used by the heat pump. In this unit we have chosen to simplify the system and stay with an air handler that holds the emergency heat, electric strip heaters, and all of the heat pump control system.

Here is how the emergency heat works:

- The user must place the heat pump in the emergency heat mode for it to operate. The emergency heat light comes on.
- The emergency heaters act on a signal from the indoor thermostat and the **W2** terminal.
- When placed in the emergency heating mode, the heat pump mechanical heating is disabled.
- From there, the signal is sent to the emergency heat relay, which is a time-delay relay. The relay will only allow one strip of electric heat to come on at a time. This prevents an inrush of amperage that would be necessary if they were all brought on at once.
- In some systems, the circuit board will monitor and signal to the emergency heat relay how much heat is required. In this way the amount of heat is matched to the need, which saves energy.

24.12 HEAT PUMP CYCLES

Now that we have covered the heat pump components not normally found in an air conditioning system, we briefly discuss the refrigeration cycle. Even though our emphasis is on electrical systems, it is important to know the refrigeration cycle. For example, you are troubleshooting a heat pump but it will not shift from the cooling to the heating cycle. The first jump to judgment is a bad or stuck reversing valve. Is the solenoid coil good? Does the solenoid coil have voltage? Not a bad start to your thinking, but if the system is low on charge, there is not enough pressure to cause the reversing valve to shift position. A 100-psi differential between the high side and low side is needed to shift the valve slider. The point is that you need a basic understanding of the refrigeration cycle to troubleshoot the complex heat pump cycle.

SEQUENCE OF OPERATION 24.3: COOLING CYCLE

We will trace the heat pump cooling cycle, as shown in Figure 24-33. Here is the sequence of operations:

1. Starting at point 1, the discharge of the compressor goes to an external muffler and into the bottom side of the reversing valve.
2. At point 2 the hot, high-pressure gas leaves the upper right side of the reversing valve and is directed to the outdoor coil, which serves as the condenser in the cooling mode.

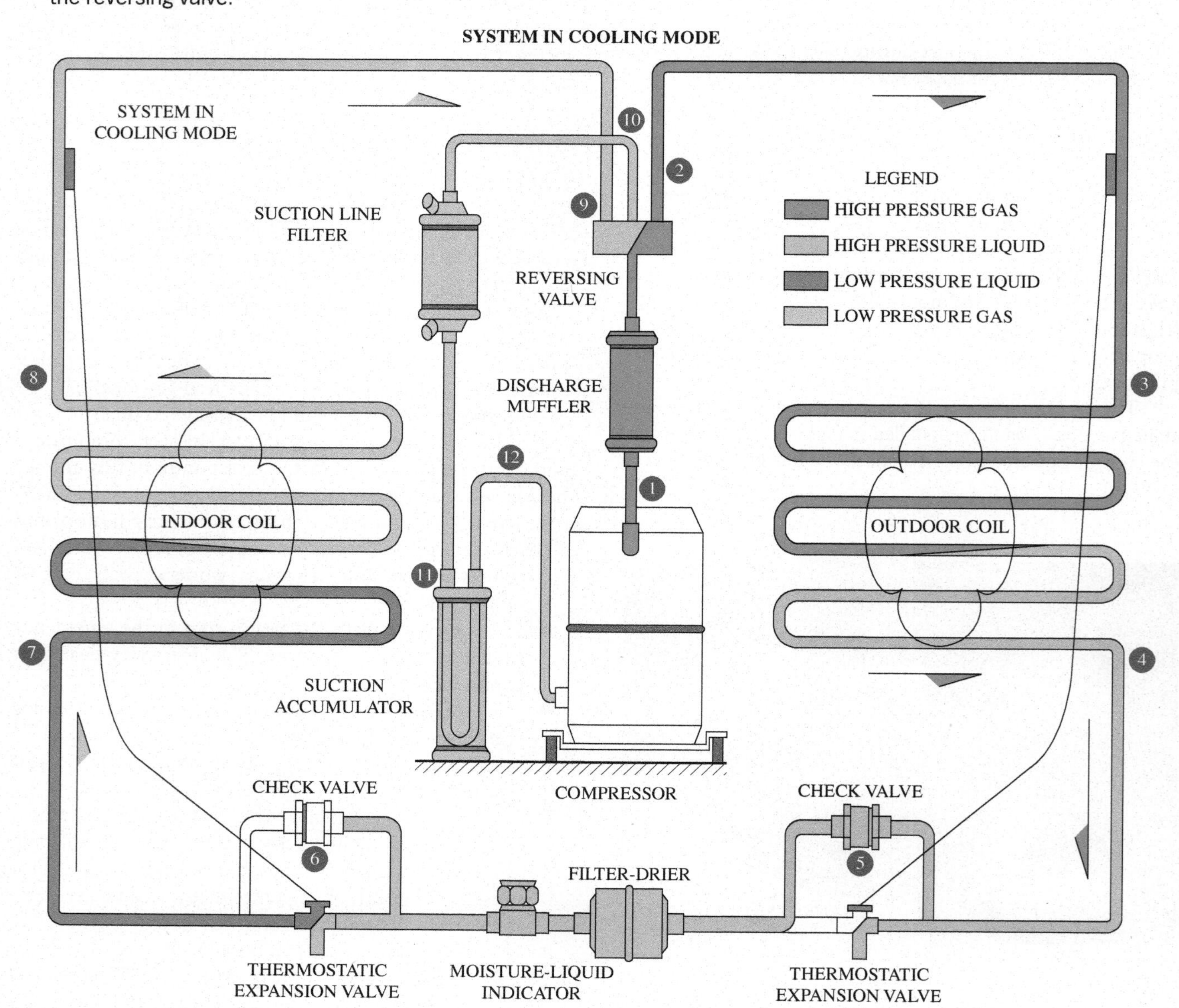

Figure 24-33 Heat pump cooling cycle.

3 At point 3 the discharge gas enters the condenser. The superheated vapor is cooled, condensed, and subcooled.

4 Point 4 shows the high-pressure, subcooled liquid leaving the condenser and going to the TXV through the liquid line.

5 The refrigerant bypasses the TXV and goes through the check valve at point 5, the filter drier and moisture-liquid indicator.

6 At point 6 the high-pressure, subcooled refrigerant goes through the TXV. The check valve above it is closed.

7 A mixture of cold, low-pressure refrigerant leaves the meter device and enters the indoor coil (evaporator) at point 7. The refrigerant picks up heat from the indoor air. This air will cause the refrigerant to boil off and convert to a superheated vapor.

8 At point 8 the cool, low-pressure superheated vapor leaves the evaporator in the return (suction line).

9 At point 9 on the reversing valve, the return gas enters the valve and is directed toward the compressor.

10 At point 10 the refrigerant goes through the middle port of the reversing valve then to the suction line filter drier.

11 The return gas enters the accumulator at point 11.

12 Any entering liquid refrigerant will fall to the bottom of the accumulator. Only vapor will leave the accumulator at point 12. The superheated gas returns to the compressor suction side to be compressed again.

Defrost Cycle

The defrost cycle is really the cooling cycle without the condenser fan operation. Refer back to Figure 24-33 as you read the following description:

- To start the defrost cycle, the heat pump electrical circuit will switch from the heating to cooling cycle. This means that the reversing valve shifts position, diverting the discharge gas at point 1 in Figure 24-33 to the outdoor coil. This superheated vapor will clear the frost from the outdoor coil.
- The outdoor fan will shut off, preventing the cold outside air from entering the coil.
- The indoor coil will receive cool refrigerant. Auxiliary heat will be activated to temper the cool air coming off the indoor coil.
- Once the frost clears from the coil or the defrost time expires, the heat pump switches to the heating mode.

TECH TIP

Typically, a sensor cannot be adjusted, but a thermostat can. Sensors are purchased by their temperature (or pressure, for pressure sensors) range and response. These things are part of the sensor manufacturer's data sheet. A thermostat is selected by range and differential, but the range is adjustable. The differential may also be adjustable, depending on manufacturer.

SEQUENCE OF OPERATION 24.4: HEATING CYCLE

We now trace the heat pump heating cycle, as shown in Figure 24-34. Here is the sequence of operations:

1 Starting at point 1, the discharge of the compressor goes to an external muffler and into the bottom side of the reversing valve.

2 At point 2 the hot, high-pressure gas leaves the upper left side of the reversing valve and is directed to the indoor coil, which serves as the condenser in the heating mode.

(continued)

3. At point 3 the discharge gas enters the indoor coil. The indoor air passes over the coil and heats the air. The superheated vapor inside the indoor coil is cooled, condensed, and subcooled.
4. Point 4 shows the high-pressure, subcooled liquid leaving the indoor coil and directed through the check valve, bypassing the TXV.
5. At point 5, the refrigerant bypasses the TXV and goes through the check valve, moisture-liquid indicator, and biflow drier.
6. At point 6 the high-pressure, subcooled refrigerant goes through the TXV. The check valve above it is closed.
7. A mixture of cold, low-pressure refrigerant leaves the meter device and enters the outdoor coil (evaporator) at point 7. The refrigerant picks up heat from the outdoor air. The outdoor air will cause the refrigerant to boil off and convert to a superheated vapor.
8. At point 8 the cool, low-pressure superheated vapor leaves the outdoor coil evaporator and flows in the return line to the reversing valve.
9. At point 9 on the reversing valve the return gas enters the reversing valve.
10. At point 10 the refrigerant goes through the valve and to the suction line filter drier.
11. The return gas enters the accumulator at point 11.
12. Any liquid refrigerant entering will fall to the bottom of the accumulator. Only vapor will leave the accumulator at point 12. The superheated gas returns to the compressor suction side to be compressed again.

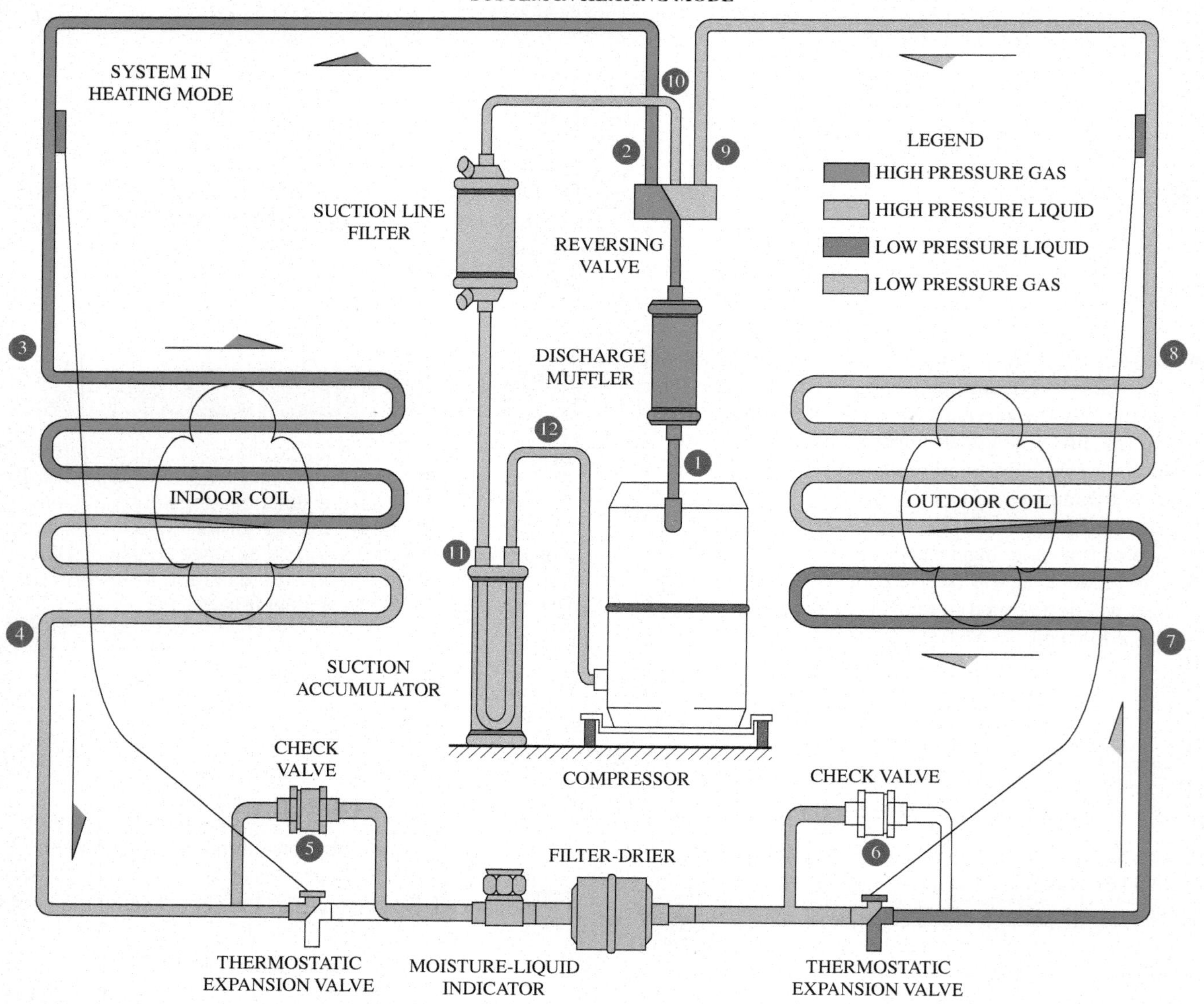

Figure 24-34 Heat pump heating cycle.

24.13 HEAT PUMP WIRING

We use enlarged Electrical Diagram ED-11 provided with this textbook and also Figure 24-35, which shows a portion of Electrical Diagram ED-11, to explain the operation of a simple heat pump cycle. Let's break down the enlarged diagram into its essential components. In Electrical Diagram ED-11 the connection diagram is found in the upper left side. The connection diagram is useful because it shows the color of the factory wiring. It also shows how the components are connected.

The connection diagram and schematic ladder diagram shown in Electrical Diagram ED-11 are used in conjunction with each other to troubleshoot this heat pump system. Actually, the connection diagram is valuable for wiring the system and the schematic diagram is valuable for troubleshooting. While the schematic diagram is more useful in troubleshooting, the wires colors and connection points of the connection diagram are helpful aids in identifying the components.

Below the connection diagram in Electrical Diagram ED-11 is the manufacturer's **NOTES section**. The notes have 10 pieces of information that relate primarily to the installation

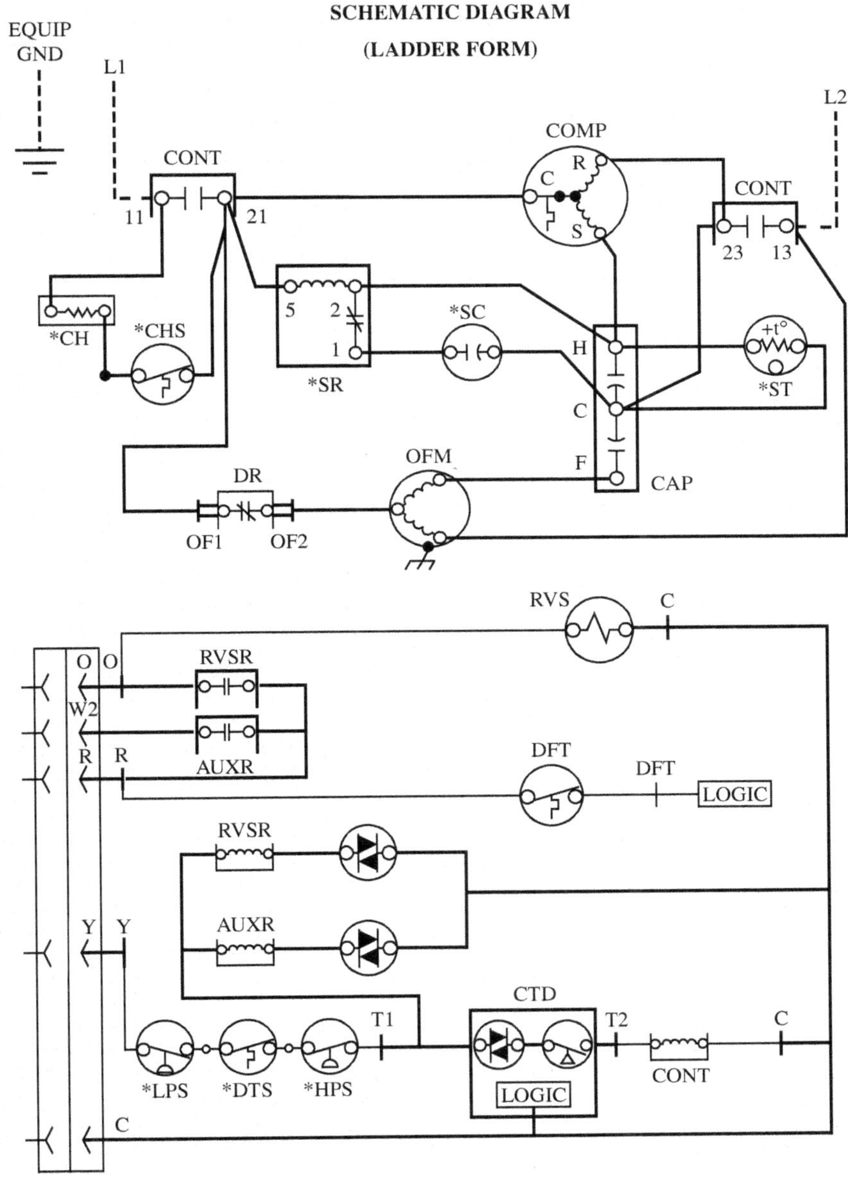

Figure 24-35 This is the schematic ladder diagram that comes from Electrical Diagram ED-11, which is part of the enlarged diagram package that relates to this unit. This diagram will be traced in red and blue in later figures.

of the heat pump. Below the notes are the DIP switch settings. The tiny switches on the circuit board have an ON or OFF position. Four options are available for setting the run time when the defrost cycle can occur if the outdoor temperature is cold enough. The options are:

- 30 minutes
- 60 minutes
- 90 minutes (factory default setting)
- 120 minutes.

The shorter time defrost option would be selected if the normal winter conditions are humid or the outdoor unit is near a moist area. The longer time period would be selected if the winter conditions are dry.

Under the schematic diagram, to the right side, there is another NOTE. It states that there is a 5-minute time delay when the unit cycles off or if it turns off unexpectedly, for example, during a short power outage. In such a case, there will not be a 5-minute time delay between the heating and defrost cycle even though the heat shifts from heating to cooling and back to heating again. Sometimes there is a short delay between the shift of cycles, but not 5 minutes.

Finally on the lower right-hand side is the LEGEND that explains the abbreviations and wiring symbols used in the diagram. Because there is no universal HVACR language of symbols, the legend is useful in identifying the components in the connection and schematic diagrams.

Heat Pump Power Circuit

Let's trace the power circuit in Figure 24-35. This circuit has high voltage and control voltage applied, but not energized to call for heating or cooling. Here is what is energized at this time:

1. Starting with the high-voltage section, upper left side at **L1**, the dashed lines mean that this is high-voltage field wiring.
2. The power goes to the crankcase heater **CH.** The asterisk ***CH** indicates that this component will be factory or field installed if present at all.
3. From the crankcase heater power goes to the crankcase heater switch **CHS**. The switch closes as the ambient temperature drops. This saves operating costs and extends the life of **CH**, especially in the cooling mode. The **CHS** may be attached to the discharge line, which will terminate the heater during compressor operation.
4. The path of electron flow continues to the normally closed contacts of **DR**, through the run winding of the outdoor fan motor **OFM**, and completing the cycle back to **L2**.
5. The room thermostat block, lower left side of the diagram, will have 24 volts applied at **R** and **C**. The voltage from the transformer is not shown.
6. Since the t'stat is open, there is no current flow but control voltage is supplied from t'stat terminal **C** to the lower LOGIC board, right side of contactor coil CONT, circuit board triacs, and reversing valve solenoid RVS . Voltage is provided from t'stat terminal **R** and to the open **RVSR** and **AUXR** contacts and to the defrost thermostat **DFT**.

Figures 24-36 and 24-37 show the actual circuit board represented by the diagram shown in Figure 24-35. The circuit board in Figure 24-36 shows the defrost timer options and the **SPEEDUP** pins, labeled as "SPEED UP TERMINALS." When the **SPEEDUP** pins are shorted together, this sends the defrost cycle through an accelerated cycle so the technician can check the complete defrost cycle operation in a minute or two. Figure 24-37 shows the circuit board with the 90-minute defrost cycle jumper selected. The circuit board will check for a possible defrost cycle every 90 minutes of run time in the heating mode.

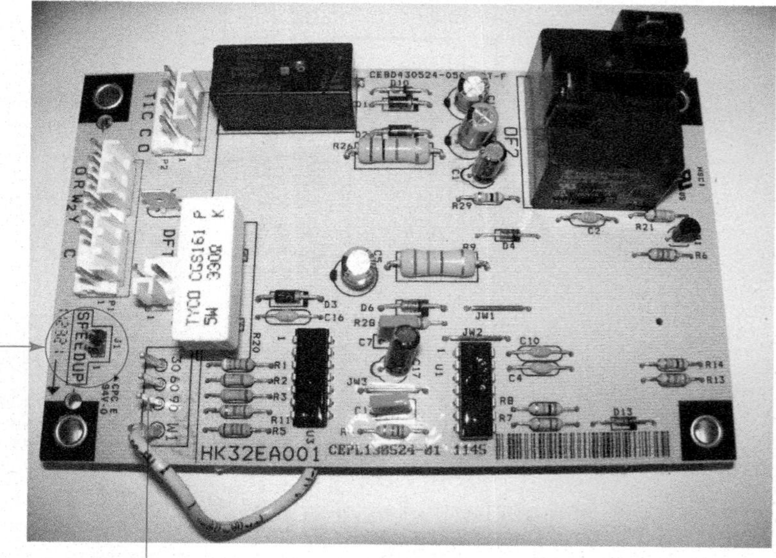

Figure 24-36 This heat pump circuit board is the actual board from the diagram shown in Figure 24-35. Notice the defrost timing options of 30, 60, and 90 minutes. The SPEEDUP terminals can be shorted so the technician can perform a quick defrost operation, usually less than 1 minute.

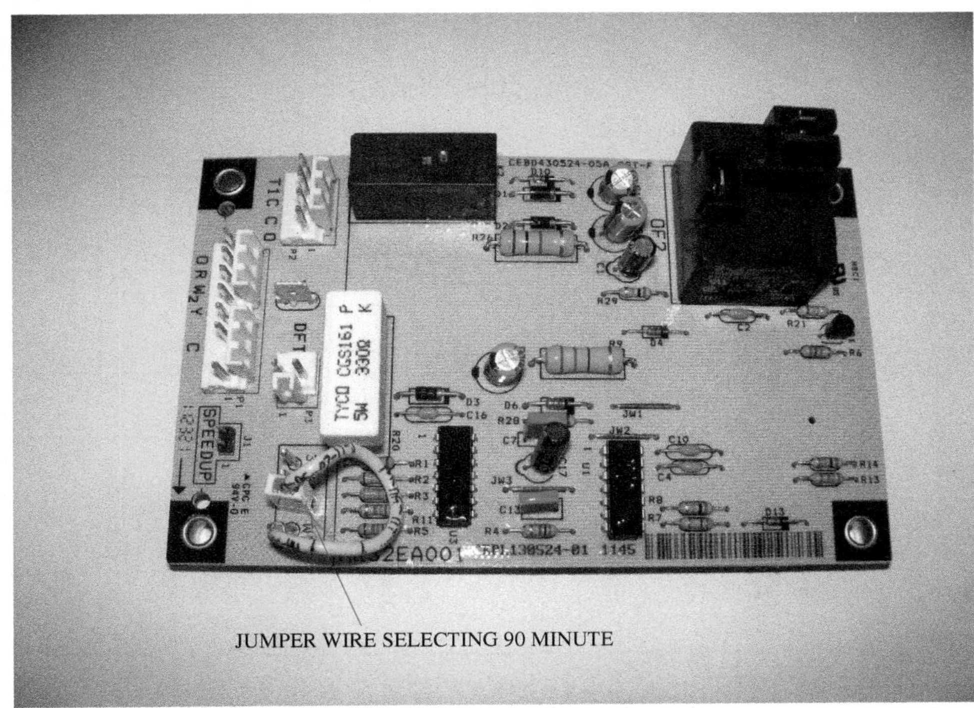

Figure 24-37 This is the same defrost circuit board as in Figure 24-36, but the 90-minute timer jumper has been selected.

Heat Pump Cooling Control Circuit Energized

Note: This section discusses the low-voltage operation of the heat pump cooling cycle. As a quick tip, when calling for cooling, the thermostat will close contacts between **C** to **R**, **C** to **O**, and **C** to **Y**. This makes a complete 24-V circuit to **R**, **C**, and **O**. The control voltage will be highlighted in red and the high-voltage section in blue. Now let's discuss the heat pump cooling operation.

1. When power is supplied to the heat pump, 24 volts will be supplied to **C** and **R**, as shown in Figure 24-38. To energize the cooling mode, the thermostat must close contacts between **C** and **Y** and between **C** and **R**. The C or common line will return the voltage to the transformer. The closed t'stat will supply 24 volts to operate the control circuit as represented in the bottom half of diagram by the red highlighting.
2. To operate the cooling circuit, power from **C** connects to **Y** then through the terminal strip **Y** to the right of the thermostat. From this point 24 volts goes through **LPS**, **DTS**, and **HPS** feeding power to the **CTD** LOGIC circuit. The components marked with an asterisk (*) are optional.
3. If the circuit in Figure 24-39 has been de-energized for 5 minutes, the LOGIC circuit will allow the voltage to continue to the contactor coil at point 1. The power circuit is completed by returning to **C** on the terminal strip and thermostat.
4. Next, the **O** circuit is energized. The thermostat closes the path between **C** and **O**. From **O** on, the thermostat power goes to the right to the terminal strip **O**.
5. From **O**, 24 volts is supplied to the left side of the reversing valve solenoid **RVS** as seen at point 2.
6. The **RVS** solenoid circuit is completed by voltage returning to the **C** or common voltage return line. The reversing valve shifts position to the cooling mode.

Heat Pump Cooling Power Circuit Energized

1. As highlighted in blue in Figure 24-40, the energized contactor coil closes the contacts between terminals **11** and **21**, and **23** and **13**.
2. Starting at **L1**, the 240 volts travels through the closed contacts and splits power to the starting relay **SR** and through the closed contacts of defrost relay **DR** as seen at point 1. Power will not be going through the crankcase heater **CH** if the crankcase heater switch, **CHR**, is open.
3. **SR** is the potential relay used to start the compressor in conjunction with the start capacitor **SC**.
4. The right side of the split power goes through closed contacts **DR** and operates the outdoor fan motor **OFM**.

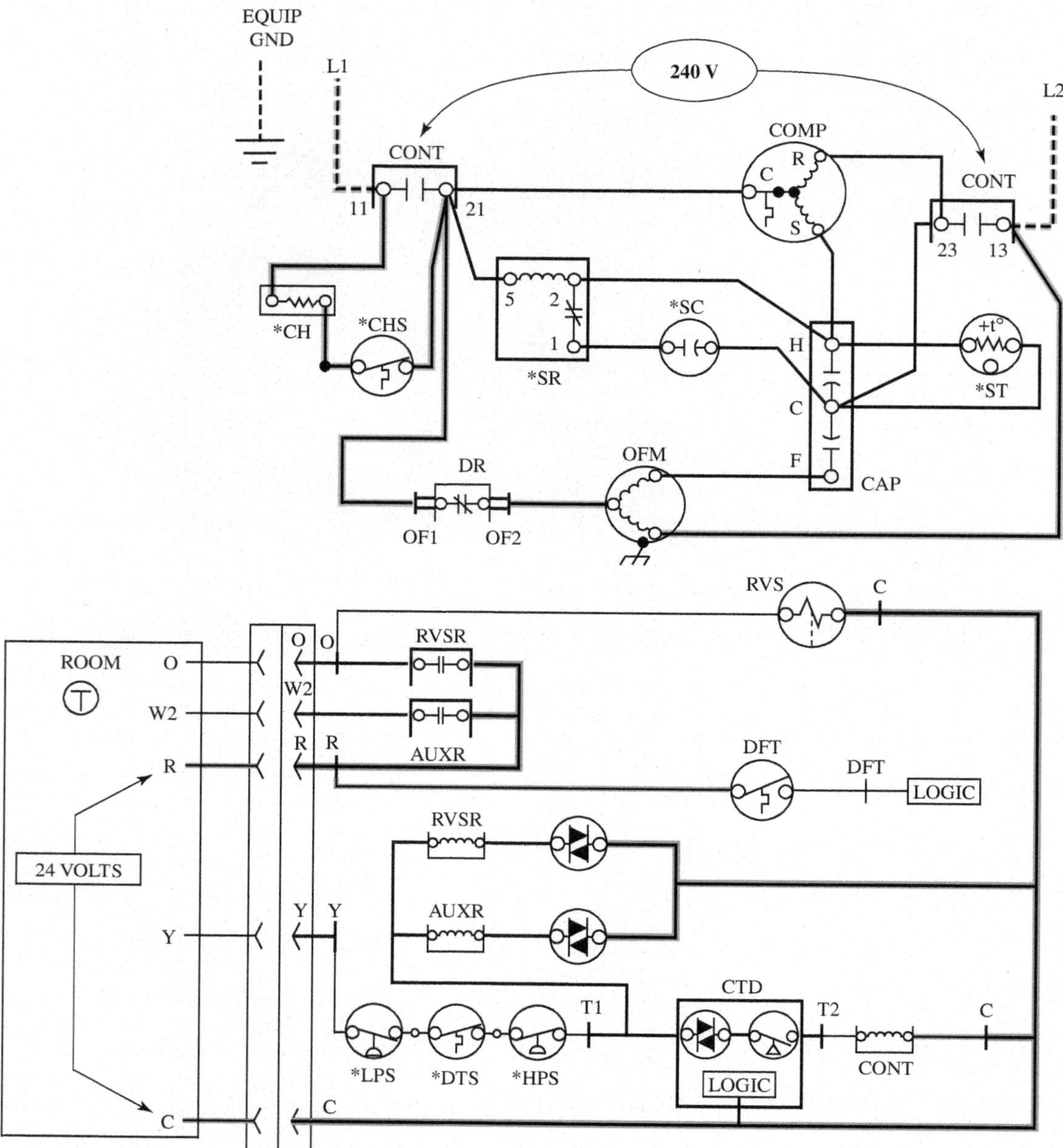

Figure 24-38 Energized circuit prior to cooling or heating operations.

5. The split power comes back together and completes the circuit through the closed contacts between **23** and **13** and back to **L2**.
6. In summary, the compressor and outdoor fan motor are operating in the cooling mode.

Heat Pump Heating Control and Power Circuits Energized

You will notice in the diagram of Figure 24-41 that the unit is now in the heating mode. The only difference between the heating mode and cooling mode is that the **O** circuit is not energized in the heating mode. In other words, the reversing valve solenoid coil **(RVS)** is not energized. This routes the refrigerant discharge from the outdoor coil to the indoor coil for heating. The outdoor coil now becomes the evaporator, absorbing heat from the cold air.

Heat Pump Defrost Control Circuit Energized

We will divide the explanation of the defrost cycle into two parts. This section will explain the controls circuit, and the next section will explain the high-voltage power section. The

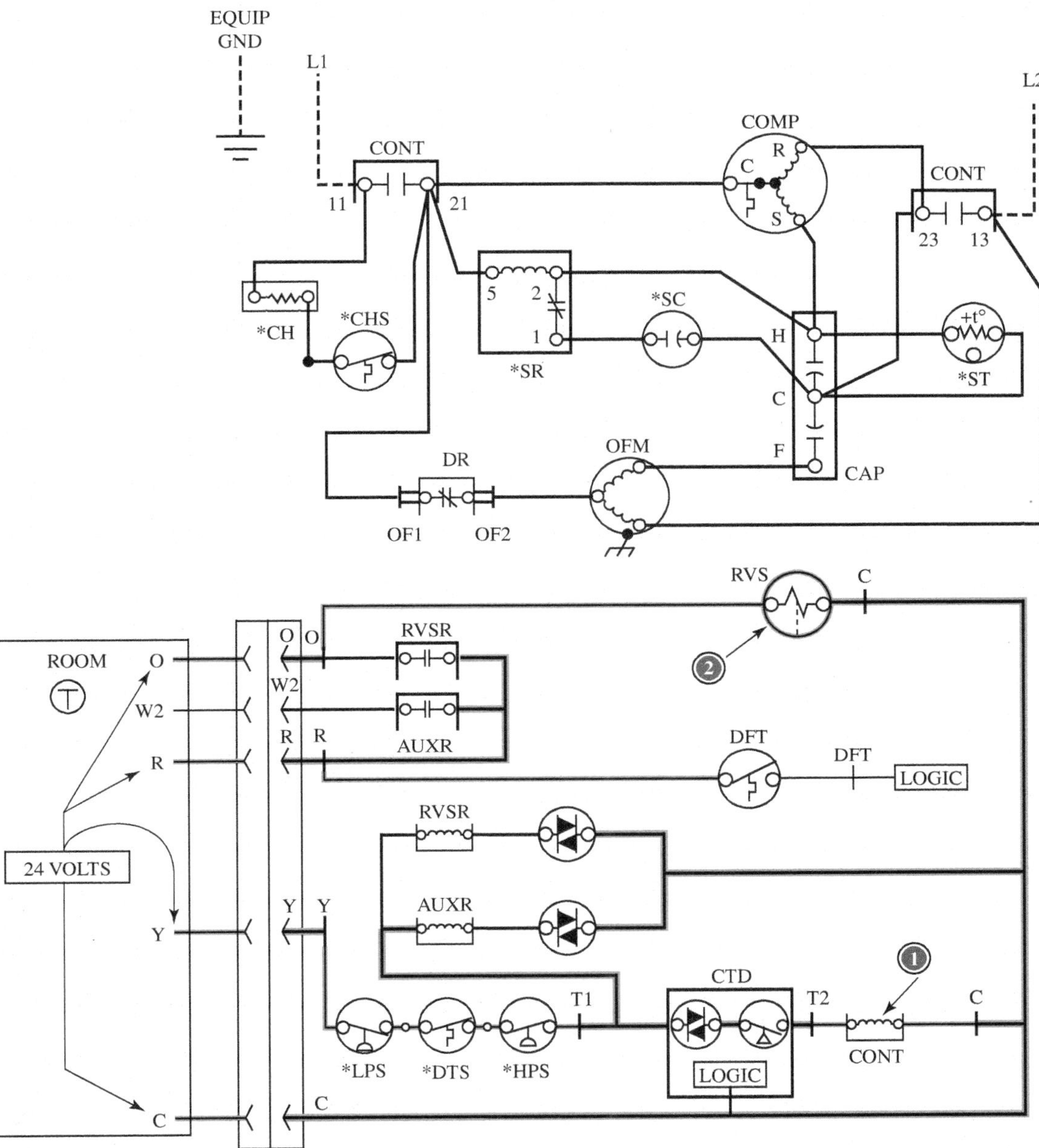

Figure 24-39 Energizing the control voltage in the cooling mode.

defrost circuit is designed to initiate the defrost cycle using the time **and** temperature method. The defrost cycle will be terminated by again using the time **or** temperature method. Here is how the control voltage section as shown in Figure 24-42 works:

1. The defrost cycle is initiated using the time and temperature method. The circuit board **LOGIC** defrost timer is selected when the equipment is installed. As we discussed earlier, the timer options are based on a run time of 30, 60, 90, or 120 minutes. The timing logic does not consider run time unless the coil temperature is below freezing. Operational time above freezing is not counted. This can be called **demand defrost** because defrosting is not automatic; it is instead energized when it is really needed. This design is considered a green feature because the demand defrost cycle does not waste energy by unnecessarily cooling the space and bringing on electric heat. Demand defrost works when there is frost to be cleared. Some of these demand defrost designs have a "learning logic" built into the circuit board. It learns how much and how often defrost is needed. This learning logic adjusts the initiation and length of the defrost cycle so that it does not operate longer than needed. Demand defrost creates fewer defrost cycles, saving energy and creating better comfort for the customer.

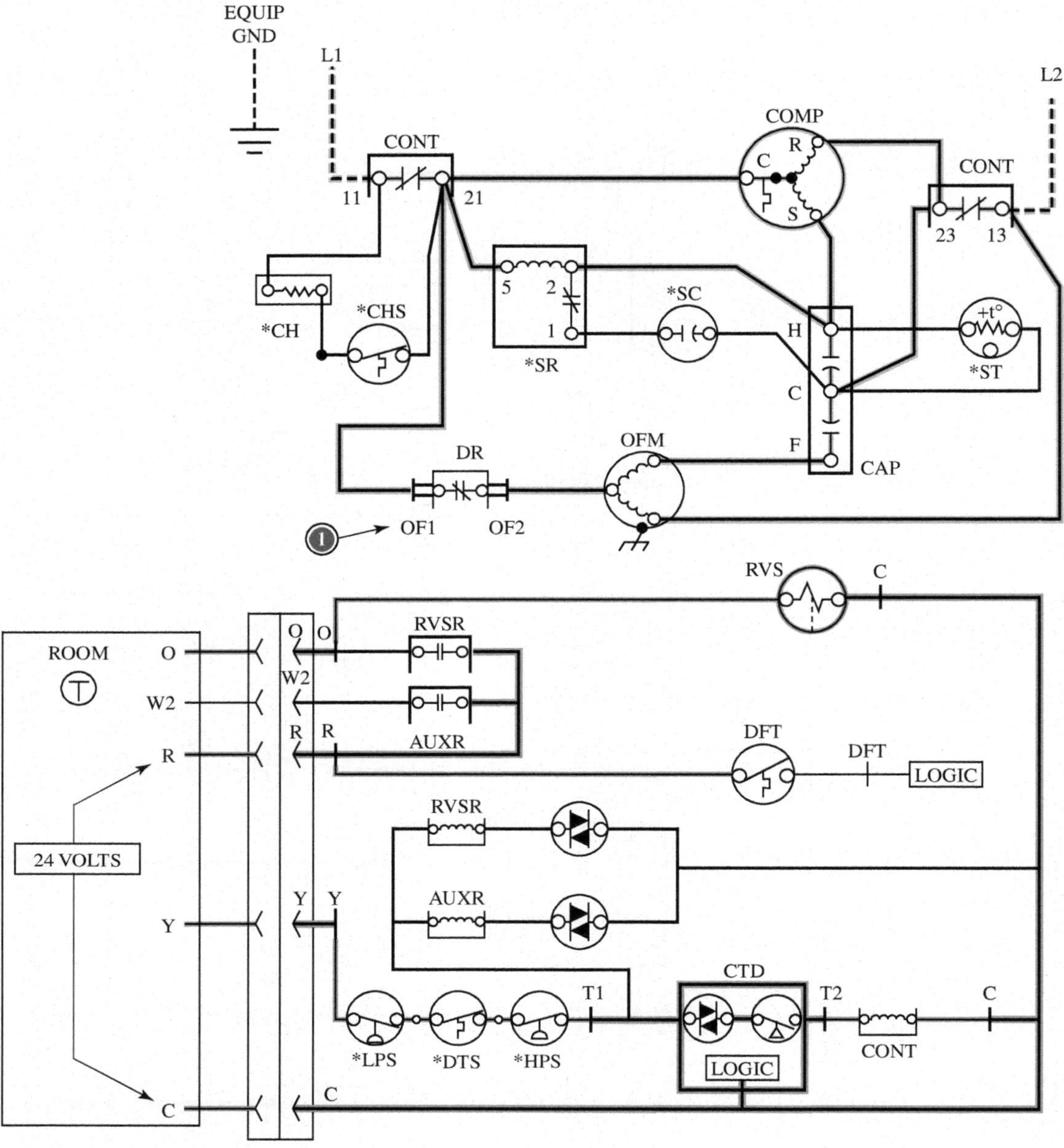

Figure 24-40 High-voltage and control voltage cooling circuit energized.

2. The defrost cycle will start when the time requirements have been met and the defrost thermostat **DFT** on the outdoor coil is closed because it is below freezing. The **RVSR** coil and **AUXR** coils will be energized (on the circuit board), closing contacts **RVSR** and **AUXR**.
3. When the **RVSR** contacts are closed, 24 volts is provided to the **RVS** coil at point 1. The **RVS** shifts the solenoid into the cooling cycle. The outdoor coil will start to defrost because hot discharge gas is going through it.
4. When the **RVSR** and **AUXR** contacts are closed (point 2), a 24-V path is provided to **W2**, which energizes the electric heat to temper the cold air now coming from the indoor coil.

Heat Pump Defrost Power Circuit Energized

Look at Figure 24-43 as we discuss the high-voltage section of the defrost operation.

1. You will notice that everything is the same in the high-voltage section except that the **DR** contacts are open, which prevents the outdoor fan from operating.

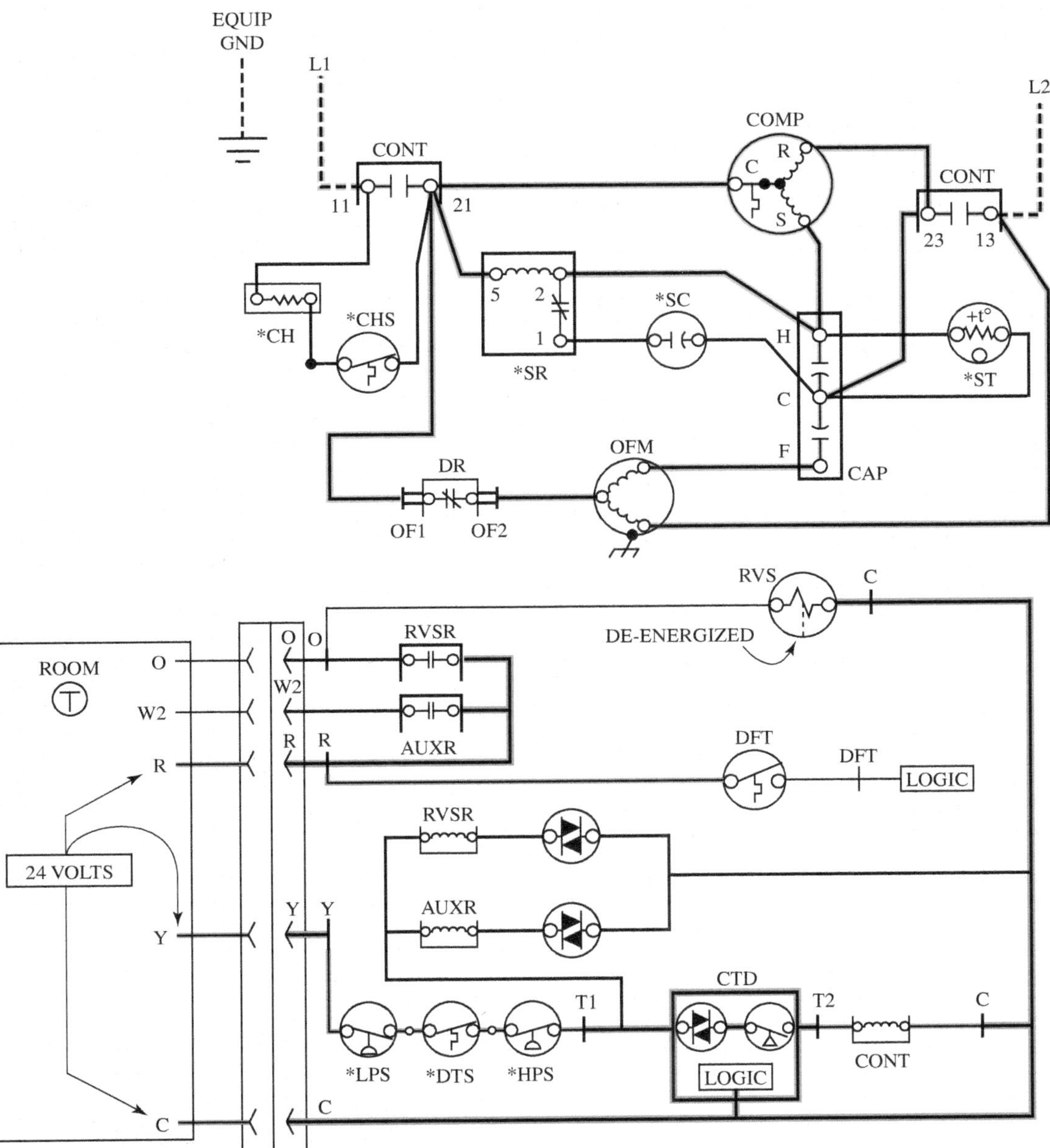

Figure 24-41 Heating circuit energized.

2. The supplemental heat source **W2** is energized. This is usually electric heat strips, but can be gas, oil, or other heat source.

Terminating the Defrost Cycle

The defrost cycle will be terminated using the time or temperature method. The defrost time is around 10 to 12 minutes. If the defrost cycle is still operating after termination time, the **LOGIC** will revert back to the heating mode, which de-energizes the **RVS** and closes the **DR** contacts to activate the outdoor fan motor. The **W2** circuit will be de-energized if the room temperature is within a degree or two of the thermostat setpoint.

SAFETY TIP

Line voltage power wiring for both indoor and outdoor components generally comes from the same panel box of breakers. Whether they come from the same panel or not, however, line voltage power supplies require manual disconnects (these are not shown in diagrams to conserve on space). Manual disconnects are for the protection of the technician and are used to perform service and maintenance operations. The line voltage connections in all diagrams in this text should be considered to be connected to manual disconnects for both the indoor and outdoor power.

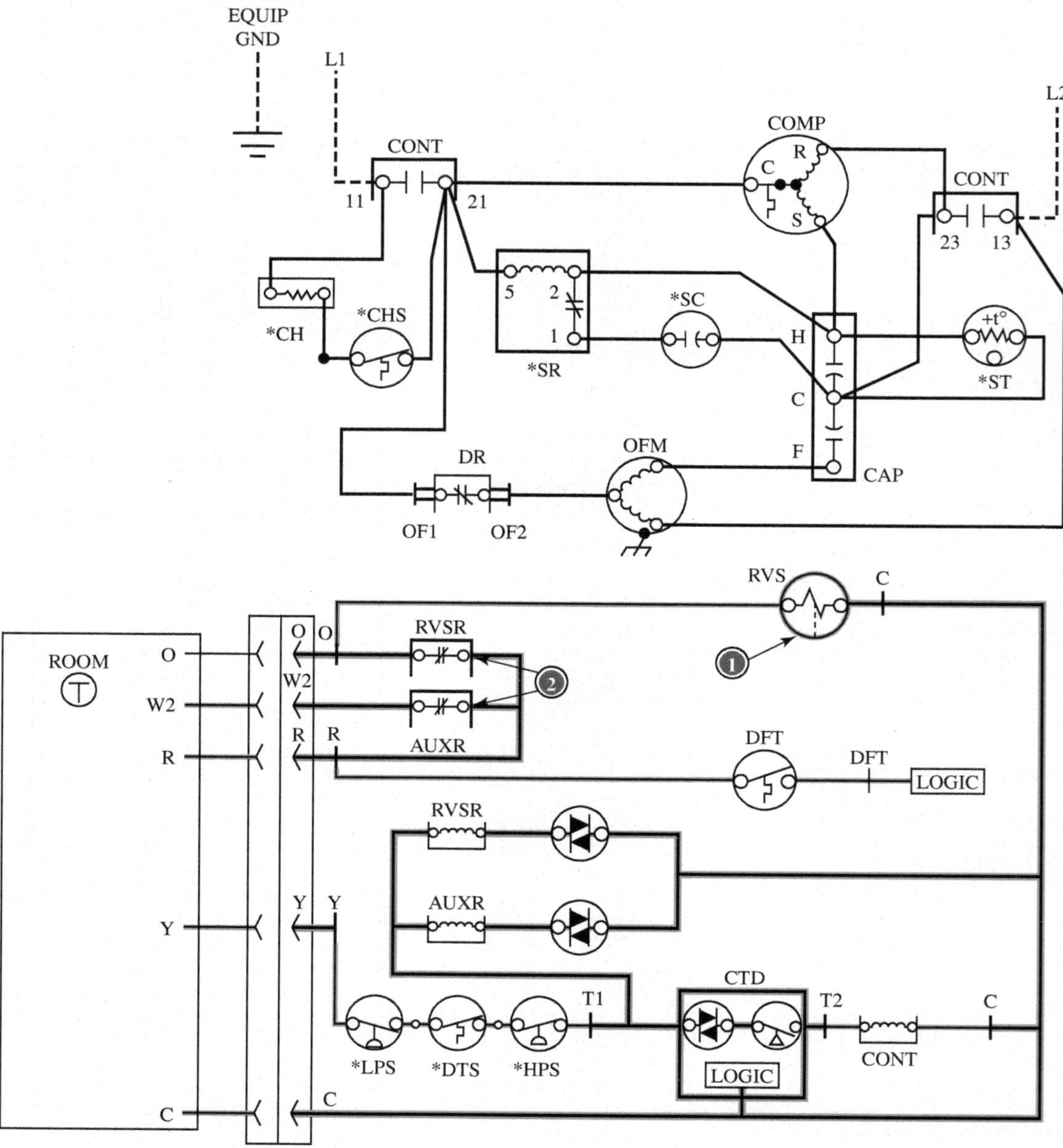

Figure 24-42 Control voltage section of the defrost circuit energized.

24.14 CONSUMER CHOICES

Consumers are often faced with the decision of changing out their heat pump after it useful life. The useful life of a heat pump is considered to be 15 years. This is the same life expectancy of a conventional air conditioning system. In some cases the consumer may decide to change the heat pump or conventional air conditioner before its normal 15 years of use have passed.

> **GREEN TIP**
>
> The **Energy Star** program provides suggestions to help the consumer and professional contractor change heat pump or air conditioning equipment prior to the end of its 15 years of useful life. Energy Star is an energy efficiency program sponsored by the U.S. Department of Energy and the U.S. Environmental Protection Agency. The purpose of the program is to promote energy conservation products.

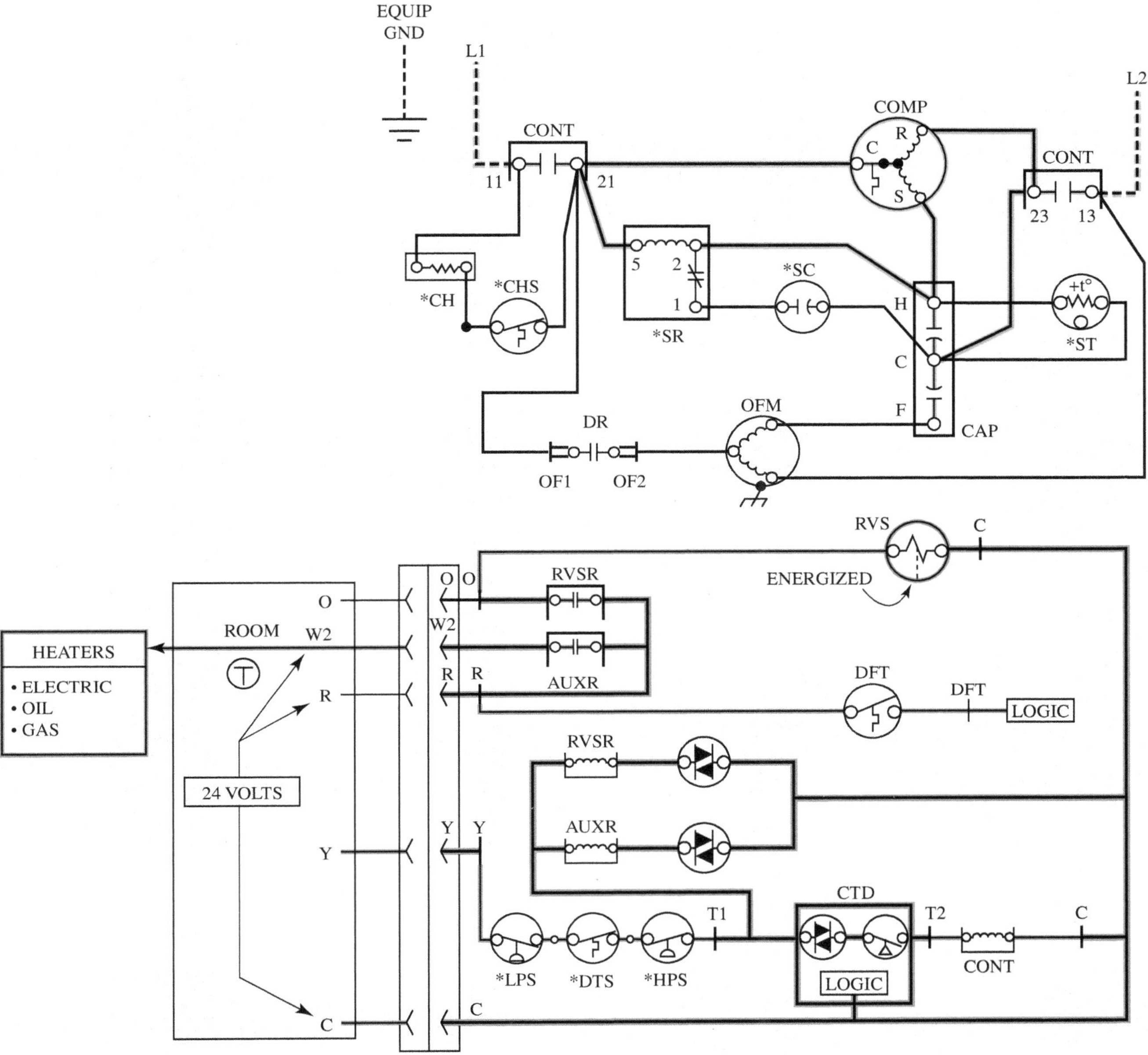

Figure 24-43 The total operation of the energized defrost circuit.

Energy Star realizes that certain telltale signs indicate when it is time to replace cooling equipment or improve parts of the system to enhance performance and reduce operating costs. A consumer may decide to replace equipment in the following situations:

1. *If the central air conditioner or heat pump is 10 years or older.* New, Energy Star–labeled equipment uses 25% to 40% less energy than typical 10-year-old models.
2. *If the equipment needs frequent repairs and the consumer's bills are increasing due to rate hikes.* This could mean that the cooling/heating equipment is becoming less efficient and needs a thorough checkup or replacement.
3. *If the system turns on and off frequently.* This can indicate that the cooling or heating system is the wrong size. Oversized cooling equipment leads to poor dehumidification and less comfort due to short cycling. An undersized system will not adequately cool or heat during extreme temperatures.
4. *If some of the rooms are too hot or too cold.* Improper equipment operation or duct design could be the problem.
5. *If the building has humidity problems.* Poor equipment operation, inadequate equipment, or leaky ductwork can cause the air to be too dry in winter or too humid in summer.

6. *If the building has excessive dust.* Leaky ducts can pull particles and air from attics, crawlspaces, and basements and distribute it throughout the structure. Sealing the ducts may be the solution.
7. *If the cooling system is noisy.* This could be the result of undersized registers or duct system. Short-run air plenums will also cause noise problems.

These guidelines are used to help the customer and contractor decide when it is time to make that investment to change the total system or make repairs on part of the system such as the ductwork. Table 24-1 lists the minimum Energy Star efficiency recommendations. These are the current efficiency standards at the time of printing. These efficiency standards will increase soon.

Table 24-1 Energy Efficiency Criteria for Energy Star–Qualified Residential Air-Source Heat Pumps and Central Air Conditioners

Product Type	SEER	EER	HSPF (For heat pumps only)
Split Systems	≥ 13	≥ 11	≥ 8.0
Single Package Equipment (including gas/electric package units)	≥ 12	≥ 10.5	≥ 7.6

Symbol ≥ Means equal to or greater than.

24.15 BASIC TROUBLESHOOTING CHECKLIST

Heat pumps are one of the most complex HVAC systems to troubleshoot. The technician's ability to troubleshoot these systems will get better with experience. In addition to the electrical system, heat pumps have a mechanical side that greatly influences sensors and controls. The technician is required to understand both the mechanical and electrical systems in order to effectively troubleshoot heat pumps.

Table 24-2 is a generic electrical system troubleshooting guide. A similar guide will be found in the manufacturer's installation manual. The manufacturer's expanded guide will cover all proprietary fault indicators and information about putting the microprocessor (MPC) into diagnostic mode. The technician will need an accurate volt meter, ammeter,

Table 24-2 Troubleshooting Guide for Narrowing Down and Isolating a Problem

Customer Complaint/Fault Indicator	First Check	Possible Cause
No heating or cooling No power—power LED is not lit	Thermostat setting and breakers	Check thermostat setting, breaker, and disconnect Check for line voltage on the contactor Check for 24 VAC between R and C Check for primary/secondary voltage on transformer
System short cycling High-pressure fault LED	Pump or blower operation	Check for pump or valve operation Check fan motor operation Check operational relays
System short cycling, but cooling Coil temperature fault LED	Check sensor	Check sensor operation for correct resistance values per manufacturer
System not running Coil temperature low limit	Check sensor	Check sensor Check pump or fan motor
System continues to lock out System reset has been used	Check voltage	Check line voltage input Check power supply wire size Check compressor starting amperage and voltage Check low-voltage output
System is off Condensate light is on	Check condensate drain pan	Check for moisture shorting the condensate switch
Fan on, but no cooling/heating	Check compressor operation	Check thermostat setting Check refrigerant pressures Check compressor contactor Check for fault indicators Check pump/fan operation

and temperature probe, since some measurements may require accuracy to decimals.

Table 24-3 gives the technician more helpful hints on air-source and water-source heat pumps. Look at the symptoms and possible solutions. The table is divided into air-source and water-source heat pumps. An air-source heat pump uses an outdoor unit. Outdoor air adds to or removes heat from the refrigerant. A water-source heat pump uses water or other fluids to add or dissipate heat from the refrigerant.

SERVICE TICKET

You are sent out to do service calls on a couple of heat pumps. The first example is a cooling problem. The second problem is a heating problem. Use the flowcharts provided to find solutions to these problems.

The first problem refers to cooling issues. Let's use Figure 24-44 to get some troubleshooting tips. You start the troubleshooting process and notice that the compressor runs but there is insufficient cooling. You think that this is a cooling problem. Point 1 on the flowchart directs you to three common problems. They are:

- Low suction pressure
- High suction pressure and low head pressure
- High suction pressure and low superheat.

After checking the thermostat setting, indoor airflow, gauges, and temperature, you find that there is high suction pressure and a low superheat condition as shown at point 2. Continuing to follow the far right side of the flowchart:

- At point 3 the flowchart directs you to check for an overcharge. You can correct an overcharge by recovering some refrigerant.
- At point 4 the flowchart directs you to check the size of the metering device piston. An oversized piston will create high suction pressure and low superheat.
- At point 5 the flowchart suggests a failed TXV. This would be a TXV that is stuck open, creating high-pressure and low-superheat conditions. A TXV sensing bulb that has lost its pressure will open the valve also and cause these conditions.

One of these suggestions is most likely to lead to the solution to the problem.

A few months later you have a service call to the same address but the problem this time is a heating problem.

Use the troubleshooting flowchart in Figure 24-45 to solve a heat pump heating problem. You start at point 1 on the flowchart because the compressor will not run. This is most likely an electrical problem, but a low charge could also cause the compressor not to operate. Next, you go to point 2 and work your way down the flowchart until you find the problem. Starting at point 2, you check the following:

- Contact(s) on the contractor open.
- Defective low-voltage transformer.
- Remote control center defective. No signal.
- Contactor coil open or shorted. If shorted, it may have burned out the transformer or blown the transformer fuse.
- Open indoor thermostat circuit including t'stat wiring.
- Liquid line pressure switch open.
- Loss of charge switch open. Lack of charge or defective switch.
- Open control circuit. You suspect wiring or component failure if the charge is adequate.

This service call practice is used to get you accustomed to using flowcharts. Manufacturers often provide troubleshooting flowcharts that give you a checklist of the most common problems.

SUMMARY

In this unit we discussed the wiring and electrical diagram for a generic heat pump. We note that there are common components in every heat pump:

1. System control (thermostat, microprocessor, etc.)
2. Compressor and a starting system
3. System reversing valve
4. Safety shutdown controls
5. Fluid system motors (pumps/blowers/fans)
6. Emergency/backup heat.

The operation of a generic heat pump was described as it would operate in the cooling mode, heating mode, and defrost mode. When in the heating mode, an air-source system would need to go into defrost mode to shed moisture accumulation on the outside coil. A geothermal system would not need a defrost mode, but would need to monitor the temperature of the ground loop with a freeze sensor.

Troubleshooting was described in a general way. Remember that the manufacturer's installation manual should be consulted for an expanded troubleshooting table. The manufacturer may also have instructions on putting the microprocessor into diagnostic mode to aid the technician in solving the service problem.

Table 24-3 Common Symptoms and Problems for Heat Pump Systems

Symptom	Possible Problem
Air-Source Heat Pump	
1. Frequent cycling, loss of temperature and humidity control.	1. Oversized equipment, thermostat needs adjustment or replacement.
2. Low indoor air flow, high compressor suction and discharge pressure, high pumping rates and compressor failure.	2. Inadequately sized ductwork, dirty air filters, registers closed or blocked, fan belts slipping.
3. Frequent fuse failure (fire and safety hazard).	3. Undersized power wiring (especially for supplemental heat), improper refrigerant charge.
4. Liquid slugging, compressor mechanical failure.	4. Oversized liquid refrigerant tubing, improper refrigerant charge.
5. Heat pump operating but indoor temperature less than thermostat setting	5. Thermostat setting or switches may not be at correct setting (i.e., fan is on but heat is not). Outdoor unit clogged by leaves, snow, ice, or other obstruction. Improper refrigerant charge. Ducts blocked. Filter clogged with dirt. Fuse blown, relay failure on supplementary heater.
6. Supplemental heat does not turn off when balance point temperature is reached.	6. Outdoor or indoor thermostat may not be set at correct temperature, thermostats may need repair.
7. Unit does not turn on.	7. Switches on both heat pump unit and thermostat may not be in the "on" position, fuse or circuit breaker may be open.
8. Excessive vibration in compressor unit.	8. Unit may not be properly anchored to concrete slab or stilts, stilt mounting may not be properly anchored in ground, piping may be installed incorrectly
9. Compressor does not run, or shuts off and will not restart.	9. Compressor, transformer burned out, electrical failure in power supply or circuit, compressor stuck.
10. Compressor runs, but cooling is insufficient.	10. Blocked ducts or registers, improper refrigerant charge.
11. Compressor cycles on and off.	11. Improper refrigerant charge, refrigerant system dirty, faulty timer or defrost controls, fan motor intermittent.
12. Compressor runs, but heating is insufficient.	12. Electric resistance heat malfunctioning, fan failure.
13. Compressor runs, but cycles on internal overload.	13. Refrigerant charge needs to be checked, defective fan motor, restrictions in refrigerant or discharge lines.
Water-Source Heat Pump	
1. Loss in system capacity, increased compressor head pressure, increased water flow to compensate for increased head pressure, reduction in outside water discharge temperature.	1. Scaling or encrustation, obstruction to water flow.
2. Loss of performance due to temperature drop of intake water.	2. Lack of separation between cold water discharge and warm water supply-source wells, excessively long pipe runs, pipes not below frost level, clogged well or filter, pump wear.
3. Water backup or loss of flow in system (system with storage tank).	3. Lack of overflow mechanism in storage tank.
4. Water backup at ground level.	4. Undersizing or clogging of disposal well.

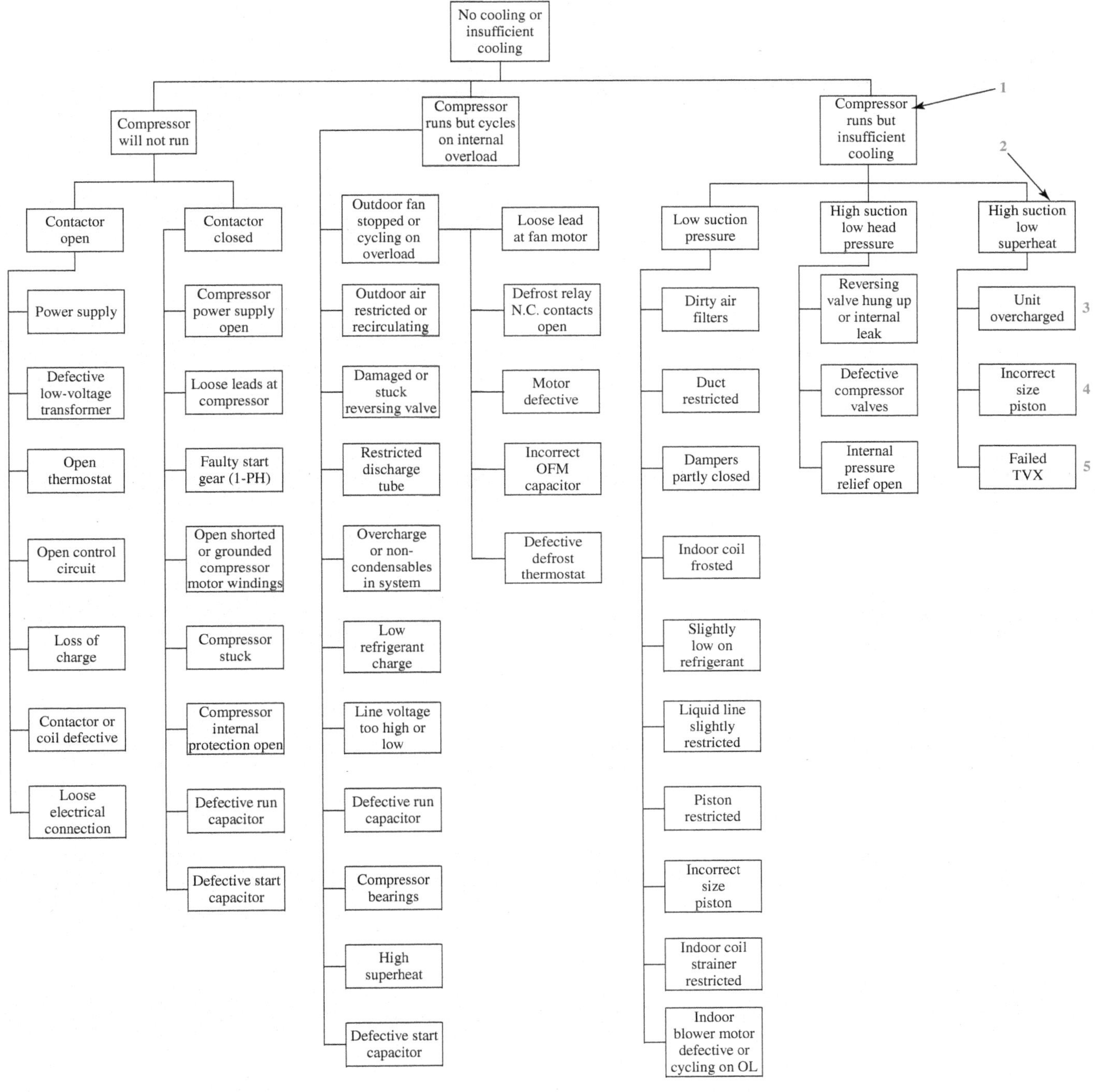

Figure 24-44 Cooling cycle troubleshooting chart.

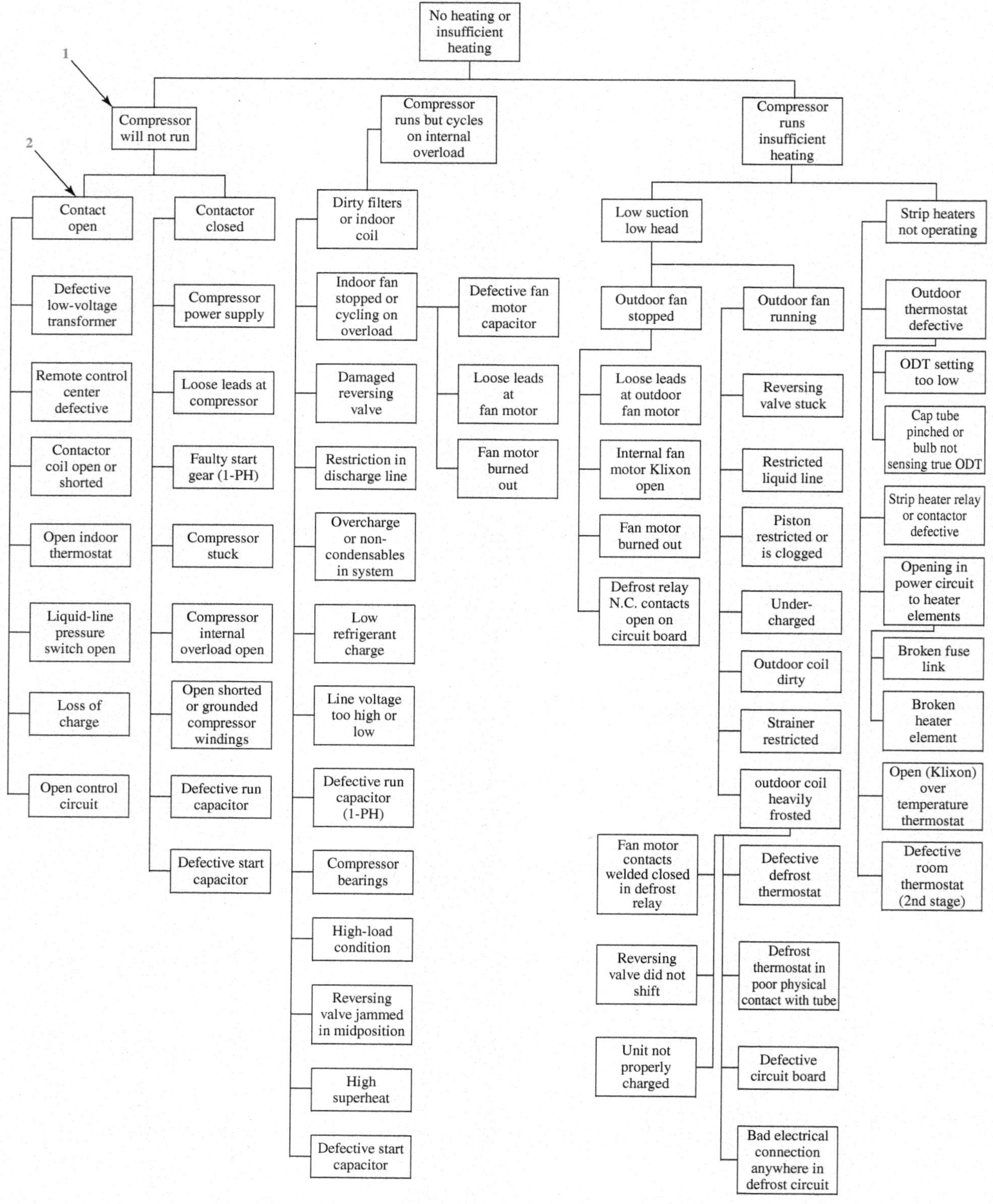

Figure 24-45 Heating cycle troubleshooting chart.

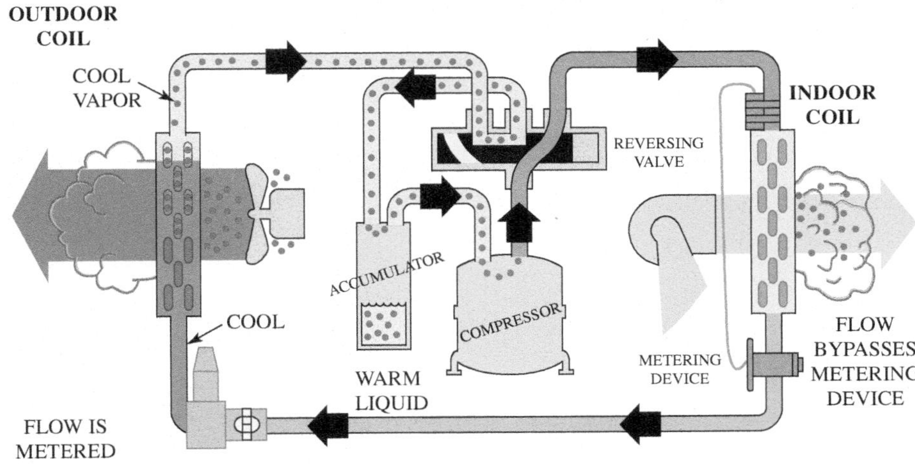

Figure 24-46 Use this diagram to answer Review Question 11.

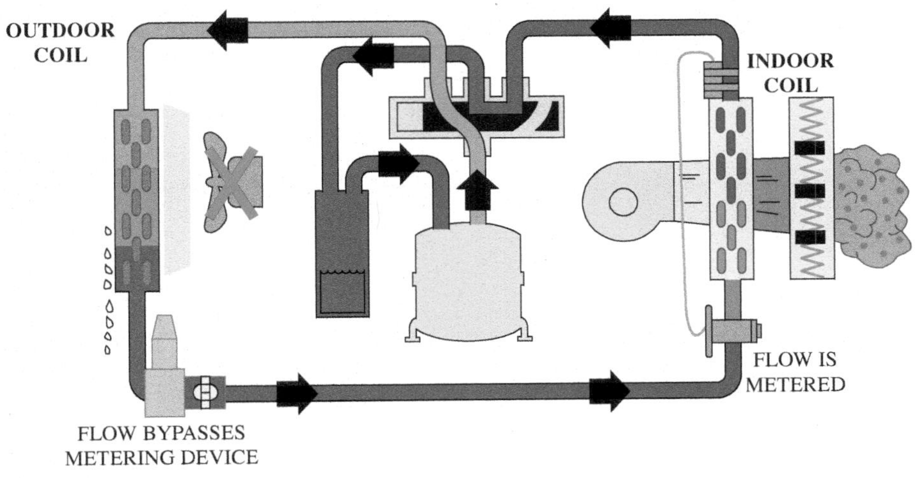

Figure 24-47 Use this diagram to answer Review Question 12.

REVIEW QUESTIONS

1. Describe what is meant by the term *electromechanical* and name components that are considered to be electromechanical.
2. Describe the function of a wiring diagram legend.
3. Name six common heat pump components.
4. Explain how inside electrical components and outside electrical components are connected to function as one unit.
5. Describe the start-up and shutdown of one mode of operation or defrost.
6. Explain why a hard start kit is used and describe its operation.
7. Describe what benefits might be derived from having a microprocessor (circuit board) as part of a system.
8. Name two ways to troubleshoot a reversing valve solenoid coil.
9. Name five electrical components found in a heat pump that are not found in a conventional air conditioning system.
10. Briefly describe the defrost cycle operation.
11. Identify the heat pump cycle shown in Figure 24-46.
12. Identify the heat pump cycle shown in Figure 24-47.
13. Refer to Electrical Diagram ED-11. Is the reversing valve energized in the heating or cooling mode?
14. List five reasons for inadequate defrost.
15. What happens when you switch the heat pump thermostat to emergency heat mode?

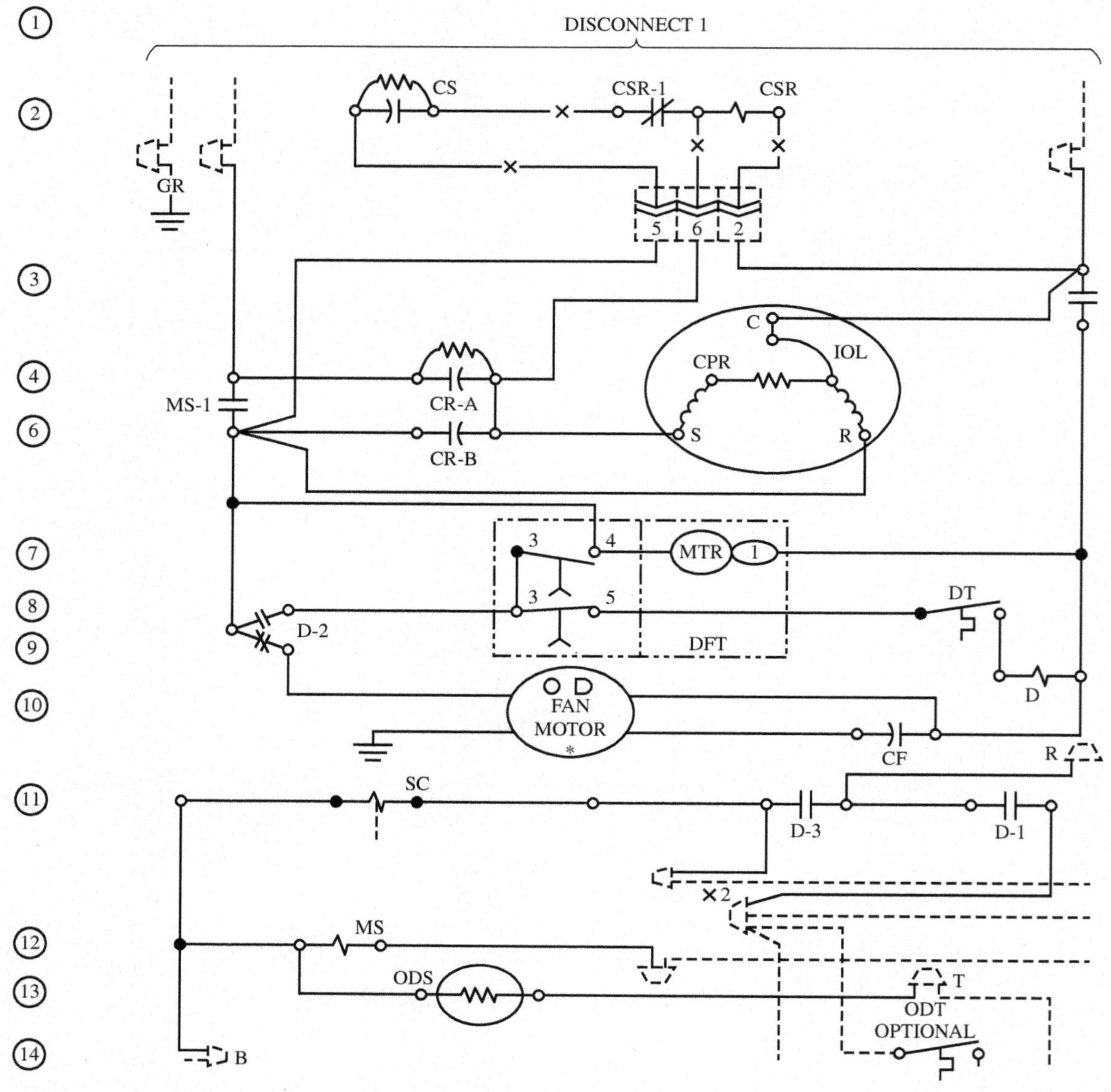

Figure 24-48 Use this diagram to answer Review Questions 21 and 22.

16. You are servicing a unit and realize that the defrost cycle is not long enough to clear the frost off the coil. How do you increase the frequency of the defrost cycle?
17. You are called to service a heat pump before the winter season. It operates in cooling mode, but will not shift to the heating mode when the thermostat is changed to heat. The R-410A pressure is 190 psi on the high side and 100 psi on the low side. What is the problem?
18. When would a heat pump, rather than some other type of heating equipment, be selected for installation?
19. True or false: The thermistor used in the defrost cycle measures pressure.
20. What are the three refrigerant cycles used in a heat pump?
21. What type of defrost system is used on the heat pump shown in Figure 24-48?
22. Does the heat pump diagram of Figure 24-48 show the outdoor section, indoor section, thermostat, or any combination of these?

UNIT 25

How to Start Electrical Troubleshooting

WHAT YOU SHOULD KNOW

After studying this unit, you will be able to:

1. Define troubleshooting.
2. State the steps of troubleshooting.
3. Describe a logical sequence for starting troubleshooting.
4. Use the human senses to detect problems.
5. Name seven types of troubleshooting techniques.

It is difficult to teach any type of electrical troubleshooting. The reason it is so difficult to teach and learn troubleshooting is because there are many problems that are not clear-cut. Is it an electrical problem? Mechanical problem? Or a combination of both electrical and mechanical?

The technician cannot see electricity and must use electrical meters to narrow down and solve the problem. Technicians need to thoroughly understand how to use all of their electric meters' functions in order to fully troubleshoot electrical problems.

Another reason it is difficult to teach troubleshooting is because problem-solving techniques vary from person to person. In many cases there is more than one way to solve a problem(s). The troubleshooting units in this book will offer different ways of solving these problems. Choose the process that works best for you.

Next, problems can blend together. There can be more than one problem. Mechanical and electrical problems can occur simultaneously. A mechanical problem can cause an electrical problem. For example, a refrigerant-flooded compressor will reduce the lubrication of moving parts and the motor bearings will wear out due to a lack of lubrication. The worn motor bearings will allow the motor rotor (the part of the motor that rotates) to drag on the stator (the stationary windings). The dragging rotor will short out the stator windings and it will appear to be an electrical problem. If the reason for the electrical failure is not corrected, the replacement compressor will soon face the same type of electrical failure, even though the original cause of the problem was a flooded compressor.

Finally, individuals learn differently. One person will learn and put into practice what is presented in this unit. Another person will read this material and totally ignore the value of using a logical approach to troubleshooting. Everyone wants hands-on practice and experience. This unit is designed to give the learner the tools, the logic, and the thought processes that can be used in isolating a problem on the job. If it is an electrical problem, the information provided here will aid in determining the area of the defective component. This unit prepares you for the units that follow. Basic troubleshooting is discussed in Unit 26, advanced troubleshooting in Unit 27, and practical troubleshooting in Unit 28.

The first part of this unit defines troubleshooting and discusses how to narrow down the problem quickly. Then we discuss the steps needed to help find the specific problem.

25.1 WHAT IS TROUBLESHOOTING?

Any operating mechanical or electrical equipment will at some time require service repair. Providing the diagnostic and repair work necessary to place the equipment back on line is one of the most important functions of the technician. The basic term that describes this type of work is *troubleshooting*.

Troubleshooting is the process of determining the cause of an equipment malfunction and performing corrective measures to bring the equipment back to normal operating parameters. Depending on the problem, this may require a high degree of knowledge, experience, and skill.

Basically, there are three types of problems:

- Electrical problems
- Mechanical problems
- Combined electrical/mechanical problems.

Whatever the nature of the problem, it is a good practice to follow a logical, structured, and systematic approach to solving it. By using a logical sequence when troubleshooting, the correct solution is usually found in the shortest possible time.

25.2 QUICK CHECKS FOR AIR CONDITIONING PROBLEMS

Figuring things out: A seasoned technician will use a set of quick checks to determine what is happening in an air conditioning system. One way to organize your troubleshooting thought process is to develop a *plan of action*. For example, use the word **ACT** as a reminder of what to check on an air conditioning system. It is important to make these checks so that you do not end up on a wild-goose chase, embarrassing yourself and your company.

The abbreviation **ACT** is used to determine if the major components in a HVACR system are operating:

A is for airflow.

C is for compressor.

T is for thermostat.

Let us discuss how the **ACT** method is used as a quick way to check the vital signs of an air conditioning or refrigeration system. The order in which the vital signs are checked is not as important as checking everything quickly before developing a diagnosis. For example, after you introduce yourself to the customer, check the thermostat setting or the air filter if it is nearby.

A Is for Airflow

Check the condensing unit airflow. Is the airflow from the condenser warmer than the surrounding air? Hot air off the condenser means it is rejecting heat from the refrigerant.

Check the indoor or evaporator airflow. Is there airflow? Is the air from the supply grilles cold? Cold air means the evaporator and air distribution system are working. Figure 25-1 illustrates checking the temperature of the supply grille. Placing your hand at the grille will not give you an accurate reading of the temperature or airflow, but it will give you an initial indicator of what is happening with the supply air conditions.

Note: Warm air being rejected by the condenser or cold air coming from the supply grille does not mean that the system is charged properly. These checks are preliminary and used as a general indicator of what is functioning and what is not functioning.

Finally, if the condenser or indoor fan comes on, you know that 24 volts is present in the control system and to the thermostat.

Figure 25-1 The troubleshooting process starts with a general survey of the equipment. Determine if cold air when in the cooling mode or warm air in the heating mode is coming from the supply grille.

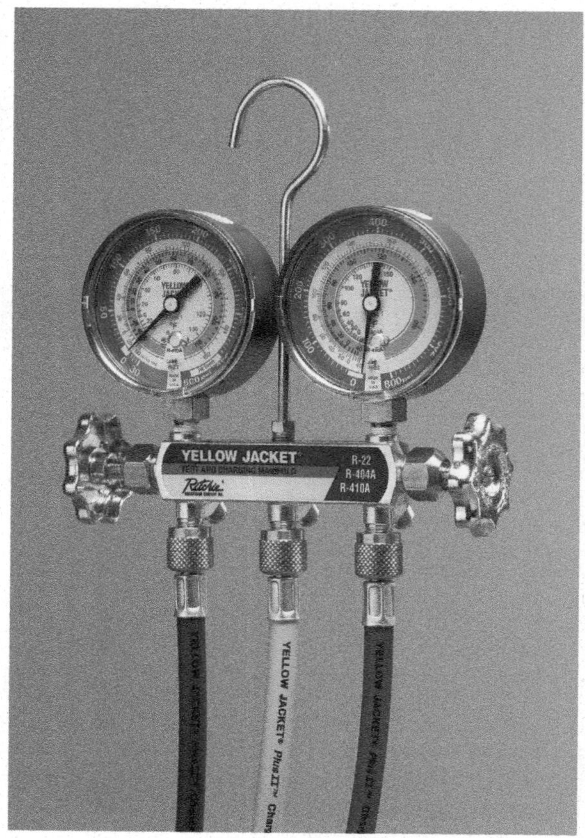

Figure 25-2 This manifold gauge set (shown not hooked up) should have a pressure differential between the low-side and high-side gauges when the compressor is operating.

C Is for Compressor

Is the compressor operating? What is the amperage draw of the compressor? Is there a pressure difference between the high side and low side? Do not assume that the compressor is running. The condenser fan noise may override the compressor operating sound. Checking the operating pressures and compressor amperage is an important part of this step. Figure 25-2 shows a manifold gauge set that can be used to determine if the compressor is operational. Because some condenser fans create noise that will prevent the technician from hearing the compressor operations, it is essential to install the gauges to determine compressor operation.

It is not advised to automatically install gauges on a system with less than 2 pounds of charge unless you are using low-loss fittings. Installation and removal of the hoses cause loss of refrigeration, which can significantly impact a system with a small charge. If you must add gauges to a system with a small charge, install the suction side hose only. There is less refrigerant loss on the low side when installing and removing hoses while the equipment is running.

Another way to determine if the compressor is running is to check the amp draw of the common wire. Any wire on a three-phase compressor will do. For an initial check, this is a little quicker than hooking up the gauges.

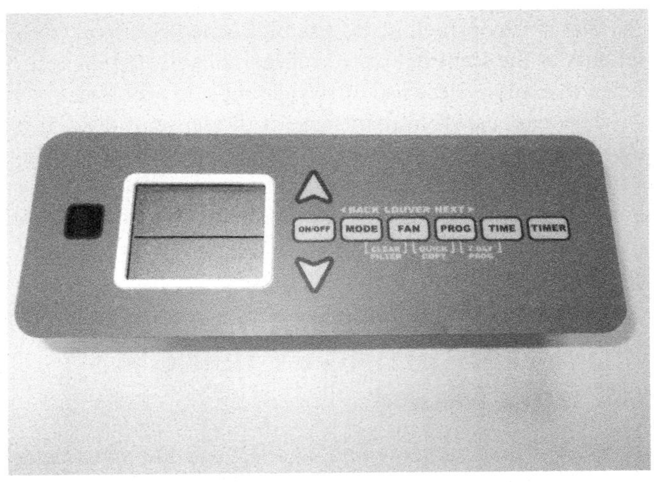

Figure 25-3 The thermostat should be checked to determine if it is set to the correct mode, heat or cooling, and that the temperature is adjusted so that it is at least 10 degrees different than the room temperature. The thermometer temperature on the thermostat is not always accurate.

T Is for Thermostat

Is the thermostat set to the cooling or refrigeration mode? Set the temperature 10°F below the cooling setpoint to ensure that the unit will begin to operate and stay operating while troubleshooting. Remember to reset the thermostat to the customer setting prior to leaving the job.

Does the digital thermostat have a display? Some systems have a time delay. Wait at least 6 minutes before reaching a conclusion. The indoor blower may be operating, but the condensing unit may be on a time-delay circuit. Figure 25-3 shows a blank screen. A blank t'stat screen indicates that there is no control voltage power. Search for the cause of the lack of control voltage, which is usually 24 volts.

In summary, the **ACT** method or another logical method you devise will eliminate parts of the system that are working and help you diagnose the problem quickly. Each job is a little different and developing a systematic way to start troubleshooting is important. Using the **ACT** method provides a fast and logical sequence for checking the system and, at the same time, familiarizes a technician with the system under scrutiny.

Because most malfunctioning problems are electrical, it is a common practice to perform electrical troubleshooting first. If the problem is mechanical, the electrical analysis will usually point the technician in that direction. Because this unit is based on electrical troubleshooting, we will discuss the specific topic of electrical troubleshooting.

Now we take a look at quick checks for a heating system.

25.3 HEATING SYSTEM ACT TROUBLESHOOTING

The **ACT** heating troubleshooting sequence is similar to the air conditioning troubleshooting sequence.

Figure 25-4 One of the ACT checks is to look for combustion (C) in a fossil fuel burning appliance. Combustion is taking place in this three-burner gas furnace.

A Is for Airflow

A is a reminder to check the airflow in the gas, heat pump, or oil furnace duct system. **A** is for airflow in handlers in electric heat, heat pumps, and hot water coils. Is the air from the supply ducts warm, indicating that the heating system is working?

C Is for Combustion

C reminds the technician to check for combustion of gas or oil heating systems. Can you see combustion? Figure 25-4 shows combustion in a gas furnace. Is heat coming from the supply ducts? There is no combustion in electrical heat, heat pumps, or hot water coils, but the technician can check for supply duct heat. Hot water coils can be part of a gas, oil, or electric system; therefore, it would be logical to check the combustion operation of a gas or oil boiler. Check the amp draw of an electric boiler.

T Is for Thermostat

Is the thermostat set for the heating mode? Set the temperature 10°F above the heating setpoint to ensure that the unit will begin to operate. Does the digital thermostat have a display? Some systems have a short time delay, allowing the heat exchanger to heat prior to beginning indoor blower operation. Remember to reset the thermostat setting to the customer setting prior to leaving the job.

25.4 USE YOUR SENSES

The troubleshooting method just discussed relies on the senses of sight, hearing, feeling, or touch as shown in Figure 25-5: Seeing that the condenser fan is on. Seeing that the thermostat is set correctly. Next, seeing that the pressures on the gauges are different, indicating that the compressor is creating a pressure difference. Finally, seeing the flames or combustion process in the case of oil or gas heat. The ACT method has the tech feeling for airflow out of the duct system.

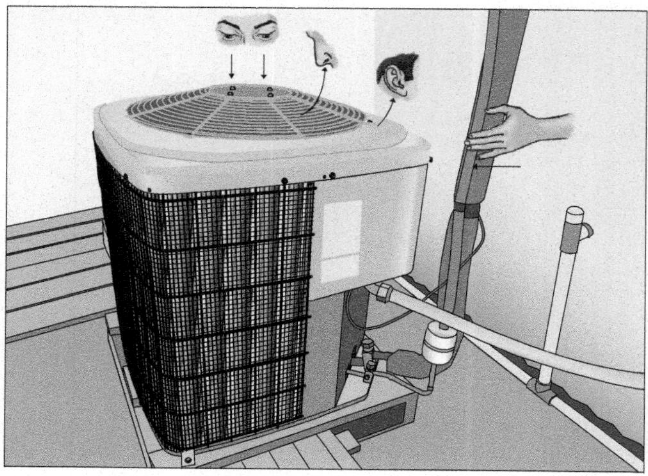

Figure 25-5 The human senses of sight, hearing, touching, and smelling are valuable troubleshooting tools. All of these senses are a valuable part of the first troubleshooting steps of gathering information.

Inspecting the system is very important. Search for burned-out parts, components, or wires. Listen for the operation of the system. Are the major system components such as blowers or compressors operating continuously? Are they cycling on and off? Are unusual noises coming from the compressor, blower, or heating system? The novice technician may not be able to pick up unusual noises, but with time and experience the sound of a correctly functioning system will be a valuable troubleshooting tool.

The sense of smell is also useful. Burned electrical components or the scent of burned refrigerant will lead the technician to explore electrical or refrigerant side problems.

The least useful human sense is the sense of taste. It is not a practice to taste damaged HVACR components, refrigerants, or other chemicals used in our profession. Using the sense of taste can be dangerous to the technician. Do not use this sense for troubleshooting.

25.5 UNDERSTANDING BASIC TROUBLESHOOTING BY USING A VOLTAGE METER

Technicians need to know how to select and use a voltmeter before trying to use this instrument for troubleshooting. Before technicians do any troubleshooting they need to know what voltage reading to expect in varying circuit conditions. Let's look at the essentials to solving electrical problems using a voltmeter.

Ways to Measure Voltage

Voltage is measured across a voltage source. This may be obvious, but a circuit without voltage will not work. One way to do this on an air conditioning system is to check the incoming voltage to the major components such as the air handler or condensing unit. If the system is a split system, as shown in Figure 25-6, it is helpful to determine if the problem is a high-voltage or low control voltage problem. Measure the voltage at the primary of the transformer. If the primary voltage is good, that means the air handler has the correct supply

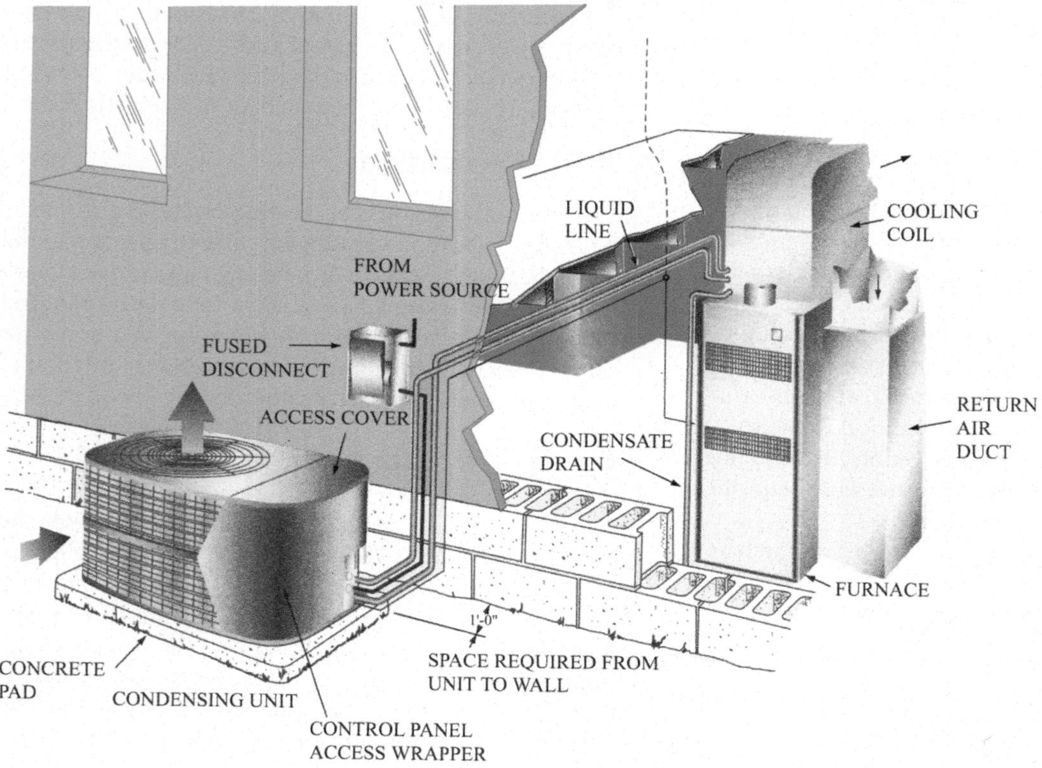

Figure 25-6 This is called a split system. This means that the condenser section is separated from the evaporator section. The condensing unit is outside the building and the air handler and evaporator are inside the building. Notice the disconnect to remove power from the outdoor section.

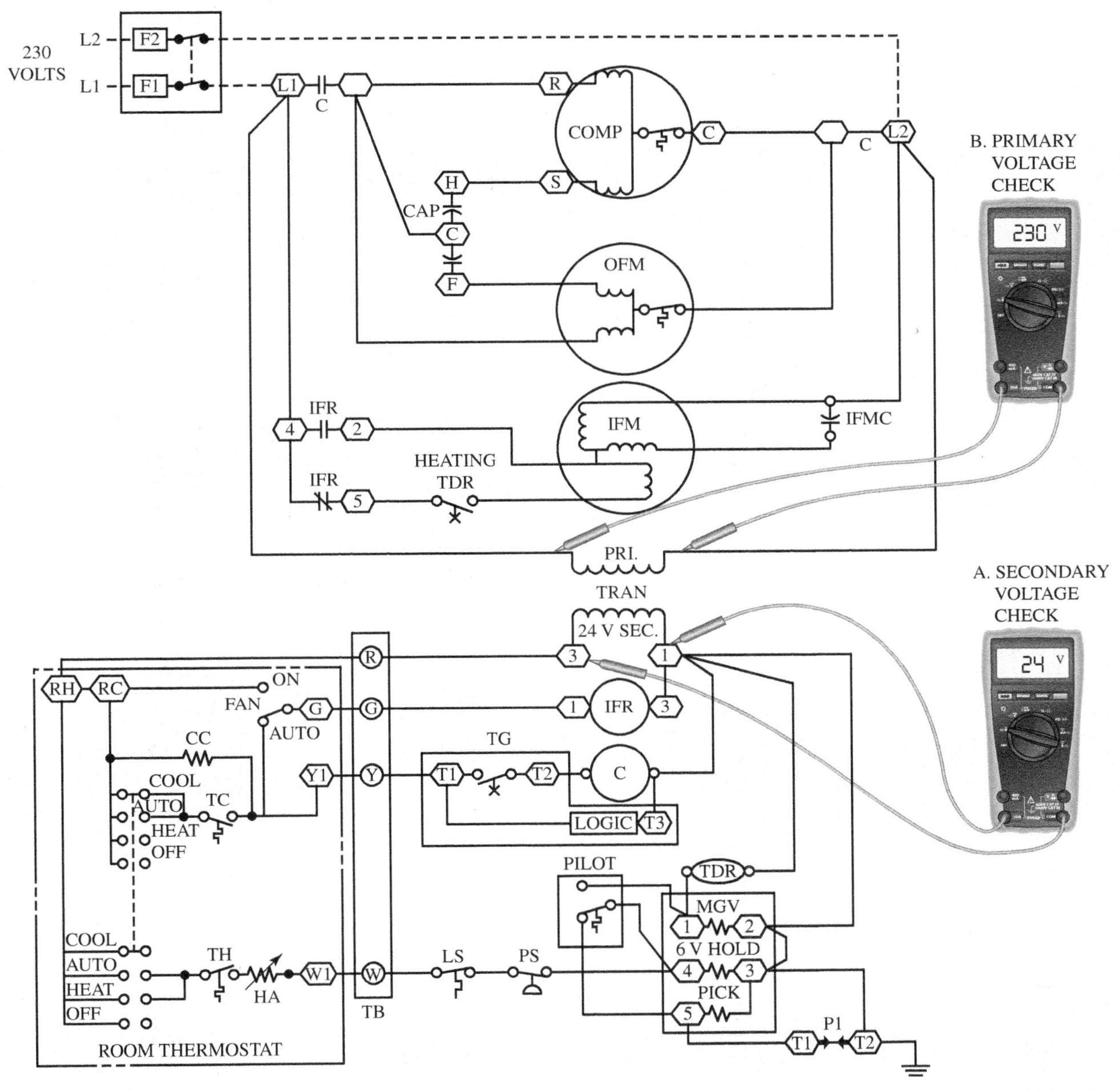

Figure 25-7 This diagram shows a quick way to determine if a system has an incoming voltage problem or control voltage problem. If the primary voltage on the transformer is good, check the voltage on the secondary side. Both of these voltages are required for the system to operate.

voltage. If the correct primary voltage is present, verify the secondary side of the transformer as shown in Figure 25-7.

Voltage will be measured across an open load. For example, you do not know that the motor windings in a simple motor are open. Assuming that everything else is working as designed, the technician will measure voltage across the motor even though it is open as illustrated in Figure 25-8. A mechanically locked-up blower motor will have supply voltage and good motor windings, yet will not rotate. Remove the power from the motor. Push-start the motor to see if it will turn freely. Seized motor bearings could prevent the motor from turning. The fan blade or blower wheel could be against its housing, preventing motor operation. ECMs will not turn freely when they are de-energized. The permanent magnets in the stator prevent the rotor from turning freely and evenly when de-energized.

Voltage will be measured across an open wire, open fuse, or open switch provided that voltage is present. The easiest way to understand this is by measuring the voltage on a plugged-in power cord. Be careful not to touch the wires together while the cord is plugged into the outlet. The voltmeter will display the wall outlet voltage on the plug prongs even though it may not operate an electrical device. Voltage will be read across an open wire because the meter is measuring

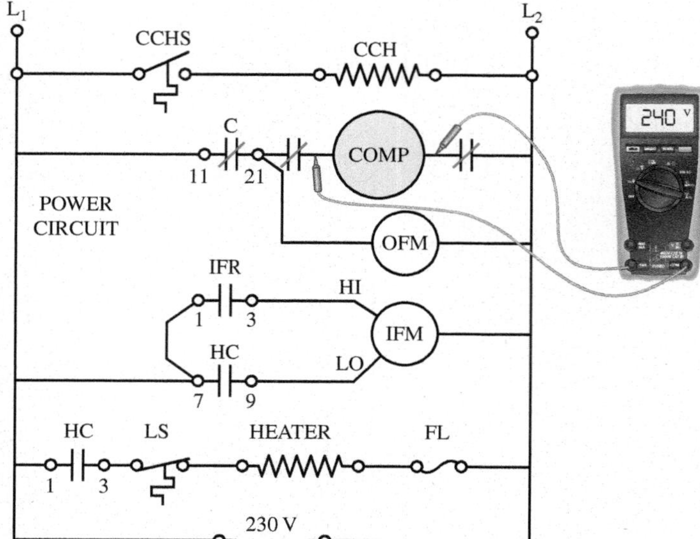

Figure 25-8 The compressor motor, COMP, may be defective even if the supply voltage is measured across it. Check the resistance of the motor windings. The motor could also be mechanically locked up.

potential difference. The voltage potential is available, but it cannot flow since the wire path is broken or open. Figure 25-9 illustrates an open switch with the voltmeter measuring the applied voltage across it.

Voltage will be measured across an open fuse. An open fuse is like an open wire and is considered to have infinite resistance. It is best to remove the suspect fuse and check it with an ohmmeter. Figure 25-10 shows how to check the fuse with an ohmmeter disconnected from a circuit. The resistance should be near 0 Ω. Measuring resistance with applied voltage will damage an ohmmeter or at a minimum blow the meter fuse. Removing the fuse also prevents a resistance reading through another component that may be in parallel with the fuse.

Where Voltage Will *Not* Be Measured

Voltage will not be measured across a piece of wire or closed set of contacts with the same potential. See the example in Figure 25-11. The voltmeter is designed to measure a voltage difference, also known as a potential difference. Measuring voltage on one bare wire with 240 volts flowing through it will not show a voltage reading since the voltmeter is measuring the same voltage potential on the wire.

Voltage will not be measured across a good fuse. The fuse is like a piece of wire and does not create resistance or a voltage drop; therefore, you will get a zero voltage reading.

Voltage will not be measured across a closed (good) overload device. An overload device, like an external

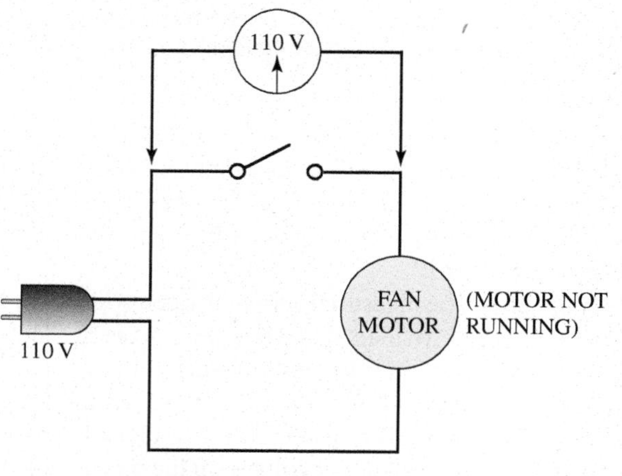

Figure 25-9 The voltmeter will measure the applied voltage across an open wire and open switch.

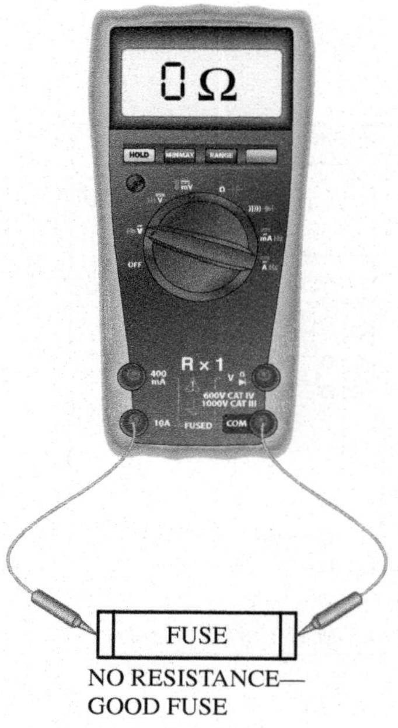

Figure 25-10 Measure resistance across a fuse while it is removed from the system. This is a safety procedure and prevents the technician from measuring something else in the system.

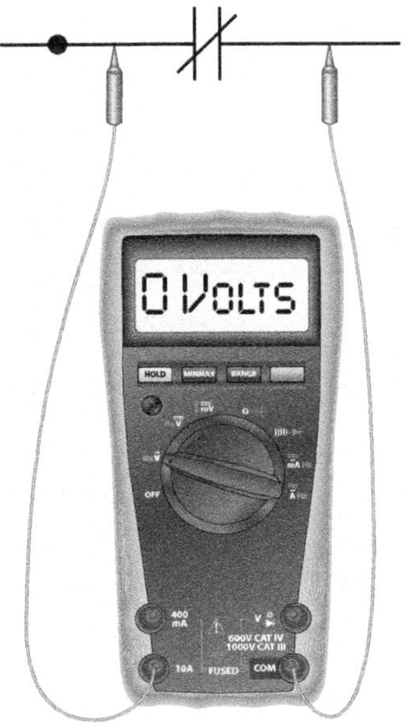

Figure 25-11 Voltage is operating the air conditioning system in this wire, but the meter measures 0 volts since there is no potential difference across the closed set of contacts. A good fuse with voltage running through it will read 0 volts. An open fuse will read the applied volts across it.

overload, is similar to a fuse. No resistance—therefore no voltage reading.

Voltage will not be measured across a closed switch. A closed switch has no resistance and will have no voltage reading.

25.6 METHODS OF ELECTRICAL TROUBLESHOOTING

There are many different methods of troubleshooting—some good, some bad. Some work in one application, yet may not work in another application. The following sections explore the many ways a technician can do troubleshooting. Read about these methods and determine which method or methods will work for you. There is no one way that is best, but there are some methods that are not acceptable. Read and decide for yourself!

Guess and Check Method

The guess and check method of troubleshooting is used by a person who has no idea of how the equipment is supposed to operate. He starts by guessing and checking anything that comes to his mind. This method is most likely used by the person who has little or no experience or the experienced technician who is asked to troubleshoot a system that is beyond her capabilities.

Part Changer Method

The part changer method of troubleshooting is used by technicians who have a low level of troubleshooting expertise. The person using this technique may or may not have an idea of how the system is supposed to operate, so he simply starts to change out parts until the system begins to operate.

One major disadvantage of this method is that there may not be anything wrong with the circuit or its components. It may be a simple thermostat operating error or a switch that has not been turned on. Changing out a part that is obviously burned out is not considered a "parts changer" behavior unless the technician does not try to find out why the component became defective. Some components burn out for a random and undeterminable reason, while other components will quickly burn out again if the reason the component was damaged in the first place is not found.

Individual Component Check Method

The individual component check method is the second stage of the guess and check method of troubleshooting. The person who uses the component check method relies on previous experience. The technician troubleshoots the components that he is most familiar with and overlook components that are unfamiliar. Again, luck is involved in finding the right defective component(s)—not a good troubleshooting method.

SAFETY TIP

Do not touch the metal part of the meter probes when measuring voltage or resistance. Touching the metal probe while measuring voltage may cause electrical shock. Touching both probes when measuring resistance will read high resistance through the body. You will think the component has a complete circuit, yet a very high resistance.

Voltage to Ground Method

Most circuits with properly supplied voltage will measure voltage to ground even if a component is open. The voltage to ground method is good to use as a safety check to ensure that no voltage is present prior to touching a bare wire or an exposed component that may be carrying voltage.

In Figure 25-12, the upper condenser motor number 2 is not operating. It is not operating because the **1R-1** contacts (circled in red) are burned open. As Figure 25-12 shows, voltage can be measured to ground with an open component to the left and right side of condenser motor 2. Using this troubleshooting method, the technician may think that the condenser motor is defective since voltage to ground is measured on each side of the condenser motor. But changing the motor would not solve the problem.

Using the voltage to ground troubleshooting method leads to confusion and misdiagnosis. It is a good safety step to ensure that no voltage is present prior to touching a component. Do not think this is a valid troubleshooting method.

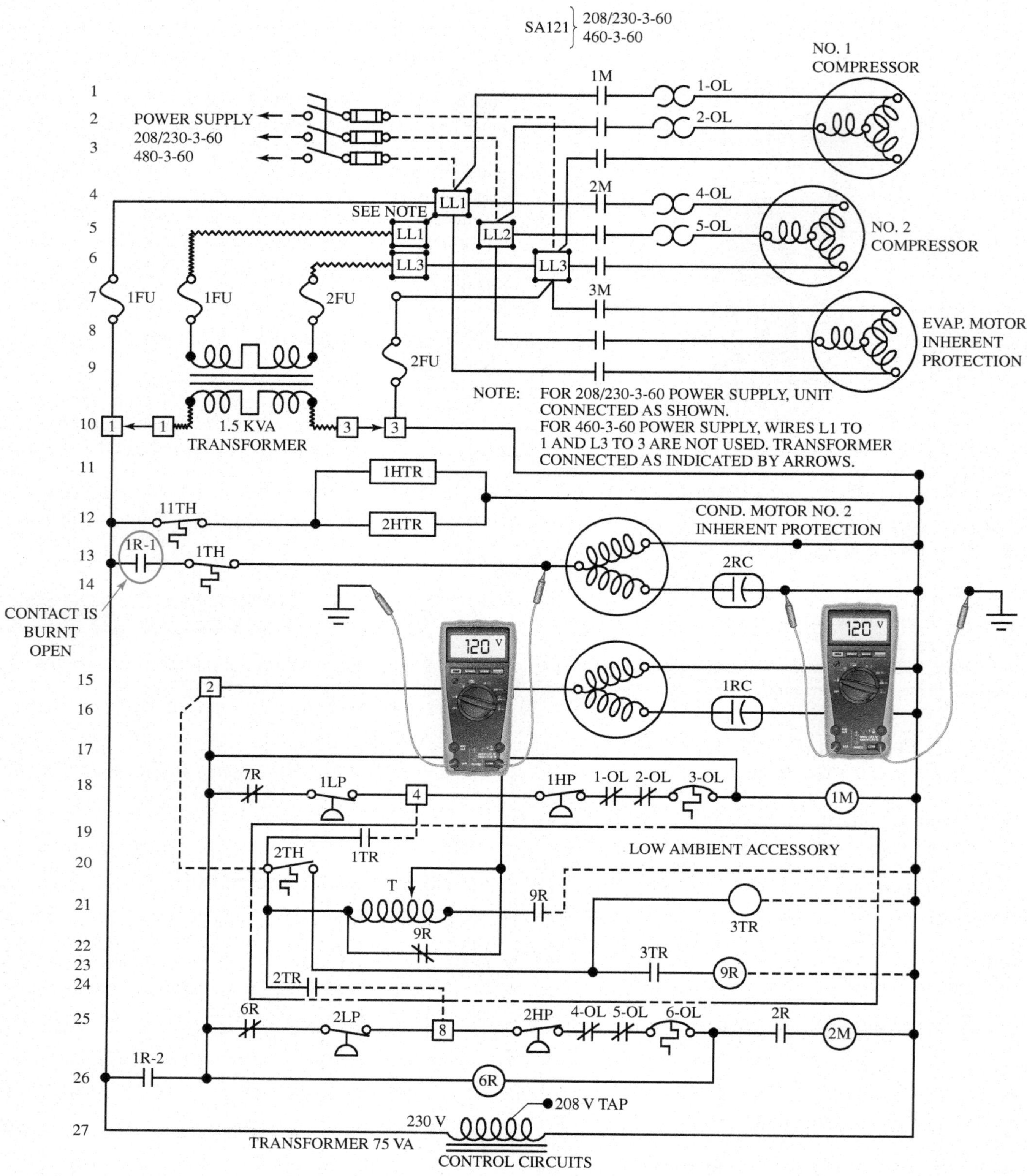

Figure 25-12 Checking voltage to ground can lead to misdiagnosis of an electrical problem. In this example **1R-1** is open, but voltage is measured to ground on both sides of the motor, leading to confusing results. Measuring to ground is a safety measure, not a good troubleshooting practice.

SAFETY TIP

Prior to using your voltmeter for voltage reading, test it against a known source of voltage such as a wall outlet or other voltage source. A defective voltmeter or open meter probe will measure zero volts and give a false and unsafe non-reading.

Hopscotch Troubleshooting Method

The **hopscotch troubleshooting** method is a recommended way to find a defective component or open wire. The hopscotch method involves measuring voltage by jumping through a circuit until the voltage is lost. Determine the **load side** of the circuit and *hopscotch* the control system to find the fault. The load side of the electrical circuit is defined as the side of the potential closest to the majority of the power-consuming devices.

The hopscotch method requires the use of a ladder wiring diagram as we previously saw in Figure 25-8. First, we examine what circuit voltage is expected when jumping or hopscotching through a circuit. Figure 25-13 shows an example of the hopscotch method of troubleshooting. When

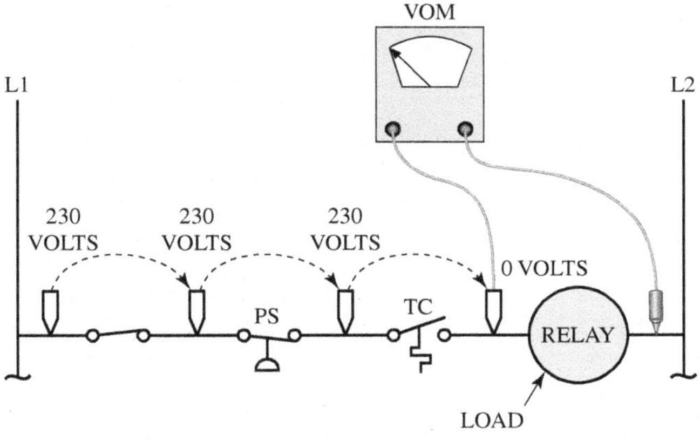

Figure 25-13 This is an example of a simple hopscotch method of troubleshooting. As shown in this diagram, the voltage will be lost when a switch or safety is open. The voltage will also be 0 volts when it jumps to the right side of the open thermostat **TC**.

the voltage is lost, the defective component, thermostat **TC**, is found. Is the thermostat open because the temperature is set too high or is the thermostat defective?

Figure 25-14, which is part of a ladder diagram, illustrates how to use the hopscotch troubleshooting method. First determine the load and **line side** of the circuit, as shown in Figure 25-15. The load side is the side of the circuit that has the component with the greatest resistance. The line side will have switches and safety devices. Connect one voltmeter probe to the right side or load side of the circuit to start the hopscotch troubleshooting procedure.

The process is simple. We measure voltage across all closed switch contacts until we cross over the load. Once the probe crosses over to the right of the load, the voltage being measured is the same, therefore the meter will have a zero-volt reading. The voltmeter only measures potential difference and having the voltmeter probes on the same side of the power supply will measure zero volts even though serious voltage is present.

Figure 25-16 shows an example of how to use the hopscotch method to troubleshoot an open compressor thermostat (**CTH1**). Attach one meter probe to the right side or load side of the ladder diagram. This can be done by securing a meter probe with an alligator clip. Start by measuring the supply voltage to the circuit with the other probe. The supplied voltage should be measured across each component until you reach the defective component.

In this case, 230 volts will be measured all the way to the left side of the open thermostat, as shown in Figures 25-17A, 25-17B, and 25-17C. Here is the hopscotch sequence of events:

- 230 volts is measured on the right side of **CR1**.
- 230 volts is measured on the left and right side of **LP1**.
- 230 volts is measured on the left and right side of **HP**.
- 230 volts is measured on the left and right sides of both **OL1** and **OL1**.
- 230 volts is measured on the left side of **CTH1**.

Measurements on the right side of the open thermostat will be zero volts, as illustrated in Figure 25-18. Verify that the thermostat is truly defective. An open **CTH1**, shown in Figure 25-19, will have the supplied voltage measured across it. Once it has been established that the thermostat is open, the technician needs to check that the thermostat is defective by removing it from the circuit. Using an ohmmeter to check the thermostat, the thermostat will be infinity or open, as shown in Figure 25-20. If it is an open thermostat, allow the device, in this case a

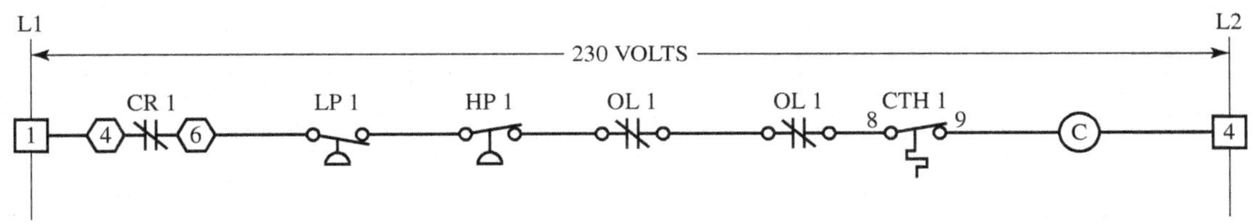

Figure 25-14 Circuit extracted from a ladder diagram.

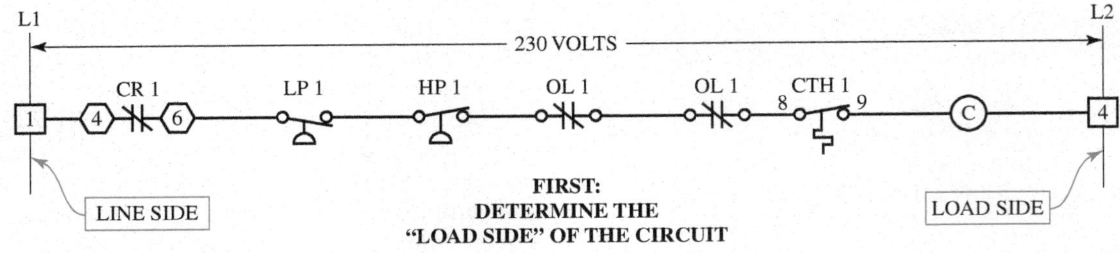

Figure 25-15 When using the hopscotch method of troubleshooting, first determine the load and line sides of the circuit.

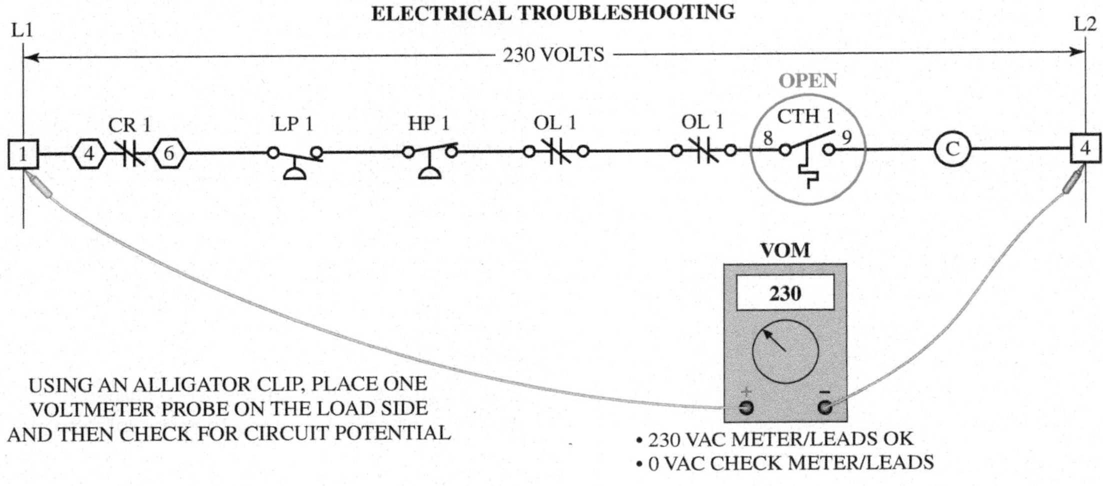

Figure 25-16 To start the hopscotch procedure, attach one meter probe to the load side of the line. Starting at the left side of the diagram, you would measure voltage across each switch until the voltage is lost. Losing the voltage indicates that a switch or safety device is open. The problem in this example is an open compressor thermostat, **CTH1.**

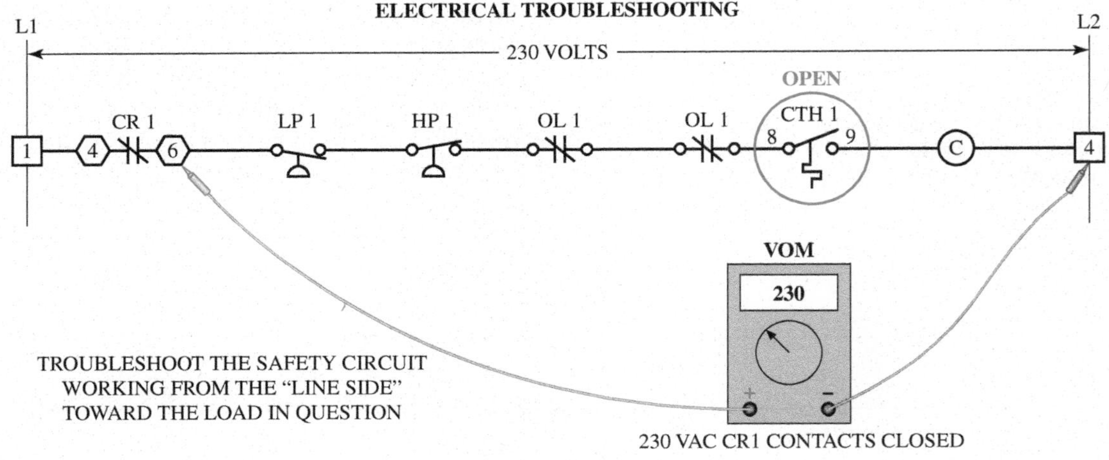

Figure 25-17A This is the second stage of hopscotch troubleshooting: the actual jumping on each side of the switches and safeties until the voltage is lost. If the circuit is good, you will lose voltage once you go to the right side of the load.

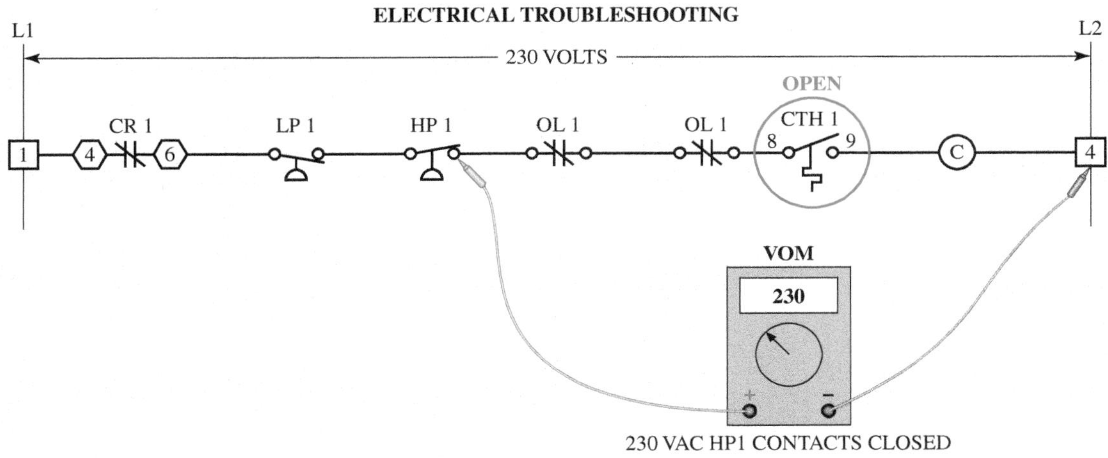

Figure 25-17B The hopscotch procedure continues toward the load.

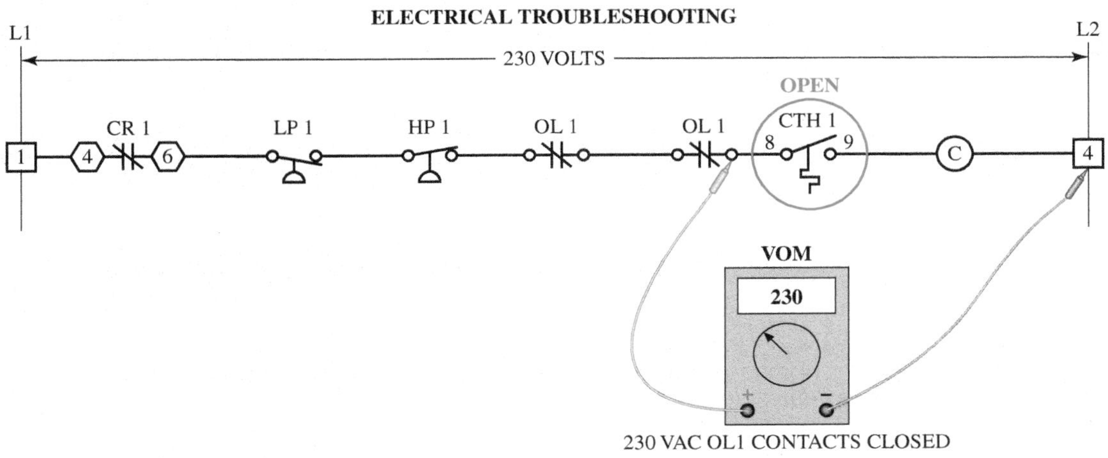

Figure 25-17C The hopscotch procedure is used across **OL1** and **OL1** to the left side of **CTH1**. Voltage is measured up to the open compressor thermostat **CTH1**.

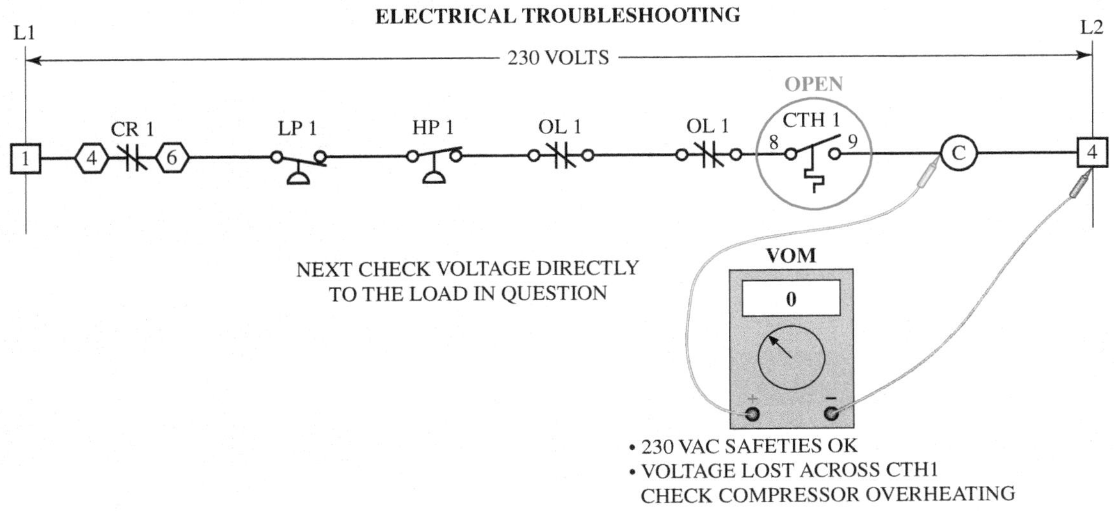

Figure 25-18 Because **CTH1** is open, the voltage is lost once the probe is placed on the right side of this compressor thermostat. This is the indication that the compressor thermostat (**CTH1**) is open.

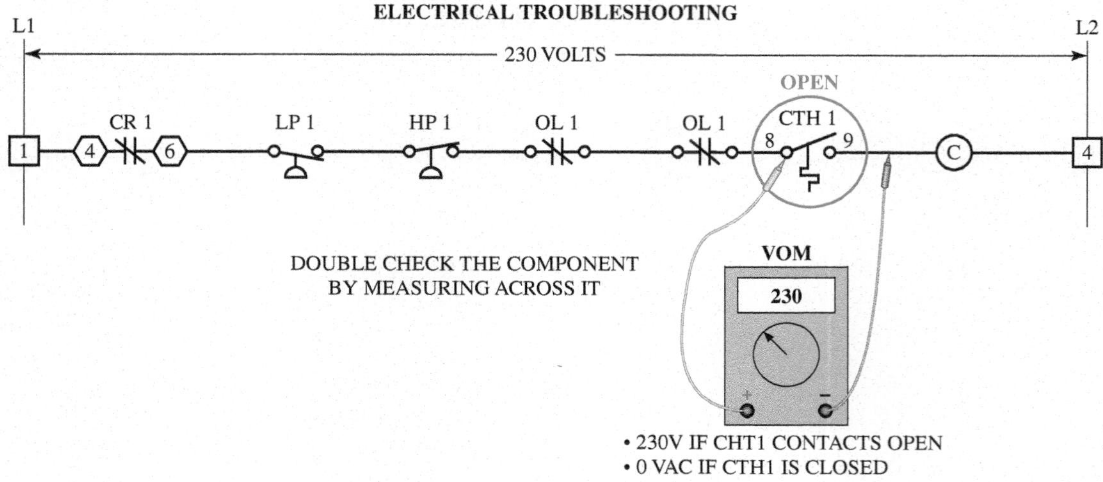

Figure 25-19 Voltage will be measured across an open switch. Is **CTH1** open because it is hot or because it is defective? Most likely because it is hot. Allow plenty of time to cool and reset. It may take hours to cool unless you take action to reduce the temperature.

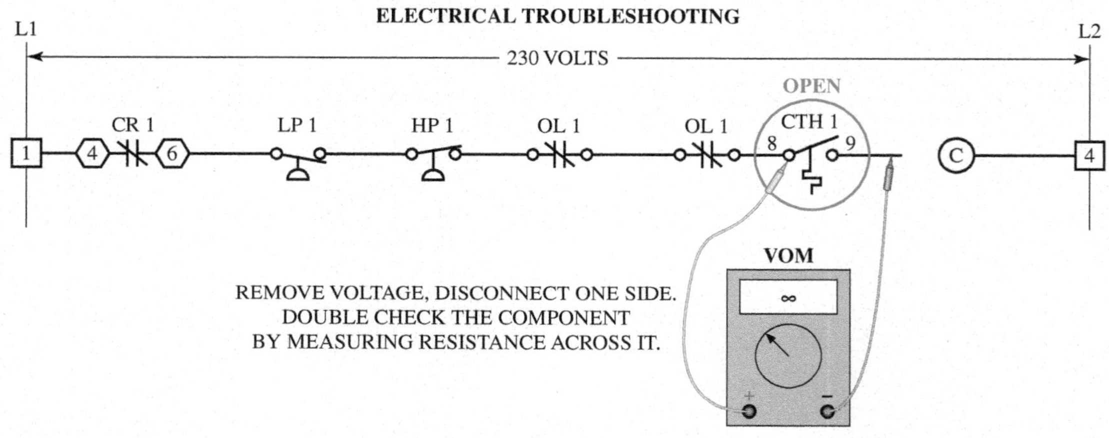

Figure 25-20 The final step in the hopscotch troubleshooting is to remove the power from the circuit. Remove the power and disconnect the component from the circuit and measure the resistance. In this case the thermostat is open, reading infinity. Is it open because it is hot? Or is it defective?

compressor, to cool before condemning it. Some compressors can take hours to cool and reset the thermostat. Manufacturers report that 24% of compressors returned under warranty have no obvious problem. Not allowing the motor to cool could be one of those "good compressor" warranty returns.

SAFETY TIP

Remove a component from its circuit prior to checking it with an ohmmeter. Voltage in the circuit can damage the ohmmeter and shock the technician. Some inexpensive meters can explode since they do not offer protection from this unsafe use.

TECH TIP

Know the resistance limits of your ohmmeter! Some multipurpose clamp-on ammeters have a limited resistance range—as low as 1,000 ohms. Some relay coils have a resistance rating higher than 1 kΩ and these limited ohmmeters will measure over limit, overload, or OL, indicating an open coil when in fact the coil is good, but the meter cannot measure a resistance higher than 1 kΩ.

Know the resistance limit of your ohmmeter no matter what brand it is or its range setting. When checking coils or resistance to ground, the technician should have an ohmmeter that can measure at least 1 megohm (MΩ) or 1 million ohms. Meters measuring up to 5 MΩ or greater would be best.

"Draw the Schematic" Method

This method can be used when no diagram is readily available or cannot be obtained in a reasonable amount of time from the manufacturer or Internet sources. The "drawing the schematic" method is used on a difficult or complex diagram. Drawing a diagram can assist the technician in understanding how the system is supposed to operate. Drawing a diagram also assists the tech in understanding the layout of the various components and, once drawn, will speed up the troubleshooting process.

"Call Someone Else" Method

Finally, the "call someone else" method is used by all smart technicians. One technician cannot know it all, so asking for help is understandable. When calling for help or seeking ideas from the Internet, several important pieces of information will be required:

- Manufacturer's name of each piece of equipment that needs troubleshooting
- Model number of all pieces of equipment
- Serial number of all pieces of equipment
- Any distinct modifications that make it different from the standard manufacturer's model
- Type of metering device. This is not electrical, but may be a requirement to solve a problem.

Take notes before seeking help. Record as much information as you can prior to obtaining assistance. This will give the person trying to help you more ammunition for solving the problem. Additionally, having information such as the refrigerant type, application, voltage and amperage readings, and pressure and temperature readings available is useful data when requesting assistance.

TECH TIP

The technician uses all of human senses except taste to troubleshoot a system. For example, when the tech is troubleshooting an air conditioning system, he can touch the suction line to feel for cold. Feel cold air from the ducts. Listen for the sound of the compressor. See that the condenser fan is blowing. Smell the air inside the building. The air should not have the odor of mildew or moisture or smell stale.

SERVICE TICKET

The A-1 Air Conditioning service department was overwhelmed by the number of air conditioning service calls in the middle of a hot Texas summer. As a new and eager technician with limited experience, you are asked to go to a long-time customer's home to survey the problem and report back information to the service manager with your findings. An experienced technician will not be able to get to the house for hours. The service manager hopes you can find the problem and leave the job with the excuse of locating parts while sending the more experienced technician in a few hours. Anyway, you are asked to make an appearance to keep and satisfy a good customer.

Upon arrival, you use the ACT method to check the pulse of the air conditioning system. Since the customer greets you and invites you inside the house, you check the following items first: There is warm airflow (**A**) coming from the supply duct system. The airflow seems to be adequate as evidenced by a quick touch of a couple of grille outlets. The thermostat (**T**) is set for 70°F, but the room temperature is 80°F.

Finally, checking the condensing unit (**C**), you notice that it is silent. While investigating the condensing unit, the compressor comes on, but the condenser fan does not. You report to the office that there is a problem with the condenser fan circuit and you report the model number of the condensing unit. The customer is notified that parts will be needed to repair the problem and that the unit will be cooling again in a few hours.

The experienced technician returns with a condenser fan motor and run capacitor. The problem was a defective run capacitor, so the problem was fixed quickly after the technician arrived. A-1 Air Conditioning's plan for dealing with a busy day worked: The customer did not call another company. Even the least experienced technician can help at crunch time if she has a logical sequence for troubleshooting.

SUMMARY

Electrical troubleshooting is the process of identifying and eliminating problems with the ultimate result of finding and correcting the fault. The troubleshooting process is very individual and there are many avenues to obtaining the final result of getting a system back on line. This unit outlined some recommended methods to first do a quick review of the job and see what is working and what is not working. Some problems can be very simple, whereas other problems can be very complex.

Each technician must develop a comfortable troubleshooting technique. The ACT method was recommended as a way of organizing a step-by-step sequence for troubleshooting. The ACT method gives the technician a way to conduct a quick check of the major components while getting familiar with the system under inspection.

The unit discussed many effective and ineffective ways of troubleshooting. The hopscotch method was explained as one of the better troubleshooting tools. The technician was encouraged to call or seek help for more difficult problems.

Prior to seeking help, the technician should document important operating and model information on the system.

REVIEW QUESTIONS

1. What is troubleshooting?
2. What are three types of troubleshooting problems?
3. The **ACT** troubleshooting method is one useful method of starting a troubleshooting process. What does each letter of **ACT** mean?
4. When doing an airflow check, what component(s) is (are) checked?
5. What is one way to verify that a compressor is operating?
6. What is checked on a thermostat to verify that it is set properly?
7. What human senses are used to assist in troubleshooting?
8. When using a voltmeter to troubleshoot a problem, what is the first voltage source that should be checked?
9. Voltage is applied to a load, like a motor, and the motor is open. In this example, will voltage be measured across a load that is open?
10. Will voltage be measured across an open fuse that has voltage applied?
11. Name three conditions in which voltage will not be measured.
12. This unit discusses seven electrical troubleshooting methods. Which method do you prefer? Discuss the reasons why you selected this method.
13. When using the hopscotch troubleshooting method, how do you locate a switch that is open?
14. The tech can use the human senses to do quick checks. What condition would a technician expect when touching a suction line that has the correct charge and no system problems?
15. The sequence of troubleshooting is a very individual process among technicians. After reading this unit, what is *your* sequence for troubleshooting a cooling problem?

UNIT 26

Basic Troubleshooting Techniques

WHAT YOU NEED TO KNOW

After studying this unit, you will be able to:

1. Further develop your troubleshooting techniques.
2. Describe basic methods of troubleshooting air conditioning systems.
3. Describe basic methods of troubleshooting electric heating systems.
4. Describe basic methods of troubleshooting heat pump systems.
5. Describe basic methods of troubleshooting fossil fuel heating systems.
6. Use troubleshooting flowcharts and troubleshooting tables.

This unit is designed to give you some basic troubleshooting tools and build on what you learned in Unit 25. We learned in the previous unit that there are basic ways in which we can organize our troubleshooting practices to be more successful and find the problem quicker.

There is no one "magic troubleshooting trick" that works for everyone and in every case. You will need to develop a technique that is comfortable for you. The troubleshooting steps that you use today will be different from those you use a year from now. Your approach will change as you learn the best way to tackle a problem. The evolutionary process of practicing troubleshooting will develop into a comfort level that will instill confidence and skill in what you do. Doctors and lawyers "practice" to improve. It is called practicing medicine or practicing law. You will practice to improve your skills as an HVACR technician. Other professionals improve their tools and instruments to become successful. You also will find better tools and instruments to help you succeed as you advance in your career. Understanding this unit is one step to your professional success. Unit 27 will expand the information learned in these first two units on troubleshooting. Unit 28 will reinforce your skills by giving you practical examples of problems and solutions. Each unit in this book that discusses specific components, such as compressors and relays, has offered a troubleshooting section on the individual components discussed.

26.1 KNOW WHAT IS HAPPENING

Prior to troubleshooting ask yourself the following questions: What should the unit be doing? What is the unit doing? From this analysis you will be able to determine: What is the unit *not* doing?

Knowing what is supposed to happen in a HVACR circuit is important to the troubleshooting process. Unfortunately, knowing exactly how the circuit is supposed to operate is not always possible. The technician needs to have a general understanding of the sequence of operations, but does not need to grasp the operation of each circuit to be successful in finding and repairing most problems. If you have a basic understanding of part of the equipment, you can "figure out" the rest of the circuit. However, there are limits to this knowledge transfer. For example, a technician who has experience with small appliances will not be able to repair a chiller without training and experience. The small appliance technician should be able to repair many residential air conditioning problems since they are closer in operation to smaller appliances than large chillers. Residential heating equipment is complex; the same small appliance technician may not be able to repair gas or oil furnaces without the training or experience required to understand these systems. By the same token, the chiller technician will be challenged to repair a small appliance, residential air conditioner, or furnace.

Training and experience are important for successful troubleshooting. Limited knowledge of a system should not stop the technician from trying to find the problem. There is always help from the office or the manufacturer's tech support line.

In summary, electrical troubleshooting is a logical process of elimination. To effectively troubleshoot, you must know the system electrically and mechanically. You must know:

- What the unit is supposed to do and how it is supposed to do it
- The function of each component used in the electrical circuit
- The physical location of each component used in the machine
- The interrelationship of components
- Electrical symbols.

You must also be able to read wiring diagrams to determine the sequence of operation.

TECH TIP

Do not waste time troubleshooting circuits that are functioning. Instead, you should:

> Eliminate the good circuits by starting with the major loads and working backwards through the control system.

(continued)

The chances of more than one failure occurring at the same time are remote, therefore you should:

Determine what is common with the major loads that is not functioning. From this you will be able to determine the most likely cause of the failure and pinpoint the component or circuit that is malfunctioning.

SAFETY TIP

Never stand in front of or look at compressor wiring terminals when starting a compressor. On rare occasions the terminal blows out, shooting high-pressure refrigerant, hot oil, and fire through the blown terminal opening. The compressor terminal cover may not be able to restrain this violent eruption. Many technicians have been injured by this occurrence. Most technicians see the aftermath of this type of blowout, because the terminals often blow out prior to the service call.

In the previous unit, we discussed the preliminary steps to troubleshooting, such as doing quick checks to determine what is working and what is not working. Now let us explore some specific examples of how to proceed with troubleshooting. The first case will be an air conditioning troubleshooting problem, followed by heating problems.

26.2 TROUBLESHOOTING AN AIR CONDITIONING SYSTEM

You have used the **ACT** method of troubleshooting to determine that the condensing unit is not operating. Now take control of the part of the system you are troubleshooting. In this case you will take control of the condensing unit by turning it off at the disconnect as shown in Figure 26-1. The disconnect is off when the red handle lever is in the down position. Be careful, however; do not assume the power has been removed. One or more knives can get stuck in the disconnect slot, meaning that the power has not been removed even though the disconnect arm is in the off position. Open the disconnect box as seen here and do a visual check to ensure the circuit is open, safely breaking the power supply on all legs.

If the disconnect is not near the condenser, remove one of the control wires to the contactor. Control wires (24 V) are shown in Figure 26-2. Install an alligator lead on the control wire to quickly connect or disconnect the control voltage to the contactor. Figure 26-3 illustrates how the alligator leads are connected in line with one side of the control voltage. This allows for control at the condensing unit. The quick connect on the contactor coil could also be used for this same control.

The important action is to take control of the condensing unit that you are troubleshooting. You do not want to rely on the thermostat for this control. You are in charge; therefore, take command!

Next, with the system off, the technician hooks up the manifold gauge hoses and the places the clamp-on ammeter to the common wire on the compressor. Locate the electrical diagram and charging information. The technician installs a manifold gauge set, since the charge will need to be checked even if it is not a charging-related problem. It would be embarrassing to repair an electrical problem and be called back in a couple days to charge a system because the technician was in a hurry and failed to check the charge.

There are exceptions to installing the manifold gauge hoses, however:

- If it is strictly an electrical problem, do not check the charge on a system that has less than 2 pounds of refrigerant. The act of adding and removing hoses will

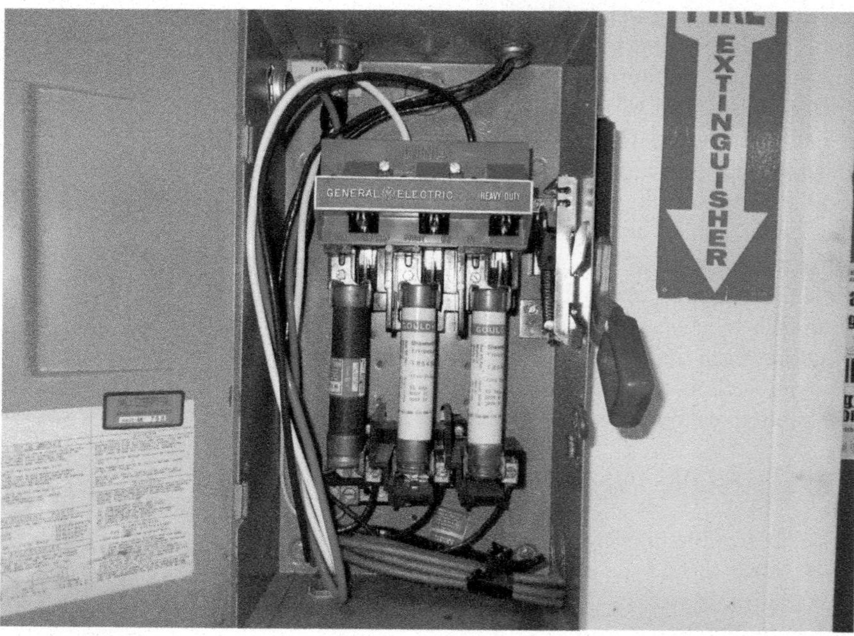

Figure 26-1 The disconnect is off when the red handle lever is in the down position. Do not assume that the knife connection in the disconnect box opens the circuit.

Figure 26-2 Control voltage (24 V) is going to the contactor. The two wires are not always on the same side of the contactor.

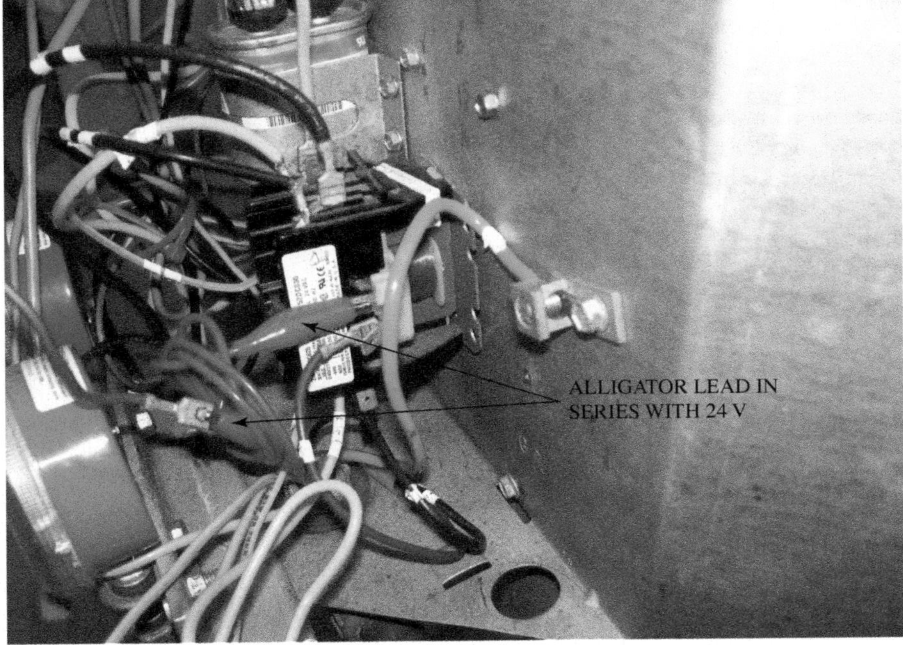

Figure 26-3 One of the control wires is removed and an alligator clip is placed between the control wire and contactor coil. The alligator clip makes it easier to control the condensing unit if the service is to take place outside.

significantly change the charge and affect the operation of the system. If gauges need to be installed on a system with a small refrigerant charge, install only the low-side hose. The suction side pressure is lower when the system is running. Less refrigerant pressure is lost if the low-side hose is installed when the system is operating.

- If it is strictly an electrical problem on a small refrigeration system without access valves, do not install a temporary or permanent access valve. Temporary valves tend to leak and need to be replaced with permanent access valves. The time involved in recovery, evacuation, leak detection, and charging will more than triple the time on the job. One way to do a quick check on the charge is to block the airflow through the evaporator by disconnecting the evaporator fan or covering the evaporator with a plastic bag. The suction line should frost up to the compressor. This is not an absolute test of the refrigerant charge, but a frost line to the compressor is a good indicator that the system has a sizable refrigerant charge.

It is important to check the compressor amperage on the common motor winding before leaving an air conditioning

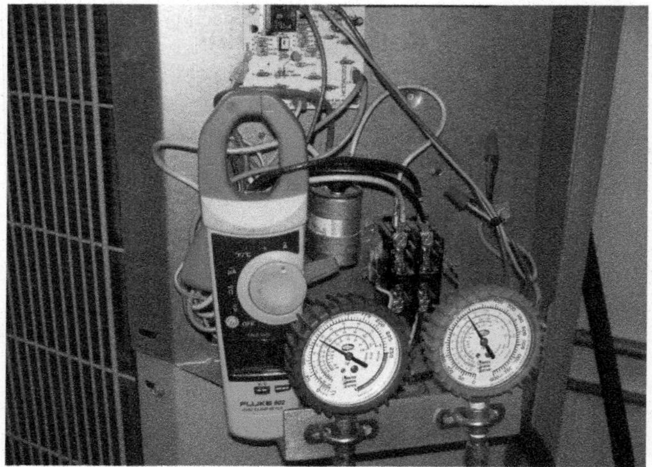

Figure 26-4 The refrigeration gauges are hooked up and the clamp-on ammeter is set up to record the compressor common winding. The technician can easily see what is happening on start-up of the condensing unit.

or refrigeration system job. This should be done at the beginning of the troubleshooting process along with installing the manifold gauge set. The purpose of doing this at the outset is to read the amperage and pressure readings as the system starts. As shown in Figure 26-4, place the clamp-on ammeter and gauges close together for simultaneous viewing when the system energizes. This process may aid in determining a problem. For example, a compressor is cutting out on its internal overload (internal thermostat and amperage protector) due to excessive operating amperage. It will only read this high-amperage condition for a second or two at start-up. There is little time for the technician to see that a high-current situation has occurred. A compressor motor cycling on and off on its internal overload will overheat the motor the more it cycles on and off. If the motor has been short cycling for a period of time, it will take longer to cool and reset. Missing this high-amperage reading may mean that the compressor will come back on for a long period of time. This slows down the troubleshooting process. Not having the clamp-on ammeter and gauges hooked up at the beginning of the troubleshooting process will prevent the technician from immediately knowing why the compressor short cycled. Was it an overamperage compressor condition? Was it a bad connection that caused the compressor to short cycle? Did it cut out on a high- or low-pressure switch? As you can see, having the instruments hooked up prior to starting is important.

Finally, the important thing to remember is that you need to take control of the condensing unit so that the unit can be turned on and off at your command and when you are ready. The first few seconds of operation can tell you about the behavior of the equipment and lead to quick diagnostic information. Allowing the system to control itself using its thermostat or safety devices will prolong the time required to find the final solution.

BLADES OR
KNIVES MAY
NOT RESET
WHEN CLOSED
OR RELEASE
WHEN OPENED

Figure 26-5 The disconnect is shown in the open position. Be careful—sometimes one or more knife blades stick in the closed position or do not reset.

SAFETY TIP

The disconnect is off when the lever is in the down position, as shown in Figure 26-5. This is a tip worth repeating: Do not assume that the knife connections in the disconnect box open the circuit. Sometimes one or more of the knife connections stick in the closed position and do not remove power from the circuit. Do a visual check to ensure that the circuit is truly open.

TECH TIP

It is always important to listen to the customer! When arriving on a service call, ask the customer to explain the problem. This is an icebreaking measure to get the customer comfortable with you, plus it allows the customer to explain the problem and any history behind the problem. Keep in mind, however, that totally relying on the customer's outlook may lead you down the wrong path. Be cautious about believing everything they say. Double check and verify a customer's story by reviewing old invoices from previous service calls. If your company serviced the equipment, check the service records. Customers are trying to be helpful, but sometimes they are misinformed or lack the understanding to communicate the whole story to you.

26.3 USING EQUIPMENT DIAGNOSTICS

The use of system diagnostics is commonplace in today's modern HVACR systems. The diagnostics can be found by counting blinking lights on a circuit board, reading digital displays on thermometers, or using external diagnostics engineered and sold especially for a piece of equipment.

CONDENSER MODULE ABC
"THE TROUBLESHOOTER"

10

TROUBLESHOOTING CODE. A PROBLEM IS INDICATED WHEN THE LIGHT BLINKS. THE FOLLOWING IS THE AREA SUSPECTED TO BE A PROBLEM. THE LIGHT WILL BLINK FOLLOWED BY A 10-SECOND PAUSE.

1 = NO TRANSFORMER
⋮
10 = LOW-PRESSURE CONDITION

Figure 26-8 Some newer condensers and air handlers have technology that assists a tech in troubleshooting. A number, which refers to a problem, appears on a module or circuit board. There will also be a legend that explains what the number means.

Figure 26-9 The total current draw of an electric heating system can be found by measuring amperage at the disconnect.

check the total amperage of the heat strips combined. The total amperage can be checked at the disconnect (see Figure 26-9), which will include a couple of amps for the blower motor. The motor wattage is also converted to heat, which is supplied to heat the space.

Electric heating systems use contactors or sequencers to activate the electric strip heaters. The contactors will activate the heat strips immediately upon being energized. The sequence device uses a time delay to bring on the heat strips and to deactivate the heat strip. A sequencer is shown in Figure 26-10. The sequencer allows the heat strips to come on one at a time or in pairs. This time-delay sequence prevents an electrical surge on the power supply provided to the furnace and is required by electrical codes. If there is adequate amp draw on all present heat strips, then the system is performing to its maximum heating capability.

Remember: Give the sequencer time to energize all heat strips. Some sequencers take a couple minutes to close the contacts and provide power to the heaters. The sequencer is used to sequence in the heat strips. The sequencer will turn on the blower and the first heat strip. The second set of heat strips will energize in about 1 minute. The sequence goes in reverse when the sequence is de-energized, with the last heat strip on becoming the first heat strip off. The fan motor is the last circuit to open to ensure cooling of the heat strips.

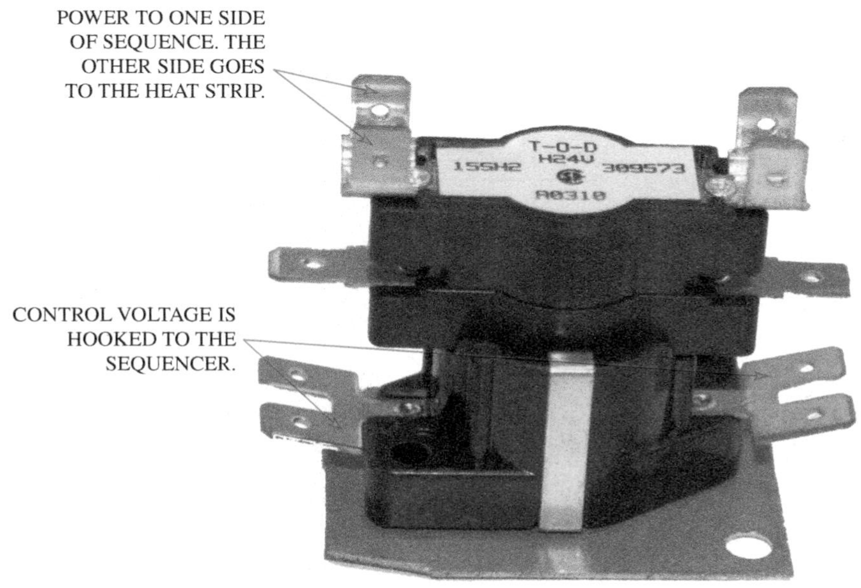
POWER TO ONE SIDE OF SEQUENCE. THE OTHER SIDE GOES TO THE HEAT STRIP.
CONTROL VOLTAGE IS HOOKED TO THE SEQUENCER.

Figure 26-10 The sequencer is used to stage in the heat strips.

SEQUENCE OF OPERATION 26.1: ELECTRIC STRIP HEAT

Figure 26-11 is a diagram showing the sequence of operations for a heating system. (For a larger view of this figure, see Electrical Diagram ED-12, which appears with the Electrical Diagrams package that accompanies this book.) Here is how this heating circuit works:

1. The t'stat calls for heat, closing the thermostat **TH** contacts (far left side of line 6). This provides a 24-V path to the sequencer coil **SEQ 1**. There may be a short time delay to close sequencer contacts **1** and **2**, **3** and **4**, and **9** and **7.** All three sets of sequencer contacts are identified

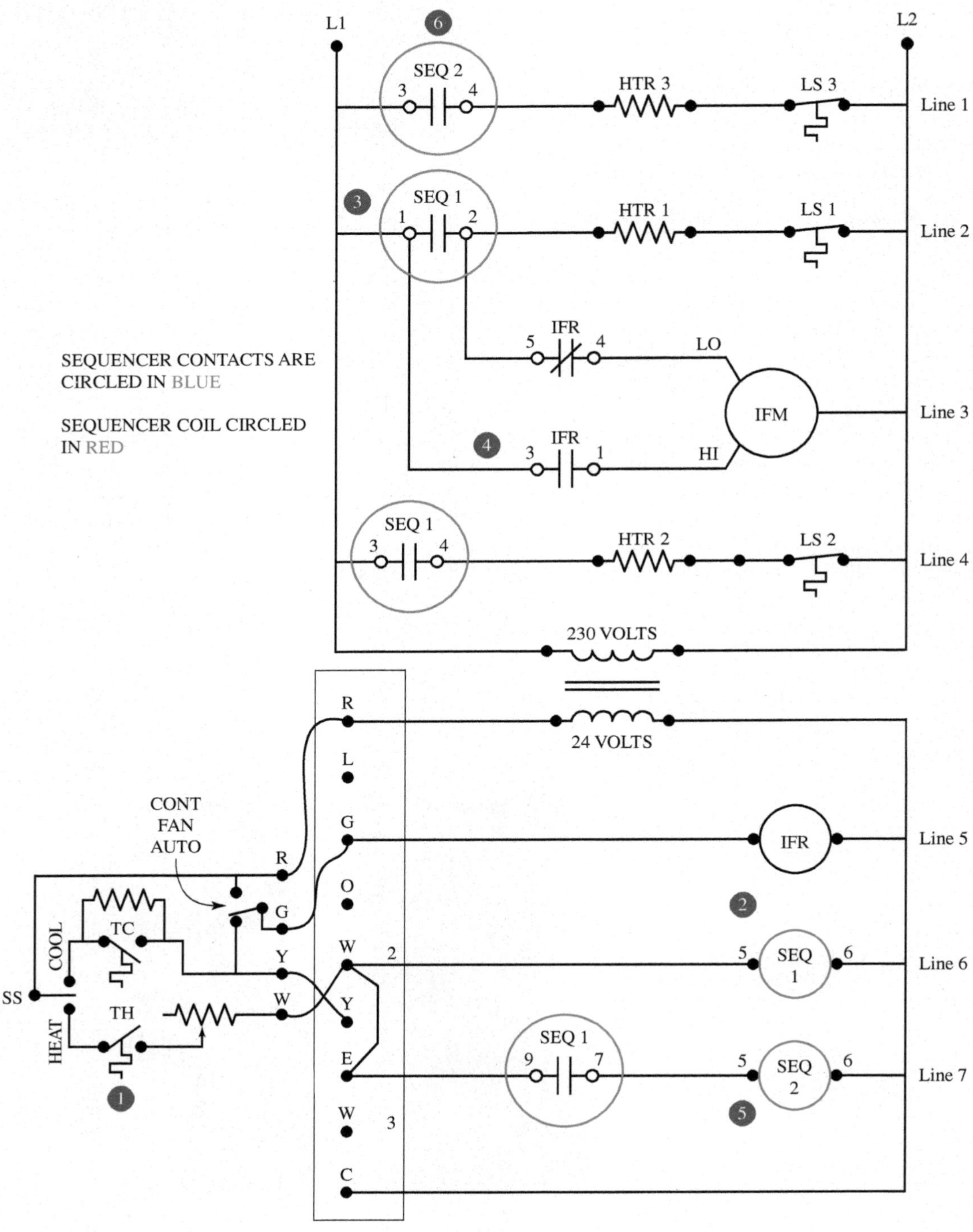

Figure 26-11 This is a sequence time-delay electric heating system.

by a blue circle. The sequencer coils are identified by a red circle.

2 This electric heating system uses two sequencers to control the indoor fan motor (IFM) and three heat strips (**HTR**). The two sequencer coils are labeled (SEQ 1) and (SEQ 2). Line 7 identifies the (SEQ 2) coil found between points **5** and **6**.

When the (SEQ 1) coil is energized, it will close the three contacts identified as **SEQ 1**. These contacts are found on the far left sides of lines 2, 4, and 7.

3 On line 2, the first sequencer, **SEQ 1**, closes first and operates fan motor (IFM) through normally closed contacts **5** and **4** on line 3. Simultaneously, two heat strips, **HTR1** and **HTR2** on lines 2 and 4, are energized. The fan needs to operate when the first heat strip is energized. This action also closes a third set of contacts in line 7. Notice that line 7 is in series with the (SEQ 2) coil.

4 Note that the normally closed **IFR** contacts do not change position. The fan will operate on low speed through the normally closed contacts.

5 Now 24 V is applied to (SEQ 2) coil.

6 The second sequencer, **SEQ 2,** closes it contacts **3** and **4** (circled in red on line 1) after a time delay. **SEQ 2** contacts **3** and **4** close and energize **HTR 3**. With these events the heating system should be fully operational.

Note: The (IFR) coil in line 5 will not be energized unless the thermostat is placed in the **Fan On** mode. The **Fan On** mode keeps the indoor blower on at all times and is not controlled by the cooling or heating thermostat setpoint.

If there is no heating, use the following troubleshooting steps:

- Check the amp draw to the heat strips. Each heat strip should draw about 20 amps for a single-phase service. The amount of amp draw will depend on supply voltage: lower voltage, lower amp draw.
- Is there supply voltage? The supply voltage is usually 208 to 240 volts. Checking the amp draw first or measuring the voltage first is a matter of preference and convenience. They can both be done very quickly.
- Check the control voltage, which is normally 24 V. If there is no control voltage, check the transformer, circuit board fuse, thermostat, and control wiring.
- If there is control voltage, in this case 24 volts on the sequence coil, check to see if contacts **1** and **2** and **3** and **4** are open. This can be done by using the voltmeter.
- With voltage applied to the sequence contacts, zero volts will be measure across a closed set of contacts.
- If the applied voltage is measured across the sequencer contacts, the contacts are open. This could mean that the sequence coil is open or the sequence contacts are open. This can be verified by an ohmmeter. The sequencer coil should measure around 100 Ω and the contacts, when energized and closed, less than one 1 Ω. Remove power to the sequencer contacts when measuring resistance across the contacts and coil voltage.
- Remember to remove power from the circuit when checking the resistance of the coil and contacts. Due to the timer-delay nature of the sequencer, the coil will need to be energized for a minute or two before the contacts close.

In summary, troubleshooting electric strip heat is not difficult. Check the amperage draw on each heat strip. Check the input and control voltages. Use the ohmmeter to verify if the sequencer coil and heat strips have continuity. These simple steps will narrow down the problem.

TROUBLESHOOTING 26.1: ELECTRIC HEAT

Troubleshooting an electric strip heat is not difficult. Check the amperage draw on each heat strip. Check the input and control voltages. Use the ohmmeter to verify if the sequencer coil and heat strips have continuity. These simple steps will narrow down the problem:

1 Check the amperage of each heat strip, which should be around 20 A. See Figure 26-12.

2 Measure supply voltage to the heating system. See Figure 26-13.

3 Measure control voltage on the sequencer coil. There is 24-V control voltage on the sequencer coil to close the contacts. See Figure 26-14.

4 Measure applied voltage across the sequencer contacts. Voltage across the sequencer contacts means that the contacts are open and no heating will occur. See Figure 26-15.

(continued)

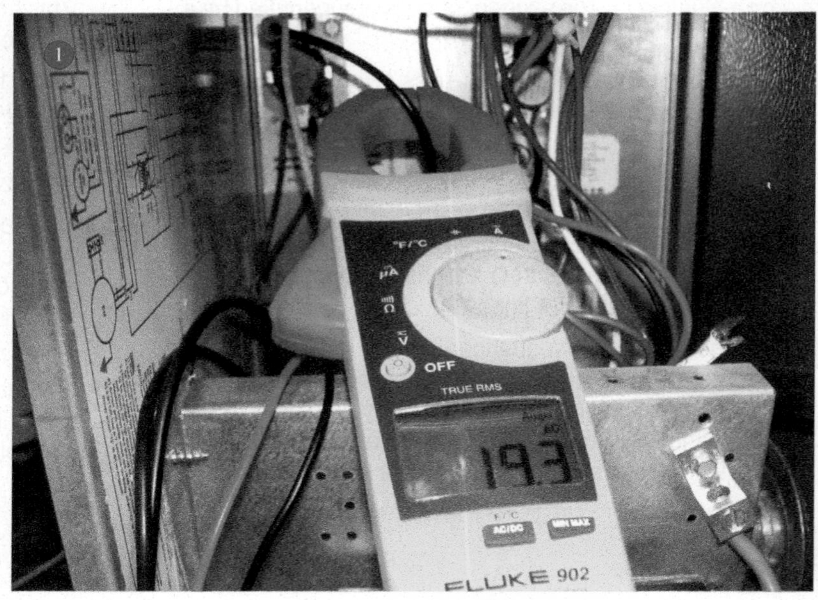

Figure 26-12 Measuring amperage.

Figure 26-13 Measuring voltage.

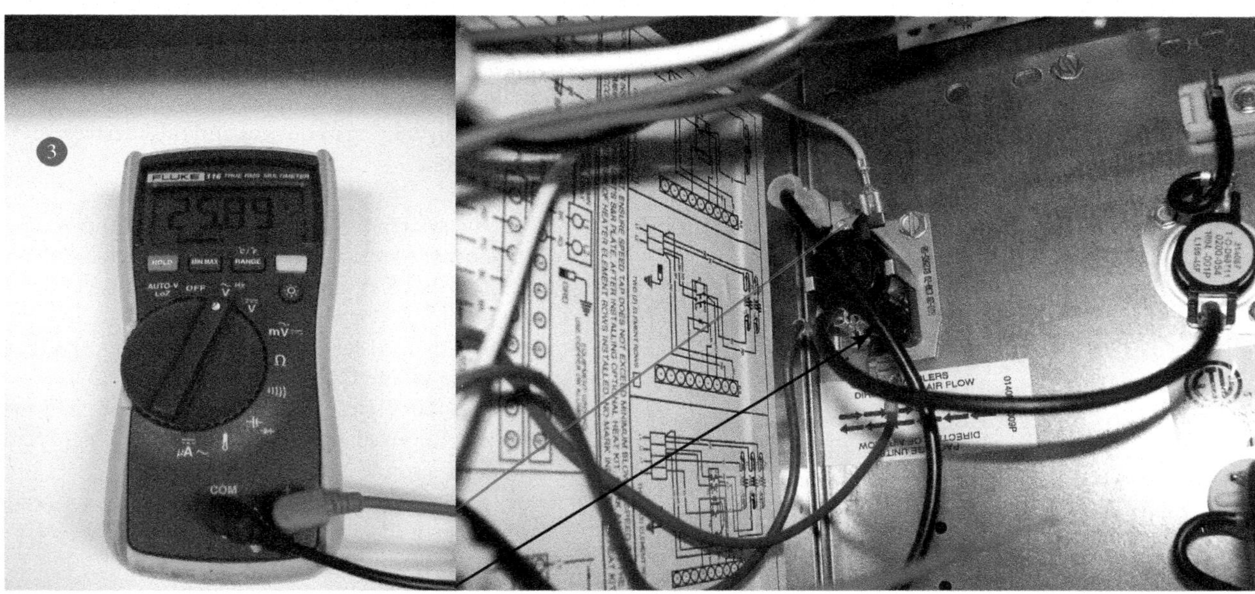

Figure 26-14 Measuring control voltage on the sequencer coil.

Figure 26-15 Measuring applied voltage across the sequencer contacts.

26.5 TROUBLESHOOTING HEAT PUMP HEATING SYSTEMS

If a heat pump system has a heating problem, use the ACT method to establish what is working. Take control of the system by turning off the disconnect, as shown in Figure 26-1 or removing one of the control voltage wires to the contactor, as shown in Figure 26-3. Like the air conditioning system troubleshooting process, hook up the gauges and clamp-on ammeter to the common of the compressor. If it is an outdoor unit heating problem (i.e., if it is the outdoor section that does not operate), use the steps discussed in the previous air conditioning section to find the problem.

If the indoor section does not operate and the outdoor does, concentrate on the indoor section including the thermostat.

Other quick checks should include the following:

- Is there at least a 100-psi high to low pressure differential? A 100-psi high to low pressure differential is needed in order to shift the reversing valve. When the heat pump is operating, there should at least a 100-psi difference between the high side and the low side.
- Is the reversing valve energized in the heating or cooling mode? Check the wiring diagram or thermostat. The **B** connection designates that the reversing valve is energized in the heating mode. Check to see if there is a **B** connection on the wiring diagram or if the thermostat **B** connection is hooked up to energize the reversing valve in the heating mode. The **O** connection designates that the reversing valve is energized in the cooling mode. The **O** connection on the wiring diagram or thermostat is hooked up to energize the reversing valve in the cooling mode. Both **B** and **O** connections are not wired on the same unit; a heat pump will use either a **B** or **O** terminal.
- Does the heat pump system operate in the cooling mode?
- Does the cooling cycle operate when the thermostat is placed in the cooling mode? The pressures will be lower than normal in the cooling mode since it is operating in winter temperature conditions.
- Is the outdoor coil covered with a thicker than normal frost, indicating a defrost cycle problem? Place the heat pump in the cooling mode to remove the frost on the outdoor coil. Disconnecting the condenser fan will reduce the defrost time.
- Does the circuit board have all voltage inputs? Check 120- or 240-V input. Check the 24-V output of the transformer.
- Does the circuit board have voltage outputs? Voltage to fan motor?
- Does the circuit board have a test sequence? Use the test sequence to check heat pump operation. The test sequence runs the heat pump through an accelerated set of operational checks including the defrost cycle.
- Are all safety devices closed? Detailed operation of heat pumps is discussed Unit 24.

26.6 TROUBLESHOOTING FOSSIL FUEL HEATING SYSTEMS

The primary fossil fuel heating systems are gas and oil. Coal and wood heating are also used as fossil fuels, but are not covered in this book. Use the ACT method to troubleshoot and determine what is operating and not operating. Fossil fuel heating systems are more complex compared to electric heat systems or the operation of conventional air conditioning systems.

Fossil Fuel Heating System Checklist

In addition to using the ACT method of troubleshooting for a quick overview of the problem, several items should be checked when a fossil fuel furnace is not working. In this case we are primarily talking about gas and oil furnaces. Here are some suggestions:

- Does the system have the correct applied power?
- Is the furnace service access door tightly closed? The furnace door has a switch that will kill the power to the furnace if it is open or not tightly closed.
- Does the thermostat have 24 volts between the thermostat on the **W** and **C** (common)?
- Does the 120-V circuit have the correct hot and neutral polarity? Reversed polarity can cause erratic operation or no operation.
- Does the furnace have a good ground? The lack of good ground can lead to erratic operation.
- Does the furnace have a good fuel supply?
- Is the main fuel valve open? Is the furnace gas valve on?
- Is the fuel pressure (gas or oil) adequate?
- Is the furnace locked out?
- Is the circuit board indicating a problem via flashing lights or a digital display?
- Is the circuit board indicating that there is a heat exchanger overheating problem? If so, check the air filter. Is it dirty? Check for restricted ductwork or a restricted or dirty evaporator coil. Check the blower motor and motor capacitor. Check anything that restricts the airflow and causes the heat exchanger to overheat. An overheated heat exchanger is a serious event and may lock out the circuit and prevent it from operating.
- Is the circuit registering a venting problem? If so, check the venting system for restrictions. Restrictions in the vent system are difficult to find. First remove the vent at the furnace and inspect the complete venting system. The restriction could be at the top of the venting system. Check the inducer draft motor for operation. A venting problem is a serious problem. It will cause the system to lock out and prevent furnace operation.
- Does the circuit board have all voltage inputs? Check 120- or 240-V input. Check the 24-V output of the transformer.
- Does the circuit board have voltage outputs? Voltage to the fan motor? Twenty-four volts to the gas valve?
- Does the circuit board have a test sequence? Use the test sequence to check furnace operation.
- Is the venting system clear and free of obstructions?
- Are all safety devices closed?

26.7 USING A FLOWCHART

The flowchart in Figure 26-16 is used to guide the technician through a number of steps to determine the problem in a simple air handler. It is a useful tool because it is designed for

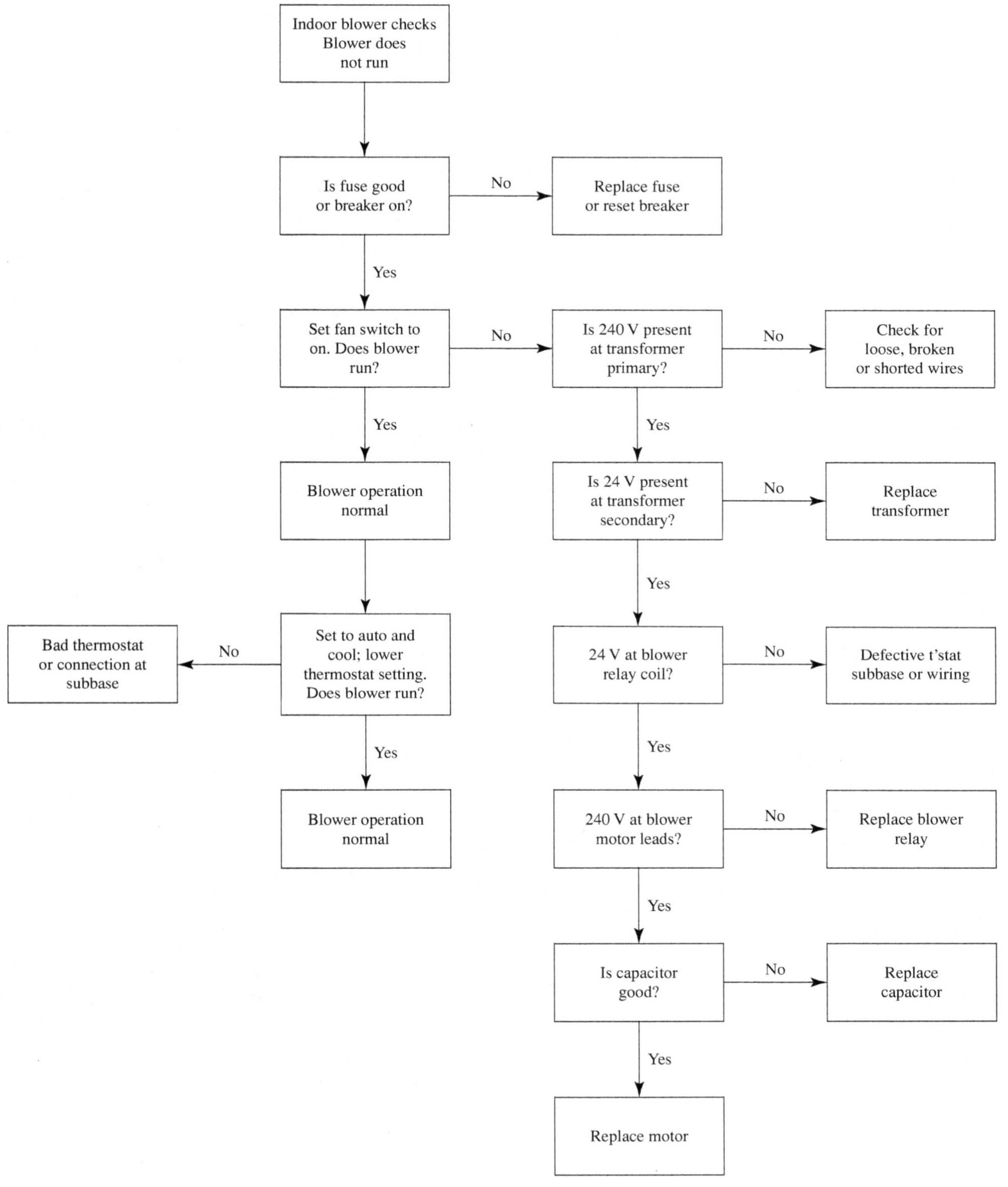

Figure 26-16 Troubleshooting flowchart for a simple air handler. This is a fault-isolation diagram that starts with an initial question. The chart guides the user through a logical troubleshooting process as the questions in each box are answered yes or no.

SYMPTOM	PROBABLE CAUSE	REMEDY
1. IFM NOT RUNNING	1. LINE FUSE(S) BAD; IFC BAD; IFR NOT PICKING UP	1. REPLACE FUSE(S); CHECK IFC; CHECK –A/C SWITCH –THERMOSTAT
2. ETC.	2. ETC.	2. ETC.

Figure 26-17 The troubleshooting table is used to assist the technician in finding a problem in an organized manner.

a specific piece of equipment. Some flowcharts are general in nature and can be used with numerous units. Other flowcharts are specific and are only useful for a specific manufacturer's model number.

26.8 TROUBLESHOOTING TABLE

Another way to organize troubleshooting is to use a manufacturer's table that is specific to a model number they produce. The table gives the technician the symptom, the probable cause, and a suggested remedy. Figure 26-17 shows a sample troubleshooting table that will be useful to technicians. These tables are normally found in the installation instructions and are not found on the job after the installation crew finishes its job.

SERVICE TICKET

You arrive on a "no cooling" service call at a small business office. The office is warm and the customer is anxious to get the cooling system working again to keep her employees comfortable. You quickly realize that the condensing unit part of the package unit is not operating. The indoor air handler is moving air and the thermostat is set correctly.

The next thing you do is determine why the condenser section is not operating. You take control of the system by turning off the disconnect to the package unit while doing a visual inspection. A visual inspection does not reveal any problems. The disconnect is turned on so that additional troubleshooting can be conducted. The unit comes on immediately, runs for a minute, then the condenser section shuts down. You wait about 10 minutes thinking the unit might come back on, but it does not. You turn the package unit disconnect off a second time, look around, but do not find anything unusual. After resetting the disconnect, the unit comes back on for about 2 minutes, then shuts down. Looking at the electrical diagram (see Figure 26-18), you realize that the unit has a lockout circuit, which stops the operation of the system if one of the following safety devices is open:

- Low-pressure switch
- High-pressure switch
- Compressor start overload contacts.

The lockout circuit was preventing the system from operating since it was low on charge. You could have saved some troubleshooting time if you had installed the manifold gauge set when arriving on the job. After starting the unit, you would have observed the suction pressure dropping within the first few moments of operation. Installing the gauges and clamp-on ammeter while taking power control of the unit is an important first step to troubleshooting.

SUMMARY

When troubleshooting define the problem, locate the problem, and fix it. Prior to troubleshooting ask yourself the following questions:

- What should the unit be doing?
- What is the unit doing?
- From this you will be able to determine: What is the unit *not* doing?

Knowing what is supposed to happen in a HVACR circuit is important to the troubleshooting process. Unfortunately, knowing exactly how the circuit is supposed to operate is not always possible.

It is important to realize that most problems are not difficult to solve. By following the steps in this unit, a technician can quickly check many of the common problems found in HVACR systems. Even the most experienced tech requires help from time to time. First, try to solve the problem without help through the use of the logical troubleshooting sequence you have chosen. Your "logical sequence" may be different from others' logical sequence. You need to select a system that is right for you and one that will help you figure things out.

REVIEW QUESTIONS

1. What are three questions you need to ask yourself prior to troubleshooting?
2. Why should you not stand in front of a compressor's terminals when starting it?
3. What is the first troubleshooting step a technician should take when working on a condensing unit that is not operating?
4. What are the second and third steps in the troubleshooting of a nonoperating condensing unit?
5. What are two exceptions to automatically installing gauges on a nonworking air conditioning or refrigeration system?

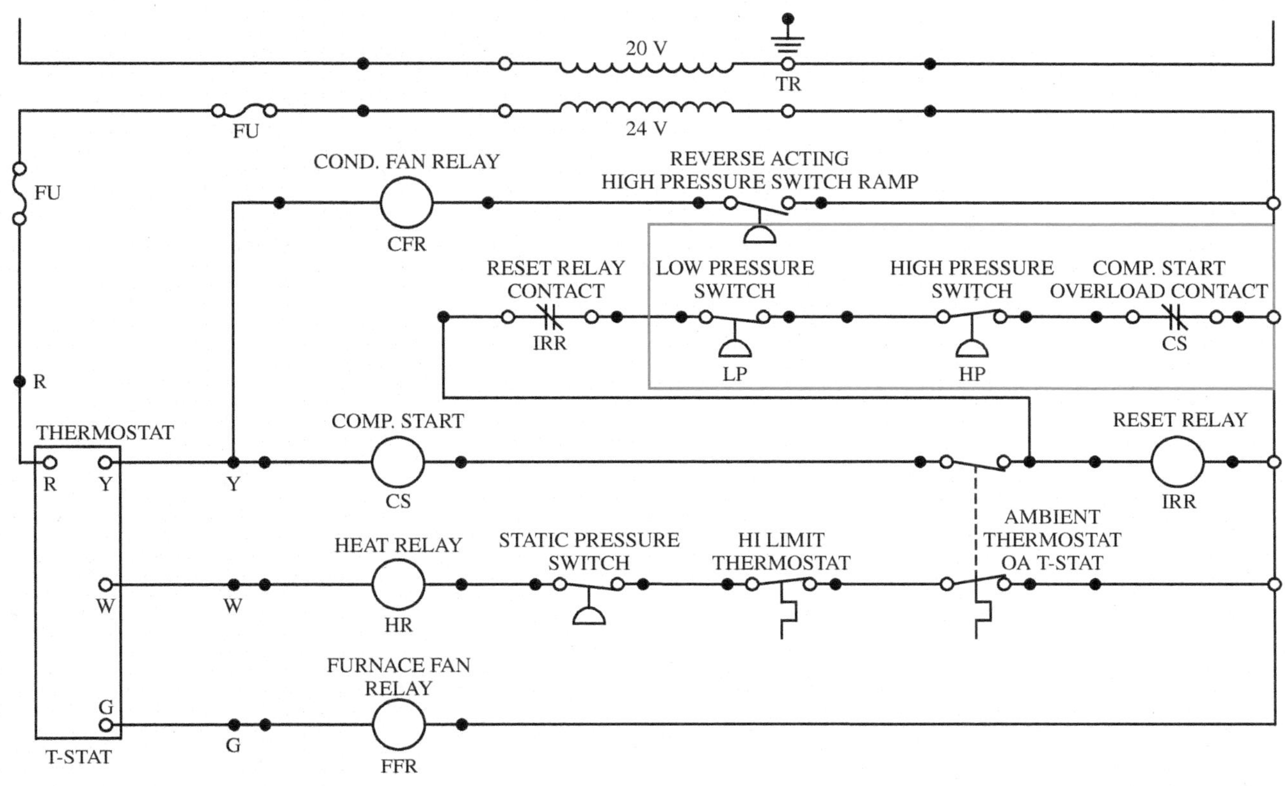

Figure 26-18 The lockout circuit is composed of the reset relay coil (**IRR**) and the normally closed reset relay contacts. The lockout circuit prevents the system from operating when one of the safety devices is open. The safety devices in this diagram, highlighted inside the red box, are the low-pressure switch (**LP**), the high-pressure switch (**HP**), and the compressor start overload contact (**CS**).

6. When troubleshooting, what instruments does the technician need to be watching when a compressor is first started?
7. The tech notices that a furnace is in the operating mode. Opening the furnace panel, a flashing light is seen that is flashing five times, followed by a delay, followed by five more flashes, etc. In general, what has the tech found?
8. What is the amperage range of a single-phase heat strip operating around 240 volts?
9. An electrical heater has four 5-kW heat strips. The customer has asked you to check its operation. The complaint is that when the heater cuts on, the lights in the house dim.
10. What basic steps are used to troubleshoot electric heat strips?
11. You are doing a service call on a heat pump that has a heating problem. What steps should you take to set up the troubleshooting process?
12. What are nine quick things you can check on a heat pump in the heating mode?
13. You are troubleshooting a fossil fuel furnace with a heating problem. What 17 items should be checked to narrow down or determine the problem?
14. What is the purpose of a troubleshooting flowchart?
15. What is the purpose of a manufacturer's troubleshooting table?
16. What control voltage would you expect in Figure 26-19?
17. You are using Figure 26-19 to troubleshoot a cooling problem. The compressor and condenser fan are not operating. The indoor fan motor is operating. At the beginning of the troubleshooting process, you attach the manifold gauge set and clamp-on ammeter to the compressor common terminal wire. The system pressure is flat, no refrigerant. What is the problem and what is keeping the system compressor from operating?

Use Figure 26-19 for the following questions:

18. What type of motor is the compressor motor?
19. Redraw the compressor motor and capacitor. Label C, S and R.
20. CFM is not identified in the Legend. What is the CFM?

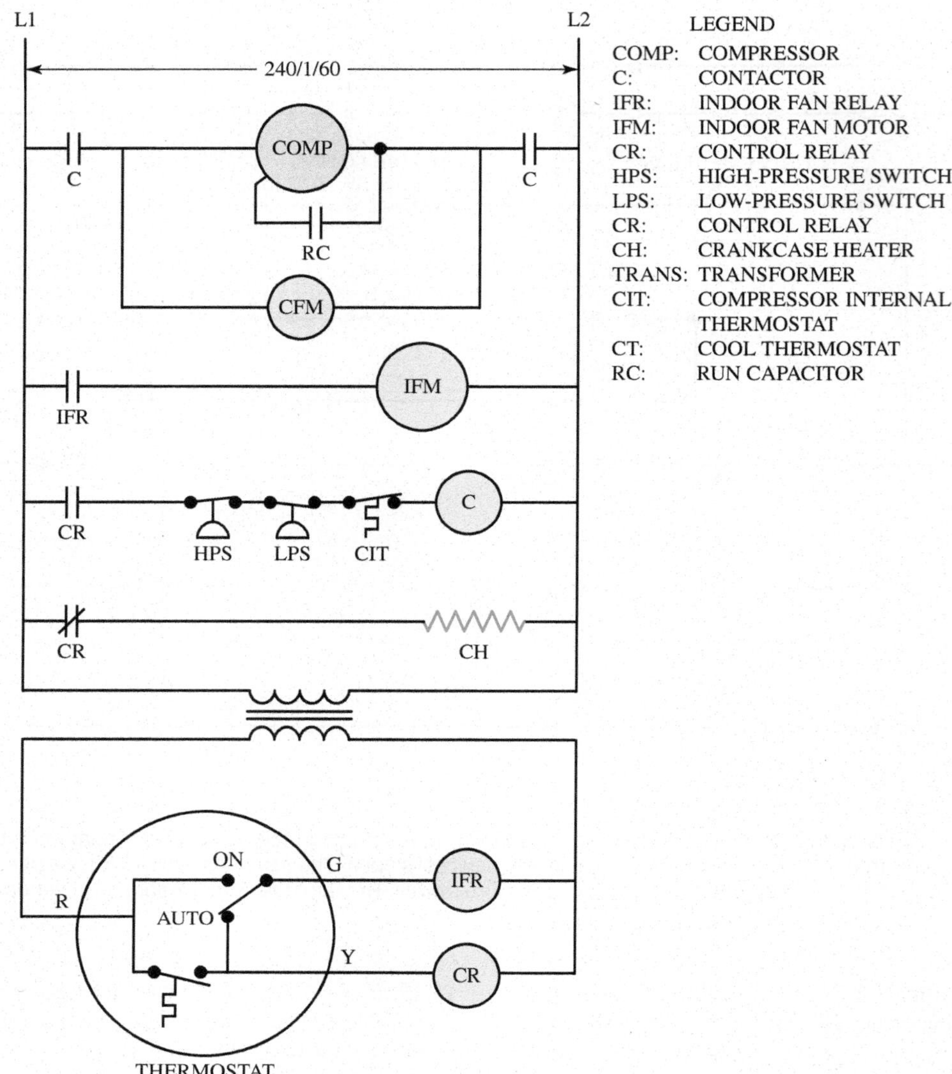

Figure 26-19 Use this diagram to answer Review Questions 16 and 17.

UNIT 27
Advanced Troubleshooting

WHAT YOU NEED TO KNOW

After studying this unit, you will be able to:

1. Explain the purpose of using the ACT method of troubleshooting.
2. Troubleshoot a compressor.
3. Determine how to troubleshoot a short circuit.
4. Use manufacturers' compressor tables to determine amp draw.
5. Use a hard start table to select a start capacitor and potential relay.
6. Calculate current imbalances and voltage imbalances.

This unit is an extension of the previous units on troubleshooting. As you may have noticed, this book uses the building block process of learning a skill such as troubleshooting. Early units discussed the individual components and how to troubleshoot them. Units 25 and 26 gave you tools to advance you from the "component level" to the "system level" of the troubleshooting process.

This unit starts with the important point of beginning the troubleshooting process. Again, a brief review of the ACT process is appropriate. Troubleshooting is a very individual process and you may prefer some other way of determining the vital signs of a system. There is no one right way to start the troubleshooting process. Your method of troubleshooting will evolve with your experience and the type of equipment you come across during your field service learning experiences.

This unit also discusses the importance of the manufacturer's notes usually found on electrical diagrams or installation manuals. It is best to sit down and read through all installation instructions when you are exposed to a piece of equipment you have not seen before.

This unit goes into detailed information on how to troubleshoot compressors. This is an important troubleshooting tool because 24% to 30% of compressors are returned under warranty with no problem found. Compressor tables are discussed. These tables provide useful information for determining compressor amperage at different operating conditions.

Selecting the right start components is important if the compressor is to start under unequal pressures or unstable voltages. This unit discusses how to use a hard start selection table to determine the correct start capacitor and potential relay for a specific compressor motor.

Finally, the unit discusses how to measure and calculate voltage and current imbalances.

This unit is packed with troubleshooting tips and ideas. Your troubleshooting skills will be tested in the next unit on practical troubleshooting.

27.1 REVIEW OF TROUBLESHOOTING STEPS

Prior to starting troubleshooting we will review common practices that will aid in the solution process. The first step is to use the **ACT** troubleshooting method to determine what is working and what is not working.

The abbreviation **ACT** is used to determine if the major components in a HVACR system are operating. The letters in the abbreviation **ACT** are used as a reminder to check for the following:

- **A** is for airflow. Is there airflow from the air handlers and condensing unit? Is the airflow from the register vents cold (cooling) or warm (heating)? Is the condenser rejecting heat to the air? Heat should be coming off the outlet of the condenser coil because it is removing heat from the refrigerant. This is a sign that the cooling circuit is working properly.
- **C** is for compressor. Is the compressor on? Is there a pressure difference between the high side and low side? Is the compressor drawing amps?
- **T** is for thermostat. Is the thermostat set to what you are troubleshooting? For example, if it is an air conditioning problem is the t'stat set for cooling? Is the temperature set 5°F below or above the room setpoint so that the unit will come on?

The order in which you do the **ACT** sequence is not important. You can start with **T** or **C** first if it makes sense for the job you are doing. The **ACT** troubleshooting starting point will help you quickly check key points on a cooling or heating system.

TROUBLESHOOTING 27.1: THE ACT METHOD

Follow these steps, in any order, to determine what is working and what is not working.

1. Check the supply airflow system. Is there airflow? Is it cool or warm? Is the condenser expelling warm air? See Figure 27-1.
2. Check compressor operations. The gauges are hooked to the low side and high side of the system. In Figure 27-2 the gauge pressures are equalized at 85 psi, therefore the compressor is not operating.
3. Set the thermostat for the heating or cooling system you are checking. The thermostat should be set 5°F below or above the room temperature. See Figure 27-3.

Figure 27-2 Check compressor operations.

Figure 27-1 Check the supply airflow system.

Figure 27-3 Set the thermostat.

The next troubleshooting steps are:

1. Diagnosis
2. Isolation
3. Correction of the fault.

Or stated another way:

- Define the problem.
- Locate the problem.
- Fix the problem.

In summary, electrical troubleshooting is a logical process of elimination. To effectively diagnose a problem, you must understand most of the system electrically and mechanically.

In the ideal world of HVACR troubleshooting, you should know the following:

- What is the system supposed to do and how is it supposed to do it?
- What is the system doing?
- What is the system *not* doing?
- The function of each component used in the electrical circuit.
- The physical location of each component used in the equipment.
- The interrelationship of components.
- Electrical symbols.

You should also be able to read a wiring diagram to determine the sequence of operations.

Even if you know only some of these indicators, you will be able to make some decisions about the problem that will lead to the solution or help in getting the solution.

27.2 MANUFACTURERS' NOTES EXPLAINED

This section explains what the manufacturer means by the "notes" listed on their electrical diagrams. The notes are usually found at the bottom of the diagram near the legend. Notes are found on Electrical Diagrams ED-10 and ED-11, which can be found with the Electrical Diagrams package that accompanies this text. Listed here are "notes" found on some common cooling diagrams. The **bold sentence** is used in the Notes section. An explanation of the meaning of each note follows the **bold sentence**. Let's start with the first note that you might find on a diagram:

1. **Symbols are electrical representations only.**
   The symbols are not drawings of the components. A pictorial diagram would show an image of what the component actually looks like.
2. **Inherent thermal protection means that the motor has an internal overload.**
   If the motor windings get too hot or if the amp draw is excessive, the overload will open. On single-phase motors the overload is located in series with the common line, inside the compressor. The overload will close when the motor windings cool or the amp draw is reduced to a safe level.
3. **The National Electrical Code (NEC) is an electrical standard used for safe equipment operation.**
   The NEC is a minimum standard code. Some local code officials modify the NEC to meet local requirements. Local code modifications usually allow another safe way of doing an electrical task.
4. **This note states that the minimum 24-V transformer rating is 40 VA. If a liquid line solenoid (LLS) is installed in the control circuit, a 60-VA transformer is required.**
   Standard air conditioning controls do not use a liquid line solenoid. The solenoid adds an extra load or amp draw to the transformer, thus requiring the larger 60-VA transformer. An undersized transformer will overheat and burn out. A control transformer should not run hot. Touch it when it is energized after 5 minutes. A hot transformer will soon burn out.
5. **Only use copper conductors, never aluminum conductors.**
   The temperature of a wire is commonly measured in degrees Celsius. The 75°C (167°F) requirement will safely allow a wire to be in conduit and travel through warm spaces such as an attic area. Wires in conduit get warmer than those exposed to open air. Hot wires cannot carry as much current.
6. **This note only has information on cooling installation.**
   Refer to the furnace installation instructions for heating or other accessories' installations.
7. **If indoor section has a transformer with a grounded secondary, connect the grounded side to the BRN/YEL lead.**
   This is a specific instruction for the grounded secondary of a transformer. Using a ground secondary side of the transformer requires all grounds to be tight or the control voltage will operate erratically or not at all. Loose grounds present an interesting problem because one time a control will work, but the next time it will not.
8. **When start relay and start capacitor are used, start thermistor is not used.**
   The start relay or potential relay and start capacitor are used as a hard start kit to start compressors. A start thermistor is a solid-state device used as an assist to start compressors. The thermistor does not create as much torque as a correctly sized hard start kit. The thermistor and start kit are not used in conjunction with one another. It is one or the other. A manufacturer's hard start kit is best.
9. **The CH means crankcase heater.**
   Not all equipment has these compressor oil heaters that keep refrigerant from condensing in the crankcase. They can be added in the field.
10. **Replace damaged or burned wire with the same gauge and same temperature conductors.**
    A larger wire size can be used. This would be a lower gauge number. For example, if a 12 AWG is needed, a larger 10 AWG can be used instead. Try to replace the wire with the same insulation color to make it easier for future troubleshooting. If one end of the wire is burned or damaged, replace the whole piece of wire, not just the burned end. Many times the overheated wire has internal damage that is not detected and will create further overheating problems.
11. **Electrical connections can become loose in shipment.**
    Check the integrity of all connections. Loose connections cause overheating and high amp draw in the wire. This will burn out the connection.
12. **Operating the compressor with closed suction and/or liquid service valves will damage the compressor.**
    Leaving the service valves closed while operating the compressor will do valve damage. Without the flow of refrigerant to cool the motor windings, the motor windings may overheat and burn. At best they will open on internal overload due to excessive winding heat.
13. **Starting the compressor without time to equalize pressures will create high starting amperage conditions, which will shorten the compressor winding life.**
    Every time a motor starts, it draws high amperage known as locked rotor amps or LRA. High amperage creates heat, which shortens the compressor motor life. Compressors started with unequal pressures normally have hard start kits or they are three-phase motors. This design will reduce high starting current conditions.
14. **Wire not present if HPS, LPS, or CTD is used.**
    According to the notes these safety devices are not factory wired. The high-pressure switch, low-pressure

TROUBLESHOOTING 27.2: FAST TRACK TO TROUBLESHOOTING COMPRESSORS

Oftentimes it is difficult to determine if a compressor is defective. About 30% of compressors replaced under warranty are found to have no problem when returned to their manufacturer. Here are some indicators to determine if a compressor is defective:

1. When the compressor is operating, do the high- and low-side pressures seem to be within range, as seen on the R-410A unit shown in Figure 27-4? If so, the compressor is most likely not the problem.

2. Is the compressor amperage equal to or lower than the rated load amps? If so, the compressor is most likely not the problem. Very low amp draw indicates compressor mechanical problems. Make certain that the compressor is not unloaded, which means operating at reduced capacity. The compressor in Figure 27-5 has a 10-amp RLA rating and is drawing adequate amperage, but not excessively low. A low amp draw would be about 60% lower than RLA. Low amp draw can also

Figure 27-4 A pressure differential indicates that the compressor is running.

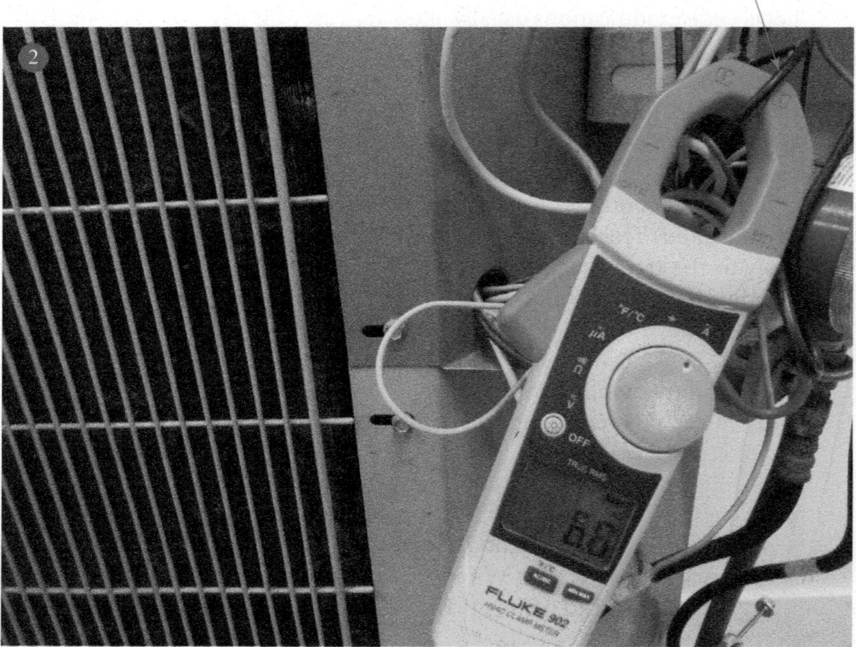

Figure 27-5 A 6.8-amp reading on the common lead indicates the small compressor is operating.

mean a low charge condition, but not a compressor problem.

3 Is the operating compressor supplied voltage within 10% of the required voltage? Low voltage will cause the motor to draw high amperage. The operating air conditioning system shown in Figure 27-6 is reading within ±10% of the rated voltage.

4 Use the manufacturer's specification sheet to determine if the compressor is operating within the designed-for pressure and amperage ratings. Figure 27-7 shows the inside cover of a condensing unit. The "spec" sheet shows the electrical diagram, charging recommendation, and amp rating. Electrical voltage and ampacity can also be found on the unit's nameplate.

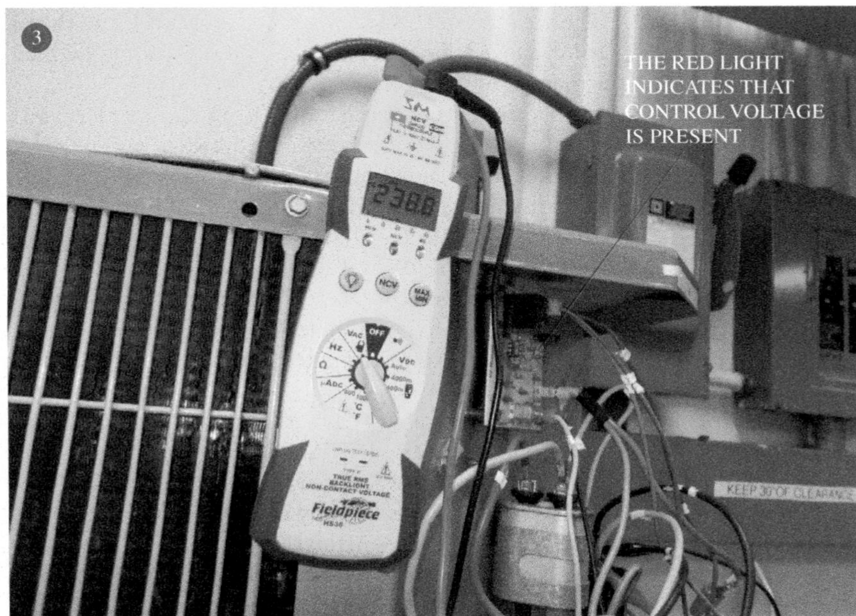

Figure 27-6 The VOM 238.8-V reading is within the ±10% input voltage tolerance.

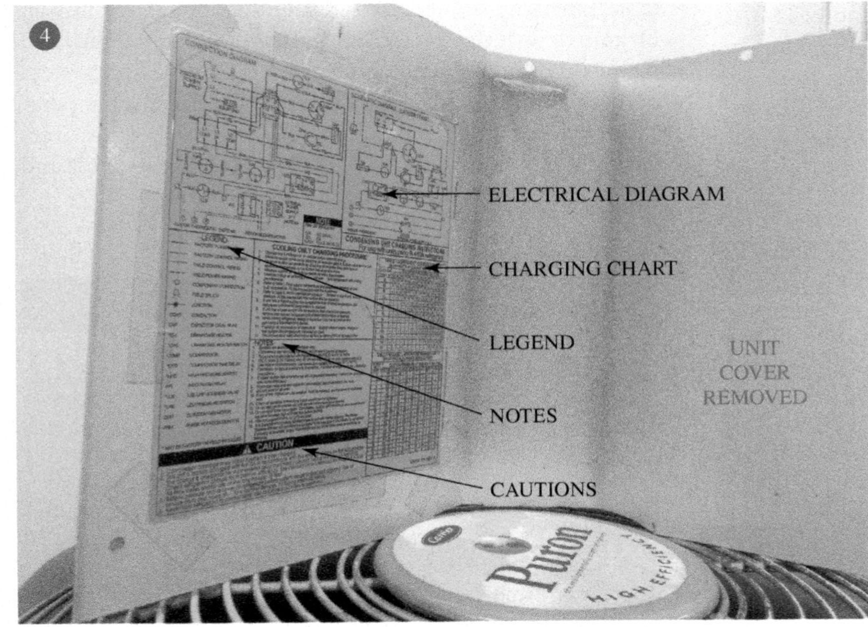

Figure 27-7 Information inside the unit cover is valuable when servicing and troubleshooting a unit.

switch, or compressor temperature discharge switch will need to be field wired. Sometimes the manufacturer will place an asterisk (*) on the diagram to indicate when something is an optional component.

Now let's practice our troubleshooting techniques. Even if you do not get any of these service calls correct, you will learn from the problems and their answers. You will start improving your troubleshooting skills and techniques.

The first three indicators should be used to determine if a compressor has an electrical or mechanical problem. The fourth indicator is the most accurate troubleshooting method, but it is less likely to be used because the manufacturer's spec sheet is not always readily available. In some instances it can be found on the Internet, but the technician is usually in a hurry to come up with a diagnosis and will not spend the time searching for the information. The compressor spec sheet database compares the amperage reading at the measured pressures. Because of voltage variations and measurement inaccuracies, the measured amperage should be compared to the actual curve sheet values within ±15%. This can vary among manufacturers.

Copeland Compressors states not to close off the suction valve to see how low the suction pressure will go. The company states that this "might cause damage to the compressor because of heat buildup" due to a lack of cooling refrigerant. Overheating the motor will cause the motor to open on internal overload or damage the motor windings. Table 27-1 reproduces a manufacturer's compressor data sheet. Some, but not all of this information may be found on the condensing unit or compressor nameplate. Let's review the data circled in red in Table 27-1 for Model No. ZP34K5E-PFV:

- This R-410A model has a 34,500-Btuh capacity.
- Current (amps) is 15.6 amps.
- The LRA is 112 amps.

The RLA listed with the MCC standards is used by Underwriters Laboratories (UL) to test the motor. MCC means maximum continuous current. It is the amount of current that a compressor may pull as it is tested by lowering the supplied voltage until the compressor trips on the protector. It is more of a UL-based safety statistic than a usable service-oriented checkpoint.

Table 27-2 illustrates how the RLA listed in Table 27-1 will vary with the condensing and evaporating temperatures. Let's compare a couple of examples. The **rating conditions** for compressor Model No. ZP34K5E-PFV are found at the upper left side of Table 27-2. They are:

- 20°F superheat
- 15°F subcooling
- 95°F outside ambient air entering the condenser coil.

The top right side of Table 27-2 provides this information:

- 208–230 volts, single phase, and 60 cycle.

The meanings of the letters **C, P, A, M,** and **E** and the **%** symbol in the left-hand column of Table 27-2 are identified at the bottom of the table.

With a condensing temperature of 130°F and an evaporating temperature of 40°F, the amp draw, **A,** will be 15.6, which is the nameplate rating as found in Table 27-1 and the same information at the same rating conditions is found in the red circle in Table 27-2.

Another example using Table 27-2, circled in blue, finds a 12.2 amp draw when the condensing temperature drops to 110°F, at the same evaporating temperature of 40°F. With a 20-psig drop in the condensing temperature, the RLA drops 3.4 amps or about 20%.

The conclusion you should draw from this table is that you should **never** charge the system using the RLA as a major indicator of a good charge. Unless you have this manufacturer's compressor chart as seen here, the amp draw should be lower than the nameplate amp rating—in this case, less than the nameplate rating of 15.6 amps—unless the outdoor temperature and operating head and suction pressure conditions are met.

In summary, you will notice when you use Table 27-2 that as the condensing temperature (far left-hand column) increases, the compressor amp draw increases. As the evaporating temperature increases (columns 2 through 10 in the table), the amp draw decreases slightly. The biggest impact on amp draw is the condensing pressure, which is influenced by the outdoor ambient temperature. An overcharge system or dirty condenser can also increase the condensing temperature, thus causing higher amp draw. High amp draw overheats the motor winding, which shortens compressor life and causes higher operating costs.

TECH TIP

As required by Underwriters Laboratories, the term *rated load amps* (RLA) is a value sometimes shown on a compressor's nameplate. The value is complicated in that it is affected by many factors such as the refrigerant charge, operating voltage, and application. It is best to keep the RLA under the nameplate rating. Never use amp draw as the sole indicator of compressor performance and never use amp draw to determine a charge method. It should be checked to determine if there is a problem.

SAFETY TIP

Check all electrical connections for a tight fit. This includes connections on the compressor terminals, capacitors, start relays, contactor terminals, and disconnect and main electrical panel boxes. Loose connections create high resistance, heat, and high amperage draw conditions. This can cause the compressor to cut out on its overload or trip a circuit breaker. Overheating a motor will shorten its life. Loose connections can also cause a burned terminal connection.

Table 27-1 Manufacturer's Compressor Data Sheet

Model	Refrig.	HP	Comp. Capacity (Btu/hr)	Comp. EER (Btu/Wh)	Evap. Capacity (Btu/hr)	Evap. EER (Btu/wh)	Return Gas Temp. (°F)	Total Subcooling (F)	Volts	ph.	Hz	Current (Amps)	RLA (MCC/1.4) (Amps)	RLA (MCC/1.56) (Amps)	MCC (Amps)	LRA (Amps)	Modulation	Application
ZP34K5E-PFV	R-410A	NA	34,500	10.24	34,500	10.24	65.0	15.0	208-230	1	60	15.6	20.0	17.9	28.0	112	.	Air Conditioning (AC)
ZP36K3E-PFV	R-410A	NA	36,200	10.00	36,200	10.00	65.0	15.0	208-230	1	60	16.6	20.7	18.6	29.0	105	.	Air Conditioning (AC)
ZP36K5E-PFV	R-410A	NA	36,000	10.17	36,000	10.17	65.0	15.0	208-230	1	60	16.3	20.0	17.9	28.0	112	.	Air Conditioning (AC)
ZP38K3E-PFV	R-410A	NA	37,800	10.00	37,800	10.00	65.0	15.0	208-230	1	60	17.3	21.4	19.2	30.0	104	.	Air Conditioning (AC)
ZP38K5E-PFV	R-410A	NA	37,500	10.14	37,500	10.14	65.0	15.0	208-230	1	60	16.9	22.1	19.9	31.0	109	.	Air Conditioning (AC)
ZP39K5E-PFV	R-410A	NA	39,000	10.29	39,000	10.29	65.0	15.0	208-230	1	60	17.5	22.1	19.9	31.0	109	.	Air Conditioning (AC)

(Source Emerson Climate Technologies.)

Table 27-2 Btuh Capacity, Power Use, Current Use, Mass Flow Rate, EER, and Percent of Isentropic Efficiency (%) for a Specific Condensing and Evaporating Temperature.

This is a good table for ensuring that the amp draw of the compressor is correct at the designed conditions.

Rating Conditions
20 °F Superheat
15 F Subcooling
95 °F Ambient Air Over

60 Hz Operation

AIR CONDITIONING

ZP34K5E-PFV
HFC-410A
COPELAND SCROLL®
PFV 208/230-1-60

Condensing Temperature °F (Sat. Dew Pt. Pressure, psig)		Evaporating Temperature °F (Sat. Dew Pt. Pressure, psig)								
		−10.0(36)	0.0(48)	10.0(62)	20.0(78)	30.0(97)	40.0(118)	45.0(130)	50.0(142)	55.0(155)
150.0	C						25,600	28,500	31,600	34,900
(611)	P						4,420	4,390	4,370	4,350
	A						20.2	20.1	20.0	19.9
	M						446	492	540	595
	E						5.8	6.5	7.2	8.0
	%						58.5	61.0	63.2	65.1
	C					22,900	28,400	31,500	34,900	38,400
140.0	P					3,930	3,880	3,850	3,830	3,820
	A					18.0	17.8	17.7	17.6	17.5
(540)	M					371	455	500	550	600
	E					5.8	7.3	8.2	9.1	10.1
	%					57.0	62.5	64.8	66.6	68.1
	C				20,100	25,300	31,200	34,500	38,100	41,800
130.0	P				3,500	3,440	3,400	3,370	3,360	3,340
	A				16.1	15.8	15.6	15.6	15.5	15.4
(475)	M				306	380	463	510	560	610
	E				5.7	7.3	9.2	10.2	11.3	12.5
	%				55.2	61.3	66.2	68.0	69.5	70.4
	C			17,300	22,100	27,500	33,900	37,400	41,100	45,100
120.0	P			3,110	3,060	3,010	2,970	2,950	2,930	2,920
	A			14.4	14.2	14.0	13.8	13.7	13.6	13.6
(417)	M			251	315	388	471	515	565	615
	E			5.6	7.2	9.1	11.4	12.7	14.0	15.5
	%			53.1	59.7	65.2	69.2	70.5	71.4	71.6
	C		14,700	19,000	24,000	29,700	36,400	40,100	44,100	48,300
110.0	P		2,760	2,720	2,670	2,630	2,590	2,580	2,560	2,540
	A		12.9	12.7	12.5	12.3	12.2	12.1	12.0	12.0
(364)	M		203	259	322	394	477	525	570	625
	E		5.3	7.0	9.0	11.3	14.0	15.6	17.2	19.0
	%		50.7	57.7	63.6	68.2	71.1	71.8	71.8	71.2
	C	12,300	16,100	20,500	25,700	31,800	38,800	42,700	46,900	51,500
100.0	P	2,430	2,400	2,370	2,340	2,300	2,270	2,250	2,230	2,210
	A	11.5	11.4	11.2	11.1	10.9	10.8	10.7	10.6	10.6
(316)	M	163	210	265	327	399	482	530	575	630
	E	5.0	6.7	8.7	11.0	13.8	17.1	19.0	21.0	23.2
	%	48.1	55.4	61.6	66.6	70.0	71.4	71.1	70.1	68.3
	C	13,400	17,300	21,900	27,300	33,700	41,000	45,100	49,500	54,000
90.0	P	2,120	2,100	2,080	2,050	2,020	1,990	1,970	1,950	1,930
	A	10.1	10.1	10.0	9.9	9.8	9.6	9.5	9.5	9.4
(273)	M	169	215	269	331	403	486	530	580	635
	E	6.3	8.2	10.6	13.3	16.7	20.6	22.9	25.4	28.0
	%	52.7	59.1	64.3	68.0	69.8	69.2	67.7	65.4	62.0
	C	14,400	18,400	23,200	28,800	35,400	43,100	47,400	52,000	57,000
80.0	P	1,840	1,840	1,830	1,810	1,780	1,750	1,730	1,710	1,690
	A	9.0	9.0	9.0	8.9	8.8	8.6	8.6	8.5	8.4
(235)	M	173	218	271	333	405	488	535	585	635
	E	7.8	10.0	12.7	16.0	19.9	24.6	27.4	30.4	33.6
	%	56.3	61.4	65.3	67.4	67.2	64.0	61.0	56.9	51.5

C: Capacity (Btu/hr), P: Power (W), A: Current (Amps), M: Mass Flow (lb/hr), E: EER (Btu/Wh), %: Isentropic Efficiency (%)

Nominal Performance Values (±5%) based on 72 hours run-in. Subject to change without notice. Current @ 230 V

(Courtesy Emerson Climate Technologies.)

SAFETY TIP

The compressor terminal block can be a weak part of the compressor shell. There is a possibility that the terminals could blow out on the technician when removing or installing wires. Check the compressor wiring away from the terminals. For example, with the power supply disconnected, check the motor winding resistance on the contactor that supplies voltage to the compressor. Check the start winding wire on the compressor side of the run cap. Be aware that the wires feeding the compressor may be shorted, thus causing a problem. Use the following procedure if the compressor windings seem to be shorted:

- Recover the refrigerant in the compressor.
- Remove the wires from the compressor terminals.
- Check the compressor windings again directly on the terminals.
- If the motor windings are not grounded, the wiring leading to the compressor terminals is probably shorted. Check the run cap for a shorted condition also.

This is the safest procedure and most accurate method for checking winding resistance.

27.3 CHECK THE CAPACITOR

Determine the capacitance of the run and start capacitors. Use a capacitor checker. Using the needle deflection of an analog or digital ohmmeter will not determine the correct capacitance of a capacitor. A multimeter will show deflection of a weak capacitor. A weak capacitor may not start a motor. A weak cap will cause the motor to draw higher running amperage.

A run capacitor should be within ±10% of its microfarad rating. Some manufacturers use a lower ±5% range. Most start caps have a microfarad range, for example, 88 to 108 MFD at 250 volts. If the measured rating is outside the rating, change it. If a compressor has been changed, a different capacitor rating may be required. This is sometimes overlooked when a compressor or other motor is changed. Any time a motor is changed, the capacitor should automatically be changed even if the capacitor rating is the same. Many warranties require installing a new capacitor and new start components with every compressor or motor replacement.

27.4 CHECK THE VOLTAGE

Check the voltage with the compressor operating at full-load conditions. Most motors will operate within ±10% of the nameplate voltage. Having a lower voltage will cause the motor to draw excess amperage. Excessive amperage causes motor overheating. For example, a motor rated for 240 volts will have a ±10% operating tolerance of ±24 volts. This will allow the motor to operate within a range of 216 to 264 volts. Many motors are rated for dual operating voltages of 208/240 volts. In this instance the acceptable voltage range will be even wider. The ±10% includes the lower and upper voltage ranges. For a 208/240-volt motor, the voltage range is 187 to 264 volts.

27.5 LOCKED ROTOR AMPS (LRA)

A locked rotor (also known as a locked motor) means the motor rotor will not move. This can be caused by mechanical or electrical problems. An indication of a locked rotor is excessive amp draw on the common terminal of a single-phase motor or any winding of a three-phase motor. The motor draws locked rotor current or LRA. LRA is usually four to six times the rated load amps (RLA) of the motor. This is merely a rule of thumb, so it could be lower or higher.

Some motors will draw locked rotor current when the pressure between the discharge and suction side is not equalized across the compressor. In this case the compressor cannot rotate because of the unequal pressures. Installing a short-cycle timer can solve this problem. This timer allows the pressures 5 minutes to equalize between off cycles.

A final indication of locked rotor is the humming of a compressor without the low- and high-side pressures changing. Without compressor rotor operation the differential pressures cannot develop across the compressor.

TROUBLESHOOTING 27.3: SOLUTION TO A STUCK COMPRESSOR

Before we discuss the troubleshooting options for a stuck compressor, you must check the voltage inputs, wire connections, compressor windings, and compressor components such as the capacitor. Most stuck compressors will be condemned as defective. Here are a few tips to try before giving up:

1. Strike the compressor with a rubber mallet or reduce blow hammer, as shown in Figure 27-8. Take care not to hit the compressor shell so hard that it jars the refrigerant line, causing a leak. In rare cases the compressor gets into a bound position and a little tap will get it moving again. This is worth a try, but rarely works.
2. Install a hard start kit, as shown in Figure 27-9. This includes the correctly matched start cap and potential relay.

(continued)

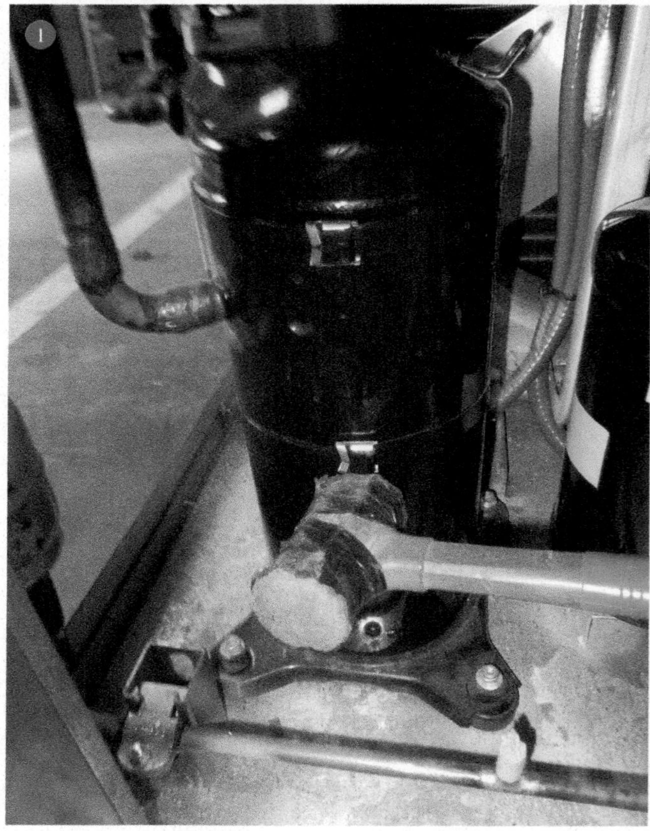

Figure 27-8 Give the side of the compressor a swift, firm strike with a rubber mallet.

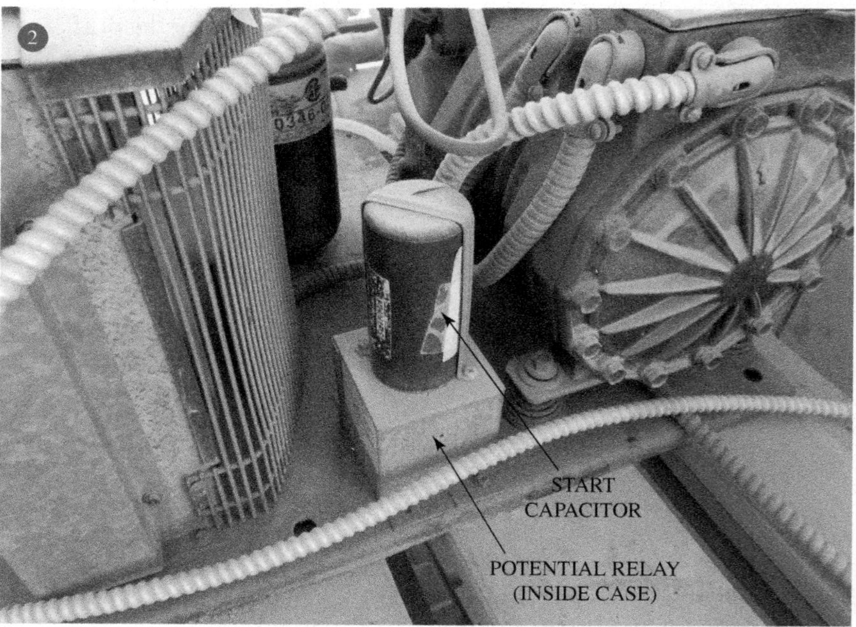

Figure 27-9 Add a correctly sized hard start kit to get a stuck compressor operating again.

27.6 INSTALLING A HARD START KIT

A hard start kit is a properly matched potential relay and start capacitor. Figure 27-10 shows the potential relay of a hard start kit. As shown, the relay coil is between terminals 5 and 2. The normally closed contacts are between terminals 1 and 2. The coil resistance is high, in the thousands of ohms. The contacts should be near 0 Ω. If it is not matched to the compressor, it may not be helpful. Use of the correct potential relay is critical. Installing the wrong potential relay will drop the capacitor out too quickly or leave the capacitor in the circuit too long, damaging the start cap.

Installing a start capacitor that is too small or too large will not improve the starting torque.

The exact size of the start capacitor and relay is obtained from the manufacturers' technical literature or from a supply house that sells that specific compressor. Table 27-3 shows the required start components for various compressor models. Here is some information we can derive from this table for this compressor Model No. ZPS20K4E-PFV (see red arrow):

- This compressor requires an 88–108-MFD start capacitor with a 330-volt rating.
- The potential relay requires a 140–150 pickup voltage and a 40–90 dropout voltage. (The pickup voltage is the voltage that opens the potential relay, which removes the start cap. The dropout voltage is the voltage at which the potential relay closes its contacts.)
- The maximum coil voltage is 332. (Voltage higher than 332 volts will burn out the potential relay coil.)

Another option is a soft start kit. Figure 27-11 shows a diagram of a soft start kit (PTC) that is found in the form of a solid relay, sized for the motor. Compared to the properly sized start capacitor and start relay, the soft start kit is not as likely to solve a starting problem. This diagram has a PTC solid-state device that changes resistance as it heats up from current flow. This is a starting device. PTC mean *positive thermal coefficient* device. At normal ambient temperature the resistance of the device is very low, 5 to 50 Ω. This allows full current flow through the start winding, starting the motor. Current flow through the PTC device instantly causes the resistance to increase and safely reduces the current flow through the start winding. The rest of the circuit operates as a PSC motor.

Figure 27-10 This is an example of a potential relay used in a hard start kit. Terminal 5 goes to the compressor common. Connection 2 goes to the start winding, and terminal 1 is connected to the start cap.

Figure 27-12 is a universal solid-state start device with a start capacitor. The common, so-called "universal" hard start kit may or may not help the motor start. On these universal start kits, the size of the start capacitor may be too small or

Table 27-3 Hard Start Kit Selection Table

	Start Components							
	Start Capacitors			Relays				
Model	MFD	Volts	Part Number	G.E. p/n	Copeland p/n	Pick-up Volts	Drop-out Volts	Coil Voltage
→ ZPS20K4E-PFV	88-108	330	014-0036-03	3ARR3KC3LP	040-0140-08	140-150	40-90	332
ZPS26K4E-PFV	88-108	330	014-0036-03	3ARR3KC3P5	040-0001-79	170-180	40-90	332
ZPS30K4E-PFV	88-108	330	014-0036-03	3ARR3KC5M5	040-0001-62	150-160	35-77	253
ZPS35K4E-PFV	88-108	330	014-0036-03	3ARR3KC3P5	040-0001-79	170-180	40-90	332
ZPS40K4E-PFV	88-108	330	014-0036-03	3ARR3KC3P5	040-0001-79	170-180	40-90	332

(Source Emerson Climate Technologies AE1311.pdf.)

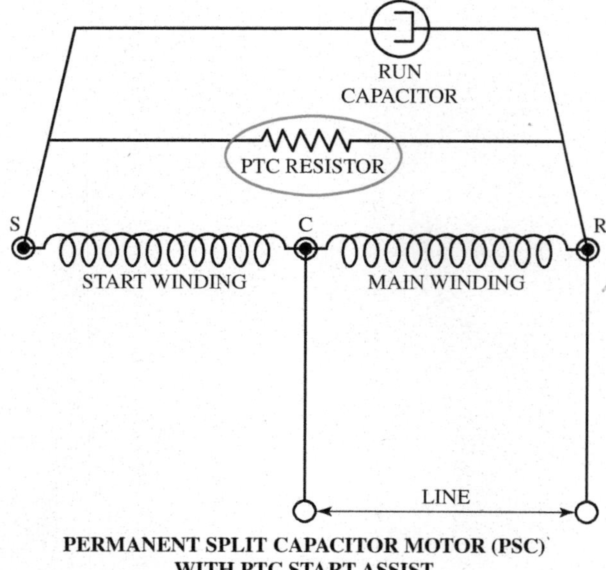

Figure 27-11 This diagram has a PTC solid-state resistor that changes resistance as it heats from current flow. This is a starting device.

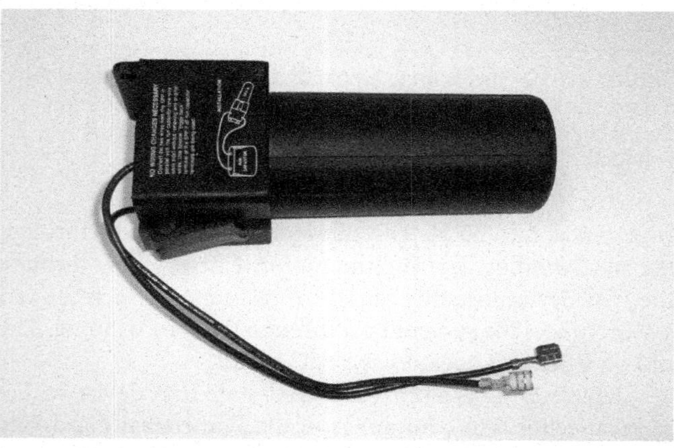

Figure 27-12 Generic solid-state relay and start capacitor. This may or may not start a stuck compressor motor since it may not be the proper microfarad rating to create the optimal starting torque.

too large to develop the optimal starting torque required to rotate the motor. The compressor may also be mechanically stuck and not movable.

A properly sized hard start kit will also help solve the problem of dimming lights when the compressor is energized. Dimming lights can also be caused by loose electrical connections in the electrical panel box, contactor, or compressor terminals.

27.7 OVERHEATING PROBLEMS

Compressor motor overheating problems have a variety of sources. If a system is undercharged, it will overheat and cut off on its overload protection device. Chronic overheating will shorten the life of the motor windings.

If a compressor overheats due to lubrication problems, the following should be checked:

- Low suction pressure
- Low charge
- High condensing pressures
- High compression ratios.

These four problems will be indicated by high discharge temperatures and the possibility of overload tripping.

Low pressure is normally the result of:

- An undercharge condition
- Low airflow or water flow across the evaporator
- An incorrect pressure switch setting
- A drop in suction line pressure
- Light load operating conditions
- A restricted evaporator coil circuit.

High condenser pressures can be caused by:

- Inadequate airflow or water flow through the condenser
- Undersized discharge line
- Undersized condenser coil
- Overcharging of refrigerant
- Noncondensables in the refrigeration system.

High compression ratios are the combination of low suction pressure and high condensing pressure.

A lack of lubrication can also cause overheating. Generally, a discharge line temperature of 225°F or lower will ensure long compressor life. Higher temperatures causes lubrication breakdown.

27.8 BEFORE CONDEMNING

Before condemning a compressor, take one more resistance readings. Check the resistance between the windings and check the windings to ground. In a single-phase compressor, the start winding's resistance should be three to five times higher than that of the run winding. A shorted winding to ground will have a very low resistance reading. Resistance readings to ground near zero or even several thousand ohms should be considered shorted. Compressor manufacturers state that a good rule of thumb is that the resistance to ground should be greater than 1,000 Ω per operating volt. For example, a motor that operates on 480 volts should have a minimum resistance of 480,000 Ω to ground. In most cases, it will be millions of ohms to ground.

Moisture or contaminated refrigerant or oil can create low resistance to ground readings. Change the refrigerant, oil, and driers to improve this condition. If no improvement is obtained, then the motor winding may soon short to ground and prove fatal to the compressor. If this compressor is involved in a critical operation, the compressor can be scheduled to be replaced prior to failure. Another option is to purchase a compressor that would be available for a change-out at the time of failure. Some compressors are not readily available and at certain times of the year compressor stock is low, taking days to obtain a replacement model.

Talk to the customer about ordering a standby compressor. Having a compressor in stock is important if a compressor failure would adversely affect a customer's critical mission. Alternate plans include renting a portable air conditioning or refrigeration system, or in the case of refrigeration, being able to purchase a lot of ice.

Check the start components such as capacitors, start relays, and contactors. These could be causing the problem.

Finally, if the windings are good and the compressor will not start, use a separate, external power source to try to start the compressor. This is considered an independent check because it does not use the condensing unit power as part of the final check. If the compressor starts, it is most likely good. Search for an external electrical problem. Copeland Compressor's website (www.copeland-corp.com) provides valuable information on compressors.

TECH TIP

Check line voltage at the load center first with the motor off and then with the motor on. The voltage should be within 10% of the motor rating.

Reduced voltage causes a motor to draw more current than normal. A severe undervoltage condition leads to overheating and premature failure.

High line voltage causes excessive inrush current at motor start. This eventually breaks down winding insulation, leading to motor failure.

27.9 VOLTAGE AND CURRENT IMBALANCES

Voltage and current imbalances are measured on three-phase motors. An imbalance means that the voltage or current between the phases is not close together.

Voltage Imbalance

Here is what a **voltage imbalance** can do to a motor:

- A voltage imbalance in a three-phase motor causes high current in the motor windings. The high-current condition degrades the winding insulation.
- A 10°F rise in winding temperature can reduce motor life by half.
- The voltage imbalance for three-phase motors should not exceed 2%.
- A voltage imbalance is usually caused by adding excessive single-phase loads on one side of a three-phase circuit, but it can also be caused by the power supplied to the motor or by loose connections or dirty contactor contacts.

EXAMPLE 27.1 CHECKING FOR A VOLTAGE IMBALANCE

To measure a voltage imbalance, the motor should be running. Figure 27-13 illustrates an example of gathering the three voltage readings to calculate a voltage imbalance.

To calculate the percentage of voltage imbalance, use this formula:

$$\%\ \text{imbalance} = \frac{\text{maximum deviation from average voltage}}{\text{average voltage}} \times 100$$

1. Find the average voltage of each phase by adding the three voltages from Figure 27-13 together and dividing by 3:

$$\begin{array}{r} 449 \\ 470 \\ +462 \\ \hline 1{,}381 \end{array} \; / \; 3 = 460 \text{ average voltage}$$

Given voltages of 449, 470, and 462, the average voltage is 460.

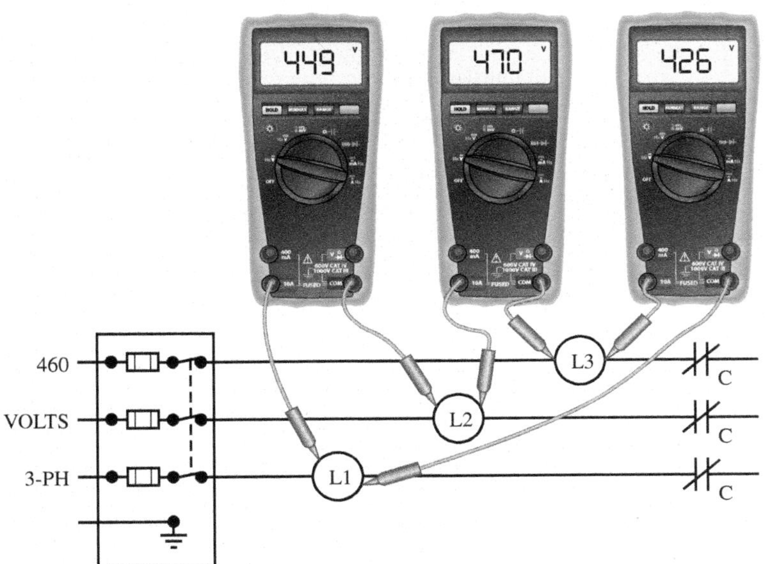

Figure 27-13 Checking voltage imbalance. The motor should be running to obtain an accurate voltage measurement.

2. Find the maximum voltage deviation from the average voltage. It does not matter if it is a negative or positive deviation. Select the greatest deviation from the average voltage:

$$449 - 460 = -11 \text{ volts}$$
$$470 - 460 = 10 \text{ volts}$$
$$462 - 460 = 2 \text{ volts}$$

The maximum deviation from the average is 11 volts.

3. Calculate the voltage imbalance:

$$\% \text{ imbalance} = \frac{\text{maximum deviation from average voltage}}{\text{average voltage}} \times 100$$

$$\% \text{ voltage imbalance} = \frac{11 \text{ volts}}{460 \text{ volts}} \times 100$$

$$\% \text{ voltage imbalance} = 0.0239 \times 100$$

The percent imbalance is 2.39%.

In this example you would have a voltage imbalance problem since it is above 2%.

Note that the 100 in the formula automatically converts the answer from a decimal to a percentage. The percentage makes the answer easier to understand.

Figure 27-14 The total current draw of a system can be found by measuring the amperage at the disconnect when the system is operating. Check each leg of a three-phase circuit.

Current Imbalance

Current should be measured to ensure that the continuous load rating on the motor's nameplate is not exceeded and that all three phases are balanced. If the measured load current exceeds the nameplate rating or the current is unbalanced, the life of the motor will be reduced by the resulting high operating temperature. Imbalanced current may be caused by a voltage imbalance between phases, a shorted motor winding, or a high-resistance connection. The **current imbalance** for three-phase motors should not exceed 10 %.

EXAMPLE 27.2 CHECKING FOR A CURRENT IMBALANCE

Figure 27-14 demonstrates how to measure amperage at the disconnect that feeds the condensing unit. Measure each power input: L_1, L_2, and L_3.

To calculate the current imbalance, use the voltage imbalance formula:

$$\% \text{current imbalance} = \frac{\text{maximum deviation from average current}}{\text{average current}} \times 100$$

Figure 27-15 shows how to measure current for the purpose of calculating a current imbalance.

1. Find the average current of each phase by adding the three amperages together and dividing by 3:

$$\begin{array}{r} 30 \\ 35 \\ +30 \\ \hline 95 \end{array}$$

95 / 3 = 31.66 average amperage, rounded to 31.7 amps

Given amperage readings of 30, 35, and 30, the average current is 31.7 amps.

2. Find the maximum amperage deviation from the average amperage. It does not matter if it is a negative or positive deviation. Select the greatest deviation from the average amperage:

$$30 - 31.7 = -1.7 \text{ amps}$$
$$35 - 31.7 = 3.3 \text{ amps}$$
$$30 - 31.7 = -1.7 \text{ amps}$$

The maximum deviation from the average is 3.3 amps.

3. Calculate the current imbalance:

$$\% \text{current imbalance} = \frac{\text{maximum deviation from average current}}{\text{average current}} \times 100$$

$$\% \text{ current imbalance} = \frac{3.3 \text{ amps}}{31.7 \text{ amps}} \times 100$$

$$\% \text{ current imbalance} = \frac{3.3 \text{ amps}}{31.7 \text{ amps}} \times 100$$

$$\% \text{ current imbalance} = 0.104 \times 100$$

The percent of current imbalance is 10.4%.

In this example you would have a current imbalance problem since it is above 10%. You can have a current and voltage problem at the same time, but they do not always occur together.

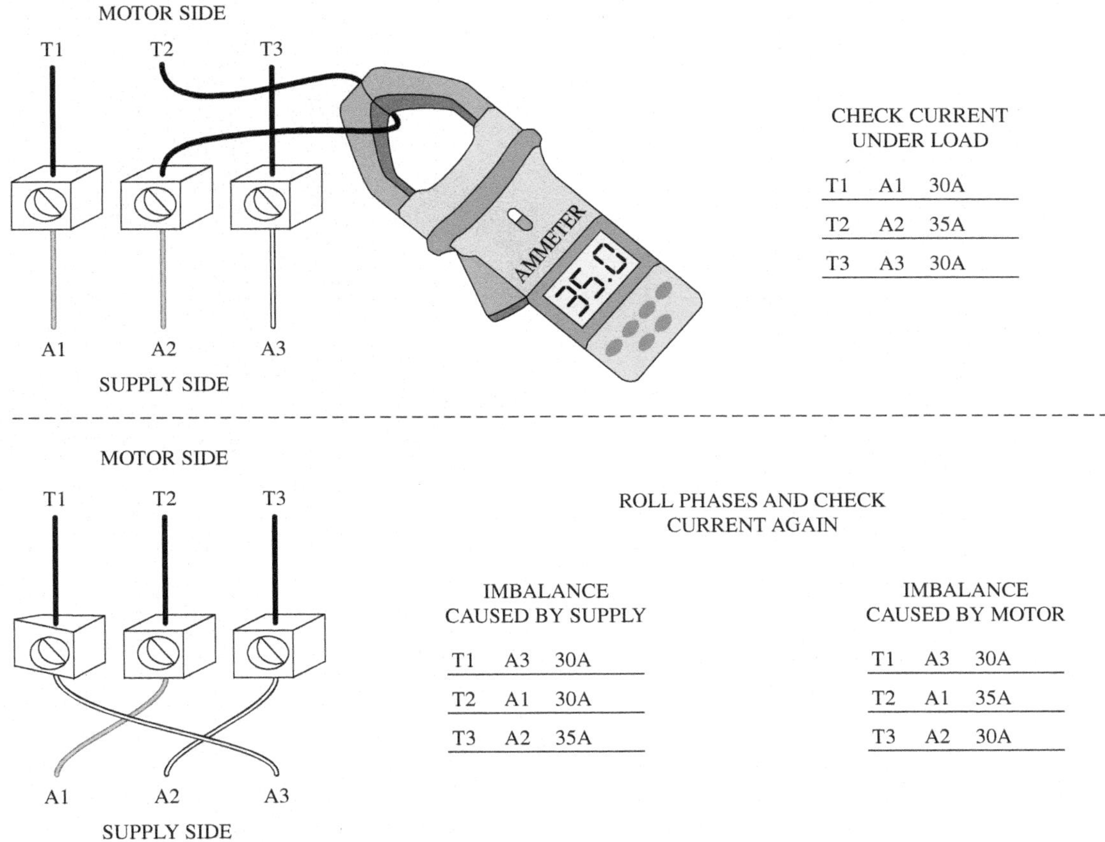

Figure 27-15 Measuring current imbalance. The upper picture shows the amperage reading of T2 or L2. The lower picture shows what to do if there is a current imbalance. It illustrates how to determine if the current imbalance is in the supply voltage or caused by the motor.

27.10 HOW TO LOCATE THE SOURCE OF CURRENT IMBALANCE

The current imbalance can be caused by the motor or the supply voltage. Measure the current on each phase and calculate the amperage imbalance. Use the following steps to isolate the current imbalance problem:

1. Determine if the motor is the problem by turning off the power, then moving all three-phase wires to a different terminal. This will not change the rotation of the motor. Some compressor motors are rotation sensitive (i.e., they turn clockwise or counterclockwise); this will keep them rotating in the same direction. Most three-phase scroll compressor motors are rotation sensitive. Few semi-hermetic compressors are rotation sensitive. Record which phase has the highest amp draw. If the same supply voltage phase has the highest amp draw as before, the supply is the problem. The lower left side of Figure 27-15 illustrates how to do this diagnosis.
2. If the highest amp draw is found on a different supply phase, the imbalance may be the result of a shorted winding in the motor or a shorted conductor feeding the compressor. Remember, when switching wiring or measuring the resistance of wires, measure from the closest disconnect, not on the compressor connection that is refrigerant charged. The terminal may blow out, severely injuring you.

27.11 MEGOHM TESTING AS A TROUBLESHOOTING TOOL

Megohm testing, also called **megging**, is a motor winding testing procedure. The test procedure is also known as motor insulation testing. A **megohmmeter**, as shown in Figure 27-16, is used to test for high resistance between the motor windings and ground. The test is conducted from the motor terminal to the suction or discharge line, as illustrated in Figure 27-17. The resistance of a compressor motor winding to ground should be a minimum of 1,000 Ω per operating voltage. For example, a 460-volt compressor should have a minimum of 460,000 Ω from any compressor terminal to ground. In most cases, the motor winding resistance to ground is a million, even tens of millions, of ohms to ground. The higher the resistance to ground, the better.

Figure 27-16 The megohmmeter is used to test the motor insulation to ground. The meter develops 250, 500, or 1,000 V to test the motor at the operating voltage. This meter has additional meter functions of low and normal resistance, voltmeter, and battery check.

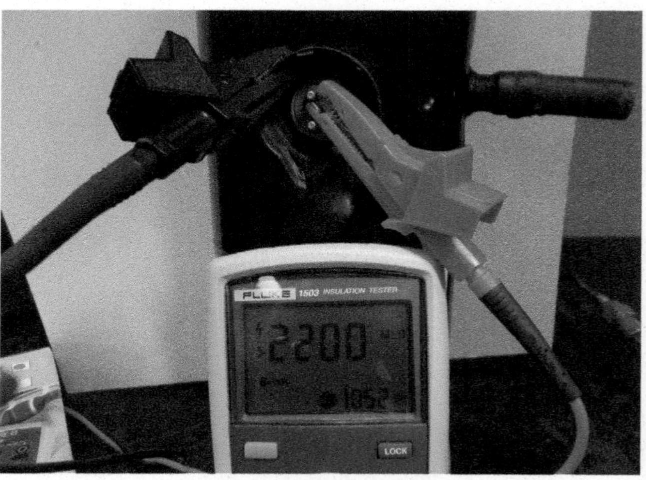

Figure 27-18 This compressor is measuring 2,200 megohms (2.2 giga-ohms) from the motor winding to ground.

An ohmmeter can be used to test motor winding resistance to ground, but a megohmmeter is the instrument of choice when testing motor insulation. When using an ohmmeter its resistance should be in excess of 1 MΩ (1,000,000 Ω). Some multipurpose clamp-on ammeters have a resistance measuring option. The maximum resistance option for this meter is usually low and not desirable for measuring high resistance. A resistance reading beyond its range will indicate infinity resistance, indicated by the symbol ∞. A digital multimeter (DMM) has a 9-volt circuit to test resistance from winding to ground. The megohmmeter generates 500 or 1,000 volts or more to test the winding to ground. This high resistance represents the higher voltages found operating these motors, therefore the megger is the meter of choice when insulation testing. Figure 27-18 demonstrates a

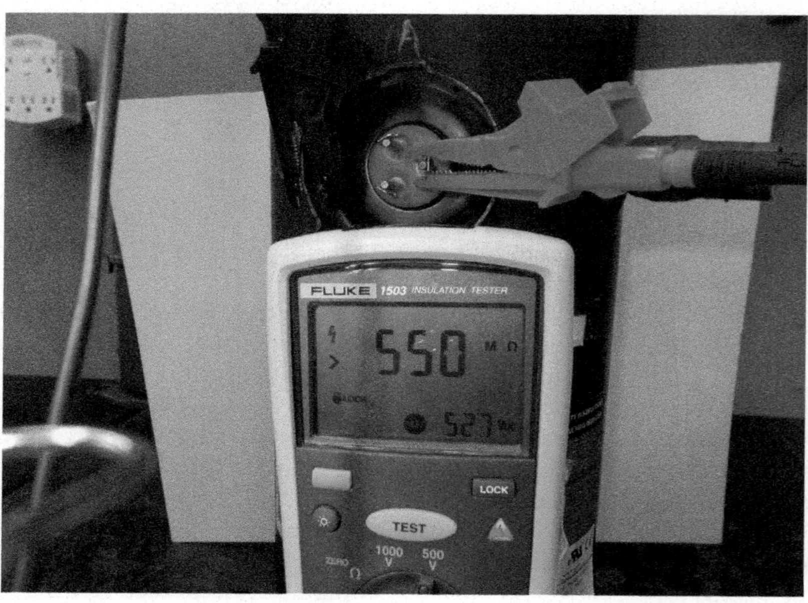

Figure 27-17 The megohmmeter checks an uncharged compressor from a terminal to a convenient refrigerant line. The reading is 550 megohms, also expressed as 550,000,000 Ω. This is excellent resistance to ground.

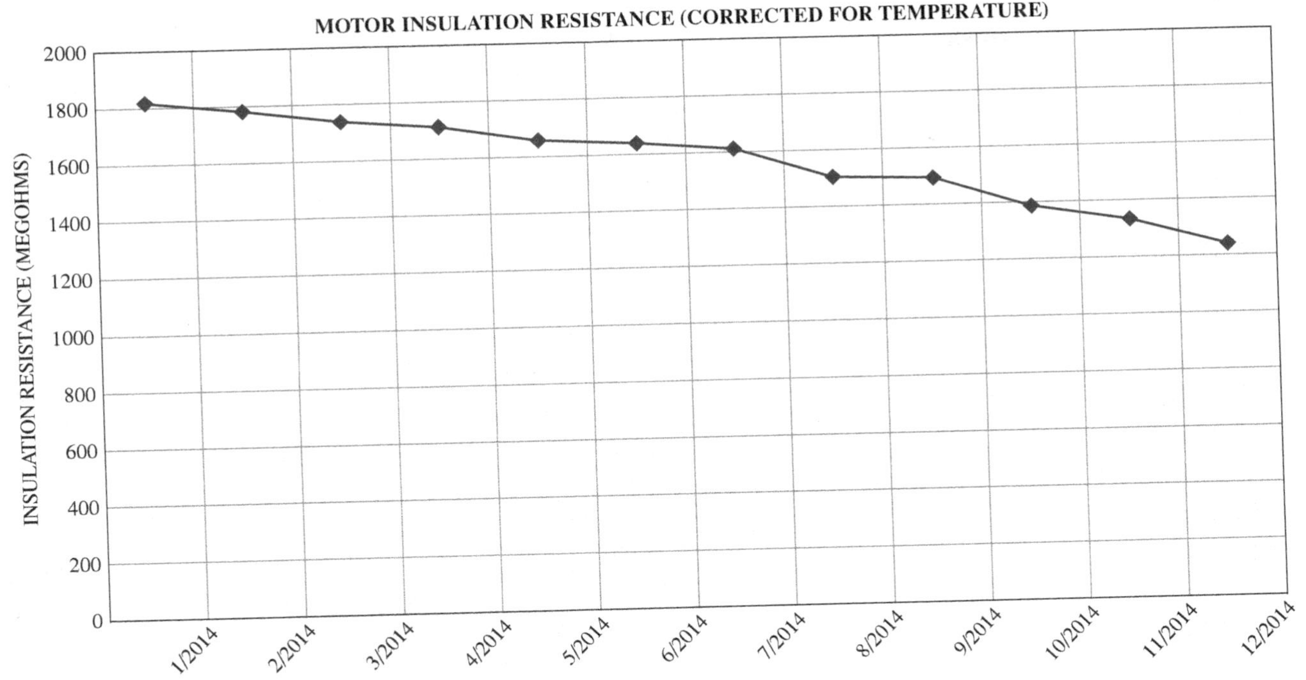

Figure 27-19 This graph shows reductions in resistance over time. A rapid downturn in resistance is an indicator that the insulation is going to fail. Having this knowledge will help technicians with plans to replace the motor.

motor with a 2.2 giga-ohm (G Ω) reading. This is an excellent resistance to ground.

The value of using a megger is to develop a history of resistance deterioration. The standard ohmmeter or the DMM can measure a shorted winding. The megohmmeter goes a step further. The megger can measure insulation breakdown in small increments, as shown in Figure 27-19. Over time, the resistance to ground will be reduced due to normal operation of the motor. As shown in this figure, the resistance declines over time. If knowing when a motor will fail is important, track the decline in resistance. When there is a sharp drop off in resistance, the unit can be replaced before it fails. Even if this is not practiced, a standby motor can be purchased and available for change out at any time.

27.12 NOT ALL ELECTRICAL FAILURES ARE ELECTRICAL PROBLEMS

Liquid refrigerant can wash out compressor motor bearings and reduce lubrication on these bearing surfaces. The bearing clearance decreases and the rotor begins to drag on the stator winding. At first the dragging rotor will cause a higher amperage draw. Eventually, the dragging rotor will short out the stator or damage the rotor. The technician will determine that the motor failed due to an electrical problem.

The real cause of the failure was liquid refrigerant washing out the bearing surfaces. Liquid refrigerant in the compressor can be caused by:

- An overcharged system
- Low airflow over the evaporator
- Low water flow through a chiller barrel
- Poor piping practices
- No crankcase heater
- Overfeeding of the expansive valve
- Low oil level
- Low load conditions.

Unfortunately, if these problems are not identified when the compressor is replaced, the replacement compressor will not last long.

TECH TIP

Inspect the oil level by looking at the compressor sight glass while it is operating. If there is no sight glass, use an infrared (IR) thermometer or a multimeter with an IR temperature probe to determine the oil level in the sump of the compressor housing. The crankcase housing temperature will change at the oil level line. The oil will be warmer. Remember that the IR device is not an accurate temperature measuring device but it will indicate a vali temperature change, which means on oil level chang

SAFETY TIP

1. Do not test insulation when the compressor is in a vacuum. Electricity can travel easily in a vacuum. The motor windings can be shorted. Be sure to use the lockout/tagout procedure with the compressor.
2. Isolate the compressor by shutting off the discharge and suction valves. Recover the refrigerant. Remove wires from the terminals before testing.
3. Remove compressor bars that are placed across compressor terminals on large tonnage motors. This will allow you to measure each set of windings in a multiwinding compressor separately.
4. Shunt compressor terminals together when checking from motor terminals to ground.
5. Clean the compressor piping of oxidation prior to placing the megger probe on the suction or discharge line.

27.13 THE MYSTERY NUISANCE TRIP

Sometimes a motor "trips" or opens on its internal or thermal overload. The overload is in series with the common connection inside the motor windings. Restarting the compressor with gauges in place and the clamp-on ammeter measuring current draw may reveal a high-amperage condition when the motor starts. Working from this condition the technician can find the cause of the motor trip.

This troubleshooting section covers the nuisance trip. The nuisance trip can be described as a one-time or random overload condition that may be difficult to replicate when arriving on the service call. This section explores what to check when it comes to tracking down nuisance trip issues.

A nuisance trip occurs when the motor experiences a high-amperage condition. To be classified as a nuisance, the overload will reset automatically or manually, but does not occur when the tech is on the job troubleshooting the problem. The technician can do little more than tighten connections and do general inspections on the equipment while monitoring the motor operation.

The first tendency is to blame the thermal overload, diagnosing it as weak, which means it opens at a lower than rated amperage. In most cases the overload is not weak and not the problem. Changing out the motor may not solve the problem either.

Here are tips to reduce a return call for nuisance trips and return calls to the job site:

1. Establish with the customer when the nuisance trip occurs. For instance, the trip may occur in the heat of the day or when the unit first comes on in the morning.
2. Determine if the system is charged correctly. An improperly charged system can underload or overload the compressor. A lack of charge reduces the refrigerant gas that is used to cool the motor. Too much charge overloads the compressing section, causing higher amperage draw.
3. Determine if the condenser, evaporator, and air filter are clean. Most condensers appear clean, but cleaning them will reveal dirt that is hidden in the coils. When in doubt, clean the condenser coils. Flush in the opposite direction of airflow.
4. Install a time-delay device. Short cycling causes a high-amperage condition and will cause a motor to trip. When the compressor restarts, the voltage may be low, which causes a high-amperage starting condition. The time delay allows the pressures to equalize and the voltage to stabilize.
5. Tighten all connections on the condensing unit, the disconnect, and the electrical panel box that feeds the condenser.
6. Add a hard start kit to a single-phase motor.
7. Determine if the wire feeding the condensing unit is the correct size.
8. Check the voltage drop directly across the closed contactor contacts. A new contactor under a full load will measure 0 volts across the closed contacts. Contacts with a carbon buildup will create resistance, leading to a voltage drop. The voltage drop should be 0 volts. If the voltage drop is 3 volts or higher, the contactor should be replaced.

TECH TIP

Ten reasons why motors fails:

1. Overloaded
2. Restricted ventilation
3. Short cycling
4. Moisture
5. Vibration
6. Voltage or current imbalance
7. Single phasing
8. High or low voltage
9. Surge voltage
10. Bearing failure.

SAFETY TIP

Using a jumper wire or a set of alligator clips can be a valuable tool when troubleshooting. Caution must be taken when using a jumper. Turn the power off prior to jumping a switch. You can jumper across a seemly open switch like a pressure switch. The wire gauge (AWG) of the jumper must be adequate to handle the amperage draw across a closed switch. Figure 27-20 demonstrates the correct way to use a jumper for troubleshooting. Do not jumper across a load. This will place a short across the power supply.

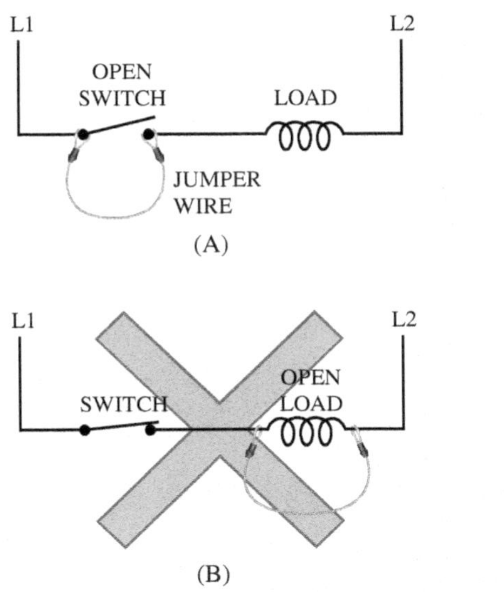

Figure 27-20 Using a jumper wire or a set of alligator clips can be a valuable tool when troubleshooting. Only jumper switches—do *not* jumper loads.

SERVICE TICKET

You are sent out on a job to do an initial diagnosis of a cooling problem. The more experienced tech is scheduled to arrive about 15 minutes later.

Here is what you found without tools or instruments: The thermostat display was set 5°F lower than the room temperature. The t'stat was set to Cool and Auto. The return air grille had air being pulled into the duct and the supply air had good airflow. Going outside, the condensing unit was located. It was operating. The air temperature coming off the condenser seemed to be about the same temperature as the outside ambient air. The suction line and liquid line were at the same temperature.

Question: Based on this limited information, where do you start looking for the problem? What will you tell the experienced tech, who has just arrived?

Answer: The compressor or refrigeration circuit does not seem to be operating. A good diagnostic starting point is to install a manifold gauge set and clamp-on ammeter on the common wire of the compressor. Further electrical troubleshooting may be required.

SUMMARY

The purpose of this unit was to give you troubleshooting practice. The unit applies some of the skills that you have learned in other sections of this book. After talking with the customer, it is important to do a quick system survey using the ACT method or some logical sequence for determining what is operating. There is really no one good way to troubleshoot. The sequence you choose for troubleshooting will differ from that of other techs.

Every tech has different ways of approaching problems or "figuring things out." The ultimate goal of troubleshooting is to find the problem and solution in a reasonable amount of time. As you do troubleshooting you will develop ways to speed up the process, create your own shortcuts, and pass your tips on to the newer techs who enter our profession. You will learn from others as they learn from you. We are in this together to help each other become better technicians.

This unit discussed the troubleshooting of various types of equipment. We spent a considerable amount of time discussing compressor troubleshooting. For your convenience, Figure 27-21 summarizes these compressor problems into one image.

REVIEW QUESTIONS

1. Refer to Figure 27-22. The heating or cooling system does not operate. List two quick checks that will narrow down the problem.
2. Refer to Figure 27-22. What prevents the operation of the heat strip when the cooling system is in operation?
3. Refer to Figure 27-22. Describe the hopscotch troubleshooting method for finding a problem with the high-voltage heating circuit. The heat strip is open.
4. Refer to Figure 27-22. Under normal cooling operation, what voltage is read across the compressor contacts **C**?
5. Refer to Figure 27-23. The figure shows two diagrams that are essentially the same, but drawn differently. The schematic wiring diagram or ladder diagram has one side of the low-pressure switch connected to the contactor coil. Use the connection diagram to determine the color of that wire.
6. Refer to Figure 27-23. Use the hopscotch method to troubleshoot a compressor problem. What is the problem after measuring the following voltage readings in the contactor coil circuit?:
   - 240 volts on the cool switch **SW**
   - 240 volts on point 4 on the left side of thermostat
   - 240 volts on point 2 on the left side of **HP**
   - 0 volts on the left side of **LP**
   - 0 volts on the left side of the contactor coil Ⓒ.
7. Refer to Figure 27-23. Does the compressor have an internal overload (sometimes referred to as an inherent overload protector)?
8. Refer to Figure 27-23. Does this unit have a start capacitor? A run capacitor? A crankcase heater?
9. Refer to Figure 27-23. Which diagram would you use for hopscotch troubleshooting? Which diagram would you use to rewire this cooling unit?

LACK OF COOLING
LOW SUCTION PRESSURE
LOW CHARGE
ICED COIL
AIRFLOW RESTRICTION
DIRTY AIR FILTER
DIRTY EVAPORATOR COIL
HIGH SUCTION PRESSURE
OVERCHARGED
BAD COMPRESSOR VALVES
WILL NOT RESTART
NO POWER
BLOWN FUSE
OPEN RELAY COIL
NO CONTROL VOLTAGE
INTERNAL OVERLOAD OPEN
FAULTY RUN OR START CAPACITOR
STUCK COMPRESSOR
THERMOSTAT OFF OR SET HIGH
COMPRESSOR WILL NOT RUN
CONTACTOR SHORT CYCLES
• PRESSURE SWITCH OPENS
• LOW CONTROL VOLTAGE
• LOW VOLTAGE CONNECTION
CONTACTOR CLOSED
• LOOSE CONNECTIONS
• MOTOR WINDING OPEN
• INTERNAL LOAD OPEN
• NO VOLTAGE
CONTACTOR OPEN
• LOOSE CONNECTIONS
• CONTACTOR COIL OPEN
• OVERLOAD OPEN
• THERMOSTAT OPEN
• OPEN TRANSFORMER
NO POWER TO THE CONTACTOR
• NO VOLTAGE
• BLOWN FUSE
COMPRESSOR CYCLES ON OVERLOAD, PRESSURE SWITCH OR INTERNAL PRESSURE RELIEF
BAD CAPACITOR
LOOSE ELECTRICAL CONNECTIONS
LOW LINE VOLTAGE
OVERCHARGED
NONCONDENSABLES IN CONDENSER
CONDENSER AIR RESTRICTION
DEFECTIVE CONDENSER MOTOR
DEFECTIVE CONDENSER MOTOR CAP
CONDENSER MOTOR BEARING TIGHT

Figure 27-21 This is a summary of conditions that will cause compressor problems.

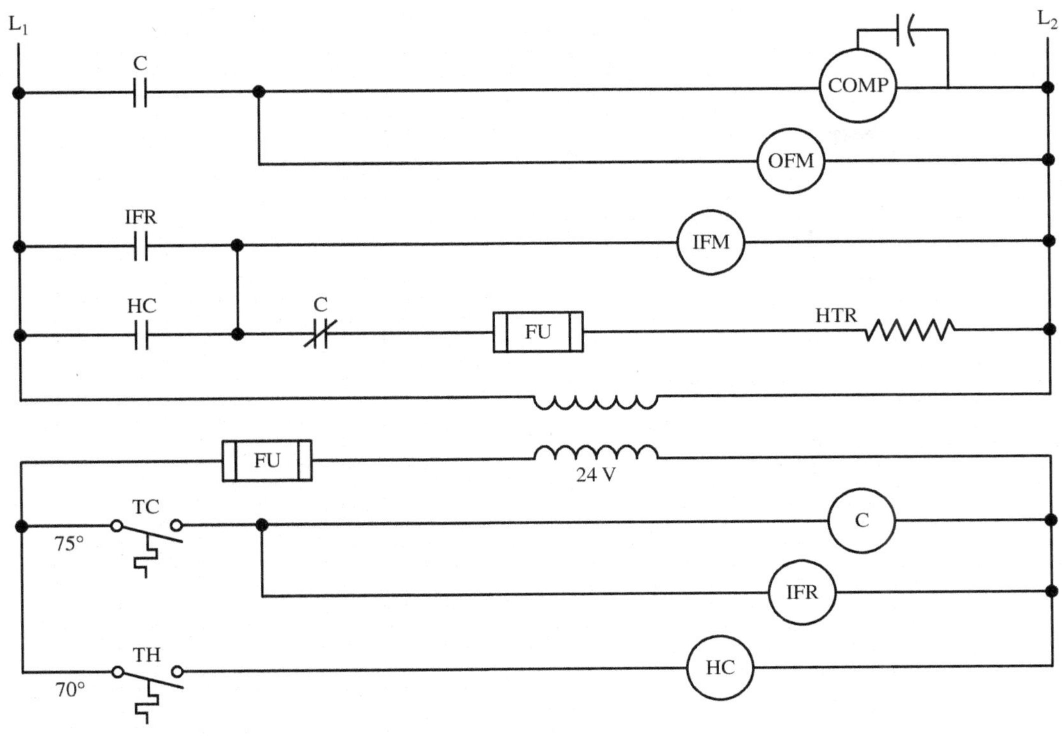

Figure 27-22 Use this diagram to answer Review Questions 1 through 4.

10. Refer to Figure 27-24 on page 541. Use the hopscotch method to troubleshoot a compressor problem. What is the problem after measuring the following voltage readings in the contactor coil circuit?:
    - 24 volts on the left and right sides of **CR**
    - 24 volts on the left and right sides of **HP** and **LP**
    - 24 volts on the left and right sides of **IT**
    - 24 volts on the left and right sides of **OL1** and **OL2**
    - 24 volts is measured across Ⓒ.
11. Refer to Figure 27-24. What system component is the **IT** sensing?
12. Refer to Figure 27-24. When this circuit is energized and operating normally, are **OL1** and **OL2** closed or open?
13. What are two advantages of installing the correctly sized hard start kit?
14. What is the purpose of the ACT procedure of troubleshooting?
15. What are the steps in the ACT process of troubleshooting?
16. What precautions should you take when unscrewing a defective high-pressure switch from a high-pressure refrigerant line?
17. From which single-phase compressor terminal will you measure RLA?
18. You wish to select a multimeter with an ohmmeter. What resistance range is required to measure high resistance?
19. What is another term for inherent thermal protection?
20. What percentage of compressors returned under warranty are found to have no defects?
21. Use Table 27-1 to answer this question: What is the compressor amperage and LRA of compressor Model No. ZP39K5E-PFV?
22. Use Table 27-2 to answer this question: What is the amp draw of the compressor model mentioned in Question 21 when the condensing temperature is 130°F and the evaporating temperature is 40°F?
23. Use Table 27-2 to answer this question: What is the amp draw of the compressor model mentioned in Question 21 when the condensing temperature is 100°F and the evaporating temperature is 40°F?
24. Use Table 27-2 to answer this question: Does the change in condensing temperature or evaporating temperature have the biggest effect on the compressor amperage difference?
25. What safety precaution should you follow when checking the resistance on compressor terminals?
26. What are two things you can do to free a stuck compressor?
27. What should be checked if a compressor is overheated due to lubrication problems?
28. Low suction pressure is normally the result of what factors?
29. What is the voltage imbalance if you measure the following voltages: 230, 245, and 250 V? Is this percentage acceptable?
30. What is the current imbalance if you measure the following current: 10, 12, and 14 A? Is this percentage acceptable?

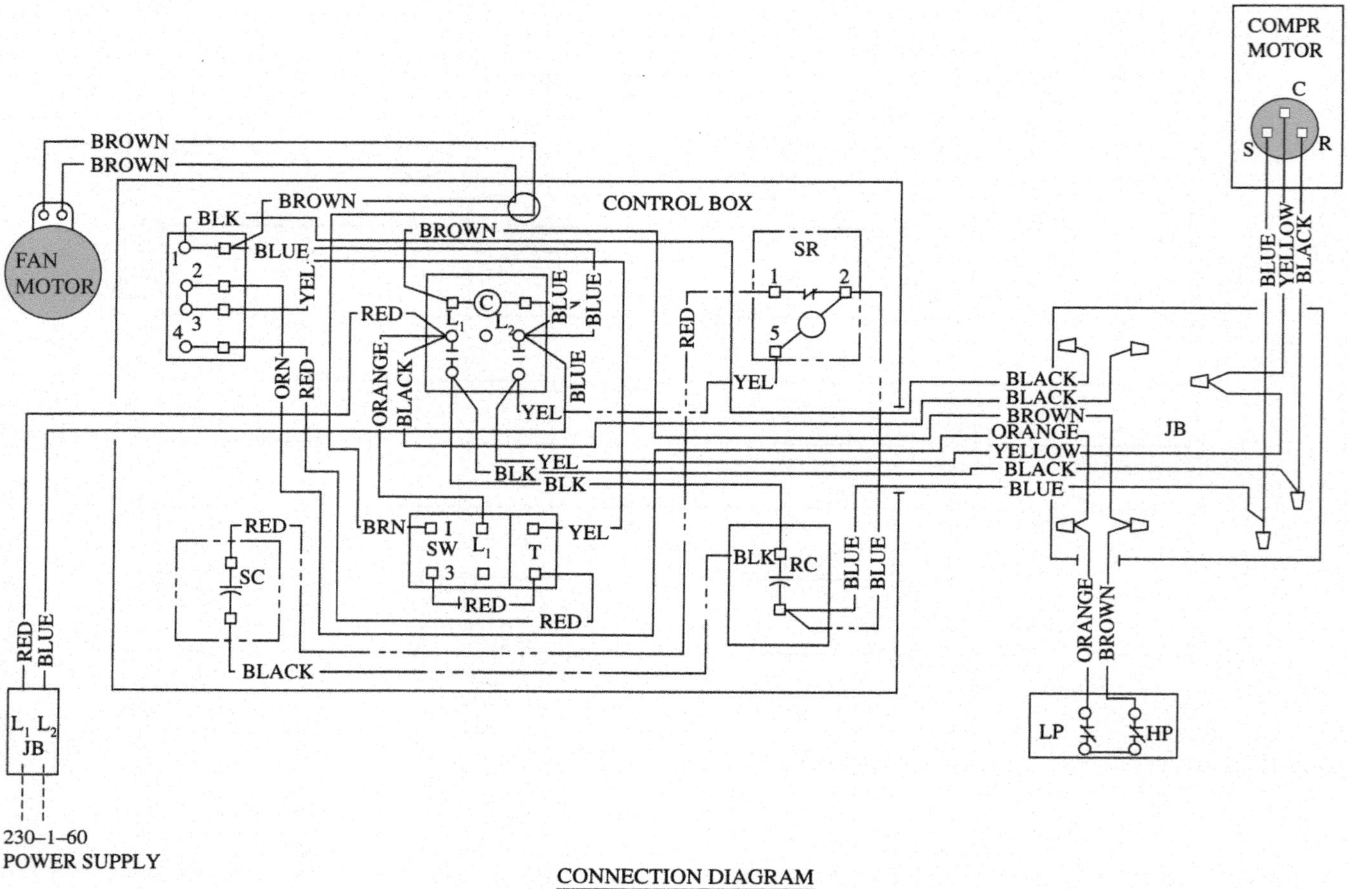

230–1–60
POWER SUPPLY

CONNECTION DIAGRAM

LEGEND

C CONTACTOR
RC RUN CAPACITOR
SC START CAPACITOR
SR START RELAY
T THERMOSTAT
SW SWITCH
HP HIGH-PRESSURE SWITCH
LP LOW-PRESSURE SWITCH
JB JUNCTION BOX

S START
R RUN
C COMMON

---- FIELD WIRING
—— FACTORY WIRING
– -- – ALTERNATE CSR WIRING

NOTES

1. FAN MOTOR PROVIDED WITH INHERENT THERMAL PROTECTOR.
2. COMPR. MOTOR PROVIDED WITH INHERENT OVERLOAD PROTECTOR.
3. MAX. FUSE SIZE 30-AMP DUAL ELEMENT.

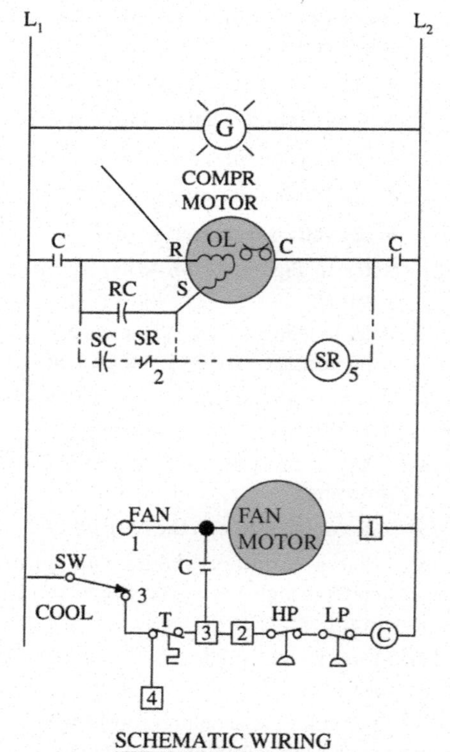

SCHEMATIC WIRING

Figure 27-23 Use this diagram to answer Review Questions 5 through 9. This pictorial wiring diagram has labeled components, a connection diagram, a legend, and notes. This is both a connection and a wiring diagram.

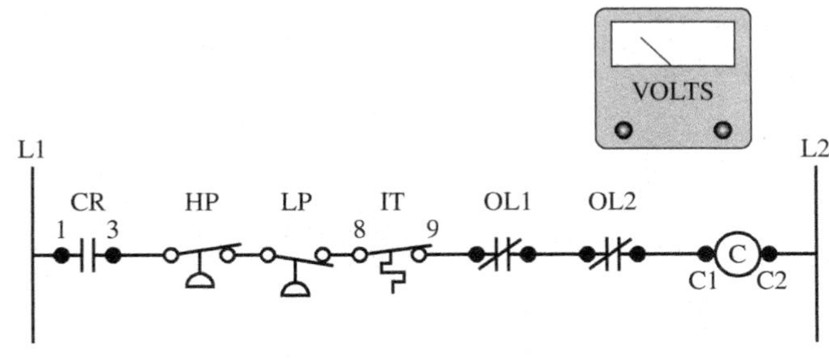

Figure 27-24 Use this diagram to answer Review Questions 10 through 12 and 35.

31. What is the purpose of megging a motor winding?
32. What is the minimum resistance to ground that should be measured on a compressor motor?
33. Liquid refrigerant in the compressor can be caused by what factors?
34. List eight steps that you should take to reduce nuisance trips callbacks.
35. Using Figure 27-24, describe how to use the hopscotch troubleshooting method if the OL2 is open.

UNIT 28
Practical Troubleshooting

WHAT YOU NEED TO KNOW

After studying this unit, you will be able to:

1. Trace wiring diagrams.
2. Use a logical sequence for troubleshooting wiring diagrams.
3. Use the hopscotch method for troubleshooting.
4. Troubleshoot a circuit using the hopscotch method.
5. Use the process of elimination to isolate electrical problems.
6. Determine how to troubleshoot a short circuit.

The goal of this unit is to help you practice electrical troubleshooting using wiring diagrams and to help you learn in-depth troubleshooting of some basic components found in HVACR systems. This unit presents some troubleshooting problems, provides their solutions, and describes how to use a logical sequence to solve the problems.

An HVACR problem will be presented and several ways to solve the problem will be explored. There may be many different ways to solve a problem, but usually there is one best way to derive a correct solution. This unit is not so much about finding the most correct way of solving a problem, but about solving it in a reasonable amount of time. Solving the problem in a reasonable amount of time includes calling for help when advice is needed instead of wasting time on the job. Everyone needs help sometime. Your troubleshooting skills will improve with experience.

Since this is an electrical textbook, the practical troubleshooting problems presented here are electrically based. Most problems in HVACR are electrical. Reading and understanding the problems in this unit will help you develop your own troubleshooting skills. Use the diagrams given or other resources to solve the problems before you look at the solutions. Even if your solution is incorrect, you will start to develop an approach to troubleshooting issues. You can learn from your mistakes.

After we discuss problems using diagrams, we will troubleshoot some of the major component parts of an air conditioning system, such as a compressor. Many parts of this book have troubleshooting tips built into the units. This section will refresh and expand on those troubleshooting techniques. For example, the hopscotch troubleshooting method was discussed in detail in earlier units. It will be used to solve some of the diagram problems presented.

Learning how to troubleshoot is the most important skill you will need as a service technician. Lesser skilled techs can change out a contactor or compressor. But determining that those components are defective takes more knowledge, skill, and experience. Troubleshooting skills come with practice and listening. Listen and learn from other techs who have found solutions to HVACR problems. It seems that one common trait of HVACR techs is their willingness to share a success story. You can learn from others' experiences without being on the job; of course, the best learning tool is figuring out the problem yourself. As stated earlier, there is not one right way to troubleshoot as long as you come up with the correct answer in a reasonable amount of time. Use your "lifelines" when you need help. Do not call for advice in front of your customer. Sometimes it is a good idea to step away from the problem for a short while. Go on a short break or get gas for your vehicle. Stepping away may clear the clouds that we all get from time to time in the troubleshooting process. You will be presented with some problems here. Step back and evaluate the information. In the problems presented in this unit, determine what you think is wrong before reading the solution.

SERVICE CALL 1: HOPSCOTCH PRACTICE

This service call problem is used to review your skills using the hopscotch troubleshooting method and give you an additional tool for troubleshooting intermittent problems.

The first service call of the day involves an air conditioning unit that occasionally cycles its pressure switch off and on, known as *short cycling*. This seems to be the only problem. For this example the problem has been isolated to the energized circuit shown in Figure 28-1. It is a simple high-voltage control circuit with **CR** contacts and low-pressure and high-pressure switches in series with ©, the compressor contactor coil. **CR**, **LP**, and **HP** will be closed when the circuit is energized and there is adequate refrigerant pressure in the system.

What should the voltage reading be across each component? Determine your answer before proceeding.

Using the hopscotch method of troubleshooting, you should measure 120 volts up to the contactor windings. The power supply is represented by **L1** and **N**. The **L1** and **N** power supply is 120 volts. When you see the terms **L1** and **L2** on a diagram, it would most likely be a 208/240-V power supply. When using the hopscotch troubleshooting method, a loss of voltage means that one of the switches is open. Figure 28-2 shows the voltage reading when the system is operating properly.

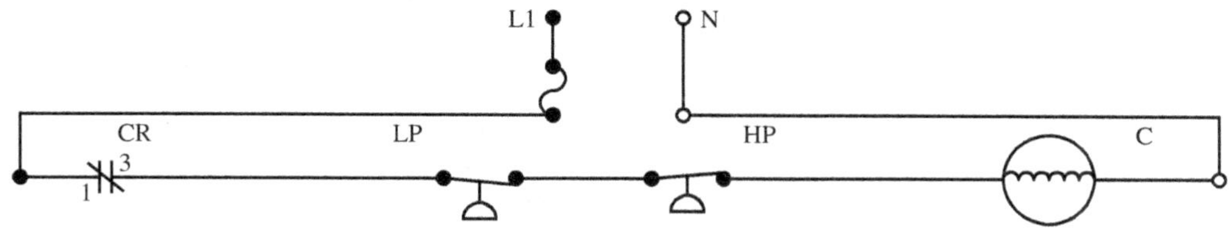

Figure 28-1 The problem for Service Call 1 has been isolated to the circuit shown here. This circuit has control relay contacts **CR 1** and **3**, low-pressure switch **LP**, high-pressure switch **HP**, and compressor contactor coil ©.

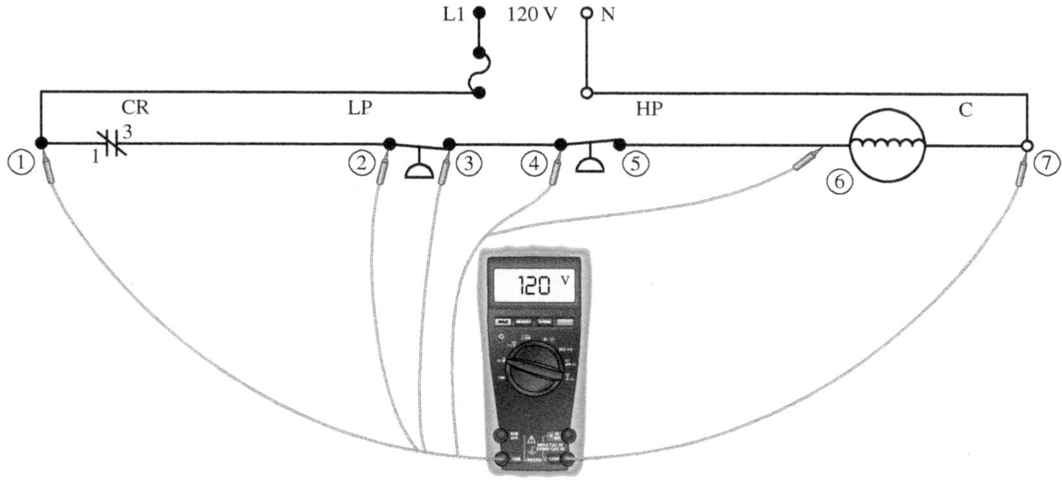

Figure 28-2 Hopscotch sequence: When the circuit is operating properly, attach one voltmeter probe to the right side of the circuit. With the second voltmeter probe, start measuring voltage on the left side of **CR**. The voltage will drop to 0 volts when it crosses the motor contactor ©.

Referring to Figure 28-2, follow the hopscotch sequence when the circuit is operating normally:

Step 1: When the circuit is operating properly, attach one voltmeter probe to the right side of the circuit, point 7.

Step 2: With the second voltmeter probe, start measuring voltage on the left side of **CR** (point 1).

Step 3: Supply voltage is measured on each side of the switch devices such as **CR**, **LP**, and **HP**, (points 2, 3, 4, 5, and 6).

Step 4: The voltage will drop to 0 volts when it crosses the motor contactor coil © at point 7.

TROUBLESHOOTING 28.1: HOPSCOTCHING A SHORT CYCLING SYSTEM

In going back to the short cycling problem, let's now investigate how to find the solution:

1. Figure 28-3 illustrates the reason why the compressor contactor is not energized continuously.
2. The low-pressure switch is open, across points 2 and 3, therefore no voltage can travel to the left side of the contactor coil ©.
3. Using hopscotch troubleshooting, you notice that the 120-V supply voltage is lost once the meter probe jumps to the right side of the low-pressure switch at point 3. The right side of the low pressure switch reads 0 volts, at point 3.
4. Since the **LPS** is open, 0 volts will be measured at remaining points 4 and 5. There is no need to make any more measurements when 0 volts is found to the right of an open component.
5. In Figure 28-4, you verify a 120-volt reading across the low-pressure switch at points 2 and 3. This voltage measurement indicates that the **LPS** is open. Is the switch open because it is defective or because the system pressure is low? The remaining steps will help you solve the specific problem.

Before we can isolate the problem hook up the manifold gauge set to determine if the system is low on charge.

(continued)

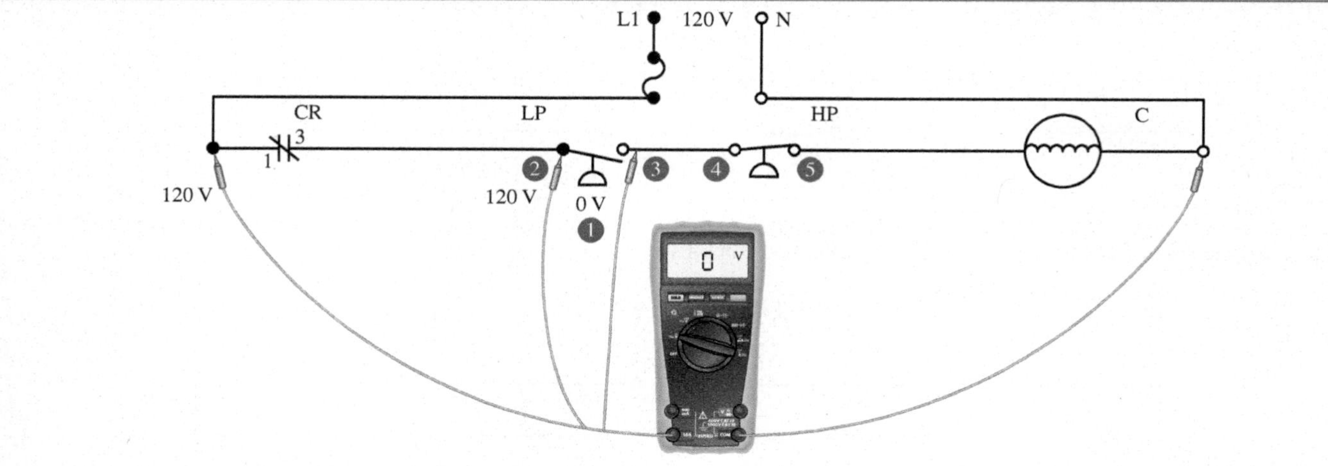

Figure 28-3 The hopscotch method of troubleshooting measures voltage starting at the left side of the diagram and measuring voltage as it jumps to the right side of the diagram. When voltage is lost, as is the case with the open low-pressure switch, the voltage on the right side of the pressure switch drops to zero volts.

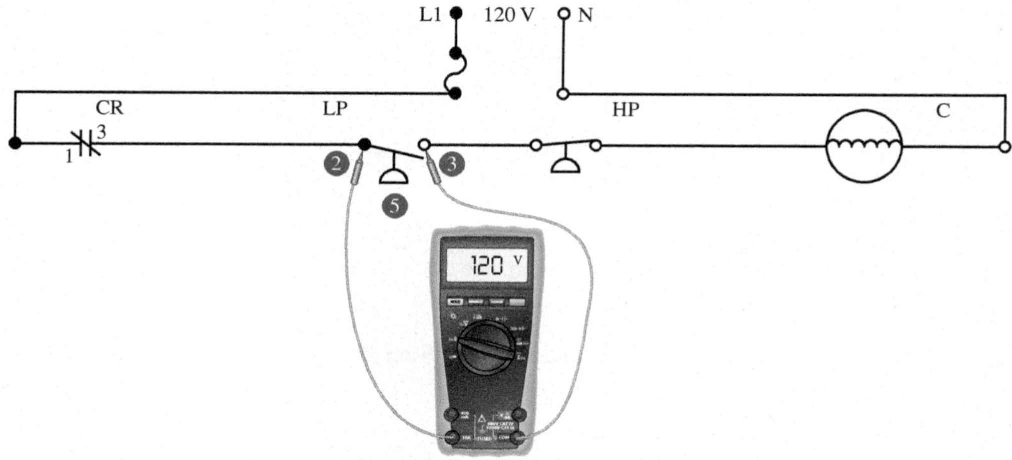

Figure 28-4 After determining that the problem is an open low-pressure switch, it is important to verify that this is really the problem.

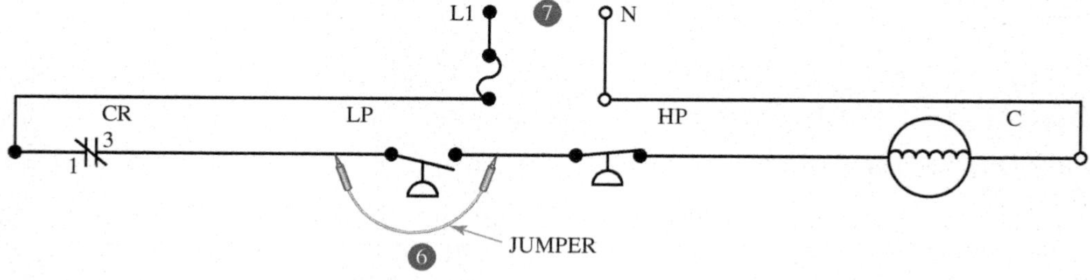

Figure 28-5 Jumper placed across an open low-pressure switch to verify it is defective.

If the suction pressure is high enough, then the low-pressure switch is defective.

6 To verify a defective low-pressure switch, first turn off the power. Then use a set of alligator clips or jumper wire to short across the defective switch as illustrated in Figure 28-5.

7 Turn the power back on.

The system should operate, unless there is another problem.

Watch the low-pressure gauge when the condenser starts to ensure that the suction pressure does not drop below it normal operating pressure.

The jumper should only be used as a temporary fix until you can get a replacement part. Too many times this temporary fix becomes permanent.

TECH TIP

This Tech Tip relates to the Troubleshooting 28.1 exercise. Once it has been determined that the problem is an open low-pressure switch, it is important to verify that this is the problem. Voltage will be measured across an open switch when voltage is applied to both sides of the circuit. This verifies that the low-pressure switch is open. It can also be verified by turning off the power to the circuit. Disconnect the low-pressure switch leads and check the switch with an ohmmeter. An open switch will measure infinity. If the switch is good and the pressure increases, the low-pressure switch will reset and the resistance will be near 0 Ω.

Finally, some problems are difficult to solve unless the event occurs when the technician is at the job site. For example, the low- or high-pressure switch may open due to random conditions that may occur periodically but do not occur when the technician is on the job. A good procedure for finding the culprit follows.

One way to determine if a safety device opens during operation is illustrated in Figure 28-6. Use a 1/10-amp fuse or smaller across any safety device that may open and reset when the adverse condition is created. When that safety component opens, the power will be directed through the fuse for a short time period until it blows. That is why a low-amperage fuse is used: It will blow because the current flow is higher than its rating. When the technician returns to the job, the blown fuse will be an indicator of the problem. It will not solve the problem, but it will narrow it down to the device that has the open fuse. The technician can then focus on the problems that would make that specific safety device open.

TECH TIP

Have a fuse holder and spare fuses available for troubleshooting. A fuse holder is shown in Figure 28-7. Use a glass fuse holder or automotive fuse holder. You will need very low amperage fuses to install across opening safeties and a 5-amp inline fuse to protect the secondary side of a 24-V, 40-VA control transformer. These are all installed on a temporary basis. The 5-amp transformer fuse is a good idea because a short circuit or a high load condition may have caused the transformer to open.

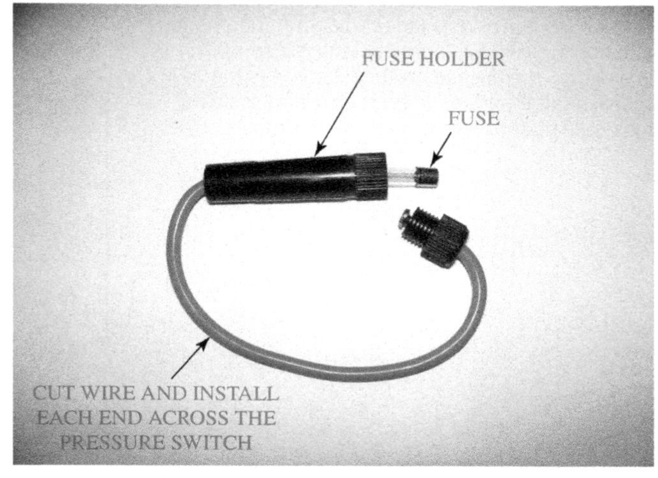

Figure 28-7 Techs should have a fuse and fuse holder with connection wires available for troubleshooting. The red wire is cut, insulation stripped, and then placed in series with the circuit it is to protect.

Time on the job: Initial time on this job is 1 hour looking for a problem that does not occur when the tech is there. Many times the problem is not solved if it is intermittent. There will be a return ticket to solve the problem if it occurs again. Total time on the job is undetermined at this time. It will depend on how quickly the intermittent problem occurs and can be identified.

SERVICE CALL 2: INADEQUATE HEATING

You are on a winter service call. The customer complaint is a lack of heating, especially at night. Use Figure 28-8 to determine the solution to this problem. Enlarged Electrical Diagram ED-1, found in the Electrical Diagrams package that accompanies this text, may also be used in this exercise. Here is the troubleshooting sequence of events:

- You arrive on the job and the heating system is working.
- Warm air is felt coming from the supply grille.
- Next, you check the amp draw of the heat strip. The heat strip is drawing about 20 amps, which is shown by the

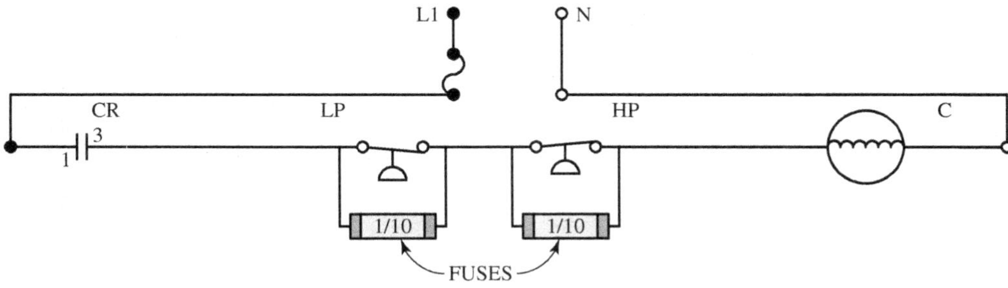

Figure 28-6 One way to determine which safety device opens during operation is to temporarily install a 1/10-amp fuse across any safety switch.

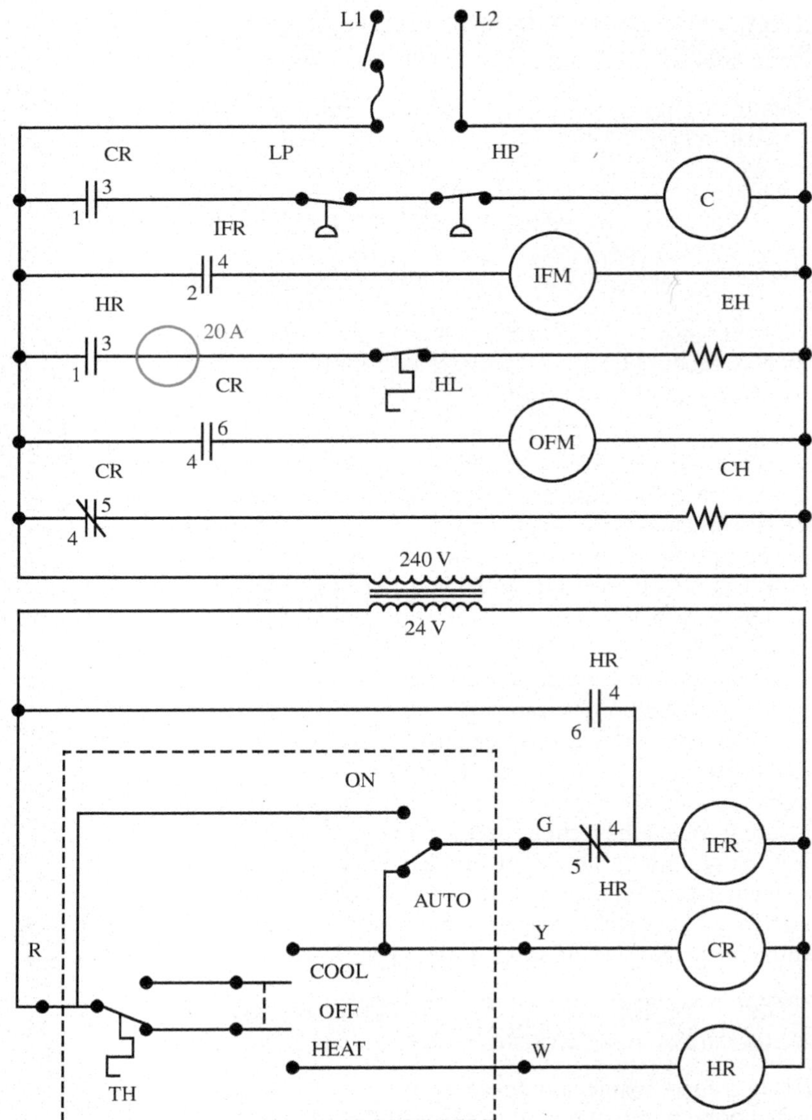

Figure 28-8 This is a ladder diagram for a heating and cooling package unit. The heating source is electric. The dashed rectangular box is the thermostat. Diagrams are not drawn to scale as indicated by the large size of the thermostat.

red circle on the diagram. This is in the normal range for a single-phase heat strip operating at 240 volts.

- After the electric furnace has been operating for about 15 minutes, the amp draw drops to zero.
- You check the thermometer on the thermostat and it is several degrees cooler than the temperature setting, therefore the heating system should still be operating.
- The blower is moving air, but it is getting cooler.

You begin to generate reasons why the heat strip is cutting out. What are the possibilities? Look at the diagram and see what you think. What are some reasons the heating system is cutting out?

From previous experiences the most likely problem is the high limit, **HL**, is opening. It may be opening at a lower than normal temperature or possibly low airflow is causing the heat strip to overheat and open the high limit.

Low airflow or short cycling could be caused by:

- ✓ A dirty filter
- ✓ A dirty evaporator
- ✓ Other duct restrictions
- ✓ The blower speed could be set too low or the blower could be slowing down after it warms up. Sometimes bearing wear causes the bearings to drag and slow down the motor. Eventually the drag will overheat the motor and the motor will cut out on its internal overload.
- ✓ Another problem could be the thermostat opening prematurely and stopping the heating process.

To find the exact problem you should use the hopscotch troubleshooting method to determine which component is opening in the electric heat **EH** line in Figure 28-8. Voltage will be lost across the open component. Doing hopscotch troubleshooting you determine that the high limit, **HL**, is opening. Here are the voltage readings:

- The left side of **HL** has 240 volts.
- The right side has 0 volts; 240 volts is measured across the open **HL**.

- Now you need to determine if the high limit is defective or if it is an airflow problem.
- The temperature rise across an electric strip heat is about 50° to 60°F. In other words, if the return air is 70°F, then the supply air should be in the range of 120° to 130°F. Higher supply air temperatures indicate that the airflow may be low. You measure 145°F from the supply grille nearest the electric furnace.

This is an indication of a lack of airflow. What are some conditions that will cause low airflow? Here are some steps to take when investigating low airflow conditions:

1. Check the air filter. Inspection reveals that it is fairly clean.
2. Inspect the evaporator. Inspection reveals a mat of dirt.
3. After cleaning the evaporator, you inspect for other airflow restrictions.
4. The blower wheel is caked with dirt. The blower wheel is removed from the motor shaft and taken outside and cleaned. No other duct restrictions are noticed.

After doing this service, measure the rise in temperature again. The supply air temperature has dropped from 145° to 120°F. Verify that this solved the problem by watching the furnace operate until it has satisfied the thermostat setting of 70°F. The amperage on the heat strip was measured while in operation. The heat strip did not cut out.

The service work you performed will also help the performance of the cooling system next spring. You recommend an annual clean and check program so that this will not happen again. *Time on the job:* Total time on the job was 2 hours. Most of the time involved cleaning and maintenance of the system. This is routine maintenance and is something you should encourage the customer to schedule with your company.

TECH TIP

When changing out a transformer that controls cooling and uses gas heat, be sure to check the polarity of the secondary of the transformer. The secondary has a hot side and common side. For example, let's say another tech changed a defective transformer during the cooling season. When the heating season arrived, the gas furnace acted erratically. Sometimes it would heat and other times it would not. The circuit board was blamed. The thermostat was blamed. Bad connections were blamed. It turns out that the tech did not observe the transformer polarity when changing it in the summer. Switching the position of the two secondary wires did the trick.

SERVICE CALL 3: "NO COOLING" CALL FOR A 5-TON PACKAGE UNIT

This service call sends you to a small office space with a 5-ton package unit, as shown in Figure 28-8. The system is silent, nothing is operating. The "silent treatment" is usually the easiest type of problem to solve. After meeting with the building manager, you notice that the thermostat is set for cooling and the thermostat setting is 10°F below the room temperature of 85°F. You place the thermostat in the Fan On mode and the blower begins to operate. What does this tell you?

This should tell you that there is supply voltage and that the secondary of the transformer also has 24 volts. Without 24 V on the secondary side of the transformer, the (IFR) coil in the control circuit would not energize the **IFR** contacts to operate the indoor fan motor (IFM). This simple check will show you that both supply and control voltage are present in the package unit. Here is the sequence for solving this problem:

1. Using the diagram in Figure 28-8, you hopscotch the thermostat circuit by attaching one voltmeter probe to the right side of the 24-V transformer.
2. Next you measure 24 volts on the **R** connection. The **R** connection will be on the t'stat or the red wire feeding the t'stat.
3. Voltage is lost when jumping over the thermostat to the **Y** connection. This indicates that the thermostat is open.
4. Measuring across **R** and **Y**, you find 24 volts, again verifying that the t'stat is open.
5. Before replacing the thermostat you want to be absolutely sure that the problem is the thermostat. You place a jumper wire across the **R** and **Y** terminals and the system begins to operate normally.

You report the problem to the building manager who approves the repair. You notify the manager that you will leave the jumper connected until returning from the supply house with a replacement thermostat in 30 minutes. This will start the cooling process while you are gone.

On your return, the replacement is installed and system operation verified. While waiting for the system to cycle, you confirm that the refrigerant charge is correct. It would be an embarrassment if the system were low on charge, requiring an additional service call scheduled for this overlooked problem. A complete system survey is important. Your doctor does a complete check-up even if you only have a sore throat. The HVACR "doctor" should do the same.

TECH TIP

When going on service calls, always check in with the owner, building manager, or person requesting the service. Ask them to explain the problem and show you where the equipment is located, including both inside and outside units. Ask for a brief history of the problem and review any service tickets from work previously completed on the equipment. This will ensure that you are working on the correct system, plus you will have a better understanding of the problem. The service calls discussed in this unit recommend that you complete these steps prior to beginning the troubleshooting process. This is part of what we like to call "customer service." Discuss the cost and outcome of your troubleshooting before doing any work and also when completing the work.

SERVICE CALL 4: INADEQUATE COOLING

Your last call of the day involves inadequate cooling at a residence. The compressor is short cycling (cutting on and off) about every 3 to 4 minutes. You find that the thermostat setting is okay. The airflow is good. The filter is clean. The condenser coil looks clean. It seems to be an electrical problem. Troubleshooting 28.2 follows the steps that will help you determine what is happening.

TROUBLESHOOTING 28.2: SHORT CYCLING COMPRESSOR

Use Figure 28-9 or enlarged Electrical Diagram ED-8, found in the Electrical Diagrams package that accompanies this text, for this exercise. Using the hopscotch procedure, you measure the following voltages:

1. 240 volts is measured on the left and right sides of the **CR** contacts (point 1).
2. If the system is energized, **CR** will be closed between points 1 and 2.
3. 240 volts is measured on the left of the high-pressure switch **HPS** (point 3).

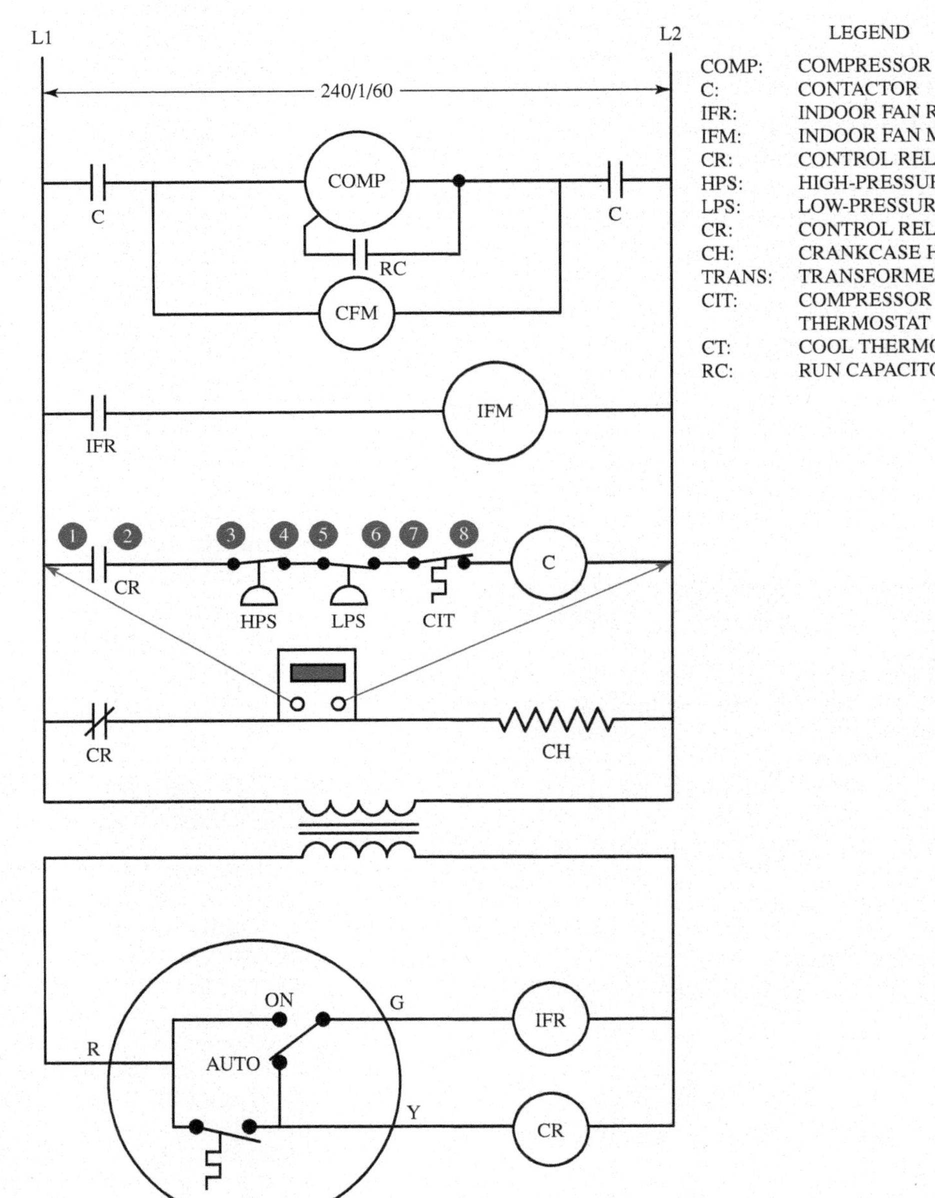

Figure 28-9 The ladder diagram is used for hopscotch troubleshooting. The VOM is hooked up for the troubleshooting procedure. This is a simplified cooling diagram without heat.

4 240 volts is measured on the right side of the high-pressure switch **HPS** (point 4).

5 240 volts is measured on the left side of the low-pressure switch **LPS** (point 5).

6 0 volts is found on the right side of the low-pressure switch **LPS** (point 6).

7 0 volts is found on the left side of the compressor internal thermostat **CIT** (point 7).

8 0 volts is found on the right side of the compressor internal thermostat **CIT** (point 8).

What is the answer to this problem?

One important troubleshooting tool left out in this exercise is the pressure readings on the manifold gauge set. This is an open low-pressure switch (**LPS**) problem. Is the pressure switch open because of low pressure or is the low-pressure switch defective, opening at a higher than normal pressure setting? Was the low-pressure switch recently changed? If so, was the correct one installed? For instance, the low-pressure switch cut-out for systems using R-410A is much higher than the low-pressure switch for R-22 systems. The R-410A low-pressure switch will get to opening pressures quicker if installed on a R-22 system. Another consideration is the state of the evaporator coil and air filter—are they clean? The coil may appear clean but it should be flushed anyway. Dirt could be embedded between the coil fins.

Before replacing the low-pressure switch, confirm that the switch is the problem. Jumper the switch and check the low-side pressures. The system may be low on charge and/or have low airflow. After the low-pressure switch is replaced, create a low-pressure condition and test the operation of the pressure switch. This will check the operation of the switch and your wiring job. You can also test the operation pressure of the switch using nitrogen pressure hooked to the manifold gauge set and power switch.

SAFETY TIP

Many high-pressure switches do not have a Schrader valve fitting under the screw-on high-pressure connection point. Refrigerant must be recovered prior to removing the high-pressure connection or you will lose the refrigerant.

SERVICE CALL 5: LACK OF COOLING

Refer to the package unit shown in Figure 28-10 (or enlarged Electrical Diagram ED-5 packaged with this text) as you try to solve this cooling problem. After a quick 10-minute survey of the job, you determine that the compressor is short cycling about every 4 or 5 minutes. The indoor blower section and condenser fan are working properly. You take control of the system at the unit disconnect by shutting the system down and installing the gauges and clamp-on ammeter on the compressor common lead of the single-phase compressor.

Once the test instruments have been connected, you close the disconnect, which will supply power to the package unit. The system has a time-delay circuit controlled by **TM**, therefore the system will not come on for about 5 minutes. Once the system comes on, the pressures stabilize to their normal readings. The clamp-on ammeter, which is connected to the common on the compressor, is running a couple of amps higher than RLA. After the system operates for about 10 minutes, the compressor shuts down. You wait for the timing circuit to finish, and the compressor comes back on again with normal pressures and the higher than normal compressor amperage. During this cycle the compressor cuts out a little quicker. What is causing this problem?

The focus seems to be the high amperage draw. What safety device is causing the compressor to shut down? Reviewing the diagram you notice that the compressor has two overload devices labeled $\mathbf{OL_1}$ and $\mathbf{OL_2}$. $\mathbf{OL_1}$ is used to protect the compressor by monitoring amperage draw through the common terminal. $\mathbf{OL_2}$ protects the start winding from a high-current condition.

(*Note:* The $\mathbf{OL_1}$ and $\mathbf{OL_2}$ thermal sensors located in the high-voltage area around the compressor have contacts located in the high-voltage control section near the timer motor **TM**.) The normally closed contacts associated with $\mathbf{OL_1}$ and $\mathbf{OL_2}$ are found in the high-voltage circuit between the high-pressure switch (**HPS**) and the timer circuit connection $\mathbf{B_2}$. You want to be certain that the $\mathbf{OL_1}$ contacts are opening before continuing the diagnosis. The **HPS** and $\mathbf{OL_2}$ are also in that line. It is not likely that either of these components is the problem since the system head pressure is not high and the start winding is essentially out of the circuit a second after the motor starts. You should confirm this, however, before continuing.

To verify the problem, you set up your DMM to hopscotch the circuit. Use Figure 28-11 to focus in on this part of the troubleshooting sequence. One probe is attached to the right side of the diagram, $\mathbf{L_2}$. With the unit operating and preparing to use the hopscotch troubleshooting method, you record the following voltages:

- 230 volts on the left side of **C**
- 230 volts on the right side of **C**
- 230 volts on the left side of contact $\mathbf{OL_1}$
- 230 volts on the right side of contact $\mathbf{OL_1}$
- 230 volts on the left side of contact $\mathbf{OL_2}$
- 230 volts on the right side of contact $\mathbf{OL_2}$.

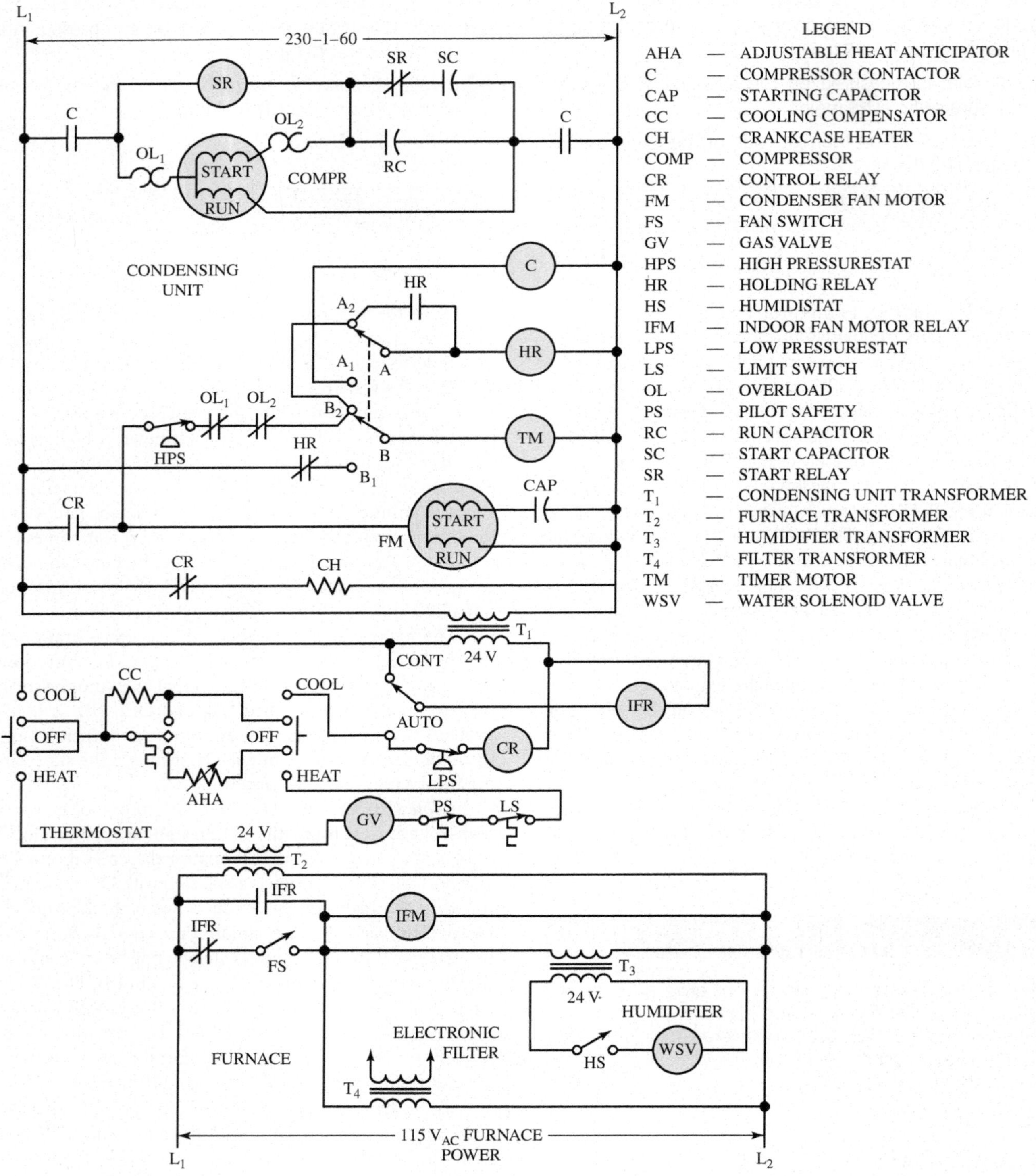

Figure 28-10 This is a diagram of a gas cooling and standard air conditioning system. The legend and ladder diagram format make it a valuable tool for troubleshooting.

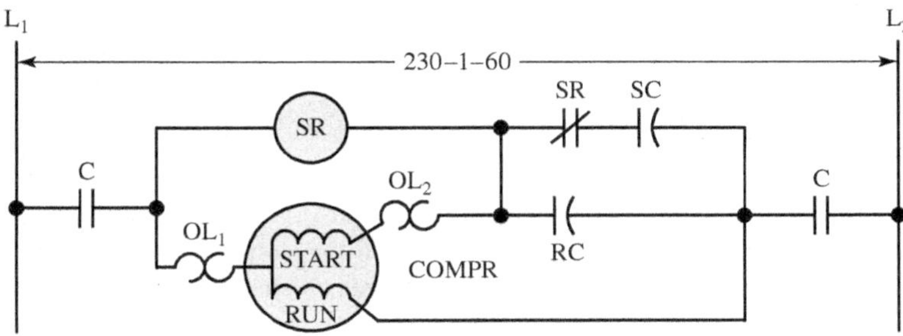

Figure 28-11 This diagram is an excerpt from Figure 28-10 used for Service Call 5.

After 10 minutes of operation, the compressor shuts down. You notice the clamp-on ammeter dropping to zero and that the high- and low-side pressures on the manifold gauge set have begun to equalize. You go back to the hopscotch measurements again and record the following readings:

- 230 volts on the left side of **C**
- 230 volts on the right side of **C**
- 230 volts on the left side of contact $\mathbf{OL_1}$
- 0 volts on the right side of contact $\mathbf{OL_1}$
- 0 volts on the left side of contact $\mathbf{OL_2}$
- 0 volts on the right side of contact $\mathbf{OL_2}$.

Once you read 0 volts on the right side of $\mathbf{OL_1}$, you can stop the hopscotch troubleshooting process and measure directly across the $\mathbf{OL_1}$. The supply voltage, 230 volts, will be measured directly across the open safety. This safety device is a thermal or temperature-sensitive device, therefore when it cools it will close the contacts and reset to its normally closed state.

After feeling assured that the problem is that a high-amperage condition is causing $\mathbf{OL_1}$ to open, you must decide what the reasons are for the high amperage draw. What do you think?

Here is a list of possible problems:

1. *Low operating voltage.* This could be caused by the energy provider. Also, carbonized contactor points create resistance and lower voltage supply to the compressor. Check the voltage on the output of the contactor when the compressor is operating. Check the voltage drop across the closed contacts with operation. The voltage drop across the closed contacts should be zero volts, but no more than 3 volts.
2. *Tight compressor motor bearings.* These will cause high-amperage operation because the motor is turning under a lot of resistance. This problem is not possible to verify without a compressor teardown and autopsy.
3. *Defective run capacitor.* An open or weak run capacitor will cause the motor to draw more amperage. Check the capacitor microfarad rating with a capacitor checker. If it is off by more than ±10%, replace the capacitor. Check the replacement capacity to verify it is within range of the required run capacitor. Some caps only allow a ±5% deviation from the microfarad rating.
4. *Undercharged system.* An undercharged system will cause the compressor motor to overheat and trip on its thermal overload. Even if the pressures appear good, it is recommended that you check superheat and subcooling. If it is undercharged the superheat will be high and the subcooling will be low. The operating pressure could be close to normal even if the charge is low.
5. *Bad connection or partially burned wire.* Either of these will create a high-amperage condition.

SERVICE CALL 6: NO COOLING—SHORT CYCLING COMPRESSOR

Your next service call is a "no cooling" problem and it is a real challenge. When calling to report the problem, the owner of the small, leased retail space said, "Nothing is working, not even the display on the digital thermostat." Arriving at the job, you verify that the package unit is not working. We will use the diagram in Figure 28-10 again to troubleshoot this service call. You find that there is no line voltage input to $\mathbf{L_1}$ and $\mathbf{L_2}$. You turn off the disconnect to take control of the situation. Next, you locate the circuit breaker and notice that the breaker is tripped. Turning your face away and stepping to the side of the electrical panel you safely reset the breaker. The breaker stays set. Energizing the disconnect causes the breaker to trip again. How are you going to solve this problem?

The system is not going to operate long enough to take any readings. Here are steps to succeed in finding the solution:

1. The first thing to do is to isolate circuits to find out what is working. This can be done by turning the thermostat to the off position. If the breaker trips with the thermostat off, the most likely problems are the transformer or a short in the $\mathbf{L_1}$ or $\mathbf{L_2}$ line. A short crankcase **CH** could also trip a breaker since it is energized when the unit is not in the cooling mode. All of these components would be energized with the thermostat set to the off position and the power applied.

2. The t'stat is off. In this case the breaker does not trip with the thermostat in the off position. First switch the thermostat to continuous fan position or **CONT**. The breaker stays closed. No problem.
3. Next, try the heating circuit. It is not likely to be in the heating circuit since the unit is operating in the cooling mode when the problem occurs. The heating circuit is working and does not trip the breaker. Turn the cooling thermostat up several degrees above the room temperature so that the cooling system will not come on. Turn the thermostat to the cooling position. The breaker does not trip.
4. Finally, setting the thermostat below the room temperature trips the breaker. You have isolated the problem to the compressor circuit. With the power off, you measure the resistance of the compressor windings. The compressor start and run windings are correct. To safely measure the windings, the windings are measured on wires running to common, start, and run. Resistance is not measured directly on the terminals. Resistance measurements are 5 Ω on the run winding and 12 Ω on the start winding. Between the run and start windings, the resistance is 18 Ω. This is close enough to be considered normal. Measuring from the common compressor terminal to ground reveals a problem. The resistance to ground is 120 Ω. Resistance to ground should be at least 1,000 ohms per volt or in this case 230,000 ohms or higher to ground. Most compressors measure millions of ohms to ground. Is the compressor shorted? If so, should you replace it? Or should you take another step?
5. After replacing the compressor, the filter drier is changed. The system is leak checked. A deep vacuum is pulled and the system is recharging with new refrigerant. The system is started up with the same results—a blown breaker. The new compressor terminals were not checked prior to or after installation. It turns out the wire feeding power to the common side of the compressor is shorted to ground. It was not obvious, since the wire was shorted as it passed through a sheetmetal partition in the condenser section.

Lesson learned: After condemning a compressor, recover the refrigerant. Measure common, start, and run terminals to ground from the compressor terminals with the wires removed, not through wires on the contactors or other switches.

In summary, it is a good safety practice to measure **CSR** away from the compressor terminals since they may blow out. If there is a problem, first recover the refrigerant. Next, remove the compressor wires and measure resistance again.

SERVICE CALL 7: NO HEATING

To troubleshoot a heating circuit, you must understand its basic operation. Now we will trace through the gas heating circuit shown in Figure 28-10 to understand how to solve the problem in this circuit.

Place the two thermostat switches to **HEAT** (left and center, on the diagram). Turn the t'stat setting up so that the heating contacts close to bring on the furnace. In this example the bimetal element will drop and complete the circuit to **AHA**, the adjustable heat anticipator. With this action the 24-V transformer will provide power to the gas valve (GV), and through pilot safety **PS**, limit switch **LS**, and back through the thermostat. The gas valve should open and provide gas to the burners. The problem in this case is an open (GV) coil. An open coil will prevent the gas valve from opening, meaning no gas to the burners. It may be easier to use Figure 28-12 for this explanation, since it is extracted and isolated from Figure 28-10. A recommended troubleshooting sequence to find the solution follows.

TROUBLESHOOTING 28.3: NO GAS HEAT

1. Start troubleshooting by attaching one DDM probe to the right side of **LS**, blue point 1. A 24-V reading will be measured on the right side of the transformer, **T2**.
2. Jumping over to the right of (GV) you will read 0 V.
3. Next, you measure 24 volts across the gas valve to verify that the voltage is applied directly to the gas valve (points 2 and 3).

Turn off the power to the unit. Finally, disconnect the control wires to the gas valve at points 2 and 3 and ohm out the gas valve. The valve coil is open and is thus defective.

Caution: Some gas valve coils have very high resistance coils. If you do not obtain a resistance reading, switch the ohmmeter to the highest resistance and measure again. At a minimum your ohmmeter should be able to read a couple of million ohms.

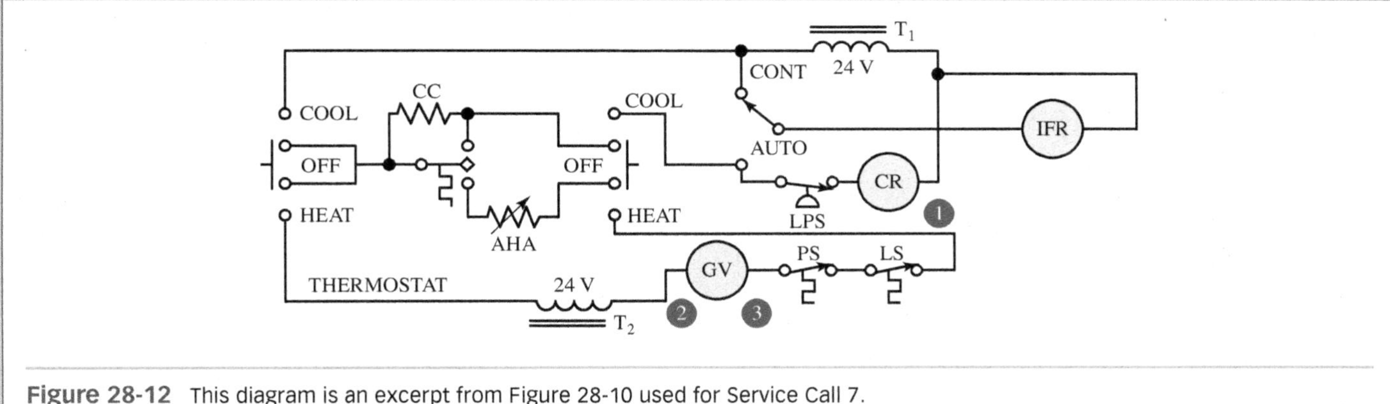

Figure 28-12 This diagram is an excerpt from Figure 28-10 used for Service Call 7.

TECH TIP

Some gas valve coils have very high resistance. Gas valve coil resistances in excess of 1 megohm can be found. It is important that your meter be able to read this high resistance or you will be condemning many good gas valves because you think they are open. A gas valve can have good resistance, yet be mechanically closed and therefore defective. Select an ohmmeter with megohm capabilities.

SERVICE CALL 8: NO CHILLED WATER FROM COMPRESSOR 1

We will use Figure 28-13 (or refer to enlarged Electrical Diagram ED-13 packaged with this text) for this service call.

This problem relates to a small chilled water system that has two separate compressor circuits. Compressor circuit 1 is off, while compressor circuit 2 is trying to keep the office building cool. The space temperature has increased a couple of degrees after an alarm notified the building manager that compressor 1 was down. Upon arriving at the call site, an inspection reveals that compressor 1 is off and the water flow in this circuit is good. You hook up the manifold gauge set and place the clamp-on ammeter on one of the leads to the compressor. Reset (turn off and on) compressor circuit 1. The compressor operates for almost 3 minutes then shuts down again. The pressures were coming up to near-normal conditions prior to shutting down. The amp draw was not excessive. What is the problem?

Let's explore how some voltage readings can help us. Figure 28-14 is an enlarged excerpt of Figure 28-13 and will useful in pinpointing sections of the diagram during this explanation.

Attach the DMM probe to the right side of the 120-V transformer labeled point 2. Start the hopscotch troubleshooting process at point 5, the far left side of the diagram, in the rectangle labeled **SS1**. You restart the compressor 1 circuit a second time and it fails to operate for more than a few minutes. You then decide to see what information can be determined from the circuit in the failed or "off" condition. Here are the voltage readings in the control circuit for **COMP1**:

- Starting at the left side (point 5) of **COMP1**, 120 V is present. The red arrow points to this section.
- Right side (point 9) of **COMP1**, 120 V.
- Both sides (points 9 and 11) of **CWFS**, 120 V.

Use Figure 28-15, an excerpt from Figure 28-13, to continue tracing the following points in the troubleshooting process:

- Both sides (points 11 and 13) of **F/S1**, 120 V.
- Both sides (points 13 and 15) of **T/S1**, 120 V.
- Both sides (points 15 and 17) of **LPS1** and (points 17 and 19) **HPS1**, 120 V.
- Both sides (points 19 and 21) of **TMR1**, 120 V.

Use Figure 28-16, excerpted from Figure 28-13, to continue tracing the following points:

- Both sides (points 21 and 23) of **C1-NO**, 120 V.
- 120 V at **L** (point 25) of **OFS1**.
- 0 V at **M** (point 27).
- Both sides (points 27 and 2) of **C1-OL**, 0 V.

Based on this information, what is the problem?

The contacts between **L** and **M** are open (points 25 and 27). This is the contact for the oil pressure switch #1. The oil pressure safety switch works on a pressure differential between the suction pressure and oil pump pressure. The oil safety switch has a start-up time delay of 2 to 3 minutes. When the compressor starts the suction pressure and oil pump pressure are nearly equal. The time delay allows the

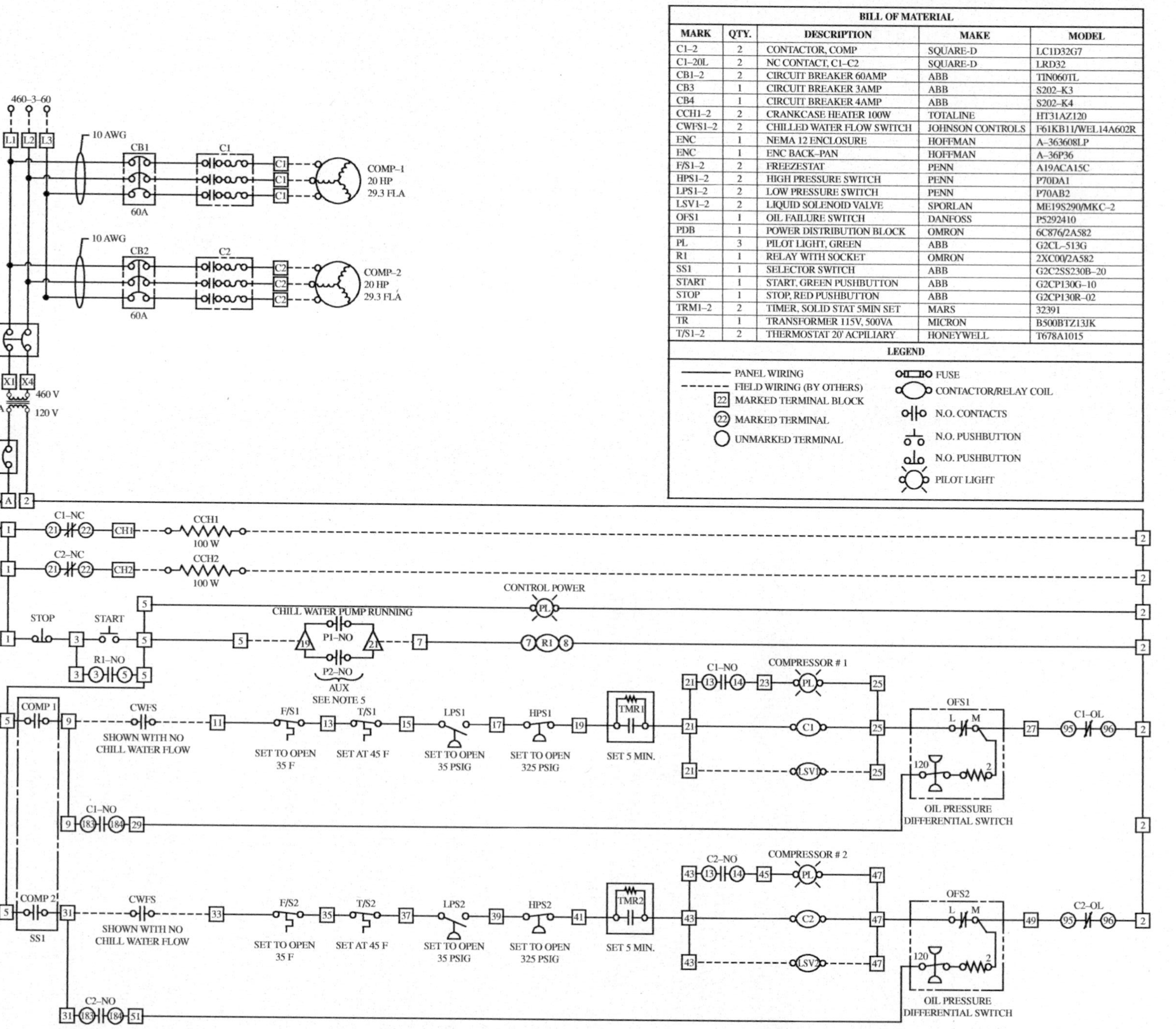

BILL OF MATERIAL				
MARK	QTY.	DESCRIPTION	MAKE	MODEL
C1–2	2	CONTACTOR, COMP	SQUARE-D	LC1D32G7
C1–20L	2	NC CONTACT, C1–C2	SQUARE-D	LRD32
CB1–2	2	CIRCUIT BREAKER 60AMP	ABB	TIN060TL
CB3	1	CIRCUIT BREAKER 3AMP	ABB	S202–K3
CB4	1	CIRCUIT BREAKER 4AMP	ABB	S202–K4
CCH1–2	2	CRANKCASE HEATER 100W	TOTALINE	HT31AZ120
CWFS1–2	2	CHILLED WATER FLOW SWITCH	JOHNSON CONTROLS	F61KB11/WEL14A602R
ENC	1	NEMA 12 ENCLOSURE	HOFFMAN	A–363608LP
ENC	1	ENC BACK–PAN	HOFFMAN	A–36P36
F/S1–2	2	FREEZESTAT	PENN	A19ACA15C
HPS1–2	2	HIGH PRESSURE SWITCH	PENN	P70DA1
LPS1–2	2	LOW PRESSURE SWITCH	PENN	P70AB2
LSV1–2	2	LIQUID SOLENOID VALVE	SPORLAN	ME19S290/MKC–2
OFS1	1	OIL FAILURE SWITCH	DANFOSS	P5292410
PDB	1	POWER DISTRIBUTION BLOCK	OMRON	6C876/2A582
PL	3	PILOT LIGHT, GREEN	ABB	G2CL–513G
R1	1	RELAY WITH SOCKET	OMRON	2XC00/2A582
SS1	1	SELECTOR SWITCH	ABB	G2C2SS230B–20
START	1	START, GREEN PUSHBUTTON	ABB	G2CP130G–10
STOP	1	STOP, RED PUSHBUTTON	ABB	G2CP130R–02
TRM1–2	2	TIMER, SOLID STAT 5MIN SET	MARS	32391
TR	1	TRANSFORMER 115V, 500VA	MICRON	B500BTZ13JK
T/S1–2	2	THERMOSTAT 20' ACPILIARY	HONEYWELL	T678A1015

Figure 28-13 This diagram is for a two-stage chilled water system. Each compressor is controlled by a separate control circuit.

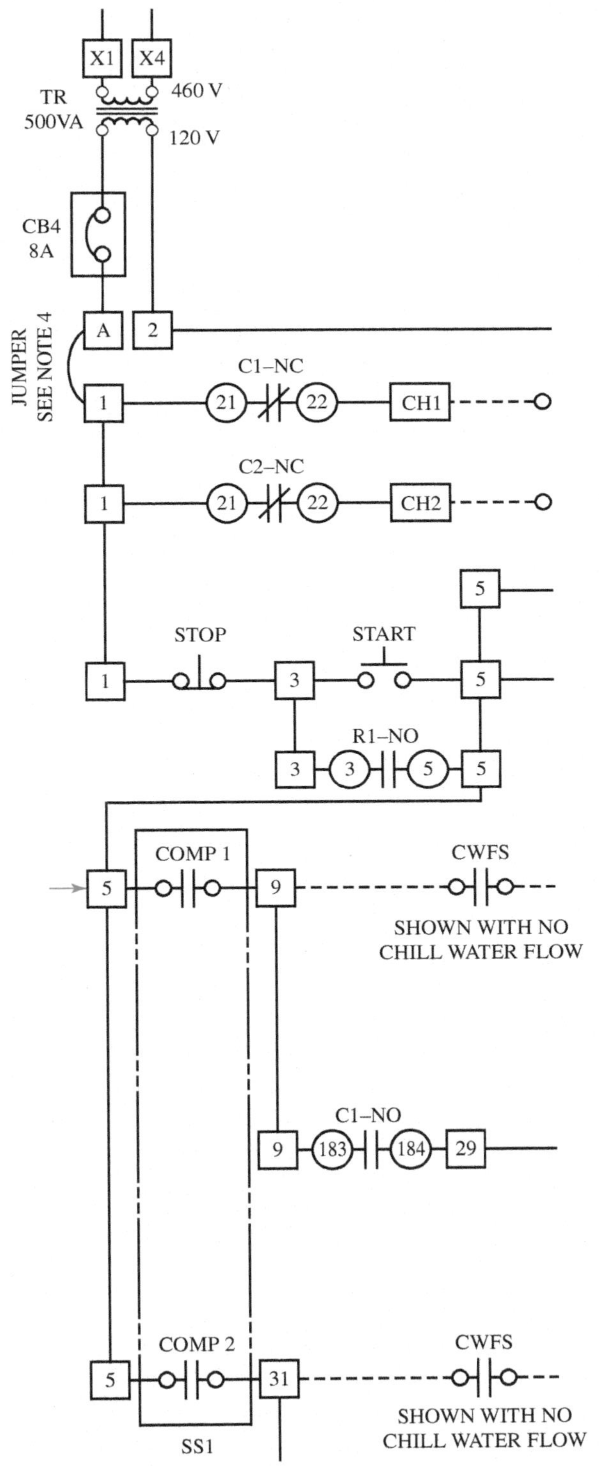

Figure 28-14 Excerpt from Figure 28-13, used for Service Call 8.

oil pressure differential to develop. On start-up, the suction pressure will drop and the oil pressure will rise. This will develop the pressure differential needed to keep the safety device closed. Low differential pressure will open the contacts between **L** and **M**.

Now you ask yourself, "What causes a low-pressure differential?"

- The compressor crankcase could be low on oil. This is not a likely problem unless the oil is trapped in the evaporator or stuck in the oil separator. Oil leaking from a compressor will be obvious. Leaking oil is very unlikely and would be accompanied by a low refrigerant charge.
- Flooded starts can create a low pressure differential. Refrigerant in the compressor crankcase will be picked up by the oil pump. The oil pick-up is at the bottom of the compressor crankcase. Refrigerant is heavier than oil and settles down to the oil pick-up tube located near the bottom of the crankcase. The liquid refrigerant will boil off to a vapor as it is being pumped. This prevents the oil pressure from developing. The oil pump is not a vapor pump. A solution to this problem includes using an operational crankcase heater and ensuring that liquid refrigerant does not return to the compressor on start-up. Looking at the piping design is important. Refrigerant draining from the evaporator during the off cycle will create flooded starts. Dips in a horizontal suction line will trap refrigeration and oil and cause a flooded start.
- Refrigerant side problems can create problems that manifest as an electrical problem. For example, liquid flooding or liquid dilution of the oil will cause bearing wear. The worn bearing will cause the rotor to fall and start grinding into the stator winding. This will short the stator winding, which will be interpreted as an electrical failure.

In this case the problem is not electrical even though the control circuit is opened by an electrical safety device. Simultaneously check both suction pressure and oil pump pressure on start-up to be sure it is a refrigerant-side problem and not an electrical problem. In this case, you find that the crankcase heater **CCH1** is open, which allows refrigerant to migrate to the compressor during the off cycle, thus creating a flooded start and a lack of oil pressure differential.

TECH TIP

Figure 28-13 has an interesting safety device. It is called the chilled water flow switch (**CWFS**) between points 9 and 11. The purpose of the flow switch is to ensure that water flows in this chilled water system before the refrigeration circuit begins to operate. Slow water flow or no water flow will cause low suction pressure with a possibility of freezing the water in the chilled water system. The frozen water will expand and cause damage to the chiller tubing. If the tubes rupture, the water and refrigerant will mix, damaging the compressor. Low water flow will open the flow switch **CWFS** and stop that compressor circuit operation.

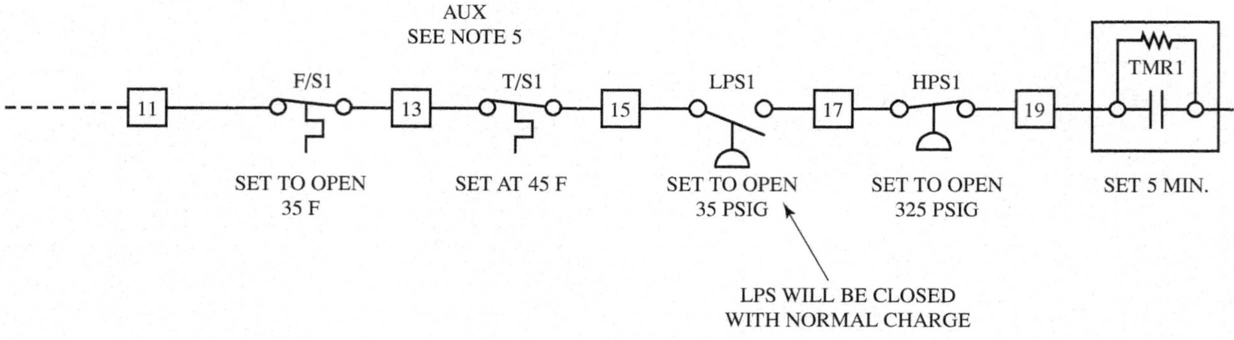

Figure 28-15 Excerpt from Figure 28-13, used for Service Call 8.

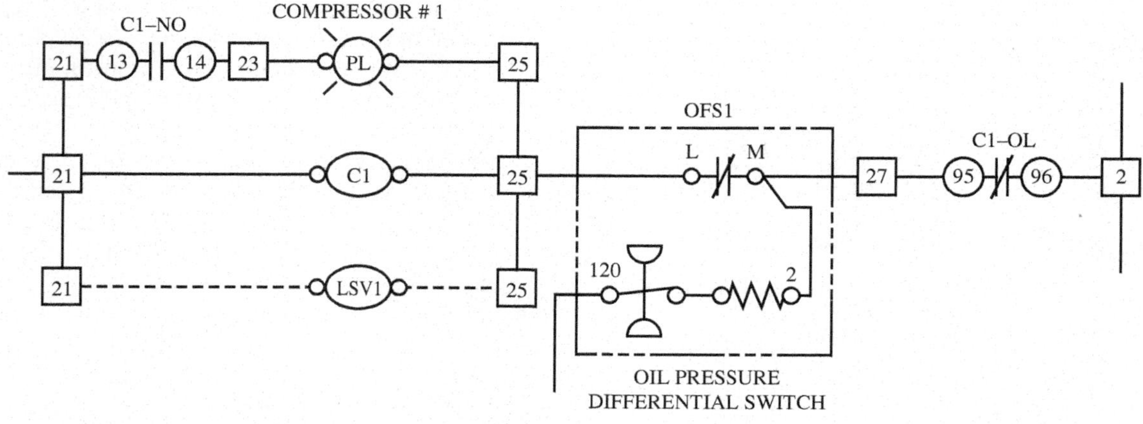

Figure 28-16 Excerpt from Figure 28-13, used for Service Call 8.

SERVICE CALL 9: NO CHILLED WATER FROM COMPRESSOR 2

The owner of a medium-size office building has complained that the air conditioning system does not cool very well in the afternoon. The owner stated that the temperature in his building begins to drift higher around 1 p.m. and continues to rise a degree every 2 hours until the office closes at 6 p.m.

After a quick system survey, you notice that the second compressor circuit is off. You install your gauge set on the nonoperating circuit and the clamp-on ammeter on one of the leads on the compressor. The nameplate reveals that the unit uses R-22. About every 5 minutes the second compressor circuit comes on for about 1 minute and then shuts down. Observing the gauges on the manifold, you notice that the low-side gauge drops below 35 psi and the high-side pressure rises to about 225 psi. At this instance circuit compressor 2 cuts out. The amp draw is below the RLA. What is the problem? Use Figure 28-13 or enlarged Electrical Diagram ED-13 to determine your answer.

1. The low-pressure switch is opening on the low-pressure cut-out of 35 psi. When the low-pressure switch is open, 120 volts will be measured across it.
2. The low-pressure switch will stay open for about 1 minute or until the suction pressure rises high enough to close the low-pressure switch.
3. After a 5-minute time delay, compressor circuit 2 will return to operation.
4. The solution is to bypass the low-pressure switch and charge up the refrigeration circuit. Remove the bypass before leaving the job. Check and repair refrigerant leaks. Let the customer know what you did and explain the invoice. It is best to give the owner an estimate prior to doing a repair or providing service.

Other than low pressure, what else could cause low suction pressure?

Low pressure on a chilled water system could be caused by low water flow in the chilled water system. Like low airflow, low water flow in the evaporator section will cause low pressure. It is best to cut off the compressor rather than allow the chilled water to freeze and then crush, expand, or crack the water tubing.

SERVICE CALL 10: NO CHILLED WATER PUMP OPERATION

You are called to a large building with a chilled water system. The chilled water pump diagram shown in Figure 28-17 and enlarged Electrical Diagram ED-14, packaged with this text, is required to operate prior to the chilled water system starting. You

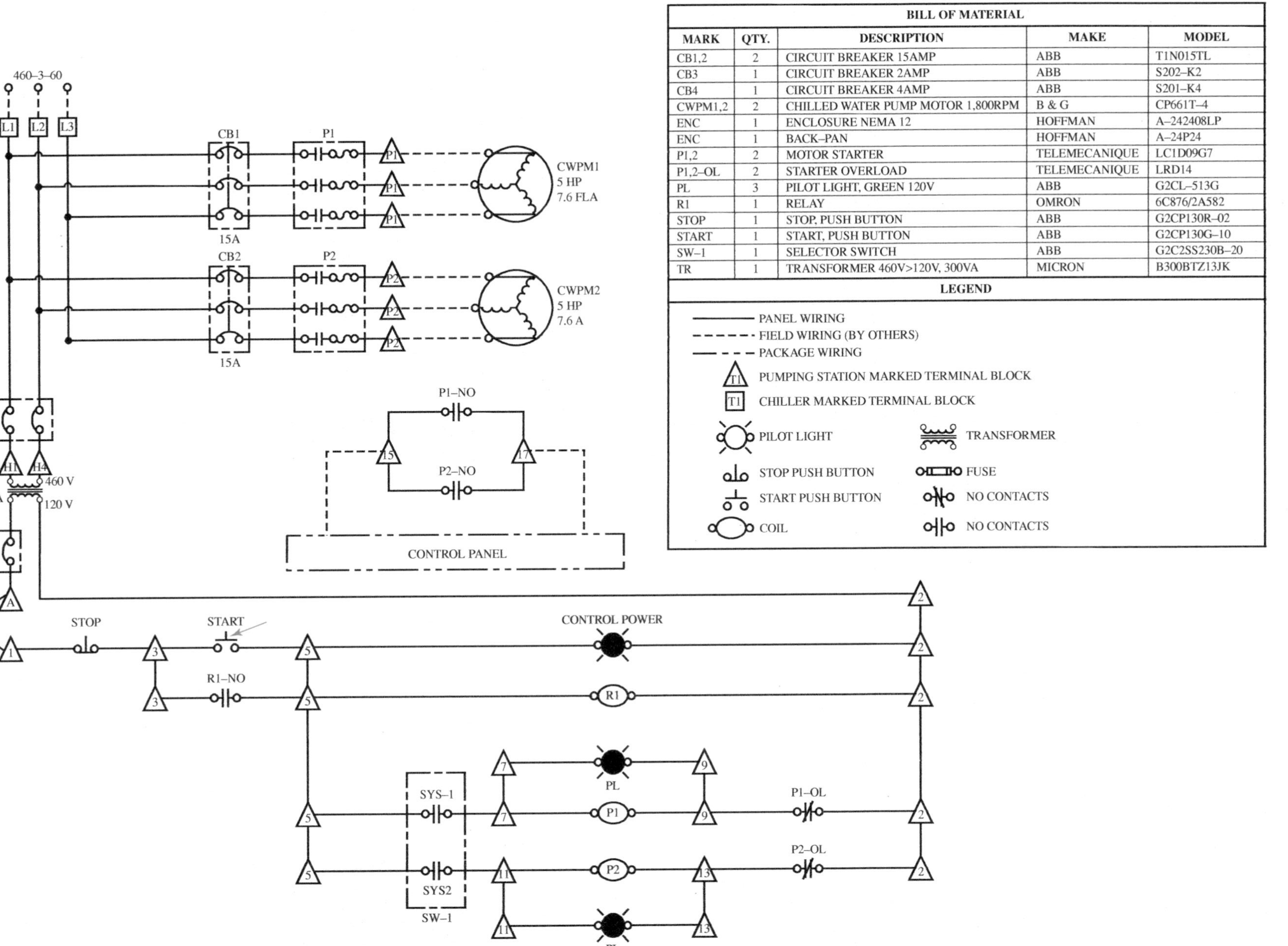

BILL OF MATERIAL

MARK	QTY.	DESCRIPTION	MAKE	MODEL
CB1,2	2	CIRCUIT BREAKER 15AMP	ABB	T1N015TL
CB3	1	CIRCUIT BREAKER 2AMP	ABB	S202–K2
CB4	1	CIRCUIT BREAKER 4AMP	ABB	S201–K4
CWPM1,2	2	CHILLED WATER PUMP MOTOR 1,800RPM	B & G	CP661T–4
ENC	1	ENCLOSURE NEMA 12	HOFFMAN	A–242408LP
ENC	1	BACK–PAN	HOFFMAN	A–24P24
P1,2	2	MOTOR STARTER	TELEMECANIQUE	LC1D09G7
P1,2–OL	2	STARTER OVERLOAD	TELEMECANIQUE	LRD14
PL	3	PILOT LIGHT, GREEN 120V	ABB	G2CL–513G
R1	1	RELAY	OMRON	6C876/2A582
STOP	1	STOP, PUSH BUTTON	ABB	G2CP130R–02
START	1	START, PUSH BUTTON	ABB	G2CP130G–10
SW–1	1	SELECTOR SWITCH	ABB	G2C2SS230B–20
TR	1	TRANSFORMER 460V>120V, 300VA	MICRON	B300BTZ13JK

Figure 28-17 This is a diagram for two chilled water pumps. Each pump is used to circulate water in a two-chilled-water-pump cooling system. It is important to ensure that water is flowing so that the chilled water does not freeze and do damage to the piping system. Frozen water expands, crushing tubing and creating restrictions or leaks.

push the **START** button between terminals 3 and 5 (pointed out by the red arrow). This type of **START** switch is a momentarily closed, spring-loaded device. After pushing the switch, the chilled water pumps do not operate. Therefore, the chillers do not operate. Look at the diagram. What do you check?

1. Check the input voltage, which measures 460 volts at **L1**, **L2**, and **L3**.
2. The voltage is measured across each leg: **L1** to **L2**, **L1** to **L3**, and **L2** to **L3**.
3. To speed up the troubleshooting process, you measure voltage on the secondary side of the transformer. It measures 120 V.
4. Next, you begin hopscotch troubleshooting by attaching one DMM probe to the right side of the 120-V circuit at any point **2**. You measure 120 volts at the circuit breaker **CB4**; 120 V on the left and right sides of the **STOP** button and the left side of the **START** button. Zero volts is measured on the right side of the **START** button.
5. You push the **START** button while measuring voltage on the right side of the **START** button. No voltage is measured. Finally, you measure 120 volts directly across the **START** button. Pushing the **START** button you still measure 120 volts across the **START** button. The 120-volt measurement indicates that the switch is open and will not close even when you push it closed.

Momentarily bypassing the **START** switch starts the chilled water pumps. After making sure that all pumps are started, you start the chilled water system and determine that it is functioning properly. You inform the building manager of your finding and get permission to pursue the needed repair of replacing the **START** switch.

SERVICE CALL 11: NO COOLING, SHORT CIRCUIT

Troubleshooting any unit with a short is always a challenge. Refer back to Figure 28-9 or ED-8. There is an electrical short in the unit represented by this diagram. How do you figure out which circuit or component is shorted?

✓ First, inspect all of the wiring and loads.
✓ Look for components or wiring that is burned.

If nothing is obviously burned, continue with the following steps to solve the problem:

1. The easiest way to do this is to check the resistance of each circuit in the ladder diagram. First, disconnect the wires from the power supply.
2. Hook an ohmmeter to the system **L1** and **L2** power wires that feed power to the lower circuits, taking care to ensure that no power is connected.
3. If there is a short circuit, the ohmmeter will read near zero ohms (0 Ω). A low resistance of 5, 10, or 15 ohms could be considered a short circuit if a component is supposed to have a much higher resistance.
4. You should be measuring the resistance of the primary of the transformer, which will be around 50 to 100 Ω. The crankcase heater is also in the circuit through the normally closed contacts **CR.**
5. In summary, with the system powered down the only resistance in the high voltage side will be the primary of the transformer and the crankcase heater. Both of these loads in parallel with have a significant resistance way above zero ohms.

Now this is where it gets a little tricky. When the power is removed, all of the switches on the left side of the diagram will be open. The only exception is the **CR** closed contacts, which keep the crankcase energized when the compressor is off. You will need to close the contacts that are open. You can close them by installing a jumper wire across each set of open contacts. Use the following sequence to narrow down the short circuit:

1. Add the jumper wire to one circuit or one line at a time to check the resistance. If the resistance is low (around 0 Ω), that circuit should be suspect for a short-circuit condition. For example, let's start with an easy circuit that has the **CR** contacts and © coil circuit. Jump the open **CR** contacts. The resistance as measured across **L1** and **L2** may change a little since this is a parallel circuit, but it should not drop to a low resistance.
2. If the **CR** line read good resistance check the next parallel circuit. Remove the jump and move to the **IFM** circuit.
3. If the **IFM** circuit has good resistance, jumper both sides of the **C** contacts and test the compressor and condenser fan circuit. This circuit has several potential shorted components: the compressor, condenser fan, run capacitor, and the wiring that controls these circuits

In summary, jumping from one circuit at a time will eventually decrease the resistance. When the resistance decreases, you have found the circuit with the shorted component(s). You can ohm out each component in that line to determine the shorted component.

If this does not work, remove all wires from the left side of the diagram and apply power. A short should not occur with all wires removed and power reapplied. If the short is still present and the breaker trips, then the short is on the right side of the diagram. Turn off the power. Next, install and check the power to each line one circuit at a time.

1. First, install power to the transformer line and apply power. Does the fuse or breaker blow?
2. If not, remove power and reconnect the crankcase heater section. Does the fuse or breaker blow?
3. If not, continue this sequence until all of the circuits have been tested individually.

Finding a shorted component or shorted wire can be a challenge. Eliminate as many circuits as possible by disconnecting the circuit, then reattaching one circuit at a time. Apply power after each circuit is hooked up until the breaker or fuse opens.

TECH TIP

One way to reduce the amount of shorted wiring conditions is to check wire insulation with a megohmmeter or "megger." A megger can measure tens of millions of ohms, much higher than an ohmmeter. This measurement would be part of a routine preventive maintenance (PM) check. Everything has resistance even if it is in the millions of ohms.

The insulation on wiring should have a high resistance value. By tracking the resistance over a period of time you may notice a gradual then sudden decrease in resistance. The sudden drop in resistance indicates impending insulation failure. Knowing that certain insulation failure is impending allows you time to schedule a replacement or repair job. You can individually check the insulation on motor windings or the wiring used to power various pieces of equipment. To know is to be prepared!

SERVICE CALL 12: FIRST STEPS

You are sent out on a job to do an initial diagnosis of a gas heating problem. The experienced tech is scheduled to arrive about 15 minutes later.

Here is what you found without tools or instruments: The thermostat display was set 5°F higher than the cool room temperature. The t'stat was set to Heat and Auto. The return air grille was pulling air into the duct and the supply air had good airflow. The furnace was located in the attic. The blower was operating. The air temperature coming from the supply air seemed to be the same temperature as the indoor air.

Question: Based on this limited information, where do you start looking for the problem? What will you tell the experienced tech, who has just arrived?

Answer: The gas furnace does not seem to be operating. A good diagnostic starting point is to open the furnace panel, bypass the door switch, and see if the gas is igniting. Feel the furnace. Is it warm? If so, that means it has been on at some point. Further gas side or electrical troubleshooting will be required.

SUMMARY

You will learn from others as they learn from you. We are in this together to help each other become better technicians. E-mail me with your troubleshooting tips and troubleshooting sequences so that I can share them with others: zmoravek@aol.com. Questions are welcome.

REVIEW QUESTIONS

1. Describe the hopscotch troubleshooting procedure.
2. What are three causes of heating short cycling?
3. What is the approximate amp draw of one heat strip operating on 240 volts?
4. Briefly discuss a general way to track down a short circuit.
5. Use Figure 28-8 to answer this question: The main power is supplied. When the system is off, what is the voltage reading across the high-pressure switch? When the system is on, what is the voltage reading across the low-pressure switch?
6. Use Figure 28-8 to answer this question: The main power is supplied and the t'stat is closed. The high-pressure switch is permanently open. Describe the full hopscotch troubleshooting sequence to find the solution to this problem.
7. Use Figure 28-8 to answer this question: The main power is supplied and the t'stat is closed. You assume that the high-pressure switch is open by using the hopscotch troubleshooting procedure. What are two ways to verify this finding?
8. Use Figure 28-9 to answer this question: Specifically, where is the symbol **CIT** located?
9. Use Figure 28-9 to answer this question: The crankcase heater is not operating when the condensing unit is operating. Is this a problem?
10. Use Figure 28-10 to answer this question: What type of compressor motor is found in this diagram?
11. Use Figure 28-13 to answer this question: What is the control voltage in this chiller?
12. Use Figure 28-13 to answer this question: Does this chiller have an oil pressure switch (oil safety switch)?
13. Use Figure 28-17 to answer this question: What type of motors are the chilled water pumps?
14. Use Figure 28-17 to answer this question: You found a defective transformer on this part of the chiller water system. After checking for shorts you did not find any problem and will replace the transformer. What voltages and VA rating will you select?

GLOSSARY

ACT Acronym used to help a technician start troubleshooting by checking **A**irflow in the condenser and ductwork, **C**ompressor operation, and the **T**hermostat setting.
adjustable thermal overload An adjustable overcurrent protection device normally used to protect a motor.
air handler Blower used to move air in a cooling and/or heating system.
air-source heat pump Heat pump that uses the ambient air to exchange heat in the outdoor coil.
alternating current (AC) Current that varies as positive and negative above and below a zero-volt reference point.
American National Standards Institute (ANSI) Safety standards used to regulate gas appliances and other equipment used in HVACR.
American Wire Gauge (AWG) Standard numbering system used to determine the diameter of a wire. The higher the gauge number, the smaller the wire.
ammeter Meter used to measure amperage.
ampacity The current, in amperes, that a conductor (wire) can carry continuously under conditions of use without exceeding its temperature rating.
ampere Current flow or movement of electrons.
amps full load (AFL) Highest current draw when a motor is operating under its highest design load. Also known as *full load amps (FLA)*.
amps locked rotor High current draw when a motor is locked or not moving. Also known as *locked rotor amps (LRA)*.
analog multimeter Volt-ohm-milliammeter (VOM) that uses a needle movement to indicate a meter reading.
annual fuel utilization efficiency (AFUE) Used to determine the energy efficiency of a gas furnace. AFUE is the heating output divided by the heat equivalent input.
arc flash Arc flash is a voltage across the resistance of air that results in an arc or voltage flash; an unexpected voltage to ground or a lower voltage between two voltage probes. Arc flash is more likely to occur with voltages above 480 V.
article A section in the National Electrical Code book. It is similar to a chapter.
auto feature Option found on some DMMs. Select the volt or ohm scale and the range is automatically determined.
auxiliary contacts Extra contacts found on a motor starter. The auxiliary contacts may have a lower amperage capacity compared to the main contacts.
auxiliary heat Secondary heat found in a heat pump. It is usually electric heat found in the second stage of heating.
babbitt lined bearings Rigid steel bearings that are quiet, have a long life, and are capable of carrying heavy direct-drive air-moving loads. The bearings are pressed into a cast metal end shield and therefore dissipate heat better than self-aligning bearings.
ball bearings Steel ball or rollers installed in a raceway on which the rotor moves in a circular pattern.
baseboard heaters Heat located in the perimeter of a room at baseboard level. Can use electric or hydronic heat.
black box Device with inputs and outputs that works without a tech needing to know how it works.
bleed resistor High-resistance resistor installed across a start capacitor terminal; used to remove a stored capacitor charge when the system is shut down.
bolt A threaded hardware device that mechanically joins or affixes two or more objects together.
bonding jumper A wire that connects a set of equipment together and to a ground rod.
branch circuit Conductors that carry power from the main or subpanel box to the HVACR equipment. It is power that branches off from a central power source or breaker box.
bridge rectifier Four rectifiers wired in a square to create DC voltages from an AC sine wave. Also called a *full wave rectifier*.
British thermal unit (Btu) Measurement of heat; the amount of heat added or removed from a pound of water to cause it to change 1°F.
butt connector Tubular piece that can be crimped and used to couple two wires together.
Canadian Standards Association (CSA) Canadian organization that regulates electrical standards similar to UL in the United States.
capacitance rating Rating of a capacitor in microfarads.
capacitive load Load that is made of up of one or more capacitors.
capacitive reactance An electrical circuit that has capacitors and is affected by the use of this device.
capacitor Device used to store electrons. A capacitor creates a phase shift between the current and voltage. Can be a run or start capacitor.
capacitor start/capacitor run motor Motor that uses a start capacitor and run capacitor to improve starting torque and running efficiency. Also known as a *CSCR motor* or *hard start kit*.
capacitor start/induction run motor Motor that uses a start capacitor for improved starting torque. The start capacitor must be removed from the circuit when the motor reaches 75% of motor speed. Also known as a *CSR* or *CSIR motor*.
carbon or wire-wound resistor Resistor that uses carbon or resistive wire to reduce current flow.
carriage bolt A fastener that has a domed or countersunk head and a shank that is topped by a short square section under the head. The square section grips into the part being fixed (typically wood), preventing the bolt from turning when the nut is tightened.
chatter Rapid opening and closing of a contactor or relay contacts due to a low-voltage condition.
circuit board Device with solid-state and mechanical components that is used to control HVACR equipment. The nonconducting board has flat, copper raceways to connect the various components.
clamp-on ammeter Electrical instrument used to measure amperage in a wire. It has an open jaw that surrounds an electrical conductor.
code jurisdiction The state, county, or city that enforces safety codes in a particular area.
combination circuit Electrical circuit that has series and parallel circuits in one diagram.
commercial transformer Larger VA transformer used in commercial or industrial systems.
complete circuit Electrical circuit with a power supply, load, and interconnecting wires. The power supply must have a return path for a circuit to be completed.

conductors Wires or metallic materials used to conduct electricity in a complete circuit.
connected load Energy-using device that drops voltage and reduces current flow.
constant torque motor An ECM-type motor that is programmed to provide a constant level of torque to the motor. This torque or power keeps the design CFM flowing through the air handler or furnace.
contactor High-current switching device used to control large loads. Has a coil to close one or more contacts. Usually has current loads in excess of 20 amps, but the contacts have current limits.
contacts Highly conducting points that open and close in the operation of relays, contactors, and overloads.
control voltage Voltage used to operate controls going through a thermostat and switch controls such as relays or contactors.
cooling anticipator Fixed resistor found in a thermostat that creates heat during the cooling-off cycle. The heat causes the thermostat to start a degree early, preventing overheating of the space when the unit is operating in the cooling mode.
Cool/Off/Heat Thermostat options to operate an air conditioning or heating system.
copper-clad aluminum conductor Aluminum conductors that are covered with a thin coating of copper.
crankcase heater Resistive heating device used to keep compressor crankcase oil warm, which reduces refrigerant condensing in the oil.
crimping tool Tool used to pinch soft tubular metal to bond a wire connection inside a tube.
current flow Flow of electrons in a circuit. Also known as *amp draw*.
cut-in When a certain pressure or temperature is reached, a circuit is closed.
cut-out When a certain pressure or temperature is reached, a circuit is opened.
cycle In reference to electricity, the term means one positive and negative peak. In reference to air conditioning, it means the movement of refrigerant from the compressor, condenser, metering device, and evaporator, then returning to the compressor to be recirculated.
DC millivolts One one-thousandth of a volt (1/1,000 or 0.001); a DC meter setting that will measure very low DC voltage.
deadband Temperature difference between cooling and heating operations. The system will be dead or off.
de-energized state State in which no power is being applied to a circuit.
defrost control Control used to remove ice from a frosted evaporator or heat pump outdoor coil.
defrost thermostat A temperature control that is part of a circuit that initiates and terminates the defrost cycle in a heat pump or low-temperature refrigeration system.
delta three-phase motor A three-phase motor with three windings in the shape of a Greek delta (Δ) or triangle configuration.
delta transformer A three-phase power supply shaped in a delta (Δ) or triangle configuration. One of the windings in a delta transformer may be center tapped to generate a higher voltage since the winding will also include the winding to the adjacent leg.
demand defrost Frost removal circuit used in a heat pump heating circuit. Defrost only occurs when it is needed or "demanded."
digital multimeter (DMM) Meter with a digital readout that measures voltage, amperage, and resistance. Some digital meters have other options such as a capacitor checker and temperature tester.
digital thermostat Thermostat with a digital display of the room temperature and selected settings.
diode Solid-state device that allows current flow in one direction only. Like an electronic check valve.
direct current (DC) Positive or negative voltage that does not change; straight-line voltage.
disconnect External, manual switching device used to remove power from a circuit or piece of equipment.
disconnect means A way to remove power from a circuit.
double-male, single-female connector Fastener used to join wires. It has two male spade-type connectors and one female connector.
dual capacitor Two capacitors in one housing; normally used to control two different components.
duct heater Electric heaters that are installed in a duct system usually for supplemental heat.
electric backup heat Electric strip heat used as auxiliary heat for a heat pump system.
electric furnace Heating system that uses electric heat strips. It usually includes a blower section.
electric heat Heating system that uses electric heat strips.
electric heat strip High resistance wire that is used for heating air.
electric unloader Electrical solenoid or mechanical valve that reduces the pumping capacity of a compressor.
electrical diagram Drawing of the electrical pathways used to operate an HVACR system; an electrical map.
electrical legend A guide to the meanings of abbreviations and symbols used on an electrical diagram.
electrical symbols Representative characters used in electrical diagrams.
electrical tape UL-rated tape used to wrap and isolate an exposed electrical connection.
electricity Movement of electrons.
electron Part of an atom, which also has protons and neutrons. The electron is the current-transferring component in an atom.
electronic controls Controls that use designed solid-state components to control HVACR equipment.
electronic expansion valve (EEV) Used to control the flow of refrigerant from a liquid line to an evaporator; an accurate way of controlling superheat and refrigerant flow.
electronically commutated motor (ECM) A device that converts electrical energy into rotary motion; a brushless DC motor that uses electronic switching to control speed and direction of rotation.
electrons A subatomic particle found in the outer orbit(s) of an atom. The electrons transfer electrical energy between atoms.
emergency heat Heat associated with a heat pump system; is designed to operate when the heat pump is off and is usually electric heat.
Energy Star Energy efficiency standards promoted by the U.S. Department of Energy. Has also been adopted by other countries.
entrance panel Electrical panel that supplies power to a piece of air conditioning equipment.
equal resistance method Method of calculating resistance that is used in circuits containing equal branch resistance values. Divide the resistance of one resistor by the total number of resistors in the parallel branch.
explosion-proof systems Electrical components that are sealed to prevent internal operating sparks from igniting a combustible environment.
external static pressure (ESP) Used to measure pressure drop and airflow in an air conditioning system.
fan Mechanical device used to move air.

fan cycling switch Switch that closes when a heat exchanger gets hot enough to blow warm air into the duct system.
farad High measure of capacitance. Most capacitors are measured in microfarads or thousands of a farad.
fastener Mechanical device used to bond two or more surfaces together.
female slip-on connector Quick connector that has on open, flat design that accepts a male slip-on connector.
ferrule Narrow circular ring of metal or, less commonly, plastic that contains stranded wire within. Is found between tubing and a fastening nut and is used as a gasket-like material to seal a joint.
flag connector Tubular wire connect that looks like a flag. Wire is crimped in the tubular end. The flag slips over a male connection.
flame sensor Device used to determine if a furnace flame is present.
fork connector Tubular wire connect that looks like a two-pronged fork. Wire is crimped in the tubular end. The fork slips around a threaded screw shaft connection and is usually fastened down tightly over the fork.
four-pole contactor Mechanical electrical switching device with up to four power inputs.
frame size Numbering and lettering system used to represent the size of a motor. Knowing the frame size makes it easier to select a replacement motor of the same size.
frequency Change in cycles or hertz per second; 60 cycle is the most common frequency in the United States.
fuse size Amperage and voltage rating of a fuse.
fusible link Device that opens when a certain amperage rating is exceeded. Used to protect one part of a system.
gas control Valve that controls the flow of gas to burners.
gas heating Furnace that uses a combustible gas to heat a heat exchanger.
gate controlled Electronic switch found in solid-state circuits.
gauge number See *American Wire Gauge (AWG).*
generator Equipment that makes electricity for general distribution.
geothermal-source heat pump Heat pump that uses water to transfer heat to or from the refrigerant. Also called *geo-source heat pump.*
ghost voltage Weak voltage that is generated by a magnetic field-inducing voltage between conductors. The voltage does not have enough current to operate a load but can be measured by a digital voltmeter.
green installation Installation that uses less energy and is, therefore, environmentally friendly.
ground circuit Wiring used to create a safety ground for equipment or other device.
ground fault current interrupter (GFCI) Used to protect electricity users from electrical shock.
ground loop A loop of water piping used in a heat pump system.
grounding Bonding a circuit to an earth ground.
groundwater system Heat pump system that uses groundwater to reject or absorb heat.
hard start kit A specifically sized potential relay and start capacitor used to provide starting torque for a compressor motor.
heat anticipator Small resistive heater located in series with the heating circuit during a call for heat. It provides a small amount of heat to the thermostat during the heating cycle to open the thermostat a degree early so that the heating system does not overheat the space.
heat pump Reverse cycle cooling and heating system. The reversing valve shifts the cooling and heating operation between the indoor and outdoor coils.
hermetic (herm) Usually found on a run capacitor, it identifies the terminal that goes to the start winding on a single-phase compressor.
hermetic compressor Compressor that has been sealed by being welded shut; cannot be opened or serviced.
hertz Another electrical term for cycles.
hex head Six-sided head of a screw.
hex screw driver Tool used to drive or remove hex head screws.
high leg The higher voltage area found on a delta transformer between one of the sets of winding.
high pressure The discharge or liquid line side of a system. Also, excessive pressure above the normal operating high-side pressures.
high-pressure switch (HPS) Safety device that opens when a certain setting on the switch has been exceeded.
holding relay Relay designed to hold a circuit in place after it has been energized.
hopscotch troubleshooting Troubleshooting method of jumping across a circuit until the voltage being measured is lost. The loss of voltage indicates that the component just tested is open.
horsepower The measure of how much work a motor can be expected to do. This value is based on the motor's full-load torque at full-load speed ratings.
hot surface ignition (HSI) Device used in gas heating to ignite the gas supply.
humidistat A moisture-sensing device that is used to control a humidifier or a dehumidifier.
ignition system Controls used to start a gas furnace's burner process.
impedance formula Formula used to find the resistance in a circuit that uses a coil or capacitor.
impedance relay High-resistance relay coil.
incandescent light bulbs Light bulb that uses a tungsten or other highly resistive element to create light output.
inductive loads Loads that have lengths of wound wiring such as motors, transformers, and relay coils.
industrial transformer Large transformer with a high VA rating.
insulation class Standard way to categorize insulation types. A letter designates the amount of allowable temperature rise based on the insulation system and the motor service factor.
insulation stripping tool Tool used to remove the insulation coating from wire.
insulator Material such as glass, plastic, or rubber that restricts electron flow.
integrated circuit (IC) Device that contains many transistors and other solid-state devices installed in one component and mounted on a circuit board. Also known as a *chip.*
integrated control module (ICM) Older name for electronically commutated motor (ECM).
inverter Device that changes DC to AC.
J-bolts Bolt head with hook or "J" appearance.
kilovolt amps (kVA) 1,000 volt-amps.
kilowatt (kW) 1,000 watts.
kilowatt-hour (kWh) 1,000 watts of power used per hour.
L1, L2, L3 The three power lines of a three-phase circuit. L1 and L2 would represent a single-phase power supply.
ladder diagram Electrical diagram drawn to look like a ladder. The rungs or sides of the diagram represent the power supply. The steps or lines between the power supply show the component symbols.
line diagram An electrical diagram that uses lines to represent connecting wires and symbols to represent components.
line side The side of an electrical diagram that has power.

line voltage monitors Device used to detect the presence of the correct voltage; will shut a system down if a low- or high-voltage condition is detected.

line voltage thermostat Thermostat that can handle high-voltage and high-current operation.

liquefied petroleum Butane or propane gas used for heating, water heating, or cooking purposes. Also known as *LP gas*.

liquid line service valve Refrigerant valve that is installed in the liquid line of a refrigeration system; can be used to check system pressures and block the flow of refrigerant.

liquid line solenoid valve Electrical valve that opens or closes a refrigerant liquid line. The valve has a coil that, when energized, will cause the valve to switch positions. The normally closed (NC) solenoid valves are the most common type found in HVACR equipment.

listed Term used to indicate that the "listed" item is approved for use by a nationally recognized agency such as Underwriters Laboratories or the National Electrical Code.

load Device that reduces electron flow such as a motor, transformer, coil, or heat strips.

locked rotor amps (LRA) High-amperage condition drawn by a motor when it is mechanically stuck and cannot move. Locked rotor amps are also drawn for a moment when a motor starts.

lockout Electric circuit designed to prevent the equipment from operating if there is a safety problem.

lockout relay Relay used to shut down a system when a safety switch opens; prevents the system from restarting even if the safety device is reset.

lockout/tagout (LOTO) Safety procedure in which a tech opens power to a piece of equipment and places a lock on the means of disconnect so that power cannot be turned on while equipment is being serviced. A tag is placed on the lock to identify who locked out the power supply.

loss of charge pressure switch A low pressure that opens near 0 psig; sometimes found in heat pumps.

low-pressure switch (LPS) Switch located in the suction side that opens when the pressure drops below a set level. The open switch causes the system to stop. The system starts when the pressure rises above a prescribed setpoint.

low-side gauge Refrigeration gauge use to check low-side pressure; has a pressure and vacuum reading.

machine screws Fastener characterized by a helical ridge, or external thread, wrapped around a cylinder. Also called a *screw* or *bolt*.

magnetic chuck Drill fitting that uses magnetism to hold screws in place.

magnetism Invisible force of a magnetic field created in or around metal such as copper or steel.

male slip-on connector Similar to a flag connector except the flat connector is in the same direction as the wire tubular connection. The male connector slips into a female connector.

manual reset flame rollout switch Gas furnace safety device that will open if burner flames spill out of the combustion area; will not reset until it is manually pushed closed.

maximum circuit ampacity (MCA) Calculation used to size the wire used on a piece of equipment.

maximum overcurrent protection (MOP) Used to size the highest fuse or breaker that should be used on a piece of equipment.

mechanical thermostat (t'stat) Older style thermostat that uses moving mechanisms to open and close contacts.

mechanical unloader Pressure-activated device that reduces the pumping capacity of a compressor by blocking refrigerant flow to the cylinder or opening a valve to prevent compression.

megawatt (MW) One million watts.

megawatt-hour (MWh) One million watts per hour.

metal oxide varistor (MOV) A two-element semiconductor with reverse resistance in which the resistance drops as the applied voltage increases.

meter loop The opening in and extension of a customer's service entrance conductors provided for installation on the electric company's meter.

microcontroller A small computer on a single integrated circuit containing a processor core, memory, and programmable input/output peripherals.

microfarad Capacitor rating that is 1/1,000 (0.001) of a farad.

microprocessor Incorporates the functions of a computer's central processing unit on a single integrated circuit or at most a few integrated circuits.

milliammeter One one-thousandth of an amp (1/1,000 or 0.001). Meter setting that will measure very low amperage.

motor nameplate Motor data found on a metal or plastic label affixed to the motor shell.

motor starter Device similar to a contactor that is used to provide power to a motor; usually has overload protection and auxiliary contacts.

multispeed fan motor Fan motor that has more than one operating speed.

multispeed motor A motor that has more than one operating speed.

multi-tap transformer Transformer with several different input voltages.

nails Sharp, pointed metal shaft with a head used as a fastener.

National Electrical Code (NEC) National standard developed by the NFPA to ensure electrical safety.

National Electrical Manufacturers Association (NEMA) Group of electrical manufacturers with the goal of standards and safety compliance.

National Fire Protection Association (NFPA) National safety group that sets standards for fire and electrical safety; controls the National Electrical Code standards.

National Fuel Gas Code Code that regulates safe use of combustible gases used in residential, commercial, and industrial applications.

NEC Article 440 National Electrical Code (NEC) article that covers air conditioning and refrigeration equipment. Other code material related to HVACR is found scattered in other sections of the NEC.

net oil pressure The difference between a compressor's oil pump pressure and suction pressure when the compressor is operating.

NFPA 70 The National Electrical Code regulated by the National Fire Protection Association.

noncontact voltage detector Electrical instrument used to measure voltage without touching a bare wire. Voltage will be measured through wire insulation or in an energized electrical component.

nonmetallic sheathed cable Several separate insulated conductors wrapped together within another insulated sheath. The bundle may have a bare, uninsulated ground conductor.

notes Information on an electrical diagram.

nut drivers Tool used to install or remove a hex head screw. Common sizes for residential equipment are 1/4 and 5/16 inches.

nylon strap Fastener made of nylon that is strong and flexible. Used to tie or bundle wires together.

nylon ties See *nylon strap*.

ohmmeter Electrical instrument used to measure resistance, shorts, or a complete circuit.

Ohm's law Voltage = amperage × resistance.
oil safety switch Compressor safety device that compares the difference in oil pump pressure and suction pressure. A low pressure differential will cause the safety device to open the control circuit.
open circuit A break in a complete circuit. In an open circuit, voltage does not have a complete path over which current can flow.
open wiring Break in a wire; NEC considers it to be wiring that is not encased in conduit.
overcurrent protection Device, such as fuses, breakers, or thermostats, that will open if an excessive current condition occurs.
overhead wiring Electric power coming into a building from aerial poles.
overload protection Device, such as fuses, breakers, or thermostats, that will open if an excessive current condition occurs.
package unit A complete air conditioning system including the condensing unit, evaporator, and indoor blower. A package unit may have a heating system such as gas, electric, or a heat pump.
parallel circuit Parallel arrangement of electrical circuits in ladder paths so that when one circuit is open, the others can be energized.
penny nail Nail size indicator. The larger the number, the longer and greater the diameter of the nail.
permanent split capacitor (PSC) motor Motor that has a run and start winding and can be identified by a run capacitor.
personal protective equipment (PPE) Safety equipment used when servicing HVACR equipment; includes, but is not limited to, safety glasses, gloves, and protective clothing.
pictorial diagram Electrical diagram that shows the images of components as installed in a circuit.
plenum rated cable Electrical conductors that are permitted for use in duct systems. The flame spread and smoke development of this type of cable are lower than those for other types.
positive thermal coefficient (PTC) A resistive device that increases resistance as the temperature increases.
power Power = volts × amps. Also known as *watts*.
power distribution Network of electricity distribution from the power plant to the end user.
power factor (PF) The ratio of power flowing in a circuit versus actual power used. This value is used as part of billing considerations in commercial systems.
power formula Power = volts × amps.
power generation The use of generators to produce electricity.
pressure switch Switch that operates on refrigerant pressure; can open on low or high pressure.
pressure transducer Device used to change pressure into an electrical measurement.
product over sum method Formula for calculating the resistance of two resistors. Multiply the resistance together, then divide that answer by the sum of the two resistors.
programmable thermostat Thermostat that can be set to automatically change cooling and heating temperatures.
quick connectors Electrical connectors that separate easily without the use of tools.
R Power designation on a transformer to the thermostat.
range Scale setting starting at a 0 reference point.
rated load current Highest current draw when a motor is operating under its highest design load.
rated voltage Required voltage that is used with a piece of equipment. The voltage rating allows a ±10% variation.
readily accessible The ability to walk directly up to a piece of equipment without having impediments in the way.
reciprocal method Formula used to calculate the resistance of two or more resistors.
rectification Changing from alternating current to direct current.
rectifier Changes AC to DC.
relay Electromechanical device used to switch low-amperage loads, usually below 15 amps. The relay has a coil and one or more sets of normally open and/or normally closed contacts.
reset relay Relay that is normally energized when a safety device opens. The relay prevents the equipment from operating after the safety device closes.
resistance Any material that reduces current flow.
resistive load Device that reduces electron flow, such as a resistor or electric heat strips.
resistor Device used to drop the voltage in an electrical circuit.
resistor tolerance Percentage of variation above and below a listed resistor's resistance.
resolution Refers to how accurate or fine a measurement a meter can deliver. By knowing the resolution of a meter, a technician can determine if it is possible to see a change in the measured signal.
reversing valve Valve used in a heat pump system to shift refrigerant flow between the indoor and outdoor coils.
reversing valve solenoid (RVS) A magnetic coil that, when energized, will cause the reversing valve to shift position.
revolutions per minute (RPM) The number of times a device will rotate in 1 minute. Usually refers to a motor's RPMs.
right-hand thread Fastener with a shaft and thread designed to go into a female thread when turned in a clockwise direction.
ring connector Wire connector with a ring on one end and a tubular opening for a crimp wire connection on the other end.
Romex Name brand of one type of nonmetallic sheathed cable.
root means square (RMS) Voltage or current level at 0.707% of its positive and negative peaks.
rotor The moving part of a motor with at least one shaft.
round-tipped screw Screw that has a pan or smooth head and a threaded shaft with a point.
run capacitor Electron storing device used to improve the running efficiency of a motor.
SAE J429 Defines the bolt grades for inch-system sized bolts and screws. Grades ranges from 0 to 8, with 8 being the strongest.
safety devices Components used to make HVACR equipment safer on the electrical and pressure side.
schematic diagram Electrical diagram showing wiring and component connections; an electrical road map.
screw Fastener with a slotted head, threaded shaft, and sharp point.
seasonal energy efficiency ratio (SEER) The efficiency rating of an air conditioning system during an entire cooling season. It is approximately the number of Btu's removed during a cooling season divided by the watts required during that cooling season.
self-aligning sleeve bearing A bearing that fits like a tube around a rotating shaft.
self-contained electric heating Complete electric heating system with blower and controls.
self-drilling Screw that has a tip sharp enough that it acts like a drill bit to start the hole and pull in the rest of the screw shaft.
self-starting screw Screw that does not require a pilot hole.
self-tapping screw Screw that has a drill bit–like tip to help start the metal penetration.
self-threading screw Screw that has a drill bit–like tip to help start the metal penetration.
semiconductor Solid-state device that has electrical conductivity intermediate to that of a conductor and an insulator.
sequencer Electric heating control device that stages in the strip heat so that the strips do not all come on at the same time.

series circuit An electrical circuit in which all of the wiring and components are in line with the power supply.
series-parallel circuits An electrical circuit that has series and parallel circuits.
service drop Power source brought into a building overhead or underground.
service factor The number by which a horsepower rating is multiplied to determine the maximum safe load that a motor may be expected to carry continuously without exceeding the allowable temperature rise of its insulation class.
set screw Fastener that is headless, meaning that the screw is fully threaded and has no head projecting past the major diameter of the screw thread; usually has a female hex opening to fasten the screw.
setback temperature Automatic temperature change that lowers the temperature setting in the winter and raises the setting in the summer to save energy.
setpoint Temperature at which a thermostat setting is reached.
setpoint temperature Temperature at which a thermostat setting is reached. Also called simply *setpoint.*
shaded pole motor Type of AC single-phase induction motor. It is basically a small squirrel-cage motor in which the auxiliary winding is composed of a copper ring or bar surrounding a portion of each pole.
sheetmetal screws Fasteners used to join sheet metal.
sheetrock screw Fasteners used to secure drywall to a wood or metal frame.
short circuit Circuit that has no load and maximum current flow.
short cycling Turning off and on in rapid succession.
silicone controlled rectifiers (SCR) A solid-state semiconductor controlled rectifier is a four-layer, solid-state current controlling device. Also known as a *silicone gate controlled rectifier.*
single-phase motor Motor that uses single-phase power.
single-pole contactor Contactor with one set of changeable contacts. The second part of the contactor has a bar to complete the circuit.
slip The difference between the rotating magnetic field and the speed at which the rotor is turning. The rotor is trying to keep up with the motor rotation magnetic field in the stator.
snap action thermostat Mechanical thermostat that closes quickly, by magnetic force, when it reaches its setpoint.
solenoid valve Electrically operated valve that opens or closes a gas, vapor, or liquid line.
solid state Circuits or devices built entirely from solid materials and in which the electrons or other charge carriers are confined entirely within the solid material.
solid-state circuits A circuit that uses solid-state components to make an operating circuit.
solid-state thermostat Electronic device that controls the room temperature in a building.
solid-state timer Electronic device that creates a time delay after air conditioning equipment cycles off; prevents short cycling.
split system Air conditioning system that separates the condensing unit and evaporator by use of a suction line and liquid line.
start capacitor Electron-storing device used to improve the starting torque of a motor.
starting torque The force or twisting motion used to get a motor moving.
stator The stationary winding of a motor. The windings seen in an electrical diagram are the stator windings.
stranded wire A conductor with numerous smaller gauge wires twisted together. It is thought that electricity travels on the surface of a wire; therefore, a stranded conductor offers more surface area for electron flow.
subbase The part of a thermostat that is installed on the wall; has switches such as cool, heat, off, and fan. The thermostat is screwed or clipped on to the subbase.
symbols Representative characters used in electrical diagrams.
synchronous speed The RPM or rotation of the magnetic field in a stator winding. The stator speed is a little faster than the rotor speed.
systems A set of components that have been designed and organized to operate together to function as a cooling or heating apparatus.
temperature tester Electronic instrument used to measure temperature.
test pins Two points on a circuit board that, once shorted across, will cause an action; used to test the operation of a circuit.
thermistor Solid-state device that changes resistance with a change of temperature.
thermostat Temperature-sensing device used to control cooling and/or heating systems.
three-phase motors Motor that has three sets of windings; wound in a delta or wye configuration.
three-pole contactor High-current control device that supplies three-phase power to a load, such as a three-phase compressor.
time-delay circuit A solid-state or mechanical device that delays the operation of a circuit when it is first turned off. The time delay can be a delay on break or a time delay when the circuit "makes" (i.e., closes).
transformer A power converter that transfers electrical energy from one circuit to another through inductively coupled conductors (i.e., the transformer coils).
transistor Semiconductor device used to amplify and switch electronic signals and electrical power.
transmission lines High-voltage conductors that start at an electrical generating station and end near the end user's facility.
troubleshooting Procedure used to figure out an electrical or refrigeration problem.
two-pole contactor High-current control device that supplies two sides of power to a load such as a single-phase compressor.
Type B venting A gas venting system that has two layers of piping. The outer piping is usually galvanized and the inner piping aluminum.
under load Power supplied to an energy-consuming device.
underground wiring Wiring that runs underground from a power distribution system to the customer. Also known as an *underground feeder.*
Underwriters Laboratories (UL) National standard rating for electrical equipment.
unit bearings Motor bearing assembly.
unknown soldier A motor that has characteristics that are difficult to identify or no nameplate.
unloader Electrical or mechanical device used to reduce the load of a compressor.
varistor Solid-state device used to protect a circuit against excessive transient voltages by placing them into the circuit in such a way that, when triggered, they will bypass the current created by the high voltage away from the sensitive components.
venting Piping used to remove flue gases burned in a gas furnace.
volt Pressure used to push electrons through a wire or load. Also called *voltage.*
volt-amps (VA) Used to give a transformer a power rating.
voltage Pressure used to push electrons through a wire or load. Also called *volt.*

voltage drops Loads that drop the voltage in a circuit. The higher the resistance, the greater the voltage drop.
voltmeter Electrical instrument used to measure volts.
watt (W) Measurement of power. Watt = volts × amps.
watts formula Power formula. Watts = volts × amps.
window units Package air conditioning or heat pump device installed in an open window.
wire cutter Tool used to cut wire.
wire gauge See *American Wire Gauge (AWG).*
wire nuts Plastic cone-shaped cap with a threaded interior; used to join two or more wires.
wire-wound resistor Resistor that uses a high-resistance wire to drop voltage and/or reduce current flow.
wiring diagram legend A guide to the meanings of abbreviations and symbols used on a wiring diagram.
wiring diagrams Electrical drawing that represents a circuit's wiring and components. This type of electrical "map" is used to trace or troubleshoot the operation of a circuit.
wood screw Fastener used to join wood material.
wye three-phase motor Three-phase motor with three sets of stator windings shaped like the letter "Y." Each set of motor windings has two sets of coils.
wye transformer Three-phase transformer with three sets of windings shaped like the letter "Y." Each phase has two sets of windings.

APPENDIX A

Abbreviations and Acronyms

Term	Definition	Unit
AC	Alternating current	1
ACT	Initials used to help a technician start troubleshooting by checking **A**irflow in condenser and ductwork, **C**ompressor operation, and **T**hermostat setting	12
AFL	Amps, full load. Also known as full load amps or FLA	9
AFUE	Annual fuel utilization efficiency	7
ALR	Amps, locked rotor. Also known as locked rotor amps or LRA	9
ANSI	American National Standards Institute	22
AWG	American Wire Gauge	6
Btu	British thermal unit	2
Btu/h	British thermal units transferred per hour	2
CCWLE	Abbreviation meaning that the motor shaft rotation will be counterclockwise from the end of the motor where the leads enter	15
CCWSE	Abbreviation meaning that the motor shaft rotation will be counterclockwise from the shaft end of the motor	15
CH	Crankcase heater	1
COM	Common; refers to a common connection	20
CPU	Central processing unit	20
CSA	Canadian Standards Association	4
CWLE	Abbreviation meaning that the motor shaft will rotate clockwise as seen from the end of the motor where the leads enter	15
CWSE	Abbreviation meaning that the motor shaft will rotate in the clockwise direction as seen from the shaft end of the motor	15
DC	Direct current	1
DMM	Digital multimeter	3
E	Letter designation for voltage in Ohm's law: $E = I \times R$	1
ECM	Electronically commutated motor	15
EEV	Electronic expansion valve	12
EMF	Electromotive force or voltage	1
ESP	External static pressure	16
FLA	Full load amps	7
GFCI	Ground fault current interrupter	12
GND	Ground	20
HERM	Hermetic	10
HPS	High-pressure switch	1
HSI	Hot surface ignition	20
I	Intensity; used to designate amp or current flow	1
kVA	Kilovolt-amps	5
kW	Kilowatt	2
kWh	Kilowatt-hour	5
LED	Light-emitting diode	20
LOTO	Lockout/tagout	3
LPS	Low-pressure switch	1
LRA	Locked rotor amps	6
MCA	Maximum circuit ampacity	7
MOP	Maximum overcurrent protection	7
MOV	Metal oxide varistor	19
NC	Normally closed contacts	21
NEC	National Electrical Code	6
NEMA	National Electrical Manufacturers Association	14
NFPA	National Fire Protection Association	3
NO	Normally open contacts	17
OL	Overload	3
PF	Power factor	5
PPE	Personal protective equipment	3
PSC	Permanent split capacitor motor	14
PTC	Positive thermal coefficient	15
R	Resistance	1
RMS	Root mean square	3
RPM	Revolutions per minute	14
RVS	Reversing valve solenoid	24
SAE J429	Society of Automotive Engineers; standard for mechanical and material requirements for externally threaded fasteners	4
SEER	Seasonal energy efficiency ratio	7
TD	Temperature difference (between the return air and supply air)	21
UL	Underwriters Laboratories	4

Term	Definition	Unit
V	Volts	1
VA	Volt-amps	8
W	Watts	1
W2	Letter-number designation for the second stage of heating on the thermostat	11
Y	Letter designation for the cooling terminal on a thermostat	11
Ω	Greek symbol for resistance or ohms	1

APPENDIX B

HVACR Formulas

BASIC FORMULAS

Btu's = 500 × GPM × ΔT

1 watt = 3.413 Btu

1 kW = 3,413 Btu

1 ton = 12,000 Btu

Motor kW = V × A × 1.73 × PF ÷ 1000

Motor tons = kW × 3,413 ÷ 12,000

KEY FORMULAS

- Volts = amps × ohms ($V = I \times R$)
- Watts = volts × amps ($P = V \times I$)
- Power factor = $\dfrac{\text{watts}}{\text{volts} \times \text{amps}}$
- Resistance in series = $R_1 + R_2$
- Two resistors in parallel = $\dfrac{R_1 \times R_2}{R_1 + R_2}$
- Multiple resistors in parallel =

  $$\frac{1}{R_T} = \frac{1}{R_1 + R_2 + R_3 + \cdots}$$
- Capacitance in parallel = $C_1 + C_2$
- Capacitance in series = $\dfrac{C_1 \times C_2}{C_1 + C_2}$

OHM'S LAW AND WATT'S LAW WHEEL

Use the Ohm's law and Watt's law wheel to help determine values of amps (I), volts (E), ohms (R), or watts (W).

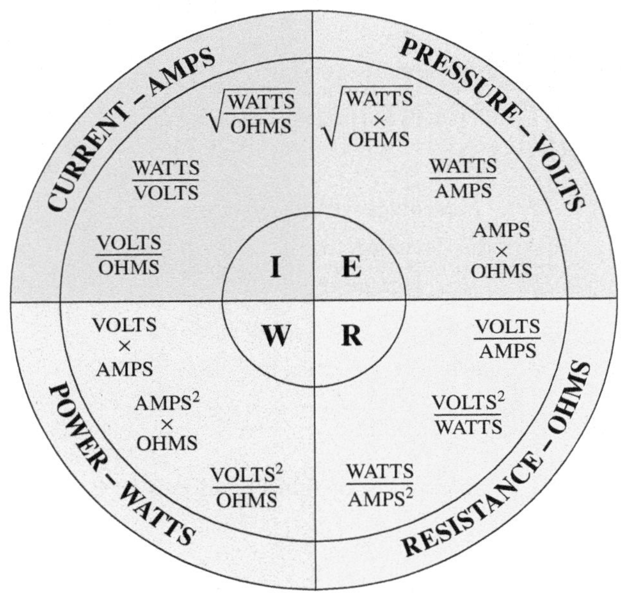

Current-Amps

$$I = \frac{\text{volts}}{\text{ohms}} = \frac{\text{watts}}{\text{volts}} = \sqrt{\frac{\text{watts}}{\text{ohms}}} = I$$

Pressure-Volts

$$E = \sqrt{\text{watts} \times \text{ohms}} = \frac{\text{watts}}{\text{amps}} = \text{amps} \times \text{ohms} = E$$

Resistance-Ohms

$$R = \frac{\text{volts}}{\text{amps}} = \frac{\text{volts}^2}{\text{watts}} = \frac{\text{watts}}{\text{amps}^2} = R$$

Power-Watts

$$\text{W} = \text{volts} \times \text{amps} = \text{amps}^2 \times \text{ohms} = \frac{\text{volts}^2}{\text{ohms}} = \text{W}$$

THREE-PHASE VOLTAGE

$$\text{Horsepower} = \frac{1.73 \times \text{amps} \times \text{volts} \times \text{motor efficiency}}{746} \times \text{power factor} \quad \textbf{(Actual)}$$

SINGLE-PHASE VOLTAGE

$$\text{Horsepower} = \frac{\text{volts} \times \text{amps} \times \text{efficiency}}{746} \times \text{power factor} \quad \textbf{(Actual)}$$

$$\text{Power factor} = \frac{\text{watts (read on meter)}}{\text{measured volts} \times \text{measured amps}}$$

THREE-PHASE VOLTAGE IMBALANCE AND CURRENT IMBALANCE FORMULAS

$$\text{Percentage voltage deviation} = \frac{\text{maximum deviation from average voltage}}{\text{average voltage}} \times 100$$

The maximum voltage deviation is 2%

$$\text{Percentage current deviation} = \frac{\text{maximum deviation from average current}}{\text{average current}} \times 100$$

The maximum current deviation is 10%

THREE-PHASE DELTA LOADS

Three-phase balanced loads $= P1 + P2 + P3$

Total line current = total power (balanced load)

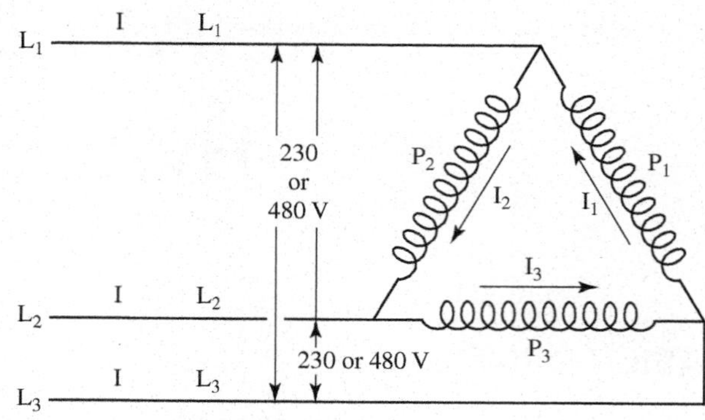

If the phases are unbalanced, each of the phases will differ from the others:

Formulas:

$$IL_1 = \sqrt{I_2^2 + I_1^2 + (I_1 \times I_2)}$$
$$IL_2 = \sqrt{I_2^2 + I_3^2 + (I_2 \times I_3)}$$
$$IL_3 = \sqrt{I_3^2 + I_1^2 + (I_1 \times I_3)}$$

USING WATTS TO MEASURE AIRFLOW IN ELECTRIC FURNACES

The upper formula is for single-phase furnaces. The lower formula is for three-phase, delta wired furnaces.

$$\text{CFM} = \frac{\text{volts} \times \text{amps} \times 3.413}{1.08 \times \text{temperature rise}^*}$$

$$\text{CFM} = \frac{1.73 \times \text{volts} \times \text{amps} \times 3.413}{1.08 \times \text{temperature rise}^*}$$

*Difference between return and supply air temperatures.

ELECTRICAL FORMULAS FOR CALCULATING AMPERES, HORSEPOWER, KILOWATTS, AND KVA

TO FIND	DIRECT CURRENT	ALTERNATING CURRENT		
		SINGLE PHASE	TWO PHASE-FOUR WIRE	THREE PHASE
AMPERES WHEN "HP" IS KNOWN	$\frac{HP \times 746}{E \times \%EFF}$	$\frac{HP \times 746}{E \times \%EFF \times PF}$	$\frac{HP \times 746}{E \times \%EPF \times PF \times 2}$	$\frac{HP \times 746}{E \times \%EFF \times PF \times 1.73}$
AMPERES WHEN "kW" IS KNOWN	$\frac{kW \times 1000}{E}$	$\frac{kW \times 1000}{E \times PF}$	$\frac{kW \times 1000}{E \times PF \times 2}$	$\frac{kW \times 1000}{E \times PF \times 1.73}$
AMPERES WHEN "kVA" IS KNOWN	———	$\frac{kVA \times 1000}{E}$	$\frac{kVA \times 1000}{E \times 2}$	$\frac{kVA \times 1000}{E \times 1.73}$
KILOWATTS	$\frac{E \times I}{1000}$	$\frac{E \times 1 \times PF}{1000}$	$\frac{E \times I \times PF \times 2}{1000}$	$\frac{E \times I \times PF \times 1.73}{1000}$
KILOVOLT-AMPERES "kVA"	———	$\frac{E \times I}{1000}$	$\frac{E \times I \times 2}{1000}$	$\frac{E \times I \times 1.73}{1000}$
HORSEPOWER	$\frac{E \times I \times \%EFF}{746}$	$\frac{E \times I \times \%EFF \times PF}{746}$	$\frac{E \times I \times \%EFF \times PF \times 2}{746}$	$\frac{E \times I \times \%EFF \times PF \times 1.73}{746}$

$$\text{Percent Efficiency} = \%\ EFF = \frac{\text{OUTPUT (WATTS)}}{\text{INPUT (WATTS)}} \qquad \text{Power Factor} = PF = \frac{\text{POWER USED (WATTS)}}{\text{APPARENT POWER}} = \frac{kW}{kVA}$$

E = VOLTS
I = AMPERES
W = WATTS

SINGLE-PHASE FULL LOAD CURRENT IN AMPERES

HP	115 V	200 V	208 V	230 V
1/6	4.4	2.5	2.4	2.2
1/4	5.8	3.3	3.2	2.9
1/3	7.2	4.1	4.0	3.6
1/2	9.8	5.6	5.4	4.9
3/4	13.8	7.9	7.6	6.9
1	16	9.2	8.8	8.0
1 1/2	20	11.5	11	10
2	24	13.8	13.2	12
3	34	19.6	18.7	17
5	56	32.2	30.8	28
7 1/2	80	46	44	40
10	100	57.5	55	50

THREE-PHASE FULL LOAD CURRENT IN AMPERES

HP	115v	200v	208v	230	460
1/2	4.4	2.5	2.4	2.2	1.1
3/4	6.4	3.7	3.5	3.2	1.6
1	8.4	4.8	4.6	4.2	2.1
1 1/2	12	6.9	6.6	6	3
2	13.6	7.8	7.5	6.8	3.4
3	–	11	10.6	9.6	4.8
5	–	17.5	16.7	15.2	7.6
7 1/2	–	25.3	24.2	22	11
10	–	32.2	30.8	28	14
15	–	48.3	46.2	42	21
20	–	62.1	59.4	54	27
25	–	78.2	74.8	68	34
30	–	92	88	80	40
40	–	120	114	104	52
50	–	150	143	130	65
60	–	177	169	154	77
75	–	221	211	192	96
100	–	285	273	248	124
125	–	359	343	312	156
150	–	414	396	360	180
200	–	552	528	480	240
250	–	–	–	–	302
300	–	–	–	–	361
350	–	–	–	–	414
400	–	–	–	–	477
450	–	–	–	–	515
500	–	–	–	–	590

PHOTO CREDITS

UNIT 1

Fig. 1-1, Figs. 1-3–1-6: STANFIELD, CARTER; SKAVES, DAVID, FUNDAMENTALS OF HVAC/R, 1st Ed., ©2010. Reprinted and Electronically reproduced by permission of Pearson Education, Inc., Upper Saddle River, New Jersey. Figs. 1-9–1-14: AHRI; MORAVEK, JOSEPH, AIR CONDITIONING SYSTEMS: PRINCIPLES, EQUIPMENT, AND SERVICE, 1st Ed., ©2001. Reprinted and Electronically reproduced by permission of Pearson Education, Inc., Upper Saddle River, New Jersey. Figs. 1-15–1-18: Reproduced with Permission, Fluke Corporation. Fig. 1-20: Courtesy Carrier Corporation. Fig. 1-21: Courtesy Giroux AC Training & Consulting.

UNIT 2

Fig. 2-2: Courtesy Richard Stockman, AmRad Engineering, Palm Coast, Florida. Fig. 2-4: AHRI; MORAVEK, JOSEPH, AIR CONDITIONING SYSTEMS: PRINCIPLES, EQUIPMENT, AND SERVICE, 1st Ed., ©2001. Reprinted and Electronically reproduced by permission of Pearson Education, Inc., Upper Saddle River, New Jersey. Figs. 2-5–2-7: STANFIELD, CARTER; SKAVES, DAVID, FUNDAMENTALS OF HVAC/R, 1st Ed., ©2010. Reprinted and Electronically reproduced by permission of Pearson Education, Inc., Upper Saddle River, New Jersey. Figs. 2-8–2-24: AHRI; MORAVEK, JOSEPH, AIR CONDITIONING SYSTEMS: PRINCIPLES, EQUIPMENT, AND SERVICE, 1st Ed., ©2001. Reprinted and Electronically reproduced by permission of Pearson Education, Inc., Upper Saddle River, New Jersey. Figs. 2-27–2-29: AHRI; MORAVEK, JOSEPH, AIR CONDITIONING SYSTEMS: PRINCIPLES, EQUIPMENT, AND SERVICE, 1st Ed., ©2001. Reprinted and Electronically reproduced by permission of Pearson Education, Inc., Upper Saddle River, New Jersey.

UNIT 3

Fig. 3-1: Reproduced with Permission, Fluke Corporation. Fig. 3-2: Printed with permission, Fluke Corporation. Figs. 3-3– 3-6: Reproduced with Permission, Fluke Corporation. Fig. 3-7: Printed with permission, Fluke Corporation. Figs. 3-8–3-9: AHRI; MORAVEK, JOSEPH, AIR CONDITIONING SYSTEMS: PRINCIPLES, EQUIPMENT, AND SERVICE, 1st Ed., ©2001. Reprinted and Electronically reproduced by permission of Pearson Education, Inc., Upper Saddle River, New Jersey. Figs. 3-10–3-13: Reproduced with Permission, Fluke Corporation. Fig. 3-15A: Printed with permission, Fluke Corporation. Figs. 3-15B–3-31: Reproduced with Permission, Fluke Corporation. Fig. 3-34: Courtesy of Ideal Industries, Inc. UNFIG1, UNFIG2, Fig. 3-36: Reproduced with Permission, Fluke Corporation.

UNIT 4

Fig. 4-1: Images provided courtesy of IDEAL INDUSTRIES, Inc. Fig. 4-2: Courtesy Diversitech. Figs. 4-5–4-9: STANFIELD, CARTER; SKAVES, DAVID, FUNDAMENTALS OF HVAC/R, 1st Ed., ©2010. Reprinted and Electronically reproduced by permission of Pearson Education, Inc., Upper Saddle River, New Jersey. Figs. 4-12–4-14: STANFIELD, CARTER; SKAVES, DAVID, FUNDAMENTALS OF HVAC/R, 1st Ed., ©2010. Reprinted and Electronically reproduced by permission of Pearson Education, Inc., Upper Saddle River, New Jersey. Fig. 4-15: Courtesy 3M Products. Fig. 4-16–4-18: Courtesy Diversitech. Figs. 4-20–4-22: Courtesy Diversitech. Figs. 4-27–4-31: Courtesy 3M Products.

UNIT 5

Fig. 5-1: © Edison Electric Institute. All rights reserved. Fig. 5-2: STANFIELD, CARTER; SKAVES, DAVID, FUNDAMENTALS OF HVAC/R, 1st Ed., ©2010. Reprinted and Electronically reproduced by permission of Pearson Education, Inc., Upper Saddle River, New Jersey. Fig. 5-3: © Edison Electric Institute. All rights reserved. Fig. 5-4: U.S. Energy Information Administration, Electric Power Annual (2010). Fig. 5-5: Courtesy DOE Energy Information Administration. Fig. 5-6: U.S. Energy Information Administration. Fig. 5-7: KISSELL, THOMAS E., ELECTRICITY, ELECTRONICS, AND CONTROL SYSTEMS FOR HVAC, 4th Ed., ©2008. Reprinted and Electronically reproduced by permission of Pearson Education, Inc., Upper Saddle River, New Jersey. Figs. 5-8–5-15: STANFIELD, CARTER; SKAVES, DAVID, FUNDAMENTALS OF HVAC/R, 1st Ed., ©2010. Reprinted and Electronically reproduced by permission of Pearson Education, Inc., Upper Saddle River, New Jersey. Fig. 5-16: Reprinted with permission from Siemens Industry, Inc. Fig. 5-17: KISSELL, THOMAS E., ELECTRICITY, ELECTRONICS, AND CONTROL SYSTEMS FOR HVAC, 4th Ed., ©2008. Reprinted and Electronically reproduced by permission of Pearson Education, Inc., Upper Saddle River, New Jersey. Fig. 5-18: © TheSupe87/Fotolia. Fig. 5-27: Reprinted with permission from Siemens Industry, Inc. Fig. 5-28: U.S. Energy Information Administration, Residential Energy Consumption Surveys 1980 and 2005. Fig. 5-29: U.S. Energy Information Administration, Residential Energy Consumption Survey: Household Consumption and Expenditure. Figs. 5-30–5-38: © Edison Electric Institute. All rights reserved. Figs. 5-40–5-41: Reprinted with permission from Siemens Industry, Inc.

UNIT 7

Fig. 7-4: Courtesy of Allied Air Enterprises, LLC, a Lennox International Inc. company. Figs. 7-5–7-6: Courtesy Carrier Corporation. Figs. 7-16–7-18: STANFIELD, CARTER; SKAVES, DAVID, FUNDAMENTALS OF HVAC/R, 1st Ed., ©2010. Reprinted and Electronically reproduced by permission of Pearson Education, Inc., Upper Saddle River, New Jersey. Fig. 7-19: © WorkSafeBC. Used with permission from Working Safely Around Electricity. Figs. 7-24–7-26: Courtesy Carrier Corporation. Fig. 7-29: STANFIELD, CARTER; SKAVES, DAVID, FUNDAMENTALS OF HVAC/R, 1st Ed., ©2010. Reprinted and Electronically reproduced by permission of Pearson Education, Inc., Upper Saddle River, New Jersey. Fig. 7-30: Courtesy HVAC/R Productions. Fig. 7-31: STANFIELD, CARTER; SKAVES, DAVID, FUNDAMENTALS OF HVAC/R, 1st Ed., ©2010. Reprinted and Electronically reproduced by permission of Pearson Education, Inc., Upper Saddle River, New Jersey.

UNIT 8

Fig. 8-2, Fig. 8-5, Fig. 8-7: Courtesy HVAC/R Productions. Fig. 8-9: Courtesy Motors & Armatures, Inc. (MARS), Fig. 8-12: KISSELL, THOMAS E., ELECTRICITY, ELECTRONICS, AND CONTROL SYSTEMS FOR HVAC, 4th Ed., ©2008. Reprinted and Electronically reproduced by permission of Pearson Education, Inc., Upper Saddle River, New Jersey. Fig. 8-14: Printed with permission, Fluke Corporation. Fig. 8-15: AHRI; MORAVEK, JOSEPH, AIR CONDITIONING SYSTEMS: PRINCIPLES, EQUIPMENT, AND SERVICE, 1st Ed., ©2001. Reprinted and Electronically reproduced by permission of Pearson Education, Inc., Upper Saddle River, New Jersey. Figs. 8-17–8-18: Courtesy Carrier Corporation. Figs. 8-20–8-21: Courtesy Motors & Armatures, Inc. (MARS), Figs. 8-22–8-23: Reprinted with permission from Siemens Industry, Inc.

UNIT 9

Figs. 9-1A–9-1B, Fig 9-5: STANFIELD, CARTER; SKAVES, DAVID, FUNDAMENTALS OF HVAC/R, 1st Ed., ©2010. Reprinted and Electronically reproduced by permission of Pearson Education, Inc., Upper Saddle River, New Jersey. Figs. 9-6–9-7: AHRI; MORAVEK, JOSEPH, AIR CONDITIONING SYSTEMS: PRINCIPLES, EQUIPMENT, AND SERVICE, 1st Ed., ©2001. Reprinted and Electronically reproduced by permission of Pearson Education, Inc., Upper Saddle River, New Jersey. Fig. 9-16, Fig. 9-20: Photo courtesy of Rockwell Automation, Inc. All Rights Reserved. Fig. 9-23–9-26: Original Material Courtesy of Schneider Electric. Fig. 9-27: Courtesy Tecumseh Products Company, Fig. 9-28: Reprinted with permission from Siemens Industries, Inc., Fig. 9-29: Courtesy Automation Direct. Fig. 9-30: Original Material Courtesy of Schneider Electric, Fig. 9-31: Courtesy Regal Beloit.

UNIT 10

Fig. 10-1: AHRI; MORAVEK, JOSEPH, AIR CONDITIONING SYSTEMS: PRINCIPLES, EQUIPMENT, AND SERVICE, 1st Ed., ©2001. Reprinted and Electronically reproduced by permission of Pearson Education, Inc., Upper Saddle River, New Jersey. Fig. 10-2: Courtesy Motors & Armatures, Inc. (MARS). Fig. 10-3: STANFIELD, CARTER; SKAVES, DAVID, FUNDAMENTALS OF HVAC/R, 1st Ed., ©2010. Reprinted and Electronically reproduced by permission of Pearson Education, Inc., Upper Saddle River, New Jersey. Fig. 10-4: Courtesy Richard Stockman, AmRad Engineering. Fig. 10-5: STANFIELD, CARTER; SKAVES, DAVID, FUNDAMENTALS OF HVAC/R, 1st Ed., ©2010. Reprinted and Electronically reproduced by permission of Pearson Education, Inc., Upper Saddle River, New Jersey. Figs. 10-6–10-7: Courtesy Richard Stockman, AmRad Engineering. Figs. 10-9–10-10: Courtesy Motors & Armatures, Inc. (MARS). Fig. 10-11: Reproduced with Permission, Fluke Corporation. Fig. 10-12: Courtesy Richard Stockman, AmRad Engineering. Fig 10-13–10-14: AHRI; MORAVEK, JOSEPH, AIR CONDITIONING SYSTEMS: PRINCIPLES, EQUIPMENT, AND SERVICE, 1st Ed., ©2001. Reprinted and Electronically reproduced by permission of Pearson Education, Inc., Upper Saddle River, New Jersey.

UNIT 11

Figs. 11-4–11-6A, Fig. 11-20: STANFIELD, CARTER; SKAVES, DAVID, FUNDAMENTALS OF HVAC/R, 1st Ed., ©2010. Reprinted and Electronically reproduced by permission of Pearson Education, Inc., Upper Saddle River, New Jersey. Figs. 11-21–11-22: Courtesy www.refrigerationbasics.com. Figs. 11-23–11-24: STANFIELD, CARTER; SKAVES, DAVID, FUNDAMENTALS OF HVAC/R, 1st Ed., ©2010. Reprinted and Electronically reproduced by permission of Pearson Education, Inc., Upper Saddle River, New Jersey. Figs. 11-25–11-26: Courtesy of Allied Air Enterprises, LLC, a Lennox International Inc. company. Fig 11-27: © Steve Cukrov/Fotolia. Fig. 11-28: STANFIELD, CARTER; SKAVES, DAVID, FUNDAMENTALS OF HVAC/R, 1st Ed., ©2010. Reprinted and Electronically reproduced by permission of Pearson Education, Inc., Upper Saddle River, New Jersey. Fig. 11-29: © Sinisa Botas/Fotolia. Fig. 11-30: Courtesy Venstar, Inc. Fig. 11-32: STANFIELD, CARTER; SKAVES, DAVID, FUNDAMENTALS OF HVAC/R, 1st Ed., ©2010. Reprinted and Electronically reproduced by permission of Pearson Education, Inc., Upper Saddle River, New Jersey. Fig. 11-38: Courtesy Thermostat Recycling Corp. Fig. 11-39: AHRI; MORAVEK, JOSEPH, AIR CONDITIONING SYSTEMS: PRINCIPLES, EQUIPMENT, AND SERVICE, 1st Ed., ©2001. Reprinted and Electronically reproduced by permission of Pearson Education, Inc., Upper Saddle River, New Jersey.

UNIT 12

Fig. 12-2: STANFIELD, CARTER; SKAVES, DAVID, FUNDAMENTALS OF HVAC/R, 1st Ed., ©2010. Reprinted and Electronically reproduced by permission of Pearson Education, Inc., Upper Saddle River, New Jersey. Fig. 12-7: Courtesy HVAC/R Productions. Fig. 12-8: Courtesy of Allied Air Enterprises, LLC, a Lennox International Inc. company. Fig. 12-13: STANFIELD, CARTER; SKAVES, DAVID, FUNDAMENTALS OF HVAC/R, 1st Ed., ©2010. Reprinted and Electronically reproduced by permission of Pearson Education, Inc., Upper Saddle River, New Jersey. Fig. 12-15: STANFIELD, CARTER; SKAVES, DAVID, FUNDAMENTALS OF HVAC/R, 1st Ed., ©2010. Reprinted and Electronically reproduced by permission of Pearson Education, Inc., Upper Saddle River, New Jersey.

UNIT 13

Fig. 13-7, Fig. 13-9: Courtesy of Sporlan Division of Park Hannifin. Figs. 13-13–13-15: Courtesy Hampden Engineering. Fig. 13-18: Courtesy of ICM Controls Corp. Fig. 13-23A: Courtesy Tecumseh Products Company. Fig. 13-23B: STANFIELD, CARTER; SKAVES, DAVID, FUNDAMENTALS OF HVAC/R, 1st Ed., ©2010. Reprinted and Electronically reproduced by permission of Pearson Education, Inc., Upper Saddle River, New Jersey. Fig. 13-24: Courtesy Tecumseh Products Company.

UNIT 14

Fig. 14-2–14-3: Reproduced with Permission, Fluke Corporation. Fig. 14-4: STANFIELD, CARTER; SKAVES, DAVID, FUNDAMENTALS OF HVAC/R, 1st Ed., ©2010. Reprinted and Electronically reproduced by permission of Pearson Education, Inc., Upper Saddle River, New Jersey. Fig. 14-7, Fig. 14-8: Courtesy Regal Beloit. Fig 14-9: STANFIELD, CARTER; SKAVES, DAVID, FUNDAMENTALS OF HVAC/R, 1st Ed., ©2010. Reprinted and Electronically reproduced by permission of Pearson Education, Inc., Upper Saddle River, New Jersey. Fig. 14-10: Courtesy Hampden Engineering Corporation. Fig. 14-12: Photo courtesy of Rockwell Automation, Inc. All rights reserved. Fig. 14-16: Courtesy Regal Beloit. Fig. 14-18: AHRI; MORAVEK, JOSEPH, AIR CONDITIONING SYSTEMS: PRINCIPLES, EQUIPMENT, AND SERVICE, 1st Ed., ©2001. Reprinted and Electronically reproduced by permission of Pearson Education, Inc., Upper Saddle River, New Jersey.

UNIT 15

Fig. 15-1: Courtesy Regal Beloit. Fig. 15-7: STANFIELD, CARTER; SKAVES, DAVID, FUNDAMENTALS OF HVAC/R, 1st Ed., ©2010. Reprinted and Electronically reproduced by permission of Pearson Education, Inc., Upper Saddle River, New Jersey. Fig. 15-8, Figs. 15-10–15-11: Courtesy of Emerson Climate Technologies, Inc. Fig. 15-12: STANFIELD, CARTER; SKAVES, DAVID, FUNDAMENTALS OF HVAC/R, 1st Ed., ©2010. Reprinted and Electronically reproduced by permission of Pearson Education, Inc., Upper Saddle River, New Jersey. Fig. 15-13: Courtesy Regal Beloit. Figs. 15-14–15-15: Courtesy of Emerson Climate Technologies, Inc. Fig. 15-24: Courtesy Regal Beloit. Figs. 15-25–15-26: Courtesy of Emerson Climate Technologies, Inc. Figs. 15-32–15-33: STANFIELD, CARTER; SKAVES, DAVID, FUNDAMENTALS OF HVAC/R, 1st Ed., ©2010. Reprinted and Electronically reproduced by permission of Pearson Education, Inc., Upper Saddle River, New Jersey. Fig. 15-34: Courtesy Carrier Corporation. Figs. 15-36–15-38: Material courtesy of Trane. Figs. 15-39–15-40: Courtesy of Emerson Climate Technologies, Inc. Fig. 15-41: Reproduced with Permission, Fluke Corporation. Fig. 15-42: STANFIELD, CARTER; SKAVES, DAVID, FUNDAMENTALS OF HVAC/R, 1st Ed., ©2010. Reprinted and Electronically reproduced by permission of Pearson Education, Inc., Upper Saddle River, New Jersey. Fig. 15-43: Courtesy Regal Beloit. Fig. 15-46: Courtesy WEG Electric Corp. Fig. 15-48: Courtesy Nidec Motor Corp. Fig. 15-49: Courtesy Carrier Corporation. Figs. 15-50–15-54: Courtesy Regal Beloit. Fig. 15-55: AHRI; MORAVEK, JOSEPH, AIR CONDITIONING SYSTEMS: PRINCIPLES, EQUIPMENT, AND SERVICE, 1st Ed., ©2001. Reprinted and Electronically reproduced by permission of Pearson Education, Inc., Upper Saddle River, New Jersey. Fig. 15-59: Reprinted with permission from Siemens Industry, Inc.

UNIT 16

Fig. 16-1, Fig. 16-4, Fig. 16-5: Courtesy Regal Beloit. Fig. 16-7: Courtesy Carrier Corporation. Fig. 16-8, Fig. 16-10, Figs. 16-12–16-13: Courtesy Regal Beloit. Fig. 16-15: Courtesy Regal Beloit. Fig. 16-17: Courtesy of Regal Beloit. Figs. 16-18–16-21, Figs. 16-23–16-25: Courtesy Regal Beloit. Fig. 16-26: Courtesy of Zebra Instrument. Figs. 16-27–16-31, Figs 16-33–16-34: Courtesy Regal Beloit.

UNIT 17

Fig. 17-1: STANFIELD, CARTER; SKAVES, DAVID, FUNDAMENTALS OF HVAC/R, 1st Ed., ©2010. Reprinted and Electronically reproduced by permission of Pearson Education, Inc., Upper Saddle River, New Jersey. Figs. 17-5–17-7: Courtesy HVAC/R Productions. Figs. 17-8A–17-22: AHRI; MORAVEK, JOSEPH, AIR CONDITIONING SYSTEMS: PRINCIPLES, EQUIPMENT, AND SERVICE, 1st Ed., ©2001. Reprinted and Electronically reproduced by permission of Pearson Education, Inc., Upper Saddle River, New Jersey. Fig. 17-23: AHRI; MAHONEY, EDWARD, ELECTRICITY, ELECTRONICS, AND WIRING DIAGRAMS FOR HVAC/R, 2nd Ed., ©2006. Reprinted and Electronically reproduced by permission of Pearson Education, Inc., Upper Saddle River, New Jersey. Figs. 17-24–17-26: Courtesy Giroux AC Training & Consulting. Fig. 17-27: STANFIELD, CARTER; SKAVES, DAVID; AHRI, FUNDAMENTALS OF HVACR, 2nd Ed., ©2013. Reprinted and Electronically reproduced by permission of Pearson Education, Inc., Upper Saddle River, New Jersey. Fig. 17-28: AHRI; MORAVEK, JOSEPH, AIR CONDITIONING SYSTEMS: PRINCIPLES, EQUIPMENT, AND SERVICE, 1st Ed., ©2001. Reprinted and Electronically reproduced by permission of Pearson Education, Inc., Upper Saddle River, NewJersey. Figs.17-29–17-30: STANFIELD, CARTER; SKAVES, DAVID, FUNDAMENTALS OF HVAC/R, 1st Ed., ©2010. Reprinted and Electronically reproduced by permission of Pearson Education, Inc., Upper Saddle River, New Jersey. Fig. 17-32–17-37: AHRI; MORAVEK, JOSEPH, AIR CONDITIONING SYSTEMS: PRINCIPLES, EQUIPMENT, AND SERVICE, 1st Ed., ©2001. Reprinted and Electronically reproduced by permission of Pearson Education, Inc., Upper Saddle River, New Jersey. Fig. 17-38: Drawing courtesy of Johnson Controls, Inc. Figs. 17-39–17-43: Courtesy Carrier Corporation. Fig. 17-44: Courtesy of Goodman Global Group, Inc. Fig. 17-45: STANFIELD, CARTER; SKAVES, DAVID, FUNDAMENTALS OF HVAC/R, 1st Ed., ©2010. Reprinted and Electronically reproduced by permission of Pearson Education, Inc., Upper Saddle River, New Jersey.

UNIT 18

Fig. 18-1: © Maximus/Fotolia, Fig. 18-2: Courtesy Quest Components. Fig. 18-3: © dexns/Fotolia, Fig. 18-4: KISSELL, THOMAS E., ELECTRICITY, ELECTRONICS, AND CONTROL SYSTEMS FOR HVAC, 4th Ed., ©2008. Reprinted and Electronically reproduced by permission of Pearson Education, Inc., Upper Saddle River, New Jersey. Fig. 18-5: © ans007/Fotolia, Fig 18-6: © Leo Blanchette/Fotolia.

UNIT 19

Fig. 19-4: Used with permission from SCILLC dba ON Semiconductor. Fig. 19-7: OSHA.

UNIT 20

Fig. 20-1: © Nomad_Soul/Fotolia, Fig. 20-3: AHRI; MORAVEK, JOSEPH, AIR CONDITIONING SYSTEMS: PRINCIPLES, EQUIPMENT, AND SERVICE, 1st Ed., ©2001. Reprinted and Electronically reproduced by permission of Pearson Education, Inc., Upper

Saddle River, New Jersey. Figs. 20-4–20-5: Courtesy of ICM Controls Corp. Fig. 20-6: Courtesy of Allied Air Enterprises, LLC, a Lennox International Inc. company. Figs. 20-7–20-10, Fig. 20-12–20-16: Courtesy Carrier Corporation.

UNIT 21

Figs. 21-1–21-6: AHRI; MAHONEY, EDWARD, ELECTRICITY, ELECTRONICS, AND WIRING DIAGRAMS FOR HVAC/R, 2nd Ed., ©2006. Reprinted and Electronically reproduced by permission of Pearson Education, Inc., Upper Saddle River, New Jersey. Figs. 21-7–21-15: Courtesy Carrier Corporation. Figs. 21-16–21-17: AHRI; MAHONEY, EDWARD, ELECTRICITY, ELECTRONICS, AND WIRING DIAGRAMS FOR HVAC/R, 2nd Ed., ©2006. Reprinted and Electronically reproduced by permission of Pearson Education, Inc., Upper Saddle River, New Jersey. Fig. 21-18: Courtesy Carrier Corporation.

UNIT 22

Figs. 22-1–22-2: US Dept of Energy. Fig. 22-8: Courtesy Carrier Corporation. Fig. 22-13: Material courtesy of Trane. Fig. 22-16: Courtesy Carrier Corporation. Fig. 22-29: Courtesy John Hohman. Figs. 22-30A–22-30B: Material courtesy of Trane. Fig. 22-31: Courtesy of Goodman Global Group, Inc. Figs. 22-32–22-33: Material courtesy of Trane. Figs. 22-34–22-43: Courtesy Carrier Corporation. Fig. 22-47: AHRI; MAHONEY, EDWARD, ELECTRICITY, ELECTRONICS, AND WIRING DIAGRAMS FOR HVAC/R, 2nd Ed., ©2006. Reprinted and Electronically reproduced by permission of Pearson Education, Inc., Upper Saddle River, New Jersey.

UNIT 23

Fig. 23-6: Courtesy Cadet Manufacturing. Fig. 23-8: Courtesy Dr. John Hohman. Fig. 23-9, Fig. 23-12: STANFIELD, CARTER; SKAVES, DAVID, FUNDAMENTALS OF HVAC/R, 1st Ed., ©2010. Reprinted and Electronically reproduced by permission of Pearson Education, Inc. Upper Saddle River, New Jersey. Fig. 23-13: Courtesy Dr. John Hohman. Fig. 23-14: Courtesy Sealed Unit Parts Co. Inc. Fig. 23-15: STANFIELD, CARTER; SKAVES, DAVID, FUNDAMENTALS OF HVAC/R, 1st Ed., ©2010. Reprinted and Electronically reproduced by permission of Pearson Education, Inc., Upper Saddle River, New Jersey. Fig. 23-22: Used with permission of HVAC/R Productions. Fig. 23-23: Courtesy Carrier Corporation. Fig. 23-24: Used with permission of HVAC/R Productions.

UNIT 24

Fig. 24-1, Figs. 24-9–24-10, Fig. 24-13: STANFIELD, CARTER; SKAVES, DAVID, FUNDAMENTALS OF HVAC/R, 1st Ed., ©2010. Reprinted and Electronically reproduced by permission of Pearson Education, Inc., Upper Saddle River, New Jersey. Fig. 24-14: Courtesy Carrier Corporation. Fig. 24-16: STANFIELD, CARTER; SKAVES, DAVID, FUNDAMENTALS OF HVAC/R, 1st Ed., ©2010. Reprinted and Electronically reproduced by permission of Pearson Education, Inc., Upper Saddle River, New Jersey. Fig. 24-22: Material courtesy of Trane. Fig. 24-23B: Material courtesy of Trane. Fig. 24-24, Figs. 24-26–24-31: Courtesy Carrier Corporation. Fig. 24-32: Courtesy Dr. John Hohman. Fig. 24-35, Figs. 24-38–24-47: Courtesy Carrier Corporation. Fig. 24-48: Material courtesy of Trane.

UNIT 25

Fig. 25-2: Courtesy of Ritchie Engineering Co.-YELLOW JACKET Products Division. Fig. 25-6: STANFIELD, CARTER; SKAVES, DAVID, FUNDAMENTALS OF HVAC/R, 1st Ed., ©2010. Reprinted and Electronically reproduced by permission of Pearson Education, Inc., Upper Saddle River, New Jersey. Figs. 25-7–25-12, AHRI; MORAVEK, JOSEPH, AIR CONDITIONING SYSTEMS: PRINCIPLES, EQUIPMENT, AND SERVICE, 1st Ed., ©2001. Reprinted and Electronically reproduced by permission of Pearson Education, Inc., Upper Saddle River, New Jersey. Fig. 25-13: Courtesy Carrier Corporation. Figs. 25-14–25-16: Courtesy Carrier Corp. Figs. 25-17A–25-20: Courtesy Carrier Corporation.

UNIT 26

Fig. 26-7: Courtesy of Goodman Company, L.P. Fig. 26-8: AHRI; MORAVEK, JOSEPH, AIR CONDITIONING SYSTEMS: PRINCIPLES, EQUIPMENT, AND SERVICE, 1st Ed., ©2001. Reprinted and Electronically reproduced by permission of Pearson Education, Inc., Upper Saddle River, New Jersey. Fig. 26-10: STANFIELD, CARTER; SKAVES, DAVID; AHRI, FUNDAMENTALS OF HVACR, 2nd Ed., © 2013, p. 584. Reprinted and Electronically reproduced by permission of Pearson Education, Inc., Upper Saddle River, New Jersey. Fig. 26-11: Courtesy Giroux AC Training & Consulting. Figs. 26-16–26-17: AHRI; MORAVEK, JOSEPH, AIR CONDITIONING SYSTEMS: PRINCIPLES, EQUIPMENT, AND SERVICE, 1st Ed., ©2001. Reprinted and Electronically reproduced by permission of Pearson Education, Inc., Upper Saddle River, New Jersey. Fig. 26-19: Courtesy Carrier Corporation.

UNIT 27

Fig. 27-10: Courtesy Sealed Unit Parts Co. Inc. Fig. 27-11: Courtesy of Emerson Climate Technologies, Inc. Fig. 27-13: Courtesy Carrier Corporation. Fig. 27-15–27-16, Fig. 27-19: Reproduced with Permission, Fluke Corporation. Fig. 27-20, Fig. 27-22: AHRI; MORAVEK, JOSEPH, AIR CONDITIONING SYSTEMS: PRINCIPLES, EQUIPMENT, AND SERVICE, 1st Ed., ©2001. Reprinted and Electronically reproduced by permission of Pearson Education, Inc., Upper Saddle River, New Jersey. Fig. 27-23: STANFIELD, CARTER; SKAVES, DAVID, FUNDAMENTALS OF HVAC/R, 1st Ed., ©2010. Reprinted and Electronically reproduced by permission of Pearson Education, Inc., Upper Saddle River, New Jersey. Fig. 27-24: Courtesy Giroux AC Training & Consulting.

UNIT 28

Fig. 28-1–28-6, Fig. 28-8: Courtesy HVAC/R Productions. Fig. 28-9: Courtesy Carrier Corporation. Figs. 28-10–28-12: AHRI; MORAVEK, JOSEPH, AIR CONDITIONING SYSTEMS: PRINCIPLES, EQUIPMENT, AND SERVICE, 1st Ed., ©2001. Reprinted and Electronically reproduced by permission of Pearson Education, Inc., Upper Saddle River, New Jersey.

INSIDE BACK COVER

Fig. IBC-01: STANFIELD, CARTER; SKAVES, DAVID; AHRI, FUNDAMENTALS OF HVACR, 2nd Ed., © 2013, p. 584. Reprinted and Electronically reproduced by permission of Pearson Education, Inc., Upper Saddle River, New Jersey.

ELECTRICAL DIAGRAMS

Fig. ED1: Courtesy HVAC/R Productions. Fig. ED4: Courtesy Hampden Engineering. Fig. ED6: Courtesy Carrier Corporation. Fig. ED7: STANFIELD, CARTER; SKAVES, DAVID, FUNDAMENTALS OF HVAC/R, 1st Ed., ©2010. Reprinted and Electronically reproduced by permission of Pearson Education, Inc., Upper Saddle River, New Jersey. Fig. ED9-ED10, Fig. ED11: Courtesy Carrier Corporation. Fig. ED12: Courtesy Giroux AC Training & Consulting.

INDEX

D

N

O

U

V

W

X